STUDENT SOLUTIONS MANUAL

C TRIMBLE & ASSOCIATES

INTERMEDIATE ALGEBRA

SEVENTH EDITION

Elayn Martin-Gay

University of New Orleans

PEARSON

Boston Columbus Indianapolis New York San Francisco
Amsterdam Cape Town Dubai London Madrid Milan Munich Paris Montreal Toronto
Delhi Mexico City São Paulo Sydney Hong Kong Seoul Singapore Taipei Tokyo

Reproduced by Pearson from electronic files supplied by the author.

Publishing as Pearson, 501 Boylston Street, Boston, MA 02116.

ISBN-13: 978-0-13-420882-4
ISBN-10: 0-13-420882-X

1 16

www.pearsonhighered.com

PEARSON

Contents

Chapter 1

Section 1.2 Practice Exercises

1. Let $b = 3.5$ and $h = 8$.
$A = \frac{1}{2}bh$
$A = \frac{1}{2}(3.5)(8) = 14$
The area is 14 square centimeters.

2. Let $p = 17$ and $q = 3$.
$2p - q = 2(17) - 3 = 34 - 3 = 31$

3. a. $\{6, 7, 8, 9\}$

b. $\{41, 42, 43, ...\}$

4. a. True, since 7 is a natural number and therefore an element of the set.

b. True, since 6 is not an element of the set $\{1, 3, 5, 7\}$.

5. a. True; every integer is a real number.

b. False; $\sqrt{8}$ is an irrational number.

c. True; every whole number is a rational number.

d. False; since the element 2 in the first set is not an element of the second set.

6. a. $|4| = 4$ since 4 is located 4 units from 0 on a number line.

b. $\left|-\frac{1}{2}\right| = \frac{1}{2}$ since $-\frac{1}{2}$ is $\frac{1}{2}$ unit from 0 on a number line.

c. $|1| = 1$ since 1 is 1 unit from 0 on a number line.

d. $-|6.8| = -6.8$
The negative sign outside the absolute value bars means to take the opposite of the absolute value of 6.8.

e. $-|-4| = -4$
Since $|-4| = 4$, $-|-4| = -4$.

7. a. The opposite of 5.4 is -5.4.

b. The opposite of $-\frac{3}{5}$ is $\frac{3}{5}$.

c. The opposite of 18 is -18.

8. a. $3 \cdot x$ or $3x$

b. $2x - 5$

c. $3\frac{5}{8} + x$

d. $x \div 2$ or $\frac{x}{2}$

e. $x - 14$

f. $5(x + 10)$

Vocabulary, Readiness & Video Check 1.2

1. Letters that represent numbers are called variables.

2. Finding the value of an expression means evaluating the expression.

3. The absolute value of a number is that number's distance from 0 on a number line.

4. An expression is formed by numbers and variables connected by operations such as addition, subtraction, multiplication, division, raising to powers, and/or taking roots.

5. The natural numbers are $\{1, 2, 3, ...\}$.

6. The whole numbers are $\{0, 1, 2, 3, ...\}$.

7. The integers are $\{..., -3, -2, -1, 0, 1, 2, 3, ...\}$.

8. The number $\sqrt{5}$ is an irrational number.

9. The number $\frac{5}{7}$ is a rational number.

10. The opposite of a is $-a$.

11. This is an application, so the answer needs to be put in context. The decimal actually represents money.

12. Every real number is either a rational number or an irrational number. There are no numbers that are both rational and irrational.

13. The absolute value of a number is that number's <u>distance</u> from zero on a number line. Or, more formally, $|a| = a$ if a is <u>0</u> or a <u>positive</u> number. Also, $|a| = -a$ if a is a <u>negative</u> number.

14. The absolute value of a number is its *distance* from 0 on a number line, regardless of direction—distance must be positive or zero. The opposite of a number is the *number* that lies the same distance from 0 on the number line as the original number, but on the other side of 0—the opposite of a number can be negative.

15. Order is important in subtraction so you must read the phrase carefully to determine the order of subtraction in your algebraic expression.

Exercise Set 1.2

1. $5x = 5(7) = 35$

3. $9.8z = 9.8(3.1) = 30.38$

5. $ab = \left(\frac{1}{2}\right)\left(\frac{3}{4}\right) = \frac{3}{8}$

7. $3x + y = 3(6) + (4) = 18 + 4 = 22$

9. $400t = 400(5) = 2000$
The B737-400 travels 2000 miles in 5 hours.

11. $lw = (5.1)(4) = 20.4$
The display needs 20.4 sq ft of floor space.

13. $2948t = 2948(3.6) = 10{,}612.80$
It costs $10,612.80 to operate the B737-400 for 3.6 hours.

15. $\{1, 2, 3, 4, 5\}$

17. $\{11, 12, 13, 14, 15, 16\}$

19. $\{0\}$

21. $\{0, 2, 4, 6, 8\}$

23. 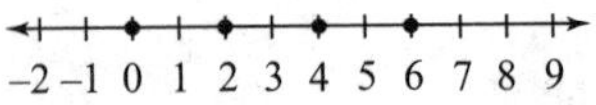

25. 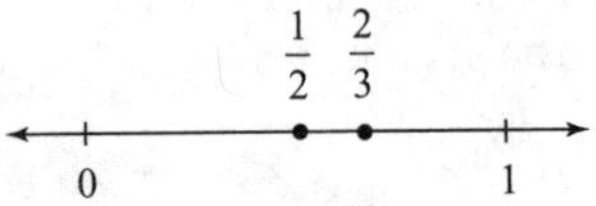

27. 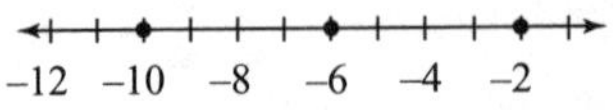

29. 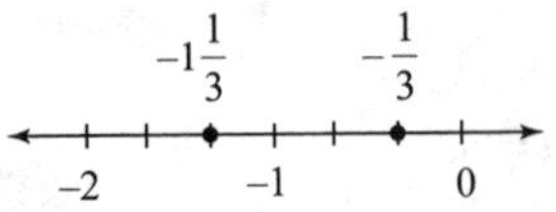

31. $\{3, 0, \sqrt{36}\}$

33. $\{3, \sqrt{36}\}$

35. $\{\sqrt{7}\}$

37. $-11 \in \{x|x \text{ is an integer}\}$

39. $-6 \notin \{2, 4, 6, \ldots\}$

41. $12 \notin \{1, 3, 5, \ldots\}$

43. $\frac{1}{2} \notin \{x|x \text{ is an irrational number}\}$

45. True; every integer is a real number.

47. True; −1 is an integer.

49. False; 0 is not a natural number.

51. False; $\sqrt{5}$ is an irrational number.

53. True; every natural number is an integer.

55. False; the number $\sqrt{7}$, for example, is a real number, but it is not a rational number.

57. $-|2| = -2$ (the opposite of $|2|$)

59. $|-4| = 4$ since −4 is located 4 units from 0 on a number line.

61. $|0| = 0$ since 0 is located 0 units from 0 on a number line.

63. $-|-3| = -3$ (the opposite of $|-3|$)

65. The opposite of −6.2 is $-(-6.2) = 6.2$.

67. The opposite of $\frac{4}{7}$ is $-\frac{4}{7}$.

69. The opposite of $-\frac{2}{3}$ is $\frac{2}{3}$.

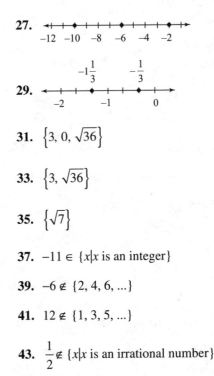

71. The opposite of 0 is 0.

73. $2x$

75. $2x + 5$

77. $x - 10$

79. $x + 2$

81. $\frac{x}{11}$

83. $12 - 3x$

85. $x + 2.3$ or $x + 2\frac{3}{10}$

87. $1\frac{1}{3} - x$

89. $\frac{5}{4-x}$

91. $2(x + 3)$

93. The height of the bar representing China is about 137. Therefore, 137 million tourists are predicted for China.

95. The height of the bar representing Spain is about 69. Therefore, 69 million tourists are predicted for Spain.

97. answers may vary

99. answers may vary

101. answers may vary

Section 1.3 Practice Exercises

1. a. $-6 + (-2) = -(6 + 2) = -8$

b. $5 + (-8) = -3$

c. $-4 + 9 = 5$

d. $-3.2 + (-4.9) = -8.1$

e. $-\frac{3}{5}+\frac{2}{3}=-\frac{3}{5}\cdot\frac{3}{3}+\frac{2}{3}\cdot\frac{5}{5}=-\frac{9}{15}+\frac{10}{15}=\frac{1}{15}$

f. $-\frac{5}{11}+\frac{3}{22}=-\frac{5}{11}\cdot\frac{2}{2}+\frac{3}{22}=-\frac{10}{22}+\frac{3}{22}=-\frac{7}{22}$

2. a. $3 - 11 = 3 + (-11) = -8$

b. $-6 - (-3) = -6 + (3) = -3$

c. $-7 - 5 = -7 + (-5) = -12$

d. $4.2 - (-3.5) = 4.2 + 3.5 = 7.7$

e.
$$\begin{aligned}-\frac{5}{7}-\frac{1}{3}&=-\frac{5\cdot 3}{7\cdot 3}-\frac{1\cdot 7}{3\cdot 7}\\&=-\frac{15}{21}+\left(-\frac{7}{21}\right)\\&=-\frac{22}{21}\end{aligned}$$

f. $3 - 1.2 = 3 + (-1.2) = 1.8$

g. $2 - 9 = 2 + (-9) = -7$

3. a. $13 + 5 - 6 = 18 - 6 = 12$

b. $-6 - 2 + 4 = -8 + 4 = -4$

4. a. Since the signs of the two numbers are different or unlike, the product is negative.
$(-5)(3) = -15$

b. Since the signs of the two numbers are the same, the product is positive.
$(-7)\left(-\frac{1}{14}\right)=\frac{7}{14}=\frac{1}{2}$

c. $5.1(-2) = -10.2$

d. $14(0) = 0$

e. $\left(-\frac{1}{4}\right)\left(\frac{8}{13}\right)=-\frac{8}{52}=-\frac{2}{13}$

f. $6(-1)(-2)(3) = -6(-2)(3) = 12(3) = 36$

g. $5(-2.3) = -11.5$

5. a. Since the signs are different or unlike, the quotient is negative.
$\frac{-16}{8}=-2$

b. Since the signs are the same, the quotient is positive.
$\frac{-15}{-3}=5$

c. $-\frac{2}{3}\div 4=-\frac{2}{3}\cdot\frac{1}{4}=-\frac{1}{6}$

d. $\frac{54}{-9}=-6$

e. $-\frac{1}{12}\div\left(-\frac{3}{4}\right)=-\frac{1}{12}\cdot-\frac{4}{3}=\frac{1}{9}$

f. $\frac{0}{-7}=0$

6. a. $2^3=2\cdot 2\cdot 2=8$

b. $\left(\frac{1}{3}\right)^2=\left(\frac{1}{3}\right)\left(\frac{1}{3}\right)=\frac{1}{9}$

c. $-12^2=-(12\cdot 12)=-144$

d. $(-12)^2=(-12)(-12)=144$

e. $-4^3=-(4\cdot 4\cdot 4)=-64$

f. $(-4)^3=(-4)(-4)(-4)=-64$

7. a. $\sqrt{49}=7$ since 7 is positive and $7^2=49$.

b. $\sqrt{\frac{1}{16}}=\frac{1}{4}$ since $\left(\frac{1}{4}\right)^2=\frac{1}{16}$.

c. $-\sqrt{64}=-8$

d. $\sqrt{-64}$ is not a real number.

e. $\sqrt{100}=10$ since $10^2=100$.

8. a. $\sqrt[3]{64}=4$ since $4^3=64$.

b. $\sqrt[5]{-1}=-1$ since $(-1)^5=-1$.

c. $\sqrt[4]{10{,}000}=10$ since $10^4=10{,}000$.

9. a. $14-3\cdot 4=14-12=2$

b. $3(5-8)^2=3(-3)^2=3(9)=27$

c. $\frac{|-5|^2+4}{\sqrt{4}-3}=\frac{5^2+4}{2-3}=\frac{25+4}{-1}=\frac{29}{-1}=-29$

10.
$$\begin{aligned}5-[(3-5)+6(2-4)]&=5-[-2+6(-2)]\\&=5-[-2+(-12)]\\&=5-[-14]\\&=5+14\\&=19\end{aligned}$$

11.
$$\begin{aligned}\frac{-2\sqrt{12+4}-(-3)^2}{6^2+|1-9|}&=\frac{-2\sqrt{16}-(-3)^2}{6^2+|-8|}\\&=\frac{-2(4)-9}{36+8}\\&=\frac{-8-9}{44}\\&=-\frac{17}{44}\end{aligned}$$

12. For each expression, replace x with 16 and y with −5.

a. $2x-7y=2(16)-7(-5)=32+35=67$

b. $-4y^2=-4(-5)^2=-4(25)=-100$

c.
$$\begin{aligned}\frac{\sqrt{x}}{y}-\frac{y}{x}&=\frac{\sqrt{16}}{-5}-\frac{-5}{16}\\&=-\frac{4}{5}+\frac{5}{16}\\&=-\frac{4}{5}\cdot\frac{16}{16}+\frac{5}{16}\cdot\frac{5}{5}\\&=-\frac{64}{80}+\frac{25}{80}\\&=-\frac{39}{80}\end{aligned}$$

13. When $x=-5$,
$\frac{9}{5}x+32=\frac{9}{5}(-5)+32=-9+32=23.$
When $x=10$,
$\frac{9}{5}x+32=\frac{9}{5}(10)+32=18+32=50.$
When $x=25$,
$\frac{9}{5}x+32=\frac{9}{5}(25)+32=45+32=77.$
The completed table is

Degrees Celsius	x	-5	10	25
Degrees Fahrenheit	$\frac{9}{5}x+32$	23	50	77

Vocabulary, Readiness & Video Check 1.3

1. $0 \cdot a = \underline{0}$

2. $\frac{0}{4}$ simplifies to $\underline{0}$ while $\frac{4}{0}$ is undefined.

3. The reciprocal of the nonzero number b is $\frac{1}{b}$.

4. The fraction $-\frac{a}{b} = \underline{\frac{-a}{b}} = \underline{\frac{a}{-b}}$.

5. An exponent is a shorthand notation for repeated multiplication of the same number.

6. In $(-5)^2$, the 2 is the exponent and the -5 is the base.

7. The opposite of squaring a number is taking the square root of a number.

8. Using order of operations, $9 \div 3 \cdot 3 = \underline{9}$.

9. addition

10. The signs of the numbers determine the sign of the product or quotient. Same signs mean a positive product or quotient and different signs mean a negative product or quotient.

11. The parentheses, or lack of them, determine the base of the expression. In Example 7, -7^2, the base is 7 and only 7 is squared. In Example 8, $(-7)^2$, the base is -7 and all of -7 is squared.

12. the positive or principal square root

13. It allows each expression to evaluate to a single number.

14. The 2 is part of a multiplication which must happen before addition in the order of operations.

Exercise Set 1.3

1. $-3+8=5$

3. $-14+(-10)=-24$

5. $-4.3-6.7=-11$

7. $13-17=-4$

9. $\frac{11}{15}-\left(-\frac{3}{5}\right)=\frac{11}{15}+\frac{9}{15}=\frac{20}{15}=\frac{4}{3}$

11. $19-10-11=9-11=-2$

13. $-\frac{4}{5}-\left(-\frac{3}{10}\right)=-\frac{4}{5}+\frac{3}{10}=-\frac{8}{10}+\frac{3}{10}=-\frac{5}{10}=-\frac{1}{2}$

15. $8-14=-6$

17. $-5 \cdot 12=-60$

19. $-7 \cdot 0=0$

21. $\frac{0}{-2}=0$

23. $\frac{-9}{3}=-3$

25. $\frac{-12}{-4}=3$

27. $3\left(-\frac{1}{18}\right)=-\frac{3}{18}=-\frac{1}{6}$

29. $(-0.7)(-0.8)=0.56$

31. $9.1 \div -1.3=\frac{9.1}{1} \cdot \frac{1}{-1.3}=-7$

33. Multiplying from left to right gives $(-4)(-2)(-1)=8(-1)=-8$.

35. $-7^2=-(7 \cdot 7)=-49$

37. $(-6)^2=(-6)(-6)=36$

39. $(-2)^3=(-2)(-2)(-2)=4(-2)=-8$

41. $\left(-\frac{1}{3}\right)^3 = \left(-\frac{1}{3}\right)\left(-\frac{1}{3}\right)\left(-\frac{1}{3}\right) = -\frac{1}{27}$

43. $\sqrt{49} = 7$ since 7 is positive and $7^2 = 49$.

45. $-\sqrt{\frac{4}{9}} = -\frac{2}{3}$ since $\left(\frac{2}{3}\right)^2 = \frac{4}{9}$.

47. $\sqrt[3]{64} = 4$ since $4^3 = 64$.

49. $\sqrt[4]{81} = 3$ since $3^4 = 81$.

51. $\sqrt{-100}$ is not a real number.

53. $3(5-7)^4 = 3(-2)^4 = 3(16) = 48$

55. $-3^2 + 2^3 = -9 + 8 = -1$

57. $\frac{3.1-(-1.4)}{-0.5} = \frac{3.1+1.4}{-0.5} = \frac{4.5}{-0.5} = -9$

59. $(-3)^2 + 2^3 = 9 + 8 = 17$

61. $-8 \div 4 \cdot 2 = -2 \cdot 2 = -4$

63. $-8\left(-\frac{3}{4}\right) - 8 = 6 - 8 = -2$

65.
$$\begin{aligned} 2-[(7-6)+(9-19)] &= 2-[1+(-10)] \\ &= 2-(-9) \\ &= 11 \end{aligned}$$

67. $\frac{(-9+6)(-1^2)}{-2-2} = \frac{(-3)(-1)}{-4} = \frac{3}{-4} = -\frac{3}{4}$

69.
$$\begin{aligned} \left(\sqrt[3]{8}\right)(-4) - \left(\sqrt{9}\right)(-5) &= (2)(-4)-(3)(-5) \\ &= -8-(-15) \\ &= -8+15 \\ &= 7 \end{aligned}$$

71.
$$\begin{aligned} 25-[(3-5)+(14-18)]^2 &= 25-[(-2)+(-4)]^2 \\ &= 25-(-6)^2 \\ &= 25-36 \\ &= -11 \end{aligned}$$

73.
$$\begin{aligned} \frac{\left(3-\sqrt{9}\right)-(-5-1.3)}{-3} &= \frac{(3-3)-(-6.3)}{-3} \\ &= \frac{0+6.3}{-3} \\ &= \frac{6.3}{-3} \\ &= -2.1 \end{aligned}$$

75. $\frac{|3-9|-|-5|}{-3} = \frac{6-5}{-3} = \frac{1}{-3} = -\frac{1}{3}$

77.
$$\begin{aligned} \frac{3(-2+1)}{5} - \frac{-7(2-4)}{1-(-2)} &= \frac{3(-1)}{5} - \frac{-7(-2)}{1+2} \\ &= \frac{-3}{5} - \frac{14}{3} \\ &= -\frac{9}{15} - \frac{70}{15} \\ &= -\frac{79}{15} \end{aligned}$$

79. $\frac{\frac{1}{3}\cdot 9 - 7}{3+\frac{1}{2}\cdot 4} = \frac{3-7}{3+2} = \frac{-4}{5} = -\frac{4}{5}$

81.
$$\begin{aligned} 3\{-2+5[1-2(-2+5)]\} &= 3\{-2+5[1-2(3)]\} \\ &= 3\{-2+5[1-6]\} \\ &= 3\{-2+5[-5]\} \\ &= 3\{-2+[-25]\} \\ &= 3\{-27\} \\ &= -81 \end{aligned}$$

83.
$$\begin{aligned} \frac{-4\sqrt{80+1}+(-4)^2}{3^3+|-2(3)|} &= \frac{-4\sqrt{81}+(-4)^2}{3^3+|-6|} \\ &= \frac{-4(9)+16}{27+6} \\ &= \frac{-36+16}{27+6} \\ &= \frac{-20}{33} \\ &= -\frac{20}{33} \end{aligned}$$

85. Let $x = 9$, $y = -2$.
$9x - 6y = 9(9) - 6(-2) = 81 + 12 = 93$

87. Let $y = -2$.
$-3y^2 = -3(-2)^2 = -3(4) = -12$

89. Let $x = 9$, $y = -2$.

$$\begin{aligned}\frac{\sqrt{x}}{y}-\frac{y}{x}&=\frac{\sqrt{9}}{-2}-\frac{-2}{9}\\&=\frac{3}{-2}+\frac{2}{9}\\&=-\frac{27}{18}+\frac{4}{18}\\&=-\frac{23}{18}\end{aligned}$$

91. Let $x = 9$, $y = -2$.

$$\begin{aligned}\frac{3+2|x-y|}{x+2y}&=\frac{3+2|9-(-2)|}{9+2(-2)}\\&=\frac{3+2|9+2|}{9+2(-2)}\\&=\frac{3+2|11|}{9+(-4)}\\&=\frac{3+22}{5}\\&=\frac{25}{5}\\&=5\end{aligned}$$

93. Let $x = 9$, $y = -2$.

$$\begin{aligned}\frac{y^3+\sqrt{x-5}}{|4x-y|}&=\frac{(-2)^3+\sqrt{9-5}}{|4\cdot 9-(-2)|}\\&=\frac{-8+\sqrt{4}}{|36+2|}\\&=\frac{-8+2}{|38|}\\&=\frac{-6}{38}\\&=-\frac{3}{19}\end{aligned}$$

95. a. $y = 5$: $8 + 2y = 8 + 2(5) = 8 + 10 = 18$
$y = 7$: $8 + 2y = 8 + 2(7) = 8 + 14 = 22$
$y = 10$: $8 + 2y = 8 + 2(10) = 8 + 20 = 28$
$y = 100$: $8 + 2y = 8 + 2(100) = 8 + 200 = 208$
The completed table is:

Length	y	5	7	10	100
Perimeter	$8 + 2y$	18	22	28	208

b. The perimeter increases as length increases; answers may vary.

97. a. $x = 10$: $\frac{100x+5000}{x} = \frac{100(10)+5000}{10}$
$= \frac{1000+5000}{10}$
$= \frac{6000}{10}$
$= 600$

$x = 100$: $\frac{100x+5000}{x} = \frac{100(100)+5000}{100}$
$= \frac{10,000+5000}{100}$
$= \frac{15,000}{100}$
$= 150$

$x = 1000$: $\frac{100x+5000}{x} = \frac{100(1000)+5000}{1000}$
$= \frac{100,000+5000}{1000}$
$= \frac{105,000}{1000}$
$= 105$

The completed table is:

Number of Bookshelves	x	10	100	1000
Cost per Bookshelf	$\frac{100x+5000}{x}$	600	150	105

b. The cost per bookshelf decreases as the number of bookshelves increases; answers may vary.

99. $-\frac{1}{7} = \frac{-1}{7} = \frac{1}{-7}$; b, c

101. $\frac{5}{-(x+y)} = \frac{-5}{(x+y)} = -\frac{5}{(x+y)}$; b, d

103. $\frac{-9x}{-2y} = \frac{9x}{2y}$; b

105. Let $x_1 = 2$, $x_2 = 4$, $y_1 = -3$, $y_2 = 2$.
$\frac{y_2 - y_1}{x_2 - x_1} = \frac{2-(-3)}{4-2} = \frac{2+3}{4-2} = \frac{5}{2}$

107. $1 - \frac{1}{5} - \frac{3}{7} = \frac{35}{35} - \frac{7}{35} - \frac{15}{35} = \frac{13}{35}$

109. 10,203 – 5998 = 4205 meters

111. $(2 + 7) \cdot (1 + 3) = 9 \cdot 4 = 36$

113. From the graph the population that was over 65 in 1970 was about 20 million.

115. From the graph, the predicted population over 65 in 2030 will be 70 million.

117. The population over 65 is increasing; answers may vary.

119. answers may vary

121. $\sqrt{10} \approx 3.1623$

123. $\sqrt{7.9} \approx 2.8107$

125. $\frac{-1.682 - 17.895}{(-7.102)(-4.691)} \approx -0.5876$

Integrated Review

1. Let $z = -4$.
$z^2 = (-4)^2 = (-4)(-4) = 16$

2. Let $z = -4$.
$-z^2 = -(-4)^2 = -(-4)(-4) = -16$

3. Let $x = -1$, $y = 3$, $z = -4$.
$\frac{4x - z}{2y} = \frac{4(-1)-(-4)}{2(3)} = \frac{-4+4}{6} = \frac{0}{6} = 0$

4. Let $x = -1$, $y = 3$, $z = -4$.
$x(y - 2z) = -1[3 - 2(-4)]$
$= -1[3+8]$
$= -1[11]$
$= -11$

5. $-7 - (-2) = -7 + 2 = -5$

6. $\frac{9}{10} - \frac{11}{12} = \frac{9}{10} \cdot \frac{6}{6} - \frac{11}{12} \cdot \frac{5}{5} = \frac{54}{60} - \frac{55}{60} = -\frac{1}{60}$

7. $\frac{-13}{2-2} = \frac{-13}{0}$ is undefined.

8. $(1.2)^2 - (2.1)^2 = 1.44 - 4.41 = -2.97$

9. $\sqrt{64} - \sqrt[3]{64} = 8 - 4 = 4$

10. $-5^2 - (-5)^2 = -25 - 25 = -50$

11. $9+2[(8-10)^2+(-3)^2]=9+2[(-2)^2+(-3)^2]$
$=9+2(4+9)$
$=9+2(13)$
$=9+26$
$=35$

12. $8-6\left[\sqrt[3]{8}(-2)+\sqrt{4}(-5)\right]=8-6[2(-2)+2(-5)]$
$=8-6[(-4)+(-10)]$
$=8-6(-14)$
$=8+84$
$=92$

13. $-15-2x$

14. $3x+5$

15. 0 is the whole number that is not a natural number.

16. true

Section 1.4 Practice Exercises

1. $\underbrace{\text{The product of } -4 \text{ and } x}_{-4x}\ \underset{=}{\text{is}}\ \underset{20}{20.}$

$-4x = 20$

2. $\underbrace{\text{Three times}}_{3}\ \underbrace{\text{the difference of } z \text{ and } 3}_{(z-3)}\ \underset{=}{\text{equals}}\ \underset{9}{9.}$

$3(z-3)=9$

3. $\underbrace{\text{The sum of } x \text{ and } 5}_{x+5}\ \underbrace{\text{is the same as}}_{=}\ \underbrace{3 \text{ less than twice } x.}_{2x-3}$

$x+5=2x-3$

4. $\underbrace{\text{The sum of } y \text{ and } 2}_{y+2}\ \underset{=}{\text{is}}\ \underbrace{4 \text{ more than the quotient of } z \text{ and } 8.}_{4+\frac{z}{8}}$

$y+2=4+\dfrac{z}{8}$

5. a. $-6<-5$ since -6 lies to the left of -5 on the number line.

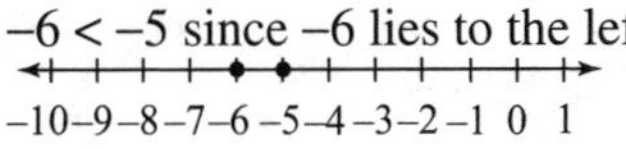

b. $\dfrac{24}{3}=8$

c. $0>-7$ since 0 lies to the right of -7 on the number line.

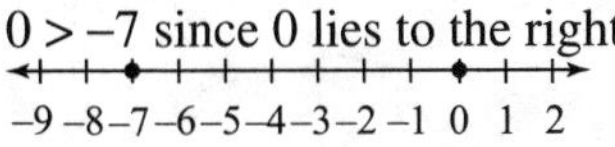

d. $-2.76<-2.67$ since -2.76 lies to the left of -2.67 on the number line.

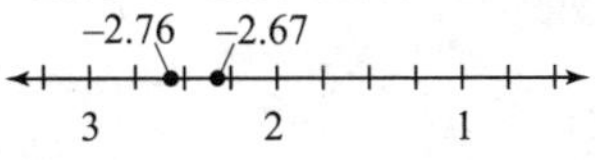

e. $\frac{9}{10} > \frac{7}{10}$

The denominators are the same, so $\frac{9}{10} > \frac{7}{10}$ since $9 > 7$.

f. $\frac{2}{3} < \frac{7}{9}$

By dividing, we see that $\frac{2}{3} = 0.666...$ and $\frac{7}{9} = 0.777....$ Thus, $\frac{2}{3} < \frac{7}{9}$ since $0.666... < 0.777....$

6. a. $x - 3 \le 5$

b. $y \ne -4$

c. $2 < 4 + \frac{1}{2}z$

7. a. The opposite of −7 is $-(-7) = 7$.

b. The opposite of 4.7 is −4.7.

c. The opposite of $-\frac{3}{8}$ is $-\left(-\frac{3}{8}\right) = \frac{3}{8}$.

8. a. The reciprocal of $-\frac{5}{3}$ is $-\frac{3}{5}$ because $-\frac{5}{3}\left(-\frac{3}{5}\right) = 1$.

b. The reciprocal of 14 is $\frac{1}{14}$.

c. The reciprocal of −2 is $-\frac{1}{2}$.

9. $8 + 13x = 13x + 8$

10. $3 \cdot (11b) = (3 \cdot 11)b = 33b$

11. a. $4(x + 5y) = 4 \cdot x + 4 \cdot 5y = 4x + 20y$

b.
$$\begin{aligned} -(3-2z) &= -1(3-2z) \\ &= -1 \cdot 3 + (-1)(-2z) \\ &= -3 + 2z \end{aligned}$$

c.
$$\begin{aligned} 0.3x(y-3) &= 0.3x \cdot y - 0.3x \cdot 3 \\ &= 0.3xy - 0.9x \end{aligned}$$

12. a.

In words:	Value of a dime	·	number of dimes	
	↓		↓	
Translate:	0.10		x	, or $0.10x$

b. In words: number of grams of carbohydrates in one cookie · number of cookies

$\downarrow$ $\downarrow$

Translate: 26 y , or $26y$

c. In words: cost of one birthday card · number of cards

$\downarrow$ $\downarrow$

Translate: 1.75 z , or $1.75z$

d. In words: Discount · purchase price

$\downarrow$ $\downarrow$

Translate: 0.15 t , or $0.15t$

13. a. If two numbers have a sum of 16 and one number is x, the other number is the rest of 16.

In words: Sixteen minus x

$\downarrow$ $\downarrow$ $\downarrow$

Translate: 16 $-$ x

b. In words: One hundred eighty minus one angle, x

$\downarrow$ $\downarrow$ $\downarrow$

Translate: 180 $-$ x

c. The next consecutive even integer is always two more than the previous even integer.

In words: first integer plus two

$\downarrow$ $\downarrow$ $\downarrow$

Translate: x $+$ 2

d. In words: younger brother's age plus nine

$\downarrow$ $\downarrow$ $\downarrow$

Translate: x $+$ 9

14. a. $6ab - ab = 6ab - 1ab = (6-1)ab = 5ab$

b. $$\begin{aligned} 4x-5+6x &= 4x+6x-5 \\ &= (4+6)x-5 \\ &= 10x-5 \end{aligned}$$

c. $17p - 9$ cannot be simplified further since $17p$ and -9 are not like terms.

15. a.
$$\begin{aligned} 5pq-2pq-11-4pq+18 &= 5pq-2pq-4pq-11+18 \\ &= (5-2-4)pq+(-11+18) \\ &= -1pq+(7) \\ &= -pq+7 \end{aligned}$$

b.
$$\begin{aligned} 3x^2+7-2(x^2-6) &= 3x^2+7-2x^2+12 \\ &= 3x^2-2x^2+7+12 \\ &= x^2+19 \end{aligned}$$

c.
$$\begin{aligned} (3.7x+2.5)-(-2.1x-1.3) &= 3.7x+2.5+2.1x+1.3 \\ &= 3.7x+2.1x+2.5+1.3 \\ &= 5.8x+3.8 \end{aligned}$$

d.
$$\begin{aligned} \frac{1}{5}(15c-25d)-\frac{1}{2}(8c+6d+1)+\frac{3}{4} &= 3c-5d-4c-3d-\frac{1}{2}+\frac{3}{4} \\ &= -c-8d+\frac{1}{4} \end{aligned}$$

Vocabulary, Readiness & Video Check 1.4

	Symbol	Meaning
1.	$<$	is less than
2.	$>$	is greater than
3.	$\neq$	is not equal to
4.	$=$	is equal to
5.	$\geq$	is greater than or equal to
6.	$\leq$	is less than or equal to

7. The opposite of nonzero number a is $\underline{-a}$.

8. The reciprocal of nonzero number a is $\underline{\frac{1}{a}}$.

9. The <u>commutative</u> property has to do with "order."

10. The <u>associative</u> property has to do with "grouping."

11. $a(b + c) = ab + ac$ illustrates the <u>distributive</u> property.

12. The <u>terms</u> of an expression are the addends of the expression.

13. $=, \neq, <, \leq, >, \geq$

14. *Reciprocal* is the same as <u>multiplicative</u> inverse and *opposite* is the same as <u>additive</u> inverse.

15. order; grouping

16. Understand the difference between the *number* of coins you have and the value the coins represent.

17. by combining like terms; distributive property

Exercise Set 1.4

1. The sum of 10 and x is -12.

$10 + x = -12$

or $10 + x = -12$

3. Twice x, plus 5, is the same as -14.

$2x + 5 = -14$

or $2x + 5 = -14$

5. The quotient of n and 5 is 4 times n.

$\frac{n}{5} = 4n$

or $\frac{n}{5} = 4n$

7. The difference of z and one-half is the same as the product of z and one-half.

$z - \frac{1}{2} = \frac{1}{2}z$

or $z - \frac{1}{2} = \frac{1}{2}z$

9. The product of 7 and x is less than or equal to -21.

$7x \le -21$

or $7x \le -21$

11. Twice the difference of x and 6 is greater than the reciprocal of 11.

$2(x - 6) > \frac{1}{11}$

or $2(x - 6) > \frac{1}{11}$

13. Twice the difference of x and 6 is -27.

$2(x - 6) = -27$

or $2(x - 6) = -27$

15. $-16 > -17$ since -16 is to the right of -17 on a number line.

17. $7.4 = 7.40$

19. $\frac{7}{11} < \frac{9}{11}$ since $7 < 9$.

21. $\frac{1}{2} < \frac{5}{8}$ since $\frac{1}{2}$ is to the left of $\frac{5}{8}$ on a number line.

23. $-7.9 < -7.09$ since -7.9 is to the left of -7.09 on a number line.

	Number	Opposite	Reciprocal
25.	5	-5	$\frac{1}{5}$
27.	-8	8	$-\frac{1}{8}$
29.	$-\frac{1}{7}$	$\frac{1}{7}$	-7
31.	0	0	Undefined
33.	$\frac{7}{8}$	$-\frac{7}{8}$	$\frac{8}{7}$

35. $7x + y = y + 7x$

37. $z \cdot w = w \cdot z$

39. $\frac{1}{3} \cdot \frac{x}{5} = \frac{x}{5} \cdot \frac{1}{3}$

41. $5 \cdot (7x) = (5 \cdot 7)x$

43. $(x + 1.2) + y = x + (1.2 + y)$

45. $(14z) \cdot y = 14(z \cdot y)$

47. $3(x + 5) = 3 \cdot x + 3 \cdot 5 = 3x + 15$

49.
$$\begin{aligned} -(2a+b) &= -1(2a+b) \\ &= -1 \cdot 2a + (-1) \cdot b \\ &= -2a - b \end{aligned}$$

51.
$$\begin{aligned} 2(6x+5y+2z) &= 2 \cdot 6x + 2 \cdot 5y + 2 \cdot 2z \\ &= 12x + 10y + 4z \end{aligned}$$

53.
$$\begin{aligned} -4(x-2y+7) &= -4 \cdot x + (-4)(-2y) + (-4)(7) \\ &= -4x + 8y - 28 \end{aligned}$$

55. $0.5x(6y-3) = 0.5x \cdot 6y - 0.5x \cdot 3 = 3xy - 1.5x$

57. $3x + 6 = 6 + 3x$

59. $\frac{2}{3} + \left(-\frac{2}{3}\right) = 0$

61. $7 \cdot 1 = 7$

63. $10(2y) = (10 \cdot 2)y$

65. In words: Value of a dime · Number of dimes
Translate: $0.1 \cdot d$ or $0.1d$

67. If two numbers have a sum of 112 and one number is x, then the other number is the "rest of 112." So, in other words, we have
One hundred twelve − x
Translate: $112 - x$

69. In words: One hundred eighty − x
Translate: $180 - x$

71. In words:
Cost of compact disc · Number of discs
Translate: $\$6.49 \cdot x$ or $\$6.49x$

73. The next odd integer would be 2 more than the given odd integer.
In words: Odd integer + Two
Translate: $x + 2$

75.
$$\begin{aligned} -9+4x+18-10x &= 4x-10x+18-9 \\ &= (4-10)x+(18-9) \\ &= -6x+9 \end{aligned}$$

77.
$$\begin{aligned} 5k-(3k-10) &= 5k-3k+10 \\ &= (5-3)k+10 \\ &= 2k+10 \end{aligned}$$

79.
$$\begin{aligned} (3x+4)-(6x-1) &= 3x+4-6x+1 \\ &= 3x-6x+4+1 \\ &= (3-6)x+5 \\ &= -3x+5 \end{aligned}$$

81.
$$\begin{aligned} 3(xy-2)+xy+15-x^2 &= 3xy-6+xy+15-x^2 \\ &= 4xy+9-x^2 \\ &= -x^2+4xy+9 \end{aligned}$$

83.
$$\begin{aligned} -(n+5)+(5n-3) &= -1(n+5)+(5n-3) \\ &= -n-5+5n-3 \\ &= 4n-8 \end{aligned}$$

85. $4(6n^2-3)-3(8n^2+4)=24n^2-12-24n^2-12$
$=24n^2-24n^2-12-12$
$=(24-24)n^2-24$
$=0n^2-24$
$=-24$

87. $3x-2(x-5)+x=3x-2x+10+x$
$=3x-2x+x+10$
$=(3-2+1)x+10$
$=2x+10$

89. $1.5x+2.3-0.7x-5.9=(1.5-0.7)x+(2.3-5.9)$
$=0.8x-3.6$

91. $\frac{3}{4}b-\frac{1}{2}+\frac{1}{6}b-\frac{2}{3}=\left(\frac{3}{4}+\frac{1}{6}\right)b+\left(-\frac{1}{2}-\frac{2}{3}\right)$
$=\frac{11}{12}b-\frac{7}{6}$

93. $2(3x+7)=6x+14$

95. $\frac{1}{2}(10x-2)-\frac{1}{6}(60x-5y)=5x-1-10x+\frac{5}{6}y$
$=5x-10x+\frac{5}{6}y-1$
$=-5x+\frac{5}{6}y-1$

97. $\frac{1}{6}(24a-18b)-\frac{1}{7}(7a-21b-2)-\frac{1}{5}$
$=4a-3b-a+3b+\frac{2}{7}-\frac{1}{5}$
$=3a+\frac{3}{35}$

99. $3(x+4)=3\cdot x+3\cdot 4=3x+12$

101. $5(7y)=(5\cdot 7)y$

103. $a(b+c)=ab+ac$

105. zero; answers may vary

107. no; answers may vary

109. answers may vary

111. $8.1z+7.3(z+5.2)-6.85$
$=8.1z+7.3z+37.96-6.85$
$=15.4z+31.11$

Chapter 1 Vocabulary Check

1. An algebraic expression is formed by numbers and variables connected by the operations of addition, subtraction, multiplication, division, raising to powers, and/or taking roots.

2. The opposite of a number a is $-a$.

3. $3(x-6)=3x-18$ by the distributive property.

4. The absolute value of a number is the distance between the number and 0 on a number line.

5. An exponent is a shorthand notation for repeated multiplication of the same factor.

6. A letter that represents a number is called a variable.

7. The symbols < and > are called inequality symbols.

8. If a is not 0, then a and $\frac{1}{a}$ are called reciprocals.

9. A + B = B + A by the commutative property.

10. (A + B) + C = A + (B + C) by the associative property.

11. The numbers 0, 1, 2, 3, ... are called whole numbers.

12. If a number corresponds to a point on a number line, we know that number is a real number.

Chapter 1 Review

1. $7x=7(3)=21$

2. $st=(1.6)(5)=8$

3. One minute = 60 seconds.
$70t=70(60)=4200$
4200 wingbeats per minute.

4. One hour is 60(60) = 3600 seconds.
$70t=70(3600)=252{,}000$
252,000 wingbeats per hour.

5. $\{x|x$ is an odd integer between -2 and $4\}$
$=\{-1, 1, 3\}$

6. $\{x|x$ is an even integer between -3 and $7\}$
$=\{-2, 0, 2, 4, 6\}$

7. There are no whole numbers that are negative.
$\varnothing$

8. All natural numbers are rational numbers.
$\varnothing$

9. $\{x|x$ is a whole number greater than $5\}$
$= \{6, 7, 8, ...\}$

10. $\{x|x$ is an integer less than $3\} = \{..., -1, 0, 1, 2\}$

11. Since $D = \{2, 4, 6, 8, 10, ..., 16\}$, $10 \in D$ is true.

12. Since $B = \{5, 9, 11\}$, $59 \in B$ is false.

13. $\sqrt{169} = 13$, which is a rational number. So $\sqrt{169} \notin G$ is true.

14. Since $F = \{\ \}$ and 0 is not an element of the empty set, then $0 \notin F$ is true.

15. Since $E = \{x|x$ is a rational number$\}$ and π is irrational, then $\pi \in E$ is false.

16. Since $H = \{x|x$ is a real number$\}$ and π is a real number, then $\pi \in H$ is true.

17. Since $\sqrt{4} = 2$ and $G = \{x|x$ is an irrational number$\}$, and 2 is a rational number, then $\sqrt{4} \in G$ is false.

18. Since $E = \{x|x$ is a rational number$\}$ and -9 is a rational number, then $-9 \in E$ is true.

19. Since $A = \{6, 10, 12\}$ and $D = \{2, 4, 6, 8, 10, 12, 14, 16\}$, then $A \subseteq D$ is true.

20. Since $C = \{..., -3, -2, -1, 0, 1, 2, 3, ...\}$ and $B = \{5, 9, 11\}$, then $C \nsubseteq B$ is true.

21. Since $C = \{..., -3, -2, -1, 0, 1, 2, 3, ...\}$ and $E = \{x|x$ is a rational number$\}$, and all integers are rational numbers, then $C \nsubseteq E$ is false.

22. Since $F = \{\ \}$ and $H = \{x|x$ is a real number$\}$, and the empty set is a subset of all sets, then $F \subseteq H$ is true.

23. Whole numbers: $\left\{5, \frac{8}{2}, \sqrt{9}\right\}$

24. Natural numbers: $\left\{5, \frac{8}{2}, \sqrt{9}\right\}$

25. Rational numbers: $\left\{5, -\frac{2}{3}, \frac{8}{2}, \sqrt{9}, 0.3, 1\frac{5}{8}, -1\right\}$

26. Irrational numbers: $\left\{\sqrt{7}, \pi\right\}$

27. Real numbers:
$\left\{5, -\frac{2}{3}, \frac{8}{2}, \sqrt{9}, 0.3, \sqrt{7}, 1\frac{5}{8}, -1, \pi\right\}$

28. Integers: $\left\{5, \frac{8}{2}, \sqrt{9}, -1\right\}$

29. The opposite of $-\frac{3}{4}$ is $-\left(-\frac{3}{4}\right) = \frac{3}{4}$.

30. The opposite of 0.6 is -0.6.

31. The opposite of 0 is $-0 = 0$.

32. The opposite of 1 is -1.

33. The reciprocal of $-\frac{3}{4}$ is $\frac{1}{\left(-\frac{3}{4}\right)} = -\frac{4}{3}$.

34. The reciprocal of 0.6 is $\frac{1}{0.6}$.

35. The reciprocal of 0 is $\frac{1}{0}$ which is undefined.

36. The reciprocal of 1 is $\frac{1}{1} = 1$.

37. $-7 + 3 = -4$

38. $-10 + (-25) = -35$

39. $5(-0.4) = -2$

40. $(-3.1)(-0.1) = 0.31$

41. $-7 - (-15) = -7 + 15 = 8$

42. $9 - (-4.3) = 9 + 4.3 = 13.3$

43. $(-6)(-4)(0)(-3) = 0$

44. $(-12)(0)(-1)(-5) = 0$

45. $-24 \div 0$ is undefined.

46. $0 \div (-45) = 0$

47. $-36 \div (-9) = 4$

48. $60 \div (-12) = -5$

49. $-\frac{4}{5}-\left(-\frac{2}{3}\right) = -\frac{4}{5}+\frac{2}{3} = -\frac{12}{15}+\frac{10}{15} = -\frac{2}{15}$

50. $\frac{5}{4}-\left(-2\frac{3}{4}\right) = \frac{5}{4}+\frac{11}{4} = \frac{16}{4} = 4$

51. $1-\frac{1}{4}-\frac{1}{3} = \frac{12}{12}-\frac{3}{12}-\frac{4}{12} = \frac{5}{12}$

52. $31{,}441 - 1589 = 29{,}852$
The elevation relative to sea level is 29,852 feet below sea level.

53. $-5+7-3-(-10) = 2-3+10 = -1+10 = 9$

54.
$$\begin{aligned} 8-(-3)+(-4)+6 &= 8+3-4+6 \\ &= 11-4+6 \\ &= 7+6 \\ &= 13 \end{aligned}$$

55. $3(4-5)^4 = 3(-1)^4 = 3(1) = 3$

56. $6(7-10)^2 = 6(-3)^2 = 6(9) = 54$

57. $\left(-\frac{8}{15}\right)\cdot\left(-\frac{2}{3}\right)^2 = -\frac{8}{15}\cdot\frac{4}{9} = -\frac{32}{135}$

58. $\left(-\frac{3}{4}\right)^2\cdot\left(-\frac{10}{21}\right) = \left(\frac{9}{16}\right)\left(-\frac{10}{21}\right) = -\frac{15}{56}$

59. $-\frac{6}{15}\div\frac{8}{25} = -\frac{6}{15}\cdot\frac{25}{8} = -\frac{150}{120} = -\frac{5}{4}$

60. $\frac{4}{9}\div\left(-\frac{8}{45}\right) = \frac{4}{9}\cdot\left(-\frac{45}{8}\right) = -\frac{180}{72} = -\frac{5}{2}$

61. $-\frac{3}{8}+3(2)\div 6 = -\frac{3}{8}+6\div 6 = -\frac{3}{8}+1 = -\frac{3}{8}+\frac{8}{8} = \frac{5}{8}$

62.
$$\begin{aligned} 5(-2)-(-3)-\frac{1}{6}+\frac{2}{3} &= -10+3-\frac{1}{6}+\frac{2}{3} \\ &= -7-\frac{1}{6}+\frac{2}{3} \\ &= -\frac{42}{6}-\frac{1}{6}+\frac{4}{6} \\ &= -\frac{39}{6} \\ &= -6\frac{1}{2} \end{aligned}$$

63.
$$\begin{aligned} \left|2^3-3^2\right|-|5-7| &= |8-9|-|-2| \\ &= |-1|-2 \\ &= 1-2 \\ &= -1 \end{aligned}$$

64.
$$\begin{aligned} \left|5^2-2^2\right|+|9\div(-3)| &= |25-4|+|-3| \\ &= |21|+3 \\ &= 21+3 \\ &= 24 \end{aligned}$$

65. $(2^3-3^2)-(5-7) = (8-9)-(-2) = -1+2 = 1$

66.
$$\begin{aligned} (5^2-2^4)+[9\div(-3)] &= (25-16)+(-3) \\ &= 9+(-3) \\ &= 6 \end{aligned}$$

67.
$$\begin{aligned} \frac{(8-10)^3-(-4)^2}{2+8(2)\div 4} &= \frac{(-2)^3-16}{2+16\div 4} \\ &= \frac{-8-16}{2+4} \\ &= \frac{-24}{6} \\ &= -4 \end{aligned}$$

68.
$$\begin{aligned} \frac{(2+4)^2+(-1)^5}{12\div 2\cdot 3-3} &= \frac{(6)^2+(-1)}{6\cdot 3-3} \\ &= \frac{36-1}{18-3} \\ &= \frac{35}{15} \\ &= \frac{7}{3} \end{aligned}$$

69. $\dfrac{(4-9)+4-9}{10-12\div 4\cdot 8}=\dfrac{(-5)+4-9}{10-3\cdot 8}$
$=\dfrac{-1-9}{10-24}$
$=\dfrac{-10}{-14}$
$=\dfrac{5}{7}$

70. $\dfrac{3-7-(7-3)}{15+30\div 6\cdot 2}=\dfrac{-4-(4)}{15+5\cdot 2}=\dfrac{-8}{15+10}=\dfrac{-8}{25}=-\dfrac{8}{25}$

71. $\dfrac{\sqrt{25}}{4+3\cdot 7}=\dfrac{5}{4+21}=\dfrac{5}{25}=\dfrac{1}{5}$

72. $\dfrac{\sqrt{64}}{24-8\cdot 2}=\dfrac{8}{24-16}=\dfrac{8}{8}=1$

73. Let $x = 0$, $y = 3$, $z = -2$.
$x^2-y^2+z^2=(0)^2-(3)^2+(-2)^2$
$=0-9+4$
$=-5$

74. Let $x = 0$, $y = 3$, $z = -2$.
$\dfrac{5x+z}{2y}=\dfrac{5(0)+(-2)}{2(3)}=\dfrac{0-2}{6}=\dfrac{-2}{6}=-\dfrac{1}{3}$

75. Let $y = 3$, $z = -2$.
$\dfrac{-7y-3z}{-3}=\dfrac{-7(3)-3(-2)}{-3}=\dfrac{-21+6}{-3}=\dfrac{-15}{-3}=5$

76. Let $x = 0$, $y = 3$, $z = -2$.
$(x-y+z)^2=(0-3+(-2))^2$
$=(-3-2)^2$
$=(-5)^2$
$=25$

77. When $r = 1$, $2\pi r = 2\pi(1) = 2(3.14) = 6.28$.
When $r = 10$,
$2\pi r = 2\pi(10) = 20(3.14) = 62.8$.
When $r = 100$,
$2\pi r = 2\pi(100) = 200(3.14) = 628$.

r	1	10	100
$2\pi r$	6.28	62.8	628

78. As the radius increases, the circumference increases.

79. $5xy-7xy+3-2+xy=5xy-7xy+xy+3-2$
$=(5-7+1)xy+(3-2)$
$=(-1)xy+1$
$=-xy+1$

80. $4x+10x-19x+10-19$
$=(4+10-19)x+(10-19)$
$=-5x+(-9)$
$=-5x-9$

81. $6x^2+2-4(x^2+1)=6x^2+2-4x^2-4$
$=6x^2-4x^2+2-4$
$=(6-4)x^2+(2-4)$
$=2x^2+(-2)$
$=2x^2-2$

82. $-7(2x^2-1)-x^2-1=-14x^2+7-x^2-1$
$=-14x^2-x^2+7-1$
$=(-14-1)x^2+(7-1)$
$=-15x^2+6$

83. $(3.2x-1.5)-(4.3x-1.2)$
$=3.2x-1.5-4.3x+1.2$
$=3.2x-4.3x-1.5+1.2$
$=(3.2-4.3)x-0.3$
$=-1.1x-0.3$

84. $(7.6x+4.7)-(1.9x+3.6)$
$=7.6x+4.7-1.9x-3.6$
$=7.6x-1.9x+4.7-3.6$
$=(7.6-1.9)x+4.7-3.6$
$=5.7x+1.1$

85. $\underbrace{\text{Twelve}}\ \underset{\downarrow}{\text{is}}\ \underbrace{\text{the product of } x \text{ and negative } 4}.$
$12 \quad = \quad -4x$
or $12 = -4x$

86. $\underbrace{\text{The sum of } n \text{ and twice } n}\ \underset{\downarrow}{\text{is}}\ \underbrace{\text{negative fifteen}}.$
$n+2n \quad = \quad -15$

87. $\underbrace{\text{Four}}\ \underbrace{\text{times}}\ \underbrace{\text{the sum of } y \text{ and three}}\ \underset{\downarrow}{\text{is}}\ \underset{\downarrow}{-1}.$
$4 \quad \cdot \quad (y+3) \quad = -1$
or $4(y + 3) = -1$

88. $\underbrace{\text{The difference of } t \text{ and } 5}_{(t-5)}, \underbrace{\text{multiplied by}}_{\cdot} \underset{6}{\text{six}} \underset{=}{\text{is}} \underset{4}{\text{four}}.$

or $6(t - 5) = 4$

89. $\underbrace{\text{Seven subtracted from } z}_{z-7} \underset{=}{\text{is}} \underset{6}{\text{six}}.$

or $z - 7 = 6$

90. $\underbrace{\text{Ten less than the product of } x \text{ and nine}}_{9x-10} \underset{=}{\text{is}} \underset{5}{\text{five}}.$

or $9x - 10 = 5$

91. $\underbrace{\text{The difference of } x \text{ and } 5}_{x-5} \underbrace{\text{is at least}}_{\geq} \underset{12}{12}.$

or $x - 5 \geq 12$

92. $\underbrace{\text{The opposite of four}}_{-4} \underbrace{\text{is less than}}_{<} \underbrace{\text{the product of } y \text{ and seven}}_{7y}.$

or $-4 < 7y$

93. $\underbrace{\text{Two-thirds}}_{\frac{2}{3}} \underbrace{\text{is not equal to}}_{\neq} \underbrace{\text{twice}}_{2\cdot} \underbrace{\text{the sum of } n \text{ and one-fourth}}_{\left(n+\frac{1}{4}\right)}.$

or $\frac{2}{3} \neq 2\left(n+\frac{1}{4}\right)$

94. $\underbrace{\text{The sum of } t \text{ and six}}_{t+6} \underbrace{\text{is not more than}}_{\leq} \underbrace{\text{negative twelve}}_{-12}.$

or $t + 6 \leq -12$

95. $(M + 5) + P = M + (5 + P)$: associative property of addition

96. $5(3x - 4) = 15x - 20$: distributive property

97. $(-4) + 4 = 0$: additive inverse property

98. $(3 + x) + 7 = 7 + (3 + x)$: commutative property of addition

99. associative and commutative properties of multiplication
To see this: $(XY)Z = X(YZ) = (YZ)X$

100. $\left(-\frac{3}{5}\right)\cdot\left(-\frac{5}{3}\right) = 1$: multiplicative inverse property

101. $T \cdot 0 = 0$: multiplication property of zero

102. $(ab)c = a(bc)$: associative property of multiplication

103. $A + 0 = A$: additive identity property

104. $8 \cdot 1 = 8$: multiplicative identity property

105. $5x - 15z = 5(x - 3z)$

106. $(7 + y) + (3 + x) = (3 + x) + (7 + y)$

107. $0 = 2 + (-2)$, for example

108. $1 = 2 \cdot \frac{1}{2}$, for example

109. $[(3.4)(0.7)]5 = (3.4)[(0.7)(5)]$

110. $7 = 7 + 0$

111. $-9 > -12$, since -9 is to the right of -12 on a number line.

112. $0 > -6$, since 0 is to the right of -6 on a number line.

113. $-3 < -1$, since -3 is to the left of -1 on a number line.

114. $7 = |-7|$

115. $-5 < -(-5)$, since $-(-5) = 5$.

116. $-(-2) > -2$, since $-(-2) = 2$.

117. The opposite of $-\frac{3}{4}$ is $\frac{3}{4}$.

The reciprocal of $-\frac{3}{4}$ is $-\frac{4}{3}$.

118. If the opposite of the number is -5, then the number is $-(-5) = 5$. The reciprocal of 5 is $\frac{1}{5}$.

119.
$$\begin{aligned} -2\left(5x+\frac{1}{2}\right)+7.1 &= -2\cdot 5x+(-2)\cdot\frac{1}{2}+7.1 \\ &= -10x-1+7.1 \\ &= -10x+6.1 \end{aligned}$$

120. $\sqrt{36} \div 2 \cdot 3 = 6 \div 2 \cdot 3 = 3 \cdot 3 = 9$

121. $-\frac{7}{11} - \left(-\frac{1}{11}\right) = -\frac{7}{11} + \frac{1}{11} = -\frac{6}{11}$

122.
$$\begin{aligned} 10-(-1)+(-2)+6 &= 10+1+(-2)+6 \\ &= 11+(-2)+6 \\ &= 9+6 \\ &= 15 \end{aligned}$$

123.
$$\begin{aligned} \left(-\frac{2}{3}\right)^3 \div \frac{10}{9} &= -\frac{8}{27} \div \frac{10}{9} \\ &= -\frac{8}{27} \cdot \frac{9}{10} \\ &= -\frac{2\cdot 4\cdot 9}{3\cdot 9\cdot 2\cdot 5} \\ &= -\frac{4}{15} \end{aligned}$$

124.
$$\begin{aligned} \frac{(3-5)^2+(-1)^3}{1+2(3-(-1))^2} &= \frac{(-2)^2+(-1)^3}{1+2(3+1)^2} \\ &= \frac{4+(-1)}{1+2(4)^2} \\ &= \frac{3}{1+2(16)} \\ &= \frac{3}{1+32} \\ &= \frac{3}{33} \\ &= \frac{1}{11} \end{aligned}$$

125.
$$\begin{aligned} &\frac{1}{3}(9x-3y)-(4x-1)+4y \\ &= \frac{1}{3}\cdot 9x-\frac{1}{3}\cdot 3y-4x+1+4y \\ &= 3x-y-4x+1+4y \\ &= 3x-4x-y+4y+1 \\ &= -x+3y+1 \end{aligned}$$

126. $1266.4z = 1266.4(7.5) = 9498$
The cost is estimated at $9498.

	Year	Increase in Life Expectancy (in years) from 10 Years Earlier
127.	1964	73.3 – 71.1 = 2.2
128.	1974	75.9 – 73.3 = 2.6
129.	1984	78.2 – 75.9 = 2.3
130.	1994	79.0 –78.2 = 0.8
131.	2004	80.4 – 79.0 = 1.4
132.	2014	81.2 – 80.4 = 0.8

Chapter 1 Getting Ready for the Test

1. Since the opposite of $\frac{2}{5}$ is $-\frac{2}{5}$, the correct directions are choice A.

2. $|7-10|+(4-6)^3=|-3|+(-2)^3=3+(-8)=-5$
Since the expression evaluates to −5, the correct directions are choice C.

3. Since the reciprocal of $\frac{2}{5}$ is $\frac{5}{2}$, the correct directions are choice B.

4. Since the opposite of 0 is 0, the correct directions are choice A. Also, 0 does not have a reciprocal, nor is it an expression that can be evaluated or simplified.

5. $\sqrt{5}$ is an irrational number; choice C.

6. $\frac{7}{8}$ is a rational number; choice B.

7. $0=\frac{0}{1}$ is a whole number and also a rational number; choices B and D.

8. $-12=\frac{-12}{1}$ is a negative integer and also a rational number; choices A and B.

9. In $7 + (x + y) = (7 + x) + y$, the grouping of the addends is changed. This illustrates the associative property of addition; choice B.

10. $-2(x - 7) = -2x + 14$ illustrates the distributive property; choice C.

11. In $20+y^3=y^3+20$, the order of addends is changed. This illustrates the commutative property of addition; choice A.

12. $z \cdot 1 = z$ illustrates the multiplicative identity property; choice E.

13. In $a(xy) = (ax)y$, the grouping of the factors is changed. This illustrates the associative property of multiplication; choice B.

14. $0+x^2=x^2$ illustrates the additive identity property; choice D.

15. $$\begin{aligned}&-5(2x^2-3)-(2-x^2)+1\\&=-5(2x^2)-(-5)(3)-2+x^2+1\\&=-10x^2+15-2+x^2+1\\&=-9x^2+14\end{aligned}$$
Choice B

16. $\underbrace{\text{Nine}}_{9}\ \underbrace{\text{times}}_{\cdot}\ \underbrace{\text{the difference of } x \text{ and } 7}_{(x-7)}\ \underset{=}{\text{is}}\ \underset{-14}{-14.}$
or $9(x-7) = -14$; choice C.

Chapter 1 Test

1. True; −2.3 lies to the right of −2.33 on a number line.

2. False; $-6^2=-36$, while $(-6)^2=36$.

3. False; $-5 - 8 = -13$, while $-(5 - 8) = -(-3) = 3$.

4. False; $(-2)(-3)(0) = 0$, while $\frac{(-4)}{0}$ is undefined.

5. True

6. False; for example, $\frac{1}{2}$ is a rational number that is not an integer.

7. $5 - 12 \div 3(2) = 5 - 4(2) = 5 - 8 = -3$

8. $5^2-3^4=25-81=-56$

9. $$\begin{aligned}(4-9)^3-|-4-6|^2&=(-5)^3-|-10|^2\\&=-125-10^2\\&=-125-100\\&=-225\end{aligned}$$

10. $$\begin{aligned}12+\{6-[5-2(-5)]\}&=12+\{6-[5+10]\}\\&=12+(6-15)\\&=12+(-9)\\&=12-9\\&=3\end{aligned}$$

11. $\dfrac{6(7-9)^3+(-2)}{(-2)(-5)(-5)}=\dfrac{6(-2)^3-2}{10(-5)}$
$=\dfrac{6(-8)-2}{-50}$
$=\dfrac{-48-2}{-50}$
$=\dfrac{-50}{-50}$
$=1$

12. $\dfrac{\left(4-\sqrt{16}\right)-(-7-20)}{-2(1-4)^2}=\dfrac{(4-4)-(-27)}{-2(-3)^2}$
$=\dfrac{0+27}{-2(9)}$
$=\dfrac{27}{-18}$
$=-\dfrac{3}{2}$

13. Let $q=4$ and $r=-2$.
$q^2-r^2=(4)^2-(-2)^2=16-4=12$

14. Let $q=4$, $r=-2$, and $t=1$.
$\dfrac{5t-3q}{3r-1}=\dfrac{5(1)-3(4)}{3(-2)-1}=\dfrac{5-12}{-6-1}=\dfrac{-7}{-7}=1$

15. **a.** When $x=1$, $8.75x=8.75(1)=8.75$.
When $x=3$, $8.75x=8.75(3)=26.25$.
When $x=10$, $8.75x=8.75(10)=87.50$.
When $x=20$, $8.75x=8.75(20)=175.00$.

x	1	3	10	20
$8.75x$	8.75	26.25	87.50	175.00

b. As the number of adults increases the total cost increases.

16. Twice the sum of x and five is 30.
$2\cdot(x+5)=30$
or $2(x+5)=30$

17. The square of the difference of six and y divided by seven is less than -2.
$(6-y)^2 \div 7 < -2$
or $\dfrac{(6-y)^2}{7}<-2$

18. The product of nine and z, divided by the absolute value of -12 is not equal to 10.

$$9z \quad \div \quad |-12| \quad \neq \quad 10$$

or $\dfrac{9z}{|-12|} \neq 10$

19. Three times the quotient of n and five is the opposite of n.

$$3 \quad \cdot \quad \frac{n}{5} \quad = \quad -n$$

or $3\left(\dfrac{n}{5}\right) = -n$

20. Twenty is equal to 6 subtracted from twice x.

$$20 \quad = \quad 2x-6$$

or $20 = 2x - 6$

21. Negative two is equal to x divided by the sum of x and five.

$$-2 \quad = \quad x \quad \div \quad (x+5)$$

or $-2 = \dfrac{x}{x+5}$

22. $6(x - 4) = 6x - 24$ illustrates the distributive property

23. $(4 + x) + z = 4 + (x + z)$ illustrates the associative property of addition

24. $(-7) + 7 = 0$ illustrates the additive inverse property

25. $(-18)(0) = 0$ illustrates the multiplication property of zero

26. The value of one nickel is 0.05 dollar; the value of one dime is 0.10 dollar. If there are n nickels and d dimes then the total amount of money is $0.05n + 0.1d$.

27. The reciprocal of $-\dfrac{7}{11}$ is $-\dfrac{11}{7}$.

The opposite of $-\dfrac{7}{11}$ is $\dfrac{7}{11}$.

28.
$$\begin{aligned}\frac{1}{3}a-\frac{3}{8}+\frac{1}{6}a-\frac{3}{4} &= \frac{1}{3}a+\frac{1}{6}a-\frac{3}{8}-\frac{3}{4}\\ &= \left(\frac{1}{3}+\frac{1}{6}\right)a-\frac{3}{8}-\frac{3}{4}\\ &= \left(\frac{2}{6}+\frac{1}{6}\right)a-\frac{3}{8}-\frac{6}{8}\\ &= \left(\frac{3}{6}\right)a-\frac{9}{8}\\ &= \frac{1}{2}a-\frac{9}{8}\end{aligned}$$

29. $\begin{aligned} 4y+10-2(y+10) &= 4y+10-2y-20 \\ &= 4y-2y+10-20 \\ &= (4-2)y-10 \\ &= 2y-10 \end{aligned}$

30. $\begin{aligned} (8.3x-2.9)-(9.6x-4.8) &= 8.3x-2.9-9.6x+4.8 \\ &= 8.3x-9.6x-2.9+4.8 \\ &= (8.3-9.6)x+1.9 \\ &= -1.3x+1.9 \end{aligned}$

Chapter 2

Section 2.1 Practice Exercises

1.
$$\begin{aligned}3x+7&=22\\3x+7-7&=22-7\\3x&=15\\\frac{3x}{3}&=\frac{15}{3}\\x&=5\end{aligned}$$

2.
$$\begin{aligned}2.5&=3-2.5t\\2.5-3&=3-2.5t-3\\-0.5&=-2.5t\\\frac{-0.5}{-2.5}&=\frac{-2.5t}{-2.5}\\0.2&=t\end{aligned}$$

3.
$$\begin{aligned}-8x-4+6x&=5x+11-4x\\-2x-4&=x+11\\-2x-4-x&=x+11-x\\-3x-4&=11\\-3x-4+4&=11+4\\-3x&=15\\\frac{-3x}{-3}&=\frac{15}{-3}\\x&=-5\end{aligned}$$

4.
$$\begin{aligned}3(x-5)&=6x-3\\3x-15&=6x-3\\3x-15-6x&=6x-3-6x\\-3x-15&=-3\\-3x-15+15&=-3+15\\-3x&=12\\\frac{-3x}{-3}&=\frac{12}{-3}\\x&=-4\end{aligned}$$

5.
$$\begin{aligned}\frac{y}{2}-\frac{y}{5}&=\frac{1}{4}\\20\left(\frac{y}{2}-\frac{y}{5}\right)&=20\left(\frac{1}{4}\right)\\20\left(\frac{y}{2}\right)-20\left(\frac{y}{5}\right)&=5\\10y-4y&=5\\6y&=5\\\frac{6y}{6}&=\frac{5}{6}\\y&=\frac{5}{6}\end{aligned}$$

6.
$$\begin{aligned}x-\frac{x-2}{12}&=\frac{x+3}{4}+\frac{1}{4}\\12\left(x-\frac{x-2}{12}\right)&=12\left(\frac{x+3}{4}+\frac{1}{4}\right)\\12\cdot x-12\left(\frac{x-2}{12}\right)&=12\left(\frac{x+3}{4}\right)+12\cdot\frac{1}{4}\\12x-(x-2)&=3(x+3)+3\\12x-x+2&=3x+9+3\\11x+2&=3x+12\\11x+2-3x&=3x+12-3x\\8x+2&=12\\8x+2-2&=12-2\\8x&=10\\\frac{8x}{8}&=\frac{10}{8}\\x&=\frac{5}{4}\end{aligned}$$

7.
$$\begin{aligned}0.15x-0.03&=0.2x+0.12\\100(0.15x-0.03)&=100(0.2x+0.12)\\100(0.15x)-100(0.03)&=100(0.2x)+100(0.12)\\15x-3&=20x+12\\15x-20x&=12+3\\-5x&=15\\\frac{-5x}{-5}&=\frac{15}{-5}\\x&=-3\end{aligned}$$

8.
$$\begin{aligned}4x-3&=4(x+5)\\4x-3&=4x+20\\4x-3-4x&=4x+20-4x\\-3&=20\end{aligned}$$

This equation is false no matter what value the variable x might have. Thus, there is no solution. The solution set is $\{\ \}$ or $\varnothing$.

9.
$$\begin{aligned}5x-2&=3+5(x-1)\\5x-2&=3+5x-5\\5x-2&=-2+5x\\5x-2+2&=-2+5x+2\\5x&=5x\\5x-5x&=5x-5x\\0&=0\end{aligned}$$

Since 0 = 0 is a true statement for every value of x, all real numbers are solutions. The solution set is the set of all real numbers or $\{x|x \text{ is a real number}\}$.

Vocabulary, Readiness & Video Check 2.1

1. Equations with the same solution set are called equivalent equations.

2. A value for the variable in an equation that makes the equation a true statement is called a solution of the equation.

3. By the addition property of equality, $y = -3$ and $y - 7 = -3 - 7$ are equivalent equations.

4. By the multiplication property of equality, $2y = -3$ and $\frac{2y}{2} = \frac{-3}{2}$ are equivalent equations.

5. $\frac{1}{3}x - 5$ expression

6. $2(x - 3) = 7$ equation

7. $\frac{5}{9}x + \frac{1}{3} = \frac{2}{9} - x$ equation

8. $\frac{5}{9}x + \frac{1}{3} - \frac{2}{9} - x$ expression

9. The addition property of equality allows us to add the same number to (or subtract the same number from) both sides of an equation and have an equivalent equation. The multiplication property of equality allows us to multiply (or divide) both sides of an equation by the same nonzero number and have an equivalent equation.

10. distributive property

11. to make the calculations less tedious

12. When solving a linear equation and all variable terms subtract out and:

    a. you have a true statement, then the equation has all real numbers for which the equation is defined as solutions.

    b. you have a false statement, then the equation has no solution.

Exercise Set 2.1

1. $$-5x = -30$$
$$\frac{-5x}{-5} = \frac{-30}{-5}$$
$$x = 6$$
Check: $$-5x = -30$$
$$-5(6) \stackrel{?}{=} -30$$
$$-30 = -30 \quad \text{True}$$
The solution is 6.

3. $$-10 = x + 12$$
$$-10 - 12 = x + 12 - 12$$
$$-22 = x$$
Check: $$-10 = x + 12$$
$$-10 \stackrel{?}{=} -22 + 12$$
$$-10 = -10 \quad \text{True}$$
The solution is -22.

5. $$x - 2.8 = 1.9$$
$$x - 2.8 + 2.8 = 1.9 + 2.8$$
$$x = 4.7$$
Check: $$x - 2.8 = 1.9$$
$$4.7 - 2.8 \stackrel{?}{=} 1.9$$
$$1.9 = 1.9 \quad \text{True}$$
The solution is 4.7.

7. $$5x - 4 = 26 + 2x$$
$$5x - 2x = 26 + 4$$
$$3x = 30$$
$$\frac{3x}{3} = \frac{30}{3}$$
$$x = 10$$
Check: $$5x - 4 = 26 + 2x$$
$$5(10) - 4 \stackrel{?}{=} 26 + 2(10)$$
$$50 - 4 \stackrel{?}{=} 26 + 20$$
$$46 = 46 \quad \text{True}$$
The solution is 10.

9. $$-4.1 - 7z = 3.6$$
$$-4.1 - 7z + 4.1 = 3.6 + 4.1$$
$$-7z = 7.7$$
$$\frac{-7z}{-7} = \frac{7.7}{-7}$$
$$z = -1.1$$
Check: $$-4.1 - 7z = 3.6$$
$$-4.1 - 7(-1.1) \stackrel{?}{=} 3.6$$
$$-4.1 + 7.7 \stackrel{?}{=} 3.6$$
$$3.6 = 3.6 \quad \text{True}$$
The solution is -1.1.

11.
$$\begin{aligned}5y+12&=2y-3\\5y+12-2y&=2y-3-2y\\3y+12&=-3\\3y+12-12&=-3-12\\3y&=-15\\\frac{3y}{3}&=\frac{-15}{3}\\y&=-5\end{aligned}$$
Check:
$$\begin{aligned}5y+12&=2y-3\\5(-5)+12&\stackrel{?}{=}2(-5)-3\\-25+12&\stackrel{?}{=}-10-3\\-13&=-13\quad\text{True}\end{aligned}$$
The solution is –5.

13.
$$\begin{aligned}3x-4-5x&=x+4+x\\-4-2x&=2x+4\\-2x-2x&=4+4\\-4x&=8\\x&=-2\end{aligned}$$
Check:
$$\begin{aligned}3x-4-5x&=x+4+x\\3(-2)-4-5(-2)&\stackrel{?}{=}-2+4-2\\-6-4+10&\stackrel{?}{=}0\\0&=0\quad\text{True}\end{aligned}$$
The solution is –2.

15.
$$\begin{aligned}8x-5x+3&=x-7+10\\3x+3&=x+3\\2x&=0\\x&=0\end{aligned}$$
Check:
$$\begin{aligned}8x-5x+3&=x-7+10\\8(0)-5(0)+3&\stackrel{?}{=}(0)-7+10\\0+3&\stackrel{?}{=}3\\3&=3\quad\text{True}\end{aligned}$$
The solution is 0.

17.
$$\begin{aligned}5x+12&=2(2x+7)\\5x+12&=4x+14\\x+12&=14\\x&=2\end{aligned}$$
Check:
$$\begin{aligned}5x+12&=2(2x+7)\\5(2)+12&\stackrel{?}{=}2(2(2)+7)\\10+12&\stackrel{?}{=}2(4+7)\\22&\stackrel{?}{=}2(11)\\22&=22\quad\text{True}\end{aligned}$$
The solution is 2.

19.
$$\begin{aligned}3(x-6)&=5x\\3x-18&=5x\\-18&=2x\\-9&=x\end{aligned}$$
Check:
$$\begin{aligned}3(x-6)&=5x\\3((-9)-6)&\stackrel{?}{=}5(-9)\\3(-15)&\stackrel{?}{=}-45\\-45&=-45\quad\text{True}\end{aligned}$$
The solution is –9.

21.
$$\begin{aligned}-2(5y-1)-y&=-4(y-3)\\-10y+2-y&=-4y+12\\-7y+2&=12\\-7y&=10\\y&=-\frac{10}{7}\end{aligned}$$
Check:
$$\begin{aligned}-2(5y-1)-y&=-4(y-3)\\-2\left(5\left(-\frac{10}{7}\right)-1\right)-\left(-\frac{10}{7}\right)&\stackrel{?}{=}-4\left(\left(-\frac{10}{7}\right)-3\right)\\-2\left(-\frac{50}{7}-1\right)+\frac{10}{7}&\stackrel{?}{=}-4\left(-\frac{31}{7}\right)\\-2\left(-\frac{57}{7}\right)+\frac{10}{7}&\stackrel{?}{=}\frac{124}{7}\\\frac{114}{7}+\frac{10}{7}&\stackrel{?}{=}\frac{124}{7}\\\frac{124}{7}&=\frac{124}{7}\quad\text{True}\end{aligned}$$
The solution is $-\frac{10}{7}$.

23.
$$\begin{aligned}\frac{x}{2}+\frac{x}{3}&=\frac{3}{4}\\12\left(\frac{x}{2}+\frac{x}{3}\right)&=12\left(\frac{3}{4}\right)\\6x+4x&=9\\10x&=9\\x&=\frac{9}{10}\end{aligned}$$
Check:
$$\begin{aligned}\frac{x}{2}+\frac{x}{3}&=\frac{3}{4}\\\frac{9}{10}\cdot\frac{1}{2}+\frac{9}{10}\cdot\frac{1}{3}&\stackrel{?}{=}\frac{3}{4}\\\frac{9}{20}+\frac{3}{10}&\stackrel{?}{=}\frac{3}{4}\\\frac{3}{4}&=\frac{3}{4}\quad\text{True}\end{aligned}$$
The solution is $\frac{9}{10}$.

25. $\frac{3t}{4} - \frac{t}{2} = 1$

$4\left(\frac{3t}{4} - \frac{t}{2}\right) = 4(1)$

$3t - 2t = 4$

$t = 4$

Check: $\frac{3t}{4} - \frac{t}{2} = 1$

$\frac{3(4)}{4} - \frac{(4)}{2} \stackrel{?}{=} 1$

$3 - 2 \stackrel{?}{=} 1$

$1 = 1$ True

The solution is 4.

27. $\frac{n-3}{4} + \frac{n+5}{7} = \frac{5}{14}$

$28\left(\frac{n-3}{4}\right) + 28\left(\frac{n+5}{7}\right) = 28\left(\frac{5}{14}\right)$

$7(n-3) + 4(n+5) = 2(5)$

$7n - 21 + 4n + 20 = 10$

$11n - 1 = 10$

$11n = 11$

$n = 1$

Check: $\frac{n-3}{4} + \frac{n+5}{7} = \frac{5}{14}$

$\frac{(1)-3}{4} + \frac{(1)+5}{7} \stackrel{?}{=} \frac{5}{14}$

$\frac{-2}{4} + \frac{6}{7} \stackrel{?}{=} \frac{5}{14}$

$\frac{-1}{2} + \frac{6}{7} \stackrel{?}{=} \frac{5}{14}$

$-\frac{7}{14} + \frac{12}{14} \stackrel{?}{=} \frac{5}{14}$

$\frac{5}{14} = \frac{5}{14}$ True

The solution is 1.

29. $0.6x - 10 = 1.4x - 14$

$-0.8x - 10 = -14$

$-0.8x = -4$

$x = 5$

Check: $0.6x - 10 = 1.4x - 14$

$0.6(5) - 10 \stackrel{?}{=} 1.4(5) - 14$

$3 - 10 \stackrel{?}{=} 7 - 14$

$-7 = -7$ True

The solution is 5.

31. $\frac{3x-1}{9} + x = \frac{3x+1}{3} + 4$

$9\left(\frac{3x-1}{9} + x\right) = 9\left(\frac{3x+1}{3} + 4\right)$

$(3x-1) + 9x = 3(3x+1) + 36$

$3x - 1 + 9x = 9x + 3 + 36$

$12x = 9x + 40$

$3x = 40$

$x = \frac{40}{3}$

Check: $\frac{3x-1}{9} + x = \frac{3x+1}{3} + 4$

$\frac{3\left(\frac{40}{3}\right) - 1}{9} + \frac{40}{3} \stackrel{?}{=} \frac{3\left(\frac{40}{3}\right) + 1}{3} + 4$

$\frac{39}{9} + \frac{120}{9} \stackrel{?}{=} \frac{41}{3} + \frac{12}{3}$

$\frac{53}{3} = \frac{53}{3}$ True

The solution is $\frac{40}{3}$.

33. $1.5(4 - x) = 1.3(2 - x)$

$10[1.5(4 - x)] = 10[1.3(2 - x)]$

$15(4 - x) = 13(2 - x)$

$60 - 15x = 26 - 13x$

$-2x = -34$

$x = 17$

Check: $1.5(4 - x) = 1.3(2 - x)$

$1.5(4 - 17) \stackrel{?}{=} 1.3(2 - 17)$

$1.5(-13) \stackrel{?}{=} 1.3(-15)$

$-19.5 = -19.5$ True

The solution is 17.

35. $4(n + 3) = 2(6 + 2n)$

$4n + 12 = 12 + 4n$

$0 = 0$

This is true for all x. Therefore, all real numbers are solutions.

37. $3(x + 1) + 5 = 3x + 2$

$3x + 3 + 5 = 3x + 2$

$3x + 8 = 3x + 2$

$8 = 2$

This is false for any x. Therefore, no solution exists, $\varnothing$.

39. $2(x-8)+x=3(x-6)+2$
$2x-16+x=3x-18+2$
$3x-16=3x-16$
$0=0$
This is true for all x. Therefore, all real numbers are solutions.

41. $4(x+5)=3(x-4)+x$
$4x+20=3x-12+x$
$4x+20=4x-12$
$20=-12$
This is false for any x. Therefore, no solution exists, $\varnothing$.

43. $\frac{3}{8}+\frac{b}{3}=\frac{5}{12}$
$24\left(\frac{3}{8}\right)+24\left(\frac{b}{3}\right)=24\left(\frac{5}{12}\right)$
$9+8b=10$
$8b=1$
$b=\frac{1}{8}$

45. $x-10=-6x-10$
$x+6x=-10+10$
$7x=0$
$x=0$

47. $5(x-2)+2x=7(x+4)-38$
$5x-10+2x=7x+28-38$
$7x-10=7x-10$
$0=0$
This is true for all x. Therefore, all real numbers are solutions.

49. $y+0.2=0.6(y+3)$
$y+0.2=0.6y+1.8$
$0.4y=1.6$
$y=4$

51. $\frac{1}{4}(a+2)=\frac{1}{6}(5-a)$
$\frac{1}{4}a+\frac{1}{2}=\frac{5}{6}-\frac{1}{6}a$
$\frac{1}{4}a+\frac{1}{6}a=\frac{5}{6}-\frac{1}{2}$
$\frac{3}{12}a+\frac{2}{12}a=\frac{5}{6}-\frac{3}{6}$
$\frac{5}{12}a=\frac{2}{6}$
$a=\frac{2}{6}\cdot\frac{12}{5}$
$a=\frac{4}{5}$

53. $2y+5(y-4)=4y-2(y-10)$
$2y+5y-20=4y-2y+20$
$7y-20=2y+20$
$7y-2y=20+20$
$5y=40$
$y=8$

55. $6x-2(x-3)=4(x+1)+4$
$6x-2x+6=4x+4+4$
$4x+6=4x+8$
$4x-4x=8-6$
$0=2$
This is false for any x. Therefore, the solution set is $\varnothing$.

57. $\frac{m-4}{3}-\frac{3m-1}{5}=1$
$15\left(\frac{m-4}{3}\right)-15\left(\frac{3m-1}{5}\right)=15(1)$
$5(m-4)-3(3m-1)=15$
$5m-20-9m+3=15$
$-4m-17=15$
$-4m=32$
$m=-8$

59. $8x-12-3x=9x-7$
$5x-12=9x-7$
$5x-9x=-7+12$
$-4x=5$
$x=-\frac{5}{4}$

61. $-(3x-5)-(2x-6)+1=-5(x-1)-(3x+2)+3$
$-3x+5-2x+6+1=-5x+5-3x-2+3$
$-5x+12=-8x+6$
$3x=-6$
$x=-2$

63. $\frac{1}{3}(y+4)+6=\frac{1}{4}(3y-1)-2$
$12\left[\frac{1}{3}(y+4)+6\right]=12\left[\frac{1}{4}(3y-1)-2\right]$
$4(y+4)+12\cdot 6=3(3y-1)-12\cdot 2$
$4y+16+72=9y-3-24$
$4y+88=9y-27$
$-5y=-115$
$y=23$

65. $2[7-5(1-n)]+8n=-16+3[6(n+1)-3n]$
$2[7-5+5n]+8n=-16+3[6n+6-3n]$
$2(2+5n)+8n=-16+3(3n+6)$
$4+10n+8n=-16+9n+18$
$4+18n=2+9n$
$9n=-2$
$n=-\frac{2}{9}$

67. Quotient means to divide. The quotient of 8 and a number: $\frac{8}{x}$

69. Product means to multiply. The product of 8 and a number: $8x$

71. Five subtracted from twice a number: $2x - 5$

73. Subtract 19 instead of adding;
$3x+19=13$
$2x=-6$
$\frac{2x}{2}=\frac{-6}{2}$
$x=-3$

75. $0.4 - 1.6 = -1.2$, not 1.2;
$9x+1.6=4x+0.4$
$5x=-1.2$
$\frac{5x}{5}=\frac{-1.2}{5}$
$x=-0.24$

77. $2x + 3 = 2x + 3$
Since the two sides of the equation are identical, the equation is true for any value of x. All real numbers are solutions.

79. $2x + 1 = 2x + 3$
Adding 1 to a number and adding 3 to the same number will not result in equal numbers for any value of x. There is no solution.

81. **a.** $4(x + 1) + 1 = 4x + 4 + 1 = 4x + 5$

b. $4(x+1)+1=-7$
$4x+4+1=-7$
$4x+5=-7$
$4x=-12$
$x=-3$
The solution is -3.

c. answers may vary

83. answers may vary

85. $3.2x+4=5.4x-7$
$3.2x+4-4=5.4x-7-4$
$3.2x=5.4x-11$
From this we see that $K=-11$.

87. $\frac{7}{11}x+9=\frac{3}{11}x-14$
$\frac{7}{11}x+9-9=\frac{3}{11}x-14-9$
$\frac{7}{11}x=\frac{3}{11}x-23$
From this we see that $K=-23$.

89. answers may vary

91. $x(x-6)+7=x(x+1)$
$x^2-6x+7=x^2+x$
$-6x+7=x$
$7=7x$
$1=x$

93. $3x(x+5)-12=3x^2+10x+3$
$3x^2+15x-12=3x^2+10x+3$
$15x-12=10x+3$
$5x=15$
$x=3$

95. $2.569x = -12.48534$

$\frac{2.569x}{2.569} = \frac{-12.48534}{2.569}$

$x = -4.86$

Check: $2.569x = -12.48534$

$2.569(-4.86) \stackrel{?}{=} -12.48534$

$-12.48534 = -12.48534$ True

The solution is –4.86.

97. $2.86z - 8.1258 = -3.75$

$2.86z = 4.3758$

$\frac{2.86z}{2.86} = \frac{4.3758}{2.86}$

$z = 1.53$

Check: $2.86z - 8.1258 = -3.75$

$2.86(1.53) - 8.1258 \stackrel{?}{=} -3.75$

$-3.75 = -3.75$ True

The solution is 1.53.

Section 2.2 Practice Exercises

1. a.

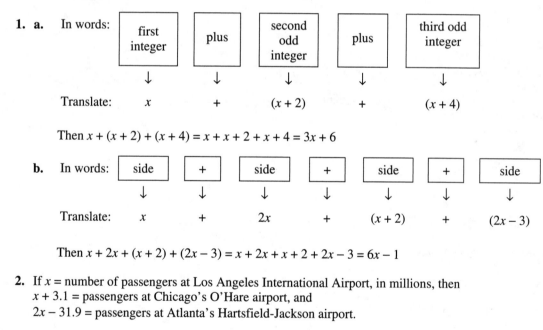

In words:	first integer	plus	second odd integer	plus	third odd integer
	↓	↓	↓	↓	↓
Translate:	x	$+$	$(x+2)$	$+$	$(x+4)$

Then $x + (x + 2) + (x + 4) = x + x + 2 + x + 4 = 3x + 6$

b.

In words:	side	+	side	+	side	+	side
	↓	↓	↓	↓	↓	↓	↓
Translate:	x	$+$	$2x$	$+$	$(x+2)$	$+$	$(2x-3)$

Then $x + 2x + (x + 2) + (2x - 3) = x + 2x + x + 2 + 2x - 3 = 6x - 1$

2. If x = number of passengers at Los Angeles International Airport, in millions, then
$x + 3.1$ = passengers at Chicago's O'Hare airport, and
$2x - 31.9$ = passengers at Atlanta's Hartsfield-Jackson airport.

In words:	passengers at Los Angeles	+	passengers at O'Hare	+	passengers at Hartsfield-Jackson
	↓	↓	↓	↓	↓
Translate:	x	$+$	$(x+3.1)$	$+$	$(2x-31.9)$

Then $x + (x + 3.1) + (2x - 31.9) = x + x + 3.1 + 2x - 31.9 = 4x - 28.8$.

3. Let x = the first number, then $3x - 8$ = the second number, and $5x$ = the third number.
The sum of the three numbers is 118.

$$\begin{aligned} x+(3x-8)+5x &= 118 \\ x+3x+5x-8 &= 118 \\ 9x-8 &= 118 \\ 9x &= 126 \\ x &= 14 \end{aligned}$$

The numbers are 14, $3x - 8 = 3(14) - 8 = 34$, and $5x = 5(14) = 70$.

4. Let x = the original price. Then $0.4x$ = the discount. The original price, minus the discount, is equal to \$270.

$$\begin{aligned} x-0.4x &= 270 \\ 0.6x &= 270 \\ x &= \frac{270}{0.6} = 450 \end{aligned}$$

The original price was \$450.

5. Let x = width, then $2x - 16$ = length.
The perimeter is 160 inches.

$$\begin{aligned} 2(x)+2(2x-16) &= 160 \\ 2x+4x-32 &= 160 \\ 6x-32 &= 160 \\ 6x &= 192 \\ x &= 32 \end{aligned}$$

$2x - 16 = 2(32) - 16 = 48$
The width is 32 inches and the length is 48 inches.

6. Let x = first odd integer, then $x + 2$ = second odd integer, and $x + 4$ = third odd integer.
The sum of the integers is 81.

$$\begin{aligned} x+(x+2)+(x+4) &= 81 \\ 3x+6 &= 81 \\ 3x &= 75 \\ x &= 25 \end{aligned}$$

$x + 2 = 27$
$x + 4 = 29$
The integers are 25, 27, and 29.

Vocabulary, Readiness & Video Check 2.2

1. 130% of a number $\underline{>}$ the number.

2. 70% of a number $\underline{<}$ the number.

3. 100% of a number $\underline{=}$ the number.

4. 200% of a number $\underline{>}$ the number.

	First Integer	All Described Integers
5. Four consecutive integers	31	31, 32, 33, 34
6. Three consecutive odd integers	31	31, 33, 35
7. Three consecutive even integers	18	18, 20, 22
8. Four consecutive even integers	92	92, 94, 96, 98
9. Three consecutive integers	y	$y, y + 1, y + 2$
10. Three consecutive even integers	z (z is even)	$z, z + 2, z + 4$
11. Four consecutive integers	p	$p, p + 1, p + 2, p + 3$
12. Three consecutive odd integers	s (s is odd)	$s, s + 2, s + 4$

13. distributive property

14. The original application asks you to find three numbers. The solution $x = 45$ only gives you the first number. You need to INTERPRET this result.

Exercise Set 2.2

1. The perimeter is the sum of the lengths of the four sides.
$y + y + y + y = 4y$

3. Let z = first integer, then $z + 1$ = second integer, and $z + 2$ = third integer.
$z+(z+1)+(z+2) = z+z+z+1+2 = 3z+3$

5. Find the sum of x nickels worth 5¢ each, and $(x + 3)$ dimes worth 10¢ each, and $2x$ quarters worth 25¢ each.
$$\begin{aligned}5x+10(x+3)+25(2x) &= 5x+10x+30+50x\\ &= 65x+30\end{aligned}$$
The total amount is $(65x + 30)$ cents.

7. $4x + 3(2x + 1) = 4x + 6x + 3 = 10x + 3$

9. The length of the side denoted by ? is $10 - 2 = 8$. Similarly, the length of the unmarked side is $(x - 3) - (x - 10) = x - 3 - x + 10 = 7$.
Thus the perimeter of the floor plan is given by $(x - 10) + 2 + 7 + 8 + (x - 3) + 10 = 2x + 14$.

11. Let x = the number.
$$\begin{aligned}4(x-2) &= 2+4x+2x\\ 4x-8 &= 2+6x\\ -2x &= 10\\ x &= -5\end{aligned}$$
The number is −5.

13. Let x = the first number, then
$5x$ = the second number, and
$x + 100$ = the third number.

$$x+5x+(x+100)=415$$
$$7x+100=415$$
$$7x=315$$
$$x=45$$

$5x = 225$
$x + 100 = 145$
The numbers are 45, 225, and 145.

15. 29% of 2271 = $0.29 \cdot 2271 = 658.59$;
$2271 - 658.59 = 1612.41$.
Approximately 1612.41 million acres are not federally owned.

17. 89.6% of 16,674 = $0.896 \cdot 16{,}674 \approx 14{,}940$
Approximately 14,940 minor earthquakes occurred world wide in 2014.

19. 15% of 1500 = $0.15 \cdot 1500 = 225$
$1500 - 225 = 1275$
1275 are willing to do business with any size retailer.

21. From the circle graph, 39% of an average worker's time at work is spent on role-specific tasks.
$100\% - 39\% = 61\%$
61% of an average worker's time at work is spent on tasks other than role-specific tasks.

23. The percents in the circle graph sum to 100%.

$$39+2x+19+x=100$$
$$3x+58=100$$
$$3x=42$$
$$x=14$$

14% of an average worker's time at work is spent on collaborating internally.

25.

$$x+4x+(x+6)=180$$
$$6x+6=180$$
$$6x=174$$
$$x=29$$

$4x = 4(29) = 116$
$x + 6 = 29 + 6 = 35$
The angles measure 29°, 35°, and 116°.

27.

$$(4x)+(5x+1)+(5x+3)=102$$
$$14x+4=102$$
$$14x=98$$
$$x=7$$

$4x = 4(7) = 28$
$5x + 1 = 5(7) + 1 = 36$
$5x + 3 = 5(7) + 3 = 38$
The sides measure 28 meters, 36 meters, and 38 meters.

29.

$$x+(2.5x-9)+x+1.5x=99$$
$$6x-9=99$$
$$6x=108$$
$$x=18$$

$1.5x = 1.5(18) = 27$
$2.5x - 9 = 2.5(18) - 9 = 36$
The sides measure 18 inches, 18 inches, 27 inches, and 36 inches.

31. Let x = first integer; then
$x + 1$ = next integer and
$x + 2$ = third integer.

$$x+(x+1)+(x+2)=228$$
$$3x+3=228$$
$$3x=225$$
$$x=75$$

$x + 1 = 75 + 1 = 76$
$x + 2 = 75 + 2 = 77$
The integers are 75, 76, and 77.

33. Let x = first even integer, then
$x + 2$ = second even integer, and
$x + 4$ = third even integer.

$$2x+(x+4)=268{,}222$$
$$3x+4=268{,}222$$
$$3x=268{,}218$$
$$x=89{,}406$$

$x + 2 = 89{,}408$
$x + 4 = 89{,}410$
Fallon's ZIP code is 89406, Fernley's ZIP code is 89408, and Gardnerville Ranchos's ZIP code is 89410.

35.

$$2x+(2x+170)+(3x-650)=6380$$
$$2x+2x+170+3x-650=6380$$
$$7x-480=6380$$
$$7x=6860$$
$$x=980$$

$2x = 2(980) = 1960$
$2x + 170 = 2(980) + 170 = 2130$
$3x - 650 = 3(980) - 650 = 2290$

Year	Increase in Social Network Users	Predicted Number
2015	$2x$	1960 million
2016	$2x + 170$	2130 million
2017	$3x - 650$	2290 million
Total	6380 million	

37. Let x be the growth in the number of food service jobs (in thousands). Then $2x - 317$ is the growth in the number of registered nurse jobs and $\frac{1}{2}x+88$ is the growth in the number of customer service jobs.

$$x+(2x-317)+\left(\frac{1}{2}x+88\right)=1248$$
$$x+2x-317+\frac{1}{2}x+88=1248$$
$$\frac{7}{2}x-229=1248$$
$$\frac{7}{2}x=1477$$
$$x=422$$

$2x - 317 = 2(422) - 317 = 527$

$\frac{1}{2}x+88=\frac{1}{2}(422)+88=299$

The growth predictions are:
food service jobs: 422 thousand;
registered nurse jobs: 527 thousand;
customer service jobs: 299 thousand

39. Let x be the number of seats in a B737-200 aircraft. Then the number of seats in a B767-300ER is $x + 88$, and the number of seats in a F-100 is $x - 32$.

$$x+(x+88)+(x-32)=413$$
$$3x+56=413$$
$$3x=357$$
$$x=119$$

$x + 88 = 119 + 88 = 207$

$x - 32 = 119 - 32 = 87$

The B737-200 has 119 seats, the B767-300ER has 207 seats, and the F-100 has 87 seats.

41. Let x be the price of the fax machine before tax.

$$x+0.08x=464.40$$
$$1.08x=464.40$$
$$x=430$$

The fax machine cost $430 before tax.

43. The new salary is 1.036 times the current salary.
$479{,}000(1.036) = 496{,}244$
The salary after the raise is $496,244.

45. Let x be the expected population of Angola in 2050. Then x is 22,400,000 plus the increase of 272% of 22,400,000.

$$x=22{,}400{,}000+2.72(22{,}400{,}000)$$
$$=22{,}400{,}000+60{,}928{,}000$$
$$=83{,}328{,}000$$

The population of Angola in 2050 is predicted to be 83,328,000.

47. Let x = measure of the angle; then $180 - x$ = measure of its supplement.

$$x=3(180-x)+20$$
$$x=540-3x+20$$
$$4x=560$$
$$x=140$$

$180 - x = 180 - 140 = 40$

The angles measure 140° and 40°.

49. Let x = measure of second angle; then $2x$ = measure of first angle and $3x - 12$ = measure of third angle.

$$x+2x+(3x-12)=180$$
$$6x-12=180$$
$$6x=192$$
$$x=32$$

$2x = 2(32) = 64$

$3x - 12 = 3(32) - 12 = 84$

The angles measure 64°, 32°, and 84°.

51. Let x = the length of a side of the square. Then $x + 6$ = the length of a side of the triangle.

$$4x=3(x+6)$$
$$4x=3x+18$$
$$x=18$$

The sides of the square are 18 cm and the sides of the triangle are 24 cm.

53. Let x = first even integer, then $x + 2$ = second even integer, and $x + 4$ = third even integer.

$$x+(x+4)=156$$
$$2x+4=156$$
$$2x=152$$
$$x=76$$

$x + 2 = 78$

$x + 4 = 80$

The integers are 76, 78, and 80.

55. Let x be the number of grandstand seats at Darlington Motor Raceway. Then $2x + 37{,}000$ is the number of grandstand seats at Daytona International Speedway.

$$\begin{aligned} x+(2x+37{,}000) &= 220{,}000 \\ 3x+37{,}000 &= 220{,}000 \\ 3x &= 183{,}000 \\ x &= 61{,}000 \end{aligned}$$

$2x + 37{,}000 = 2(61{,}000) + 37{,}000 = 159{,}000$

Darlington has 61,000 grandstand seats and Daytona has 159,000 grandstand seats.

57. Let x be the population of the urban area of New York, in millions. Then $x + 4.6$ is the population of Delhi's urban area and $2x - 3.4$ is the population of Tokyo-Yokohama.

$$\begin{aligned} x+(x+4.6)+(2x-3.4) &= 83.6 \\ 4x+1.2 &= 83.6 \\ 4x &= 82.4 \\ x &= 20.6 \end{aligned}$$

$x + 4.6 = 20.6 + 4.6 = 25.2$

$2x - 3.4 = 2(20.6) - 3.4 = 37.8$

The populations are:
New York urban area: 20.6 million;
Delhi urban area: 25.2 million
Tokyo-Yokohama urban area: 37.8 million

59. $$\begin{aligned} x+5x+(6x-3) &= 483 \\ 12x-3 &= 483 \\ 12x &= 486 \\ x &= 40.5 \end{aligned}$$

$5x = 5(40.5) = 202.5$

$6x - 3 = 6(40.5) - 3 = 240$

The sides measure 40.5 feet, 202.5 feet, and 240 feet.

61. Let x be the number of bulb hours for a halogen bulb. Then $25x$ is the number of bulb hours for a fluorescent bulb, and $x - 2500$ is the number of bulb hours for an incandescent bulb.

$$\begin{aligned} x+25x+(x-2500) &= 105{,}500 \\ 27x-2500 &= 105{,}500 \\ 27x &= 108{,}000 \\ x &= 4000 \end{aligned}$$

$25x = 25(4000) = 100{,}000$

$x - 2500 = 4000 - 2500 = 1500$

A halogen bulb lasts 4000 bulb hours. A fluorescent bulb lasts 100,000 bulb hours. An incandescent bulb lasts 1500 bulb hours.

63. Let x be the number of wins for the Milwaukee Brewers. Then $x + 1$ is the number of wins for the Toronto Blue Jays, and $x + 2$ is the number of wins for the New York Yankees.

$$\begin{aligned} x+(x+1)+(x+2) &= 249 \\ 3x+3 &= 249 \\ 3x &= 246 \\ x &= 82 \end{aligned}$$

$x + 1 = 82 + 1 = 83$

$x + 2 = 82 + 2 = 84$

The Milwaukee Brewers had 82 wins, the Toronto Blue Jays had 83 wins, and the New York Yankees had 84 wins.

65. Let x = height of Galter Pavilion; then $x + 67$ = height of Guy's Tower and $x + 47$ = height of Queen Mary

$$\begin{aligned} x+(x+67)+(x+47) &= 1320 \\ 3x+114 &= 1320 \\ 3x &= 1206 \\ x &= 402 \end{aligned}$$

$x + 67 = 402 + 67 = 469$

$x + 47 = 402 + 47 = 449$

Galter Pavilion: 402 ft
Guy's Tower: 469 ft
Queen Mary: 449 ft

67. $$\begin{aligned} 4ab-3bc &= 4(-5)(-8)-3(-8)(2) \\ &= 160+48 \\ &= 208 \end{aligned}$$

69. $n^2 - m^2 = (-3)^2 - (-8)^2 = 9 - 64 = -55$

71. $P + PRT = 3000 + 3000(0.0325)(2) = 3195$

73. yes; answers may vary

75. Let $x°$ be the measure of an angle. Then its complement measures $(90 - x)°$ and its supplement measures $(180 - x)°$.

$$\begin{aligned} 180-x &= 10(90-x) \\ 180-x &= 900-10x \\ 180+9x &= 900 \\ 9x &= 720 \\ x &= 80 \end{aligned}$$

The angle measures 80°.

77. $$\begin{aligned} y &= -80.6x+2054 \\ 0 &= -80.6x+2054 \\ 80.6x &= 2054 \\ x &\approx 25.5 \end{aligned}$$

The average number of cigarettes smoked will be zero 25.5 years after 2000, or in 2025.

79. $y = -80.6(20) + 2054 = 442$

The average number of cigarettes smoked by an American adult in 2020 is predicted to be 442.

81. Let x be the first integer. Then $x + 1$ and $x + 2$ are the next two consecutive integers.

$$x+(x+1)+(x+2)=3(x+1)$$
$$3x+3=3x+3$$
$$3=3 \quad \text{True}$$

The sum of any three consecutive integers is equal to three times the second integer.

83.

$$R=C$$
$$24x=100+20x$$
$$4x=100$$
$$x=25$$

25 skateboards are needed to break even.

85. The company loses money if it makes and sells fewer products than the break-even number.

Section 2.3 Practice Exercises

1.

$$I=PRT$$
$$\frac{I}{PR}=\frac{PRT}{PR}$$
$$\frac{I}{PR}=T \text{ or } T=\frac{I}{PR}$$

2.

$$7x-2y=5$$
$$7x-2y-7x=5-7x$$
$$-2y=5-7x$$
$$\frac{-2y}{-2}=\frac{5-7x}{-2}$$
$$y=\frac{7}{2}x-\frac{5}{2}$$

3.

$$A=P+Prt$$
$$A-P=P+Prt-P$$
$$A-P=Prt$$
$$\frac{A-P}{Pt}=\frac{Prt}{Pt}$$
$$\frac{A-P}{Pt}=r \text{ or } r=\frac{A-P}{Pt}$$

4. Let $P = 8000$, $r = 6\% = 0.06$, $t = 4$, $n = 2$.

$$A=P\left(1+\frac{r}{n}\right)^{nt}$$
$$A=8000\left(1+\frac{0.06}{2}\right)^{2\cdot 4}$$
$$A=8000(1.03)^8$$
$$A\approx 8000(1.266770081)$$
$$A\approx 10{,}134.16$$

Russ will have $10,134.16 in his account.

5. Let $d = 190$ and $r = 7.5$.

$$d=rt$$
$$190=7.5t$$
$$\frac{190}{7.5}=\frac{7.5t}{7.5}$$
$$25\frac{1}{3}=t$$

They spent $25\frac{1}{3}$ hours cycling, or 25 hours 20 minutes.

Vocabulary, Readiness & Video Check 2.3

1.

$$2x+y=5$$
$$y=5-2x$$

2.

$$7x-y=3$$
$$-y=3-7x$$
$$y=-3+7x \text{ or } y=7x-3$$

3.

$$a-5b=8$$
$$a=5b+8$$

4.

$$7r+s=10$$
$$s=10-7r$$

5.

$$5j+k-h=6$$
$$5j+k=h+6$$
$$k=h-5j+6$$

6.

$$w-4y+z=0$$
$$w+z=4y$$
$$z=4y-w$$

7. That the specified variable will equal some expression and that this expression should not contain the specified variable.

8. The only way to check the solution is in the formula used, because if the wrong formula is used, a wrong answer may seem to check correctly.

Exercise Set 2.3

1.

$$d=rt$$
$$\frac{d}{r}=\frac{rt}{r}$$
$$\frac{d}{r}=t$$
$$t=\frac{d}{r}$$

3.
$$I = PRT$$
$$\frac{I}{PT} = \frac{PRT}{PT}$$
$$\frac{I}{PT} = R$$
$$R = \frac{I}{PT}$$

5.
$$9x - 4y = 16$$
$$9x - 4y - 9x = 16 - 9x$$
$$-4y = 16 - 9x$$
$$\frac{-4y}{-4} = \frac{16-9x}{-4}$$
$$y = \frac{9x-16}{4}$$

7.
$$P = 2L + 2W$$
$$P - 2L = 2W$$
$$\frac{P-2L}{2} = \frac{2W}{2}$$
$$\frac{P-2L}{2} = W$$
$$W = \frac{P-2L}{2}$$

9.
$$J = AC - 3$$
$$J + 3 = AC$$
$$\frac{J+3}{C} = \frac{AC}{C}$$
$$\frac{J+3}{C} = A$$
$$A = \frac{J+3}{C}$$

11.
$$W = gh - 3gt^2$$
$$W = g(h - 3t^2)$$
$$\frac{W}{h-3t^2} = \frac{g(h-3t^2)}{h-3t^2}$$
$$\frac{W}{h-3t^2} = g$$
$$g = \frac{W}{h-3t^2}$$

13.
$$T = C(2 + AB)$$
$$T = 2C + ABC$$
$$T - 2C = 2C + ABC - 2C$$
$$T - 2C = ABC$$
$$\frac{T-2C}{AC} = \frac{ABC}{AC}$$
$$\frac{T-2C}{AC} = B$$
$$B = \frac{T-2C}{AC}$$

15.
$$C = 2\pi r$$
$$\frac{C}{2\pi} = \frac{2\pi r}{2\pi}$$
$$\frac{C}{2\pi} = r$$
$$r = \frac{C}{2\pi}$$

17.
$$E = I(r + R)$$
$$E = Ir + IR$$
$$E - IR = Ir + IR - IR$$
$$E - IR = Ir$$
$$\frac{E-IR}{I} = \frac{Ir}{I}$$
$$\frac{E-IR}{I} = r$$
$$r = \frac{E-IR}{I}$$

19.
$$s = \frac{n}{2}(a + L)$$
$$2s = 2 \cdot \frac{n}{2}(a + L)$$
$$2s = n(a + L)$$
$$2s = na + nL$$
$$2s - na = na + nL - na$$
$$2s - na = nL$$
$$\frac{2s-na}{n} = \frac{nL}{n}$$
$$\frac{2s-na}{n} = L$$
$$L = \frac{2s-na}{n}$$

21.
$$N = 3st^4 - 5sv$$
$$N - 3st^4 = 3st^4 - 5sv - 3st^4$$
$$N - 3st^4 = -5sv$$
$$\frac{N - 3st^4}{-5s} = \frac{-5sv}{-5s}$$
$$\frac{3st^4 - N}{5s} = v$$
$$v = \frac{3st^4 - N}{5s}$$

23.
$$S = 2LW + 2LH + 2WH$$
$$S - 2LW = 2LW + 2LH + 2WH - 2LW$$
$$S - 2LW = 2LH + 2WH$$
$$S - 2LW = H(2L + 2W)$$
$$\frac{S - 2LW}{2L + 2W} = \frac{H(2L + 2W)}{2L + 2W}$$
$$\frac{S - 2LW}{2L + 2W} = H$$
$$H = \frac{S - 2LW}{2L + 2W}$$

25. $A = P\left(1 + \frac{r}{n}\right)^{nt} = 3500\left(1 + \frac{0.03}{n}\right)^{10n}$

n	1	2	4
A	\$4703.71	\$4713.99	\$4719.22

n	12	365
A	\$4722.74	\$4724.45

27. $A = P\left(1 + \frac{r}{n}\right)^{nt} = 6000\left(1 + \frac{0.04}{n}\right)^{5n}$

a. $n = 2$

$$A = 6000\left(1 + \frac{0.04}{2}\right)^{5 \cdot 2} \approx 7313.97$$

\$7313.97

b. $n = 4$

$$A = 6000\left(1 + \frac{0.04}{4}\right)^{5 \cdot 4} \approx 7321.14$$

\$7321.14

c. $n = 12$

$$A = 6000\left(1 + \frac{0.04}{12}\right)^{5 \cdot 12} \approx 7325.98$$

\$7325.98

29. Roundtrip distance = 90 + 90 = 180 miles
$$d = r \cdot t$$
$$180 = 50t$$
$$\frac{180}{50} = t$$
$$t = 3.6$$
3.6 hours = 3 hours 36 minutes
The trip will take 3.6 hours or 3 hours 36 minutes.

31.
$$C = \frac{5}{9}(F - 32)$$
$$C = \frac{5}{9}(104 - 32)$$
$$C = \frac{5}{9}(72)$$
$$C = 40$$
The day's high temperature was 40°C.

33. $A = s^2 = (64)^2 = 4096;\ \frac{4096}{24} \approx 171$

171 packages of tiles should be purchased.

35.
$$A = \frac{1}{2}bh$$
$$18 = \frac{1}{2}(4)h$$
$$18 = 2h$$
$$9 = h$$
The height is 9 feet.

37. The area of one pair of walls is $2 \cdot 14 \cdot 8 = 224\text{ ft}^2$ and the area of the other walls is $2 \cdot 16 \cdot 8 = 256\text{ ft}^2$ for a total of 480 ft^2. Multiplying by 2, the number of coats, yields 960 ft^2. Dividing this by 500 yields 1.92. Thus, 2 gallons should be purchased.

39.
$$V = \pi r^2 h$$
$$980\pi = \pi(7)^2 h$$
$$980\pi = 49\pi h$$
$$980 = 49h$$
$$20 = h$$
The height is 20 meters.

41. a. $V = \frac{4}{3}\pi r^3;\ r = \frac{d}{2} = \frac{12}{2} = 6$

$V = \frac{4}{3}\pi(6)^3$

$V = \frac{4}{3}\pi(216)$

$V = 288\pi$

The volume is 288π cubic mm.

b. $V = 288\pi \approx 904.78$ cubic mm.

43. a. $V = \pi r^2 h$

$V = \pi(4.2)^2(2.12)$

$V \approx 1174.86$

The volume of the cylinder is 1174.86 cubic meters.

b. $V = \frac{4}{3}\pi r^3$

$V = \frac{4}{3}\pi(4.2)^3$

$V \approx 310.34$

The volume of the sphere is 310.34 cubic meters.

c. $V = 1174.86 + 310.34 = 1485.20$

The volume of the tank is 1485.20 cubic meters.

45. 19 hours 5 minutes $= 19\frac{5}{60}$ hours

$d = rt$

$2447.8 = r\left(19\frac{5}{60}\right)$

$128.3 \approx r$

Her average speed was 128.3 miles per hour.

47. $V = \pi r^2 h$

1 mile = 5280 feet

1.3 miles = 6864 feet

$3800 = \pi r^2(6864)$

$0.42 \approx r$

The radius of the hole is 0.42 feet.

49. $C = \pi d = \pi(41.125) = 41.125\pi \text{ ft} \approx 129.1325 \text{ ft}$

The circumference of Eartha is $41.125\pi \approx 129.1325$ feet.

51. $A = P\left(1+\frac{r}{n}\right)^{nt}$

$= 10{,}000\left(1+\frac{0.085}{4}\right)^{4\cdot 2}$

$= 10{,}000(1+0.02125)^8$

$\approx \$11{,}831.96$

$\$11{,}831.96 - \$10{,}000 = \$1831.96$

53. $C = 4h + 9f + 4p$

$C - 4h - 4p = 9f$

$\frac{C-4h-4p}{9} = f$

$f = \frac{C-4h-4p}{9}$

55. $C = 4h + 9f + 4p$

$C = 4(7) + 9(14) + 4(6)$

$C = 178$

There are 178 calories in this serving.

57. $C = 4h + 9f + 4p$

$130 = 4(31) + 9(0) + 4p$

$130 = 124 + 4p$

$6 = 4p$

$\frac{6}{4} = p$

$p = 1.5$

There are 1.5 g of protein provided by this serving of raisins.

59. $-3, -2, -1$ satisfy $x < 0$.

61. $-3, -2, -1, 0, 1$ satisfy $x + 5 \le 6$ or $x \le 1$.

63. answers may vary

	Planet	AU from Sun
	Planet	AU from Sun
65.	Venus	$\frac{67.2}{92.9} = 0.723$
67.	Mars	$\frac{141.5}{92.9} = 1.523$
69.	Saturn	$\frac{886.1}{92.9} \approx 9.538$
71.	Neptune	$\frac{279.3}{92.9} \approx 30.065$

73. answers may vary

75. answers may vary;

$$W = gh - 3gt^2$$
$$W = g(h - 3t^2)$$
$$\frac{W}{h-3t^2} = \frac{g(h-3t^2)}{h-3t^2}$$
$$\frac{W}{h-3t^2} = g \text{ or } g = \frac{W}{h-3t^2}$$

77. Only one of the 8 sectors is green.

$$P(\text{green}) = \frac{1}{8}$$

79. Only one of the 8 sectors is black.

$$P(\text{black}) = \frac{1}{8}$$

81. Four of the sectors are green or blue.

$$P(\text{green or blue}) = \frac{4}{8} = \frac{1}{2}$$

83. Three of the sectors are red, green, or black.

$$P(\text{red, green, or black}) = \frac{3}{8}$$

85. None of the sectors is white.

$$P(\text{white}) = \frac{0}{8} = 0$$

87. $P(\text{impossible event}) = 0$

Section 2.4 Practice Exercises

1. a. $\{x|x < 3.5\}$ $(-\infty, 3.5)$

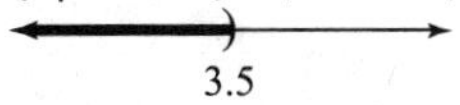

b. $\{x|x \geq -3\}$ $[-3, \infty)$

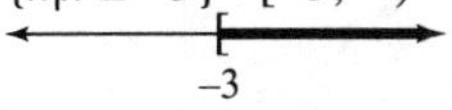

c. $\{x|-1 \leq x < 4\}$ $[-1, 4)$

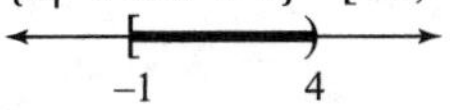

2.
$$x + 5 > 9$$
$$x + 5 - 5 > 9 - 5$$
$$x > 4$$
$(4, \infty)$

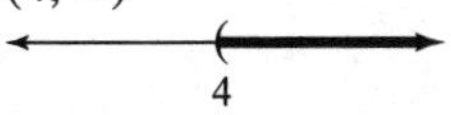

3.
$$3x + 1 \leq 2x - 3$$
$$3x + 1 - 2x \leq 2x - 3 - 2x$$
$$x + 1 \leq -3$$
$$x + 1 - 1 \leq -3 - 1$$
$$x \leq -4$$
$(-\infty, -4]$

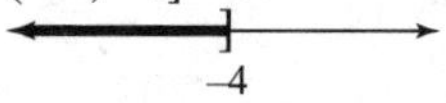

4. a.
$$\frac{2}{5}x \geq \frac{4}{15}$$
$$\frac{5}{2} \cdot \frac{2}{5}x \geq \frac{5}{2} \cdot \frac{4}{15}$$
$$x \geq \frac{2}{3}$$
$\left[\frac{2}{3}, \infty\right)$

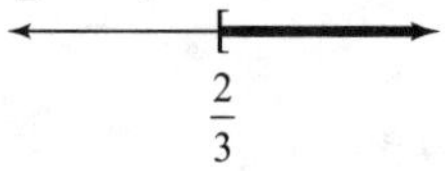

b.
$$-2.4x < 9.6$$
$$\frac{-2.4x}{-2.4} > \frac{9.6}{-2.4}$$
$$x > -4$$
$(-4, \infty)$

5.
$$-(4x + 6) \leq 2(5x + 9) + 2x$$
$$-4x - 6 \leq 10x + 18 + 2x$$
$$-4x - 6 \leq 12x + 18$$
$$-4x - 6 + 4x \leq 12x + 18 + 4x$$
$$-6 \leq 16x + 18$$
$$-6 - 18 \leq 16x + 18 - 18$$
$$-24 \leq 16x$$
$$\frac{-24}{16} \leq \frac{16x}{16}$$
$$-\frac{3}{2} \leq x$$
$\left[-\frac{3}{2}, \infty\right)$

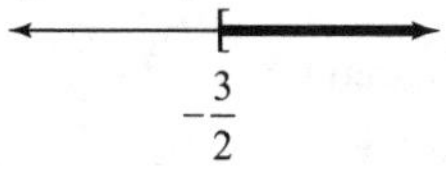

6. $\frac{3}{5}(x-3) \geq x-7$

$$5\left[\frac{3}{5}(x-3)\right] \geq 5(x-7)$$
$$3(x-3) \geq 5(x-7)$$
$$3x-9 \geq 5x-35$$
$$3x-9-5x \geq 5x-35-5x$$
$$-2x-9 \geq -35$$
$$-2x-9+9 \geq -35+9$$
$$-2x \geq -26$$
$$\frac{-2x}{-2} \leq \frac{-26}{-2}$$
$$x \leq 13$$

$(-\infty, 13]$

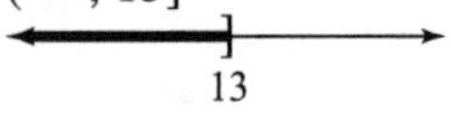

7. $4(x-2) < 4x+5$

$$4x-8 < 4x+5$$
$$4x-8-4x < 4x+5-4x$$
$$-8 < 5$$

This is a true statement for all values of x. The solution set is $\{x|x \text{ is a real number}\}$ or $(-\infty, \infty)$.

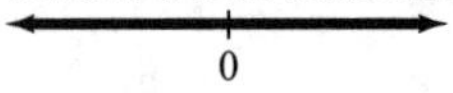

8.

In words:	900	+	commission (15% of sales)	≥	2400
	↓	↓	↓	↓	↓
Translate:	900	+	$0.15x$	≥	2400

$$900+0.15x \geq 2400$$
$$900+0.15x-900 \geq 2400-900$$
$$0.15x \geq 1500$$
$$x \geq 10{,}000$$

Sales must be greater than or equal to \$10,000 per month.

9. $-11.8t+390 < 175$

$$-11.8t < -215$$
$$t > \text{approximately } 18.2$$

The annual consumption of cigarettes will be less than 175 billion more than 18.2 years after 2004, or in approximately 18 + 2004 = 2022 and after.

Vocabulary, Readiness & Video Check 2.4

1. d. $(-\infty, -5)$

2. c. $[-11, \infty)$

3. b. $\left(-2.5, \frac{7}{4}\right]$

4. a. $\left[-\frac{10}{3}, 0.2\right)$

5. The set $\{x|x \geq -0.4\}$ written in interval notation is $\underline{[-0.4, \infty)}$.

6. The set $\{x|x < -0.4\}$ written in interval notation is $\underline{(-\infty, -0.4)}$.

7. The set $\{x|x \leq -0.4\}$ written in interval notation is $\underline{(-\infty, -0.4]}$.

8. The set $\{x|x > -0.4\}$ written in interval notation is $\underline{(-0.4, \infty)}$.

9. The graph of Example 1 is shaded from $-\infty$ to, but not including, -3, as indicated by a parenthesis. To write interval notation, write down what is shaded for the inequality from left to right. A parenthesis is always used with $-\infty$, so from the graph, the interval notation is $(-\infty, -3)$.

10. We can add the same number to (or subtract the same number from) both sides of a linear inequality in one variable and have an equivalent inequality; addition property of equality.

11. If you multiply or divide both sides of an inequality by the <u>same</u> nonzero negative number, you must <u>reverse</u> the direction of the inequality symbol.

12. maximum, or less

Exercise Set 2.4

1. $\{x|x < -3\}$
$(-\infty, -3)$

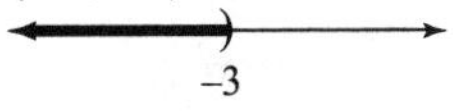

3. $\{x|x \geq 0.3\}$
$[0.3, \infty)$

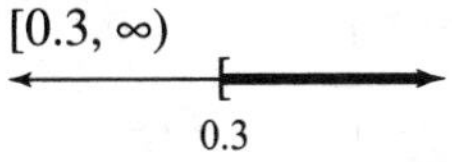

5. $\{x|-7 \leq x\}$
$[-7, \infty)$

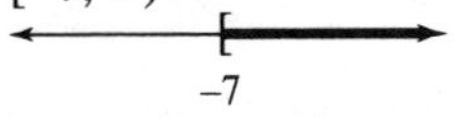

7. $\{x|-2 < x < 5\}$
$(-2, 5)$

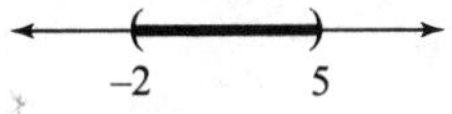

9. $\{x|5 \geq x > -1\}$
$(-1, 5]$

11.
$$x - 7 \geq -9$$
$$x \geq -2$$
$[-2, \infty)$

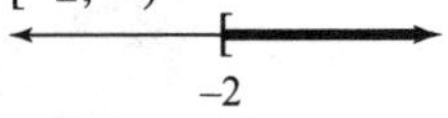

13.
$$7x < 6x + 1$$
$$x < 1$$
$(-\infty, 1)$

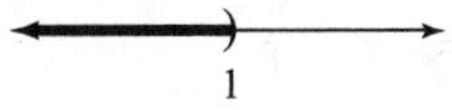

15.
$$8x - 7 \leq 7x - 5$$
$$x - 7 \leq -5$$
$$x \leq 2$$
$(-\infty, 2]$

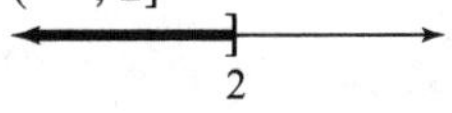

17.
$$\frac{3}{4}x \geq 6$$
$$\frac{4}{3} \cdot \frac{3}{4}x \geq \frac{4}{3} \cdot 6$$
$$x \geq 8$$
$[8, \infty)$

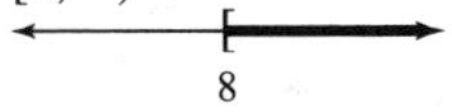

19.
$$5x < -23.5$$
$$x < -4.7$$
$(-\infty, -4.7)$

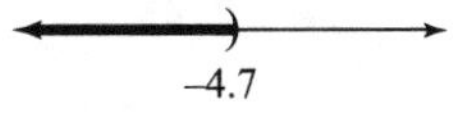

21.
$$-3x \geq 9$$
$$x \leq -3$$
$(-\infty, -3]$

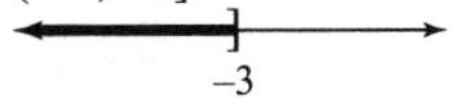

23.
$$-2x + 7 \geq 9$$
$$-2x \geq 2$$
$$x \leq -1$$
$(-\infty, -1]$

25.
$$15 + 2x \geq 4x - 7$$
$$15 \geq 2x - 7$$
$$22 \geq 2x$$
$$11 \geq x \text{ or } x \leq 11$$
$(-\infty, 11]$

27. $4(2x+1)>4$
$8x+4>4$
$8x>0$
$x>0$
$(0, \infty)$

29. $3(x-5)<2(2x-1)$
$3x-15<4x-2$
$-15<x-2$
$-13<x$
$x>-13$
$(-13, \infty)$

31. $\frac{5x+1}{7}-\frac{2x-6}{4}\ge -4$
$28\left(\frac{5x+1}{7}-\frac{2x-6}{4}\right)\ge 28(-4)$
$4(5x+1)-7(2x-6)\ge -112$
$20x+4-14x+42\ge -112$
$6x+46\ge -112$
$6x\ge -158$
$x\ge -\frac{79}{3}$
$\left[-\frac{79}{3}, \infty\right)$

33. $-3(2x-1)<-4[2+3(x+2)]$
$-6x+3<-4(2+3x+6)$
$-6x+3<-4(8+3x)$
$-6x+3<-32-12x$
$6x+3<-32$
$6x<-35$
$x<-\frac{35}{6}$
$\left(-\infty, -\frac{35}{6}\right)$

35. $x+9<3$
$x+9-9<3-9$
$x<-6$
$(-\infty, -6)$

37. $-x<-4$
$\frac{-x}{-1}>\frac{-4}{-1}$
$x>4$
$(4, \infty)$

39. $-7x\le 3.5$
$\frac{-7x}{-7}\ge \frac{3.5}{-7}$
$x\ge -0.5$
$[-0.5, \infty)$

41. $\frac{1}{2}+\frac{2}{3}\ge \frac{x}{6}$
$6\left(\frac{1}{2}+\frac{2}{3}\right)\ge 6\left(\frac{x}{6}\right)$
$3+4\ge x$
$7\ge x$
$x\le 7$
$(-\infty, 7]$

43. $-5x+4\le -4(x-1)$
$-5x+4\le -4x+4$
$-x\le 0$
$x\ge 0$
$[0, \infty)$

45. $\frac{3}{4}(x-7)\ge x+2$
$4\left[\frac{3}{4}(x-7)\right]\ge 4(x+2)$
$3(x-7)\ge 4(x+2)$
$3x-21\ge 4x+8$
$-x-21\ge 8$
$-x\ge 29$
$x\le -29$
$(-\infty, -29]$

47. $0.8x+0.6x\ge 4.2$
$1.4x\ge 4.2$
$x\ge 3$
$[3, \infty)$

49. $4(x-6)+2x-4\ge 3(x-7)+10x$
$4x-24+2x-4\ge 3x-21+10x$
$6x-28\ge 13x-21$
$-28\ge 7x-21$
$-7\ge 7x$
$-1\ge x$
$x\le -1$
$(-\infty, -1]$

51. $14-(5x-6)\ge -6(x+1)-5$
$$14-5x+6\ge -6x-6-5$$
$$-5x+20\ge -6x-11$$
$$x+20\ge -11$$
$$x\ge -31$$
$[-31, \infty)$

53. $4(x-1)\ge 4x-8$
$$4x-4\ge 4x-8$$
$$-4\ge -8 \quad \text{(True for all } x\text{)}$$
All real numbers
$(-\infty, \infty)$

55. $3x+1<3(x-2)$
$$3x+1<3x-6$$
$$1<-6 \quad \text{False}$$
$\varnothing$

57. $0.4(4x-3)<1.2(x+2)$
$$10[0.4(4x-3)]<10[1.2(x+2)]$$
$$4(4x-3)<12(x+2)$$
$$16x-12<12x+24$$
$$4x-12<24$$
$$4x<36$$
$$x<9$$
$(-\infty, 9)$

59. $\frac{2}{5}x-\frac{1}{4}\le \frac{3}{10}x-\frac{4}{5}$
$$20\left[\frac{2}{5}x-\frac{1}{4}\right]\le 20\left[\frac{3}{10}x-\frac{4}{5}\right]$$
$$4\cdot 2x-5\le 2\cdot 3x-4\cdot 4$$
$$8x-5\le 6x-16$$
$$2x-5\le -16$$
$$2x\le -11$$
$$x\le -\frac{11}{2}$$
$\left(-\infty, -\frac{11}{2}\right]$

61. $\frac{1}{2}(3x-4)\le \frac{3}{4}(x-6)+1$
$$4\left[\frac{1}{2}(3x-4)\right]\le 4\left[\frac{3}{4}(x-6)+1\right]$$
$$2(3x-4)\le 3(x-6)+4$$
$$6x-8\le 3x-18+4$$
$$6x-8\le 3x-14$$
$$3x-8\le -14$$
$$3x\le -6$$
$$x\le -2$$
$(-\infty, -2]$

63. $\frac{-x+2}{2}-\frac{1-5x}{8}<-1$
$$8\left(\frac{-x+2}{2}-\frac{1-5x}{8}\right)<8(-1)$$
$$4(-x+2)-(1-5x)<-8$$
$$-4x+8-1+5x<-8$$
$$x+7<-8$$
$$x<-15$$
$(-\infty, -15)$

65. $\frac{x+5}{5}-\frac{3+x}{8}\ge -\frac{3}{10}$
$$40\left(\frac{x+5}{5}-\frac{3+x}{8}\right)\ge 40\left(-\frac{3}{10}\right)$$
$$8(x+5)-5(3+x)\ge -12$$
$$8x+40-15-5x\ge -12$$
$$3x+25\ge -12$$
$$3x\ge -37$$
$$x\ge -\frac{37}{3}$$
$\left[-\frac{37}{3}, \infty\right)$

67. $\frac{x+3}{12}+\frac{x-5}{15}<\frac{2}{3}$
$$60\left(\frac{x+3}{12}+\frac{x-5}{15}\right)<60\left(\frac{2}{3}\right)$$
$$5(x+3)+4(x-5)<20(2)$$
$$5x+15+4x-20<40$$
$$9x-5<40$$
$$9x<45$$
$$x<5$$
$(-\infty, 5)$

69. a. Let x be the final exam score.

$$\frac{72+67+82+79+x+x}{6} \ge 77$$
$$6\left(\frac{72+67+82+79+2x}{6}\right) \ge 6(77)$$
$$72+67+82+79+2x \ge 462$$
$$300+2x \ge 462$$
$$300+2x-300 \ge 462-300$$
$$2x \ge 162$$
$$\frac{2x}{2} \ge \frac{162}{2}$$
$$x \ge 81$$

The solution is $\{x|x \ge 81\}$.

b. A final exam grade of 81 or higher will result in an average of 77 or higher.

71. a. Let x be the number of whole boxes on the elevator.

$$147+66x \le 1500$$
$$147+66x-147 \le 1500-147$$
$$66x \le 1353$$
$$x \le 20.5$$

The solution is $\{x|x \le 20.5\}$ or since x is a whole number $\{x|x \le 20\}$.

b. She can move at most 20 whole boxes at one time.

73. a. Let x be the weight of luggage and cargo. The six passengers have a total weight of 6(160) pounds.

$$x+6(160) \le 2000$$
$$x+960 \le 2000$$
$$x \le 1040$$

The solution is $\{x|x \le 1040\}$.

b. The luggage and cargo must weigh no more than 1040 pounds.

75. a. Let x be the number of daily miles driven.

$$36 < 24+0.15x$$
$$12 < 0.15x$$
$$\frac{12}{0.15} < \frac{0.15x}{0.15}$$
$$80 < x$$
$$\{x|x > 80\}$$

b. If you drive more than 80 miles a day, plan A is more economical.

77. $F \ge \frac{9}{5}C+32$

$$F \ge \frac{9}{5}(500)+32$$
$$F \ge 932$$
$$\{F|F \ge 932\}$$

Glass is a liquid at temperatures of 932°F or higher.

79. a.

$$2464t+40,067 > 70,000$$
$$2464t > 29,933$$
$$t > \text{approximately } 12.1$$

t is more than 12.1 so $t \ge 13$.
2011 + 13 = 2024
Salaries will be greater than $70,000 in 2024.

b. answers may vary

81. Consumption of whole milk is decreasing over time; answers may vary.

83. 2024 is 20 years after 2004, so 2024 corresponds to $t = 20$.

$w = -1.9t + 59$
$w = -1.9(20) + 59 = -38 + 59 = 21$
The average consumption of whole milk is predicted to be 21 pounds per person per year in 2024.

85.

$$-1.9t+59 < 35$$
$$-1.9t < -24$$
$$t > \text{approximately } 12.6$$

12.6 years after 2004 corresponds to some time during the year 2016.
During 2016, the consumption of whole milk will be less than 35 pounds per person per year.

87. answers may vary

89. a.

$$s > w$$
$$-0.22t+27.4 > -1.9t+59$$
$$1.68t > 31.6$$
$$t > \text{approximately } 18.8$$

The solution is $\{t|t > 18.8\}$.

b. 2004 + 19 = 2023
2023 is the first year when the consumption of skim milk is predicted to be greater than the consumption of whole milk.

91. $x < 5$ and $x > 1$
The integers are 2, 3, and 4.

93. $x \ge -2$ and $x \ge 2$
The integers are 2, 3, 4, ...

95.
$$\begin{aligned} 2x-6&=4 \\ 2x-6+6&=4+6 \\ 2x&=10 \\ \frac{2x}{2}&=\frac{10}{2} \\ x&=5 \end{aligned}$$

97.
$$\begin{aligned} -x+7&=5x-6 \\ -x-5x&=-6-7 \\ -6x&=-13 \\ \frac{-6x}{-6}&=\frac{-13}{-6} \\ x&=\frac{13}{6} \end{aligned}$$

99. $\{x|x \ge 2\}$, $[2, \infty)$

101.
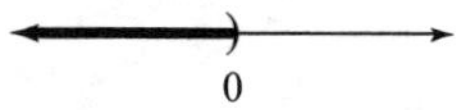

$(-\infty, 0)$

103. $\{x|-2 < x \le 1.5\}$

105. To solve $-3x \le 14$, both sides must be divided by -3, so the inequality symbol will be reversed.

107. To solve $-x \ge -23$, both sides must be divided by -1, so the inequality symbol will be reversed.

109.
$$\begin{aligned} 2x-3&=5 \\ 2x&=8 \\ x&=4 \end{aligned}$$
The solution set is $\{4\}$.

111.
$$\begin{aligned} 2x-3&<5 \\ 2x&<8 \\ x&<4 \end{aligned}$$
The solution set is $(-\infty, 4)$.

113. answers may vary

115. answers may vary

Integrated Review

1.
$$\begin{aligned} -4x&=20 \\ \frac{-4x}{-4}&=\frac{20}{-4} \\ x&=-5 \end{aligned}$$

2.
$$\begin{aligned} -4x&<20 \\ \frac{-4x}{-4}&>\frac{20}{-4} \\ x&>-5 \end{aligned}$$
$(-5, \infty)$

3.
$$\begin{aligned} \frac{3x}{4}&\ge 2 \\ 4\left(\frac{3x}{4}\right)&\ge 4(2) \\ 3x&\ge 8 \\ x&\ge \frac{8}{3} \end{aligned}$$
$\left[\frac{8}{3}, \infty\right)$

4.
$$\begin{aligned} 5x+3&\ge 2+4x \\ x+3&\ge 2 \\ x&\ge -1 \end{aligned}$$
$[-1, \infty)$

5.
$$\begin{aligned} 6(y-4)&=3(y-8) \\ 6y-24&=3y-24 \\ 3y&=0 \\ y&=0 \end{aligned}$$

6.
$$\begin{aligned} -4x&\le \frac{2}{5} \\ -20x&\le 2 \\ x&\ge -\frac{1}{10} \end{aligned}$$
$\left[-\frac{1}{10}, \infty\right)$

7.
$$\begin{aligned} -3x&\ge \frac{1}{2} \\ 2(-3x)&\ge 2\left(\frac{1}{2}\right) \\ -6x&\ge 1 \\ x&\le -\frac{1}{6} \end{aligned}$$
$\left(-\infty, -\frac{1}{6}\right]$

8.
$$\begin{aligned} 5(y+4)&=4(y+5) \\ 5y+20&=4y+20 \\ y&=0 \end{aligned}$$

9. $7x < 7(x-2)$
$7x < 7x - 14$
$0 < -14$ (False)
No solution; $\varnothing$

10. $\dfrac{-5x+11}{2} \le 7$
$2\left(\dfrac{-5x+11}{2}\right) \le 2(7)$
$-5x + 11 \le 14$
$-5x \le 3$
$x \ge -\dfrac{3}{5}$
$\left[-\dfrac{3}{5}, \infty\right)$

11. $-5x + 1.5 = -19.5$
$-5x + 1.5 - 1.5 = -19.5 - 1.5$
$-5x = -21$
$\dfrac{-5x}{-5} = \dfrac{-21}{-5}$
$x = 4.2$

12. $-5x + 4 = -26$
$-5x = -30$
$x = 6$

13. $5 + 2x - x = -x + 3 - 14$
$5 + x = -x - 11$
$5 + 2x = -11$
$2x = -16$
$x = -8$

14. $12x + 14 < 11x - 2$
$x + 14 < -2$
$x < -16$
$(-\infty, -16)$

15. $\dfrac{x}{5} - \dfrac{x}{4} = \dfrac{x-2}{2}$
$20\left(\dfrac{x}{5} - \dfrac{x}{4}\right) = 20\left(\dfrac{x-2}{2}\right)$
$4x - 5x = 10(x-2)$
$-x = 10x - 20$
$-11x = -20$
$x = \dfrac{20}{11}$

16. $12x - 12 = 8(x-1)$
$12x - 12 = 8x - 8$
$4x - 12 = -8$
$4x = 4$
$x = 1$

17. $2(x-3) > 70$
$2x - 6 > 70$
$2x > 76$
$x > 38$
$(38, \infty)$

18. $-3x - 4.7 = 11.8$
$-3x - 4.7 + 4.7 = 11.8 + 4.7$
$-3x = 16.5$
$\dfrac{-3x}{-3} = \dfrac{16.5}{-3}$
$x = -5.5$

19. $-2(b-4) - (3b-1) = 5b + 3$
$-2b + 8 - 3b + 1 = 5b + 3$
$-5b + 9 = 5b + 3$
$-10b = -6$
$b = \dfrac{-6}{-10} = \dfrac{3}{5}$

20. $8(x+3) < 7(x+5) + x$
$8x + 24 < 7x + 35 + x$
$8x + 24 < 8x + 35$
$24 < 35$ (True for all x)
All real numbers; $(-\infty, \infty)$

21. $\dfrac{3t+1}{8} = \dfrac{5+2t}{7} + 2$
$56\left(\dfrac{3t+1}{8}\right) = 56\left(\dfrac{5+2t}{7}\right) + 56(2)$
$7(3t+1) = 8(5+2t) + 112$
$21t + 7 = 40 + 16t + 112$
$21t + 7 = 16t + 152$
$5t = 145$
$t = 29$

22. $4(x-6) - x = 8(x-3) - 5x$
$4x - 24 - x = 8x - 24 - 5x$
$3x - 24 = 3x - 24$
$-24 = -24$ (True for all x)
The solution is all real numbers.

23. $$\frac{x}{6}+\frac{3x-2}{2}<\frac{2}{3}$$
$$6\left(\frac{x}{6}+\frac{3x-2}{2}\right)<6\left(\frac{2}{3}\right)$$
$$x+3(3x-2)<4$$
$$x+9x-6<4$$
$$10x-6<4$$
$$10x<10$$
$$x<1$$
$(-\infty, 1)$

24. $$\frac{y}{3}+\frac{y}{5}=\frac{y+3}{10}$$
$$30\left(\frac{y}{3}\right)+30\left(\frac{y}{5}\right)=30\left(\frac{y+3}{10}\right)$$
$$10y+6y=3(y+3)$$
$$16y=3y+9$$
$$13y=9$$
$$y=\frac{9}{13}$$

25. $$5(x-6)+2x>3(2x-1)-4$$
$$5x-30+2x>6x-3-4$$
$$7x-30>6x-7$$
$$x>23$$
$(23, \infty)$

26. $$14(x-1)-7x\le 2(3x-6)+4$$
$$14x-14-7x\le 6x-12+4$$
$$7x-14\le 6x-8$$
$$x\le 6$$
$(-\infty, 6]$

27. $$\frac{1}{4}(3x+2)-x\ge\frac{3}{8}(x-5)+2$$
$$8\left[\frac{1}{4}(3x+2)-x\right]\ge 8\left[\frac{3}{8}(x-5)+2\right]$$
$$2(3x+2)-8x\ge 3(x-5)+16$$
$$6x+4-8x\ge 3x-15+16$$
$$-2x+4\ge 3x+1$$
$$3\ge 5x$$
$$\frac{3}{5}\ge x \quad\text{or}\quad x\le\frac{3}{5}$$
$\left(-\infty, \frac{3}{5}\right]$

28. $$\frac{1}{3}(x-10)-4x>\frac{5}{6}(2x+1)-1$$
$$6\left[\frac{1}{3}(x-10)-4x\right]>6\left[\frac{5}{6}(2x+1)-1\right]$$
$$2(x-10)-24x>5(2x+1)-6$$
$$2x-20-24x>10x+5-6$$
$$-22x-20>10x-1$$
$$-19>32x$$
$$-\frac{19}{32}>x \quad\text{or}\quad x<-\frac{19}{32}$$
$\left(-\infty, -\frac{19}{32}\right)$

Section 2.5 Practice Exercises

1. $A = \{1, 3, 5, 7, 9\}$ and $B = \{1, 2, 3, 4\}$
The numbers 1 and 3 are in sets A and B.
The intersection is $\{1, 3\}$. $A \cap B = \{1, 3\}$.

2. $x+3<8$ *and* $2x-1<3$
$x<5$ *and* $2x<4$
$x<5$ *and* $x<2$

$\{x|x<5\}$, $(-\infty, 5)$

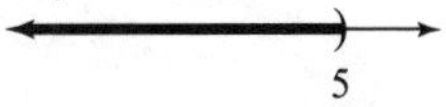

$\{x|x<2\}$, $(-\infty, 2)$

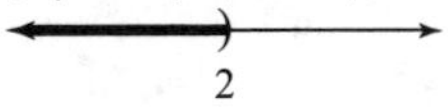

$\{x|x<5 \text{ and } x<2\} = \{x|x<2\}$

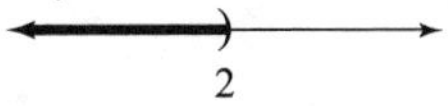

The solution set is $(-\infty, 2)$.

3. $4x\le 0$ *and* $3x+2>8$
$x\le 0$ *and* $3x>6$
$x\le 0$ *and* $x>2$

$\{x|x\le 0\}$, $(-\infty, 0]$

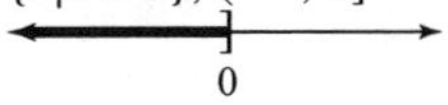

$\{x|x>2\}$, $(2, \infty)$

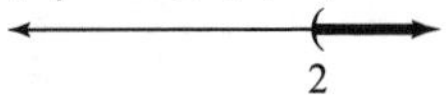

$\{x|4x\le 0 \text{ and } 3x+2>8\} = \{\ \}$ or $\varnothing$

4. $$3<5-x<9$$
$$3-5<5-x-5<9-5$$
$$-2<-x<4$$
$$\frac{-2}{-1}>\frac{-x}{-1}>\frac{4}{-1}$$
$$2>x>-4$$
or $-4<x<2$
The solution set is $(-4, 2)$.

5. $-4 \le \frac{x}{2} - 1 \le 3$

$2(-4) \le 2\left(\frac{x}{2} - 1\right) \le 2(3)$

$-8 \le x - 2 \le 6$

$-8 + 2 \le x - 2 + 2 \le 6 + 2$

$-6 \le x \le 8$

The solution set is $[-6, 8]$.

6. $A = \{1, 3, 5, 7, 9\}$ and $B = \{2, 3, 4, 5, 6\}$.
The numbers that are in either set or both sets are $\{1, 2, 3, 4, 5, 6, 7, 9\}$. This set is the union, $A \cup B$.

7. $8x + 5 \le 8$ *or* $x - 1 \ge 2$

$8x \le 3$ *or* $x \ge 3$

$x \le \frac{3}{8}$ *or* $x \ge 3$

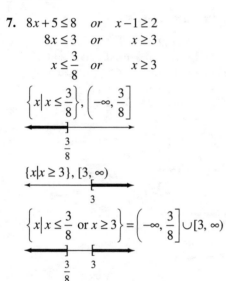

$\left\{x \middle| x \le \frac{3}{8}\right\}, \left(-\infty, \frac{3}{8}\right]$

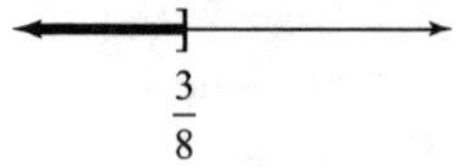

$\{x|x \ge 3\}, [3, \infty)$

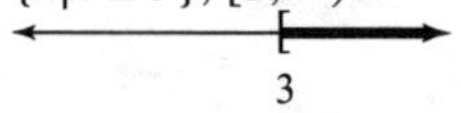

$\left\{x \middle| x \le \frac{3}{8} \text{ or } x \ge 3\right\} = \left(-\infty, \frac{3}{8}\right] \cup [3, \infty)$

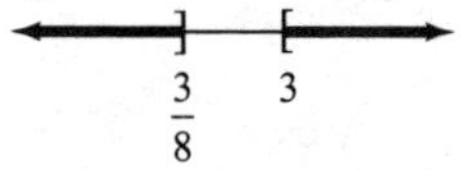

The solution set is $\left(-\infty, \frac{3}{8}\right] \cup [3, \infty)$.

8. $-3x - 2 > -8$ *or* $5x > 0$

$-3x > -6$ *or* $x > 0$

$x < 2$ *or* $x > 0$

$\{x|x < 2\}, (-\infty, 2)$

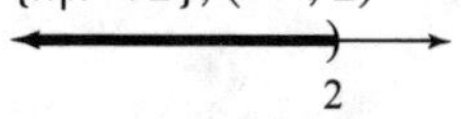

$\{x|x > 0\}, (0, \infty)$

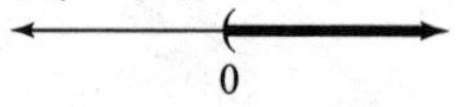

$\{x|x < 2 \text{ or } x > 0\}, (-\infty, \infty)$

The solution set is $(-\infty, \infty)$.

Vocabulary, Readiness & Video Check 2.5

1. Two inequalities joined by the words "and" or "or" are called <u>compound</u> inequalities.

2. The word <u>and</u> means intersection.

3. The word <u>or</u> means union.

4. The symbol <u>$\cap$</u> means intersection.

5. The symbol <u>$\cup$</u> represents union.

6. The symbol <u>$\varnothing$</u> is the empty set.

7. For an element to be in the intersection of sets *A* and *B*, the element must be in set *A* <u>and</u> in set *B*.

8. Graph the two intervals, each on its own number line, so you can see their intersection. Graph this intersection on the third number line—this intersection is the solution set.

9. For an element to be in the union of sets *A* and *B*, the element must be in set *A* <u>or</u> in set *B*.

10. Graph the two intervals, each on its own number line, so you can see their union. Graph this union on the third number line—this union is the solution set.

Exercise Set 2.5

1. $C \cup D = \{2, 3, 4, 5, 6, 7\}$

3. $A \cap D = \{4, 6\}$

5. $A \cup B = \{..., -2, -1, 0, 1, ...\}$

7. $B \cap D = \{5, 7\}$

9. $B \cup C = \{x|x \text{ is an odd integer or } x = 2 \text{ or } x = 4\}$

11. $A \cap C = \{2, 4\}$

13. $x < 1$ *and* $x > -3$

$-3 < x < 1$

$(-3, 1)$

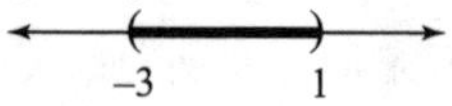

15. $x \le -3$ *and* $x \ge -2$

$\varnothing$

17. $x < -1$ *and* $x < 1$

$x < -1$

$(-\infty, -1)$

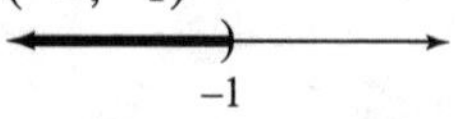

19. $x+1\ge 7$ *and* $3x-1\ge 5$
$x\ge 6$ *and* $3x\ge 6$
$x\ge 2$

$x\ge 6$
$[6, \infty)$

21. $4x+2\le -10$ *and* $2x\le 0$
$4x\le -12$ *and* $x\le 0$
$x\le -3$

$x\le -3$
$(-\infty, -3]$

23. $-2x<-8$ *and* $x-5<5$
$x>4$ *and* $x<10$
$(4, 10)$

25. $5<x-6<11$
$11<x<17$
$(11, 17)$

27. $-2\le 3x-5\le 7$
$3\le 3x\le 12$
$1\le x\le 4$
$[1, 4]$

29. $1\le \frac{2}{3}x+3\le 4$
$-2\le \frac{2}{3}x\le 1$
$-3\le x\le \frac{3}{2}$
$\left[-3, \frac{3}{2}\right]$

31. $-5\le \frac{-3x+1}{4}\le 2$
$4(-5)\le 4\left(\frac{-3x+1}{4}\right)\le 4(2)$
$-20\le -3x+1\le 8$
$-21\le -3x\le 7$
$7\ge x\ge -\frac{7}{3}$
$-\frac{7}{3}\le x\le 7$
$\left[-\frac{7}{3}, 7\right]$

33. $x<4$ *or* $x<5$
$(-\infty, 5)$

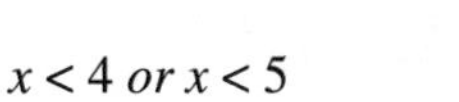

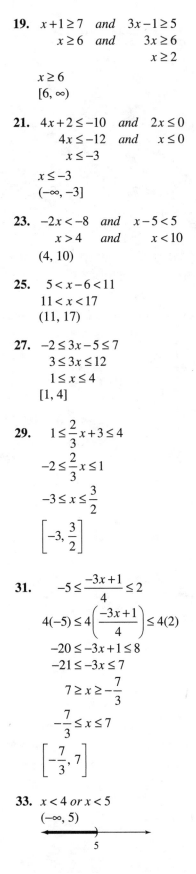

35. $x\le -4$ *or* $x\ge 1$
$(-\infty, -4]\cup[1, \infty)$

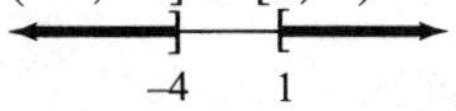

37. $x>0$ *or* $x<3$
$(-\infty, \infty)$

39. $-2x\le -4$ *or* $5x-20\ge 5$
$x\ge 2$ *or* $5x\ge 25$
$x\ge 5$

$x\ge 2$
$[2, \infty)$

41. $x+4<0$ *or* $6x>-12$
$x<-4$ *or* $x>-2$
$(-\infty, -4)\cup(-2, \infty)$

43. $3(x-1)<12$ *or* $x+7>10$
$x-1<4$ *or* $x>3$
$x<5$
$(-\infty, \infty)$

45. $x<\frac{2}{3}$ *and* $x>-\frac{1}{2}$
$-\frac{1}{2}<x<\frac{2}{3}$
$\left(-\frac{1}{2}, \frac{2}{3}\right)$

47. $x<\frac{2}{3}$ *or* $x>-\frac{1}{2}$
$(-\infty, \infty)$

49. $0\le 2x-3\le 9$
$3\le 2x\le 12$
$\frac{3}{2}\le x\le 6$
$\left[\frac{3}{2}, 6\right]$

51. $\frac{1}{2} < x - \frac{3}{4} < 2$

$4\left(\frac{1}{2}\right) < 4\left(x - \frac{3}{4}\right) < 4(2)$

$2 < 4x - 3 < 8$

$5 < 4x < 11$

$\frac{5}{4} < x < \frac{11}{4}$

$\left(\frac{5}{4}, \frac{11}{4}\right)$

53. $x + 3 \geq 3$ *and* $x + 3 \leq 2$

$x \geq 0$ *and* $x \leq -1$

No solution exists.

$\varnothing$

55. $3x \geq 5$ *or* $-\frac{5}{8}x - 6 > 1$

$x \geq \frac{5}{3}$ *or* $-\frac{5}{8}x > 7$

$x < -\frac{56}{5}$

$\left(-\infty, -\frac{56}{5}\right) \cup \left[\frac{5}{3}, \infty\right)$

57. $0 < \frac{5 - 2x}{3} < 5$

$0 < 5 - 2x < 15$

$\frac{-5}{-2} > \frac{-2x}{-2} > \frac{10}{-2}$

$\frac{5}{2} > x > -5$

$-5 < x < \frac{5}{2}$

$\left(-5, \frac{5}{2}\right)$

59. $-6 < 3(x - 2) \leq 8$

$-6 < 3x - 6 \leq 8$

$0 < 3x \leq 14$

$0 < x < \frac{14}{3}$

$\left(0, \frac{14}{3}\right]$

61. $-x + 5 > 6$ *and* $1 + 2x \leq -5$

$-x > 1$ *and* $2x \leq -6$

$x < -1$ *and* $x \leq -3$

$x \leq -3$

$(-\infty, -3]$

63. $3x + 2 \leq 5$ *or* $7x > 29$

$3x \leq 3$ *or* $x > \frac{29}{7}$

$x \leq 1$ *or* $x > \frac{29}{7}$

$(-\infty, 1] \cup \left(\frac{29}{7}, \infty\right)$

65. $5 - x > 7$ *and* $2x + 3 \geq 13$

$-x > 2$ *and* $2x \geq 10$

$x < -2$ *and* $x \geq 5$

No solution exists.

$\varnothing$

67. $-\frac{1}{2} \leq \frac{4x - 1}{6} < \frac{5}{6}$

$6\left(-\frac{1}{2}\right) \leq 6\left(\frac{4x - 1}{6}\right) < 6\left(\frac{5}{6}\right)$

$-3 \leq 4x - 1 < 5$

$-2 \leq 4x < 6$

$-\frac{1}{2} \leq x < \frac{3}{2}$

$\left[-\frac{1}{2}, \frac{3}{2}\right)$

69. $\frac{1}{15} < \frac{8 - 3x}{15} < \frac{4}{5}$

$15\left(\frac{1}{15}\right) < 15\left(\frac{8 - 3x}{15}\right) < 15\left(\frac{4}{5}\right)$

$1 < 8 - 3x < 12$

$-7 < -3x < 4$

$-\frac{4}{3} < x < \frac{7}{3}$

$\left(-\frac{4}{3}, \frac{7}{3}\right)$

71. $0.3 < 0.2x - 0.9 < 1.5$

$1.2 < 0.2x < 2.4$

$6 < x < 12$

$(6, 12)$

73. $|-7| - |19| = 7 - 19 = -12$

75. $-(-6) - |-10| = 6 - 10 = -4$

77. $|x| = 7$
$x = -7, 7$

79. $|x| = 0$
$x = 0$

81. Both lines are above the level representing 1500 for the years 2004 and 2005.

83. answers may vary

85. $2x - 3 < 3x + 1 < 4x - 5$
$2x - 3 < 3x + 1$ *and* $3x + 1 < 4x - 5$
$-x < 4$ *and* $-x < -6$
$x > -4$ *and* $x > 6$
$x > 6$
$(6, \infty)$

87. $-3(x-2) \le 3 - 2x \le 10 - 3x$
$-3x + 6 \le 3 - 2x$ *and* $3 - 2x \le 10 - 3x$
$-x \le -3$ *and* $x \le 7$
$x \ge 3$
$3 \le x \le 7$
$[3, 7]$

89. $5x - 8 < 2(2 + x) < -2(1 + 2x)$
$5x - 8 < 4 + 2x$ *and* $4 + 2x < -2 - 4x$
$3x < 12$ *and* $6x < -6$
$x < 4$ *and* $x < -1$
$x < -1$
$(-\infty, -1)$

91.
$$-29 \le C \le 35$$
$$-29 \le \frac{5}{9}(F - 32) \le 35$$
$$-52.5 \le F - 32 \le 63$$
$$-20.2 \le F \le 95$$
$$-20.2° \le F \le 95°$$

93.
$$70 \le \frac{68 + 65 + 75 + 78 + 2x}{6} \le 79$$
$$420 \le 286 + 2x \le 474$$
$$134 \le 2x \le 188$$
$$67 \le x \le 94$$
If Christian scores between 67 and 94 inclusive on his final exam, he will receive a C in the course.

Section 2.6 Practice Exercises

1. $|q| = 13$
$q = 13$ or $q = -13$
The solution set is $\{-13, 13\}$.

2. $|2x - 3| = 5$
$2x - 3 = 5$ or $2x - 3 = -5$
$2x = 8$ or $2x = -2$
$x = 4$ or $x = -1$
The solution set is $\{-1, 4\}$.

3. $\left|\frac{x}{5} + 1\right| = 15$
$\frac{x}{5} + 1 = 15$ or $\frac{x}{5} + 1 = -15$
$\frac{x}{5} = 14$ or $\frac{x}{5} = -16$
$x = 70$ or $x = -80$
The solutions are -80 and 70.

4. $|3x| + 8 = 14$
$|3x| = 6$
$3x = 6$ or $3x = -6$
$x = 2$ or $x = -2$
The solutions are -2 and 2.

5. $|z| = 0$
The solution is 0.

6. $3|z| + 9 = 7$
$3|z| = -2$
$|z| = -\frac{2}{3}$
The absolute value of a number is never negative, so there is no solution. The solution set is { } or ∅.

7. $\left|\frac{5x + 3}{4}\right| = -8$
The absolute value of a number is never negative, so there is no solution. The solution set is { } or ∅.

8. $|2x + 4| = |3x - 1|$

$$\begin{aligned} 2x+4 &= 3x-1 \quad \text{or} \quad & 2x+4 &= -(3x-1) \\ -x+4 &= -1 & 2x+4 &= -3x+1 \\ -x &= -5 & 5x+4 &= 1 \\ x &= 5 & 5x &= -3 \\ & & x &= -\frac{3}{5} \end{aligned}$$

The solutions are $-\frac{3}{5}$ and 5.

9. $|x - 2| = |8 - x|$

$$\begin{aligned} x-2 &= 8-x \quad \text{or} \quad & x-2 &= -(8-x) \\ 2x-2 &= 8 & x-2 &= -8+x \\ 2x &= 10 & -2 &= -8 \quad \text{False} \\ x &= 5 \end{aligned}$$

The solution is 5.

Vocabulary, Readiness & Video Check 2.6

1. $|x - 2| = 5$
C. $x - 2 = 5$ or $x - 2 = -5$

2. $|x - 2| = 0$
A. $x - 2 = 0$

3. $|x - 2| = |x + 3|$
B. $x - 2 = x + 3$ or $x - 2 = -(x + 3)$

4. $|x + 3| = 5$
E. $x + 3 = 5$ or $x + 3 = -5$

5. $|x + 3| = -5$
D. $\varnothing$

6. If a is negative, $|X| = a$ has no solution. (Also, if a is 0, we solve $X = 0$.)

Exercise Set 2.6

1. $|x| = 7$
$x = 7$ or $x = -7$

3. $|3x| = 12.6$

$$\begin{aligned} 3x &= 12.6 \quad \text{or} \quad & 3x &= -12.6 \\ x &= 4.2 \quad \text{or} \quad & x &= -4.2 \end{aligned}$$

5. $|2x - 5| = 9$

$$\begin{aligned} 2x-5 &= 9 \quad \text{or} \quad & 2x-5 &= -9 \\ 2x &= 14 \quad \text{or} \quad & 2x &= -4 \\ x &= 7 \quad \text{or} \quad & x &= -2 \end{aligned}$$

7. $\left|\frac{x}{2} - 3\right| = 1$

$$\begin{aligned} \frac{x}{2}-3 &= 1 \quad \text{or} \quad & \frac{x}{2}-3 &= -1 \\ 2\left(\frac{x}{2}-3\right) &= 2(1) \quad \text{or} \quad & 2\left(\frac{x}{2}-3\right) &= 2(-1) \\ x-6 &= 2 \quad \text{or} \quad & x-6 &= -2 \\ x &= 8 \quad \text{or} \quad & x &= 4 \end{aligned}$$

9. $|z| + 4 = 9$

$$\begin{aligned} |z| &= 5 \\ z = -5 \quad &\text{or} \quad z = -5 \end{aligned}$$

11. $|3x| + 5 = 14$

$$\begin{aligned} |3x| &= 9 \\ 3x &= 9 \quad \text{or} \quad & 3x &= -9 \\ x &= 3 \quad \text{or} \quad & x &= -3 \end{aligned}$$

13. $|2x| = 0$

$$\begin{aligned} 2x &= 0 \\ x &= 0 \end{aligned}$$

15. $|4n + 1| + 10 = 4$
$|4n + 1| = -6$ which is impossible.
The solution set is $\varnothing$.

17. $|5x - 1| = 0$

$$\begin{aligned} 5x-1 &= 0 \\ 5x &= 1 \\ x &= \frac{1}{5} \end{aligned}$$

19. $|5x - 7| = |3x + 11|$

$$\begin{aligned} 5x-7 &= 3x+11 \quad \text{or} \quad & 5x-7 &= -(3x+11) \\ 2x &= 18 \quad \text{or} \quad & 5x-7 &= -3x-11 \\ x &= 9 \quad \text{or} \quad & 8x &= -4 \\ & & x &= -\frac{1}{2} \end{aligned}$$

21. $|z + 8| = |z - 3|$

$$\begin{aligned} z+8 &= z-3 \quad \text{or} \quad & z+8 &= -(z-3) \\ 8 &= -3 \quad \text{or} \quad & z+8 &= -z+3 \\ & & 2z &= -5 \\ & & z &= -\frac{5}{2} \end{aligned}$$

The only solution is $-\frac{5}{2}$.

23. $|x| = 4$
$x = 4$ or $x = -4$

25. $|y| = 0$; $y = 0$

27. $|z| = -2$ is impossible. The solution set is $\varnothing$.

29. $|7 - 3x| = 7$
$7 - 3x = 7$ or $7 - 3x = -7$
$-3x = 0$ or $-3x = -14$
$x = 0$ or $x = \frac{14}{3}$

31. $|6x| - 1 = 11$
$|6x| = 12$
$6x = 12$ or $6x = -12$
$x = 2$ or $x = -2$

33. $|4p| = -8$ is impossible. The solution set is $\varnothing$.

35. $|x - 3| + 3 = 7$
$|x - 3| = 4$
$x - 3 = 4$ or $x - 3 = -4$
$x = 7$ or $x = -1$

37. $\left|\frac{z}{4} + 5\right| = -7$ is impossible. The solution set is $\varnothing$.

39. $|9v - 3| = -8$ is impossible. The solution set is $\varnothing$.

41. $|8n + 1| = 0$
$8n + 1 = 0$
$8n = -1$
$n = -\frac{1}{8}$

43. $|1 - 6c| - 7 = -3$
$|1 - 6c| = 4$
$1 - 6c = 4$ or $1 - 6c = -4$
$6c = 3$ or $6c = -5$
$c = \frac{1}{2}$ or $c = -\frac{5}{6}$

45. $|5x + 1| = 11$
$5x + 1 = 11$ or $5x + 1 = -11$
$5x = 10$ or $5x = -12$
$x = 2$ or $x = -\frac{12}{5}$

47. $|4x - 2| = |-10|$
$|4x - 2| = 10$
$4x - 2 = 10$ or $4x - 2 = -10$
$4x = 12$ or $4x = -8$
$x = 3$ or $x = -2$

49. $|5x + 1| = |4x - 7|$
$5x + 1 = 4x - 7$ or $5x + 1 = -(4x - 7)$
$x = -8$ or $5x + 1 = -4x + 7$
$9x = 6$
$x = \frac{2}{3}$

51. $|6 + 2x| = -|-7|$
$|6 + 2x| = -7$ which is impossible. The solution set is $\varnothing$.

53. $|2x - 6| = |10 - 2x|$
$2x - 6 = 10 - 2x$ or $2x - 6 = -(10 - 2x)$
$4x = 16$ or $2x - 6 = -10 + 2x$
$x = 4$ or $-6 = -10$
$-6 = -10$ is impossible. The only solution is 4.

55. $\left|\frac{2x - 5}{3}\right| = 7$
$\frac{2x - 5}{3} = 7$ or $\frac{2x - 5}{3} = -7$
$2x - 5 = 21$ or $2x - 5 = -21$
$2x = 26$ or $2x = -16$
$x = 13$ or $x = -8$

57. $2 + |5n| = 17$
$|5n| = 15$
$5n = 15$ or $5n = -15$
$n = 3$ or $n = -3$

59. $\left|\frac{2x - 1}{3}\right| = |-5|$
$\left|\frac{2x - 1}{3}\right| = 5$
$\frac{2x - 1}{3} = 5$ or $\frac{2x - 1}{3} = -5$
$2x - 1 = 15$ or $2x - 1 = -15$
$2x = 16$ or $2x = -14$
$x = 8$ or $x = -7$

61. $|2y-3| = |9-4y|$

$$\begin{aligned} 2y-3 &= 9-4y &\text{ or }&& 2y-3 &= -(9-4y) \\ 6y &= 12 &\text{ or }&& 2y-3 &= -9+4y \\ y &= 2 &\text{ or }&& -2y &= -6 \\ &&&& y &= 3 \end{aligned}$$

63. $\left|\frac{3n+2}{8}\right| = |-1|$

$$\left|\frac{3n+2}{8}\right| = 1$$

$$\begin{aligned} \frac{3n+2}{8} &= 1 &\text{ or }&& \frac{3n+2}{8} &= -1 \\ 3n+2 &= 8 &\text{ or }&& 3n+2 &= -8 \\ 3n &= 6 &\text{ or }&& 3n &= -10 \\ n &= 2 &\text{ or }&& n &= -\frac{10}{3} \end{aligned}$$

65. $|x+4| = |7-x|$

$$\begin{aligned} x+4 &= 7-x &\text{ or }&& x+4 &= -(7-x) \\ 2x &= 3 &\text{ or }&& x+4 &= -7+x \\ x &= \frac{3}{2} &\text{ or }&& 4 &= -7 \end{aligned}$$

$4 = -7$ is impossible. The only solution is $\frac{3}{2}$.

67. $\left|\frac{8c-7}{3}\right| = -|-5|$

$\left|\frac{8c-7}{3}\right| = -5$ which is impossible.

The solution set is $\varnothing$.

69. In 2014, 29.1% of cheese production came from cheddar cheese.

71. 9.2% of 360° is 0.092(360°) = 33.12°

73. answers may vary

75. answers may vary

77. Since absolute value is never negative, the solution set is $\varnothing$.

79. All numbers whose distance from 0 is 5 units is written as $|x| = 5$.

81. answers may vary

83. $|x-1| = 5$

85. answers may vary

87. $|x| = 6$

89. $|x-2| = |3x-4|$

Section 2.7 Practice Exercises

1. $|x| < 5$

The solution set of this inequality contains all numbers whose distance from 0 is less than 5. The solution set is $(-5, 5)$.

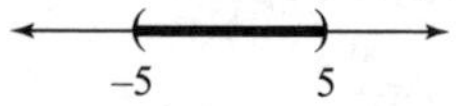

2. $|b+1| < 3$

$$\begin{aligned} -3 &< b+1 < 3 \\ -3-1 &< b+1-1 < 3-1 \\ -4 &< b < 2 \end{aligned}$$

$(-4, 2)$

3. $|3x-2| + 5 \le 9$

$$\begin{aligned} |3x-2| &\le 9-5 \\ |3x-2| &\le 4 \\ -4 \le 3x&-2 \le 4 \\ -4+2 \le 3x&-2+2 \le 4+2 \\ -2 \le &\,3x \le 6 \\ -\frac{2}{3} \le &\,x \le 2 \end{aligned}$$

$\left[-\frac{2}{3}, 2\right]$

4. $\left|3x + \frac{5}{8}\right| < -4$

The absolute value of a number is always nonnegative and can never be less than –4. The solution set is { } or $\varnothing$.

5. $\left|\frac{3(x-2)}{5}\right| \le 0$

$$\begin{aligned} \frac{3(x-2)}{5} &= 0 \\ 5\left[\frac{3(x-2)}{5}\right] &= 5(0) \\ 3(x-2) &= 0 \\ 3x-6 &= 0 \\ 3x &= 6 \\ x &= 2 \end{aligned}$$

The solution set is $\{2\}$.

6. $|y+4| \ge 6$

$$y+4 \le -6 \quad \text{or} \quad y+4 \ge 6$$
$$y+4-4 \le -6-4 \quad \text{or} \quad y+4-4 \ge 6-4$$
$$y \le -10 \quad \text{or} \quad y \ge 2$$

$(-\infty, -10] \cup [2, \infty)$

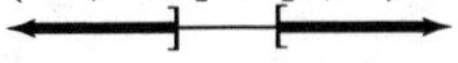

7.
$$|4x+3|+5 > 3$$
$$|4x+3|+5-5 > 3-5$$
$$|4x+3| > -2$$

The absolute value of any number is always nonnegative and thus is always greater than −2.

$(-\infty, \infty)$

8.
$$\left|\frac{x}{2}-3\right|-5 > -2$$
$$\left|\frac{x}{2}-3\right|-5+5 > -2+5$$
$$\left|\frac{x}{2}-3\right| > 3$$

$$\frac{x}{2}-3 < -3 \quad \text{or} \quad \frac{x}{2}-3 > 3$$
$$2\left(\frac{x}{2}-3\right) < 2(-3) \quad \text{or} \quad 2\left(\frac{x}{2}-3\right) > 2(3)$$
$$x-6 < -6 \quad \text{or} \quad x-6 > 6$$
$$x < 0 \quad \text{or} \quad x > 12$$

$(-\infty, 0) \cup (12, \infty)$

Vocabulary, Readiness & Video Check 2.7

1. D
2. E
3. C
4. B
5. A
6. The left side of the inequality is an absolute value, which must be nonnegative—it must be 0 or positive. Therefore, there is no value of x that can make the value of this absolute value be less than the negative value on the right side of the inequality.
7. The solution set involves "or" and "or" means "union."

Exercise Set 2.7

1. $|x| \le 4$

$-4 \le x \le 4$

$[-4, 4]$

3. $|x-3| < 2$

$-2 < x-3 < 2$

$1 < x < 5$

$(1, 5)$

5. $|x+3| < 2$

$-2 < x+3 < 2$

$-5 < x < -1$

$(-5, -1)$

7. $|2x+7| \le 3$

$-13 \le 2x+7 \le 13$

$-20 \le 2x \le 6$

$-10 \le x \le 3$

$[-10, 3]$

9. $|x|+7 \le 12$

$|x| \le 5$

$-5 \le x \le 5$

$[-5, 5]$

11. $|3x-1| < -5$

No real solutions; ∅

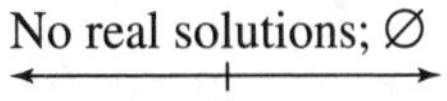

13. $|x-6|-7 \le -1$

$|x-6| \le 6$

$-6 \le x-6 \le 6$

$0 \le x \le 12$

$[0, 12]$

15. $|x| > 3$

$x < -3$ or $x > 3$

$(-\infty, -3) \cup (3, \infty)$

17. $|x+10| \geq 14$

$x+10 \leq -14$ or $x+10 \geq 14$

$x \leq -24$ or $x \geq 4$

$(-\infty, -24] \cup [4, \infty)$

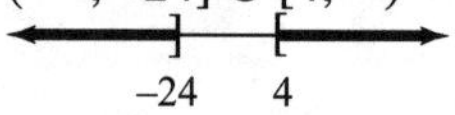

19. $|x|+2>6$

$|x|>4$

$x<-4$ or $x>4$

$(-\infty, -4) \cup (4, \infty)$

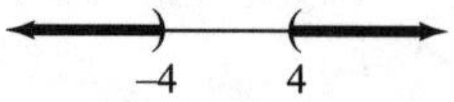

21. $|5x|>-4$

All real numbers

$(-\infty, \infty)$

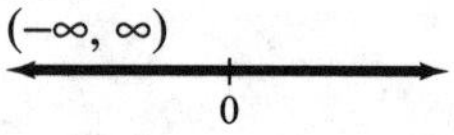

23. $|6x-8|+3>7$

$|6x-8|>4$

$6x-8<-4$ or $6x-8>4$

$6x<4$ or $6x>12$

$x<\frac{2}{3}$ or $x>2$

$\left(-\infty, \frac{2}{3}\right) \cup (2, \infty)$

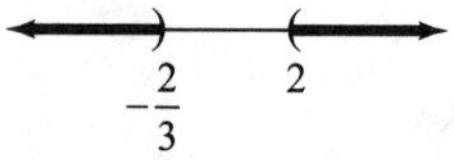

25. $|x| \leq 0$

$|x|=0$

$x=0$

27. $|8x+3|>0$ only excludes $|8x+3|=0$

$8x+3=0$

$8x=-3$

$x=-\frac{3}{8}$

All real numbers except $-\frac{3}{8}$.

$\left(-\infty, -\frac{3}{8}\right) \cup \left(-\frac{3}{8}, \infty\right)$

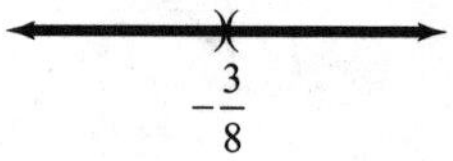

29. $|x| \leq 2$

$-2 \leq x \leq 2$

$[-2, 2]$

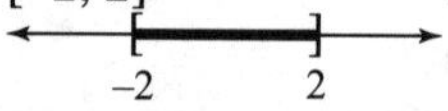

31. $|y|>1$

$y<-1$ or $y>1$

$(-\infty, -1) \cup (1, \infty)$

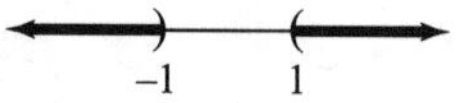

33. $|x-3|<8$

$-8<x-3<8$

$-5<x<11$

$(-5, 11)$

35. $|0.6x-3|>0.6$

$0.6x-3<-0.6$ or $0.6x-3>0.6$

$0.6x<2.4$ or $0.6x>3.6$

$x<4$ or $x>6$

$(-\infty, 4) \cup (6, \infty)$

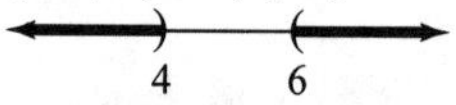

37. $5+|x| \leq 2$

$|x| \leq -3$

No real solution

$\varnothing$

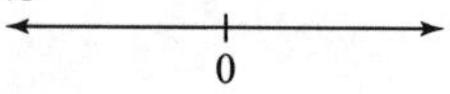

39. $|x|>-4$

All real numbers

$(-\infty, \infty)$

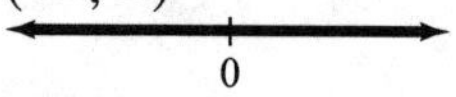

41. $|2x-7| \leq 11$

$-11 \leq 2x-7 \leq 11$

$-4 \leq 2x \leq 18$

$-2 \leq x \leq 9$

$[-2, 9]$

43. $|x+5|+2 \geq 8$

$|x+5| \geq 6$

$x+5 \leq -6$ or $x+5 \geq 6$

$x \leq -11$ or $x \geq 1$

$(-\infty, -11] \cup [1, \infty)$

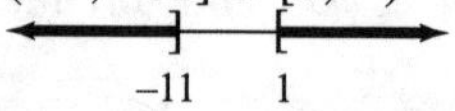

45. $|x| > 0$ only excludes $|x| = 0$, or $x = 0$.
All real numbers except $x = 0$
$(-\infty, 0) \cup (0, \infty)$

47. $9 + |x| > 7$
$|x| > -2$
All real numbers
$(-\infty, \infty)$

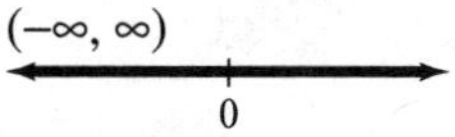

49. $6 + |4x - 1| \le 9$
$|4x - 1| \le 3$
$-3 \le 4x - 1 \le 3$
$-2 \le 4x \le 4$
$-\frac{1}{2} \le x \le 1$
$\left[-\frac{1}{2}, 1\right]$

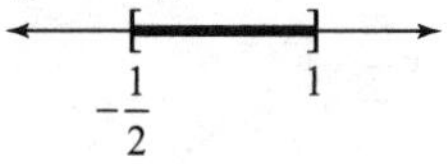

51. $\left|\frac{2}{3}x + 1\right| > 1$
$\frac{2}{3}x + 1 < -1$ or $\frac{2}{3}x + 1 > 1$
$\frac{2}{3}x < -2$ or $\frac{2}{3}x > 0$
$x < -3$ or $x > 0$
$(-\infty, -3) \cup (0, \infty)$

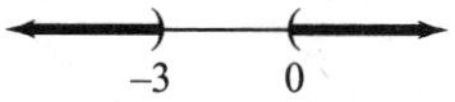

53. $|5x + 3| < -6$
No real solution
$\varnothing$

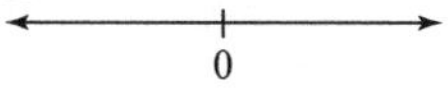

55. $\left|\frac{8x - 3}{4}\right| \le 0$
$\frac{8x - 3}{4} = 0$
$8x - 3 = 0$
$8x = 3$
$x = \frac{3}{8}$
$\left\{\frac{3}{8}\right\}$

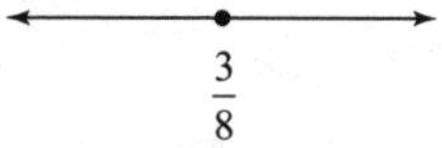

57. $|1 + 3x| + 4 < 5$
$|1 + 3x| < 1$
$-1 < 1 + 3x < 1$
$-2 < 3x < 0$
$-\frac{2}{3} < x < 0$
$\left(-\frac{2}{3}, 0\right)$

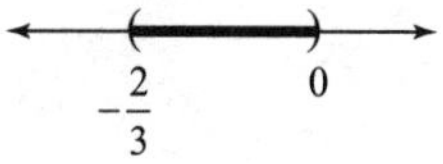

59. $\left|\frac{x + 6}{3}\right| > 2$
$\frac{x + 6}{3} < -2$ or $\frac{x + 6}{3} > 2$
$x + 6 < -6$ or $x + 6 > 6$
$x < -12$ or $x > 0$
$(-\infty, -12) \cup (0, \infty)$

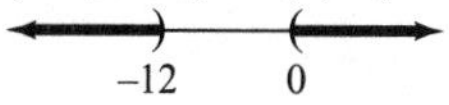

61. $-15 + |2x - 7| \le -6$
$|2x - 7| \le 9$
$-9 \le 2x - 7 \le 9$
$-2 \le 2x \le 16$
$-1 \le x \le 8$
$[-1, 8]$

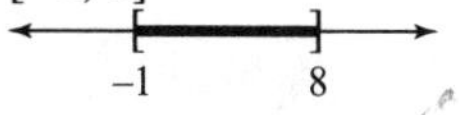

63. $\left|2x+\frac{3}{4}\right|-7\le -2$
$$\left|2x+\frac{3}{4}\right|\le 5$$
$$-5\le 2x+\frac{3}{4}\le 5$$
$$-20\le 8x+3\le 20$$
$$-23\le 8x\le 17$$
$$-\frac{23}{8}\le x\le \frac{17}{8}$$
$$\left[-\frac{23}{8},\frac{17}{8}\right]$$

65. $|2x-3|<7$
$$-7<2x-3<7$$
$$-4<2x<10$$
$$-2<x<5$$
$(-2, 5)$

67. $|2x-3|=7$
$$2x-3=7 \quad \text{or} \quad 2x-3=-7$$
$$2x=10 \quad \text{or} \quad 2x=-4$$
$$x=5 \quad \text{or} \quad x=-2$$

69. $|x-5|\ge 12$
$$x-5\le -12 \quad \text{or} \quad x-5\ge 12$$
$$x\le -7 \quad \text{or} \quad x\ge 17$$
$(-\infty, -7]\cup[17, \infty)$

71. $|9+4x|=0$
$$9+4x=0$$
$$4x=-9$$
$$x=-\frac{9}{4}$$

73. $|2x+1|+4<7$
$$|2x+1|<3$$
$$-3<2x+1<3$$
$$-4<2x<2$$
$$-2<x<1$$
$(-2, 1)$

75. $|3x-5|+4=5$
$$|3x-5|=1$$
$$3x-5=1 \quad \text{or} \quad 3x-5=-1$$
$$3x=6 \quad \text{or} \quad 3x=4$$
$$x=2 \quad \text{or} \quad x=\frac{4}{3}$$

77. $|x+11|=-1$ is impossible. The solution set is $\varnothing$.

79. $\left|\frac{2x-1}{3}\right|=6$
$$\frac{2x-1}{3}=6 \quad \text{or} \quad \frac{2x-1}{3}=-6$$
$$2x-1=18 \quad \text{or} \quad 2x-1=-18$$
$$2x=19 \quad \text{or} \quad 2x=-17$$
$$x=\frac{19}{2} \quad \text{or} \quad x=-\frac{17}{2}$$

81. $\left|\frac{3x-5}{6}\right|>5$
$$\frac{3x-5}{6}<-5 \quad \text{or} \quad \frac{3x-5}{6}>5$$
$$3x-5<-30 \quad \text{or} \quad 3x-5>30$$
$$3x<-25 \quad \text{or} \quad 3x>35$$
$$x<-\frac{25}{3} \quad \text{or} \quad x>\frac{35}{3}$$
$$\left(-\infty, -\frac{25}{3}\right)\cup\left(\frac{35}{3}, \infty\right)$$

83. $P(\text{rolling a 2})=\frac{1}{6}$

85. $P(\text{rolling a 7})=0$

87. $P(\text{rolling a 1 or 3})=\frac{1}{3}$

89.
$$3x-4y=12$$
$$3(2)-4y=12$$
$$6-4y=12$$
$$-4y=6$$
$$y=-\frac{3}{2}=-1.5$$

91.
$$3x-4y=12$$
$$3x-4(-3)=12$$
$$3x+12=12$$
$$3x=0$$
$$x=0$$

93. $|x| < 7$

95. $|x| \le 5$

97. answers may vary

99. $|3.5 - x| < 0.05$

$$\begin{aligned} -0.05 < 3.5 - x < 0.05 \\ -3.55 < -x < -3.45 \\ 3.55 > x > 3.45 \\ 3.45 < x < 3.55 \end{aligned}$$

Chapter 2 Vocabulary Check

1. The statement "$x < 5$ *or* $x > 7$" is called a compound inequality.
2. An equation in one variable that has no solution is called a contradiction.
3. The intersection of two sets is the set of all elements common to both sets.
4. The union of two sets is the set of all elements that belong to either of the sets.
5. An equation in one variable that has every number (for which the equation is defined) as a solution is called an identity.
6. The equation $d = rt$ is also called a formula.
7. A number's distance from 0 is called its absolute value.
8. When a variable in an equation is replaced by a number and the resulting equation is true, then that number is called a solution of the equation.
9. The integers 17, 18, 19 are examples of consecutive integers.
10. The statement $5x - 0.2 < 7$ is an example of a linear inequality in one variable.
11. The statement $5x - 0.2 = 7$ is an example of a linear equation in one variable.

Chapter 2 Review

1.
$$\begin{aligned} 4(x-5) &= 2x-14 \\ 4x-20 &= 2x-14 \\ 2x &= 6 \\ x &= 3 \end{aligned}$$

2.
$$\begin{aligned} x+7 &= -2(x+8) \\ x+7 &= -2x-16 \\ 3x &= -23 \\ x &= -\frac{23}{3} \end{aligned}$$

3.
$$\begin{aligned} 3(2y-1) &= -8(6+y) \\ 6y-3 &= -48-8y \\ 14y &= -45 \\ y &= -\frac{45}{14} \end{aligned}$$

4.
$$\begin{aligned} -(z+12) &= 5(2z-1) \\ -z-12 &= 10z-5 \\ -11z &= 7 \\ z &= -\frac{7}{11} \end{aligned}$$

5.
$$\begin{aligned} n-(8+4n) &= 2(3n-4) \\ n-8-4n &= 6n-8 \\ -3n &= 6n \\ -9n &= 0 \\ n &= 0 \end{aligned}$$

6.
$$\begin{aligned} 4(9v+2) &= 6(1+6v)-10 \\ 36v+8 &= 6+36v-10 \\ 36v+8 &= 36v-4 \\ 8 &= -4 \end{aligned}$$

No solution, or $\varnothing$

7.
$$\begin{aligned} 0.3(x-2) &= 1.2 \\ 10[0.3(x-2) &= 10(1.2) \\ 3(x-2) &= 12 \\ 3x-6 &= 12 \\ 3x &= 18 \\ x &= 6 \end{aligned}$$

8.
$$\begin{aligned} 1.5 &= 0.2(c-0.3) \\ 1.5 &= 0.2c-0.06 \\ 100(1.5) &= 100(0.2c-0.06) \\ 150 &= 20c-6 \\ 156 &= 20c \\ 7.8 &= c \end{aligned}$$

9.
$$\begin{aligned} -4(2-3x) &= 2(3x-4)+6x \\ -8+12x &= 6x-8+6x \\ -8+12x &= 12x-8 \\ -8 &= -8 \end{aligned}$$

All real numbers

10. $6(m-1)+3(2-m)=0$
$6m-6+6-3m=0$
$3m=0$
$m=0$

11. $6-3(2g+4)-4g=5(1-2g)$
$6-6g-12-4g=5-10g$
$-6-10g=5-10g$
$-6=5$
No solution, $\varnothing$

12. $20-5(p+1)+3p=-(2p-15)$
$20-5p-5+3p=-2p+15$
$15-2p=-2p+15$
$15=15$
All real numbers

13. $\frac{x}{3}-4=x-2$
$3\left(\frac{x}{3}-4\right)=3(x-2)$
$x-12=3x-6$
$-2x=6$
$x=-3$

14. $\frac{9}{4}y=\frac{2}{3}y$
$12\left(\frac{9}{4}y\right)=12\left(\frac{2}{3}y\right)$
$27y=8y$
$19y=0$
$y=0$

15. $\frac{3n}{8}-1=3+\frac{n}{6}$
$24\left(\frac{3n}{8}-1\right)=24\left(3+\frac{n}{6}\right)$
$9n-24=72+4n$
$5n=96$
$n=\frac{96}{5}$

16. $\frac{z}{6}+1=\frac{z}{2}+2$
$6\left(\frac{z}{6}+1\right)=6\left(\frac{z}{2}+2\right)$
$z+6=3z+12$
$-2z=6$
$z=-3$

17. $\frac{y}{4}-\frac{y}{2}=-8$
$4\left(\frac{y}{4}-\frac{y}{2}\right)=4(-8)$
$y-2y=-32$
$-y=-32$
$y=32$

18. $\frac{2x}{3}-\frac{8}{3}=x$
$2x-8=3x$
$-8=x$

19. $\frac{b-2}{3}=\frac{b+2}{5}$
$5(b-2)=3(b+2)$
$5b-10=3b+6$
$2b=16$
$b=8$

20. $\frac{2t-1}{3}=\frac{3t+2}{15}$
$15\left(\frac{2t-1}{3}\right)=15\left(\frac{3t+2}{15}\right)$
$5(2t-1)=3t+2$
$10t-5=3t+2$
$7t=7$
$t=1$

21. $\frac{2(t+1)}{3}=\frac{2(t-1)}{3}$
$3\left[\frac{2(t+1)}{3}\right]=3\left[\frac{2(t-1)}{3}\right]$
$2(t+1)=2(t-1)$
$2t+2=2t-2$
$2=-2$
No solution, $\varnothing$

22. $\frac{3a-3}{6}=\frac{4a+1}{15}+2$
$30\left(\frac{3a-3}{6}\right)=30\left(\frac{4a+1}{15}+2\right)$
$5(3a-3)=2(4a+1)+30(2)$
$15a-15=8a+2+60$
$15a-15=8a+62$
$7a=77$
$a=11$

23. Let x = the number.
$2(x-3)=3x+1$
$2x-6=3x+1$
$-7=x$
The number is −7.

24. Let x = smaller number, then
$x + 5$ = larger number.
$x+x+5=285$
$2x=280$
$x=140$
$x + 5 = 145$
The numbers are 140 and 145.

25. $40\% \cdot 130 = 0.40 \cdot 130 = 52$

26. $1.5\% \cdot 8 = 0.015 \cdot 8 = 0.12$

27. Let x = width of the playing field, then
$2x - 5$ = length of the playing field.
$2x+2(2x-5)=230$
$2x+4x-10=230$
$6x=240$
$x=40$
Then $2x - 5 = 2(40) - 5 = 75$. The field is 75 meters long and 40 meters wide.

28. Let x be the median weekly earnings for a young adult with an associate's degree in 2013.
$x+0.43x=1108$
$1.43x=1108$
$x\approx 775$
The median weekly earnings for a young adult with an associate's degree in 2013 was $775.

29. Let n = the first integer, then
$n + 1$ = the second integer,
$n + 2$ = the third integer, and
$n + 3$ = the fourth integer.
$(n+1)+(n+2)+(n+3)-2n=16$
$n+6=16$
$n=10$
Therefore, the integers are 10, 11, 12, and 13.

30. Let x = smaller odd integer, then
$x + 2$ = larger odd integer.
$5x=3(x+2)+54$
$5x=3x+6+54$
$2x=60$
$x=30$
Since this is not odd, no such consecutive odd integers exist.

31. Let m = number of miles of driven.
$2(19.95)+0.12(m-200)=46.86$
$39.90+0.12m-24=46.86$
$0.12m+15.90=46.86$
$0.12m=30.96$
$m=258$
He drove 258 miles.

32. Solve $R = C$.
$16.50x=4.50x+3000$
$12x=3000$
$x=250$
Thus, 250 calculators must be produced and sold in order to break even.

33. $V=lwh$
$w=\frac{V}{lh}$

34. $C=2\pi r$
$\frac{C}{2\pi}=r$

35. $5x-4y=-12$
$5x+12=4y$
$y=\frac{5x+12}{4}$

36. $5x-4y=-12$
$5x=4y-12$
$x=\frac{4y-12}{5}$

37. $y-y_1=m(x-x_1)$
$m=\frac{y-y_1}{x-x_1}$

38. $y-y_1=m(x-x_1)$
$y-y_1=mx-mx_1$
$y-y_1+mx_1=mx$
$\frac{y-y_1+mx_1}{m}=x$

39. $E=I(R+r)$
$E=IR+Ir$
$I-IR=Ir$
$\frac{E-IR}{I}=r$

40.
$$S = vt + gt^2$$
$$S - vt = gt^2$$
$$\frac{S - vt}{t^2} = g$$

41.
$$T = gr + gvt$$
$$T = g(r + vt)$$
$$g = \frac{T}{r + vt}$$

42.
$$I = Prt + P$$
$$I = P(rt + 1)$$
$$\frac{I}{rt + 1} = P$$

43. $A = P\left(1 + \frac{r}{n}\right)^{nt} = 3000\left(1 + \frac{0.03}{n}\right)^{7n}$

a. $A = 3000\left(1 + \frac{0.03}{2}\right)^{14} \approx \$3695.27$

b. $A = 3000\left(1 + \frac{0.03}{52}\right)^{364} \approx \$3700.81$

44.
$$C = \frac{5}{9}(F - 32)$$
$$C = \frac{5}{9}(90 - 32)$$
$$C = \frac{5}{9}(58)$$
$$C = \frac{290}{9} \approx 32.2$$

90°F is $\left(\frac{290}{9}\right)$°C ≈ 32.2°C.

45. Let x = original width, then
$x + 2$ = original length.
$$(x + 4)(x + 2 + 4) = x(x + 2) + 88$$
$$(x + 4)(x + 6) = x^2 + 2x + 88$$
$$x^2 + 10x + 24 = x^2 + 2x + 88$$
$$8x = 64$$
$$x = 8$$
$x + 2 = 10$
The original width is 8 in. and the original length is 10 in.

46. Area $= 18 \times 21 = 378$ ft^2

Packages $= \frac{378}{24} = 15.75$

There are 16 packages needed.

47.
$$3(x - 5) > -(x + 3)$$
$$3x - 15 > -x - 3$$
$$4x > 12$$
$$x > 3$$
$(3, \infty)$

48.
$$-2(x + 7) \geq 3(x + 2)$$
$$-2x - 14 \geq 3x + 6$$
$$-5x \geq 20$$
$$x \leq -4$$
$(-\infty, -4]$

49.
$$4x - (5 + 2x) < 3x - 1$$
$$4x - 5 - 2x < 3x - 1$$
$$2x - 5 < 3x - 1$$
$$-x < 4$$
$$x > -4$$
$(-4, \infty)$

50.
$$3(x - 8) < 7x + 2(5 - x)$$
$$3x - 24 < 7x + 10 - 2x$$
$$3x - 24 < 5x + 10$$
$$-2x < 34$$
$$x > -17$$
$(-17, \infty)$

51.
$$24 \geq 6x - 2(3x - 5) + 2x$$
$$24 \geq 6x - 6x + 10 + 2x$$
$$24 \geq 10 + 2x$$
$$14 \geq 2x$$
$$7 \geq x$$
$(-\infty, 7]$

52.
$$\frac{x}{3} + \frac{1}{2} > \frac{2}{3}$$
$$6\left(\frac{x}{3} + \frac{1}{2}\right) > 6\left(\frac{2}{3}\right)$$
$$2x + 3 > 4$$
$$2x > 1$$
$$x > \frac{1}{2}$$
$\left(\frac{1}{2}, \infty\right)$

53. $x+\frac{3}{4}<-\frac{x}{2}+\frac{9}{4}$
$4\left(x+\frac{3}{4}\right)<4\left(-\frac{x}{2}+\frac{9}{4}\right)$
$4x+3<-2x+9$
$6x<6$
$x<1$
$(-\infty, 1)$

54. $\frac{x-5}{2}\le\frac{3}{8}(2x+6)$
$8\left(\frac{x-5}{2}\right)\le 8\left[\frac{3}{8}(2x+6)\right]$
$4(x-5)\le 3(2x+6)$
$4x-20\le 6x+18$
$-2x\le 38$
$x\ge -19$
$[-19, \infty)$

55. Let n = number of pounds of laundry.
$15<0.5(10)+0.4(n-10)$
$15<5+0.4n-4$
$15<1+0.4n$
$14<0.4n$
$35<n$
It is more economical to use the housekeeper for more than 35 pounds of laundry per week.

56. Let x = the score from the last judge.
$\frac{9.5+9.7+9.9+9.7+9.7+9.6+9.5+x}{8}\ge 9.65$
$67.6+x\ge 77.2$
$x\ge 9.6$
The last judge must give Nana at least a 9.6 for her to win the silver medal.

57. $1\le 4x-7\le 3$
$8\le 4x\le 10$
$2\le x\le\frac{5}{2}$
$\left[2, \frac{5}{2}\right]$

58. $-2\le 8+5x<-1$
$-10\le 5x\le -9$
$-2\le x\le -\frac{9}{5}$
$\left[-2, \frac{9}{5}\right)$

59. $-3<4(2x-1)<12$
$-3<8x-4<12$
$1<8x<16$
$\frac{1}{8}<x<2$
$\left(\frac{1}{8}, 2\right)$

60. $-6<x-(3-4x)<-3$
$-6<x-3+4x<-3$
$-6<5x-3<-3$
$-3<5x<0$
$-\frac{3}{5}<x<0$
$\left(-\frac{3}{5}, 0\right)$

61. $\frac{1}{6}<\frac{4x-3}{3}\le\frac{4}{5}$
$30\left(\frac{1}{6}\right)<30\left(\frac{4x-3}{3}\right)\le 30\left(\frac{4}{5}\right)$
$5<10(4x-3)\le 24$
$5<40x-30\le 24$
$35<40x<54$
$\frac{7}{8}<x\le\frac{27}{20}$
$\left(\frac{7}{8}, \frac{27}{20}\right]$

62. $x\le 2$ *and* $x>-5$
$-5<x\le 2$
$(-5, 2]$

63. $3x-5>6$ *or* $-x<-5$
$3x>11$ *or* $x>5$
$x>\frac{11}{3}$ *or* $x>5$
$x>\frac{11}{3}$
$\left(\frac{11}{3}, \infty\right)$

64. $500 \le F \le 1000$
$500 \le \frac{9}{5}C + 32 \le 1000$
$468 \le \frac{9}{5}C \le 968$
$260 \le C \le 538$
Rounded to the nearest degree, firing temperatures range from 260°C to 538°C.

65. Let x = the amount saved each summer.
$4000 \le 2x + 500 \le 8000$
$3500 \le 2x \le 7500$
$1750 \le x \le 3750$
She must save between \$1750 and \$3750 each summer.

66. $|x - 7| = 9$
$x - 7 = 9$ or $x - 7 = -9$
$x = 16$ or $x = -2$

67. $|8 - x| = 3$
$8 - x = 3$ or $8 - x = -3$
$-x = -5$ or $-x = -11$
$x = 5$ or $x = 11$

68. $|2x + 9| = 9$
$2x + 9 = 9$ or $2x + 9 = -9$
$2x = 0$ or $2x = -18$
$x = 0$ or $x = -9$

69. $|-3x + 4| = 7$
$-3x + 4 = 7$ or $-3x + 4 = -7$
$-3x = 3$ or $-3x = -11$
$x = -1$ or $x = \frac{11}{3}$

70. $|3x - 2| + 6 = 10$
$|3x - 2| = 4$
$3x - 2 = 4$ or $3x - 2 = -4$
$3x = 6$ or $3x = -2$
$x = 2$ or $x = -\frac{2}{3}$

71. $5 + |6x + 1| = 5$
$|6x + 1| = 0$
$6x + 1 = 0$
$6x = -1$
$x = -\frac{1}{6}$

72. $-5 = |4x - 3|$
The solution set is $\varnothing$.

73. $|5 - 6x| + 8 = 3$
$|5 - 6x| = -5$
The solution set is $\varnothing$.

74. $-8 = |x - 3| - 10$
$2 = |x - 3|$
$x - 3 = 2$ or $x - 3 = -2$
$x = 5$ or $x = 1$

75. $\left|\frac{3x - 7}{4}\right| = 2$
$\frac{3x - 7}{4} = 2$ or $\frac{3x - 7}{4} = -2$
$3x - 7 = 8$ or $3x - 7 = -8$
$3x = 15$ or $3x = -1$
$x = 5$ or $x = -\frac{1}{3}$

76. $|6x + 1| = |15 + 4x|$
$6x + 1 = 15 + 4x$ or $6x + 1 = -(15 + 4x)$
$2x = 14$ or $6x + 1 = -15 - 4x$
$x = 7$ or $10x = -16$
$x = -\frac{8}{5}$

77. $|5x - 1| < 9$
$-9 < 5x - 1 < 9$
$-8 < 5x < 10$
$-\frac{8}{5} < x < 2$
$\left(-\frac{8}{5}, 2\right)$

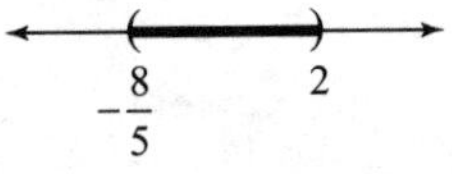

78. $|6 + 4x| \ge 10$
$6 + 4x \le -10$ or $6 + 4x \ge 10$
$4x \le -16$ or $4x \ge 4$
$x \le -4$ or $x \ge 1$
$(-\infty, -4] \cup [1, \infty)$

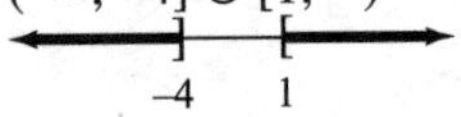

79. $|3x| - 8 > 1$

$|3x| > 9$

$3x < -9$ or $3x > 9$

$x < -3$ or $x > 3$

$(-\infty, -3) \cup (3, \infty)$

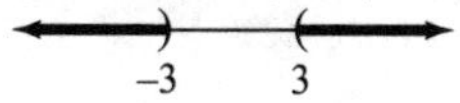

80. $9 + |5x| < 24$

$|5x| < 15$

$-15 < 5x < 15$

$-3 < x < 3$

$(-3, 3)$

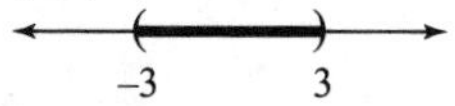

81. $|6x - 5| \le -1$

The solution set is $\varnothing$.

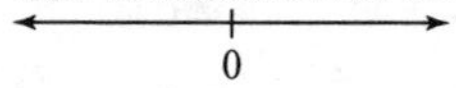

82. $\left|3x + \frac{2}{5}\right| \ge 4$

$3x + \frac{2}{5} \le -4$ or $3x + \frac{2}{5} \ge 4$

$5\left(3x + \frac{2}{5}\right) \le 5(-4)$ or $5\left(3x + \frac{2}{5}\right) \ge 5(4)$

$15x + 2 \le -20$ or $15x + 2 \ge 20$

$15x \le -22$ or $15x \ge 18$

$x \le -\frac{22}{15}$ or $x \ge \frac{6}{5}$

$\left(-\infty, -\frac{22}{15}\right] \cup \left[\frac{6}{5}, \infty\right)$

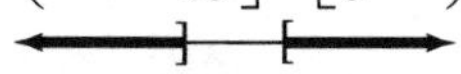

83. $\left|\frac{x}{3} + 6\right| - 8 > -5$

$\left|\frac{x}{3} + 6\right| > 3$

$\frac{x}{3} + 6 < -3$ or $\frac{x}{3} + 6 > 3$

$\frac{x}{3} < -9$ or $\frac{x}{3} > -3$

$x < -27$ or $x > -9$

$(-\infty, -27) \cup (-9, \infty)$

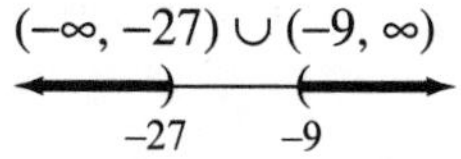

84. $\left|\frac{4(x-1)}{7}\right| + 10 < 2$

$\left|\frac{4(x-1)}{7}\right| < -8$

The solution set is $\varnothing$.

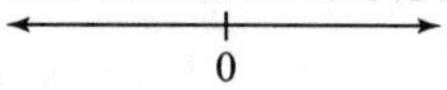

85. $\frac{x-2}{5} + \frac{x+2}{2} = \frac{x+4}{3}$

$30\left(\frac{x-2}{5} + \frac{x+2}{2}\right) = 30\left(\frac{x+4}{3}\right)$

$6(x-2) + 15(x+2) = 10(x+4)$

$6x - 12 + 15x + 30 = 10x + 40$

$21x + 18 = 10x + 40$

$11x = 22$

$x = 2$

86. $\frac{2z-3}{4} - \frac{4-z}{2} = \frac{z+1}{3}$

$12\left(\frac{2z-3}{4} - \frac{4-z}{2}\right) = 12\left(\frac{z+1}{3}\right)$

$3(2z-3) - 6(4-z) = 4(z+1)$

$6z - 9 - 24 + 6z = 4z + 4$

$12z - 33 = 4z + 4$

$8z = 37$

$z = \frac{37}{8}$

87. $A = \frac{h}{2}(B + b)$

$2A = hB + hb$

$2A - hb = hB$

$\frac{2A - hb}{h} = B$

88. $V = \frac{1}{3}\pi r^2 h$

$3V = \pi r^2 h$

$\frac{3V}{\pi r^2} = h$

89. Let x = number of tourists for France, then $x + 9$ = number of tourists for United States, and $x + 44$ = number of tourists for China.

$x + (x+9) + (x+44) = 332$

$3x + 53 = 332$

$3x = 279$

$x = 93$

$x + 9 = 102$
$x + 44 = 137$
China is predicted to have 137 million tourists, whereas the United States is predicted to have 102 million and France, 93 million.

90. $d = rt$ or $r = \frac{d}{t}$
11:00 a.m. to 1:15 p.m. is 2.25 hours.
$r = \frac{130}{2.25} \approx 58$
His average speed was 58 mph.

91. $V_{\text{box}} = lwh = 8 \cdot 5 \cdot 3 = 120 \text{ in}^3$, while
$V_{\text{cyl}} = \pi r^2 h = \pi \cdot 3^2 \cdot 6 = 54\pi \approx 170 \text{ in}^3$
Therefore, the cylinder holds more ice cream.

92. $48 + x \geq 5(2x+4) - 2x$
$48 + x \geq 10x + 20 - 2x$
$48 + x \geq 8x + 20$
$28 \geq 7x$
$4 \geq x$
$(-\infty, 4]$

93. $\frac{3(x-2)}{5} > \frac{-5(x-2)}{3}$
$15\left[\frac{3(x-2)}{5}\right] > 15\left[\frac{-5(x-2)}{3}\right]$
$9(x-2) > -25(x-2)$
$9x - 18 > -25x + 50$
$34x > 68$
$x > 2$
$(2, \infty)$

94. $0 \leq \frac{2(3x+4)}{5} \leq 3$
$5(0) \leq 5\left[\frac{2(3x+4)}{5}\right] \leq 5(3)$
$0 \leq 2(3x+4) \leq 15$
$0 \leq 6x + 8 \leq 15$
$-8 \leq 6x \leq 7$
$-\frac{4}{3} \leq x \leq \frac{7}{6}$
$\left[-\frac{4}{3}, \frac{7}{6}\right]$

95. $x \leq 2$ *or* $x > -5$
$(-\infty, \infty)$

96. $-2x \leq 6$ *and* $-2x + 3 < -7$
$x \geq -3$ *and* $-2x < -10$
$x \geq -3$ *and* $x > 5$
$x > 5$
$(5, \infty)$

97. $|7x| - 26 = -5$
$|7x| = 21$
$7x = 21$ or $7x = -21$
$x = 3$ or $x = -3$

98. $\left|\frac{9-2x}{5}\right| = -3$
The solution set is $\varnothing$.

99. $|x - 3| = |7 + 2x|$
$x - 3 = 7 + 2x$ or $x - 3 = -(7 + 2x)$
$-10 = x$ or $x - 3 = -7 - 2x$
$3x = -4$
$x = -\frac{4}{3}$

100. $|6x - 5| \geq -1$
Since $|6x - 5|$ is nonnegative for all numbers x, the solution set is $(-\infty, \infty)$.

101. $\left|\frac{4x-3}{5}\right| < 1$
$-1 < \frac{4x-3}{5} < 1$
$-5 < 4x - 3 < 5$
$-2 < 4x < 8$
$-\frac{1}{2} < x < 2$
$\left(-\frac{1}{2}, 2\right)$

Chapter 2 Getting Ready for the Test

1. $x - 9 = -6x - 9$
$7x = 0$
$x = 0$
The solution is 0; C.

2. $4x + 8 = 2(x + 4)$
$4x + 8 = 2x + 8$
$2x = 0$
The solution is 0; C.

3. $5(2x-4)=10(x-2)$
$10x-20=10x-20$
Both sides of the equation are identical, so all real numbers are solutions; A.

4. $3(x+2)+x=4(x-1)+1$
$3x+6+x=4x-4+1$
$4x+6=4x-3$
$6=-3$ False
$6=-3$ is false for all values of x, so the equation has no solution; B.

5. $\{x|x\leq -11\}$ is $(-\infty, -11]$; A.

6. $\{x|-5<x\}$ is $(-5, \infty)$; B.

7. **A.** $|-7-3| \stackrel{?}{=} 7$, $|-10| \stackrel{?}{=} 7$, $10\neq 7$ $\qquad$ $|7-3| \stackrel{?}{=} 7$, $|4| \stackrel{?}{=} 7$, $4\neq 7$

B. $|-10-3| \stackrel{?}{=} 7$, $|-13| \stackrel{?}{=} 7$, $13\neq 7$ $\qquad$ $|10-3| \stackrel{?}{=} 7$, $|7| \stackrel{?}{=} 7$, $7=7$

C. $|4-3| \stackrel{?}{=} 7$, $|1| \stackrel{?}{=} 7$, $1\neq 7$ $\qquad$ $|10-3| \stackrel{?}{=} 7$, $|7| \stackrel{?}{=} 7$, $7=7$

D. $|-4-3| \stackrel{?}{=} 7$, $|-7| \stackrel{?}{=} 7$, $7=7$ $\qquad$ $|10-3| \stackrel{?}{=} 7$, $|7| \stackrel{?}{=} 7$, $7=7$

D gives the correct solutions.

8. $|5x-2|\leq 4$ is equivalent to $-4\leq 5x-2\leq 4$; C.

9. $|5x-2|=4$ is equivalent to $5x-2=4$ or $5x-2=-4$; A.

10. $|5x-2|\geq 4$ is equivalent to $5x-2\geq 4$ or $5x-2\leq -4$; E.

11. $|5x|-2=4$ or $|5x|=6$ is equivalent to $5x=6$ or $5x=-6$; B.

12. An absolute value will never be negative, so $|x+3|=-9$ has no solution, or $\varnothing$; A.

13. An absolute value will never be negative, so $|x+3|<-9$ has no solution, or $\varnothing$; A.

14. An absolute value will always be greater than equal to 0, so $|x+3|>-9$ has all real numbers as solutions, or $(-\infty, \infty)$; B.

Chapter 2 Test

1. $8x+14=5x+44$
$3x=30$
$x=10$

2. $9(x+2)=5[11-2(2-x)+3]$
$9x+18=5[11-4+2x+3]$
$9x+18=5[10+2x]$
$9x+18=50+10x$
$-x=32$
$x=-32$

3. $3(y-4)+y=2(6+2y)$
$3y-12+y=12+4y$
$4y-12=12+4y$
$-12=12$
No solution, $\varnothing$

4. $7n-6+n=2(4n-3)$
$8n-6=8n-6$
$-6=-6$
All real numbers

5. $\frac{7w}{4}+5=\frac{3w}{10}+1$
$20\left(\frac{7w}{4}+5\right)=20\left(\frac{3w}{10}+1\right)$
$35w+100=6w+20$
$29w=-80$
$w=-\frac{80}{29}$

6. $\frac{z+7}{9}+1=\frac{2z+1}{6}$
$18\left(\frac{z+7}{9}+1\right)=18\left(\frac{2z+1}{6}\right)$
$2(z+7)+18=3(2z+1)$
$2z+14+18=6z+3$
$2z+32=6z+3$
$2z-6z=3-32$
$-4z=-29$
$z=\frac{29}{4}$

7. $|6x-5|-3=-2$
$|6x-5|=1$
$6x-5=1$ or $6x-5=-1$
$6x=6$ or $6x=4$
$x=1$ or $x=\frac{2}{3}$

8. $|8-2t|=-6$
No solution, $\varnothing$

9. $|2x-3|=|4x+5|$
$2x-3=4x+5$ or $2x-3=-(4x+5)$
$2x-4x=5+3$ or $2x-3=-4x-5$
$-2x=8$ or $2x+4x=-5+3$
$x=-4$ or $6x=-2$
$x=-4$ or $x=-\frac{1}{3}$

10. $|x-5|=|x+2|$
$x-5=x+2$ or $x-5=-(x+2)$
$-5=2$ False or $x-5=-x-2$
$2x=3$
$x=\frac{3}{2}$

Since $-5=2$ is not possible, the only solution is $\frac{3}{2}$.

11. $3x-4y=8$
$3x-8=4y$
$y=\frac{3x-8}{4}$

12. $S=gt^2+gvt$
$S=g(t^2+vt)$
$g=\frac{S}{t^2+vt}$

13. $F=\frac{9}{5}C+32$
$F-32=\frac{9}{5}C$
$C=\frac{5}{9}(F-32)$

14. $3(2x-7)-4x>-(x+6)$
$6x-21-4x>-x-6$
$2x-21>-x-6$
$3x>15$
$x>5$
$(5, \infty)$

15. $\frac{3x-2}{3}-\frac{5x+1}{4}\ge 0$
$12\left[\frac{3x-2}{3}-\frac{5x+1}{4}\right]\ge 12(0)$
$4(3x-2)-3(5x+1)\ge 0$
$12x-8-15x-3\ge 0$
$-3x-11\ge 0$
$-3x\ge 11$
$x\le -\frac{11}{3}$
$\left(-\infty, -\frac{11}{3}\right]$

16. $-3<2(x-3)\le 4$
$-3<2x-6\le 4$
$3<2x\le 10$
$\frac{3}{2}<x\le 5$
$\left(\frac{3}{2}, 5\right]$

17. $|3x+1|>5$
$3x+1<-5$ or $3x+1>5$
$3x<-6$ or $3x>4$
$x<-2$ or $x>\frac{4}{3}$
$(-\infty, -2)\cup\left(\frac{4}{3}, \infty\right)$

18. $|x-5|-4<-2$
$|x-5|<2$
$-2<x-5<2$
$3<x<7$
$(3, 7)$

19. $x\ge 5$ *and* $x\ge 4$
$[5, \infty)$

20. $x\ge 5$ *or* $x\ge 4$
$[4, \infty)$

21. $-1 \le \frac{2x-5}{3} < 2$

$3(-1) \le 3\left(\frac{2x-5}{3}\right) < 3(2)$

$-3 \le 2x-5 < 6$

$-3+5 \le 2x-5+5 < 6+5$

$2 \le 2x < 11$

$\frac{2}{2} \le \frac{2x}{2} < \frac{11}{2}$

$1 \le x < \frac{11}{2}$

$\left[1, \frac{11}{2}\right)$

22. $6x+1 > 5x+4$ *or* $1-x > -4$

$x > 3$ *or* $5 > x$

$(-\infty, \infty)$

23. $12\% \cdot 80 = 0.12 \cdot 80 = 9.6$

24. Let x be the number of new vehicles sold by Ford in 2010. The number of new vehicles sold is increased by 29.1%, or by $0.291x$.

$x + 0.291x = 2,480,942$

$1.291x = 2,480,942$

$x \approx 1,922,000$

Ford sold approximately 1,922,000 new vehicles in 2010.

25. Recall that $C = 2\pi r$. Here $C = 78.5$.

$78.5 = 2\pi r$

$r = \frac{78.5}{2\pi} = \frac{39.25}{\pi}$

Also recall that $A = \pi r^2$.

$A = \pi\left(\frac{39.25}{\pi}\right)^2 = \frac{39.25^2}{\pi} \approx \frac{39.25^2}{3.14} \approx 491$

The area of the pen is about 491 square feet. Each dog requires at least 60 square feet of space, and $\frac{491}{60} \approx 8.18$. At most 8 dogs could be kept in the pen.

26. Let x be the number of people employed as registered nurses in 2012. The number of people employed in this field in 2022 is x increased by 19%.

$x + 0.19x = 3,240,000$

$1.19x = 3,240,000$

$x \approx 2,723,000$

In 2012, there were 2,723,000 registered nurses employed.

27. Use $A = P\left(1+\frac{r}{n}\right)^{nt}$ where $P = 2500$, $r = 3.5\% = 0.035$, $t = 10$, and $n = 4$.

$A = 2500\left(1+\frac{0.035}{4}\right)^{4\cdot 10}$

$A = 2500(1.00875)^{40}$

$A \approx \$3542.27$

28. Let x be the amount of money international travelers spend in New York. Then $x + 4$ is the amount of money international travelers spend in California and $2x - 1$ is the amount of money international travelers spend in Florida.

$x + (x+4) + (2x-1) = 39$

$4x + 3 = 39$

$4x = 36$

$x = 9$

$x + 4 = 9 + 4 = 13$

$2x - 1 = 2(9) - 1 = 18 - 1 = 17$

International travelers spend \$9 billion in New York, \$13 billion in California, and \$17 billion in Florida.

Chapter 2 Cumulative Review

1. a. $\{101, 102, 103, \ldots\}$

b. $\{2, 3, 4, 5\}$

2. a. $\{-2, -1, 0, 1, 2, 3, 4\}$

b. $\{4\}$

3. a. $|3| = 3$

b. $\left|-\frac{1}{7}\right| = \frac{1}{7}$

c. $-|2.7| = -2.7$

d. $-|-8| = -8$

e. $|0| = 0$

4. a. The opposite of $\frac{2}{3}$ is $-\frac{2}{3}$.

b. The opposite of -9 is 9.

c. The opposite of 1.5 is -1.5.

5. a. $-3+(-11)=-14$

b. $3+(-7)=-4$

c. $-10+15=5$

d. $-8.3+(-1.9)=-10.2$

e. $-\frac{1}{4}+\frac{1}{2}=-\frac{1}{4}+\frac{2}{4}=\frac{1}{4}$

f. $-\frac{2}{3}+\frac{3}{7}=-\frac{14}{21}+\frac{9}{21}=-\frac{5}{21}$

6. a. $-2-(-10)=-2+10=8$

b. $1.7-8.9=-7.2$

c. $-\frac{1}{2}-\frac{1}{4}=-\frac{2}{4}-\frac{1}{4}=-\frac{3}{4}$

7. a. $\sqrt{9}=3$ since $3^2=9$.

b. $\sqrt{25}=5$ since $5^2=25$.

c. $\sqrt{\frac{1}{4}}=\frac{1}{2}$ since $\left(\frac{1}{2}\right)^2=\frac{1}{4}$.

d. $-\sqrt{36}=-6$ since $6^2=36$.

e. $\sqrt{-36}$ is not a real number.

8. a. $-3(-2)=6$

b. $-\frac{3}{4}\left(-\frac{4}{7}\right)=\frac{3}{7}$

c. $\frac{0}{-2}=0$

d. $\frac{-20}{-2}=10$

9. Let $x=4$, $y=-3$.

a. $3x-7y=3(4)-7(-3)=12+21=33$

b. $-2y^2=-2(-3)^2=-2(9)=-18$

c.
$$\frac{\sqrt{x}}{y}-\frac{y}{x}=\frac{\sqrt{4}}{-3}-\frac{-3}{4}$$
$$=-\frac{2}{3}+\frac{3}{4}$$
$$=-\frac{8}{12}+\frac{9}{12}$$
$$=\frac{1}{12}$$

10. a. $\sqrt[4]{1}=1$ since $1^4=1$.

b. $\sqrt[3]{8}=2$ since $2^3=8$.

c. $\sqrt[4]{81}=3$ since $3^4=81$.

11. a. $x+5=20$

b. $2(3+y)=4$

c. $x-8=2x$

d. $\frac{z}{9}=9+z$

12. a. $-3>-5$ since -3 is to the right of -5 on a number line.

b. $\frac{-12}{-4}=3$

c. $0>-2$ since 0 is to the right of -2 on a number line.

13. $7x+5=5+7x$

14. $5\cdot(7x)=(5\cdot 7)x=35x$

15.
$$2x+5=9$$
$$2x=4$$
$$x=2$$

16.
$$11.2=1.2-5x$$
$$10=-5x$$
$$-2=x$$

17.
$$6x-4=2+6(x-1)$$
$$6x-4=2+6x-6$$
$$6x-4=6x-4$$
$$-4=-4, \text{ which is always true.}$$
All real numbers

18. $2x+1.5=-0.2+1.6x$
$0.4x=-1.7$
$x=-4.25$

19. a. Let x = the first integer. Then $x+1$ = the second integer and $x+2$ = the third integer.
$x+(x+1)+(x+2)=3x+3$

b. $x+(5x)+(6x-3)=12x-3$

20. a. Let x = the first integer. Then $x+2$ = the second even integer and $x+4$ = the third even integer.
$x+(x+2)+(x+4)=3x+6$

b. $4(3x+1)=12x+4$

21. Let x = first number, then $2x+3$ = second number and $4x$ = third number.

$$x+(2x+3)+4x=164$$
$$7x+3=164$$
$$7x=161$$
$$x=23$$

$2x+3=2(23)+3=49$
$4x=4(23)=92$
The three numbers are 23, 49 and 92.

22. Let x = first number, then $3x+2$ = second number.

$$(3x+2)-x=24$$
$$2x+2=24$$
$$2x=22$$
$$x=11$$

$3x+2=3(11)+2=35$
The two numbers are 11 and 35.

23. $3y-2x=7$

$$3y=2x+7$$
$$y=\frac{2x+7}{3},\text{ or } y=\frac{2x}{3}+\frac{7}{3}$$

24. $7x-4y=10$

$$7x=4y+10$$
$$x=\frac{4y+10}{7},\text{ or } x=\frac{4y}{7}+\frac{10}{7}$$

25. $A=\frac{1}{2}(B+b)h$

$$2A=(B+b)h$$
$$2A=Bh+bh$$
$$2A-Bh=bh$$
$$\frac{2A-Bh}{h}=b$$

26. $P=2l+2w$

$$P-2w=2l$$
$$\frac{P-2w}{2}=l$$

27. a. $\{x|x\ge 2\}$

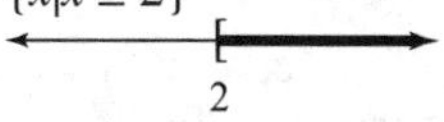

$[2, \infty)$

b. $\{x|x<-1\}$

$(-\infty, -1)$

c. $\{x|0.5<x\le 3\}$

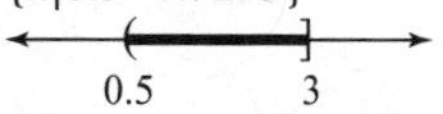

$(0.5, 3]$

28. a. $\{x|x\le -3\}$

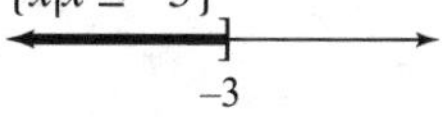

$(-\infty, -3]$

b. $\{x|-2\le x<0.1\}$

$[-2, 0.1)$

29. $-(x-3)+2\le 3(2x-5)+x$

$$-x+3+2\le 6x-15+x$$
$$-x+5\le 7x-15$$
$$20\le 8x$$
$$\frac{5}{2}\le x$$

$\left[\frac{5}{2}, \infty\right)$

30. $2(7x-1)-5x>-(-7x)+4$

$$14x-2-5x>7x+4$$
$$9x-2>7x+4$$
$$2x>6$$
$$x>3$$

$(3, \infty)$

31. $2(x+3) > 2x+1$
$2x+6 > 2x+1$
$6 > 1$; True for all real numbers x.
$(-\infty, \infty)$

32. $4(x+1)-3 < 4x+1$
$4x+4-3 < 4x+1$
$4x+1 < 4x+1$
$1 < 1$ Never true
$\varnothing$

33. $A = \{2, 4, 6, 8\}$, $B = \{3, 4, 5, 6\}$; the numbers 4 and 6 are in both sets so the intersection of A and B is $\{4, 6\}$.

34. The elements in either set or both sets are $-2, -1, 0, 1, 2, 3, 4$, and 5, so the union is $\{-2, -1, 0, 1, 2, 3, 4, 5\}$.

35. $x-7 < 2$ *and* $2x+1 < 9$
$x < 9$ *and* $2x < 8$
$x < 4$
$x < 4$
$(-\infty, 4)$

36. $x+3 \le 1$ *or* $3x-1 < 8$
$x \le -2$ *or* $3x < 9$
$x < 3$
$x < 3$
$(-\infty, 3)$

37. $A = \{2, 4, 6, 8\}$ and $B = \{3, 4, 5, 6\}$, so the union of A and B is $\{2, 3, 4, 5, 6, 8\}$.

38. $\varnothing$; there are no elements in common.

39. $-2x-5 < -3$ *or* $6x < 0$
$-2x < 2$ *or* $x < 0$
$x > -1$
All real numbers
$(-\infty, \infty)$

40. $-2x-5 < -3$ *and* $6x < 0$
$-2x < 2$ *and* $x < 0$
$x > -1$
$-1 < x < 0$
$(-1, 0)$

41. $|p| = 2$
$p = 2$ or $p = -2$

42. $|x| = 5$
$x = 5$ or $x = -5$

43. $\left|\frac{x}{2} - 1\right| = 11$
$\frac{x}{2} - 1 = 11$ or $\frac{x}{2} - 1 = -11$
$\frac{x}{2} = 12$ or $\frac{x}{2} = -10$
$x = 24$ or $x = -20$

44. $\left|\frac{y}{3} + 2\right| = 10$
$\frac{y}{3} + 2 = 10$ or $\frac{y}{3} + 2 = -10$
$\frac{y}{3} = 8$ or $\frac{y}{3} = -12$
$y = 24$ or $y = -36$

45. $|x-3| = |5-x|$
$x-3 = 5-x$ or $x-3 = -(5-x)$
$2x = 8$ or $x-3 = -5+x$
$x = 4$ or $-3 = -5$
Since $-3 = -5$ is not possible, the only solution is 4.

46. $|x+3| = |7-x|$
$x+3 = 7-x$ or $x+3 = -(7-x)$
$2x = 4$ or $x-3 = -7+x$
$x = 2$ or $-3 = -7$
Since $-3 = -7$ is not possible, the only solution is 2.

47. $|x| \le 3$
$-3 \le x \le 3$
$[-3, 3]$

48. $|x| > 1$
$x < -1$ or $x > 1$
$(-\infty, -1) \cup (1, \infty)$

49. $|2x+9| + 5 > 3$
$|2x+9| > -2$
Since $|2x + 9|$ is nonnegative for all numbers x, the solution set is $(-\infty, \infty)$.

50. $|3x+1| + 9 < 1$
$|3x+1| < -8$
The solution set is $\varnothing$.

Chapter 3

Section 3.1 Practice Exercises

1. The six points are graphed as shown.

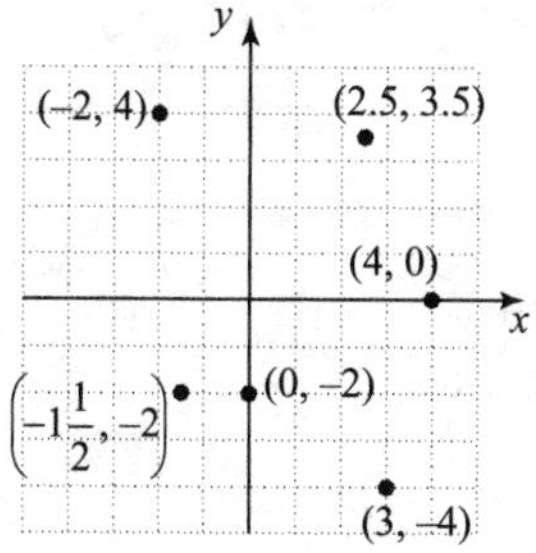

a. (3, −4) lies in quadrant IV.

b. (0, −2) is on the y-axis.

c. (−2, 4) lies in quadrant II.

d. (4, 0) is on the x-axis.

e. $\left(-1\frac{1}{2},\ -2\right)$ is in quadrant III.

f. (2.5, 3.5) is in quadrant I.

2. Let $x = 1$ and $y = 4$.

$$4x + y = 8$$
$$4(1) + 4 \overset{?}{=} 8$$
$$4 + 4 \overset{?}{=} 8$$
$$8 = 8 \quad \text{True}$$

Let $x = 0$ and $y = 6$.

$$4x + y = 8$$
$$4(0) + 6 \overset{?}{=} 8$$
$$0 + 6 \overset{?}{=} 8$$
$$6 = 8 \quad \text{False}$$

Let $x = 3$ and $y = -4$.

$$4x + y = 8$$
$$4(3) + (-4) \overset{?}{=} 8$$
$$12 - 4 \overset{?}{=} 8$$
$$8 = 8 \quad \text{True}$$

Thus, (0, 6) is not a solution, but both (1, 4) and (3, −4) are solutions.

3. a. Since x is products sold, find 6000 along the x-axis and move vertically up until you reach a point on the line. From this point on the line, move horizontally to the left until you reach the y-axis. Its value on the y-axis is 4200, which means if \$6000 worth of products is sold, the salary for the month is \$4200.

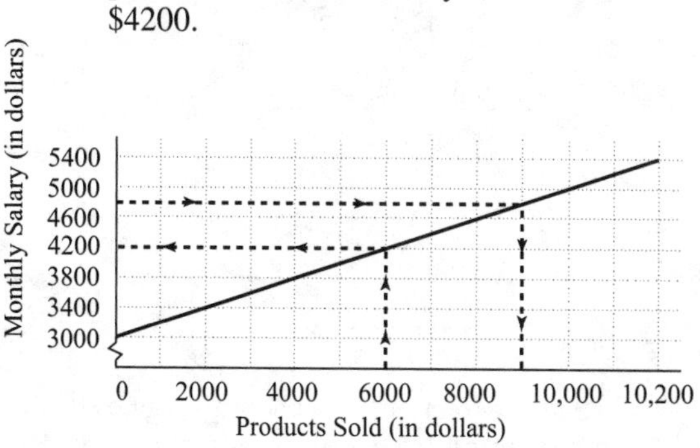

b. Since y is monthly salary, find 4800 along the y-axis and move horizontally to the right until you reach a point on the line. Move vertically downward until you reach the x-axis. The corresponding x-value is 9000. This means that \$9000 worth of products sold gives a salary of \$4800 for the month. For the salary to be greater than \$4800, products sold must be greater than \$9000.

4. $y = -3x - 2$

This is a linear equation. (In standard form, it is $3x + y = -2$.) Since the equation is solved for y, we choose three x-values.

Let $x = 0$.

$$y = -3x - 2$$
$$y = -3 \cdot 0 - 2$$
$$y = -2$$

Let $x = -1$.

$$y = -3x - 2$$
$$y = -3(-1) - 2$$
$$y = 1$$

Let $x = -2$.

$$y = -3x - 2$$
$$y = -3(-2) - 2$$
$$y = 4$$

The three ordered pairs (0, −2), (−1, 1), and (−2, 4) are listed in the table.

x	y
0	−2
−1	1
−2	4

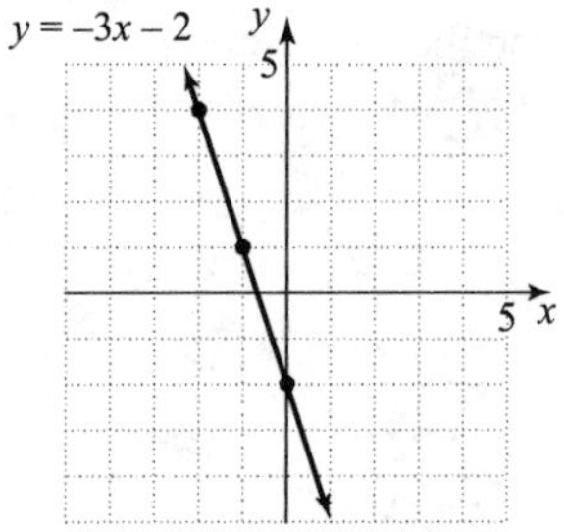

5. $y = -\frac{1}{2}x$

To avoid fractions, we choose x-values that are multiples of 2. To find the y-intercept, we let $x = 0$.

If $x = 0$, then $y = -\frac{1}{2}(0)$, or 0.

If $x = 2$, then $y = -\frac{1}{2}(2)$, or −1.

If $x = -2$, then $y = -\frac{1}{2}(-2)$, or 1.

x	y
0	0
2	−1
−2	1

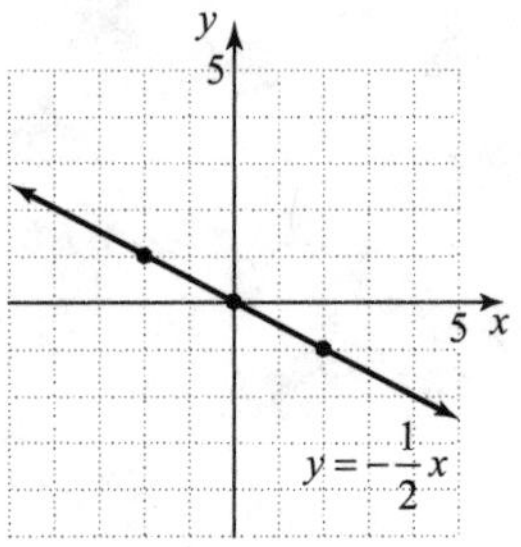

6. $y = 4x^2$

This equation is not linear because of the x^2 term. Its graph is not a line.

If $x = -3$, then $y = 4(-3)^2$, or 36.

If $x = -2$, then $y = 4(-2)^2$, or 16.

If $x = -1$, then $y = 4(-1)^2$, or 4.

If $x = 0$, then $y = 4(0)^2$, or 0.

If $x = 1$, then $y = 4(1)^2$, or 4.

If $x = 2$, then $y = 4(2)^2$, or 16.

If $x = 3$, then $y = 4(3)^2$, or 36.

x	y
−3	36
−2	16
−1	4
0	0
1	4
2	16
3	36

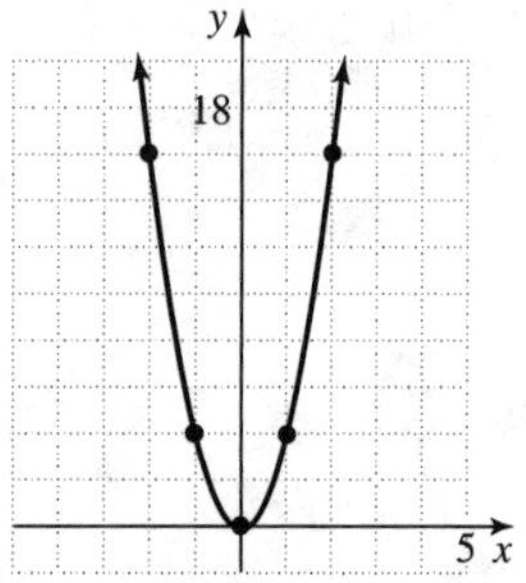

7. $y = -|x|$

This equation is not linear because it cannot be written in the form $Ax + By = C$. Its graph is not a line.

If $x = -3$, then $y = -|-3|$, or −3.

If $x = -2$, then $y = -|-2|$, or −2.

If $x = -1$, then $y = -|-1|$, or −1.

If $x = 0$, then $y = -|0|$, or 0.

If $x = 1$, then $y = -|1|$, or −1.

If $x = 2$, then $y = -|2|$, or −2.

If $x = 3$, then $y = -|3|$, or −3.

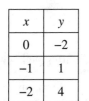

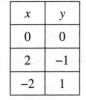

x	y
−3	−3
−2	−2
−1	−1
0	0
1	−1
2	−2
3	−3

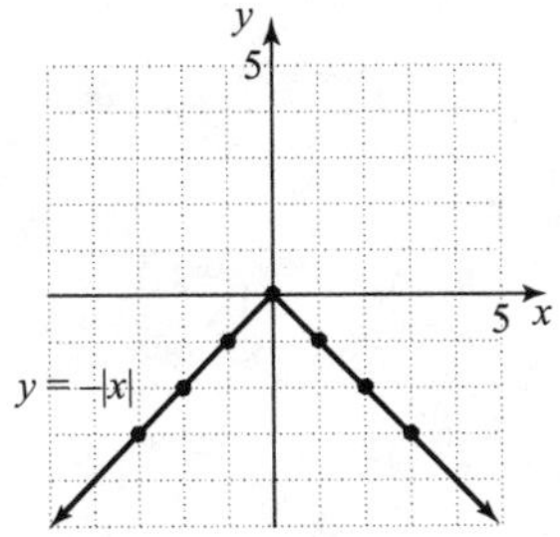

Graphing Calculator Explorations

1.

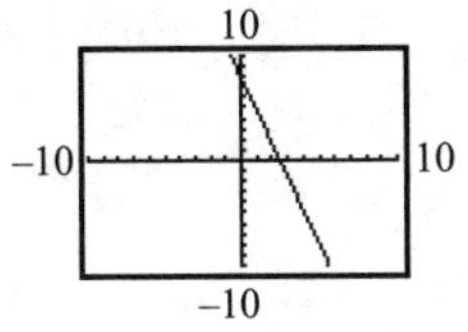

2.

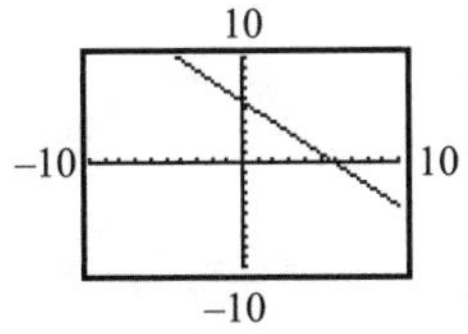

3.

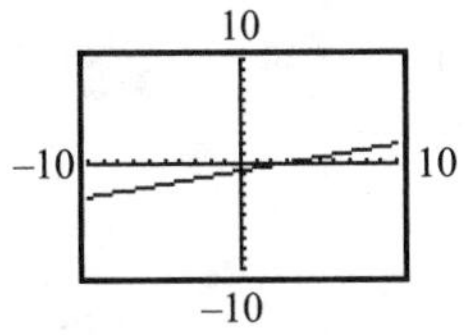

4.

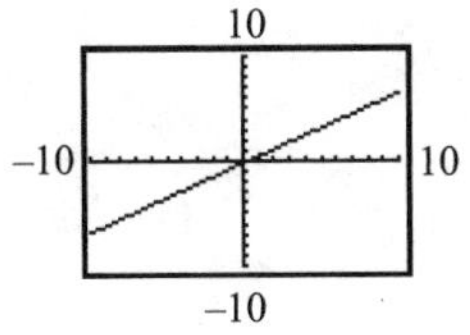

5.

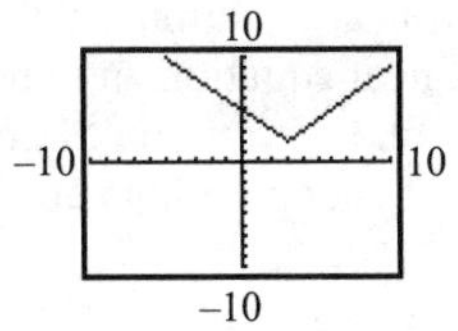

6.

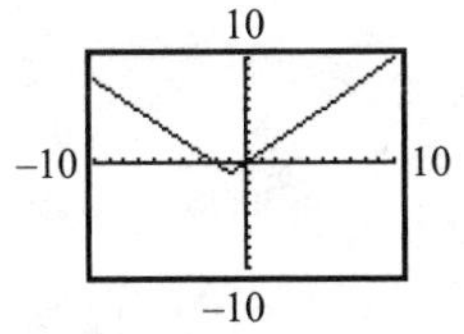

7.

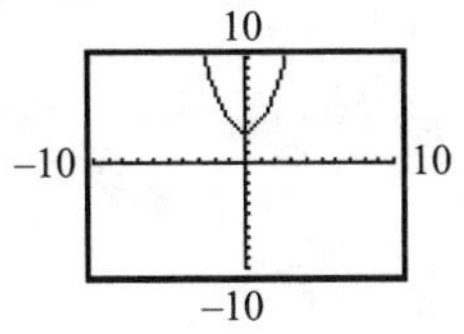

8.

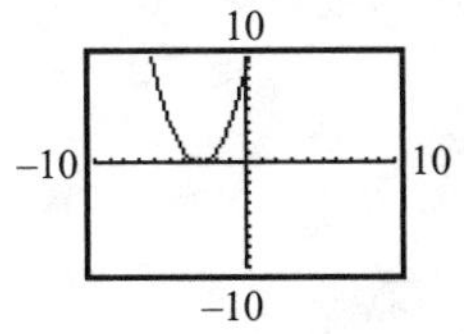

Vocabulary, Readiness & Video Check 3.1

1. The intersection of the x-axis and y-axis is a point called the origin.

2. The rectangular coordinate system has 4 quadrants and 2 axes.

3. The graph of a single ordered pair of numbers is how many points? 1

4. The graph of $Ax + By = C$, where A and B are not both 0 is a line.

5. The graph of $y = |x|$ looks V-shaped.

6. The graph of $y = x^2$ is a parabola.

7. origin; left (if negative) or right (if positive); up (if positive) or down (if negative)

8. An ordered pair is a solution of an equation in two variables if, when the variables are replaced with their ordered pair values, a true statement results.

9. The graph of an equation is a picture of its ordered pair solutions; a third point is found as a check to make sure the points line up and a mistake hasn't been made.

10. When graphing a nonlinear equation, first recognize it as a nonlinear equation and know that the graph is <u>not</u> a line. If you don't know the <u>shape</u> of the graph, plot enough points until you see a pattern.

Exercise Set 3.1

1. (3, 2) is in quadrant I

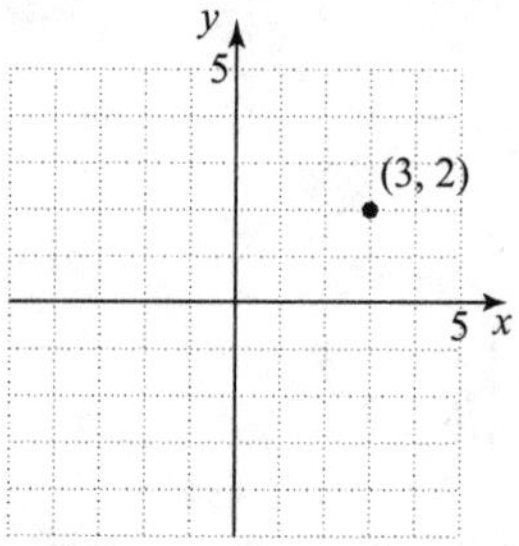

3. (−5, 3) is in quadrant II.

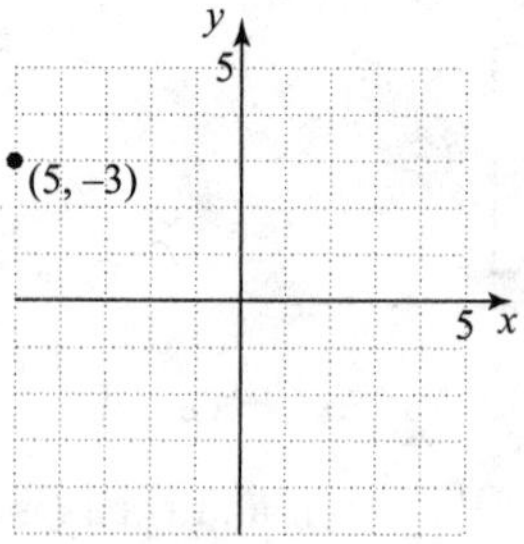

5. $\left(5\frac{1}{2}, -4\right)$ is in quadrant IV.

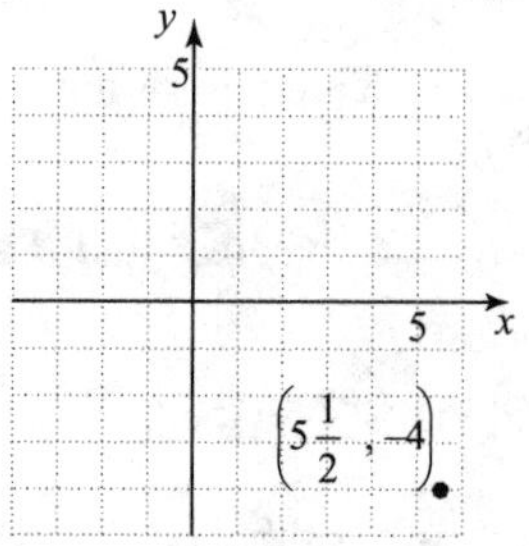

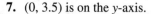

7. (0, 3.5) is on the *y*-axis.

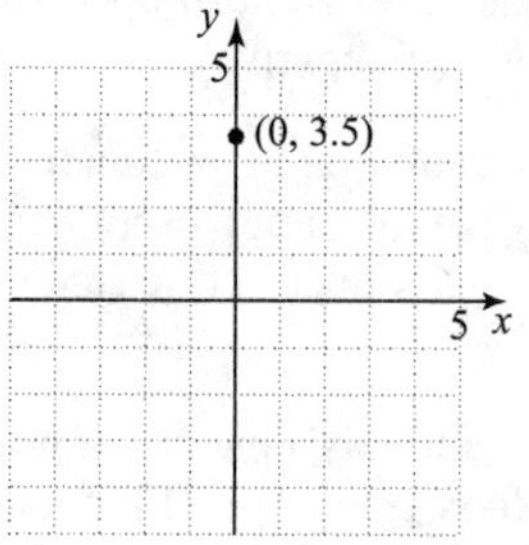

9. (−2, −4) is in quadrant III.

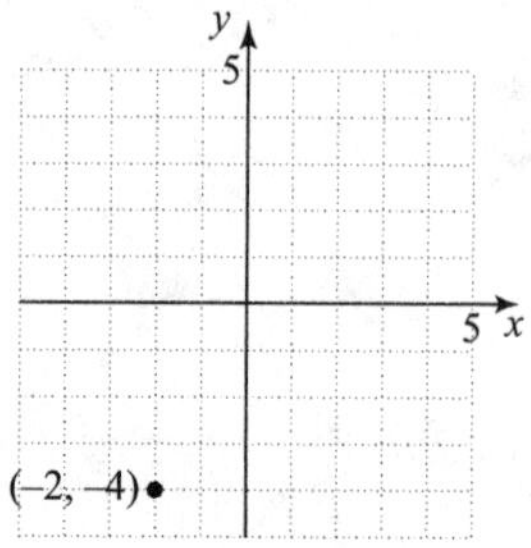

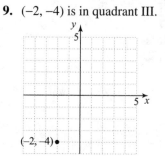

11. Point *C* is (−5, −2).

13. Point *E* is (−1, 0).

15. Point *B* is (3, 0).

17. Let $x = 0$, $y = 5$.

$$y = 3x - 5$$
$$5 = 3 \cdot 0 - 5$$
$$5 = -5$$

False; no

Let $x = -1$, $y = -8$.

$$y = 3x - 5$$
$$-8 = 3 \cdot (-1) - 5$$
$$-8 = -8$$

True; yes

19. Let $x = 1$, $y = 0$.

$$-6x + 5y = -6$$
$$-6(1) + 5(0) = -6$$
$$-6 = -6$$

True; yes

Let $x = 2$, $y = \frac{6}{5}$.

$$-6x + 5y = -6$$
$$-6(2) + 5\left(\frac{6}{5}\right) = -6$$
$$-6 = -6$$

True; yes

21. Let $x = 1$, $y = 2$.

$$y = 2x^2$$
$$2 = 2(1)^2$$
$$2 = 2$$

True; yes

Let $x = 3$, $y = 18$.
$y = 2x^2$
$18 = 2(3)^2$
$18 = 18$
True; yes

23. Let $x = 2$, $y = 8$.
$y = x^3$
$8 = (2)^3$
$8 = 8$
True; yes

Let $x = 3$, $y = 9$,
$y = x^3$
$9 = (3)^3$
$9 = 27$
False; no

25. Let $x = 1$, $y = 3$.
$y = \sqrt{x} + 2$
$3 = \sqrt{1} + 2$
$3 = 3$
True; yes
Let $x = 4$, $y = 4$.
$y = \sqrt{x} + 2$
$4 = \sqrt{4} + 2$
$4 = 4$
True; yes

27. $x + y = 3$
Linear

x	y
0	3
3	0
1	2

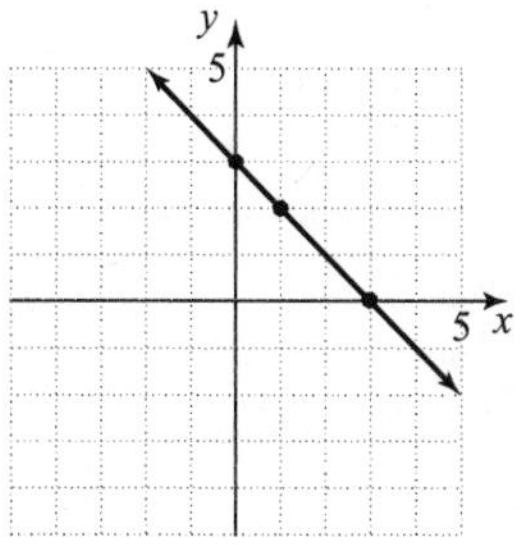

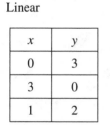

29. $y = 4x$
Linear

x	y
−1	−4
0	0
1	4

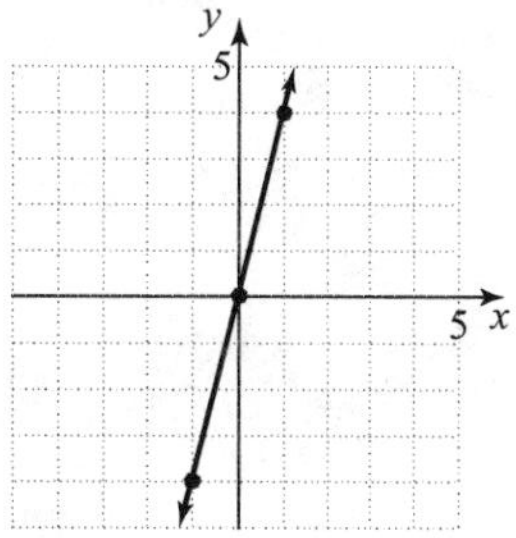

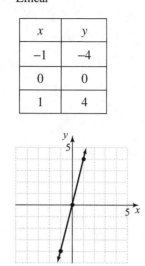

31. $y = 4x - 2$
Linear

x	y
0	−2
$\frac{1}{2}$	0
1	2

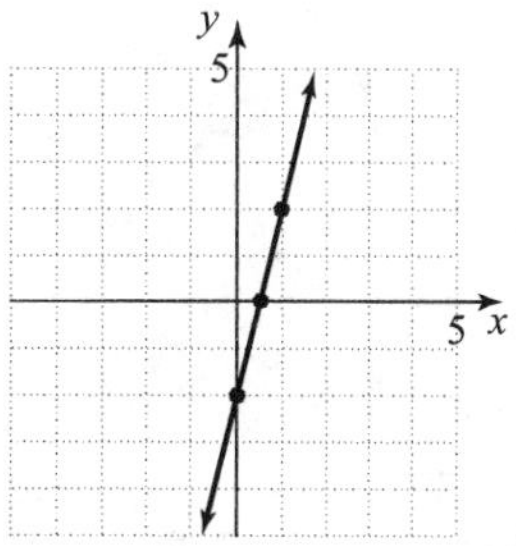

33. $y = |x| + 3$
Not linear

x	y
−2	5
−1	4
0	3
1	4
2	5

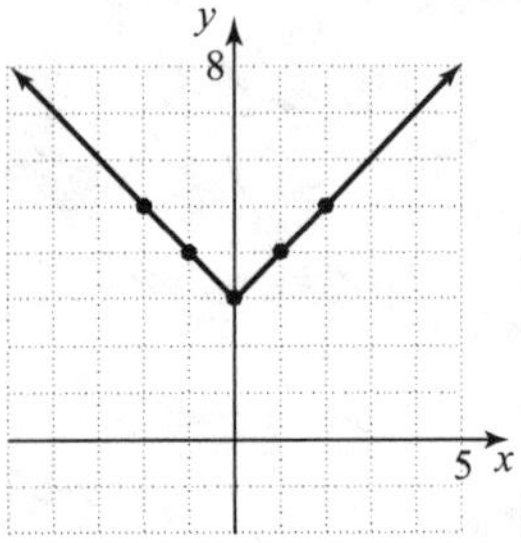

35. $2x - y = 5$
Linear

x	y
$2\frac{1}{2}$	0
0	−5
1	−3

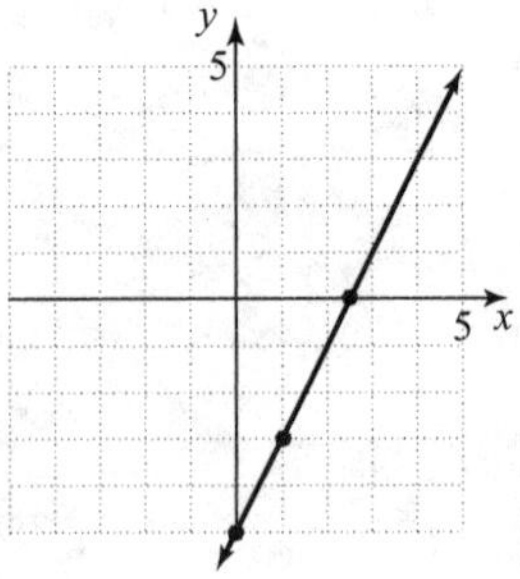

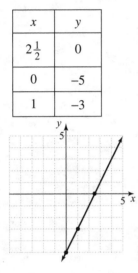

37. $y = 2x^2$
Not linear

x	y
−2	8
−1	2
0	0
1	2
2	8

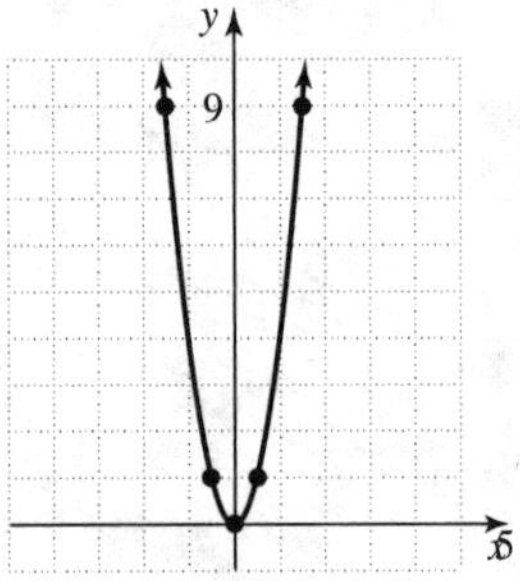

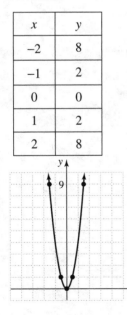

39. $y = x^2 - 3$
Not linear

x	y
−2	1
−1	−2
0	−3
1	−2
2	1

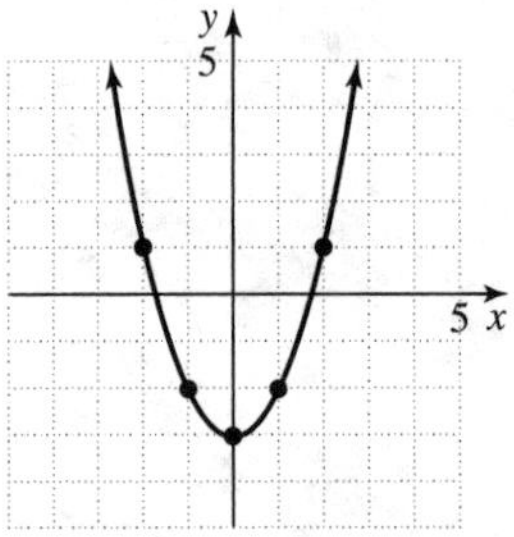

41. $y = -2x$
Linear

x	y
−1	2
0	0
1	−2

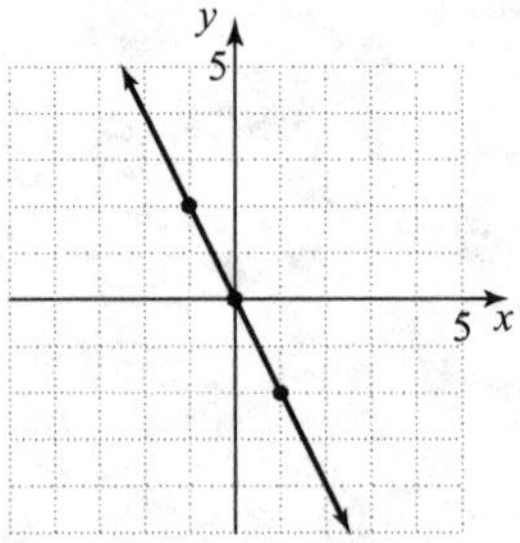

43. $y = -2x + 3$
Linear

x	y
−1	5
0	3
1	1

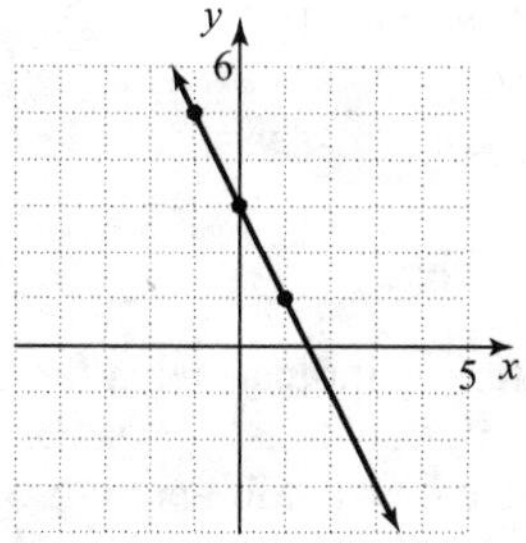

45. $y = |x + 2|$
Not linear

x	y
−4	2
−3	1
−2	0
−1	1
0	2

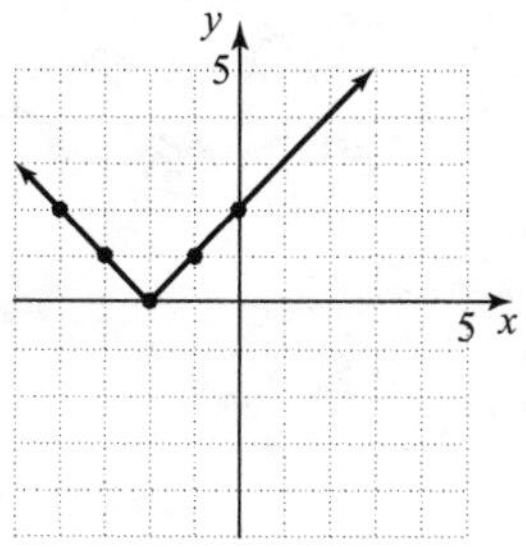

47. $y = x^3$
Not linear

x	y
−3	−27
−2	−8
−1	−1
0	0
1	1
2	8

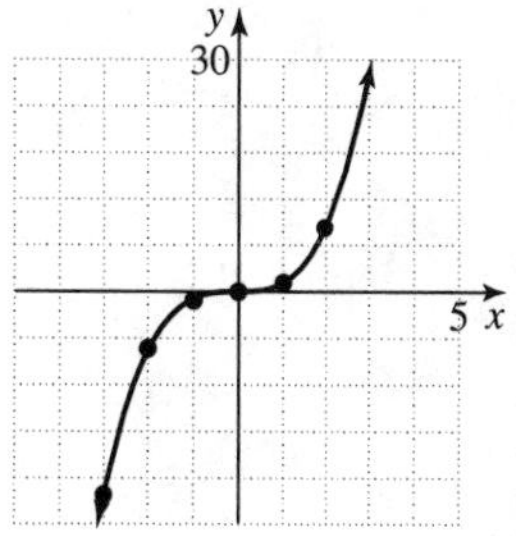

49. $y = -|x|$
Not linear

x	y
−2	−2
−2	−1
0	0
1	−1
2	−2

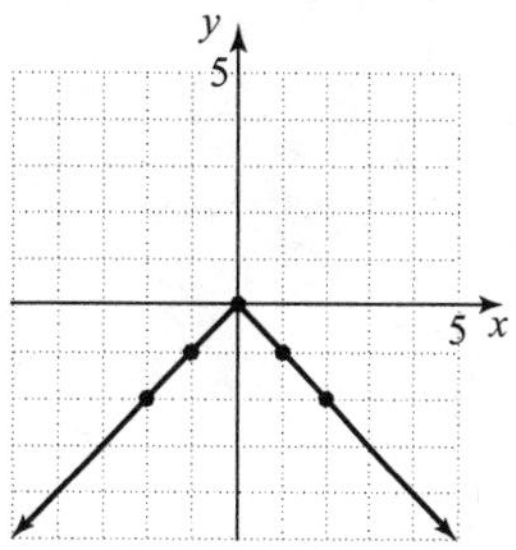

51. $y = \frac{1}{3}x - 1$
Linear

x	y
−3	−2
0	−1
3	0

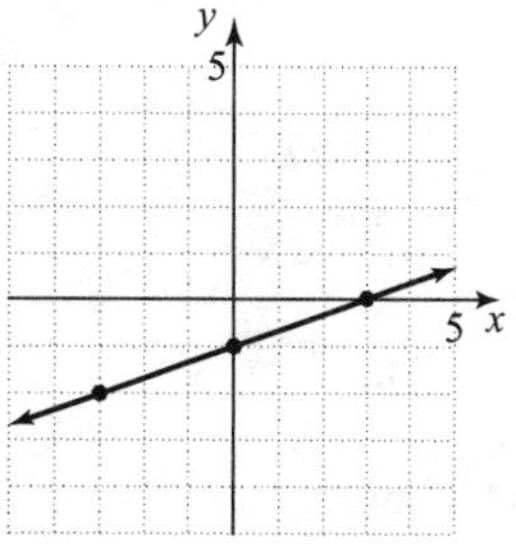

53. $x = -\frac{3}{2}x + 1$

Linear

x	y
−2	4
0	1
2	−2

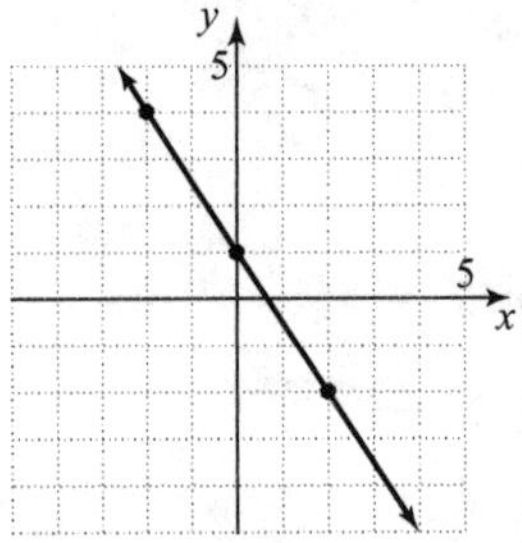

55. $3(x-2)+5x = 6x-16$

$$3x-6+5x = 6x-16$$
$$8x-6 = 6x-16$$
$$2x = -10$$
$$x = -5$$

The solution is -5.

57. $3x + \frac{2}{5} = \frac{1}{10}$

$$30x+4 = 1$$
$$30x = -3$$
$$x = -\frac{1}{10}$$

The solution is $-\frac{1}{10}$.

59. $3x \le -15$

$$x \le -5$$

$(-\infty, -5]$

61. $2x-5 > 4x+3$

$$-2x > 8$$
$$x < -4$$

$(-\infty, -4)$

63. (4, −2) is in quadrant IV.

65. (0, −100) is on the y-axis.

67. (−10, −30) is in quadrant III.

69. $(x, -y)$ lies in quadrant IV.

71. $(x, 0)$ lies on the x-axis.

73. $(-x, -y)$ lies in quadrant III.

75. The first coordinate, −1, indicates that the point is 1 unit to the left of the y-axis. The second coordinate, 5.3, indicates that the point is 5.3 units above the x-axis. The answer is b.

77. Look for the graph where the only nonzero y-values are 40 and 60. The answer is B.

79. Look for the graph where all the y-values are between 10 and 30. The answer is C.

81. The first segment in the graph with y-coordinate greater than 5.0 begins in 1997. Thus, 1997 is the first year that the minimum hourly wage rose above $5.00.

83. answers may vary

85. $y = x^2 - 4x + 7$

x	y
0	7
1	4
2	3
3	4
4	7

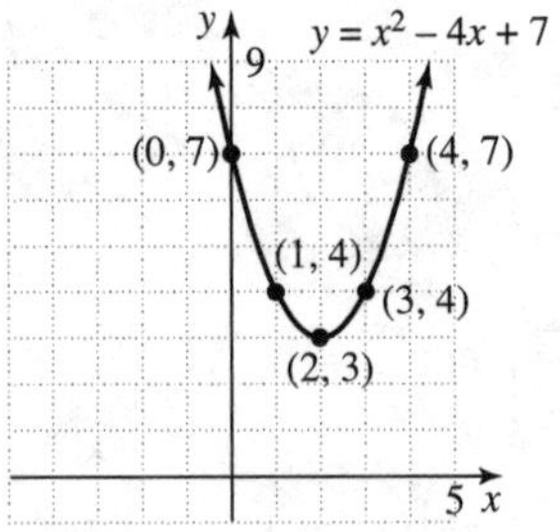

87. a. $y = 2x + 6$

x	y
0	6
1	8
2	10

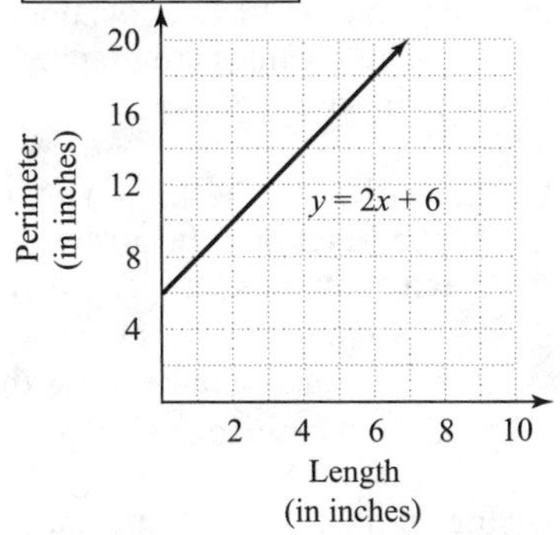

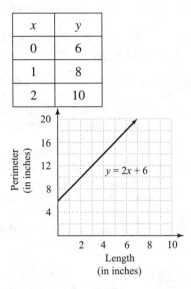

b. When x is 4, y is 14. Thus, when the length is 4 inches, the perimeter is 14 inches.

89. When $x = 0$, $y = 7$. Thus, the purchase price was \$7000.

91. 7000 − 6500 = \$500

93. Depreciation is the same from year to year.

95. They are parallel.

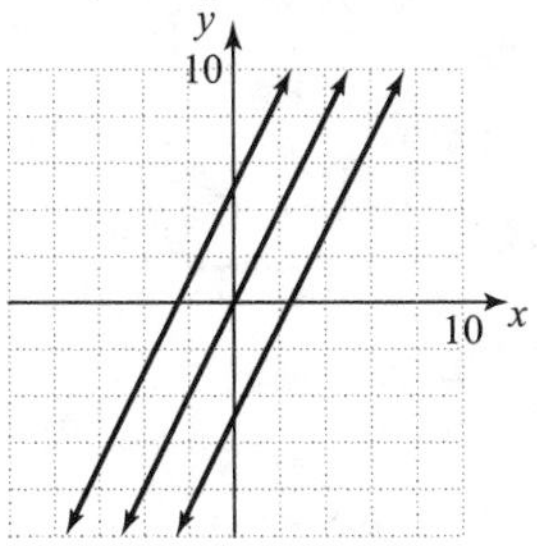

97. "The y-value is 5 more than three times the x-value" is written as $y = 3x + 5$.

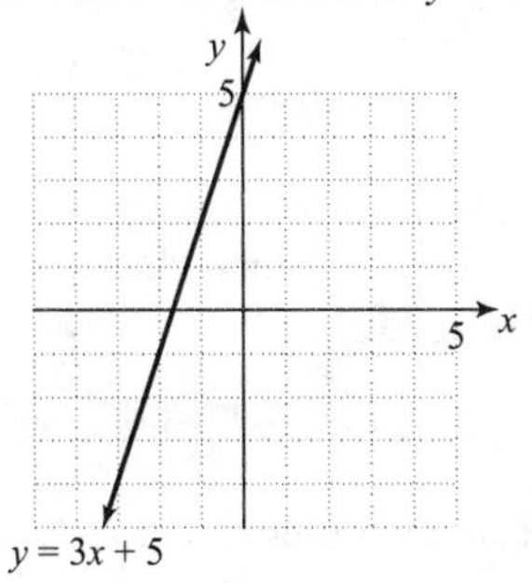

99. "The y-value is 2 more than the square of the x-value" is written as $y = x^2 + 2$.

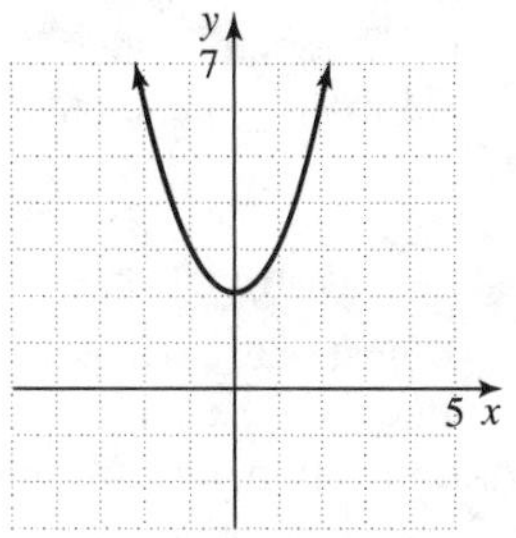

101. The graphs are the same.

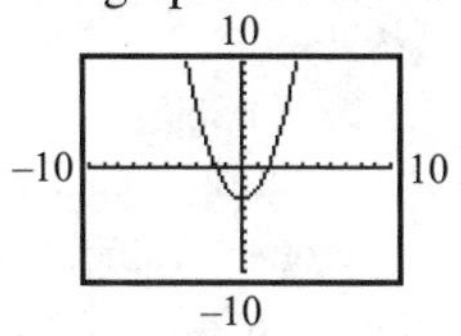

103. The graphs are the same.

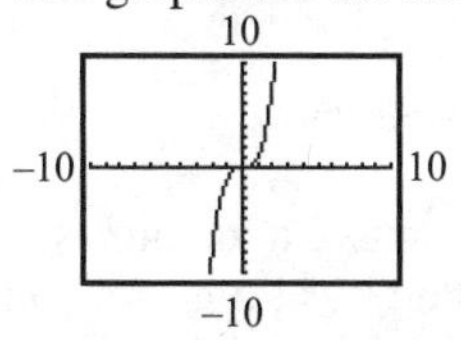

Section 3.2 Practice Exercises

1. a. The domain is the set of all first coordinates, {4, 5}. The range is the set of all second coordinates, {1, −3, −2, 6}.

b. Ordered pairs are not listed here but are given in graph form. The relation is {(3, −4), (3, −3), (3, −2), (3, −1), (3, 0), (3, 1), (3, 2), (3, 3), (3, 4)}. The domain is {3}. The range is {−4, −3, −2, −1, 0, 1, 2, 3, 4}.

c. The domain is the set of inputs, {Accountant, Computer Programmer, Engineer, Social Worker, Paralegal}. The range is the numbers in the set of outputs that correspond to elements in the set of inputs {51, 73, 92, 81, 48}.

2. a. Although the ordered pairs (3, 1) and (9, 1) have the same y-value, each x-value is assigned to only one y-value, so this set of ordered pairs is a function.

b. The x-value −2 is assigned to two y-values, −3 and 4, in this graph, so this relation does not define a function.

c. This relation is a function because although two different people may have the same birth date, each person has only one birth date. This means that each element in the first set is assigned to only one element in the second set.

3. The relation $y = -3x + 5$ is a function if each x-value corresponds to just one y-value. For each x-value substituted into the equation $y = -3x + 5$, the multiplication and addition performed on each gives a single result, so only one y-value will be associated with each x-value. Thus, $y = -3x + 5$ is a function.

4. The relation $y = -x^2$ is a function if each x-value corresponds to just one y-value. For each x-value substituted into the equation $y = -x^2$, squaring each gives a single result, so only one y-value will be associated with each x-value. Thus, $y = -x^2$ is a function.

5. a. Yes, this is the graph of a function since no vertical line will intersect this graph more than once.

b. Yes, this is the graph of a function since no vertical line will intersect this graph more than once.

c. No, this is not the graph of a function. Note that vertical lines can be drawn that intersect the graph in two points.

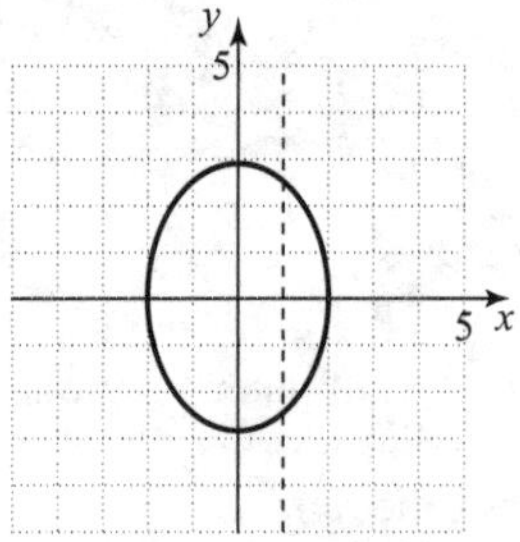

d. Yes, this is the graph of a function since no vertical line will intersect this graph more than once.

e. No, this is not the graph of a function. A vertical line can be drawn that intersects this line at every point.

6. a. By the vertical line test, the graph is the graph of a function. The x-values are graphed from −1 to 2, so the domain is [−1, 2]. The y-values are graphed from −2 to 9, so the range is [−2, 9].

b. By the vertical line test, the graph is not the graph of a function. The x-values are graphed from −1 to 1, so the domain is [−1, 1]. The y-values are graphed from −4 to 4, so the range is [−4, 4].

c. By the vertical line test, the graph is the graph of a function. The arrows indicate that the graph continues forever. All x-values are graphed, so the domain is $(-\infty, \infty)$. The y-values for 4 and numbers less than 4 are graphed, so the range is $(-\infty, 4]$.

d. By the vertical line test, the graph is the graph of a function. The arrows indicate that the graph continues forever. All x-values and all y-values are graphed, so the domain is $(-\infty, \infty)$ and the range is $(-\infty, \infty)$.

7. a. Substitute 1 for x in $f(x)$.

$$f(x) = 3x - 2$$
$$f(1) = 3(1) - 2 = 3 - 2 = 1$$

b. Substitute 1 for x in $g(x)$.

$$g(x) = 5x^2 + 2x - 1$$
$$g(1) = 5(1)^2 + 2(1) - 1$$
$$= 5 + 2 - 1 = 6$$

c. Substitute 0 for x in $f(x)$.

$$f(x) = 3x - 2$$
$$f(0) = 3(0) - 2 = 0 - 2 = -2$$

d. Substitute −2 for x in $g(x)$.

$$g(x) = 5x^2 + 2x - 1$$
$$g(-2) = 5(-2)^2 + 2(-2) - 1$$
$$= 5(4) - 4 - 1$$
$$= 20 - 4 - 1 = 15$$

8. a. To find $f(1)$, find the y-value when $x = 1$. We see from the graph that when $x = 1$, y or $f(x) = -3$. Thus, $f(1) = -3$.

b. $f(0) = -2$ from the ordered pair (0, −2).

c. $g(-2) = 3$ from the ordered pair (−2, 3).

d. $g(0) = 1$ from the ordered pair (0, 1).

e. To find x-values such that $f(x) = 1$, we are looking for any ordered pairs on the graph of f whose $f(x)$ or y-value is 1. They are $(-1, 1)$ and $(3, 1)$. Thus, $f(-1) = 1$ and $f(3) = 1$. The x-values are -1 and 3.

f. Find ordered pairs on the graph of g whose $g(x)$ or y-value is -2. There is one such ordered pair, $(-3, -2)$. Thus, $g(-3) = -2$. The only x-value is -3.

9. Find the semester Fall 2010 and move upward until you reach the graph representing 2-year enrollment. From the point on the graph, move horizontally to the left until the vertical axis is reached. In fall 2010, approximately 7.6 million or 7,600,000 students were enrolled.

10. 2015 is 15 years after 2000, so find $f(15)$.

$$f(x) = -0.05x^2 + 1.1x + 2.2$$

$$f(15) = -0.05(15)^2 + 1.1(15) + 2.2 = 7.45$$

The 2-year institution enrollment in Fall 2015 was 7.45 million or 7,450,000.

Graphing Calculator Explorations

1.

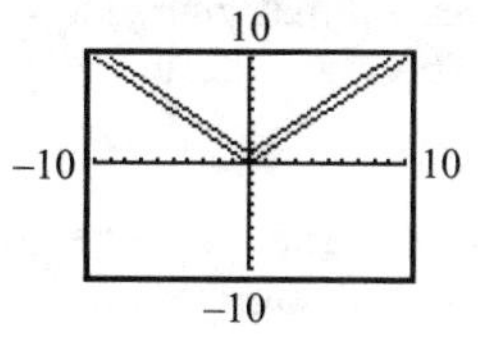

2.

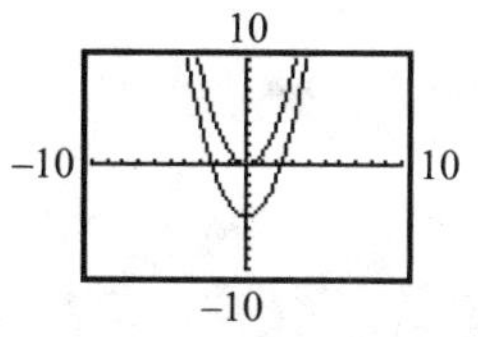

3.

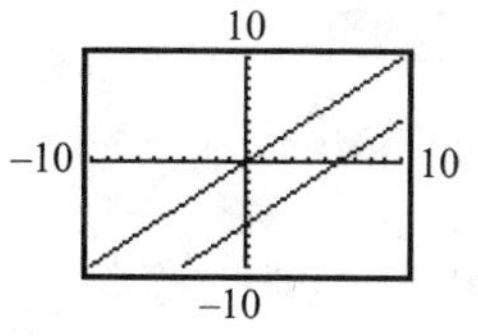

4.

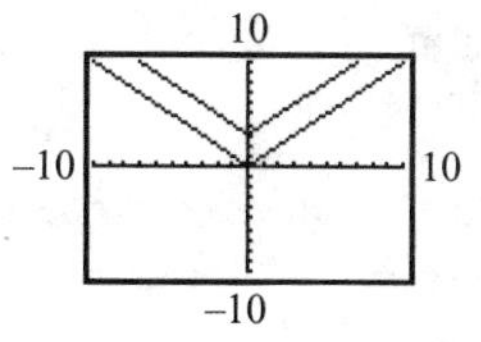

5.

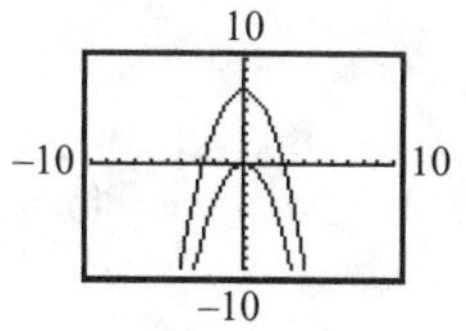

6.

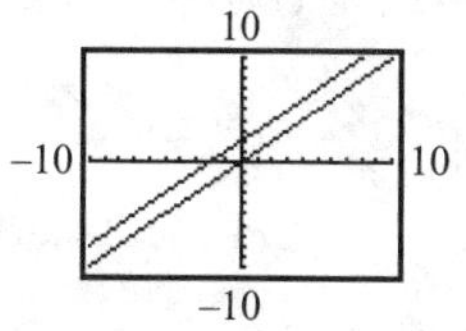

Vocabulary, Readiness & Video Check 3.2

1. A <u>relation</u> is a set of ordered pairs.

2. The <u>range</u> of a relation is the set of all second components of the ordered pairs.

3. The <u>domain</u> of a relation is the set of all first components of the ordered pairs.

4. A <u>function</u> is a relation in which each first component in the ordered pairs corresponds to *exactly* one second component.

5. By the vertical line test, all linear equations are functions except those whose graphs are <u>vertical</u> lines.

6. If $f(-2) = 1.7$, the corresponding ordered pair is <u>$(-2, 1.7)$</u>.

7. An equation is one way of writing down a correspondence that produces ordered pair solutions; a relation between two sets of coordinates, x's and y's.

8. Yes. The function definition restricts each first component to correspond to exactly one second component, but it makes no such restriction on each second component.

9. If a vertical line intersects a graph two (or more) times, then there's an x-value corresponding to two (or more) different y-values and is thus not a function.

10. No, equations of the form $x = c$, whose graphs are vertical lines, are not functions.

11. Using function notation, the replacement value for x and the resulting $f(x)$ or y-value corresponds to an ordered pair (x, y) solution to the function.

Exercise Set 3.2

1. The domain is the set of all first coordinates and the range is the set of all second coordinates.
Domain = {−1, 0, −2, 5}
Range = {7, 6, 2}
Function since each x-value corresponds to exactly one y-value.

3. The domain is the set of all first coordinates and the range is the set of all second coordinates.
Domain = {−2, 6, −7}
Range = {4, −3, −8}
The relation is not a function since −2 is paired with both 4 and −3.

5. The domain is the set of all first coordinates and the range is the set of all second coordinates.
Domain = {1}
Range = {1, 2, 3, 4}
The relation is not a function since 1 is paired with both 1 and 2 for example.

7. The domain is the set of all first coordinates and the range is the set of all second coordinates.
$\text{Domain} = \left\{\frac{3}{2}, 0\right\}$
$\text{Range} = \left\{\frac{1}{2}, -7, \frac{4}{5}\right\}$
The relation is not a function since $\frac{3}{2}$ is paired with both $\frac{1}{2}$ and −7.

9. The domain is the set of all first coordinates and the range is the set of all second coordinates.
Domain = {−3, 0, 3}
Range = {−3, 0, 3}
Function since each x-value corresponds to exactly one y-value.

11. Points on graph: (−1, 2), (1, 1), (2, 1), (3, 1)
Domain = {−1, 1, 2, 3}
Range = {2, 1}
Function since each x-value corresponds to exactly one y-value.

13. Domain
= {1994, 1998, 2002, 2006, 2010, 2014}
Range = {6, 9, 10}
Function since each input corresponds to exactly one output.

15. Domain = {32°, 104°, 212°, 50°}
Range = {0°, 40°, 10°, 100°}
Function since each input corresponds to exactly one output.

17. Domain = {0}
Range = {2, −1, 5, 100}
Not a function since the input 0 corresponds to more than one output.

19. This relation is a function because although two different students may have the same final grade average, each student has only one final grade average. This means that each element in the first set is assigned to only one element in the second set.

21. This relation is not a function because more than one person in Cincinnati has blue eyes. This means that one element in the first set is assigned to more than one element in the second set.

23. Yes, this is the graph of a function since no vertical line will intersect this graph more than once.

25. No, this is not the graph of a function. Note that vertical lines can be drawn that intersect the graph in two points. The y-axis is such a vertical line.

27. Yes, this is the graph of a function since no vertical line will intersect this graph more than once.

29. The x-values are graphed from 0 to $+\infty$;
domain = $[0, \infty)$
The arrows indicate the y-values continue forever; range = $(-\infty, \infty)$
The relation is not a function since it fails the vertical line test (try $x = 1$).

31. The x-values are graphed from −1 to 1;
domain = $[-1, 1]$
The arrows indicate the y-values continue forever; range = $(-\infty, \infty)$
The relation is not a function since it fails the vertical line test (try $x = 0$).

33. The arrows indicate the x-values continue forever; domain = $(-\infty, \infty)$
The y-values do not include $(-3, 3)$;
range = $(-\infty, -3] \cup [3, \infty)$
The relation is not a function since it fails the vertical line test (try $x = 2$).

35. The x-values are graphed from 2 to 7; domain = [2, 7]
The y-values are graphed from 1 to 6; range = [1, 6]
The relation is not a function since it fails the vertical line test (try $x = 4$).

37. The only x-value is 2; domain = {−2}
The y-values continue forever; range = $(-\infty, \infty)$
The relation is not a function since it fails the vertical line test (try $x = -2$).

39. The x-values continue forever; domain = $(-\infty, \infty)$
The y-values are 3 and less; range = $(-\infty, 3]$
Function since it passes the vertical line test.

41. $y = x + 1$
For each x-value substituted into the equation $y = x + 1$, the addition performed gives a single result, so only one y-value will be associated with each x-value. Thus, $y = x + 1$ is a function.

43. $x = 2y^2$
The x-value 8 is associated with two y-values, −2 and 2. Thus, $x = 2y^2$ is not a function.

45. $y - x = 7$
For each x-value substituted into the equation $y - x = 7$, the process of solving for y gives a single result, so only one y-value will be associated with each x-value. Thus, $y - x = 7$ is a function.

47. $y = \frac{1}{x}$
For each x-value substituted into the equation $y = \frac{1}{x}$, the division performed gives a single result, so only one y-value will be associated with each x-value. Thus, $y = \frac{1}{x}$ is a function.

49. $y = 5x - 12$
For each x-value substituted into the equation $y = 5x - 12$, the multiplication and addition performed on each gives a single result, so only one y-value will be associated with each x-value. Thus, $y = 5x - 12$ is a function.

51. $x = y^2$
The x-value 4 is associated with two y-values, −2 and 2. Thus, $x = y^2$ is not a function.

53. $f(x) = 3x + 3$
$f(4) = 3(4) + 3 = 12 + 3 = 15$

55. $h(x) = 5x^2 - 7$

$$\begin{aligned} h(-3) &= 5(-3)^2 - 7 \\ &= 5(9) - 7 \\ &= 45 - 7 \\ &= 38 \end{aligned}$$

57. $g(x) = 4x^2 - 6x + 3$

$$\begin{aligned} g(2) &= 4(2)^2 - 6(2) + 3 \\ &= 4(4) - 12 + 3 \\ &= 16 - 12 + 3 \\ &= 7 \end{aligned}$$

59. $g(x) = 4x^2 - 6x + 3$

$$\begin{aligned} g(0) &= 4(0)^2 - 6(0) + 3 \\ &= 4(0) - 0 + 3 \\ &= 0 - 0 + 3 \\ &= 3 \end{aligned}$$

61. $f(x) = \frac{1}{2}x$

a. $f(0) = \frac{1}{2}(0) = 0$

b. $f(2) = \frac{1}{2}(2) = 1$

c. $f(-2) = \frac{1}{2}(-2) = -1$

63. $g(x) = 2x^2 + 4$

a.
$$\begin{aligned} g(-11) &= 2(-11)^2 + 4 \\ &= 2(121) + 4 \\ &= 242 + 4 \\ &= 246 \end{aligned}$$

b.
$$\begin{aligned} g(-1) &= 2(-1)^2 + 4 \\ &= 2(1) + 4 \\ &= 2 + 4 \\ &= 6 \end{aligned}$$

c. $g\left(\frac{1}{2}\right) = 2\left(\frac{1}{2}\right)^2 + 4$
$= 2\left(\frac{1}{4}\right) + 4$
$= \frac{1}{2} + \frac{8}{2}$
$= \frac{9}{2}$

65. $f(x) = -5$

a. $f(2) = -5$

b. $f(0) = -5$

c. $f(606) = -5$

67. $f(x) = 1.3x^2 - 2.6x + 5.1$

a. $f(2) = 1.3(2)^2 - 2.6(2) + 5.1$
$= 1.3(4) - 5.2 + 5.1$
$= 5.2 - 5.2 + 5.1$
$= 5.1$

b. $f(-2) = 1.3(-2)^2 - 2.6(-2) + 5.1$
$= 1.3(4) + 5.2 + 5.1$
$= 5.2 + 5.2 + 5.1$
$= 15.5$

c. $f(3.1) = 1.3(3.1)^2 - 2.6(3.1) + 5.1$
$= 1.3(9.61) - 8.06 + 5.1$
$= 12.493 - 8.06 + 5.1$
$= 9.533$

69. If $f(1) = -10$, then $y = -10$ when $x = 1$. The ordered pair is $(1, -10)$.

71. If $g(4) = 56$, then $y = 56$ when $x = 4$. The ordered pair is $(4, 56)$.

73. The ordered pair $(-1, -2)$ is on the graph of f. Thus, $f(-1) = -2$.

75. The ordered pair $(2, 0)$ is on the graph of g. Thus, $g(2) = 0$.

77. There are two ordered pairs on the graph of f with a y-value of -5, $(-4, -5)$ and $(0, -5)$. The x-values are -4 and 0.

79. To the right of the y-axis, there is one ordered pair on the graph of g with a y-value of 4, $(3, 4)$. The x-value is 3.

81. $f(x) = 0.48x + 87.8$
$f(6) = 0.48(6) + 87.8 = 2.88 + 87.8 = 90.68$
In 2014, about 90.68% of Americans 25 and older had completed at least high school.

83. $g(x) = 0.008x^2 + 0.43x + 87.8$

$g(12) = 0.008(12)^2 + 0.43(12) + 87.8$
$= 1.152 + 5.16 + 87.8$
$= 94.112$
In 2020, we predict that 94.112% of Americans 25 and older will have completed at least high school.

85. answers may vary

87. Since 2012 is 12 years after 2000, find $f(12)$.
$f(x) = 0.42x + 10.5$
$f(12) = 0.42(12) + 10.5$
$= 15.54$
We predict that diamond production was \$15.54 billion in 2012.

89. $f(x) = x + 7$

91. $A(r) = \pi r^2$
$A(5) = \pi(5)^2 = 25\pi$ square centimeters

93. $V(x) = x^3$
$V(14) = (14)^3 = 2744$ cubic inches

95. $H(f) = 2.59f + 47.24$
$H(46) = 2.59(46) + 47.24$
$= 166.38$ centimeters

97. $D(x) = \frac{136}{25}x$
$D(30) = \frac{136}{25}(30) = 163.2$ milligrams

99. $C(x) = -1.17x + 61.7$

a. $C(6) = -1.17(6) + 61.7$
$= -7.02 + 61.7$
$= 54.68$
Per capita consumption of beef was 54.68 pounds in 2014.

b. $2019 - 2008 = 11$

$$C(11) = -1.17(11) + 61.7$$
$$= -12.87 + 61.7$$
$$= 48.83$$

It is predicted that per capita consumption of beef in the United States will be 48.83 pounds in 2019.

101. $x - y = -5$

x	0	–5	1
y	5	0	6

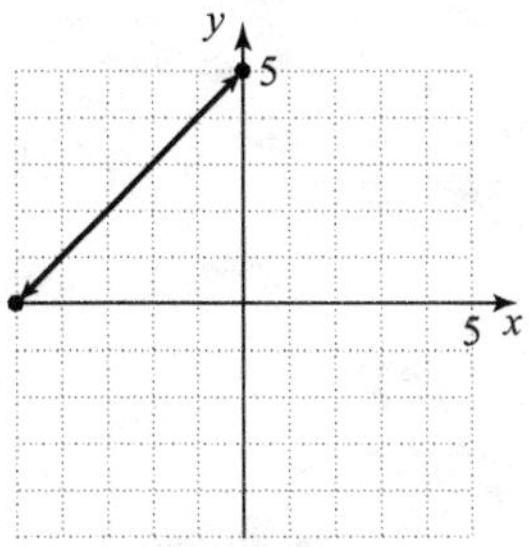

103. $7x + 4y = 8$

x	0	$\frac{8}{7}$	$\frac{12}{7}$
y	2	0	–1

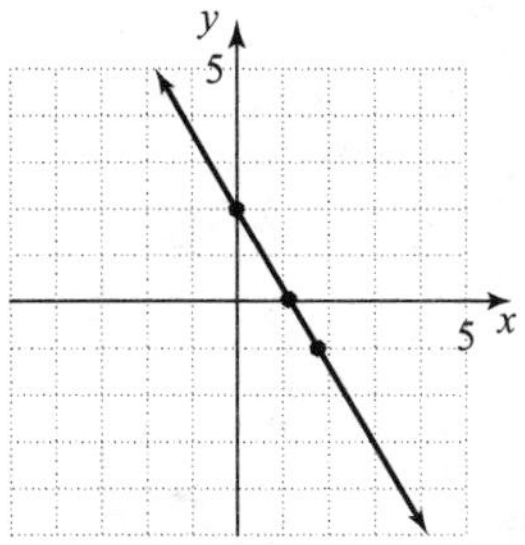

105. $y = 6x$

x	0	0	–1
y	0	0	–6

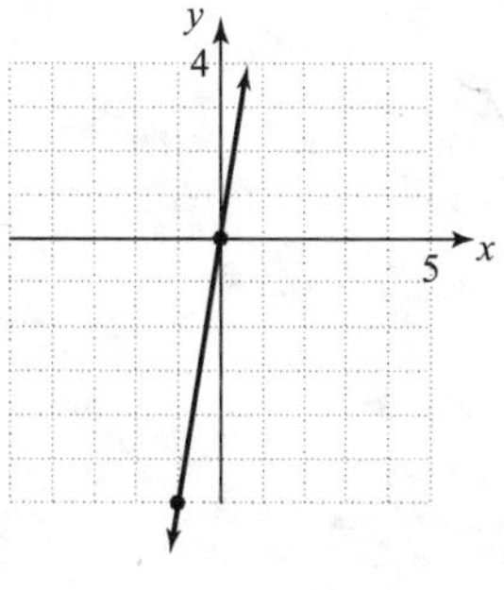

107. Yes, it is possible to find the perimeter. The sum of the lengths of the two sides parallel to the side measuring 45 meters is also 45 meters. The sum of the lengths of the two sides parallel to the side measuring 40 meters is also 40 meters. The perimeter is 45 + 45 + 40 + 40 = 170 meters.

109. $f(7) = 50$ means that $y = 50$ when $x = 7$. Thus, (7, 50) is an ordered pair of the function. The given statement is true.

111. Since $f(7) = 50$ when $f(x) = x^2 + 1$, the statement is true.

113. $h(x) = x^2 + 7$

a. $h(3) = (3)^2 + 7 = 16$

b. $h(a) = a^2 + 7$

115. $f(x) = 3x - 12$

a. $f(4) = 3(4) - 12 = 12 - 12 = 0$

b. $f(a) = 3a - 12$

c. $f(-x) = 3(-x) - 12 = -3x - 12$

d. $f(x + h) = 3(x + h) - 12 = 3x + 3h - 12$

117. infinitely many

119. answers may vary

121. answers may vary

Section 3.3 Practice Exercises

1. $f(x) = 4x,\ g(x) = 4x - 3$

x	$f(x)$	$g(x)$
0	0	–3
–1	–4	–7
1	4	1

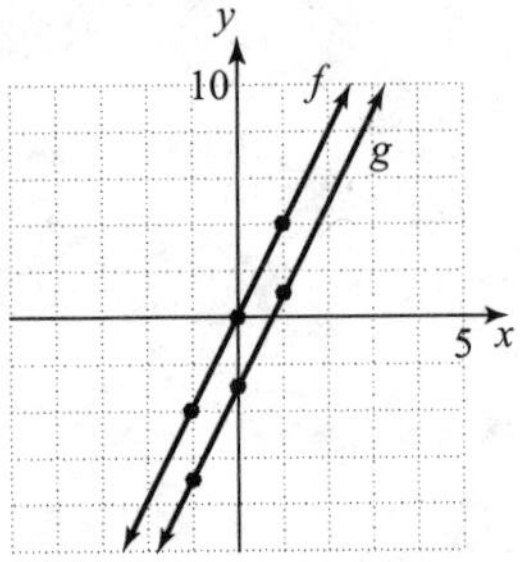

2. $f(x) = -2x,\ g(x) = -2x + 5$

x	$f(x)$	$g(x)$
0	0	5
−1	2	7
1	−2	3
2	−4	1

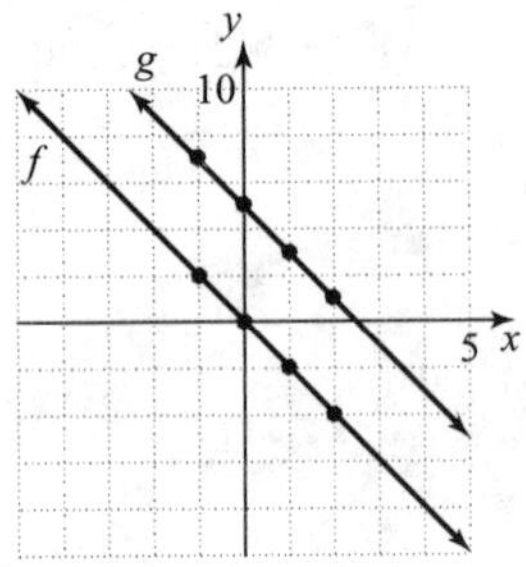

3. **a.** The y-intercept of $f(x) = \frac{3}{4}x - \frac{2}{5}$ is $\left(0,\ -\frac{2}{5}\right)$.

b. The y-intercept of $y = 2.6x + 4.1$ is (0, 4.1).

4. $4x - 5y = -20$

Let $x = 0$.

$$\begin{aligned} 4x - 5y &= -20 \\ 4 \cdot 0 - 5y &= -20 \\ -5y &= -20 \\ y &= 4 \end{aligned}$$

Let $y = 0$.

$$\begin{aligned} 4x - 5y &= -20 \\ 4x - 5 \cdot 0 &= -20 \\ 4x &= -20 \\ x &= -5 \end{aligned}$$

Let $x = -2$.

$$\begin{aligned} 4x - 5y &= -20 \\ 4(-2) - 5y &= -20 \\ -8 - 5y &= -20 \\ -5y &= -12 \\ y &= \frac{12}{5} = 2\frac{2}{5} \end{aligned}$$

The ordered pairs are in the table.

x	y
0	4
−5	0
−2	$2\frac{2}{5}$

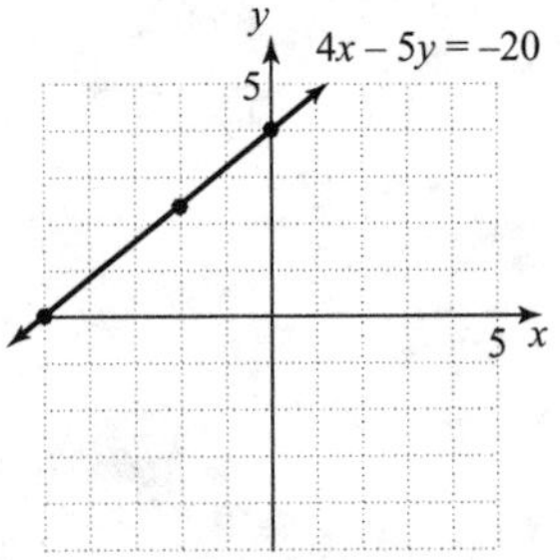

5. $y = -3x$

If $x = 0$, then $y = -3(0) = 0$.

If $x = 1$, then $y = -3(1) = -3$.

If $x = -1$, then $y = -3(-1) = 3$.

The ordered pairs are in the table.

x	y
0	0
1	−3
−1	3

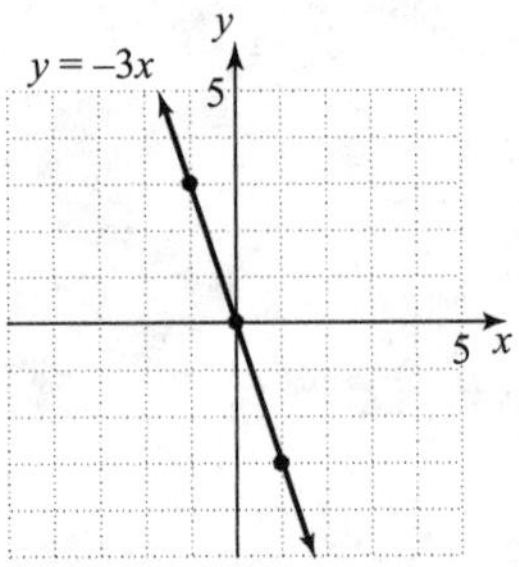

6. $x = -4$
The equation can be written as $x + 0y = -4$. For any y-value chosen, notice that x is -4.

x	y
−4	−3
−4	0
−4	3

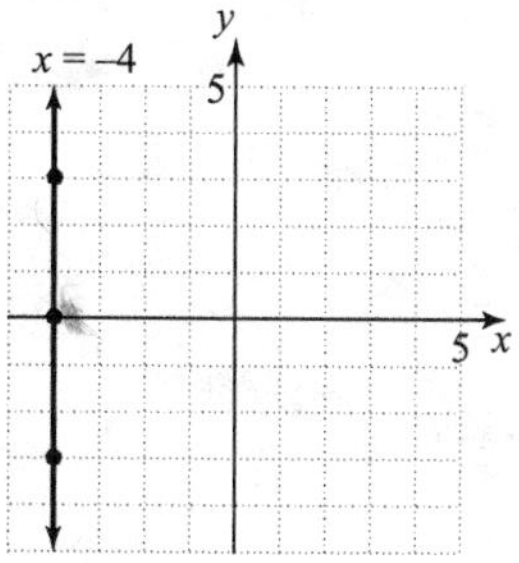

7. $y = 4$
The equation can be written as $0x + y = 4$. For any x-value chosen, notice that y is 4.

x	y
−3	4
0	4
3	4

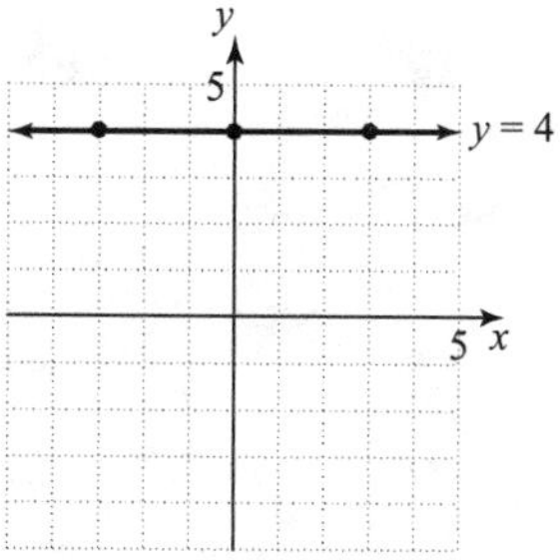

Graphing Calculator Explorations

1. $x = 3.5y$

$$y = \frac{x}{3.5}$$

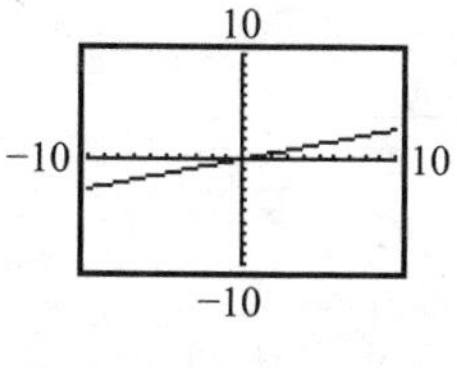

2. $-2.7y = x$

$$y = \frac{x}{-2.7} = -\frac{x}{2.7}$$

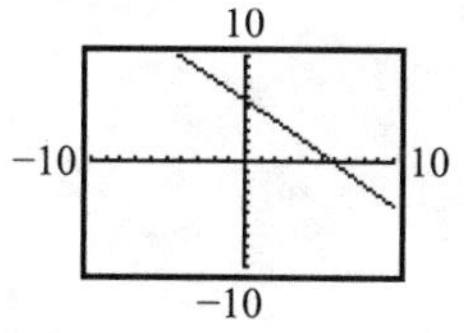

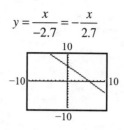

3. $5.78x + 2.31y = 10.98$

$$2.31y = -5.78x + 10.98$$
$$y = -\frac{5.78}{2.31}x + \frac{10.98}{2.31}$$

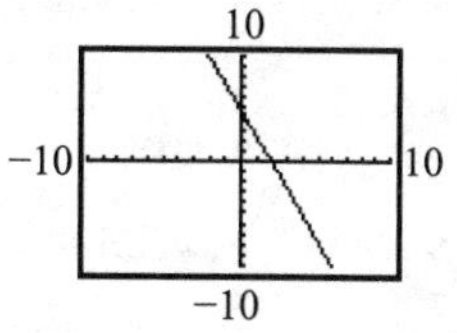

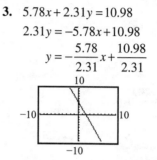

4. $-7.22x + 3.89y = 12.57$

$$3.89y = 7.22x + 12.57$$
$$y = \frac{7.22}{3.89}x + \frac{12.57}{3.89}$$

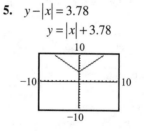

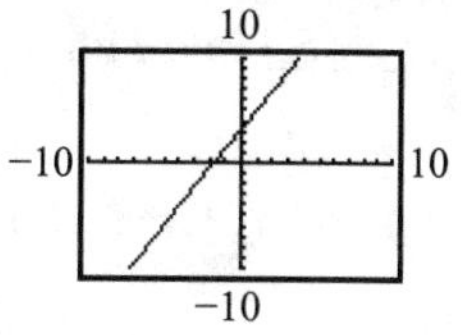

5. $y - |x| = 3.78$

$$y = |x| + 3.78$$

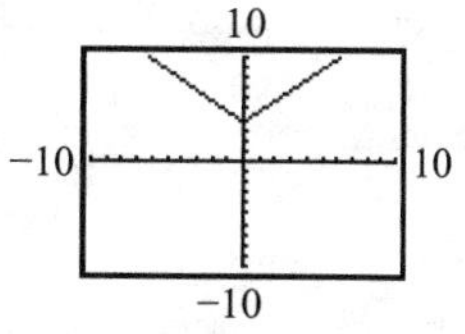

6. $3y - 5x^2 = 6x - 4$

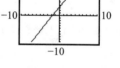

$$3y = 5x^2 + 6x - 4$$
$$y = \frac{5}{3}x^2 + 2x - \frac{4}{3}$$

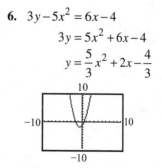

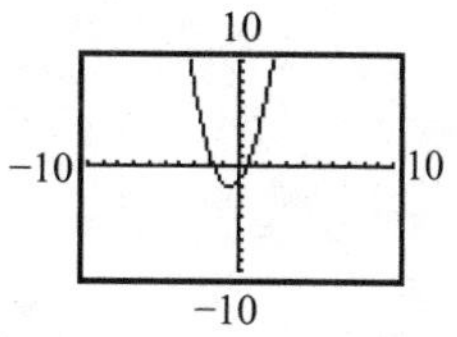

7. $y - 5.6x^2 = 7.7x + 1.5$

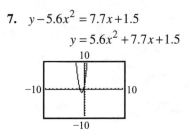

$y = 5.6x^2 + 7.7x + 1.5$

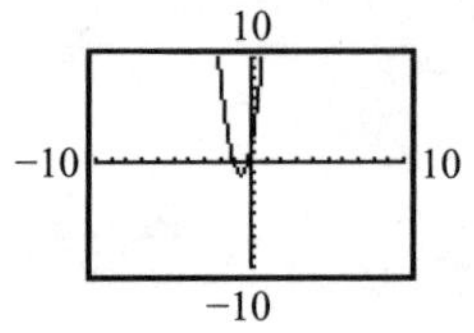

8. $y + 2.6|x| = -3.2$

$y = -2.6|x| - 3.2$

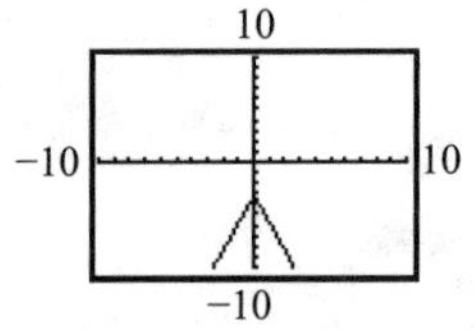

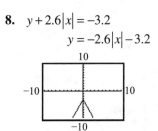

Vocabulary, Readiness & Video Check 3.3

1. A linear function can be written in the form $f(x) = mx + b$.
2. In the form $f(x) = mx + b$, the y-intercept is $(0, b)$.
3. The graph of $x = c$ is a vertical line with x-intercept $(c, 0)$.
4. The graph of $y = c$ is a horizontal line with y-intercept $(0, c)$.
5. To find an x-intercept, let $y = 0$ or $f(x) = 0$ and solve for x.
6. To find a y-intercept, let $x = 0$ and solve for y.
7. $f(x) = mx + b$, or slope-intercept form
8. A third point is found as a check that the points lie along the same line.
9. For a horizontal line, the coefficient of x is 0; for a vertical line, the coefficient of y is 0.

Exercise Set 3.3

1. $f(x) = -2x$

x	0	−1	1
y	0	2	−2

Plot the points to obtain the graph.

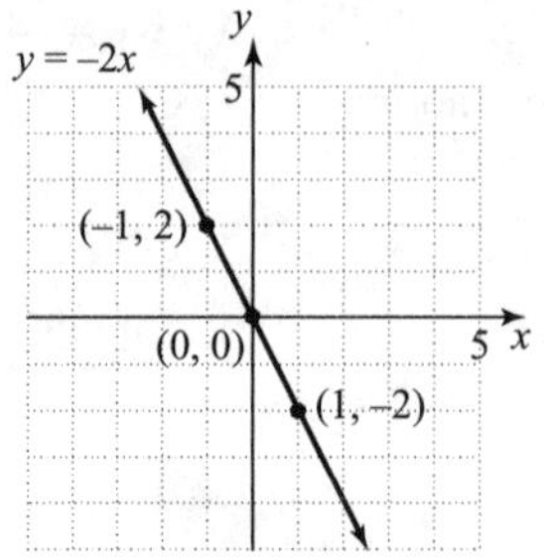

3. $f(x) = -2x + 3$

x	0	1	−1
y	3	1	5

Plot the points to obtain the graph.

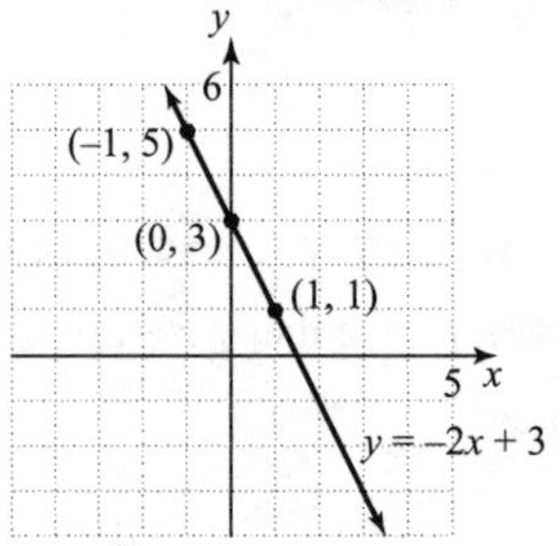

5. $f(x) = \frac{1}{2}x$

x	0	2	−2
y	0	1	−1

Plot the points to obtain the graph.

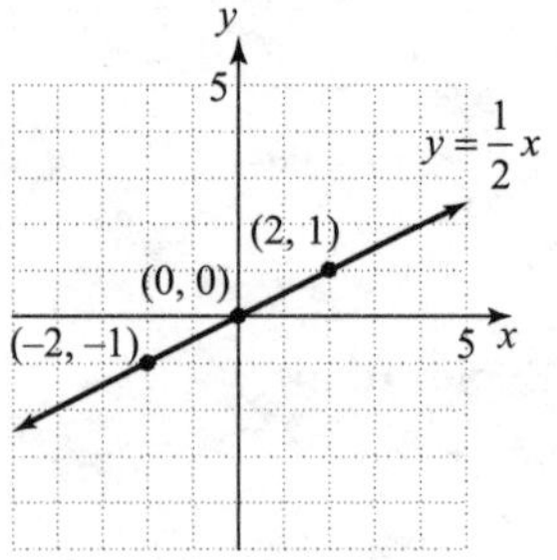

7. $f(x) = \frac{1}{2}x - 4$

x	0	2	4
y	−4	−3	−2

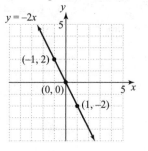

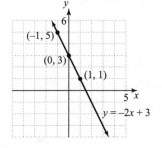

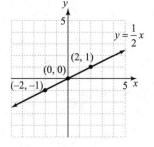

Plot the points to obtain the graph.

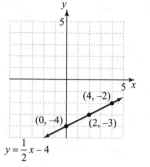

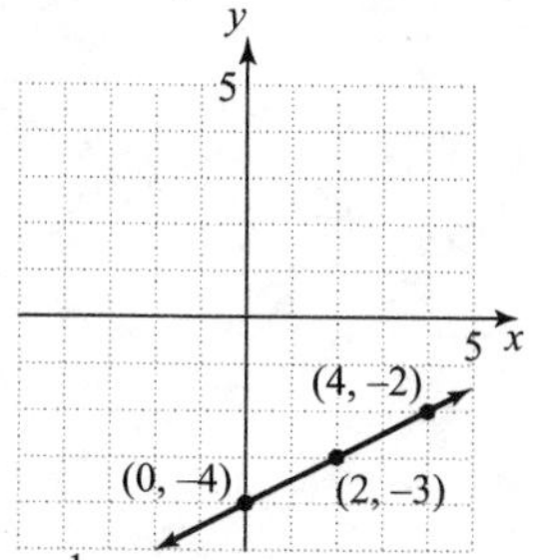

$y = \frac{1}{2}x - 4$

9. The graph of $f(x) = 5x - 3$ is the graph of $f(x) = 5x$ shifted down 3 units. The correct graph is C.

11. The graph of $f(x) = 5x + 1$ is the graph of $f(x) = 5x$ shifted up 1 unit. The correct graph is D.

13. $x - y = 3$

Let $x = 0$. $\quad$ Let $y = 0$. $\quad$ Let $x = 2$.
$0 - y = 3$ $\quad$ $x - 0 = 3$ $\quad$ $2 - y = 3$
$y = -3$ $\quad$ $x = 3$ $\quad$ $y = -1$

x	0	3	2
y	–3	0	–1

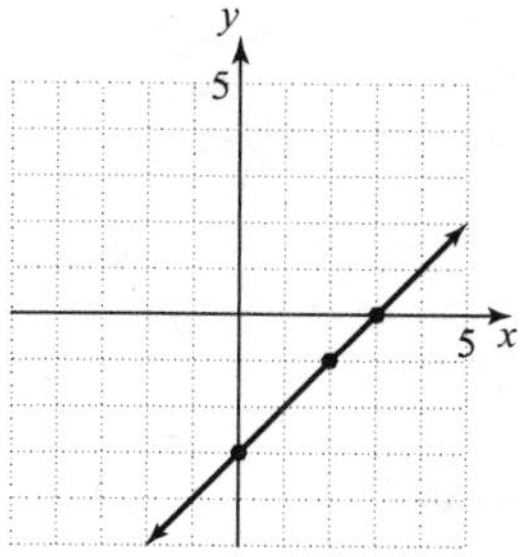

$f(x) = x - 3$

15. $x = 5y$

Let $x = 0$. $\quad$ Let $x = 5$. $\quad$ Let $x = -5$.
$0 = 5y$ $\quad$ $5 = 5y$ $\quad$ $-5 = 5y$
$y = 0$ $\quad$ $y = 1$ $\quad$ $y = -1$

x	0	5	–5
y	0	1	–1

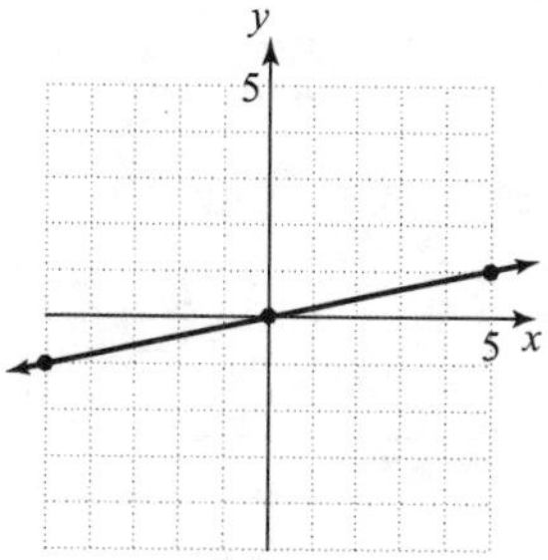

$f(x) = \frac{x}{5}$

17. $-x + 2y = 6$

Let $x = 0$. $\quad$ Let $y = 0$. $\quad$ Let $x = 2$.
$-0 + 2y = 6$ $\quad$ $-x + 2(0) = 6$ $\quad$ $-2 + 2y = 6$
$y = 3$ $\quad$ $x = -6$ $\quad$ $y = 4$

x	0	–6	2
y	3	0	4

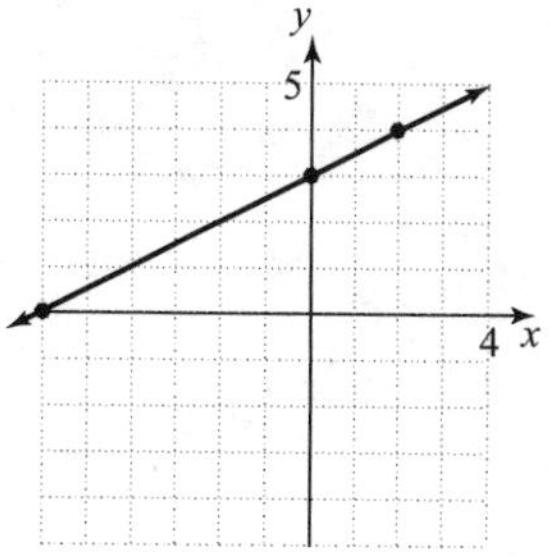

$f(x) = \frac{1}{2}x + 3$

19. $2x - 4y = 8$

Let $x = 0$. $\quad$ Let $y = 0$.
$2(0) - 4y = 8$ $\quad$ $2x - 4(0) = 8$
$y = -2$ $\quad$ $x = 4$
Let $x = 2$.
$2(2) - 4y = 8$
$y = -1$

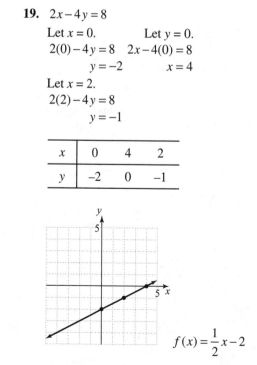

x	0	4	2
y	–2	0	–1

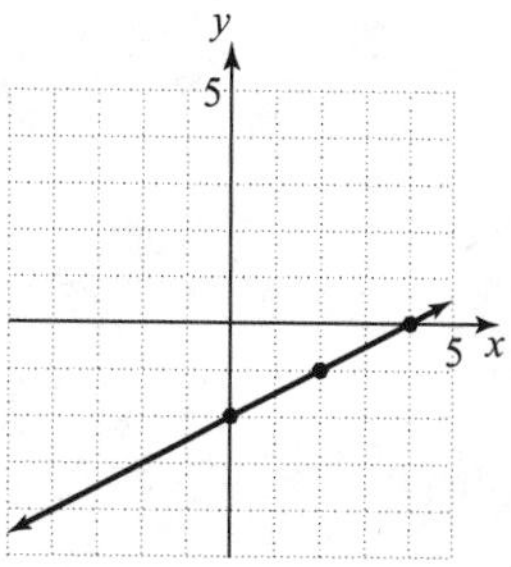

$f(x) = \frac{1}{2}x - 2$

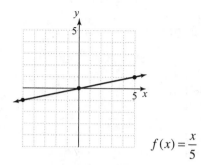

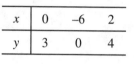

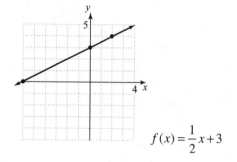

21. $x = -1$
Vertical line with x-intercept at -1

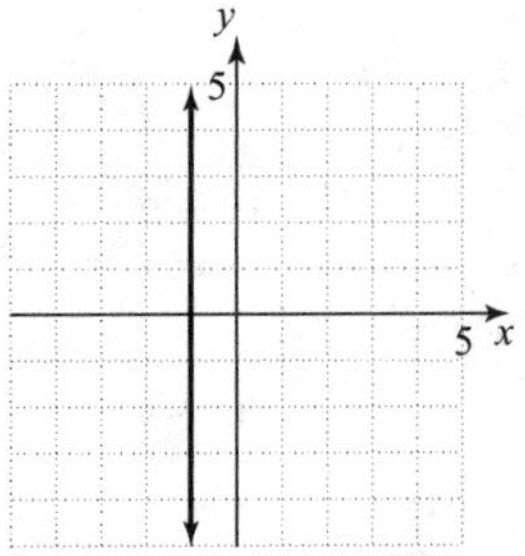

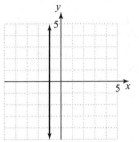

23. $y = 0$
Horizontal line with y-intercept at 0

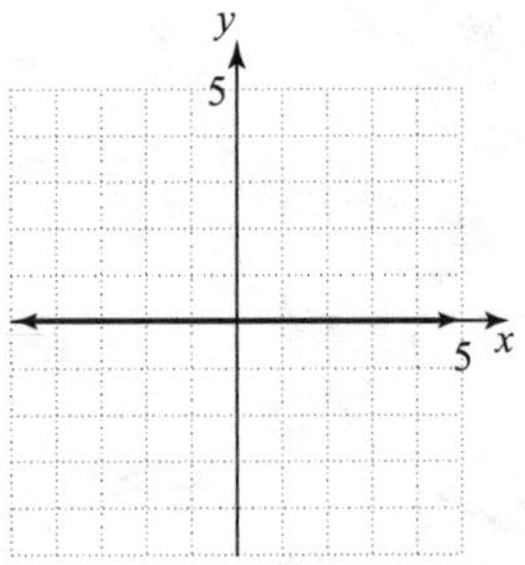

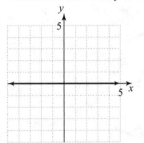

25. $y + 7 = 0$
$y = -7$
Horizontal line with y-intercept at -7

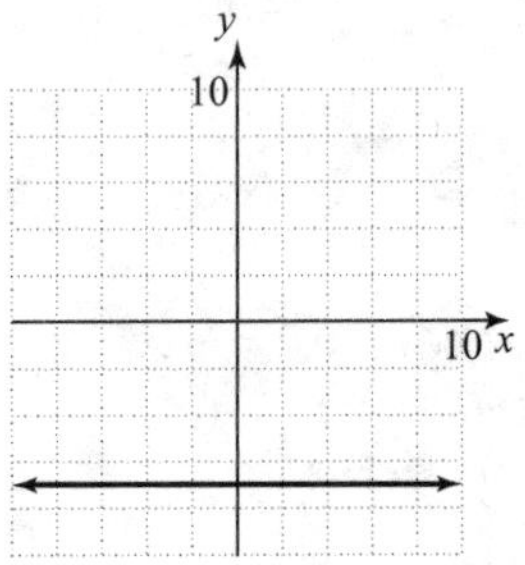

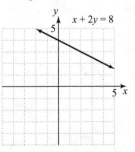

27. The graph of $y = 2$ is a horizontal line with y-intercept (0, 2). The correct graph is C.

29. The graph of $x - 2 = 0$ or $x = 2$ is a vertical line with x-intercept (2, 0). The correct graph is A.

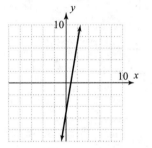

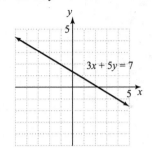

31. $x + 2y = 8$

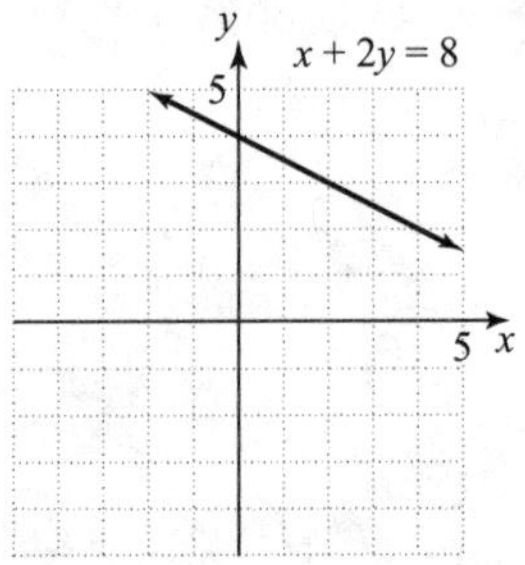

33. $3x + 5y = 7$

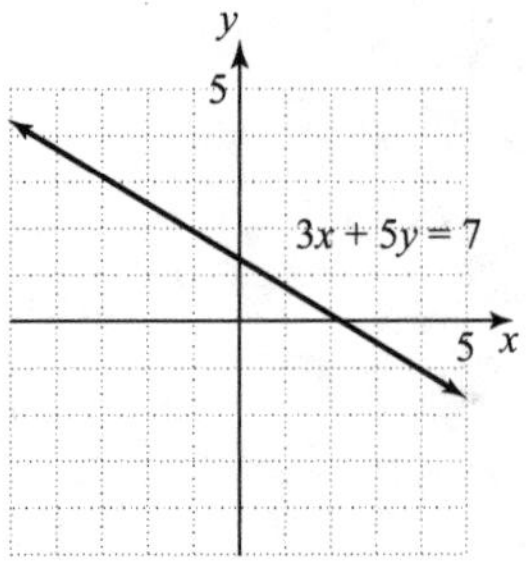

35. $x + 8y = 8$

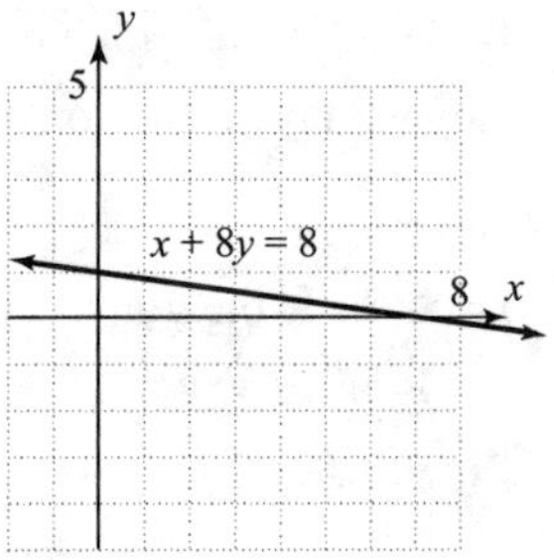

37. $5 = 6x - y$

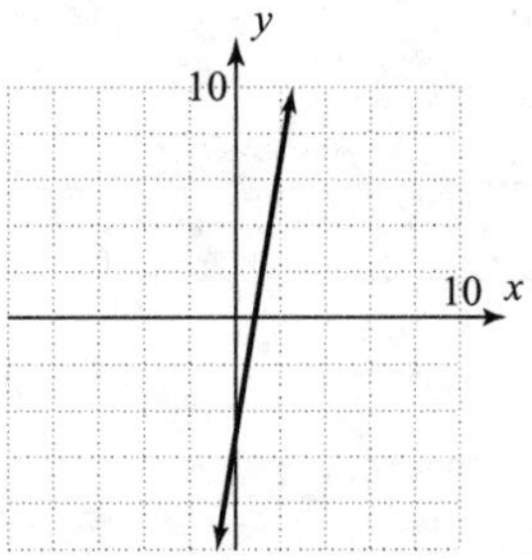

39. $-x + 10y = 11$

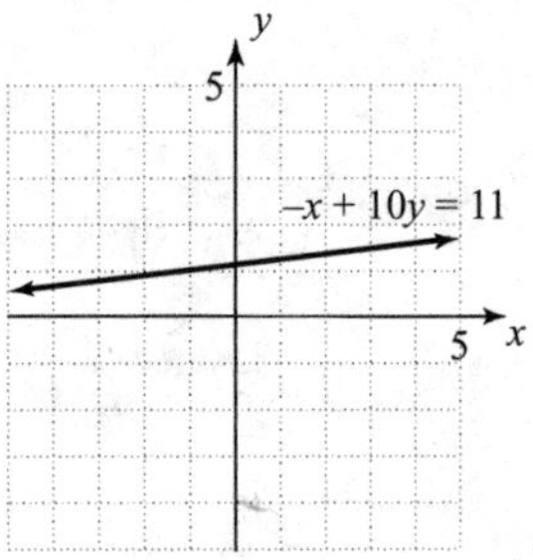

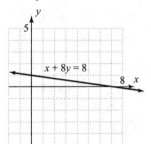

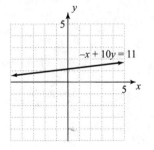

41. $y = \frac{3}{2}$

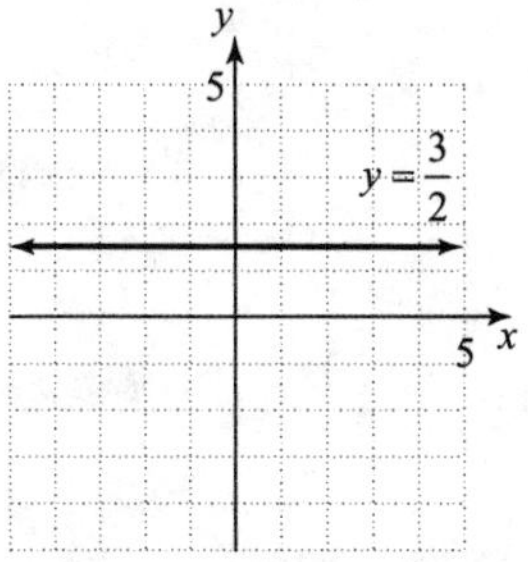

43. $2x + 3y = 6$

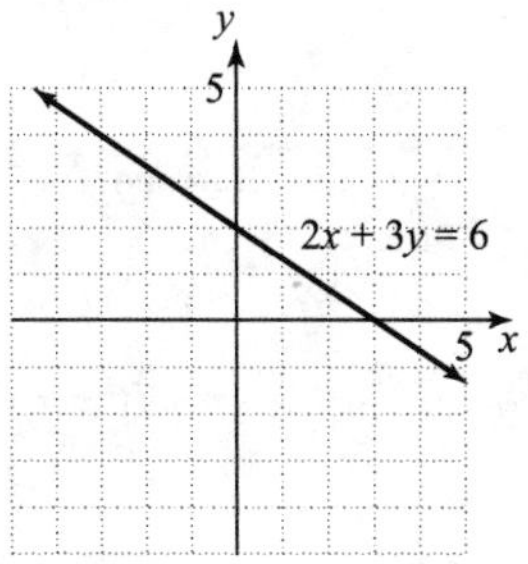

45. $x + 3 = 0$, or $x = -3$

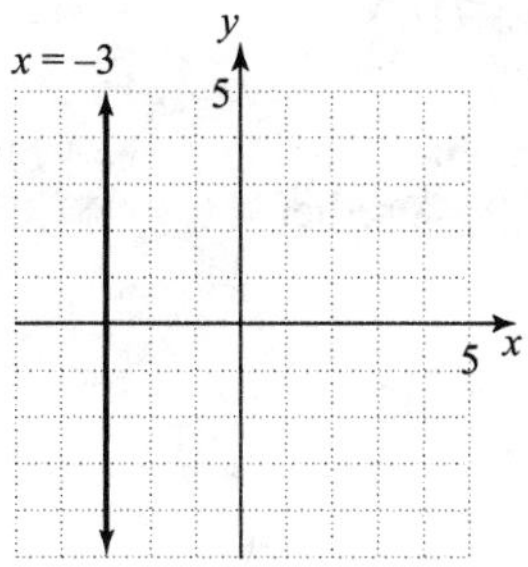

47. $f(x) = \frac{3}{4}x + 2$

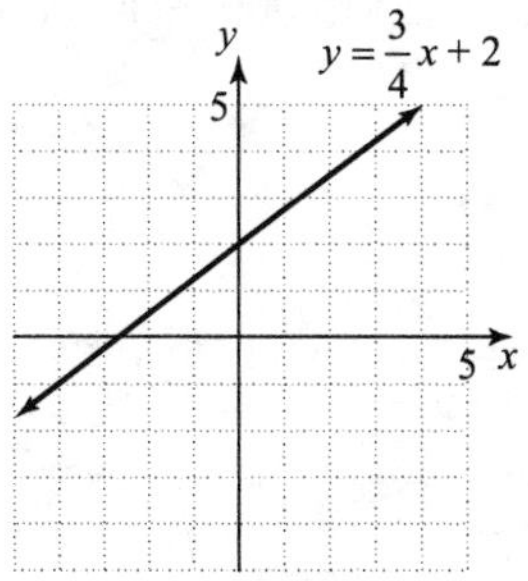

49. $f(x) = x$

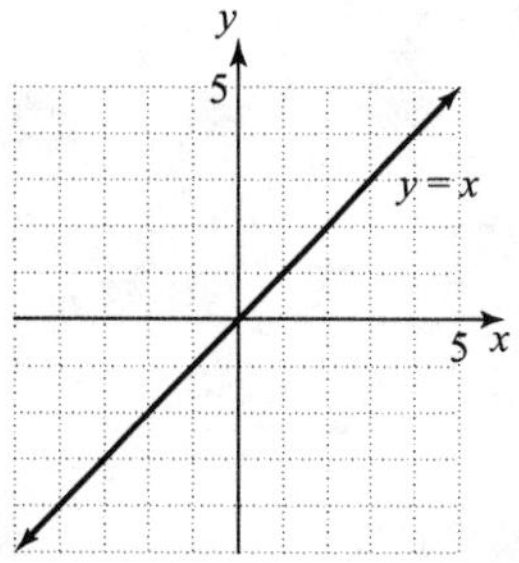

51. $f(x) = \frac{1}{2}x$

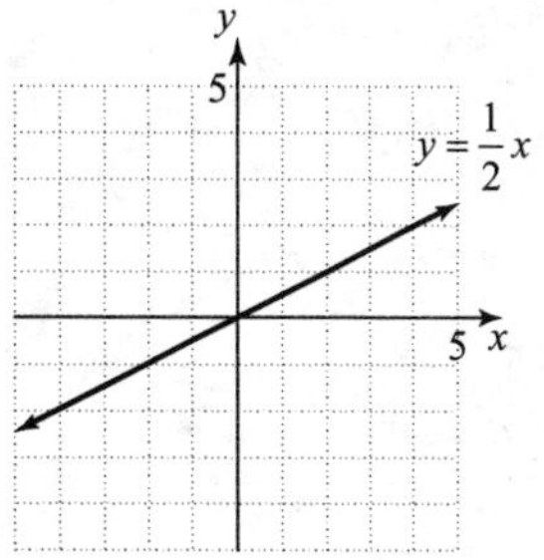

53. $f(x) = 4x - \frac{1}{3}$

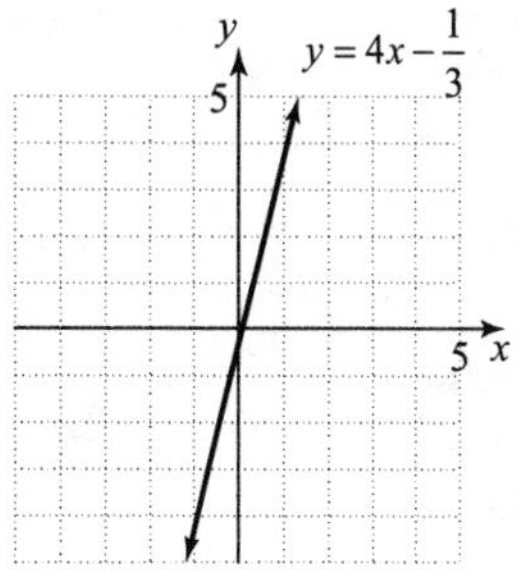

55. $x = -3$

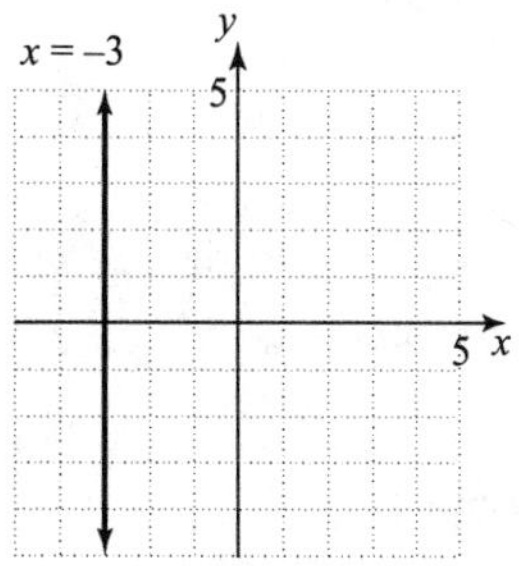

57. $|x - 3| = 6$

$x - 3 = 6$ or $x - 3 = -6$

$x = 9$ or $x = -3$

59. $|2x+5|>3$
$2x+5<-3$ or $2x+5>3$
$2x<-8$ or $2x>-2$
$x<-4$ or $x>-1$
$(-\infty,-4)\cup(-1,\infty)$

61. $|3x-4|\le 2$
$-2\le 3x-4\le 2$
$2\le 3x\le 6$
$\frac{2}{3}\le x\le 2$
$\left[\frac{2}{3},2\right]$

63. $\frac{-6-3}{2-8}=\frac{-9}{-6}=\frac{3}{2}$

65. $\frac{-8-(-2)}{-3-(-2)}=\frac{-8+2}{-3+2}=\frac{-6}{-1}=6$

67. $\frac{0-6}{5-0}=\frac{-6}{5}=-\frac{6}{5}$

69. no; answers may vary

71. yes; answers may vary

73. $2x+3y=1500$

a. $2(0)+3y=1500$
$3y=1500$
$y=500$
(0, 500); If no tables are produced, 500 chairs can be produced.

b. $2x+3(0)=1500$
$2x=1500$
$x=750$
(750, 0); If no chairs are produced, 750 tables can be produced.

c. $2(50)+3y=1500$
$100+3y=1500$
$3y=1400$
$y=466.7$
466 chairs

75. $C(x)=0.2x+24$

a. $C(200)=0.2(200)+24$
$=40+24$
$=64$
\$64

b.

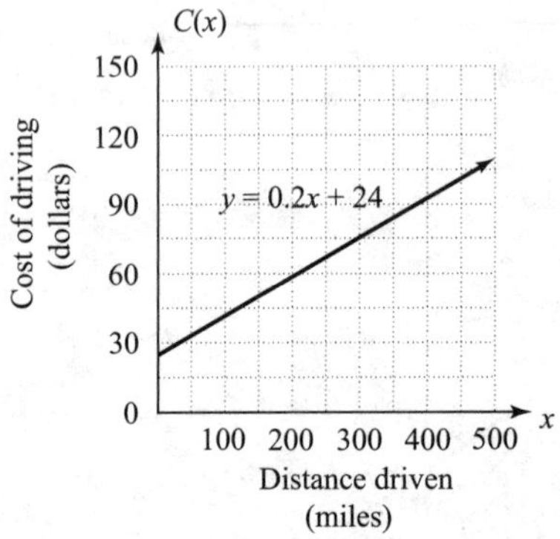

c. The line moves upward from left to right.

77. $f(x) = 502x + 7180$

a. $f(13)=502(13)+7180$
$=6526+7180$
$=13,706$
The yearly cost of attending a two-year college in 2020 is predicted to be \$13,706.

b. $f(x) = 15,000$
$15,000=502x+7180$
$7820=502x$
$15.6\approx x$
Round up to 16.
2007 + 16 = 2023
The yearly cost of attending a two-year college will exceed \$15,000 in 2023.

c. answers may vary

79. answers may vary

81. The vertical line $x=0$ has y-intercepts.

83. $y=-4x+2$ is a line parallel to $y=-4x$ but with y-intercept (0, 2).

85. The graph shows the graph of $y=|x|$ shifted up 3 units. Its equation is $y=|x|+3$. The correct answer is D.

87. The graph shows the graph of $y = |x|$ shifted down 3 units. Its equation is $y = |x| - 3$. The correct answer is C.

89.

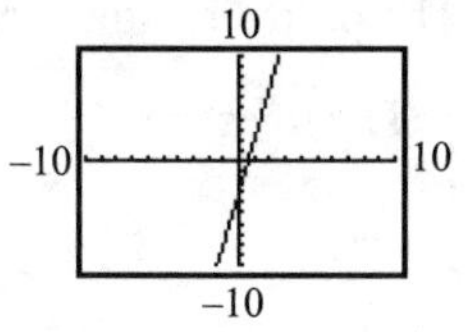

91.

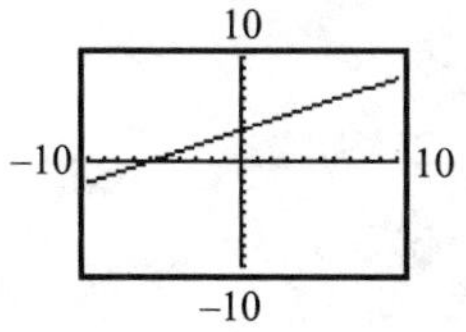

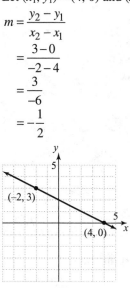

Section 3.4 Practice Exercises

1. Let $(x_1, y_1) = (4, 0)$ and $(x_2, y_2) = (-2, 3)$.

$$m = \frac{y_2 - y_1}{x_2 - x_1} = \frac{3-0}{-2-4} = \frac{3}{-6} = -\frac{1}{2}$$

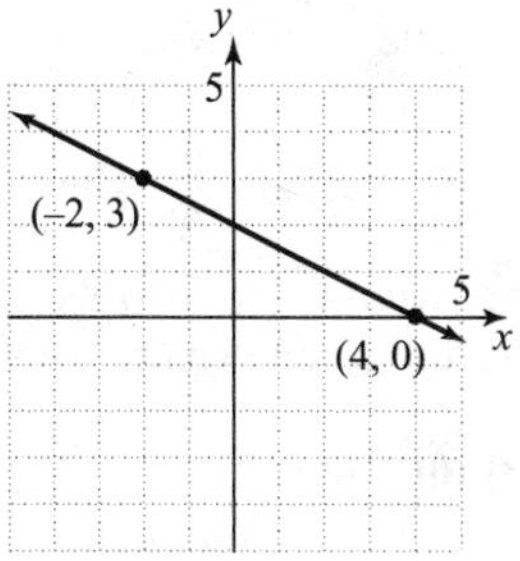

2. Let $(x_1, y_1) = (-5, -4)$ and $(x_2, y_2) = (5, 2)$.

$$m = \frac{y_2 - y_1}{x_2 - x_1} = \frac{2-(-4)}{5-(-5)} = \frac{6}{10} = \frac{3}{5}$$

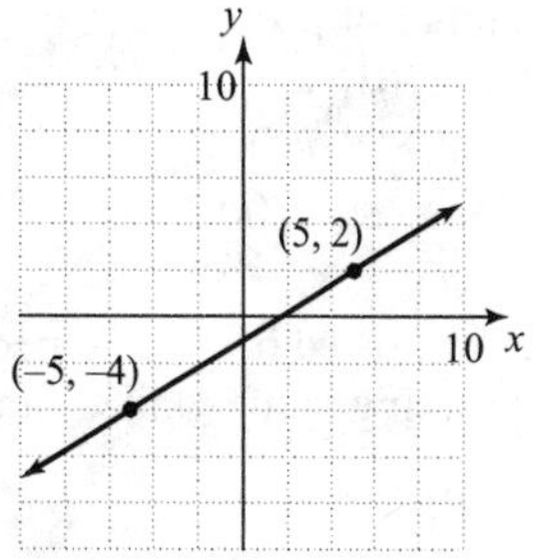

3. $f(x) = -4x + 6$ or $y = -4x + 6$

We need two points. We find the intercepts.

Let $x = 0$.

$y = -4 \cdot 0 + 6 = 6$

Let $y = 0$.

$$0 = -4x + 6$$
$$4x = 6$$
$$x = \frac{6}{4} = \frac{3}{2}$$

The intercepts are (0, 6) and $\left(\frac{3}{2}, 0\right)$. Let $(x_1, y_1) = (0, 6)$ and $(x_2, y_2) = \left(\frac{3}{2}, 0\right)$.

$$m = \frac{y_2 - y_1}{x_2 - x_1} = \frac{0-6}{\frac{3}{2}-0} = \frac{-6}{\frac{3}{2}} = -\frac{6}{1} \cdot \frac{2}{3} = -\frac{12}{3} = -4$$

4. $2x - 3y = 9$

Write the equation in slope-intercept form by solving for y.

$$2x - 3y = 9$$
$$-3y = -2x + 9$$
$$\frac{-3y}{-3} = \frac{-2x}{-3} + \frac{9}{-3}$$
$$y = \frac{2}{3}x - 3$$

The coefficient of x, $\frac{2}{3}$, is the slope, and the y-intercept is (0, −3).

5. a. To find the number of people in 2015, find y when x is $2015 - 2010 = 5$.

$$\begin{aligned} y &= 269{,}370x + 4{,}869{,}700 \\ &= 269{,}370(5) + 4{,}869{,}700 \\ &= 6{,}216{,}550 \end{aligned}$$

There were an estimated 6,216,550 people age 85 or older living in the United States in 2015.

b. The number of people age 85 or older in the United States increases at a rate of 269,370 per year.

c. At year $x = 0$, or 2010, there were 4,869,700 people age 85 or older in the United States.

6. $x = 4$

The graph of $x = 4$ is a vertical line. We choose two points on the line, (4, 0) and (4, 3). Let $(x_1, y_1) = (4, 0)$ and $(x_2, y_2) = (4, 3)$.

$$\begin{aligned} m &= \frac{y_2 - y_1}{x_2 - x_1} \\ &= \frac{3-0}{4-4} \\ &= \frac{3}{0} \end{aligned}$$

Since $\frac{3}{0}$ is undefined, the slope of the vertical line $x = 4$ is undefined.

7. $y = -3$

The graph of $y = -3$ is a horizontal line. We choose two points on the line, $(0, -3)$ and $(4, -3)$. Let $(x_1, y_1) = (0, -3)$ and $(x_2, y_2) = (4, -3)$.

$$\begin{aligned} m &= \frac{y_2 - y_1}{x_2 - x_1} \\ &= \frac{-3-(-3)}{4-0} \\ &= \frac{0}{4} \\ &= 0 \end{aligned}$$

The slope of the horizontal line $y = -3$ is 0.

8. a. Find the slope of each line.

$$\begin{aligned} x - 2y &= 3 \\ -2y &= -x + 3 \\ \frac{-2y}{-2} &= \frac{-x}{-2} + \frac{3}{-2} \\ y &= \frac{1}{2}x - \frac{3}{2} \end{aligned}$$

The slope is $\frac{1}{2}$.

$$\begin{aligned} 2x + y &= 3 \\ y &= -2x + 3 \end{aligned}$$

The slope is -2. The product of the slopes is $-1\left[\frac{1}{2}(-2) = -1\right]$. The lines are perpendicular.

b. Find the slope of each line.

$$\begin{aligned} 4x - 3y &= 2 \\ -3y &= -4x + 2 \\ \frac{-3y}{-3} &= \frac{-4x}{-3} + \frac{2}{-3} \\ y &= \frac{4}{3}x - \frac{2}{3} \end{aligned}$$

The slope is $\frac{4}{3}$. The y-intercept is $\left(0, -\frac{2}{3}\right)$.

$$\begin{aligned} -8x + 6y &= -6 \\ 6y &= 8x - 6 \\ \frac{6y}{6} &= \frac{8x}{6} - \frac{6}{6} \\ y &= \frac{4}{3}x - 1 \end{aligned}$$

The slope is $\frac{4}{3}$. The y-intercept is $(0, -1)$.

The slopes of both lines are $\frac{4}{3}$. The y-intercepts are different, so the lines are not the same. Therefore, the lines are parallel.

Graphing Calculator Explorations

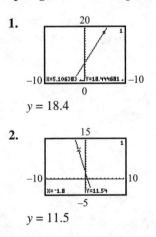

1.

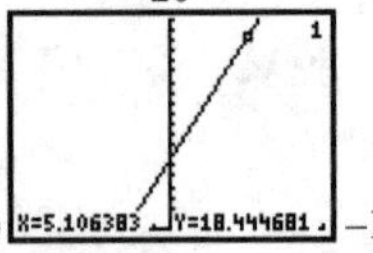

$y = 18.4$

2.

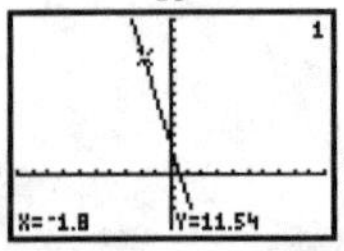

$y = 11.5$

3.

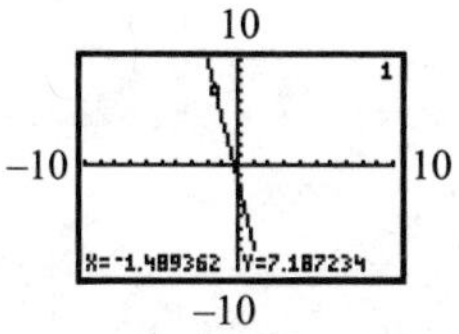

$x = -1.5$

4.

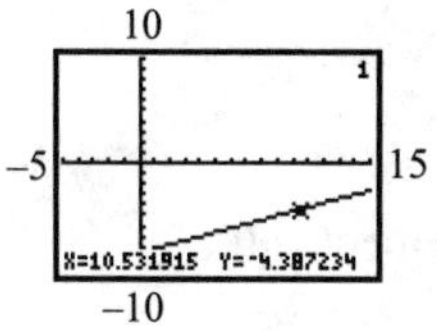

$x = 10.5$

5.

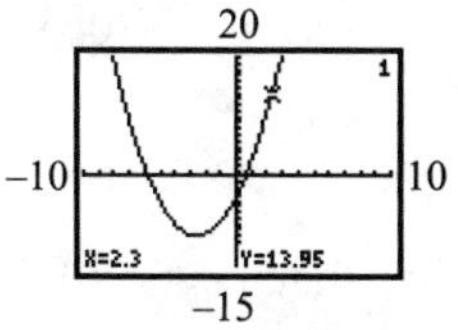

$y = 14.0$

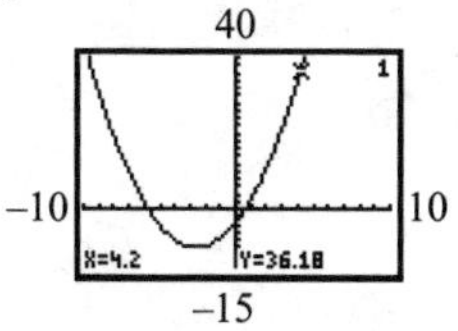

$x = 4.2$

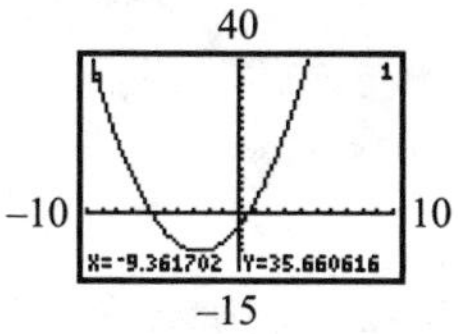

$x = -9.4$

6.

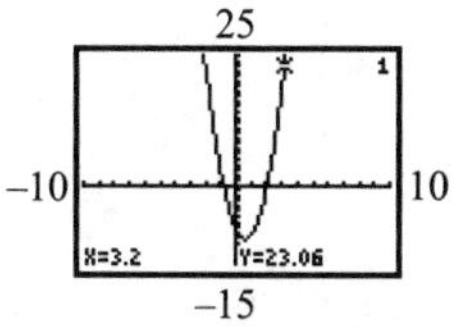

$y = 23.1$

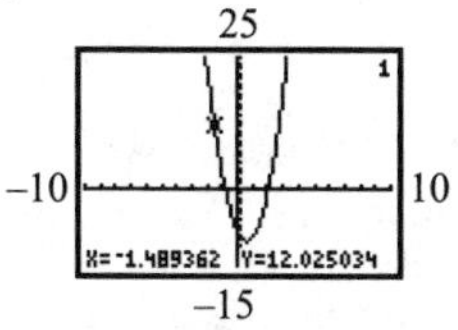

$x = -1.5$

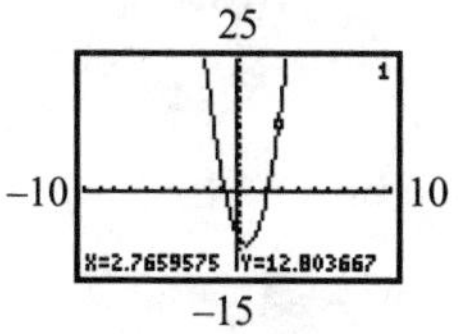

$x = 2.8$

Vocabulary, Readiness & Video Check 3.4

1. The measure of the steepness or tilt of a line is called slope.

2. The slope of a line through two points is measured by the ratio of vertical change to horizontal change.

3. If a linear equation is in the form $y = mx + b$, or $f(x) = mx + b$, the slope of the line is m and the y-intercept is $(0, b)$.

4. The form $y = mx + b$ or $f(x) = mx + b$ is the slope-intercept form.

5. The slope of a horizontal line is 0.

6. The slope of a vertical line is undefined.

7. Two nonvertical perpendicular lines have slopes whose product is -1.

8. Two nonvertical lines are parallel if they have the same slope and different y-intercepts.

9. A positive slope means the line increases from left to right. A negative slope means the line decreases from left to right.

10. Solve the equation for y; the slope is the coefficient of x.

11. Slope-intercept form allows us to see the slope and y-intercept so we can discuss them in the context of the application. Slope is a rate of change and tells us the rate of increase of registered nurses (52.7 thousand per year). The y-intercept tells us the number of registered nurses at "zero" year (2711 thousand in 2012).

12. Slope is 0 indicates $m = 0$ and a horizontal line; undefined slope indicates m is undefined and a vertical line; no slope refers to an undefined slope.

13. The slopes of vertical lines are undefined and we can't mathematically work with undefined slopes.

Exercise Set 3.4

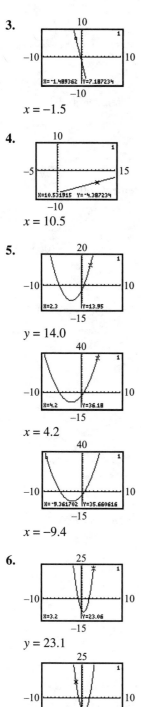

1. $m = \dfrac{11-2}{8-3} = \dfrac{9}{5}$

3. $m=\dfrac{8-1}{1-3}=\dfrac{7}{-2}=-\dfrac{7}{2}$

5. $m=\dfrac{3-8}{4-(-2)}=\dfrac{-5}{6}=-\dfrac{5}{6}$

7. $m=\dfrac{-4-(-6)}{4-(-2)}=\dfrac{-4+6}{4+2}=\dfrac{2}{6}=\dfrac{1}{3}$

9. $m=\dfrac{11-(-1)}{-12-(-3)}=\dfrac{12}{-9}=-\dfrac{4}{3}$

11. $m=\dfrac{5-5}{3-(-2)}=\dfrac{0}{5}=0$

13. $m=\dfrac{-5-1}{-1-(-1)}=\dfrac{-6}{0}$
undefined slope

15. $m=\dfrac{0-6}{-3-0}=\dfrac{-6}{-3}=2$

17. $m=\dfrac{4-2}{-3-(-1)}=\dfrac{2}{-2}=-1$

19. A line with slope $m=\dfrac{7}{6}$ slants upward.

21. Since $m = 0$, the line is horizontal.

23. The slope of l_1 is negative, and the slope of l_2 is positive. Since a positive number is greater than any negative number, l_2 has the greater slope.

25. The slope of l_1 is negative, and the slope of l_2 is 0. Since 0 is greater than any negative number, l_2 has the greater slope.

27. Both lines have positive slope. Since l_2 is steeper, it has the greater slope.

29. $f(x)=5x-2$
$m = 5$, $b = -2$ so y-intercept is $(0, -2)$.

31. $2x+y=7$
$y=-2x+7$
$m = -2$, $b = 7$ so y-intercept is $(0, 7)$.

33. $2x-3y=10$
$-3y=-2x+10$
$y=\dfrac{2}{3}x-\dfrac{10}{3}$
$m=\dfrac{2}{3}$, $b=-\dfrac{10}{3}$ so y-intercept is $\left(0, -\dfrac{10}{3}\right)$.

35. $f(x)=\dfrac{1}{2}x$
$m=\dfrac{1}{2}$, $b=0$ so y-intercept is $(0, 0)$.

37. $f(x)=2x+3$
The slope is 2, and the y-intercept is $(0, 3)$. The correct graph is A.

39. $f(x)=-2x+3$
The slope is -2, and the y-intercept is $(0, 3)$. The correct graph is B.

41. $x=1$ is a vertical line.
m is undefined.

43. $y=-3$ is a horizontal line.
$m=0$

45. $x+2=0$
$x=-2$
This is a vertical line. m is undefined.

47. $f(x) = -x + 5$ or $y = -1x + 5$
$m = -1$, $b = 5$ so y-intercept is $(0, 5)$.

49. $-6x+5y=30$
$5y=6x+30$
$y=\dfrac{6}{5}x+6$
$m=\dfrac{6}{5}$, $b=6$ so y-intercept is $(0, 6)$.

51. $3x+9=y$
$y=3x+9$
$m = 3$, $b = 9$ so y-intercept is $(0, 9)$.

53. $y=4$
$m = 0$, $b = 4$ so y-intercept is $(0, 4)$.

55. $f(x)=7x$
$m = 7$, $b = 0$ so y-intercept is $(0, 0)$.

57. $6+y=0$
$y=-6$
$m = 0, b = -6$ so y-intercept is $(0, -6)$.

59. $2-x=3$
$x=-1$
m is undefined. There is no y-intercept.

61. $y = 12x + 6$, $m = 12$; $y = 12x - 2$, $m = 12$
Parallel, since they have the same slope.

63. $y=-9x+3$, $m=-9$; $y=\frac{3}{2}x-7$, $m=\frac{3}{2}$
Neither, since their slopes are not equal nor does their product equal -1.

65. $f(x)=-3x+6$, $m=-3$; $g(x)=\frac{1}{3}x+5$, $m=\frac{1}{3}$
Perpendicular, since the product of their slopes is -1.

67. $-4x+2y=5$
$y=2x+\frac{5}{2}$
$m=2$
$2x-y=7$
$y=2x-7$
$m=2$
Parallel, since they have the same slope.

69. $-2x+3y=1$
$y=\frac{2}{3}x+\frac{1}{3}$
$m=\frac{2}{3}$
$3x+2y=12$
$y=-\frac{3}{2}x+6$
$m=-\frac{3}{2}$
Perpendicular, since the product of their slopes is -1.

71. Two points on the line: (0, 0), (2, 3)
$$m=\frac{3-0}{2-0}=\frac{3}{2}$$

73. Two points on the line: (4, 0), (0, 2)
$$m=\frac{2-0}{0-4}=\frac{2}{-4}=-\frac{1}{2}$$

75. $m=\frac{8}{12}=\frac{2}{3}$

77.
$$m=\frac{-1600 \text{ ft}}{2.5 \text{ mi}}=\frac{-1600 \text{ ft}}{2.5(5280 \text{ ft})}=\frac{-1600}{13,200}\approx -0.12$$

79. a. Use $y = 0.15x + 71.5$ with
$x = 1990 - 1950 = 40$.
$y = 0.15(40) + 71.5 = 77.5$
The life expectancy of an American female born in 1990 is 77.5 years.

b. From $y = 0.15x + 71.5$, $m = 0.15$.
The life expectancy of a female born in the United States is approximately 0.15 year more than a female born one year before.

c. From $y = 0.15x + 71.5$, the y-intercept is $(0, 71.5)$. At year = 0, or 1950, the life expectancy of an American female was 71.5 years.

81. a.
$$527x-10y=-27,110$$
$$-10y=-527x-27,110$$
$$y=52.7x+2711$$
$m = 52.7$, y-intercept $(0, 2711)$

b. The slope means that the number of people employed as registered nurses increases by 52.7 thousand for every 1 year.

c. The y-intercept means that there were 2711 thousand registered nurses in 2012.

83. a. Use $y = 0.18x + 0.95$ with
$x = 2014 - 2010 = 4$.
$y = 0.18(4) + 0.95 = 1.67$
The annual sales of Converse in 2014 were \$1.67 billion.

b.
$$2=0.18x+0.95$$
$$1.05=0.18x$$
$$5.8\approx x$$
$2010 + 5.8 = 2015.8$
Annual sales of Converse were above \$2 billion in 2015.

c. answers may vary

85.
$$y-2=5(x+6)$$
$$y-2=5x+30$$
$$y=5x+32$$

87. $y-(-1)=2(x-0)$

$y+1=2x$

$y=2x-1$

89. The denominator in the first fraction should be $7-(-2)$.

$m=\frac{-14-6}{7-(-2)}=\frac{-20}{9}=-\frac{20}{9}$

91. The numerator in the first fraction should be $-10-(-5)$. The denominator in the first fraction should be $-8-(-11)$.

$m=\frac{-10-(-5)}{-8-(-11)}=\frac{-5}{3}=-\frac{5}{3}$

93. $f(x)=-\frac{7}{2}x-6$ or $y=-\frac{7}{2}x-6$

$m=-\frac{7}{2}$

The slope of a parallel line is $-\frac{7}{2}$.

95. $f(x)=-\frac{7}{2}x-6$ or $y=-\frac{7}{2}x-6$

$m=-\frac{7}{2}$

The slope of a perpendicular line is $\frac{2}{7}$.

97. $5x-2y=6 \Rightarrow y=\frac{5}{2}x-3$

$m=\frac{5}{2}$

The slope of a parallel line is $\frac{5}{2}$.

99. $5x-2y=6 \Rightarrow y=\frac{5}{2}x-3$

$m=\frac{5}{2}$

The slope of a perpendicular line is $-\frac{2}{5}$.

101. $m_1=\frac{-2-4}{2-(-1)}=\frac{-6}{3}=-2$

$m_2=\frac{2-6}{-4-(-8)}=\frac{-4}{4}=-1$

$m_3=\frac{0-(-4)}{-6-0}=\frac{4}{-6}=-\frac{2}{3}$

$m_1=-2,\ m_2=-1,\ m_3=-\frac{2}{3}$

103. The coordinates of point *B* are (6, 20).

105. $m=\frac{13-20}{10-6}=\frac{-7}{4}=-\frac{7}{4}=-1.75$

The rate of change is –1.75 yd per sec.

107. answers may vary

109. answers may vary

111. $m=\frac{0.25 \text{ inch}}{1 \text{ foot}}$

$=0.25\frac{\text{inch}}{\text{foot}}$

$=\frac{1}{4}\cdot\frac{1 \text{ inch}}{12 \text{ inches}}$

$=\frac{1}{48}$

113. $-4x+2y=5$ or $y=2x+\frac{5}{2}$

$2x-y=7$ or $y=2x-7$

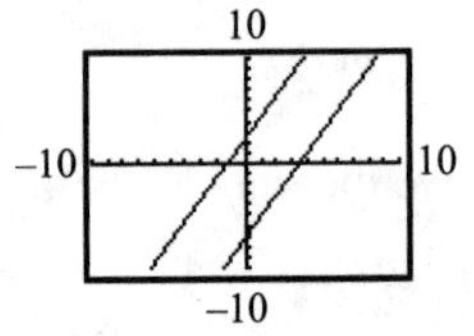

Section 3.5 Practice Exercises

1. $y=\frac{3}{4}x+2$

The slope is $\frac{3}{4}$, and the *y*-intercept is (0, 2). Plot (0, 2). Then plot a second point by starting at (0, 2), rising 3 units up, and running 4 units to the right. The second point is (4, 5).

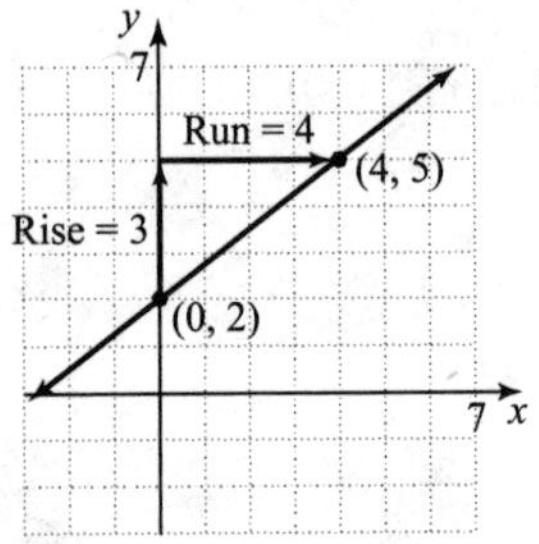

2. $x+2y=6$
Solve for y.
$$x+2y=6$$
$$2y=-x+6$$
$$y=-\frac{1}{2}x+3$$
The slope is $-\frac{1}{2}$, and the y-intercept is (0, 3).
Plot (0, 3). Then plot a second point by starting at (0, 3), moving 1 unit down, and moving 2 units to the right. The second point is (2, 2).

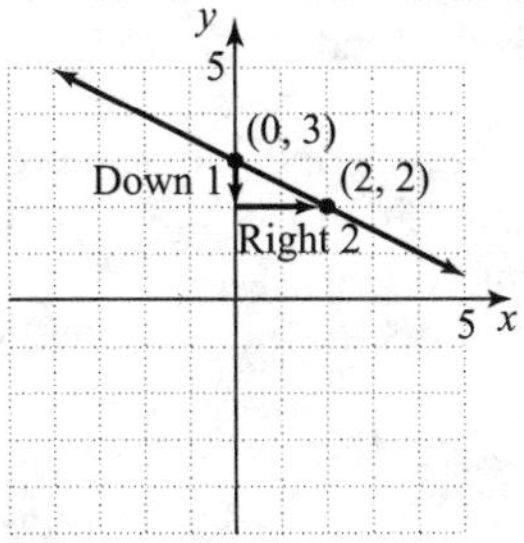

3. We are given the slope, $-\frac{3}{4}$, and the y-intercept, (0, 4).
Let $m=-\frac{3}{4}$ and $b = 4$.
$$y=mx+b$$
$$y=-\frac{3}{4}x+4$$

4. Use the point-slope form with $m = -4$ and $(x_1, y_1) = (-2, 5)$.
$$y-y_1=m(x-x_1)$$
$$y-5=-4[x-(-2)]$$
$$y-5=-4(x+2)$$
$$y-5=-4x-8$$
$$y=-4x-3$$

5. First find the slope.
$$m=\frac{0-2}{2-(-1)}=\frac{-2}{3}=-\frac{2}{3}$$
Use the slope and one of the points in the point-slope form. We use (2, 0).
$$y-y_1=m(x-x_1)$$
$$y-0=-\frac{2}{3}(x-2)$$
$$y=-\frac{2}{3}x+\frac{4}{3}$$
$$f(x)=-\frac{2}{3}x+\frac{4}{3}$$

6. The points on the graph have coordinates (–2, 3) and (1, 1). Find the slope.
$$m=\frac{1-3}{1-(-2)}=\frac{-2}{3}=-\frac{2}{3}$$
Use the slope and one of the points in the point-slope form. We use (1, 1).
$$y-y_1=m(x-x_1)$$
$$y-1=-\frac{2}{3}(x-1)$$
$$3(y-1)=-2(x-1)$$
$$3y-3=-2x+2$$
$$2x+3y=5$$

7. Let x = the number of years after 2000 and y = home sales in the fourth quarter of the year corresponding to x.
We have two ordered pairs, (9, 15,710) and (14, 12,680). Find the slope.
$$m=\frac{12{,}680-15{,}710}{14-9}=\frac{-3030}{5}=-606$$
Use the slope and the point (9, 15,710) in the point-slope form.
$$y-y_1=m(x-x_1)$$
$$y-15{,}710=-606(x-9)$$
$$y-15{,}710=-606x+5454$$
$$y=-606x+21{,}164$$
The year 2020 corresponds to $x = 20$.
$$y=-606(20)+21{,}164$$
$$=-12{,}120+21{,}164$$
$$=9044$$
We predict that there will be 9044 home sales in the fourth quarter of 2020.

8. A horizontal line has an equation of the form $y = b$. Since the line contains the point (6, –2), the equation is $y = -2$.

9. Since the line has undefined slope, the line must be vertical. A vertical line has an equation of the form $x = c$. Since the line contains the point (6, –2), the equation is $x = 6$.

10. Solve the given equation for y.
$$3x+4y=1$$
$$4y=-3x+1$$
$$y=-\frac{3}{4}x+\frac{1}{4}$$
The slope of this line is $-\frac{3}{4}$, so the slope of any line parallel to it is also $-\frac{3}{4}$. Use this slope and the point (8, –3) in the point-slope form.

$$y - y_1 = m(x - x_1)$$
$$y - (-3) = -\frac{3}{4}(x - 8)$$
$$4(y + 3) = -3(x - 8)$$
$$4y + 12 = -3x + 24$$
$$3x + 4y = 12$$

11. Solve the given equation for y.

$$3x + 4y = 1$$
$$4y = -3x + 1$$
$$y = -\frac{3}{4}x + \frac{1}{4}$$

The slope of this line is $-\frac{3}{4}$, so the slope of any line perpendicular to it is the negative reciprocal of $-\frac{3}{4}$, or $\frac{4}{3}$. Use this slope and the point $(8, -3)$ in the point-slope form.

$$y - y_1 = m(x - x_1)$$
$$y - (-3) = \frac{4}{3}(x - 8)$$
$$3(y + 3) = 4(x - 8)$$
$$3y + 9 = 4x - 32$$
$$3y = 4x - 41$$
$$y = \frac{4}{3}x - \frac{41}{3}$$
$$f(x) = \frac{4}{3}x - \frac{41}{3}$$

Vocabulary, Readiness & Video Check 3.5

1. $m = -4$, $b = 12$ so y-intercept is $(0, 12)$.

2. $m = \frac{2}{3}$, $b = -\frac{7}{2}$ so y-intercept is $\left(0, -\frac{7}{2}\right)$.

3. $m = 5$, $b = 0$ so y-intercept is $(0, 0)$.

4. $m = -1$, $b = 0$ so y-intercept is $(0, 0)$.

5. $m = \frac{1}{2}$, $b = 6$ so y-intercept is $(0, 6)$.

6. $m = -\frac{2}{3}$, $b = 5$ so y-intercept is $(0, 5)$.

7. The lines both have slope 12 and they have different y-intercepts, $(0, 6)$ and $(0, -2)$, so they are parallel.

8. The lines both have slope -5 and they have different y-intercepts, $(0, 8)$ and $(0, -8)$, so they are parallel.

9. The line have slopes -9 and $\frac{3}{2}$. The slopes are not equal and their product is not -1, so the lines are neither parallel nor perpendicular.

10. The line have slopes 2 and $\frac{1}{2}$. The slopes are not equal and their product is not -1, so the lines are neither parallel nor perpendicular.

11. To graph a line using its slope and y-intercept, first write the equation in slope-intercept form. Graph the one point you now know, the y-intercept. Use the slope to find a second point.

12. The y-intercept is of the form $(0, b)$, so the b value is the y-coordinate of the y-intercept.

13. if one of the two points given is the y-intercept

14. Example 5: $y = 3$; Example 6: $x = -1$

15. $f(x) = \frac{1}{3}x - \frac{17}{3}$

Exercise Set 3.5

1. $y = 5x - 2$

   possible points: $(0, -2)$, $(1, 3)$

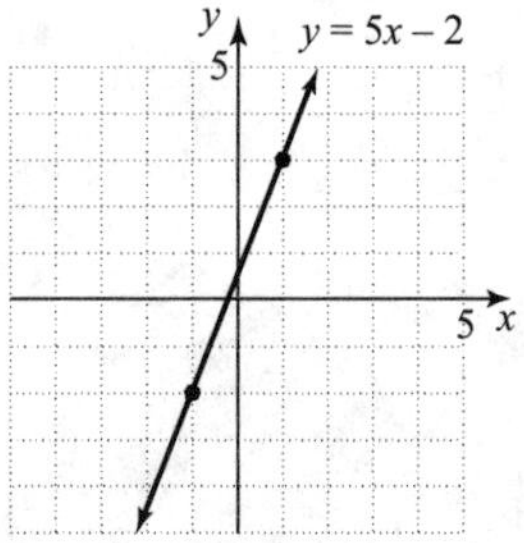

3. $4x + y = 7$

   $y = -4x + 7$

   possible points: $(0, 7)$, $(1, 3)$

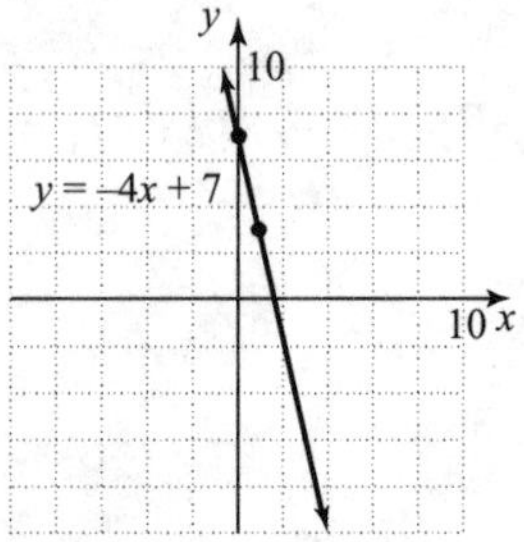

5. $-3x+2y=3$

$2y=3x+3$

$y=\frac{3}{2}x+\frac{3}{2}$

possible points: $\left(0,\frac{3}{2}\right),\left(2,\frac{9}{2}\right)$

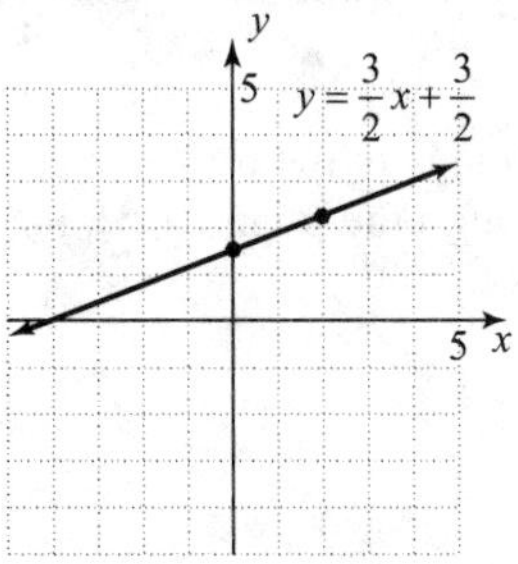

7. $m=-1, b=1$

$y=mx+b$

$y=-1x+1$

$y=-x+1$

9. $m=2,\ b=\frac{3}{4}$

$y=mx+b$

$y=2x+\frac{3}{4}$

11. $m=\frac{2}{7},\ b=0$

$y=mx+b$

$y=\frac{2}{7}x+0$

$y=\frac{2}{7}x$

13. $y-y_1=m(x-x_1)$

$y-2=3(x-1)$

$y-2=3x-3$

$y=3x-1$

15. $y-y_1=m(x-x_1)$

$y-(-3)=-2(x-1)$

$y+3=-2x+2$

$y=-2x-1$

17. $y-y_1=m(x-x_1)$

$y-2=\frac{1}{2}[x-(-6)]$

$y-2=\frac{1}{2}(x+6)$

$y-2=\frac{1}{2}x+3$

$y=\frac{1}{2}x+5$

19. $y-y_1=m(x-x_1)$

$y-0=-\frac{9}{10}[x-(-3)]$

$y=-\frac{9}{10}(x+3)$

$y=-\frac{9}{10}x-\frac{27}{10}$

21. $m=\frac{6-0}{4-2}=\frac{6}{2}=3$

$y-0=3(x-2)$

$y=3x-6$

$f(x)=3x-6$

23. $m=\frac{13-5}{-6-(-2)}=\frac{8}{-4}=-2$

$y-5=-2[x-(-2)]$

$y-5=-2(x+2)$

$y-5=-2x-4$

$y=-2x+1$

$f(x)=-2x+1$

25. $m=\frac{-3-(-4)}{-4-(-2)}=\frac{1}{-2}=-\frac{1}{2}$

$y-(-4)=-\frac{1}{2}[x-(-2)]$

$y+4=-\frac{1}{2}(x+2)$

$2y+8=-(x+2)$

$2y+8=-x-2$

$2y=-x-10$

$y=-\frac{1}{2}x-5$

$f(x)=-\frac{1}{2}x-5$

27. $m=\frac{-9-(-8)}{-6-(-3)}=\frac{-1}{-3}=\frac{1}{3}$

$$\begin{aligned} y-(-8) &= \frac{1}{3}[x-(-3)] \\ y+8 &= \frac{1}{3}(x+3) \\ 3y+24 &= x+3 \\ 3y &= x-21 \\ y &= \frac{1}{3}x-7 \\ f(x) &= \frac{1}{3}x-7 \end{aligned}$$

29. $m=\frac{\frac{7}{10}-\frac{4}{10}}{-\frac{1}{5}-\frac{3}{5}}=\frac{\frac{3}{10}}{-\frac{4}{5}}=\frac{3}{10}\left(-\frac{5}{4}\right)=-\frac{3}{8}$

$$\begin{aligned} y-\frac{4}{10} &= -\frac{3}{8}\left(x-\frac{3}{5}\right) \\ y-\frac{4}{10} &= -\frac{3}{8}x+\frac{9}{40} \\ y &= -\frac{3}{8}x+\frac{5}{8} \\ f(x) &= -\frac{3}{8}x+\frac{5}{8} \end{aligned}$$

31. (0, 3), (1, 1)

$m=\frac{1-3}{1-0}=\frac{-2}{1}=-2$

$b=3$

$y=-2x+3$

$2x+y=3$

33. (−2, 1), (4, 5)

$m=\frac{5-1}{4-(-2)}=\frac{4}{6}=\frac{2}{3}$

$$\begin{aligned} y-1 &= \frac{2}{3}(x+2) \\ 3y-3 &= 2(x+2) \\ 3y-3 &= 2x+4 \\ 2x-3y &= -7 \end{aligned}$$

35. $f(0)=-2$

37. $f(2)=2$

39. $f(x)=-6$

$f(-2)=-6$

$x=-2$

41. $y=mx+b$

$-4=0(-2)+b$

$-4=b$

$y=-4$

43. Every vertical line is in the form $x = c$. Since the line passes through the point (4, 7), its equation is $x = 4$.

45. Every horizontal line is in the form $y = c$. Since the line passes through the point (0, 5), its equation is $y = 5$.

47. $y=4x-2$ so $m=4$

$$\begin{aligned} y-8 &= 4(x-3) \\ y-8 &= 4x-12 \\ y &= 4x-4 \\ f(x) &= 4x-4 \end{aligned}$$

49. $3y=x-6$ or $y=\frac{1}{3}x-2$ so

$m=\frac{1}{3}$ and $m_{\perp}=-3$

$$\begin{aligned} y-(-5) &= -3(x-2) \\ y+5 &= -3x+6 \\ y &= -3x+1 \\ f(x) &= -3x+1 \end{aligned}$$

51. $3x+2y=5$

$2y=-3x+5$

$y=-\frac{3}{2}x+\frac{5}{2}$ so $m=-\frac{3}{2}$

$$\begin{aligned} y-(-3) &= -\frac{3}{2}[x-(-2)] \\ 2(y+3) &= -3(x+2) \\ 2y+6 &= -3(x+2) \\ 2y+6 &= -3x-6 \\ y &= -\frac{3}{2}x-6 \\ f(x) &= -\frac{3}{2}x-6 \end{aligned}$$

53.

$$\begin{aligned} y-3 &= 2[x-(-2)] \\ y-3 &= 2(x+2) \\ y-3 &= 2x+4 \\ 2x-y &= -7 \end{aligned}$$

55. $m = \frac{2-6}{5-1} = \frac{-4}{4} = -1$

$y-6 = -1(x-1)$
$y-6 = -x+1$
$y = -x+7$
$f(x) = -x+7$

57. $y = -\frac{1}{2}x+11$
$2y = -x+22$
$x+2y = 22$

59. $m = \frac{-6-(-4)}{0-(-7)} = \frac{-2}{7} = -\frac{2}{7}$

$y = -\frac{2}{7}x-6$
$7y = -2x-42$
$2x+7y = -42$

61. $y-0 = -\frac{4}{3}[x-(-5)]$
$3y = -4(x+5)$
$3y = -4x-20$
$4x+3y = -20$

63. Every vertical line is in the form $x = c$. Since the line passes through the point (−2, −10), its equation is $x = -2$.

65. $2x+4y = 9$
$4y = -2x+9$
$y = -\frac{1}{2}x+\frac{9}{4}$ so $m = -\frac{1}{2}$

$y-(-2) = -\frac{1}{2}(x-6)$
$2(y+2) = -(x-6)$
$2y+4 = -x+6$
$x+2y = 2$

67. Lines with slopes of 0 are horizontal. Every horizontal line is in the form $y = c$. Since the line passes through (−9, 12), its equation is $y = 12$.

69. $8x-y = 9$
$y = 8x-9$ so $m = 8$
$y-1 = 8(x-6)$
$y-1 = 8x-48$
$8x-y = 47$

71. A line perpendicular to $y = 9$ will have the form $x = c$. Since the line passes through the point (5, −6), its equation is $x = 5$.

73. $m = \frac{-5-(-8)}{-6-2} = \frac{3}{-8} = -\frac{3}{8}$

$y-(-8) = -\frac{3}{8}(x-2)$
$8(y+8) = -3(x-2)$
$8y+64 = -3x+6$
$y = -\frac{3}{8}x-\frac{29}{4}$
$f(x) = -\frac{3}{8}x-\frac{29}{4}$

75. a. Use the points (1, 32) and (3, 96).
$m = \frac{96-32}{3-1} = \frac{64}{2} = 32$
$y-y_1 = m(x-x_1)$
$y-32 = 32(x-1)$
$y-32 = 32x-32$
$y = 32x$

b. Let $x = 4$. $y = 32x = 32(4) = 128$
The rock is traveling 128 feet per second at 4 seconds.

77. a. Use the points (6, 2000) and (8, 1500).
$m = \frac{1500-2000}{8-6} = \frac{-500}{2} = -250$
$y-y_1 = m(x-x_1)$
$y-2000 = -250(x-6)$
$y-2000 = -250x+1500$
$y = -250x+3500$

b. Let $x = 7.50$.
$y = -250x + 3500$
$y = -250(7.50) + 3500 = 1625$
The daily sales of Frisbees at \$7.50 each is predicted to be 1625 Frisbees.

79. a. Use the points (0, 2711) and (10, 3238).
$m = \frac{3238-2711}{10-0} = \frac{527}{10} = 52.7$
$y = mx + b$
$y = 52.7x + 2711$

b. Let $x = 2017 - 2012 = 5$.
$y = 52.7(5) + 2711 = 2974.5$
The number of people employed as registered nurses in 2017 is predicted to be 2974.5 thousand, or 2,974,500.

81. a. Use the points (0, 268,200) and (5, 342,800).

$$m = \frac{342,800 - 268,200}{5 - 0} = \frac{74,600}{5} = 14,920$$

$y = mx + b$

$y = 14,920x + 268,200$

b. Let $x = 2018 - 2009 = 9$.

$y = 14,920(9) + 268,200 = 402,480$

The average price of a new home in 2018 is predicted to be \$402,480.

83.
$$\begin{aligned} 2x - 7 &\le 21 \\ 2x &\le 28 \\ x &\le 14 \end{aligned}$$
$(-\infty, 14]$

85.
$$\begin{aligned} 5(x-2) &\ge 3(x-1) \\ 5x - 10 &\ge 3x - 3 \\ 2x &\ge 7 \\ x &\ge \frac{7}{2} \end{aligned}$$
$\left[\frac{7}{2}, \infty\right)$

87.
$$\begin{aligned} \frac{x}{2} + \frac{1}{4} &< \frac{1}{8} \\ 8\left(\frac{x}{2} + \frac{1}{4}\right) &< 8\left(\frac{1}{8}\right) \\ 4x + 2 &< 1 \\ 4x &< -1 \\ x &< -\frac{1}{4} \end{aligned}$$
$\left(-\infty, -\frac{1}{4}\right)$

89. Since any vertical line intersects any horizontal line in a right angle, the statement is true.

91. $m = \frac{1-(-1)}{-5-3} = \frac{2}{-8} = -\frac{1}{4}$ so $m_{\perp} = 4$

$$M((3, -1), (5, 1)) = \left(\frac{3-5}{2}, \frac{-1+1}{2}\right) = (1, 0)$$

$$\begin{aligned} y - 0 &= 4[x - (-1)] \\ y &= 4(x+1) \\ y &= 4x + 4 \\ -4x + y &= 4 \end{aligned}$$

93. $m = \frac{-4-6}{-22-(-2)} = \frac{-10}{-20} = \frac{1}{2}$ so $m_{\perp} = -2$

$$M((-2, 6), (-22, -4)) = \left(\frac{-2-22}{2}, \frac{6-4}{2}\right) = (-12, 1)$$

$$\begin{aligned} y - 1 &= -2[x - (-12)] \\ y - 1 &= -2(x+12) \\ y - 1 &= -2x - 24 \\ 2x + y &= -23 \end{aligned}$$

95. $m = \frac{7-3}{-4-2} = \frac{4}{-6} = -\frac{2}{3}$ so $m_{\perp} = \frac{3}{2}$

$$M((2, 3), (-4, 7)) = \left(\frac{2-4}{2}, \frac{3+7}{2}\right) = (-1, 5)$$

$$\begin{aligned} y - 5 &= \frac{3}{2}[x - (-1)] \\ 2(y-5) &= 3(x+1) \\ 2y - 10 &= 3x + 3 \\ 3x - 2y &= -13 \end{aligned}$$

97. answers may vary

99. $f(x) = -x + 7$

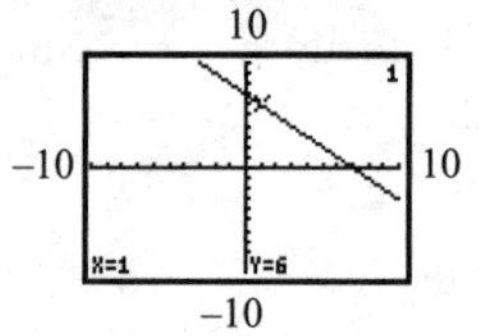

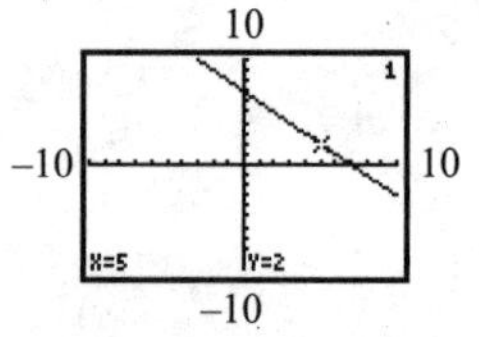

101. $4x + 3y = -20$

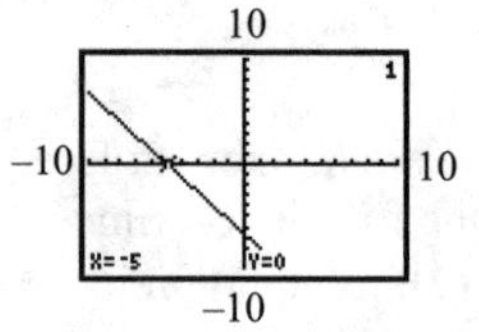

103. $y = 4x - 2$
$y = 4x - 4$

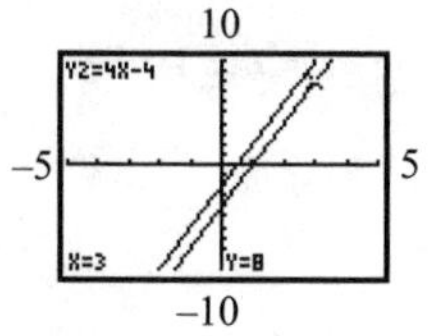

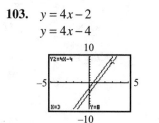

Integrated Review

1. $y = -2x$

x	-1	0	1
y	2	0	-2

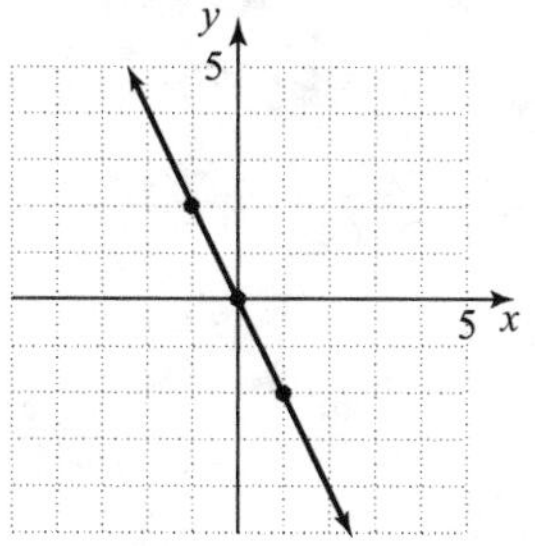

2. $3x - 2y = 6$

x	0	2	4
y	-3	0	3

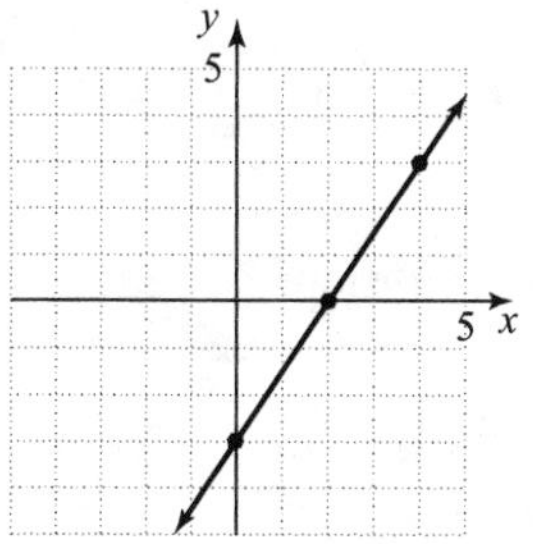

3. $x = -3$
The graph of $x = -3$ is a vertical line.

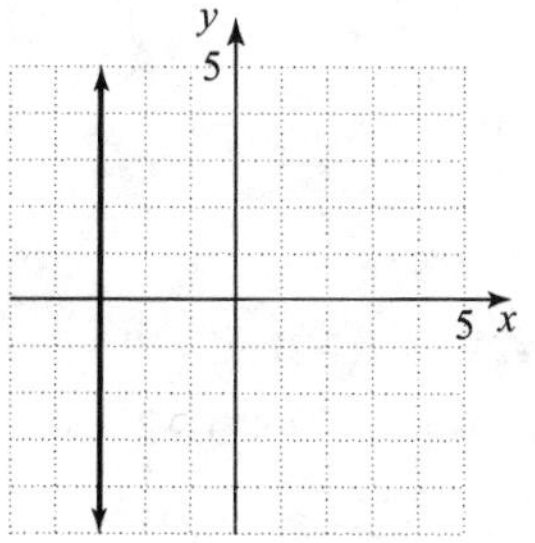

4. $y = 1.5$

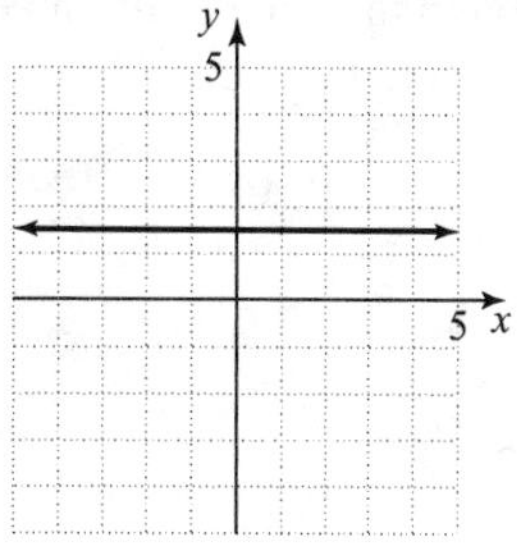

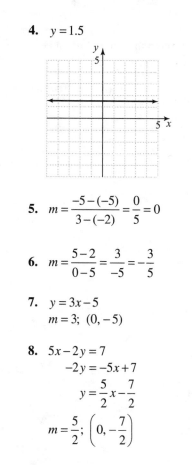

5. $m = \dfrac{-5-(-5)}{3-(-2)} = \dfrac{0}{5} = 0$

6. $m = \dfrac{5-2}{0-5} = \dfrac{3}{-5} = -\dfrac{3}{5}$

7. $y = 3x - 5$
$m = 3;\ (0, -5)$

8. $5x - 2y = 7$
$-2y = -5x + 7$
$y = \dfrac{5}{2}x - \dfrac{7}{2}$
$m = \dfrac{5}{2};\ \left(0, -\dfrac{7}{2}\right)$

9. $y = 8x - 6 \quad y = 8x + 6$
$m = 8 \qquad m = 8$
Parallel, since their slopes are equal.

10. $y = \dfrac{2}{3}x + 1$
$m = \dfrac{2}{3}$

$2y + 3x = 1$
$2y = -3x + 1$
$y = -\dfrac{3}{2}x + \dfrac{1}{2}$
$m = -\dfrac{3}{2}$

Perpendicular, since the product of their slopes is -1.

11. $m = \dfrac{2-6}{5-1} = \dfrac{-4}{4} = -1$
$y - 6 = -1(x - 1)$
$y - 6 = -x + 1$
$y = -x + 7$

12. Every vertical line is in the form $x = c$. Since the line passes through the point $(-2, -10)$, its equation is $x = -2$.

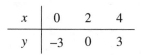

13. Every horizontal line is in the form $y = c$. Since the line passes through the point (1, 0), its equation is $y = 0$.

14. $m = \frac{-5-(-9)}{-6-2} = \frac{4}{-8} = -\frac{1}{2}$

$$y-(-9) = -\frac{1}{2}(x-2)$$
$$2(y+9) = -1(x-2)$$
$$2y+18 = -x+2$$
$$2y = -x-16$$
$$y = -\frac{1}{2}x-8$$
$$f(x) = -\frac{1}{2}x-8$$

15.
$$y-4 = -5[x-(-2)]$$
$$y-4 = -5(x+2)$$
$$y-4 = -5x-10$$
$$y = -5x-6$$
$$f(x) = -5x-6$$

16.
$$y = -4x+\frac{1}{3}$$
$$f(x) = -4x+\frac{1}{3}$$

17.
$$y = \frac{1}{2}x-1$$
$$f(x) = \frac{1}{2}x-1$$

18.
$$y-0 = 3\left(x-\frac{1}{2}\right)$$
$$y = 3x-\frac{3}{2}$$

19.
$$3x-y = 5$$
$$y = 3x-5$$
$$m = 3$$
$$y-(-5) = 3[x-(-1)]$$
$$y+5 = 3(x+1)$$
$$y+5 = 3x+3$$
$$y = 3x-2$$

20.
$$4x-5y = 10$$
$$-5y = -4x+10$$
$$y = \frac{4}{5}x-2;\ m = \frac{4}{5} \text{ so } m_{\perp} = -\frac{5}{4}$$

Therefore, $y = -\frac{5}{4}x+4$.

21.
$$4x+y = \frac{2}{3}$$
$$y = -4x+\frac{2}{3};\ m = -4 \text{ so } m_{\perp} = \frac{1}{4}$$
$$y-(-3) = \frac{1}{4}(x-2)$$
$$4(y+3) = x-2$$
$$4y+12 = x-2$$
$$4y = x-14$$
$$y = \frac{1}{4}x-\frac{7}{2}$$

22.
$$5x+2y = 2$$
$$2y = -5x+2$$
$$y = -\frac{5}{2}x+1$$
$$m = -\frac{5}{2}$$
$$y-0 = -\frac{5}{2}[x-(-1)]$$
$$y = -\frac{5}{2}(x+1)$$
$$2y = -5(x+1)$$
$$2y = -5x-5$$
$$y = -\frac{5}{2}x-\frac{5}{2}$$

23. A line having undefined slope is vertical. Therefore, the equation is $x = -1$.

24.
$$y-3 = 0[x-(-1)]$$
$$y-3 = 0$$
$$y = 3$$

Section 3.6 Practice Exercises

1. $f(x) = \begin{cases} -4x-2 & \text{if } x \le 0 \\ x+1 & \text{if } x > 0 \end{cases}$

Since $4 > 0$, $f(4) = 4+1 = 5$.

Since $-2 \le 0$, $f(-2) = -4(-2)-2 = 8-2 = 6$.

Since $0 \le 0$, $f(0) = -4(0)-2 = 0-2 = -2$.

2. $f(x)=\begin{cases}-4x-2 & \text{if } x\le 0\\ x+1 & \text{if } x>0\end{cases}$

For $x \le 0$:

x	$f(x)$
−2	6
−1	2
0	−2

For $x > 0$:

x	$f(x)$
1	2
2	3
3	4

Graph a closed circle at (0, −2). Graph an open circle at (0, 1), which is found by substituting 0 for x in $f(x)=x+1$.

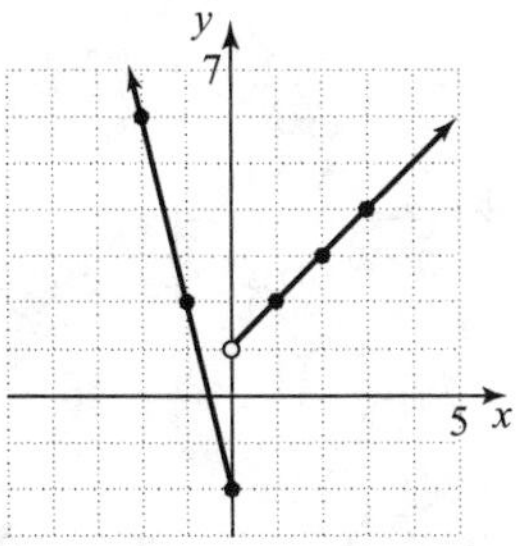

3. $f(x)=x^2$ and $g(x)=x^2-3$

The graph of $g(x)=x^2-3$ is the graph of $f(x)=x^2$ moved downward 3 units.

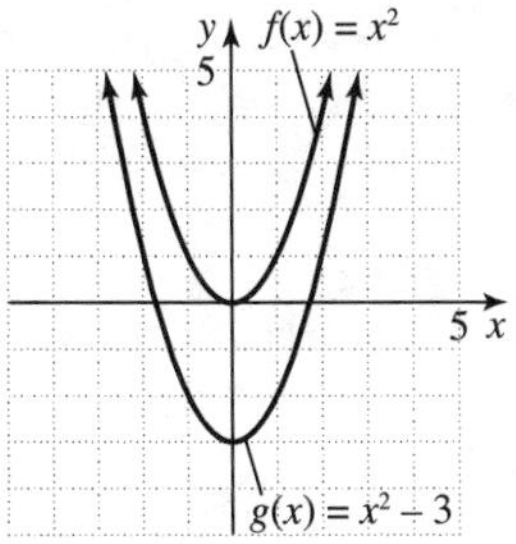

4. $f(x)=\sqrt{x}$ and $g(x)=\sqrt{x}+1$

The graph of $g(x)=\sqrt{x}+1$ is the graph of $f(x)=\sqrt{x}$ moved upward 1 unit.

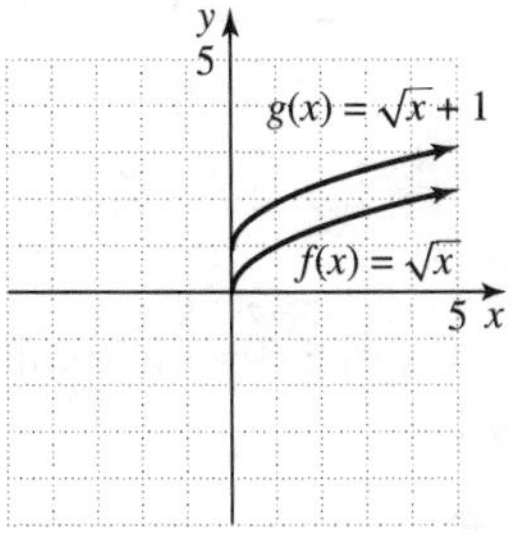

5. $f(x)=|x|$ and $g(x)=|x-3|$

x	$f(x)$	$g(x)$
−2	2	5
−1	1	4
0	0	3
1	1	2
2	2	1
3	3	0
4	4	1
5	5	2

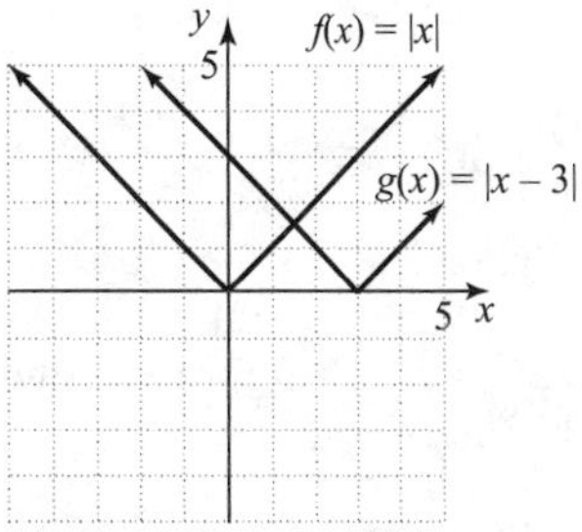

6. $f(x)=|x|$ and $g(x)=|x-2|+3$

The graph of $g(x)$ is the same as the graph of $f(x)$ shifted 2 units to the right and 3 units up.

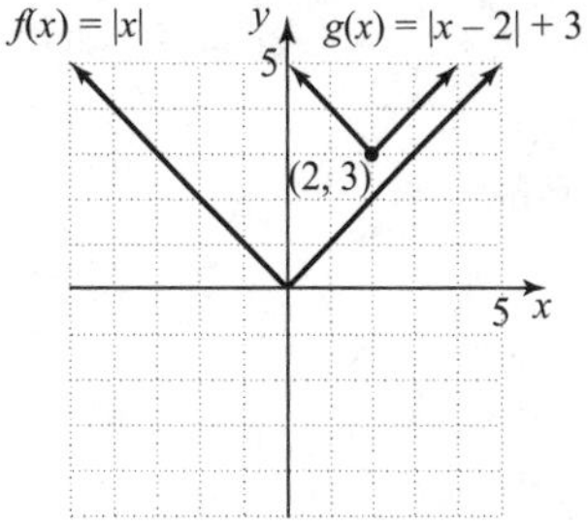

7. $h(x)=-(x+2)^2-1$

The graph of $h(x)=-(x+2)^2-1$ is the same as the graph of $f(x)=x^2$ reflected about the x-axis, then moved 2 units to the left and 1 unit downward.

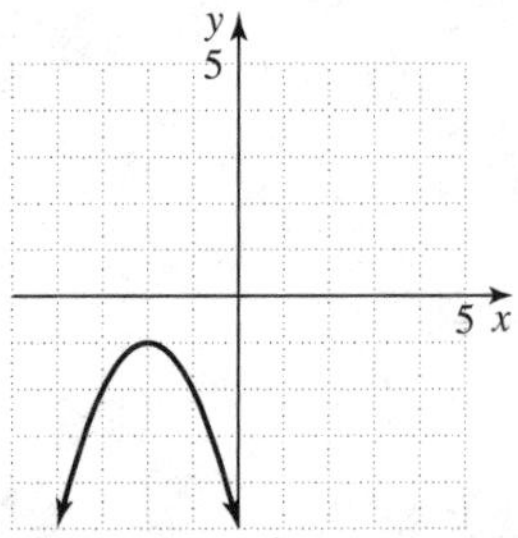

$h(x) = -(x + 2)^2 - 1$

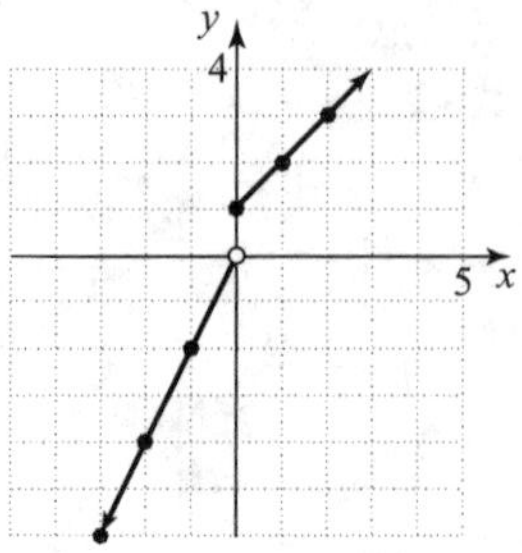

Vocabulary, Readiness & Video Check 3.6

1. The graph that corresponds to $y = \sqrt{x}$ is C.

2. The graph that corresponds to $y = x^2$ is B.

3. The graph that corresponds to $y = x$ is D.

4. The graph that corresponds to $y = |x|$ is A.

5. Although $f(x) = x + 3$ isn't defined for $x = -1$, we need to clearly indicate the point where this piece of the graph ends. Therefore, we find this point and graph it as an open circle.

6. Once you know the shapes, use the shifting rules to tell you where to move the vertex or starting point for each graph, then you can easily draw in the appropriate basic shape.

7. The graph of $f(x) = -\sqrt{x+6}$ has the same shape as the graph of $f(x) = \sqrt{x+6}$ but it is reflected about the <u>x-axis</u>.

Exercise Set 3.6

1. $f(x) = \begin{cases} 2x & \text{if } x < 0 \\ x+1 & \text{if } x \geq 0 \end{cases}$

For $x < 0$:

x	$f(x)$
−3	−6
−2	−4
−1	−2

For $x \geq 0$:

x	$f(x)$
0	1
1	2
2	3

Graph a closed circle at (0, 1). Graph an open circle at (0, 0), which is found by substituting 0 for x in $f(x) = 2x$.

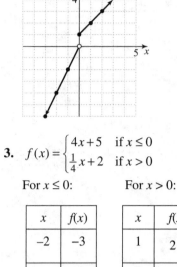

3. $f(x) = \begin{cases} 4x+5 & \text{if } x \leq 0 \\ \frac{1}{4}x+2 & \text{if } x > 0 \end{cases}$

For $x \leq 0$:

x	$f(x)$
−2	−3
−1	1
0	5

For $x > 0$:

x	$f(x)$
1	$2\frac{1}{4}$
2	$2\frac{1}{2}$
4	3

Graph a closed circle at (0, 5). Graph an open circle at (0, 2), which is found by substituting 0 for x in $f(x) = \frac{1}{4}x + 2$.

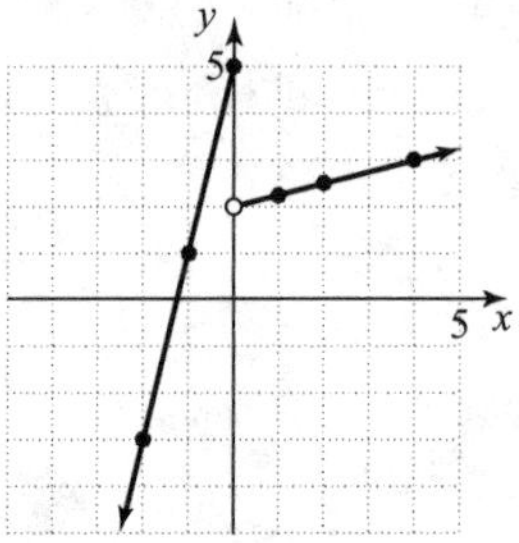

5. $g(x) = \begin{cases} -x & \text{if } x \leq 1 \\ 2x+1 & \text{if } x > 1 \end{cases}$

For $x \leq 1$:

x	$g(x)$
−1	1
0	0
1	−1

For $x > 1$:

x	$g(x)$
2	5
3	7
4	9

Graph a closed circle at (1, −1). Graph an open circle at (1, 3), which is found by substituting 1 for x in $g(x) = 2x + 1$.

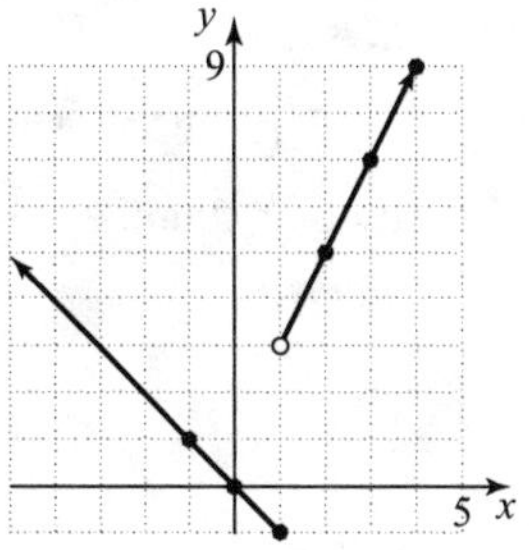

7. $f(x)=\begin{cases}5 & \text{if } x<-2\\ 3 & \text{if } x\geq -2\end{cases}$

For $x<-2$: For $x\geq -2$:

x	f(x)
–5	5
–4	5
–3	5

x	f(x)
–2	3
–1	3
0	3

Graph a closed circle at (–2, 3). Graph an open circle at (–2, 5), which is found by substituting –2 for *x* in $f(x)=5$.

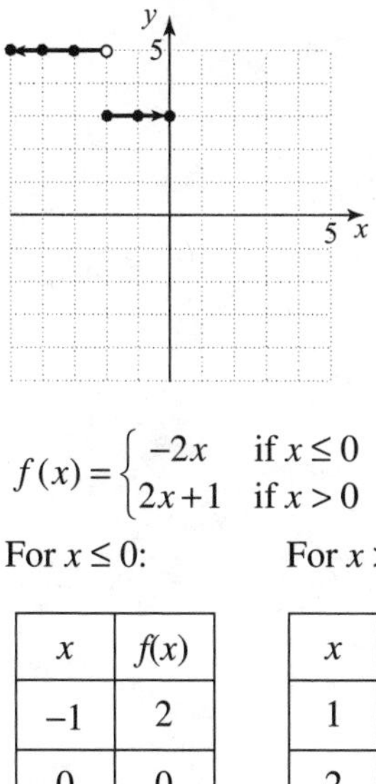

9. $f(x)=\begin{cases}-2x & \text{if } x\leq 0\\ 2x+1 & \text{if } x>0\end{cases}$

For $x\leq 0$: For $x>0$:

x	f(x)
–1	2
0	0

x	f(x)
1	3
2	5

Graph a closed circle at (0, 0). Graph an open circle at (0, 1), which is found by substituting 0 for *x* in $f(x)=2x+1$.

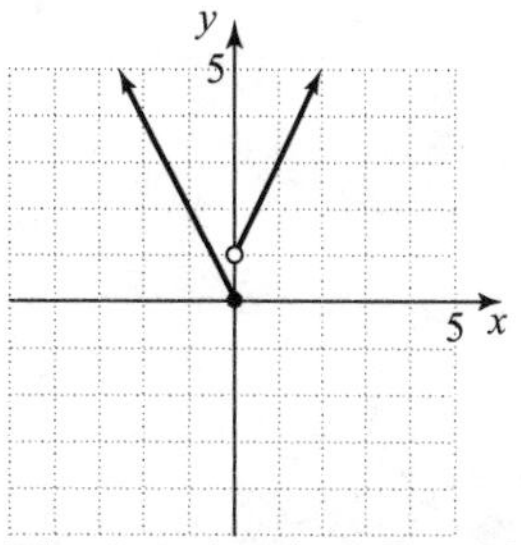

The function is defined for all real numbers, so the domain is $(-\infty, \infty)$. The function takes on all *y*-values greater than or equal to 0, so the range is $[0, \infty)$.

11. $h(x)=\begin{cases}5x-5 & \text{if } x<2\\ -x+3 & \text{if } x\geq 2\end{cases}$

For $x<2$: For $x\geq 2$:

x	h(x)
0	–5
1	0

x	h(x)
2	1
3	0

Graph a closed circle at (2, 1). Graph an open circle at (2, 5), which is found by substituting 2 for *x* in $h(x)=5x-5$.

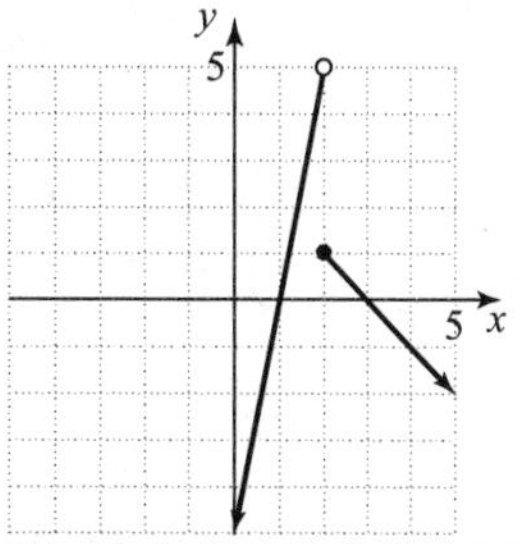

The function is defined for all real numbers, so the domain is $(-\infty, \infty)$. The function takes on all *y*-values less than 5, so the range is $(-\infty, 5)$.

13. $f(x)=\begin{cases}x+3 & \text{if } x<-1\\ -2x+4 & \text{if } x\geq -1\end{cases}$

For $x<-1$: For $x\geq -1$:

x	f(x)
–4	–1
–3	0
–2	1

x	f(x)
–1	6
0	4
1	2

Graph a closed circle at (–1, 6). Graph an open circle at (–1, 2), which is found by substituting

−1 for *x* in $f(x) = x + 3$.

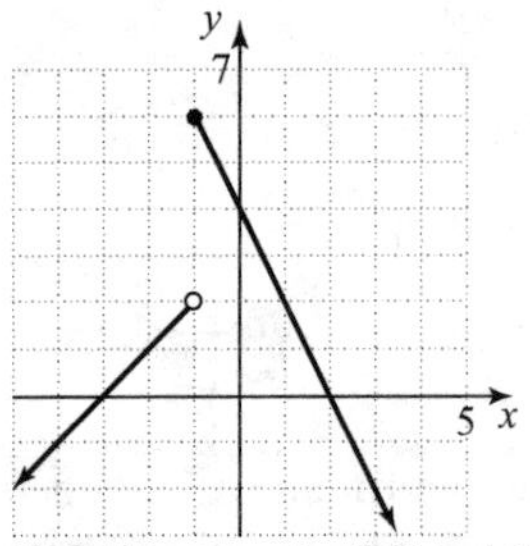

The function is defined for all real numbers, so the domain is (−∞, ∞). The function takes on all *y*-values less than or equal to 6, so the range is (−∞, 6].

15. $g(x) = \begin{cases} -2 & \text{if } x \le 0 \\ -4 & \text{if } x \ge 1 \end{cases}$

For $x \le 0$:

x	$g(x)$
−2	−2
−1	−2
0	−2

For $x \ge 1$:

x	$g(x)$
1	−4
2	−4
3	−4

Graph closed circles at (0, −2) and (1, −4).

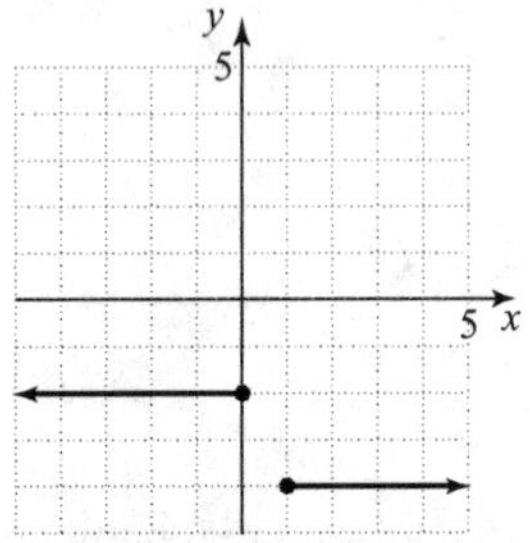

The function is defined for $x \le 0$ or $x \ge 1$, so the domain is (−∞, 0] ∪ [1, ∞). The function takes on two *y*-values, −2 and −4, so the range is {−2, −4}.

17. $f(x) = |x| + 3$

The graph of $f(x) = |x| + 3$ is the same as the graph of $y = |x|$ shifted up 3 units.

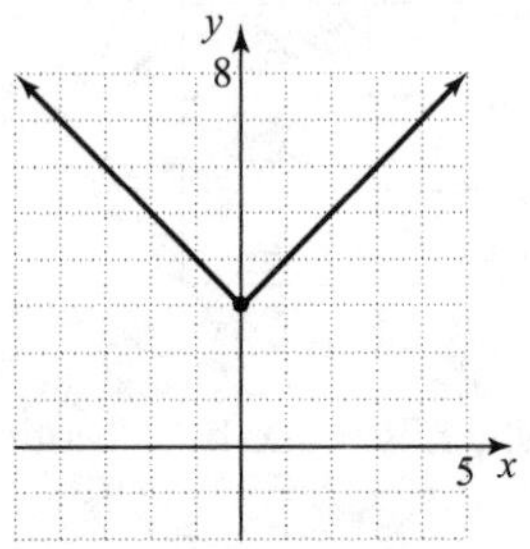

19. $f(x) = \sqrt{x} - 2$

The graph of $f(x) = \sqrt{x} - 2$ is the same as the graph of $y = \sqrt{x}$ shifted down 2 units.

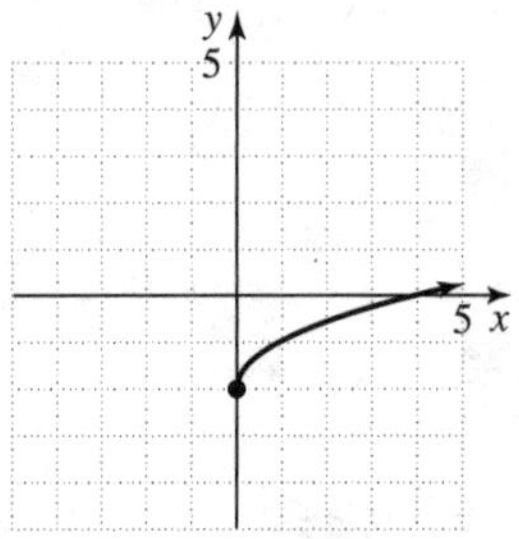

21. $f(x) = |x - 4|$

The graph of $f(x) = |x - 4|$ is the same as the graph of $y = |x|$ shifted right 4 units.

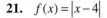

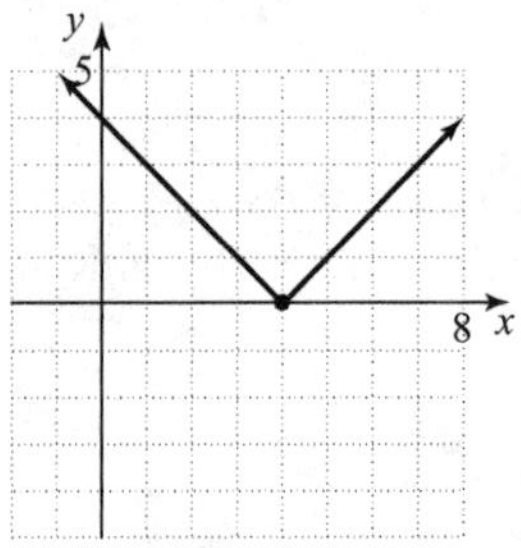

23. $f(x) = \sqrt{x + 2}$

The graph of $f(x) = \sqrt{x + 2}$ is the same as the graph of $y = \sqrt{x}$ shifted left 2 units.

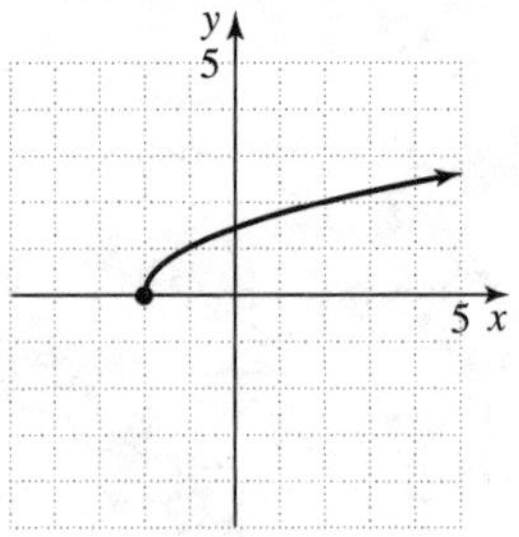

25. $y = (x-4)^2$

The graph of $y = (x-4)^2$ is the same as the graph of $y = x^2$ shifted right 4 units.

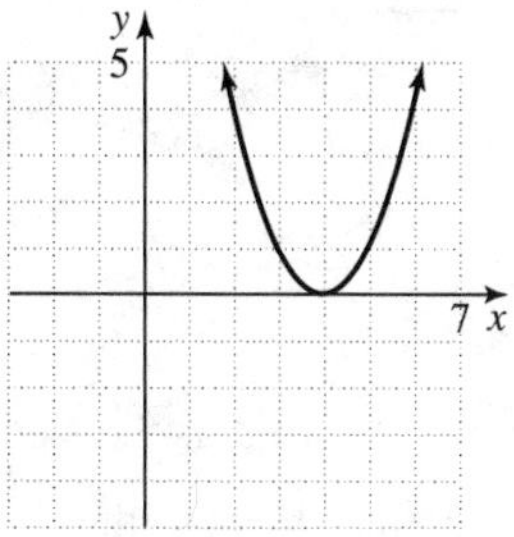

27. $f(x) = x^2 + 4$

The graph of $f(x) = x^2 + 4$ is the same as the graph of $y = x^2$ shifted up 4 units.

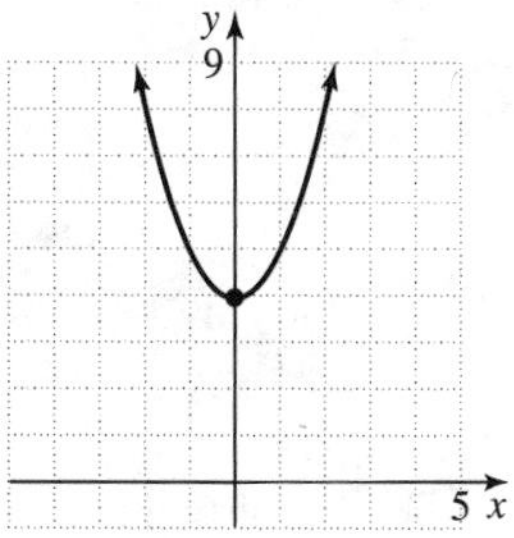

29. $f(x) = \sqrt{x-2} + 3$

The graph of $f(x) = \sqrt{x-2} + 3$ is the same as the graph of $y = \sqrt{x}$ shifted right 2 units and up 3 units.

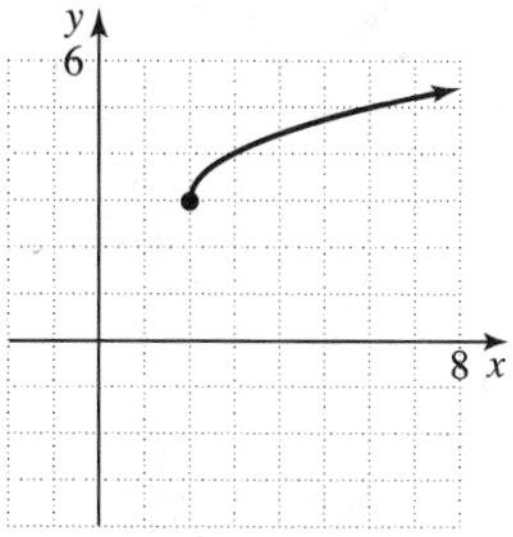

31. $f(x) = |x-1| + 5$

The graph of $f(x) = |x-1| + 5$ is the same as the graph of $y = |x|$ shifted right 1 unit and up 5 units.

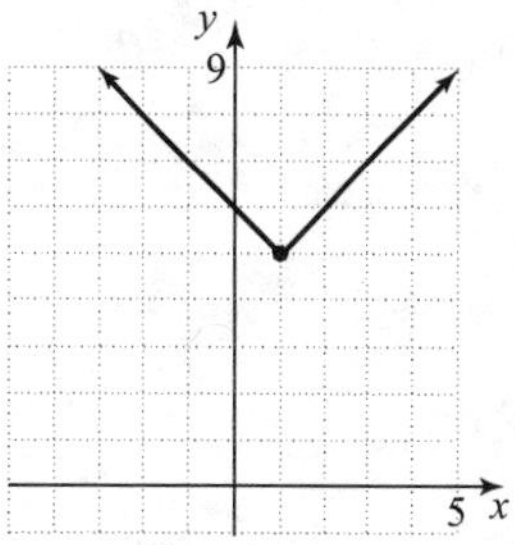

33. $f(x) = \sqrt{x+1} + 1$

The graph of $f(x) = \sqrt{x+1} + 1$ is the same as the graph of $y = \sqrt{x}$ shifted left 1 unit and up 1 unit.

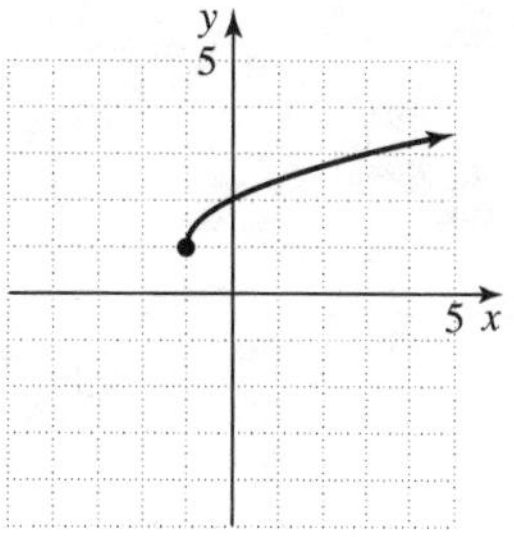

35. $f(x) = |x+3| - 1$

The graph of $f(x) = |x+3| - 1$ is the same as the graph of $y = |x|$ shifted left 3 units and down 1 unit.

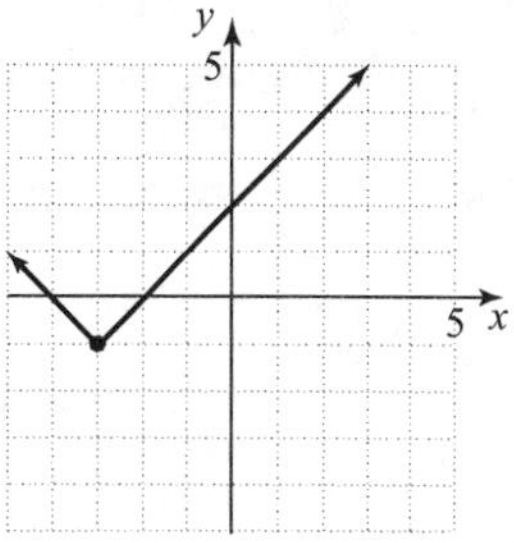

37. $g(x) = (x-1)^2 - 1$

The graph of $g(x) = (x-1)^2 - 1$ is the same as the graph of $y = x^2$ shifted right 1 unit and down 1 unit.

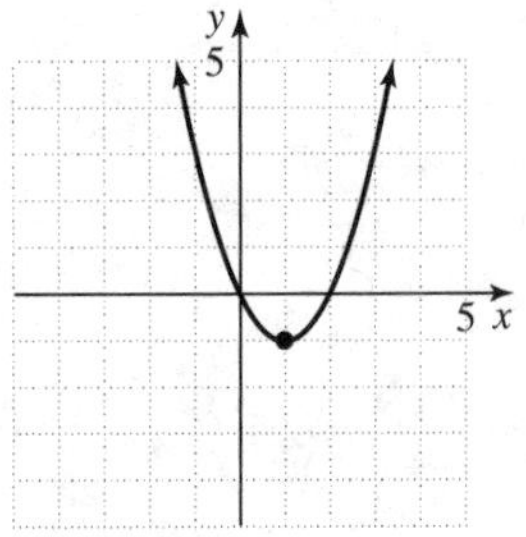

39. $f(x) = (x+3)^2 - 2$

The graph of $f(x) = (x+3)^2 - 2$ is the same as the graph of $y = x^2$ shifted left 3 units and down 2 units.

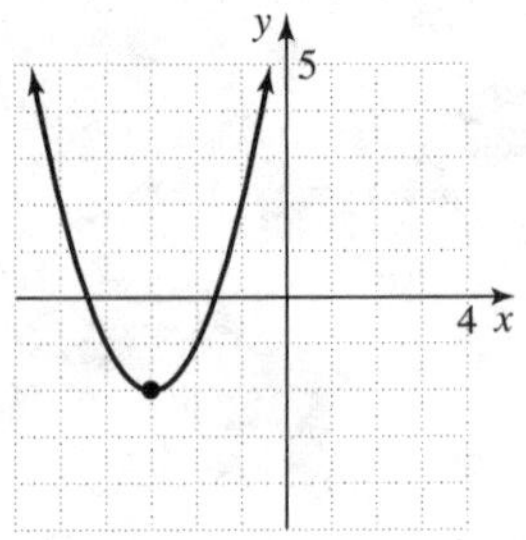

41. $f(x) = -(x-1)^2$

The graph of $f(x) = -(x-1)^2$ is the same as the graph of $y = x^2$ reflected about the x-axis and then shifted right 1 unit.

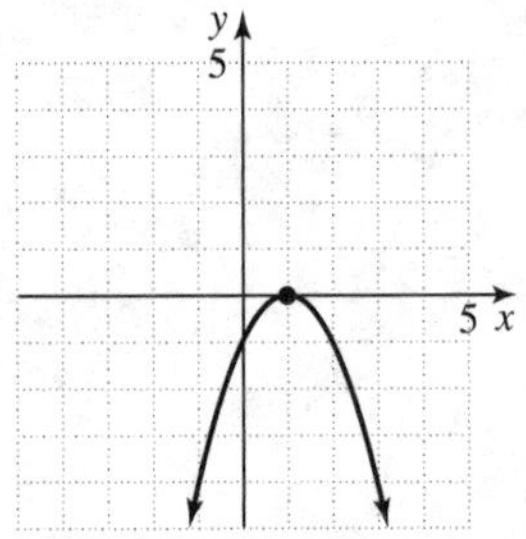

43. $h(x) = -\sqrt{x} + 3$

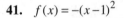

The graph of $h(x) = -\sqrt{x} + 3$ is the same as the graph of $y = \sqrt{x}$ reflected about the x-axis and then shifted up 3 units.

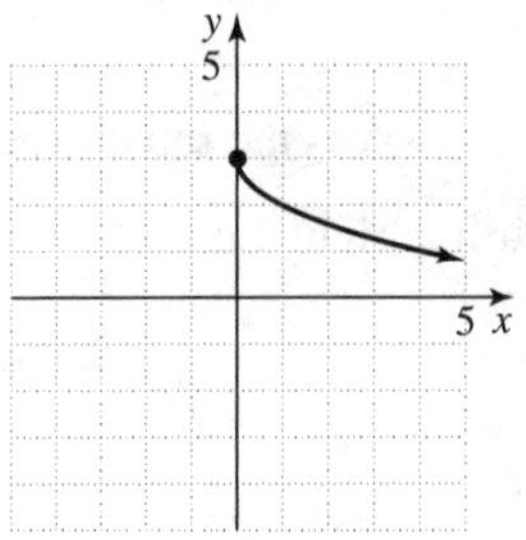

45. $h(x) = -|x+2| + 3$

The graph of $h(x) = -|x+2| + 3$ is the same as the graph of $y = |x|$ reflected about the x-axis and then shifted left 2 units and up 3 units.

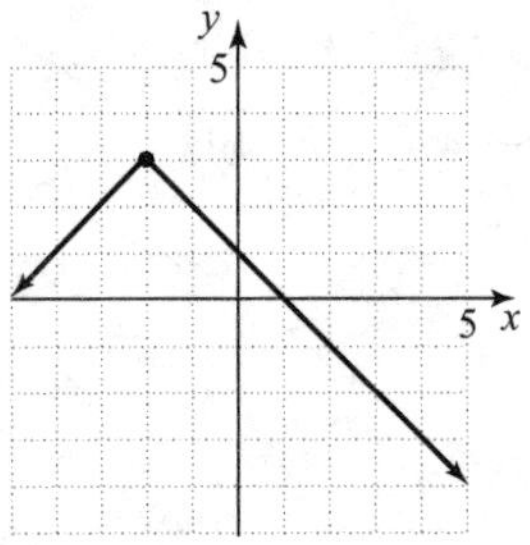

47. $f(x) = (x-3) + 2$

Since the function can be simplified to $f(x) = x - 1$, we see that its graph is a line with slope $m = 1$ and y-intercept $(0, -1)$.

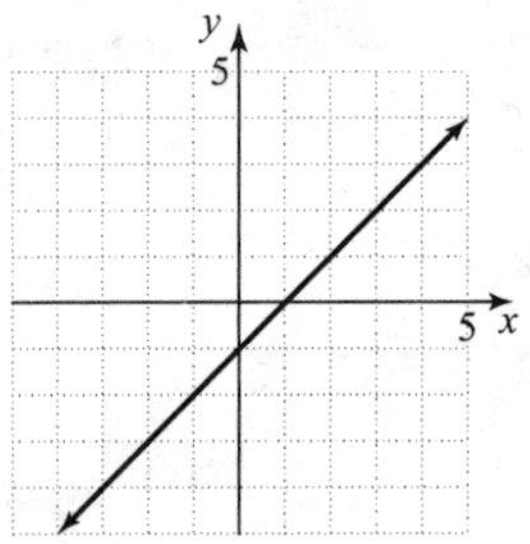

49. The graph of $y = -1$ is a horizontal line with y-intercept $(0, -1)$. The correct graph is A.

51. The graph of $x = 3$ is a vertical line with x-intercept $(3, 0)$. The correct graph is D.

53. answers may vary

55. $f(x) = \begin{cases} -\frac{1}{2}x & \text{if } x \le 0 \\ x+1 & \text{if } 0 < x \le 2 \\ 2x-1 & \text{if } x > 2 \end{cases}$

Some points for $x \le 0$: $(-4, 2)$, $(-2, 1)$, $(0, 0)$
Closed dot at $(0, 0)$

Some points for $0 < x \le 2$: (1, 2), (2, 3)
Open dot at (0, 1), closed dot at (2, 3)
Some points for $x > 2$: (3, 5), (4, 7)
There would be an open dot at (2, 3) except that it gets filled by the middle piece of the graph.

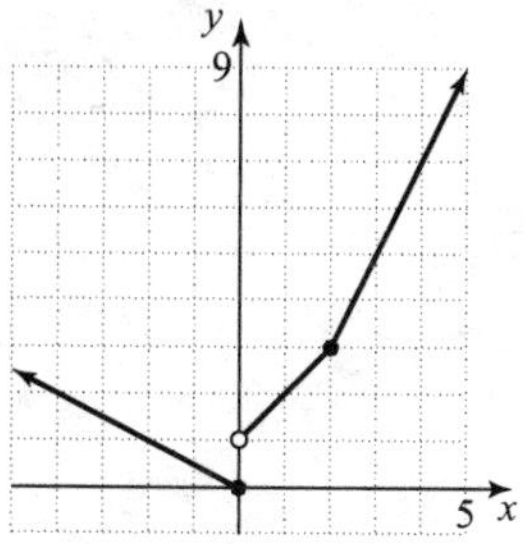

57. $f(x) = \sqrt{x-2}+3$
The function is defined when $x-2 \ge 0$, or $x \ge 2$, so the domain is $[2, \infty)$. The function takes on all y-values greater than or equal to 3, so the range is $[3, \infty)$.

59. $h(x) = -|x+2|+3$
The function is defined for all real numbers, so the domain is $(-\infty, \infty)$. The function takes on all y-values less than or equal to 3, so the range is $(-\infty, 3]$.

61. $f(x) = 5\sqrt{x-20}+1$
The function is defined when $x-20 \ge 0$, or $x \ge 20$, so the domain is $[20, \infty)$.

63. $h(x) = 5|x-20|+1$
The function is defined for all real numbers, so the domain is $(-\infty, \infty)$.

65. $g(x) = 9-\sqrt{x+103}$
The function is defined when $x+103 \ge 0$, or $x \ge -103$, so the domain is $[-103, \infty)$.

67. $f(x) = \begin{cases} |x| & \text{if } x \le 0 \\ x^2 & \text{if } x > 0 \end{cases}$

For $x \le 0$:

x	$f(x)$
−2	2
−1	1
0	0

For $x > 0$:

x	$f(x)$
1	1
2	4
3	9

Graph a closed circle at (0, 0). The graph of $f(x) = x^2$ for $x > 0$ also approaches the point (0, 0).

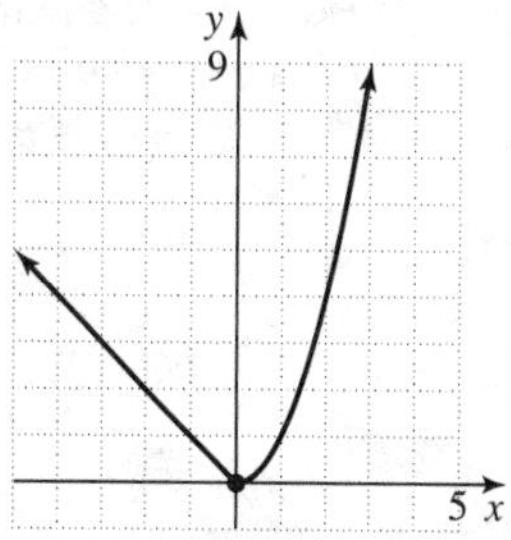

The function is defined for all real numbers, so the domain is $(-\infty, \infty)$. The function takes on all y-values greater than or equal to 0, so the range is $[0, \infty)$.

69. $g(x) = \begin{cases} |x-2| & \text{if } x < 0 \\ -x^2 & \text{if } x \ge 0 \end{cases}$

For $x < 0$:

x	$g(x)$
−3	5
−2	4
−1	3

For $x \ge 0$:

x	$g(x)$
0	0
1	−1
2	−4

Graph an open circle at (0, 2). Graph a closed circle at (0, 0).

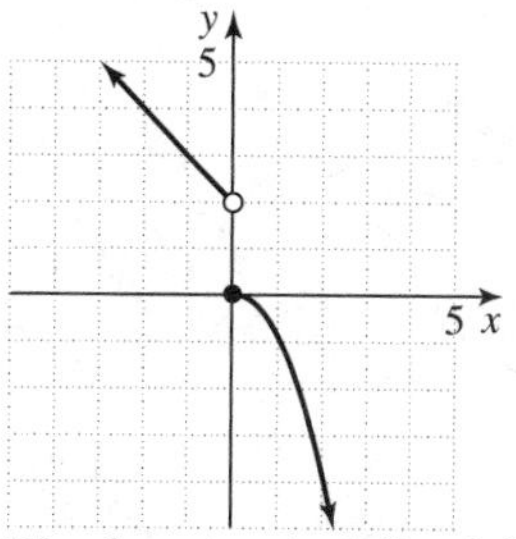

The function is defined for all real numbers, so the domain is $(-\infty, \infty)$. The function takes on all y-values such $y > 2$ or $y \le 0$, so the range is $(-\infty, 0] \cup (2, \infty)$.

Section 3.7 Practice Exercises

1. $3x + y < 8$
The boundary line is $3x + y = 8$. Graph a dashed boundary line because the inequality symbol is <. The point (0, 0) is not on the boundary line, so we use it as a test point. Replace x with 0 and y with 0 in the original inequality.

$3x + y < 8$
$3(0) + 0 < 8$
$0 < 8$ True

Since (0, 0) satisfies the inequality, shade the half-plane that contains (0, 0). Every point in the shaded half-plane satisfies the original inequality.

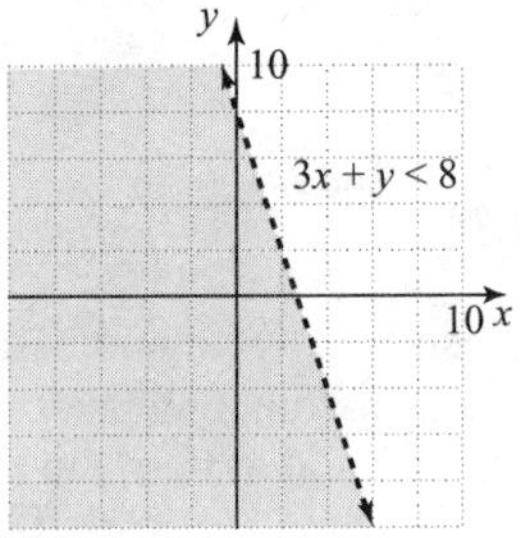

2. $x \geq 3y$

First graph the boundary line $x = 3y$. Graph a solid boundary line because the inequality symbol is $\geq$. We choose (0, 1) as a test point.

$x \geq 3y$
$0 \geq 3(1)$
$0 \geq 3$ False

Since this point does not satisfy the inequality, shade the half-plane on the opposite side of the boundary line from (0, 1). The graph of $x \geq 3y$ is the boundary line together with the shaded region.

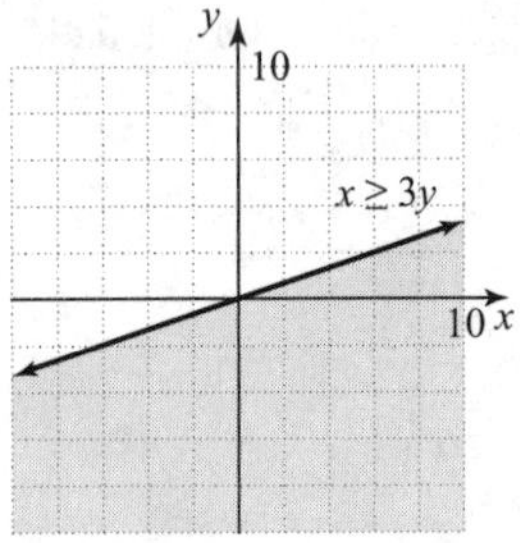

3. $x \leq 3$ and $y \leq x - 2$

Graph each inequality. The intersection of the two graphs is all points common to both regions, as shown by the darker shading in the graph.

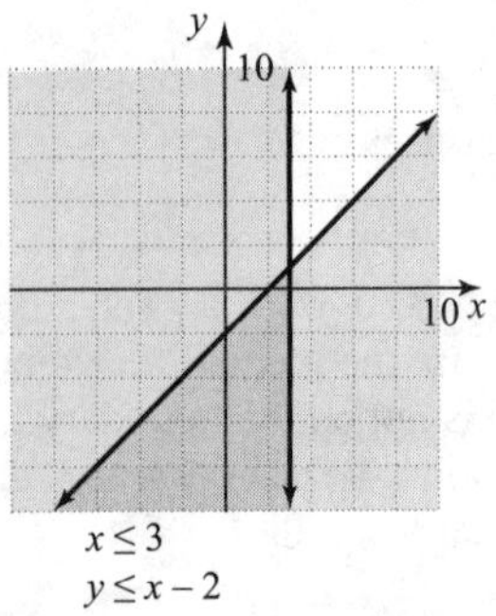

4. $2x - 3y \leq -2$ or $y \geq 1$

Graph each inequality. The union of the two inequalities is both shaded regions, including the boundary lines, as shown in the graph.

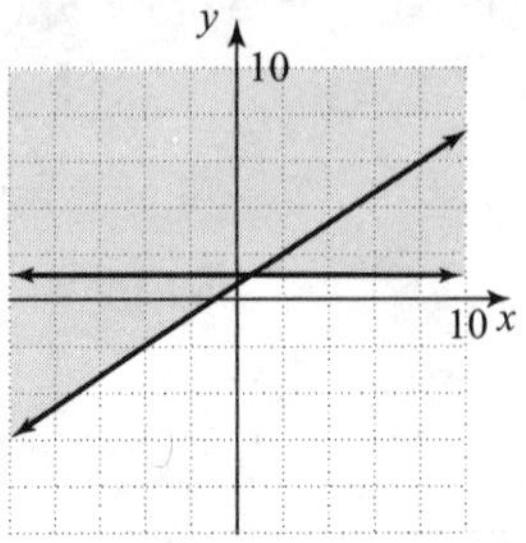

Vocabulary, Readiness & Video Check 3.7

1. We find the boundary line equation by replacing the inequality symbol with =. The points on this line are solutions (line is solid) if the inequality is $\geq$ or $\leq$; the points on this line are not solutions (line is dashed) if the inequality is > or <.

2. Graph each inequality separately on the same rectangular coordinate system. The intersection is where their solution regions overlap—that is, where you've shaded twice.

Exercise Set 3.7

1. $x < 2$

Graph $x = 2$ as a dashed line. Shade to the left of the line.

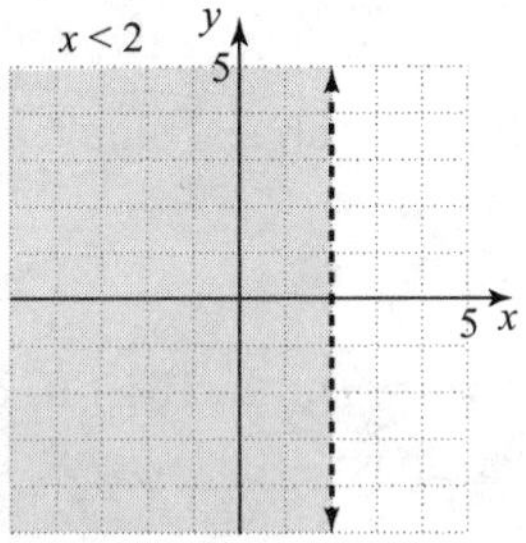

3. $x - y \geq 7$

Graph $x - y = 7$ as a solid line.
Test: (0, 0)
$0 - 0 \geq 7$
$0 \geq 7$ False

Shade the half-plane that does not contain (0, 0).

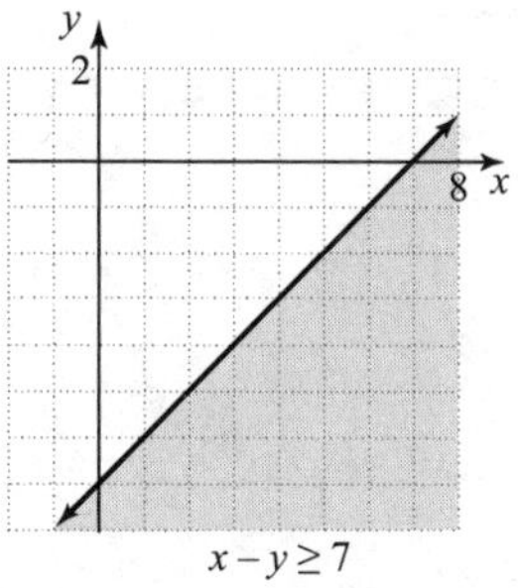

5. $3x + y > 6$

Graph $3x + y = 6$ as a dashed line.
Test: (0, 0)
$3(0) + 0 > 6$
$0 > 6$ False
Shade the half-plane that does not contain (0, 0).

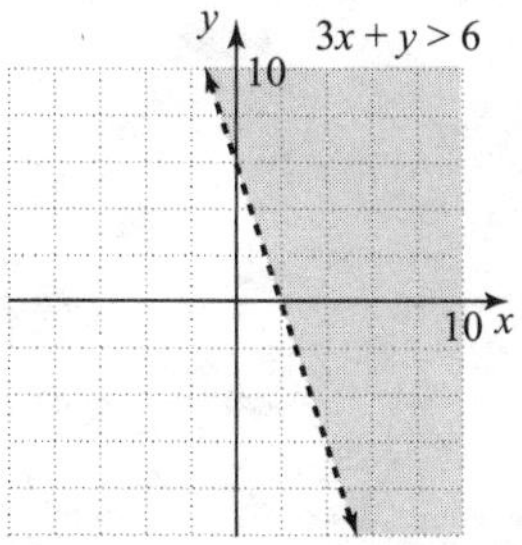

7. $y \le -2x$

Graph $y = -2x$ as a solid line.
Test: (1, 1)
$1 \le -2(1)$
$1 \le -2$ False
Shade the half-plane that does not contain (1, 1).

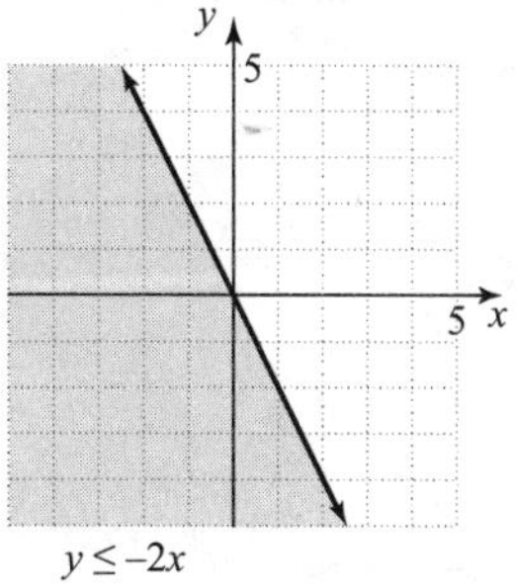

9. $2x + 4y \ge 8$

Graph $2x + 4y = 8$ as a solid line.
Test: (0, 0)
$2(0) + 4(0) \ge 8$
$0 \ge 8$ False
Shade the half-plane that does not contain (0, 0).

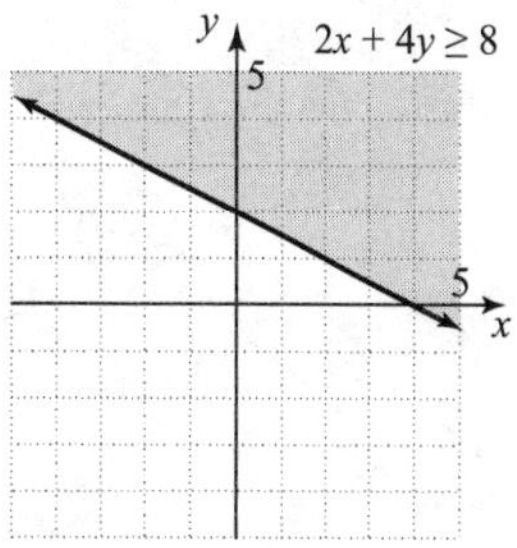

11. $5x + 3y > -15$

Graph $5x + 3y = -15$ as a dashed line.
Test: (0, 0)
$5(0) + 3(0) > -15$
$0 > -15$ True
Shade the half-plane that contains (0, 0).

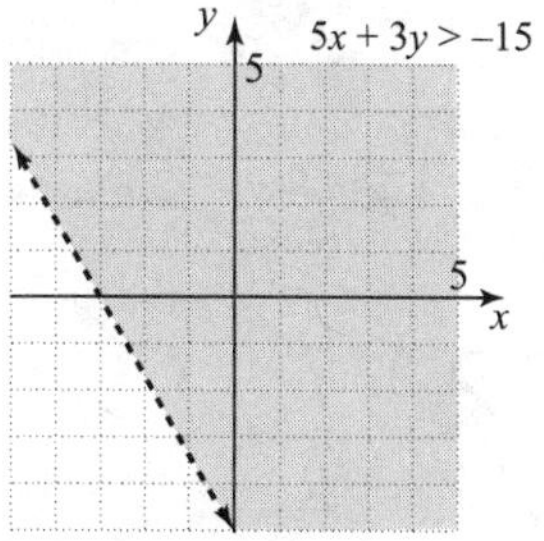

13. $x \ge 3$ and $y \le -2$

The intersection is shown by the darker shading in the graph.

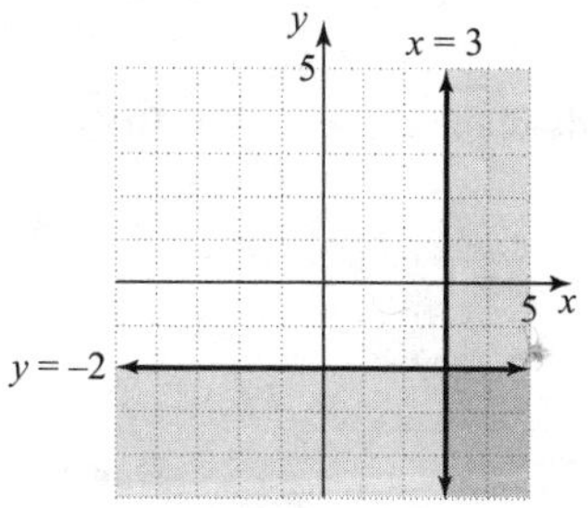

15. $x \le -2$ or $y \ge 4$

The union is both shaded regions, including the boundary line.

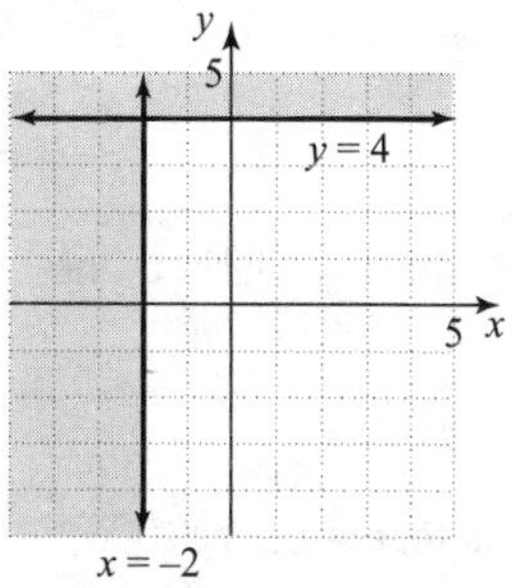

17. $x - y < 3$ and $x > 4$

The intersection is shown by the darker shading in the graph.

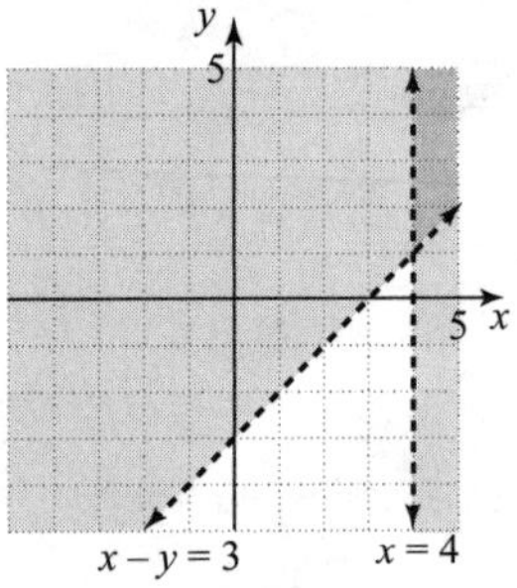

19. $x + y \le 3$ or $x - y \ge 5$

The union is both shaded regions, including the boundary lines.

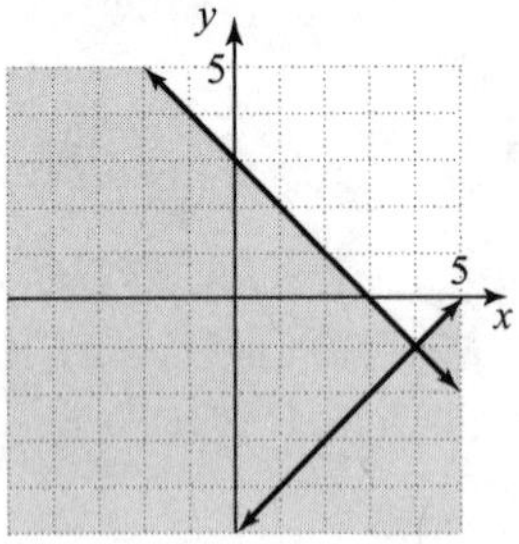

21. $y \ge -2$

Graph $y = -2$ as a solid line.
Test: (0, 0)
$0 \ge -2$ True
Shade the half-plane that contains (0, 0).

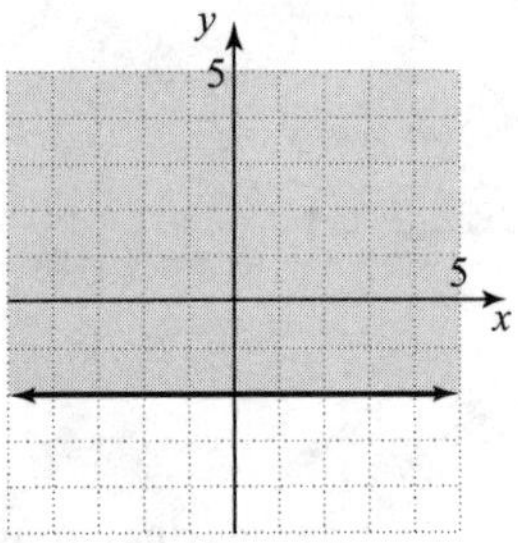

23. $x - 6y < 12$

$$-6y < 12 - x$$
$$y > \frac{1}{6}x - 2$$

Graph the boundary line as a dashed line.
Test: (0, 0)

$$0 > \frac{1}{6}(0) - 2 \quad \text{True}$$

Shade the half-plane that contains (0, 0).

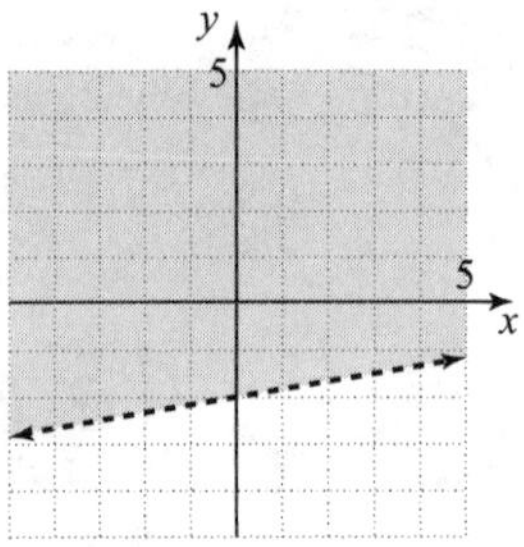

25. $x > 5$

Graph $x = 5$ as a dashed line.
Test: (0, 0)
$0 > 5$ False
Shade the half-plane that does not contain (0, 0).

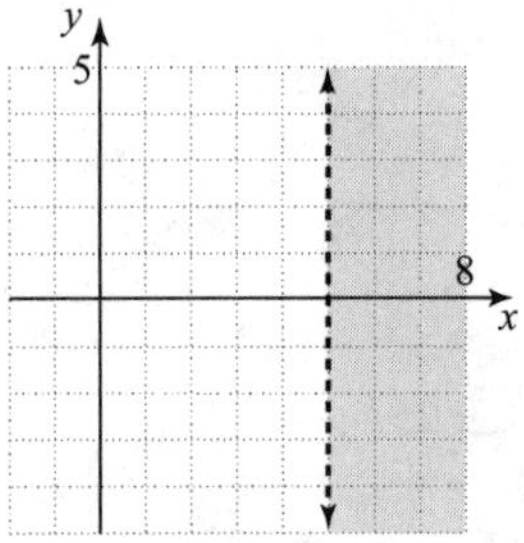

27. $-2x + y \le 4$

$$y \le 2x + 4$$

Graph the boundary line as a solid line.
Test: (0, 0)
$0 \le 2(0) + 4$ True
Shade the half-plane that contains (0, 0).

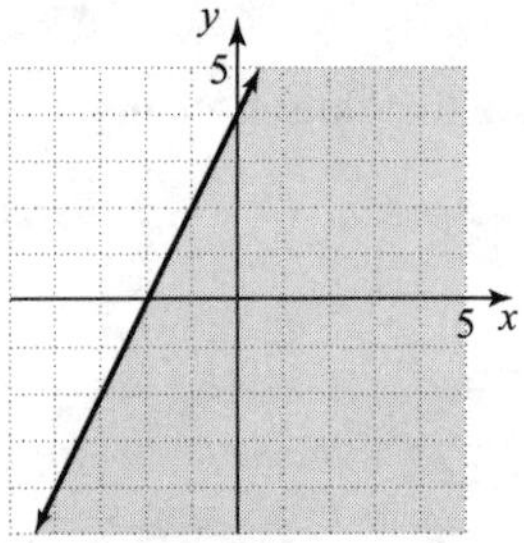

29. $x - 3y < 0$

$$-3y < -x$$
$$y > \frac{x}{3}$$

Graph the boundary line as a dashed line.
Test: (0, 1)

$$1 > \frac{0}{3} \quad \text{True}$$

Shade the half-plane that contains (0, 1).

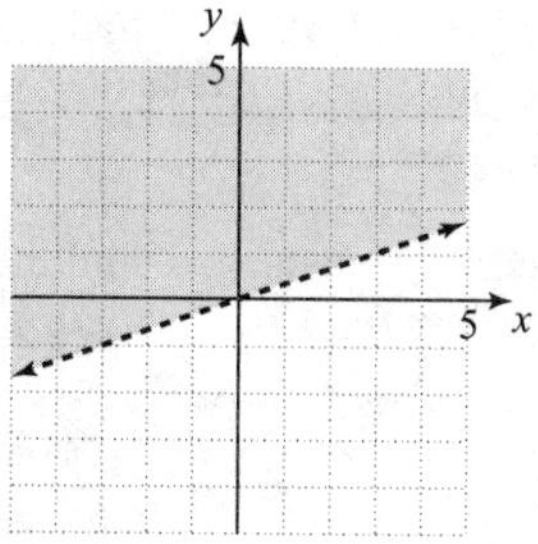

31. $3x - 2y \le 12$

$-2y \le -3x + 12$

$y \ge \frac{3}{2}x - 6$

Graph the boundary line as a solid line.
Test: (0, 0)
$3(0) - 2(0) \le 12$ True
Shade the half-plane that contains (0, 0).

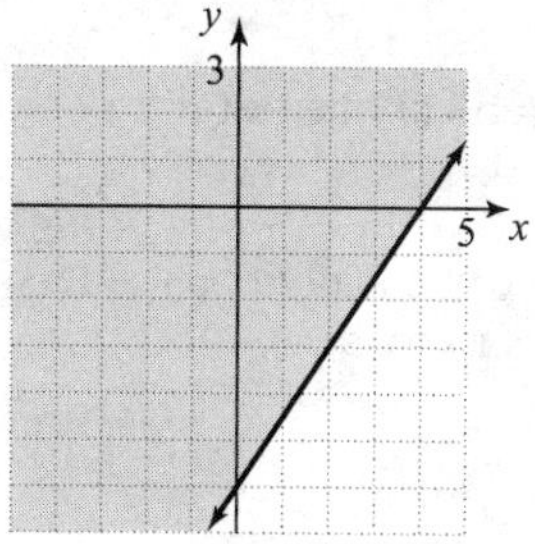

33. $x - y > 2$ or $y < 5$

$y < x - 2$ or $y < 5$

Graph each inequality. The union of the two inequalities is both shaded regions, as shown by the shading in the graph below.

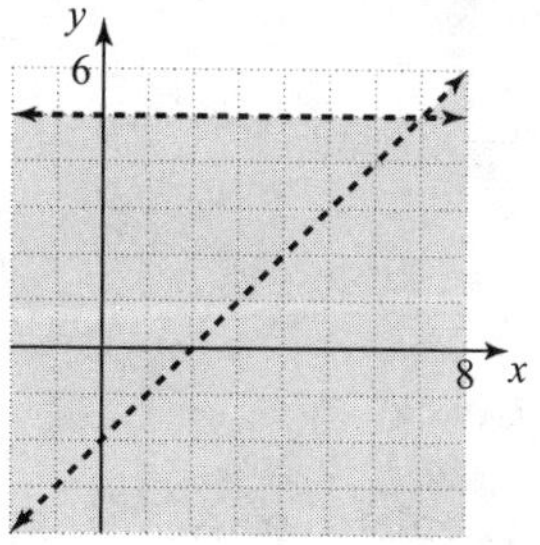

35. $x + y \le 1$ and $y \le -1$

$y \le -x + 1$ and $y \le -1$

Graph each inequality. The intersection of the two inequalities is all points common to both regions, as shown by the shading in the graph below.

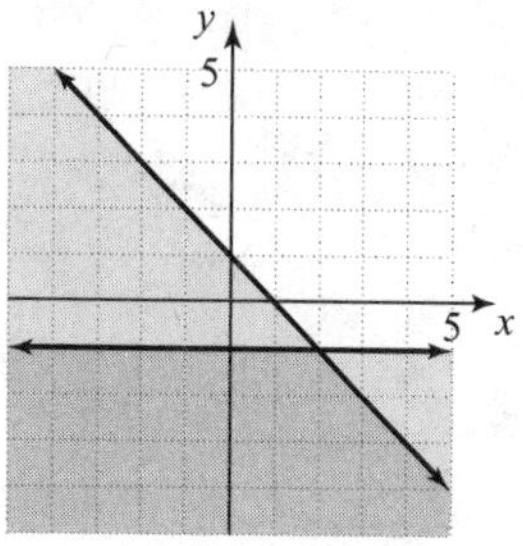

37. $2x + y > 4$ or $x \ge 1$

$y > -2x + 4$ or $x \ge 1$

Graph each inequality. The union of the two inequalities is both shaded regions, as shown by the shading in the graph below.

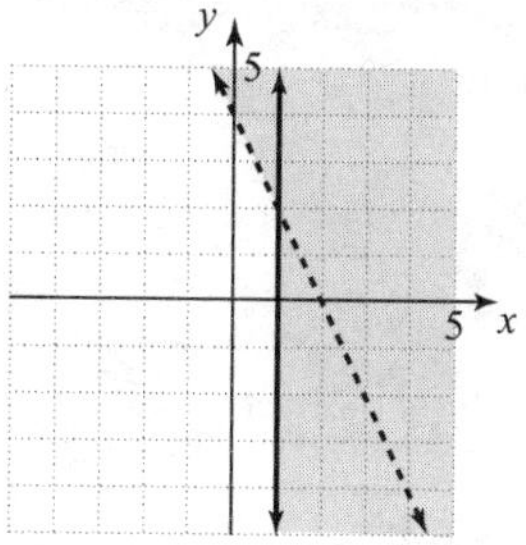

39. $x \ge -2$ and $x \le 1$

$-2 \le x \le 1$

Graph each inequality. The intersection of the two inequalities is all points common to both regions, as shown by the shading in the graph below.

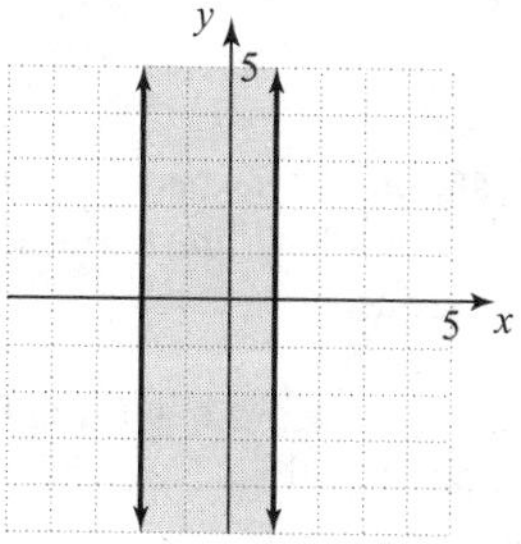

41. $x+y \le 0$ or $3x-6y \ge 12$

$y \le -x$ or $-6y \ge -3x+12$

$y \le -x$ or $y \le \dfrac{1}{2}x-2$

Graph each inequality. The union of the two inequalities is both shaded regions, as shown by the shading in the graph below.

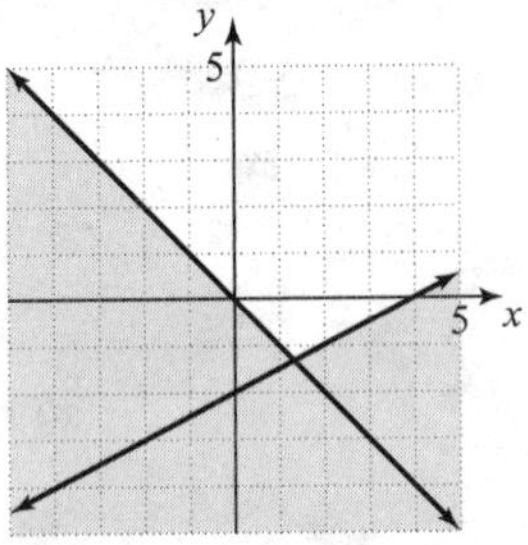

43. $2x-y>3$ and $x>0$

$y<2x-3$ and $x>0$

Graph each inequality. The intersection of the two inequalities is all points common to both regions, as shown by the shading in the graph below.

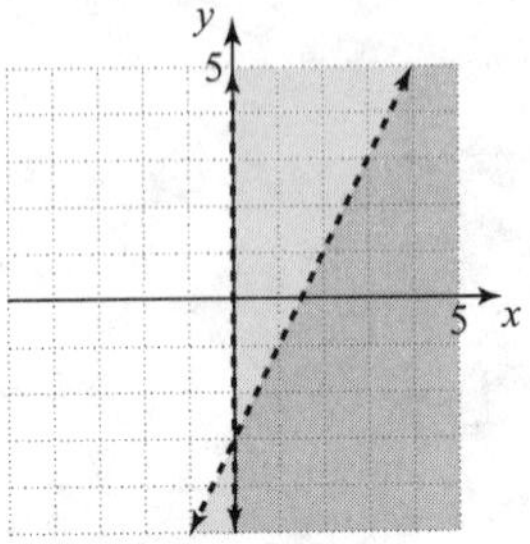

45. $y \le 2x+3$

The boundary line should be solid, and the half-plane below the boundary line should be shaded. The correct graph is D.

47. $y > 2x+3$

The boundary line should be dashed, and the half-plane above the boundary line should be shaded. The correct graph is A.

49. The boundary line, $x = 2$, is solid, and the half-plane with x-values greater than 2 is shaded. The inequality is $x \ge 2$.

51. The boundary line, $y = -3$, is solid, and the half-plane with y-values less than -3 is shaded. The inequality is $y \le -3$.

53. The boundary line, $y = 4$, is dashed, and the half-plane with y-values greater than 4 is shaded. The inequality is $y > 4$.

55. The boundary line, $x = 1$, is dashed, and the half-plane with x-values less than 1 is shaded. The inequality is $x < 1$.

57. $2^3 = 2 \cdot 2 \cdot 2 = 8$

59. $-5^2 = -(5 \cdot 5) = -25$

61. $(-2)^4 = (-2)(-2)(-2)(-2) = 16$

63. $\left(\dfrac{3}{5}\right)^3 = \dfrac{3^3}{5^3} = \dfrac{27}{125}$

65. Domain: [1, 5]
Range: [1, 3]
Since it fails the vertical line test it is not a function.

67. A dashed boundary line should be used when the inequality contains a < or >.

69. $\begin{cases} x \ge 0 \\ y \ge 0 \\ 2x+4y \le 40 \end{cases}$

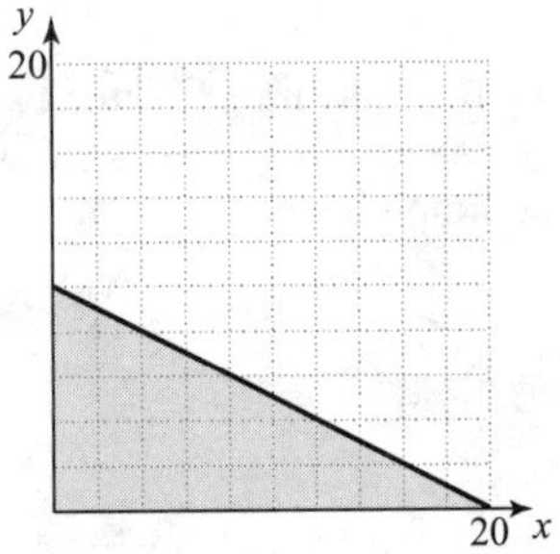

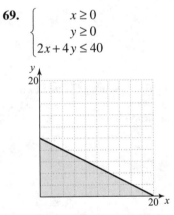

Chapter 3 Vocabulary Check

1. A <u>relation</u> is a set of ordered pairs.
2. The graph of every linear equation in two variables is a <u>line</u>.
3. The statement $-x + 2y > 0$ is called a <u>linear inequality</u> in two variables.
4. <u>Standard</u> form of linear equation in two variables is $Ax + By = C$.
5. The <u>range</u> of a relation is the set of all second components of the ordered pairs of the relation.

6. Parallel lines have the same slope and different y-intercepts.

7. Slope-intercept form of a linear equation in two variables is $y = mx + b$.

8. A function is a relation in which each first component in the ordered pairs corresponds to exactly one second component.

9. In the equation $y = 4x - 2$, the coefficient of x is the slope of its corresponding graph.

10. Two lines are perpendicular if the product of their slopes is -1.

11. To find the x-intercept of a linear equation, let $y = 0$ and solve for the other variable.

12. The domain of a relation is the set of all first components of the ordered pairs of the relation.

13. A linear function is a function that can be written in the form $f(x) = mx + b$.

14. To find the y-intercept of a linear equation, let $x = 0$ and solve for the other variable.

15. The equation $y - 8 = -5(x + 1)$ is written in point-slope form.

Chapter 3 Review

1. $A(2, -1)$, quadrant IV
$B(-2, 1)$, quadrant II
$C(0, 3)$, y-axis
$D(-3, -5)$, quadrant III

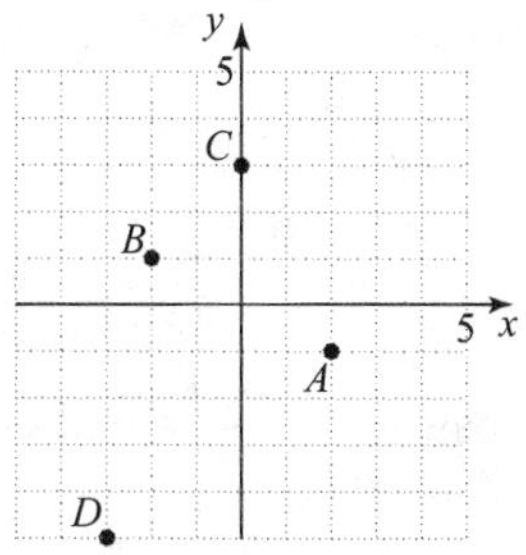

2. $A(-3, 4)$, quadrant II
$B(4, -3)$, quadrant IV
$C(-2, 0)$, x-axis
$D(-4, 1)$, quadrant II

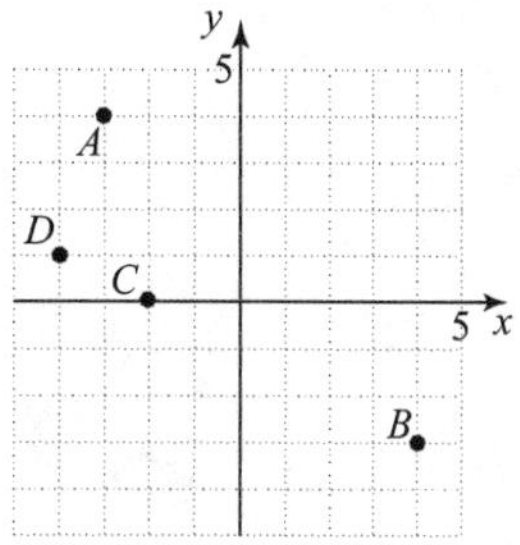

3. $7x - 8y = 56$
(0, 56); No
$7(0) - 8(56) \stackrel{?}{=} 56$
$-448 = 56$, False
(8, 0); Yes
$7(8) - 8(0) \stackrel{?}{=} 56$
$56 = 56$, True

4. $-2x + 5y = 10$
(−5, 0); Yes
$-2(-5) + 5(0) \stackrel{?}{=} 10$
$10 = 10$, True
(1, 1), No
$-2(1) + 5(1) \stackrel{?}{=} 10$
$3 = 10$, False

5. $x = 13$
(13, 5); Yes
$13 = 13$, True
(13, 13); Yes
$13 = 13$, True

6. $y = 2$
(7, 2); Yes
$2 = 2$, True
(2, 7); No
$7 = 2$, False

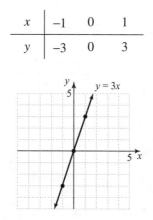

7. $y = 3x$; Linear

x	−1	0	1
y	−3	0	3

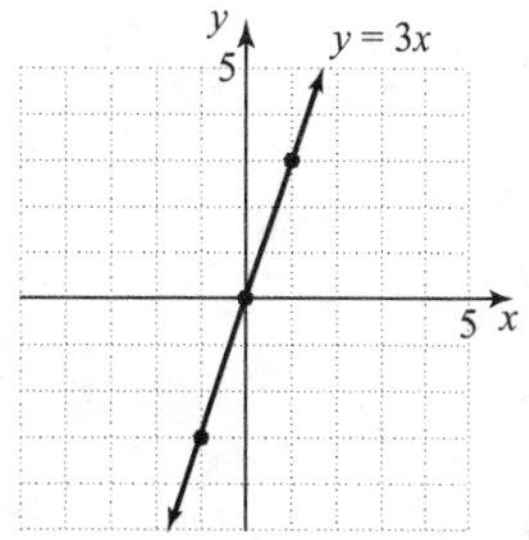

8. $y = 5x$; Linear

x	-1	0	1
y	-5	0	5

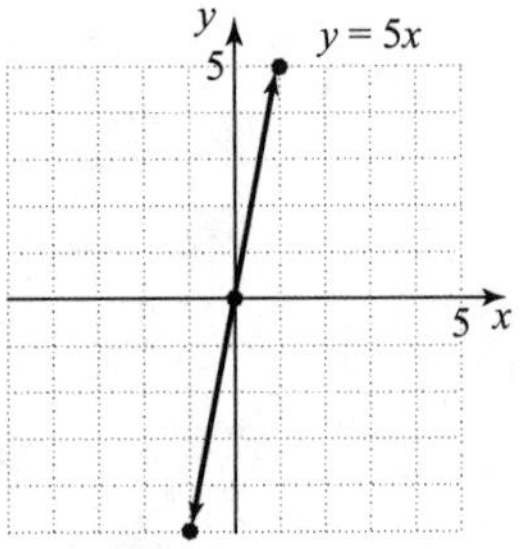

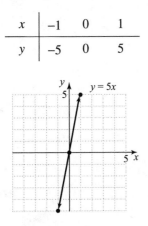

9. $3x - y = 4$; Linear
Find three ordered pair solutions, or find x- and y-intercepts, or find m and b.

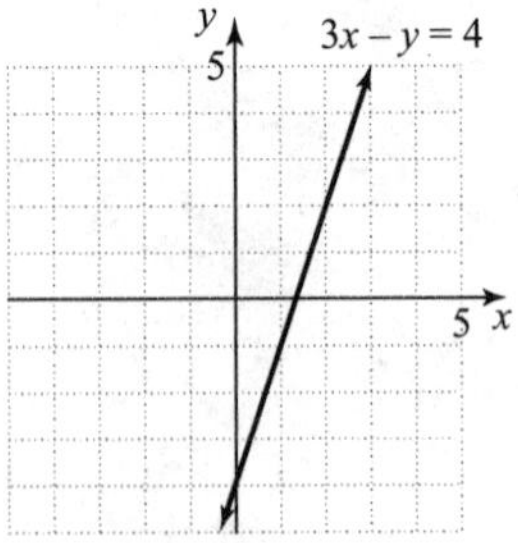

10. $x - 3y = 2$; Linear
Find three ordered pair solutions, or find x- and y-intercepts, or find m and b.

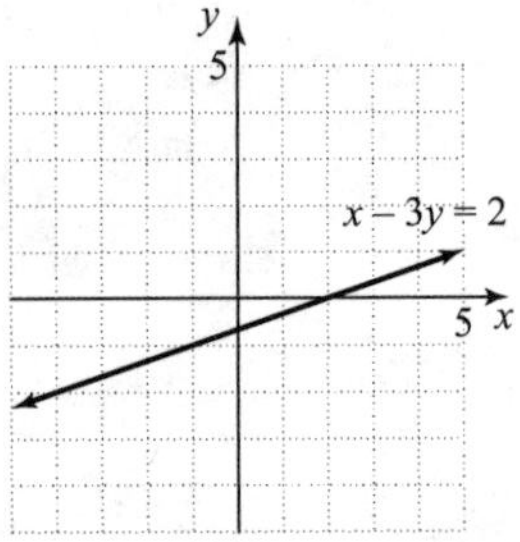

11. $y = |x| + 4$; Nonlinear

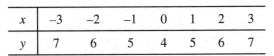

x	-3	-2	-1	0	1	2	3
y	7	6	5	4	5	6	7

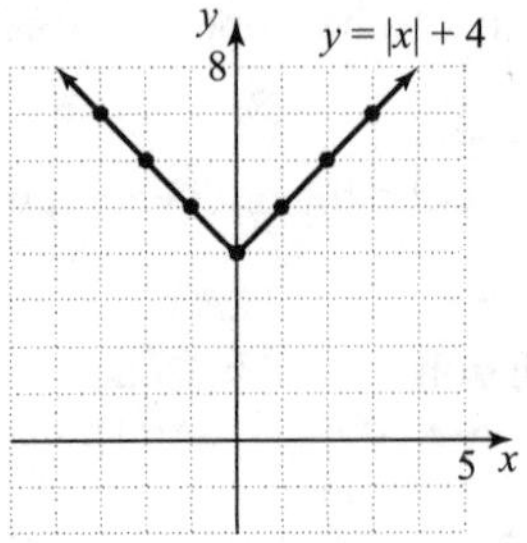

12. $y = x^2 + 4$; Nonlinear

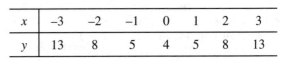

x	-3	-2	-1	0	1	2	3
y	13	8	5	4	5	8	13

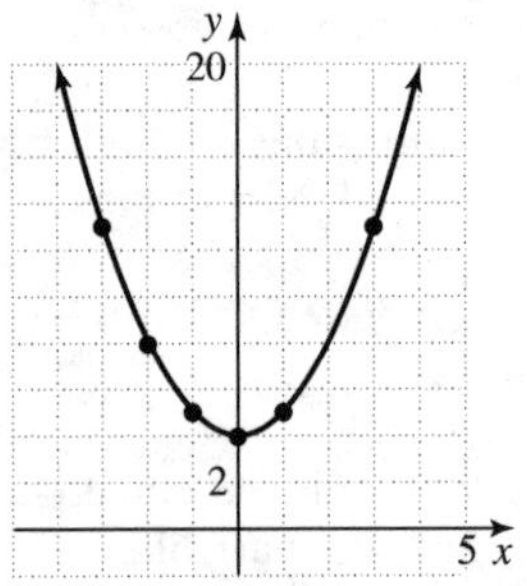

13. $y = -\frac{1}{2}x + 2$; Linear

Find three ordered pair solutions, or find x- and y-intercepts, or find m and b.

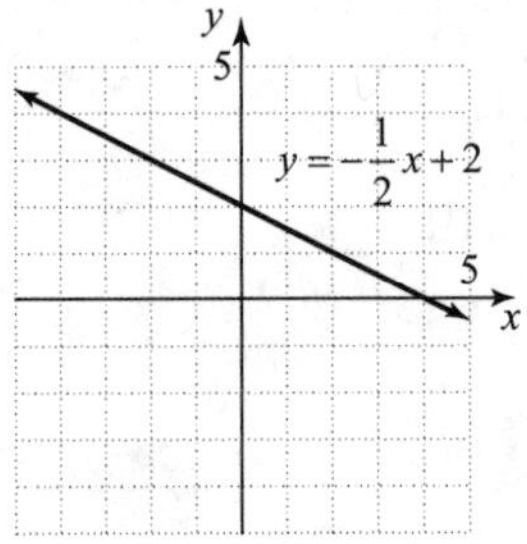

14. $y = -x + 5$; Linear
Find three ordered pair solutions, or find x- and y-intercepts, or find m and b.

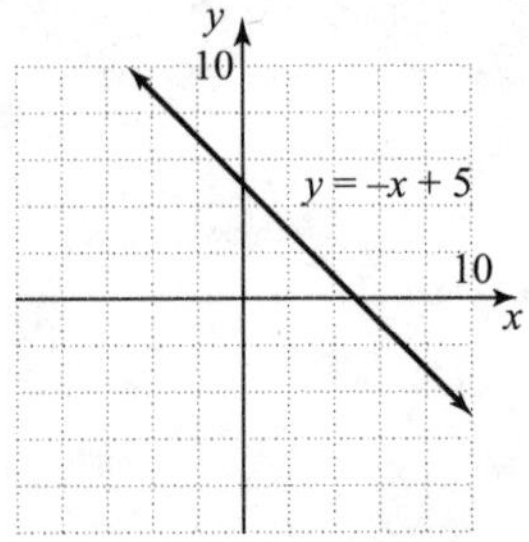

15. $y = 2x - 1$; Linear
Find three ordered pair solutions, or find x- and y-intercepts, or find m and b.

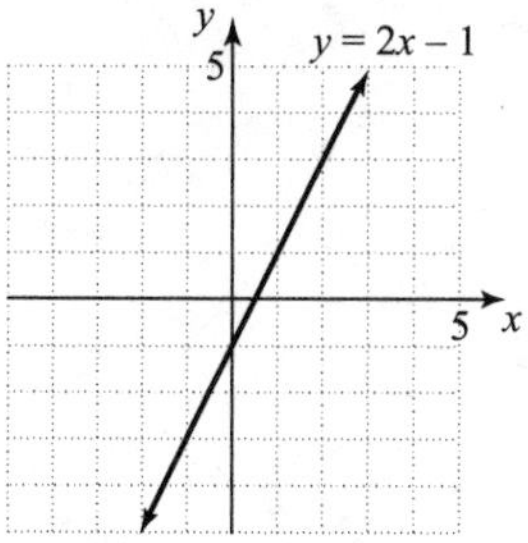

16. $y = \frac{1}{3}x + 1$; Linear

Find three ordered pair solutions, or find x- and y-intercepts, or find m and b.

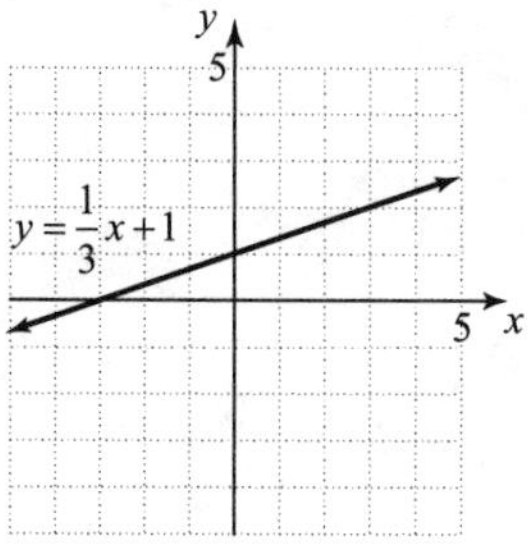

17. $y = -1.36x$; Linear
Find three ordered pair solutions, or find x- and y-intercepts, or find m and b.

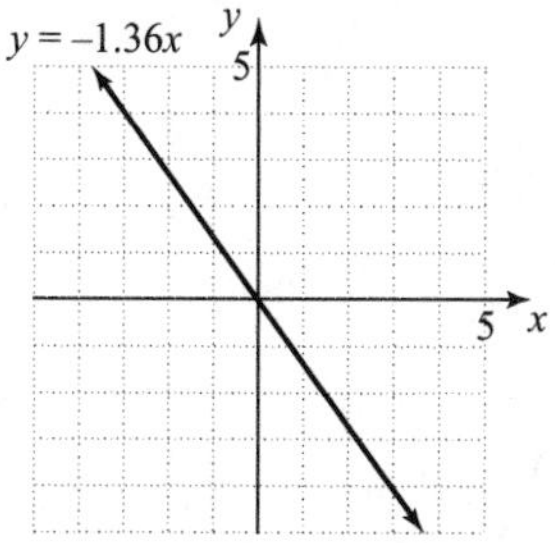

18. $y = 2.1x + 5.9$

Find three ordered pair solutions, or find x- and y-intercepts, or find m and b.

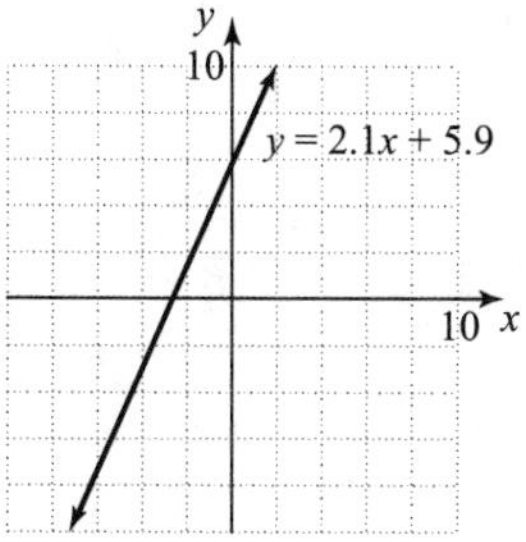

19. The domain is the set of all first coordinates (or inputs) and the range is the set of all second coordinates (or outputs).

Domain: $\left\{-\frac{1}{2}, 6, 0, 25\right\}$

Range: $\left\{\frac{3}{4}, -12, 25\right\}$

Function since each x-value corresponds to exactly one y-value.

20. The domain is the set of all first coordinates (or inputs) and the range is the set of all second coordinates (or outputs).

Domain: $\left\{\frac{3}{4}, -12, 25\right\}$

Range: $\left\{-\frac{1}{2}, 6, 0, 25\right\}$

Not a function since $\frac{3}{4}$ (or 0.75) is paired with both $-\frac{1}{2}$ and 6.

21. The domain is the set of all first coordinates (or inputs) and the range is the set of all second coordinates (or outputs).
Domain: {2, 4, 6, 8}
Range: {2, 4, 5, 6}
Not a function since 2 is paired with both 2 and 4.

22. The domain is the set of all first coordinates (or inputs) and the range is the set of all second coordinates (or outputs).
Domain:
{Triangle, Square, Rectangle, Parallelogram}
Range: {3, 4}
Function since each input is paired with exactly one output.

23. Domain: $(-\infty, \infty)$

Range: $(-\infty, -1] \cup [1, \infty)$

Not a function since it fails the vertical line test.

24. Domain: $\{-3\}$
Range: $(-\infty, \infty)$

Not a function since it fails the vertical line test.

25. Domain: $(-\infty, \infty)$

Range: $\{4\}$
Function since it passes the vertical line test.

26. Domain: [–1, 1]
Range: [–1, 1]
Not a function since it fails the vertical line test.

27. $f(x) = x - 5$
$f(2) = (2) - 5 = -3$

28. $g(x) = -3x$
$g(0) = -3(0) = 0$

29. $g(x) = -3x$
$g(-6) = -3(-6) = 18$

30. $h(x) = 2x^2 - 6x + 1$
$h(-1) = 2(-1)^2 - 6(-1) + 1$
$= 2(1) + 6 + 1$
$= 9$

31. $h(x) = 2x^2 - 6x + 1$
$h(1) = 2(1)^2 - 6(1) + 1 = 2 - 6 + 1 = -3$

32. $f(x) = x - 5$
$f(5) = (5) - 5 = 0$

33. $J(x) = 2.54x$
$J(150) = 2.54(150) = 381$ pounds

34. $J(x) = 2.54x$
$J(2000) = 2.54(2000) = 5080$ pounds

35. The point (–1, 0) is on the graph, so $f(-1) = 0$.

36. The point (1, –2) is on the graph, so $f(1) = -2$.

37. $f(x) = 1$
$f(-2) = f(4) = 1$
$x = -2, 4$

38. $f(x) = -1$
$f(0) = f(2) = -1$
$x = 0, 2$

39. $f(x) = \frac{1}{5}x$ or $y = \frac{1}{5}x$
$m = \frac{1}{5}$, $b = 0$

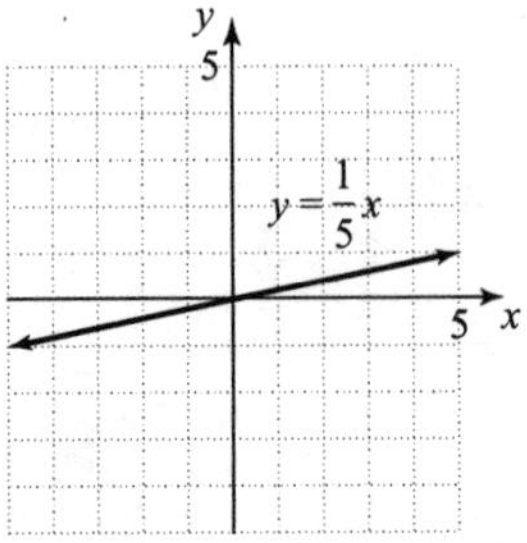

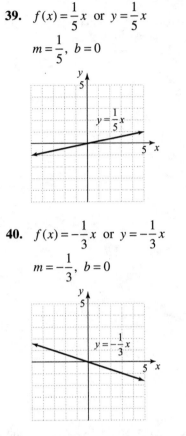

40. $f(x) = -\frac{1}{3}x$ or $y = -\frac{1}{3}x$
$m = -\frac{1}{3}$, $b = 0$

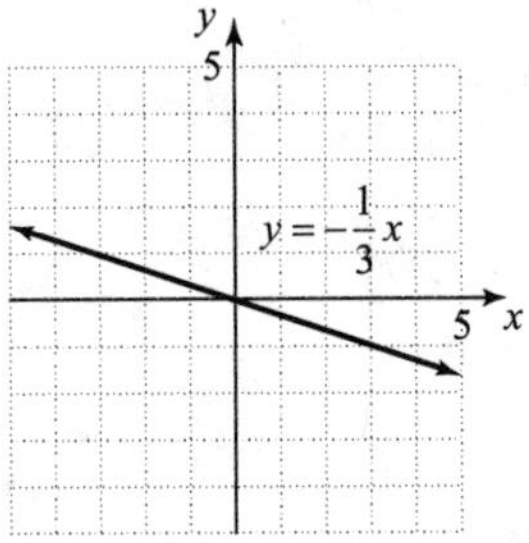

41. $g(x) = -2x + 3$ or $y = -2x + 3$
$m = -2$, $b = 3$

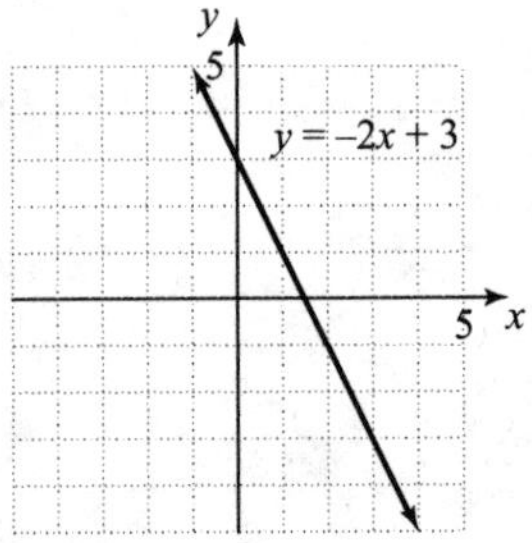

42. $g(x) = 4x - 1$ or $y = 4x - 1$
$m = 4$, $b = -1$

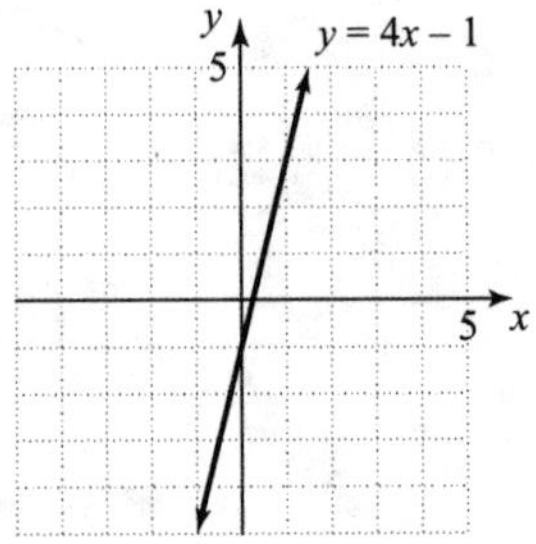

43. $f(x) = 3x + 1$

The y-intercept should be (0, 1). The correct graph is C.

44. $f(x) = 3x - 2$

The y-intercept should be (0, −2). The correct graph is A.

45. $f(x) = 3x + 2$

The y-intercept should be (0, 2). The correct graph is B.

46. $f(x) = 3x - 5$

The y-intercept should be (0, −5). The correct graph is D.

47. $4x + 5y = 20$

Let $x = 0$	Let $y = 0$
$4(0) + 5y = 20$	$4x + 5(0) = 20$
$y = 4$	$x = 5$
$(0, 4)$	$(5, 0)$

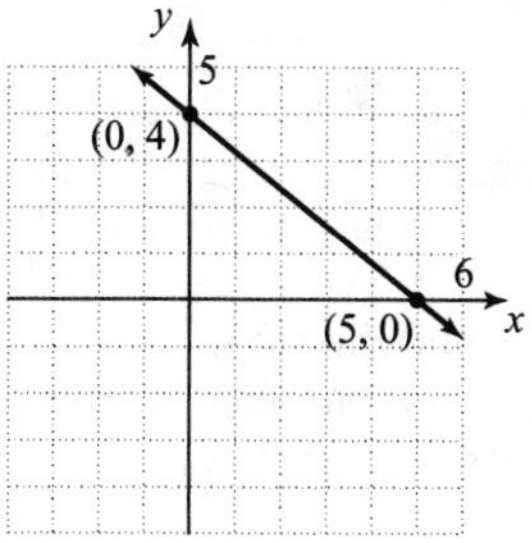

48. $3x - 2y = -9$

Let $x = 0$	Let $y = 0$
$3(0) - 2y = -9$	$3x - 2(0) = -9$
$y = \frac{9}{2}$	$x = -3$
$\left(0, \frac{9}{2}\right)$	$(-3, 0)$

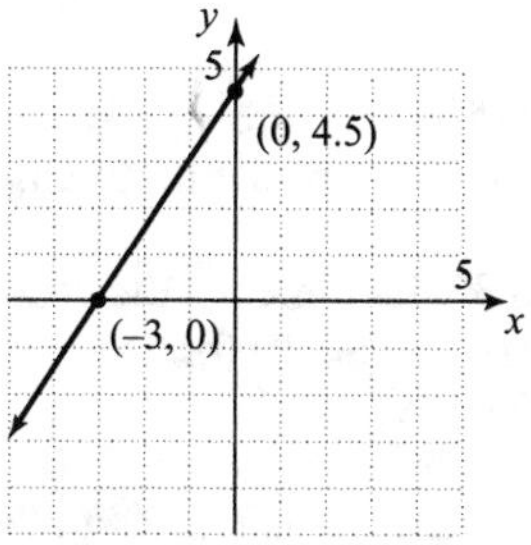

49. $4x - y = 3$

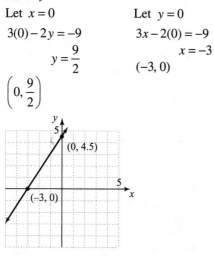

Let $x = 0$	Let $y = 0$
$4(0) - y = 3$	$4x - (0) = 3$
$y = -3$	$x = \frac{3}{4}$
$(0, -3)$	$\left(\frac{3}{4}, 0\right)$

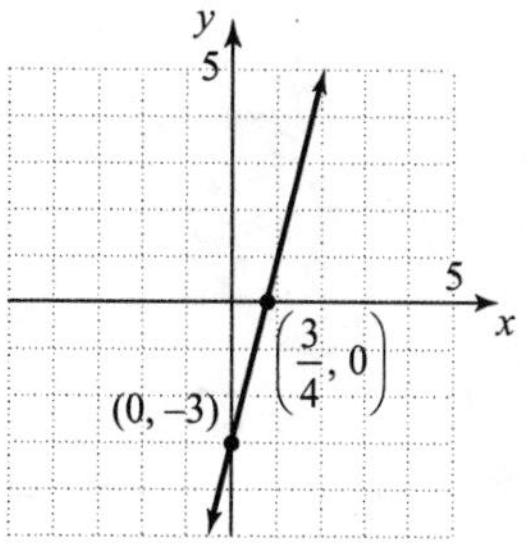

50. $2x + 6y = 9$

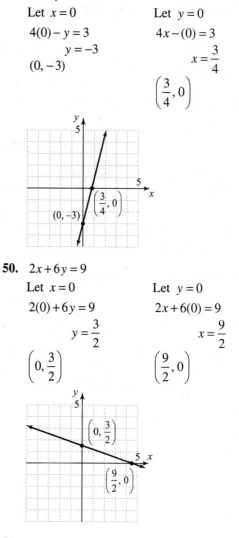

Let $x = 0$	Let $y = 0$
$2(0) + 6y = 9$	$2x + 6(0) = 9$
$y = \frac{3}{2}$	$x = \frac{9}{2}$
$\left(0, \frac{3}{2}\right)$	$\left(\frac{9}{2}, 0\right)$

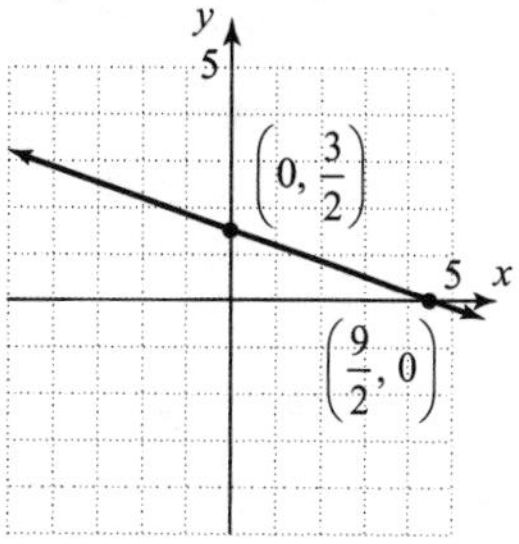

51. $y = 5$

Horizontal line with y-intercept 5.

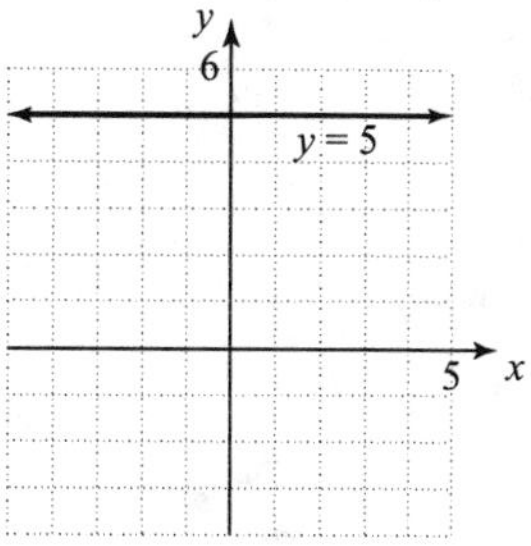

52. $x = -2$
Vertical line with x-intercept -2.

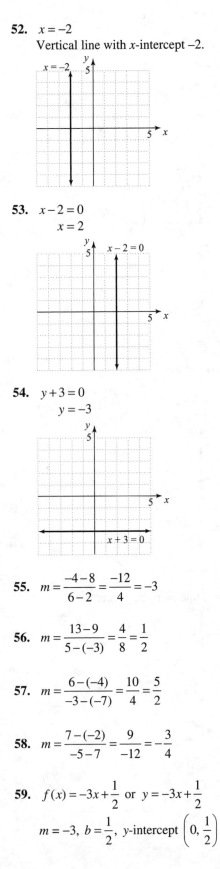

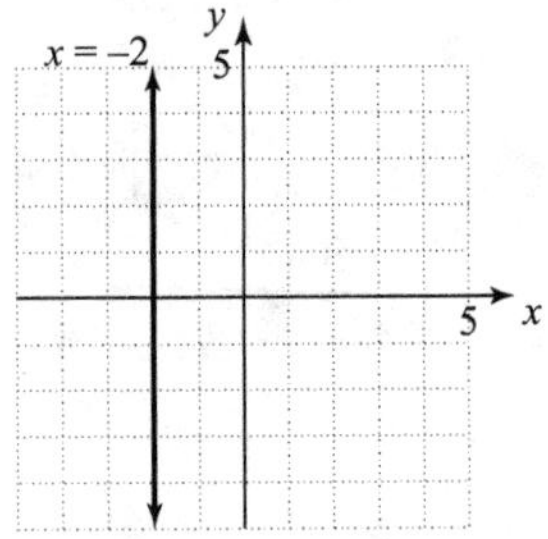

53. $x - 2 = 0$
$x = 2$

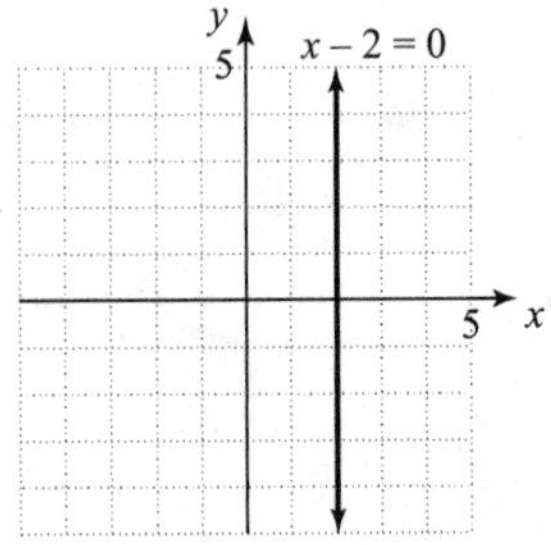

54. $y + 3 = 0$
$y = -3$

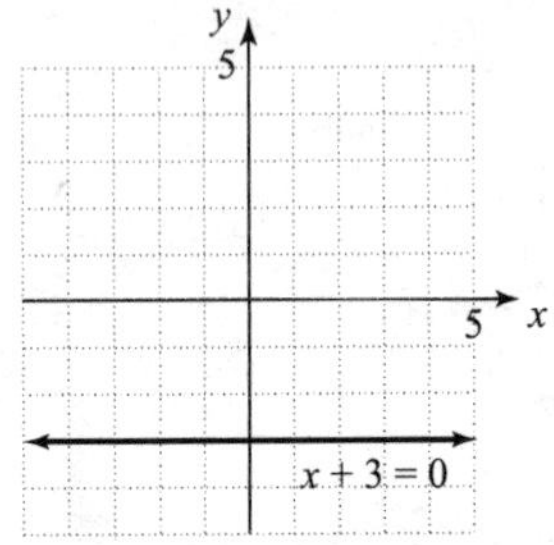

55. $m = \dfrac{-4-8}{6-2} = \dfrac{-12}{4} = -3$

56. $m = \dfrac{13-9}{5-(-3)} = \dfrac{4}{8} = \dfrac{1}{2}$

57. $m = \dfrac{6-(-4)}{-3-(-7)} = \dfrac{10}{4} = \dfrac{5}{2}$

58. $m = \dfrac{7-(-2)}{-5-7} = \dfrac{9}{-12} = -\dfrac{3}{4}$

59. $f(x) = -3x + \dfrac{1}{2}$ or $y = -3x + \dfrac{1}{2}$

$m = -3,\ b = \dfrac{1}{2},$ y-intercept $\left(0, \dfrac{1}{2}\right)$

60. $g(x) = 2x + 4$ or $y = 2x + 4$
$m = 2,\ b = 4,$ y-intercept $(0, 4)$

61. $6x - 15y = 20$

$$-15y = -6x + 20$$
$$y = \frac{2}{5}x - \frac{4}{3}$$

$m = \dfrac{2}{5},\ b = -\dfrac{4}{3},$ y-intercept $\left(0, -\dfrac{4}{3}\right)$

62. $4x + 14y = 21$

$$14y = -4x + 21$$
$$y = -\frac{2}{7}x + \frac{3}{2}$$

$m = -\dfrac{2}{7},\ b = \dfrac{3}{2},$ y-intercept $\left(0, \dfrac{3}{2}\right)$

63. $y - 3 = 0$
$y = 3;$ Slope $= 0$

64. $x = -5$; Vertical line
Slope is undefined.

65. The slope of l_1 is negative, and the slope of l_2 is positive. Since a positive number is greater than any negative number, l_2 has the greater slope.

66. The slope of l_1 is 0, and the slope of l_2 is positive. Since a positive number is greater than 0, l_2 has the greater slope.

67. The slope of l_1 and the slope of l_2 are both positive. Since l_2 is steeper, it has the greater slope.

68. The slope of l_1 is 0, and the slope of l_2 is negative. Since a negative number is less than 0, l_1 has the greater slope.

69. $C = 0.3x + 42$

a. Let $x = 150$.
$C(150) = 0.3(150) + 42 = 87$
It would cost \$87 to rent the minivan.

b. The slope is 0.3 which means the cost of renting increases by \$0.30 for each additional mile driven.

c. The y-intercept is $(0, 42)$. The cost of renting for 0 miles driven is \$42.

70. $C = 0.4x + 19$

a. Let $x = 325$.
$C = 0.4(325) + 19 = 149$
It will cost \$149 to rent the car.

b. The slope is 0.4; the cost increases by \$0.40 for each additional mile driven.

c. The y-intercept is (0, 19); the cost for 0 miles driven is \$19.

71. $f(x) = -2x + 6$ $\quad$ $g(x) = 2x - 1$
$m = -2$ $\quad$ $m = 2$
Neither; The slopes are not the same and their product is not −1.

72. $y = \frac{3}{4}x + 1$ $\quad$ $y = -\frac{4}{3}x + 1$
$m = \frac{3}{4}$ $\quad$ $m = -\frac{4}{3}$
The lines are perpendicular since the product of their slopes is −1.

73. $-x + 3y = 2$ $\quad$ $6x - 18y = 3$
$y = \frac{1}{3}x + \frac{2}{3}$ $\quad$ $y = \frac{1}{3}x - \frac{1}{6}$
$m = \frac{1}{3}$ $\quad$ $m = \frac{1}{3}$
Parallel, since their slopes are equal.

74. $x - 2y = 6$ $\quad$ $4x + y = 8$
$x - 6 = 2y$ $\quad$ $y = -4x + 8$
$\frac{1}{2}x - 3 = y$
$m = \frac{1}{2}$ $\quad$ $m = -4$
Neither, since their slopes are not equal, nor does their product equal −1.

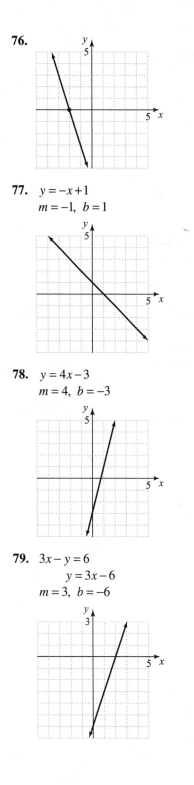

75.

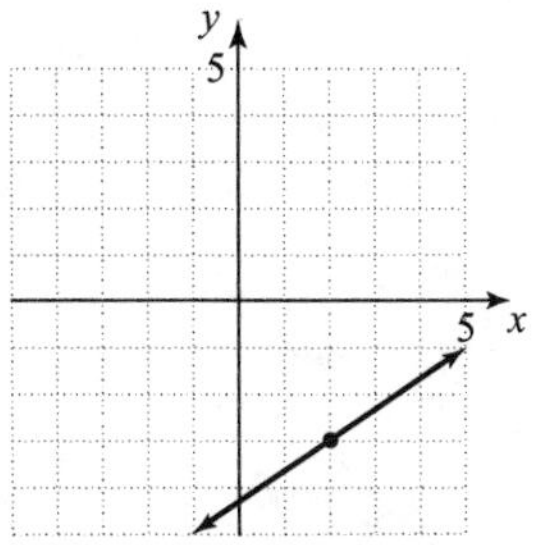

76.

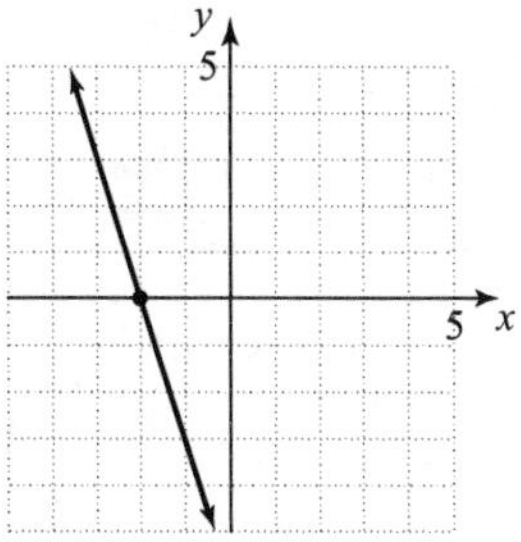

77. $y = -x + 1$
$m = -1,\ b = 1$

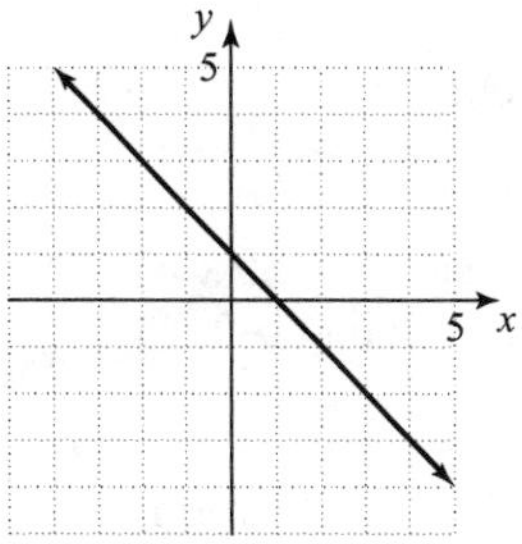

78. $y = 4x - 3$
$m = 4,\ b = -3$

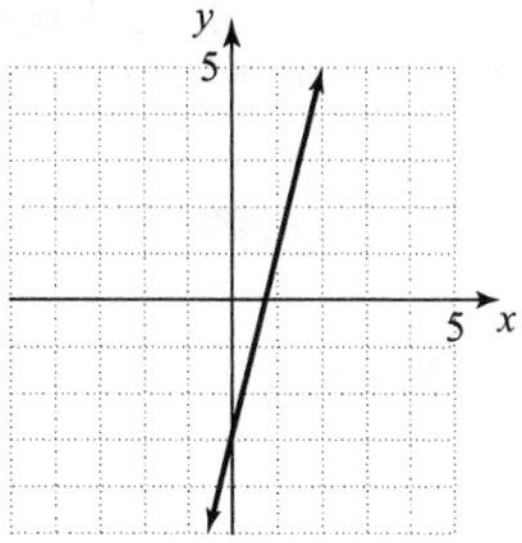

79. $3x - y = 6$
$y = 3x - 6$
$m = 3,\ b = -6$

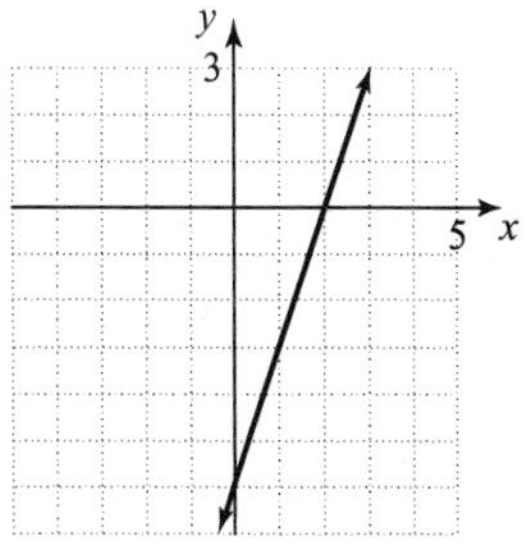

80. $y = -5x$
$m = -5,\ b = 0$

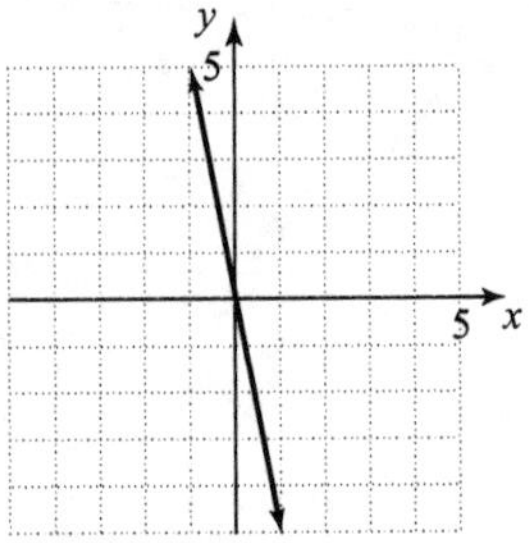

81. Every horizontal line is in the form $y = c$. Since the line passes through the point (3, −1), its equation is $y = -1$.

82. Every vertical line has the form $x = c$. Since the line passes through the point (−2, −4), its equation is $x = -2$.

83. A line parallel to $x = 6$ has the form $x = c$. Since the line passes through (−4, −3), its equation is $x = -4$.

84. Lines with slope 0 are horizontal, and have the form $y = c$. Since it passes through (2, 5), its equation is $y = 5$.

85.
$$\begin{aligned} y - y_1 &= m(x - x_1) \\ y - 5 &= 3[x - (-3)] \\ y - 5 &= 3(x + 3) \\ y - 5 &= 3x + 9 \\ 3x - y &= -14 \end{aligned}$$

86.
$$\begin{aligned} y - y_1 &= m(x - x_1) \\ y - (-2) &= 2(x - 5) \\ y + 2 &= 2x - 10 \\ 2x - y &= 12 \end{aligned}$$

87. $m = \dfrac{-2-(-1)}{-4-(-6)} = \dfrac{-1}{2} = -\dfrac{1}{2}$
$$\begin{aligned} y - y_1 &= m(x - x_1) \\ y - (-1) &= -\frac{1}{2}[x - (-6)] \\ 2(y + 1) &= -(x + 6) \\ 2y + 2 &= -x - 6 \\ x + 2y &= -8 \end{aligned}$$

88. $m = \dfrac{-8-3}{-4-(-5)} = \dfrac{-11}{1} = -11$
$$\begin{aligned} y - y_1 &= m(x - x_1) \\ y - 3 &= -11[x - (-5)] \\ y - 3 &= -11(x + 5) \\ y - 3 &= -11x - 55 \\ 11x + y &= -52 \end{aligned}$$

89. $x = 4$ has undefined slope.
A line perpendicular to $x = 4$ has slope = 0 and is therefore horizontal.
$y = 3$

90. $y = 8$ has slope = 0
A line parallel to $y = 8$ has slope = 0.
$y = -5$

91.
$$\begin{aligned} y &= mx + b \\ y &= -\frac{2}{3}x + 4 \\ f(x) &= -\frac{2}{3}x + 4 \end{aligned}$$

92.
$$\begin{aligned} y &= mx + b \\ y &= -x - 2 \\ f(x) &= -x - 2 \end{aligned}$$

93. $6x + 3y = 5$
$$\begin{aligned} 3y &= -6x + 5 \\ y &= -2x + \frac{5}{3} \quad \text{so } m = -2 \end{aligned}$$
$$\begin{aligned} y - y_1 &= m(x - x_1) \\ y - (-6) &= -2(x - 2) \\ y + 6 &= -2x + 4 \\ y &= -2x - 2 \\ f(x) &= -2x - 2 \end{aligned}$$

94. $3x+2y=8$

$$2y=-3x+8$$
$$y=-\frac{3}{2}x+4 \text{ so } m=-\frac{3}{2}$$
$$y-y_1=m(x-x_1)$$
$$y-(-2)=-\frac{3}{2}[x-(-4)]$$
$$2(y+2)=-3(x+4)$$
$$2y+4=-3x-12$$
$$2y=-3x-16$$
$$y=-\frac{3}{2}x-8$$
$$f(x)=-\frac{3}{2}x-8$$

95. $4x+3y=5$

$$3y=-4x+5$$
$$y=-\frac{4}{3}x+\frac{5}{3}$$
$$\text{so } m=-\frac{4}{3} \text{ and } m_{\perp}=\frac{3}{4}$$
$$y-y_1=m(x-x_1)$$
$$y-(-1)=\frac{3}{4}[x-(-6)]$$
$$4(y+1)=3(x+6)$$
$$4y+4=3x+18$$
$$4y=3x+14$$
$$y=\frac{3}{4}x+\frac{7}{2}$$
$$f(x)=\frac{3}{4}x+\frac{7}{2}$$

96. $2x-3y=6$

$$-3y=-2x+6$$
$$y=\frac{2}{3}x-2$$
$$\text{so } m=\frac{2}{3} \text{ and } m_{\perp}=-\frac{3}{2}$$
$$y-y_1=m(x-x_1)$$
$$y-5=-\frac{3}{2}[x-(-4)]$$
$$2(y-5)=-3(x+4)$$
$$2y-10=-3x-12$$
$$2y=-3x-2$$
$$y=-\frac{3}{2}x-1$$
$$f(x)=-\frac{3}{2}x-1$$

97. a. Use the points (7, 210,000) and (12, 270,000).

$$m=\frac{270{,}000-210{,}000}{12-7}=\frac{60{,}000}{5}=12{,}000$$
$$y-y_1=m(x-x_1)$$
$$y-210{,}000=12{,}000(x-7)$$
$$y-210{,}000=12{,}000x-84{,}000$$
$$y=12{,}000x+126{,}000$$

b. Let $x = 18$ since 2018 is 18 years after 2000.

$$y=12{,}000x+126{,}000$$
$$y=12{,}000(18)+126{,}000$$
$$=216{,}000+126{,}000$$
$$=342{,}000$$

In 2018, the value of the building is estimated to be $342,000.

98. a. Use the points (2, 20,600) and (4, 14,600).

$$m=\frac{14{,}600-20{,}600}{4-2}=\frac{-6000}{2}=-3000$$
$$y-y_1=m(x-x_1)$$
$$y-20{,}600=-3000(x-2)$$
$$y-20{,}600=-3000x+6000$$
$$y=-3000x+26{,}600$$

b. Let $x = 2018 - 2012 = 6$.

$y = -3000(6) + 26{,}600 = 8600$

the value of the car in 2018 is estimated to be $8600.

99. $f(x)=\begin{cases}-3x & \text{if } x<0\\ x-3 & \text{if } x\ge 0\end{cases}$

For $x < 0$:

x	$f(x)$
−3	9
−2	6
−1	3

For $x \ge 0$:

x	$f(x)$
0	−3
1	−2
2	−1

Graph a closed circle at (0, −3). Graph an open circle at (0, 0), which is found by substituting 0 for x in $f(x)=-3x$.

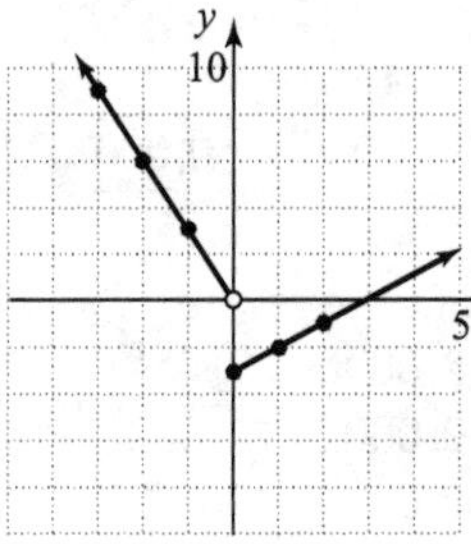

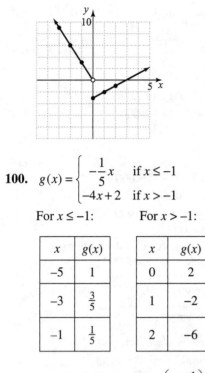

100. $g(x)=\begin{cases}-\frac{1}{5}x & \text{if } x\le -1\\ -4x+2 & \text{if } x>-1\end{cases}$

For $x \le -1$:

x	$g(x)$
–5	1
–3	$\frac{3}{5}$
–1	$\frac{1}{5}$

For $x > -1$:

x	$g(x)$
0	2
1	–2
2	–6

Graph a closed circle at $\left(-1, \frac{1}{5}\right)$. Graph an open circle at (–1, 6), which is found by substituting –1 for x in $g(x) = -4x + 2$.

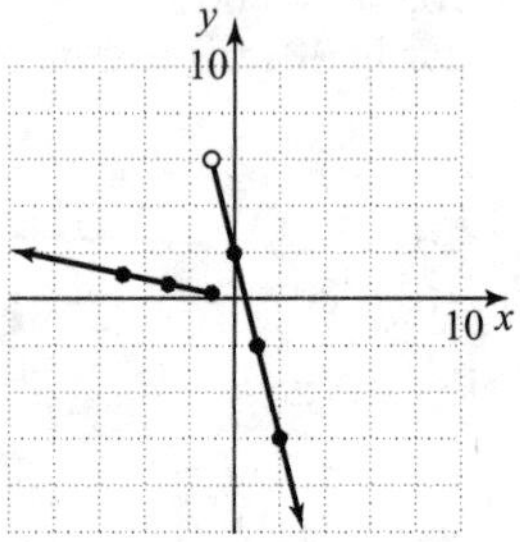

101. $y = \sqrt{x} - 4$

The graph of $f(x) = \sqrt{x} - 4$ is the same as the graph of $y = \sqrt{x}$ shifted down 4 units.

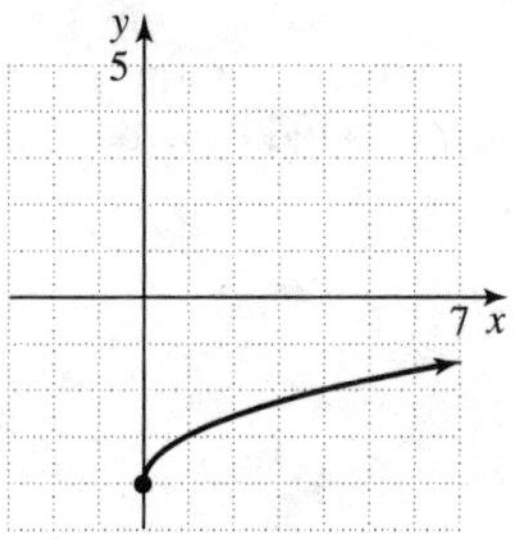

102. $f(x) = \sqrt{x-4}$

The graph of $f(x) = \sqrt{x-4}$ is the same as the graph of $y = \sqrt{x}$ shifted right 4 units.

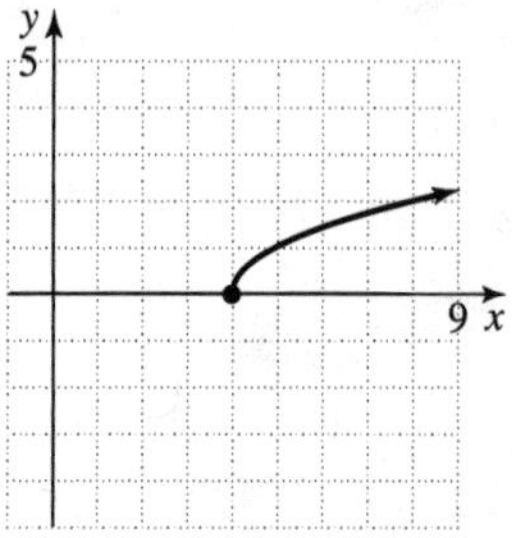

103. $g(x) = |x-2| - 2$

The graph of $g(x) = |x-2| - 2$ is the same as the graph of $y = |x|$ shifted right 2 units and down 2 units.

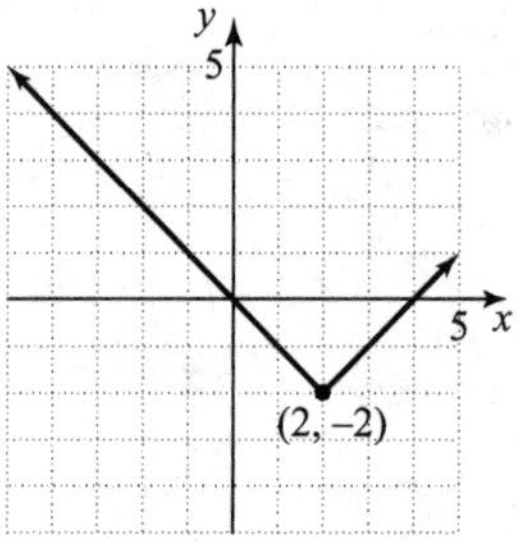

104. $h(x) = -(x+3)^2 - 1$

The graph of $h(x) = -(x+3)^2 - 1$ is the same as the graph of $y = x^2$ reflected about the x-axis and then shifted left 3 units and down 1 unit.

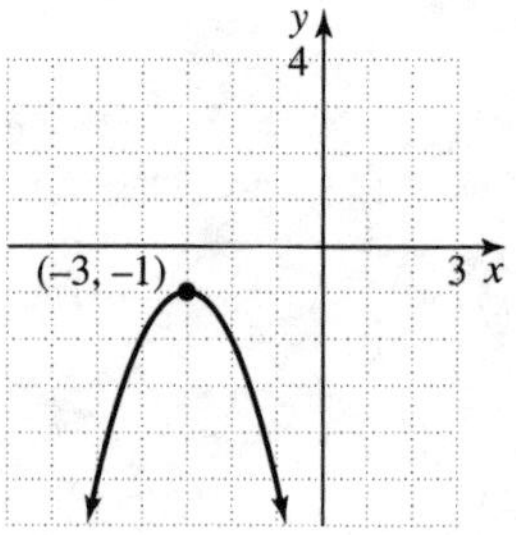

105. $3x + y > 4$

$y > -3x + 4$

Graph the boundary line as dashed.
Test: (0, 0)
$3(0) + 0 > 4$ False
Shade the half-plane that does not include (0, 0).

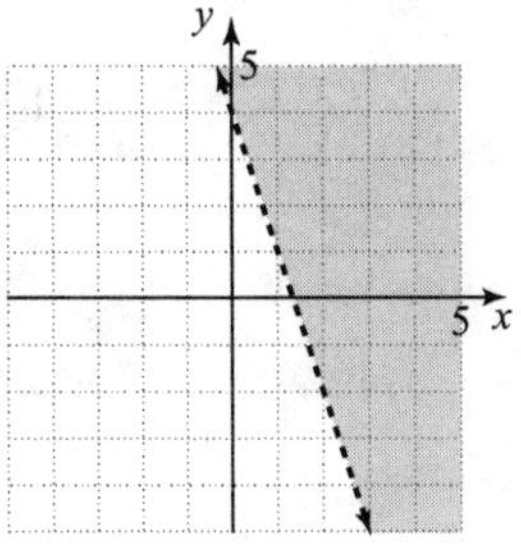

106. $\frac{1}{2}x - y < 2$

$y > \frac{1}{2}x - 2$

Graph the boundary line as dashed.
Test: (0, 0)

$\frac{1}{2}(0) - 0 < 2$ True

Shade the half-plane that contains (0, 0).

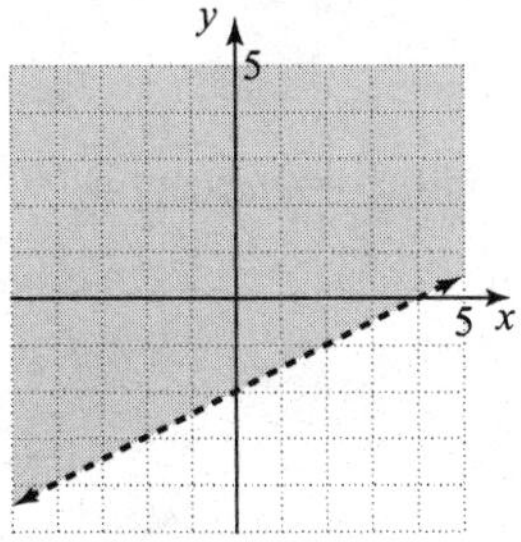

107. $5x - 2y \le 9$

$-2y \le -5x + 9$

$y \ge \frac{5}{2}x - \frac{9}{2}$

Graph the boundary line as solid.
Test: (0, 0)
$5(0) - 2(0) \le 9$ True
Shade the half-plane that contains (0, 0).

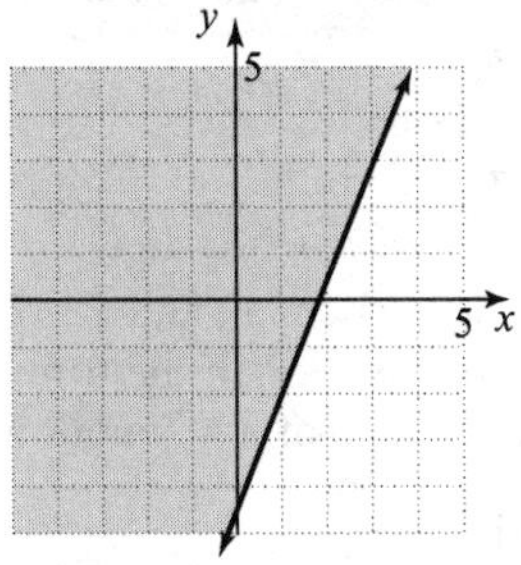

108. $2x \le 6y$

$\frac{x}{3} \le y$ or $y \ge \frac{x}{3}$

Graph the boundary line as solid.
Test: (0, 1)
$3(1) \ge 0$ True

Shade the half-plane that contains (0, 1).

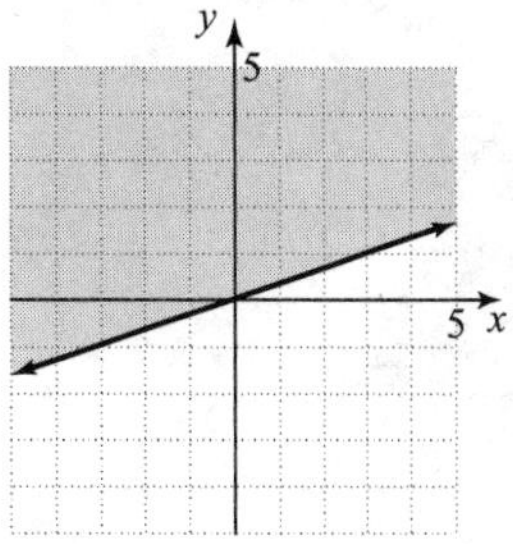

109. $y < 1$

Graph the boundary line as dashed. Shade the half-plane below $y = 1$.

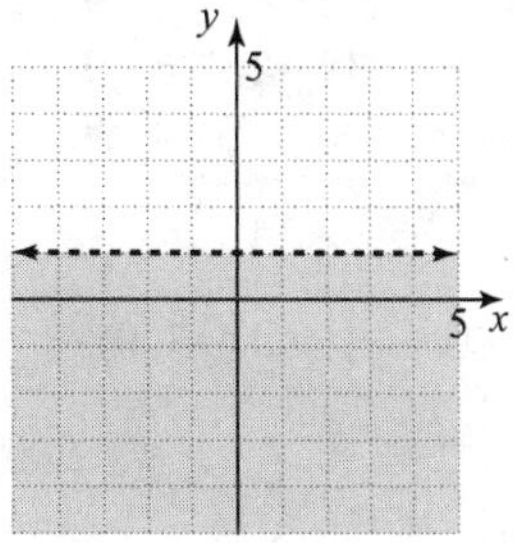

110. $x > -2$

Graph the boundary line as dashed. Shade the half-plane to the right of $x = -2$.

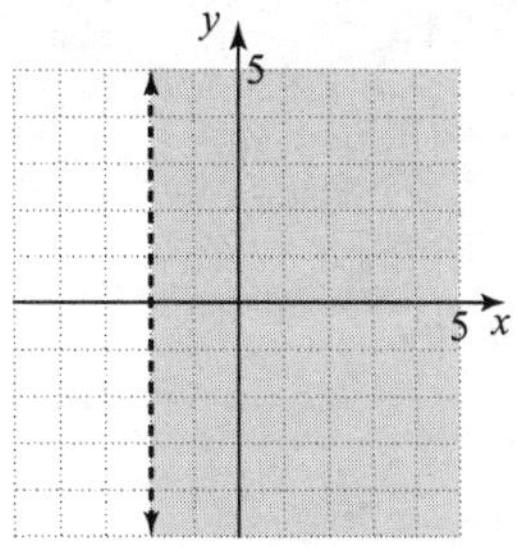

111. $y > 2x + 3$ or $x \le -3$

Graph each inequality. The union of the two inequalities is both shaded regions, as shown by the graph below.

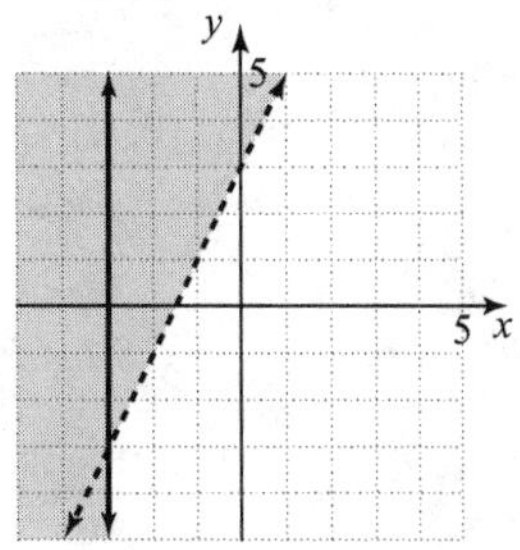

112. $2x < 3y + 8$ and $y \ge -2$

Graph each inequality. The intersection of the two inequalities is all points common to both regions, as shown by the shading in the graph below.

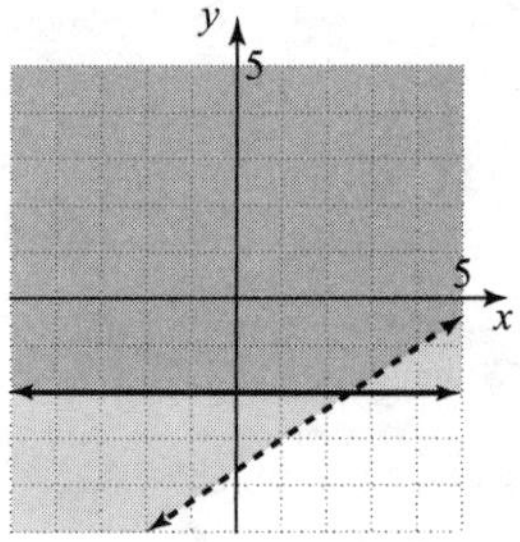

113. $3x - 2y = -9$

Let $x = 0$.

$$3x - 2y = -9$$
$$3(0) - 2y = -9$$
$$-2y = -9$$
$$y = \frac{9}{2}$$

Let $y = 0$.

$$3x - 2y = -9$$
$$3x - 2(0) = -9$$
$$3x = -9$$
$$x = -3$$

The intercepts are $\left(0, \frac{9}{2}\right)$ and $(-3, 0)$.

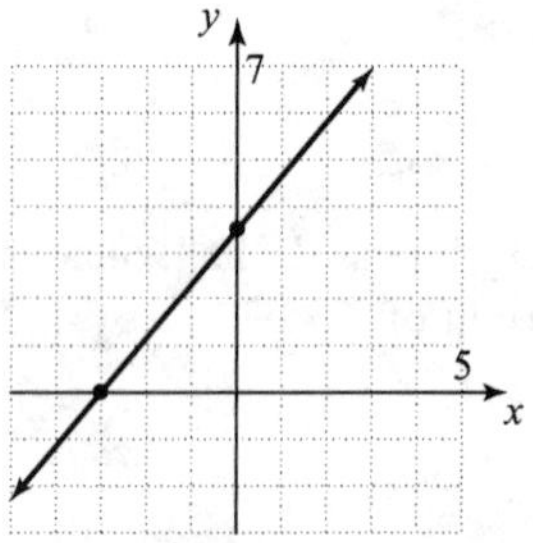

114. $5x - 3y < 10$

Graph the boundary line

$$5x - 3y = 10$$
$$-3y = -5x + 10$$
$$y = \frac{5}{3}x - \frac{10}{3}$$

as a dashed line.

Test (0, 0).

$$5x - 3y < 10$$
$$5(0) - 3(0) < 10$$
$$0 < 10 \quad \text{True}$$

Shade the half-plane that contains (0, 0).

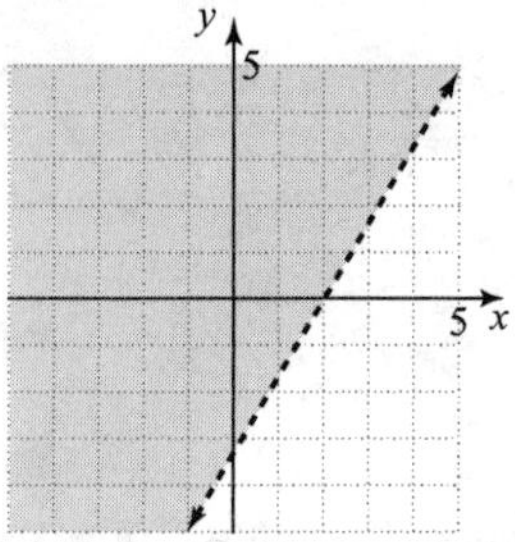

115. $3y \geq x$

Graph the boundary line $3y = x$ or $y = \frac{1}{3}x$ as a solid line.

Test (0, 1).

$$3y \geq x$$
$$3(1) \geq 0$$
$$3 \geq 0 \quad \text{True}$$

Shade the half-plane that contains (0, 1).

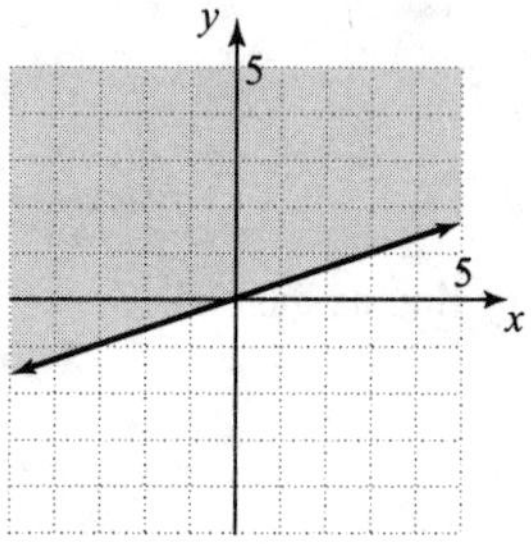

116. $x = -4y$ or $y = -\frac{1}{4}x$

The slope is $-\frac{1}{4}$, and the *y*-intercept is (0, 0).

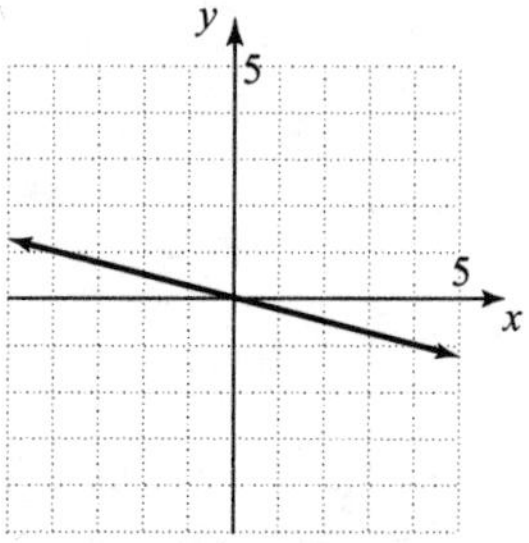

117. Vertical; through $\left(-7, -\frac{1}{2}\right)$

A vertical line has an equation of the form $x = a$, where a is the *x*-coordinate of any point on the line. The equation is $x = -7$.

118. Slope 0; through $\left(-4, \frac{9}{2}\right)$

A line with slope 0 is horizontal, and a horizontal line has an equation of the form $y = b$, where b is the *y*-coordinate of any point on the line. The equation is $y = \frac{9}{2}$.

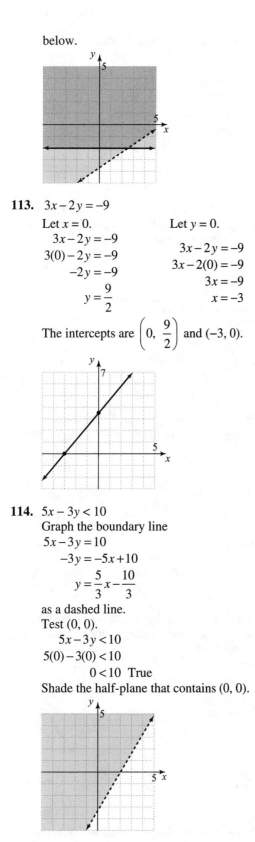

119. Slope $\frac{3}{4}$; through (–8, –4)

$$y - y_1 = m(x - x_1)$$
$$y-(-4) = \frac{3}{4}(x-(-8))$$
$$y+4 = \frac{3}{4}(x+8)$$
$$4(y+4) = 3(x+8)$$
$$4y+16 = 3x+24$$
$$4y = 3x+8$$
$$y = \frac{3}{4}x+2$$

120. Through (–3, 8) and (–2, 3)
Find the slope.

$$m = \frac{3-8}{-2-(-3)} = \frac{-5}{1} = -5$$

Use the slope and one of the points in the point-slope form. We use (–2, 3).

$$y - y_1 = m(x - x_1)$$
$$y-3 = -5(x-(-2))$$
$$y-3 = -5(x+2)$$
$$y-3 = -5x-10$$
$$y = -5x-7$$

121. Through (–6, 1); parallel to $y = -\frac{3}{2}x+11$

The slope of a line parallel to $y = -\frac{3}{2}x+11$ will have the same slope, $-\frac{3}{2}$.

$$y - y_1 = m(x - x_1)$$
$$y-1 = -\frac{3}{2}(x-(-6))$$
$$y-1 = -\frac{3}{2}(x+6)$$
$$2(y-1) = -3(x+6)$$
$$2y-2 = -3x-18$$
$$2y = -3x-16$$
$$y = -\frac{3}{2}x-8$$

122. Through (–5, 7); perpendicular to $5x-4y=10$
Find the slope of $5x-4y=10$.

$$5x-4y = 10$$
$$-4y = -5x+10$$
$$y = \frac{5}{4}x-\frac{5}{2}$$

The slope is $\frac{5}{4}$. The slope of any line perpendicular to this line is the negative reciprocal of $\frac{5}{4}$, or $-\frac{4}{5}$.

$$y - y_1 = m(x - x_1)$$
$$y-7 = -\frac{4}{5}(x-(-5))$$
$$y-7 = -\frac{4}{5}(x+5)$$
$$5(y-7) = -4(x+5)$$
$$5y-35 = -4x-20$$
$$5y = -4x+15$$
$$y = -\frac{4}{5}x+3$$

123. $f(x) = \begin{cases} x-2 & \text{if } x \le 0 \\ -\frac{x}{3} & \text{if } x \ge 3 \end{cases}$

For $x \le 0$:

x	$f(x)$
–2	–4
–1	–3
0	–2

For $x \ge 3$:

x	$f(x)$
3	–1
4	$-\frac{4}{3}$
6	–2

Graph closed circles at (0, –2) and (3, –1).

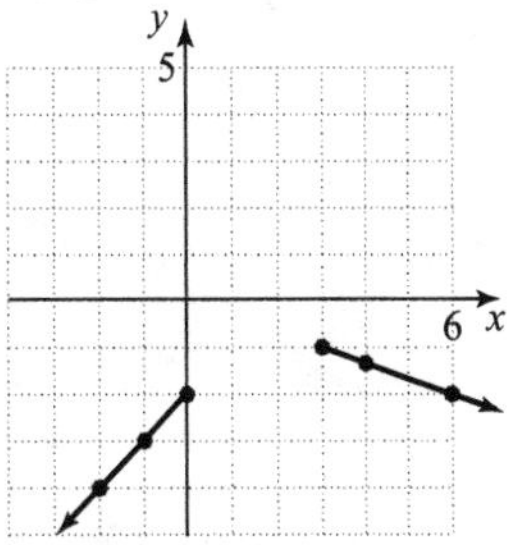

124. $g(x) = \begin{cases} 4x-3 & \text{if } x \le 1 \\ 2x & \text{if } x > 1 \end{cases}$

For $x \le 1$:

x	$g(x)$
–1	–7
0	–3
1	1

For $x > 1$:

x	$g(x)$
2	4
3	6
4	8

Graph a closed circle at (1, 1). Graph an open circle at (1, 2), which is found by substituting 1 for x in $g(x) = 2x$.

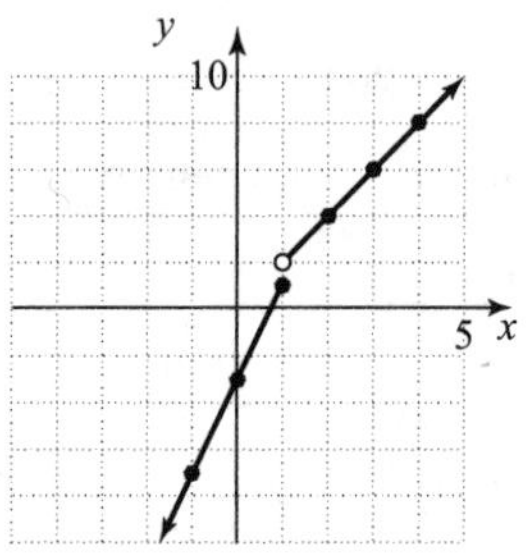

125. $f(x) = \sqrt{x-2}$

The graph of $f(x) = \sqrt{x-2}$ is the same as the graph of $y = \sqrt{x}$ shifted right 2 units.

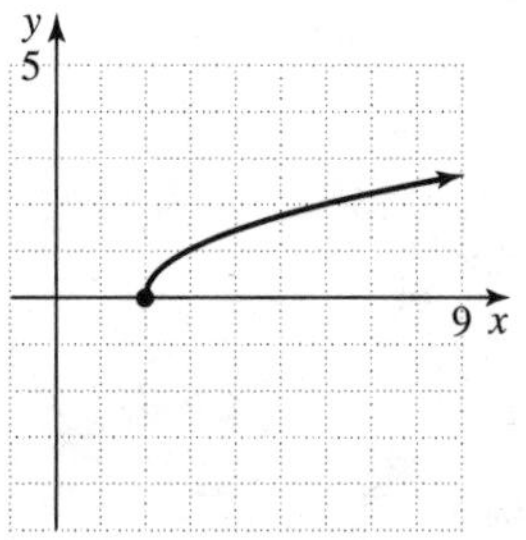

126. $f(x) = |x+1| - 3$

The graph of $f(x) = |x+1| - 3$ is the same as the graph of $y = |x|$ shifted left 1 unit and down 3 units.

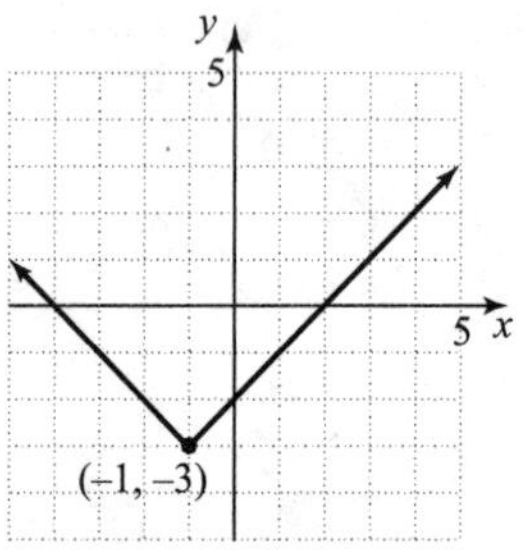

Chapter 3 Getting Ready for the Test

1. For $a > 0$ and $b > 0$, (a, b) is in Quadrant I; A.

2. For $a > 0$ and $b > 0$, $(-a, -b)$ is in Quadrant III; C.

3. For $a > 0$ and $b > 0$, $(0, b)$ is on the y-axis; F.

4. For $a > 0$ and $b > 0$, $(a, 0)$ is on the x-axis; E.

5. For $a > 0$ and $b > 0$, $(-a, b)$ is in Quadrant II; B.

6. For $a > 0$ and $b > 0$, $(a, -b)$ is in Quadrant IV; D.

7. In the diagram, each input is paired with exactly one output, so the relation is a function; A.

8. The range of the relation is the set of outputs that correspond to inputs of the relation, which is $\{1, 3\}$; B.

9. $f(x) = -x^2$

   $f(-3) = -(-3)^2 = -9$

   D

10. $f(x) = 2x - 3$

    $f(a + h) = 2(a + h) - 3 = 2a + 2h - 3$

    C

11. If $(-6, 0)$ is an ordered pair solution for $g(x)$, then the x-value -6 corresponds to the y-value, or $g(x)$-value, 0. In function notation, this is $g(-6) = 0$; B.

12. The given information is two points and the final answer is an equation of a line written using function notation. The correct directions are choice C.

13. Every linear equation, except the equation for a vertical line, represents a function. Therefore, $y = -3x + 13$ represents a function; A.

14. Every linear equation, except the equation for a vertical line, represents a function. Therefore, $y = 5$ (a horizontal line) represents a function; A.

15. Every linear equation, except the equation for a vertical line, represents a function. Therefore, $x = -4$ (a vertical line) does not represent a function; B.

16. Every linear equation, except the equation for a vertical line, represents a function. Therefore, $y = 4x$ represents a function; A.

17. The vertex is the highest point on the graph. This is $(4, 3)$, which is $f(4) = 3$, written in function notation; B.

18. The point $(0, -9)$ is on the graph. Thus, $f(0) = -9$; D.

19. The points on the graph with y-value of 0 are $(2, 0)$ and $(6, 0)$. Thus, $f(2) = 0$ and $f(6) = 0$; C and E.

20. The point $(8, -9)$ is on the graph. Thus, $f(8) = -9$; D.

21. All real numbers can be used as x-values in the function. Therefore, the domain of $f(x)$ is $(-\infty, \infty)$; G.

22. The largest y-value on the graph is $y = 3$. The range of $f(x)$ is $(-\infty, 3]$; I.

23. The line with the greatest slope is the line that tilts most steeply upward from left to right. This is line D.

24. The line with the least slope is the line that tilts most steeply downward from left to right. This is line A.

Chapter 3 Test

1.

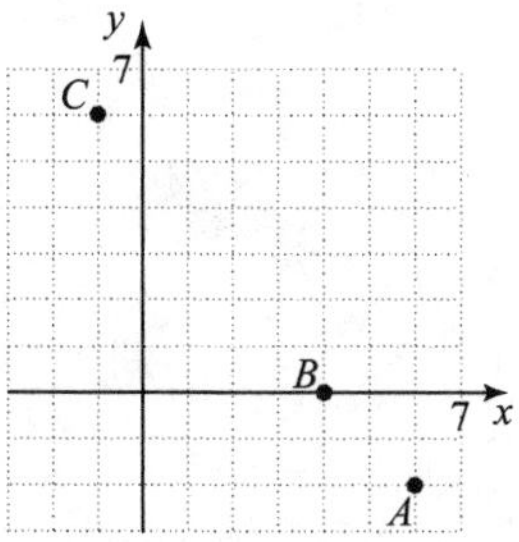

A is in quadrant IV.
B is on the x-axis, no quadrant.
C is in quadrant II.

2. $2x-3y=-6$

$$-3y=-2x-6$$
$$y=\frac{2}{3}x+2$$
$$m=\frac{2}{3},\ b=2$$

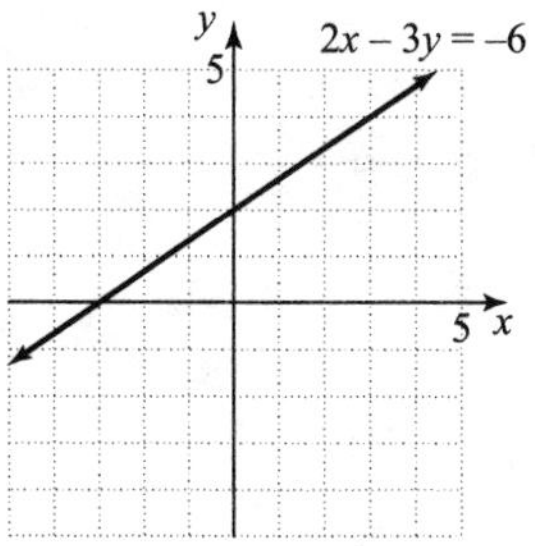

3. $4x+6y=7$

$$6y=-4x+7$$
$$y=-\frac{2}{3}x+\frac{7}{6}$$
$$m=-\frac{2}{3},\ b=\frac{7}{6}$$

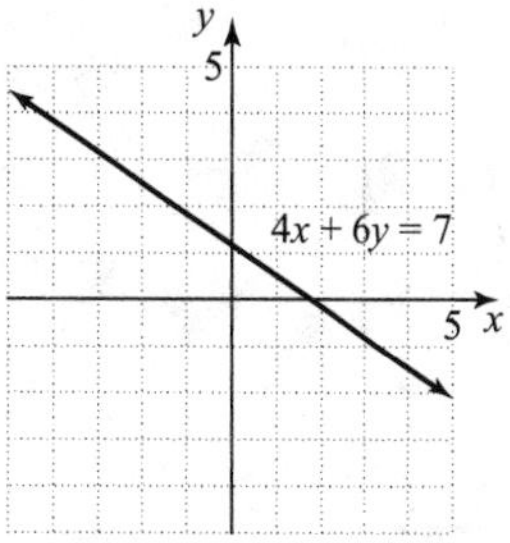

4. $f(x)=\frac{2}{3}x$ or $y=\frac{2}{3}x$

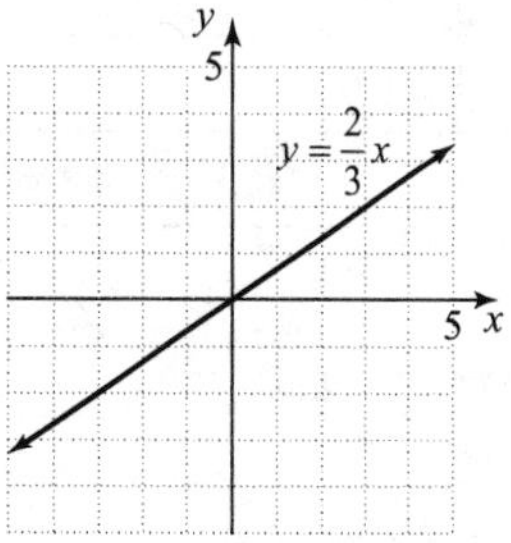

5. $y=-3$

Horizontal line with y-intercept at -3.

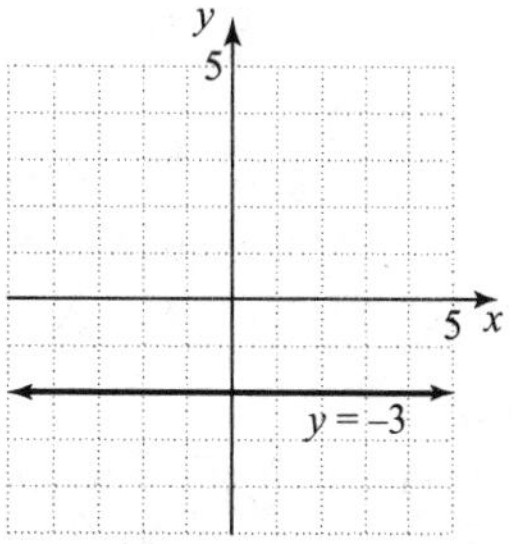

6. $m=\frac{10-(-8)}{-7-5}=\frac{18}{-12}=-\frac{3}{2}$

7. $3x+12y=8$

$$12y=-3x+8$$
$$y=-\frac{1}{4}x+\frac{2}{3}$$
$$m=-\frac{1}{4},\ b=\frac{2}{3},\ y\text{-intercept }\left(0,\frac{2}{3}\right)$$

8. $f(x)=(x-1)^2$

x	−2	−1	0	1	2	3	4
y	9	4	1	0	1	4	9

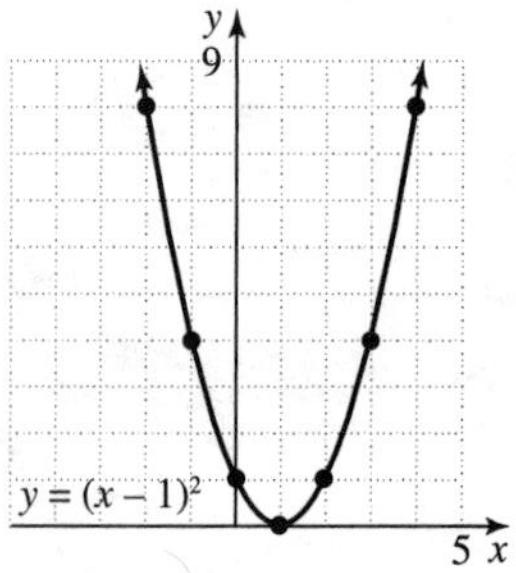

9. $g(x)=|x|+2$

x	−3	−2	−1	0	1	2	3
y	5	4	3	2	3	4	5

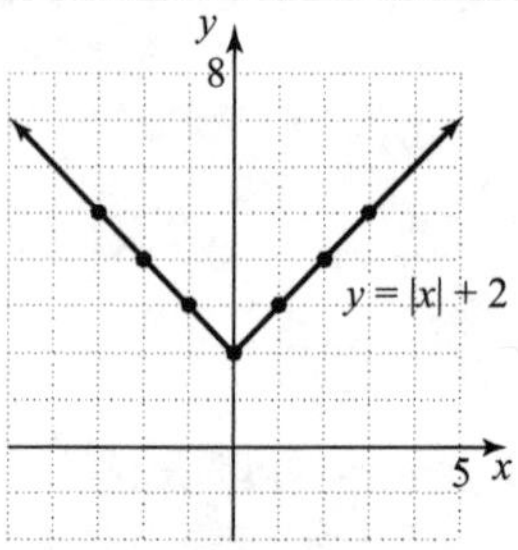

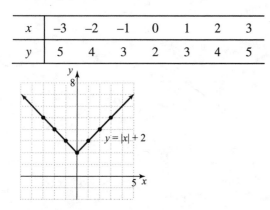

10. Horizontal; through (2, −8)
A horizontal line has an equation of the form $y = b$, where b is the y-coordinate of any point on the line. The equation is $y = -8$.

11. Vertical; through (−4, −3)
A vertical line has an equation of the form $x = a$, where a is the x-coordinate of any point on the line. The equation is $x = -4$.

12. Perpendicular to $x = 5$; through (3, −2)
The line $x = 5$ is vertical, so any line perpendicular to it is horizontal. A horizontal line has an equation of the form $y = b$, where b is the y-coordinate of any point on the line. The equation is $y = -2$.

13.
$$\begin{aligned} y-y_1&=m(x-x_1)\\ y-(-1)&=-3(x-4)\\ y+1&=-3x+12\\ 3x+y&=11 \end{aligned}$$

14.
$$\begin{aligned} y-y_1&=m(x-x_1)\\ y-(-2)&=5(x-0)\\ y+2&=5x\\ 5x-y&=2 \end{aligned}$$

15. $m=\dfrac{-3-(-2)}{6-4}=\dfrac{-1}{2}=-\dfrac{1}{2}$
$$\begin{aligned} y-y_1&=m(x-x_1)\\ y-(-2)&=-\frac{1}{2}(x-4)\\ 2(y+2)&=-(x-4)\\ 2y+4&=-x+4\\ 2y&=-x\\ y&=-\frac{1}{2}x\\ f(x)&=-\frac{1}{2}x \end{aligned}$$

16.
$$\begin{aligned} 3x-y&=4\\ y&=3x-4 \end{aligned}$$
$m=3$ so $m_\perp=-\dfrac{1}{3}$
$$\begin{aligned} y-y_1&=m(x-x_1)\\ y-2&=-\frac{1}{3}[x-(-1)]\\ 3(y-2)&=-(x+1)\\ 3y-6&=-x-1\\ 3y&=-x+5\\ y&=-\frac{1}{3}x+\frac{5}{3}\\ f(x)&=-\frac{1}{3}x+\frac{5}{3} \end{aligned}$$

17.
$$\begin{aligned} 2y+x&=3\\ 2y&=-x+3 \end{aligned}$$
$y=-\dfrac{1}{2}x+3$ so $m=-\dfrac{1}{2}$
$$\begin{aligned} y-y_1&=m(x-x_1)\\ y-(-2)&=-\frac{1}{2}(x-3)\\ 2(y+2)&=-(x-3)\\ 2y+4&=-x+3\\ 2y&=-x-1\\ y&=-\frac{1}{2}x-\frac{1}{2}\\ f(x)&=-\frac{1}{2}x-\frac{1}{2} \end{aligned}$$

18. $2x-5y=8$

$-5y=-2x+8$

$y=\frac{2}{5}x-\frac{8}{5}$ so $m_1=\frac{2}{5}$

$m_2=\frac{-1-4}{-1-1}=\frac{-5}{-2}=\frac{5}{2}$

Therefore, lines L_1 and L_2 are neither parallel nor perpendicular since their slopes are not equal and the product of their slopes is not −1.

19. $x \le -4$

Graph a solid boundary line and shade the half-plane to the left of $x = -4$.

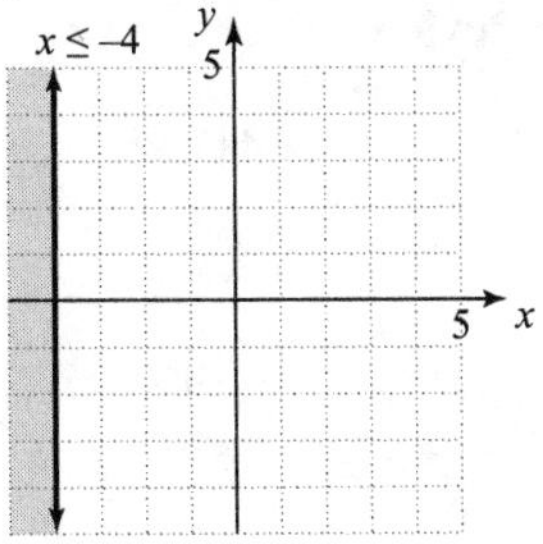

20. $2x - y > 5$

$y < 2x - 5$

Graph the boundary line as a dashed line.

Test: (0, 0)

$2(0) - 0 > 5$ False

Shade the half-plane that does not contain the point (0, 0).

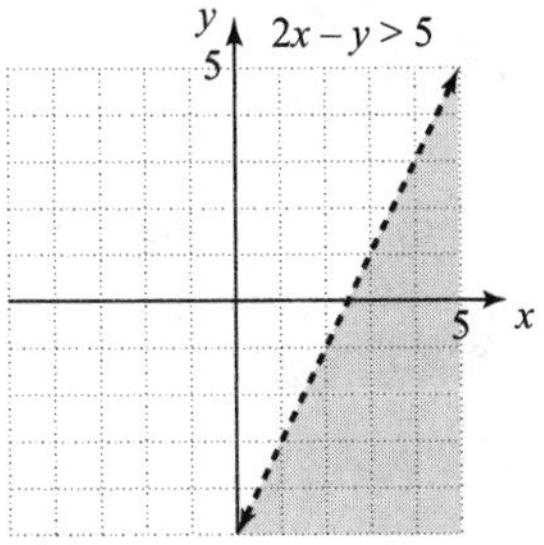

21. $2x+4y<6$ and $y\le 4$

$4y<-2x+6$ and $y\le 4$

$y<-\frac{1}{2}x+\frac{3}{2}$ and $y\le 4$

Graph each inequality. The intersection of the two inequalities is all points common to both regions, as shown by the shading in the graph below.

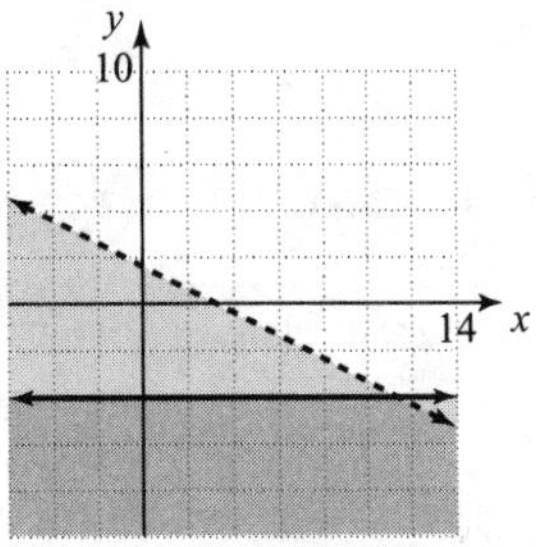

22. Domain: $(-\infty, \infty)$

Range: $\{5\}$

Function since it passes the vertical line test.

23. Domain: $\{-2\}$

Range: $(-\infty, \infty)$

Not a function since it fails the vertical line test.

24. Domain: $(-\infty, \infty)$

Range: $[0, \infty)$

Function since it passes the vertical line test.

25. Domain: $(-\infty, \infty)$

Range: $(-\infty, \infty)$

Function since it passes the vertical line test.

26. $y = 0.024x + 79.44$

a. Let $x = 90$.

$y = 0.024(90) + 79.44 = 81.6$

A team with a payroll of \$90 million would be expected to win 82 games.

b. Let $x = 204$.

$y = 0.024(204) + 79.44 = 84.336$

The Yankees would be expected to have won 84 games.

c. Let $y = 95$ and solve for x.

$95=0.024x+79.44$

$15.56=0.024x$

$648\approx x$

A payroll of \$648 million would be needed to win 95 games.

d. The slope is 0.024. Every million dollars spent on payroll increases winnings by 0.024 game.

27. $f(x)=\begin{cases}-\frac{1}{2}x & \text{if } x\le 0\\ 2x-3 & \text{if } x>0\end{cases}$

For $x \le 0$:

x	f(x)
–4	2
–2	1
0	0

For $x > 0$:

x	f(x)
1	–1
2	1
3	3

Graph a closed circle at (0, 0). Graph an open circle at (0, –3), which is found by substituting 0 for x in $f(x)=2x-3$.

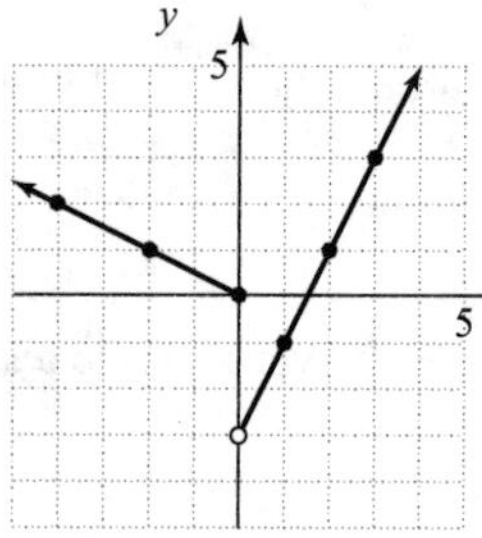

28. $f(x)=(x-4)^2$

The graph of $f(x)=(x-4)^2$ is the same as the graph of $y=x^2$ shifted right 4 units.

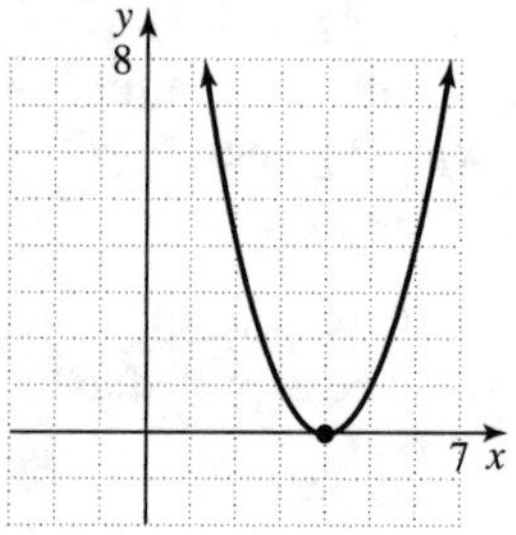

29. $g(x)=-|x+2|-1$

The graph of $g(x)=-|x+2|-1$ is the same as the graph of $y=|x|$ reflected about the x-axis and then shifted left 2 units and down 1 unit.

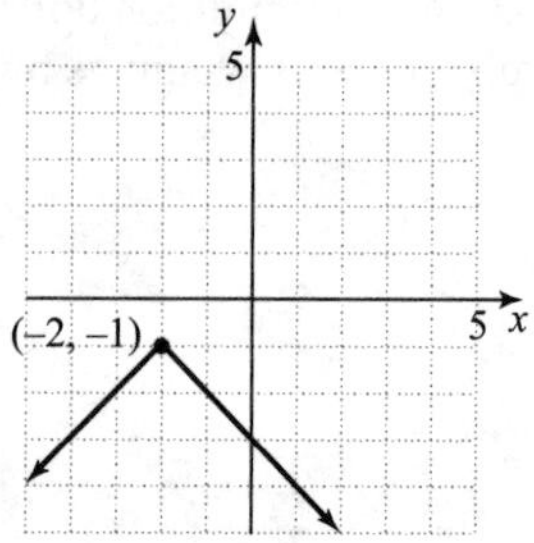

30. $h(x)=\sqrt{x}-1$

The graph of $h(x)=\sqrt{x}-1$ is the same as the graph of $y=\sqrt{x}$ shifted down 1 unit.

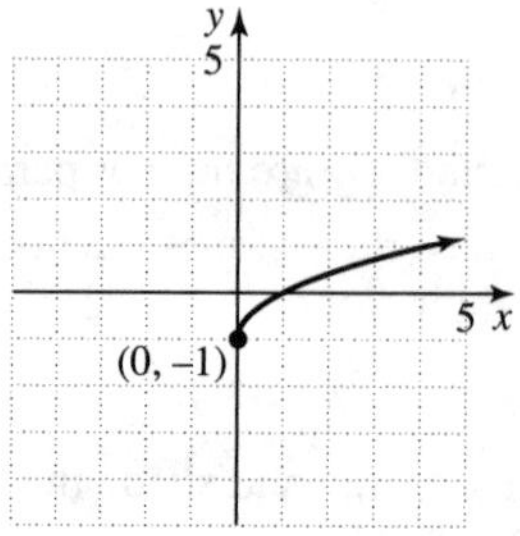

Chapter 3 Cumulative Review

1. $3x-y=3(15)-(4)=45-4=41$

2. a. $-4+(-3)=-7$

b. $\frac{1}{2}-\left(-\frac{1}{3}\right)=\frac{3}{6}+\frac{2}{6}=\frac{5}{6}$

c. $7-20=-13$

3. a. True, 3 is a real number

b. False, $\frac{1}{5}$ is not an irrational number.

c. False, every rational number is not an integer, for example, $\frac{2}{3}$.

d. False, since 1 is not in the second set.

4. a. The opposite of –7 is 7.

b. The opposite of 0 is 0.

c. The opposite of $\frac{1}{4}$ is $-\frac{1}{4}$.

5. a. $2-8=-6$

b. $-8-(-1)=-8+1=-7$

c. $-11-5=-16$

d. $10.7-(-9.8)=10.7+9.8=20.5$

e. $-\frac{2}{3}-\frac{1}{2}=-\frac{4}{6}-\frac{3}{6}=-\frac{7}{6}$

f. $1-0.06=0.94$

g. $4-7=-3$

6. a. $\frac{-42}{-6}=7$

b. $\frac{0}{14}=0$

c. $-1(-5)(-2)=5(-2)=-10$

7. a. $3^2=3\cdot 3=9$

b. $\left(\frac{1}{2}\right)^4=\frac{1}{2^4}=\frac{1}{16}$

c. $-5^2=-(5\cdot 5)=-25$

d. $(-5)^2=(-5)(-5)=25$

e. $-5^3=-(5\cdot 5\cdot 5)=-125$

f. $(-5)^3=(-5)(-5)(-5)=-125$

8. a. distributive property

b. commutative property of addition

9. a. $-1>-2$ since -1 is to the right of -2 on a number line.

b. $\frac{12}{4}=3$

c. $-5<0$ since -5 is to the left of 0 on a number line.

d. $-3.5<-3.05$ since -3.5 is to the left of -3.05 on a number line.

e. $\frac{5}{8}>\frac{3}{8}$ since 5 is to the right of 3 on a number line.

f. $\frac{2}{3}<\frac{3}{4}$ since $\frac{8}{12}<\frac{9}{12}$ and 8 is to the left of 9 on a number line.

10. $2x^2$

a. $2(7)^2=2(49)=98$

b. $2(-7)^2=2(49)=98$

11. a. The reciprocal of 11 is $\frac{1}{11}$.

b. The reciprocal of -9 is $-\frac{1}{9}$.

c. The reciprocal of $\frac{7}{4}$ is $\frac{4}{7}$.

12. $$\begin{aligned}-2+3[5-(7-10)]&=-2+3[5-(-3)]\\&=-2+3(8)\\&=-2+24\\&=22\end{aligned}$$

13. $$\begin{aligned}0.6&=2-3.5c\\-1.4&=-3.5c\\\frac{-1.4}{-3.5}&=\frac{-3.5c}{-3.5}\\0.4&=c\end{aligned}$$

14. $$\begin{aligned}2(x-3)&=-40\\2x-6&=-40\\2x&=-34\\x&=-17\end{aligned}$$

15. $$\begin{aligned}3x+5&=3(x+2)\\3x+5&=3x+6\\5&=6 \quad \text{False}\end{aligned}$$

The solution is { } or ∅.

16. $$\begin{aligned}5(x-7)&=4x-35+x\\5x-35&=5x-35\\-35&=-35 \quad \text{True for any number}\end{aligned}$$

The solution is all real numbers.

17. a. If x is the first integer, then the next two consecutive integers are $x+1$ and $x+2$. The sum is $x+(x+1)+(x+2)=3x+3$.

b. The perimeter is found by adding the lengths of the sides.
$x+5x+(6x-3)=12x-3$

18. 25% of $16=0.25(16)=4$

19. Let x = the lowest of the scores. Then the other two scores are $x + 2$ and $x + 4$.
$$x+(x+2)+(x+4)=264$$
$$3x+6=264$$
$$3x=258$$
$$x=86$$
$x + 2 = 86 + 2 = 88$
$x + 4 = 86 + 4 = 90$
The scores are 86, 88, and 90.

20. Let x = first odd integer, then
$x + 2$ = next odd integer and
$x + 4$ = third odd integer.
$$x+(x+2)+(x+4)=213$$
$$3x+6=213$$
$$3x=207$$
$$x=69$$
$x+2=69+2=71$
$x+4=69+4=73$
The integers are 69, 71, and 73.

21. $V=lwh$
$$\frac{V}{lw}=\frac{lwh}{lw}$$
$$\frac{V}{lw}=h$$

22. $7x+3y=21$
$$3y=-7x+21$$
$$y=-\frac{7}{3}x+7$$

23. $x-2<5$
$$x<7$$
$(-\infty, 7)$ or $\{x|x<7\}$

24. $-x-17\geq 9$
$$-x\geq 26$$
$$x\leq -26$$
$(-\infty, -26]$ or $\{x|x\leq -26\}$

25. $\frac{2}{5}(x-6)\geq x-1$
$$5\left[\frac{2}{5}(x-6)\right]\geq 5[x-1]$$
$$2(x-6)\geq 5x-5$$
$$2x-12\geq 5x-5$$
$$-3x\geq 7$$
$$x\leq -\frac{7}{3}$$
$\left(-\infty, -\frac{7}{3}\right]$

26. $3x+10>\frac{5}{2}(x-1)$
$$2(3x+10)>2\left[\frac{5}{2}(x-1)\right]$$
$$6x+20>5(x-1)$$
$$6x+20>5x-5$$
$$x>-25$$
$(-25, \infty)$

27. $2x\geq 0$ *and* $4x-1\leq -9$
$x\geq 0$ *and* $4x\leq -8$
$x\geq 0$ *and* $x\leq -2$
The solution set is $\varnothing$.

28. $x-2<6$ *and* $3x+1>1$
$x<8$ *and* $3x>0$
$x<8$ *and* $x>0$
$0<x<8$
$(0, 8)$

29. $5x-3\leq 10$ *or* $x+1\geq 5$
$5x\leq 13$ *or* $x\geq 4$
$x\leq \frac{13}{5}$ *or* $x\geq 4$
$\left(-\infty, \frac{13}{5}\right]\cup[4, \infty)$

30. $x-2<6$ *or* $3x+1>1$
$x<8$ *or* $3x>0$
$x<8$ *or* $x>0$
$(-\infty, \infty)$

31. $|5w+3|=7$
$5w+3=7$ or $5w+3=-7$
$5w=4$ or $5w=-10$
$w=\frac{4}{5}$ or $w=-2$

32. $|5x-2|=3$
$5x-2=3$ or $5x-2=-3$
$5x=5$ or $5x=-1$
$x=1$ or $x=-\frac{1}{5}$

33. $|3x+2|=|5x-8|$
$3x+2=5x-8$ or $3x+2=-(5x-8)$
$-2x=-10$ or $3x+2=-5x+8$
$x=5$ or $8x=6$
$x=\frac{3}{4}$

34. $|7x-2|=|7x+4|$
$7x-2=7x+4$ or $7x-2=-(7x+4)$
$-2=4$ or $7x-2=-7x-4$
False or $14x=-2$
$x=-\frac{1}{7}$

The only solution is $-\frac{1}{7}$.

35. $|5x+1|+1\le 10$
$|5x+1|\le 9$
$-9\le 5x+1\le 9$
$-10\le 5x\le 8$
$-2\le x\le \frac{8}{5}$
$\left[-2, \frac{8}{5}\right]$

36. $|-x+8|-2\le 8$
$|-x+8|\le 10$
$-10\le -x+8\le 10$
$-18\le -x\le 2$
$18\ge x\ge -2$
$-2\le x\le 18$
$[-2, 18]$

37. $|y-3|>7$
$y-3<-7$ or $y-3>7$
$y<-4$ or $y>10$
$(-\infty, -4)\cup(10, \infty)$

38. $|x+3|>1$
$x+3<-1$ or $x+3>1$
$x<-4$ or $x>-2$
$(-\infty, -4)\cup(-2, \infty)$

39. $3x-y=12$
$3(0)-(-12)\stackrel{?}{=}12$
$12=12$ True
(0, −12) is a solution.
$3(1)-9\stackrel{?}{=}12$
$-6=12$ False
(1, 9) is not a solution.
$3(2)-(-6)\stackrel{?}{=}12$
$6+6\stackrel{?}{=}12$
$12=12$ True
(2, −6) is a solution.

40. $7x+2y=10$
$2y=-7x+10$
$y=-\frac{7}{2}x+5$
$m=-\frac{7}{2}$, *y*-intercept = (0, 5)

41. Yes, $y=2x+1$ is a function (graph the function and use the vertical line test).

42. No, it is not a function (by the vertical line test).

43. a. $f(x)=\frac{1}{2}x+\frac{3}{7}$
$y = mx + b$
$b=\frac{3}{7}$
y-intercept = $\left(0, \frac{3}{7}\right)$

b. $y=-2.5x-3.2$
$y = mx + b$
$b=-3.2$
y-intercept = (0, −3.2)

44. $m=\frac{y_2-y_1}{x_2-x_1}=\frac{9-6}{0-(-1)}=\frac{3}{1}=3$

45. $f(x)=\frac{2}{3}x+4$
$y = mx + b$
The slope of the line is *m*, the coefficient of *x*, $\frac{2}{3}$.

46. Vertical; through $\left(-2, -\frac{3}{4}\right)$

A vertical line has an equation of the form $x = a$, where a is the x-coordinate of any point on the line. The equation is $x = -2$.

47. y-intercept = (0, –3) means that $b = -3$. Using the equation $y = mx + b$, we have $y = \frac{1}{4}x - 3$.

48. Horizontal; through $\left(-2, -\frac{3}{4}\right)$

A horizontal line has an equation of the form $y = b$, where b is the y-coordinate of any point on the line. The equation is $y = -\frac{3}{4}$.

49. $2x - y < 6$

Graph the boundary line as a dashed line.
Test: (0, 0)
$2(0) - 0 < 6$ True
Shade the half-plane that contains the point (0, 0).

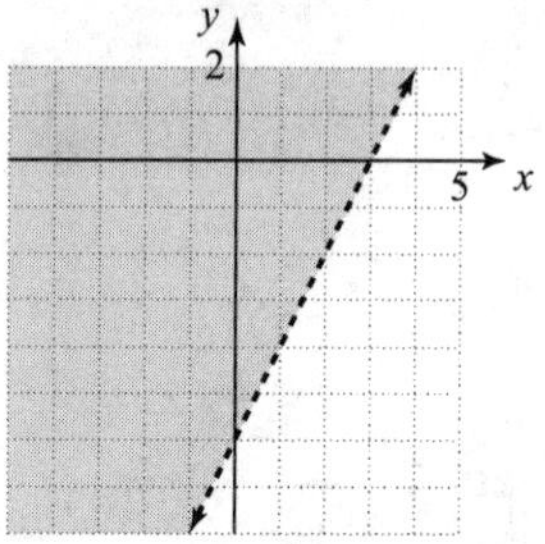

50. $m = \frac{7-5}{-4-(-2)} = \frac{2}{-4+2} = \frac{2}{-2} = -1$

$$y - y_1 = m(x - x_1)$$
$$y - 5 = -1[x - (-2)]$$
$$y - 5 = -(x + 2)$$
$$y - 5 = -x - 2$$
$$x + y = 3$$

Chapter 4

Section 4.1 Practice Exercises

1. a. $\begin{cases} -x - 4y = 1 \\ 2x + y = 5 \end{cases}$

Replace x with 3 and y with -1 in each equation.

$$\begin{aligned} -x - 4y &= 1 \\ -3 - 4(-1) &\stackrel{?}{=} 1 \\ -3 + 4 &\stackrel{?}{=} 1 \\ 1 &= 1 \quad \text{True} \end{aligned}$$

$$\begin{aligned} 2x + y &= 5 \\ 2(3) + (-1) &\stackrel{?}{=} 5 \\ 6 - 1 &\stackrel{?}{=} 5 \\ 5 &= 5 \quad \text{True} \end{aligned}$$

Since (3, −1) makes both equations true, it is a solution.

b. $\begin{cases} 4x + y = -4 \\ -x + 3y = 8 \end{cases}$

Replace x with −2 and y with 4 in each equation.

$$\begin{aligned} 4x + y &= -4 \\ 4(-2) + 4 &\stackrel{?}{=} -4 \\ -8 + 4 &\stackrel{?}{=} -4 \\ -4 &= -4 \quad \text{True} \end{aligned}$$

$$\begin{aligned} -x + 3y &= 8 \\ -(-2) + 3(4) &\stackrel{?}{=} 8 \\ 2 + 12 &\stackrel{?}{=} 8 \\ 14 &= 8 \quad \text{False} \end{aligned}$$

Since (−2, 4) does not make both equations true, it is not a solution.

2. $\begin{cases} y = 5x \\ 2x + y = 7 \end{cases}$

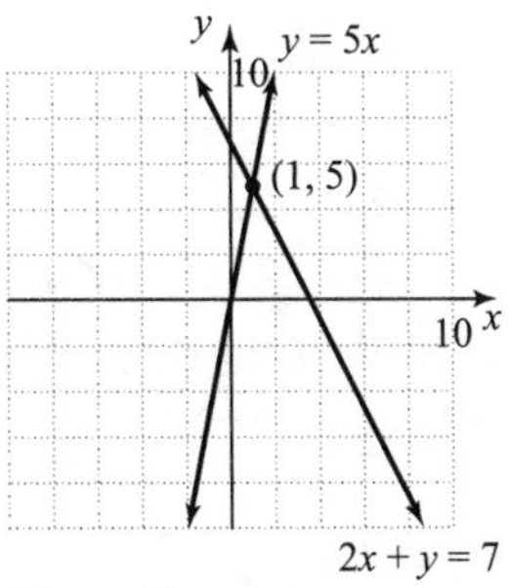

These lines intersect at one point as shown in the graph. The coordinates of the point appear to be (1, 5). Check by replacing x with 1 and y with 5 in both equations.

$$\begin{aligned} y &= 5x \\ 5 &\stackrel{?}{=} 5(1) \\ 5 &= 5 \quad \text{True} \end{aligned} \qquad \begin{aligned} 2x + y &= 7 \\ 2(1) + 5 &\stackrel{?}{=} 7 \\ 2 + 5 &\stackrel{?}{=} 7 \\ 7 &= 7 \quad \text{True} \end{aligned}$$

Since (1, 5) satisfies both equations, we conclude that (1, 5) is the solution of the system.

3. $\begin{cases} y = \frac{3}{4}x + 1 \\ 3x - 4y = 12 \end{cases}$

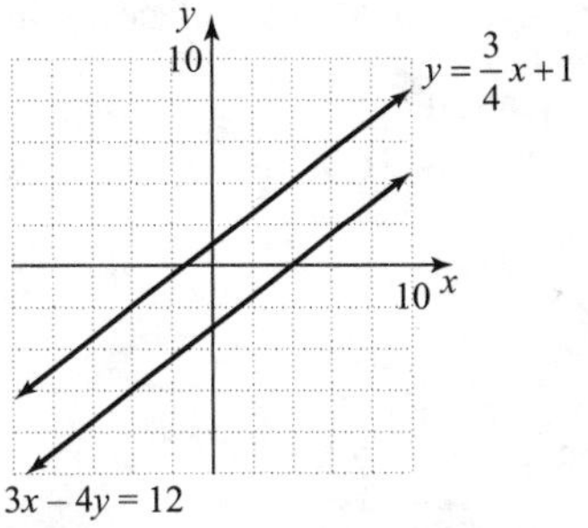

The lines appear to be parallel. The first equation is in point-slope. Write the second equation in point-slope form.

$$\begin{aligned} 3x - 4y &= 12 \\ -4y &= -3x + 12 \\ y &= \frac{3}{4}x - 3 \end{aligned}$$

The graphs have the same slope, $\frac{3}{4}$, but different y-intercepts, so the lines are parallel. The system has no solution.

4. $\begin{cases} 3x - 2y = 4 \\ -9x + 6y = -12 \end{cases}$

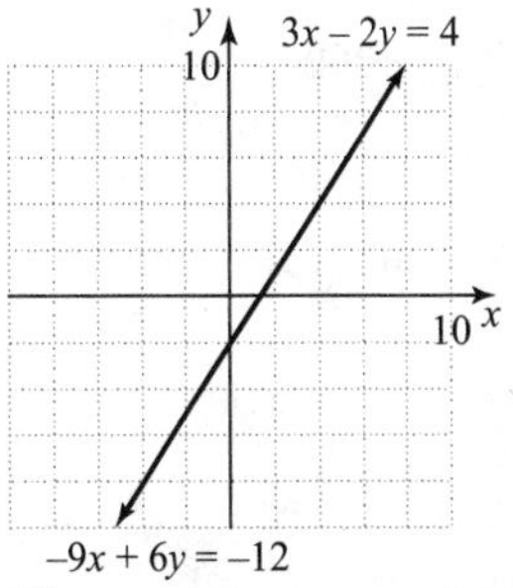

The graphs appear to be the same line. Notice that if both sides of the first equation are multiplied by −3, the result is the second equation. Any solution of one equation satisfies the other equation as well. There is an infinite number of solutions in the form $\{(x, y) | 3x - 2y = 4\}$ or $\{(x, y) | -9x + 6y = -12\}$.

5. $\begin{cases} y=4x+7 \\ 2x+y=4 \end{cases}$

In the first equation, we are told that y is equal to $4x + 7$, so we substitute $4x + 7$ for y in the second equation and solve for x.

$$\begin{aligned} 2x+y&=4 \\ 2x+(4x+7)&=4 \\ 6x+7&=4 \\ 6x&=-3 \\ x&=\frac{-3}{6}=-\frac{1}{2} \end{aligned}$$

To find the y-coordinate, we replace x with $-\frac{1}{2}$ in the first equation.

$$\begin{aligned} y&=4x+7 \\ y&=4\left(-\frac{1}{2}\right)+7 \\ y&=-2+7=5 \end{aligned}$$

The solution is $\left(-\frac{1}{2}, 5\right)$.

6. $\begin{cases} -\frac{x}{3}+\frac{y}{4}=\frac{1}{2} \\ \frac{x}{4}-\frac{y}{2}=-\frac{1}{4} \end{cases}$

Multiply each equation by its LCD to clear fractions.

$$\begin{cases} 12\left(-\frac{x}{3}+\frac{y}{4}\right)=12\left(\frac{1}{2}\right) \\ 4\left(\frac{x}{4}-\frac{y}{2}\right)=4\left(-\frac{1}{4}\right) \end{cases}$$

$$\begin{cases} -4x+3y=6 \\ x-2y=-1 \end{cases}$$

Solve the second equation for x.

$$\begin{aligned} x-2y&=-1 \\ x&=2y-1 \end{aligned}$$

Replace x with $2y-1$ in the first equation.

$$\begin{aligned} -4x+3y&=6 \\ -4(2y-1)+3y&=6 \\ -8y+4+3y&=6 \\ -5y+4&=6 \\ -5y&=2 \\ y&=-\frac{2}{5} \end{aligned}$$

To find the x-coordinate, replace y with $-\frac{2}{5}$ in $x = 2y - 1$.

$$x=2\left(-\frac{2}{5}\right)-1=-\frac{4}{5}-1=-\frac{4}{5}-\frac{5}{5}=-\frac{9}{5}$$

The solution is $\left(-\frac{9}{5}, -\frac{2}{5}\right)$.

7. $\begin{cases} 3x-y=5 \\ 5x+y=11 \end{cases}$

We add the equations.

$$\begin{aligned} 3x-y&=5 \\ 5x+y&=11 \\ \hline 8x\quad&=16 \\ x&=2 \end{aligned}$$

Replace x with 2 in the second equation to find y.

$$\begin{aligned} 5x+y&=11 \\ 5(2)+y&=11 \\ 10+y&=11 \\ y&=1 \end{aligned}$$

The solution is (2, 1).

8. $\begin{cases} 3x-2y=-6 \\ 4x+5y=-8 \end{cases}$

We can eliminate y if we multiply both sides of the first equation by 5 and both sides of the second equation by 2.

$$\begin{cases} 5(3x-2y)=5(-6) \\ 2(4x+5y)=2(-8) \end{cases}$$

$$\begin{aligned} 15x-10y&=-30 \\ 8x+10y&=-16 \\ \hline 23x\quad&=-46 \\ x&=-2 \end{aligned}$$

To find y, replace x with -2 in either equation.

$$\begin{aligned} 4x+5y&=-8 \\ 4(-2)+5y&=-8 \\ -8+5y&=-8 \\ 5y&=0 \\ y&=0 \end{aligned}$$

The solution is (−2, 0).

9. $\begin{cases} 8x+y=6 \\ 2x+\frac{y}{4}=-2 \end{cases}$

If we multiply the second equation by −4, the coefficients of x will be opposites.

$$\begin{cases} 8x+y=6 \\ -4\left(2x+\frac{y}{4}\right)=-4(-2) \end{cases}$$

$$\begin{cases} 8x + y = 6 \\ -8x - y = 8 \end{cases}$$
$$0 = 14 \quad \text{False}$$

The system has no solution. The solution set is { } or ∅.

10. $\begin{cases} -3x + 2y = -1 \\ 9x - 6y = 3 \end{cases}$

To eliminate x, we multiply both sides of the first equation by 3.

$$\begin{cases} 3(-3x + 2y) = 3(-1) \\ 9x - 6y = 3 \end{cases}$$

$$\begin{cases} -9x + 6y = -3 \\ 9x - 6y = 3 \end{cases}$$
$$0 = 0 \quad \text{True}$$

There is an infinite number of solutions. The solution set is $\{(x, y) | -3x + 2y = -1\}$ or $\{(x, y) | 9x - 6y = 3\}$.

Graphing Calculator Explorations

1.

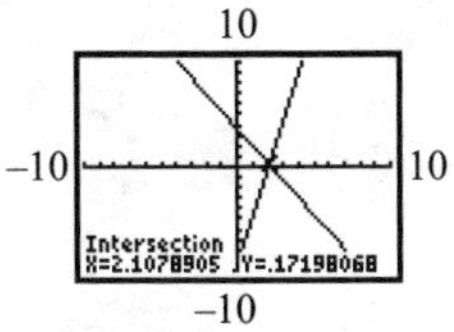

(2.11, 0.17)

2.

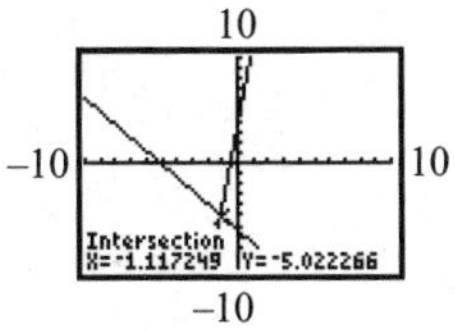

(−1.12, −5.02)

3.

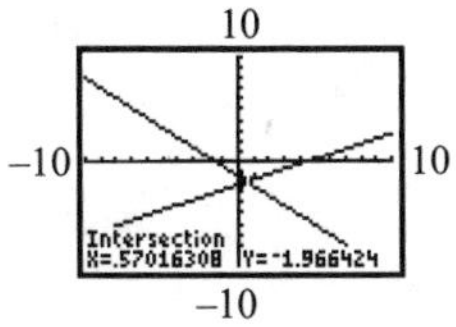

(0.57, −1.97)

4.

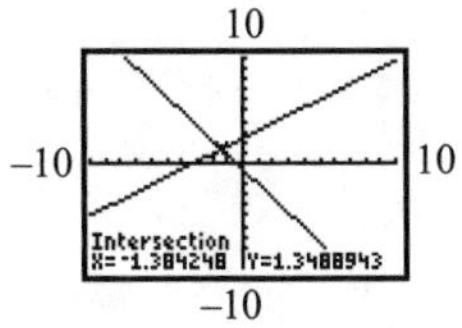

(−1.38, 1.35)

Vocabulary, Readiness & Video Check 4.1

1. A system with no solution has lines that are parallel. The correct graph is B.

2. A system with an infinite number of solutions has lines that are the same. The correct graph is C.

3. A system with solution (1, −2) has lines that intersect at (1, −2). The correct graph is A.

4. A system with solution (−3, 0) has lines that intersect at (−3, 0). The correct graph is D.

5. The ordered pair must be a solution of *both* equations of the system in order to be a solution of the system.

6. If the system contains two equations in two variables and has a single solution, this solution is the intersection of two lines. An intersection point with noninteger coordinates would be very difficult to graph/guess correctly.

7. Solve one equation for a variable. Next be sure to substitute this expression for the variable into the *other* equation.

8. Step 2 says to multiply one or both equations by a nonzero number to get opposite coefficients on a variable. However, once we eliminate fractions, the variable x already has opposite coefficients, so we proceed to Step 3.

Exercise Set 4.1

1. $\begin{cases} x - y = 3 \\ 2x - 4y = 8 \end{cases}$

$$x - y = 3$$
$$2 - (-1) \stackrel{?}{=} 3$$
$$3 = 3 \quad \text{True}$$
$$2x - 4y = 8$$
$$2(2) - 4(-1) \stackrel{?}{=} 8$$
$$4 + 4 \stackrel{?}{=} 8$$
$$8 = 8 \quad \text{True}$$

Yes, (2, −1) is a solution.

3. $\begin{cases} 2x-3y=-9 \\ 4x+2y=-2 \end{cases}$

$$\begin{aligned} 2x-3y&=-9 \\ 2(3)-3(5)&\stackrel{?}{=}-9 \\ 6-15&\stackrel{?}{=}-9 \\ -9&=-9 \quad \text{True} \\ 4x+2y&=-2 \\ 4(3)+2(5)&\stackrel{?}{=}-2 \\ 12+10&\stackrel{?}{=}-2 \\ 22&=-2 \quad \text{False} \end{aligned}$$

No, (3, 5) is not a solution.

5. $\begin{cases} y=-5x \\ x=-2 \end{cases}$

$$\begin{aligned} y&=-5x \\ 10&\stackrel{?}{=}-5(-2) \\ 10&=10 \quad \text{True} \end{aligned} \qquad \begin{aligned} x&=-2 \\ -2&=-2 \quad \text{True} \end{aligned}$$

Yes, (–2, 10) is a solution of the system.

7. $\begin{cases} 3x+7y=-19 \\ -6x=5y+8 \end{cases}$

$$\begin{aligned} 3x+7y&=-19 \\ 3\left(\frac{2}{3}\right)+7(-3)&\stackrel{?}{=}-19 \\ 2+(-21)&\stackrel{?}{=}-19 \\ -19&=-19 \quad \text{True} \\ -6x&=5y+8 \\ -6\left(\frac{2}{3}\right)&\stackrel{?}{=}5(-3)+8 \\ -4&\stackrel{?}{=}-15+8 \\ -4&=-7 \quad \text{False} \end{aligned}$$

No, $\left(\frac{2}{3}, -3\right)$ is not a solution.

9. $\begin{cases} x+y=1 \\ x-2y=4 \end{cases}$

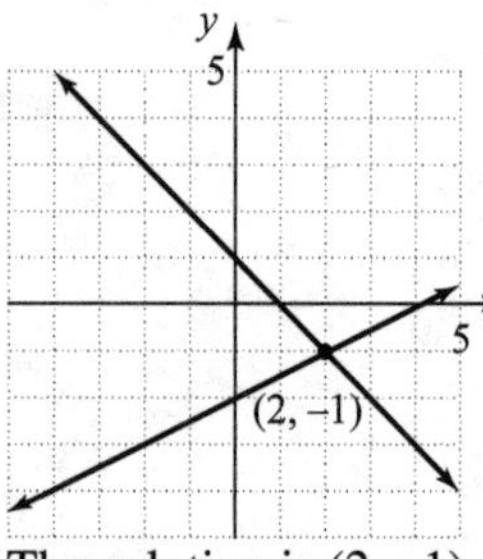

The solution is (2, –1).

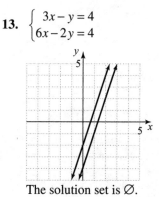

11. $\begin{cases} 2y-4x=0 \\ x+2y=5 \end{cases}$

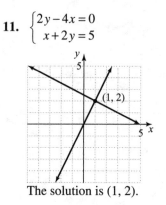

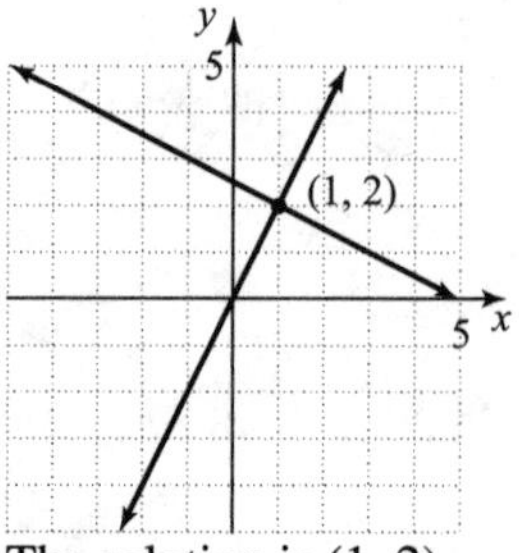

The solution is (1, 2).

13. $\begin{cases} 3x-y=4 \\ 6x-2y=4 \end{cases}$

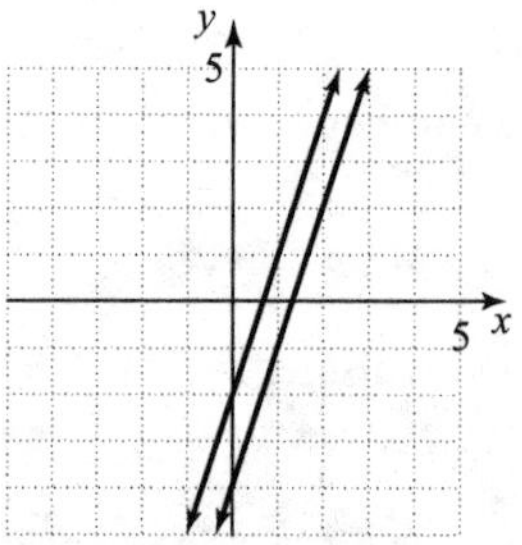

The solution set is $\varnothing$.

15. $\begin{cases} x+y=10 \\ y=4x \end{cases}$

Replace y with $4x$ in E1.

$$\begin{aligned} x+(4x)&=10 \\ 5x&=10 \\ x&=2 \end{aligned}$$

Replace x with 2 in E2.

$$\begin{aligned} y&=4(2) \\ y&=8 \end{aligned}$$

The solution is (2, 8).

17. $\begin{cases} 4x-y=9 \\ 2x+3y=-27 \end{cases}$

Solve E1 for y.

$$\begin{aligned} 4x-y&=9 \\ y&=4x-9 \end{aligned}$$

Replace y with $4x-9$ in E2.

$$\begin{aligned} 2x+3(4x-9)&=-27 \\ 2x+12x-27&=-27 \\ 14x&=0 \\ x&=0 \end{aligned}$$

Replace x with 0 in E1.

$$\begin{aligned} 4(0)-y&=9 \\ y&=-9 \end{aligned}$$

The solution is (0, –9).

19. $\begin{cases} \frac{1}{2}x + \frac{3}{4}y = -\frac{1}{4} \\ \frac{3}{4}x - \frac{1}{4}y = 1 \end{cases}$

Clear fractions by multiplying each equation by 4.

$\begin{cases} 2x + 3y = -1 \\ 3x - y = 4 \end{cases}$

Now solve E2 for y.

$3x - y = 4$
$y = 3x - 4$

Replace y with $3x - 4$ in E1.

$2x + 3(3x - 4) = -1$
$2x + 9x - 12 = -1$
$11x = 11$
$x = 1$

Replace x with 1 in equation $y = 3x - 4$.

$y = 3(1) - 4$
$y = -1$

The solution is (1, –1).

21. $\begin{cases} \frac{x}{3} + y = \frac{4}{3} \\ -x + 2y = 11 \end{cases}$

Clear fractions by multiplying the first equation by 3.

$\begin{cases} x + 3y = 4 \\ -x + 2y = 11 \end{cases}$

Solve E2 for x.

$2y - 11 = x$
$x = 2y - 11$

Replace x with $2y - 11$ in E1.

$(2y - 11) + 3y = 4$
$5y = 15$
$y = 3$

Replace y with 3 in equation $x = 2y - 11$.

$x = 2(3) - 11$
$x = -5$

The solution is (–5, 3).

23. $\begin{cases} -x + 2y = 0 \\ x + 2y = 5 \end{cases}$

$-x + 2y = 0$
$x + 2y = 5$
E1 + E2: $4y = 5$

$y = \frac{5}{4}$

Replace y with $\frac{5}{4}$ in E1.

$-x + 2\left(\frac{5}{4}\right) = 0$

$-x + \frac{5}{2} = 0$

$\frac{5}{2} = x$

The solution is $\left(\frac{5}{2}, \frac{5}{4}\right)$.

25. $\begin{cases} 5x + 2y = 1 \\ x - 3y = 7 \end{cases}$

Multiply E2 by –5.

$\begin{cases} 5x + 2y = 1 \\ -5x + 15y = -35 \end{cases}$

E1 + E2: $5x + 2y = 1$
$-5x + 15y = -35$
$17y = -34$
$y = -2$

Replace y with –2 in E2.

$x - 3(-2) = 7$
$x + 6 = 7$
$x = 1$

The solution is (1, –2).

27. $\begin{cases} \frac{3}{4}x + \frac{5}{2}y = 11 \\ \frac{1}{16}x - \frac{3}{4}y = -1 \end{cases}$

Clear fractions by multiplying E1 by 4 and E2 by 16.

$\begin{cases} 3x + 10y = 44 \\ x - 12y = -16 \end{cases}$

Multiply E2 by –3.

$\begin{cases} 3x + 10y = 44 \\ -3x + 36y = 48 \end{cases}$

E1 + E2: $3x + 10y = 44$
$-3x + 36y = 48$
$46y = 92$
$y = 2$

Replace y with 2 in the equation $x - 12y = -16$.

$x - 12(2) = -16$
$x - 24 = -16$
$x = 8$

The solution is (8, 2).

29. $\begin{cases} 3x - 5y = 11 \\ 2x - 6y = 2 \end{cases}$

Multiply E1 by 2 and E2 by –3.

$$\begin{cases} 6x - 10y = 22 \\ -6x + 18y = -6 \end{cases}$$

E1 + E2:
$$\begin{array}{r} 6x - 10y = 22 \\ \underline{-6x + 18y = -6} \\ 8y = 16 \\ y = 2 \end{array}$$

Replace y with 2 in E2.

$$\begin{aligned} 2x - 6(2) &= 2 \\ 2x - 12 &= 2 \\ 2x &= 14 \\ x &= 7 \end{aligned}$$

The solution is (7, 2).

31. $\begin{cases} x - 2y = 4 \\ 2x - 4y = 4 \end{cases}$

Multiply E1 by –2.

$$\begin{cases} -2x + 4y = -8 \\ 2x - 4y = 4 \end{cases}$$

E1 + E2:
$$\begin{array}{r} -2x + 4y = -8 \\ \underline{2x - 4y = 4} \\ 0 = -4 \quad \text{False} \end{array}$$

Inconsistent system; the solution set is ∅.

33. $\begin{cases} 3x + y = 1 \\ 2y = 2 - 6x \end{cases}$

$$\begin{cases} 3x + y = 1 \\ 6x + 2y = 2 \end{cases}$$

Multiply E1 by –2.

$$\begin{cases} -6x - 2y = -2 \\ 6x + 2y = 2 \end{cases}$$

E1 + E2:
$$\begin{array}{r} -6x - 2y = -2 \\ \underline{6x + 2y = 2} \\ 0 = 0 \quad \text{True} \end{array}$$

Dependent system; the solution set is $\{(x, y) | 3x + y = 1\}$.

35. $\begin{cases} 2x + 5y = 8 \\ 6x + y = 10 \end{cases}$

Multiply E1 by –3.

$$\begin{cases} -6x - 15y = -24 \\ 6x + y = 10 \end{cases}$$

E1 + E2:
$$\begin{array}{r} -6x - 15y = -24 \\ \underline{6x + y = 10} \\ -14y = -14 \\ y = 1 \end{array}$$

Replace y with 1 in E2.

$$\begin{aligned} 6x + 1 &= 10 \\ 6x &= 9 \\ x &= \frac{9}{6} = \frac{3}{2} \end{aligned}$$

The solution is $\left(\frac{3}{2}, 1\right)$.

37. $\begin{cases} 2x + 3y = 1 \\ x - 2y = 4 \end{cases}$

Multiply E1 by –1 and E2 by 2.

$$\begin{cases} -2x - 3y = -1 \\ 2x - 4y = 8 \end{cases}$$

E1 + E2:
$$\begin{array}{r} -2x - 3y = -1 \\ \underline{2x - 4y = 8} \\ -7y = 7 \\ y = -1 \end{array}$$

Replace y with –1 in E2.

$$\begin{aligned} x - 2(-1) &= 4 \\ x + 2 &= 4 \\ x &= 2 \end{aligned}$$

The solution is (2, –1).

39. $\begin{cases} \frac{1}{3}x + y = \frac{4}{3} \\ -\frac{1}{4}x - \frac{1}{2}y = -\frac{1}{4} \end{cases}$

Clear fractions by multiplying E1 by 3 and E2 by 4.

$$\begin{cases} x + 3y = 4 \\ -x - 2y = -1 \end{cases}$$

E1 + E2:
$$\begin{array}{r} x + 3y = 4 \\ \underline{-x - 2y = -1} \\ y = 3 \end{array}$$

Replace y with 3 in the equation $x + 3y = 4$.

$$\begin{aligned} x + 3(3) &= 4 \\ x + 9 &= 4 \\ x &= -5 \end{aligned}$$

The solution is (–5, 3).

41. $\begin{cases} 2x + 6y = 8 \\ 3x + 9y = 12 \end{cases}$

Multiply E1 by –3 and E2 by 2.

$$\begin{cases} -6x - 18y = -24 \\ 6x + 18y = 24 \end{cases}$$

E1 + E2:
$$\begin{array}{r} -6x - 18y = -24 \\ \underline{6x + 18y = 24} \\ 0 = 0 \quad \text{True} \end{array}$$

Dependent system; the solution is $\{(x, y) | 3x + 9y = 12\}$.

43. $\begin{cases} 4x+2y=5 \\ 2x+y=-1 \end{cases}$

Multiply E2 by –2.

$\begin{cases} 4x+2y=5 \\ -4x-2y=2 \end{cases}$

E1 + E2: $\begin{array}{r} 4x+2y=5 \\ \underline{-4x-2y=2} \\ 0=7 \end{array}$ False

Inconsistent system; the solution set is ∅.

45. $\begin{cases} 10y-2x=1 \\ 5y=4-6x \end{cases}$

$\begin{cases} 10y-2x=1 \\ 5y+6x=4 \end{cases}$

Multiply E2 by –2.

$\begin{cases} 10y-2x=1 \\ -10y-12x=-8 \end{cases}$

E1 + E2: $\begin{array}{r} 10y-2x=1 \\ \underline{-10y-12x=-8} \\ -14x=-7 \\ x=\frac{1}{2} \end{array}$

Replace x with $\frac{1}{2}$ in the equation $5y = 4 - 6x$.

$5y=4-6\left(\frac{1}{2}\right)$
$5y=4-3$
$5y=1$
$y=\frac{1}{5}$

The solution is $\left(\frac{1}{2}, \frac{1}{5}\right)$.

47. $\begin{cases} 5x-2y=27 \\ -3x+5y=18 \end{cases}$

Multiply E1 by 3 and E2 by 5.

$\begin{cases} 15x-6y=81 \\ -15x+25y=90 \end{cases}$

E1 + E2: $\begin{array}{r} 15x-6y=81 \\ \underline{-15x+25y=90} \\ 19y=171 \\ y=9 \end{array}$

Replace y with 9 in E1.

$5x-2(9)=27$
$5x-18=27$
$5x=45$
$x=9$

The solution is (9, 9).

49. $\begin{cases} x=3y+2 \\ 5x-15y=10 \end{cases}$

Replace x with $3y+2$ in E2.

$5(3y+2)-15y=10$
$15y+10-15y=10$
$10=10$ True

The system is dependent. The solution set is $\{(x, y)|x = 3y + 2\}$.

51. $\begin{cases} 2x-y=-1 \\ y=-2x \end{cases}$

Replace y with $-2x$ in E1.

$2x-(-2x)=-1$
$4x=-1$
$x=-\frac{1}{4}$

Replace x with $-\frac{1}{4}$ in E2.

$y=-2\left(-\frac{1}{4}\right)$
$y=\frac{1}{2}$

The solution is $\left(-\frac{1}{4}, \frac{1}{2}\right)$.

53. $\begin{cases} 2x=6 \\ y=5-x \end{cases}$

E1 yields $x = 3$.
Replace x with 3 in E2.

$y=5-3$
$y=2$

The solution is (3, 2).

55. $\begin{cases} \frac{x+5}{2}=\frac{6-4y}{3} \\ \frac{3x}{5}=\frac{21-7y}{10} \end{cases}$

Multiply E1 by 6 and E2 by 10.

$\begin{cases} 3x+15=12-8y \\ 6x=21-7y \end{cases}$

$\begin{cases} 3x+8y=-3 \\ 6x+7y=21 \end{cases}$

Multiply E1 by –2.

$$\begin{cases} -6x-16y=6 \\ 6x+7y=21 \end{cases}$$

E1 + E2: $\begin{array}{r} -6x-16y=6 \\ 6x+7y=21 \\ \hline -9y=27 \\ y=-3 \end{array}$

Replace y with –3 in the equation $3x + 8y = -3$.

$$\begin{aligned} 3x+8(-3)&=-3 \\ 3x-24&=-3 \\ 3x&=21 \\ x&=7 \end{aligned}$$

The solution is (7, –3).

57. $\begin{cases} 4x-7y=7 \\ 12x-21y=24 \end{cases}$

Multiply E1 by –3.

$$\begin{array}{r} \begin{cases} -12x+21y=-21 \\ 12x-21y=24 \end{cases} \\ \hline 0=3 \quad \text{False} \end{array}$$

Inconsistent system; the solution set is ∅.

59. $\begin{cases} \frac{2}{3}x-\frac{3}{4}y=-1 \\ -\frac{1}{6}x+\frac{3}{8}y=1 \end{cases}$

Multiply E1 by 12 and E2 by 24.

$$\begin{cases} 8x-9y=-12 \\ -4x+9y=24 \end{cases}$$

E1 + E2: $\begin{array}{r} 8x-9y=-12 \\ -4x+9y=24 \\ \hline 4x \quad =12 \\ x=3 \end{array}$

Replace x with 3 in the equation $-4x + 9y = 24$.

$$\begin{aligned} -4(3)+9y&=24 \\ -12+9y&=24 \\ 9y&=36 \\ y&=4 \end{aligned}$$

The solution is (3, 4).

61. $\begin{cases} 0.7x-0.2y=-1.6 \\ 0.2x-y=-1.4 \end{cases}$

Multiply both equations by 10.

$$\begin{cases} 7x-2y=-16 \\ 2x-10y=-14 \end{cases}$$

Multiply E1 by –5.

$$\begin{cases} -35x+10y=80 \\ 2x-10y=-14 \end{cases}$$

E1 + E2: $\begin{array}{r} -35x+10y=80 \\ 2x-10y=-14 \\ \hline -33x \quad =66 \\ x=-2 \end{array}$

Replace x with –2 in the equation $7x - 2y = -16$.

$$\begin{aligned} 7(-2)-2y&=-16 \\ -14-2y&=-16 \\ -2y&=-2 \\ y&=1 \end{aligned}$$

The solution is (–2, 1).

63. $\begin{cases} 4x-1.5y=10.2 \\ 2x+7.8y=-25.68 \end{cases}$

Multiply E2 by –2.

$$\begin{cases} 4x-1.5y=10.2 \\ -4x-15.6y=51.36 \end{cases}$$

E1 + E2: $\begin{array}{r} 4x-1.5y=10.2 \\ -4x-15.6y=51.36 \\ \hline -17.1y=61.56 \\ y=-3.6 \end{array}$

Replace y with –3.6 in E1.

$$\begin{aligned} 4x-1.5(-3.6)&=10.2 \\ 4x+5.4&=10.2 \\ 4x&=4.8 \\ x&=1.2 \end{aligned}$$

The solution is (1.2, –3.6).

65.

$$\begin{aligned} 3x-4y+2z&=5 \\ 3(1)-4(2)+2(5)&\stackrel{?}{=}5 \\ 3-8+10&\stackrel{?}{=}5 \\ 5&=5 \end{aligned}$$

True

67.

$$\begin{aligned} -x-5y+3z&=15 \\ -(0)-5(-1)+3(5)&\stackrel{?}{=}15 \\ 0+5+15&\stackrel{?}{=}15 \\ 20&=15 \end{aligned}$$

False

69. $\begin{array}{r} 3x+2y-5z=10 \\ -3x+4y+z=15 \\ \hline 6y-4z=25 \end{array}$

71. $\begin{array}{r} 10x+5y+6z=14 \\ -9x+5y-6z=-12 \\ \hline x+10y \quad =2 \end{array}$

73. $\begin{cases} y = 2x - 5 \\ y = 2x + 1 \end{cases}$

The lines have the same slope and different y-intercepts, so they are parallel. There is no solution.

75. $\begin{cases} x + y = 3 \\ 5x + 5y = 15 \end{cases}$

The result of multiplying the first equation by 5 is $5x + 5y = 15$, which is the second equation. The system is dependent, so it has an infinite number of solutions.

77. no; answers may vary

79. The lines intersect at the point (5, 21), which corresponds to 5000 DVDs at a price of $21 per DVD.

81. The graph for supply is higher than that for demand when x is greater than 6. Therefore, supply is greater than demand.

83. $\begin{cases} y = 2.5x \\ y = 0.9x + 3000 \end{cases}$

Replace y with $2.5x$ in the second equation.

$$\begin{aligned} y &= 0.9x + 3000 \\ 2.5x &= 0.9x + 3000 \\ 1.6x &= 3000 \\ x &= 1875 \end{aligned}$$

Replace x with 1875 in the first equation.

$y = 2.5x = 2.5(1875) = 4687.5$

The break-even point is (1875, 4687.5).

85. At $x = 2000$, the graph for revenue is higher than that for cost, so the company makes money.

87. The revenue graph is above the cost graph to the right of the intersection point. Thus, for x-values greater than 1875, the company will make a profit.

89. answers may vary; one possibility:

$\begin{cases} -2x + y = 1 \\ x - 2y = -8 \end{cases}$

91. a. Since the slope of the equation for mobile phone subscriptions is positive, the number of mobile phone subscriptions is increasing. Since the slope of the equation for landline subscriptions is negative, the number of landline subscriptions is decreasing.

b. $\begin{cases} y = 3.9x + 44.1 \\ y = -1.8x + 67.3 \end{cases}$

Replace y with $3.9x + 44.1$ in the second equation.

$$\begin{aligned} 3.9x + 44.1 &= -1.8x + 67.3 \\ 3.9x &= -1.8x + 23.2 \\ 5.7x &= 23.2 \\ 4 \approx 4.07 &\approx 4 \end{aligned}$$

Replace x with 4 in the first equation.

$y = 3.9(4) + 44.1 = 59.7 \approx 60$

The solution is approximately (4, 60).

c. $2000 + 4 = 2004$

In the year 2004, the number of mobile phone subscriptions per hundred people was about the same as the number of landline subscriptions.

93. $\begin{cases} \dfrac{1}{x} + y = 12 \\ \dfrac{3}{x} - y = 4 \end{cases}$

Replacing $\dfrac{1}{x}$ with a, we have

$\begin{cases} a + y = 12 \\ 3a - y = 4 \end{cases}$

Adding the two new equations we get

$$\begin{aligned} 4a &= 16 \\ a &= 4 \end{aligned}$$

Replace a with 4 in the equation $a + y = 12$.

$$\begin{aligned} 4 + y &= 12 \\ y &= 8 \end{aligned}$$

Since $a = 4$, $x = \dfrac{1}{4}$.

The solution is $\left(\dfrac{1}{4}, 8\right)$.

95. $\begin{cases} \dfrac{1}{x} + \dfrac{1}{y} = 5 \\ \dfrac{1}{x} - \dfrac{1}{y} = 1 \end{cases}$

Replace $\dfrac{1}{x}$ with a and $\dfrac{1}{y}$ with b.

$\begin{cases} a + b = 5 \\ a - b = 1 \end{cases}$

Adding the two new equations we get

$$\begin{aligned} 2a &= 6 \\ a &= 3 \end{aligned}$$

Replace a with 3 in the equation $a + b = 5$.

$$\begin{aligned}3+b&=5\\ b&=2\end{aligned}$$

Since $a = 3$, $x=\frac{1}{3}$. Similarly, $y=\frac{1}{2}$.

The solution is $\left(\frac{1}{3},\frac{1}{2}\right)$.

97. $\begin{cases}\frac{2}{x}-\frac{4}{y}=5\\ \frac{1}{x}-\frac{2}{y}=\frac{3}{2}\end{cases}$

Replace $\frac{1}{x}$ with a and $\frac{1}{y}$ with b.

$$\begin{cases}2a-4b=5\\ a-2b=\frac{3}{2}\end{cases}$$

Multiply E2 by 2.

$$\begin{cases}2a-4b=5\\ 2a-4b=3\end{cases}$$

This system is inconsistent. The solution set is $\varnothing$.

99. $\begin{cases}\frac{3}{x}-\frac{2}{y}=-18\\ \frac{2}{x}+\frac{3}{y}=1\end{cases}$

Replace $\frac{1}{x}$ with a and $\frac{1}{y}$ with b.

$$\begin{cases}3a-2b=-18\\ 2a+3b=1\end{cases}$$

Multiply E1 by 3 and E2 by 2, then add.

$$\begin{cases}9a-6b=-54\\ 4a+6b=2\end{cases}$$

$$\begin{aligned}13a\quad &=-52\\ a&=-4\end{aligned}$$

Replace a with -4 in the equation $2a + 3b = 1$.

$$\begin{aligned}2(-4)+3b&=1\\ 3b&=9\\ b&=3\end{aligned}$$

Since $a=-4$, $x=-\frac{1}{4}$. Also, since $b = 3$, $y=\frac{1}{3}$.

The solution is $\left(-\frac{1}{4},\frac{1}{3}\right)$.

Section 4.2 Practice Exercises

1. $\begin{cases}3x+2y-z=0 & (1)\\ x-y+5z=2 & (2)\\ 2x+3y+3z=7 & (3)\end{cases}$

Multiply equation (2) by 2 and add to equation (1) to eliminate y.

$$\begin{cases}3x+2y-z=0\\ 2(x-y+5z)=2(2)\end{cases}$$

$$\begin{cases}3x-2y\quad -z=0\\ 2x-2y+10z=4\end{cases}$$

$$5x\qquad +9z=4\quad (4)$$

Multiply equation (2) by 3 and add to equation (3) to eliminate y again.

$$\begin{cases}3(x-y+5z)=3(2)\\ 2x+3y+3z=7\end{cases}$$

$$\begin{cases}3x-3y+15z=6\\ 2x+3y\quad +3z=7\end{cases}$$

$$5x+\qquad 18z=13\quad (5)$$

Multiply equation (4) by -1 and add to equation (5) to eliminate x.

$$\begin{cases}-1(5x+9z)=-1(4)\\ 5x+18z=13\end{cases}$$

$$\begin{cases}-5x\quad -9z=-4\\ 5x+18z=13\end{cases}$$

$$\begin{aligned}9z&=9\\ z&=1\end{aligned}$$

Replace z with 1 in equation (4) or (5).

$$\begin{aligned}5x+9z&=4\\ 5x+9(1)&=4\\ 5x&=-5\\ x&=-1\end{aligned}$$

Replace x with -1 and z with 1 in equation (1), (2), or (3).

$$\begin{aligned}x-y+5z&=2\\ -1-y+5(1)&=2\\ -y+4&=2\\ -y&=-2\\ y&=2\end{aligned}$$

The solution is $(-1, 2, 1)$. To check, let $x = -1$, $y = 2$, and $z = 1$ in all three original equations of the system.

2. $\begin{cases}6x-3y+12z=4 & (1)\\ -6x+4y-2z=7 & (2)\\ -2x+y-4z=3 & (3)\end{cases}$

Multiply equation (3) by 3 and add to equation (1) to eliminate x.

$$\begin{cases} 6x-3y+12z=4 \\ 3(-2x+y-4z)=3(3) \end{cases}$$

$$\begin{cases} 6x-3y+12z=4 \\ -6x+3y-12z=9 \end{cases}$$

$$0=13 \quad \text{False}$$

Since the statement is false, this system is inconsistent and has no solution. The solution set is { } or ∅.

3. $\begin{cases} 3x+4y \quad\quad =0 & (1) \\ 9x \quad\quad -4z=6 & (2) \\ \quad -2y+7z=1 & (3) \end{cases}$

Equation (2) has no term containing the variable y. Eliminate y using equations (1) and (3). Multiply equation (3) by 2 and add to equation (1).

$$\begin{cases} 3x+4y=0 \\ 2(-2y+7z)=2(1) \end{cases}$$

$$\begin{cases} 3x+4y \quad\quad =0 \\ \quad -4y+14z=2 \end{cases}$$

$$3x \quad +14z=2 \quad (4)$$

Multiply equation (4) by −3 and add to equation (2) to eliminate x.

$$\begin{cases} 9x-4z=6 \\ -3(3x+14z)=-3(2) \end{cases}$$

$$\begin{cases} 9x \; -4z=6 \\ -9x-52z=-6 \end{cases}$$

$$-56z=0$$
$$z=0$$

Replace z with 0 in equation (2) and solve for x.

$$9x-4z=6$$
$$9x-4(0)=6$$
$$9x=6$$
$$x=\frac{6}{9}=\frac{2}{3}$$

Replace z with 0 in equation (3) and solve for y.

$$-2y+7z=1$$
$$-2y+7(0)=1$$
$$-2y=1$$
$$y=-\frac{1}{2}$$

The solution is $\left(\frac{2}{3}, -\frac{1}{2}, 0\right)$.

4. $\begin{cases} 2x+y-3z=6 & (1) \\ x+\frac{1}{2}y-\frac{3}{2}z=3 & (2) \\ -4x-2y+6z=-12 & (3) \end{cases}$

Multiply both sides of equation (2) by 2 to eliminate fractions, and multiply both sides of equation (3) by $-\frac{1}{2}$ since all coefficients in equation (3) are divisible by 2 and the coefficient of x is negative. The resulting system is

$$\begin{cases} 2x+y-3z=6 \\ 2x+y-3z=6 \\ 2x+y-3z=6 \end{cases}$$

Since the three equations are identical, there are infinitely many solutions of the system. The equations are dependent. The solution set can be written as $\{(x, y, z) | 2x + y - 3z = 6\}$.

5. $\begin{cases} x+2y+4z=16 & (1) \\ x \quad\quad +2z=-4 & (2) \\ \quad\quad y-3z=30 & (3) \end{cases}$

Solve equation (2) for x and equation (3) for y.

$x+2z=-4$ $\qquad\qquad$ $y-3z=30$

$x=-2z-4$ $\qquad\qquad$ $y=3z+30$

Substitute $-2z-4$ for x and $3z+30$ for y in equation (1) and solve for z.

$$x+2y+4z=16$$
$$(-2z-4)+2(3z+30)+4z=16$$
$$-2z-4+6z+60+4z=16$$
$$8z+56=16$$
$$8z=-40$$
$$z=-5$$

Use $x=-2z-4$ to find x:
$x=-2(-5)-4=10-4=6$.
Use $y=3z+30$ to find y:
$y=3(-5)+30=-15+30=15$.
The solution is (6, 15, −5).

Vocabulary, Readiness & Video Check 4.2

1. a. $x+y+z=3$

$$-1+3+1 \stackrel{?}{=} 3$$
$$3=3 \quad \text{True}$$

b. $-x+y+z=5$

$$-(-1)+3+1 \stackrel{?}{=} 5$$
$$1+3+1 \stackrel{?}{=} 5$$
$$5=5 \quad \text{True}$$

c. $-x+y+2z=0$
$-(-1)+3+2(1) \stackrel{?}{=} 0$
$1+3+2 \stackrel{?}{=} 0$
$6=0$ False

d. $x+2y-3z=2$
$-1+2(3)-3(1) \stackrel{?}{=} 2$
$-1+6-3 \stackrel{?}{=} 2$
$2=2$ True

(−1, 3, 1) is a solution to the equations a, b, and d.

2. a. $x+y+z=-1$
$2+1+(-4) \stackrel{?}{=} -1$
$-1=-1$ True

b. $x-y-z=-3$
$2-1-(-4) \stackrel{?}{=} -3$
$2-1+4 \stackrel{?}{=} -3$
$5=-3$ False

c. $2x-y+z=-1$
$2(2)-1+(-4) \stackrel{?}{=} -1$
$4-1-4 \stackrel{?}{=} -1$
$-1=-1$ True

d. $-x-3y-z=-1$
$-2-3(1)-(-4) \stackrel{?}{=} -1$
$-2-3+4 \stackrel{?}{=} -1$
$-1=-1$ True

(2, 1, −4) is a solution to the equations a, c, and d.

3. yes; answers may vary

4. no; answers may vary

5. Once we have one equation in two variables, we need to get another equation in the *same* two variables, giving us a system of two equations in two variables. We solve this new system to find the values of two variables. We then substitute these values into an original equation to find the value of the third.

Exercise Set 4.2

1. $\begin{cases} x-y+z=-4 & (1) \\ 3x+2y-z=5 & (2) \\ -2x+3y-z=15 & (3) \end{cases}$

Add the equations (1) and (2) to eliminate z.

$$\begin{array}{l} x-y+z=-4 \\ \underline{3x+2y-z=5} \\ 4x+y\quad=1 \quad (4) \end{array}$$

Add equations (1) and (3) to eliminate z.

$$\begin{array}{l} x-y+z=-4 \\ \underline{-2x+3y-z=15} \\ -x+2y\quad=11 \quad (5) \end{array}$$

Multiply equation (5) by 4 and add it to equation (4).

$$\begin{array}{r} 4x+y=1 \\ \underline{-4x+8y=44} \\ 9y=45 \\ y=5 \end{array}$$

Replace y with 5 in equation (4).

$4x+y=1$
$4x+5=1$
$4x=-4$
$x=-1$

Replace x with −1 and y with 5 in equation (1).

$x-y+z=-4$
$-1-5+z=-4$
$-6+z=-4$
$z=2$

The solution of the system is (−1, 5, 2).

3. $\begin{cases} x+y\quad=3 & (1) \\ 2y\quad=10 & (2) \\ 3x+2y-3z=1 & (3) \end{cases}$

Solve equation (2) for y.

$2y=10$
$y=5$

Replace y with 5 in equation (1).

$x+y=3$
$x+5=3$
$x=-2$

Replace x with −2 and y with 5 in equation (3).

$3x+2y-3z=1$
$3(-2)+2(5)-3z=1$
$-6+10-3z=1$
$4-3z=1$
$-3z=-3$
$z=1$

The solution of the system is (−2, 5, 1).

5. $\begin{cases} 2x+2y+z=1 & (1) \\ -x+y+2z=3 & (2) \\ x+2y+4z=0 & (3) \end{cases}$

Add equations (2) and (3) to eliminate x.

$$\begin{array}{r} -x+y+2z=3 \\ \underline{x+2y+4z=0} \\ 3y+6z=3 \\ y+2z=1 \quad (4) \end{array}$$

Multiply equation (2) by 2 and add it to equation (1) to eliminate x.

$$\begin{array}{r} 2x+2y+z=1 \\ \underline{-2x+2y+4z=6} \\ 4y+5z=7 \quad (5) \end{array}$$

Multiply equation (4) by –4 and add it to equation (5).

$$\begin{array}{r} -4y-8z=-4 \\ \underline{4y+5z=7} \\ -3z=3 \\ z=-1 \end{array}$$

Replace z with –1 in equation (4).

$$\begin{aligned} y+2z&=1 \\ y+2(-1)&=1 \\ y-2&=1 \\ y&=3 \end{aligned}$$

Replace y with 3 and z with –1 in equation (3).

$$\begin{aligned} x+2y+4z&=0 \\ x+2(3)+4(-1)&=0 \\ x+6-4&=0 \\ x+2&=0 \\ x&=-2 \end{aligned}$$

The solution of the system is (–2, 3, –1).

7. $\begin{cases} x-2y+z=-5 & (1) \\ -3x+6y-3z=15 & (2) \\ 2x-4y+2z=-10 & (3) \end{cases}$

Multiply equation (1) by –3.

$-3x + 6y - 3z = 15$

This is equation (2).

Multiply equation (1) by 2.

$2x - 4y + 2z = -10$

This is equation (3).

The system is dependent and the solution set is $\{(x, y, z)|x - 2y + z = -5\}$.

9. $\begin{cases} 4x-y+2z=5 & (1) \\ 2y+z=4 & (2) \\ 4x+y+3z=10 & (3) \end{cases}$

Multiply equation (1) by –1 and add it to equation (3) to eliminate x.

$$\begin{array}{r} -4x+y-2z=-5 \\ \underline{4x+y+3z=10} \\ 2y+z=5 \quad (4) \end{array}$$

Multiply equation (2) by –1 and add it to equation (4).

$$\begin{array}{r} -2y-z=-4 \\ \underline{2y+z=5} \\ 0=1 \end{array}$$

The statement 0 = 1 is false, so the system has no solution.

11. $\begin{cases} x+5z=0 & (1) \\ 5x+y=0 & (2) \\ y-3z=0 & (3) \end{cases}$

Solve equation (1) for x.

$$\begin{aligned} x+5z&=0 \\ x&=-5z \end{aligned}$$

Solve equation (3) for y.

$$\begin{aligned} y-3z&=0 \\ y&=3z \end{aligned}$$

Replace x with $-5z$ and y with $3z$ in equation (2).

$$\begin{aligned} 5x+y&=0 \\ 5(-5z)+3z&=0 \\ -25z+3z&=0 \\ -22z&=0 \\ z&=0 \end{aligned}$$

Replace z with 0 in equation (1).

$$\begin{aligned} x+5z&=0 \\ x+5(0)&=0 \\ x+0&=0 \\ x&=0 \end{aligned}$$

Replace x with 0 in equation (2).

$$\begin{aligned} 5x+y&=0 \\ 5(0)+y&=0 \\ 0+y&=0 \\ y&=0 \end{aligned}$$

The solution of the system is (0, 0, 0).

13. $\begin{cases} 6x-5z=17 & (1) \\ 5x-y+3z=-1 & (2) \\ 2x+y=-41 & (3) \end{cases}$

Add equations (2) and (3) to eliminate y.

$$\begin{array}{r} 5x-y+3z=-1 \\ \underline{2x+y=-41} \\ 7x+3z=-42 \quad (4) \end{array}$$

Multiply equation (1) by 3 and equation (4) by 5, then add the resulting equations.

$$\begin{array}{r} 18x-15z=51 \\ \underline{35x+15z=-210} \\ 53x=-159 \\ x=-3 \end{array}$$

Replace x with −3 in equation (1).

$$\begin{aligned} 6x-5z&=17\\ 6(-3)-5z&=17\\ -18-5z&=17\\ -5z&=35\\ z&=-7 \end{aligned}$$

Replace x with −3 in equation (3).

$$\begin{aligned} 2x+y&=-41\\ 2(-3)+y&=-41\\ -6+y&=-41\\ y&=-35 \end{aligned}$$

The solution of the system is (−3, −35, −7).

15. $\begin{cases} x+y+z=8 & (1)\\ 2x-y-z=10 & (2)\\ x-2y-3z=22 & (3) \end{cases}$

Add equations (1) and (2) to eliminate y and z.

$$\begin{aligned} x+y+z&=8\\ 2x-y-z&=10\\ \hline 3x\qquad&=18\\ x&=6 \end{aligned}$$

Multiply equation (1) by 2 and add it to equation (3) to eliminate y.

$$\begin{aligned} 2x+2y+2z&=16\\ x-2y-3z&=22\\ \hline 3x\qquad-z&=38 \quad (4) \end{aligned}$$

Replace x with 6 in equation (4).

$$\begin{aligned} 3x-z&=38\\ 3(6)-z&=38\\ 18-z&=38\\ z&=-20 \end{aligned}$$

Replace x with 6 and z with −20 in equation (1).

$$\begin{aligned} x+y+z&=8\\ 6+y+(-20)&=8\\ y-14&=8\\ y&=22 \end{aligned}$$

The solution of the system is (6, 22, −20).

17. $\begin{cases} x+2y-z=5 & (1)\\ 6x+y+z=7 & (2)\\ 2x+4y-2z=5 & (3) \end{cases}$

Add equations (1) and (2) to eliminate z.

$$\begin{aligned} x+2y-z&=5\\ 6x+y+z&=7\\ \hline 7x+3y\qquad&=12 \quad (4) \end{aligned}$$

Multiply equation (2) by 2 and add to equation (3) to eliminate z.

$$\begin{aligned} 12x+2y+2z&=14\\ 2x+4y-2z&=5\\ \hline 14x+6y\qquad&=19 \quad (5) \end{aligned}$$

Multiply equation (4) by −2 and add to equation (5).

$$\begin{aligned} -14x-6y&=-24\\ 14x+6y&=19\\ \hline 0&=-5 \end{aligned}$$

The statement 0 = −5 is false, so the system has no solution.

19. $\begin{cases} 2x-3y+z=2 & (1)\\ x-5y+5z=3 & (2)\\ 3x+y-3z=5 & (3) \end{cases}$

Multiply equation (1) by −5 and add it to equation (2) to eliminate z.

$$\begin{aligned} -10x+15y-5z&=-10\\ x-5y+5z&=3\\ \hline -9x+10y\qquad&=-7 \quad (4) \end{aligned}$$

Multiply equation (1) by 3 and add it to equation (3) to eliminate z.

$$\begin{aligned} 6x-9y+3z&=6\\ 3x+y-3z&=5\\ \hline 9x-8y\qquad&=11 \quad (5) \end{aligned}$$

Add equations (4) and (5) to eliminate x.

$$\begin{aligned} -9x+10y&=-7\\ 9x-8y&=11\\ \hline 2y&=4\\ y&=2 \end{aligned}$$

Replace y with 2 in equation (5).

$$\begin{aligned} 9x-8y&=11\\ 9x-8(2)&=11\\ 9x-16&=11\\ 9x&=27\\ x&=3 \end{aligned}$$

Replace x with 3 and y with 2 in equation (1).

$$\begin{aligned} 2x-3y+z&=2\\ 2(3)-3(2)+z&=2\\ 6-6+z&=2\\ z&=2 \end{aligned}$$

The solution of the system is (3, 2, 2).

21. $\begin{cases} -2x-4y+6z=-8 & (1)\\ x+2y-3z=4 & (2)\\ 4x+8y-12z=16 & (3) \end{cases}$

Multiply equation (2) by −2.

$-2x-4y+6z=-8$

This is equation (1).

Multiply equation (2) by 4.

$4x+8y-12z=16$

This is equation (3).

The system is dependent and the solution set is $\{(x, y, z)|x+2y-3z=4\}$.

23. $\begin{cases} 2x+2y-3z=1 & (1) \\ y+2z=-14 & (2) \\ 3x-2y=-1 & (3) \end{cases}$

Multiply equation (1) by −3 and equation (3) by 2, then add the results to eliminate *x*.

$$\begin{aligned} -6x-6y+9z&=-3 \\ 6x-4y&=-2 \\ \hline -10y+9z&=-5 \quad (4) \end{aligned}$$

Multiply equation (2) by 10 and add the result to equation (4) to eliminate *y*.

$$\begin{aligned} 10y+20z&=-140 \\ -10y+9z&=-5 \\ \hline 29z&=-145 \\ z&=-5 \end{aligned}$$

Replace *z* with −5 in equation (2).

$$\begin{aligned} y+2z&=-14 \\ y+2(-5)&=-14 \\ y-10&=-14 \\ y&=-4 \end{aligned}$$

Replace *y* with −4 in equation (3).

$$\begin{aligned} 3x-2y&=-1 \\ 3x-2(-4)&=-1 \\ 3x+8&=-1 \\ 3x&=-9 \\ x&=-3 \end{aligned}$$

The solution of the system is (−3, −4, −5).

25. $\begin{cases} x+2y-z=5 & (1) \\ -3x-2y-3z=11 & (2) \\ 4x+4y+5z=-18 & (3) \end{cases}$

Add equations (1) and (2) to eliminate *y*.

$$\begin{aligned} x+2y-z&=5 \\ -3x-2y-3z&=11 \\ \hline -2x-4z&=16 \\ x+2z&=-8 \quad (4) \end{aligned}$$

Multiply equation (2) by 2 and add to equation (3) to eliminate *y*.

$$\begin{aligned} -6x-4y-6z&=22 \\ 4x+4y+5z&=-18 \\ \hline -2x-z&=4 \quad (5) \end{aligned}$$

Multiply equation (5) by 2 and add to equation (4).

$$\begin{aligned} x+2z&=-8 \\ -4x-2z&=8 \\ \hline -3x&=0 \\ x&=0 \end{aligned}$$

Replace *x* with 0 in equation (4).

$$\begin{aligned} x+2z&=-8 \\ 0+2z&=-8 \\ 2z&=-8 \\ z&=-4 \end{aligned}$$

Replace *x* with 0 and *z* with −4 in equation (1).

$$\begin{aligned} x+2y-z&=5 \\ 0+2y-(-4)&=5 \\ 2y+4&=5 \\ 2y&=1 \\ y&=\frac{1}{2} \end{aligned}$$

The solution of the system is $\left(0, \frac{1}{2}, -4\right)$.

27. $\begin{cases} \frac{3}{4}x-\frac{1}{3}y+\frac{1}{2}z=9 & (1) \\ \frac{1}{6}x+\frac{1}{3}y-\frac{1}{2}z=2 & (2) \\ \frac{1}{2}x-y+\frac{1}{2}z=2 & (3) \end{cases}$

Multiply equation (1) by 12, equation (2) by 6, and equation (3) by 2 to clear fractions.

$\begin{cases} 9x-4y+6z=108 & (4) \\ x+2y-3z=12 & (5) \\ x-2y+z=4 & (6) \end{cases}$

Multiply equation (5) by 2 and add it to equation (4) to eliminate *y* and *z*.

$$\begin{aligned} 9x-4y+6z&=108 \\ 2x+4y-6z&=24 \\ \hline 11x&=132 \\ x&=12 \end{aligned}$$

Add equations (5) and (6) to eliminate *y*.

$$\begin{aligned} x+2y-3z&=12 \\ x-2y+z&=4 \\ \hline 2x-2z&=16 \quad (7) \end{aligned}$$

Replace *x* with 12 in equation (7).

$$\begin{aligned} 2x-2z&=16 \\ 2(12)-2z&=16 \\ 24-2z&=16 \\ -2z&=-8 \\ z&=4 \end{aligned}$$

Replace *x* with 12 and *z* with 4 in equation (6).

$$\begin{aligned} x-2y+z&=4 \\ 12-2y+4&=4 \\ -2y+16&=4 \\ -2y&=-12 \\ y&=6 \end{aligned}$$

The solution of the system is (12, 6, 4).

29. Let x = the first number, then
$2x$ = the second number.

$$x + 2x = 45$$
$$3x = 45$$
$$x = 15$$
$$2x = 2(15) = 30$$

The numbers are 15 and 30.

31.
$$2(x-1) - 3x = x - 12$$
$$2x - 2 - 3x = x - 12$$
$$-x - 2 = x - 12$$
$$-2x = -10$$
$$x = 5$$

33.
$$-y - 5(y+5) = 3y - 10$$
$$-y - 5y - 25 = 3y - 10$$
$$-6y - 25 = 3y - 10$$
$$-9y = 15$$
$$y = -\frac{15}{9} = -\frac{5}{3}$$

35. answers may vary

37. answers may vary

39.
$$\begin{cases} x + y + z = 1 & (1) \\ 2x - y + z = 0 & (2) \\ -x + 2y + 2z = -1 & (3) \end{cases}$$

Add E1 and E3.
$3y + 3z = 0$ or $y + z = 0$ (4)

Add –2 times E1 to E2.
$$\begin{array}{r} -2x - 2y - 2z = -2 \\ 2x - y + z = 0 \\ \hline -3y - z = -2 \quad (5) \end{array}$$

Add E4 and E5.
$$-2y = -2$$
$$y = 1$$

Replace y with 1 in E4.
$$1 + z = 0$$
$$z = -1$$

Replace y with 1 and z with –1 in E1.
$$x + 1 + (-1) = 1$$
$$x = 1$$

The solution is (1, 1, –1), and
$$\frac{x}{8} + \frac{y}{4} + \frac{z}{3} = \frac{1}{8} + \frac{1}{4} - \frac{1}{3}$$
$$= \frac{3}{24} + \frac{6}{24} - \frac{8}{24}$$
$$= \frac{1}{24}.$$

41.
$$\begin{cases} x + y \quad - w = 0 & (1) \\ y + 2z + w = 3 & (2) \\ x \quad - z \quad = 1 & (3) \\ 2x - y \quad - w = -1 & (4) \end{cases}$$

Add E1 and E2.
$x + 2y + 2z = 3$ (5)

Add E2 and E4.
$2x + 2z = 2$ or $x + z = 1$ (6)

Add E3 and E6.
$$\begin{array}{l} x - z = 1 \\ x + z = 1 \\ \hline 2x \quad = 2 \end{array}$$

Replace x with 1 in E3.
$$1 - z = 1$$
$$z = 0$$

Replace x with 1 and z with 0 in E5.
$$1 + 2y + 2(0) = 3$$
$$1 + 2y = 3$$
$$2y = 2$$
$$y = 1$$

Replace y with 1, and z with 0 in E2.
$$1 + 2(0) + w = 3$$
$$1 + w = 3$$
$$w = 2$$

The solution is (1, 1, 0, 2).

43.
$$\begin{cases} x + y + z + w = 5 & (1) \\ 2x + y + z + w = 6 & (2) \\ x + y + z \quad = 2 & (3) \\ x + y \quad = 0 & (4) \end{cases}$$

Add –1 times E4 to E3.
$$\begin{array}{r} -x - y \quad = 0 \\ x + y + z = 2 \\ \hline z = 2 \end{array}$$

Replace z with 2 in E1 and E2.
$$\begin{cases} x + y + w = 3 & (5) \\ 2x + y + w = 4 & (6) \end{cases}$$

Add –1 times E5 to E6.
$$\begin{array}{l} -x - y - w = -3 \\ 2x + y + w = 4 \\ \hline x \quad = 1 \end{array}$$

Replace x with 1 in E4.
$$1 + y = 0$$
$$y = -1$$

Replace x with 1, y with –1, and z with 2 in E1.
$$1 + (-1) + 2 + w = 5$$
$$2 + w = 5$$
$$w = 3$$

The solution is (1, –1, 2, 3).

45. answers may vary

Section 4.3 Practice Exercises

1. a. $\begin{cases} y = -0.91x + 110.45 \\ y = 0.11x + 98.6 \end{cases}$

Use substitution.

$$\begin{aligned} -0.91x + 110.45 &= 0.11x + 98.6 \\ -1.02x &= -11.85 \\ x &\approx 11.6 \approx 12 \end{aligned}$$

The predicted year is 12 years after 2005 or 2017.

b. yes; answers may vary

2. Let x = first number
y = second number
"A first number is five more than a second number" is translated as $x = y + 5$. "Twice the first number is 2 less than 3 times the second number" is translated as $2x = 3y - 2$.

We solve the following system.

$\begin{cases} x = y + 5 \\ 2x = 3y - 2 \end{cases}$

Since the first equation is solved for x, we use substitution. Substitute $y + 5$ for x in the second equation.

$$\begin{aligned} 2(y+5) &= 3y - 2 \\ 2y + 10 &= 3y - 2 \\ 12 &= y \end{aligned}$$

Replace y with 12 in the equation $x = y + 5$ and solve for x.

$x = 12 + 5 = 17$

The numbers are 12 and 17.

3. Let x = speed of the V150
y = speed of the Atlantique
We summarize the information in a chart. Both trains have traveled two hours.

	Rate	•	Time	=	Distance
V150	x		2		$2x$
Atlantique	y		2		$2y$

The trains are 2150 kilometers apart, so the sum of the distances is 2150: $2x + 2y = 2150$.

The V150 is 75 kph faster than the Atlantique: $x = y + 75$.

We solve the following system.

$\begin{cases} 2x + 2y = 2150 \\ x = y + 75 \end{cases}$

Since the second equation is solved for x, we use substitution. Substitute $y + 75$ for x in the first equation.

$$\begin{aligned} 2(y+75) + 2y &= 2150 \\ 2y + 150 + 2y &= 2150 \\ 4y + 150 &= 2150 \\ 4y &= 2000 \\ y &= 500 \end{aligned}$$

To find x, we replace y with 500 in the second equation.

$x = 500 + 75 = 575$

The speed of the V150 is 575 kph, and the speed of the Atlantique is 500 kph.

4. Let x = amount of 99% acid
y = amount of water (0%)
Both x and y are measured in liters. We use a table to organize the given data.

	Amount	Acid Strength	Amount of Pure Acid
99% acid	x	99%	$0.99x$
Water	y	0%	$0y$

The amount of 99% acid and water combined must equal 1 liter, so $x + y = 1$.
The amount of pure acid in the mixture must equal the sum of the amounts of pure acid in the 99% acid and in the water, so
$0.99x + 0y = 0.05(1)$, which simplifies to $0.99x = 0.05$.
We solve the following system.

$\begin{cases} x + y = 1 \\ 0.99x = 0.05 \end{cases}$

Since the second equation does not contain y, we solve it for x.

$$\begin{aligned} 0.99x &= 0.05 \\ x &= \frac{0.05}{0.99} \approx 0.05 \end{aligned}$$

To find y, we replace x with 0.05 in the first equation.

$$\begin{aligned} x + y &= 1 \\ 0.05 + y &= 1 \\ y &= 0.95 \end{aligned}$$

The teacher should use 0.05 liter of the 99% HCL solution and 0.95 liter of water.

5. Let x = the number of packages.
The firm charges the customer \$4.50 for each package, so the revenue equation is $R(x) = 4.5x$.
Each package costs \$2.50 to produce and the equipment costs \$3000, so the cost equation is

$C(x) = 2.5x + 3000.$
Since the break-even point is when $R(x) = C(x)$, we solve the equation $4.5x = 2.5x + 3000$.

$$4.5x = 2.5x + 3000$$
$$2x = 3000$$
$$x = 1500$$

The company must sell 1500 packages to break even.

6. Let x = measure of smallest angle
y = measure of largest angle
z = measure of third angle
The sum of the measures is 180°:
$x + y + z = 180.$

The measure of the largest angle is 40° more than the measure of the smallest angle:
$y = x + 40.$

The measure of the remaining angle is 20° more than the measure of the smallest angle:
$y = x + 20.$

We solve the following system.

$$\begin{cases} x + y + z = 1180 \\ y = x + 40 \\ z = x + 20 \end{cases}$$

We substitute $x + 40$ for y and $x + 20$ for z in the first equation.

$$x + (x + 40) + (x + 20) = 180$$
$$3x + 60 = 180$$
$$3x = 120$$
$$x = 40$$

Then $y = x + 40 = 40 + 40 = 80$ and $z = x + 20 = 40 + 20 = 60$.
The angle measures are 40°, 60°, and 80°.

Vocabulary, Readiness & Video Check 4.3

1. Up to now we've been choosing one variable/unknown and translating into one equation. To solve by a system of equations, we'll choose two variables to represent two unknowns and translate into two equations.

2. The break-even point occurs when revenue equals cost—money has not been lost or made; set the revenue function equal to the cost function and solve for the variable.

3. The ordered triple still needs to be interpreted in the context of the application. Each value actually represents the angle measure of a triangle, in degrees.

Exercise Set 4.3

1. Let x = the first number, y = the second number.

$$\begin{cases} x = y + 2 \\ 2x = 3y - 4 \end{cases}$$

Substitute $x = y + 2$ in the second equation.

$$2(y + 2) = 3y - 4$$
$$2y + 4 = 3y - 4$$
$$y = 8$$

Replace y with 8 in the first equation.
$x = 8 + 2 = 10$
The numbers are 10 and 8.

3. a. Let e = length of the Enterprise class, n = length of the Nimitz class.

$$\begin{cases} e + n = 2193 \\ e - n = 9 \end{cases}$$

Add the equations.

$$2e = 2202$$
$$e = 1101$$

Replace e with 1101 in the first equation.

$$1101 + n = 2193$$
$$n = 1092$$

The Enterprise class is 1101 feet and the Nimitz class is 1092 feet.

b. There are 3 feet in each yard, so there are 300 feet in 100 yards.

$$\frac{1101}{300} = 3.67$$

The length of the Enterprise class carrier is 3.67 football fields.

5. With the wind, the plane was moving at $\frac{2520}{4.5} = 560$ mph. Against the wind, the plane was moving at $\frac{2160}{4.5} = 480$ mph.
Let p = speed of the plane in still air, w = speed of the wind.

$$\begin{cases} p + w = 560 \\ p - w = 480 \end{cases}$$

Add the equations.

$$2p = 1040$$
$$p = 520$$

Replace p with 520 in the first equation.

$$520 + w = 560$$
$$w = 40$$

The speed of the plane is 520 mph and the speed of the wind is 40 mph.

7. Let x = amount of 4% butterfat milk, and y = amount of 1% butterfat milk.

qt	strength	amount of butterfat
x	4%	$0.04x$
y	1%	$0.01y$
60	2%	$0.02(60) = 1.2$

$$\begin{cases} x+y=60 \\ 0.04x+0.01y=1.2 \end{cases}$$

Multiply the second equation by −100 and add the result to the first equation.

$$\begin{array}{r} x+y=60 \\ \underline{-4x-y=-120} \\ -3x=-60 \\ x=20 \end{array}$$

Replace x with 20 in the first equation.

$$\begin{aligned} 20+y&=60 \\ y&=40 \end{aligned}$$

Thus, mix 20 quarts of 4% butterfat milk with 40 quarts of 1% butterfat milk.

9. Let x be the number of students studying in the United Kingdom, and y be the number studying in Italy.

$$\begin{cases} x+y=66{,}564 \\ x=y+8682 \end{cases}$$

Replace x with $y + 8682$ in the first equation.

$$\begin{aligned} x+y&=66{,}564 \\ y+8682+y&=66{,}564 \\ 2y+8682&=66{,}564 \\ 2y&=57{,}882 \\ y&=28{,}941 \end{aligned}$$

Replace y with 28,941 in the second equation.

$x = y + 8682 = 28{,}941 + 8682 = 37{,}623$

37,623 students studied in the United Kingdom, and 28,941 studied in Italy.

11. Let l be the number of large frames and s be the number of small frames.

$$\begin{cases} l+s=22 \\ 15l+8s=239 \end{cases}$$

Solve the first equation for l.

$l = 22 - s$

Replace l with $22 - s$ in the second equation.

$$\begin{aligned} 15(22-s)+8s&=239 \\ 330-15s+8s&=239 \\ -7s&=-91 \\ s&=13 \end{aligned}$$

Replace s with 13 in the first equation.

$$\begin{aligned} l+13&=22 \\ l&=9 \end{aligned}$$

She bought 9 large frames and 13 small frames.

13. Let m = the first number, n = the second number.

$$\begin{cases} m=n-2 \\ 2m=3n+4 \end{cases}$$

Substitute $m = n - 2$ in the second equation.

$$\begin{aligned} 2(n-2)&=3n+4 \\ 2n-4&=3n+4 \\ -8&=n \end{aligned}$$

Replace n with −8 in the first equation.

$m = -8 - 2 = -10$

The numbers are −10 and −8.

15. a. answers may vary

b. $\begin{cases} y=-0.24x+10.6 \\ y=0.21x+10.5 \end{cases}$

Replace y with $-0.24x + 10.6$ in the second equation.

$$\begin{aligned} -0.24x+10.6&=0.21x+10.5 \\ -0.45x+10.6&=10.5 \\ -0.45x&=-0.1 \\ x&\approx 0.2\approx 0 \end{aligned}$$

$2008 + 0 = 2008$

The pounds of cheddar cheese consumed equaled the pounds of mozzarella cheese consumed in 2008.

17. Let p be the price of a pen and w be the price of a writing tablet.

$$\begin{cases} 7w+4p=6.40 \\ 2w+19p=5.40 \end{cases}$$

Multiply the first equation by −2 and the second equation by 7 and add the resulting equations.

$$\begin{array}{r} -14w\quad -8p=-12.8 \\ \underline{14w+133p=37.8} \\ 125p=25 \\ p=0.2 \end{array}$$

Replace p with 0.2 in the first equation.

$$\begin{aligned} 7w+4p&=6.40 \\ 7w+4(0.2)&=6.40 \\ 7w+0.8&=6.4 \\ 7w&=5.6 \\ w&=0.8 \end{aligned}$$

The price of a writing tablet is $0.80 and the price of a pen is $0.20.

19. Let p be the speed of the plane and w be the speed of the wind.

$$\begin{cases} 3p+3w=2160 \\ 4p-4w=2160 \end{cases}$$

Multiply the first equation by $\frac{1}{3}$ and the second equation by $\frac{1}{4}$, and add the results.

$$\begin{aligned} p+w&=720 \\ p-w&=540 \\ \hline 2p\quad &=1260 \\ p&=630 \end{aligned}$$

Replace p with 630 in the first equation.

$$\begin{aligned} 3(630)+3w&=2160 \\ 1890+3w&=2160 \\ 3w&=270 \\ w&=90 \end{aligned}$$

The plane's speed is 630 mph and the wind's speed is 90 mph.

21. a. $\begin{cases} y=0.04x+4.27 \\ y=0.67x+3.01 \end{cases}$

Replace y with $0.04x + 4.27$ in the second equation.

$$\begin{aligned} 0.04x+4.27&=0.67x+3.01 \\ -0.63x+4.27&=3.01 \\ -0.63x&=-1.26 \\ x&=2 \end{aligned}$$

$2010 + 2 = 2012$

The average American adult spent the same number of hours on television viewing as on using digital media in 2012.

b. answers may vary

23. Let x be the length of each of the equal sides and y be the length of the third side.

$$\begin{cases} 2x+y=93 \\ y=x+9 \end{cases}$$

Replace y with $x + 9$ in the first equation.

$$\begin{aligned} 2x+y&=93 \\ 2x+x+9&=93 \\ 3x&=84 \\ x&=28 \end{aligned}$$

Replace x with 28 in the second equation.

$y = x + 9 = 28 + 9 = 37$

The lengths of the sides are 28 cm, 28 cm, and 37 cm.

25. Let m be the number of miles.

Hertz = $25 + 0.10m$

Budget = $20 + 0.25m$

Using Budget = 2 · Hertz gives

$$\begin{aligned} 20+0.25m&=2(25+0.10m) \\ 20+0.25m&=50+0.20m \\ 0.25m&=30+0.20m \\ 0.05m&=30 \\ m&=\frac{30}{0.05}=600 \end{aligned}$$

The Budget charge is twice the Hertz charge for a daily mileage of 600 miles.

27. $\begin{cases} x=y-30 \\ x+y=180 \end{cases}$

Replace x with $y - 30$ in the second equation.

$$\begin{aligned} x+y&=180 \\ y-30+y&=180 \\ 2y&=210 \\ y&=105 \end{aligned}$$

Replace y with 105 in the first equation.

$x = y - 30 = 105 - 30 = 75$

The values are $x = 75$ and $y = 105$.

29. The break-even point is where $C(x) = R(x)$.

$$\begin{aligned} 30x+10{,}000&=46x \\ 10{,}000&=16x \\ 625&=x \end{aligned}$$

625 units must be sold to break even.

31. The break-even point is where $C(x) = R(x)$.

$$\begin{aligned} 1.2x+1500&=1.7x \\ 1500&=0.5x \\ 3000&=x \end{aligned}$$

3000 units must be sold to break even.

33. The break-even point is where $C(x) = R(x)$.

$$\begin{aligned} 75x+160{,}000&=200x \\ 160{,}000&=125x \\ 1280&=x \end{aligned}$$

1280 units must be sold to break even.

35. a. Let x be the number of desks. The revenue from each desk is \$450, so $R(x) = 450x$.

b. The cost is \$6000 plus \$200 for each desk, so $C(x) = 200x + 6000$.

c.

$$\begin{aligned} R(x)&=C(x) \\ 450x&=200x+6000 \\ 250x&=6000 \\ x&=24 \end{aligned}$$

The break-even point is 24 desks.

37. Let x = number of units of Mix A,
y = number of units of Mix B,
z = number of units of Mix C.

$$\begin{cases} 4x+6y+4z=30 & (1) \\ 6x+y+z=16 & (2) \\ 3x+2y+12z=24 & (3) \end{cases}$$

Multiply equation (2) by −6 and add to equation (1).

$-32x - 2z = -66$ or

$16x + z = 33$ (4)

Multiply equation (2) by −2 and add to equation (3).

$-9x + 10z = -8$ (5)

Multiply equation (4) by −10 and add to equation (5).

$$-169x = -338$$
$$x = 2$$

Replace x with 2 in equation (4).

$$16(2) + z = 33$$
$$32 + z = 33$$
$$z = 1$$

Replace x with 2 and z with 1 in equation (2).

$$6(2) + y + 1 = 16$$
$$12 + y + 1 = 16$$
$$y + 13 = 16$$
$$y = 3$$

Combine 2 units of Mix A, 3 units of Mix B, and 1 unit of Mix C.

39. Let x = length of shortest side,
y = length of equal sides,
z = length of longest side.

$$\begin{cases} x+2y+z=29 \\ z=2x \\ y=x+2 \end{cases}$$

Replace y with $x + 2$ and z with $2x$ in the first equation.

$$x + 2(x+2) + 2x = 29$$
$$x + 2x + 4 + 2x = 29$$
$$5x = 25$$
$$x = 5$$

Replace x with 5 in the third equation.

$y = 5 + 2 = 7$

Replace x with 5 in the second equation.

$z = 2(5) = 10$

The four sides measure 5 inches, 7 inches, 7 inches, and 10 inches.

41. Let x = the first number, y = the second number, and z = the third number.

$$\begin{cases} x+y+z=40 \\ x=y+5 \\ x=2z \end{cases}$$

$$\begin{cases} x+y+z=40 \\ y=x-5 \\ z=\frac{1}{2}x \end{cases}$$

Substitute $y = x - 5$ and $z = \frac{1}{2}x$ in the first equation.

$$x + x - 5 + \frac{1}{2}x = 40$$
$$\frac{5}{2}x - 5 = 40$$
$$\frac{5}{2}x = 45$$
$$x = \frac{2}{5}(45) = 18$$

$y = x - 5 = 18 - 5 = 13$

$z = \frac{1}{2}x = \frac{1}{2}(18) = 9$

The three numbers are 18, 13, and 9.

43. Let x = number of free throws,
y = number of two-point field goals, and
z = number of three-point field goals.

$$\begin{cases} x+2y+3z=2593 \\ x=3z+127 \\ y=x-46 \end{cases}$$

Substitute $3z + 127$ for x in the third equation.

$y = x - 46 = 3z + 127 - 46 = 3z + 81$

Substitute $3z + 127$ for x and $3z + 81$ for y in the first equation.

$$(3z+127) + 2(3z+81) + 3z = 2593$$
$$3z + 127 + 6z + 162 + 3z = 2593$$
$$12z + 289 = 2593$$
$$12z = 2304$$
$$z = 192$$

$x = 3z + 127 = 3(192) + 127 = 703$

$y = x - 46 = 703 - 46 = 657$

He made 703 free throws, 657 2-point field goals, and 192 3-point field goals.

45. $\begin{cases} x+y+z=180 \\ 2x+5+y=180 \\ 2x-5+z=180 \end{cases}$

$z = 185 - 2x$
$y = 175 - 2x$
Replace y with $175 - 2x$ and z with $185 - 2x$ in the first equation.

$$\begin{aligned} x+(175-2x)+(185-2x) &= 180 \\ 360-3x &= 180 \\ -3x &= -180 \\ x &= 60 \end{aligned}$$

$z = 185 - 2(60) = 185 - 120 = 65$
$y = 175 - 2(60) = 175 - 120 = 55$
The values are $x = 60$, $y = 55$, and $z = 65$.

47.
$$\begin{array}{r} 6x-2y+2z=4 \\ -x+2y+3z=6 \\ \hline 5x \qquad +5z=10 \end{array}$$

49.
$$\begin{array}{r} -3x-6y+3z=0 \\ 3x+y-z=2 \\ \hline -5y+2z=2 \end{array}$$

51. Let x = number filed in 2007 and y = number filed in 2014.

$\begin{cases} y=2x-623,200 \\ y=x+156,800 \end{cases}$

Use substitution.

$$\begin{aligned} 2x-623,200 &= x+156,800 \\ x-623,200 &= 156,800 \\ x &= 780,000 \end{aligned}$$

$y = x + 156,800 = 780,000 + 156,800 = 936,800$
There were 780,000 personal bankruptcy petitions filed in 2007 and 936,800 filed in 2014.

53. $y=ax^2+bx+c$
For (1, 6), use $x = 1$ and $y = 6$.
$6 = a + b + c$ (1)
For (−1, −2), use $x = -1$ and $y = -2$.
$-2 = a - b + c$ (2)
For (0, −1), use $x = 0$ and $y = -1$.
$-1 = a \cdot 0 + b \cdot 0 + c$
$-1 = c$ (3)
The system is

$$\begin{aligned} 6&=a+b+c && (1) \\ -2&=a-b+c && (2) \\ -1&=c && (3) \end{aligned}$$

From equation (3), we see that $c = -1$. Multiply equation (2) by −1 and add to equation (1).
$8 = 2b$
$4 = b$
Replace b with 4 and c with −1 in equation (1).

$$\begin{aligned} 6&=a+4-1 \\ 6&=a+3 \\ 3&=a \end{aligned}$$

The solution is $a = 3$, $b = 4$, and $c = -1$.

55. $y=ax^2+bx+c$
For (3, 927), $x = 3$ and $y = 927$.
$927=a(3)^2+b(3)+c$
$927=9a+3b+c$ (1)
For (11, 1179), $x = 11$ and $y = 1179$.
$1179=a(11)^2+b(11)+c$
$1179=121a+11b+c$ (2)
For (19, 1495), $x = 19$ and $y = 1495$.
$1495=a(19)^2+b(19)+c$
$1495=361a+19b+c$ (3)
The system is

$\begin{cases} 927=9a+3b+c & (1) \\ 1179=121a+11b+c & (2) \\ 1495=361a+19b+c & (3) \end{cases}$

Multiply equation (1) by −1 and add to equation (2).
$252 = 112a + 8b$ (4)
Multiply equation (2) by −1 and add to equation (3).
$316 = 240a + 8b$ (5)
Multiply equation (4) by −1 and add to equation (5).

$$\begin{array}{r} -252=-112a-8b \\ 316=240a+8b \\ \hline 64=128a \\ \frac{1}{2}=a \end{array}$$

Replace a with $\frac{1}{2}=0.5$ in equation (4).

$$\begin{aligned} 252&=112(0.5)+8b \\ 252&=56+8b \\ 196&=8b \\ b&=\frac{49}{2}=24.5 \end{aligned}$$

Replace a with 0.5 and b with 24.5 in equation (1).

$$\begin{aligned} 927&=9(0.5)+3(24.5)+c \\ 927&=4.5+73.5+c \\ 927&=78+c \\ 849&=c \end{aligned}$$

The solution is $a = 0.5$, $b = 24.5$, and $c = 849$.
For 2015, $x = 25$.

$y = 0.5(25)^2 + 24.5(25) + 849 = 1774$

According to the model, 1774 thousand students will take the ACT in 2015.

57. $f(x) = 0.42x + 8.4$
$f(x) = -0.17x + 14.2$
Use substitution.

$$0.42x + 8.4 = -0.17x + 14.2$$
$$0.59x + 8.4 = 14.2$$
$$0.59x = 5.8$$
$$x \approx 9.8 \approx 10$$

$f(10) = 0.42(10) + 8.4 = 12.6 \approx 13$

The solution is (10, 13).

Integrated Review

1. A system with solution (1, 2) has lines that intersect at (1, 2). The correct graph is C.

2. A system with solution (−2, 3) has lines that intersect at (−2, 3). The correct graph is D.

3. A system with no solution has lines that are parallel. The correct graph is A.

4. A system with an infinite number of solutions has lines that are the same. The correct graph is B.

5. $\begin{cases} x + y = 4 & (1) \\ y = 3x & (2) \end{cases}$

Substitute $y = 3x$ in E1.

$$x + (3x) = 4$$
$$4x = 4$$
$$x = 1$$

Replace x with 1 in E2.

$y = 3x = 3(1) = 3$

The solution is (1, 3).

6. $\begin{cases} x - y = -4 & (1) \\ y = 4x & (2) \end{cases}$

Substitute $y = 4x$ in E1.

$$x - (4x) = -4$$
$$-3x = -4$$
$$x = \frac{4}{3}$$

Replace x with $\frac{4}{3}$ in E2.

$$y = 4x = 4\left(\frac{4}{3}\right) = \frac{16}{3}$$

The solution is $\left(\frac{4}{3}, \frac{16}{3}\right)$.

7. $\begin{cases} x + y = 1 & (1) \\ x - 2y = 4 & (2) \end{cases}$

Multiply E1 by −1 and add to E2.

$$\begin{array}{r} -x - y = -1 \\ \underline{x - 2y = 4} \\ -3y = 3 \\ y = -1 \end{array}$$

Replace y with −1 in E1.

$$x + (-1) = 1$$
$$x - 1 = 1$$
$$x = 2$$

The solution is (2, −1).

8. $\begin{cases} 2x - y = 8 & (1) \\ x + 3y = 11 & (2) \end{cases}$

Multiply E1 by 3 and add to E2.

$$\begin{array}{r} 6x - 3y = 24 \\ \underline{x + 3y = 11} \\ 7x \quad\quad = 35 \\ x = 5 \end{array}$$

Replace x with 5 in E1.

$$2(5) - y = 8$$
$$10 - y = 8$$
$$y = 2$$

The solution is (5, 2).

9. $\begin{cases} 2x + 5y = 8 & (1) \\ 6x + y = 10 & (2) \end{cases}$

Multiply E2 by −5 and add to E1.

$$\begin{array}{r} 2x + 5y = 8 \\ \underline{-30x - 5y = -50} \\ -28x \quad\quad = -42 \\ x = \frac{3}{2} \end{array}$$

Replace x with $\frac{3}{2}$ in E2.

$$6\left(\frac{3}{2}\right) + y = 10$$
$$9 + y = 8$$
$$y = 1$$

The solution is $\left(\frac{3}{2}, 1\right)$.

10. $\begin{cases} \frac{1}{8}x - \frac{1}{2}y = -\frac{5}{8} & (1) \\ -3x - 8y = 0 & (2) \end{cases}$

Multiply E1 by −16 and add to E2.

$$\begin{array}{r} -2x+8y=10 \\ -3x-8y=0 \\ \hline -5x \quad =10 \end{array}$$
$$x=-2$$
Replace x with –2 in E2.
$$-3(-2)-8y=0$$
$$6-8y=0$$
$$-8y=-6$$
$$y=\frac{-6}{-8}=\frac{3}{4}$$
The solution is $\left(-2, \frac{3}{4}\right)$.

11. $\begin{cases} 4x-7y=7 & (1) \\ 12x-21y=24 & (2) \end{cases}$

Multiply E1 by –3 and add to E2.
$$\begin{array}{r} -12x+21y=-21 \\ 12x-21y=24 \\ \hline 0=3 \end{array} \quad \text{False}$$
The system is inconsistent. The solution set is ∅.

12. $\begin{cases} 2x-5y=3 & (1) \\ -4x+10y=-6 & (2) \end{cases}$

Multiply E1 by 2 and add to E2.
$$\begin{array}{r} 4x-10y=6 \\ -4x+10y=-6 \\ \hline 0=0 \end{array} \quad \text{True}$$
The system is dependent. The solution set is $\{(x, y)|2x-5y=3\}$.

13. $\begin{cases} y=\frac{1}{3}x \\ 5x-3y=4 \end{cases}$

Substitute $\frac{1}{3}x$ for y in E2.
$$5x-3\left(\frac{1}{3}x\right)=4$$
$$5x-x=4$$
$$4x=4$$
$$x=1$$
Replace x with 1 in E1.
$$y=\frac{1}{3}(1)=\frac{1}{3}$$
The solution is $\left(1, \frac{1}{3}\right)$.

14. $\begin{cases} y=\frac{1}{4}x \\ 2x-4y=3 \end{cases}$

Substitute $\frac{1}{4}x$ for y in E2.
$$2x-4\left(\frac{1}{4}x\right)=3$$
$$2x-x=3$$
$$x=3$$
Replace x with 3 in E1.
$$y=\frac{1}{4}(3)=\frac{3}{4}$$
The solution is $\left(3, \frac{3}{4}\right)$.

15. $\begin{cases} x+y \quad =2 & (1) \\ -3y+z=-7 & (2) \\ 2x+y-z=-1 & (3) \end{cases}$

Add E2 and E3.
$$2x-2y=-8 \text{ or } x-y=-4 \quad (4)$$
Add E1 and E4.
$$2x=-2$$
$$x=-1$$
Replace x with –1 in E1.
$$-1+y=2$$
$$y=3$$
Replace y with 3 in E2.
$$-3(3)+z=-7$$
$$-9+z=-7$$
$$z=2$$
The solution is (–1, 3, 2).

16. $\begin{cases} y+2z=-3 & (1) \\ x-2y \quad =7 & (2) \\ 2x-y+z=5 & (3) \end{cases}$

Multiply E2 by –2 and add to E3.
$$\begin{array}{r} -2x+4y \quad =-14 \\ 2x-y+z=5 \\ \hline 3y+z=-9 \end{array} \quad (4)$$
Multiply E4 by –2 and add to E1.
$$\begin{array}{r} -6y-2z=18 \\ y+2z=-3 \\ \hline -5y \quad =15 \end{array}$$
$$y=-3$$
Replace y with –3 in E4.
$$3(-3)+z=-9$$
$$z=0$$
Replace y with –3 in E2.

$$x-2(-3)=7$$
$$x+6=7$$
$$x=1$$

The solution is (1, –3, 0).

17. $\begin{cases} 2x+4y-6z=3 & (1) \\ -x+\ \ y-\ z=6 & (2) \\ x+2y-3z=1 & (3) \end{cases}$

Multiply E3 by –2 and add to E1.

$$-2x-4y+6z=-2$$
$$\underline{2x+4y-6z=3}$$
$$0=1 \text{ False}$$

The system is inconsistent. The solution set is ∅.

18. $\begin{cases} x-y+3z=2 & (1) \\ -2x+2y-6z=-4 & (2) \\ 3x-3y+9z=6 & (3) \end{cases}$

Multiply E1 by 2 and add to E2.

$$2x-2y+6z=4$$
$$\underline{-2x+2y-6z=-4}$$
$$0=0 \text{ True}$$

The system is dependent. The solution set is $\{(x, y)|x-y+3z=2\}$.

19. $\begin{cases} x+y-4z=5 & (1) \\ x-y+2z=-2 & (2) \\ 3x+2y+4z=18 & (3) \end{cases}$

Add E1 and E2.

$2x-2z=3$ (4)

Multiply E2 by 2 and add to E3.

$$2x-2y+4z=-4$$
$$\underline{3x+2y+4z=18}$$
$$5x \qquad +8z=14 \quad (5)$$

Multiply E4 by 4 and add to E5.

$$8x-8z=12$$
$$\underline{5x+8z=14}$$
$$13x \qquad =26$$
$$x=2$$

Replace x with 2 in E4.

$$2(2)-2z=3$$
$$-2z=-1$$
$$z=\frac{1}{2}$$

Replace x with 2 and z with $\frac{1}{2}$ in E1.

$$2+y-4\left(\frac{1}{2}\right)=5$$
$$2+y-2=5$$
$$y=5$$

The solution is $\left(2, 5, \frac{1}{2}\right)$.

20. $\begin{cases} 2x-y+3z=2 & (1) \\ x+y-6z=0 & (2) \\ 3x+4y-3z=6 & (3) \end{cases}$

Add E1 and E3.

$5x+3y=8$ (4)

Multiply E1 by 2 and add to E2.

$$4x-2y+6z=4$$
$$\underline{x+\ \ y-6z=0}$$
$$5x-y \qquad =4 \quad (5)$$

Multiply E5 by 3 and add to E4.

$$15x-3y=12$$
$$\underline{5x+3y=8}$$
$$20x \qquad =20$$
$$x=1$$

Replace x with 1 in E5.

$$5(1)-y=4$$
$$-y=-1$$
$$y=1$$

Replace both x and y with 1 in E1.

$$2(1)-1+3z=2$$
$$1+3z=2$$
$$3z=1$$
$$z=\frac{1}{3}$$

The solution is $\left(1, 1, \frac{1}{3}\right)$.

21. Let x = the first number and y = the second number.

$\begin{cases} x=y-8 & (1) \\ 2x=y+11 & (2) \end{cases}$

Substitute $x=y-8$ in E2.

$$2(y-8)=y+11$$
$$2y-16=y+11$$
$$y=27$$

Replace y with 27 in E1.

$x=27-8=19$

The numbers are 19 and 27.

22. Let x = measure of the two smallest angles, y = measure of the third angle, and z = measure of the fourth angle.

$$\begin{cases} 2x+y+z=360 \\ y=x+30 \\ z=x+50 \end{cases}$$

Substitute $y=x+30$ and $z=x+50$ in the first equation.

$$\begin{aligned} 2x+(x+30)+(x+50)&=360 \\ 4x+80&=360 \\ 4x&=280 \\ x&=70 \end{aligned}$$

so $y=70+30=100$ and $z=70+50=120$

The two smallest angles are 70°, the third angle is 100°, and the fourth angle is 120°.

Section 4.4 Practice Exercises

1. $\begin{cases} x+4y=-2 \\ 3x-y=7 \end{cases}$

The corresponding matrix is $\left[\begin{array}{cc|c} 1 & 4 & -2 \\ 3 & -1 & 7 \end{array}\right]$. The element in the first row, first column is already 1. Multiply row 1 by −3 and add to row 2 to get a 0 below the 1.

$$\left[\begin{array}{cc|c} 1 & 4 & -2 \\ -3(1)+3 & -3(4)+(-1) & -3(-2)+7 \end{array}\right]$$

$$\left[\begin{array}{cc|c} 1 & 4 & -2 \\ 0 & -13 & 13 \end{array}\right]$$

We change −13 to a 1 by dividing row 2 by −13.

$$\left[\begin{array}{cc|c} 1 & 4 & -2 \\ 0 & \frac{-13}{-13} & \frac{13}{-13} \end{array}\right]$$

$$\left[\begin{array}{cc|c} 1 & 4 & -2 \\ 0 & 1 & -1 \end{array}\right]$$

The last matrix corresponds to $\begin{cases} x+4y=-2 \\ y=-1 \end{cases}$

To find x, we let $y=-1$ in the first equation.

$$\begin{aligned} x+4y&=-2 \\ x+4(-1)&=-2 \\ x-4&=-2 \\ x&=2 \end{aligned}$$

The solution is (2, −1).

2. $\begin{cases} x-3y=3 \\ -2x+6y=4 \end{cases}$

The corresponding matrix is $\left[\begin{array}{cc|c} 1 & -3 & 3 \\ -2 & 6 & 4 \end{array}\right]$. The element in the first row, first column is already 1. Multiply row 1 by 2 and add to row 2 to get a 0 below the 1.

$$\left[\begin{array}{cc|c} 1 & -3 & 3 \\ 2(1)+(-2) & 2(-3)+6 & 2(3)+4 \end{array}\right]$$

$$\left[\begin{array}{cc|c} 1 & -3 & 3 \\ 0 & 0 & 10 \end{array}\right]$$

The corresponding system is $\begin{cases} x-3y=3 \\ 0=10 \end{cases}$

The equation 0 = 10 is false. Hence, the system is inconsistent and has no solution. The solution set is ∅.

3. $\begin{cases} x+3y-z=0 \\ 2x+y+3z=5 \\ -x-2y+4z=7 \end{cases}$

The corresponding matrix is $\left[\begin{array}{ccc|c} 1 & 3 & -1 & 0 \\ 2 & 1 & 3 & 5 \\ -1 & -2 & 4 & 7 \end{array}\right]$.

The element in the first row, first column is already 1. Multiply row 1 by −2 and add to row 2 to get a 0 below the 1 in row 2. Add row 1 to row 3 to get a 0 below the 1 in row 3.

$$\left[\begin{array}{ccc|c} 1 & 3 & -1 & 0 \\ -2(1)+2 & -2(3)+1 & -2(-1)+3 & -2(0)+5 \\ 1+(-1) & 3+(-2) & -1+4 & 0+7 \end{array}\right]$$

$$\left[\begin{array}{ccc|c} 1 & 3 & -1 & 0 \\ 0 & -5 & 5 & 5 \\ 0 & 1 & 3 & 7 \end{array}\right]$$

Now we want a 1 where the −5 is now. Interchange rows 2 and 3.

$$\left[\begin{array}{ccc|c} 1 & 3 & -1 & 0 \\ 0 & 1 & 3 & 7 \\ 0 & -5 & 5 & 5 \end{array}\right]$$

Now we want a 0 below the 1. Multiply row 2 by 5 and add to row 3.

$$\left[\begin{array}{ccc|c} 1 & 3 & -1 & 0 \\ 0 & 1 & 3 & 7 \\ 5(0)+0 & 5(1)+(-5) & 5(3)+5 & 5(7)+5 \end{array}\right]$$

$$\left[\begin{array}{ccc|c} 1 & 3 & -1 & 0 \\ 0 & 1 & 3 & 7 \\ 0 & 0 & 20 & 40 \end{array}\right]$$

Finally, divide row 3 by 20.

$$\left[\begin{array}{ccc|c} 1 & 3 & -1 & 0 \\ 0 & 1 & 3 & 7 \\ 0 & 0 & \frac{20}{20} & \frac{40}{20} \end{array}\right]$$

$$\left[\begin{array}{ccc|c} 1 & 3 & -1 & 0 \\ 0 & 1 & 3 & 7 \\ 0 & 0 & 1 & 2 \end{array}\right]$$

This matrix corresponds to the system

$$\begin{cases} x+3y-z=0 \\ y+3z=7 \\ z=2 \end{cases}$$

The z-coordinate is 2. Replace z with 2 in the second equation and solve for y.

$$y+3z=7$$
$$y+3(2)=7$$
$$y+6=7$$
$$y=1$$

To find x, we let $z=2$ and $y=1$ in the first equation.

$$x+3y-z=0$$
$$x+3(1)-2=0$$
$$x+1=0$$
$$x=-1$$

The solution is (−1, 1, 2).

Vocabulary, Readiness & Video Check 4.4

1. A matrix is a rectangular array of numbers.

2. Each of the numbers in a matrix is called an element.

3. The numbers aligned horizontally in a matrix are in the same row.

4. The numbers aligned vertically in a matrix are in the same column.

5. Any two columns may be interchanged. false

6. Any two rows may be interchanged. true

7. The elements in a row may be added to their corresponding elements in another row. true

8. The elements of a column may be multiplied by any nonzero number. false

9. Two rows may be interchanged, the elements of any row may be multiplied/divided by the same nonzero number, the elements of any row may be multiplied/divided by the same nonzero number and added to their corresponding elements in any other row; rows were not interchanged in Example 1.

10. Consider the possible confusion or errors that might occur if you're careless and it's unclear which row or column numbers belong in, especially when working with larger matrices and/or with a lot of calculations.

Exercise Set 4.4

1. $\begin{cases} x+y=1 \\ x-2y=4 \end{cases}$

$$\left[\begin{array}{cc|c} 1 & 1 & 1 \\ 1 & -2 & 4 \end{array}\right]$$

Multiply R1 by −1 and add to R2.

$$\left[\begin{array}{cc|c} 1 & 1 & 1 \\ 0 & -3 & 3 \end{array}\right]$$

Divide R2 by −3.

$$\left[\begin{array}{cc|c} 1 & 1 & 1 \\ 0 & 1 & -1 \end{array}\right]$$

This corresponds to $\begin{cases} x+y=1 \\ y=-1 \end{cases}$.

$$x+(-1)=1$$
$$x-1=1$$
$$x=2$$

The solution is (2, −1).

3. $\begin{cases} x+3y=2 \\ x+2y=0 \end{cases}$

$$\left[\begin{array}{cc|c} 1 & 3 & 2 \\ 1 & 2 & 0 \end{array}\right]$$

Multiply R1 by −1 and add to R2.

$$\left[\begin{array}{cc|c} 1 & 3 & 2 \\ 0 & -1 & -2 \end{array}\right]$$

Multiply R2 by −1.

$$\left[\begin{array}{cc|c} 1 & 3 & 2 \\ 0 & 1 & 2 \end{array}\right]$$

This corresponds to $\begin{cases} x+3y=2 \\ y=2 \end{cases}$.

$$x+3(2)=2$$
$$x+6=2$$
$$x=-4$$

The solution is (−4, 2).

5. $\begin{cases} x-2y=4 \\ 2x-4y=4 \end{cases}$

$$\left[\begin{array}{cc|c} 1 & -2 & 4 \\ 2 & -4 & 4 \end{array}\right]$$

Multiply R1 by −2 and add to R2.

$$\left[\begin{array}{cc|c} 1 & -2 & 4 \\ 0 & 0 & -4 \end{array}\right]$$

This corresponds to $\begin{cases} x-2y=4 \\ 0=-4 \end{cases}$.

This is an inconsistent system. The solution is ∅.

7. $\begin{cases} 3x-3y=9 \\ 2x-2y=6 \end{cases}$

$$\left[\begin{array}{cc|c} 3 & -3 & 9 \\ 2 & -2 & 6 \end{array}\right]$$

Divide R1 by 3.

$$\left[\begin{array}{cc|c} 1 & -1 & 3 \\ 2 & -2 & 6 \end{array}\right]$$

Multiply R1 by –2 and add to R2.

$$\left[\begin{array}{cc|c} 1 & -1 & 3 \\ 0 & 0 & 0 \end{array}\right]$$

This corresponds to $\begin{cases} x-y=3 \\ 0=0 \end{cases}$.

This is a dependent system. The solution is $\{(x, y)|x-y=3\}$.

9. $\begin{cases} x+y=3 \\ 2y=10 \\ 3x+2y-4z=12 \end{cases}$

$$\left[\begin{array}{ccc|c} 1 & 1 & 0 & 3 \\ 0 & 2 & 0 & 10 \\ 3 & 2 & -4 & 12 \end{array}\right]$$

Multiply R1 by –3 and add to R3.

$$\left[\begin{array}{ccc|c} 1 & 1 & 0 & 3 \\ 0 & 2 & 0 & 10 \\ 0 & -1 & -4 & 3 \end{array}\right]$$

Divide R2 by 2.

$$\left[\begin{array}{ccc|c} 1 & 1 & 0 & 3 \\ 0 & 1 & 0 & 5 \\ 0 & -1 & -4 & 3 \end{array}\right]$$

Add R2 to R3.

$$\left[\begin{array}{ccc|c} 1 & 1 & 0 & 3 \\ 0 & 1 & 0 & 5 \\ 0 & 0 & -4 & 8 \end{array}\right]$$

Divide R3 by –4.

$$\left[\begin{array}{ccc|c} 1 & 1 & 0 & 3 \\ 0 & 1 & 0 & 5 \\ 0 & 0 & 1 & -2 \end{array}\right]$$

This corresponds to $\begin{cases} x+y=3 \\ y=5 \\ z=-2 \end{cases}$.

$$x+5=3$$
$$x=-2$$

The solution is (–2, 5, –2).

11. $\begin{cases} 2y-z=-7 \\ x+4y+z=-4 \\ 5x-y+2z=13 \end{cases}$

$$\left[\begin{array}{ccc|c} 0 & 2 & -1 & -7 \\ 1 & 4 & 1 & -4 \\ 5 & -1 & 2 & 13 \end{array}\right]$$

Interchange R1 and R2.

$$\left[\begin{array}{ccc|c} 1 & 4 & 1 & -4 \\ 0 & 2 & -1 & -7 \\ 5 & -1 & 2 & 13 \end{array}\right]$$

Multiply R1 by –5 and add to R3.

$$\left[\begin{array}{ccc|c} 1 & 4 & 1 & -4 \\ 0 & 2 & -1 & -7 \\ 0 & -21 & -3 & 33 \end{array}\right]$$

Divide R2 by 2.

$$\left[\begin{array}{ccc|c} 1 & 4 & 1 & -4 \\ 0 & 1 & -\frac{1}{2} & -\frac{7}{2} \\ 0 & -21 & -3 & 33 \end{array}\right]$$

Multiply R2 by 21 and add to R3.

$$\left[\begin{array}{ccc|c} 1 & 4 & 1 & -4 \\ 0 & 1 & -\frac{1}{2} & -\frac{7}{2} \\ 0 & 0 & -\frac{27}{2} & -\frac{81}{2} \end{array}\right]$$

Multiply R2 by $-\frac{2}{27}$.

$$\left[\begin{array}{ccc|c} 1 & 4 & 1 & -4 \\ 0 & 1 & -\frac{1}{2} & -\frac{7}{2} \\ 0 & 0 & 1 & 3 \end{array}\right]$$

This corresponds to $\begin{cases} x+4y+z=-4 \\ y-\frac{1}{2}z=-\frac{7}{2} \\ z=3 \end{cases}$.

$$y-\frac{1}{2}(3)=-\frac{7}{2}$$
$$y-\frac{3}{2}=-\frac{7}{2}$$
$$y=-2$$
$$x+4(-2)+3=-4$$
$$x-8+3=-4$$
$$x=1$$

The solution is (1, –2, 3).

13. $\begin{cases} x-4=0 \\ x+y=1 \end{cases}$ or $\begin{cases} x \quad\; =4 \\ x+y=1 \end{cases}$

$\left[\begin{array}{cc|c} 1 & 0 & 4 \\ 1 & 1 & 1 \end{array}\right]$

Multiply R1 by –1 and add to R2.

$\left[\begin{array}{cc|c} 1 & 0 & 4 \\ 0 & 1 & -3 \end{array}\right]$

This corresponds to $\begin{cases} x=4 \\ y=-3 \end{cases}$

The solution is (4, –3).

15. $\begin{cases} x+y+z=2 \\ 2x \quad\; -z=5 \\ \quad 3y+z=2 \end{cases}$

$\left[\begin{array}{ccc|c} 1 & 1 & 1 & 2 \\ 2 & 0 & -1 & 5 \\ 0 & 3 & 1 & 2 \end{array}\right]$

Multiply R1 by –2 and add to R2.

$\left[\begin{array}{ccc|c} 1 & 1 & 1 & 2 \\ 0 & -2 & -3 & 1 \\ 0 & 3 & 1 & 2 \end{array}\right]$

Divide R2 by –2.

$\left[\begin{array}{ccc|c} 1 & 1 & 1 & 2 \\ 0 & 1 & \frac{3}{2} & -\frac{1}{2} \\ 0 & 3 & 1 & 2 \end{array}\right]$

Multiply R2 by –3 and add to R3.

$\left[\begin{array}{ccc|c} 1 & 1 & 1 & 2 \\ 0 & 1 & \frac{3}{2} & -\frac{1}{2} \\ 0 & 0 & -\frac{7}{2} & \frac{7}{2} \end{array}\right]$

Multiply R3 by $-\frac{2}{7}$.

$\left[\begin{array}{ccc|c} 1 & 1 & 1 & 2 \\ 0 & 1 & \frac{3}{2} & -\frac{1}{2} \\ 0 & 0 & 1 & -1 \end{array}\right]$

This corresponds to $\begin{cases} x+y+z=2 \\ y+\frac{3}{2}z=-\frac{1}{2}. \\ z=-1 \end{cases}$

$$y+\frac{3}{2}(-1)=-\frac{1}{2}$$
$$y-\frac{3}{2}=-\frac{1}{2}$$
$$y=1$$

$$x+1+(-1)=2$$
$$x=2$$

The solution is (2, 1, –1).

17. $\begin{cases} 5x-2y=27 \\ -3x+5y=18 \end{cases}$

$\left[\begin{array}{cc|c} 5 & -2 & 27 \\ -3 & 5 & 18 \end{array}\right]$

Divide R1 by 5.

$\left[\begin{array}{cc|c} 1 & -\frac{2}{5} & \frac{27}{5} \\ -3 & 5 & 18 \end{array}\right]$

Multiply R1 by 3 and add to R2.

$\left[\begin{array}{cc|c} 1 & -\frac{2}{5} & \frac{27}{5} \\ 0 & \frac{19}{5} & \frac{171}{5} \end{array}\right]$

Multiply R2 by $\frac{5}{19}$.

$\left[\begin{array}{cc|c} 1 & -\frac{2}{5} & \frac{27}{5} \\ 0 & 1 & 9 \end{array}\right]$

This corresponds to $\begin{cases} x-\frac{2}{5}y=\frac{27}{5}. \\ y=9 \end{cases}$

$$x-\frac{2}{5}(9)=\frac{27}{5}$$
$$x-\frac{18}{5}=\frac{27}{5}$$
$$x=9$$

The solution is (9, 9).

19. $\begin{cases} 4x-7y=7 \\ 12x-21y=24 \end{cases}$

$\left[\begin{array}{cc|c} 4 & -7 & 7 \\ 12 & -21 & 24 \end{array}\right]$

Divide R1 by 4.

$\left[\begin{array}{cc|c} 1 & -\frac{7}{4} & \frac{7}{4} \\ 12 & -21 & 24 \end{array}\right]$

Multiply R1 by –12 and add to R2.

$\left[\begin{array}{cc|c} 1 & -\frac{7}{4} & \frac{7}{4} \\ 0 & 0 & 3 \end{array}\right]$

This corresponds to $\begin{cases} x-\frac{7}{4}y=\frac{7}{4}. \\ 0=3 \end{cases}$

This is an inconsistent system. The solution set is ∅.

21. $\begin{cases} 4x-y+2z=5 \\ \quad 2y+z=4 \\ 4x+y+3z=10 \end{cases}$

$\left[\begin{array}{ccc|c} 4 & -1 & 2 & 5 \\ 0 & 2 & 1 & 4 \\ 4 & 1 & 3 & 10 \end{array}\right]$

Divide R1 by 4.

$\left[\begin{array}{ccc|c} 1 & -\frac{1}{4} & \frac{1}{2} & \frac{5}{4} \\ 0 & 2 & 1 & 4 \\ 4 & 1 & 3 & 10 \end{array}\right]$

Multiply R1 by –4 and add to R3.

$\left[\begin{array}{ccc|c} 1 & -\frac{1}{4} & \frac{1}{2} & \frac{5}{4} \\ 0 & 2 & 1 & 4 \\ 0 & 2 & 1 & 5 \end{array}\right]$

Divide R2 by 2.

$\left[\begin{array}{ccc|c} 1 & -\frac{1}{4} & \frac{1}{2} & \frac{5}{4} \\ 0 & 1 & \frac{1}{2} & 2 \\ 0 & 2 & 1 & 5 \end{array}\right]$

Multiply R2 by –2 and add to R3.

$\left[\begin{array}{ccc|c} 1 & -\frac{1}{4} & \frac{1}{2} & \frac{5}{4} \\ 0 & 1 & \frac{1}{2} & 2 \\ 0 & 0 & 0 & 1 \end{array}\right]$

This corresponds to $\begin{cases} x-\frac{1}{4}y+\frac{1}{2}z=\frac{5}{4} \\ \quad y+\frac{1}{2}z=2 \\ \quad 0=1 \end{cases}$.

This is an inconsistent system. The solution set is $\varnothing$.

23. $\begin{cases} 4x+y+z=3 \\ -x+y-2z=-11 \\ x+2y+2z=-1 \end{cases}$

$\left[\begin{array}{ccc|c} 4 & 1 & 1 & 3 \\ -1 & 1 & -2 & -11 \\ 1 & 2 & 2 & -1 \end{array}\right]$

Interchange R1 and R3.

$\left[\begin{array}{ccc|c} 1 & 2 & 2 & -1 \\ -1 & 1 & -2 & -11 \\ 4 & 1 & 1 & 3 \end{array}\right]$

Add R1 to R2. Multiply R1 by –4 and add to R3.

$\left[\begin{array}{ccc|c} 1 & 2 & 2 & -1 \\ 0 & 3 & 0 & -12 \\ 0 & -7 & -7 & 7 \end{array}\right]$

Divide R2 by 3.

$\left[\begin{array}{ccc|c} 1 & 2 & 2 & -1 \\ 0 & 1 & 0 & -4 \\ 0 & -7 & -7 & 7 \end{array}\right]$

Multiply R2 by 7 and add to R3.

$\left[\begin{array}{ccc|c} 1 & 2 & 2 & -1 \\ 0 & 1 & 0 & -4 \\ 0 & 0 & -7 & -21 \end{array}\right]$

Divide R3 by –7.

$\left[\begin{array}{ccc|c} 1 & 2 & 2 & -1 \\ 0 & 1 & 0 & -4 \\ 0 & 0 & 1 & 3 \end{array}\right]$

This corresponds to $\begin{cases} x+2y+2z=-1 \\ \quad y=-4 \\ \quad z=3 \end{cases}$.

$$x+2(-4)+2(3)=-1$$
$$x-8+6=-1$$
$$x=1$$

The solution is (1, –4, 3).

25. No vertical line intersects the graph more than once. It is the graph of a function.

27. The *y*-axis is a vertical line that intersects the graph more than once. It is not the graph of a function.

29. $(-1)(-5)-(6)(3)=5-18=-13$

31. $(4)(-10)-(2)(-2)=-40+4=-36$

33. $(-3)(-3)-(-1)(-9)=9-9=0$

35. The matrix should have four columns, so (a) is not the correct matrix. The matrix should have a 0 in the first column, second row since the coefficient of *x* in the second equation is 0, so (b) is not the correct matrix. The correct matrix is (c).

37. a. $\begin{cases} y=1.46x+14.05 \\ 4.24x-y=0.65 \end{cases}$ or $\begin{cases} -1.46x+y=14.05 \\ 4.24x-y=0.65 \end{cases}$

$\left[\begin{array}{cc|c} -1.46 & 1 & 14.05 \\ 4.24 & -1 & 0.65 \end{array}\right]$

Multiply row 1 by $-\frac{1}{1.46}$.

$$\begin{bmatrix} 1 & -0.68 & | & -9.62 \\ 4.24 & -1 & | & 0.65 \end{bmatrix}$$

Multiply row 1 by −4.24 and add to row 2.

$$\begin{bmatrix} 1 & -0.68 & | & -9.62 \\ 0 & 1.90 & | & 41.45 \end{bmatrix}$$

Multiply row 2 by $\frac{1}{1.90}$.

$$\begin{bmatrix} 1 & -0.68 & | & -9.62 \\ 0 & 1 & | & 21.82 \end{bmatrix}$$

This corresponds to $\begin{cases} x - 0.68y = -9.62 \\ y = 21.82 \end{cases}$.

$x - 0.68(21.82) = -9.62$

$x \approx 5.2 \approx 5$

2000 + 5 = 2005

During 2005, geothermal sources and wind power generated the same amount of electricity.

b. no; answers may vary

c. no; it has a positive slope

d. answers may vary

39. answers may vary

Section 4.5 Practice Exercises

1. $\begin{cases} 4x \geq y \\ x + 3y \geq 6 \end{cases}$

Graph both inequalities on the same set of axes. The solution is the intersection of the solution regions.

For $4x \geq y$, the boundary line is the graph of $4x = y$. The boundary line is solid since the inequality means $4x > y$ or $4x = y$. The test point (1, 0) satisfies the inequality, so we shade the half-plane that includes (1, 0).

For $x + 3y \geq 6$, sketch the solid boundary line $x + 3y = 6$. The test point (0, 0) does not satisfy the inequality, so shade the half-plane that does not include (0, 0).

The solution of the system is the darker shaded region. This solution includes parts of both boundary lines.

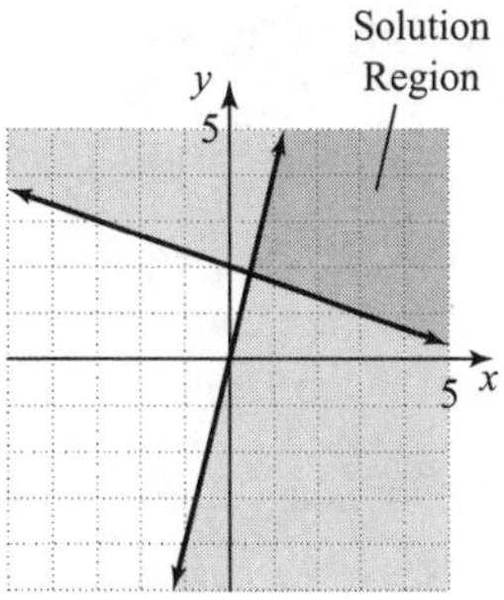

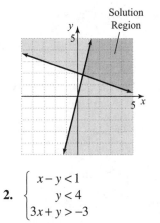

2. $\begin{cases} x - y < 1 \\ y < 4 \\ 3x + y > -3 \end{cases}$

Graph all three inequalities on the same set of axes. All boundary lines are dashed since the inequality symbols are < and >. The solution set of the system is the shaded region. The boundary lines are not a part of the solution.

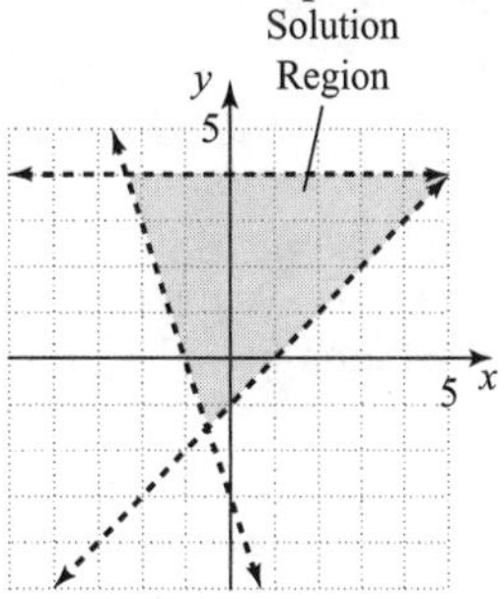

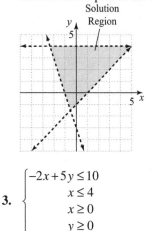

3. $\begin{cases} -2x + 5y \leq 10 \\ x \leq 4 \\ x \geq 0 \\ y \geq 0 \end{cases}$

Graph the inequalities on the same set of axes. The intersection of the inequalities is the solution region. It is the only shaded region in this graph and includes the portion of all four boundary lines that border the shaded region.

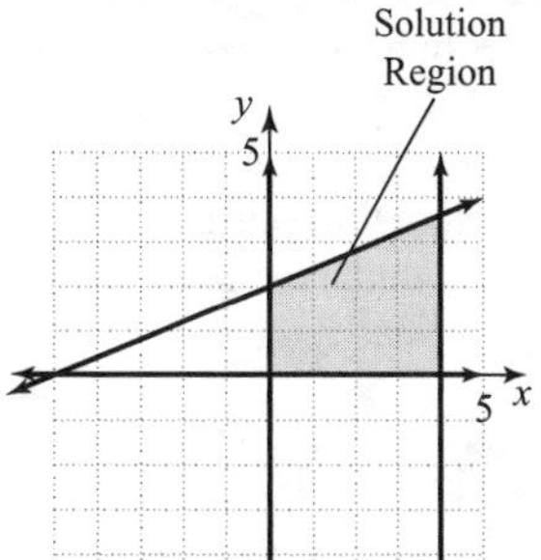

Vocabulary, Readiness & Video Check 4.5

1. Two or more linear inequalities form a system of linear inequalities.

2. An ordered pair that satisfies each inequality in a system is a solution of the system.

3. The point where two boundary lines intersect is a corner point.

4. The solution region of a system of inequalities consists of the intersection of the solution regions of the inequalities in the system.

5. No; we can choose any test point except a point on the second inequality's own boundary line.

Exercise Set 4.5

1. $\begin{cases} y \geq x+1 \\ y \geq 3-x \end{cases}$

Graph both inequalities on the same set of axes. The solution is the intersection of the solution regions. The solution of the system is the darker shaded region. This solution includes parts of both boundary lines.

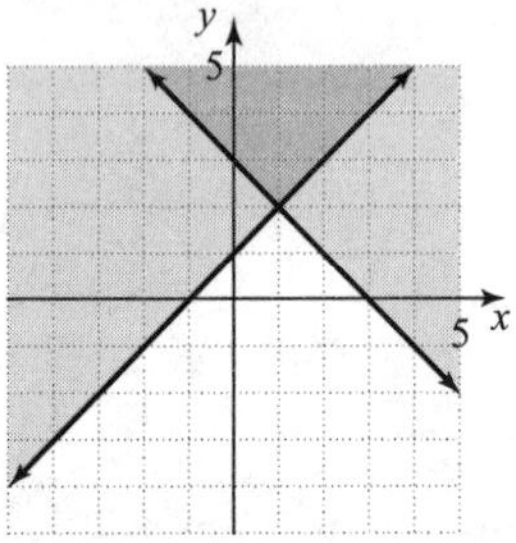

3. $\begin{cases} y < 3x-4 \\ y \leq x+2 \end{cases}$

Graph both inequalities on the same set of axes. The solution is the intersection of the solution regions. The solution of the system is the darker shaded region. This solution includes the part of the solid boundary line that borders the region but not the dashed boundary line or the point where the boundary lines intersect.

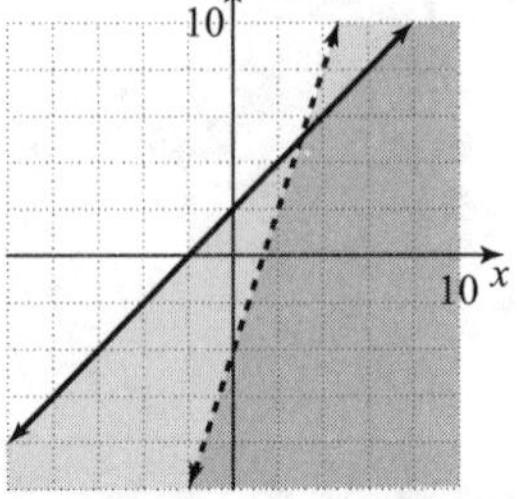

5. $\begin{cases} y < -2x-2 \\ y > x+4 \end{cases}$

Graph both inequalities on the same set of axes. The solution is the intersection of the solution regions. The solution of the system is the darker shaded region. The boundary lines are not a part of the solution.

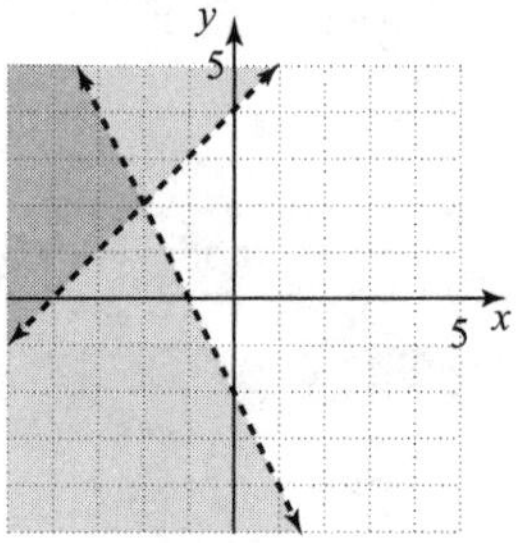

7. $\begin{cases} y \geq -x+2 \\ y \leq 2x+5 \end{cases}$

Graph both inequalities on the same set of axes. The solution is the intersection of the solution regions. The solution of the system is the darker shaded region. This solution includes parts of both boundary lines.

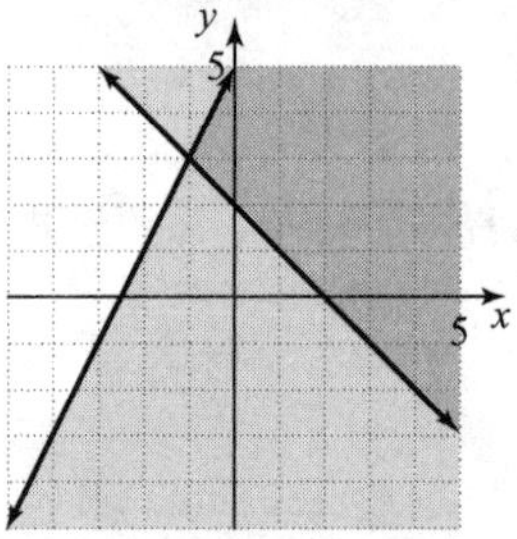

9. $\begin{cases} x \geq 3y \\ x+3y \leq 6 \end{cases}$

Graph both inequalities on the same set of axes. The solution is the intersection of the solution regions. The solution of the system is the darker shaded region. This solution includes parts of both boundary lines.

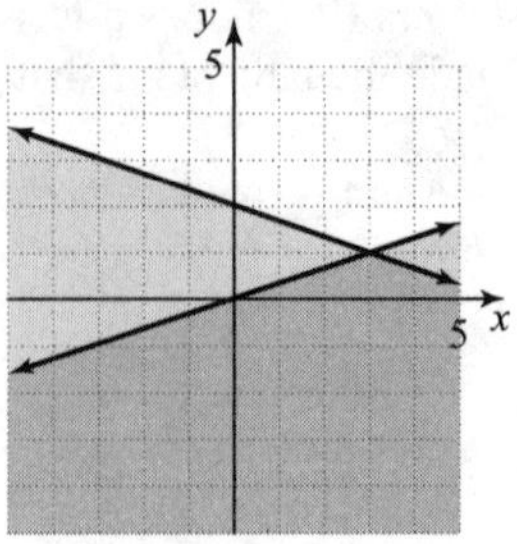

11. $\begin{cases} x \le 2 \\ y \ge -3 \end{cases}$

Graph both inequalities on the same set of axes. The solution is the intersection of the solution regions. The solution of the system is the darker shaded region. This solution includes parts of both boundary lines.

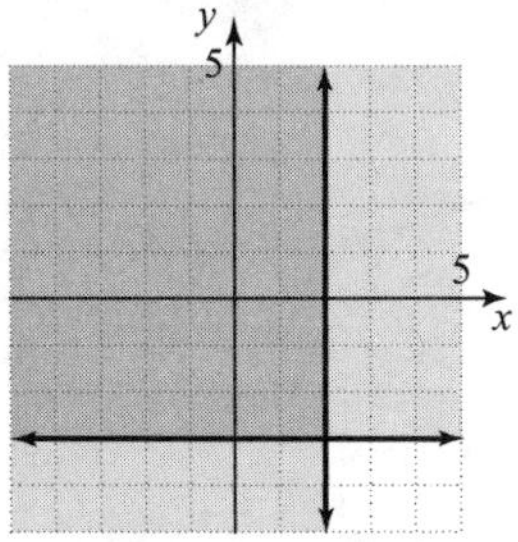

13. $\begin{cases} y \ge 1 \\ x < -3 \end{cases}$

Graph both inequalities on the same set of axes. The solution is the intersection of the solution regions. The solution of the system is the darker shaded region. This solution includes the part of the solid boundary line that borders the region but not the dashed boundary line or the point where the boundary lines intersect.

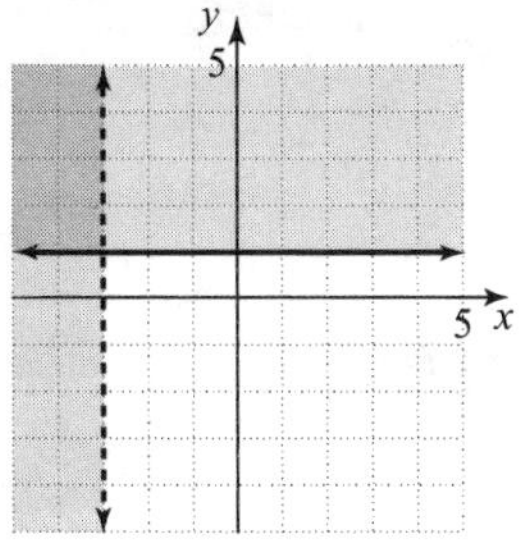

15. $\begin{cases} y + 2x \ge 0 \\ 5x - 3y \le 12 \\ y \le 2 \end{cases}$

Graph all three inequalities on the same set of axes. The solution set of the system is the shaded region. The parts of the boundary lines that border the shaded region are part of the solution.

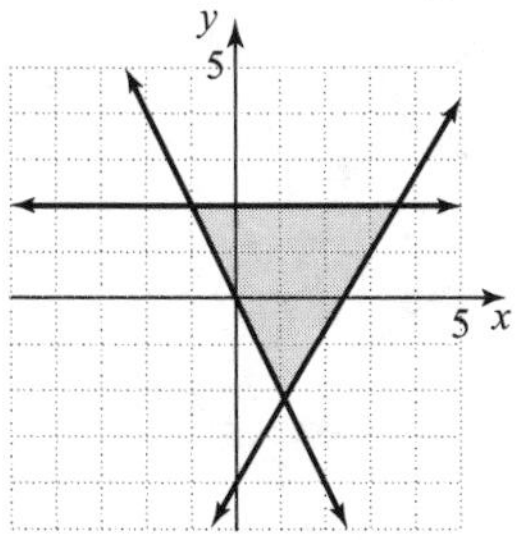

17. $\begin{cases} 3x - 4y \ge -6 \\ 2x + y \le 7 \\ y \ge -3 \end{cases}$

Graph all three inequalities on the same set of axes. The solution set of the system is the shaded region. The parts of the boundary lines that border the shaded region are part of the solution.

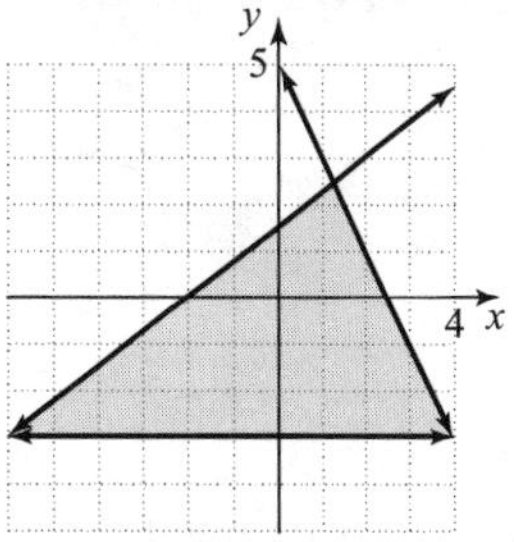

19. $\begin{cases} 2x + y \le 5 \\ x \le 3 \\ x \ge 0 \\ y \ge 0 \end{cases}$

Graph the inequalities on the same set of axes. The intersection of the inequalities is the solution region. It is the only shaded region in this graph and includes the portion of all four boundary lines that border the shaded region.

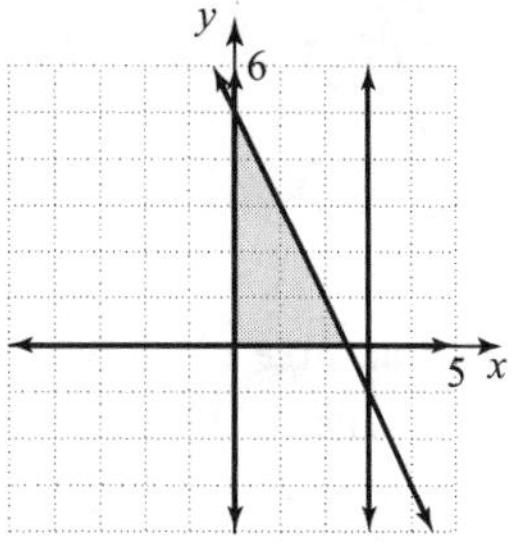

21. $\begin{cases} y < 5 \\ x > 3 \end{cases}$

Both boundary lines should be dashed. The region including (5, 0) should be shaded since (5, 0) satisfies both inequalities. The correct graph is C.

23. $\begin{cases} y \le 5 \\ x < 3 \end{cases}$

The boundary line $y = 5$ should be solid. The correct graph is D.

25. $(-3)^2 = (-3)(-3) = 9$

27. $\left(\frac{2}{3}\right)^2 = \frac{2}{3} \cdot \frac{2}{3} = \frac{2 \cdot 2}{3 \cdot 3} = \frac{4}{9}$

29. $(-2)^2-(-3)+2(-1)=4-(-3)+2(-1)$
$=4-(-3)+(-2)$
$=4+3-2$
$=7-2$
$=5$

31. $8^2+(-13)-4(-2)=64+(-13)-4(-2)$
$=64+(-13)-(-8)$
$=64-13+8$
$=51+8$
$=59$

33. $\begin{cases} y \le 3 \\ y \ge 3 \end{cases}$

The only y-values that satisfy both inequalities are those that equal 3. The solution of the system is the line $y = 3$.

35. answers may vary.

Chapter 4 Vocabulary Check

1. Two or more linear equations in two variables form a <u>system of equations</u>.
2. A <u>solution</u> of a system of two equations in two variables is an ordered pair that makes both equations true.
3. A <u>consistent</u> system of equations has at least one solution.
4. A solution of a system of three equations in three variables is an ordered <u>triple</u> that makes all three equations true.
5. An <u>inconsistent</u> system of equations has no solution.
6. A <u>matrix</u> is a rectangular array of numbers.
7. Each of the numbers in a matrix is called an <u>element</u>.
8. The numbers aligned horizontally in a matrix are in the same <u>row</u>.
9. The numbers aligned vertically in a matrix are in the same <u>column</u>.

Chapter 4 Review

1. $\begin{cases} 3x+10y=1 & (1) \\ x+\ 2y=-1 & (2) \end{cases}$

a.

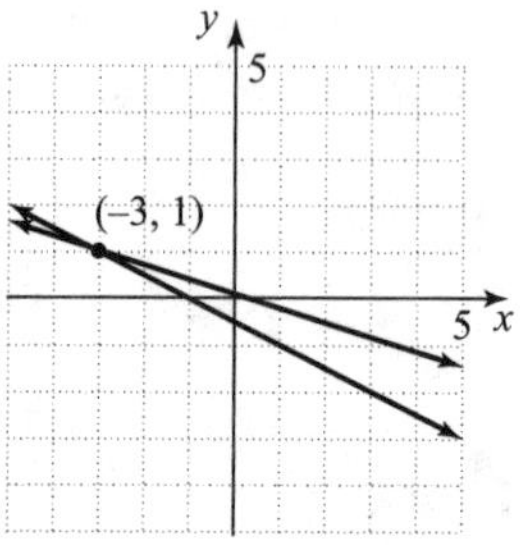

b. From E2: $x=-2y-1$

Replace x with $-2y-1$ in E1.

$3(-2y-1)+10y=1$
$-6y-3+10y=1$
$4y=4$
$y=1$

Replace y with 1 in the equation $x = -2y - 1$.

$x=-2(1)-1=-3$

The solution is (−3, 1).

c. Multiply E2 by −3 and add to E1.

$3x+10y=1$
$\underline{-3x-6y=3}$
$4y=4$
$y=1$

Replace y with 1 in E2.

$x+2(1)=-1$
$x+2=-1$
$x=-3$

The solution is (−3, 1).

2. $\begin{cases} y=\frac{1}{2}x+\frac{2}{3} & (1) \\ 4x+6y=4 & (2) \end{cases}$

a.

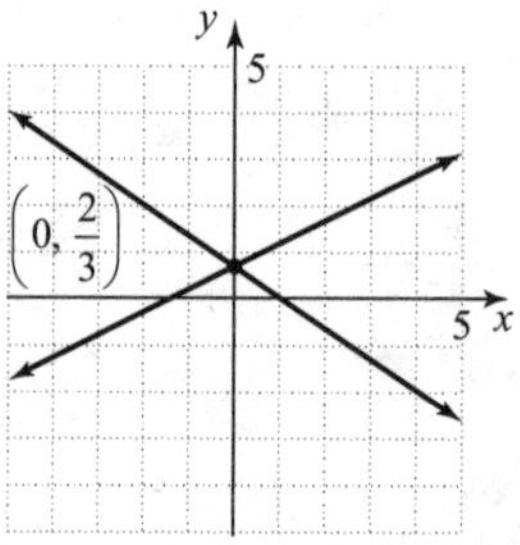

b. Replace y with $\frac{1}{2}x+\frac{2}{3}$ in E2.

$$4x+6\left(\frac{1}{2}x+\frac{2}{3}\right)=4$$
$$4x+3x+4=4$$
$$x=0$$

Replace x with 0 in E1.

$$y=\frac{1}{2}(0)+\frac{2}{3}=\frac{2}{3}$$

The solution is $\left(0, \frac{2}{3}\right)$.

c. Rewrite the system: $\begin{cases} -\frac{1}{2}x+y=\frac{2}{3} \\ 4x+6y=4 \end{cases}$.

Multiply the first equation by –6.

$$\begin{cases} 3x-6y=-4 \\ 4x+6y=4 \end{cases}$$

Add these equations.

$$7x=0$$
$$x=0$$

Replace x with 0 in second equation.

$$4(0)+6y=4$$
$$6y=4$$
$$y=\frac{4}{6}=\frac{2}{3}$$

The solution is $\left(0, \frac{2}{3}\right)$.

3. $\begin{cases} 2x-4y=22 & (1) \\ 5x-10y=15 & (2) \end{cases}$

a.

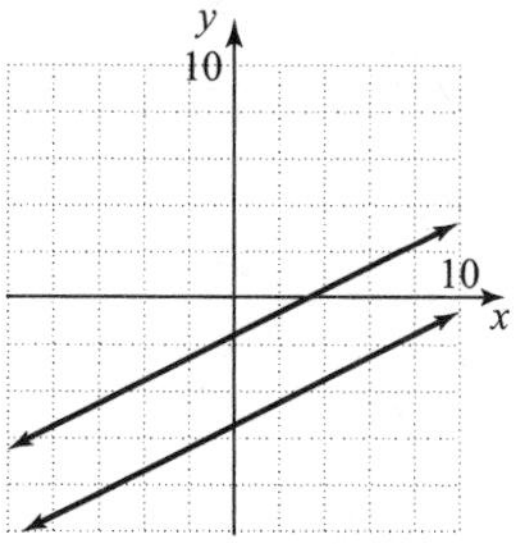

b. Solve E1 for x.

$$2x-4y=22$$
$$2x=4y+22$$
$$x=2y+11$$

Replace x with $2y+11$ in E2.

$$5(2y+11)-10y=15$$
$$10y+55-10y=15$$
$$55=15 \text{ False}$$

This is an inconsistent system. The solution is $\varnothing$.

c. Multiply E1 by 5 and E2 by –2.

$$\begin{cases} 10x-20y=110 \\ -10x+20y=-30 \end{cases}$$

Add these equations.

$$10x-20y=110$$
$$\underline{-10x+20y=-30}$$
$$0=80 \quad \text{False}$$

This is an inconsistent system. The solution is $\varnothing$.

4. $\begin{cases} 3x-6y=12 & (1) \\ 2y=x-4 & (2) \end{cases}$

a.

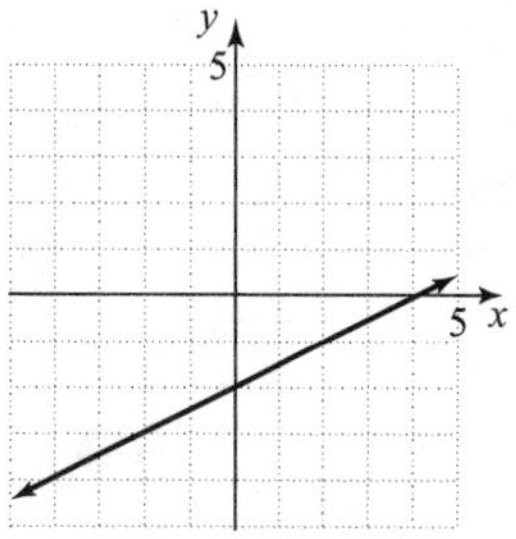

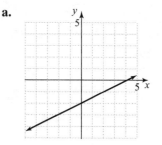

b. Solve E2 for x.

$$x=2y+4$$

Replace x with $2y+4$ in E1.

$$3(2y+4)-6y=12$$
$$6y+12-6y=12$$
$$12=12 \text{ True}$$

This is a dependent system. The solution is $\{(x, y)|3x-6y=12\}$.

c. $\begin{cases} 3x-6y=12 & (1) \\ -x+2y=-4 & (2) \end{cases}$

Multiply E2 by 3.

$$\begin{cases} 3x-6y=12 \\ -3x+6y=-12 \end{cases}$$

Add these equations.

$$3x-6y=12$$
$$\underline{-3x+6y=-12}$$
$$0=0 \quad \text{True}$$

This is a dependent system. The solution is $\{(x, y)|3x-6y=12\}$.

5. $\begin{cases} \frac{1}{2}x - \frac{3}{4}y = -\frac{1}{2} & (1) \\ \frac{1}{8}x + \frac{3}{4}y = \frac{19}{8} & (2) \end{cases}$

a.

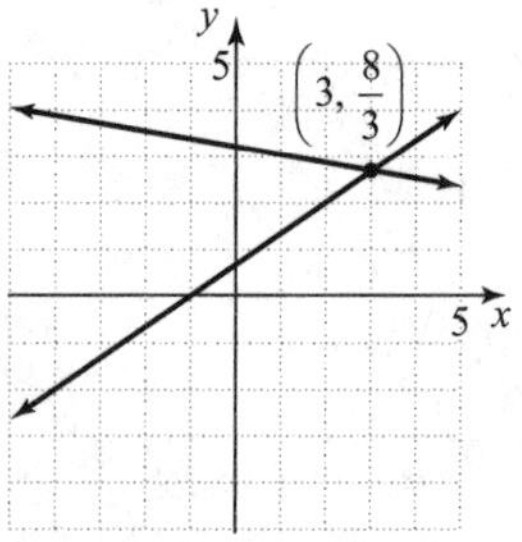

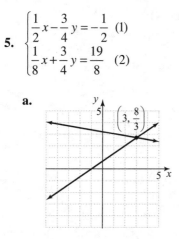

b. Clear fractions by multiplying E1 by 4 and E2 by 8.

$\begin{cases} 2x - 3y = -2 & (1) \\ x + 6y = 19 & (2) \end{cases}$

Solve the new E2 for *x*.

$x = -6y + 19$

Replace *x* with $-6y+19$ in new E1.

$2(-6y+19) - 3y = -2$
$-12y + 38 - 3y = -2$
$-15y = -40$
$y = \frac{-40}{-15} = \frac{8}{3}$

Replace *y* with $\frac{8}{3}$ in the equation

$x = -6y + 19.$
$x = -6\left(\frac{8}{3}\right) + 19$
$x = -16 + 19$
$x = 3$

The solution is $\left(3, \frac{8}{3}\right)$.

c. Add the equations.

$\frac{1}{2}x - \frac{3}{4}y = -\frac{1}{2}$
$\frac{1}{8}x + \frac{3}{4}y = \frac{19}{8}$
$\frac{5}{8}x = \frac{15}{8}$
$5x = 15$
$x = 3$

Replace *x* with 3 in E1.

$\frac{1}{2}(3) - \frac{3}{4}y = -\frac{1}{2}$
$-\frac{3}{4}y = -2$
$-3y = -8$
$y = \frac{8}{3}$

The solution is $\left(3, \frac{8}{3}\right)$.

6. $\begin{cases} y = 32x & (1) \\ y = 15x + 25{,}500 & (2) \end{cases}$

Multiply E1 by –1 and add to E2.

$-y = -32$
$y = 15x + 25{,}500$
$0 = -17x + 25{,}500$
$17x = 25{,}500$
$x = 1500$

Replace *x* with 1500 in E1.

$y = 32(1500) = 48{,}000$

The number of backpacks that the company must sell is 1500.

7. $\begin{cases} x + z = 4 & (1) \\ 2x - y = 4 & (2) \\ x + y - z = 0 & (3) \end{cases}$

Adding E2 and E3 gives $3x - z = 4$ (4)
Adding E1 and E4 gives $4x = 8$ or $x = 2$
Replace *x* with 2 in E1.

$2 + z = 4$
$z = 2$

Replace *x* with 2 and *z* with 2 in E3.

$2 + y - 2 = 0$
$y = 0$

The solution is (2, 0, 2).

8. $\begin{cases} 2x + 5y = 4 & (1) \\ x - 5y + z = -1 & (2) \\ 4x - z = 11 & (3) \end{cases}$

Add E2 and E3.

$5x - 5y = 10$ (4)

Add E1 and E4.

$7x = 14$
$x = 2$

Replace *x* with 2 in E1.

$2(2) + 5y = 4$
$4 + 5y = 4$
$5y = 0$
$y = 0$

Replace x with 2 in E3.

$$\begin{aligned}4(2)-z&=11\\8-z&=11\\z&=-3\end{aligned}$$

The solution is (2, 0, –3).

9. $\begin{cases}4y+2z=5 & (1)\\2x+8y=5 & (2)\\6x+4z=1 & (3)\end{cases}$

Multiply E1 by –2 and add to E2.

$$\begin{array}{l}-8y-4z=-10\\ \underline{2x+8y=5}\\ 2x-4z=-5\ (4)\end{array}$$

Add E3 and E4.

$$\begin{aligned}8x&=-4\\x&=-\frac{1}{2}\end{aligned}$$

Replace x with $-\frac{1}{2}$ in E2.

$$\begin{aligned}2\left(-\frac{1}{2}\right)+8y&=5\\-1+8y&=5\\8y&=6\\y&=\frac{3}{4}\end{aligned}$$

Replace x with $-\frac{1}{2}$ in E3.

$$\begin{aligned}6\left(-\frac{1}{2}\right)+4z&=1\\-3+4z&=1\\4z&=4\\z&=1\end{aligned}$$

The solution is $\left(-\frac{1}{2}, \frac{3}{4}, 1\right)$.

10. $\begin{cases}5x+7y=9 & (1)\\14y-z=28 & (2)\\4x+2z=-4 & (3)\end{cases}$

Dividing E3 by 2 gives $2x+z=-2$.
Add this equation to E2.

$$\begin{array}{l}2x+z=-2\\ \underline{14y-z=28}\\ 2x+14y=26 \text{ or } x+7y=13\ (4)\end{array}$$

Multiply E4 by –1 and add to E1.

$$\begin{array}{l}-x-7y=-13\\ \underline{5x+7y=9}\\ 4x=-4\\ x=-1\end{array}$$

Replace x with –1 in E4.

$$\begin{aligned}-1+7y&=13\\7y&=14\\y&=2\end{aligned}$$

Replace x with –1 in E3.

$$\begin{aligned}4(-1)+2z&=-4\\-4+2z&=-4\\2z&=0\\z&=0\end{aligned}$$

The solution is (–1, 2, 0).

11. $\begin{cases}3x-2y+2z=5 & (1)\\-x+6y+z=4 & (2)\\3x+14y+7z=20 & (3)\end{cases}$

Multiply E2 by 3 and add to E1.

$$\begin{array}{l}3x-2y+2z=5\\ \underline{-3x+18y+3z=12}\\ 16y+5z=17\ (4)\end{array}$$

Multiply E3 by –1 and add to E1.

$$\begin{array}{l}3x-2y+2z=5\\ \underline{-3x-14y-7z=-20}\\ -16y-5z=-15\ (5)\end{array}$$

Add E4 and E5.

$$\begin{array}{l}16y+5z=17\\ \underline{-16y-5z=-15}\\ 0=2 \quad \text{False}\end{array}$$

The system is inconsistent. The solution is ∅.

12. $\begin{cases}x+2y+3z=11 & (1)\\y+2z=3 & (2)\\2x+2z=10 & (3)\end{cases}$

Multiply E2 by –2 and add to E1.

$$\begin{array}{l}x+2y+3z=11\\ \underline{-2y-4z=-6}\\ x-z=5\ (4)\end{array}$$

Multiply E4 by 2 and add to E3.

$$\begin{array}{l}2x+2z=10\\ \underline{2x-2z=10}\\ 4x=20\\ x=5\end{array}$$

Replace x with 5 in E3.

$$\begin{aligned}2(5)+2z&=10\\10+2z&=10\\2z&=0\\z&=0\end{aligned}$$

Replace z with 0 in E2.

$$\begin{aligned}y+2(0)&=3\\y+0&=3\\y&=3\end{aligned}$$

The solution is (5, 3, 0).

13. $\begin{cases} 7x-3y+2z=0 & (1) \\ 4x-4y-\ z=2 & (2) \\ 5x+2y+3z=1 & (3) \end{cases}$

Multiply E2 by 2 and add to E1.

$$\begin{array}{r} 7x-3y+2z=0 \\ 8x-8y-2z=4 \\ \hline 15x-11y\quad=4\ (4) \end{array}$$

Multiply E2 by 3 and add to E3.

$$\begin{array}{r} 12x-12y-3z=6 \\ 5x+2y+3z=1 \\ \hline 17x-10y\quad=7\ (5) \end{array}$$

Solve the new system.

$\begin{cases} 15x-11y=4 & (4) \\ 17x-10y=7 & (5) \end{cases}$

Multiply E4 by –10, multiply E5 by 11, and add.

$$\begin{array}{r} -150x+110y=-40 \\ 187x-110y=77 \\ \hline 37x\quad=37 \\ x=1 \end{array}$$

Replace x with 1 in E4.

$$\begin{aligned} 15(1)-11y&=4 \\ 15-11y&=4 \\ -11y&=-11 \\ y&=1 \end{aligned}$$

Replace x with 1 and y with 1 in E1.

$$\begin{aligned} 7(1)-3(1)+2z&=0 \\ 4+2z&=0 \\ 2z&=-4 \\ z&=-2 \end{aligned}$$

The solution is (1, 1, –2).

14. $\begin{cases} x-3y-5z=-5 & (1) \\ 4x-2y+3z=13 & (2) \\ 5x+3y+4z=22 & (3) \end{cases}$

Multiply E1 by –4 and add to E2.

$$\begin{array}{r} -4x+12y+20z=20 \\ 4x-\ 2y+\ 3z=13 \\ \hline 10y+23z=33\ (4) \end{array}$$

Multiply E1 by –5 and add to E3.

$$\begin{array}{r} -5x+15y+25z=25 \\ 5x+\ 3y+\ 4z=22 \\ \hline 18y+29z=47\ (5) \end{array}$$

Solve the new system.

$\begin{cases} 10y+23z=33 & (4) \\ 18y+29z=47 & (5) \end{cases}$

Multiply E4 by 9, multiply E5 by –5 and add.

$$\begin{array}{r} 90y+207z=297 \\ -90y-145z=-235 \\ \hline 62z=62 \\ z=1 \end{array}$$

Replace z with 1 in E4.

$$\begin{aligned} 10y+23(1)&=33 \\ 10y&=10 \\ y&=1 \end{aligned}$$

Replace y with 1 and z with 1 in E1.

$$\begin{aligned} x-3(1)-5(1)&=-5 \\ x-8&=-5 \\ x&=3 \end{aligned}$$

The solution is (3, 1, 1).

15. Let x = the first number, y = the second number, and z = the third number.

$\begin{cases} x+y+z=98 & (1) \\ x+y=z+2 & (2) \\ y=4x & (3) \end{cases}$

Replace y with $4x$ in E1 and E2.

$$\begin{aligned} x+4x+z&=98 \\ 5x+z&=98 \quad (4) \\ x+4x&=z+2 \\ 5x-z&=2 \quad (5) \end{aligned}$$

Add E4 and E5.

$$\begin{array}{r} 5x+z=98 \\ 5x-z=2 \\ \hline 10x\quad=100 \\ x=10 \end{array}$$

Replace x with 10 in E3.

$y=4(10)=40$

Replace x with 10 and y with 40 in E2.

$$\begin{aligned} 10+40&=z+2 \\ 50&=z+2 \\ 48&=z \end{aligned}$$

The numbers are 10, 40, and 48.

16. Let x = the first number and y = the second number.

$\begin{cases} x=3y & (1) \\ 2(x+y)=168 & (2) \end{cases}$

Replace x with $3y$ in E2.

$$\begin{aligned} 2(3y+y)&=168 \\ 8y&=168 \\ y&=21 \end{aligned}$$

Replace y with 21 in E1.

$x=3(21)=63$

The numbers are 63 and 21.

17. Let x = speed of first car and y = speed of the second car.

$$\begin{cases} 4x+4y=492 & (1) \\ y=x+7 & (2) \end{cases}$$

Replace y with $x+7$ in E1.

$$\begin{aligned} 4x+4(x+7)&=492 \\ 8x+28&=492 \\ 8x&=464 \\ x&=58 \end{aligned}$$

Replace x with 58 in E2.

$y=58+7=65$

The cars are going 58 and 65 miles per hour.

18. Let w = the width of the foundation and l = the length of the foundation.

$$\begin{cases} l=3w & (1) \\ 2w+2l=296 & (2) \end{cases}$$

Replace l with $3w$ in E2.

$$\begin{aligned} 2w+2(3w)&=296 \\ 2w+6w&=296 \\ 8w&=296 \\ w&=37 \end{aligned}$$

Replace w with 37 in E1.

$l=3(37)=111$

The foundation is 37 feet wide and 111 feet long.

19. Let x = liters of 10% solution and y = liters of 60% solution.

$$\begin{cases} x+y=50 & (1) \\ 0.10x+0.60y=0.40(50) & (2) \end{cases}$$

Solve E1 for y.

$y=50-x$

Replace y with $50-x$ in E2.

$$\begin{aligned} 0.10x+0.60(50-x)&=0.40(50) \\ 10[0.10x+0.60(50-x)]&=10[0.40(50)] \\ x+6(50-x)&=4(50) \\ x+300-6x&=200 \\ -5x&=-100 \\ x&=20 \end{aligned}$$

Replace x with 20 in the equation $y = 50 - x$.

$y=50-20=30$

He should use 20 liters of 10% solution and 30 liters of 60% solution.

20. Let c = pounds of chocolate used, n = pounds of nuts used, and r = pounds of raisins used.

$$\begin{cases} r=2n & (1) \\ c+n+r=45 & (2) \\ 3.00c+2.70n+2.25r=2.80(45) & (3) \end{cases}$$

Replace r with $2n$ in E2.

$$\begin{aligned} c+n+2n&=45 \\ c+3n&=45 \\ c&=-3n+45 \end{aligned}$$

Replace r with $2n$ and c with $-3n+45$ in E3.

$$\begin{aligned} 3.00(-3n+45)+2.70n+2.25(2n)&=126 \\ -9n+135+2.7n+4.5n&=126 \\ -1.8n+135&=126 \\ -1.8n&=-9 \\ n&=5 \end{aligned}$$

Replace n with 5 in E1.

$r=2(5)=10$

Replace n with 5 and r with 10 in E2.

$$\begin{aligned} c+5+10&=45 \\ c+15&=45 \\ c&=30 \end{aligned}$$

She should use 30 pounds of creme-filled chocolates, 5 pounds of chocolate-covered nuts, and 10 pounds of chocolate-covered raisins.

21. Let x = the number of pennies, y = the number of nickels, and z = the number dimes.

$$\begin{cases} x+y+z=53 & (1) \\ 0.01x+0.05y+0.10z=2.77 & (2) \\ y=z+4 & (3) \end{cases}$$

Clear the decimals from E2 by multiplying by 100.

$x+5y+10z=277$ (4)

Replace y with $z+4$ in E1.

$$\begin{aligned} x+z+4+z&=53 \\ x+2z&=49 \quad (5) \end{aligned}$$

Replace y with $z+4$ in E4.

$$\begin{aligned} x+5(z+4)+10z&=277 \\ x+15z&=257 \quad (6) \end{aligned}$$

Solve the new system.

$$\begin{cases} x+2z=49 & (5) \\ x+15z=257 & (6) \end{cases}$$

Multiply E5 by –1 and add to E6.

$$\begin{aligned} -x-2z&=-49 \\ x+15z&=257 \\ \hline 13z&=208 \\ z&=16 \end{aligned}$$

Replace z with 16 in E3.

$$\begin{aligned} x+2(16)&=49 \\ x+32&=49 \\ x&=17 \end{aligned}$$

Replace z with 16 in E3.

$y=16+4=20$

He has 17 pennies, 20 nickels, and 16 dimes in his jar.

22. Let l = rate of interest on the larger investment and s = the rate of interest on the smaller investment, both expressed as decimals.

$$\begin{cases} 10{,}000l + 4000s = 1250 & (1) \\ l = s + 0.02 & (2) \end{cases}$$

Replace l with $s + 0.02$ in E1.

$$10{,}000(s+0.02) + 4000s = 1250$$
$$10{,}000s + 200 + 4000s = 1250$$
$$14{,}000s = 1050$$
$$s = \frac{1050}{14{,}000} = 0.075$$

and $l = 0.075 + 0.02 = 0.095$.

The interest rate on the larger investment is 9.5% and the rate on the smaller investment is 7.5%.

23. Let x = length of the equal sides and y = length of the third side.

$$\begin{cases} 2x + y = 73 & (1) \\ y = x + 7 & (2) \end{cases}$$

Replace y with $x + 7$ in E1.

$$2x + x + 7 = 73$$
$$3x = 66$$
$$x = 22$$

Replace x with 22 in E2.

$$y = 22 + 7 = 29$$

Two sides of the triangle have length 22 cm and the third side has length 29 cm.

24. Let f = the first number, s = the second number, and t = the third number.

$$\begin{cases} f + s + t = 295 & (1) \\ f = s + 5 & (2) \\ f = 2t & (3) \end{cases}$$

Solve E2 for s and E3 for t.

$$s = f - 5$$
$$t = \frac{f}{2}$$

Replace s with $f - 5$ and t with $\frac{f}{2}$ in E1.

$$f + f - 5 + \frac{f}{2} = 295$$
$$\frac{5}{2}f = 300$$
$$f = 120$$

Replace f with 300 in the equation $s = f - 5$.

$$s = 120 - 5 = 115$$

Replace f with 120 the equation $\frac{f}{2}$.

$$t = \frac{120}{2} = 60$$

The first number is 120, the second number is 115, and the third number is 60.

25. $\begin{cases} 3x + 10y = 1 \\ x + 2y = -1 \end{cases}$

$$\left[\begin{array}{cc|c} 3 & 10 & 1 \\ 1 & 2 & -1 \end{array}\right]$$

Interchange R1 and R2.

$$\left[\begin{array}{cc|c} 1 & 2 & -1 \\ 3 & 10 & 1 \end{array}\right]$$

Multiply R1 by –3 and add to R2.

$$\left[\begin{array}{cc|c} 1 & 2 & -1 \\ 0 & 4 & 4 \end{array}\right]$$

Divide R2 by 4.

$$\left[\begin{array}{cc|c} 1 & 2 & -1 \\ 0 & 1 & 1 \end{array}\right]$$

This corresponds to $\begin{cases} x + 2y = -1 \\ y = 1 \end{cases}$.

$$x + 2(1) = -1$$
$$x = -3$$

The solution is (–3, 1).

26. $\begin{cases} 3x - 6y = 12 \\ 2y = x - 4 \end{cases}$, or $\begin{cases} 3x - 6y = 12 \\ -x + 2y = -4 \end{cases}$

$$\left[\begin{array}{cc|c} 3 & -6 & 12 \\ -1 & 2 & -4 \end{array}\right]$$

Divide R1 by 3.

$$\left[\begin{array}{cc|c} 1 & -2 & 4 \\ -1 & 2 & -4 \end{array}\right]$$

Add R1 to R2.

$$\left[\begin{array}{cc|c} 1 & -2 & 4 \\ 0 & 0 & 0 \end{array}\right]$$

This corresponds to $\begin{cases} x - 2y = 4 \\ 0 = 0 \end{cases}$.

This is a dependent system. The solution is $\{(x, y) | 3x - 6y = 12\}$.

27. $\begin{cases} 3x - 2y = -8 \\ 6x + 5y = 11 \end{cases}$

$$\left[\begin{array}{cc|c} 3 & -2 & -8 \\ 6 & 5 & 11 \end{array}\right]$$

Divide R1 by 3.

$$\left[\begin{array}{cc|c} 1 & -\frac{2}{3} & -\frac{8}{3} \\ 6 & 5 & 11 \end{array}\right]$$

Multiply R1 by –6 and add to R2.

$$\left[\begin{array}{cc|c} 1 & -\frac{2}{3} & -\frac{8}{3} \\ 0 & 9 & 27 \end{array}\right]$$

Divide R2 by 9.

$$\left[\begin{array}{cc|c} 1 & -\frac{2}{3} & -\frac{8}{3} \\ 0 & 1 & 3 \end{array}\right]$$

This corresponds to $\begin{cases} x-\frac{2}{3}y=-\frac{8}{3} \\ y=3 \end{cases}$.

$$x-\frac{2}{3}(3)=-\frac{8}{3}$$
$$x-2=-\frac{8}{3}$$
$$x=-\frac{2}{3}$$

The solution is $\left(-\frac{2}{3}, 3\right)$.

28. $\begin{cases} 6x-6y=-5 \\ 10x-2y=1 \end{cases}$

$$\left[\begin{array}{cc|c} 6 & -6 & -5 \\ 10 & -2 & 1 \end{array}\right]$$

Divide R1 by 6.

$$\left[\begin{array}{cc|c} 1 & -1 & -\frac{5}{6} \\ 10 & -2 & 1 \end{array}\right]$$

Multiply R1 by –10 and add to R2.

$$\left[\begin{array}{cc|c} 1 & -1 & -\frac{5}{6} \\ 0 & 8 & \frac{28}{3} \end{array}\right]$$

Divide R2 by 8.

$$\left[\begin{array}{cc|c} 1 & -1 & -\frac{5}{6} \\ 0 & 1 & \frac{7}{6} \end{array}\right]$$

Add R2 to R1.

$$\left[\begin{array}{cc|c} 1 & 0 & \frac{1}{3} \\ 0 & 1 & \frac{7}{6} \end{array}\right]$$

This corresponds to $\begin{cases} x=\frac{1}{3} \\ y=\frac{7}{6} \end{cases}$. The solution is $\left(\frac{1}{3}, \frac{7}{6}\right)$.

29. $\begin{cases} 3x-6y=0 \\ 2x+4y=5 \end{cases}$

$$\left[\begin{array}{cc|c} 3 & -6 & 0 \\ 2 & 4 & 5 \end{array}\right]$$

Divide R1 by 3.

$$\left[\begin{array}{cc|c} 1 & -2 & 0 \\ 2 & 4 & 5 \end{array}\right]$$

Multiply R1 by –2 and add to R2.

$$\left[\begin{array}{cc|c} 1 & -2 & 0 \\ 0 & 8 & 5 \end{array}\right]$$

Divide R2 by 8.

$$\left[\begin{array}{cc|c} 1 & -2 & 0 \\ 0 & 1 & \frac{5}{8} \end{array}\right]$$

This corresponds to $\begin{cases} x-2y=0 \\ y=\frac{5}{8} \end{cases}$.

$$x-2\left(\frac{5}{8}\right)=0$$
$$x-\frac{5}{4}=0$$
$$x=\frac{5}{4}$$

The solution is $\left(\frac{5}{4}, \frac{5}{8}\right)$.

30. $\begin{cases} 5x-3y=10 \\ -2x+y=-1 \end{cases}$

$$\left[\begin{array}{cc|c} 5 & -3 & 10 \\ -2 & 1 & -1 \end{array}\right]$$

Divide R1 by 5.

$$\left[\begin{array}{cc|c} 1 & -\frac{3}{5} & 2 \\ -2 & 1 & -1 \end{array}\right]$$

Multiply R1 by 2 and add to R2.

$$\left[\begin{array}{cc|c} 1 & -\frac{3}{5} & 2 \\ 0 & -\frac{1}{5} & 3 \end{array}\right]$$

Multiply R2 by –5.

$$\left[\begin{array}{cc|c} 1 & -\frac{3}{5} & 2 \\ 0 & 1 & -15 \end{array}\right]$$

This corresponds to $\begin{cases} x-\frac{3}{5}y=2 \\ y=-15 \end{cases}$.

$$x - \frac{3}{5}(-15) = 2$$
$$x + 9 = 2$$
$$x = -7$$

The solution is (–7, –15).

31. $\begin{cases} 0.2x - 0.3y = -0.7 \\ 0.5x + 0.3y = 1.4 \end{cases}$

$$\left[\begin{array}{cc|c} 0.2 & -0.3 & -0.7 \\ 0.5 & 0.3 & 1.4 \end{array}\right]$$

Multiply both rows by 10 to clear decimals.

$$\left[\begin{array}{cc|c} 2 & -3 & -7 \\ 5 & 3 & 14 \end{array}\right]$$

Divide R1 by 2.

$$\left[\begin{array}{cc|c} 1 & -\frac{3}{2} & -\frac{7}{2} \\ 5 & 3 & 14 \end{array}\right]$$

Multiply R1 by –5 and add to R2.

$$\left[\begin{array}{cc|c} 1 & -\frac{3}{2} & -\frac{7}{2} \\ 0 & \frac{21}{2} & \frac{63}{2} \end{array}\right]$$

Multiply R2 by $\frac{2}{21}$.

$$\left[\begin{array}{cc|c} 1 & -\frac{3}{2} & -\frac{7}{2} \\ 0 & 1 & 3 \end{array}\right]$$

This corresponds to $\begin{cases} x - \frac{3}{2}y = -\frac{7}{2} \\ y = 3 \end{cases}$.

$$x - \frac{3}{2}(3) = -\frac{7}{2}$$
$$x - \frac{9}{2} = -\frac{7}{2}$$
$$x = 1$$

The solution is (1, 3).

32. $\begin{cases} 3x + 2y = 8 \\ 3x - y = 5 \end{cases}$

$$\left[\begin{array}{cc|c} 3 & 2 & 8 \\ 3 & -1 & 5 \end{array}\right]$$

Divide R1 by 3.

$$\left[\begin{array}{cc|c} 1 & \frac{2}{3} & \frac{8}{3} \\ 3 & -1 & 5 \end{array}\right]$$

Multiply R1 by –3 and add to R2.

$$\left[\begin{array}{cc|c} 1 & \frac{2}{3} & \frac{8}{3} \\ 0 & -3 & -3 \end{array}\right]$$

Divide R2 by –3.

$$\left[\begin{array}{cc|c} 1 & \frac{2}{3} & \frac{8}{3} \\ 0 & 1 & 1 \end{array}\right]$$

This corresponds to $\begin{cases} x + \frac{2}{3}y = \frac{8}{3} \\ y = 1 \end{cases}$.

$$x + \frac{2}{3}(1) = \frac{8}{3}$$
$$x = 2$$

The solution is (2, 1).

33. $\begin{cases} x \quad + z = 4 \\ 2x - y \quad = 0 \\ x + y - z = 0 \end{cases}$

$$\left[\begin{array}{ccc|c} 1 & 0 & 1 & 4 \\ 2 & -1 & 0 & 0 \\ 1 & 1 & -1 & 0 \end{array}\right]$$

Multiply R1 by –2 and add to R2. Multiply R1 by –1 and add to R3.

$$\left[\begin{array}{ccc|c} 1 & 0 & 1 & 4 \\ 0 & -1 & -2 & -8 \\ 0 & 1 & -2 & -4 \end{array}\right]$$

Multiply R2 by –1.

$$\left[\begin{array}{ccc|c} 1 & 0 & 1 & 4 \\ 0 & 1 & 2 & 8 \\ 0 & 1 & -2 & -4 \end{array}\right]$$

Multiply R2 by –1 and add to R3.

$$\left[\begin{array}{ccc|c} 1 & 0 & 1 & 4 \\ 0 & 1 & 2 & 8 \\ 0 & 0 & -4 & -12 \end{array}\right]$$

Divide R3 by –4.

$$\left[\begin{array}{ccc|c} 1 & 0 & 1 & 4 \\ 0 & 1 & 2 & 8 \\ 0 & 0 & 1 & 3 \end{array}\right]$$

This corresponds to $\begin{cases} x + z = 4 \\ y + 2z = 8 \\ z = 3 \end{cases}$.

$$y + 2(3) = 8$$
$$y + 6 = 8$$
$$y = 2$$
$$x + 3 = 4$$
$$x = 1$$

The solution is (1, 2, 3).

34. $\begin{cases} 2x+5y = 4 \\ x-5y+z=-1 \\ 4x - z = 11 \end{cases}$

$\left[\begin{array}{ccc|c} 2 & 5 & 0 & 4 \\ 1 & -5 & 1 & -1 \\ 4 & 0 & -1 & 11 \end{array}\right]$

Interchange R1 and R2.

$\left[\begin{array}{ccc|c} 1 & -5 & 1 & -1 \\ 2 & 5 & 0 & 4 \\ 4 & 0 & -1 & 11 \end{array}\right]$

Multiply R1 by –2 and add to R2. Multiply R1 by –4 and add to R3.

$\left[\begin{array}{ccc|c} 1 & -5 & 1 & -1 \\ 0 & 15 & -2 & 6 \\ 0 & 20 & -5 & 15 \end{array}\right]$

Divide R2 by 15.

$\left[\begin{array}{ccc|c} 1 & -5 & 1 & -1 \\ 0 & 1 & -\frac{2}{15} & \frac{2}{5} \\ 0 & 20 & -5 & 15 \end{array}\right]$

Multiply R2 by –20 and add to R3.

$\left[\begin{array}{ccc|c} 1 & -5 & 1 & -1 \\ 0 & 1 & -\frac{2}{15} & \frac{2}{5} \\ 0 & 0 & -\frac{7}{3} & 7 \end{array}\right]$

Multiply R3 by $-\frac{3}{7}$.

$\left[\begin{array}{ccc|c} 1 & -5 & 1 & -1 \\ 0 & 1 & -\frac{2}{15} & \frac{2}{5} \\ 0 & 0 & 1 & -3 \end{array}\right]$

This corresponds to $\begin{cases} x-5y+z=-1 \\ y-\frac{2}{15}z=\frac{2}{5} \\ z=-3 \end{cases}$.

$y-\frac{2}{15}(-3)=\frac{2}{5}$

$y+\frac{2}{5}=\frac{2}{5}$

$y=0$

$x-5(0)+(-3)=-1$

$x-3=-1$

$x=2$

The solution is (2, 0, –3).

35. $\begin{cases} 3x-y = 11 \\ x + 2z = 13 \\ y - z = -7 \end{cases}$

$\left[\begin{array}{ccc|c} 3 & -1 & 0 & 11 \\ 1 & 0 & 2 & 13 \\ 0 & 1 & -1 & -7 \end{array}\right]$

Interchange R1 and R2.

$\left[\begin{array}{ccc|c} 1 & 0 & 2 & 13 \\ 3 & -1 & 0 & 11 \\ 0 & 1 & -1 & -7 \end{array}\right]$

Interchange R2 and R3.

$\left[\begin{array}{ccc|c} 1 & 0 & 2 & 13 \\ 0 & 1 & -1 & -7 \\ 3 & -1 & 0 & 11 \end{array}\right]$

Multiply R1 by –3 and add to R3.

$\left[\begin{array}{ccc|c} 1 & 0 & 2 & 13 \\ 0 & 1 & -1 & -7 \\ 0 & -1 & -6 & -28 \end{array}\right]$

Add R2 to R3.

$\left[\begin{array}{ccc|c} 1 & 0 & 2 & 13 \\ 0 & 1 & -1 & -7 \\ 0 & 0 & -7 & -35 \end{array}\right]$

Divide R3 by –7.

$\left[\begin{array}{ccc|c} 1 & 0 & 2 & 13 \\ 0 & 1 & -1 & -7 \\ 0 & 0 & 1 & 5 \end{array}\right]$

This corresponds to $\begin{cases} x+2z=13 \\ y-z=-7 \\ z=5 \end{cases}$.

$y-5=-7$

$y=-2$

$x+2(5)=13$

$x=3$

The solution is (3, –2, 5).

36. $\begin{cases} 5x+7y+3z=9 \\ 14y-z=28 \\ 4x + 2z=-4 \end{cases}$

$\left[\begin{array}{ccc|c} 5 & 7 & 3 & 9 \\ 0 & 14 & -1 & 28 \\ 4 & 0 & 2 & -4 \end{array}\right]$

Divide R1 by 5.

$\left[\begin{array}{ccc|c} 1 & \frac{7}{5} & \frac{3}{5} & \frac{9}{5} \\ 0 & 14 & -1 & 28 \\ 4 & 0 & 2 & -4 \end{array}\right]$

Multiply R1 by –4 and add to R3.

$$\left[\begin{array}{ccc|c} 1 & \frac{7}{5} & \frac{3}{5} & \frac{9}{5} \\ 0 & 14 & -1 & 28 \\ 0 & -\frac{28}{5} & -\frac{2}{5} & -\frac{56}{5} \end{array}\right]$$

Divide R2 by 14.

$$\left[\begin{array}{ccc|c} 1 & \frac{7}{5} & \frac{3}{5} & \frac{9}{5} \\ 0 & 1 & -\frac{1}{14} & 2 \\ 0 & -\frac{28}{5} & -\frac{2}{5} & -\frac{56}{5} \end{array}\right]$$

Multiply R2 by $\frac{28}{5}$ and add to R3.

$$\left[\begin{array}{ccc|c} 1 & \frac{7}{5} & \frac{3}{5} & \frac{9}{5} \\ 0 & 1 & -\frac{1}{14} & 2 \\ 0 & 0 & -\frac{4}{5} & 0 \end{array}\right]$$

Multiply R3 by $-\frac{5}{4}$.

$$\left[\begin{array}{ccc|c} 1 & \frac{7}{5} & \frac{3}{5} & \frac{9}{5} \\ 0 & 1 & -\frac{1}{14} & 2 \\ 0 & 0 & 1 & 0 \end{array}\right]$$

This corresponds to $\begin{cases} x+\frac{7}{5}y+\frac{3}{5}z=\frac{9}{5} \\ y-\frac{1}{14}z=2. \\ z=0 \end{cases}$

$$y-\frac{1}{14}(0)=2$$
$$y=2$$
$$x+\frac{7}{5}(2)+\frac{3}{5}(0)=\frac{9}{5}$$
$$x+\frac{14}{5}=\frac{9}{5}$$
$$x=-1$$

The solution is (–1, 2, 0).

37. $\begin{cases} 7x-3y+2z=0 \\ 4x-4y-z=2 \\ 5x+2y+3z=1 \end{cases}$

$$\left[\begin{array}{ccc|c} 7 & -3 & 2 & 0 \\ 4 & -4 & -1 & 2 \\ 5 & 2 & 3 & 1 \end{array}\right]$$

Interchange R1 and R2.

$$\left[\begin{array}{ccc|c} 4 & -4 & -1 & 2 \\ 7 & -3 & 2 & 0 \\ 5 & 2 & 3 & 1 \end{array}\right]$$

Divide R1 by 4.

$$\left[\begin{array}{ccc|c} 1 & -1 & -\frac{1}{4} & \frac{1}{2} \\ 7 & -3 & 2 & 0 \\ 5 & 2 & 3 & 1 \end{array}\right]$$

Multiply R1 by –7 and add to R2. Multiply R1 by –5 and add to R3.

$$\left[\begin{array}{ccc|c} 1 & -1 & -\frac{1}{4} & \frac{1}{2} \\ 0 & 4 & \frac{15}{4} & -\frac{7}{2} \\ 0 & 7 & \frac{17}{4} & -\frac{3}{2} \end{array}\right]$$

Divide R2 by 4.

$$\left[\begin{array}{ccc|c} 1 & -1 & -\frac{1}{4} & \frac{1}{2} \\ 0 & 1 & \frac{15}{16} & -\frac{7}{8} \\ 0 & 7 & \frac{17}{4} & -\frac{3}{2} \end{array}\right]$$

Multiply R2 by –7 and add to R3.

$$\left[\begin{array}{ccc|c} 1 & -1 & -\frac{1}{4} & \frac{1}{2} \\ 0 & 1 & \frac{15}{16} & -\frac{7}{8} \\ 0 & 0 & -\frac{37}{16} & -\frac{37}{8} \end{array}\right]$$

Multiply R3 by $-\frac{16}{37}$.

$$\left[\begin{array}{ccc|c} 1 & -1 & -\frac{1}{4} & \frac{1}{2} \\ 0 & 1 & \frac{15}{16} & -\frac{7}{8} \\ 0 & 0 & 1 & -2 \end{array}\right]$$

This corresponds to $\begin{cases} x-y-\frac{1}{4}z=\frac{1}{2} \\ y+\frac{15}{16}z=-\frac{7}{8}. \\ z=-2 \end{cases}$

$$y+\frac{15}{16}(-2)=-\frac{7}{8}$$
$$y-\frac{15}{8}=-\frac{7}{8}$$
$$y=1$$
$$x-1-\frac{1}{4}(-2)=\frac{1}{2}$$
$$x-1+\frac{1}{2}=\frac{1}{2}$$
$$x=1$$

The solution is (1, 1, –2).

38. $\begin{cases} x+2y+3z=14 \\ \quad y+2z=3 \\ 2x \quad -2z=10 \end{cases}$

$$\left[\begin{array}{ccc|c} 1 & 2 & 3 & 14 \\ 0 & 1 & 2 & 3 \\ 2 & 0 & -2 & 10 \end{array}\right]$$

Multiply R1 by –2 and add to R3.

$$\left[\begin{array}{ccc|c} 1 & 2 & 3 & 14 \\ 0 & 1 & 2 & 3 \\ 0 & -4 & -8 & -18 \end{array}\right]$$

Multiply R2 by 4 and add to R3.

$$\left[\begin{array}{ccc|c} 1 & 2 & 3 & 14 \\ 0 & 1 & 2 & 3 \\ 0 & 0 & 0 & -6 \end{array}\right]$$

This corresponds to $\begin{cases} x+2y+3z=14 \\ \quad y+2z=3 \\ \quad 0=-6 \end{cases}$.

This system is inconsistent. The solution is $\varnothing$.

39. $\begin{cases} y \ge 2x-3 \\ y \le -2x+1 \end{cases}$

Graph both inequalities on the same set of axes. The solution is the intersection of the solution regions. The solution of the system is the darker shaded region. This solution includes parts of both boundary lines.

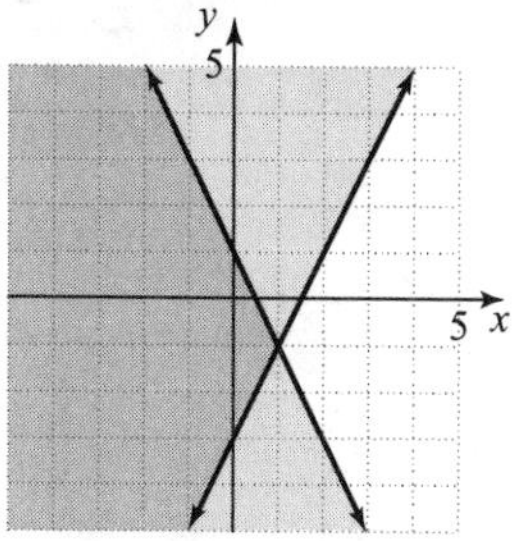

40. $\begin{cases} y \le -3x-3 \\ y \le 2x+7 \end{cases}$

Graph both inequalities on the same set of axes. The solution is the intersection of the solution regions. The solution of the system is the darker shaded region. This solution includes parts of both boundary lines.

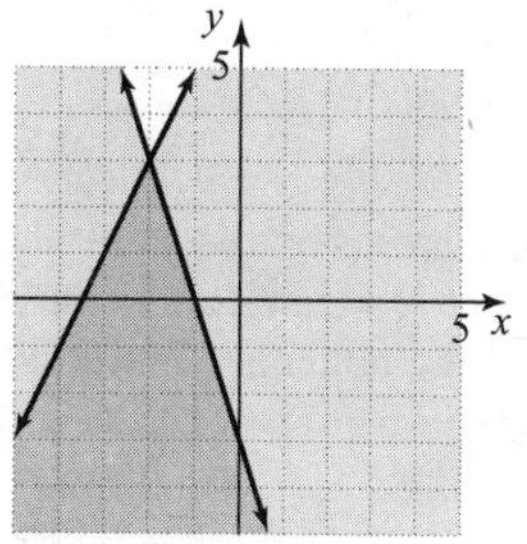

41. $\begin{cases} x+2y>0 \\ x-y \le 6 \end{cases}$

Graph both inequalities on the same set of axes. The solution is the intersection of the solution regions. The solution of the system is the darker shaded region. This solution includes the part of the solid boundary line that borders the region but not the dashed boundary line or the point where the boundary lines intersect.

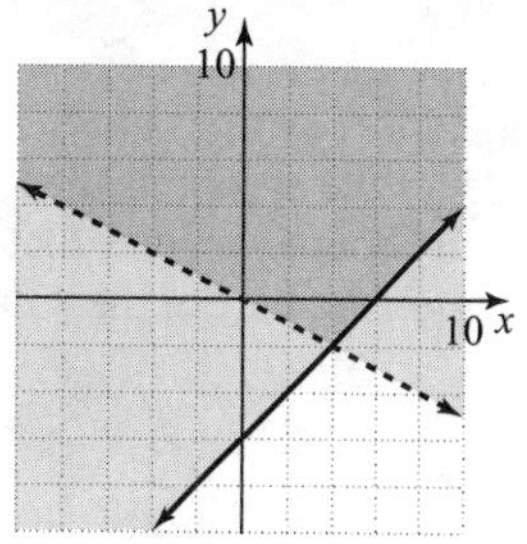

42. $\begin{cases} x-2y \ge 7 \\ x+y \le -5 \end{cases}$

Graph both inequalities on the same set of axes. The solution is the intersection of the solution regions. The solution of the system is the darker shaded region. This solution includes parts of both boundary lines.

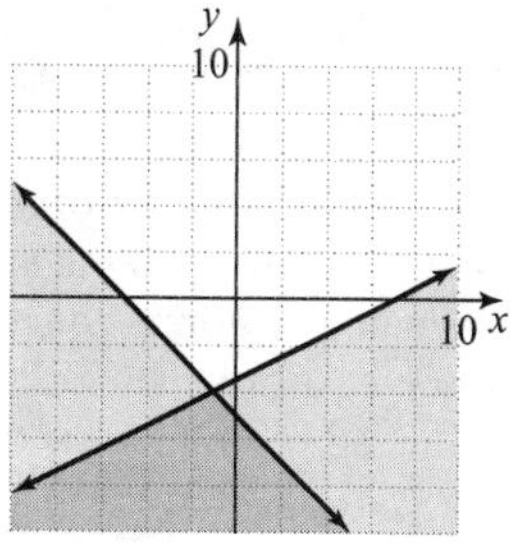

43. $\begin{cases} 3x-2y \le 4 \\ 2x+y \ge 5 \\ y \le 4 \end{cases}$

Graph all three inequalities on the same set of axes. The solution set of the system is the shaded region. The parts of the boundary lines that border the shaded region are part of the solution.

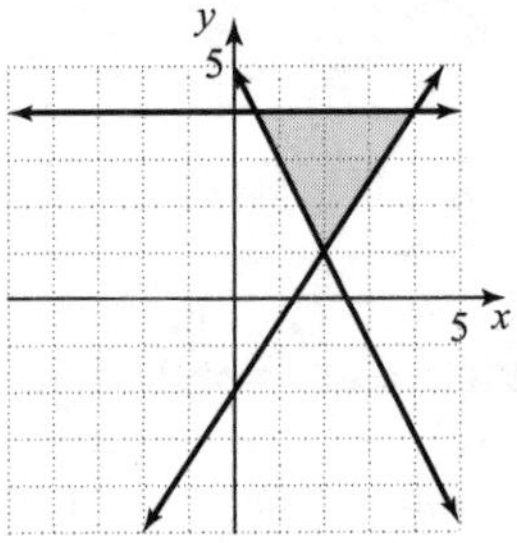

44. $\begin{cases} 4x-y \le 0 \\ 3x-2y \ge -5 \\ y \ge -4 \end{cases}$

Graph all three inequalities on the same set of axes. The solution set of the system is the shaded region. The parts of the boundary lines that border the shaded region are part of the solution.

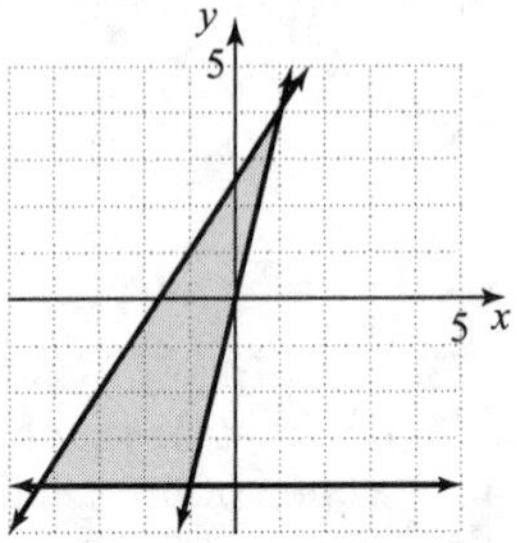

45. $\begin{cases} x+2y \le 5 \\ x \le 2 \\ x \ge 0 \\ y \ge 0 \end{cases}$

Graph the inequalities on the same set of axes. The intersection of the inequalities is the solution region. It is the only shaded region in this graph and includes the portion of all four boundary lines that border the shaded region.

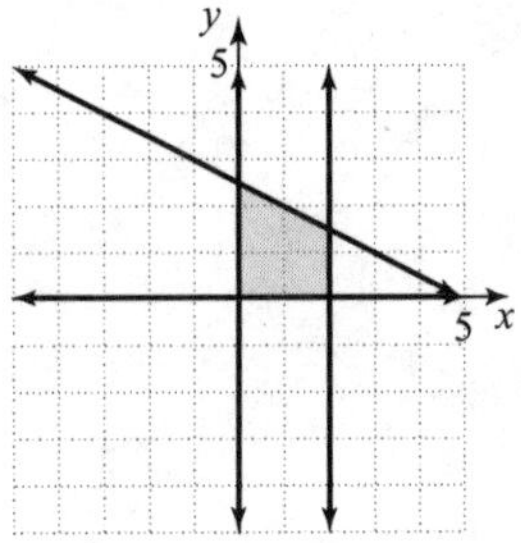

46. $\begin{cases} x+3y \le 7 \\ y \le 5 \\ x \ge 0 \\ y \ge 0 \end{cases}$

Graph the inequalities on the same set of axes. The intersection of the inequalities is the solution region. It is the only shaded region in this graph and includes the portion of all four boundary lines that border the shaded region.

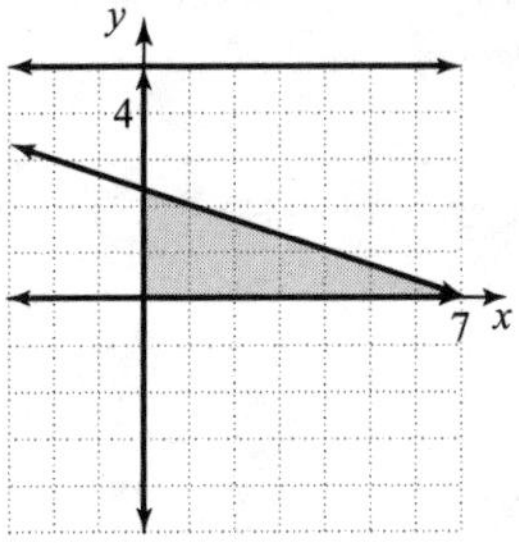

47. $\begin{cases} y = x-5 \\ y = -2x+2 \end{cases}$

We substitute $x - 5$ for y in the second equation.

$$x-5 = -2x+2$$
$$3x = 7$$
$$x = \frac{7}{3}$$

Replace x with $\frac{7}{3}$ in the first equation.

$$y = \frac{7}{3} - 5 = \frac{7}{3} - \frac{15}{3} = -\frac{8}{3}$$

The solution is $\left(\frac{7}{3}, -\frac{8}{3}\right)$.

48. $\begin{cases} \frac{2}{5}x + \frac{3}{4}y = 1 \\ x+3y = -2 \end{cases}$

Multiply both sides of the first equation by 20 to eliminate fractions.

$$\begin{cases} 20\left(\frac{2}{5}x + \frac{3}{4}y\right) = 20(1) \\ x+3y = -2 \end{cases}$$

$$\begin{cases} 8x+15y = 20 \\ x+3y = -2 \end{cases}$$

Multiply both sides of the second equation by -5 and add to the first equation to eliminate y.

$$\begin{cases} 8x+15y = 20 \\ -5x-15y = 10 \end{cases}$$
$$3x = 30$$
$$x = 10$$

To find y, replace x with 10 in the second equation.

$$10+3y=-2$$
$$3y=-12$$
$$y=-4$$

The solution is (10, –4).

49. $\begin{cases} 5x-2y=10 \\ x=\frac{2}{5}y+2 \end{cases}$

Multiply both sides of the second equation by 5.

$$\begin{cases} 5x-2y=10 \\ 5x=5\left(\frac{2}{5}y+2\right) \end{cases}$$

$$\begin{cases} 5x-2y=10 \\ 5x=2y+10 \end{cases}$$

Subtract $2y$ from both sides of the second equation.

$$\begin{cases} 5x-2y=10 \\ 5x-2y=10 \end{cases}$$

The equations are the same. The system has an infinite number of solutions. The solution set can be written as $\{(x, y)|5x-2y=10\}$.

50. $\begin{cases} x-4y=4 \\ \frac{1}{8}x-\frac{1}{2}y=3 \end{cases}$

Multiply the second by –8 and add to the first equation to eliminate x.

$$\begin{cases} x-4y=4 \\ -x+4y=-24 \end{cases}$$

The equation $0=-20$ is false. The system has no solution. The solution set is { } or ∅.

51. $\begin{cases} x-3y+2z=0 \\ 9y-z=22 \\ 5x+3z=10 \end{cases}$

The corresponding matrix is $\left[\begin{array}{ccc|c} 1 & -3 & 2 & 0 \\ 0 & 9 & -1 & 22 \\ 5 & 0 & 3 & 10 \end{array}\right]$

Multiply row 1 by –5 and add to row 3.

$$\left[\begin{array}{ccc|c} 1 & -3 & 2 & 0 \\ 0 & 9 & -1 & 22 \\ 0 & 15 & -7 & 10 \end{array}\right]$$

Divide row 2 by 9.

$$\left[\begin{array}{ccc|c} 1 & -3 & 2 & 0 \\ 0 & 1 & -\frac{1}{9} & \frac{22}{9} \\ 0 & 15 & -7 & 10 \end{array}\right]$$

Multiply row 2 by –15 and add to row 3.

$$\left[\begin{array}{ccc|c} 1 & -3 & 2 & 0 \\ 0 & 1 & -\frac{1}{9} & \frac{22}{9} \\ 0 & 0 & -\frac{48}{9} & -\frac{240}{9} \end{array}\right]$$

Multiply row 3 by $-\frac{9}{48}$.

$$\left[\begin{array}{ccc|c} 1 & -3 & 2 & 0 \\ 0 & 1 & -\frac{1}{9} & \frac{22}{9} \\ 0 & 0 & 1 & 5 \end{array}\right]$$

This matrix represents the system

$$\begin{cases} x-3y+2z=0 \\ y-\frac{1}{9}z=\frac{22}{9}. \\ z=5 \end{cases}$$

Replace z with 5 in the second equation to find y.

$$y-\frac{1}{9}(5)=\frac{22}{9}$$
$$y=\frac{22}{9}+\frac{5}{9}=\frac{27}{9}=3$$

Replace y with 3 and z with 5 in the first equation to find x.

$$x-3(3)+2(5)=0$$
$$x-9+10=0$$
$$x+1=0$$
$$x=-1$$

The solution is (–1, 3, 5).

52. Let x = the first number

y = the second number

$$\begin{cases} x=3y-5 \\ x+y=127 \end{cases}$$

Substitute $3y-5$ for x in the second equation.

$$(3y-5)+y=127$$
$$4y-5=127$$
$$4y=132$$
$$y=33$$

Replace y with 33 in the first equation to find x.

$$x=3(33)-5=99-5=94$$

The numbers are 94 and 33.

53. Let x = length of the shortest side

y = length of the second side

z = length of the third side

$$\begin{cases} x+y+z=126 \\ y=2x \\ z=x+14 \end{cases}$$

Substitute $2x$ for y and $x+14$ for z in the first equation.

$$x + 2x + (x + 14) = 126$$
$$4x + 14 = 126$$
$$4x = 112$$
$$x = 28$$
$y = 2x = 2(28) = 56$
$z = x + 14 = 28 + 14 = 42$
The lengths are 28 units, 42 units, and 56 units.

54. $\begin{cases} y \le 3x - \dfrac{1}{2} \\ 3x + 4y \ge 6 \end{cases}$

Graph both inequalities on the same set of axes. The solution is the intersection of the solution regions. The solution of the system is the darker shaded region. This solution includes parts of both boundary lines.

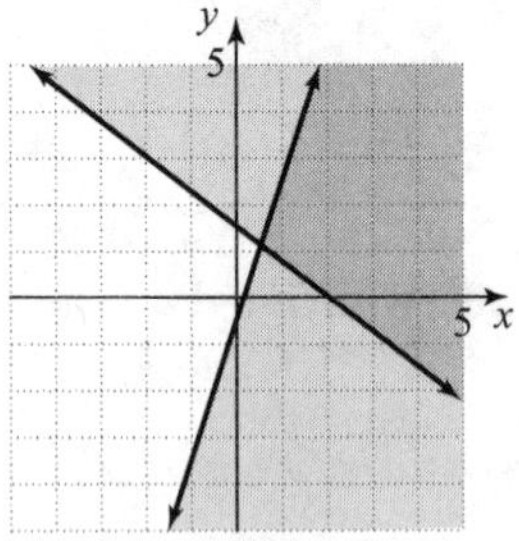

55. $\begin{cases} y = 16.0x + 666 \\ y = -24.8x + 872 \end{cases}$

Substitute $16x + 666$ for y in the second equation.
$$16x + 666 = -24.8x + 872$$
$$40.8x + 666 = 872$$
$$40.8x = 206$$
$$x \approx 5.0 \approx 5$$
$1995 + 5 = 2000$
The number of newspapers printed as morning editions was the same as the number printed as evening editions in 2000.

Chapter 4 Getting Ready for the Test

1. Since the equation $-2 = 0$ is a false statement, the original system of equations has no solution; C.

2. **A.** $5(1) - 0 - (-2) \stackrel{?}{=} -7$
$7 = -7$ False
Since (1, 0, −2) is not a solution of $5x - y - z = -7$, it cannot be a solution of the system.

B. $3(1) - 0 - 2(-2) \stackrel{?}{=} 7$
$7 = 7$ True
$1 + 0 + (-2) \stackrel{?}{=} -1$
$-1 = -1$ True
$5(1) + (-2) \stackrel{?}{=} 3$
$3 = 3$ True
(1, 0, −2) is a solution of system B.

C. In (1, 0, −2), $x = 1$, $y = 0$, and $z = -2$. Thus, (1, 0, −2) cannot be a solution of the equation $x = -2$. Thus (1, 0, −2) cannot be a solution of the system.

D. In (1, 0, −2), $x = 1$, $y = 0$, and $z = -2$. Thus, (1, 0, −2) cannot be a solution of the equation $y = 1$. Thus, (1, 0, −2) cannot be a solution of the system.

(1, 0, −2) is a solution of system B.

3. $\begin{cases} 4x - y = 0 \\ x + z = 1 \\ 3y - z = 12 \end{cases}$ or $\begin{cases} 4x - 1y + 0z = 0 \\ 1x + 0y + 1z = 1 \\ 0x + 3y - 1z = 12 \end{cases}$

The correct matrix is $\left[\begin{array}{ccc|c} 4 & -1 & 0 & 0 \\ 1 & 0 & 1 & 1 \\ 0 & 3 & -1 & 12 \end{array}\right]$; C.

Use the following system and choices for Exercises 4–6.

$\begin{cases} 3x - y + 2z = 3 & \text{Equation (1)} \\ 4x + y - z = 5 & \text{Equation (2)} \\ -x + 5y + 3z = 12 & \text{Equation (3)} \end{cases}$

A. [Equation (1)] + [Equation (2)]
$$\begin{array}{r} 3x - y + 2z = 3 \\ \underline{4x + y - z = 5} \\ 7x + z = 8 \end{array}$$

B. [Equation (1)] + 3 · [Equation (3)]
$$\begin{array}{r} 3x - y + 2z = 3 \\ \underline{-3x + 15y + 9z = 36} \\ 14y + 11z = 39 \end{array}$$

C. [Equation (1)] + 2 · [Equation (2)]
$$\begin{array}{r} 3x - y + 2z = 3 \\ \underline{8x + 2y - 2z = 10} \\ 11x + y = 13 \end{array}$$

D. [Equation (1)] + [Equation (3)]

$$\begin{array}{r} 3x - y + 2z = 3 \\ -x + 5y + 3z = 12 \\ \hline 2x + 4y + 5z = 15 \end{array}$$

4. The variable x is eliminated in choice B.

5. The variable y is eliminated in choice A.

6. The variable z is eliminated in choice C.

7. $\begin{cases} x \le -3 \\ y \le 3 \end{cases}$

The solution area is both to the left of $x = -3$ $(x \le -3)$ and below $y = 3$ $(y \le 3)$; C.

8. $\begin{cases} x \le -3 \\ y \ge 3 \end{cases}$

The solution area is both to the left of $x = -3$ $(x \le -3)$ and above $y = 3$ $(y \ge 3)$; D.

9. $\begin{cases} x \ge -3 \\ y \ge 3 \end{cases}$

The solution area is both to the right of $x = -3$ $(x \ge -3)$ and above $y = 3$ $(y \ge 3)$; B.

10. $\begin{cases} x \ge -3 \\ y \le 3 \end{cases}$

The solution area is both to the right of $x = -3$ $(x \ge -3)$ and below $y = 3$ $(y \le 3)$; A.

Chapter 4 Test

1. $\begin{cases} 2x - y = -1 & (1) \\ 5x + 4y = 17 & (2) \end{cases}$

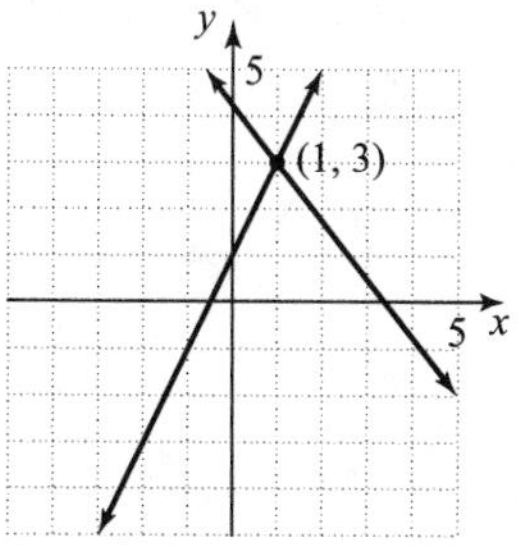

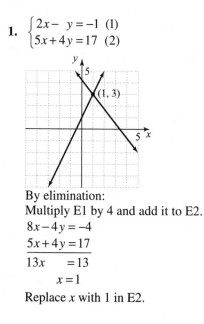

By elimination:
Multiply E1 by 4 and add it to E2.

$$\begin{array}{r} 8x - 4y = -4 \\ 5x + 4y = 17 \\ \hline 13x \quad = 13 \\ x = 1 \end{array}$$

Replace x with 1 in E2.

$$5(1) + 4y = 17$$
$$4y = 12$$
$$y = 3$$

The solution is (1, 3).

2. $\begin{cases} 7x - 14y = 5 & (1) \\ x = 2y & (2) \end{cases}$

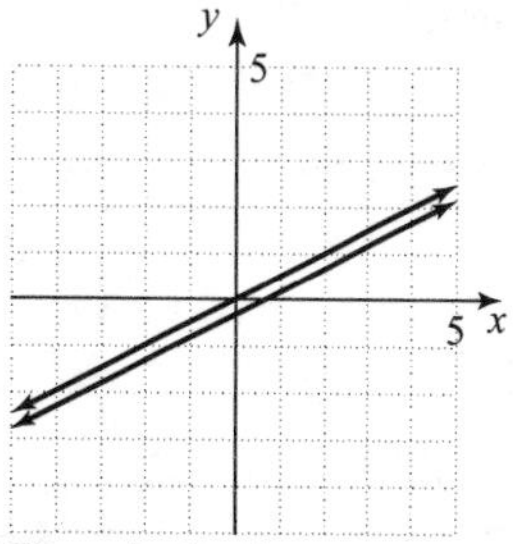

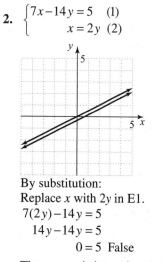

By substitution:
Replace x with $2y$ in E1.

$$7(2y) - 14y = 5$$
$$14y - 14y = 5$$
$$0 = 5 \quad \text{False}$$

The system is inconsistent. The solution set is $\varnothing$.

3. $\begin{cases} 4x - 7y = 29 \\ 2x + 5y = -11 \end{cases}$

Multiply E2 by –2 and add to E1.

$$\begin{array}{r} -4x - 10y = 22 \\ 4x - 7y = 29 \\ \hline -17y = 51 \\ y = -3 \end{array}$$

Replace y with –3 in E1.

$$4x - 7(-3) = 29$$
$$4x + 21 = 29$$
$$4x = 8$$
$$x = 2$$

The solution is (2, –3).

4. $\begin{cases} 15x + 6y = 15 \\ 10x + 4y = 10 \end{cases}$

Divide E1 by 3 and E2 by 2.

$\begin{cases} 5x + 2y = 5 \\ 5x + 2y = 5 \end{cases}$

The system is dependent. The solution is $\{(x, y) | 10x + 4y = 10\}$.

5. $\begin{cases} 2x - 3y \quad = 4 & (1) \\ \quad 3y + 2z = 2 & (2) \\ x \quad - z = -5 & (3) \end{cases}$

Add E1 and E2.

$2x + 2z = 6$ or $x + z = 3$ (4)

Add E3 and E4.

$$\begin{array}{l} x+z=3 \\ \underline{x-z=-5} \\ 2x \quad =-2 \\ \quad x=-1 \end{array}$$

Replace x with –1 in E3.

$$\begin{array}{l} -1-z=-5 \\ -z=-4 \text{ so } z=4 \end{array}$$

Replace x with –1 in E1.

$$\begin{array}{l} 2(-1)-3y=4 \\ -2-3y=4 \\ -3y=6 \\ y=-2 \end{array}$$

The solution is (–1, –2, 4).

6. $\begin{cases} 3x-2y-z=-1 & (1) \\ 2x-2y \quad =4 & (2) \\ 2x \quad -2z=-12 & (3) \end{cases}$

Multiply E2 by –1 and add to E1.

$$\begin{array}{l} 3x-2y-z=-1 \\ \underline{-2x+2y \quad =-4} \\ x \quad -z=-5 \quad (4) \end{array}$$

Multiply E4 by –2 and add to E3.

$$\begin{array}{l} 2x-2z=-12 \\ \underline{-2x+2z=10} \\ 0=-2 \quad \text{False} \end{array}$$

The system is inconsistent. The solution set is ∅.

7. $\begin{cases} \frac{x}{2}+\frac{y}{4}=-\frac{3}{4} \\ x+\frac{3}{4}y=-4 \end{cases}$

Clear fractions by multiplying both equations by 4.

$\begin{cases} 2x+\ y=-3 & (1) \\ 4x+3y=-16 & (2) \end{cases}$

Multiply E1 by –2 and add to E2.

$$\begin{array}{l} -4x-2y=6 \\ \underline{4x+3y=-16} \\ y=-10 \end{array}$$

Replace y with –10 in E1.

$$\begin{array}{l} 2x+(-10)=-3 \\ 2x=7 \quad \text{so } x=\frac{7}{2} \end{array}$$

The solution is $\left(\frac{7}{2}, -10\right)$.

8. $\begin{cases} x-\ y=-2 \\ 3x-3y=-6 \end{cases}$

$$\left[\begin{array}{cc|c} 1 & -1 & -2 \\ 3 & -3 & -6 \end{array}\right]$$

Multiply R1 by –3 and add to R2.

$$\left[\begin{array}{cc|c} 1 & -1 & -2 \\ 0 & 0 & 0 \end{array}\right]$$

This corresponds to $\begin{cases} x-y=-2 \\ 0=0 \end{cases}$.

This is a dependent system. The solution is $\{(x, y) | x-y=-2\}$.

9. $\begin{cases} x+2y=-1 \\ 2x+5y=-5 \end{cases}$

$$\left[\begin{array}{cc|c} 1 & 2 & -1 \\ 2 & 5 & -5 \end{array}\right]$$

Multiply R1 by –2 and add to R2.

$$\left[\begin{array}{cc|c} 1 & 2 & -1 \\ 0 & 1 & -3 \end{array}\right]$$

This corresponds to $\begin{cases} x+2y=-1 \\ y=-3 \end{cases}$.

$$\begin{array}{l} x+2(-3)=-1 \\ x-6=-1 \\ x=5 \end{array}$$

The solution is (5, –3).

10. $\begin{cases} x-y-\ z=0 \\ 3x-y-5z=-2 \\ 2x+3y \quad =-5 \end{cases}$

$$\left[\begin{array}{ccc|c} 1 & -1 & -1 & 0 \\ 3 & -1 & -5 & -2 \\ 2 & 3 & 0 & -5 \end{array}\right]$$

Multiply R1 by –3 and add to R2. Multiply R1 by –2 and add to R3.

$$\left[\begin{array}{ccc|c} 1 & -1 & -1 & 0 \\ 0 & 2 & -2 & -2 \\ 0 & 5 & 2 & -5 \end{array}\right]$$

Divide R2 by 2.

$$\left[\begin{array}{ccc|c} 1 & -1 & -1 & 0 \\ 0 & 1 & -1 & -1 \\ 0 & 5 & 2 & -5 \end{array}\right]$$

Multiply R2 by –5 and add to R3.

$$\left[\begin{array}{ccc|c} 1 & -1 & -1 & 0 \\ 0 & 1 & -1 & -1 \\ 0 & 0 & 7 & 0 \end{array}\right]$$

Divide R3 by 7.

$$\left[\begin{array}{ccc|c} 1 & -1 & -1 & 0 \\ 0 & 1 & -1 & -1 \\ 0 & 0 & 1 & 0 \end{array}\right]$$

This corresponds to $\begin{cases} x-y-z=0 \\ y-z=-1. \\ z=0 \end{cases}$

$$y-0=-1$$
$$y=-1$$
$$x-(-1)-0=0$$
$$x+1=0$$
$$x=-1$$

The solution is (–1, –1, 0).

11. Let x = double occupancy rooms and y = single occupancy rooms.

$$\begin{cases} x+\quad y=80 & (1) \\ 90x+80y=6930 & (2) \end{cases}$$

Multiply E1 by –80 and add to E2.

$$-80x-80y=-6400$$
$$90x+80y=6930$$
$$\overline{10x\qquad =530}$$
$$x=53$$

Replace x with 53 in E1.

$$53+y=80$$
$$y=27$$

53 double-occupancy and 27 single-occupancy rooms are occupied.

12. Let x = gallons of 10% solution and y = gallons of 20% solution.

$$\begin{cases} x+\qquad y=20 & (1) \\ 0.10x+0.20y=0.175(20) & (2) \end{cases}$$

Multiply E1 by –0.10 add to E2.

$$-0.10x-0.10y=-2.0$$
$$0.10x+0.20y=3.5$$
$$\overline{0.10y=1.5}$$
$$y=15$$

Replace y with 15 in E1.

$$x+15=20$$
$$x=5$$

They should use 5 gallons of 10% fructose solution and 15 gallons of the 20% solution.

13. $R(x)=4x$ and $C(x)=1.5x+2000$

The break-even point occurs when $R(x)=C(x)$.

$$4x=1.5x+2000$$
$$2.5x=2000$$
$$x=800$$

The company must sell 800 packages to break even.

14. Let x = measure of the smallest angle. Then the largest angle has a measure of $5x - 3$, and the remaining angle has a measure of $2x - 1$.The sum of the three angles must add to 180°:

$$a+b+c=180$$
$$x+(5x-3)+(2x-1)=180$$
$$x+5x-3+2x-1=180$$
$$8x-4=180$$
$$8x=184$$
$$x=23$$
$$5x-3=5(23)-3=115-3=112$$
$$2x-1=2(23)-1=46-1=45$$

The angle measures are 23°, 45°, and 112°.

15. $\begin{cases} 2y-x\ge 1 \\ x+y\ge -4 \\ y\le 2 \end{cases}$

Graph all three inequalities on the same set of axes. The solution set of the system is the shaded region. The parts of the boundary lines that border the shaded region are part of the solution.

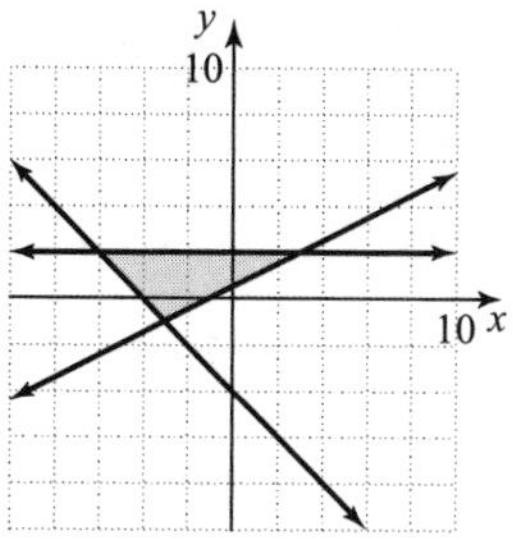

Chapter 4 Cumulative Review

1. a. Since 3 is a natural number, the statement is true.

b. Since 7 is not one of the three numbers listed in the set, the statement is true.

2. a. Since 7 is not an element of the second set, the first set is not a subset of the second set. The statement is false.

b. Since all three numbers in the first set are also elements of the second set, the first set is a subset of the second set. The statement is true.

3. a. $11 + 2 - 7 = 13 - 7 = 6$

b. $-5 - 4 + 2 = -9 + 2 = -7$

4. a. $-7-(-2)=-7+2=-5$

b. $14-38=-24$

5. a. The opposite of 4 is –4.

b. The opposite of $\frac{3}{7}$ is $-\frac{3}{7}$.

c. The opposite of –11.2 is 11.2.

6. a. The reciprocal of 5 is $\frac{1}{5}$.

b. The reciprocal of $-\frac{2}{3}$ is $-\frac{3}{2}$.

7. a. $3(2x+y)=6x+3y$

b. $-(3x-1)=-3x+1$

c. $0.7a(b-2)=0.7ab-1.4a$

8. a. $7(3x-2y+4)=21x-14y+28$

b. $-(-2s-3t)=2s+3t$

9. a. $3x-5x+4=(3-5)x+4=-2x+4$

b. $7yz+yz=(7+1)yz=8yz$

c. $4z+6.1=4z+6.1$

10. a.
$$\begin{aligned}5y^2-1+2(y^2+2)&=5y^2-1+2y^2+4\\&=7y^2+3\end{aligned}$$

b.
$$\begin{aligned}&(7.8x-1.2)-(5.6x-2.4)\\&=7.8x-1.2-5.6x+2.4\\&=2.2x+1.2\end{aligned}$$

11.
$$\begin{aligned}-4x-1+5x&=9x+3-7x\\x-1&=2x+3\\-x&=4\\x&=-4\end{aligned}$$

12.
$$\begin{aligned}8y-14&=6y-14\\2y&=0\\y&=0\end{aligned}$$

13.
$$\begin{aligned}0.3x+0.1&=0.27x-0.02\\0.03x&=-0.12\\x&=-4\end{aligned}$$

14.
$$\begin{aligned}2(m-6)-m&=4(m-3)-3m\\2m-12-m&=4m-12-3m\\m-12&=m-12\\0&=0 \quad \text{Always True}\end{aligned}$$
The solution is all real numbers.

15. Let x = length of the third side, then $2x + 12$ = length of the two equal sides.
$$\begin{aligned}x+(2x+12)+(2x+12)&=149\\5x+24&=149\\5x&=125\\x&=25\end{aligned}$$
$2(25)+12=50+12=62$
The sides are 25 cm, 62 cm, and 62 cm.

16. Let x = measure of the equal angles, $x + 10$ = measure of the third angle, and $\frac{1}{2}x$ = measure of the fourth angle.
$$\begin{aligned}x+x+(x+10)+\frac{1}{2}x&=360\\\frac{7}{2}x+10&=360\\\frac{7}{2}x&=350\\7x&=700\\x&=100\end{aligned}$$
$x+10=100+10=110$
$\frac{1}{2}x=\frac{1}{2}(100)=50$
The measure of the angles are 100°, 100°, 110°, and 50°.

17.
$$\begin{aligned}3x+4&\ge 2x-6\\x&\ge -10\end{aligned}$$
$[-10, \infty)$

18.
$$\begin{aligned}5(2x-1)&>-5\\10x-5&>-5\\10x&>0\\x&>0\end{aligned}$$
$(0, \infty)$

19. $2 < 4 - x < 7$
$-2 < -x < 3$
$2 > x > -3$
$-3 < x < 2$
$(-3, 2)$

20. $-1 < \dfrac{-2x-1}{3} < 1$
$3(-1) < 3\left[\dfrac{-2x-1}{3}\right] < 3(1)$
$-3 < -2x - 1 < 3$
$-2 < -2x < 4$
$1 > x > -2$
$-2 < x < 1$
$(-2, 1)$

21. $|2x| + 5 = 7$
$|2x| = 2$
$2x = 2$ or $2x = -2$
$x = 1$ or $x = -1$

22. $|x-5| = 4$
$x - 5 = 4$ or $x - 5 = -4$
$x = 9$ or $x = 1$

23. $|m-6| < 2$
$-2 < m - 6 < 2$
$4 < m < 8$
$(4, 8)$

24. $|2x+1| > 5$
$2x + 1 < -5$ or $2x + 1 > 5$
$2x < -6$ or $2x > 4$
$x < -3$ or $x > 2$
$(-\infty, -3) \cup (2, \infty)$

25.

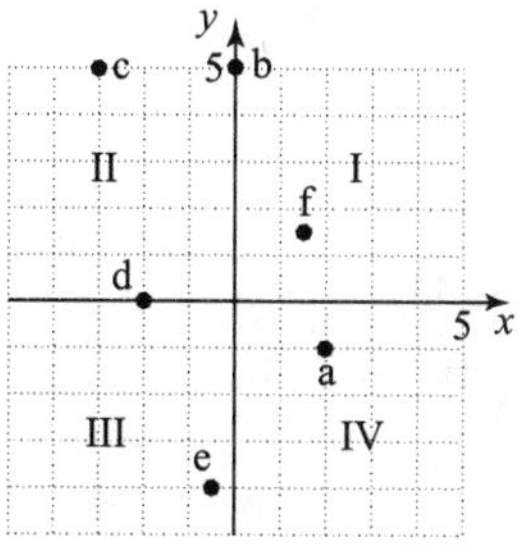

a. (2, –1) is in quadrant IV.

b. (0, 5) is on the y-axis.

c. (–3, 5) is in quadrant II.

d. (–2, 0) is on the x-axis.

e. $\left(-\dfrac{1}{2}, -4\right)$ is in quadrant III.

f. (1.5, 1.5) is in quadrant I.

26. a. (–1, –5) is in quadrant III.

b. (4, –2) is in quadrant IV.

c. (0, 2) is on the y-axis.

27. No; for the input $x = 4$, there are two outputs, $y = \pm 2$.

28. $-2x + \dfrac{1}{2}y = -2$, or $y = 4x - 4$

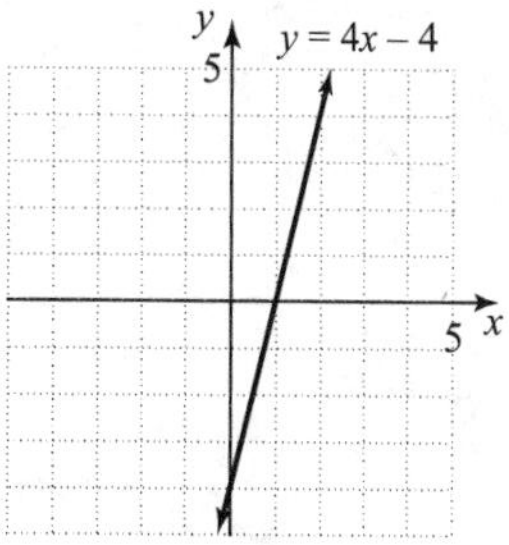

29. $f(x) = 7x^2 - 3x + 1$, $g(x) = 8x - 2$

a. $f(1) = 7(1)^2 - 3(1) + 1 = 7 - 3 + 1 = 5$

b. $g(1) = 8(1) - 2 = 8 - 2 = 6$

c. $f(-2) = 7(-2)^2 - 3(-2) + 1$
$= 7(4) + 6 + 1$
$= 28 + 6 + 1$
$= 35$

d. $g(0) = 8(0) - 2 = 0 - 2 = -2$

30. $f(x) = 3x^2$

a. $f(5) = 3(5)^2 = 3(25) = 75$

b. $f(-2) = 3(-2)^2 = 3(4) = 12$

31. $g(x) = 2x + 1$ and $f(x) = 2x$

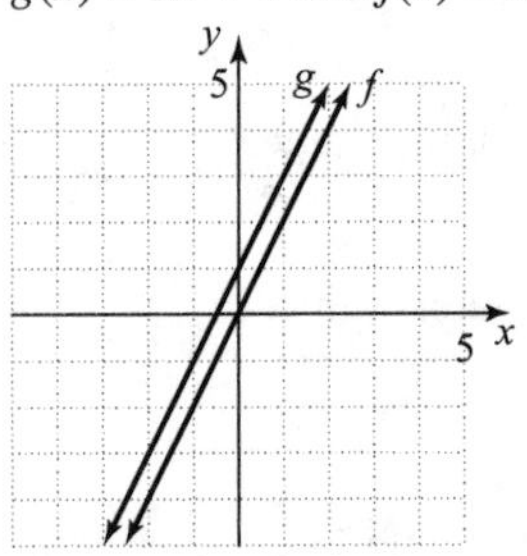

The graph of g is the graph of f shifted 1 unit up.

32. $m = \frac{9-6}{0-(-2)} = \frac{3}{2}$

33. $3x - 4y = 4$
$-4y = -3x + 4$
$y = \frac{3}{4}x - 1$

$m = \frac{3}{4}$, y-intercept = $(0, -1)$

34. $y = 2$
$m = 0$, y-intercept = $(0, 2)$

35. a. $3x + 7y = 4$
$7y = -3x + 4$
$y = -\frac{3}{7}x + \frac{4}{7}$
$m = -\frac{3}{7}$

$6x + 14y = 7$
$14y = -6x + 7$
$y = -\frac{3}{7}x + \frac{1}{2}$
$m = -\frac{3}{7}$

Parallel, since the slopes are equal.

b. $-x + 3y = 2$
$3y = x + 2$
$y = \frac{1}{3}x + \frac{2}{3}$
$m = \frac{1}{3}$
$2x + 6y = 5$
$6y = -2x + 5$
$y = -\frac{1}{3}x + \frac{5}{6}$
$m = -\frac{1}{3}$

Neither, since the slopes are not equal and their product is not -1.

36. $y - (-9) = \frac{1}{5}(x - 0)$
$y + 9 = \frac{1}{5}x$
$y = \frac{1}{5}x - 9$

37. $m = \frac{-5-0}{-4-4} = \frac{-5}{-8} = \frac{5}{8}$
$y - 0 = \frac{5}{8}(x - 4)$
$y = \frac{5}{8}x - \frac{5}{2}$
$f(x) = \frac{5}{8}x - \frac{5}{2}$

38. $f(x) = \frac{1}{2}x - \frac{1}{3}$ or $y = \frac{1}{2}x - \frac{1}{3}$
$m = \frac{1}{2}$ so $m_{\perp} = -2$
$y - 6 = -2[x - (-2)]$
$y - 6 = -2(x + 2)$
$y - 6 = -2x - 4$
$y = -2x + 2$

39. $3x \geq y$, or $y \leq 3x$
Graph the boundary line $y = 3x$ with a solid line because the inequality symbol is $\leq$.
Test: (0, 1)
$3x \geq y$
$3(0) \geq 1$
$0 \geq 1$ False
Shade the half-plane that does not contain (0, 1).

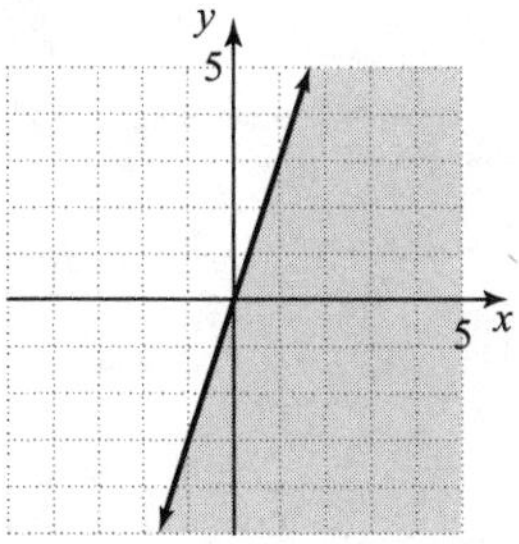

40. $x \geq 1$

Graph the boundary line $x = 1$ with a solid line because the inequality symbol is ≥.

Shade the half-plane that does not contain (0, 0).

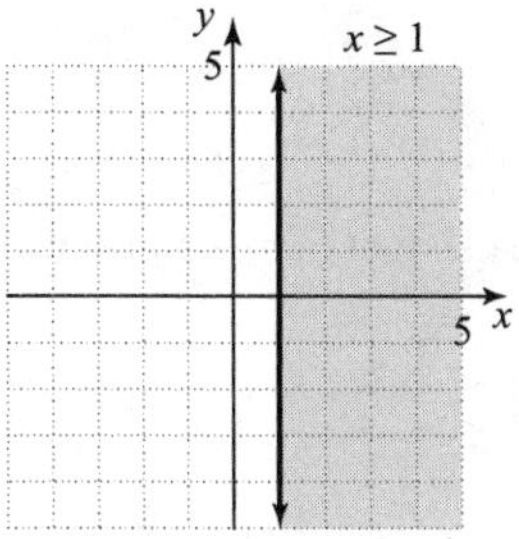

41. a. $\begin{cases} -x + y = 2 \\ 2x - y = -3 \end{cases}$

$$-(-1) + 1 = 2$$
$$1 + 1 = 2$$
$$2 = 2 \text{ True}$$
$$2(-1) - (1) = -3$$
$$-2 - 1 = -3$$
$$-3 = -3 \text{ True}$$

Yes, (–1, 1) is a solution.

b. $\begin{cases} 5x + 3y = -1 \\ x - y = 1 \end{cases}$

$$5(-2) + 3(3) = -1$$
$$-10 + 9 = -1$$
$$-1 = -1 \text{ True}$$
$$-2 - 3 = -1$$
$$-5 = -1 \text{ False}$$

No, (–2, 3) is not a solution.

42. $\begin{cases} 5x + y = -2 & (1) \\ 4x - 2y = -10 & (2) \end{cases}$

Multiply E1 by 2 and add to E2.

$$10x + 2y = -4$$
$$4x - 2y = -10$$
$$14x = -14$$
$$x = -1$$

Replace x with –1 in E1.

$$5(-1) + y = -2$$
$$-5 + y = -2$$
$$y = 3$$

The solution is (–1, 3).

43. $\begin{cases} 3x - y + z = -15 & (1) \\ x + 2y - z = 1 & (2) \\ 2x + 3y - 2z = 0 & (3) \end{cases}$

Add E1 and E2.

$$4x + y = -14 \quad (4)$$

Multiply E1 by 2 and add to E3.

$$6x - 2y + 2z = -30$$
$$2x + 3y - 2z = 0$$
$$8x + y = -30 \quad (5)$$

Solve the new system:

$\begin{cases} 4x + y = -14 & (4) \\ 8x + y = -30 & (5) \end{cases}$

Multiply E4 by –1 and add to E5.

$$-4x - y = 14$$
$$8x + y = -30$$
$$4x = -16$$
$$x = -4$$

Replace x with –4 in E4.

$$4(-4) + y = -14$$
$$-16 + y = -14$$
$$y = 2$$

Replace x with –4 and y with 2 in E1.

$$3(-4) - (2) + z = -15$$
$$-12 - 2 + z = -15$$
$$-14 + z = -15$$
$$z = -1$$

The solution is (–4, 2, –1).

44. $\begin{cases} x - 2y + z = 0 & (1) \\ 3x - y - 2z = -15 & (2) \\ 2x - 3y + 3z = 7 & (3) \end{cases}$

Multiply E1 by 2 and add to E2.

$$2x - 4y + 2z = 0$$
$$3x - y - 2z = -15$$
$$5x - 5y = -15 \text{ or } x - y = -3 \quad (4)$$

Multiply E1 by –3 and add to E3.

$$-3x + 6y - 3z = 0$$
$$2x - 3y + 3z = 7$$
$$-x + 3y = 7 \quad (5)$$

Add E4 and E5.

$$2y = 4$$
$$y = 2$$

Replace y with 2 in E4.

$x-2=-3$
$x=-1$
Replace x with –1 and y with 2 in E1.
$-1-2(2)+z=0$
$-5+z=0$
$z=5$
The solution is (–1, 2, 5).

45. $\begin{cases} x+3y=5 \\ 2x-y=-4 \end{cases}$

$\left[\begin{array}{cc|c} 1 & 3 & 5 \\ 2 & -1 & -4 \end{array}\right]$

Multiply R1 by –2 and add to R2.

$\left[\begin{array}{cc|c} 1 & 3 & 5 \\ 0 & -7 & -14 \end{array}\right]$

Divide R2 by –7.

$\left[\begin{array}{cc|c} 1 & 3 & 5 \\ 0 & 1 & 2 \end{array}\right]$

This corresponds to $\begin{cases} x+3y=5 \\ y=2 \end{cases}$.

$x+3(2)=5$
$x+6=5$
$x=-1$
The solution is (–1, 2).

46. $\begin{cases} -6x+8y=0 \\ 9x-12y=2 \end{cases}$

$\left[\begin{array}{cc|c} -6 & 8 & 0 \\ 9 & -12 & 2 \end{array}\right]$

Divide R1 by –6.

$\left[\begin{array}{cc|c} 1 & -\frac{4}{3} & 0 \\ 9 & -12 & 2 \end{array}\right]$

Multiply R1 by –9 and add to R2.

$\left[\begin{array}{cc|c} 1 & -\frac{4}{3} & 0 \\ 0 & 0 & 2 \end{array}\right]$

This corresponds to the system.

$\begin{cases} x-\frac{4}{3}y=0 \\ 0=2 \end{cases}$.

The system is inconsistent. The solution set is ∅.

Chapter 5

Section 5.1 Practice Exercises

1. a. $3^4 \cdot 3^2 = 3^{4+2} = 3^6$

b. $x^5 \cdot x^2 = x^{5+2} = x^7$

c. $y \cdot y^3 \cdot y^5 = (y^1 \cdot y^3) \cdot y^5 = y^4 \cdot y^5 = y^9$

2. a. $(5z^3)(7z) = 5(7)z^3z^1 = 35z^4$

b. $(-4.1t^5q^3)(5tq^5) = -4.1(5)t^5t^1q^3q^5$
$= -20.5t^6q^8$

3. a. $5^0 = 1$

b. $-5^0 = -(5^0) = -(1) = -1$

c. $(3x-8)^0 = 1$

d. $3x^0 = 3(1) = 3$

4. a. $\dfrac{z^8}{z^3} = z^{8-3} = z^5$

b. $\dfrac{3^9}{3^3} = 3^{9-3} = 3^6$

c. $\dfrac{45x^7}{5x^3} = 9x^{7-3} = 9x^4$

d. $\dfrac{24a^{14}b^6}{18a^7b^6} = \dfrac{4}{3}a^{14-7}b^{6-6}$
$= \dfrac{4}{3}a^7b^0$
$= \dfrac{4}{3}a^7 \text{ or } \dfrac{4a^7}{3}$

5. a. $6^{-2} = \dfrac{1}{6^2} = \dfrac{1}{36}$

b. $(-2)^{-6} = \dfrac{1}{(-2)^6} = \dfrac{1}{64}$

c. $3x^{-5} = 3 \cdot \dfrac{1}{x^5} = \dfrac{3}{x^5}$

d. $(5y)^{-1} = \dfrac{1}{(5y)^1} = \dfrac{1}{5y}$

e. $\dfrac{k^4}{k^{11}} = k^{4-11} = k^{-7} = \dfrac{1}{k^7}$

f. $\dfrac{5^3}{5^5} = 5^{3-5} = 5^{-2} = \dfrac{1}{5^2} = \dfrac{1}{25}$

g. $5^{-1} + 2^{-2} = \dfrac{1}{5^1} + \dfrac{1}{2^2} = \dfrac{1}{5} + \dfrac{1}{4} = \dfrac{4}{20} + \dfrac{5}{20} = \dfrac{9}{20}$

h. $\dfrac{1}{z^{-8}} = \dfrac{1}{\frac{1}{z^8}} = 1 \div \dfrac{1}{z^8} = 1 \cdot \dfrac{z^8}{1} = z^8$

6. a. $\dfrac{z^{-8}}{z^3} = z^{-8-3} = z^{-11} = \dfrac{1}{z^{11}}$

b. $\dfrac{7t^3}{t^{-5}} = 7 \cdot t^{3-(-5)} = 7t^8$

c. $\dfrac{3^{-2}}{3^{-4}} = 3^{-2-(-4)} = 3^{-2+4} = 3^2 = 9$

d. $\dfrac{5a^{-5}b^3}{15a^2b^{-4}} = \dfrac{a^{-5-2}b^{3-(-4)}}{3} = \dfrac{a^{-7}b^7}{3} = \dfrac{b^7}{3a^7}$

e. $\dfrac{(2x^{-5})(x^6)}{x^5} = \dfrac{2x^{-5+6}}{x^5}$
$= \dfrac{2x^1}{x^5}$
$= 2x^{1-5}$
$= 2x^{-4}$
$= \dfrac{2}{x^4}$

7. a. $x^{3a} \cdot x^4 = x^{3a+4}$

b. $\dfrac{x^{3t-2}}{x^{t-3}} = x^{(3t-2)-(t-3)} = x^{3t-2-t+3} = x^{2t+1}$

8. a. Move the decimal point until the number is between 1 and 10. The decimal point is moved 4 places and the original number is 10 or greater, so the count is positive 4.
$65,000 = 6.5\times10^4$

b. Move the decimal point until the number is between 1 and 10. The decimal point is moved 5 places and the original number is less than 1, so the count is –5.
$0.000038 = 3.8\times10^{-5}$

9. a. Since the exponent is positive, move the decimal point 5 places to the right.
$6.2\times10^5 = 620,000$

b. Since the exponent is negative, more the decimal point 2 places to the left.
$3.109\times10^{-2} = 0.03109$

Graphing Calculator Explorations

1. $(3\times10^{11})(2\times10^{32}) = 6\times10^{43}$

2. $(6\times10^{14})\div(3\times10^9) = 2\times10^5$

3. $(5.2\times10^{23})(7.3\times10^4) = 3.796\times10^{28}$

4. $(4.38\times10^{41})\div(3\times10^{17}) = 1.46\times10^{24}$

Vocabulary, Readiness & Video Check 5.1

1. $9x^5$; base x

2. yz^5; base z

3. -3^5; base 3

4. $(-3)^5$; base –3

5. $(y^7)^5$; base y^7

6. $9\cdot2^5$; base 2

7. These properties allow us to reorder and regroup factors to put those with the same bases together so that we may apply the product rule.

8. Since there are no parentheses, the negative is not part of the base and is therefore not raised to the zero power. The evaluation is –1, not 1.

9. Subtract the exponents on like bases when applying the quotient rule.

10. A negative exponent will not make an expression evaluate to a negative number. A negative exponent moves an expression or factor from numerator to denominator or from denominator to numerator.

11. When you move the decimal point to the left, the sign of the exponent is positive; when you move the decimal point to the right, the sign of the exponent is negative.

Exercise Set 5.1

1. $5x^{-1}y^{-2} = \dfrac{5}{xy^2}$

3. $a^2b^{-1}c^{-5} = \dfrac{a^2}{bc^5}$

5. $\dfrac{y^{-2}}{x^{-4}} = \dfrac{x^4}{y^2}$

7. $4^2\cdot4^3 = 4^{2+3} = 4^5$

9. $x^5\cdot x^3 = x^{5+3} = x^8$

11. $m\cdot m^7\cdot m^6 = m^{1+7+6} = m^{14}$

13. $(4xy)(-5x) = -20x^{1+1}y = -20x^2y$

15. $(-4x^3p^2)(4y^3x^3) = -16x^{3+3}y^3p^2$
$= -16x^6y^3p^2$

17. $-8^0 = -(8^0) = -1$

19. $(4x+5)^0 = 1$

21. $-x^0 = -(x^0) = -(1) = -1$

23. $4x^0 + 5 = 4(1) + 5 = 4 + 5 = 9$

25. $\frac{a^5}{a^2} = a^{5-2} = a^3$

27. $-\frac{26z^{11}}{2z^7} = -13z^{11-7} = -13z^4$

29. $\frac{x^9y^6}{x^8y^6} = x^{9-8}y^{6-6} = x^1y^0 = x$

31. $\frac{12x^4y^7}{9xy^5} = \frac{4x^{4-1}y^{7-5}}{3} = \frac{4}{3}x^3y^2$

33. $\frac{-36a^5b^7c^{10}}{6ab^3c^4} = -6a^{5-1}b^{7-3}c^{10-4} = -6a^4b^4c^6$

35. $4^{-2} = \frac{1}{4^2} = \frac{1}{16}$

37. $(-3)^{-3} = \frac{1}{(-3)^3} = \frac{1}{-27} = -\frac{1}{27}$

39. $\frac{x^7}{x^{15}} = x^{7-15} = x^{-8} = \frac{1}{x^8}$

41. $5a^{-4} = \frac{5}{a^4}$

43. $\frac{x^{-7}}{y^{-2}} = \frac{y^2}{x^7}$

45. $\frac{x^{-2}}{x^5} = x^{-2-5} = x^{-7} = \frac{1}{x^7}$

47. $\frac{8r^4}{2r^{-4}} = 4r^{4-(-4)} = 4r^8$

49. $\frac{x^{-9}x^4}{x^{-5}} = \frac{x^{-9+4}}{x^{-5}} = \frac{x^{-5}}{x^{-5}} = x^{-5-(-5)} = x^0 = 1$

51. $\frac{2a^{-6}b^2}{18ab^{-5}} = \frac{a^{-6-1}b^{2-(-5)}}{9} = \frac{a^{-7}b^7}{9} = \frac{b^7}{9a^7}$

53. $\frac{(24x^8)(x)}{20x^{-7}} = \frac{6x^{8+1}}{5x^{-7}} = \frac{6x^{9-(-7)}}{5} = \frac{6x^{16}}{5}$

55. $-7x^3 \cdot 20x^9 = -7 \cdot 20x^{3+9} = -140x^{12}$

57. $x^7 \cdot x^8 \cdot x = x^{7+8+1} = x^{16}$

59. $2x^3 \cdot 5x^7 = 2 \cdot 5x^{3+7} = 10x^{10}$

61. $(5x)^0 + 5x^0 = 1 + 5 \cdot 1 = 1 + 5 = 6$

63. $\frac{z^{12}}{z^{15}} = z^{12-15} = z^{-3} = \frac{1}{z^3}$

65. $3^0 - 3t^0 = 1 - 3 \cdot 1 = 1 - 3 = -2$

67. $\frac{y^{-3}}{y^{-7}} = y^{-3-(-7)} = y^4$

69.
$$\begin{aligned} 4^{-1} + 3^{-2} &= \frac{1}{4^1} + \frac{1}{3^2} \\ &= \frac{1}{4} + \frac{1}{9} \\ &= \frac{9}{36} + \frac{4}{36} \\ &= \frac{13}{36} \end{aligned}$$

71. $3x^{-1} = \frac{3}{x^1} = \frac{3}{x}$

73. $\frac{r^4}{r^{-4}} = r^{4-(-4)} = r^8$

75. $\frac{x^{-7}y^{-2}}{x^2y^2} = x^{-7-2}y^{-2-2} = x^{-9}y^{-4} = \frac{1}{x^9y^4}$

77.
$$\begin{aligned} &(-4x^2y)(3x^4)(-2xy^5) \\ &= -4(3)(-2)x^2 \cdot x^4 \cdot x \cdot y \cdot y^5 \\ &= 24x^{2+4+1} \cdot y^{1+5} \\ &= 24x^7y^6 \end{aligned}$$

79. $2^{-4} \cdot x = \frac{x}{2^4} = \frac{x}{16}$

81. $\frac{5^{17}}{5^{13}} = 5^{17-13} = 5^4 = 625$

83. $\dfrac{8^{-7}}{8^{-6}} = 8^{-7-(-6)} = 8^{-7+6} = 8^{-1} = \dfrac{1}{8}$

85. $\dfrac{9^{-5}a^4}{9^{-3}a^{-1}} = 9^{-5-(-3)}a^{4-(-1)}$
$= 9^{-5+3}a^{4+1}$
$= 9^{-2}a^5$
$= \dfrac{a^5}{9^2}$
$= \dfrac{a^5}{81}$

87. $\dfrac{14x^{-2}yz^{-4}}{2xyz} = \dfrac{14}{2} \cdot x^{-2-1}y^{1-1}z^{-4-1}$
$= 7x^{-3}y^0z^{-5}$
$= \dfrac{7}{x^3z^5}$

89. $x^5 \cdot x^{7a} = x^{5+7a}$ or x^{7a+5}

91. $\dfrac{x^{3t-1}}{x^t} = x^{3t-1-t} = x^{2t-1}$

93. $y^{2p} \cdot y^{9p} = y^{2p+9p} = y^{11p}$

95. $\dfrac{z^{6x}}{z^7} = z^{6x-7}$

97. $\dfrac{x^{3t} \cdot x^{4t-1}}{x^t} = \dfrac{x^{3t+(4t-1)}}{x^t} = x^{7t-1-t} = x^{6t-1}$

99. $31,250,000 = 3.125 \times 10^7$

101. $0.016 = 1.6 \times 10^{-2}$

103. $67,413 = 6.7413 \times 10^4$

105. $0.0125 = 1.25 \times 10^{-2}$

107. $0.000053 = 5.3 \times 10^{-5}$

109. $778,300,000 = 7.783 \times 10^8$

111. $16,773,000,000,000 = 1.6773 \times 10^{13}$

113. $124,000,000,000 = 1.24 \times 10^{11}$

115. $0.001 = 1.0 \times 10^{-3}$

117. $3.6 \times 10^{-9} = 0.0000000036$

119. $9.3 \times 10^7 = 93,000,000$

121. $1.278 \times 10^6 = 1,278,000$

123. $7.35 \times 10^{12} = 7,350,000,000,000$

125. $4.03 \times 10^{-7} = 0.000000403$

127. $3.0 \times 10^8 = 300,000,000$

129. $9.5 \times 10^{-3} = 0.0095$

131. $6.6683 \times 10^{10} = 66,683,000,000$

133. $(5 \cdot 2)^2 = (10)^2 = 100$

135. $\left(\dfrac{3}{4}\right)^3 = \left(\dfrac{3}{4}\right)\left(\dfrac{3}{4}\right)\left(\dfrac{3}{4}\right) = \dfrac{3 \cdot 3 \cdot 3}{4 \cdot 4 \cdot 4} = \dfrac{27}{64}$

137. $(2^3)^2 = 8^2 = 64$

139. answers may vary

141. answers may vary

143. **a.** $x^a \cdot x^a = x^{a+a} = x^{2a}$

b. $x^a + x^a = 2x^a$

c. $\dfrac{x^a}{x^b} = x^{a-b}$

d. $x^a \cdot x^b = x^{a+b}$

e. $x^a + x^b = x^a + x^b$

145. 7^{13}

147. 7^{-11}

Section 5.2 Practice Exercises

1. a. $(z^3)^5 = z^{3\cdot 5} = z^{15}$

b. $(5^2)^2 = 5^{2\cdot 2} = 5^4 = 625$

c. $(3^{-1})^3 = 3^{-1\cdot 3} = 3^{-3} = \frac{1}{3^3} = \frac{1}{27}$

d. $(x^{-4})^{-6} = x^{-4(-6)} = x^{24}$

2. a. $(2x^3)^5 = 2^5 \cdot (x^3)^5 = 2^5 \cdot x^{3\cdot 5} = 32x^{15}$

b. $\left(\frac{3}{5}\right)^2 = \frac{3^2}{5^2} = \frac{9}{25}$

c. $\left(\frac{2a^5}{b^7}\right)^4 = \frac{(2a^5)^4}{(b^7)^4} = \frac{2^4 \cdot (a^5)^4}{(b^7)^4} = \frac{16a^{20}}{b^{28}}$

d. $\left(\frac{3^{-2}}{x}\right)^{-1} = \frac{(3^{-2})^{-1}}{x^{-1}} = \frac{3^2}{x^{-1}} = 9x$

e.
$$\begin{aligned}(a^{-2}b^{-5}c^4)^{-2} &= (a^{-2})^{-2} \cdot (b^{-5})^{-2} \cdot (c^4)^{-2}\\ &= a^4 b^{10} c^{-8}\\ &= \frac{a^4 b^{10}}{c^8}\end{aligned}$$

3. a.
$$\begin{aligned}(3ab^{-5})^{-3} &= 3^{-3}a^{-3}(b^{-5})^{-3}\\ &= 3^{-3}a^{-3}b^{15}\\ &= \frac{b^{15}}{3^3a^3}\\ &= \frac{b^{15}}{27a^3}\end{aligned}$$

b.
$$\begin{aligned}\left(\frac{y^{-7}}{y^{-4}}\right)^{-5} &= \frac{(y^{-7})^{-5}}{(y^{-4})^{-5}}\\ &= \frac{y^{35}}{y^{20}}\\ &= y^{35-20}\\ &= y^{15}\end{aligned}$$

c. $\left(\frac{3}{8}\right)^{-2} = \frac{3^{-2}}{8^{-2}} = \frac{8^2}{3^2} = \frac{64}{9}$

d.
$$\begin{aligned}\frac{9^{-2}a^{-4}b^3}{a^2b^{-5}} &= 9^{-2}\left(\frac{a^{-4}}{a^2}\right)\left(\frac{b^3}{b^{-5}}\right)\\ &= 9^{-2}a^{-4-2}b^{3-(-5)}\\ &= 9^{-2}a^{-6}b^8\\ &= \frac{b^8}{9^2a^6}\\ &= \frac{b^8}{81a^6}\end{aligned}$$

4. a.
$$\begin{aligned}\left(\frac{5a^4b}{a^{-8}c}\right)^{-3} &= \left(\frac{5a^{12}b}{c}\right)^{-3}\\ &= \frac{5^{-3}a^{-36}b^{-3}}{c^{-3}}\\ &= \frac{c^3}{5^3a^{36}b^3}\\ &= \frac{c^3}{125a^{36}b^3}\end{aligned}$$

b.
$$\begin{aligned}\left(\frac{2x^4}{5y^{-2}}\right)^3\left(\frac{x^{-4}}{10y^{-2}}\right)^{-1} &= \frac{8x^{12}}{125y^{-6}} \cdot \frac{x^4}{10^{-1}y^2}\\ &= \frac{8\cdot 10\cdot x^{12}x^4y^6}{125y^2}\\ &= \frac{16x^{16}y^4}{25}\end{aligned}$$

5. a.
$$\begin{aligned}x^{-2a}(3x^a)^3 &= x^{-2a} \cdot 3^3 \cdot x^{a\cdot 3}\\ &= 27x^{-2a+3a}\\ &= 27x^a\end{aligned}$$

b.
$$\begin{aligned}\frac{(y^{3b})^3}{y^{4b-3}} &= \frac{y^{9b}}{y^{4b-3}}\\ &= y^{9b-(4b-3)}\\ &= y^{9b-4b+3}\\ &= y^{5b+3}\end{aligned}$$

6. a.
$$\begin{aligned}&(3.4\times 10^4)(5\times 10^{-7})\\ &= 3.4\times 5\times 10^4\times 10^{-7}\\ &= 17.0\times 10^{-3}\\ &= (1.7\times 10^1)\times 10^{-3}\\ &= 1.7\times 10^{-2}\end{aligned}$$

b. $\frac{1.6\times10^8}{4\times10^{-2}}=\left(\frac{1.6}{4}\right)\left(\frac{10^8}{10^{-2}}\right)$
$=0.4\times10^{8-(-2)}$
$=0.4\times10^{10}$
$=(4\times10^{-1})\times10^{10}$
$=4\times10^9$

7. $\frac{2400\times0.0000014}{800}$
$=\frac{(2.4\times10^3)(1.4\times10^{-6})}{8\times10^2}$
$=\frac{2.4(1.4)}{8}\cdot\frac{10^3\cdot10^{-6}}{10^2}$
$=0.42\times10^{-5}$
$=(4.2\times10^{-1})\times10^{-5}$
$=4.2\times10^{-6}$

Vocabulary, Readiness & Video Check 5.2

1. $(x^4)^5=x^{4(5)}=x^{20}$

2. $(5^6)^2=5^{6(2)}=5^{12}$

3. $x^4\cdot x^5=x^{4+5}=x^9$

4. $x^7\cdot x^8=x^{7+8}=x^{15}$

5. $(y^6)^7=y^{6(7)}=y^{42}$

6. $(x^3)^4=x^{3(4)}=x^{12}$

7. $(z^4)^9=z^{4(9)}=z^{36}$

8. $(z^3)^7=z^{3(7)}=z^{21}$

9. $(z^{-6})^{-3}=z^{-6(-3)}=z^{18}$

10. $(y^{-4})^{-2}=y^{-4(-2)}=y^8$

11. The power rule involves a power of a base raised to a power and exponents are multiplied; the product rule involves a product of like bases and exponents are added.

12. power of a quotient, power of a product, power rule, negative exponent, quotient rule

13. We are asked to write the answer in scientific notation. The first product isn't because the number multiplied by the power of 10 is not between 1 and 10.

Exercise Set 5.2

1. $(3^{-1})^2=3^{-2}=\frac{1}{3^2}=\frac{1}{9}$

3. $(x^4)^{-9}=x^{-36}=\frac{1}{x^{36}}$

5. $(3x^2y^3)^2=3^2(x^2)^2(y^3)^2=9x^4y^6$

7. $\left(\frac{2x^5}{y^{-3}}\right)^4=(2x^5y^3)^4$
$=2^4(x^5)^4(y^3)^4$
$=16x^{20}y^{12}$

9. $(2a^2bc^{-3})^{-6}=2^{-6}(a^2)^{-6}b^{-6}(c^{-3})^{-6}$
$=2^{-6}a^{-12}b^{-6}c^{18}$
$=\frac{c^{18}}{2^6a^{12}b^6}$
$=\frac{c^{18}}{64a^{12}b^6}$

11. $\left(\frac{x^7y^{-3}}{z^{-4}}\right)^{-5}=\frac{x^{7(-5)}y^{-3(-5)}}{z^{-4(-5)}}$
$=\frac{x^{-35}y^{15}}{z^{20}}$
$=\frac{y^{15}}{x^{35}z^{20}}$

13. $(-2^{-2}y^{-1})^{-3}=[-1\cdot(2)^{-2}y^{-1}]^{-3}$
$=(-1)^{-3}(2^{-2})^{-3}(y^{-1})^{-3}$
$=(-1)\cdot2^6y^3$
$=-64y^3$

15. $\left(\dfrac{a^{-4}}{a^{-5}}\right)^{-2} = (a^{-4-(-5)})^{-2}$
$= (a^1)^{-2}$
$= a^{-2}$
$= \dfrac{1}{a^2}$

17. $\left(\dfrac{6p^6}{p^{12}}\right)^2 = (6p^{6-12})^2$
$= (6p^{-6})^2$
$= 6^2 p^{-12}$
$= \dfrac{36}{p^{12}}$

19. $(-8y^3xa^{-2})^{-3} = [-1\cdot 8y^3xa^{-2}]^{-3}$
$= (-1)^{-3}\cdot 8^{-3}y^{-9}x^{-3}a^6$
$= -\dfrac{a^6}{8^3x^3y^9}$
$= -\dfrac{a^6}{512x^3y^9}$

21. $\left(\dfrac{3}{4}\right)^{-3} = \dfrac{3^{-3}}{4^{-3}} = \dfrac{4^3}{3^3} = \dfrac{64}{27}$

23. $\left(\dfrac{2a^{-2}b^5}{4a^2b^7}\right)^{-2} = \left(\dfrac{1}{2}a^{-4}b^{-2}\right)^{-2}$
$= \left(\dfrac{1}{2}\right)^{-2} a^8b^4$
$= \dfrac{1^{-2}a^8b^4}{2^{-2}}$
$= \dfrac{2^2a^8b^4}{1^2}$
$= 4a^8b^4$

25. $\left(\dfrac{x^{-2}y^{-2}}{a^{-3}}\right)^{-7} = \dfrac{(x^{-2})^{-7}(y^{-2})^{-7}}{(a^{-3})^{-7}} = \dfrac{x^{14}y^{14}}{a^{21}}$

27. $(y^{-5})^2 = y^{-5\cdot 2} = y^{-10} = \dfrac{1}{y^{10}}$

29. $(5^{-1})^3 = 5^{-3} = \dfrac{1}{5^3} = \dfrac{1}{125}$

31. $(x^7)^{-9} = x^{7\cdot(-9)} = x^{-63} = \dfrac{1}{x^{63}}$

33. $\left(\dfrac{x^4}{y^{-3}}\right)^{-5} = \dfrac{(x^4)^{-5}}{(y^{-3})^{-5}} = \dfrac{x^{-20}}{y^{15}} = \dfrac{1}{x^{20}y^{15}}$

35. $(4x^2)^2 = 4^2(x^2)^2 = 16x^4$

37. $\left(\dfrac{4^{-4}}{y^3x}\right)^{-2} = \dfrac{(4^{-4})^{-2}}{(y^3)^{-2}x^{-2}} = \dfrac{4^8}{y^{-6}x^{-2}} = 4^8x^2y^6$

39. $\left(\dfrac{2x^{-3}}{y^{-1}}\right)^{-3} = \dfrac{2^{-3}(x^{-3})^{-3}}{(y^{-1})^{-3}}$
$= \dfrac{2^{-3}x^9}{y^3}$
$= \dfrac{x^9}{2^3y^3}$
$= \dfrac{x^9}{8y^3}$

41. $\dfrac{4^{-1}x^2yz}{x^{-2}yz^3} = 4^{-1}x^{2-(-2)}y^{1-1}z^{1-3}$
$= \dfrac{1}{4}x^4y^0z^{-2}$
$= \dfrac{x^4}{4z^2}$

43. $\left(\dfrac{3x^5}{6x^4}\right)^4 = \left(\dfrac{1}{2}x\right)^4$
$= \left(\dfrac{1}{2}\right)^4 x^4$
$= \dfrac{1^4}{2^4}x^4$
$= \dfrac{1}{16}x^4$
$= \dfrac{x^4}{16}$

45. $\dfrac{(y^3)^{-4}}{y^3} = \dfrac{y^{-12}}{y^3} = y^{-12-3} = y^{-15} = \dfrac{1}{y^{15}}$

47. $\dfrac{3^{-2}a^{-5}b^{6}}{4^{-2}a^{-7}b^{-3}}=\left(\dfrac{3^{-2}}{4^{-2}}\right)a^{-5-(-7)}b^{6-(-3)}$
$=\dfrac{4^2}{3^2}a^2b^9$
$=\dfrac{16a^2b^9}{9}$

49. $(4x^6y^5)^{-2}(6x^4y^3)$
$=4^{-2}(x^6)^{-2}(y^5)^{-2}\cdot 6x^4y^3$
$=\dfrac{1}{4^2}x^{-12}y^{-10}\cdot 6x^4y^3$
$=\dfrac{6}{16}x^{-12+4}y^{-10+3}$
$=\dfrac{3}{8}x^{-8}y^{-7}$
$=\dfrac{3}{8x^8y^7}$

51. $x^6(x^6bc)^{-6}=x^6\cdot(x^6)^{-6}(b)^{-6}(c)^{-6}$
$=x^6x^{-36}b^{-6}c^{-6}$
$=x^{6-36}b^{-6}c^{-6}$
$=x^{-30}b^{-6}c^{-6}$
$=\dfrac{1}{x^{30}b^6c^6}$

53. $\dfrac{2^{-3}x^2y^{-5}}{5^{-2}x^7y^{-1}}=\dfrac{5^2}{2^3}x^{2-7}y^{-5-(-1)}$
$=\dfrac{25}{8}x^{-5}y^{-4}$
$=\dfrac{25}{8x^5y^4}$

55. $\left(\dfrac{2x^2}{y^4}\right)^3\left(\dfrac{2x^5}{y}\right)^{-2}=\dfrac{2^3x^6}{y^{12}}\cdot\dfrac{2^{-2}x^{-10}}{y^{-2}}$
$=\dfrac{2^{3+(-2)}x^{6+(-10)}}{y^{12+(-2)}}$
$=\dfrac{2^1x^{-4}}{y^{10}}$
$=\dfrac{2}{x^4y^{10}}$

57. $(x^{3a+6})^3=x^{(3a+6)\cdot 3}=x^{9a+18}$

59. $\dfrac{x^{4a}(x^{4a})^3}{x^{4a-2}}=\dfrac{x^{4a}\cdot x^{12a}}{x^{4a-2}}$
$=\dfrac{x^{4a+12a}}{x^{4a-2}}$
$=x^{16a-(4a-2)}$
$=x^{12a+2}$

61. $(b^{5x-2})^2=b^{(5x-2)\cdot 2}=b^{10x-4}$

63. $\dfrac{(y^{2a})^8}{y^{a-3}}=\dfrac{y^{16a}}{y^{a-3}}=y^{16a-(a-3)}=y^{15a+3}$

65. $\left(\dfrac{2x^{3t}}{x^{2t-1}}\right)^4=\dfrac{2^4x^{3t(4)}}{x^{(2t-1)\cdot 4}}$
$=\dfrac{16x^{12t}}{x^{8t-4}}$
$=16x^{12t-(8t-4)}$
$=16x^{4t+4}$

67. $\dfrac{25x^{2a+1}y^{a-1}}{5x^{3a+1}y^{2a-3}}=\left(\dfrac{25}{5}\right)\left(\dfrac{x^{2a+1}}{x^{3a+1}}\right)\left(\dfrac{y^{a-1}}{y^{2a-3}}\right)$
$=5x^{2a+1-(3a+1)}y^{a-1-(2a-3)}$
$=5x^{2a+1-3a-1}y^{a-1-2a+3}$
$=5x^{-a}y^{-a+2}$

69. $(5\times10^{11})(2.9\times10^{-3})=5\times2.9\times10^{11+(-3)}$
$=14.5\times10^8$
$=1.45\times10^1\times10^8$
$=1.45\times10^9$

71. $(2\times10^5)^3=2^3\times10^{5(3)}=8\times10^{15}$

73. $\dfrac{3.6\times10^{-4}}{9\times10^2}=\dfrac{3.6}{9}\times10^{-4-2}$
$=0.4\times10^{-6}$
$=4\times10^{-1}\times10^{-6}$
$=4\times10^{-7}$

75. $\dfrac{0.0069}{0.023}=\dfrac{6.9\times10^{-3}}{2.3\times10^{-2}}$
$=\dfrac{6.9}{2.3}\times10^{-3-(-2)}$
$=3\times10^{-1}$

77. $\dfrac{18,200\times 100}{91,000}=\dfrac{1.82\times 10^4\times 1\times 10^2}{9.1\times 10^4}$
$=\dfrac{1.82\times 10^6}{9.1\times 10^4}$
$=0.2\times 10^{6-4}$
$=2\times 10^{-1}\times 10^2$
$=2\times 10^{-1+2}$
$=2\times 10^1$

79. $\dfrac{6000\times 0.006}{0.009\times 400}=\dfrac{6\times 10^3\times 6\times 10^{-3}}{9\times 10^{-3}\times 4\times 10^2}$
$=\dfrac{36\times 10^0}{36\times 10^{-1}}$
$=1\times 10^{0-(-1)}$
$=1\times 10^1$

81. $\dfrac{0.00064\times 2000}{16,000}=\dfrac{6.4\times 10^{-4}\times 2\times 10^3}{1.6\times 10^4}$
$=\dfrac{12.8\times 10^{-1}}{1.6\times 10^4}$
$=8\times 10^{-1-4}$
$=8\times 10^{-5}$

83. $\dfrac{66,000\times 0.001}{0.002\times 0.003}=\dfrac{6.6\times 10^4\times 1\times 10^{-3}}{2\times 10^{-3}\times 3\times 10^{-3}}$
$=\dfrac{6.6\times 10^1}{6\times 10^{-6}}$
$=1.1\times 10^{1-(-6)}$
$=1.1\times 10^7$

85. $\dfrac{9.24\times 10^{15}}{(2.2\times 10^{-2})(1.2\times 10^{-5})}$
$=\dfrac{9.24}{(2.2)(1.2)}\cdot 10^{15-(-2)-(-5)}$
$=3.5\times 10^{22}$

87. $(6.452\times 10^{-4})(4\times 10^{-2})=25.808\times 10^{-6}$
$=(2.5808\times 10^1)\times 10^{-6}$
$=2.5808\times 10^{-5}$

The area of the square is 2.5808×10^{-5} square meter.

89. $-5y+4y-18-y=-2y-18$

91. $-3x-(4x-2)=-3x-4x+2=-7x+2$

93. $3(z-4)-2(3z+1)=3z-12-6z-2$
$=-3z-14$

95. $\left(\dfrac{2x^{-2}}{y}\right)^3=\dfrac{2^3x^{-2(3)}}{y^3}=\dfrac{8x^{-6}}{y^3}=\dfrac{8}{x^6y^3}$

The volume is $\dfrac{8}{x^6y^3}$ cubic meters.

97. $D=\dfrac{500,000}{250}$
$=\dfrac{5\times 10^5}{2.5\times 10^2}$
$=2\times 10^{5-2}$
$=2\times 10^3$

Its density is 2×10^3 pounds per cubic feet.

99. $a^{-1}=a^1$
$\dfrac{1}{a}=a$
$1=a^2$
$(1)^2=1$ and $(-1)^2=1$
Yes; $a = 1$ and $a = -1$.

101. no; answers may vary

103. $\dfrac{3.19\times 10^8}{3.536\times 10^6}=\dfrac{3.19}{3.536}\times\dfrac{10^8}{10^6}\approx 0.90\times 10^2=90$

The population density for the United States in 2015 was 90 people per square mile.

105. $\dfrac{5.07\times 10^8}{3.19\times 10^8}=\dfrac{5.07}{3.19}\times\dfrac{10^8}{10^8}\approx 1.6$

The population of the European Union was approximately 1.6 times the population of the United States in 2015.

Section 5.3 Practice Exercises

1. a. $4x^5$: The exponent on x is 5, so the degree of the term is 5.

b. -4^3y^3: The exponent on y is 3, so the degree of the term is 3.

c. The degree of z, or z^1, is 1.

d. $65a^3b^7c$: The degree is the sum of the exponents on the variables, or $3 + 7 + 1 = 11$.

e. The degree of 36, which can be written as $36x^0$, is 0.

	Polynomial	Degree	Classification
2. a.	$3x^4 + 2x^2 - 3$	4	trinomial
b.	$9abc^3$	5	monomial
c.	$8x^5 + 5x^3$	5	binomial

3.

Term	Degree
$2x^3y$	4
$-3x^3y^2$	5
$-9y^5$	5
9.6	0

The largest degree of any term is 5, so the degree of this polynomial is 5.

4. a. Substitute −1 for *x* and simplify.

$$P(x) = -5x^2 + 2x - 8$$
$$\begin{aligned} P(-1) &= -5(-1)^2 + 2(-1) - 8 \\ &= -5 - 2 - 8 \\ &= -15 \end{aligned}$$

b. Substitute 3 for *x* and simplify.

$$P(x) = -5x^2 + 2x - 8$$
$$\begin{aligned} P(3) &= -5(3)^2 + 2(3) - 8 \\ &= -45 + 6 - 8 \\ &= -47 \end{aligned}$$

5. $P(t) = -16t^2 + 290$

Let $t = 0$: $P(0) = -16(0)^2 + 290 = 290$

Let $t = 2$: $P(2) = -16(2)^2 + 290 = 226$

The height of the object at $t = 0$ second is 290 feet and the height at $t = 2$ seconds is 226 feet.

6. a. $$\begin{aligned} 8x^4 - 5x^4 - 5x &= (8-5)x^4 - 5x \\ &= 3x^4 - 5x \end{aligned}$$

b. $$\begin{aligned} &4ab - 5b + 3ab + 2b \\ &= 4ab + 3ab - 5b + 2b \\ &= (4+3)ab + (-5+2)b \\ &= 7ab - 3b \end{aligned}$$

7. $$\begin{aligned} &(5x^3 - 3x^2 - 9x - 8) + (x^3 + 9x^2 + 2x) \\ &= 5x^3 + x^3 - 3x^2 + 9x^2 - 9x + 2x - 8 \\ &= 6x^3 + 6x^2 - 7x - 8 \end{aligned}$$

8. a. $$\begin{aligned} &(3a^4b - 5ab^2 + 7) + (9ab^2 - 12) \\ &= 3a^4b - 5ab^2 + 7 + 9ab^2 - 12 \\ &= 3a^4b - 5ab^2 + 9ab^2 + 7 - 12 \\ &= 3a^4b + 4ab^2 - 5 \end{aligned}$$

b. $$\begin{aligned} &(2x^5 - 3y + x - 6) + (4y - 2x - 3) \\ &= 2x^5 - 3y + x - 6 + 4y - 2x - 3 \\ &= 2x^5 - 3y + 4y + x - 2x - 6 - 3 \\ &= 2x^5 + y - x - 9 \end{aligned}$$

9. $$\begin{aligned} &(13a^4 - 7a^3 - 9) - (-2a^4 + 8a^3 - 12) \\ &= 13a^4 - 7a^3 - 9 + 2a^4 - 8a^3 + 12 \\ &= 13a^4 + 2a^4 - 7a^3 - 8a^3 - 9 + 12 \\ &= 15a^4 - 15a^3 + 3 \end{aligned}$$

10. $$\begin{aligned} &(11x^2y^2 - 7xy^2) - (5x^2y^2 - 3xy^2 + 5y^3) \\ &= 11x^2y^2 - 7xy^2 - 5x^2y^2 + 3xy^2 - 5y^3 \\ &= 6x^2y^2 - 4xy^2 - 5y^3 \end{aligned}$$

11. The degree of *f*(*x*) is 3, which means that its graph has the shape of **C**.

Graphing Calculator Explorations

1. $$\begin{aligned} &(2x^2 + 7x + 6) + (x^3 - 6x^2 - 14) \\ &= x^3 - 4x^2 + 7x - 8 \end{aligned}$$

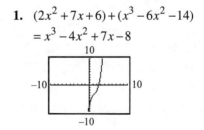

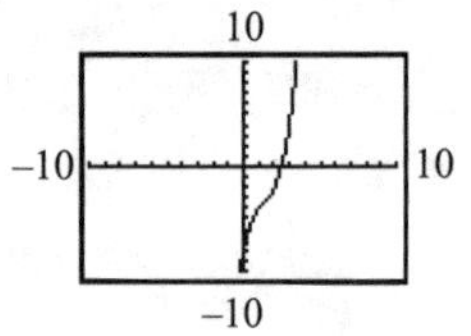

2. $(-14x^3 - x + 2) + (-x^3 + 3x^2 + 4x)$
$= -15x^3 + 3x^2 + 3x + 2$

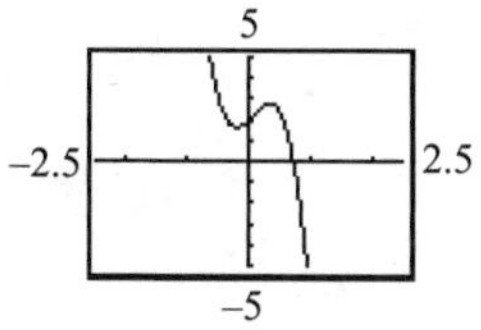

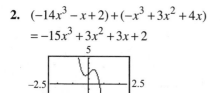

3. $(1.8x^2 - 6.8x - 1.7) - (3.9x^2 - 3.6x)$
$= -2.1x^2 - 3.2x - 1.7$

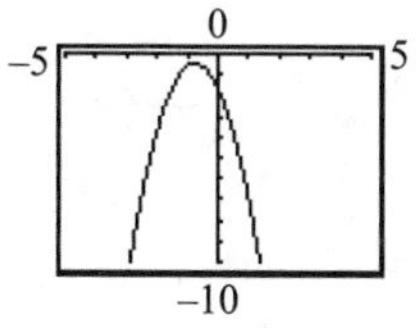

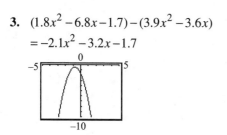

4. $(-4.8x^2 + 12.5x - 7.8) - (3.1x^2 - 7.8x)$
$= -7.9x^2 + 20.3x - 7.8$

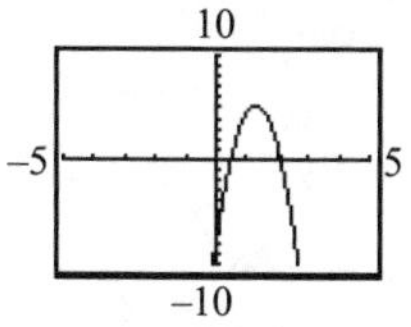

5. $(1.29x - 5.68) + (7.69x^2 - 2.55x + 10.98)$
$= 7.69x^2 - 1.26x + 5.3$

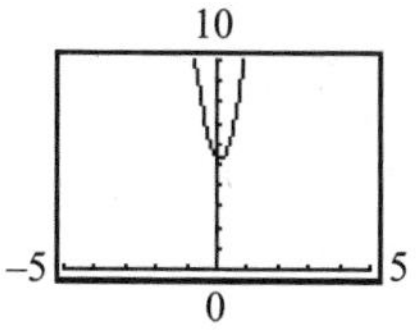

6. $(-0.98x^2 - 1.56x + 5.57) + (4.36x - 3.71)$
$= -0.98x^2 + 2.8x + 1.86$

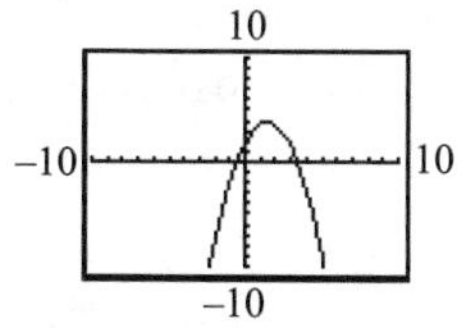

Vocabulary, Readiness & Video Check 5.3

1. The numerical factor of a term is the coefficient.

2. A polynomial is a finite sum of terms in which all variables are raised to nonnegative integer powers and no variables appear in any denominator.

3. A binomial is a polynomial with 2 terms.

4. A monomial is a polynomial with 1 term.

5. A trinomial is a polynomial with 3 terms.

6. The degree of a term is the sum of the exponents on the variables in the term.

7. The degree of a polynomial is the largest degree of all its terms.

8. Like terms contain the same variables raised to the same powers.

9. We are finding the degree of the polynomial, which is the greatest degree of any of its terms, so we find the degree of each term first.

10. evaluating a polynomial at a given replacement value

11. by combining like terms

12. combine like terms

13. Change the operation from subtraction to addition and add the opposite of the polynomial that is being subtracted—that is, change the signs of all terms of the polynomial being subtracted and add.

14. C; degree 2 tells you it's a parabola and its negative coefficient tells you the parabola opens downward.

Exercise Set 5.3

1. 4 has degree 0.

3. $5x^2$ has degree 2.

5. $-3xy^2$ has degree 1 + 2 = 3.

7. -8^7y^3 has degree 3 (note: the degree on the *variable* is 3).

9. $3.78ab^3c^5$ has degree 1 + 3 + 5 = 9.

11. $6x + 0.3$ has degree 1 and is a binomial.

13. $3x^2 - 2x + 5$ has degree 2 and is a trinomial.

15. -3^4xy^2 has degree 1 + 2 = 3 and is a monomial.

17.

Term	Degree
x^2y	3
$-4xy^2$	3
$5x$	1
y^4	4

$x^2y-4xy^2+5x+y^4$ has degree 4 and is none of these.

19. $P(x)=x^2+x+1$
$P(7)=7^2+7+1=49+7+1=57$

21. $Q(x)=5x^2-1$
$$\begin{aligned}Q(-10)&=5(-10)^2-1\\&=5(100)-1\\&=500-1\\&=499\end{aligned}$$

23. $Q(x)=5x^2-1$
$$\begin{aligned}Q\left(\frac{1}{4}\right)&=5\left(\frac{1}{4}\right)^2-1\\&=5\left(\frac{1}{16}\right)-1\\&=\frac{5}{16}-\frac{16}{16}\\&=-\frac{11}{16}\end{aligned}$$

25. $P(t)=-16t^2+1125$
$P(2)=-16(2)^2+1125=1061$
After 2 seconds, the height is 1061 feet.

27. $P(t)=-16t^2+1125$
$P(6)=-16(6)^2+1125=549$
After 6 seconds, the height is 549 feet.

29. $$\begin{aligned}5y+y-6y^2-y^2&=(5+1)y+(-6-1)y^2\\&=6y-7y^2 \text{ or } -7y^2+6y\end{aligned}$$

31. $$\begin{aligned}&4x^2y+2x-3x^2y-\frac{1}{2}-7x\\&=4x^2y-3x^2y+2x-7x-\frac{1}{2}\\&=x^2y-5x-\frac{1}{2}\end{aligned}$$

33. $$\begin{aligned}&7x^2-2xy+5y^2-x^2+xy+11y^2\\&=7x^2-x^2-2xy+xy+5y^2+11y^2\\&=(7-1)x^2+(-2+1)xy+(5+11)y^2\\&=6x^2-xy+16y^2\end{aligned}$$

35. $$\begin{aligned}&(9y^2+y-8)+(9y^2-y-9)\\&=9y^2+y-8+9y^2-y-9\\&=9y^2+9y^2+y-y-8-9\\&=18y^2-17\end{aligned}$$

37. $$\begin{aligned}&(x^2+xy-y^2)+(2x^2-4xy+7y^2)\\&=x^2+xy-y^2+2x^2-4xy+7y^2\\&=x^2+2x^2+xy-4xy-y^2+7y^2\\&=3x^2-3xy+6y^2\end{aligned}$$

39. $$\begin{array}{r}x^2-6x+3\\+\quad(2x+5)\\\hline x^2-4x+8\end{array}$$

41. $$\begin{aligned}&(7x^3y-4xy+8)+(5x^3y+4xy+8x)\\&=7x^3y-4xy+8+5x^3y+4xy+8x\\&=7x^3y+5x^3y-4xy+4xy+8x+8\\&=12x^3y+8x+8\end{aligned}$$

43. $$\begin{aligned}&(0.6x^3+1.2x^2-4.5x+9.1)+(3.9x^3-x^2+0.7x)\\&=0.6x^3+1.2x^2-4.5x+9.1+3.9x^3-x^2+0.7x\\&=0.6x^3+3.9x^3+1.2x^2-x^2-4.5x+0.7x+9.1\\&=4.5x^3+0.2x^2-3.8x+9.1\end{aligned}$$

45. $$\begin{aligned}&(9y^2-7y+5)-(8y^2-7y+2)\\&=9y^2-7y+5-8y^2+7y-2\\&=9y^2-8y^2-7y+7y+5-2\\&=y^2+3\end{aligned}$$

47. $$\begin{aligned}(4x^2+2x)-(6x^2-3x)&=4x^2+2x-6x^2+3x\\&=4x^2-6x^2+2x+3x\\&=-2x^2+5x\end{aligned}$$

49. $\begin{array}{r} 6y^2-6y+4 \\ \underline{-(-y^2+6y+7)} \end{array} \rightarrow \begin{array}{r} 6y^2-6y+4 \\ \underline{+\ y^2-6y-7} \\ 7y^2-12y-3 \end{array}$

51. $(9x^3-2x^2+4x-7)-(2x^3-6x^2-4x+3)$
$=9x^3-2x^2+4x-7-2x^3+6x^2+4x-3$
$=9x^3-2x^3-2x^2+6x^2+4x+4x-7-3$
$=7x^3+4x^2+8x-10$

53. $\left(-19y^2+7yx+\frac{1}{7}\right)-\left(y^2+4yx+\frac{1}{7}\right)$
$=-19y^2+7yx+\frac{1}{7}-y^2-4yx-\frac{1}{7}$
$=-19y^2-y^2+7yx-4yx+\frac{1}{7}-\frac{1}{7}$
$=-20y^2+3yx$

55. $(-3x+8)+(-3x^2+3x-5)$
$=-3x+8-3x^2+3x-5$
$=-3x^2-3x+3x+8-5$
$=-3x^2+3$

57. $(5y^4-7y^2+x^2-3)+(-3y^4+2y^2+4)$
$=5y^4-7y^2+x^2-3-3y^4+2y^2+4$
$=5y^4-3y^4-7y^2+2y^2+x^2-3+4$
$=2y^4-5y^2+x^2+1$

59. $(4x^2-6x+2)-(-x^2+3x+5)$
$=4x^2-6x+2+x^2-3x-5$
$=4x^2+x^2-6x-3x+2-5$
$=5x^2-9x-3$

61. $(5x^2+x+9)-(2x^2-9)=5x^2+x+9-2x^2+9$
$=5x^2-2x^2+x+9+9$
$=3x^2+x+18$

63. $(5x-11)+(-x-2)=5x-11-x-2$
$=5x-x-11-2$
$=4x-13$

65. $(3x^3-b+2a-6)+(-4x^3+b+6a-6)$
$=3x^3-b+2a-6-4x^3+b+6a-6$
$=3x^3-4x^3-b+b+2a+6a-6-6$
$=-x^3+8a-12$

67. $(14ab-10a^2b+6b^2)-(18a^2-20a^2b-6b^2)$
$=14ab-10a^2b+6b^2-18a^2+20a^2b+6b^2$
$=14ab-10a^2b+20a^2b-18a^2+6b^2+6b^2$
$=14ab+10a^2b-18a^2+12b^2$

69. $\begin{array}{r} 3x^2+15x\ +8 \\ \underline{+\ (2x^2\ +7x\ +8)} \\ 5x^2+22x+16 \end{array}$

71. $(7x^2-5)+(-3x^2-2)-(4x^2-7)$
$=7x^2-5-3x^2-2-4x^2+7$
$=7x^2-3x^2-4x^2-5-2+7$
$=0$

73. $(-3+4x^2+7xy^2)+(2x^3-x^2+xy^2)$
$=-3+4x^2+7xy^2+2x^3-x^2+xy^2$
$=7xy^2+xy^2+2x^3+4x^2-x^2-3$
$=8xy^2+2x^3+3x^2-3$

75. $\begin{array}{r} 3x^2-4x\ +8 \\ \underline{-\qquad (5x\ -7)} \end{array}$
Is equivalent to:
$\begin{array}{r} 3x^2-4x\ +8 \\ \underline{+\quad (-5x\ +7)} \\ 3x^2-9x+15 \end{array}$

77. $(7x^2+4x+9)+(8x^2+7x-8)-(3x+7)$
$=7x^2+4x+9+8x^2+7x-8-3x-7$
$=7x^2+8x^2+4x+7x-3x+9-8-7$
$=15x^2+8x-6$

79. $\left(\frac{2}{3}x^2-\frac{1}{6}x+\frac{5}{6}\right)-\left(\frac{1}{3}x^2+\frac{5}{6}x-\frac{1}{6}\right)$
$=\frac{2}{3}x^2-\frac{1}{6}x+\frac{5}{6}-\frac{1}{3}x^2-\frac{5}{6}x+\frac{1}{6}$
$=\frac{2}{3}x^2-\frac{1}{3}x^2-\frac{1}{6}x-\frac{5}{6}x+\frac{5}{6}+\frac{1}{6}$
$=\frac{1}{3}x^2-x+1$

81. If $L=5$, $W=4$, and $H=9$, then
$2HL+2LW+2HW$
$=2(9)(5)+2(5)(4)+2(9)(4)$
$=90+40+72$
$=202$
The surface area is 202 square inches.

83. $P(t) = -16t^2 + 300t$

a. $P(1) = -16(1)^2 + 300(1) = 284$
After 1 second, the height is 284 feet.

b. $P(2) = -16(2)^2 + 300(2) = 536$
After 2 seconds, the height is 536 feet.

c. $P10) = -16(10)^2 + 300(10) = 1400$
After 10 seconds, the height is 1400 feet.

d. $P(14) = -16(14)^2 + 300(14) = 1064$
After 14 seconds, the height is 1064 feet.

e. answers may vary

f. $P(18) = -16(18)^2 + 300(18) = 216$
$P(19) = -16(19)^2 + 300(19) = -76$
The object hits the ground after approximately 19 seconds.

85. $P(x) = 45x - 100{,}000$
$P(4000) = 45(4000) - 100{,}000$
$= 180{,}000 - 100{,}000$
$= 80{,}000$
The profit from selling 4000 computer briefcases is \$80,000.

87. $R(x) = 11x$
$R(1500) = 11(1500) = 16{,}500$
The revenue from selling 1500 boxes is \$16,500.

89. A: The degree of $f(x)$ is 2, so the graph has the shape of A or C. The coefficient of x^2 is a positive number, so the graph has the shape of A.

91. D: The degree of $g(x)$ is 3, so the graph has the shape of B or D. The coefficient of x^3 is a negative number, so the graph has the shape of D.

93. $5(3x-2) = 15x - 10$

95. $-2(x^2 - 5x + 6) = -2x^2 + 10x - 12$

97. The opposite of $8x - 6$ is $-(8x - 6)$ or $-8x + 6$; a and c.

99. $(12x - 1.7) - (15x + 6.2)$
$= 12x - 1.7 - 15x - 6.2$
$= -3x - 7.9$

101. answers may vary

103. answers may vary

105. $(4x^{2a} - 3x^a + 0.5) - (x^{2a} - 5x^a - 0.2)$
$= 4x^{2a} - 3x^a + 0.5 - x^{2a} + 5x^a + 0.2$
$= 3x^{2a} + 2x^a + 0.7$

107. $(8x^{2y} - 7x^y + 3) + (-4x^{2y} + 9x^y - 14)$
$= 8x^{2y} - 7x^y + 3 - 4x^{2y} + 9x^y - 14$
$= 4x^{2y} + 2x^y - 11$

109. $P = 2l + 2w$
$= 2(3x^2 - x + 2y) + 2(x + 5y)$
$= 6x^2 - 2x + 4y + 2x + 10y$
$= 6x^2 + 14y$
The perimeter is $P = (6x^2 + 14y)$ units.

111. $P(x) + Q(x) = (3x + 3) + (4x^2 - 6x + 3)$
$= 3x + 3 + 4x^2 - 6x + 3$
$= 4x^2 - 3x + 6$

113. $Q(x) - R(x) = (4x^2 - 6x + 3) - (5x^2 - 7)$
$= 4x^2 - 6x + 3 - 5x^2 + 7$
$= -x^2 - 6x + 10$

115. $2[Q(x)] - R(x)$
$= 2(4x^2 - 6x + 3) - (5x^2 - 7)$
$= 8x^2 - 12x + 6 - 5x^2 + 7$
$= 3x^2 - 12x + 13$

117. $3[R(x)] + 4[P(x)] = 3(5x^2 - 7) + 4(3x + 3)$
$= 15x^2 - 21 + 12x + 12$
$= 15x^2 + 12x - 9$

119. $P(x) = 2x - 3$

a. $P(a) = 2a - 3$

b. $P(-x) = 2(-x) - 3 = -2x - 3$

c. $P(x + h) = 2(x + h) - 3 = 2x + 2h - 3$

121. $P(x) = 4x$

a. $P(a) = 4a$

b. $P(-x) = 4(-x) = -4x$

c. $P(x+h) = 4(x+h) = 4x+4h$

123. $P(x) = 4x-1$

a. $P(a) = 4a-1$

b. $P(-x) = 4(-x)-1 = -4x-1$

c. $P(x+h) = 4(x+h)-1 = 4x+4h-1$

125. $f(x) = 0.24x^2 + 8.41x + 122.8$

a. 2005 is 25 years after 1980, so $x = 25$.

$$\begin{aligned} f(25) &= 0.24(25)^2 + 8.41(25) + 122.8 \\ &= 483.05 \end{aligned}$$

Restaurant food-and-drink sales were $483.05 billion in 2005.

b. 2010 is 30 years after 1980, so $x = 30$.

$$\begin{aligned} f(30) &= 0.24(30)^2 + 8.41(30) + 122.8 \\ &= 591.1 \end{aligned}$$

Restaurant food-and-drink sales were $591.1 billion in 2010.

c. 2020 is 40 years after 1980, so $x = 40$.

$$\begin{aligned} f(40) &= 0.24(40)^2 + 8.41(40) + 122.8 \\ &= 843.2 \end{aligned}$$

Restaurant food-and-drink sales are estimated to be $843.2 billion in 2020.

d. answers may vary

Section 5.4 Practice Exercises

1. a. $(3x^4)(2x^2) = 3(2)(x^4)(x^2) = 6x^6$

b. $$\begin{aligned} &(-5m^4np^3)(-8mnp^5) \\ &= -5(-8)(m^4m)(n \cdot n)(p^3p^5) \\ &= 40m^5n^2p^8 \end{aligned}$$

2. a. $$\begin{aligned} 3x(7x-1) &= 3x(7x) + 3x(-1) \\ &= 21x^2 - 3x \end{aligned}$$

b. $$\begin{aligned} &-5a^2(3a^2 - 6a + 5) \\ &= -5a^2(3a^2) + (-5a^2)(-6a) \\ &\quad + (-5a^2)(5) \\ &= -15a^4 + 30a^3 - 25a^2 \end{aligned}$$

c. $$\begin{aligned} &-mn^3(5m^2n^2 + 2mn - 5m) \\ &= -mn^3(5m^2n^2) + (-mn^3)(2mn) \\ &\quad + (-mn^3)(-5m) \\ &= -5m^3n^5 - 2m^2n^4 + 5m^2n^3 \end{aligned}$$

3. a. $$\begin{aligned} (x+5)(2x+3) &= x(2x+3) + 5(2x+3) \\ &= 2x^2 + 3x + 10x + 15 \\ &= 2x^2 + 13x + 15 \end{aligned}$$

b. $$\begin{aligned} &(3x-1)(x^2 - 6x + 2) \\ &= 3x(x^2 - 6x + 2) + (-1)(x^2 - 6x + 2) \\ &= 3x^3 - 18x^2 + 6x - x^2 + 6x - 2 \\ &= 3x^3 - 19x^2 + 12x - 2 \end{aligned}$$

4. $$\begin{array}{r} x^2 - 4x - 5 \\ 3x^2 + 2 \\ \hline 2x^2 - 8x - 10 \\ 3x^4 - 12x^3 - 15x^2 \qquad\qquad \\ \hline 3x^4 - 12x^3 - 13x^2 - 8x - 10 \end{array}$$

5. $$\begin{aligned} &(x-5)(x+3) \\ &= x \cdot x + 3 \cdot x + (-5)x + (-5)(3) \\ &= x^2 + 3x - 5x - 15 \\ &= x^2 - 2x - 15 \end{aligned}$$

6. a. $$\begin{aligned} &(3x-5)(2x-7) \\ &= 3x(2x) + 3x(-7) + (-5)(2x) \\ &\quad + (-5)(-7) \\ &= 6x^2 - 21x - 10x + 35 \\ &= 6x^2 - 31x + 35 \end{aligned}$$

b. $$\begin{aligned} &(2x^2 - 3y)(4x^2 + y) \\ &= 8x^4 + 2x^2y - 12x^2y - 3y^2 \\ &= 8x^4 - 10x^2y - 3y^2 \end{aligned}$$

7. a. $$\begin{aligned} (x+6)^2 &= x^2 + 2 \cdot x \cdot 6 + 6^2 \\ &= x^2 + 12x + 36 \end{aligned}$$

b. $(x-2)^2 = x^2 - 2\cdot x\cdot 2 + 2^2$
$= x^2 - 4x + 4$

c. $(3x+5y)^2$
$= (3x)^2 + 2(3x)(5y) + (5y)^2$
$= 9x^2 + 30xy + 25y^2$

d. $(3x^2 - 8b)^2$
$= (3x^2)^2 - 2(3x^2)(8b) + (8b)^2$
$= 9x^4 - 48x^2b + 64b^2$

8. a. $(x-7)(x+7) = x^2 - 7^2 = x^2 - 49$

b. $(2a+5)(2a-5) = (2a)^2 - 5^2$
$= 4a^2 - 25$

c. $\left(5x^2 + \frac{1}{4}\right)\left(5x^2 - \frac{1}{4}\right) = (5x^2)^2 - \left(\frac{1}{4}\right)^2$
$= 25x^4 - \frac{1}{16}$

d. $(a^3 - 4b^2)(a^3 + 4b^2) = (a^3)^2 - (4b^2)^2$
$= a^6 - 16b^4$

9. $[2 + (3x-y)]^2$
$= 2^2 + 2(2)(3x-y) + (3x-y)^2$
$= 4 + 4(3x-y) + (3x)^2 - 2(3x)\cdot y + y^2$
$= 4 + 12x - 4y + 9x^2 - 6xy + y^2$

10. $[(3x-y)-5][(3x-y)+5]$
$= (3x-y)^2 - 5^2$
$= (3x)^2 - 2(3x)(y) + y^2 - 25$
$= 9x^2 - 6xy + y^2 - 25$

11. $(x+4)(x-4)(x^2-16)$
$= (x^2-16)(x^2-16)$
$= (x^2-16)^2$
$= x^4 - 32x^2 + 256$

12. $f(x) = x^2 - 3x + 5$
$f(h+1) = (h+1)^2 - 3(h+1) + 5$
$= h^2 + 2h + 1 - 3h - 3 + 5$
$= h^2 - h + 3$

Graphing Calculator Explorations

1. $(x+4)(x-4) = x^2 - 16$

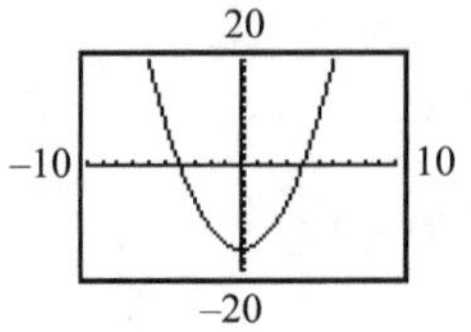

2. $(x+3)(x+3) = x^2 + 6x + 9$

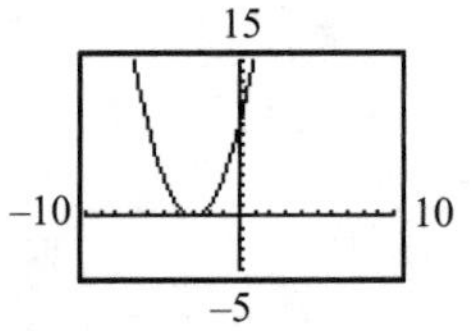

3. $(3x-7)^2 = 9x^2 - 42x + 49$

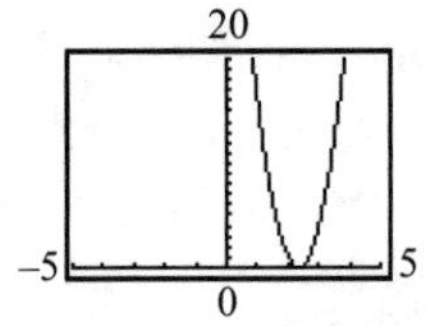

4. $(5x-2)^2 = 25x^2 - 20x + 4$

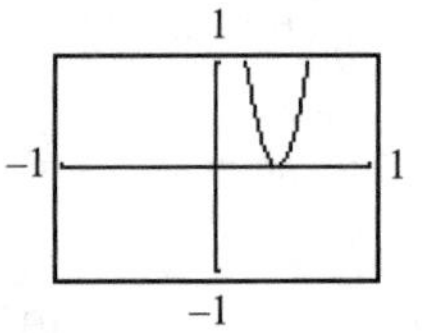

5. $(5x+1)(x^2 - 3x - 2) = 5x^3 - 14x^2 - 13x - 2$

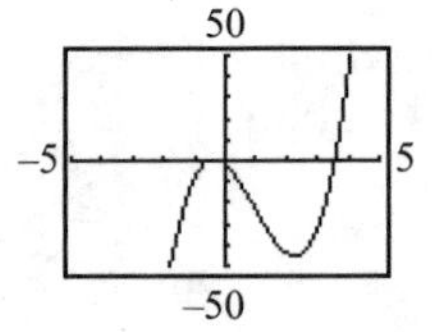

6. $(7x+4)(2x^2 + 3x - 5) = 14x^3 + 29x^2 - 23x - 20$

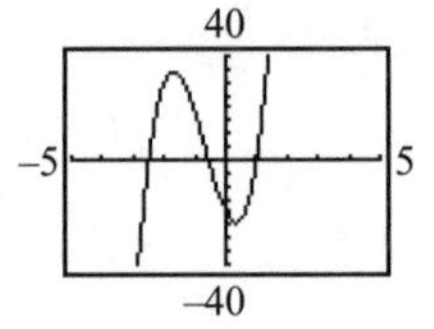

Vocabulary, Readiness & Video Check 5.4

1. $(6x^3)\left(\frac{1}{2}x^3\right) = \underline{3x^6}$; choice b

2. $(x+7)^2 = \underline{x^2+14x+49}$; choice c

3. $(x+7)(x-7) = \underline{x^2-49}$; choice b

4. The product of $(3x-1)(4x^2-2x+1)$ is a polynomial of degree 3; choice a.

5. If $f(x) = x^2+1$ then $f(a+1) = \underline{(a+1)^2+1}$; choice d.

6. $[x+(2y+1)]^2 = \underline{[x+(2y+1)][x+(2y+1)]}$; choice b

7. distributive property, product rule

8. multiplying two binomials

9. FOIL order or distributive property

10. Multiplying using the FOIL order gives you four terms and the two middle terms will be opposites and subtract out.

11. Multiply two at a time and multiply as usual, using patterns if you recognize them. Keep multiplying until you are through.

12. No, a variable can also be replaced by a variable expression, so a function can also be evaluated for a variable expression.

Exercise Set 5.4

1. $$\begin{aligned}(-4x^3)(3x^2) &= -4(3)(x^3)(x^2)\\ &= -12x^{3+2}\\ &= -12x^5\end{aligned}$$

3. $$\begin{aligned}(8.6a^4b^5c)(10ab^3c^2) &= 8.6(10)(a^4a)(b^5b^3)(cc^2)\\ &= 86a^5b^8c^3\end{aligned}$$

5. $3x(4x+7) = 3x(4x)+3x(7) = 12x^2+21x$

7. $$\begin{aligned}-6xy(4x+y) &= -6xy(4x)+(-6xy)(y)\\ &= -24x^2y-6xy^2\end{aligned}$$

9. $$\begin{aligned}-4ab(xa^2+ya^2-3) &= -4ab(xa^2)+(-4ab)(ya^2)+(-4ab)(-3)\\ &= -4a^3bx-4a^3by+12ab\end{aligned}$$

11. $$\begin{aligned}(x-3)(2x+4) &= x(2x+4)-3(2x+4)\\ &= 2x^2+4x-6x-12\\ &= 2x^2-2x-12\end{aligned}$$

13. $(2x+3)(x^3-x+2) = 2x(x^3-x+2)+3(x^3-x+2)$
$= 2x^4-2x^2+4x+3x^3-3x+6$
$= 2x^4+3x^3-2x^2+x+6$

15.
$$\begin{array}{r} 3x-2 \\ \times\ 5x+1 \\ \hline 3x-2 \\ 15x^2-10x \\ \hline 15x^2-7x-2 \end{array}$$

17.
$$\begin{array}{r} 3m^2+2m-1 \\ \times\ \qquad 5m+2 \\ \hline 6m^2+4m-2 \\ 15m^3+10m^2-5m \\ \hline 15m^3+16m^2-m-2 \end{array}$$

19. $-6a^2b^2(5a^2b^2-6a-6b) = -6a^2b^2(5a^2b^2)+(-6a^2b^2)(-6a)+(-6a^2b^2)(-6b)$
$= -30a^4b^4+36a^3b^2+36a^2b^3$

21. $(x-3)(x+4) = x^2+4x-3x-12 = x^2+x-12$

23. $(5x-8y)(2x-y) = 10x^2-5xy-16xy+8y^2$
$= 10x^2-21xy+8y^2$

25. $\left(4x+\frac{1}{3}\right)\left(4x-\frac{1}{2}\right) = 16x^2-2x+\frac{4}{3}x-\frac{1}{6}$
$= 16x^2-\frac{2}{3}x-\frac{1}{6}$

27. $(5x^2-2y^2)(x^2-3y^2) = 5x^4-15x^2y^2-2x^2y^2+6y^4$
$= 5x^4-17x^2y^2+6y^4$

29. $(x+4)^2 = x^2+2\cdot x\cdot 4+4^2 = x^2+8x+16$

31. $(6y-1)(6y+1) = (6y)^2-1^2 = 36y^2-1$

33. $(3x-y)^2 = (3x)^2-2\cdot 3x\cdot y+y^2$
$= 9x^2-6xy+y^2$

35. $(7ab+3c)(7ab-3c) = (7ab)^2-(3c)^2$
$= 49a^2b^2-9c^2$

37. $\left(3x+\frac{1}{2}\right)\left(3x-\frac{1}{2}\right) = (3x)^2-\left(\frac{1}{2}\right)^2 = 9x^2-\frac{1}{4}$

39. $[3+(4b+a)]^2 = 3^2+2\cdot 3\cdot(4b+a)+(4b+a)^2$
$= 9+6(4b+a)+(4b)^2+2\cdot 4b\cdot a+a^2$
$= 9+24b+6a+16b^2+8ab+a^2$

41. $[(2s-3)-1][(2s-3)+1] = (2s-3)^2-1^2$
$= (2s)^2-2\cdot 2s\cdot 3+3^2-1$
$= 4s^2-12s+9-1$
$= 4s^2-12s+8$

43. $[(x^2+4x)-6]^2$
$= (x^2+4x)^2-2(x^2+4x)(6)+6^2$
$= x^4+8x^3+16x^2-12x^2-48x+36$
$= x^4+8x^3+4x^2-48x+3$

45. $(x+y)(x-y)(x^2-y^2) = (x^2-y^2)(x^2-y^2)$
$= (x^2)^2-2(x^2)(y^2)+(y^2)^2$
$= x^4-2x^2y^2+y^4$

47. $(x-2)^4 = (x-2)(x-2)(x-2)(x-2)$
$= [(x-2)^2][(x-2)^2]$
$= [x^2-2\cdot x\cdot 2+2^2][x^2-2\cdot x\cdot 2+2^2]$
$= (x^2-4x+4)(x^2-4x+4)$
$= x^2(x^2-4x+4)-4x(x^2-4x+4)+4(x^2-4x+4)$
$= x^4-4x^3+4x^2-4x^3+16x^2-16x+4x^2-16x+16$
$= x^4-8x^3+24x^2-32x+16$

49. $(x-5)(x+5)(x^2+25) = (x^2-25)(x^2+25)$
$= (x^2)^2-25^2$
$= x^4-625$

51. $-8a^2b(3b^2-5b+20) = -24a^2b^3+40a^2b^2-160a^2b$

53. $(6x+1)^2 = (6x)^2+2\cdot 6x\cdot 1+1^2$
$= 36x^2+12x+1$

55. $(5x^3+2y)(5x^3-2y) = (5x^3)^2-(2y)^2$
$= 25x^6-4y^2$

57. $(2x^3+5)(5x^2+4x+1) = 2x^3(5x^2+4x+1)+5(5x^2+4x+1)$
$= 10x^5+8x^4+2x^3+25x^2+20x+5$

59. $(3x^2+2x-1)^2 = (3x^2+2x-1)(3x^2+2x-1)$
$= 3x^2(3x^2+2x-1)+2x(3x^2+2x-1)-1(3x^2+2x-1)$
$= 9x^4+6x^3-3x^2+6x^3+4x^2-2x-3x^2-2x+1$
$= 9x^4+12x^3-2x^2-4x+1$

61. $(3x-1)(x+3) = 3x(x+3)-1(x+3)$
$= 3x^2+9x-x-3$
$= 3x^2+8x-3$

63. $(3x^4+1)(3x^2+5) = 9x^6+15x^4+3x^2+5$

65. $(3x+1)^2 = (3x)^2+2\cdot 3x\cdot 1+1^2 = 9x^2+6x+1$

67. $(3b-6y)(3b+6y) = (3b)^2-(6y)^2 = 9b^2-36y^2$

69. $(7x-3)(7x+3) = (7x)^2-3^2 = 49x^2-9$

71.
$$\begin{array}{r} 3x^2+4x-4 \\ \times\qquad\qquad 3x+6 \\ \hline 18x^2+24x-24 \\ 9x^3+12x^2-12x\qquad \\ \hline 9x^3+30x^2+12x-24 \end{array}$$

73. $(4x^2-2x+5)(3x+1) = 4x^2(3x+1)-2x(3x+1)+5(3x+1)$
$= 12x^3+4x^2-6x^2-2x+15x+5$
$= 12x^3-2x^2+13x+5$

75. $[(x+y)-7]^2 = (x+y)^2-14(x+y)+49$
$= x^2+2xy+y^2-14x-14y+49$

77. $(11a^2+1)(2a+1) = 22a^3+11a^2+2a+1$

79. $\left(\frac{2}{3}n-2\right)\left(\frac{1}{2}n-9\right) = \frac{1}{3}n^2-6n-n+18$
$= \frac{1}{3}n^2-7n+18$

81. $(3x+1)(3x-1)(2y+5x) = [(3x)^2-1^2](2y+5x)$
$= (9x^2-1)(2y+5x)$
$= 18x^2y+45x^3-2y-5x$
or $45x^3+18x^2y-5x-2y$

83. $f(x) = x^2-3x$
$f(a) = a^2-3a$

85. $f(x) = x^2 - 3x$

$f(a+h) = (a+h)^2 - 3(a+h)$
$= a^2 + 2ah + h^2 - 3a - 3h$

87. $f(x) = x^2 - 3x$

$f(b-2) = (b-2)^2 - 3(b-2)$
$= b^2 - 4b + 4 - 3b + 6$
$= b^2 - 7b + 10$

89. $y = -2x + 7$
$m = -2$

91. $3x - 5y = 14$
$-5y = -3x + 14$
$y = \frac{3}{5}x - \frac{14}{5}$
$m = \frac{3}{5}$

93. Since any vertical line crosses the graph at most once, it is a function.

95. $7y(3z - 2) + 1 = 21yz - 14y + 1$

97. answers may vary

99. $F(x) = x^2 + 3x + 2$

a. $F(a+h) = (a+h)^2 + 3(a+h) + 2$
$= a^2 + 2ah + h^2 + 3a + 3h + 2$

b. $F(a) = a^2 + 3a + 2$

c. $F(a+h) - F(a)$
$= a^2 + 2ah + h^2 + 3a + 3h + 2 - (a^2 + 3a + 2)$
$= 2ah + h^2 + 3h$

101. $5x^2y^n(6y^{n+1} - 2)$
$= 5x^2y^n(6y^{n+1}) + 5x^2y^n(-2)$
$= 30x^2y^{2n+1} - 10x^2y^n$

103. $(x^a + 5)(x^{2a} - 3)$
$= x^a \cdot x^{2a} + x^a(-3) + 5(x^{2a}) + 5(-3)$
$= x^{3a} - 3x^a + 5x^{2a} - 15$

105. Area $= \pi r^2$
$= \pi(5x-2)^2$
$= \pi(25x^2 - 20x + 4)$ square km
or $(25\pi x^2 - 20\pi x + 4\pi)$ square km

107. Area $= (3x-2)^2 - x^2$
$= (9x^2 - 12x + 4) - x^2$
$= (8x^2 - 12x + 4)$ square inches

109. One operation is addition, the other is multiplication.

a. $(3x+5) + (3x+7) = 6x + 12$

b. $(3x+5)(3x+7)$
$= 9x^2 + 21x + 15x + 35$
$= 9x^2 + 36x + 35$

111. $P(x) \cdot R(x) = (5x)(x+5)$
$= 5x \cdot x + 5x \cdot 5$
$= 5x^2 + 25x$

113. $[Q(x)]^2 = (x^2 - 2)^2$
$= (x^2)^2 - 2(x^2)(2) + 2^2$
$= x^4 - 4x^2 + 4$

115. $R(x) \cdot Q(x) = (x+5)(x^2 - 2)$
$= x^3 - 2x + 5x^2 - 10$
$= x^3 + 5x^2 - 2x - 10$

Section 5.5 Practice Exercises

1. $32x^4y^2 = 2 \cdot 2 \cdot 2 \cdot 2 \cdot 2 \cdot x^4 \cdot y \cdot y$
$48x^3y = 2 \cdot 2 \cdot 2 \cdot 2 \cdot 3 \cdot x^3 \cdot y$
$24y^2 = 2 \cdot 2 \cdot 2 \cdot 3 \cdot y \cdot y$
$\text{GCF} = 2 \cdot 2 \cdot 2 \cdot y = 8y$

2. a. The GCF of $12x^2$, 9, and $15x$ is 3.
$12x^2 + 9 + 15x = 3(4x^2) + 3(3) + 3(5x)$
$= 3(4x^2 + 3 + 5x)$

b. There is no common factor of the terms $3x$ and $-8y^3$ other than 1 (or -1).

c. The GCF of $8a^4$ and $-2a^3$ is $2a^3$.

$$8a^4 - 2a^3 = 2a^3 \cdot 4a - 2a^3 \cdot 1$$
$$= 2a^3(4a-1)$$

3. Factor out the GCF of the two terms, $8x^3y^2$.

$$64x^5y^2 - 8x^3y^2 = 8x^3y^2 \cdot 8x^2 - 8x^3y^2 \cdot 1$$
$$= 8x^3y^2(8x^2 - 1)$$

4. Factor out the GCF of the three terms, $-xy^2$.

$$-9x^4y^2 + 5x^2y^2 + 7xy^2$$
$$= -xy^2 \cdot 9x^3 - xy^2(-5x) - xy^2(-7)$$
$$= -xy^2(9x^3 - 5x - 7)$$

5. The GCF is $(x + 4)$.
$3(x + 4) + 5b(x + 4) = (x + 4)(3 + 5b)$

6. $8b(a^3 + 2y) - (a^3 + 2y)$
$= 8b(a^3 + 2y) - 1(a^3 + 2y)$
$= (a^3 + 2y)(8b - 1)$

7. $xy - 5x - 10 + 2y = (xy - 5x) + (-10 + 2y)$
$= x(y-5) + 2(-5 + y)$
$= x(y-5) + 2(y-5)$
$= (y-5)(x+2)$

8. $a^3 + 2a^2 + 5a + 10 = (a^3 + 2a^2) + (5a + 10)$
$= a^2(a+2) + 5(a+2)$
$= (a+2)(a^2 + 5)$

9. $x^2y^2 + 3y^2 - 5x^2 - 15$
$= (x^2y^2 + 3y^2) + (-5x^2 - 15)$
$= y^2(x^2 + 3) - 5(x^2 + 3)$
$= (x^2 + 3)(y^2 - 5)$

10. $pq + 3p - q - 3 = (pq + 3p) + (-q - 3)$
$= p(q+3) - 1(q+3)$
$= (q+3)(p-1)$

Vocabulary, Readiness & Video Check 5.5

1. The reverse of multiplying is factoring.

2. The greatest common factor (GCF) of x^7, x^3, x^5 is x^3.

3. In general, the GCF of a list of common variables raised to powers is the least exponent in the list.

4. Factoring means writing as a product.

5. True or false: A factored form of $2xy^3 + 10xy$ is $2xy \cdot y^2 + 2xy \cdot 5$. false

6. True or false: A factored form of $x^3 - 6x^2 + x$ is $x(x^2 - 6x)$. false

7. True or false: A factored form of $5x - 5y + x^3 - x^2y$ is $5(x-y) + x^2(x-y)$. false

8. True or false: A factored form of $5x - 5y + x^3 - x^2y$ is $(x-y)(5+x^2)$. true

9. The GCF is that common variable raised to the smallest exponent in the list.

10. Multiplication (what times the GCF gives you each term of the polynomial?) or division (what is each term of the polynomial divided by the GCF?).

11. Look for a GCF other than 1 or −1; if you have a four-term polynomial.

Exercise Set 5.5

1. a^8, a^5, and a^3; GCF = a^3

3. $x^2y^3z^3, y^2z^3$, and xy^2z^2; GCF = y^2z^2

5. $6x^3y, 9x^2y^2$, and $12x^2y$; GCF = $3x^2y$

7. $10x^3yz^3, 20x^2z^5, 45xz^3$; GCF = $5xz^3$

9. $18x - 12 = 6 \cdot 3x - 6 \cdot 2 = 6(3x - 2)$

11. $4y^2 - 16xy^3 = 4y^2 \cdot 1 - 4y^2 \cdot 4xy$
$= 4y^2(1 - 4xy)$

13. $6x^5 - 8x^4 + 2x^3 = 2x^3(3x^2) - 2x^3(4x) + 2x^3(1)$
$= 2x^3(3x^2 - 4x + 1)$

15. $8a^3b^3 - 4a^2b^2 + 4ab + 16ab^2$
$= 4ab(2a^2b^2) - 4ab(ab) + 4ab(1) + 4ab(4b)$
$= 4ab(2a^2b^2 - ab + 1 + 4b)$

17. $6(x+3) + 5a(x+3) = (x+3)(6+5a)$

19. $2x(z+7) + (z+7) = (z+7)(2x+1)$

21. $3x(6x^2+5) - 2(6x^2+5) = (6x^2+5)(3x-2)$

23. $20a^3 + 5a^2 + 1$
There is no common factor other than 1.

25. $39x^3y^3 - 26x^2y^3 = 13x^2y^3 \cdot 3x - 13x^2y^3 \cdot 2$
$= 13x^2y^3(3x-2)$

27. $ab + 3a + 2b + 6 = a(b+3) + 2(b+3)$
$= (a+2)(b+3)$

29. $ac + 4a - 2c - 8 = a(c+4) - 2(c+4)$
$= (a-2)(c+4)$

31. $2xy - 3x - 4y + 6 = x(2y-3) - 2(2y-3)$
$= (x-2)(2y-3)$

33. $12xy - 8x - 3y + 2 = 4x(3y-2) - (3y-2)$
$= (4x-1)(3y-2)$

35. $6x^3 + 9 = 3 \cdot 2x^3 + 3 \cdot 3 = 3(2x^3+3)$

37. $x^3 + 3x^2 = x^2 \cdot x + x^2 \cdot 3 = x^2(x+3)$

39. $8a^3 - 4a = 4a \cdot 2a^2 - 4a \cdot 1 = 4a(2a^2 - 1)$

41. $8m^3 + 4m^2 + 1$
There is no common factor other than 1.

43. $-20x^2y + 16xy^3 = 4xy \cdot (-5x) + 4xy \cdot 4y^2$
$= 4xy(-5x + 4y^2)$
or
$-20x^2y + 16xy^3 = -4xy \cdot 5x + (-4xy)(-4y^2)$
$= -4xy(5x - 4y^2)$

45. $10a^2b^3 + 5ab^2 - 15ab^3$
$= 5ab^2 \cdot 2ab + 5ab^2 \cdot 1 - 5ab^2 \cdot 3b$
$= 5ab^2(2ab + 1 - 3b)$

47. $9abc^2 + 6a^2bc - 6ab + 3bc$
$= 3b \cdot 3ac^2 + 3b \cdot 2a^2c + 3b \cdot (-2a) + 3b \cdot c$
$= 3b(3ac^2 + 2a^2c - 2a + c)$

49. $4x(y-2) - 3(y-2) = (y-2)(4x-3)$

51. $6xy + 10x + 9y + 15 = (6xy + 10x) + (9y + 15)$
$= 2x(3y+5) + 3(3y+5)$
$= (3y+5)(2x+3)$

53. $xy + 3y - 5x - 15 = (xy + 3y) + (-5x - 15)$
$= y(x+3) - 5(x+3)$
$= (x+3)(y-5)$

55. $6ab - 2a - 9b + 3 = (6ab - 2a) + (-9b + 3)$
$= 2a(3b-1) + (-3)(3b-1)$
$= (3b-1)(2a-3)$

57. $12xy + 18x + 2y + 3 = (12xy + 18x) + (2y + 3)$
$= 6x(2y+3) + 1(2y+3)$
$= (2y+3)(6x+1)$

59. $2m(n-8) - (n-8) = 2m(n-8) - 1(n-8)$
$= (n-8)(2m-1)$

61. $15x^3y^2 - 18x^2y^2 = 3x^2y^2 \cdot 5x - 3x^2y^2 \cdot 6$
$= 3x^2y^2(5x-6)$

63. $2x^2 + 3xy + 4x + 6y = (2x^2 + 3xy) + (4x + 6y)$
$= x(2x+3y) + 2(2x+3y)$
$= (2x+3y)(x+2)$

65. $5x^2 + 5xy - 3x - 3y = 5x(x+y) - 3(x+y)$
$= (x+y)(5x-3)$

67. $x^3 + 3x^2 + 4x + 12 = (x^3 + 3x^2) + (4x + 12)$
$= x^2(x+3) + 4(x+3)$
$= (x+3)(x^2+4)$

69. $x^3 - x^2 - 2x + 2 = (x^3 - x^2) + (-2x + 2)$
$= x^2(x-1) - 2(x-1)$
$= (x-1)(x^2-2)$

71. $(5x^2)(11x^5) = 5(11)x^2x^5 = 55x^7$

73. $(5x^2)^3 = 5^3(x^2)^3 = 125x^6$

75. $(x+2)(x-5) = x^2 - 5x + 2x - 10$
$= x^2 - 3x - 10$

77. $(x+3)(x+2) = x^2 + 3x + 2x + 6$
$= x^2 + 5x + 6$

79. $(y-3)(y-1) = y^2 - 1y - 3y + 3$
$= y^2 - 4y + 3$

81. d

a. $2(5x^2 - x + 1) = 10x^2 - 2x + 2$

b. $2(5x^2 - x) = 10x^2 - 2x$

c. $2(5x^2 - x - 2) = 10x^2 - 2x - 4$

d. $2(5x^2 - x - 1) = 10x^2 - 2x - 2$

83. $2\pi r^2 + 2\pi rh = 2\pi r(r + h)$

85. $A = 5600 + 5600RT$
$A = 5600(1) + 5600(RT)$
$A = 5600(1 + RT)$

87. answers may vary

89. a. $(2-x)(3-y) = 6 - 2y - 3x + xy$
$= xy - 3x - 2y + 6$

b. $(-2+x)(-3+y) = 6 - 2y - 3x + xy$
$= xy - 3x - 2y + 6$

c. $(y-3)(x-2) = yx - 2y - 3x + 6$
$= xy - 3x - 2y + 6$

d. $(-x+2)(-y+3) = xy - 3x - 2y + 6$

None; all are factored forms of $xy - 3x - 2y + 6$.

91. answers may vary

93. $IR_1 + IR_2 = E$
$I(R_1 + R_2) = E$

95. $x^{3n} - 2x^{2n} + 5x^n$
$= x^n \cdot x^{2n} - x^n \cdot 2x^n + x^n \cdot 5$
$= x^n(x^{2n} - 2x^n + 5)$

97. $6x^{8a} - 2x^{5a} - 4x^{3a} = 2x^{3a}(3x^{5a} - x^{2a} - 2)$

99. $h(t) = -16t^2 + 64t$

a. $h(t) = -16t(t - 4)$

b. $h(1) = -16(1)^2 + 64(1) = 48$
$h(1) = -16(1)(1 - 4) = 48$ feet

c. answers may vary

101. The greatest common factor is 4.
$f(x) = 64x^3 - 744x^2 + 13,452x + 75,784$
$f(x) = 4(16x^3 - 186x^2 + 3363x + 18,946)$

Section 5.6 Practice Exercises

1. Find two integers whose product is 6 and whose sum is 5. Since our integers must have a positive product and a positive sum, look for positive factors.

Positive Factors of 6	Sum of Factors
1, 6	1 + 6 = 7
3, 2	3 + 2 = 5 (correct)

$x^2 + 5x + 6 = (x + 3)(x + 2)$

2. Find two integers whose product is 24 and whose sum is −11. Since our integers must have a positive product and a negative sum, look for negative factors.

Negative Factors of 24	Sum of Factors
−1, −24	−1 + (−24) = −25
−2, −12	−2 + (−12) = −14
−3, −8	−3 + (−8) = −11 (correct)
−4, −6	−4 + (−6) = −10

$x^2 - 11x + 24 = (x - 3)(x - 8)$

3. $3x^3 - 9x^2 - 30x = 3x(x^2 - 3x - 10)$
Find two integers whose product is −10 and whose sum is −3. The numbers are −5 and 2.
$3x^3 - 9x^2 - 30x = 3x(x^2 - 3x - 10)$
$= 3x(x - 5)(x + 2)$

4. $2b^2 - 18b - 22 = 2(b^2 - 9b - 11)$
Find two integers whose product is −11 and whose sum is −9.

Factors	Sum
−1, 11	10
1, −11	−10

Neither of the pairs has a sum of −9, so no further factoring is possible.
$2b^2 - 18b - 22 = 2(b^2 - 9b - 11)$

5. Factors of $2x^2$: $2x \cdot x$
Factors of 6: $1 \cdot 6$ and $2 \cdot 3$
$(2x + 6)(x + 1) \Rightarrow 2x + 6x = 8x$ (incorrect middle term)
$(2x + 1)(x + 6) \Rightarrow 12x + x = 13x$ (correct middle term)
$(2x + 2)(x + 3) \Rightarrow 6x + 2x = 8x$ (incorrect middle term)
$(2x + 3)(x + 2) \Rightarrow 4x + 3x = 7x$ (incorrect middle term)
$2x^2 + 13x + 6 = (2x + 1)(x + 6)$

6. Factors of $4x^2$: $4x \cdot x$ and $2x \cdot 2x$
Factors of −6: $-6 \cdot 1, 6 \cdot -1, 2 \cdot -3, -2 \cdot 3$
$(4x - 6)(x + 1) \Rightarrow 4x - 6x = -2x$ (incorrect)
$(4x + 6)(x - 1) \Rightarrow -4x + 6x = 2x$ (incorrect)
$(4x + 2)(x - 3) \Rightarrow -12x + 2x = -10x$ (incorrect)
$(4x - 3)(x + 2) \Rightarrow 8x - 3x = 5x$ (correct)
$4x^2 + 5x - 6 = (4x - 3)(x + 2)$

7. $18b^4 - 57b^3 + 30b^2 = 3b^2(6b^2 - 19b + 10)$
Factors of $6b^2$: $2b \cdot 3b, 6b \cdot b$
Negative factors of 10: $-1 \cdot -10, -5 \cdot -2$
$(2b - 1)(3b - 10) \Rightarrow -20b - 3b = -23b$ (incorrect)
$(2b - 10)(3b - 1) \Rightarrow -2b - 30b = -32b$ (incorrect)
$(2b - 5)(3b - 2) \Rightarrow -4b - 15b = -19b$ (correct)

$$\begin{aligned} 18b^4 - 57b^3 + 30b^2 &= 3b^2(6b^2 - 19b + 10) \\ &= 3b^2(2b - 5)(3b - 2) \end{aligned}$$

8. No greatest common factor can be factored out.
Factors of $25x^2$: $25x \cdot x, 5x \cdot 5x$
Factors of $4y^2$: $4y \cdot y, 2y \cdot 2y$
Try possible combinations.

$$\begin{aligned} 25x^2 + 20xy + 4y^2 &= (5x + 2y)(5x + 2y) \\ &= (5x + 2y)^2 \end{aligned}$$

9. $20x^2 + 23x + 6$
$a = 20, b = 23, c = 6$
Find two numbers whose product is $a \cdot c = 20 \cdot 6 = 120$, and whose sum is b, 23. The two numbers are 8 and 15.

$$\begin{aligned} 20x^2 + 23x + 6 &= 20x^2 + 8x + 15x + 6 \\ &= 4x(5x + 2) + 3(5x + 2) \\ &= (5x + 2)(4x + 3) \end{aligned}$$

10. $15x^2 + 4x - 3$
$a = 15, b = 4, c = -3$
Find two numbers whose product is $a \cdot c = 15(-3) = -45$ and whose sum is b, 4. The two numbers are 9 and −5.

$$\begin{aligned} 15x^2 + 4x - 3 &= 15x^2 + 9x - 5x - 3 \\ &= 3x(5x + 3) - 1(5x + 3) \\ &= (5x + 3)(3x - 1) \end{aligned}$$

11. Let $y = x + 1$.

$$\begin{aligned} 3(x + 1)^2 - 7(x + 1) - 20 &= 3y^2 - 7y - 20 \\ &= (3y + 5)(y - 4) \end{aligned}$$

Replace y with $x + 1$.

$$\begin{aligned} (3y + 5)(y - 4) &= [3(x + 1) + 5][(x + 1) - 4)] \\ &= (3x + 3 + 5)(x + 1 - 4) \\ &= (3x + 8)(x - 3) \end{aligned}$$

Thus, $3(x + 1)^2 - 7(x + 1) - 20 = (3x + 8)(x - 3)$.

12. Let $y = x^2$.

$$\begin{aligned} 6x^4 - 11x^2 - 10 &= 6y^2 - 11y - 10 \\ &= (3y + 2)(2y - 5) \end{aligned}$$

Replace y with x^2.
$6x^4 - 11x^2 - 10 = (3x^2 + 2)(2x^2 - 5)$

Vocabulary, Readiness & Video Check 5.6

1. $10 = 2 \cdot 5$
$7 = 2 + 5$
2 and 5

2. $12 = 2 \cdot 2 \cdot 3 = 2 \cdot 6$
$8 = 2 + 6$
2 and 6

3. $24 = 2 \cdot 2 \cdot 2 \cdot 3 = 8 \cdot 3$
$11 = 8 + 3$
8 and 3

4. $30 = 2 \cdot 3 \cdot 5 = 10 \cdot 3$
$13 = 10 + 3$
10 and 3

5. Check by multiplying. If you get a middle term of $2x$, not $-2x$, switch the signs of your factors of -24.

6. 8 is positive, and since the factors need to sum to -6, both factors must be negative.

7. Write down the factors of the first and last terms. Try various combinations of these factors and look at the sum of the outer and inner products to see if you get the middle term of the trinomial. If not, try another combination of factors.

8. This gives us a four-term polynomial, which may be factored by grouping, if it can be factored.

9. Our factoring involves a substitution from the original expression, so we need to substitute back in order to get a factorization in terms of the original variable/expression, and simplify if necessary.

Exercise Set 5.6

1. $x^2 + 9x + 18 = (x + 6)(x + 3)$

3. $x^2 - 12x + 32 = (x - 4)(x - 8)$

5. $x^2 + 10x - 24 = (x + 12)(x - 2)$

7. $x^2 - 2x - 24 = (x - 6)(x + 4)$

9. Note that the GCF is 3.

$$3x^2 - 18x + 24 = 3(x^2 - 6x + 8) = 3(x - 2)(x - 4)$$

11. Note that the GCF is $4z$.

$$4x^2z + 28xz + 40z = 4z(x^2 + 7x + 10) = 4z(x + 2)(x + 5)$$

13. Note that the GCF is 2.

$$2x^2 - 24x - 64 = 2(x^2 - 12x - 32)$$

15. $5x^2 + 16x + 3 = (5x + 1)(x + 3)$

17. $2x^2 - 11x + 12 = (2x - 3)(x - 4)$

19. $2x^2 + 25x - 20$ is prime.

21. $4x^2 - 12x + 9 = (2x - 3)(2x - 3) = (2x - 3)^2$

23. Note that the GCF is 2.

$$12x^2 + 10x - 50 = 2(6x^2 + 5x - 25) = 2(3x - 5)(2x + 5)$$

25. Note that the GCF is y^2.

$$3y^4 - y^3 - 10y^2 = y^2(3y^2 - y - 10) = y^2(3y + 5)(y - 2)$$

27. Note that the GCF is $2x$.

$$6x^3 + 8x^2 + 24x = 2x(3x^2 + 4x + 12)$$

29. $2x^2 - 5xy - 3y^2 = (2x + y)(x - 3y)$

31. Note that the GCF is 2, so that $28y^2 + 22y + 4 = 2(14y^2 + 11y + 2)$.
$ac = 28$; the two numbers are 4 and 7.

$$\begin{aligned} 14y^2 + 11y + 2 &= 14y^2 + 7y + 4y + 2 \\ &= 7y(2y + 1) + 2(2y + 1) \\ &= (7y + 2)(2y + 1) \end{aligned}$$

So, $28y^2 + 22y + 4 = 2(7y + 2)(2y + 1)$.

33. $2x^2 + 15x - 27$; $ac = -54$ so the two numbers are 18 and -3.

$$\begin{aligned} 2x^2 + 15x - 27 &= 2x^2 + 18x - 3x - 27 \\ &= 2x(x + 9) - 3(x + 9) \\ &= (2x - 3)(x + 9) \end{aligned}$$

35. Let $y = x^2$. Then we have

$$x^4 + x^2 - 6 = y^2 + y - 6 = (y + 3)(y - 2).$$

This yields $(x^2 + 3)(x^2 - 2)$.

37. Let $y = 5x + 1$. Then we have

$$(5x+1)^2 + 8(5x+1) + 7 = y^2 + 8y + 7 = (y+1)(y+7).$$

This yields

$[(5x+1)+1][(5x+1)+7] = (5x+2)(5x+8).$

39. Let $y = x^3$. Then we have

$$x^6 - 7x^3 + 12 = y^2 - 7y + 12 = (y-4)(y-3).$$

This yields $(x^3 - 4)(x^3 - 3)$.

41. Let $y = a + 5$. Then we have

$$(a+5)^2 - 5(a+5) - 24 = y^2 - 5y - 24 = (y-8)(y+3).$$

This yields

$[(a+5)-8][(a+5)+3] = (a-3)(a+8).$

43. $x^2 - 24x - 81 = (x-27)(x+3)$

45. $x^2 - 15x - 54 = (x-18)(x+3)$

47. $3x^2 - 6x + 3 = 3(x^2 - 2x + 1) = 3(x-1)(x-1) = 3(x-1)^2$

49. $3x^2 - 5x - 2 = (3x+1)(x-2)$

51. $8x^2 - 26x + 15 = (4x-3)(2x-5)$

53. $18x^4 + 21x^3 + 6x^2 = 3x^2(6x^2 + 7x + 2) = 3x^2(3x+2)(2x+1)$

55. $x^2 + 8xz + 7z^2 = (x+z)(x+7z)$

57. $x^2 - x - 12$; $ac = -12$ so the two numbers are -4 and 3.

$$x^2 - x - 12 = x^2 - 4x + 3x - 12 = x(x-4) + 3(x-4) = (x+3)(x-4)$$

59. $3a^2 + 12ab + 12b^2 = 3(a^2 + 4ab + 4b^2) = 3(a+2b)(a+2b) = 3(a+2b)^2$

61. $x^2 + 4x + 5$ is prime.

63. Let $y = x + 4$. Then

$$2(x+4)^2 + 3(x+4) - 5 = 2y^2 + 3y - 5 = (2y+5)(y-1) = [2(x+4)+5][(x+4)-1] = (2x+8+5)(x+3) = (2x+13)(x+3)$$

65. $6x^2 - 49x + 30 = (3x-2)(2x-15)$

67. Let $y = x^2$. Then

$$x^4 - 5x^2 - 6 = y^2 - 5y - 6 = (y-6)(y+1) = (x^2 - 6)(x^2 + 1)$$

69. $6x^3 - x^2 - x = x(6x^2 - x - 1) = x(3x+1)(2x-1)$

71. $12a^2 - 29ab + 15b^2 = (4a - 3b)(3a - 5b)$

73. $9x^2 + 30x + 25 = (3x+5)(3x+5) = (3x+5)^2$

75. $3x^2y - 11xy + 8y = y(3x^2 - 11x + 8) = y(3x-8)(x-1)$

77. $2x^2 + 2x - 12 = 2(x^2 + x - 6) = 2(x+3)(x-2)$

79. Let $y = x - 4$. Then

$$(x-4)^2 + 3(x-4) - 18 = y^2 + 3y - 18 = (y+6)(y-3) = [(x-4)+6][(x-4)-3] = (x+2)(x-7)$$

81. Let $y = x^3$. Then

$$2x^6 + 3x^3 - 9 = 2y^2 + 3y - 9 = (2y-3)(y+3) = (2x^3 - 3)(x^3 + 3)$$

83. $72xy^4 - 24xy^2z + 2xz^2 = 2x(36y^4 - 12y^2z + z^2) = 2x(6y^2 - z)(6y^2 - z) = 2x(6y^2 - z)^2$

85. $2x^3y + 2x^2y - 12xy = 2xy(x^2 + x - 6)$
$= 2xy(x+3)(x-2)$

87. $x^2 + 6xy + 5y^2 = (x+5y)(x+y)$

89. $(x-3)(x+3) = x^2 - 3^3 = x^2 - 9$

91. $(2x+1)^2 = (2x)^2 + 2(2x)(1) + 1^2$
$= 4x^2 + 4x + 1$

93.
$$\begin{array}{r} x^2 + 2x + 4 \\ x - 2 \\ \hline -2x^2 - 4x - 8 \\ x^3 + 2x^2 + 4x \quad \\ \hline x^3 \qquad\qquad -8 \end{array}$$

95. $x^2 + bx + 6$
$6 = 2 \cdot 3$ or $6 = (-2)(-3)$
$6 = 1 \cdot 6$ or $6 = (-1)(-6)$
$(x+2)(x+3) = x^2 + 5x + 6$
$(x-2)(x-3) = x^2 - 5x + 6$
$(x+1)(x+6) = x^2 + 7x + 6$
$(x-1)(x-6) = x^2 - 7x + 6$
$b = \pm 5$ and $b = \pm 7$

97. $V(x) = x^3 + 2x^2 - 8x$
$= x(x^2 + 2x - 8)$
$= x(x+4)(x-2)$

99. $h(t) = -16t^2 + 80t + 576$

a. $h(0) = -16(0)^2 + 80(0) + 576 = 576$ ft
$h(2) = -16(2)^2 + 80(2) + 576$
$= -16(4) + 160 + 576$
$= -64 + 160 + 576$
$= 672$ ft
$h(4) = -16(4)^2 + 80(4) + 576$
$= -16(16) + 320 + 576$
$= -256 + 320 + 576$
$= 640$ ft
$h(6) = -16(6)^2 + 80(6) + 576$
$= -16(36) + 480 + 576$
$= -576 + 480 + 576$
$= 480$ ft

b. answers may vary

c. $h(t) = -16t^2 + 80t + 576$
$= -16(t^2 - 5t - 36)$
$= -16(t-9)(t+4)$

101. $x^{2n} + 10x^n + 16 = (x^n + 2)(x^n + 8)$

103. $x^{2n} - 3x^n - 18 = (x^n - 6)(x^n + 3)$

105. $2x^{2n} + 11x^n + 5 = (2x^n + 1)(x^n + 5)$

107. $4x^{2n} - 12x^n + 9 = (2x^n - 3)(2x^n - 3)$
$= (2x^n - 3)^2$

109. $x^4 + 6x^3 + 5x^2 = x^2(x^2 + 6x + 5)$
$= x^2(x+5)(x+1)$

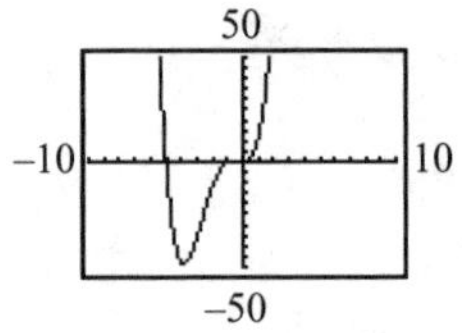

111. $30x^3 + 9x^2 - 3x = 3x(10x^2 + 3x - 1)$
$= 3x(5x-1)(2x+1)$

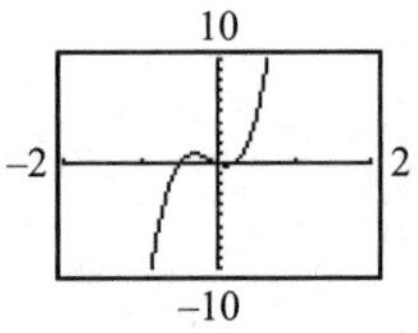

Section 5.7 Practice Exercises

1. $b^2 + 16b + 64 = b^2 + 2(b)(8) + 8^2 = (b+8)^2$

2. $45x^2b - 30xb + 5b = 5b(9x^2 - 6x + 1)$
$= 5b[(3x)^2 - 2(3x)(1) + 1^2]$
$= 5b(3x-1)^2$

3. a. $x^2 - 16 = x^2 - 4^2 = (x+4)(x-4)$

b. $25b^2 - 49 = (5b)^2 - 7^2 = (5b-7)(5b+7)$

c. $45 - 20x^2 = 5(9 - 4x^2)$
$= 5[3^2 - (2x)^2]$
$= 5(3-2x)(3+2x)$

d. $y^2 - \frac{1}{81} = y^2 - \left(\frac{1}{9}\right)^2 = \left(y - \frac{1}{9}\right)\left(y + \frac{1}{9}\right)$

4. a. $x^4 - 10,000 = (x^2)^2 - 100^2$
$= (x^2 + 100)(x^2 - 100)$
$= (x^2 + 100)(x + 10)(x - 10)$

b. $(x+2)^2 - 49 = (x+2)^2 - 7^2$
$= [(x+2)+7][(x+2)-7]$
$= (x+2+7)(x+2-7)$
$= (x+9)(x-5)$

5. $m^2 + 6m + 9 - n^2 = (m^2 + 6m + 9) - n^2$
$= (m+3)^2 - n^2$
$= [(m+3)+n][(m+3)-n]$
$= (m+3+n)(m+3-n)$

6. $x^3 + 64 = x^3 + 4^3$
$= (x+4)(x^2 - x \cdot 4 + 4^2)$
$= (x+4)(x^2 - 4x + 16)$

7. $a^3 + 8b^3 = a^3 + (2b)^3$
$= (a+2b)[a^2 - a(2b) + (2b)^2]$
$= (a+2b)(a^2 - 2ab + 4b^2)$

8. $27 - y^3 = 3^3 - y^3$
$= (3-y)(3^2 + 3 \cdot y + y^2)$
$= (3-y)(9 + 3y + y^2)$

9. $b^3x^2 - 8x^2 = x^2(b^3 - 8)$
$= x^2(b^3 - 2^3)$
$= x^2(b-2)(b^2 + b \cdot 2 + 2^2)$
$= x^2(b-2)(b^2 + 2b + 4)$

Vocabulary, Readiness & Video Check 5.7

1. $81y^2 = (9y)^2$

2. $4z^2 = (2z)^2$

3. $64x^6 = (8x^3)^2$

4. $49y^6 = (7y^3)^2$

5. $8x^3 = (2x)^3$

6. $27y^3 = (3y)^3$

7. $64y^6 = (4x^2)^3$

8. $x^3y^6 = (xy^2)^3$

9. See if the first term is a square, say a^2, and the last term is a square, say b^2. Check to see if the middle term is $2 \cdot a \cdot b$ or $-2 \cdot a \cdot b$.

10. In order to help you see that the binomial is a difference of squares and also to identify the terms to use in the special factoring formula.

11. First rewrite the original binomial so that each term is some quantity cubed. Your answers will then vary, depending on your interpretation.

Exercise Set 5.7

1. $x^2 + 6x + 9 = (x)^2 + 2 \cdot x \cdot 3 + (3)^2 = (x+3)^2$

3. $4x^2 - 12x + 9 = (2x)^2 - 2 \cdot 2x \cdot 3 + 3^2$
$= (2x-3)^2$

5. $25x^2 + 10x + 1 = (5x)^2 + 2 \cdot 5x \cdot 1 + 1^2 = (5x+1)^2$

7. $3x^2 - 24x + 48 = 3(x^2 - 8x + 16)$
$= 3(x^2 - 2 \cdot x \cdot 4 + 4^2)$
$= 3(x-4)^2$

9. $9y^2x^2 + 12yx^2 + 4x^2$
$= x^2(9y^2 + 12y + 4)$
$= x^2[(3y)^2 + 2 \cdot 3y \cdot 2 + (2)^2]$
$= x^2(3y+2)^2$

11. $16x^2 - 56xy + 49y^2 = (4x)^2 - 2 \cdot 4x \cdot 7y + (7y)^2$
$= (4x - 7y)^2$

13. $x^2 - 25 = (x)^2 - (5)^2 = (x+5)(x-5)$

15. $\frac{1}{9} - 4z^2 = \left(\frac{1}{3}\right)^2 - (2z)^2 = \left(\frac{1}{3} + 2z\right)\left(\frac{1}{3} - 2z\right)$

17. $(y+2)^2-49=(y+2)^2-(7)^2$
$=[(y+2)+7][(y+2)-7]$
$=(y+9)(y-5)$

19. $64x^2-100=4(16x^2-25)$
$=4[(4x)^2-5^2]$
$=4(4x+5)(4x-5)$

21. $(x+2y)^2-9=(x+2y)^2-(3)^2$
$=[(x+2y)+3][(x+2y)-3]$
$=(x+2y+3)(x+2y-3)$

23. $x^2+6x+9-y^2=(x+3)^2-y^2$
$=(x+3+y)(x+3-y)$

25. $x^2+16x+64-x^4=(x^2+16x+64)-x^4$
$=(x^2+2\cdot x\cdot 8+8^2)-x^4$
$=(x+8)^2-(x^2)^2$
$=[(x+8)+x^2][(x+8)-x^2]$
$=(x+8+x^2)(x+8-x^2)$

27. $x^3+27=x^3+(3)^3=(x+3)(x^2-3x+9)$

29. $z^3-1=z^3-1^3=(z-1)(z^2+z+1)$

31. $m^3+n^3=(m+n)(m^2-mn+n^2)$

33. $27y^2-x^3y^2=y^2(27-x^3)$
$=y^2(3^3-x^3)$
$=y^2(3-x)(3^2+3\cdot x+x^2)$
$=y^2(3-x)(9+3x+x^2)$

35. $8ab^3+27a^4=a(8b^3+27a^3)$
$=a[(2b)^3+(3a)^3]$
$=a(2b+3a)[(2b)^2-2b\cdot 3a+(3a)^2]$
$=a(2b+3a)(4b^2-6ab+9a^2)$

37. $250y^3-16x^3$
$=2(125y^3-8x^3)$
$=2[(5y)^3-(2x)^3]$
$=2(5y-2x)[(5y)^2+5y\cdot 2x+(2x)^2]$
$=2(5y-2x)(25y^2+10xy+4x^2)$

39. $x^2-12x+36=x^2-2\cdot x\cdot 6+6^2=(x-6)^2$

41. $18x^2y-2y=2y(9x^2-1)$
$=2y[(3x)^2-1^2]$
$=2y(3x+1)(3x-1)$

43. $9x^2-49=(3x)^2-7^2=(3x+7)(3x-7)$

45. $x^4-1=(x^2)^2-1^2$
$=(x^2+1)(x^2-1)$
$=(x^2+1)(x+1)(x-1)$

47. $x^6-y^3=(x^2)^3-y^3$
$=(x^2-y)[(x^2)^2+x^2y+y^2]$
$=(x^2-y)(x^4+x^2y+y^2)$

49. $8x^3+27y^3$
$=(2x)^3+(3y)^3$
$=(2x+3y)[(2x)^2-2x\cdot 3y+(3y)^2]$
$=(2x+3y)(4x^2-6xy+9y^2)$

51. $4x^2+4x+1-z^2=(4x^2+4x+1)-z^2$
$=[(2x)^2+2\cdot 2x\cdot 1+1^2]-z^2$
$=(2x+1)^2-z^2$
$=(2x+1+z)(2x+1-z)$

53. $3x^6y^2+81y^2=3y^2(x^6+27)$
$=3y^2[(x^2)^3+3^3]$
$=3y^2(x^2+3)[(x^2)^2-x^2\cdot 3+3^2]$
$=3y^2(x^2+3)(x^4-3x^2+9)$

55. $n^3-\frac{1}{27}=n^3-\left(\frac{1}{3}\right)^3$
$=\left(n-\frac{1}{3}\right)\left[n^2+n\cdot\frac{1}{3}+\left(\frac{1}{3}\right)^2\right]$
$=\left(n-\frac{1}{3}\right)\left(n^2+\frac{1}{3}n+\frac{1}{9}\right)$

57. $-16y^2+64=-16(y^2-4)$
$=-16(y^2-2^2)$
$=-16(y+2)(y-2)$

59. $x^2 - 10x + 25 - y^2 = (x^2 - 10x + 25) - y^2$
$= (x^2 - 2 \cdot x \cdot 5 + 5^2) - y^2$
$= (x-5)^2 - y^2$
$= (x - 5 + y)(x - 5 - y)$

61. $a^3b^3 + 125 = (ab)^3 + 5^3$
$= (ab+5)[(ab)^2 - ab \cdot 5 + 5^2]$
$= (ab+5)(a^2b^2 - 5ab + 25)$

63. $\frac{x^2}{25} - \frac{y^2}{9} = \left(\frac{x}{5}\right)^2 - \left(\frac{y}{3}\right)^2 = \left(\frac{x}{5} + \frac{y}{3}\right)\left(\frac{x}{5} - \frac{y}{3}\right)$

65. $(x+y)^3 + 125$
$= (x+y)^3 + 5^3$
$= [(x+y)+5][(x+y)^2 - (x+y) \cdot 5 + 5^2]$
$= (x+y+5)(x^2 + 2xy + y^2 - 5x - 5y + 25)$

67. $x - 5 = 0$
$x = 5$

69. $3x + 1 = 0$
$3x = -1$
$x = -\frac{1}{3}$

71. $-2x = 0$
$x = 0$

73. $-5x + 25 = 0$
$-5x = -25$
$x = 5$

75. No; $x^2 - 4$ can be factored further.
$5x(x^2 - 4) = 5x(x^2 - 2^2) = 5x(x+2)(x-2)$

77. Yes; $7y(a^2 + a + 1)$ is factored completely.

79. Area $= \pi R^2 - \pi r^2$
$= \pi(R^2 - r^2)$
$= \pi(R+r)(R-r)$ sq units

81. Volume $= x^3 - y^2x$
$= x(x^2 - y^2)$
$= x(x+y)(x-y)$ cubic units

83. $\frac{1}{2} \cdot b = \frac{1}{2} \cdot 6 = 3$ so $c = 3^2 = 9$

85. $\frac{1}{2} \cdot b = \frac{1}{2}(-14) = -7$ so $c = (-7)^2 = 49$

87. $\frac{1}{2} \cdot c = \frac{c}{2}$ so $\left(\frac{c}{2}\right)^2 = 16$
$\frac{c^2}{4} = 16$
$c^2 = 64$
$c = \pm 8$

89. $x^6 - 1$

a. $(x^3)^2 - 1^2$
$= (x^3 + 1)(x^3 - 1)$
$= (x+1)(x^2 - x + 1)(x-1)(x^2 + x + 1)$

b. $(x^2)^3 - 1^3 = (x^2 - 1)(x^4 + x^2 + 1)$
$= (x+1)(x-1)(x^4 + x^2 + 1)$

c. answers may vary

91. $x^{2n} - 25 = (x^n)^2 - 5^2 = (x^n + 5)(x^n - 5)$

93. $36x^{2n} - 49 = (6x^n)^2 - 7^2 = (6x^n + 7)(6x^n - 7)$

95. $x^{4n} - 16 = (x^{2n})^2 - 4^2$
$= (x^{2n} + 4)(x^{2n} - 4)$
$= (x^{2n} + 4)[(x^n)^2 - 2^2]$
$= (x^{2n} + 4)(x^n + 2)(x^n - 2)$

Integrated Review Practice Exercises

1. a. $12x^2y - 3xy = 3xy(4x) + 3xy(-1)$
$= 3xy(4x - 1)$

b. $49x^2 - 4 = (7x)^2 - 2^2 = (7x+2)(7x-2)$

c. $5x^2 + 2x - 3 = (5x-3)(x+1)$

d. $3x^2 + 6 + x^3 + 2x = 3(x^2 + 2) + x(x^2 + 2)$
$= (x^2 + 2)(3 + x)$

e. $4x^2+20x+25=(2x)^2+2\cdot 2x\cdot 5+5^2$
$=(2x+5)^2$

f. b^2+100 cannot be factored.

2. a. $64x^3+y^3=(4x)^3+y^3$
$=(4x+y)[(4x)^2-4x\cdot y+y^2]$
$=(4x+y)(16x^2-4xy+y^2)$

b. $7x^2y^2-63y^4=7y^2(x^2-9y^2)$
$=7y^2[x^2-(3y)^2]$
$=7y^2(x-3y)(x+3y)$

c. $3x^2+12x+12-3b^2$
$=3(x^2+4x+4-b^2)$
$=3[(x+2)^2-b^2]$
$=3(x+2+b)(x+2-b)$

d. $x^5y^4+27x^2y$
$=x^2y(x^3y^3+27)$
$=x^2y[(xy)^3+3^3]$
$=x^2y(xy+3)(x^2y^2-3xy+9)$

e. $(x+7)^2-81y^2=(x+7)^2-(9y)^2$
$=(x+7+9y)(x+7-9y)$

Integrated Review Exercise Set

1. $(-y^2+6y-1)+(3y^2-4y-10)$
$=-y^2+6y-1+3y^2-4y-10$
$=2y^2+2y-11$

2. $(5z^4-6z^2+z+1)-(7z^4-2z+1)$
$=5z^4-6z^2+z+1-7z^4+2z-1$
$=-2z^4-6z^2+3z$

3. $(x^2-6x+2)-(x-5)=x^2-6x+2-x+5$
$=x^2-7x+7$

4. $(2x^2+6x-5)+(5x^2-10x)=7x^2-4x-5$

5. $(5x-3)^2=(5x)^2-2(5x)(3)+3^2$
$=25x^2-30x+9$

6. $(5x^2-14x-3)-(5x+1)=5x^2-14x-3-5x-1$
$=5x^2-19x-4$

7. $\frac{2x^4}{x}-\frac{3x^2}{x}+\frac{5x}{x}-\frac{2}{2}=2x^3-3x+5-1$
$=2x^3-3x+4$

8.
$$\begin{array}{r} x^2-3x-2 \\ \times \qquad 4x-1 \\ \hline -x^2+3x+2 \\ 4x^3-12x^2-8x \quad \\ \hline 4x^3-13x^2-5x+2 \end{array}$$

9. $x^2-8x+16-y^2=(x-4)^2-y^2$
$=(x-4+y)(x-4-y)$

10. $12x^2-22x-20=2(6x^2-11x-10)$
$=2(3x+2)(2x-5)$

11. $x^4-x=x(x^3-1)=x(x-1)(x^2+x+1)$

12. Let $y=2x+1$. Then
$(2x+1)^2-3(2x+1)+2$
$=y^2-3y+2$
$=(y-2)(y-1)$
$=[(2x+1)-2][(2x+1)-1]$
$=(2x-1)(2x)$
$=2x(2x-1)$

13. $14x^2y-2xy=2xy(7x-1)$

14. $24ab^2-6ab=6ab(4b-1)$

15. $4x^2-16=4(x^2-4)=4(x+2)(x-2)$

16. $9x^2-81=9(x^2-9)=9(x+3)(x-3)$

17. $3x^2-8x-11=(3x-11)(x+1)$

18. $5x^2-2x-3=(5x+3)(x-1)$

19. $4x^2+8x-12=4(x^2+2x-3)$
$=4(x+3)(x-1)$

20. $6x^2-6x-12=6(x^2-x-2)$
$=6(x-2)(x+1)$

21. $4x^2+36x+81=(2x)^2+2\cdot 2x\cdot 9+9^2$
$=(2x+9)^2$

22. $25x^2+40x+16=(5x)^2+2\cdot 5x\cdot 4+4^2$
$=(5x+4)^2$

23. $8x^3+125y^3=(2x)^3+(5y)^3$
$=(2x+5y)(4x^2-10xy+25y^2)$

24. $27x^3-64y^3=(3x)^3-(4y)^3$
$=(3x-4y)(9x^2+12xy+16y^2)$

25. $64x^2y^3-8x^2=8x^2(8y^3-1)$
$=8x^2[(2y)^3-1^3]$
$=8x^2(2y-1)(4y^2+2y+1)$

26. $27x^5y^4-216x^2y$
$=27x^2y(x^3y^3-8)$
$=27x^2y[(xy)^3-2^3]$
$=27x^2y(xy-2)(x^2y^2+2xy+4)$

27. $(x+5)^3+y^3$
$=[(x+5)+y][(x+5)^2-(x+5)y+y^2]$
$=(x+y+5)(x^2+10x+25-xy-5y+y^2)$
$=(x+y+5)(x^2+10x-xy-5y+y^2+25)$

28. $(y-1)^3+27x^3$
$=(y-1)^3+(3x)^3$
$=[(y-1)+3x][(y-1)^2-(y-1)(3x)+(3x)^2]$
$=(y-1+3x)(y^2-2y+1-3xy+3x+9x^2)$

29. Let $y = 5a - 3$. Then
$(5a-3)^2-6(5a-3)+9=y^2-6y+9$
$=(y-3)(y-3)$
$=(y-3)^2$
$=[(5a-3)-3]^2$
$=(5a-6)^2$

30. Let $y = 4r + 1$. Then
$(4r+1)^2+8(4r+1)+16=y^2+8y+16$
$=(y+4)(y+4)$
$=(y+4)^2$
$=[(4r+1)+4]^2$
$=(4r+5)^2$

31. $7x^2-63x=7x(x-9)$

32. $20x^2+23x+6=(4x+3)(5x+2)$

33. $ab-6a+7b-42=a(b-6)+7(b-6)$
$=(a+7)(b-6)$

34. $20x^2-220x+600=20(x^2-11x+30)$
$=20(x-6)(x-5)$

35. $x^4-1=(x^2)^2-1^2$
$=(x^2+1)(x^2-1)$
$=(x^2+1)(x+1)(x-1)$

36. $15x^2-20x=5x(3x-4)$

37. $10x^2-7x-33=(5x-11)(2x+3)$

38. $45m^3n^3-27m^2n^2=9m^2n^2(5mn-3)$

39. $5a^3b^3-50a^3b=5a^3b(b^2-10)$

40. $x^4+x=x(x^3+1)$
$=x(x^3+1^3)$
$=x(x+1)(x^2-x+1)$

41. $16x^2+25$ is a prime polynomial.

42. $20x^3+20y^3=20(x^3+y^3)$
$=20(x+y)(x^2-xy+y^2)$

43. $10x^3-210x^2+1100x=10x(x^2-21x+110)$
$=10x(x-11)(x-10)$

44. $9y^2-42y+49=(3y)^2-2\cdot 3y\cdot 7+7^2$
$=(3y-7)^2$

45. $64a^3b^4 - 27a^3b$
$= a^3b(64b^3 - 27)$
$= a^3b[(4b)^3 - 3^3]$
$= a^3b(4b-3)(16b^2 + 12b + 9)$

46. $y^4 - 16 = (y^2)^2 - 4^2$
$= (y^2 + 4)(y^2 - 4)$
$= (y^2 + 4)(y+2)(y-2)$

47. $2x^3 - 54 = 2(x^3 - 27)$
$= 2(x^3 - 3^3)$
$= 2(x-3)(x^2 + 3x + 9)$

48. $2sr + 10s - r - 5 = 2s(r+5) - 1(r+5)$
$= (2s-1)(r+5)$

49. $3y^5 - 5y^4 + 6y - 10 = y^4(3y-5) + 2(3y-5)$
$= (y^4 + 2)(3y-5)$

50. $64a^2 + b^2$ is a prime polynomial.

51. $100z^3 + 100 = 100(z^3 + 1)$
$= 100(z+1)(z^2 - z + 1)$

52. $250x^4 - 16x = 2x(125x^3 - 8)$
$= 2x[(5x)^3 - 2^3]$
$= 2x(5x-2)(25x^2 + 10x + 4)$

53. $4b^2 - 36b + 81 = (2b)^2 - 2 \cdot 2b \cdot 9 + 9^2$
$= (2b-9)^2$

54. $2a^5 - a^4 + 6a - 3 = a^4(2a-1) + 3(2a-1)$
$= (a^4 + 3)(2a-1)$

55. Let $x = y - 6$. Then
$(y-6)^2 + 3(y-6) + 2 = x^2 + 3x + 2$
$= (x+2)(x+1)$
$= [(y-6)+2][(y-6)+1]$
$= (y-4)(y-5)$

56. Let $x = c + 2$. Then
$(c+2)^2 - 6(c+2) + 5 = x^2 - 6x + 5$
$= (x-5)(x-1)$
$= [(c+2)-5][(c+2)-1]$
$= (c-3)(c+1)$

57. Area $= 3^2 - 4x^2 = 3^2 - (2x)^2 = (3+2x)(3-2x)$

Section 5.8 Practice Exercises

1. $(x+8)(x-5) = 0$
$x + 8 = 0$ or $x - 5 = 0$
$x = -8$ or $x = 5$
Both solutions check. The solution set is $\{-8, 5\}$.

2. $3x^2 + 10x - 8 = 0$
$(3x-2)(x+4) = 0$
$3x - 2 = 0$ or $x + 4 = 0$
$3x = 2$
$x = \frac{2}{3}$ or $x = -4$
Both solutions check. The solution set is $\left\{-4, \frac{2}{3}\right\}$.

3. $x(3x+14) = -8$
$3x^2 + 14x = -8$
$3x^2 + 14x + 8 = 0$
$(3x+2)(x+4) = 0$
$3x + 2 = 0$ or $x + 4 = 0$
$3x = -2$
$x = -\frac{2}{3}$ or $x = -4$
The solutions are $-\frac{2}{3}$ and -4.

4. $8(x^2 + 3) + 4 = -8x(x+3) + 19$
$8x^2 + 24 + 4 = -8x^2 - 24x + 19$
$16x^2 + 24x + 9 = 0$
$(4x+3)(4x+3) = 0$
$4x + 3 = 0$ or $4x + 3 = 0$
$4x = -3$ or $4x = -3$
$x = -\frac{3}{4}$ or $x = -\frac{3}{4}$
The solution is $-\frac{3}{4}$.

5. $4x^2 = \frac{15}{2}x+1$

$2(4x^2) = 2\left(\frac{15}{2}x+1\right)$

$8x^2 = 15x+2$

$8x^2 - 15x - 2 = 0$

$(8x+1)(x-2) = 0$

$8x+1=0$ or $x-2=0$

$8x=-1$

$x=-\frac{1}{8}$ or $x=2$

The solutions are $-\frac{1}{8}$ and 2.

6. $x^3 = 2x^2 + 3x$

$x^3 - 2x^2 - 3x = 0$

$x(x^2 - 2x - 3) = 0$

$x(x-3)(x+1) = 0$

$x=0$ or $x-3=0$ or $x+1=0$

$x=3$ or $x=-1$

The solutions are 0, 3, and −1.

7. $x^3 - 9x = 18 - 2x^2$

$x^3 + 2x^2 - 9x - 18 = 0$

$x^2(x+2) - 9(x+2) = 0$

$(x+2)(x^2-9) = 0$

$(x+2)(x+3)(x-3) = 0$

$x+2=0$ or $x+3=0$ or $x-3=0$

$x=-2$ or $x=-3$ or $x=3$

The solutions are 3, −3, and −2.

8. Let $h = 0$.

$-16t^2 + 96t = 0$

$-16t(t-6) = 0$

$-16t = 0$ or $t-6=0$

$t=0$ or $t=6$

The rocket will return to the ground in 6 seconds.

9. Let x = first even integer, then $x + 2$ = next even integer, and $x + 4$ = third even integer.

$x^2 + (x+2)^2 = (x+4)^2$

$x^2 + x^2 + 4x + 4 = x^2 + 8x + 16$

$x^2 - 4x - 12 = 0$

$(x-6)(x+2) = 0$

$x-6=0$ or $x+2=0$

$x=6$ or $x=-2$

Discard $x = -2$ since length cannot be negative. The legs are $x = 6$ units, $x + 2 = 8$ units, and $x + 4 = 10$ units.

10. The graph of $f(x) = (x-1)(x+3)$ has two x-intercepts, (1, 0) and (−3, 0), so the graph is C. The graph of $g(x) = x(x+3)(x-2)$ has three x-intercepts, (0, 0), (−3, 0), and (2, 0), so the graph is A. The graph of $h(x) = (x-3)(x+2)(x-2)$ has three x-intercepts, (3, 0), (−2, 0), and (2, 0), so the graph is B.

Graphing Calculator Explorations

1. $y = x^2 + 3x - 2$

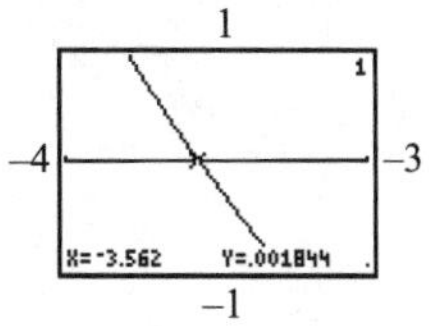

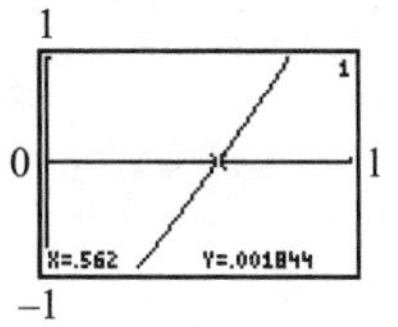

The intercepts are −3.562, 0.562.

2. $y = 5x^2 - 7x + 1$

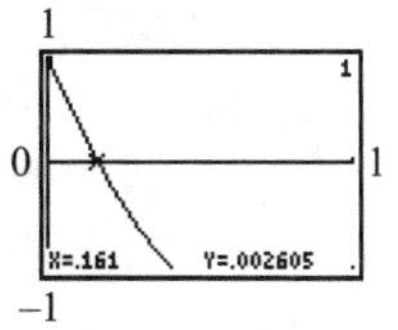

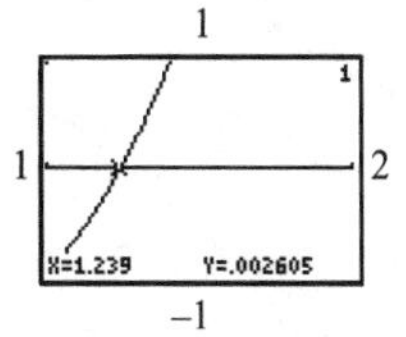

The intercepts are 0.161, 1.239.

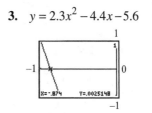

3. $y = 2.3x^2 - 4.4x - 5.6$

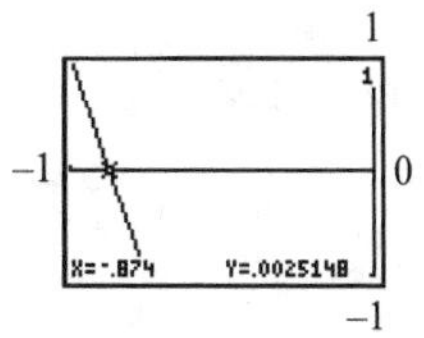

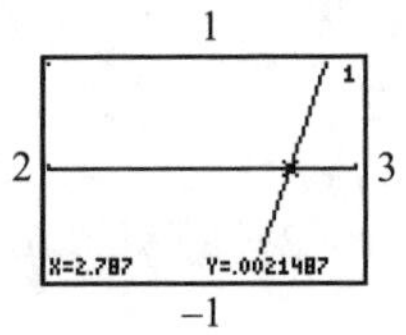

The intercepts are –0.874, 2.787.

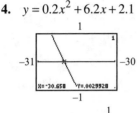

4. $y = 0.2x^2 + 6.2x + 2.1$

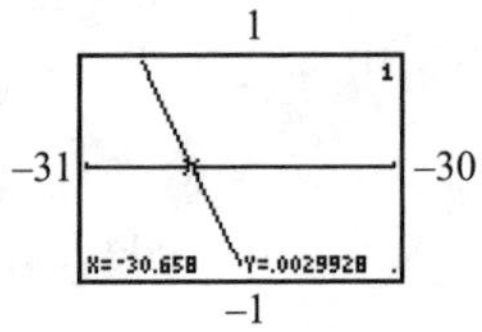

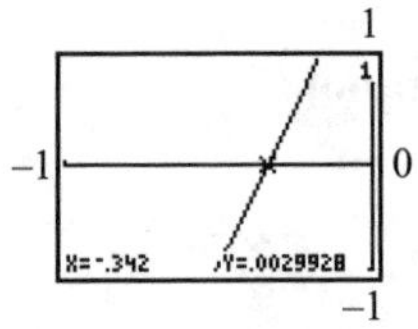

The intercepts are –30.658, –0.342.

5. $y = 0.09x^2 - 0.13x - 0.08$

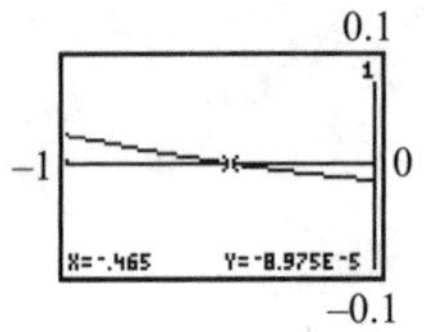

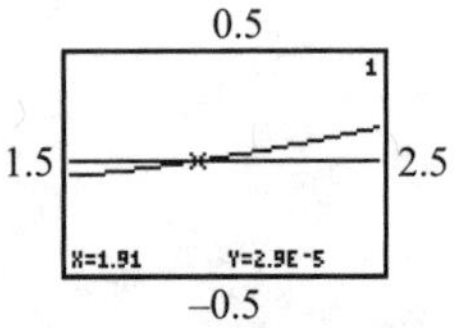

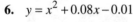

The intercepts are –0.465, 1.910.

6. $y = x^2 + 0.08x - 0.01$

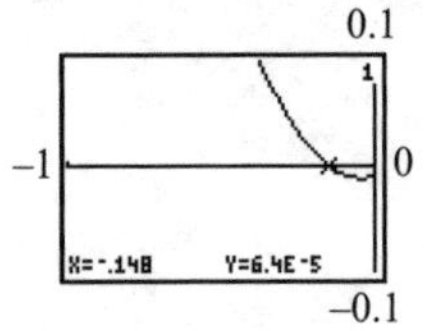

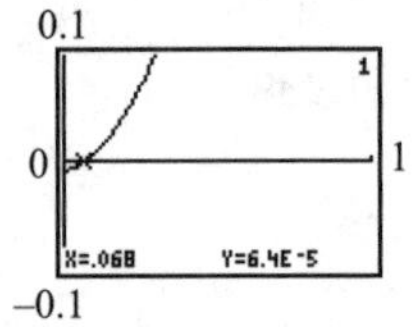

The intercepts are –0.148, 0.068.

Vocabulary, Readiness & Video Check 5.8

1. $(x-3)(x+5) = 0$
$x-3=0$ or $x+5=0$
$x=3$ or $x=-5$
The solutions are –5, 3.

2. $(y+5)(y+3) = 0$
$y+5=0$ or $y+3=0$
$y=-5$ or $y=-3$
The solutions are –5, –3.

3. $(z-3)(z+7) = 0$
$z-3=0$ or $z+7=0$
$z=3$ or $z=-7$
The solutions are –7, 3.

4. $(c-2)(c-4) = 0$
$c-2=0$ or $c-4=0$
$c=2$ or $c=4$
The solutions are 2, 4.

5. $x(x-9) = 0$
$x=0$ or $x-9=0$
$x=9$
The solutions are 0, 9.

6. $w(w+7) = 0$
$w=0$ or $w+7=0$
$w=-7$
The solutions are –7, 0.

7. One side of the equation must be a factored polynomial and the other side must be zero.

8. A negative solution makes no sense in the context of our application since we are looking for a positive length.

9. Finding the x-intercepts of any equation in two variables means you let $y = 0$ and solve for x. Doing this with a quadratic equation gives us an equation = 0, which may be solved by factoring.

Exercise Set 5.8

1. $(x+3)(3x-4)=0$
$x+3=0$ or $3x-4=0$
$x=-3$ or $3x=4$
$x=\frac{4}{3}$
The solutions are $-3, \frac{4}{3}$.

3. $3(2x-5)(4x+3)=0$
$2x-5=0$ or $4x+3=0$
$2x=5$ or $4x=-3$
$x=\frac{5}{2}$ or $x=-\frac{3}{4}$
The solutions are $-\frac{3}{4}, \frac{5}{2}$.

5. $x^2+11x+24=0$
$(x+8)(x+3)=0$
$x+8=0$ or $x+3=0$
$x=-8$ or $x=-3$
The solutions are –8, –3.

7. $12x^2+5x-2=0$
$(4x-1)(3x+2)=0$
$4x-1=0$ or $3x+2=0$
$4x=1$ or $3x=-2$
$x=\frac{1}{4}$ or $x=-\frac{2}{3}$
The solutions are $-\frac{2}{3}, \frac{1}{4}$.

9. $z^2+9=10z$
$z^2-10z+9=0$
$(z-9)(z-1)=0$
$z-9=0$ or $z-1=0$
$z=9$ or $z=1$
The solutions are 1, 9.

11. $x(5x+2)=3$
$5x^2+2x-3=0$
$(5x-3)(x+1)=0$
$5x-3=0$ or $x+1=0$
$5x=3$ or $x=-1$
$x=\frac{3}{5}$
The solutions are $-1, \frac{3}{5}$.

13. $x^2-6x=x(8+x)$
$x^2-6x=8x+x^2$
$-14x=0$
$x=0$
The solution is 0.

15. $\frac{z^2}{6}-\frac{z}{2}-3=0$
$z^2-3z-18=0$
$(z-6)(z+3)=0$
$z-6=0$ or $z+3=0$
$z=6$ or $z=-3$
The solutions are –3, 6.

17. $\frac{x^2}{2}+\frac{x}{20}=\frac{1}{10}$
$10x^2+x=2$
$10x^2+x-2=0$
$(5x-2)(2x+1)=0$
$5x-2=0$ or $2x+1=0$
$5x=2$ or $2x=-1$
$x=\frac{2}{5}$ or $x=-\frac{1}{2}$
The solutions are $-\frac{1}{2}, \frac{2}{5}$.

19. $\frac{4t^2}{5}=\frac{t}{5}+\frac{3}{10}$
$8t^2=2t+3$
$8t^2-2t-3=0$
$(4t-3)(2t+1)=0$
$4t-3=0$ or $2t+1=0$
$4t=3$ or $2t=-1$
$t=\frac{3}{4}$ or $t=-\frac{1}{2}$
The solutions are $-\frac{1}{2}, \frac{3}{4}$.

21. $(x+2)(x-7)(3x-8)=0$
$x+2=0$ or $x-7=0$ or $3x-8=0$
$x=-2$ or $x=7$ or $3x=8$
$x=\frac{8}{3}$
The solutions are $-2, 7, \frac{8}{3}$.

23. $(x^2-1)(6x+1)=0$
$(x+1)(x-1)(6x+1)=0$
$x+1=0$ or $x-1=0$ or $6x+1=0$
$x=-1$ or $x=1$ or $x=-\frac{1}{6}$
The solutions are $-1, 1, -\frac{1}{6}$.

25. $y^3=9y$
$y^3-9y=0$
$y(y^2-9)=0$
$y(y+3)(y-3)=0$
$y=0$ or $y+3=0$ or $y-3=0$
$y=-3$ or $y=3$
The solutions are $-3, 0, 3$.

27. $x^3-x=2x^2-2$
$x^3-2x^2-x+2=0$
$x^2(x-2)-1(x-2)=0$
$(x^2-1)(x-2)=0$
$(x+1)(x-1)(x-2)=0$
$x+1=0$ or $x-1=0$ or $x-2=0$
$x=-1$ or $x=1$ or $x=2$
The solutions are $-1, 1, 2$.

29. $(2x+7)(x-10)=0$
$2x+7=0$ or $x-10=0$
$2x=-7$ or $x=10$
$x=-\frac{7}{2}$
The solutions are $-\frac{7}{2}, 10$.

31. $3x(x-5)=0$
$3x=0$ or $x-5=0$
$x=0$ or $x=5$
The solutions are 0, 5.

33. $x^2-2x-15=0$
$(x-5)(x+3)=0$
$x-5=0$ or $x+3=0$
$x=5$ or $x=-3$
The solutions are $-3, 5$.

35. $12x^2+2x-2=0$
$2(6x^2+x-1)=0$
$2(3x-1)(2x+1)=0$
$3x-1=0$ or $2x+1=0$
$3x=1$ or $2x=-1$
$x=\frac{1}{3}$ or $x=-\frac{1}{2}$
The solutions are $-\frac{1}{2}, \frac{1}{3}$.

37. $w^2-5w=36$
$w^2-5w-36=0$
$(w-9)(w+4)=0$
$w-9=0$ or $w+4=0$
$w=9$ or $w=-4$
The solutions are $-4, 9$.

39. $25x^2-40x+16=0$
$(5x-4)^2=0$
$5x-4=0$
$5x=4$
$x=\frac{4}{5}$
The solution is $\frac{4}{5}$.

41. $2r^3+6r^2=20r$
$2r^3+6r^2-20r=0$
$2r(r^2+3r-10)=0$
$2r(r+5)(r-2)=0$
$2r=0$ or $r+5=0$ or $r-2=0$
$r=0$ or $r=-5$ or $r=2$
The solutions are $-5, 0, 2$.

43. $z(5z-4)(z+3)=0$
$z=0$ or $5z-4=0$ or $z+3=0$
$5z=4$ or $z=-3$
$z=\frac{4}{5}$
The solutions are $-3, 0, \frac{4}{5}$.

45. $2z(z+6)=2z^2+12z-8$
$2z^2+12z=2z^2+12z-8$
$0=-8$ False
No solutions exist; $\varnothing$.

47. $(x-1)(x+4)=24$
$x^2+3x-4=24$
$x^2+3x-28=0$
$(x+7)(x-4)=0$
$x+7=0$ or $x-4=0$
$x=-7$ or $x=4$
The solutions are –7, 4.

49. $\frac{x^2}{4}-\frac{5}{2}x+6=0$
$x^2-10x+24=0$
$(x-6)(x-4)=0$
$x-6=0$ or $x-4=0$
$x=6$ or $x=4$
The solutions are 4, 6.

51. $y^2+\frac{1}{4}=-y$
$4y^2+1=-4y$
$4y^2+4y+1=0$
$(2y+1)^2=0$
$2y+1=0$
$2y=-1$
$y=-\frac{1}{2}$
The solution is $-\frac{1}{2}$.

53. $y^3+4y^2=9y+36$
$y^3+4y^2-9y-36=0$
$y^2(y+4)-9(y+4)=0$
$(y^2-9)(y+4)=0$
$(y+3)(y-3)(y+4)=0$
$y+3=0$ or $y-3=0$ or $y+4=0$
$y=-3$ or $y=3$ or $y=-4$
The solutions are –4, –3, 3.

55. $2x^3=50x$
$2x^3-50x=0$
$2x(x^2-25)=0$
$2x(x+5)(x-5)=0$
$2x=0$ or $x+5=0$ or $x-5=0$
$x=0$ or $x=-5$ or $x=5$
The solutions are –5, 0, 5.

57. $x^2+(x+1)^2=61$
$x^2+x^2+2x+1=61$
$2x^2+2x-60=0$
$2(x^2+x-30)=0$
$2(x+6)(x-5)=0$
$x+6=0$ or $x-5=0$
$x=-6$ or $x=5$
The solutions are –6, 5.

59. $m^2(3m-2)=m$
$3m^3-2m^2=m$
$3m^3-2m^2-m=0$
$m(3m^2-2m-1)=0$
$m(3m+1)(m-1)=0$
$m=0$ or $3m+1=0$ or $m-1=0$
$3m=-1$ or $m=1$
$m=-\frac{1}{3}$
The solutions are $-\frac{1}{3}$, 0, 1.

61. $3x^2=-x$
$3x^2+x=0$
$x(3x+1)=0$
$x=0$ or $3x+1=0$
$3x=-1$
$x=-\frac{1}{3}$
The solutions are $-\frac{1}{3}$, 0.

63. $x(x-3)=x^2+5x+7$

$x^2-3x=x^2+5x+7$

$-8x=7$

$x=-\frac{7}{8}$

The solution is $-\frac{7}{8}$.

65. $3(t-8)+2t=7+t$

$3t-24+2t=7+t$

$5t-24=7+t$

$4t=31$

$t=\frac{31}{4}$

The solution is $\frac{31}{4}$.

67. $-3(x-4)+x=5(3-x)$

$-3x+12+x=15-5x$

$-2x+12=15-5x$

$3x=3$

$x=1$

The solution is 1.

69. Let n = the one number and $n+5$ = the other number.

$n(n+5)=66$

$n^2+5n-66=0$

$(n+11)(n-6)=0$

$n+11=0$ or $n-6=0$

$n=-11$ or $n=6$

The two solutions are –11 and –6 and 6 and 11.

71. Let d = amount of cable needed. Then from the Pythagorean theorem, $d^2=45^2+60^2=5625$ so $d=\sqrt{5625}=75$ ft.

73. $C(x)=x^2-15x+50$

$9500=x^2-15x+50$

$0=x^2-15x-9450$

$0=(x-105)(x+90)$

$x-105=0$ or $x+90=0$

$x=105$ or $x=-90$

Disregard the negative; 105 units.

75. Let x = one leg of a right triangle and $x-3$ = the other leg of the right triangle.

$15^2=x^2+(x-3)^2$

$225=x^2+x^2-6x+9$

$225=2x^2-6x+9$

$0=2x^2-6x-216$

$0=2(x^2-3x-108)$

$0=2(x-12)(x+9)$

$x-12=0$ or $x+9=0$

$x=12$ or $x=-9$

Disregarding the negative solution, we find that one leg of the right triangle is 12 cm and the other leg is 9 cm.

77. Note that the outer rectangle has lengths of $2x+12$ and $2x+16$. Thus, the area of the border is $(2x+12)(2x+16)-12\cdot 16$. Set this equal to 128 and solve for x.

$(2x+12)(2x+16)-12\cdot 16=128$

$4x^2+56x+192-192=128$

$4x^2+56x=128$

$4x^2+56x-128=0$

$x^2+14x-32=0$

$(x+16)(x-2)=0$

$x+16=0$ or $x-2=0$

$x=-16$ or $x=2$

Since x must be positive, we see that $x=2$ inches.

79. The sunglasses will hit the ground when $h(t)$ equals 0.

$-16t^2+1600=0$

$-16(t^2-100)=0$

$-16(t-10)(t+10)=0$

$t-10=0$ or $t+10=0$

$t=10$ or $t=-10$

The sunglasses will hit the ground 10 seconds after being dropped.

81. Let the width of the floor = w. Then the length is $2w-3$ and so the area is

$(2w-3)w=90$

$2w^2-3w-90=0$

$(2w-15)(w+6)=0$

$2w-15=0$ or $w+6=0$
$2w=15$ or $w=-6$
$w=\frac{15}{2}=7.5$
Disregard –6.
$2w-3=2(7.5)-3=15-3=12$
The width is 7.5 ft and the length is 12 ft.

83. $0.5x^2=50$
$0.5x^2-50=0$
$5x^2-500=0$
$x^2-100=0$
$(x+10)(x-10)=0$
$x+10=0$ or $x-10=0$
$x=-10$ or $x=10$
Disregard the negative solution. A 10-inch square tier is needed, provided each person has one serving.

85. The object will hit the ground when $h(t)$ equals 0.
$-16t^2+80t+576=0$
$-16(t^2-5t-36)=0$
$-16(t-9)(t+4)=0$
$t-9=0$ or $t+4=0$
$t=9$ or $t=-4$
The object will hit the ground 9 seconds after being dropped.

87. E; x-intercepts (2, 0), (–5, 0)

89. F; x-intercepts (0, 0), (–3, 0), (3, 0)

91. B; x-intercepts $\left(-\frac{1}{2}, 0\right)$, (–4, 0)

93. (–3, 0), (0, 2); function, because any vertical line will cross only once.

95. (–4, 0), (0, 2), (4, 0), (0, –2); not a function, because a vertical line can be drawn that crosses twice.

97. $(x-5)(x+2)=0$
$x-5=0$ or $x+2=0$
$x=5$ or $x=-2$

99. $y(y-5)=-6$
$y^2-5y+6=0$
$(y-2)(y-3)=0$
$y-2=0$ or $y-3=0$
$y=2$ or $y=3$

101. $(x^2+x-6)(3x^2-14x-5)=0$
$x^2+x-6=0$ or $3x^2-14x-5=0$
$(x+3)(x-2)=0$ or $(3x+1)(x-5)=0$
$x+3=0$ or $x-2=0$ or $3x+1=0$
$x=-3$ $\quad x=2$ $\quad 3x=-1$
$x=-\frac{1}{3}$
or $x-5=0$
$x=5$
The solutions are $-3, -\frac{1}{3}, 2, 5$.

103. $y=0.7x^2+4.9x+8$
$38.8=0.7x^2+4.9x+8$
$0=0.7x^2+4.9x-30.8$
$0=7x^2+49x-308$
$0=7(x^2+7x-44)$
$0=7(x+11)(x-4)$
$x+11=0$ or $x-4=0$
$x=-11$ $\quad x=4$
The solutions are –11, 4.

105. answers may vary

107. no; answers may vary

109. answers may vary

111. answers may vary

113. answers may vary

Chapter 5 Vocabulary Check

1. A polynomial is a finite sum of terms in which all variables are raised to nonnegative integer powers and no variables appear in any denominator.

2. Factoring is the process of writing a polynomial as a product.

3. Exponents are used to write repeated factors in a more compact form.

4. The degree of a term is the sum of the exponents on the variables contained in the term.

5. A monomial is a polynomial with one term.

6. If a is not 0, $a^0 = 1$.

7. A trinomial is a polynomial with three terms.

8. A polynomial equation of degree 2 is also called a quadratic equation.

9. A positive number is written in scientific notation if it is written as the product of a number a, where $1 \le a < 10$, and a power of 10.

10. The degree of a polynomial is the largest degree of all its terms.

11. A binomial is a polynomial with two terms.

12. If a and b are real numbers and $a \cdot b = 0$, then $a = 0$ and $b = 0$.

Chapter 5 Review

1. $(-2)^2 = (-2)(-2) = 4$

2. $(-3)^4 = (-3)(-3)(-3)(-3) = 81$

3. $-2^2 = -(2 \cdot 2) = -4$

4. $-3^4 = -(3 \cdot 3 \cdot 3 \cdot 3) = -81$

5. $8^0 = 1$

6. $-9^0 = -1$

7. $-4^{-2} = -\dfrac{1}{4^2} = -\dfrac{1}{16}$

8. $(-4)^2 = \dfrac{1}{(-4)^2} = \dfrac{1}{16}$

9. $-xy^2 \cdot y^3 \cdot xy^2z = -x^{1+1}y^{2+3+2}z = -x^2y^7z$

10. $(-4xy)(-3xy^2b) = (-4)(-3)x^{1+1}y^{1+2}b$
$= 12x^2y^3b$

11. $a^{-14}a^5 = a^{-14+5} = a^{-9} = \dfrac{1}{a^9}$

12. $\dfrac{a^{16}}{a^{17}} = a^{16-17} = a^{-1} = \dfrac{1}{a}$

13. $\dfrac{x^{-7}}{x^4} = x^{-7-4} = x^{-11} = \dfrac{1}{x^{11}}$

14. $\dfrac{9a(a^{-3})}{18a^{15}} = \dfrac{a^{1-3-15}}{2} = \dfrac{a^{-17}}{2} = \dfrac{1}{2a^{17}}$

15. $\dfrac{y^{6p-3}}{y^{6p+2}} = y^{(6p-3)-(6p+2)}$
$= y^{6p-3-6p-2}$
$= y^{-5}$
$= \dfrac{1}{y^5}$

16. $(3x^{2a+b}y^{-3b})^2 = 3^2x^{2(2a+b)}y^{2(-3b)}$
$= 9x^{4a+2b}y^{-6b}$ or $\dfrac{9x^{4a+2b}}{y^{6b}}$

17. $36,890,000 = 3.689 \times 10^7$

18. $0.000362 = 3.62 \times 10^{-4}$

19. $1.678 \times 10^{-6} = 0.000001678$

20. $4.1 \times 10^5 = 410,000$

21. $(8^5)^3 = 8^{5 \cdot 3} = 8^{15}$

22. $\left(\dfrac{a}{4}\right)^2 = \dfrac{a^2}{4^2} = \dfrac{a^2}{16}$

23. $(3x)^3 = 3^3x^3 = 27x^3$

24. $(-4x)^{-2} = \dfrac{1}{(-4x)^2} = \dfrac{1}{(-4)^2x^2} = \dfrac{1}{16x^2}$

25. $\left(\dfrac{6x}{5}\right)^2 = \dfrac{(6x)^2}{5^2} = \dfrac{36x^2}{25}$

26. $(8^6)^{-3} = 8^{6(-3)} = 8^{-18} = \frac{1}{8^{18}}$

27. $\left(\frac{4}{3}\right)^{-2} = \frac{4^{-2}}{3^{-2}} = \frac{3^2}{4^2} = \frac{9}{16}$

28. $(-2x^3)^{-3} = \frac{1}{(-2x^3)^3}$
$= \frac{1}{(-2)^3(x^3)^3}$
$= \frac{1}{-8x^9}$
$= -\frac{1}{8x^9}$

29. $\left(\frac{8p^6}{4p^4}\right)^{-2} = (2p^2)^{-2} = 2^{-2}p^{-4} = \frac{1}{4p^4}$

30. $(-3x^{-2}y^2)^3 = (-3)^3(x^{-2})^3(y^2)^3$
$= -27x^{-6}y^6$
$= -\frac{27y^6}{x^6}$

31. $\left(\frac{x^{-5}y^{-3}}{z^3}\right)^{-5} = \frac{x^{25}y^{15}}{z^{-15}} = x^{25}y^{15}z^{15}$

32. $\frac{4^{-1}x^3yz}{x^{-2}yx^4} = \frac{x^{3-(-2)-4}z}{4} = \frac{x^{3+2-4}z}{4} = \frac{xz}{4}$

33. $(5xyz)^{-4}(x^{-2})^{-3} = \frac{1}{(5xyz)^4}x^6$
$= \frac{x^6}{5^4x^4y^4z^4}$
$= \frac{x^2}{625y^4z^4}$

34. $\frac{2(3yz)^{-3}}{y^{-3}} = \frac{2(3)^{-3}y^{-3}z^{-3}}{y^{-3}} = \frac{2}{3^3z^3} = \frac{2}{27z^3}$

35. $x^{4a}(3x^{5a})^3 = x^{4a}(3^3x^{15a})$
$= 27x^{4a+15a}$
$= 27x^{19a}$

36. $\frac{4y^{3x-3}}{2y^{2x+4}} = 2y^{(3x-3)-(2x+4)}$
$= 2y^{3x-3-2x-4}$
$= 2y^{x-7}$

37. $\frac{(0.00012)(144,000)}{0.0003} = \frac{(1.2\times10^{-4})(1.44\times10^5)}{3\times10^{-4}}$
$= 0.576\times10^5$
$= 5.76\times10^4$

38. $\frac{(-0.00017)(0.00039)}{3000}$
$= \frac{(-1.7\times10^{-4})(3.9\times10^{-4})}{3\times10^3}$
$= -2.21\times10^{-4-4-3}$
$= -2.21\times10^{-11}$

39. $\frac{27x^{-5}y^5}{18x^{-6}y^2}\cdot\frac{x^4y^{-2}}{x^{-2}y^3} = \frac{3x^{-5+4}y^{5-2}}{2x^{-6-2}y^{2+3}}$
$= \frac{3x^{-1}y^3}{2x^{-8}y^5}$
$= \frac{3}{2}x^{-1-(-8)}y^{3-5}$
$= \frac{3}{2}x^7y^{-2}$
$= \frac{3x^7}{2y^2}$

40. $\frac{3x^5}{y^{-4}}\cdot\frac{(3xy^{-3})^{-2}}{(z^{-3})^{-4}} = \frac{3x^5\cdot3^{-2}x^{-2}y^6}{y^{-4}z^{12}}$
$= \frac{3^{1-2}x^{5-2}y^{6-(-4)}}{z^{12}}$
$= \frac{3^{-1}x^3y^{10}}{z^{12}}$
$= \frac{x^3y^{10}}{3z^{12}}$

41. The degree of the polynomial $x^2y - 3xy^3z + 5x + 7y$ is the degree of the term $-3xy^3z$ which is 5.

42. $3x+2$ has degree 1.

43. $4x+8x-6x^2-6x^2y=(4+8)x-6x^2-6x^2y$
$=12x-6x^2-6x^2y$

44. $-8xy^3+4xy^3-3x^3y=(-8+4)xy^3-3x^3y$
$=-4xy^3-3x^3y$

45. $(3x+7y)+(4x^2-3x+7)+(y-1)$
$=3x+7y+4x^2-3x+7+y-1$
$=4x^2+(3-3)x+(7+1)y+(7-1)$
$=4x^2+8y+6$

46. $(4x^2-6xy+9y^2)-(8x^2-6xy-y^2)$
$=4x^2-6xy+9y^2-8x^2+6xy+y^2$
$=(4-8)x^2+(9+1)y^2$
$=-4x^2+10y^2$

47. $(3x^2-4b+28)+(9x^2-30)-(4x^2-6b+20)$
$=3x^2-4b+28+9x^2-30-4x^2+6b-20$
$=(3+9-4)x^2+(-4+6)b+(28-30-20)$
$=8x^2+2b-22$

48. $(9xy+4x^2+18)+(7xy-4x^3-9x)$
$=9xy+4x^2+18+7xy-4x^3-9x$
$=-4x^3+4x^2+(9+7)xy-9x+18$
$=-4x^3+4x^2+16xy-9x+18$

49. $2x-(3x-5)=2x-3x+5$
$=(2-3)x+5$
$=-x+5$

50. $(3x^2y-7xy-4)+(9x^2y+x)-(x-7)$
$=3x^2y-7xy-4+9x^2y+x-x+7$
$=(3+9)x^2y-7xy+(-4+7)$
$=12x^2y-7xy+3$

51.
$$\begin{array}{r} x^2-5x+7 \\ -\quad (x+4) \\ \hline x^2-6x+3 \end{array}$$

52.
$$\begin{array}{l} x^3 \qquad +2xy^2-y \\ +\quad (x-4xy^2 \quad -7) \\ \hline x^3+x-2xy^2-y-7 \end{array}$$

53. $P(6)=9(6)^2-7(6)+8=290$

54. $P(0)=9(0)^2-7(0)+8=8$

55. $P(-2)=9(-2)^2-7(-2)+8=58$

56. $P(-3)=9(-3)^2-7(-3)+8=110$

57. $P(x)+Q(x)=(2x-1)+(x^2+2x-5)$
$=2x-1+x^2+2x-5$
$=x^2+4x-6$

58. $2\cdot P(x)-Q(x)=2(2x-1)-(x^2+2x-5)$
$=4x-2-x^2-2x+5$
$=-x^2+2x+3$

59. $2(2x^2y-6x+1)+2(x^2y+5)$
$=4x^2y-12x+2+2x^2y+10$
$=(6x^2y-12x+12)$ cm

60. $(3x+11)+(3x+11)+3x=3x+3x+3x+11+11$
$=9x+22$
The perimeter is $(9x+22)$ inches.

61. $-6x(4x^2-6x+1)=-24x^3+36x^2-6x$

62. $-4ab^2(3ab^3+7ab+1)$
$=-4ab^2(3ab^3)-4ab^2(7ab)-4ab^2(1)$
$=-12a^2b^5-28a^2b^3-4ab^2$

63. $(x-4)(2x+9)=2x^2+9x-8x-36$
$=2x^2+x-36$

64. $(-3xa+4b)^2=(-3xa)^2+2(-3xa)(4b)+(4b)^2$
$=9x^2a^2-24xab+16b^2$

65.
$$\begin{array}{r} 9x^2+4x+1 \\ 4x-3 \\ \hline -27x^2-12x-3 \\ 36x^3+16x^2\quad+4x \\ \hline 36x^3-11x^2\quad-8x-3 \end{array}$$

66. $(2x-1)(x^2+2x-5) = 2x(x^2+2x-5)-1(x^2+2x-5)$
$= 2x^3+4x^2-10x-x^2-2x+5$
$= 2x^3+3x^2-12x+5$

67. $(5x-9y)(3x+9y) = 15x^2+45xy-27xy+81y^2$
$= 15x^2+18xy-81y^2$

68. $\left(x-\frac{1}{3}\right)\left(x+\frac{2}{3}\right) = x^2+\frac{2}{3}x-\frac{1}{3}x-\frac{1}{3}\left(\frac{2}{3}\right)$
$= x^2+\frac{1}{3}x-\frac{2}{9}$

69. $(x^2+9x+1)^2 = (x^2+9x+1)(x^2+9x+1)$
$= x^2(x^2+9x+1)+9x(x^2+9x+1)+1(x^2+9x+1)$
$= x^4+9x^3+x^2+9x^3+81x^2+9x+x^2+9x+1$
$= x^4+18x^3+83x^2+18x+1$

70. $(m^2+m-2)^2 = (m^2+m-2)(m^2+m-2)$
$= m^2(m^2+m-2)+m(m^2+m-2)-2(m^2+m-2)$
$= m^4+m^3-2m^2+m^3+m^2-2m-2m^2-2m+4$
$= m^4+2m^3-3m^2-4m+4$

71. $(3x-y)^2 = (3x)^2-2(3x)y+y^2$
$= 9x^2-6xy+y^2$

72. $(4x+9)^2 = (4x)^2+2(4x)(9)+9^2$
$= 16x^2+72x+81$

73. $(x+3y)(x-3y) = x^2-(3y)^2 = x^2-9y^2$

74. $[4+(3a-b)][4-(3a-b)] = 4^2-(3a-b)^2$
$= 16-[(3a)^2-2(3a)b+b^2]$
$= 16-(9a^2-6ab+b^2)$
$= 16-9a^2+6ab-b^2$

75. Area $= lw$
$= (3y+7z)(3y-7z)$
$= (3y)^2-(7z)^2$
$= 9y^2-49z^2$

The area of the rectangle is $(9y^2-49z^2)$ square units.

76. $(11y+9)^2 = (11y)^2 + 2(11y)(9) + 9^2$
$= 121y^2 + 198y + 81$
The area of the square is
$(121y^2 + 198y + 81)$ square units.

77. $4a^b(3a^{b+2} - 7) = 4a^b(3a^{b+2}) + 4a^b(-7)$
$= 12a^{b+b+2} - 28a^b$
$= 12a^{2b+2} - 28a^b$

78. $(3x^a - 4)(3x^a + 4) = (3x^a)^2 - 4^2$
$= 3^2(x^a)^2 - 16$
$= 9x^{2a} - 16$

79. $16x^3 - 24x^2 = 8x^2(2x-3)$

80. $36y - 24y^2 = 12y(3-2y)$

81. $6ab^2 + 8ab - 4a^2b^2 = 2ab(3b + 4 - 2ab)$

82. $14a^2b^2 - 21ab^2 + 7ab = 7ab(2ab - 3b + 1)$

83. $6a(a+3b) - 5(a+3b) = (6a-5)(a+3b)$

84. $4x(x-2y) - 5(x-2y) = (4x-5)(x-2y)$

85. $xy - 6y + 3x - 18 = y(x-6) + 3(x-6)$
$= (x-6)(y+3)$

86. $ab - 8b + 4a - 32 = b(a-8) + 4(a-8)$
$= (a-8)(b+4)$

87. $pq - 3p - 5q + 15 = p(q-3) - 5(q-3)$
$= (q-3)(p-5)$

88. $x^3 - x^2 - 2x + 2 = x^2(x-1) - 2(x-1)$
$= (x-1)(x^2-2)$

89. $2x \cdot y - x \cdot x = 2xy - x^2 = x(2y - x)$

90. $9y(4y) - 4(x^2) = 36y^2 - 4x^2$
$= 4(9y^2 - x^2)$
$= 4(3y+x)(3y-x)$

91. $x^2 - 14x - 72 = (x-18)(x+4)$

92. $x^2 + 16x - 80 = (x-4)(x+20)$

93. $2x^2 - 18x + 28 = 2(x^2 - 9x + 14)$
$= 2(x-7)(x-2)$

94. $3x^2 + 33x + 54 = 3(x^2 + 11x + 18)$
$= 3(x+9)(x+2)$

95. $2x^3 - 7x^2 - 9x = x(2x^2 - 7x - 9)$
$= x(2x-9)(x+1)$

96. $3x^3 + 2x^2 - 16x = x(3x^2 + 2x - 16)$
$= x(3x+8)(x-2)$

97. $6x^2 + 17x + 10 = (6x+5)(x+2)$

98. $15x^2 - 91x + 6 = (15x-1)(x-6)$

99. $4x^2 + 2x - 12 = 2(2x^2 + x - 6)$
$= 2(2x-3)(x+2)$

100. $9x^2 - 12x - 12 = 3(3x^2 - 4x - 4)$
$= 3(3x+2)(x-2)$

101. $y^2(x+6)^2 - 2y(x+6)^2 - 3(x+6)^2$
$= (x+6)^2(y^2 - 2y - 3)$
$= (x+6)^2(y-3)(y+1)$

102. Let $y = x + 5$. Then
$(x+5)^2 + 6(x+5) + 8 = y^2 + 6y + 8$
$= (y+4)(y+2)$
$= [(x+5)+4][(x+5)+2]$
$= (x+9)(x+7)$

103. $x^4 - 6x^2 - 16 = (x^2 - 8)(x^2 + 2)$

104. $x^4 + 8x^2 - 20 = (x^2 + 10)(x^2 - 2)$

105. $x^2 - 100 = x^2 - 10^2 = (x+10)(x-10)$

106. $x^2 - 81 = x^2 - 9^2 = (x+9)(x-9)$

107. $2x^2 - 32 = 2(x^2 - 16)$
$= 2(x^2 - 4^2)$
$= 2(x+4)(x-4)$

108. $6x^2-54=6(x^2-9)$
$=6(x^2-3^2)$
$=6(x+3)(x-3)$

109. $81-x^4=9^2-(x^2)^2$
$=(9+x^2)(9-x^2)$
$=(9+x^2)(3+x)(3-x)$

110. $16-y^4=4^2-(y^2)^2$
$=(4+y^2)(4-y^2)$
$=(4+y^2)(2+y)(2-y)$

111. $(y+2)^2-25=(y+2)^2-5^2$
$=[(y+2)+5][(y+2)-5]$
$=(y+7)(y-3)$

112. $(x-3)^2-16=(x-3)^2-4^2$
$=[(x-3)+4][(x-3)-4]$
$=(x+1)(x-7)$

113. $x^3+216=x^3+6^3$
$=(x+6)(x^2-6\cdot x+6^2)$
$=(x+6)(x^2-6x+36)$

114. $y^3+512=y^3+8^3$
$=(y+8)(y^2-8\cdot y+8^2)$
$=(y+8)(y^2-8y+64)$

115. $8-27y^3=2^3-(3y)^3$
$=(2-3y)(4+2\cdot 3y+(3y)^2)$
$=(2-3y)(4+6y+9y^2)$

116. $1-64y^3=1^3-(4y)^3$
$=(1-4y)(1^2+1\cdot 4y+(4y)^2)$
$=(1-4y)(1+4y+16y^2)$

117. $6x^4y+48xy=6xy(x^3+8)$
$=6xy(x^3+2^3)$
$=6xy(x+2)(x^2-2x+2^2)$
$=6xy(x+2)(x^2-2x+4)$

118. $2x^5+16x^2y^3=2x^2(x^3+8y^3)$
$=2x^2(x^3+(2y)^3)$
$=2x^2(x+2y)(x^2-x\cdot 2y+(2y)^2)$
$=2x^2(x+2y)(x^2-2xy+4y^2)$

119. $x^2-2x+1-y^2=(x^2-2x+1)-y^2$
$=(x-1)^2-y^2$
$=[(x-1)+y][(x-1)-y]$
$=(x-1+y)(x-1-y)$

120. $x^2-6x+9-4y^2=(x^2-6x+9)-4y^2$
$=(x-3)^2-(2y)^2$
$=[(x-3)+2y][(x-3)-2y]$
$=(x-3+2y)(x-3-2y)$

121. $4x^2+12x+9=(2x+3)(2x+3)$
$=(2x+3)^2$

122. $16a^2-40ab+25b^2=(4a-5b)(4a-5b)$
$=(4a-5b)^2$

123. $\text{Volume}=\pi R^2h-\pi r^2h$
$=\pi h(R^2-r^2)$
$=\pi h(R+r)(R-r)$ cubic units

124. $\text{SA}=2\pi Rh+2\pi rh$
$=2\pi h(R+r)$ square units

125. $(3x-1)(x+7)=0$
$3x-1=0$ or $x+7=0$
$x=\frac{1}{3}$ or $x=-7$

The solutions are $-7, \frac{1}{3}$.

126. $3(x+5)(8x-3)=0$
$x+5=0$ or $8x-3=0$
$x=-5$ or $x=\frac{3}{8}$

The solutions are $-5, \frac{3}{8}$.

127. $5x(x-4)(2x-9)=0$

$5x=0$ or $x-4=0$ or $2x-9=0$

$x=0$ or $x=4$ or $x=\frac{9}{2}$

The solutions are $0, 4, \frac{9}{2}$.

128. $6(x+3)(x-4)(5x+1)=0$

$x+3=0$ or $x-4=0$ or $5x+1=0$

$x=-3$ or $x=4$ or $5x=-1$

$x=-\frac{1}{5}$

The solutions are $-\frac{1}{5}, -3, 4$.

129. $2x^2=12x$

$2x^2-12x=0$

$2x(x-6)=0$

$2x=0$ or $x-6=0$

$x=0$ or $x=6$

The solutions are 0, 6.

130. $4x^3-36x=0$

$4x(x^2-9)=0$

$4x(x+3)(x-3)=0$

$4x=0$ or $x+3=0$ or $x-3=0$

$x=0$ or $x=-3$ or $x=3$

The solutions are –3, 0, 3.

131. $(1-x)(3x+2)=-4x$

$3x+2-3x^2-2x=-4x$

$-3x^2+x+2=-4x$

$-3x^2+5x+2=0$

$3x^2-5x-2=0$

$(3x+1)(x-2)=0$

$3x+1=0$ or $x-2=0$

$3x=-1$ or $x=2$

$x=-\frac{1}{3}$

The solutions are $-\frac{1}{3}, 2$.

132. $2x(x-12)=-40$

$2x^2-24x=-40$

$2x^2-24x+40=0$

$2(x^2-12x+20)=0$

$2(x-10)(x-2)=0$

$x-10=0$ or $x-2=0$

$x=10$ or $x=2$

The solutions are 2, 10.

133. $3x^2+2x=12-7x$

$3x^2+9x-12=0$

$3(x^2+3x-4)=0$

$3(x+4)(x-1)=0$

$x+4=0$ or $x-1=0$

$x=-4$ or $x=1$

The solutions are –4, 1.

134. $2x^2+3x=35$

$2x^2+3x-35=0$

$(2x-7)(x+5)=0$

$2x-7=0$ or $x+5=0$

$2x=7$ or $x=-5$

$x=\frac{7}{2}$

The solutions are $-5, \frac{7}{2}$.

135. $x^3-18x=3x^2$

$x^3-3x^2-18x=0$

$x(x^2-3x-18)=0$

$x(x-6)(x+3)=0$

$x=0$ or $x-6=0$ or $x+3=0$

$x=6$ or $x=-3$

The solutions are –3, 0, 6.

136. $19x^2-42x=-x^3$

$x^3+19x^2-42x=0$

$x(x^2+19x-42)=0$

$x(x+21)(x-2)=0$

$x=0$ or $x+21=0$ or $x-2=0$

$x=-21$ or $x=2$

The solutions are –21, 0, 2.

137. $12x = 6x^3 + 6x^2$

$-6x^3 - 6x^2 + 12x = 0$

$-6x(x^2 + x - 2) = 0$

$-6x(x+2)(x-1) = 0$

$-6x = 0$ or $x + 2 = 0$ or $x - 1 = 0$

$x = 0$ or $x = -2$ or $x = 1$

The solutions are –2, 0, 1.

138. $8x^3 + 10x^2 = 3x$

$8x^3 + 10x^2 - 3x = 0$

$x(8x^2 + 10x - 3) = 0$

$x(4x-1)(2x+3) = 0$

$x = 0$ or $4x - 1 = 0$ or $2x + 3 = 0$

$4x = 1$ or $2x = -3$

$x = \frac{1}{4}$ or $x = -\frac{3}{2}$

The solutions are $-\frac{3}{2}$, 0, $\frac{1}{4}$.

139. Let x = the number. Then

$x + 2x^2 = 105$

$2x^2 + x - 105 = 0$

$(2x+15)(x-7) = 0$

$2x + 15 = 0$ or $x - 7 = 0$

$2x = -15$ or $x = 7$

$x = -\frac{15}{2}$

The number is $-\frac{15}{2}$ or 7.

140. Let x = width; then $2x - 5$ = length.

$x(2x-5) = 33$

$2x^2 - 5x = 33$

$2x^2 - 5x - 33 = 0$

$(2x-11)(x+3) = 0$

$2x - 11 = 0$ or $x + 3 = 0$

$2x = 11$ or $x = -3$

$x = \frac{11}{2}$

Disregard the negative.

Width = $\frac{11}{2} = 5\frac{1}{2}$ m

Length = $2\left(\frac{11}{2}\right) - 5 = 6$ m

141. $h(t) = -16t^2 + 400$

$0 = -16t^2 + 400$

$0 = -16(t^2 - 25)$

$0 = -16(t+5)(t-5)$

$t + 5 = 0$ or $t - 5 = 0$

$t = -5$ or $t = 5$

Disregard the negative. The stunt dummy will reach the ground after 5 seconds.

142. $P(t) = -16t^2 + 1053$

$P(1) = -16(1)^2 + 1053 = -16 + 1053 = 1037$

$P(8) = -16(8)^2 + 1053 = -1024 + 1053 = 29$

After 1 second, the object is at 1037 feet and after 8 seconds, the object is at 29 feet.

143. $(x+5)(3x^2 - 2x + 1)$

$= x(3x^2 - 2x + 1) + 5(3x^2 - 2x + 1)$

$= 3x^3 - 2x^2 + x + 15x^2 - 10x + 5$

$= 3x^3 + 13x^2 - 9x + 5$

144. $(3x^2 + 4x - 1.2) - (5x^2 - x + 5.7)$

$= 3x^2 + 4x - 1.2 - 5x^2 + x - 5.7$

$= -2x^2 + 5x - 6.9$

145. $(3x^2 + 4x - 1.2) + (5x^2 - x + 5.7)$

$= 3x^2 + 4x - 1.2 + 5x^2 - x + 5.7$

$= 8x^2 + 3x + 4.5$

146. $\left(7ab - \frac{1}{2}\right)^2 = (7ab)^2 - 2(7ab)\left(\frac{1}{2}\right) + \left(\frac{1}{2}\right)^2$

$= 49a^2b^2 - 7ab + \frac{1}{4}$

147. $P(x) = -x^2 + x - 4$

$P(5) = -5^2 + 5 - 4 = -25 + 5 - 4 = -24$

148. $P(x) = -x^2 + x - 4$

$P(-2) = -(-2)^2 + (-2) - 4 = -4 - 2 - 4 = -10$

149. $12y^5 - 6y^4 = 6y^4(2y) + 6y^4(-1) = 6y^4(2y - 1)$

150. $x^2y + 4x^2 - 3y - 12 = x^2(y+4) - 3(y+4)$

$= (y+4)(x^2 - 3)$

151. $6x^2 - 34x - 12 = 2(3x^2 - 17x - 6)$
$= 2(3x+1)(x-6)$

152. $y^2(4x+3)^2 - 19y(4x+3)^2 - 20(4x+3)^2$
$= (4x+3)^2(y^2 - 19y - 20)$
$= (4x+3)^2(y-20)(y+1)$

153. $4z^7 - 49z^5 = z^5(4z^2 - 49)$
$= z^5[(2z)^2 - 7^2]$
$= z^5(2z+7)(2z-7)$

154. $5x^4 + 4x^2 - 9 = (x^2 - 1)(5x^2 + 9)$
$= (x+1)(x-1)(5x^2 + 9)$

155. $8x^2 = 24x$
$8x^2 - 24x = 0$
$8x(x-3) = 0$
$8x = 0$ or $x - 3 = 0$
$x = 0$ or $x = 3$

156. $x(x-11) = 26$
$x^2 - 11x = 26$
$x^2 - 11x - 26 = 0$
$(x+2)(x-13) = 0$
$x + 2 = 0$ or $x - 13 = 0$
$x = -2$ or $x = 13$

Chapter 5 Getting Ready for the Test

1. $x^a \cdot x^a = x^{a+a} = x^{2a}$; C

2. $x^a + x^a = 2x^a$; D

3. $x^a \cdot x^b = x^{a+b}$; F

4. $x^a + x^b$ cannot be simplified; G

5. $(x^a)^b = x^{a \cdot b} = x^{ab}$; A

6. $\dfrac{x^a}{x^b} = x^{a-b}$; E

7. $3x - 5 = 3x^1 - 5$ has degree 1.
$x^3y - 5x + 2 = x^3y^1 - 5x^1 + 2$ has degree $3 + 1 = 4$.
$x^2y + 4y = x^2y^1 + 4y^1$ has degree $2 + 1 = 3$.
$3 = 3x^0$ has degree 0.
Choice C is correct.

8. $P(x) = 5x^3 - 2x + 1$
$P(-1) = 5(-1)^3 - 2(-1) + 1$
$= 5(-1) - 2(-1) + 1$
$= -5 + 2 + 1$
$= -2$
Choice B is correct.

9. $P(x) - Q(x) = (2x-3) - (4x+3)$
$= 2x - 3 - 4x - 3$
$= (2-4)x + (-3-3)$
$= -2x - 6$
Choice C is correct.

10. $5x^0 = 5 \cdot 1 = 5$
$-(5x)^0 = -1$
$(-5x)^0 = 1$
Choice C is correct.

11. $7^{-2} = \dfrac{1}{7^2} = \dfrac{1}{49}$; D

12. $(5x+1)^2 = (5x)^2 + 2(5x)(1) + 1^2$
$= 25x^2 + 10x + 1$
Choice B is correct.

13. $25x^2y^3 = 5 \cdot 5 \cdot x^2 \cdot y^3$
$-10x^2y = -2 \cdot 5 \cdot x^2 \cdot y^1$
$5x^2y^2 = 5 \cdot x^2 \cdot y^2$
$\text{GCF} = 5 \cdot x^2 \cdot y^1 = 5x^2y$; C

14. $(x-2)(x^2 + 2x + 4)$ is a factored expression; A.

15. $x(y^2 + 16) + 5(y^2 + 16)$ is a sum, not a product, so it is not a factored expression; B.

16. $x^a \cdot x^a - 10 \cdot 10$ is a difference, not a product, so it is not a factored expression; B.

17. To solve $x(12x + 25) = 50$, the expressions need to be simplified and rearranged so that one side of the equation is 0. Choice A is an incorrect step since it is neither simplification (Choice B), nor rearranging so that one side of the equation is 0 (Choice C).

Chapter 5 Test

1. $(-9x)^{-2} = \frac{1}{(-9x)^2} = \frac{1}{81x^2}$

2. $-3xy^{-2}(4xy^2)z = -12x^{1+1}y^{-2+2}z = -12x^2z$

3. $\frac{6^{-1}a^2b^{-3}}{3^{-2}a^{-5}b^2} = \frac{3^2a^{2+5}}{6^1b^{2+3}} = \frac{9a^7}{6b^5} = \frac{3a^7}{2b^5}$

4. $$\begin{aligned}\left(\frac{-xy^{-5}z}{xy^3}\right)^{-5} &= \frac{-x^{-5}y^{25}z^{-5}}{x^{-5}y^{-15}}\\ &= \frac{x^{-5+5}y^{25-(-15)}}{z^5}\\ &= -\frac{y^{40}}{z^5}\end{aligned}$$

5. $630,000,000 = 6.3\times10^8$

6. $0.01200 = 1.2\times10^{-2}$

7. $5\times10^{-6} = 0.000005$

8. $$\begin{aligned}\frac{(0.0024)(0.00012)}{0.00032} &= \frac{(2.4\times10^{-3})(1.2\times10^{-4})}{3.2\times10^{-4}}\\ &= \frac{(2.4)(1.2)}{3.2}\times10^{-3+(-4)-(-4)}\\ &= 0.9\times10^{-3}\\ &= 0.0009\end{aligned}$$

9. $$\begin{aligned}&(4x^3y-3x-4)-(9x^3y+8x+5)\\ &= 4x^3y-3x-4-9x^3y-8x-5\\ &= -5x^3y-11x-9\end{aligned}$$

10. $$\begin{aligned}-3xy(4x+y) &= -3xy(4x)-3xy(y)\\ &= -12x^2y-3xy^2\end{aligned}$$

11. $$\begin{aligned}(3x+4)(4x-7) &= 12x^2-21x+16x-28\\ &= 12x^2-5x-28\end{aligned}$$

12. $(5a-2b)(5a+2b) = (5a)^2-(2b)^2 = 25a^2-4b^2$

13. $$\begin{aligned}(6m+n)^2 &= (6m)^2+2(6m)n+n^2\\ &= 36m^2+12mn+n^2\end{aligned}$$

14. $$\begin{array}{r} x^2-6x+4\\ \times \qquad 2x-1\\ \hline -x^2+6x-4\\ 2x^3-12x^2+8x \qquad\\ \hline 2x^3-13x^2+14x-4\end{array}$$

15. $16x^3y-12x^2y^4 = 4x^2y(4x-3y^3)$

16. $x^2-13x-30 = (x-15)(x+2)$

17. $$\begin{aligned}4y^2+20y+25 &= (2y+5)(2y+5)\\ &= (2y+5)^2\end{aligned}$$

18. $$\begin{aligned}6x^2-15x-9 &= 3(2x^2-5x-3)\\ &= 3(2x+1)(x-3)\end{aligned}$$

19. $4x^2-25 = (2x)^2-5^2 = (2x+5)(2x-5)$

20. $x^3+64 = x^3+4^3 = (x+4)(x^2-4x+16)$

21. $$\begin{aligned}3x^2y-27y^3 &= 3y(x^2-9y^2)\\ &= 3y(x^2-(3y)^2)\\ &= 3y(x+3y)(x-3y)\end{aligned}$$

22. $6x^2+24 = 6(x^2+4)$

23. $$\begin{aligned}16y^3-2 &= 2(8y^3-1)\\ &= 2((2y)^3-1^3)\\ &= 2(2y-1)(4y^2+2y+1)\end{aligned}$$

24. $$\begin{aligned}x^2y-9y-3x^2+27 &= y(x^2-9)-3(x^2-9)\\ &= (x^2-9)(y-3)\\ &= (x+3)(x-3)(y-3)\end{aligned}$$

25. $$\begin{aligned}3n(7n-20) &= 96\\ 21n^2-60n &= 96\\ 21n^2-60n-96 &= 0\\ 3(7n^2-20n-32) &= 0\\ 3(7n+8)(n-4) &= 0\end{aligned}$$

$7n+8=0$ or $n-4=0$
$7n=-8$ or $n=4$
$n=-\frac{8}{7}$

The solutions are $-\frac{8}{7}$, 4.

26. $(x+2)(x-2)=5(x+4)$
$x^2-4=5x+20$
$x^2-5x-24=0$
$(x-8)(x+3)=0$
$x-8=0$ or $x+3=0$
$x=8$ or $x=-3$
The solutions are 8, –3.

27. $2x^3+5x^2-8x-20=0$
$x^2(2x+5)-4(2x+5)=0$
$(2x+5)(x^2-4)=0$
$(2x+5)(x+2)(x-2)=0$
$2x+5=0$ or $x+2=0$ or $x-2=0$
$2x=-5$ or $x=-2$ or $x=2$
$x=-\frac{5}{2}$

The solutions are $-\frac{5}{2}$, –2, 2.

28. Area $= x^2-(2y)^2$
$=(x+2y)(x-2y)$ square units

29. $h(t)=-16t^2+96t+880$

a. $-16(1)^2+96(1)+880=-16+96+880$
$=960$
After 1 second, the height is 960 feet.

b. $-16(5.1)^2+96(5.1)+880$
$=-416.16+489.6+880$
$=953.44$
After 5.1 seconds, the height is 953.44 feet.

c. $-16t^2+96t+880=-16(t^2-6t-55)$
$=-16(t-11)(t+5)$

d. $-16(t-11)(t+5)=0$
$t-11=0$ or $t+5=0$
$t=11$ or $t=-5$
Disregard the negative. The pebble will hit the ground in 11 seconds.

Chapter 5 Cumulative Review

1. a. $\sqrt[3]{-27}=-3$ since $(-3)^3=-27$.

b. $\sqrt[5]{1}=1$ since $1^5=1$.

c. $\sqrt[4]{16}=2$ since $2^4=16$.

2. a. $\sqrt[3]{64}=4$ since $4^3=64$.

b. $\sqrt[4]{81}=3$ since $3^4=81$.

c. $\sqrt[5]{32}=2$ since $2^5=32$.

3. $2(x-3)=5x-9$
$2x-6=5x-9$
$-3x=-3$
$x=1$
The solution is 1.

4. $0.3y+2.4=0.1y+4$
$10(0.3y+2.4)=10(0.1y+4)$
$3y+24=y+40$
$2y=16$
$y=8$
The solution is 8.

5. $A=10{,}000\left(1+\frac{0.05}{4}\right)^{4(3)}$
$=10{,}000(1.0125)^{12}$
$=10{,}000(1.160754518)$
$=11{,}607.54518$
There will be \$11,607.55 in the account.

6. The area of the room is
$2(14\cdot 8)+2(18\cdot 8)=512$ sq ft. Two coats means
$2\cdot 512=1024$ sq ft of wall needs paint.
$\frac{1}{400}=\frac{x}{1024}$
$1024\left(\frac{1}{400}\right)=1024\left(\frac{x}{512}\right)$
$2.56=x$
$x\approx 3$
3 gallons of paint are needed.

7. a.
$$\frac{1}{4}x \le \frac{3}{8}$$
$$8\left(\frac{1}{4}x\right) \le 8\left(\frac{3}{8}\right)$$
$$2x \le 3$$
$$x \le \frac{3}{2}$$
$$\left\{x \middle| x \le \frac{3}{2}\right\} \text{ or } \left(-\infty, \frac{3}{2}\right]$$

b.
$$-2.3x < 6.9$$
$$10(-2.3x) < 10(6.9)$$
$$-23x < 69$$
$$\frac{-23x}{-23} > \frac{69}{-23}$$
$$x > -3$$
$$\{x \mid x > -3\} \text{ or } (-3, \infty)$$

8.
$$x + 2 \le \frac{1}{4}(x - 7)$$
$$4(x+2) \le 4\left[\frac{1}{4}(x-7)\right]$$
$$4x + 8 \le x - 7$$
$$3x \le -15$$
$$x \le -5$$
$$\{x \mid x \le -5\} \text{ or } (-\infty, -5]$$

9.
$$-1 \le \frac{2x}{3} + 5 \le 2$$
$$3(-1) \le 3\left(\frac{2x}{3} + 5\right) \le 3(2)$$
$$-3 \le 2x + 15 \le 6$$
$$-18 \le 2x \le -9$$
$$-9 \le x \le -\frac{9}{2}$$
$$\left[-9, -\frac{9}{2}\right]$$

10.
$$-\frac{1}{3} < \frac{3x+1}{6} \le \frac{1}{3}$$
$$6\left(-\frac{1}{3}\right) < 6\left(\frac{3x+1}{6}\right) \le 6\left(\frac{1}{3}\right)$$
$$-2 < 3x + 1 \le 2$$
$$-3 < 3x \le 1$$
$$-1 < x \le \frac{1}{3}$$
$$\left(-1, \frac{1}{3}\right]$$

11. $|y| = 0$
$y = 0$
The solution is 0.

12. $8 + |4c| = 24$
$|4c| = 16$
$4c = 16$ or $4c = -16$
$c = 4$ or $c = -4$
The solutions are –4, 4.

13. $\left|2x - \frac{1}{10}\right| < -13$ is impossible; { } or ∅.

14. $|5x - 1| + 9 > 5$
$|5x - 1| > -4$ is always true.
$(-\infty, \infty)$

15. $y = \frac{1}{3}x$

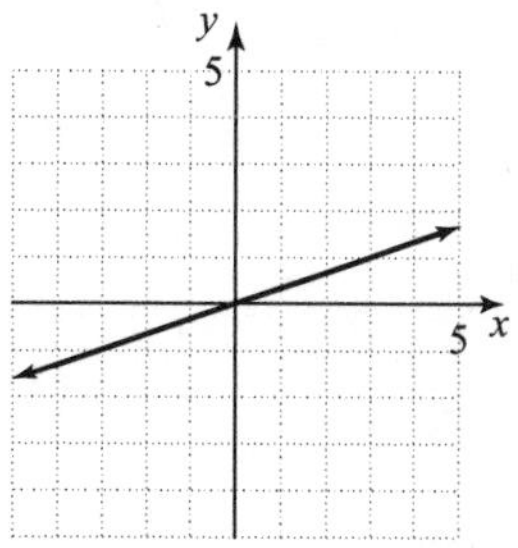

16. $y = 3x$

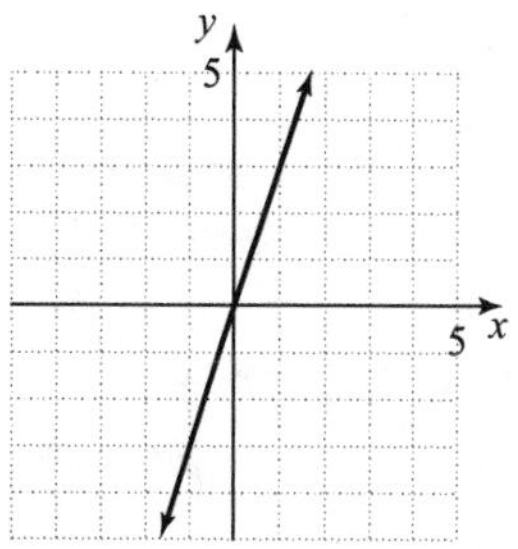

17. $f(x)=\begin{cases}2x+3 & \text{if} \quad x\le 0\\ -x-1 & \text{if} \quad x>0\end{cases}$

$f(2) = -2 - 1 = -3$: $(2, -3)$
$f(-6) = 2(-6) + 3 = -9$: $(-6, -9)$
$f(0) = 2(0) + 3 = 3$: $(0, 3)$

18. $f(x)=3x^2+2x+3$

$$\begin{aligned}f(-3)&=3(-3)^2+2(-3)+3\\&=3(9)-6+3\\&=27-6+3\\&=24\end{aligned}$$

19. $x=2$

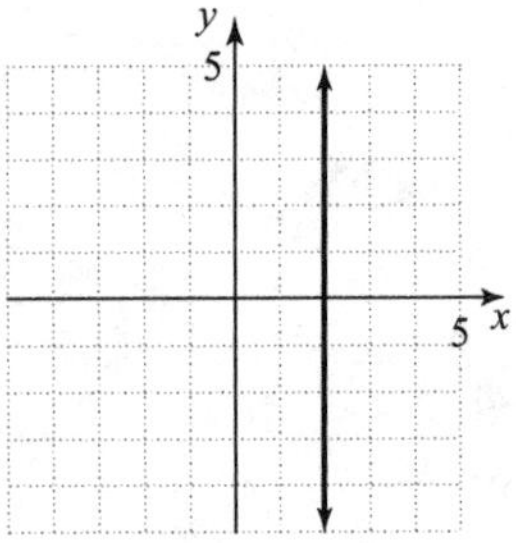

20. $y-5=0$
$y=5$

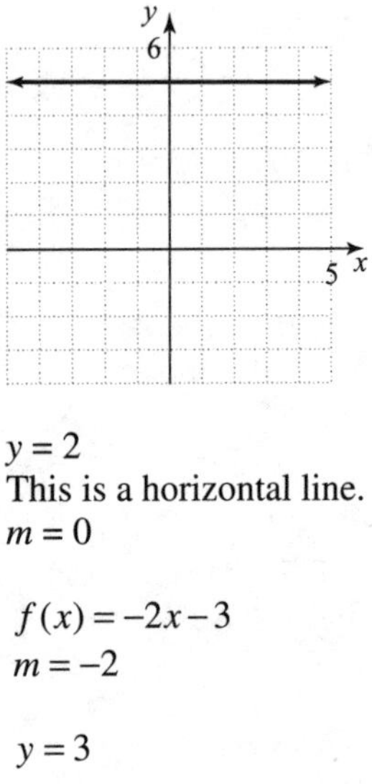

21. $y=2$
This is a horizontal line.
$m=0$

22. $f(x)=-2x-3$
$m=-2$

23. $y=3$

24. $x=-3$

25. $x+\frac{1}{2}y\ge -4$ or $y\le -2$

$\frac{1}{2}y\ge -x-4$ or $y\le -2$

$y\ge -2x-8$ or $y\le -2$

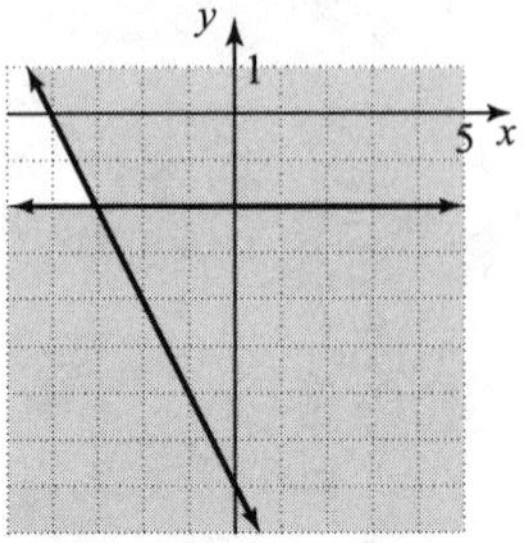

26. $y-3=0[x-(-2)]$
$y-3=0$
$y=3$

27. $\begin{cases}2x+4y=-6 & (1)\\ x=2y-5 & (2)\end{cases}$

Substitute $2y-5$ for x in E1.

$$\begin{aligned}2(2y-5)+4y&=-6\\4y-10+4y&=-6\\8y&=4\\y&=\frac{1}{2}\end{aligned}$$

Substitute $\frac{1}{2}$ for y in E2.

$x=2\left(\frac{1}{2}\right)-5=1-5=-4$

The solution is $\left(-4, \frac{1}{2}\right)$.

28. $\begin{cases}4x-2y=8 & (1)\\ y=3x-6 & (2)\end{cases}$

Substitute $3x-6$ for y in E1.

$$\begin{aligned}4x-2(3x-6)&=8\\4x-6x+12&=8\\-2x&=-4\\x&=2\end{aligned}$$

Substitute 2 for x in E2.

$y=3(2)-6=6-6=0$

The solution is (2, 0).

29. $\begin{cases}2x+4y \quad =1 & (1)\\ 4x \quad -4z=-1 & (2)\\ y-4z=-3 & (3)\end{cases}$

Multiply E2 by –1 and add to E3.

$$\begin{array}{l} -4x \quad +4z = 1 \\ \quad\quad y - 4z = -3 \\ \hline -4x + y \quad = -2 \;\; (4) \end{array}$$

Muliply E1 by 2 and add to E4.

$$\begin{array}{r} 4x + 8y = 2 \\ -4x + y = -2 \\ \hline 9y = 0 \\ y = 0 \end{array}$$

Replace y with 0 in E1.

$$2x + 4(0) = 1$$
$$2x = 1$$
$$x = \frac{1}{2}$$

Replace y with 0 in E3.

$$y - 4z = -3$$
$$-4z = -3$$
$$z = \frac{3}{4}$$

The solution is $\left(\frac{1}{2}, 0, \frac{3}{4}\right)$.

30. $\begin{cases} x + y - \frac{3}{2}z = \frac{1}{2} & (1) \\ -y - 2z = 14 & (2) \\ x - \frac{2}{3}y = -\frac{1}{3} & (3) \end{cases}$

Multiply E1 by 2 and E3 by 3 to clear fractions.

$\begin{cases} 2x + 2y - 3z = 1 & (1) \\ -y - 2z = 14 & (2) \\ 3x - 2y = -1 & (3) \end{cases}$

Add E1 and E3.

$5x - 3z = 0$ (4)

Multiply E2 by 2 and add to E1.

$$\begin{array}{l} -2y - 4z = 28 \\ 2x + 2y - 3z = 1 \\ \hline 2x \quad -7z = 29 \;\; (5) \end{array}$$

Solve the new system $\begin{cases} 5x - 3z = 0 & (4) \\ 2x - 7z = 29 & (5) \end{cases}$.

Multiply E4 by –2, multiply E5 by 5, and add.

$$\begin{array}{r} -10x + 6z = 0 \\ 10x - 35z = 145 \\ \hline -29z = 145 \\ z = -5 \end{array}$$

Replace z with –5 in E4.

$$5x - 3(-5) = 0$$
$$5x + 15 = 0$$
$$5x = -15$$
$$x = -3$$

Replace z with –5 in E2.

$$-y - 2(-5) = 14$$
$$-y + 10 = 14$$
$$-y = 4$$
$$y = -4$$

The solution is (–3, –4, –5).

31. Let x = the first number and y = the second number.

$\begin{cases} x = y - 4 & (1) \\ 4x = 2y + 6 & (2) \end{cases}$

Multiply E1 by –4 and add to E2.

$$\begin{array}{r} -4x = -4y + 16 \\ 4x = 2y + 6 \\ \hline 0 = -2y + 22 \\ 2y = 22 \\ y = 11 \end{array}$$

Replace y with 11 in E1.

$x = 11 - 4 = 7$

The numbers are 7 and 11.

32. Let x = ounces of 20% solution and y = ounces of 60% solution.

$\begin{cases} x + y = 50 & (1) \\ 0.20x + 0.60y = 50(0.30) & (2) \end{cases}$

Multiply E2 by 100 to clear decimals.

$\begin{cases} x + y = 50 & (1) \\ 20x + 60y = 1500 & (2) \end{cases}$

Multiply E1 by –20 and add to E2.

$$\begin{array}{r} -20x - 20y = -1000 \\ 20x + 60y = 1500 \\ \hline 40y = 500 \end{array}$$
$$y = \frac{500}{40} = 12.5$$

Replace y with 12.5 in E1.

$$x + 12.5 = 50$$
$$x = 37.5$$

You should mix 37.5 ounces of the 20% solution and 12.5 ounces of the 60% solution.

33. $\begin{cases} 2x - y = 3 \\ 4x - 2y = 5 \end{cases}$

$$\left[\begin{array}{rr|r} 2 & -1 & 3 \\ 4 & -2 & 5 \end{array}\right]$$

Divide R1 by 2.

$$\left[\begin{array}{rr|r} 1 & -\frac{1}{2} & \frac{3}{2} \\ 4 & -2 & 5 \end{array}\right]$$

Multiply R1 by –4 and add to R2.

$$\left[\begin{array}{rr|r} 1 & -\frac{1}{2} & \frac{3}{2} \\ 0 & 0 & -1 \end{array}\right]$$

This corresponds to $\begin{cases} x-\frac{1}{2}y=\frac{3}{2} \\ 0=-1 \end{cases}$. The last equation is impossible. The system is inconsistent. The solution set is ∅.

34. $\begin{cases} 4y=8 \\ x+y=7 \end{cases}$

$\left[\begin{array}{cc|c} 0 & 4 & 8 \\ 1 & 1 & 7 \end{array}\right]$

Interchange R1 and R2.

$\left[\begin{array}{cc|c} 1 & 1 & 7 \\ 0 & 4 & 8 \end{array}\right]$

Divide R2 by 4.

$\left[\begin{array}{cc|c} 1 & 1 & 7 \\ 0 & 1 & 2 \end{array}\right]$

This corresponds to $\begin{cases} x+y=7 \\ y=2 \end{cases}$.

Replace y with 2 in the equation $x + y = 7$.

$x+2=7$
$x=5$

The solution is (5, 2).

35. Let x = measure of smallest angle, then $x + 80$ = measure of largest angle, and $x + 10$ = measure of remaining angle.

$x+(x+80)+(x+10)=180$
$3x+90=180$
$3x=90$
$x=30$

$x + 80 = 110$
$x + 10 = 40$

The angles measure 30°, 110°, and 40°.

36. $m=\frac{1}{2}$, y-intercept (0, 5), $b = 5$

$y=mx+b$
$y=\frac{1}{2}x+5$
$f(x)=\frac{1}{2}x+5$

37. a. $730,000=7.3\times10^5$

b. $0.00000104=1.04\times10^{-6}$

38. a. $8,250,000=8.25\times10^6$

b. $0.0000346=3.46\times10^{-5}$

39. a. $(2x^0y^{-3})^{-2}=2^{-2}(1)^{-2}(y^{-3})^{-2}=\frac{y^6}{2^2}=\frac{y^6}{4}$

b. $\left(\frac{x^{-5}}{x^{-2}}\right)^{-3}=\frac{(x^{-5})^{-3}}{(x^{-2})^{-3}}=\frac{x^{15}}{x^6}=x^{15-6}=x^9$

c. $\left(\frac{2}{7}\right)^{-2}=\frac{2^{-2}}{7^{-2}}=\frac{7^2}{2^2}=\frac{49}{4}$

d. $\frac{5^{-2}x^{-3}y^{11}}{x^2y^{-5}}=\frac{x^{-3-2}y^{11-(-5)}}{5^2}$
$=\frac{x^{-5}y^{16}}{25}$
$=\frac{y^{16}}{25x^5}$

40. a. $(4a^{-1}b^0)^{-3}=4^{-3}(a^{-1})^{-3}(1)^{-3}=\frac{a^3}{4^3}=\frac{a^3}{64}$

b. $\left(\frac{a^{-6}}{a^{-8}}\right)^{-2}=\frac{(a^{-6})^{-2}}{(a^{-8})^{-2}}$
$=\frac{a^{12}}{a^{16}}$
$=a^{12-16}$
$=a^{-4}$
$=\frac{1}{a^4}$

c. $\left(\frac{2}{3}\right)^{-3}=\frac{2^{-3}}{3^{-3}}=\frac{3^3}{2^3}=\frac{27}{8}$

d. $\frac{3^{-2}a^{-2}b^{12}}{a^4b^{-5}}=\frac{a^{-2-4}b^{12-(-5)}}{3^2}$
$=\frac{a^{-6}b^{17}}{9}$
$=\frac{b^{17}}{9a^6}$

41. The degree is the degree of the term x^2y^2, which is 2 + 2 = 4.

42. $(3x^2-2x)-(5x^2+3x)=3x^2-2x-5x^2-3x$
$=-2x^2-5x$

43. **a.** $(2x^3)(5x^6) = 2(5)x^{3+6} = 10x^9$

b. $(7y^4z^4)(-xy^{11}z^5) = -7xy^{4+11}z^{4+5}$
$= -7xy^{15}z^9$

44. **a.** $(3y^6)(4y^2) = 3(4)y^{6+2} = 12y^8$

b. $(6a^3b^2)(-a^2bc^4) = -6a^{3+2}b^{2+1}c^4$
$= -6a^5b^3c^4$

45. $17x^3y^2 - 34x^4y^2 = 17x^3y^2(1-2x)$

46. $12x^3y - 3xy^3 = 3xy(4x^2 - y^2)$
$= 3xy((2x)^2 - y^2)$
$= 3xy(2x+y)(2x-y)$

47. $x^2 + 10x + 16 = (x+8)(x+2)$

48. $5a^2 + 14a - 3 = (5a-1)(a+3)$

49. $2x^2 + 9x - 5 = 0$
$(2x-1)(x+5) = 0$
$2x - 1 = 0$ or $x + 5 = 0$
$2x = 1$ or $x = -5$
$x = \frac{1}{2}$

The solution is $-5, \frac{1}{2}$.

50. $3x^2 - 10x - 8 = 0$
$(3x+2)(x-4) = 0$
$3x + 2 = 0$ or $x - 4 = 0$
$3x = -2$ or $x = 4$
$x = -\frac{2}{3}$

The solution is $-\frac{2}{3}, 4$.

Chapter 6

Section 6.1 Practice Exercises

1. a. The denominator of $f(x)$ is never 0.
Domain: $\{x|x \text{ is a real number}\}$

b. Undefined values when
$x+3=0$, or $x=-3$
Domain: $\{x|x \text{ is a real number and } x \neq -3\}$

c. Undefined values when

$$x^2-5x+6=0$$
$$(x-3)(x-2)=0$$
$$x-3=0 \quad \text{or} \quad x-2=0$$
$$x=3 \quad \text{or} \quad x=2$$

Domain:
$\{x|x \text{ is a real number and } x \neq 2,\ x \neq 3\}$

2. a. $$\frac{5z^4}{10z^5-5z^4}=\frac{5z^4\cdot 1}{5z^4(2z-1)}=1\cdot\frac{1}{2z-1}=\frac{1}{2z-1}$$

b. $$\frac{5x^2+13x+6}{6x^2+7x-10}=\frac{(5x+3)(x+2)}{(6x-5)(x+2)}=\frac{5x+3}{6x-5}\cdot 1=\frac{5x+3}{6x-5}$$

3. a. $$\frac{x+3}{3+x}=\frac{x+3}{x+3}=1$$

b. $$\frac{3-x}{x-3}=\frac{-1(-3+x)}{x-3}=\frac{-1(x-3)}{x-3}=\frac{-1}{1}=-1$$

4. $$\frac{20-5x^2}{x^2+x-6}=\frac{5(4-x^2)}{(x+3)(x-2)}=\frac{5(2+x)(2-x)}{(x+3)(x-2)}=\frac{5(2+x)\cdot(-1)(x-2)}{(x+3)(x-2)}=-\frac{5(2+x)}{x+3}$$

5. a. $$\frac{x^3+64}{4+x}=\frac{(x+4)(x^2-4x+16)}{x+4}=x^2-4x+16$$

b. $$\frac{5z^2+10}{z^3-3z^2+2z-6}=\frac{5(z^2+2)}{(z^3-3z^2)+(2z-6)}=\frac{5(z^2+2)}{z^2(z-3)+2(z-3)}=\frac{5(z^2+2)}{(z-3)(z^2+2)}=\frac{5}{z-3}$$

6. a. $$\frac{2+5n}{3n}\cdot\frac{6n+3}{5n^2-3n-2}=\frac{2+5n}{3n}\cdot\frac{3(2n+1)}{(5n+2)(n-1)}=\frac{2n+1}{n(n-1)}$$

b. $$\frac{x^3-8}{-6x+12}\cdot\frac{6x^2}{x^2+2x+4}=\frac{(x-2)(x^2+2x+4)}{-6(x-2)}\cdot\frac{6x^2}{x^2+2x+4}=\frac{(x-2)(x^2+2x+4)\cdot 6\cdot x^2}{-1\cdot 6(x-2)(x^2+2x+4)}=\frac{x^2}{-1}=-x^2$$

7. a. $$\frac{6y^3}{3y^2-27}\div\frac{42}{3-y}=\frac{6y^3}{3y^2-27}\cdot\frac{3-y}{42}=\frac{6y^3(3-y)}{3(y+3)(y-3)\cdot 42}=\frac{6y^3\cdot(-1)(y-3)}{3(y+3)(y-3)\cdot 6\cdot 7}=-\frac{y^3}{21(y+3)}$$

b. $\dfrac{10x^2+23x-5}{5x^2-51x+10} \div \dfrac{2x^2+9x+10}{7x^2-68x-20}$

$= \dfrac{10x^2+23x-5}{5x^2-51x+10} \cdot \dfrac{7x^2-68x-20}{2x^2+9x+10}$

$= \dfrac{(5x-1)(2x+5)}{(5x-1)(x-10)} \cdot \dfrac{(7x+2)(x-10)}{(2x+5)(x+2)}$

$= \dfrac{7x+2}{x+2}$

8. $\dfrac{x^2-16}{(x-4)^2} \cdot \dfrac{5x-20}{3x} \div \dfrac{x^2+x-12}{x}$

$= \dfrac{x^2-16}{(x-4)^2} \cdot \dfrac{5x-20}{3x} \cdot \dfrac{x}{x^2+x-12}$

$= \dfrac{(x+4)(x-4)}{(x-4)(x-4)} \cdot \dfrac{5(x-4)}{3x} \cdot \dfrac{x}{(x+4)(x-3)}$

$= \dfrac{5}{3(x-3)}$

9. a. $C(100) = \dfrac{3.2(100)+400}{100} = \dfrac{720}{100} = 7.2$

$7.20 per tee shirt

b. $C(1000) = \dfrac{3.2(1000)+400}{1000} = \dfrac{3600}{1000} = 3.6$

$3.60 per tee shirt

Graphing Calculator Explorations

1. $x^2-4=0$

$(x+2)(x-2)=0$

$x+2=0$ or $x-2=0$

$x=-2$ or $x=2$

Domain: $\{x|x$ is a real number and $x \neq -2, x \neq 2\}$

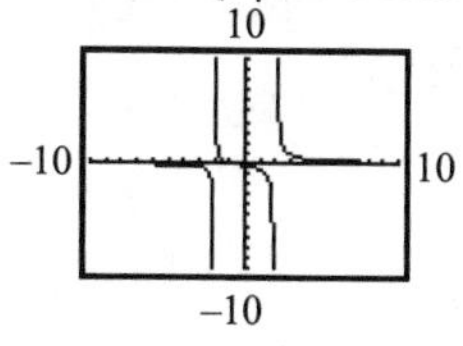

2. $x^2-9=0$

$(x+3)(x-3)=0$

$x+3=0$ or $x-3=0$

$x=-3$ or $x=3$

Domain: $\{x|x$ is a real number and $x \neq -3, x \neq 3\}$

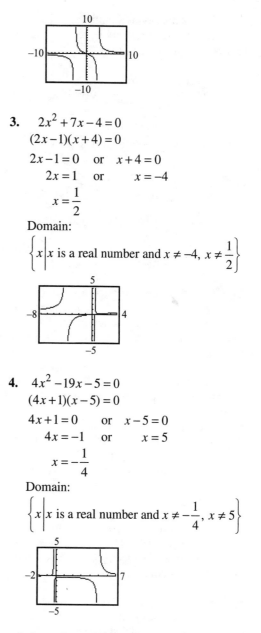

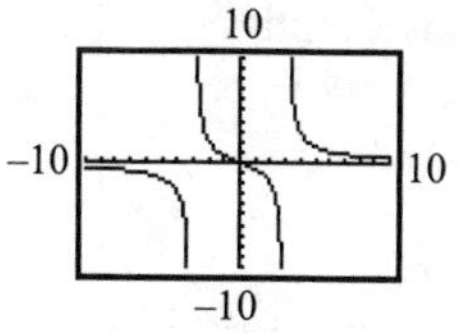

3. $2x^2+7x-4=0$

$(2x-1)(x+4)=0$

$2x-1=0$ or $x+4=0$

$2x=1$ or $x=-4$

$x=\dfrac{1}{2}$

Domain:

$\left\{x \middle| x \text{ is a real number and } x \neq -4,\ x \neq \dfrac{1}{2}\right\}$

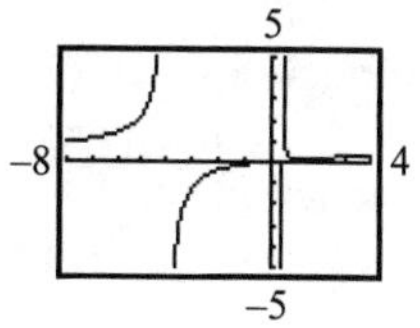

4. $4x^2-19x-5=0$

$(4x+1)(x-5)=0$

$4x+1=0$ or $x-5=0$

$4x=-1$ or $x=5$

$x=-\dfrac{1}{4}$

Domain:

$\left\{x \middle| x \text{ is a real number and } x \neq -\dfrac{1}{4},\ x \neq 5\right\}$

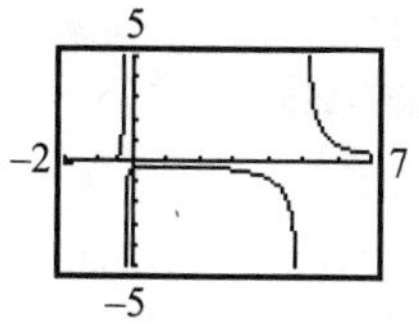

Vocabulary, Readiness & Video Check 6.1

1. A rational expression is an expression that can be written as the quotient $\dfrac{P}{Q}$ of two polynomials P and Q as long as $Q \neq 0$.

2. A rational expression is undefined if the denominator is 0.

3. The domain of the rational function $f(x)=\dfrac{2}{x}$ is $\{x|x$ is a real number and $x \neq 0\}$.

4. A rational expression is <u>simplified</u> if the numerator and denominator have no common factors other than 1 or −1.

5. The expression $\frac{x^2+2}{2+x^2}$ simplifies to <u>1</u>.

6. The expression $\frac{y-z}{z-y}$ simplifies to <u>−1</u>.

7. For a rational expression, $-\frac{a}{b}=\frac{-a}{\underline{b}}=\frac{a}{\underline{-b}}$.

8. The statement $\frac{a-6}{a+2}=\frac{-(a-6)}{-(a+2)}=\frac{-a+6}{-a-2}$ is true.

9. Rational expressions are fractions and are therefore undefined if the denominator is zero; the domain of a rational function consists of all real numbers except those for which the rational expression is undefined.

10. If the numerator and denominator of a rational expression are opposites of each other, the expression simplifies to −1.

11. Yes, multiplying and also simplifying rational expressions often require polynomial factoring. Example 6 involves factoring out a GCF and factoring a difference of squares.

12. Division of rational expressions is similar to division of <u>fractions</u>. Therefore, to divide rational expressions, it's the first rational expression times the <u>reciprocal</u> of the second rational expression.

13. Since the domain consists of all allowable x-values and x is defined as the number of years since the book was published, we do not allow x to be negative for the context of this application.

Exercise Set 6.1

1. 4 is never 0, so the domain of $f(x)=\frac{5x-7}{4}$ is $\{x|x \text{ is a real number}\}$.

3. $2t=0$
$t=0$
The domain of $s(t)=\frac{t^2+1}{2t}$ is $\{t|t \text{ is a real number and } t\neq 0\}$.

5. $7-x=0$
$7=x$
The domain of $f(x)=\frac{3x}{7-x}$ is $\{x|x \text{ is a real number and } x\neq 7\}$.

7. $3x-1=0$
$3x=1$
$x=\frac{1}{3}$
The domain of $f(x)=\frac{x}{3x-1}$ is $\left\{x\middle|x \text{ is a real number and } x\neq \frac{1}{3}\right\}$.

9. $x^3+x^2-2x=0$
$x(x^2+x-2)=0$
$x(x+2)(x-1)=0$
$x=0$ or $x+2=0$ or $x-1=0$
$x=0$ or $x=-2$ or $x=1$
The domain of $R(x)=\frac{3+2x}{x^3+x^2-2x}$ is $\{x|x \text{ is a real number and } x\neq -2, x\neq 0, x\neq 1\}$.

11. $x^2-4=0$
$(x+2)(x-2)=0$
$x+2=0$ or $x-2=0$
$x=-2$ or $x=2$
The domain of $C(x)=\frac{x+3}{x^2-4}$ is $\{x|x \text{ is a real number and } x\neq 2, x\neq -2\}$.

13. $\frac{8x-16x^2}{8x}=\frac{8x(1-2x)}{8x}=1-2x$

15. $\frac{x^2-9}{3+x}=\frac{(x+3)(x-3)}{3+x}=\frac{(x+3)(x-3)}{x+3}=x-3$

17. $\frac{9y-18}{7y-14}=\frac{9(y-2)}{7(y-2)}=\frac{9}{7}$

19. $\frac{x^2+6x-40}{x+10}=\frac{(x+10)(x-4)}{x+10}=x-4$

21. $\frac{x-9}{9-x}=\frac{-1(9-x)}{9-x}=-1$

23. $\frac{x^2-49}{7-x}=\frac{(x+7)(x-7)}{7-x}$
$=\frac{(x+7)(x-7)}{-1(-7+x)}$
$=\frac{(x+7)(x-7)}{-1(x-7)}$
$=-(x+7)$

25. $\frac{2x^2-7x-4}{x^2-5x+4}=\frac{(2x+1)(x-4)}{(x-1)(x-4)}=\frac{2x+1}{x-1}$

27. $\frac{x^3-125}{2x-10}=\frac{(x-5)(x^2+5x+25)}{2(x-5)}$
$=\frac{x^2+5x+25}{2}$

29. $\frac{3x^2-5x-2}{6x^3+2x^2+3x+1}=\frac{(3x+1)(x-2)}{2x^2(3x+1)+1(3x+1)}$
$=\frac{(3x+1)(x-2)}{(3x+1)(2x^2+1)}$
$=\frac{x-2}{2x^2+1}$

31. $\frac{9x^2-15x+25}{27x^3+125}=\frac{9x^2-15x+25}{(3x+5)(9x^2-15x+25)}$
$=\frac{1}{3x+5}$

33. $\frac{2x-4}{15}\cdot\frac{6}{2-x}=\frac{2(x-2)}{3\cdot 5}\cdot\frac{2\cdot 3}{-(x-2)}$
$=\frac{2\cdot 2}{5(-1)}$
$=-\frac{4}{5}$

35. $\frac{18a-12a^2}{4a^2+4a+1}\cdot\frac{4a^2+8a+3}{4a^2-9}$
$=\frac{6a(3-2a)}{(2a+1)(2a+1)}\cdot\frac{(2a+3)(2a+1)}{(2a+3)(2a-3)}$
$=\frac{-6a(2a-3)}{(2a+1)(2a+1)}\cdot\frac{(2a+3)(2a+1)}{(2a+3)(2a-3)}$
$=-\frac{6a}{2a+1}$

37. $\frac{9x+9}{4x+8}\cdot\frac{2x+4}{3x^2-3}=\frac{9(x+1)}{4(x+2)}\cdot\frac{2(x+2)}{3(x^2-1)}$
$=\frac{3\cdot 3(x+1)}{2\cdot 2(x+2)}\cdot\frac{2(x+2)}{3(x+1)(x-1)}$
$=\frac{3}{2(x-1)}$

39. $\frac{2x^3-16}{6x^2+6x-36}\cdot\frac{9x+18}{3x^2+6x+12}$
$=\frac{2(x^3-8)}{6(x^2+x-6)}\cdot\frac{9(x+2)}{3(x^2+2x+4)}$
$=\frac{2(x-2)(x^2+2x+4)}{2\cdot 3(x-2)(x+3)}\cdot\frac{3\cdot 3(x+2)}{3(x^2+2x+4)}$
$=\frac{x+2}{x+3}$

41. $\frac{a^3+a^2b+a+b}{5a^3+5a}\cdot\frac{6a^2}{2a^2-2b^2}$
$=\frac{a^2(a+b)+1(a+b)}{5a(a^2+1)}\cdot\frac{6a^2}{2(a^2-b^2)}$
$=\frac{(a+b)(a^2+1)}{5a(a^2+1)}\cdot\frac{2\cdot 3\cdot a\cdot a}{2(a+b)(a-b)}$
$=\frac{3a}{5(a-b)}$

43. $\frac{x^2-6x-16}{2x^2-128}\cdot\frac{x^2+16x+64}{3x^2+30x+48}$
$=\frac{(x-8)(x+2)}{2(x^2-64)}\cdot\frac{(x+8)(x+8)}{3(x^2+10x+16)}$
$=\frac{(x-8)(x+2)}{2(x+8)(x-8)}\cdot\frac{(x+8)(x+8)}{3(x+2)(x+8)}$
$=\frac{1}{2\cdot 3}$
$=\frac{1}{6}$

45. $\frac{2x}{5}\div\frac{6x+12}{5x+10}=\frac{2x}{5}\cdot\frac{5x+10}{6x+12}$
$=\frac{2x}{5}\cdot\frac{5(x+2)}{6(x+2)}$
$=\frac{2x}{5}\cdot\frac{5(x+2)}{2\cdot 3(x+2)}$
$=\frac{x}{3}$

47. $\frac{a+b}{ab} \div \frac{a^2-b^2}{4a^3b} = \frac{a+b}{ab} \cdot \frac{4a^3b}{a^2-b^2}$
$= \frac{a+b}{ab} \cdot \frac{4a^3b}{(a+b)(a-b)}$
$= \frac{4a^2}{a-b}$

49. $\frac{x^2-6x+9}{x^2-x-6} \div \frac{x^2-9}{4}$
$= \frac{x^2-6x+9}{x^2-x-6} \cdot \frac{4}{x^2-9}$
$= \frac{(x-3)^2}{(x-3)(x+2)} \cdot \frac{4}{(x+3)(x-3)}$
$= \frac{4}{(x+2)(x+3)}$

51. $\frac{x^2-6x-16}{2x^2-128} \div \frac{x^2+10x+16}{x^2+16x+64}$
$= \frac{x^2-6x-16}{2x^2-128} \cdot \frac{x^2+16x+64}{x^2+10x+16}$
$= \frac{(x-8)(x+2)}{2(x^2-64)} \cdot \frac{(x+8)(x+8)}{(x+2)(x+8)}$
$= \frac{(x-8)(x+2)}{2(x-8)(x+8)} \cdot \frac{(x+8)(x+8)}{(x+2)(x+8)}$
$= \frac{1}{2}$

53. $\frac{3x-x^2}{x^3-27} \div \frac{x}{x^2+3x+9}$
$= \frac{3x-x^2}{x^3-27} \cdot \frac{x^2+3x+9}{x}$
$= \frac{x(3-x)}{(x-3)(x^2+3x+9)} \cdot \frac{x^2+3x+9}{x}$
$= \frac{-x(x-3)}{(x-3)(x^2+3x+9)} \cdot \frac{x^2+3x+9}{x}$
$= -1$

55. $\frac{8b+24}{3a+6} \div \frac{ab-2b+3a-6}{a^2-4a+4}$
$= \frac{8b+24}{3a+6} \cdot \frac{a^2-4a+4}{ab-2b+3a-6}$
$= \frac{8(b+3)}{3(a+2)} \cdot \frac{(a-2)(a-2)}{b(a-2)+3(a-2)}$
$= \frac{8(b+3)}{3(a+2)} \cdot \frac{(a-2)(a-2)}{(a-2)(b+3)}$
$= \frac{8(a-2)}{3(a+2)}$

57. $\frac{x^2-9}{4} \cdot \frac{x^2-x-6}{x^2-6x+9}$
$= \frac{(x+3)(x-3)}{4} \cdot \frac{(x+2)(x-3)}{(x-3)^2}$
$= \frac{(x+3)(x+2)}{4}$

59. $\frac{2x^2-4x-30}{5x^2-40x-75} \div \frac{x^2-8x+15}{x^2-6x+9}$
$= \frac{2x^2-4x-30}{5x^2-40x-75} \cdot \frac{x^2-6x+9}{x^2-8x+15}$
$= \frac{2(x^2-2x-15)}{5(x^2-8x-15)} \cdot \frac{(x-3)(x-3)}{(x-3)(x-5)}$
$= \frac{2(x-5)(x+3)}{5(x^2-8x-15)} \cdot \frac{(x-3)(x-3)}{(x-3)(x-5)}$
$= \frac{2(x+3)(x-3)}{5(x^2-8x-15)}$

61. $\frac{r^3+s^3}{r+s} = \frac{(r+s)(r^2-rs+s^2)}{r+s} = r^2-rs+s^2$

63. $\frac{4}{x} \div \frac{3xy}{x^2} \cdot \frac{6x^2}{x^4} = \frac{4}{x} \cdot \frac{x^2}{3xy} \cdot \frac{6x^2}{x^4} = \frac{8}{x^2y}$

65. $\frac{3x^2-5x-2}{y^2+y-2} \cdot \frac{y^2+4y-5}{12x^2+7x+1} \div \frac{5x^2-9x-2}{8x^2-2x-1}$
$= \frac{3x^2-5x-2}{y^2+y-2} \cdot \frac{y^2+4y-5}{12x^2+7x+1} \cdot \frac{8x^2-2x-1}{5x^2-9x-2}$
$= \frac{(3x+1)(x-2)(y+5)(y-1)(4x+1)(2x-1)}{(y+2)(y-1)(4x+1)(3x+1)(5x+1)(x-2)}$
$= \frac{(y+5)(2x-1)}{(y+2)(5x+1)}$

67. $\frac{5a^2-20}{3a^2-12a} \div \frac{a^3+2a^2}{2a^2-8a} \cdot \frac{9a^3+6a^2}{2a^2-4a}$
$= \frac{5a^2-20}{3a^2-12a} \cdot \frac{2a^2-8a}{a^3+2a^2} \cdot \frac{9a^3+6a^2}{2a^2-4a}$
$= \frac{5(a^2-4)}{3a(a-4)} \cdot \frac{2a(a-4)}{a^2(a+2)} \cdot \frac{3a^2(3a+2)}{2a(a-2)}$
$= \frac{5(a+2)(a-2)}{3a(a-4)} \cdot \frac{2a(a-4)}{a^2(a+2)} \cdot \frac{3a^2(3a+2)}{2a(a-2)}$
$= \frac{5(3a+2)}{a}$

69. $\frac{5x^4+3x^2-2}{x-1} \cdot \frac{x+1}{x^4-1}$
$= \frac{(5x^2-2)(x^2+1)}{x-1} \cdot \frac{x+1}{(x^2+1)(x^2-1)}$
$= \frac{(5x^2-2)(x^2+1)}{x-1} \cdot \frac{x+1}{(x^2+1)(x+1)(x-1)}$
$= \frac{5x^2-2}{(x-1)^2}$

71. $f(x) = \frac{x+8}{2x-1}$
$f(2) = \frac{2+8}{2(2)-1} = \frac{10}{4-1} = \frac{10}{3}$
$f(0) = \frac{0+8}{2(0)-1} = \frac{8}{0-1} = \frac{8}{-1} = -8$
$f(-1) = \frac{-1+8}{2(-1)-1} = \frac{7}{-2-1} = \frac{7}{-3} = -\frac{7}{3}$

73. $g(x) = \frac{x^2+8}{x^3-25x}$
$g(3) = \frac{3^2+8}{3^3-25(3)} = \frac{9+8}{27-75} = \frac{17}{-48} = -\frac{17}{48}$
$g(-2) = \frac{(-2)^2+8}{(-2)^3-25(-2)} = \frac{4+8}{-8+50} = \frac{12}{42} = \frac{2}{7}$
$g(1) = \frac{1^2+8}{1^3-25(1)} = \frac{1+8}{1-25} = \frac{9}{-24} = -\frac{3}{8}$

75. $R(x) = \frac{1000x^2}{x^2+4}$

a. $R(1) = \frac{1000 \cdot 1^2}{1^2+4} = \frac{1000}{5} = 200$
The revenue at the end of the first year is \$200 million.

b. $R(2) = \frac{1000 \cdot 2^2}{2^2+4}$
$= \frac{1000 \cdot 4}{4+4}$
$= \frac{4000}{8}$
$= 500$
The revenue at the end of the second year is \$500 million.

c. The revenue during the second year is equal to the revenue at the end of the second year minus the revenue at the end of the first year.
$500 - 200 = 300$
The revenue during the second year is \$300 million.

d. $x^2+4=0$
$x^2=-4$
This equation has no solutions. Thus there are no values of x that would make the denominator equal to zero and the function undefined. The domain of $R(x)$ is $\{x | x \text{ is a real number}\}$.

77. $\frac{4}{5} + \frac{3}{5} = \frac{7}{5}$

79. $\frac{5}{28} - \frac{2}{21} = \frac{15}{84} - \frac{8}{84} = \frac{7}{84} = \frac{1}{12}$

81. $\frac{3}{8} + \frac{1}{2} - \frac{3}{16} = \frac{6}{16} + \frac{8}{16} - \frac{3}{16} = \frac{11}{16}$

83. a. $\frac{-x}{5-x} \neq \frac{x}{5-x}$

b. $\frac{-x}{-5+x} = \frac{x}{-(-5+x)} = \frac{x}{5-x}$

c. $\frac{x}{x-5}=\frac{x}{-5+x}\neq\frac{x}{5-x}$

d. $\frac{-x}{x-5}=\frac{x}{-(x-5)}=\frac{x}{-x+5}=\frac{x}{5-x}$

85. no; answers may vary

87. $A=l\cdot w$
$$=\left(\frac{x+2}{x}\right)\left(\frac{5x}{x^2-4}\right)$$
$$=\frac{x+2}{x}\cdot\frac{5x}{(x+2)(x-2)}$$
$$=\frac{5}{x-2}$$

The area is $\frac{5}{x-2}$ square meters.

89. Since $A = b\cdot h$, $b=\frac{A}{h}$.
$$b=\frac{\frac{x^2+x-2}{x^3}}{\frac{x^2}{x-1}}$$
$$=\frac{x^2+x-2}{x^3}\cdot\frac{x-1}{x^2}$$
$$=\frac{(x+2)(x-1)}{x^3}\cdot\frac{x-1}{x^2}$$
$$=\frac{(x+2)(x-1)^2}{x^5}$$

The length of the base is $\frac{(x+2)(x-1)^2}{x^5}$ feet.

91. answers may vary

93. a. $\frac{x+5}{5+x}=\frac{x+5}{x+5}=1$

b. $\frac{x-5}{5-x}=\frac{x-5}{-(x-5)}=-1$

c. $\frac{x+5}{x-5}$ neither

d. $\frac{-x-5}{x+5}=\frac{-(x+5)}{x+5}=-1$

e. $\frac{x-5}{-x+5}=\frac{x-5}{-(x-5)}=-1$

f. $\frac{-5+x}{x-5}=\frac{x-5}{x-5}=1$

95. answers may vary

97. $f(x)=\frac{20x}{100-x}$

x	0	10	30	50	70	90	95	99
y	0	$\frac{20}{9}$	$\frac{60}{7}$	20	$\frac{140}{3}$	180	380	1980

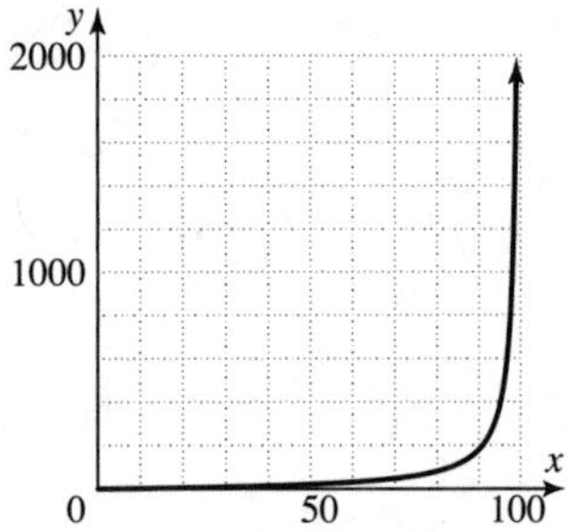

99. $\frac{p^x-4}{4-p^x}=\frac{p^x-4}{-(-4+p^x)}=\frac{p^x-4}{-(p^x-4)}=\frac{1}{-1}=-1$

101. $\frac{x^n+4}{x^{2n}-16}=\frac{x^n+4}{(x^n+4)(x^n-4)}=\frac{1}{x^n-4}$

103. $\frac{x^{2n}-4}{7x}\cdot\frac{14x^3}{x^n-2}=\frac{(x^n+2)(x^n-2)}{7x}\cdot\frac{14x^3}{x^n-2}$
$$=2x^2(x^n+2)$$

105. $\frac{y^{2n}+9}{10y}\cdot\frac{y^n-3}{y^{4n}-81}$
$$=\frac{y^{2n}+9}{10y}\cdot\frac{y^n-3}{(y^{2n}+9)(y^{2n}-9)}$$
$$=\frac{y^{2n}+9}{10y}\cdot\frac{y^n-3}{(y^{2n}+9)(y^n+3)(y^n-3)}$$
$$=\frac{1}{10y(y^n+3)}$$

107. $\dfrac{y^{2n}-y^n-2}{2y^n-4} \div \dfrac{y^{2n}-1}{1+y^n}$

$= \dfrac{y^{2n}-y^n-2}{2y^n-4} \cdot \dfrac{1+y^n}{y^{2n}-1}$

$= \dfrac{(y^n-2)(y^n+1)}{2(y^n-2)} \cdot \dfrac{1+y^n}{(y^n+1)(y^n-1)}$

$= \dfrac{1+y^n}{2(y^n-1)}$

Section 6.2 Practice Exercises

1. a. $\dfrac{9}{11z^2}+\dfrac{x}{11z^2}=\dfrac{9+x}{11z^2}$

b. $\dfrac{x}{8}+\dfrac{5x}{8}=\dfrac{x+5x}{8}=\dfrac{6x}{8}=\dfrac{3x}{4}$

c. $\dfrac{x^2}{x+4}-\dfrac{16}{x+4}=\dfrac{x^2-16}{x+4}$

$=\dfrac{(x+4)(x-4)}{x+4}$

$=x-4$

d. $\dfrac{z}{2a^2}-\dfrac{z+3}{2a^2}=\dfrac{z-(z+3)}{2a^2}$

$=\dfrac{z-z-3}{2a^2}$

$=-\dfrac{3}{2a^2}$

2. a. $\dfrac{7}{6x^3y^5}, \dfrac{2}{9x^2y^4}$

We factor each denominator.

$6x^3y^5 = 2\cdot 3\cdot x^3 \cdot y^5$

$9x^2y^4 = 3^2 \cdot x^2 \cdot y^4$

$\text{LCD} = 2\cdot 3^2 \cdot x^3 \cdot y^5 = 18x^3y^5$

b. $\dfrac{11}{x-2}, \dfrac{x}{x-3}$

The denominators $x - 2$ and $x - 3$ do not factor further.

$\text{LCD} = (x-2)(x-3)$

c. $\dfrac{b+2}{b^2-16}, \dfrac{8}{b^2-8b+16}, \dfrac{5b}{2b^2-5b-12}$

We factor each denominator.

$b^2-16=(b-4)(b+4)$

$b^2-8b+16=(b-4)(b-4)$

$2b^2-5b-12=(2b+3)(b-4)$

$\text{LCD}=(b-4)^2(b+4)(2b+3)$

d. $\dfrac{y}{y^2-9}, \dfrac{3}{12-4y}$

We factor each denominator.

$y^2-9=(y-3)(y+3)$

$12-4y=4(3-y)=4(-1)(y-3)$

$\text{LCD}=-4(y-3)(y+3)$

3. a. The LCD is $5p^4q$.

$\dfrac{4}{p^3q}+\dfrac{3}{5p^4q}=\dfrac{4\cdot 5p}{p^3q\cdot 5p}+\dfrac{3}{5p^4q}$

$=\dfrac{20p}{5p^4q}+\dfrac{3}{5p^4q}$

$=\dfrac{20p+3}{5p^4q}$

b. The LCD is the product of the two denominators: $(y + 3)(y - 3)$.

$\dfrac{4}{y+3}+\dfrac{5y}{y-3}$

$=\dfrac{4\cdot(y-3)}{(y+3)\cdot(y-3)}+\dfrac{5y\cdot(y+3)}{(y-3)\cdot(y+3)}$

$=\dfrac{4y-12}{(y+3)(y-3)}+\dfrac{5y^2+15y}{(y+3)(y-3)}$

$=\dfrac{4y-12+5y^2+15y}{(y+3)(y-3)}$

$=\dfrac{5y^2+19y-12}{(y+3)(y-3)}$

c. The LCD is either $z - 5$ or $5 - z$.

$$\frac{3z-18}{z-5}-\frac{3}{5-z}=\frac{3z-18}{z-5}-\frac{3}{-1(z-5)}$$
$$=\frac{3z-18}{z-5}-\frac{-1\cdot 3}{z-5}$$
$$=\frac{3z-18-(-3)}{z-5}$$
$$=\frac{3z-18+3}{z-5}$$
$$=\frac{3z-15}{z-5}$$
$$=\frac{3(z-5)}{z-5}$$
$$=3$$

4. $$\frac{t}{t^2-25}-\frac{3}{t^2-3t-10}$$
$$=\frac{t}{(t+5)(t-5)}-\frac{3}{(t-5)(t+2)}$$

The LCD is $(t + 5)(t - 5)(t + 2)$.

$$\frac{t}{(t+5)(t-5)}-\frac{3}{(t-5)(t+2)}$$
$$=\frac{t\cdot(t+2)}{(t+5)(t-5)\cdot(t+2)}-\frac{3\cdot(t+5)}{(t-5)(t+2)\cdot(t+5)}$$
$$=\frac{t^2+2t}{(t+5)(t-5)(t+2)}-\frac{3t+15}{(t+5)(t-5)(t+2)}$$
$$=\frac{t^2+2t-3t-15}{(t+5)(t-5)(t+2)}$$
$$=\frac{t^2-t-15}{(t+5)(t-5)(t+2)}$$

5. $$\frac{2x+3}{3x^2-5x-2}+\frac{x-6}{6x^2-13x-5}$$
$$=\frac{2x+3}{(3x+1)(x-2)}+\frac{x-6}{(3x+1)(2x-5)}$$

The LCD is $(3x + 1)(x - 2)(2x - 5)$.

$$=\frac{(2x+3)\cdot(2x-5)}{(3x+1)(x-2)\cdot(2x-5)}+\frac{(x-6)\cdot(x-2)}{(3x+1)(2x-5)\cdot(x-2)}$$
$$=\frac{4x^2-4x-15}{(3x+1)(x-2)(2x-5)}+\frac{x^2-8x+12}{(3x+1)(x-2)(2x-5)}$$
$$=\frac{4x^2-4x-15+x^2-8x+12}{(3x+1)(x-2)(2x-5)}$$
$$=\frac{5x^2-12x-3}{(3x+1)(x-2)(2x-5)}$$

6. $$\frac{2}{x-2}+\frac{3x}{x^2-x-2}-\frac{1}{x+1}$$
$$=\frac{2}{x-2}+\frac{3x}{(x-2)(x+1)}-\frac{1}{x+1}$$

The LCD is $(x - 2)(x - 1)$.

$$=\frac{2\cdot(x+1)}{(x-2)\cdot(x+1)}+\frac{3x}{(x-2)(x+1)}-\frac{1\cdot(x-2)}{(x+1)\cdot(x-2)}$$
$$=\frac{2x+2}{(x-2)(x+1)}+\frac{3x}{(x-2)(x+1)}-\frac{x-2}{(x-2)(x+1)}$$
$$=\frac{2x+2+3x-x+2}{(x-2)(x+1)}$$
$$=\frac{4x+4}{(x-2)(x+1)}$$
$$=\frac{4(x+1)}{(x-2)(x+1)}$$
$$=\frac{4}{x-2}$$

Vocabulary, Readiness & Video Check 6.2

1. The denominators must be the same before performing the operations of addition and subtraction (a, b).

2. To perform the operation of division (d), you multiply the first rational expression by the reciprocal of the second rational expression.

3. Numerator times numerator all over denominator times denominator is multiplication (c).

4. The operations of addition and multiplication (a, c) are commutative (order doesn't matter).

5. Addition: $\frac{5}{y}+\frac{7}{y}=\frac{12}{y}$

6. Subtraction: $\frac{5}{y}-\frac{7}{y}=-\frac{2}{y}$

7. Multiplication: $\frac{5}{y}\cdot\frac{7}{y}=\frac{35}{y^2}$

8. Division: $\frac{5}{y}\div\frac{7}{y}=\frac{5}{y}\cdot\frac{y}{7}=\frac{5}{7}$

9. We need to be sure we subtract the entire second numerator—that is, make sure we "distribute" the subtraction to each term in the second numerator.

10. The LCD contains a factor the greatest number of times it appears in any *one* factored denominator. $(a - b)$ appears at most two times in any one factored denominator.

11. To write an equivalent rational expression, you multiply the numerator of the expression by the exact same thing as the denominator. This is the same as multiplying the original rational expression by one, which doesn't change the value of the original expression.

Exercise Set 6.2

1. $\frac{5}{2x}-\frac{x+1}{2x}=\frac{5-(x+1)}{2x}=\frac{-x+4}{2x}$

3. $\frac{y+11}{y-2}-\frac{y-5}{y-2}=\frac{y+11-(y-5)}{y-2}=\frac{16}{y-2}$

5. $\frac{2}{xz^2}-\frac{5}{xz^2}=\frac{2-5}{xz^2}=\frac{-3}{xz^2}=-\frac{3}{xz^2}$

7. $\frac{2}{x-2}+\frac{x}{x-2}=\frac{2+x}{x-2}=\frac{x+2}{x-2}$

9. $\frac{x^2}{x+2}-\frac{4}{x+2}=\frac{x^2-4}{x+2}$
$=\frac{(x+2)(x-2)}{x+2}$
$=x-2$

11. $\frac{2x-6}{x^2+x-6}+\frac{3-3x}{x^2+x-6}=\frac{2x-6+3-3x}{x^2+x-6}$
$=\frac{-x-3}{x^2+x-6}$
$=\frac{-1(x+3)}{(x+3)(x-2)}$
$=\frac{-1}{x-2}$ or $\frac{1}{-(x-2)}$
$=-\frac{1}{x-2}$ or $\frac{1}{2-x}$

13. $\frac{x-5}{2x}-\frac{x+5}{2x}=\frac{x-5-x-5}{2x}=\frac{-10}{2x}=-\frac{5}{x}$

15. $7 = 7$
$5x = 5x$
$\text{LCD} = 7 \cdot 5x = 35x$

17. $x = x$
$x + 1 = x + 1$
$\text{LCD} = x(x + 1)$

19. $x + 7 = x + 7$
$x - 7 = x - 7$
$\text{LCD} = (x + 7)(x - 7)$

21. $3x + 6 = 3(x + 2)$
$2x - 4 = 2(x - 2)$
$\text{LCD} = 3 \cdot 2(x + 2)(x - 2) = 6(x + 2)(x - 2)$

23. $a^2-b^2=(a+b)(a-b)$
$a^2-2ab+b^2=(a-b)^2$
$\text{LCD}=(a+b)(a-b)^2$

25. $x^2-9=(x+3)(x-3)$
$x = x$
$12 - 4x = -4(x - 3)$
$\text{LCD} = -4x(x + 3)(x - 3)$

27. $\frac{4}{3x}+\frac{3}{2x}=\frac{4\cdot 2}{3x(2)}+\frac{3\cdot 3}{2x(3)}=\frac{8}{6x}+\frac{9}{6x}=\frac{17}{6x}$

29. $\frac{5}{2y^2}-\frac{2}{7y}=\frac{5\cdot 7}{2y^2\cdot 7}-\frac{2\cdot 2y}{7y\cdot 2y}$
$=\frac{35}{14y^2}-\frac{4y}{14y^2}$
$=\frac{35-4y}{14y^2}$

31. $\frac{x-3}{x+4}-\frac{x+2}{x-4}$
$=\frac{(x-3)(x-4)}{(x+4)(x-4)}-\frac{(x+2)(x+4)}{(x-4)(x+4)}$
$=\frac{x^2-7x+12}{(x+4)(x-4)}-\frac{x^2+6x+8}{(x+4)(x-4)}$
$=\frac{x^2-7x+12-x^2-6x-8}{(x+4)(x-4)}$
$=\frac{-13x+4}{(x+4)(x-4)}$

33. $\dfrac{1}{x-5}-\dfrac{19-2x}{(x-5)(x+4)}$
$=\dfrac{1\cdot(x+4)}{(x-5)\cdot(x+4)}-\dfrac{19-2x}{(x-5)(x+4)}$
$=\dfrac{x+4+19-2x}{(x-5)(x+4)}$
$=\dfrac{3x-15}{(x-5)(x+4)}$
$=\dfrac{3(x-5)}{(x-5)(x+4)}$
$=\dfrac{3}{x+4}$

35. $\dfrac{1}{a-b}+\dfrac{1}{b-a}=\dfrac{1}{a-b}+\dfrac{1}{-1(-b+a)}$
$=\dfrac{1}{a-b}+\dfrac{-1}{a-b}$
$=\dfrac{0}{a-b}$
$=0$

37. $\dfrac{x+1}{1-x}+\dfrac{1}{x-1}=\dfrac{x+1}{-(x-1)}+\dfrac{1}{x-1}$
$=\dfrac{-(x+1)}{x-1}+\dfrac{1}{x-1}$
$=\dfrac{-x-1+1}{x-1}$
$=\dfrac{-x}{x-1}$
$=-\dfrac{x}{x-1}$

39. $\dfrac{5}{x-2}+\dfrac{x+4}{2-x}=\dfrac{5}{x-2}+\dfrac{x+4}{-(-2+x)}$
$=\dfrac{5}{x-2}+\dfrac{-(x+4)}{x-2}$
$=\dfrac{5-x-4}{x-2}$
$=\dfrac{-x+1}{x-2}$

41. $\dfrac{y+1}{y^2-6y+8}-\dfrac{3}{y^2-16}$
$=\dfrac{y+1}{(y-2)(y-4)}-\dfrac{3}{(y+4)(y-4)}$
$=\dfrac{(y+1)(y+4)}{(y-2)(y-4)(y+4)}-\dfrac{3(y-2)}{(y-2)(y+4)(y-4)}$
$=\dfrac{(y+1)(y+4)-3(y-2)}{(y-2)(y-4)(y+4)}$
$=\dfrac{y^2+5y+4-3y+6}{(y-2)(y-4)(y+4)}$
$=\dfrac{y^2+2y+10}{(y-2)(y-4)(y+4)}$

43. $\dfrac{x+4}{3x^2+11x+6}+\dfrac{x}{2x^2+x-15}$
$=\dfrac{x+4}{(3x+2)(x+3)}+\dfrac{x}{(2x-5)(x+3)}$
$=\dfrac{(x+4)(2x-5)}{(3x+2)(x+3)(2x-5)}+\dfrac{x(3x+2)}{(2x-5)(x+3)(3x+2)}$
$=\dfrac{2x^2+3x-20}{(3x+2)(x+3)(2x-5)}+\dfrac{3x^2+2x}{(3x+2)(x+3)(2x-5)}$
$=\dfrac{2x^2+3x-20+3x^2+2x}{(3x+2)(x+3)(2x-5)}$
$=\dfrac{5x^2+5x-20}{(3x+2)(x+3)(2x-5)}$
$=\dfrac{5(x^2+x-4)}{(3x+2)(x+3)(2x-5)}$

45. $\dfrac{7}{x^2-x-2}-\dfrac{x-1}{x^2+4x+3}$
$=\dfrac{7}{(x-2)(x+1)}-\dfrac{x-1}{(x+3)(x+1)}$
$=\dfrac{7(x+3)}{(x-2)(x+1)(x+3)}-\dfrac{(x-1)(x-2)}{(x+3)(x+1)(x-2)}$
$=\dfrac{7(x+3)-(x-1)(x-2)}{(x-2)(x+1)(x+3)}$
$=\dfrac{7x+21-x^2+3x-2}{(x-2)(x+1)(x+3)}$
$=\dfrac{-x^2+10x+19}{(x-2)(x+1)(x+3)}$

47. $\dfrac{x}{x^2-8x+7}-\dfrac{x+2}{2x^2-9x-35}$
$=\dfrac{x}{(x-1)(x-7)}-\dfrac{x+2}{(2x+5)(x-7)}$
$=\dfrac{x(2x+5)}{(x-1)(x-7)(2x+5)}-\dfrac{(x+2)(x-1)}{(2x+5)(x-7)(x-1)}$
$=\dfrac{x(2x+5)-(x+2)(x-1)}{(2x+5)(x-7)(x-1)}$
$=\dfrac{2x^2+5x-(x^2+x-2)}{(2x+5)(x-7)(x-1)}$
$=\dfrac{2x^2+5x-x^2-x+2}{(2x+5)(x-7)(x-1)}$
$=\dfrac{x^2+4x+2}{(2x+5)(x-7)(x-1)}$

49. $\dfrac{2}{a^2+2a+1}+\dfrac{3}{a^2-1}$
$=\dfrac{2}{(a+1)^2}+\dfrac{3}{(a+1)(a-1)}$
$=\dfrac{2(a-1)}{(a+1)^2(a-1)}+\dfrac{3(a+1)}{(a+1)(a-1)(a+1)}$
$=\dfrac{2(a-1)+3(a+1)}{(a+1)^2(a-1)}$
$=\dfrac{2a-2+3a+3}{(a+1)^2(a-1)}$
$=\dfrac{5a+1}{(a+1)^2(a-1)}$

51. $\dfrac{4}{3x^2y^3}+\dfrac{5}{3x^2y^3}=\dfrac{9}{3x^2y^3}=\dfrac{3}{x^2y^3}$

53. $\dfrac{13x-5}{2x}-\dfrac{13x+5}{2x}=\dfrac{13x-5-13x-5}{2x}$
$=\dfrac{-10}{2x}$
$=-\dfrac{5}{x}$

55. $\dfrac{3}{2x+10}+\dfrac{8}{3x+15}=\dfrac{3}{2(x+5)}+\dfrac{8}{3(x+5)}$
$=\dfrac{3\cdot 3}{2(x+5)\cdot 3}+\dfrac{8\cdot 2}{3(x+5)\cdot 2}$
$=\dfrac{9}{6(x+5)}+\dfrac{16}{6(x+5)}$
$=\dfrac{25}{6(x+5)}$

57. $\dfrac{-2}{x^2-3x}-\dfrac{1}{x^3-3x^2}=\dfrac{-2}{x(x-3)}-\dfrac{1}{x^2(x-3)}$
$=\dfrac{-2x}{x^2(x-3)}-\dfrac{1}{x^2(x-3)}$
$=\dfrac{-2x-1}{x^2(x-3)}$

59. $\dfrac{ab}{a^2-b^2}+\dfrac{b}{a+b}$
$=\dfrac{ab}{(a+b)(a-b)}+\dfrac{b}{a+b}$
$=\dfrac{ab}{(a+b)(a-b)}+\dfrac{b(a-b)}{(a+b)(a-b)}$
$=\dfrac{ab}{(a+b)(a-b)}+\dfrac{ab-b^2}{(a+b)(a-b)}$
$=\dfrac{ab+ab-b^2}{(a+b)(a-b)}$
$=\dfrac{2ab-b^2}{(a+b)(a-b)}$
$=\dfrac{b(2a-b)}{(a+b)(a-b)}$

61. $\dfrac{5}{x^2-4}-\dfrac{3}{x^2+4x+4}$
$=\dfrac{5}{(x+2)(x-2)}-\dfrac{3}{(x+2)^2}$
$=\dfrac{5(x+2)}{(x+2)(x-2)(x+2)}-\dfrac{3(x-2)}{(x+2)^2(x-2)}$
$=\dfrac{5(x+2)-3(x-2)}{(x+2)^2(x-2)}$
$=\dfrac{5x+10-3x+6}{(x+2)^2(x-2)}$
$=\dfrac{2x+16}{(x+2)^2(x-2)}$
$=\dfrac{2(x+8)}{(x+2)^2(x-2)}$

63. $\frac{3x}{2x^2-11x+5}+\frac{7}{x^2-2x-15}=\frac{3x}{(2x-1)(x-5)}+\frac{7}{(x-5)(x+3)}$

$=\frac{3x(x+3)}{(2x-1)(x-5)(x+3)}+\frac{7(2x-1)}{(x-5)(x+3)(2x-1)}$

$=\frac{3x^2+9x}{(2x-1)(x-5)(x+3)}+\frac{14x-7}{(2x-1)(x-5)(x+3)}$

$=\frac{3x^2+9x+14x-7}{(2x-1)(x-5)(x+3)}$

$=\frac{3x^2+23x-7}{(2x-1)(x-5)(x+3)}$

65. $\frac{2}{x+1}-\frac{3x}{3x+3}+\frac{1}{2x+2}=\frac{2}{x+1}-\frac{3x}{3(x+1)}+\frac{1}{2(x+1)}$

$=\frac{2}{x+1}-\frac{x}{x+1}+\frac{1}{2(x+1)}$

$=\frac{2\cdot 2}{2(x+1)}-\frac{2\cdot x}{2(x+1)}+\frac{1}{2(x+1)}$

$=\frac{4-2x+1}{2(x+1)}$

$=\frac{5-2x}{2(x+1)}$

67. $\frac{3}{x+3}+\frac{5}{x^2+6x+9}-\frac{x}{x^2-9}=\frac{3}{x+3}+\frac{5}{(x+3)^2}-\frac{x}{(x+3)(x-3)}$

$=\frac{3(x+3)(x-3)}{(x+3)(x+3)(x-3)}+\frac{5(x-3)}{(x+3)^2(x-3)}-\frac{x(x+3)}{(x+3)(x-3)(x+3)}$

$=\frac{3x^2-27}{(x+3)^2(x-3)}+\frac{5x-15}{(x+3)^2(x-3)}-\frac{x^2+3x}{(x+3)^2(x-3)}$

$=\frac{3x^2-27+5x-15-x^2-3x}{(x+3)^2(x-3)}$

$=\frac{2x^2+2x-42}{(x+3)^2(x-3)}$

$=\frac{2(x^2+x-21)}{(x+3)^2(x-3)}$

69. $$\frac{x}{x^2-9}+\frac{3}{x^2-6x+9}-\frac{1}{x+3}=\frac{x}{(x+3)(x-3)}+\frac{3}{(x-3)^2}-\frac{1}{x+3}$$
$$=\frac{x(x-3)}{(x+3)(x-3)^2}+\frac{3(x+3)}{(x-3)^2(x+3)}-\frac{1(x-3)^2}{(x+3)(x-3)^2}$$
$$=\frac{x(x-3)+3(x+3)-(x-3)^2}{(x+3)(x-3)^2}$$
$$=\frac{x^2-3x+3x+9-(x^2-6x+9)}{(x+3)(x-3)^2}$$
$$=\frac{x^2+9-x^2+6x-9}{(x+3)(x-3)^2}$$
$$=\frac{6x}{(x+3)(x-3)^2}$$

71. $$\left(\frac{1}{x}+\frac{2}{3}\right)-\left(\frac{1}{x}-\frac{2}{3}\right)=\left(\frac{3}{3x}+\frac{2x}{3x}\right)-\left(\frac{3}{3x}-\frac{2x}{3x}\right)$$
$$=\left(\frac{3+2x}{3x}\right)-\left(\frac{3-2x}{3x}\right)$$
$$=\frac{3+2x-3+2x}{3x}$$
$$=\frac{4x}{3x}$$
$$=\frac{4}{3}$$

73. $$\left(\frac{2}{3}-\frac{1}{x}\right)\cdot\left(\frac{3}{x}+\frac{1}{2}\right)=\left(\frac{2x}{3x}-\frac{3}{3x}\right)\cdot\left(\frac{3\cdot 2}{2x}+\frac{x}{2x}\right)$$
$$=\left(\frac{2x-3}{3x}\right)\cdot\left(\frac{6+x}{2x}\right)$$
$$=\frac{(2x-3)(x+6)}{6x^2}\text{ or }\frac{2x^2+9x-18}{6x^2}$$

75. $$\left(\frac{2a}{3}\right)^2\div\left(\frac{a^2}{a+1}-\frac{1}{a+1}\right)=\frac{4a^2}{9}\div\frac{a^2-1}{a+1}$$
$$=\frac{4a^2}{9}\div\frac{(a+1)(a-1)}{a+1}$$
$$=\frac{4a^2}{9}\div(a-1)$$
$$=\frac{4a^2}{9}\cdot\frac{1}{a-1}$$
$$=\frac{4a^2}{9(a-1)}$$

77. $\left(\frac{2x}{3}\right)^2 \div \left(\frac{x}{3}\right)^2 = \left(\frac{2^2x^2}{3^2}\right) \div \left(\frac{x^2}{3^2}\right)$
$= \frac{2^2x^2}{3^2} \cdot \frac{3^2}{x^2}$
$= 2^2$
$= 4$

79. $\left(\frac{x}{x+1} - \frac{x}{x-1}\right) \div \frac{x}{2x+2}$
$= \left(\frac{x \cdot (x-1)}{(x+1)(x-1)} - \frac{x \cdot (x+1)}{(x-1)(x+1)}\right) \div \frac{x}{2(x+1)}$
$= \frac{x(x-1) - x(x+1)}{(x+1)(x-1)} \div \frac{x}{2(x+1)}$
$= \frac{x^2 - x - x^2 - x}{(x+1)(x-1)} \div \frac{x}{2(x+1)}$
$= \frac{-2x}{(x+1)(x-1)} \cdot \frac{2(x+1)}{x}$
$= -\frac{4}{x-1}$

81. $\frac{4}{x} \cdot \left(\frac{2}{x+2} - \frac{2}{x-2}\right)$
$= \frac{4}{x} \cdot \left(\frac{2(x-2)}{(x+2)(x-2)} - \frac{2(x+2)}{(x-2)(x+2)}\right)$
$= \frac{4}{x} \cdot \left(\frac{2x-4}{(x+2)(x-2)} - \frac{2x+4}{(x+2)(x-2)}\right)$
$= \frac{4}{x} \cdot \left(\frac{2x-4-2x-4}{(x+2)(x-2)}\right)$
$= \frac{4}{x}\left(\frac{-8}{(x+2)(x-2)}\right)$
$= -\frac{32}{x(x+2)(x-2)}$

83. $12\left(\frac{2}{3} + \frac{1}{6}\right) = 12 \cdot \frac{2}{3} + 12 \cdot \frac{1}{6} = \frac{24}{3} + \frac{12}{6} = 8 + 2 = 10$

85. $x^2\left(\frac{4}{x^2} + 1\right) = x^2 \cdot \frac{4}{x^2} + x^2 \cdot 1 = 4 + x^2$

87. $\sqrt{100} = 10$ because $10^2 = 100$.

89. $\sqrt[3]{8} = 2$ because $2^3 = 8$.

91. $\sqrt[4]{81} = 3$ because $3^4 = 81$.

93. $a^2 + b^2 = c^2$
$3^2 + 4^2 = c^2$
$9 + 16 = c^2$
$25 = c^2$
$c = 5$ meters

95. $\frac{2x-3}{x^2+1} - \frac{x-6}{x^2+1} = \frac{2x-3-(x-6)}{x^2+1}$
$= \frac{2x-3-x+6}{x^2+1}$
$= \frac{x+3}{x^2+1}$

97. $P = 4s$
$P = 4 \cdot \left(\frac{x}{x+5}\right) = \frac{4x}{x+5}$
$A = s^2$
$A = \left(\frac{x}{x+5}\right)^2$
$A = \left(\frac{x}{x+5}\right)\left(\frac{x}{x+5}\right) = \frac{x^2}{(x+5)^2}$

The perimeter is $\frac{4x}{x+5}$ feet, and the area is $\frac{x^2}{(x+5)^2}$ square feet.

99. answers may vary

101. answers may vary

103. answers may vary

105. $x^{-1} + (2x)^{-1} = \frac{1}{x} + \frac{1}{2x}$
$= \frac{1 \cdot 2}{x \cdot 2} + \frac{1}{2x}$
$= \frac{2+1}{2x}$
$= \frac{3}{2x}$

107. $4x^{-2} - 3x^{-1} = \frac{4}{x^2} - \frac{3}{x} = \frac{4}{x^2} - \frac{3x}{x^2} = \frac{4-3x}{x^2}$

109.

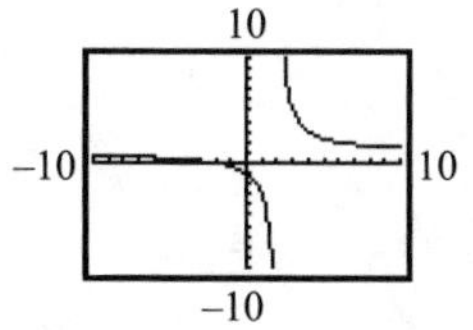

Section 6.3 Practice Exercises

1. a. $\dfrac{\frac{5k}{36m}}{\frac{15k}{9}} = \dfrac{5k}{36m} \div \dfrac{15k}{9}$
$= \dfrac{5k}{36m} \cdot \dfrac{9}{15k}$
$= \dfrac{5k \cdot 9}{36m \cdot 15k}$
$= \dfrac{1}{12m}$

b. $\dfrac{\frac{8x}{x-4}}{\frac{3}{x+4}} = \dfrac{8x}{x-4} \div \dfrac{3}{x+4}$
$= \dfrac{8x}{x-4} \cdot \dfrac{x+4}{3}$
$= \dfrac{8x(x+4)}{3(x-4)}$

c. $\dfrac{\frac{5}{a}+\frac{b}{a^2}}{\frac{5a}{b^2}+\frac{1}{b}} = \dfrac{\frac{5 \cdot a}{a \cdot a}+\frac{b}{a^2}}{\frac{5a}{b^2}+\frac{1 \cdot b}{b \cdot b}}$
$= \dfrac{\frac{5a+b}{a^2}}{\frac{5a+b}{b^2}}$
$= \dfrac{5a+b}{a^2} \cdot \dfrac{b^2}{5a+b}$
$= \dfrac{b^2(5a+b)}{a^2(5a+b)}$
$= \dfrac{b^2}{a^2}$

2. a. The LCD is $(x-4)(x+4)$.
$\dfrac{\frac{8x}{x-4}}{\frac{3}{x+4}} = \dfrac{\left(\frac{8x}{x-4}\right) \cdot (x-4)(x+4)}{\left(\frac{3}{x+4}\right) \cdot (x-4)(x+4)}$
$= \dfrac{8x(x+4)}{3(x-4)}$

b. The LCD is a^2b^2.
$\dfrac{\frac{b}{a^2}+\frac{1}{a}}{\frac{a}{b^2}+\frac{1}{b}} = \dfrac{\left(\frac{b}{a^2}+\frac{1}{a}\right) \cdot a^2b^2}{\left(\frac{a}{b^2}+\frac{1}{b}\right) \cdot a^2b^2}$
$= \dfrac{\frac{b}{a^2} \cdot a^2b^2 + \frac{1}{a} \cdot a^2b^2}{\frac{a}{b^2} \cdot a^2b^2 + \frac{1}{b} \cdot a^2b^2}$
$= \dfrac{b^3+ab^2}{a^3+a^2b}$
$= \dfrac{b^2(b+a)}{a^2(a+b)}$
$= \dfrac{b^2}{a^2}$

3. $\dfrac{3x^{-1}+x^{-2}y^{-1}}{y^{-2}+xy^{-1}} = \dfrac{\frac{3}{x}+\frac{1}{x^2y}}{\frac{1}{y^2}+\frac{x}{y}}$

The LCD is x^2y^2.
$= \dfrac{\left(\frac{3}{x}+\frac{1}{x^2y}\right) \cdot x^2y^2}{\left(\frac{1}{y^2}+\frac{x}{y}\right) \cdot x^2y^2}$
$= \dfrac{\frac{3}{x} \cdot x^2y^2 + \frac{1}{x^2y} \cdot x^2y^2}{\frac{1}{y^2} \cdot x^2y^2 + \frac{x}{y} \cdot x^2y^2}$
$= \dfrac{3xy^2+y}{x^2+x^3y}$ or $\dfrac{y(3xy+1)}{x^2(1+xy)}$

4. $\dfrac{(3x)^{-1}-2}{5x^{-1}+2} = \dfrac{\frac{1}{3x}-2}{\frac{5}{x}+2}$
$= \dfrac{\left(\frac{1}{3x}-2\right) \cdot 3x}{\left(\frac{5}{x}+2\right) \cdot 3x}$
$= \dfrac{\frac{1}{3x} \cdot 3x - 2 \cdot 3x}{\frac{5}{x} \cdot 3x + 2 \cdot 3x}$
$= \dfrac{1-6x}{15+6x}$ or $\dfrac{1-6x}{3(5+2x)}$

Vocabulary, Readiness & Video Check 6.3

1. $\dfrac{\frac{7}{x}}{\frac{1}{x}+\frac{z}{x}} = \dfrac{x\left(\frac{7}{x}\right)}{x\left(\frac{1}{x}\right)+x\left(\frac{z}{x}\right)} = \dfrac{7}{\underline{1+z}}$

2. $\dfrac{\frac{x}{4}}{\frac{x^2}{2}+\frac{1}{4}} = \dfrac{4\left(\frac{x}{4}\right)}{4\left(\frac{x^2}{2}\right)+4\left(\frac{1}{4}\right)} = \underline{\dfrac{x}{2x^2+1}}$

3. $x^{-2} = \underline{\dfrac{1}{x^2}}$

4. $y^{-3} = \underline{\dfrac{1}{y^3}}$

5. $2x^{-1} = \underline{\dfrac{2}{x}}$

6. $(2x)^{-1} = \underline{\dfrac{1}{2x}}$

7. $(9y)^{-1} = \underline{\dfrac{1}{9y}}$

8. $9y^{-2} = \underline{\dfrac{9}{y^2}}$

9. a single fraction in the numerator and in the denominator

10. In method 2, you find the LCD of all fractions in both the numerator and the denominator in order to clear the complex fraction of fractions in the numerator and denominator; in method 1, you find the LCD of the fractions only in the numerator and/or only in the denominator in order to get single fractions in the numerator and denominator.

11. Since a negative exponent moves its base from a numerator to a denominator of the expression, a rational expression containing negative exponents can become a complex fraction when rewritten with positive exponents.

Exercise Set 6.3

1. $\dfrac{\frac{10}{3x}}{\frac{5}{6x}} = \dfrac{10}{3x}\cdot\dfrac{6x}{5} = \dfrac{60x}{15x} = 4$

3. $\dfrac{1+\frac{2}{5}}{2+\frac{3}{5}} = \dfrac{5\left(1+\frac{2}{5}\right)}{5\left(2+\frac{3}{5}\right)} = \dfrac{5+2}{10+3} = \dfrac{7}{13}$

5. $\dfrac{\frac{4}{x-1}}{\frac{x}{x-1}} = \dfrac{4}{x-1}\cdot\dfrac{x-1}{x} = \dfrac{4}{x}$

7. $\dfrac{1-\frac{2}{x}}{x+\frac{4}{9x}} = \dfrac{9x\left(1-\frac{2}{x}\right)}{9x\left(x+\frac{4}{9x}\right)} = \dfrac{9x-18}{9x^2+4} = \dfrac{9(x-2)}{9x^2+4}$

9.
$$\dfrac{\frac{4x^2-y^2}{xy}}{\frac{2}{y}-\frac{1}{x}} = \dfrac{\left(\frac{4x^2-y^2}{xy}\right)\cdot xy}{\left(\frac{2}{y}-\frac{1}{x}\right)\cdot xy}$$
$$= \dfrac{4x^2-y^2}{2x-y}$$
$$= \dfrac{(2x-y)(2x+y)}{2x-y}$$
$$= 2x+y$$

11. $\dfrac{\frac{x+1}{3}}{\frac{2x-1}{6}} = \dfrac{x+1}{3}\cdot\dfrac{6}{2x-1} = \dfrac{2(x+1)}{2x-1}$

13. $\dfrac{\frac{2}{x}+\frac{3}{x^2}}{\frac{4}{x^2}-\frac{9}{x}} = \dfrac{\left(\frac{2}{x}+\frac{3}{x^2}\right)x^2}{\left(\frac{4}{x^2}-\frac{9}{x}\right)x^2} = \dfrac{2x+3}{4-9x}$

15.
$$\dfrac{\frac{1}{x}+\frac{2}{x^2}}{x+\frac{8}{x^2}} = \dfrac{x^2\left(\frac{1}{x}+\frac{2}{x^2}\right)}{x^2\left(x+\frac{8}{x^2}\right)}$$
$$= \dfrac{x+2}{x^3+8}$$
$$= \dfrac{x+2}{(x+2)(x^2-2x+4)}$$
$$= \dfrac{1}{x^2-2x+4}$$

17.
$$\dfrac{\frac{4}{5-x}+\frac{5}{x-5}}{\frac{2}{x}+\frac{3}{x-5}} = \dfrac{-\frac{4}{x-5}+\frac{5}{x-5}}{\frac{2(x-5)+3x}{x(x-5)}}$$
$$= \dfrac{\frac{1}{x-5}}{\frac{2x-10+3x}{x(x-5)}}$$
$$= \dfrac{1}{x-5}\cdot\dfrac{x(x-5)}{5x-10}$$
$$= \dfrac{x}{5x-10} \text{ or } \dfrac{x}{5(x-2)}$$

19. $$\frac{\frac{x+2}{x}-\frac{2}{x-1}}{\frac{x+1}{x}+\frac{x+1}{x-1}}=\frac{\frac{(x+2)(x-1)-2x}{x(x-1)}}{\frac{(x+1)(x-1)+(x+1)(x)}{x(x-1)}}$$
$$=\frac{\frac{x^2+x-2-2x}{x(x-1)}}{\frac{x^2-1+x^2+x}{x(x-1)}}$$
$$=\frac{x^2-x-2}{x(x-1)}\cdot\frac{x(x-1)}{2x^2+x-1}$$
$$=\frac{(x-2)(x+1)}{x(x-1)}\cdot\frac{x(x-1)}{(2x-1)(x+1)}$$
$$=\frac{x-2}{2x-1}$$

21. $$\frac{\frac{2}{x}+3}{\frac{4}{x^2}-9}=\frac{\left(\frac{2}{x}+3\right)\cdot x^2}{\left(\frac{4}{x^2}-9\right)\cdot x^2}$$
$$=\frac{2x+3x^2}{4-9x^2}$$
$$=\frac{x(2+3x)}{(2+3x)(2-3x)}$$
$$=\frac{x}{2-3x}$$

23. $$\frac{1-\frac{x}{y}}{\frac{x^2}{y^2}-1}=\frac{\left(1-\frac{x}{y}\right)\cdot y^2}{\left(\frac{x^2}{y^2}-1\right)\cdot y^2}$$
$$=\frac{y^2-xy}{x^2-y^2}$$
$$=\frac{y(y-x)}{(x+y)(x-y)}$$
$$=\frac{-y(x-y)}{(x+y)(x-y)}$$
$$=-\frac{y}{x+y}$$

25. $$\frac{\frac{-2x}{x-y}}{\frac{y}{x^2}}=\frac{-2x}{x-y}\cdot\frac{x^2}{y}=-\frac{2x^3}{y(x-y)}$$

27. $$\frac{\frac{2}{x}+\frac{1}{x^2}}{\frac{y}{x^2}}=\frac{\left(\frac{2}{x}+\frac{1}{x^2}\right)x^2}{\left(\frac{y}{x^2}\right)x^2}=\frac{2x+1}{y}$$

29. $$\frac{\frac{x}{9}-\frac{1}{x}}{1+\frac{3}{x}}=\frac{\left(\frac{x}{9}-\frac{1}{x}\right)\cdot 9x}{\left(1+\frac{3}{x}\right)\cdot 9x}$$
$$=\frac{x^2-9}{9x+27}$$
$$=\frac{(x+3)(x-3)}{9(x+3)}$$
$$=\frac{x-3}{9}$$

31. $$\frac{\frac{x-1}{x^2-4}}{1+\frac{1}{x-2}}=\frac{\frac{x-1}{x^2-4}}{\frac{x-2+1}{x-2}}=\frac{\frac{x-1}{x^2-4}}{\frac{x-1}{x-2}}$$
$$=\frac{x-1}{x^2-4}\cdot\frac{x-2}{x-1}$$
$$=\frac{x-1}{(x+2)(x-2)}\cdot\frac{x-2}{x-1}$$
$$=\frac{1}{x+2}$$

33. $$\frac{\frac{2}{x+5}+\frac{4}{x+3}}{\frac{3x+13}{x^2+8x+15}}=\frac{\frac{2}{x+5}+\frac{4}{x+3}}{\frac{3x+13}{(x+5)(x+3)}}$$
$$=\frac{\left(\frac{2}{x+5}+\frac{4}{x+3}\right)(x+5)(x+3)}{\frac{3x+13}{(x+5)(x+3)}(x+5)(x+3)}$$
$$=\frac{2(x+3)+4(x+5)}{3x+13}$$
$$=\frac{2x+6+4x+20}{3x+13}$$
$$=\frac{6x+26}{3x+13}$$
$$=\frac{2(3x+13)}{3x+13}$$
$$=2$$

35. $$\frac{x^{-1}}{x^{-2}+y^{-2}}=\frac{\frac{1}{x}}{\frac{1}{x^2}+\frac{1}{y^2}}$$
$$=\frac{x^2y^2\left(\frac{1}{x}\right)}{x^2y^2\left(\frac{1}{x^2}+\frac{1}{y^2}\right)}$$
$$=\frac{xy^2}{y^2+x^2}$$
$$=\frac{xy^2}{x^2+y^2}$$

37. $\dfrac{2a^{-1}+3b^{-2}}{a^{-1}-b^{-1}}=\dfrac{\frac{2}{a}+\frac{3}{b^2}}{\frac{1}{a}-\frac{1}{b}}$
$=\dfrac{ab^2\left(\frac{2}{a}+\frac{3}{b^2}\right)}{ab^2\left(\frac{1}{a}-\frac{1}{b}\right)}$
$=\dfrac{2b^2+3a}{b^2-ab}$
$=\dfrac{2b^2+3a}{b(b-a)}$

39. $\dfrac{1}{x-x^{-1}}=\dfrac{1}{x-\frac{1}{x}}$
$=\dfrac{x(1)}{x\left(x-\frac{1}{x}\right)}$
$=\dfrac{x}{x^2-1}$
$=\dfrac{x}{(x+1)(x-1)}$

41. $\dfrac{a^{-1}+1}{a^{-1}-1}=\dfrac{\frac{1}{a}+1}{\frac{1}{a}-1}=\dfrac{a\left(\frac{1}{a}+1\right)}{a\left(\frac{1}{a}-1\right)}=\dfrac{1+a}{1-a}$

43. $\dfrac{3x^{-1}+(2y)^{-1}}{x^{-2}}=\dfrac{\frac{3}{x}+\frac{1}{2y}}{\frac{1}{x^2}}$
$=\dfrac{2x^2y\left(\frac{3}{x}+\frac{1}{2y}\right)}{2x^2y\left(\frac{1}{x^2}\right)}$
$=\dfrac{6xy+x^2}{2y}$
$=\dfrac{x(x+6y)}{2y}$

45. $\dfrac{2a^{-1}+(2a)^{-1}}{a^{-1}+2a^{-2}}=\dfrac{\frac{2}{a}+\frac{1}{2a}}{\frac{1}{a}+\frac{2}{a^2}}$
$=\dfrac{2a^2\left(\frac{2}{a}+\frac{1}{2a}\right)}{2a^2\left(\frac{1}{a}+\frac{2}{a^2}\right)}$
$=\dfrac{4a+a}{2a+4}$
$=\dfrac{5a}{2(a+2)}$

47. $\dfrac{5x^{-1}+2y^{-1}}{x^{-2}y^{-2}}=\dfrac{\frac{5}{x}+\frac{2}{y}}{\frac{1}{x^2y^2}}$
$=\dfrac{x^2y^2\left(\frac{5}{x}+\frac{2}{y}\right)}{x^2y^2\left(\frac{1}{x^2y^2}\right)}$
$=5xy^2+2x^2y$
$=xy(5y+2x)$

49. $\dfrac{5x^{-1}-2y^{-1}}{25x^{-2}-4y^{-2}}=\dfrac{\frac{5}{x}-\frac{2}{y}}{\frac{25}{x^2}-\frac{4}{y^2}}$
$=\dfrac{x^2y^2\left(\frac{5}{x}-\frac{2}{y}\right)}{x^2y^2\left(\frac{25}{x^2}-\frac{4}{y^2}\right)}$
$=\dfrac{5xy^2-2x^2y}{25y^2-4x^2}$
$=\dfrac{xy(5y-2x)}{(5y+2x)(5y-2x)}$
$=\dfrac{xy}{5y+2x}$ or $\dfrac{xy}{2x+5y}$

51. $\dfrac{3x^3y^2}{12x}=\dfrac{3x\cdot x^2y^2}{3x\cdot 4}=\dfrac{x^2y^2}{4}$

53. $\dfrac{144x^5y^5}{-16x^2y}=\dfrac{16x^2y\cdot 9x^3y^4}{16x^2y\cdot(-1)}=-9x^3y^4$

55. $|x-5|=9$
$x-5=-9$ or $x-5=9$
$x=-4$ or $x=14$

57. $\dfrac{\frac{x+1}{9}}{\frac{y-2}{5}}=\dfrac{x+1}{9}\div\dfrac{y-2}{5}=\dfrac{x+1}{9}\cdot\dfrac{5}{y-2}$
Both a and c are equivalent to the original expression.

59. $\dfrac{a}{1-\frac{s}{770}}=\dfrac{770(a)}{770\left(1-\frac{s}{770}\right)}=\dfrac{770a}{770-s}$

61. $\dfrac{\frac{1}{x}}{\frac{3}{y}}=\dfrac{1}{x}\div\dfrac{3}{y}=\dfrac{1}{x}\cdot\dfrac{y}{3}$
Both a and b are equivalent to the original expression.

63. answers may vary

65. $$\frac{1}{1+(1+x)^{-1}}=\frac{1}{1+\frac{1}{1+x}}=\frac{(1+x)\cdot 1}{(1+x)\left(1+\frac{1}{1+x}\right)}=\frac{1+x}{1+x+1}=\frac{1+x}{2+x}$$

67. $$\frac{x}{1-\frac{1}{1+\frac{1}{x}}}=\frac{x}{1-\frac{1}{\frac{x+1}{x}}}=\frac{x}{1-\frac{x}{x+1}}=\frac{(x+1)(x)}{(x+1)\left(1-\frac{x}{x+1}\right)}=\frac{x(x+1)}{x+1-x}=\frac{x(x+1)}{1}=x(x+1)$$

69. $$\frac{\frac{2}{y^2}-\frac{5}{xy}-\frac{3}{x^2}}{\frac{2}{y^2}+\frac{7}{xy}+\frac{3}{x^2}}=\frac{x^2y^2\left(\frac{2}{y^2}-\frac{5}{xy}-\frac{3}{x^2}\right)}{x^2y^2\left(\frac{2}{y^2}+\frac{7}{xy}+\frac{3}{x^2}\right)}=\frac{2x^2-5xy-3y^2}{2x^2+7xy+3y^2}=\frac{(2x+y)(x-3y)}{(2x+y)(x+3y)}=\frac{x-3y}{x+3y}$$

71. $$\frac{3(a+1)^{-1}+4a^{-2}}{(a^3+a^2)^{-1}}=\frac{\frac{3}{a+1}+\frac{4}{a^2}}{\frac{1}{a^3+a^2}}=\frac{\frac{3a^2+4(a+1)}{a^2(a+1)}}{\frac{1}{a^2(a+1)}}=\frac{3a^2+4a+4}{a^2(a+1)}\cdot\frac{a^2(a+1)}{1}=3a^2+4a+4$$

73. $f(x)=\dfrac{1}{x}$

a. $f(a+h)=\dfrac{1}{a+h}$

b. $f(a)=\dfrac{1}{a}$

c. $\dfrac{f(a+h)-f(a)}{h}=\dfrac{\frac{1}{a+h}-\frac{1}{a}}{h}$

d. $$\frac{\frac{1}{a+h}-\frac{1}{a}}{h}=\frac{a(a+h)\left(\frac{1}{a+h}-\frac{1}{a}\right)}{a(a+h)\cdot h}=\frac{a-(a+h)}{ah(a+h)}=\frac{-h}{ah(a+h)}=-\frac{1}{a(a+h)}$$

75. $f(x)=\dfrac{3}{x+1}$

a. $f(a+h)=\dfrac{3}{a+h+1}$

b. $f(a)=\dfrac{3}{a+1}$

c. $\dfrac{f(a+h)-f(a)}{h}=\dfrac{\frac{3}{a+h+1}-\frac{3}{a+1}}{h}$

d. $$\frac{\frac{3}{a+h+1}-\frac{3}{a+1}}{h}=\frac{\left(\frac{3}{a+h+1}-\frac{3}{a+1}\right)\cdot(a+h+1)(a+1)}{h\cdot(a+h+1)(a+1)}=\frac{3(a+1)-3(a+h+1)}{h(a+h+1)(a+1)}=\frac{3a+3-3a-3h-3}{h(a+h+1)(a+1)}=\frac{-3h}{h(a+h+1)(a+1)}=-\frac{3}{(a+h+1)(a+1)}$$

Section 6.4 Practice Exercises

1. $\frac{18a^3-12a^2+30a}{6a}=\frac{18a^3}{6a}-\frac{12a^2}{6a}+\frac{30a}{6a}$
$=3a^2-2a+5$

2. $\frac{5a^3b^4-8a^2b^3+ab^2-8b}{ab^2}$
$=\frac{5a^3b^4}{ab^2}-\frac{8a^2b^3}{ab^2}+\frac{ab^2}{ab^2}-\frac{8b}{ab^2}$
$=5a^2b^2-8ab+1-\frac{8}{ab}$

3.
$$\begin{array}{r}
3x-2\\
x+3\overline{)3x^2+7x-6}\\
\underline{3x^2+9x}\\
-2x-6\\
\underline{-2x-6}\\
0
\end{array}$$
Answer: $3x-2$

4.
$$\begin{array}{r}
3x-2\\
2x-1\overline{)6x^2-7x+8}\\
\underline{6x^2-3x}\\
-4x+8\\
\underline{-4x+2}\\
6
\end{array}$$
Answer: $3x-2+\frac{6}{2x-1}$

5.
$$\begin{array}{r}
5x^2-6x+8\\
x+3\overline{)5x^3+9x^2-10x+30}\\
\underline{5x^3+15x^2}\\
-6x^2-10x\\
\underline{-6x^2-18x}\\
8x+30\\
\underline{8x+24}\\
6
\end{array}$$
Answer: $5x^2-6x+8+\frac{6}{x+3}$

6.
$$\begin{array}{r}
2x^2+3x-2\\
x^2+0x+1\overline{)2x^4+3x^3+0x^2-5x+2}\\
\underline{2x^4+0x^3+2x^2}\\
3x^3-2x^2-5x\\
\underline{3x^3+0x^2+3x}\\
-2x^2-8x+2\\
\underline{-2x^2+0x-2}\\
-8x+4
\end{array}$$
Answer: $2x^2+3x-2+\frac{-8x+4}{x^2+1}$

7.
$$\begin{array}{r}
16x^2+20x+25\\
4x-5\overline{)64x^3+0x^2+0x-125}\\
\underline{64x^3-80x^2}\\
80x^2+0x\\
\underline{80x^2-100x}\\
100x-125\\
\underline{100x-125}\\
0
\end{array}$$
Answer: $16x^2+20x+25$

8. Since $x-c=x-1$, c is 1.
$$\begin{array}{r|rrrr}
1 & 4 & -3 & 6 & 5\\
 & & 4 & 1 & 7\\
\hline
 & 4 & 1 & 7 & 12
\end{array}$$
$4x^2+x+7+\frac{12}{x-1}$

9. Since $x-c=x+3=x-(-3)$, c is -3.
$$\begin{array}{r|rrrrr}
-3 & 1 & 3 & -5 & 0 & 12\\
 & & -3 & 0 & 15 & -45\\
\hline
 & 1 & 0 & -5 & 15 & -33
\end{array}$$
$x^3-5x+15-\frac{33}{x+3}$

10. a. $P(x)=x^3-5x-2$
$P(2)=2^3-5(2)-2$
$=8-10-2$
$=-4$

b. Since $x-c=x-2$, c is 2.
$$\begin{array}{r|rrrr}
2 & 1 & 0 & -5 & -2\\
 & & 2 & 4 & -2\\
\hline
 & 1 & 2 & -1 & -4
\end{array}$$
The remainder is – 4.

11.
$$\begin{array}{r|rrrrrr} 3 & 2 & -18 & 0 & 90 & 59 & 0 \\ & & 6 & -36 & -108 & -54 & 15 \\ \hline & 2 & -12 & -36 & -18 & 5 & 15 \end{array}$$
$P(3) = 15$

Vocabulary, Readiness & Video Check 6.4

1. the common denominator
2. When the degree of the remainder is less than the degree of the divisor.
3. The last number is the remainder and the other numbers are the coefficients of the variables in the quotient; the degree of the quotient is one less than the degree of the dividend.
4. When the calculations for substituting in c might be tedious, that is, if $P(x)$ is a many-termed, high-degree polynomial and/or of the coefficients are very large numbers, etc.

Exercise Set 6.4

1. $\frac{4a^2+8a}{2a} = \frac{4a^2}{2a} + \frac{8a}{2a} = 2a+4$

3. $\frac{12a^5b^2+16a^4b}{4a^4b} = \frac{12a^5b^2}{4a^4b} + \frac{16a^4b}{4a^4b}$
$= 3ab+4$

5. $\frac{4x^2y^2+6xy^2-4y^2}{2x^2y}$
$= \frac{4x^2y^2}{2x^2y} + \frac{6xy^2}{2x^2y} - \frac{4y^2}{2x^2y}$
$= 2y + \frac{3y}{x} - \frac{2y}{x^2}$

7.
$$\begin{array}{r} x+1 \\ x+2\overline{)x^2+3x+2} \\ \underline{x^2+2x} \\ x+2 \\ \underline{x+2} \\ 0 \end{array}$$
$\frac{x^2+3x+2}{x+2} = x+1$

9.
$$\begin{array}{r} 2x-8 \\ x+1\overline{)2x^2-6x-8} \\ \underline{2x^2+2x} \\ -8x-8 \\ \underline{-8x-8} \\ 0 \end{array}$$
$\frac{2x^2-6x-8}{x+1} = 2x-8$

11.
$$\begin{array}{r} x-\frac{1}{2} \\ 2x+4\overline{)2x^2+3x-2} \\ \underline{2x^2+4x} \\ -x-2 \\ \underline{-x-2} \\ 0 \end{array}$$
$\frac{2x^2+3x-2}{2x+4} = x-\frac{1}{2}$

13.
$$\begin{array}{r} 2x^2-\frac{1}{2}x+5 \\ 2x+4\overline{)4x^3+7x^2+8x+20} \\ \underline{4x^3+8x^2} \\ -x^2+8x \\ \underline{-x^2-2x} \\ 10x+20 \\ \underline{10x+20} \\ 0 \end{array}$$
$\frac{4x^3+7x^2+8x+20}{2x+4} = 2x^2-\frac{1}{2}x+5$

15.
$$\begin{array}{r} 2x^2 \qquad -6 \\ 3x+1\overline{)6x^3+2x^2-18x-6} \\ \underline{6x^3+2x^2} \\ -18x-6 \\ \underline{-18x-6} \\ 0 \end{array}$$
$\frac{2x^2+6x^3-18x-6}{3x+1} = 2x^2-6$

17.
$$\begin{array}{r@{}l} & 3x^3+5x+4 \\ x^2+0x-2 & \big)\overline{3x^5+0x^4-x^3+4x^2-12x-8} \\ & \underline{3x^5+0x^4-6x^3} \\ & \qquad\quad 5x^3+4x^2-12x \\ & \qquad\quad \underline{5x^3+0x^2-10x} \\ & \qquad\qquad\quad 4x^2-2x-8 \\ & \qquad\qquad\quad \underline{4x^2+0x-8} \\ & \qquad\qquad\qquad\quad -2x \end{array}$$

$$\frac{3x^5-x^3+4x^2-12x-8}{x^2-2}=3x^3+5x+4-\frac{2x}{x^2-2}$$

19.
$$\begin{array}{r@{}l} & 2x^3+\frac{9}{2}x^2+10x+21 \\ x-2 & \big)\overline{2x^4+\frac{1}{2}x^3+x^2+x+0} \\ & \underline{2x^4-4x^3} \\ & \quad \frac{9}{2}x^3+x^2 \\ & \quad \underline{\frac{9}{2}x^3-9x^2} \\ & \qquad 10x^2+x \\ & \qquad \underline{10x^2-20x} \\ & \qquad\qquad 21x+0 \\ & \qquad\qquad \underline{21x-42} \\ & \qquad\qquad\quad 42 \end{array}$$

$$\frac{2x^4+\frac{1}{2}x^3+x^2+x}{x-2}=2x^3+\frac{9}{2}x^2+10x+21+\frac{42}{x-2}$$

21. $x-5=x-c$ where $c=5$.

$$\begin{array}{r|rrr} 5 & 1 & 3 & -40 \\ & & 5 & 40 \\ \hline & 1 & 8 & 0 \end{array}$$

$$\frac{x^2+3x-40}{x-5}=x+8$$

23. $x+6=x-c$ where $c=-6$.

$$\begin{array}{r|rrr} -6 & 1 & 5 & -6 \\ & & -6 & 6 \\ \hline & 1 & -1 & 0 \end{array}$$

$$\frac{x^2+5x-6}{x+6}=x-1$$

25. $x-2=x-c$ where $c=2$.

$$\begin{array}{r|rrrr} 2 & 1 & -7 & -13 & 5 \\ & & 2 & -10 & -46 \\ \hline & 1 & -5 & -23 & -41 \end{array}$$

$$\frac{x^3-7x^2-13x+5}{x-2}=x^2-5x-23-\frac{41}{x-2}$$

27. $x-2=x-c$ where $c=2$.

$$\begin{array}{r|rrr} 2 & 4 & 0 & -9 \\ & & 8 & 16 \\ \hline & 4 & 8 & 7 \end{array}$$

$$\frac{4x^2-9}{x-2}=4x+8+\frac{7}{x-2}$$

29.
$$\frac{4x^7y^4+8xy^2+4xy^3}{4xy^3}$$
$$=\frac{4x^7y^4}{4xy^3}+\frac{8xy^2}{4xy^3}+\frac{4xy^3}{4xy^3}$$
$$=x^6y+\frac{2}{y}+1$$

31.
$$\begin{array}{r@{}l} & 5x^2\qquad -6 \\ 2x-1 & \big)\overline{10x^3-5x^2-12x+1} \\ & \underline{10x^3-5x^2} \\ & \qquad\qquad -12x+1 \\ & \qquad\qquad \underline{-12x+6} \\ & \qquad\qquad\qquad -5 \end{array}$$

$$\frac{10x^3-5x^2-12x+1}{2x-1}=5x^2-6-\frac{5}{2x-1}$$

33. $x-4$ has the form $x-c$, where $c=4$, so synthetic division can be used.

$2x^3-6x^2-4=2x^3-6x^2+0x-4$

$$\begin{array}{r|rrrr} 4 & 2 & -6 & 0 & -4 \\ & & 8 & 8 & 32 \\ \hline & 2 & 2 & 8 & 28 \end{array}$$

$$\frac{2x^3-6x^2-4}{x-4}=2x^2+2x+8+\frac{28}{x-4}$$

35. $x - 5$ has the form $x - c$, where $c = 5$, so synthetic division can be used.

5	2	−13	16	−9	20
		10	−15	5	−20
	2	−3	1	−4	0

$$\frac{2x^4 - 13x^3 + 16x^2 - 9x + 20}{x - 5}$$
$$= 2x^3 - 3x^2 + x - 4$$

37. $x + 1$ has the form $x - c$, where $c = -1$, so synthetic division can be used.

$7x^2 - 4x + 12 + 3x^3 = 3x^3 + 7x^2 - 4x + 12$

−1	3	7	−4	12
		−3	−4	8
	3	4	−8	20

$$\frac{7x^2 - 4x + 12 + 3x^3}{x + 1} = 3x^2 + 4x - 8 + \frac{20}{x + 1}$$

39. $x - \frac{1}{3}$ has the form $x - c$, where $c = \frac{1}{3}$, so synthetic division can be used.

$\frac{1}{3}$	3	2	−4	1
		1	1	−1
	3	3	−3	0

$$\frac{3x^3 + 2x^2 - 4x + 1}{x - \frac{1}{3}} = 3x^2 + 3x - 3$$

41. $x - 1$ has the form $x - c$, where $c = 1$, so synthetic division can be used.

$x^3 - 1 = x^3 + 0x^2 + 0x - 1$

1	1	0	0	−1
		1	1	1
	1	1	1	0

$$\frac{x^3 - 1}{x - 1} = x^2 + x + 1$$

43. $$\frac{25xy^2 + 75xyz + 125x^2yz}{-5x^2y}$$
$$= \frac{25xy^2}{-5x^2y} + \frac{75xyz}{-5x^2y} + \frac{125x^2yz}{-5x^2y}$$
$$= -\frac{5y}{x} - \frac{15z}{x} - 25z$$

45. $9x^5 + 6x^4 - 6x^2 - 4x$
$= 9x^5 + 6x^4 + 0x^3 - 6x^2 - 4x + 0$

$$\begin{array}{r} 3x^4 - 2x \\ 3x+2 \overline{)\, 9x^5 + 6x^4 + 0x^3 - 6x^2 - 4x + 0} \\ \underline{9x^5 + 6x^4} \qquad\qquad\qquad\quad \\ 0 + 0x^3 - 6x^2 - 4x \\ \underline{-6x^2 - 4x} \\ 0 \end{array}$$

$(9x^5 + 6x^4 - 6x^2 - 4x) \div (3x + 2) = 3x^4 - 2x$

47.

1	1	3	−7	4
		1	4	−3
	1	4	−3	1

Thus, $P(1) = 1$.

49.

−3	3	−7	−2	5
		−9	48	−138
	3	−16	46	−133

Thus, $P(-3) = -133$.

51.

−1	4	0	1	0	−2
		−4	4	−5	5
	4	−4	5	−5	3

Thus, $P(-1) = 3$.

53.

$\frac{1}{3}$	2	0	−3	0	−2
		$\frac{2}{3}$	$\frac{2}{9}$	$-\frac{25}{27}$	$-\frac{25}{81}$
	2	$\frac{2}{3}$	$-\frac{25}{9}$	$-\frac{25}{27}$	$-\frac{187}{181}$

Thus, $P\left(\frac{1}{3}\right) = -\frac{187}{81}$.

55.

$\frac{1}{2}$	1	1	−1	0	0	3
		$\frac{1}{2}$	$\frac{3}{4}$	$-\frac{1}{8}$	$-\frac{1}{16}$	$-\frac{1}{32}$
	1	$\frac{3}{2}$	$-\frac{1}{4}$	$-\frac{1}{8}$	$-\frac{1}{16}$	$\frac{95}{32}$

Thus, $P\left(\frac{1}{2}\right) = \frac{95}{32}$.

57. $$\begin{aligned} 7x+2&=x-3\\ 7x-x&=-3-2\\ 6x&=-5\\ x&=-\frac{5}{6} \end{aligned}$$
The solution is $-\frac{5}{6}$.

59. $$\begin{aligned} x^2&=4x-4\\ x^2-4x+4&=0\\ (x-2)^2&=0\\ x-2&=0\\ x&=2 \end{aligned}$$
The solution is 2.

61. $$\begin{aligned} \frac{x}{3}-5&=13\\ 3\left(\frac{x}{3}-5\right)&=(13)\cdot 3\\ x-15&=39\\ x&=54 \end{aligned}$$
The solution is 54.

63. $x^3-1=x^3-1^3=(x-1)(x^2+x+1)$

65. $$\begin{aligned} 125z^3+8&=(5z)^3+2^3\\ &=(5z+2)(25z^2-10z+4) \end{aligned}$$

67. $$\begin{aligned} xy+2x+3y+6&=(xy+2x)+(3y+6)\\ &=x(y+2)+3(y+2)\\ &=(y+2)(x+3) \end{aligned}$$

69. $x^3-9x=x(x^2-9)=x(x+3)(x-3)$

71. $(5x^2-3x+2)\div(x+2)$ is a candidate for synthetic division since $x+2$ is in the form $x-c$, where $c=-2$.

73. $(x^7-2)\div(x^5+1)$ is not a candidate for synthetic division since x^5+1 does not have the form $x-c$.

75. The degree of the remainder must be less than that of the divisor, or 3 in this case. The choices are a or d.

77. $$\begin{aligned} \frac{3x^4+6x^2-18}{3}&=\frac{3x^4}{3}+\frac{6x^2}{3}-\frac{18}{3}\\ &=x^4+2x^2-6 \end{aligned}$$
The length of each piece is (x^4+2x^2-6) meters.

79. $$\begin{array}{r} 3x-7\\ 5x+2\overline{\big)\,15x^2-29x-14}\\ \underline{15x^2+6x}\\ -35x-14\\ \underline{-35x-14}\\ 0 \end{array}$$
The width is $(3x-7)$ inches.

81. $A=bh$ so $h=\frac{A}{b}=\frac{x^4-23x^2+9x-5}{x+5}$
$$\begin{array}{r|rrrrr} -5 & 1 & 0 & -23 & 9 & -5\\ & & -5 & 25 & -10 & 5\\ \hline & 1 & -5 & 2 & -1 & 0 \end{array}$$
The height is (x^3-5x^2+2x-1) cm.

83. $$\begin{array}{r} x^3+\frac{5}{3}x^2+\frac{5}{3}x+\frac{8}{3}\\ x-1\overline{\big)\,x^4+\frac{2}{3}x^3+0x^2+x+0}\\ \underline{x^4-x^3}\\ \frac{5}{3}x^3-0x^2\\ \underline{\frac{5}{3}x^3-\frac{5}{3}x^2}\\ \frac{5}{3}x^2+x\\ \underline{\frac{5}{3}x^2-\frac{5}{3}x}\\ \frac{8}{3}x+0\\ \underline{\frac{8}{3}x-\frac{8}{3}}\\ \frac{8}{3} \end{array}$$
Answer: $x^3+\frac{5}{3}x^2+\frac{5}{3}x+\frac{8}{3}+\frac{8}{3(x-1)}$

85.
$$\begin{array}{r} \frac{3}{2}x^3+\frac{1}{4}x^2+\frac{1}{8}x-\frac{7}{16} \\ 2x-1\overline{)3x^4-x^3+0x^2-x+\frac{1}{2}} \\ \underline{3x^4-\frac{3}{2}x^3} \\ \frac{1}{2}x^3+0x^2 \\ \underline{\frac{1}{2}x^3-\frac{1}{4}x^2} \\ \frac{1}{4}x^2-x \\ \underline{\frac{1}{4}x^2-\frac{1}{8}x} \\ -\frac{7}{8}x+\frac{1}{2} \\ \underline{-\frac{7}{8}x+\frac{7}{16}} \\ \frac{1}{16} \end{array}$$

Answer: $\frac{3}{2}x^3+\frac{1}{4}x^2+\frac{1}{8}x-\frac{7}{16}+\frac{1}{16(2x-1)}$

87.
$$\begin{array}{r} x^3 \qquad -\frac{2}{5}x \\ 5x+10\overline{)5x^4+10x^3-2x^2-4x+0} \\ \underline{5x^4+10x^3} \\ -2x^2-4x \\ \underline{-2x^2-4x} \\ 0 \end{array}$$

Answer: $x^3-\frac{2}{5}x$

89. $\frac{f(x)}{g(x)}=\frac{25x^2-5x+30}{5x}$

$=\frac{25x^2}{5x}-\frac{5x}{5x}+\frac{30}{5x}$

$=5x-1+\frac{6}{x}$

Setting the denominator equal to 0, we get

$5x=0$

$x=0.$

Thus, $x = 0$ is not in the domain of $\frac{f(x)}{g(x)}$.

91. $\frac{f(x)}{g(x)}=\frac{7x^4-3x^2+2}{x-2}$

$$\begin{array}{r} 7x^3+14x^2+25x+50 \\ x-2\overline{)7x^4+0x^3-3x^2+0x+2} \\ \underline{7x^4-14x^3} \\ 14x^3-3x^2 \\ \underline{14x^3-28x^2} \\ 25x^2+0x \\ \underline{25x^2-50x} \\ 50x+2 \\ \underline{50x-100} \\ 102 \end{array}$$

Therefore, $\frac{f(x)}{g(x)}=\frac{7x^4-3x^2+2}{x-2}$

$=7x^3+14x^2+25x+50+\frac{102}{x-2}$

Setting the denominator equal to 0, we get $x-2=0$, or $x=2$. Thus, $x = 2$ is not in the domain of $\frac{f(x)}{g(x)}$.

93. answers may vary

95. answers may vary

97.
$$\begin{array}{r|rrrr} -3 & 1 & 3 & 4 & 12 \\ & & -3 & 0 & -12 \\ \hline & 1 & 0 & 4 & 0 \end{array}$$

Remainder = 0 and

$(x+3)(x^2+4)=x^3+3x^2+4x+12$

99. Multiply $(2x^2+5x-6)$ by $(x - 5)$ and add the remainder, 3.

$(2x^2+5x-6)(x-5)+3$

$=2x^3-10x^2+5x^2-25x-6x+30+3$

$=2x^3-5x^2-31x+33$

101. a. $m(x)=\frac{P(x)}{R(x)}$

$=\frac{-0.08x^3-0.04x^2+1.06x+1.9}{2.2x+9.38}$

b. 2018 is 8 years after 2010, so $x = 8$.

$$\begin{aligned} m(8) &= \frac{-0.08(8)^3 - 0.04(8)^2 + 1.06(8) + 1.9}{2.2(8) + 9.38} \\ &= \frac{-33.14}{26.98} \\ &\approx -1.23 \end{aligned}$$

103. $P(c)$ is equal to the remainder when $P(x)$ is divided by $x - c$. Therefore, $P(c) = 0$.

Section 6.5 Practice Exercises

1. The LCD is 8.

$$\begin{aligned} \frac{5x}{4} - \frac{3}{2} &= \frac{7x}{8} \\ 8\left(\frac{5x}{4} - \frac{3}{2}\right) &= 8\left(\frac{7x}{8}\right) \\ 8 \cdot \frac{5x}{4} - 8 \cdot \frac{3}{2} &= 8 \cdot \frac{7x}{8} \\ 10x - 12 &= 7x \\ 3x &= 12 \\ x &= 4 \end{aligned}$$

2. The LCD of the denominators x, $5x$, and 5 is $5x$.

$$\begin{aligned} \frac{6}{x} - \frac{x+9}{5x} &= \frac{2}{5} \\ 5x\left(\frac{6}{x} - \frac{x+9}{5x}\right) &= 5x\left(\frac{2}{5}\right) \\ 5x \cdot \frac{6}{x} - 5x \cdot \frac{x+9}{5x} &= 5x \cdot \frac{2}{5} \\ 30 - (x+9) &= 2x \\ 30 - x - 9 &= 2x \\ 21 &= 3x \\ 7 &= x \end{aligned}$$

3. The LCD is $x + 3$.

$$\begin{aligned} \frac{x-5}{x+3} &= \frac{2(x-1)}{x+3} \\ (x+3) \cdot \frac{x-5}{x+3} &= (x+3) \cdot \frac{2(x-1)}{x+3} \\ x - 5 &= 2(x-1) \\ x - 5 &= 2x - 2 \\ -3 &= x \end{aligned}$$

The number -3 makes the denominator $x + 3$ equal to 0, so it is not a solution. The solution set is { } or ∅.

4. The LCD is $x(5x - 1)$.

$$\frac{5x}{5x-1}+\frac{1}{x}=\frac{1}{5x-1}$$

$$x(5x-1)\cdot\frac{5x}{5x-1}+x(5x-1)\cdot\frac{1}{x}=x(5x-1)\cdot\frac{1}{5x-1}$$

$$x(5x)+(5x-1)=x$$

$$5x^2+5x-1-x=0$$

$$5x^2+4x-1=0$$

$$(5x-1)(x+1)=0$$

$$5x-1=0 \text{ or } x+1=0$$

$$x=\frac{1}{5} \text{ or } x=-1$$

The number $\frac{1}{5}$ makes the denominator $5x-1$ equal 0, so it is not a solution. The solution is -1.

5. $x^2-4=(x+2)(x-2)$

The LCD is $(x + 2)(x - 2)$.

$$\frac{2}{x-2}-\frac{5+2x}{x^2-4}=\frac{x}{x+2}$$

$$(x+2)(x-2)\cdot\frac{2}{x-2}-(x+2)(x-2)\cdot\frac{5+2x}{(x+2)(x-2)}=(x+2)(x-2)\cdot\frac{x}{x+2}$$

$$2(x+2)-(5+2x)=x(x-2)$$

$$2x+4-5-2x=x^2-2x$$

$$x^2-2x+1=0$$

$$(x-1)(x-1)=0$$

$$x-1=0$$

$$x=1$$

Since 1 does not make any denominator 0, the solution is 1.

6. $2z^2-z-6=(2z+3)(z-2)$

$z^2-2z=z(z-2)$

The LCD is $3z(2z + 3)(z - 2)$.

$$\frac{z}{2z^2-z-6}-\frac{1}{3z}=\frac{2}{z^2-2z}$$

$$\frac{z}{(2z+3)(z-2)}-\frac{1}{3z}=\frac{2}{z(z-2)}$$

$$3z(2z+3)(z-2)\cdot\frac{z}{(2z+3)(z-2)}-3z(2z+3)(z-2)\cdot\frac{1}{3z}=3z(2z+3)(z-2)\cdot\frac{2}{z(z-2)}$$

$$3z(z)-(2z+3)(z-2)=2\cdot3(2z+3)$$

$$3z^2-(2z^2-z-6)=12z+18$$

$$3z^2-2z^2+z+6=12z+18$$

$$z^2+z+6=12z+18$$

$$z^2-11z-12=0$$

$$(z-12)(z+1)=0$$

$$z-12=0 \quad \text{or} \quad z+1=0$$
$$z=12 \quad \text{or} \quad z=-1$$

Neither 12 nor −1 makes any denominator 0, so they are both solutions. The solutions are 12 and −1.

Vocabulary, Readiness & Video Check 6.5

1. $\frac{x}{2}=\frac{3x}{5}+\frac{x}{6}$ is an equation.

2. $\frac{3x}{5}+\frac{x}{6}$ is an expression.

3. $\frac{x}{x-1}+\frac{2x}{x+1}$ is an expression.

4. $\frac{x}{x-1}+\frac{2x}{x+1}=5$ is an equation.

5. $\frac{y+7}{2}=\frac{y+1}{6}+\frac{1}{y}$ is an equation.

6. $\frac{y+1}{6}+\frac{1}{y}$ is an expression.

7. The LCD of $\frac{x}{7}$, $-\frac{x}{2}$, and $\frac{1}{2}$ is <u>14</u>. **c**

8. The LCD of $\frac{9}{x+1}$, $\frac{5}{(x+1)^2}$, and $\frac{x}{x+1}$ is <u>$(x+1)^2$</u>. **b**

9. The LCD of $\frac{7}{x-4}$, $\frac{x}{x^2-16}=\frac{x}{(x+4)(x-4)}$, and $\frac{1}{x+4}$ is <u>$(x+4)(x-4)$</u>. **a**

10. The LCD of $3=\frac{3}{1}$, $\frac{1}{x-5}$, and $-\frac{2}{x^2-5x}=-\frac{2}{x(x-5)}$ is <u>$x(x-5)$</u>. **d**

11. Linear and quadratic equations are solved in very different ways, so you need to determine the next correct move to make.

12. First check to see if a proposed solution makes any denominator of the original equation zero, giving you an undefined rational expression. If so, the solution is an extraneous solution and is not a solution of the equation.

Exercise Set 6.5

1.
$$\frac{x}{2}-\frac{x}{3}=12$$
$$6\left(\frac{x}{2}-\frac{x}{3}\right)=6(12)$$
$$3x-2x=72$$
$$x=72$$

3.
$$\frac{x}{3}=\frac{1}{6}+\frac{x}{4}$$
$$12\left(\frac{x}{3}\right)=12\left(\frac{1}{6}+\frac{x}{4}\right)$$
$$4x=2+3x$$
$$x=2$$

5.
$$\frac{2}{x}+\frac{1}{2}=\frac{5}{x}$$
$$2x\left(\frac{2}{x}+\frac{1}{2}\right)=2x\left(\frac{5}{x}\right)$$
$$4+x=10$$
$$x=6$$

7.
$$\frac{x^2+1}{x}=\frac{5}{x}$$
$$x\left(\frac{x^2+1}{x}\right)=x\left(\frac{5}{x}\right)$$
$$x^2+1=5$$
$$x^2-4=0$$
$$x+2=0 \quad \text{or} \quad x-2=0$$
$$x=-2 \quad \text{or} \quad x=2$$

9.
$$\frac{x+5}{x+3}=\frac{2}{x+3}$$
$$(x+3)\cdot\frac{x+5}{x+3}=(x+3)\cdot\frac{2}{x+3}$$
$$x+5=2$$
$$x=-3$$

which we disregard as extraneous. No solution, or ∅.

11.

$$\frac{5}{x-2}-\frac{2}{x+4}=-\frac{4}{x^2+2x-8}$$
$$\frac{5}{x-2}-\frac{2}{x+4}=-\frac{4}{(x+4)(x-2)}$$
$$(x-4)(x+2)\left(\frac{5}{x-2}-\frac{2}{x+4}\right)=(x-4)(x+2)\left(-\frac{4}{(x+4)(x-2)}\right)$$
$$5(x+4)-2(x-2)=-4$$
$$5x+20-2x+4=-4$$
$$3x+24=-4$$
$$3x=-28$$
$$x=-\frac{28}{3}$$

13.

$$\frac{1}{x-1}=\frac{2}{x+1}$$
$$(x+1)(x-1)\cdot\frac{1}{x-1}=(x+1)(x-1)\cdot\frac{2}{x+1}$$
$$1(x+1)=2(x-1)$$
$$x+1=2x-2$$
$$-x=-3$$
$$x=3$$

15.

$$\frac{x^2-23}{2x^2-5x-3}+\frac{2}{x-3}=\frac{-1}{2x+1}$$
$$\frac{x^2-23}{(2x+1)(x-3)}+\frac{2}{x-3}=\frac{-1}{2x+1}$$
$$(2x+1)(x-3)\left(\frac{x^2-23}{(2x+1)(x-3)}+\frac{2}{x-3}\right)=(2x+1)(x-3)\left(\frac{-1}{2x+1}\right)$$
$$(x^2-23)+2(2x+1)=-1(x-3)$$
$$x^2-23+4x+2=-x+3$$
$$x^2+5x-24=0$$
$$(x+8)(x-3)=0$$

$x+8=0$ or $x-3=0$

$x=-8$ or $x=3$

We discard 3 as extraneous.

$x=-8$

17.

$$\frac{1}{x-4}-\frac{3x}{x^2-16}=\frac{2}{x+4}$$
$$\frac{1}{x-4}-\frac{3x}{(x+4)(x-4)}=\frac{2}{x+4}$$
$$(x+4)(x-4)\left[\frac{1}{x-4}-\frac{3x}{(x+4)(x-4)}\right]=(x+4)(x-4)\left(\frac{2}{x+4}\right)$$
$$1(x+4)-3x=2(x-4)$$
$$x+4-3x=2x-8$$
$$-2x+4=2x-8$$
$$-4x=-12$$
$$x=3$$

19.

$$\frac{1}{x-4}=\frac{8}{x^2-16}$$
$$\frac{1}{x-4}=\frac{8}{(x+4)(x-4)}$$
$$(x+4)(x-4)\left(\frac{1}{x-4}\right)=(x+4)(x-4)\left(\frac{8}{(x+4)(x-4)}\right)$$
$$1(x+4)=8$$
$$x+4=8$$
$$x=-4$$

which we discard as extraneous. No solution, or $\varnothing$.

21.

$$\frac{1}{x-2}-\frac{2}{x^2-2x}=1$$
$$\frac{1}{x-2}-\frac{2}{x(x-2)}=1$$
$$x(x-2)\left[\frac{1}{x-2}-\frac{2}{x(x-2)}\right]=x(x-2)\cdot 1$$
$$x-2=x(x-2)$$
$$x-2=x^2-2x$$
$$0=x^2-3x+2$$
$$0=(x-2)(x-1)$$

$x-2=0$ or $x-1=1$
$x=2$ or $x=1$

We discard 2 as extraneous.
$x=1$

23.

$$\frac{5}{x}=\frac{20}{12}$$
$$12x\left(\frac{5}{x}\right)=12x\left(\frac{20}{12}\right)$$
$$60=20x$$
$$3=x$$

25.
$$1-\frac{4}{a}=5$$
$$a\left(1-\frac{4}{a}\right)=a(5)$$
$$a-4=5a$$
$$-4=4a$$
$$-1=a$$

27.
$$\frac{x^2+5}{x}-1=\frac{5(x+1)}{x}$$
$$x\left(\frac{x^2+5}{x}-1\right)=x\left[\frac{5(x+1)}{x}\right]$$
$$x^2+5-x=5x+5$$
$$x^2-6x=0$$
$$x(x-6)=0$$
$$x=0 \text{ or } x-6=0$$
$$x=6$$

We discard 0 as extraneous.

$x=6$

29.
$$\frac{1}{2x}-\frac{1}{x+1}=\frac{1}{3x^2+3x}$$
$$\frac{1}{2x}-\frac{1}{x+1}=\frac{1}{3x(x+1)}$$
$$6x(x+1)\left(\frac{1}{2x}-\frac{1}{x+1}\right)=6x(x+1)\left(\frac{1}{3x(x+1)}\right)$$
$$3(x+1)-6x=2$$
$$3x+3-6x=2$$
$$-3x+3=2$$
$$-3x=-1$$
$$x=\frac{1}{3}$$

31.
$$\frac{1}{x}-\frac{x}{25}=0$$
$$25x\left(\frac{1}{x}-\frac{x}{25}\right)=25x(0)$$
$$25-x^2=0$$
$$-(x^2-25)=0$$
$$-(x+5)(x-5)=0$$
$$x+5=0 \quad \text{or} \quad x-5=0$$
$$x=-5 \quad \text{or} \quad x=5$$

33.
$$5-\frac{2}{2y-5}=\frac{3}{2y-5}$$
$$(2y-5)\left(5-\frac{2}{2y-5}\right)=(2y-5)\cdot\frac{3}{2y-5}$$
$$5(2y-5)-2=3$$
$$10y-25-2=3$$
$$10y-27=3$$
$$10y=30$$
$$y=3$$

35.
$$\frac{x-1}{x+2}=\frac{2}{3}$$
$$3(x+2)\left(\frac{x-1}{x+2}\right)=3(x+2)\left(\frac{2}{3}\right)$$
$$3(x-1)=2(x+2)$$
$$3x-3=2x+4$$
$$x=7$$

37.
$$\frac{x+3}{x+2}=\frac{1}{x+2}$$
$$(x+2)\cdot\frac{x+3}{x+2}=(x+2)\cdot\frac{1}{x+2}$$
$$x+3=1$$
$$x=-2$$
which we discard as extraneous. No solution, or ∅.

39.
$$\frac{1}{a-3}+\frac{2}{a+3}=\frac{1}{a^2-9}$$
$$\frac{1}{a-3}+\frac{2}{a+3}=\frac{1}{(a+3)(a-3)}$$
$$(a+3)(a-3)\left(\frac{1}{a-3}+\frac{2}{a+3}\right)=(a+3)(a-3)\cdot\frac{1}{(a+3)(a-3)}$$
$$1(a+3)+2(a-3)=1$$
$$a+3+2a-6=1$$
$$3a-3=1$$
$$3a=4$$
$$a=\frac{4}{3}$$

41.

$$\frac{64}{x^2-16}+1=\frac{2x}{x-4}$$
$$\frac{64}{(x+4)(x-4)}+1=\frac{2x}{x-4}$$
$$(x+4)(x-4)\left[\frac{64}{(x+4)(x-4)}+1\right]=(x+4)(x-4)\cdot\frac{2x}{x-4}$$
$$64+1(x+4)(x-4)=2x(x+4)$$
$$64+(x^2-16)=2x^2+8x$$
$$x^2+48=2x^2+8x$$
$$0=x^2+8x-48$$
$$0=(x+12)(x-4)$$

$x+12=0$ or $x-4=0$
$x=-12$ or $x=4$

We discard 4 as extraneous.
$x=-12$

43.

$$\frac{-15}{4y+1}+4=y$$
$$(4y+1)\left(\frac{-15}{4y+1}+4\right)=(4y+1)y$$
$$-15+4(4y+1)=4y^2+y$$
$$-15+16y+4=4y^2+y$$
$$-11+16y=4y^2+y$$
$$0=4y^2-15y+11$$
$$0=(4y-11)(y-1)$$

$4y-11=0$ or $y-1=0$
$4y=11$ or $y=1$
$y=\frac{11}{4}$

45.

$$\frac{28}{x^2-9}+\frac{2x}{x-3}+\frac{6}{x+3}=0$$
$$\frac{28}{(x+3)(x-3)}+\frac{2x}{x-3}+\frac{6}{x+3}=0$$
$$(x+3)(x-3)\left[\frac{28}{(x+3)(x-3)}+\frac{2x}{x-3}+\frac{6}{x+3}\right]=(x+3)(x-3)\cdot 0$$
$$28+2x(x+3)+6(x-3)=0$$
$$28+2x^2+6x+6x-18=0$$
$$2x^2+12x+10=0$$
$$2(x^2+6x+5)=0$$
$$2(x+5)(x+1)=0$$

$x+5=0$ or $x+1=0$
$x=-5$ or $x=-1$

47.
$$\frac{x+2}{x^2+7x+10}=\frac{1}{3x+6}-\frac{1}{x+5}$$
$$\frac{x+2}{(x+5)(x+2)}=\frac{1}{3(x+2)}-\frac{1}{x+5}$$
$$3(x+5)(x+2)\cdot\frac{x+2}{(x+5)(x+2)}=3(x+5)(x+2)\left[\frac{1}{3(x+2)}-\frac{1}{x+5}\right]$$
$$3(x+2)=1(x+5)-1\cdot 3(x+2)$$
$$3x+6=x+5-3x-6$$
$$3x+6=-2x-1$$
$$5x=-7$$
$$x=-\frac{7}{5}$$

49. Let x = the number.
$$3x+4=19$$
$$3x=15$$
$$x=5$$
The number is 5.

51. Let w = width. Then $w+5=\text{length}$.
$$2l+2w=50$$
$$2(w+5)+2w=50$$
$$2w+10+2w=50$$
$$4w+10=50$$
$$4w=40$$
$$w=10;$$
$$w+5=10+5=15$$
The length is 15 inches and the width is 10 inches.

53. From the graph, 10.8% + 8.4% = 19.2% of state and federal prison inmates were age 45 to 54.

55. From the graph, the category that shows the highest percent of prison inmates is the 30–34-year-olds.

57. 15.3% of 1,574,742 = 0.153(1,574,742) ≈ 240,936
Approximately 240,936 25- to 29-year-old inmates would be expected at the end of 2014.

59. answers may vary

61.
$$f(x)=3.3+\frac{5400}{x}$$
$$5.10=3.3+\frac{5400}{x}$$
$$1.8=\frac{5400}{x}$$
$$1.8x=5400$$
$$x=3000$$
3000 game disks

63.
$$x^{-2}-19x^{-1}+48=0$$
$$\frac{1}{x^2}-\frac{19}{x}+48=0$$
$$x^2\left(\frac{1}{x^2}-\frac{19}{x}+48\right)=x^2\cdot 0$$
$$1-19x+48x^2=0$$
$$48x^2-19x+1=0$$
$$(16x-1)(3x-1)=0$$
$$x=\frac{1}{16} \text{ or } x=\frac{1}{3}$$

65.
$$p^{-2}+4p^{-1}-5=0$$
$$\frac{1}{p^2}+\frac{4}{p}-5=0$$
$$p^2\left(\frac{1}{p^2}+\frac{4}{p}-5\right)=p^2\cdot 0$$
$$1+4p-5p^2=0$$
$$5p^2-4p-1=0$$
$$(5p+1)(p-1)=0$$
$$p=-\frac{1}{5} \text{ or } p=1$$

67.
$$\frac{1.4}{x-2.6}=\frac{-3.5}{x+7.1}$$
$$(x-2.6)(x+7.1)\cdot\frac{1.4}{x-2.6}=(x-2.6)(x+7.1)\cdot\frac{-3.5}{x+7.1}$$
$$1.4(x+7.1)=-3.5(x-2.6)$$
$$1.4x+9.94=-3.5x+9.1$$
$$4.9x=-0.84$$
$$x\approx-0.17$$

69.
$$\frac{10.6}{y}-14.7=\frac{9.92}{3.2}+7.6$$
$$\frac{10.6}{y}-14.7=10.7$$
$$\frac{10.6}{y}=25.4$$
$$y\cdot\frac{10.6}{y}=y\cdot 25.4$$
$$10.6=25.4y$$
$$0.42\approx y$$

71. $(x-1)^2+3(x-1)+2=0$

Let $u = x - 1$.

$$u^2+3u+2=0$$
$$(u+2)(u+1)=0$$
$$u=-2 \text{ or } u=-1$$
$$x-1=-2 \text{ or } x-1=-1$$
$$x=-1 \text{ or } x=0$$

73. $\left(\frac{3}{x-1}\right)^2+2\left(\frac{3}{x-1}\right)+1=0$

Let $u=\frac{3}{x-1}$.

$$u^2+2u+1=0$$
$$(u+1)^2=0$$
$$u+1=0$$
$$u=-1$$
$$\frac{3}{x-1}=-1$$
$$3=-1(x-1)$$
$$3=-x+1$$
$$x=-2$$

75.

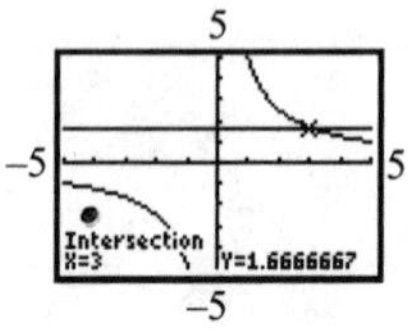

77.

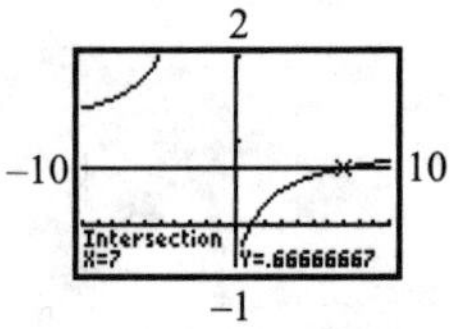

Integrated Review

1. $\frac{x}{2}=\frac{1}{8}+\frac{x}{4}$

The LCD is 8.

$$8\cdot\frac{x}{2}=8\cdot\frac{1}{8}+8\cdot\frac{x}{4}$$
$$4x=1+2x$$
$$2x=1$$
$$x=\frac{1}{2}$$

The solution is $\frac{1}{2}$.

2. $\frac{x}{4}=\frac{3}{2}+\frac{x}{10}$

The LCD is 20.

$$20\cdot\frac{x}{4}=20\cdot\frac{3}{2}+20\cdot\frac{x}{10}$$
$$5x=30+2x$$
$$3x=30$$
$$x=10$$

The solution is 10.

3. $\frac{1}{8}+\frac{x}{4}=\frac{1}{8}+\frac{x}{4}\cdot\frac{2}{2}=\frac{1}{8}+\frac{2x}{8}=\frac{1+2x}{8}$

4. $\frac{3}{2}+\frac{x}{10}=\frac{5}{5}\cdot\frac{3}{2}+\frac{x}{10}=\frac{15}{10}+\frac{x}{10}=\frac{15+x}{10}$

5.
$$\begin{aligned}\frac{4}{x+2}-\frac{2}{x-1}&=\frac{4}{x+2}\cdot\frac{x-1}{x-1}-\frac{2}{x-1}\cdot\frac{x+2}{x+2}\\&=\frac{4(x-1)}{(x+2)(x-1)}-\frac{2(x+2)}{(x+2)(x+1)}\\&=\frac{4x-4-2x-4}{(x+2)(x-1)}\\&=\frac{2x-8}{(x+2)(x-1)}\\&=\frac{2(x-4)}{(x+2)(x-1)}\end{aligned}$$

6.
$$\begin{aligned}\frac{5}{x-2}-\frac{10}{x+4}&=\frac{5}{x-2}\cdot\frac{x+4}{x+4}-\frac{10}{x+4}\cdot\frac{x-2}{x-2}\\&=\frac{5(x+4)}{(x-2)(x+4)}-\frac{10(x-2)}{(x-2)(x+4)}\\&=\frac{5x+20-10x+20}{(x-2)(x+4)}\\&=\frac{-5x+40}{(x-2)(x+4)}\\&=\frac{-5(x-8)}{(x-2)(x+4)} \text{ or } -\frac{5(x-8)}{(x-2)(x+4)}\end{aligned}$$

7. $\frac{4}{x+2}=\frac{2}{x-1}$

The LCD is $(x + 2)(x - 1)$.

$$(x+2)(x-1)\cdot\frac{4}{x+2}=(x+2)(x-1)\cdot\frac{2}{x-1}$$
$$4(x-1)=2(x+2)$$
$$4x-4=2x+4$$
$$4x=2x+8$$
$$2x=8$$
$$x=4$$

The solution is 4.

8. $\frac{5}{x-2}=\frac{10}{x+4}$

The LCD is $(x-2)(x+4)$.

$$\begin{aligned}(x-2)(x+4)\cdot\frac{5}{x-2}&=(x-2)(x+4)\cdot\frac{10}{x+4}\\ 5(x+4)&=10(x-2)\\ 5x+20&=10x-20\\ 5x&=10x-40\\ -5x&=-40\\ x&=8\end{aligned}$$

The solution is 8.

9. $x^2-4=(x+2)(x-2)$

The LCD is $(x+2)(x-2)$.

$$\begin{aligned}\frac{2}{x^2-4}&=\frac{1}{x+2}-\frac{3}{x-2}\\ (x+2)(x-2)\cdot\frac{2}{x^2-4}&=(x+2)(x-2)\cdot\frac{1}{x+2}-(x+2)(x-2)\cdot\frac{3}{x-2}\\ 2&=(x-2)-3(x+2)\\ 2&=x-2-3x-6\\ 2&=-2x-8\\ 2x&=-10\\ x&=-5\end{aligned}$$

The solution is -5.

10. $x^2-25=(x+5)(x-5)$

The LCD is $(x+5)(x-5)$.

$$\begin{aligned}\frac{3}{x^2-25}&=\frac{1}{x+5}+\frac{2}{x-5}\\ (x+5)(x-5)\cdot\frac{3}{(x+5)(x-5)}&=(x+5)(x-5)\cdot\frac{1}{x+5}+(x+5)(x-5)\cdot\frac{2}{x-5}\\ 3&=(x-5)+2(x+5)\\ 3&=x-5+2x+10\\ 3&=3x+5\\ -2&=3x\\ -\frac{2}{3}&=x\end{aligned}$$

The solution is $-\frac{2}{3}$.

11. $$\frac{5}{x^2-3x}+\frac{4}{2x-6}=\frac{5}{x(x-3)}+\frac{4}{2(x-3)}$$
$$=\frac{5}{x(x-3)}\cdot\frac{2}{2}+\frac{4}{2(x-3)}\cdot\frac{x}{x}$$
$$=\frac{10}{2x(x-3)}+\frac{4x}{2x(x-3)}$$
$$=\frac{4x+10}{2x(x-3)}$$
$$=\frac{2(2x+5)}{2x(x-3)}$$
$$=\frac{2x+5}{x(x-3)}$$

12. $$\frac{5}{x^2-3x}\div\frac{4}{2x-6}=\frac{5}{x^2-3x}\cdot\frac{2x-6}{4}$$
$$=\frac{5}{x(x-3)}\cdot\frac{2(x-3)}{4}$$
$$=\frac{5}{2x}$$

13. $x^2-1=(x-1)(x+1)$

The LCD is $(x-1)(x+1)$.
$$\frac{x-1}{x+1}+\frac{x+7}{x-1}=\frac{4}{x^2-1}$$
$$(x-1)(x+1)\cdot\frac{x-1}{x+1}+(x-1)(x+1)\cdot\frac{x+7}{x-1}=(x-1)(x+1)\cdot\frac{4}{(x-1)(x+1)}$$
$$(x-1)(x-1)+(x+1)(x+7)=4$$
$$x^2-2x+1+x^2+8x+7=4$$
$$2x^2+6x+8=4$$
$$2x^2+6x+4=0$$
$$2(x^2+3x+2)=0$$
$$2(x+1)(x+2)=0$$

$x+1=0$ or $x+2=0$
$x=-1$ or $x=-2$

The number -1 makes the denominator $x+1$ equal to 0, so it is not a solution. The solution is -2.

14. $$\left(1-\frac{y}{x}\right)\div\left(1-\frac{x}{y}\right)=\left(\frac{x}{x}-\frac{y}{x}\right)\div\left(\frac{y}{y}-\frac{x}{y}\right)$$
$$=\left(\frac{x-y}{x}\right)\div\left(\frac{y-x}{y}\right)$$
$$=\frac{x-y}{x}\cdot\frac{y}{y-x}$$
$$=\frac{x-y}{x}\cdot\frac{y}{-(x-y)}$$
$$=-\frac{y}{x}$$

15. $\frac{a^2-9}{a-6}\cdot\frac{a^2-5a-6}{a^2-a-6}=\frac{(a+3)(a-3)}{a-6}\cdot\frac{(a-6)(a+1)}{(a-3)(a+2)}$
$=\frac{(a+3)(a+1)}{a+2}$

16. $\frac{2}{a-6}+\frac{3a}{a^2-5a-6}-\frac{a}{5a+5}=\frac{2}{a-6}+\frac{3a}{(a-6)(a+1)}-\frac{a}{5(a+1)}$
$=\frac{2}{a-6}\cdot\frac{5(a+1)}{5(a+1)}+\frac{3a}{(a-6)(a+1)}\cdot\frac{5}{5}-\frac{a}{5(a+1)}\cdot\frac{a-6}{a-6}$
$=\frac{10a+10}{5(a+1)(a-6)}+\frac{15a}{5(a+1)(a-6)}-\frac{a^2-6a}{5(a+1)(a-6)}$
$=\frac{10a+10+15a-a^2+6a}{5(a+1)(a-6)}$
$=\frac{-a^2+31a+10}{5(a+1)(a-6)}$

17. $\frac{2x+3}{3x-2}=\frac{4x+1}{6x+1}$

The LCD is $(3x-2)(6x+1)$.

$$(3x-2)(6x+1)\cdot\frac{2x+3}{3x-2}=(3x-2)(6x+1)\cdot\frac{4x+1}{6x+1}$$
$$(6x+1)(2x+3)=(3x-2)(4x+1)$$
$$12x^2+18x+2x+3=12x^2+3x-8x-2$$
$$12x^2+20x+3=12x^2-5x-2$$
$$20x+3=-5x-2$$
$$25x+3=-2$$
$$25x=-5$$
$$x=-\frac{5}{25}$$
$$x=-\frac{1}{5}$$

The solution is $-\frac{1}{5}$.

18. The LCD is $2x(4x + 1)$.

$$\frac{5x-3}{2x}=\frac{10x+3}{4x+1}$$
$$2x(4x+1)\cdot\frac{5x-3}{2x}=2x(4x+1)\cdot\frac{10x+3}{4x+1}$$
$$(4x+1)(5x-3)=2x(10x+3)$$
$$20x^2-12x+5x-3=20x^2+6x$$
$$20x^2-7x-3=20x^2+6x$$
$$-7x-3=6x$$
$$-3=13x$$
$$-\frac{3}{13}=x$$

The solution is $-\frac{3}{13}$.

19. $\frac{a}{9a^2-1}+\frac{2}{6a-2}$
$$=\frac{a}{(3a-1)(3a+1)}+\frac{2}{2(3a-1)}$$
$$=\frac{a}{(3a-1)(3a+1)}\cdot\frac{2}{2}+\frac{2}{2(3a-1)}\cdot\frac{(3a+1)}{(3a+1)}$$
$$=\frac{2a}{2(3a-1)(3a+1)}+\frac{6a+2}{2(3a-1)(3a+1)}$$
$$=\frac{8a+2}{2(3a-1)(3a+1)}$$
$$=\frac{2(4a+1)}{2(3a-1)(3a+1)}$$
$$=\frac{4a+1}{(3a-1)(3a+1)}$$

20. $\frac{3}{4a-8}-\frac{a+2}{a^2-2a}=\frac{3}{4(a-2)}-\frac{a+2}{a(a-2)}$
$$=\frac{3}{4(a-2)}\cdot\frac{a}{a}-\frac{a+2}{a(a-2)}\cdot\frac{4}{4}$$
$$=\frac{3a}{4a(a-2)}-\frac{4(a+2)}{4a(a-2)}$$
$$=\frac{3a}{4a(a-2)}-\frac{4a+8}{4a(a-2)}$$
$$=\frac{3a-4a-8}{4a(a-2)}$$
$$=\frac{-a-8}{4a(a-2)}$$
$$=-\frac{a+8}{4a(a-2)}$$

21. The LCD is x^2.

$$-\frac{3}{x^2}-\frac{1}{x}+2=0$$
$$x^2\cdot-\frac{3}{x^2}-x^2\cdot\frac{1}{x}+x^2\cdot 2=0$$
$$-3-x+2x^2=0$$
$$2x^2-x-3=0$$
$$(2x-3)(x+1)=0$$
$$2x-3=0 \quad\text{or}\quad x+1=0$$
$$2x=3 \quad\text{or}\quad x=-1$$
$$x=\frac{3}{2} \quad\text{or}\quad x=-1$$

The solutions are -1, $\frac{3}{2}$.

22. $\frac{x}{2x+6}+\frac{5}{x^2-9}$
$$=\frac{x}{2(x+3)}+\frac{5}{(x-3)(x+3)}$$
$$=\frac{x}{2(x+3)}\cdot\frac{x-3}{x-3}+\frac{5}{(x-3)(x+3)}\cdot\frac{2}{2}$$
$$=\frac{x(x-3)}{2(x+3)(x-3)}+\frac{10}{2(x+3)(x-3)}$$
$$=\frac{x^2-3x}{2(x+3)(x-3)}+\frac{10}{2(x+3)(x-3)}$$
$$=\frac{x^2-3x+10}{2(x+3)(x-3)}$$

23.
$$\begin{aligned}\frac{x-8}{x^2-x-2}+\frac{2}{x-2}&=\frac{x-8}{(x-2)(x+1)}+\frac{2}{x-2}\\&=\frac{x-8}{(x-2)(x+1)}+\frac{2}{x-2}\cdot\frac{x+1}{x+1}\\&=\frac{x-8}{(x-2)(x+1)}+\frac{2x+2}{(x-2)(x+1)}\\&=\frac{x-8+2x+2}{(x-2)(x+1)}\\&=\frac{3x-6}{(x-2)(x+1)}\\&=\frac{3(x-2)}{(x-2)(x+1)}\\&=\frac{3}{x+1}\end{aligned}$$

24. $x^2-x-2=(x-2)(x+1)$

The LCD is $(x-2)(x+1)$.

$$\begin{aligned}\frac{x-8}{x^2-x-2}+\frac{2}{x-2}&=\frac{3}{x+1}\\(x-2)(x+1)\cdot\frac{x-8}{(x-2)(x+1)}+(x-2)(x+1)\cdot\frac{2}{x-2}&=(x-2)(x+1)\cdot\frac{3}{x+1}\\(x-8)+2(x+1)&=3(x-2)\\x-8+2x+2&=3x-6\\3x-6&=3x-6\\-6&=-6\quad\text{True}\end{aligned}$$

The solution set is
$\{x|x \text{ is a real number and } x\neq 2, x\neq -1\}$.

25. The LCD is a.

$$\begin{aligned}\frac{3}{a}-5&=\frac{7}{a}-1\\a\cdot\frac{3}{a}-a\cdot 5&=a\cdot\frac{7}{a}-a\cdot 1\\3-5a&=7-a\\3&=7+4a\\-4&=4a\\-1&=a\end{aligned}$$

The solution is -1.

26. $\dfrac{7}{3z-9}+\dfrac{5}{z}=\dfrac{7}{3(z-3)}+\dfrac{5}{z}$
$=\dfrac{7}{3(z-3)}\cdot\dfrac{z}{z}+\dfrac{5}{z}\cdot\dfrac{3(z-3)}{3(z-3)}$
$=\dfrac{7z}{3z(z-3)}+\dfrac{15(z-3)}{3z(z-3)}$
$=\dfrac{7z+15z-45}{3z(z-3)}$
$=\dfrac{22z-45}{3z(z-3)}$

27. a. $\dfrac{x}{5}-\dfrac{x}{4}+\dfrac{1}{10}$ is an expression.

b. The first step to simplify this expression is to write each rational expression term so that the denominator is the LCD, 20.

c. $\dfrac{x}{5}-\dfrac{x}{4}+\dfrac{1}{10}=\dfrac{x}{5}\cdot\dfrac{4}{4}-\dfrac{x}{4}\cdot\dfrac{5}{5}+\dfrac{1}{10}\cdot\dfrac{2}{2}$
$=\dfrac{4x}{20}-\dfrac{5x}{20}+\dfrac{2}{20}$
$=\dfrac{4x-5x+2}{20}$
$=\dfrac{-x+2}{20}$

28. a. $\dfrac{x}{5}-\dfrac{x}{4}=\dfrac{1}{10}$ is an equation.

b. The first step to solve this equation is to clear the equation of fractions by multiplying each term by the LCD, 20.

c. $\dfrac{x}{5}-\dfrac{x}{4}=\dfrac{1}{10}$
$20\cdot\dfrac{x}{5}-20\cdot\dfrac{x}{4}=20\cdot\dfrac{1}{10}$
$4x-5x=2$
$-x=2$
$x=-2$

The solution is –2.

29. $\dfrac{\triangle+\square}{\triangle}=\dfrac{\triangle}{\triangle}+\dfrac{\square}{\triangle}=1+\dfrac{\square}{\triangle}$

b is the correct answer.

30. $\dfrac{\triangle}{\square}+\dfrac{\square}{\triangle}=\dfrac{\triangle}{\square}\cdot\dfrac{\triangle}{\triangle}+\dfrac{\square}{\triangle}\cdot\dfrac{\square}{\square}$
$=\dfrac{\triangle\triangle}{\square\triangle}+\dfrac{\square\square}{\square\triangle}$
$=\dfrac{\triangle\triangle+\square\square}{\square\triangle}$

d is the correct answer.

31. $\dfrac{\triangle}{\square}\cdot\dfrac{\bigcirc}{\square}=\dfrac{\triangle\bigcirc}{\square\square}$

d is the correct answer.

32. $\dfrac{\triangle}{\square}\div\dfrac{\bigcirc}{\triangle}=\dfrac{\triangle}{\square}\cdot\dfrac{\triangle}{\bigcirc}=\dfrac{\triangle\triangle}{\square\bigcirc}$

a is the correct answer.

33. $\dfrac{\frac{\triangle+\square}{\bigcirc}}{\frac{\triangle}{\bigcirc}}=\dfrac{\triangle+\square}{\bigcirc}\div\dfrac{\triangle}{\bigcirc}=\dfrac{\triangle+\square}{\bigcirc}\cdot\dfrac{\bigcirc}{\triangle}=\dfrac{\triangle+\square}{\triangle}$

d is the correct answer.

Section 6.6 Practice Exercises

1. abc is the LCD.

$$\frac{1}{a}-\frac{1}{b}=\frac{1}{c}$$
$$abc\left(\frac{1}{a}-\frac{1}{b}\right)=abc\left(\frac{1}{c}\right)$$
$$abc\left(\frac{1}{a}\right)-abc\left(\frac{1}{b}\right)=abc\left(\frac{1}{c}\right)$$
$$bc-ac=ab$$
$$bc=ab+ac$$
$$bc=a(b+c)$$
$$\frac{bc}{b+c}=a$$

2. Let n = the number.

$$\frac{3+n}{11-n}=\frac{5}{2}$$
$$2(11-n)\cdot\frac{3+n}{11-n}=2(11-n)\cdot\frac{5}{2}$$
$$2(3+n)=5(11-n)$$
$$6+2n=55-5n$$
$$7n=49$$
$$n=7$$

The number is 7.

3. Let x = the number of homes heated with natural gas.

$$\frac{2}{5} = \frac{x}{36,000}$$
$$5x = 2 \cdot 36,000$$
$$x = \frac{72,000}{5}$$
$$x = 14,400$$

14,400 homes in the community are heated with natural gas.

4. Let t = the time it takes them to clean the cages together.

	Time	Part done in one hour
Elissa	3	$\frac{1}{3}$
Bill	2	$\frac{1}{2}$
Together	t	$\frac{1}{t}$

$$\frac{1}{3} + \frac{1}{2} = \frac{1}{t}$$
$$6t\left(\frac{1}{3} + \frac{1}{2}\right) = 6t\left(\frac{1}{t}\right)$$
$$2t + 3t = 6$$
$$5t = 6$$
$$t = \frac{6}{5} \text{ or } 1\frac{1}{5}$$

Elissa and Bill can clean the cages in $1\frac{1}{5}$ hours if they work together.

5. Let x = the speed of the tugboat in still water.

	Distance	Rate	Time $\left(\frac{d}{r}\right)$
Upstream	100	$x - 2$	$\frac{100}{x-2}$
Downstream	100	$x + 2$	$\frac{100}{x+2}$

The time spent traveling upstream is $1\frac{2}{3} = \frac{5}{3}$ times the time spent traveling downstream.

$$\frac{100}{x-2} = \frac{5}{3}\left(\frac{100}{x+2}\right)$$
$$3(x-2)(x+2)\left(\frac{100}{x-2}\right) = 3(x-2)(x+2)\left(\frac{5}{3}\right)\left(\frac{100}{x+2}\right)$$
$$300(x+2) = 500(x-2)$$
$$300x+600 = 500x-1000$$
$$1600 = 200x$$
$$8 = x$$

The speed of the tugboat in still water is 8 mph.

Vocabulary, Readiness & Video Check 6.6

1. After we multiply through by the LCD, we factor the specified variable out so that we can divide and get the specified variable alone.

2. $\frac{1}{x}$

3. We write the same units in the numerators and the same units in the denominators.

4. Two or more people (or machines) take different amounts of time to complete a task. When working together, they will complete the task in less time than the fastest person, so your answer will be less than the time of the fastest person.

5. $x + y$ (or $y + x$)

Exercise Set 6.6

1.
$$F = \frac{9}{5}C+32$$
$$F-32 = \frac{9}{5}C$$
$$C = \frac{5}{9}(F-32)$$

3.
$$Q = \frac{A-I}{L}$$
$$QL = L\left(\frac{A-I}{L}\right)$$
$$QL = A-I$$
$$I = A-QL$$

5.
$$\frac{1}{R} = \frac{1}{R_1}+\frac{1}{R_2}$$
$$RR_1R_2\cdot\frac{1}{R} = RR_1R_2\left(\frac{1}{R_1}+\frac{1}{R_2}\right)$$
$$R_1R_2 = RR_2+RR_1$$
$$R_1R_2 = R(R_2+R_1)$$
$$R = \frac{R_1R_2}{R_1+R_2}$$

7. $S=\frac{n(a+L)}{2}$
$2S=n(a+L)$
$n=\frac{2S}{a+L}$

9. $A=\frac{h(a+b)}{2}$
$2A=h(a+b)$
$2A=ah+bh$
$2A-ah=bh$
$b=\frac{2A-ah}{h}$

11. $\frac{P_1V_1}{T_1}=\frac{P_2V_2}{T_2}$
$T_1T_2\cdot\frac{P_1V_1}{T_1}=T_1T_2\cdot\frac{P_2V_2}{T_2}$
$P_1V_1T_2=P_2V_2T_1$
$T_2=\frac{P_2V_2T_1}{P_1V_1}$

13. $f=\frac{f_1f_2}{f_1+f_2}$
$(f_1+f_2)f=f_1f_2$
$f_1f+f_2f=f_1f_2$
$f_1f=f_1f_2-f_2f$
$f_1f=f_2(f_1-f)$
$\frac{f_1f}{f_1-f}=f_2$

15. $\lambda=\frac{2L}{n}$
$n\lambda=2L$
$\frac{n\lambda}{2}=L$

17. $\frac{\theta}{\omega}=\frac{2L}{c}$
$c\omega\cdot\frac{\theta}{\omega}=c\omega\cdot\frac{2L}{c}$
$c\theta=2L\omega$
$c=\frac{2L\omega}{\theta}$

19. Let x = the number. Then
$\frac{1}{x}$ = the reciprocal of the number.
$x+5\left(\frac{1}{x}\right)=6$
$x+\frac{5}{x}=6$
$x\left(x+\frac{5}{x}\right)=6x$
$x^2+5=6x$
$x^2-6x+5=0$
$(x-5)(x-1)=0$
$x-5=0$ or $x-1=0$
$x=5$ or $x=1$
The number is either 1 or 5.

21. Let x = the number.
$\frac{12+x}{41+2x}=\frac{1}{3}$
$3(12+x)=1(41+2x)$
$36+3x=41+2x$
$x=5$
The number is 5.

23. Let a = amount of water in 3 minutes.
$\frac{15}{10}=\frac{a}{3}$
$10a=15(3)$
$10a=45$
$a=4.5$
The camel can drink 4.5 gallons.

25. Let w = the number of women.
$\frac{16.3}{100}=\frac{w}{535,370}$
$16.3(535,370)=100\cdot w$
$8,726,531=100w$
$87,265\approx w$
There were 87,265 women.

27. Let x = number of hours needed working together.
$\frac{1}{26}+\frac{1}{39}=\frac{1}{x}$
$78x\left(\frac{1}{26}+\frac{1}{39}\right)=78x\left(\frac{1}{x}\right)$
$3x+2x=78$
$5x=78$
$x=\frac{78}{5}=15.6$
The roofers together would take 15.6 hours.

29. Let x = time to sort the stack working together.

$$\frac{1}{20}+\frac{1}{30}+\frac{1}{60}=\frac{1}{x}$$
$$60x\left(\frac{1}{20}+\frac{1}{30}+\frac{1}{60}\right)=60x\cdot\frac{1}{x}$$
$$3x+2x+x=60$$
$$6x=60$$
$$x=10$$

It takes them 10 minutes to sort the mail when all three work together.

31. Let r = speed of the car. Then $r+150$ = speed of the plane.

$$t_{\text{plane}}=t_{\text{car}}$$
$$\frac{600}{r+150}=\frac{150}{r}$$
$$600r=150(r+150)$$
$$600r=150r+22,500$$
$$450r=22,500$$
$$r=\frac{22,500}{450}=50$$
$$r+150=50+150=200$$

The speed of the plane was 200 mph.

33. Let r = speed of the boat in still water.

$$t_{\text{downstream}}=t_{\text{upstream}}$$
$$\frac{20}{r+5}=\frac{10}{r-5}$$
$$20(r-5)=10(r+5)$$
$$20r-100=10r+50$$
$$10r=150$$
$$r=15$$

The speed of the boat in still water is 15 mph.

35. Let x = the first integer. Then $x+1$ = the next integer.

$$\frac{1}{x}+\frac{1}{x+1}=-\frac{15}{56}$$
$$56x(x+1)\left[\frac{1}{x}+\frac{1}{x+1}\right]=56x(x+1)\left(-\frac{15}{56}\right)$$
$$56(x+1)+56x=-15x(x+1)$$
$$56x+56+56x=-15x^2-15x$$
$$112x+56=-15x^2-15x$$
$$15x^2+127x+56=0$$
$$(15x+7)(x+8)=0$$
$$15x+7=0 \quad\text{or}\quad x+8=0$$
$$15x=-7 \quad\text{or}\quad x=-8$$
$$x=-\frac{7}{15} \qquad x+1=-8+1=-7$$

$-\frac{7}{15}$ is not an integer.

The integers are –8 and –7.

37. Let t = time for 2nd hose to fill the pond.

$$\frac{1}{45}+\frac{1}{t}=\frac{1}{20}$$
$$180t\left(\frac{1}{45}+\frac{1}{t}\right)=180t\cdot\frac{1}{20}$$
$$4t+180=9t$$
$$180=5t$$
$$36=t$$

The second hose will take 36 minutes to fill the pond alone.

39. Let r = the speed of the first train. Then $r+15$ = the speed of the 2nd train.

$$d_{\text{train 1}}+d_{\text{train 2}}=630$$
$$6r+6(r+15)=630$$
$$6r+6r+90=630$$
$$12r=540$$
$$r=45$$
$$r+15=45+15=60$$

The speed of the trains were 45 mph and 60 mph.

41. Let t = time to travel 1 mile.

$$\frac{0.17}{1}=\frac{1}{t}$$
$$0.17t=1$$
$$100(0.17t)=100(1)$$
$$17t=100$$
$$t=\frac{100}{17}\approx 5.882352941$$

It would take 5.9 hours.

43. Let t = time to fill quota working together.

$$\frac{1}{5}+\frac{1}{6}+\frac{1}{7.5}=\frac{1}{t}$$
$$225t\left(\frac{1}{5}+\frac{1}{6}+\frac{1}{7.5}\right)=225t\cdot\frac{1}{t}$$
$$45t+37.5t+30t=225$$
$$112.5t=225$$
$$t=2$$

It would take 2 hours using all three machines.

45. Let r = the speed of plane in still air.

$$t_{\text{with}} = t_{\text{against}}$$
$$\frac{465}{r+20} = \frac{345}{r-20}$$
$$465(r-20) = 345(r+20)$$
$$465r - 9300 = 345r + 6900$$
$$120r = 16,200$$
$$r = 135$$

The planes speed in still air is 135 mph.

47. Let d = the distance of the run.

$$t_{\text{jogger 2}} = \frac{1}{2} + t_{\text{jogger 1}}$$
$$\frac{d}{6} = \frac{1}{2} + \frac{d}{8}$$
$$24\left(\frac{d}{6}\right) = 24\left(\frac{1}{2} + \frac{d}{8}\right)$$
$$4d = 12 + 3d$$
$$d = 12$$

The run was 12 miles.

49. Let n = numerator. Then $n+1$ = the denominator, and $\frac{n}{n+1}$ = the fraction. Now

$$\frac{n-3}{(n+1)-3} = \frac{4}{5}$$
$$\frac{n-3}{n-2} = \frac{4}{5}$$
$$5(n-3) = 4(n-2)$$
$$5n - 15 = 4n - 8$$
$$n = 7$$

Thus, $\frac{n}{n+1} = \frac{7}{7+1} = \frac{7}{8}$ is the fraction.

51. Let t = time to move the cans working together.

$$\frac{1}{2} + \frac{1}{6} = \frac{1}{t}$$
$$6t\left(\frac{1}{2} + \frac{1}{6}\right) = 6t \cdot \frac{1}{t}$$
$$3t + t = 6$$
$$4t = 6$$
$$t = \frac{6}{4} = 1\frac{1}{2}$$

It would take them $1\frac{1}{2}$ minutes.

53.

	distance =	rate	· time
upstream	3	$x-6$	$\frac{3}{x-6}$
downstream	9	$x+6$	$\frac{9}{x+6}$

$$\frac{3}{x-6} = \frac{9}{x+6}$$
$$3(x+6) = 9(x-6)$$
$$3x + 18 = 9x - 54$$
$$3x + 72 = 9x$$
$$72 = 6x$$
$$12 = x$$

Time upstream $= \frac{3}{12-6} = \frac{3}{6} = \frac{1}{2}$ hour.

Total time is $2\left(\frac{1}{2}\right) = 1$ hour.

55. Let t = time to complete the job working together.

$$\frac{1}{4} + \frac{1}{5} = \frac{1}{t}$$
$$20t\left(\frac{1}{4} + \frac{1}{5}\right) = 20t \cdot \frac{1}{t}$$
$$5t + 4t = 20$$
$$9t = 20$$
$$t = \frac{20}{9} = 2\frac{2}{9}$$

It would take them $2\frac{2}{9}$ hours.

57. Let t = time if they worked together.

$$\frac{1}{3} + \frac{1}{6} = \frac{1}{t}$$
$$6t\left(\frac{1}{3} + \frac{1}{6}\right) = 6t \cdot \frac{1}{t}$$
$$2t + t = 6$$
$$3t = 6$$
$$t = 2$$

It would take them 2 hours.

59. Let m = the number of movies that would be rated PG-13.

$$\frac{12}{20} = \frac{m}{692}$$
$$20m = 12 \cdot 692$$
$$20m = 8304$$
$$m = \frac{8304}{20} = 415.2$$

Approximately 415 movies would be PG-13.

61.
$$\frac{x}{5}=\frac{x+2}{3}$$
$$5(x+2)=3x$$
$$5x+10=3x$$
$$10=-2x$$
$$-5=x$$

63.
$$\frac{x-3}{2}=\frac{x-5}{6}$$
$$6(x-3)=2(x-5)$$
$$6x-18=2x-10$$
$$4x=8$$
$$x=2$$

65. 5 ft 8 in. = (5 · 12 + 8) in. = 68 in.
$$\frac{705w}{h^2}=25$$
$$\frac{705w}{68^2}=25$$
$$\frac{705w}{4624}=25$$
$$705w=25(4624)$$
$$705w=115,600$$
$$w\approx 163.9716312$$
The person should weigh 164 pounds.

67. Use $h=\dfrac{a}{1-\frac{s}{770}}$ with a = 329.63 and s = 50.
$$h=\frac{a}{1-\frac{s}{770}}=\frac{329.63}{1-\frac{50}{770}}\approx 352.52$$
The pitch that the observer hears is higher than the actual pitch—it is closest to the musical note F.

69.
$$\frac{1}{R}=\frac{1}{R_1}+\frac{1}{R_2}$$
$$\frac{1}{2}=\frac{1}{3}+\frac{1}{R_2}$$
$$6R_2\left(\frac{1}{2}\right)=6R_2\left(\frac{1}{3}+\frac{1}{R_2}\right)$$
$$3R_2=2R_2+6$$
$$R_2=6$$
6 ohms

71.
$$\frac{1}{R}=\frac{1}{R_1}+\frac{1}{R_2}+\frac{1}{R_3}$$
$$\frac{1}{R}=\frac{1}{5}+\frac{1}{6}+\frac{1}{2}$$
$$30R\left(\frac{1}{R}\right)=30R\left(\frac{1}{5}+\frac{1}{6}+\frac{1}{2}\right)$$
$$30=6R+5R+15R$$
$$30=26R$$
$$R=\frac{30}{26}=\frac{15}{13}$$
$\frac{15}{13}$ ohms

Section 6.7 Practice Exercises

1.
$$y=kx$$
$$20=k(15)$$
$$\frac{4}{3}=k$$
$$k=\frac{4}{3};\ y=\frac{4}{3}x$$

2.
$$d=kw$$
$$9=k(36)$$
$$\frac{1}{4}=k$$
$$d=\frac{1}{4}w$$
$$d=\frac{1}{4}(75)$$
$d=\frac{75}{4}$ inches or $18\frac{3}{4}$ inches

3.
$$b=\frac{k}{a}$$
$$5=\frac{k}{9}$$
$$k=45;\ b=\frac{45}{a}$$

4. $P = \frac{k}{V}$
$350 = \frac{k}{2.8}$
$980 = k$
$P = \frac{980}{V}$
$P = \frac{980}{1.5}$
$P = 653\frac{1}{3}$

The pressure is $653\frac{1}{3}$ kilopascals.

5. $A = kap$

6. $y = \frac{k}{x^3}$
$\frac{1}{2} = \frac{k}{2^3}$
$\frac{1}{2} = \frac{k}{8}$
$4 = k$
$k = 4;\ y = \frac{4}{x^3}$

7. $y = \frac{kz}{x^3}$
$15 = \frac{k \cdot 5}{3^3}$
$81 = k$
$k = 81;\ y = \frac{81z}{x^3}$

Vocabulary, Readiness & Video Check 6.7

1. $y = 5x$ represents direct variation.

2. $y = \frac{700}{x}$ represents inverse variation.

3. $y = 5xz$ represents joint variation.

4. $y = \frac{1}{2}abc$ represents joint variation.

5. $y = \frac{9.1}{x}$ represents inverse variation.

6. $y = 2.3x$ represents direct variation.

7. $y = \frac{2}{3}x$ represents direct variation.

8. $y = 3.1st$ represents joint variation.

9. linear; slope

10. When x and y vary inversely, the product of x and y is the constant of variation, k.

11. $y = ka^2b^5$

12. A combined variation problem involves combinations of direct, inverse, and/or joint variation.

Exercise Set 6.7

1. $y = kx$
$4 = k(20)$
$k = \frac{1}{5}$
$y = \frac{1}{5}x$

3. $y = kx$
$6 = k(4)$
$k = \frac{3}{2}$
$y = \frac{3}{2}x$

5. $y = kx$
$7 = k\left(\frac{1}{2}\right)$
$k = 14$
$y = 14x$

7. $y = kx$
$0.2 = k(0.8)$
$k = 0.25$
$y = 0.25x$

9. $W = kr^3$
$1.2 = k \cdot 2^3$
$k = \frac{1.2}{8} = 0.15$
$W = 0.15r^3$
$= 0.15(3)^3$
$= 0.15(27)$
$= 4.05$
The ball weighs 4.05 pounds.

11. $W = kN$
$36.5 = k(467{,}000)$
$k = \frac{36.5}{467{,}000}$
$P = \frac{36.5}{467{,}000}N$
$P = \frac{36.5}{467{,}000}(160{,}000) \approx 12.5$
We expect Springfield to use 12.5 billion gallons of water.

13. $y = \frac{k}{x}$
$6 = \frac{k}{5}$
$k = 30$
$y = \frac{30}{x}$

15. $y = \frac{k}{x}$
$100 = \frac{k}{7}$
$k = 700$
$y = \frac{700}{x}$

17. $y = \frac{k}{x}$
$\frac{1}{8} = \frac{k}{16}$
$k = 2$
$y = \frac{2}{x}$

19. $y = \frac{k}{x}$
$0.2 = \frac{k}{0.7}$
$k = 0.14$
$y = \frac{0.14}{x}$

21. $R = \frac{k}{T}$
$45 = \frac{k}{6}$
$k = 270$
$R = \frac{270}{5} = 54$
The car's speed is 54 mph.

23. $I = \frac{k}{R}$
$40 = \frac{k}{270}$
$k = 10{,}800$
$I = \frac{10{,}800}{R} = \frac{10{,}800}{150} = 72$
The current is 72 amps.

25. $I_1 = \frac{k}{d^2}$
Replace d by $2d$.
$I_2 = \frac{k}{(2d)^2} = \frac{k}{4d^2} = \frac{1}{4}I_1$
Thus, the intensity is divided by 4.

27. $x = kyz$

29. $r = kst^3$

31. $y = kx^3$
$9 = k(3)^3$
$9 = 27k$
$k = \frac{1}{3}$
$y = \frac{1}{3}x^3$

33. $y = k\sqrt{x}$
$0.4 = k\sqrt{4}$
$0.4 = 2k$
$\frac{0.4}{2} = k$
$0.2 = k$
$y = 0.2\sqrt{x}$

35.
$$y = \frac{k}{x^2}$$
$$0.052 = \frac{k}{5^2}$$
$$k = 1.3$$
$$y = \frac{1.3}{x^2}$$

37.
$$y = kxz^3$$
$$120 = k(5)(2^3)$$
$$120 = k(5)(8)$$
$$120 = 40k$$
$$3 = k$$
$$y = 3xz^3$$

39.
$$\text{Weight} = \frac{kwh^2}{l}$$
$$12 = \frac{k\left(\frac{1}{2}\right)\left(\frac{1}{3}\right)^2}{10}$$
$$120 = k \cdot \frac{1}{2} \cdot \frac{1}{9}$$
$$k = 2160$$
$$\text{Weight} = \frac{2160wh^2}{l} = \frac{2160\left(\frac{2}{3}\right)\left(\frac{1}{2}\right)^2}{16} = 22.5$$

The beam can support 22.5 tons.

41.
$$V = kr^2h$$
$$32\pi = k(4)^2(6)$$
$$32\pi = k(16)(6)$$
$$32\pi = 96k$$
$$\frac{32\pi}{96} = k$$
$$\frac{\pi}{3} = k$$
$$V = \frac{\pi}{3}r^2h$$
$$V = \frac{\pi}{3}(3)^2(5)$$
$$V = 15\pi$$

The volume is 15π cubic inches.

43.
$$I = \frac{k}{x^2}$$
$$80 = \frac{k}{2^2}$$
$$k = 320$$
$$I = \frac{320}{x^2}$$
$$5 = \frac{320}{x^2}$$
$$5x^2 = 320$$
$$x^2 = 64$$
$$x = 8$$

The source is 8 feet from the light source.

45. y varies directly as x is written as $y = kx$.

47. a varies inversely as b is written as $a = \frac{k}{b}$.

49. y varies jointly as x and z is written as $y = kxz$.

51. y varies inversely as x^3 is written as $y = \frac{k}{x^3}$.

53. y varies directly as x and inversely as p^2 is written as $y = \frac{kx}{p^2}$.

55. $r = 4$ in.
$C = 2\pi r = 2\pi(4) = 8\pi$ in.
$A = \pi r^2 = \pi(4)^2 = 16\pi$ sq in.

57. $r = 9$ cm
$C = 2\pi r = 2\pi(9) = 18\pi$ cm
$A = \pi r^2 = \pi(9)^2 = 81\pi$ sq cm

59. $\sqrt{81} = 9$

61. $\sqrt{1} = 1$

63. $\sqrt{\frac{1}{4}} = \frac{1}{2}$

65. $\sqrt{\frac{4}{9}} = \frac{2}{3}$

67. $y = \frac{2}{3}x$ is an example of direct variation; a.

69. $y = 9ab$ is an example of joint variation; c.

71. $H_1 = ks^3$
$H_2 = k(2s)^3 = 8(ks^3) = 8H_1$
It is multiplied by 8.

73. $y_1 = kx$
$y_2 = k(2x) = 2(kx) = 2y_1$
It is multiplied by 2.

75.

x	$\frac{1}{4}$	$\frac{1}{2}$	1	2	4
$y = \frac{3}{x}$	12	6	3	$\frac{3}{2}$	$\frac{3}{4}$

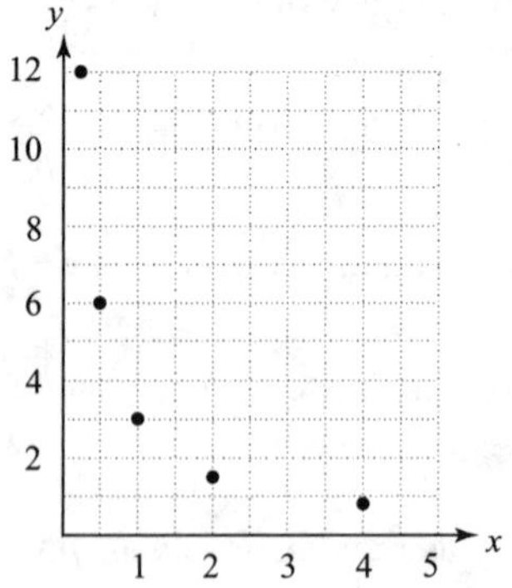

77.

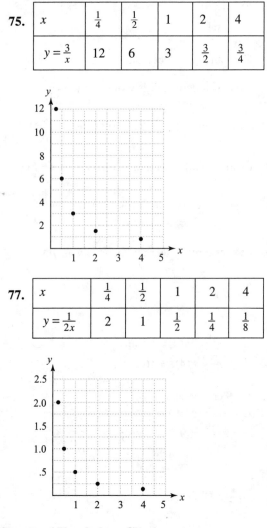

x	$\frac{1}{4}$	$\frac{1}{2}$	1	2	4
$y = \frac{1}{2x}$	2	1	$\frac{1}{2}$	$\frac{1}{4}$	$\frac{1}{8}$

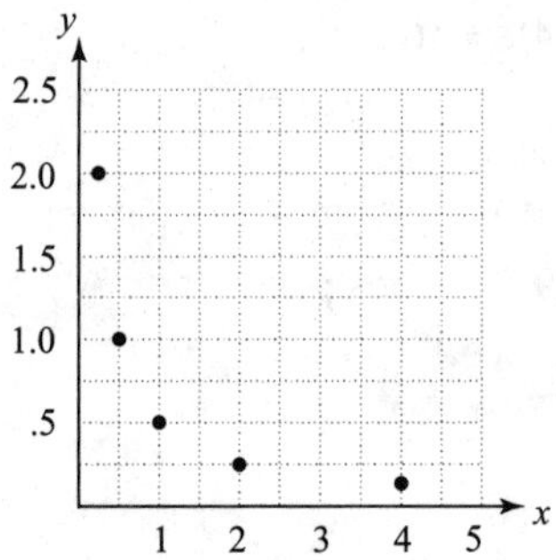

Chapter 6 Vocabulary Check

1. A rational expression whose numerator, denominator, or both contain one or more rational expressions is called a complex fraction.

2. To divide a polynomial by a polynomial other than a monomial, we use long division.

3. In the equation $y = kx$, y varies directly as x.

4. In the equation $y = \frac{k}{x}$, y varies inversely as x.

5. The least common denominator of a list of rational expressions is a polynomial of least degree whose factors include the denominator factors in the list.

6. When a polynomial is to be divided by a binomial of the form $x - c$, a shortcut process called synthetic division may be used.

7. In the equation $y = kxz$, y varies jointly as x and z.

8. The expressions $(x - 5)$ and $(5 - x)$ are called opposites.

9. A rational expression is an expression that can be written as the quotient $\frac{P}{Q}$ of two polynomials P and Q as long as Q is not 0.

10. Which is an expression and which is an equation? An example of an equation is $\frac{2}{x} + \frac{2}{x^2} = 7$, and an example of an expression is $\frac{2}{x} + \frac{5}{x^2}$.

Chapter 6 Review

1. 7 is never 0 so the domain of $f(x) = \frac{3-5x}{7}$ is $\{x|x \text{ is a real number}\}$.

2. 11 is never 0 so the domain of $g(x) = \frac{2x+4}{11}$ is $\{x|x \text{ is a real number}\}$.

3. $x - 5 = 0$
$x = 5$
The domain of $F(x) = \frac{-3x^2}{x-5}$ is $\{x|x \text{ is a real number and } x \neq 5\}$.

4. $3x - 12 = 0$
$3x = 12$
$x = 4$
The domain of $h(x) = \frac{4x}{3x-12}$ is $\{x|x \text{ is a real number and } x \neq 4\}$.

5. $x^2+8x=0$
$x(x+8)=0$
$x=0$ or $x+8=0$
$x=0$ or $x=-8$

The domain of $f(x)=\dfrac{x^3+2}{x^2+8x}$ is $\{x|x \text{ is a real number and } x \neq 0, x \neq -8\}$.

6. $3x^2-48=0$
$3(x^2-16)=0$
$3(x+4)(x-4)=0$
$x+4=0$ or $x-4=0$
$x=-4$ or $x=4$

The domain of $G(x)=\dfrac{20}{3x^2-48}$ is $\{x|x \text{ is a real number and } x \neq -4, x \neq 4\}$.

7. $\dfrac{x-12}{12-x}=\dfrac{x-12}{-(x-12)}=-1$

8. $\dfrac{5x-15}{25x-75}=\dfrac{5(x-3)}{25(x-3)}=\dfrac{5}{25}=\dfrac{1}{5}$

9. $\dfrac{2x}{2x^2-2x}=\dfrac{2x}{2x(x-1)}=\dfrac{1}{x-1}$

10. $\dfrac{x+7}{x^2-49}=\dfrac{x+7}{(x-7)(x+7)}=\dfrac{1}{x-7}$

11. $\dfrac{2x^2+4x-30}{x^2+x-20}=\dfrac{2(x^2+2x-15)}{(x+5)(x-4)}$
$=\dfrac{2(x+5)(x-3)}{(x+5)(x-4)}$
$=\dfrac{2(x-3)}{x-4}$

12. $C(x)=\dfrac{35x+4200}{x}$

a. $C(50)=\dfrac{35(50)+4200}{50}$
$=\dfrac{1750+4200}{50}$
$=\dfrac{5950}{50}$
$=119$

The average cost is \$119.

b. $C(100)=\dfrac{35(100)+4200}{100}$
$=\dfrac{3500+4200}{100}$
$=\dfrac{7700}{100}$
$=77$

The average cost is \$77.

c. It will decrease.

13. $\dfrac{4-x}{5}\cdot\dfrac{15}{2x-8}=\dfrac{4-x}{5}\cdot\dfrac{5\cdot 3}{2(x-4)}$
$=\dfrac{-1(x-4)}{5}\cdot\dfrac{5\cdot 3}{2(x-4)}$
$=-\dfrac{3}{2}$

14. $\dfrac{x^2-6x+9}{2x^2-18}\cdot\dfrac{4x+12}{5x-15}$
$=\dfrac{(x-3)(x-3)}{2(x^2-9)}\cdot\dfrac{4(x+3)}{5(x-3)}$
$=\dfrac{(x-3)(x-3)}{2(x-3)(x+3)}\cdot\dfrac{4(x+3)}{5(x-3)}$
$=\dfrac{4}{10}$
$=\dfrac{2}{5}$

15. $\dfrac{a-4b}{a^2+ab}\cdot\dfrac{b^2-a^2}{8b-2a}$
$=\dfrac{a-4b}{a(a+b)}\cdot\dfrac{(b-a)(b+a)}{2(4b-a)}$
$=\dfrac{-(4b-a)}{a(a+b)}\cdot-\dfrac{(a-b)(a+b)}{2(4b-a)}$
$=\dfrac{a-b}{2a}$

16. $\dfrac{x^2-x-12}{2x^2-32}\cdot\dfrac{x^2+8x+16}{3x^2+21x+36}$
$=\dfrac{(x-4)(x+3)}{2(x^2-16)}\cdot\dfrac{(x+4)(x+4)}{3(x^2+7x+12)}$
$=\dfrac{(x-4)(x+3)}{2(x-4)(x+4)}\cdot\dfrac{(x+4)(x+4)}{3(x+4)(x+3)}$
$=\dfrac{1}{2\cdot 3}$
$=\dfrac{1}{6}$

17. $\frac{4x+8y}{3} \div \frac{5x+10y}{9} = \frac{4x+8y}{3} \cdot \frac{9}{5x+10y}$
$= \frac{4(x+2y)}{3} \cdot \frac{3 \cdot 3}{5(x+2y)}$
$= \frac{12}{5}$

18. $\frac{x^2-25}{3} \div \frac{x^2-10x+25}{x^2-x-20}$
$= \frac{x^2-25}{3} \cdot \frac{x^2-x-20}{x^2-10x+25}$
$= \frac{(x+5)(x-5)}{3} \cdot \frac{(x-5)(x+4)}{(x-5)(x-5)}$
$= \frac{(x+5)(x+4)}{3}$

19. $\frac{a-4b}{a^2+ab} \div \frac{20b-5a}{b^2-a^2}$
$= \frac{a-4b}{a^2+ab} \cdot \frac{b^2-a^2}{20b-5a}$
$= \frac{a-4b}{a(a+b)} \cdot \frac{(b+a)(b-a)}{5(4b-a)}$
$= \frac{a-4b}{a(a+b)} \cdot \frac{-(a+b)(a-b)}{-5(a-4b)}$
$= \frac{a-b}{5a}$

20. $\frac{3x+3}{x-1} \div \frac{x^2-6x-7}{x^2-1} = \frac{3x+3}{x-1} \cdot \frac{x^2-1}{x^2-6x-7}$
$= \frac{3(x+1)}{x-1} \cdot \frac{(x+1)(x-1)}{(x-7)(x+1)}$
$= \frac{3(x+1)}{x-7}$

21. $\frac{2x-x^2}{x^3-8} \div \frac{x^2}{x^2+2x+4}$
$= \frac{2x-x^2}{x^3-8} \cdot \frac{x^2+2x+4}{x^2}$
$= \frac{x(2-x)}{(x-2)(x^2+2x+4)} \cdot \frac{x^2+2x+4}{x^2}$
$= \frac{-x(x-2)}{(x-2)(x^2+2x+4)} \cdot \frac{x^2+2x+4}{x^2}$
$= -\frac{1}{x}$

22. $\frac{5x-15}{3-x} \cdot \frac{x+2}{10x+20} \cdot \frac{x^2-9}{x^2-x-6}$
$= \frac{5(x-3)}{-(x-3)} \cdot \frac{x+2}{10(x+2)} \cdot \frac{(x+3)(x-3)}{(x-3)(x+2)}$
$= -\frac{x+3}{2(x+2)}$

23. $4x^2y^5 = 2 \cdot 2x^2y^5$
$10x^2y^4 = 2 \cdot 5x^2y^4$
$6y^4 = 2 \cdot 3y^4$
The LCD is $2 \cdot 2 \cdot 5 \cdot 3x^2y^5 = 60x^2y^5$.

24. The LCD is $2x(x-2)$.

25. The LCD is $5x(x-5)$.

26. $5x^3 = 5 \cdot x^3$
$x^2+3x-28 = (x+7)(x-4)$
$10x^2-30x = 10x(x-3) = 2 \cdot 5 \cdot x \cdot (x-3)$
The LCD is
$2 \cdot 5 \cdot x^3 \cdot (x-4)(x+7)(x-3)$
$= 10x^3(x-4)(x+7)(x-3)$.

27. $\frac{4}{x-4} + \frac{x}{x-4} = \frac{4+x}{x-4}$

28. $\frac{4}{3x^2} + \frac{2}{3x^2} = \frac{4+2}{3x^2} = \frac{6}{3x^2} = \frac{2}{x^2}$

29. $\frac{1}{x-2}-\frac{1}{4-2x}=\frac{1}{x-2}-\frac{1}{2(2-x)}$
$=\frac{1}{x-2}-\frac{-1}{2(x-2)}$
$=\frac{1}{x-2}\cdot\frac{2}{2}-\frac{-1}{2(x-2)}$
$=\frac{2}{2(x-2)}-\frac{-1}{2(x-2)}$
$=\frac{2+1}{2(x-2)}$
$=\frac{3}{2(x-2)}$

30. $\frac{1}{10-x}+\frac{x-1}{x-10}=\frac{-1}{x-10}+\frac{x-1}{x-10}$
$=\frac{-1+x-1}{x-10}$
$=\frac{x-2}{x-10}$

31. $\frac{x}{9-x^2}-\frac{2}{5x-15}=\frac{-x}{x^2-9}-\frac{2}{5x-15}$
$=\frac{-x}{(x-3)(x+3)}-\frac{2}{5(x-3)}$
$=\frac{-x}{(x-3)(x+3)}\cdot\frac{5}{5}-\frac{2}{5(x-3)}\cdot\frac{x+3}{x+3}$
$=\frac{-5x}{5(x-3)(x+3)}-\frac{2(x+3)}{5(x-3)(x+3)}$
$=\frac{-5x-2(x+3)}{5(x-3)(x+3)}$
$=\frac{-5x-2x-6}{5(x-3)(x+3)}$
$=\frac{-7x-6}{5(x-3)(x+3)}$

32. $2x+1-\frac{1}{x-3}=2x\cdot\frac{x-3}{x-3}+1\cdot\frac{x-3}{x-3}-\frac{1}{x-3}$
$=\frac{2x^2-6x}{x-3}+\frac{x-3}{x-3}-\frac{1}{x-3}$
$=\frac{2x^2-6x+x-3-1}{x-3}$
$=\frac{2x^2-5x-4}{x-3}$

33. $\dfrac{2}{a^2-2a+1}+\dfrac{3}{a^2-1}=\dfrac{2}{(a-1)(a-1)}+\dfrac{3}{(a+1)(a-1)}$

$=\dfrac{2}{(a-1)(a-1)}\cdot\dfrac{a+1}{a+1}+\dfrac{3}{(a+1)(a-1)}\cdot\dfrac{(a-1)}{(a-1)}$

$=\dfrac{2(a+1)}{(a-1)^2(a+1)}+\dfrac{3(a-1)}{(a-1)^2(a+1)}$

$=\dfrac{2a+2}{(a-1)^2(a+1)}+\dfrac{3a-3}{(a-1)^2(a+1)}$

$=\dfrac{2a+2+3a-3}{(a-1)^2(a+1)}$

$=\dfrac{5a-1}{(a-1)^2(a+1)}$

34. $\dfrac{x}{9x^2+12x+16}-\dfrac{3x+4}{27x^3-64}=\dfrac{x}{9x^2+12x+16}-\dfrac{3x+4}{(3x-4)(9x^2+12x+16)}$

$=\dfrac{x}{9x^2+12x+16}\cdot\dfrac{3x-4}{3x-4}-\dfrac{3x+4}{(3x-4)(9x^2+12x+16)}$

$=\dfrac{3x^2-4x-3x-4}{(3x-4)(9x^2+12x+16)}$

$=\dfrac{3x^2-7x-4}{(3x-4)(9x^2+12x+16)}$

35. $\dfrac{2}{x-1}-\dfrac{3x}{3x-3}+\dfrac{1}{2x-2}=\dfrac{2}{x-1}-\dfrac{3x}{3(x-1)}+\dfrac{1}{2(x-1)}$

$=\dfrac{2}{x-1}\cdot\dfrac{6}{6}-\dfrac{3x}{3(x-1)}\cdot\dfrac{2}{2}+\dfrac{1}{2(x-1)}\cdot\dfrac{3}{3}$

$=\dfrac{12}{6(x-1)}-\dfrac{6x}{6(x-1)}+\dfrac{3}{6(x-1)}$

$=\dfrac{12-6x+3}{6(x-1)}$

$=\dfrac{15-6x}{6(x-1)}$

$=\dfrac{3(5-2x)}{6(x-1)}$

$=\dfrac{5-2x}{2(x-1)}$

36. $$\text{Perimeter} = \frac{1}{x}+\frac{1}{x}+\frac{1}{x}+\frac{1}{x}+\frac{2}{x}+\frac{2}{x}+\frac{3}{2x}+\frac{5}{2x}$$
$$= \frac{7}{x}+\frac{8}{2x}$$
$$= \frac{7}{x}\cdot\frac{2}{2}+\frac{8}{2x}$$
$$= \frac{14}{2x}+\frac{8}{2x}$$
$$= \frac{14+8}{2x}$$
$$= \frac{22}{2x}$$
$$= \frac{11}{x}$$

37. $$\frac{1-\frac{3x}{4}}{2+\frac{x}{4}} = \frac{4\left(1-\frac{3x}{4}\right)}{4\left(2+\frac{x}{4}\right)} = \frac{4-3x}{8+x}$$

38. $$\frac{\frac{x^2}{15}}{\frac{x+1}{5x}} = \frac{x^2}{15} \div \frac{x+1}{5x} = \frac{x^2}{15}\cdot\frac{5x}{x+1} = \frac{x^3}{3(x+1)}$$

39. $$\frac{2-\frac{3}{2x}}{x-\frac{2}{5x}} = \frac{10x\left(2-\frac{3}{2x}\right)}{10x\left(x-\frac{2}{5x}\right)}$$
$$= \frac{20x-15}{10x^2-4}$$
$$= \frac{5(4x-3)}{2(5x^2-2)}$$

40. $$\frac{1+\frac{x}{y}}{\frac{x^2}{y^2}-1} = \frac{y^2\left(1+\frac{x}{y}\right)}{y^2\left(\frac{x^2}{y^2}-1\right)}$$
$$= \frac{y^2+xy}{x^2-y^2}$$
$$= \frac{y(y+x)}{(x+y)(x-y)}$$
$$= \frac{y}{x-y}$$

41. $$\frac{\frac{5}{x}+\frac{1}{xy}}{\frac{3}{x^2}} = \frac{x^2y\left(\frac{5}{x}+\frac{1}{xy}\right)}{x^2y\left(\frac{3}{x^2}\right)} = \frac{5xy+x}{3y} = \frac{x(5y+1)}{3y}$$

42. $$\frac{\frac{x}{3}-\frac{3}{x}}{1+\frac{3}{x}} = \frac{3x\left(\frac{x}{3}-\frac{3}{x}\right)}{3x\left(1+\frac{3}{x}\right)}$$
$$= \frac{x^2-9}{3x+9}$$
$$= \frac{(x+3)(x-3)}{3(x+3)}$$
$$= \frac{x-3}{3}$$

43. $$\frac{\frac{1}{x-1}+1}{\frac{1}{x+1}-1} = \frac{(x+1)(x-1)\left(\frac{1}{x-1}+1\right)}{(x+1)(x-1)\left(\frac{1}{x+1}-1\right)}$$
$$= \frac{x+1+(x+1)(x-1)}{x-1-(x+1)(x-1)}$$
$$= \frac{x+1+x^2-1}{x-1-x^2+1}$$
$$= \frac{x+x^2}{x-x^2}$$
$$= \frac{x(1+x)}{x(1-x)}$$
$$= \frac{1+x}{1-x}$$

44. $$\frac{\frac{x-3}{x+3}+\frac{x+3}{x-3}}{\frac{x-3}{x+3}-\frac{x+3}{x-3}} = \frac{(x-3)(x+3)\left(\frac{x-3}{x+3}+\frac{x+3}{x-3}\right)}{(x-3)(x+3)\left(\frac{x-3}{x+3}-\frac{x+3}{x-3}\right)}$$
$$= \frac{(x-3)^2+(x+3)^2}{(x-3)^2-(x+3)^2}$$
$$= \frac{x^2-6x+9+x^2+6x+9}{x^2-6x+9-(x^2+6x+9)}$$
$$= \frac{2x^2+18}{x^2-6x+9-x^2-6x-9}$$
$$= \frac{2(x^2+9)}{-12x}$$
$$= -\frac{x^2+9}{6x}$$

45. $$f(a+h) = \frac{3}{a+h}$$

46. $$f(a) = \frac{3}{a}$$

47. $$\frac{f(a+h)-f(a)}{h} = \frac{\frac{3}{a+h}-\frac{3}{a}}{h}$$

48.
$$\begin{aligned}\frac{f(a+h)-f(a)}{h} &= \frac{\frac{3}{a+h}-\frac{3}{a}}{h}\\ &= \frac{\left(\frac{3}{a+h}-\frac{3}{a}\right)\cdot a(a+h)}{h\cdot a(a+h)}\\ &= \frac{3a-3(a+h)}{h\cdot a(a+h)}\\ &= \frac{3a-3a-3h}{h\cdot a(a+h)}\\ &= \frac{-3h}{h\cdot a(a+h)}\\ &= -\frac{3}{a(a+h)}\end{aligned}$$

49.
$$\begin{aligned}\frac{4xy+2x^2-9}{4xy} &= \frac{4xy}{4xy}+\frac{2x^2}{4xy}-\frac{9}{4xy}\\ &= 1+\frac{x}{2y}-\frac{9}{4xy}\end{aligned}$$

50.
$$\begin{aligned}\frac{12xb^2+16xb^4}{4xb^3} &= \frac{12xb^2}{4xb^3}+\frac{16xb^4}{4xb^3}\\ &= \frac{3}{b}+4b\end{aligned}$$

51.
$$\begin{array}{r} 3x^3+9x^2+2x+6 \\ x-3\overline{\big)3x^4+0x^3-25x^2+0x-20} \\ \underline{3x^4-9x^3} \\ 9x^3-25x^2 \\ \underline{9x^3-27x^2} \\ 2x^2+0x \\ \underline{2x^2-6x} \\ 6x-20 \\ \underline{6x-18} \\ -2 \end{array}$$

Answer: $3x^3+9x^2+2x+6-\frac{2}{x-3}$

52.
$$\begin{array}{r} 2x^3-4x^2+7x-9 \\ x+2\overline{\big)2x^4+0x^3-x^2+5x-12} \\ \underline{2x^4+4x^3} \\ -4x^3-x^2 \\ \underline{-4x^3-8x^2} \\ 7x^2+5x \\ \underline{7x^2+14x} \\ -9x-12 \\ \underline{-9x-18} \\ 6 \end{array}$$

Answer: $2x^3-4x^2+7x-9+\frac{6}{x+2}$

53.
$$\begin{array}{r} x^2-1 \\ 2x+3\overline{\big)2x^3+3x^2-2x+2} \\ \underline{2x^3+3x^2} \\ -2x+2 \\ \underline{-2x-3} \\ 5 \end{array}$$

Answer: $x^2-1+\frac{5}{2x+3}$

54.
$$\begin{array}{r} 3x^2+2x-1 \\ x^2+x+2\overline{\big)3x^4+5x^3+7x^2+3x-2} \\ \underline{3x^4+3x^3+6x^2} \\ 2x^3+x^2+3x \\ \underline{2x^3+2x^2+4x} \\ -x^2-x-2 \\ \underline{-x^2-x-2} \\ 0 \end{array}$$

Answer: $3x^2+2x-1$

55.
$$\begin{array}{r|rrrr} 2 & 3 & 0 & 12 & -4 \\ & & 6 & 12 & 48 \\ \hline & 3 & 6 & 24 & 44 \end{array}$$

Answer: $3x^2+6x+24+\frac{44}{x-2}$

56.
$$\begin{array}{r|rrrrrr} -1 & 1 & 0 & 0 & 0 & 0 & -1 \\ & & -1 & 1 & -1 & 1 & -1 \\ \hline & 1 & -1 & 1 & -1 & 1 & -2 \end{array}$$

Answer: $x^4-x^3+x^2-x+1-\frac{2}{x+1}$

57. $\begin{array}{r|rrrr} 3 & 1 & 0 & 0 & -81 \\ & & 3 & 9 & 27 \\ \hline & 1 & 3 & 9 & -54 \end{array}$

Answer: $x^2+3x+9-\dfrac{54}{x-3}$

58. $\begin{array}{r|rrrrr} -2 & 3 & 0 & -2 & 0 & 10 \\ & & -6 & 12 & -20 & 40 \\ \hline & 3 & -6 & 10 & -20 & 50 \end{array}$

Answer: $3x^3-6x^2+10x-20+\dfrac{50}{x+2}$

59. $\begin{array}{r|rrrrrr} 4 & 3 & 0 & 0 & 0 & -9 & 7 \\ & & 12 & 48 & 192 & 768 & 3036 \\ \hline & 3 & 12 & 48 & 192 & 759 & 3043 \end{array}$

Thus, $P(4)=3043$.

60. $\begin{array}{r|rrrrrr} -5 & 3 & 0 & 0 & 0 & -9 & 7 \\ & & -15 & 75 & -375 & 1875 & -9330 \\ \hline & 3 & -15 & 75 & -375 & 1866 & -9323 \end{array}$

Thus, $P(-5)=-9323$.

61. $\begin{array}{r|rrrrrr} -\frac{1}{2} & 3 & 0 & 0 & 0 & -9 & 7 \\ & & -\frac{3}{2} & \frac{3}{4} & -\frac{3}{8} & \frac{3}{16} & \frac{141}{32} \\ \hline & 3 & -\frac{3}{2} & \frac{3}{4} & -\frac{3}{8} & -\frac{141}{16} & \frac{365}{32} \end{array}$

Thus, $P\left(-\dfrac{1}{2}\right)=\dfrac{365}{32}$.

62. $\begin{array}{r|rrrrr} 3 & 1 & -1 & -6 & -6 & 18 \\ & & 3 & 6 & 0 & -18 \\ \hline & 1 & 2 & 0 & -6 & 0 \end{array}$

length $=(x^3+2x^2-6)$ miles

63. The LCD is $3x$.

$$\frac{3}{x}+\frac{1}{3}=\frac{5}{x}$$
$$3x\left(\frac{3}{x}+\frac{1}{3}\right)=3x\left(\frac{5}{x}\right)$$
$$3x\left(\frac{3}{x}\right)+3x\left(\frac{1}{3}\right)=3x\left(\frac{5}{x}\right)$$
$$9+x=15$$
$$x=6$$

The solution is 6.

64. The LCD is $2(5x-9)$.

$$\begin{aligned}\frac{2x+3}{5x-9}&=\frac{3}{2}\\ 2(5x-9)\cdot\frac{2x+3}{5x-9}&=2(5x-9)\cdot\frac{3}{2}\\ 2(2x+3)&=(5x-9)\cdot 3\\ 4x+6&=15x-27\\ 6&=11x-27\\ 33&=11x\\ 3&=x\end{aligned}$$

The solution is 3.

65. The LCD is $(x-2)(x+2)$.

$$\begin{aligned}\frac{1}{x-2}-\frac{3x}{x^2-4}&=\frac{2}{x+2}\\ (x-2)(x+2)\left(\frac{1}{x-2}-\frac{3x}{x^2-4}\right)&=(x-2)(x+2)\left(\frac{2}{x+2}\right)\\ (x-2)(x+2)\cdot\frac{1}{x-2}-(x-2)(x+2)\cdot\frac{3x}{(x+2)(x-2)}&=(x-2)(x+2)\cdot\frac{2}{x+2}\\ (x+2)-3x&=2(x-2)\\ -2x+2&=2x-4\\ -4x+2&=-4\\ -4x&=-6\\ x&=\frac{-6}{-4}=\frac{3}{2}\end{aligned}$$

The solution is $\frac{3}{2}$.

66. The LCD is $7x$.

$$\begin{aligned}\frac{7}{x}-\frac{x}{7}&=0\\ 7x\left(\frac{7}{x}-\frac{x}{7}\right)&=7x(0)\\ 7x\left(\frac{7}{x}\right)-7x\left(\frac{x}{7}\right)&=0\\ 49-x^2&=0\\ (7-x)(7+x)&=0\end{aligned}$$

$$\begin{aligned}7-x=0 &\quad\text{or}\quad 7+x=0\\ 7=x &\quad\text{or}\quad x=-7\end{aligned}$$

The solutions are -7 and 7.

67. $\frac{5}{x^2-7x}+\frac{4}{2x-14}=\frac{5}{x(x-7)}+\frac{4}{2(x-7)}$
$=\frac{5}{x(x-7)}\cdot\frac{2}{2}+\frac{4}{2(x-7)}\cdot\frac{x}{x}$
$=\frac{10}{2x(x-7)}+\frac{4x}{2x(x-7)}$
$=\frac{10+4x}{2x(x-7)}$
$=\frac{2(5+2x)}{2x(x-7)}$
$=\frac{5+2x}{x(x-7)}$

68. $\frac{4}{3-x}-\frac{7}{2x-6}+\frac{5}{x}$
$=\frac{-4}{x-3}-\frac{7}{2(x-3)}+\frac{5}{x}$
$=\frac{-4}{x-3}\cdot\frac{2x}{2x}-\frac{7}{2(x-3)}\cdot\frac{x}{x}+\frac{5}{x}\cdot\frac{2(x-3)}{2(x-3)}$
$=\frac{-8x}{2x(x-3)}-\frac{7x}{2x(x-3)}+\frac{10x-30}{2x(x-3)}$
$=\frac{-8x-7x+10x-30}{2x(x-3)}$
$=\frac{-5x-30}{2x(x-3)}$
$=\frac{-5(x+6)}{2x(x-3)}$

69. The LCD is x^2.
$3-\frac{5}{x}-\frac{2}{x^2}=0$
$x^2\left(3-\frac{5}{x}-\frac{2}{x^2}\right)=x^2(0)$
$x^2\cdot 3-x^2\left(\frac{5}{x}\right)-x^2\left(\frac{2}{x^2}\right)=0$
$3x^2-5x-2=0$
$(3x+1)(x-2)=0$
$3x+1=0$ or $x-2=0$
$3x=-1$ or $x=2$
$x=-\frac{1}{3}$ or $x=2$

The solutions are $-\frac{1}{3}$ and 2.

70. The LCD is x^2.
$2+\frac{15}{x^2}=\frac{13}{x}$
$x^2\left(2+\frac{15}{x^2}\right)=x^2\left(\frac{13}{x}\right)$
$2x^2+15=13x$
$2x^2-13x+15=0$
$(2x-3)(x-5)=0$
$2x-3=0$ or $x-5=0$
$x=\frac{3}{2}$ or $x=5$

The solutions are $\frac{3}{2}$ and 5.

71. $A=\frac{h(a+b)}{2}$
$2A=h(a+b)$
$\frac{2A}{h}=a+b$
$a=\frac{2A}{h}-b$

72. $\frac{1}{R}=\frac{1}{R_1}+\frac{1}{R_2}$
$RR_1R_2\left(\frac{1}{R}\right)=RR_1R_2\left(\frac{1}{R_1}\right)+RR_1R_2\left(\frac{1}{R_2}\right)$
$R_1R_2=RR_2+RR_1$
$R_1R_2-RR_2=RR_1$
$R_2(R_1-R)=RR_1$
$R_2=\frac{RR_1}{R_1-R}$

73. $I=\frac{E}{R+r}$
$I(R+r)=E$
$R+r=\frac{E}{I}$
$R=\frac{E}{I}-r$

74. $A=P+Prt$
$A-P=Prt$
$\frac{A-P}{Pt}=r$ or $r=\frac{A-P}{Pt}$

75.
$$\frac{1}{x}=\frac{1}{y}-\frac{1}{z}$$
$$xyz\left(\frac{1}{x}\right)=xyz\left(\frac{1}{y}-\frac{1}{z}\right)$$
$$yz=xz-xy$$
$$yz=x(z-y)$$
$$\frac{yz}{z-y}=\frac{x(z-y)}{z-y}$$
$$\frac{yz}{z-y}=x \text{ or } x=\frac{yz}{z-y}$$

76.
$$H=\frac{kA(T_1-T_2)}{L}$$
$$HL=kA(T_1-T_2)$$
$$\frac{HL}{k(T_1-T_2)}=A \text{ or } A=\frac{HL}{k(T_1-T_2)}$$

77. Let x = the number.
$$x+2\left(\frac{1}{x}\right)=3$$
$$x\left[x+2\left(\frac{1}{x}\right)\right]=x\cdot 3$$
$$x^2+2=3x$$
$$x^2-3x+2=0$$
$$(x-1)(x-2)=0$$
$$x-1=0 \quad\text{or}\quad x-2=0$$
$$x=1 \quad\text{or}\quad x=2$$
The numbers are 1 and 2.

78. Let x = the number.
$$\frac{3+x}{7+2x}=\frac{10}{21}$$
$$21(3+x)=10(7+2x)$$
$$63+21x=70+20x$$
$$63+1x=70$$
$$x=7$$
The number is 7.

79. Let x = amount of time required for all three boys to paint the fence together.
$$\frac{1}{4}+\frac{1}{5}+\frac{1}{6}=\frac{1}{x}$$
$$60x\cdot\frac{1}{4}+60x\cdot\frac{1}{5}+60x\cdot\frac{1}{6}=60x\cdot\frac{1}{x}$$
$$15x+12x+10x=60$$
$$37x=60$$
$$x=\frac{60}{37}=1\frac{23}{37}$$
It will take $1\frac{23}{37}$ hours for all three boys to paint the fence together.

80. Let x = amount of time it takes Tom to type the mailing labels when working alone.
$$\frac{1}{6}+\frac{1}{x}=\frac{1}{4}$$
$$12x\cdot\frac{1}{6}+12x\cdot\frac{1}{x}=12x\cdot\frac{1}{4}$$
$$2x+12=3x$$
$$12=x$$
It takes Tom 12 hours to complete the task alone.

81. Let x = the speed of the current.

	distance =	rate ·	time
Upstream	72	$32-x$	$\frac{72}{32-x}$
Downstream	120	$32+x$	$\frac{120}{32+x}$

$$\frac{72}{32-x}=\frac{120}{32+x}$$
$$72(32+x)=120(32-x)$$
$$2304+72x=3840-120x$$
$$72x=1536-120x$$
$$192x=1536$$
$$x=8$$
The speed of the current is 8 mph.

82. Let x = the speed of the walker.

	distance =	rate ·	time
Jogger	14	$x+3$	$\frac{14}{x+3}$
Walker	8	x	$\frac{8}{x}$

$$\frac{14}{x+3}=\frac{8}{x}$$
$$14x=8(x+3)$$
$$14x=8x+24$$
$$6x=24$$
$$x=4$$
The speed of the walker is 4 mph.

83. $A = kB$
$6 = k(14)$
$k = \frac{6}{14} = \frac{3}{7}$
$A = \frac{3}{7}B = \frac{3}{7}(21) = 9$

84. $P = \frac{K}{V}$
$1250 = \frac{K}{2}$
$K = 2500$
$P = \frac{2500}{V}$
$800 = \frac{2500}{V}$
$800V = 2500$
$V = 3.125$
When the pressure is 800 kilopascals, the volume is 3.125 cubic meters.

85. $\frac{22x+8}{11x+4} = \frac{2(11x+4)}{11x+4} = 2$

86. $\frac{xy-3x+2y-6}{x^2+4x+4} = \frac{x(y-3)+2(y-3)}{(x+2)(x+2)}$
$= \frac{(x+2)(y-3)}{(x+2)(x+2)}$
$= \frac{y-3}{x+2}$

87. $\frac{2}{5x} \div \frac{4-18x}{6-27x} = \frac{2}{5x} \cdot \frac{6-27x}{4-18x}$
$= \frac{2}{5x} \cdot \frac{3(2-9x)}{2(2-9x)}$
$= \frac{3}{5x}$

88. $\frac{7x+28}{2x+4} \div \frac{x^2+2x-8}{x^2-2x-8}$
$= \frac{7x+28}{2x+4} \cdot \frac{x^2-2x-8}{x^2+2x-8}$
$= \frac{7(x+4)}{2(x+2)} \cdot \frac{(x-4)(x+2)}{(x+4)(x-2)}$
$= \frac{7(x-4)}{2(x-2)}$

89. $\frac{5a^2-20}{a^3+2a^2+a+2} \div \frac{7a}{a^3+a}$
$= \frac{5a^2-20}{a^3+2a^2+a+2} \cdot \frac{a^3+a}{7a}$
$= \frac{5(a^2-4)}{a^2(a+2)+1(a+2)} \cdot \frac{a(a^2+1)}{7a}$
$= \frac{5(a+2)(a-2)}{(a+2)(a^2+1)} \cdot \frac{a(a^2+1)}{7a}$
$= \frac{5(a-2)}{7}$

90. $\frac{4a+8}{5a^2-20} \cdot \frac{3a^2-6a}{a+3} \div \frac{2a^2}{5a+15}$
$= \frac{4a+8}{5a^2-20} \cdot \frac{3a^2-6a}{a+3} \cdot \frac{5a+15}{2a^2}$
$= \frac{4(a+2)}{5(a^2-4)} \cdot \frac{3a(a-2)}{a+3} \cdot \frac{5(a+3)}{2a^2}$
$= \frac{4(a+2)}{5(a-2)(a+2)} \cdot \frac{3a(a-2)}{a+3} \cdot \frac{5(a+3)}{2a^2}$
$= \frac{6}{a}$

91. $\frac{7}{2x} + \frac{5}{6x} = \frac{7}{2x} \cdot \frac{3}{3} + \frac{5}{6x}$
$= \frac{21}{6x} + \frac{5}{6x}$
$= \frac{21+5}{6x}$
$= \frac{26}{6x}$
$= \frac{13}{3x}$

92. $\frac{x-2}{x+1} - \frac{x-3}{x-1} = \frac{x-2}{x+1} \cdot \frac{x-1}{x-1} - \frac{x-3}{x-1} \cdot \frac{x+1}{x+1}$
$= \frac{(x-2)(x-1)}{(x+1)(x-1)} - \frac{(x-3)(x+1)}{(x+1)(x-1)}$
$= \frac{x^2-3x+2}{(x+1)(x-1)} - \frac{x^2-2x-3}{(x+1)(x-1)}$
$= \frac{x^2-3x+2-x^2+2x+3}{(x+1)(x-1)}$
$= \frac{-x+5}{(x+1)(x-1)}$

93. $$\frac{2x+1}{x^2+x-6}+\frac{2-x}{x^2+x-6}=\frac{2x+1+2-x}{x^2+x-6}$$
$$=\frac{x+3}{x^2+x-6}$$
$$=\frac{x+3}{(x+3)(x-2)}$$
$$=\frac{1}{x-2}$$

94. $$\frac{2}{x^2-16}-\frac{3x}{x^2+8x+16}+\frac{3}{x+4}=\frac{2}{(x+4)(x-4)}-\frac{3x}{(x+4)^2}+\frac{3}{x+4}$$
$$=\frac{2}{(x+4)(x-4)}\cdot\frac{x+4}{x+4}-\frac{3x}{(x+4)^2}\cdot\frac{x-4}{x-4}+\frac{3}{x+4}\cdot\frac{(x+4)(x-4)}{(x+4)(x-4)}$$
$$=\frac{2x+8}{(x+4)^2(x-4)}-\frac{3x^2-12x}{(x+4)^2(x-4)}+\frac{3x^2-48}{(x+4)^2(x-4)}$$
$$=\frac{2x+8-3x^2+12x+3x^2-48}{(x+4)^2(x-4)}$$
$$=\frac{14x-40}{(x+4)^2(x-4)}$$
$$=\frac{2(7x-20)}{(x+4)^2(x-4)}$$

95. $$\frac{\frac{1}{x}-\frac{2}{3x}}{\frac{5}{2x}-\frac{1}{3}}=\frac{6x\left(\frac{1}{x}-\frac{2}{3x}\right)}{6x\left(\frac{5}{2x}-\frac{1}{3}\right)}=\frac{6-4}{15-2x}=\frac{2}{15-2x}$$

96. $$\frac{2}{1-\frac{2}{x}}=\frac{x(2)}{x\left(1-\frac{2}{x}\right)}=\frac{2x}{x-2}$$

97. $$\frac{\frac{x^2+5x-6}{4x+3}}{\frac{(x+6)^2}{8x+6}}=\frac{x^2+5x-6}{4x+3}\div\frac{(x+6)^2}{8x+6}$$
$$=\frac{x^2+5x-6}{4x+3}\cdot\frac{8x+6}{(x+6)^2}$$
$$=\frac{(x+6)(x-1)}{4x+3}\cdot\frac{2(4x+3)}{(x+6)^2}$$
$$=\frac{2(x-1)}{x+6}$$

98. $\dfrac{\frac{3}{x-1}-\frac{2}{1-x}}{\frac{2}{x-1}-\frac{2}{x}}=\dfrac{\frac{3}{x-1}-\frac{-2}{x-1}}{\frac{2}{x-1}-\frac{2}{x}}$

$$=\frac{(x)(x-1)\left(\frac{3}{x-1}-\frac{-2}{x-1}\right)}{(x)(x-1)\left(\frac{2}{x-1}-\frac{2}{x}\right)}$$

$$=\frac{3x+2x}{2x-2(x-1)}$$

$$=\frac{5x}{2x-2x+2}$$

$$=\frac{5x}{2}$$

99. $4+\dfrac{8}{x}=8$

The LCD is x.

$$x(4)+x\left(\frac{8}{x}\right)=x(8)$$
$$4x+8=8x$$
$$8=4x$$
$$x=2$$

The solution is 2.

100. $\dfrac{x-2}{x^2-7x+10}=\dfrac{1}{5x-10}-\dfrac{1}{x-5}$

$$\frac{x-2}{(x-2)(x-5)}=\frac{1}{5(x-2)}-\frac{1}{x-5}$$

The LCD is $5(x-2)(x-5)$.

$$5(x-2)(x-5)\cdot\frac{x-2}{(x-2)(x-5)}=5(x-2)(x-5)\cdot\frac{1}{5(x-2)}-5(x-2)(x-5)\cdot\frac{1}{x-5}$$
$$5(x-2)=(x-5)-5(x-2)$$
$$5x-10=x-5-5x+10$$
$$5x-10=-4x+5$$
$$9x-10=5$$
$$9x=15$$
$$x=\frac{15}{9}=\frac{5}{3}$$

The solution is $\dfrac{5}{3}$.

101. Let x be the numerator of a fraction. Then $x + 2$ is the denominator.

$$\frac{x-3}{x+2+5}=\frac{2}{3}$$
$$\frac{x-3}{x+7}=\frac{2}{3}$$
$$3(x-3)=2(x+7)$$
$$3x-9=2x+14$$
$$x-9=14$$
$$x=23$$

$x + 2 = 23 + 2 = 25$

The fraction is $\frac{23}{25}$.

102. Let x be the first even integer and $x + 2$ be the next consecutive even integer.

$$\frac{1}{x}+\frac{1}{x+2}=-\frac{9}{40}$$
$$40x(x+2)\cdot\frac{1}{x}+40x(x+2)\cdot\frac{1}{x+2}=40x(x+2)\cdot\frac{-9}{40}$$
$$40(x+2)+40x=-9x(x+2)$$
$$40x+80+40x=-9x^2-18x$$
$$80x+80=-9x^2-18x$$
$$9x^2+98x+80=0$$
$$(9x+8)(x+10)=0$$

$9x+8=0$ or $x+10=0$

$9x=-8$ or $x=-10$

$x=-\frac{8}{9}$ or $x=-10$

$-\frac{8}{9}$ is not an integer.

$x + 2 = -10 + 2 = -8$

The two integers are −10 and −8.

103. Let x = time it takes to empty a full tank if both pipes are open.

$$-\frac{1}{2.5}+\frac{1}{2}=\frac{1}{x}$$
$$(2.5)(2x)\cdot-\frac{1}{2.5}+(2.5)(2x)\cdot\frac{1}{2}=(2.5)(2x)\cdot\frac{1}{x}$$
$$-2x+2.5x=5$$
$$0.5x=5$$
$$x=10$$

It takes 10 hours to empty a full tank if both pipes are open.

104. Let x = the speed of the car and
$x + 430$ = the speed of the jet.

	distance =	rate ·	time
Car	210	x	$\frac{210}{x}$
Jet	1715	$x + 430$	$\frac{1715}{x+430}$

$$\frac{210}{x}=\frac{1715}{x+430}$$
$$210(x+430)=1715x$$
$$210x+90{,}300=1715x$$
$$90{,}300=1505x$$
$$60=x$$

$x + 430 = 60 + 430 = 490$

The speed of the jet is 490 mph.

105. One train traveled 382 miles in 6 hours and the second train traveled 382 − 112 = 270 miles in 6 hours.

Recall $D = rt$ or $r = \frac{D}{t}$.

$r = \frac{382}{6} = 63\frac{2}{3}$ $r = \frac{270}{6} = 45$

The speeds of the trains are $63\frac{2}{3}$ mph and 45 mph.

106.
$$A = kr^2$$
$$36\pi = k(3)^2$$
$$36\pi = 9k$$
$$k = 4\pi$$
$$A = 4\pi r^2$$
$$A = 4\pi(4)^2 = 4\pi \cdot 16 = 64\pi$$
The surface area is 64π square inches.

107. Let p = number of smart phones.
$$\frac{29}{50} = \frac{p}{43{,}560}$$
$$50 \cdot p = 29 \cdot 43{,}560$$
$$50p = 1{,}263{,}240$$
$$p = 25{,}264.8$$
You would expect 25,265 adults to own smart phones.

108. Let d = number of Internet users who purchase digital music.
$$\frac{8}{25} = \frac{d}{2000}$$
$$25 \cdot d = 8 \cdot 2000$$
$$25d = 16{,}000$$
$$d = 640$$
In a group of 2000, it is predicted there would be 640 who purchase digital music.

109.
$$\begin{array}{r|l} & 3x^3 + 13x^2 + 51x + 204 \\ x-4 & 3x^4 + x^3 - x^2 + 0x - 2 \\ & \underline{3x^4 - 12x^3} \\ & 13x^3 - x^2 \\ & \underline{13x^3 - 52x^2} \\ & 51x^2 + 0x \\ & \underline{51x^2 - 204x} \\ & 204x - 2 \\ & \underline{204x - 816} \\ & 814 \end{array}$$
$$3x^3 + 13x^2 + 51x + 204 + \frac{814}{x-4}$$

110.
$$C = \frac{k}{D}$$
$$12 = \frac{k}{8}$$
$$96 = k$$
$$C = \frac{96}{D} = \frac{96}{24} = 4$$

Chapter 6 Getting Ready for the Test

1. $\frac{-x}{4-x} = \frac{-x}{-(-4+x)} = \frac{x}{-4+x} = \frac{x}{x-4}$; D

2. $\frac{x+3}{x-1} = \frac{-1}{-1} \cdot \frac{x+3}{x-1} = \frac{-(x+3)}{-(x-1)}$ which is choice A.

$\frac{x+3}{x-1} = \frac{-(x+3)}{-(x-1)} = \frac{-x-3}{-x+1}$ which is choice C.

$\frac{x+3}{x-1} = \frac{-x-3}{-x+1} = \frac{-x-3}{1-x}$ which is choice D.

Choice B is **not** equivalent to $\frac{x+3}{x-1}$.

3. The expression $\frac{x+3}{x^2+9}$ is undefined when $x^2 + 9 = 0$ or $x^2 = -9$. The equation $x^2 = -9$ has no real solutions so the expression is undefined for none of the values given; E.

4. $\frac{y-6}{6-y} = \frac{y-6}{-1(-6+y)} = \frac{y-6}{-1(y-6)} = \frac{1}{-1} = -1$; B

5. $\frac{y+3}{3+y} = \frac{y+3}{y+3} = 1$; A

6. $\dfrac{x-2}{-2+x}=\dfrac{x-2}{x-2}=1$; A

7. $\dfrac{m-4}{m+4}$ cannot be simplified; C

8. To multiply rational expressions, multiply the numerators and multiply the denominators.
$\dfrac{5}{x+5}\cdot\dfrac{7(x+1)}{x+5}=\dfrac{5\cdot 7(x+1)}{(x+5)\cdot(x+5)}=\dfrac{35(x+1)}{(x+5)^2}$; B

9. To subtract rational expressions with a common denominator, subtract the numerators and put the result over the common denominator.
$$\begin{aligned}\frac{x-3}{x+5}-\frac{x-1}{x+5}&=\frac{x-3-(x-1)}{x+5}\\&=\frac{x-3-x+1}{x+5}\\&=\frac{-2}{x+5}\text{; C}\end{aligned}$$

10. $4x+8=4(x+2)$
$8x-8=8(x-1)=2\cdot 4(x-1)$
The LCD of $\dfrac{5}{4x+8}$ and $\dfrac{9}{8x-8}$ is
$2\cdot 4\cdot (x+2)(x-1)=8(x+2)(x-1)$; D.

11. $\dfrac{2x^{-1}}{y^{-2}+(5x)^{-1}}=\dfrac{\frac{2}{x}}{\frac{1}{y^2}+\frac{1}{5x}}$; A

12. $4(x+1)=2\cdot 2(x+1)$
$6(x+1)=2\cdot 3(x+1)$
The LCD is $2\cdot 2\cdot 3(x+1)=12(x+1)$.
$$\frac{3}{1}-\frac{10x}{4(x+1)}=\frac{5}{6(x+1)}$$
$$12(x+1)\cdot\frac{3}{1}-12(x+1)\frac{10x}{4(x+1)}=12(x+1)\frac{5}{6(x+1)}$$
$$3\cdot 12(x+1)-3\cdot 10x=2\cdot 5$$
Choice C is correct.

13. If x is the first even integer, then $x+2$ is the next even integer, and the reciprocals of the two even integers are $\dfrac{1}{x}$ and $\dfrac{1}{x+2}$.
The phrase translates to $\dfrac{1}{x}+\dfrac{1}{x+2}=\dfrac{9}{40}$; D.

14. y varies inversely as x is written as $y=\dfrac{k}{x}$; D.

15. y varies directly as x is written as $y=kx$; B.

16. y varies jointly as x and z is written as $y=kxz$; C.

17. y varies directly as x^2 is written as $y=kx^2$; A.

18. y varies inversely as x^2 is written as $y=\dfrac{k}{x^2}$; E.

Chapter 6 Test

1. $1-x=0$
$1=x$
The domain of $f(x)=\dfrac{5x^2}{1-x}$ is
$\{x|x \text{ is a real number and } x\neq 1\}$.

2. $x^2+4x+3=0$
$(x+3)(x+1)=0$
$x+3=0$ or $x+1=0$
$x=-3$ or $x=-1$
The domain of $g(x)=\dfrac{9x^2-9}{x^2+4x+3}$ is
$\{x|x \text{ is a real number and } x\neq -3, x\neq -1\}$.

3. $\dfrac{7x-21}{24-8x}=\dfrac{7(x-3)}{8(3-x)}=\dfrac{7(x-3)}{-8(x-3)}=-\dfrac{7}{8}$

4. $\dfrac{x^2-4x}{x^2+5x-36}=\dfrac{x(x-4)}{(x+9)(x-4)}=\dfrac{x}{x+9}$

5.
$$\begin{aligned}\frac{x^3-8}{x-2}&=\frac{x^3-2^3}{x-2}\\&=\frac{(x-2)(x^2+2x+4)}{x-2}\\&=x^2+2x+4\end{aligned}$$

6.
$$\begin{aligned}&\frac{2x^3+16}{6x^2+12x}\cdot\frac{5}{x^2-2x+4}\\&=\frac{2(x^3+8)}{6x(x+2)}\cdot\frac{5}{x^2-2x+4}\\&=\frac{2(x+2)(x^2-2x+4)}{6x(x+2)}\cdot\frac{5}{x^2-2x+4}\\&=\frac{5}{3x}\end{aligned}$$

7. $\dfrac{5}{4x^3}+\dfrac{7}{4x^3}=\dfrac{5+7}{4x^3}=\dfrac{12}{4x^3}=\dfrac{3}{x^3}$

8. $\dfrac{3x^2-12}{x^2+2x-8}\div\dfrac{6x+18}{x+4}$
$=\dfrac{3x^2-12}{x^2+2x-8}\cdot\dfrac{x+4}{6x+18}$
$=\dfrac{3(x^2-4)}{(x+4)(x-2)}\cdot\dfrac{x+4}{6(x+3)}$
$=\dfrac{3(x+2)(x-2)}{(x+4)(x-2)}\cdot\dfrac{(x+4)}{6(x+3)}$
$=\dfrac{x+2}{2(x+3)}$

9. $\dfrac{4x-12}{2x-9}\div\dfrac{3-x}{4x^2-81}\cdot\dfrac{x+3}{5x+15}$
$=\dfrac{4x-12}{2x-9}\cdot\dfrac{4x^2-81}{3-x}\cdot\dfrac{x+3}{5x+15}$
$=\dfrac{4(x-3)}{2x-9}\cdot\dfrac{(2x+9)(2x-9)}{-(x-3)}\cdot\dfrac{x+3}{5(x+3)}$
$=\dfrac{4(2x+9)}{-5}$
$=-\dfrac{4(2x+9)}{5}$

10. $\dfrac{3+2x}{10-x}+\dfrac{13+x}{x-10}=\dfrac{-(3+2x)}{x-10}+\dfrac{13+x}{x-10}$
$=\dfrac{-3-2x+13+x}{x-10}$
$=\dfrac{-x+10}{x-10}$
$=-\dfrac{x-10}{x-10}$
$=-1$

11. $\dfrac{2x^2+7}{2x^4-18x^2}-\dfrac{6x+7}{2x^4-18x^2}=\dfrac{2x^2+7-6x-7}{2x^4-18x^2}$
$=\dfrac{2x^2-6x}{2x^4-18x^2}$
$=\dfrac{2x(x-3)}{2x^2(x^2-9)}$
$=\dfrac{2x(x-3)}{2x^2(x-3)(x+3)}$
$=\dfrac{1}{x(x+3)}$

12. $\dfrac{3}{x^2-x-6}+\dfrac{2}{x^2-5x+6}$
$=\dfrac{3}{(x-3)(x+2)}+\dfrac{2}{(x-3)(x-2)}$
$=\dfrac{3}{(x-3)(x+2)}\cdot\dfrac{x-2}{x-2}+\dfrac{2}{(x-3)(x-2)}\cdot\dfrac{x+2}{x+2}$
$=\dfrac{3(x-2)}{(x-3)(x+2)(x-2)}+\dfrac{2(x+2)}{(x-3)(x+2)(x-2)}$
$=\dfrac{3x-6}{(x-3)(x+2)(x-2)}+\dfrac{2x+4}{(x-3)(x+2)(x-2)}$
$=\dfrac{3x-6+2x+4}{(x-3)(x+2)(x-2)}$
$=\dfrac{5x-2}{(x-3)(x+2)(x-2)}$

13. $3x-21=3(x-7)$
$2x-14=2(x-7)$
The LCD is $3\cdot 2(x-7)=6(x-7)$.
$\dfrac{5}{x-7}-\dfrac{2x}{3x-21}+\dfrac{x}{2x-14}$
$=\dfrac{5}{x-7}\cdot\dfrac{6}{6}-\dfrac{2x}{3(x-7)}\cdot\dfrac{2}{2}+\dfrac{x}{2(x-7)}\cdot\dfrac{3}{3}$
$=\dfrac{30}{6(x-7)}-\dfrac{4x}{6(x-7)}+\dfrac{3x}{6(x-7)}$
$=\dfrac{30-4x+3x}{6(x-7)}$
$=\dfrac{30-x}{6(x-7)}$

14. $\dfrac{3x}{5}\cdot\left(\dfrac{5}{x}-\dfrac{5}{2x}\right)=\dfrac{3x}{5}\cdot\left(\dfrac{5}{x}\cdot\dfrac{2}{2}-\dfrac{5}{2x}\right)$
$=\dfrac{3x}{5}\left(\dfrac{10}{2x}-\dfrac{5}{2x}\right)$
$=\dfrac{3x}{5}\left(\dfrac{10-5}{2x}\right)$
$=\dfrac{3x}{5}\left(\dfrac{5}{2x}\right)$
$=\dfrac{3}{2}$

15. $\dfrac{\frac{5}{x}-\frac{7}{3x}}{\frac{9}{8x}-\frac{1}{x}}=\dfrac{24x\left(\frac{5}{x}-\frac{7}{3x}\right)}{24x\left(\frac{9}{8x}-\frac{1}{x}\right)}=\dfrac{120-56}{27-24}=\dfrac{64}{3}$

16. $\dfrac{\frac{x^2-5x+6}{x+3}}{\frac{x^2-4x+4}{x^2-9}} = \dfrac{x^2-5x+6}{x+3} \div \dfrac{x^2-4x+4}{x^2-9}$

$= \dfrac{x^2-5x+6}{x+3} \cdot \dfrac{x^2-9}{x^2-4x+4}$

$= \dfrac{(x-3)(x-2)}{x+3} \cdot \dfrac{(x+3)(x-3)}{(x-2)(x-2)}$

$= \dfrac{(x-3)^2}{x-2}$

17. $\dfrac{4x^2y+9x+3xz}{3xz} = \dfrac{4x^2y}{3xz} + \dfrac{9x}{3xz} + \dfrac{3xz}{3xz}$

$= \dfrac{4xy}{3z} + \dfrac{3}{z} + 1$

18.

$$\begin{array}{r} 2x^2 - x - 2 \\ 2x+1 \overline{)\, 4x^3 + 0x^2 - 5x + 0} \\ \underline{4x^3 + 2x^2} \\ -2x^2 - 5x \\ \underline{-2x^2 - x} \\ -4x + 0 \\ \underline{-4x - 2} \\ 2 \end{array}$$

Answer: $2x^2 - x - 2 + \dfrac{2}{2x+1}$

19.

$$\begin{array}{r|rrrrr} -3 & 4 & -3 & 0 & -1 & -1 \\ & & -12 & 45 & -135 & 408 \\ \hline & 4 & -15 & 45 & -136 & 407 \end{array}$$

Answer: $4x^3 - 15x^2 + 45x - 136 + \dfrac{407}{x+3}$

20.

$$\begin{array}{r|rrrrr} -2 & 4 & 0 & 7 & -2 & -5 \\ & & -8 & 16 & -46 & 96 \\ \hline & 4 & -8 & 23 & -48 & 91 \end{array}$$

Thus, $P(-2) = 91$.

21. $\dfrac{5x+3}{3x-7} = \dfrac{19}{7}$

$7(5x+3) = 19(3x-7)$

$35x+21 = 57x-133$

$154 = 22x$

$7 = x$

The solution is 7.

22. The LCD is $x - 4$.

$\dfrac{x}{x-4} = 3 - \dfrac{4}{x-4}$

$(x-4)\dfrac{x}{x-4} = (x-4)\cdot 3 - (x-4)\cdot\dfrac{4}{x-4}$

$x = 3(x-4) - 4$

$x = 3x - 12 - 4$

$x = 3x - 16$

$-2x = -16$

$x = 8$

The solution is 8.

23. $\dfrac{3}{x+2} - \dfrac{1}{5x} = \dfrac{2}{5x^2+10x}$

$\dfrac{3}{x+2} - \dfrac{1}{5x} = \dfrac{2}{5x(x+2)}$

$5x(x+2)\left(\dfrac{3}{x+2} - \dfrac{1}{5x}\right) = 5x(x+2)\cdot\dfrac{2}{5x(x+2)}$

$3(5x) - 1(x+2) = 2$

$15x - x - 2 = 2$

$14x = 4$

$x = \dfrac{4}{14} = \dfrac{2}{7}$

The solution is $\dfrac{2}{7}$.

24. $\dfrac{x^2+8}{x} - 1 = \dfrac{2(x+4)}{x}$

$x\left(\dfrac{x^2+8}{x} - 1\right) = x\left(\dfrac{2(x+4)}{x}\right)$

$(x^2+8) - x = 2(x+4)$

$x^2 - x + 8 = 2x + 8$

$x^2 - 3x = 0$

$x(x-3) = 0$

$x = 0$ or $x - 3 = 0$

$x = 3$

Discard the answer 0 as extraneous.
The solution is 3.

25. $\frac{x+b}{a} = \frac{4x-7a}{b}$

$ab\left(\frac{x+b}{a}\right) = ab\left(\frac{4x-7a}{b}\right)$

$b(x+b) = a(4x-7a)$

$xb+b^2 = 4ax-7a^2$

$b^2+7a^2 = 4ax-xb$

$b^2+7a^2 = x(4a-b)$

$x = \frac{b^2+7a^2}{4a-b}$

26. Let x = the number.

$(x+1)\cdot\frac{2}{x} = \frac{12}{5}$

$\frac{2(x+1)}{x} = \frac{12}{5}$

$\frac{2x+2}{x} = \frac{12}{5}$

$5(2x+2) = 12x$

$10x+10 = 12x$

$10 = 2x$

$5 = x$

The number is 5.

27. Let t = time to weed garden together.

Note that 1 hr and 30 min = $\frac{3}{2}$ hours.

$\frac{1}{2}+\frac{1}{\frac{3}{2}} = \frac{1}{t}$

$\frac{1}{2}+\frac{2}{3} = \frac{1}{t}$

$6t\left(\frac{1}{2}+\frac{2}{3}\right) = 6t\left(\frac{1}{t}\right)$

$3t+4t = 6$

$7t = 6$

$t = \frac{6}{7}$

It takes them $\frac{6}{7}$ hour.

28. $W = \frac{k}{V}$

$20 = \frac{k}{12}$

$k = 20(12) = 240$

$W = \frac{240}{V} = \frac{240}{15} = 16$

29. $Q = kRS^2$

$24 = k(3)(4)^2$

$24 = 48k$

$k = \frac{24}{48} = \frac{1}{2}$

$Q = \frac{1}{2}RS^2 = \frac{1}{2}(2)(3)^2 = 9$

30. $S = k\sqrt{d}$

$160 = k\sqrt{400}$

$160 = 20k$

$k = \frac{160}{20} = 8$

$S = 8\sqrt{d}$

$128 = 8\sqrt{d}$

$\sqrt{d} = \frac{128}{8}$

$\sqrt{d} = 16$

$d = 256$

The height of the cliff is 256 feet.

Chapter 6 Cumulative Review

1. a. $8x$

b. $8x+3$

c. $x\div(-7)$ or $\frac{x}{-7}$

d. $2x-1\frac{6}{10} = 2x-1.6$

e. $x-6$

f. $2(4+x)$

2. a. $x-\frac{1}{3}$

b. $5x-6$

c. $8x+3$

d. $\frac{7}{2-x}$

3.
$$\frac{y}{3}-\frac{y}{4}=\frac{1}{6}$$
$$12\left(\frac{y}{3}-\frac{y}{4}\right)=12\left(\frac{1}{6}\right)$$
$$4y-3y=2$$
$$y=2$$

4.
$$\frac{x}{7}+\frac{x}{5}=\frac{12}{5}$$
$$35\left(\frac{x}{7}+\frac{x}{5}\right)=35\left(\frac{12}{5}\right)$$
$$5x+7x=84$$
$$12x=84$$
$$x=7$$

5.
$$c<200$$
$$-11.8t+390<200$$
$$-11.8t<-190$$
$$t>\text{approximately } 16.1$$
2004 + 16 = 2020
The consumption of cigarettes will be less than 200 billion per year in 2020 and after.

6. Let x = score on final exam.
$$\frac{78+65+82+79+2x}{6}\ge 78$$
$$\frac{304+2x}{6}\ge 78$$
$$6\left(\frac{304+2x}{6}\right)\ge 6(78)$$
$$304+2x\ge 468$$
$$2x\ge 164$$
$$x\ge 82$$
The minimum score she can make on her final is 82.

7. $\left|\frac{3x+1}{2}\right|=-2$ is impossible. The solution set is $\varnothing$.

8.
$$\left|\frac{2x-1}{3}\right|+6=3$$
$\left|\frac{2x-1}{3}\right|=-3,$ which is impossible.
The solution set is $\varnothing$.

9.
$$\left|\frac{2(x+1)}{3}\right|\le 0$$
$$\frac{2(x+1)}{3}=0$$
$$2(x+1)=0$$
$$2x+2=0$$
$$2x=-2$$
$$x=-1$$
The solution is -1.

10.
$$\left|\frac{3(x-1)}{4}\right|\ge 2$$
$$\frac{3(x-1)}{4}\le -2 \quad\text{or}\quad \frac{3(x-1)}{4}\ge 2$$
$$\frac{3x-3}{4}\le -2 \quad\text{or}\quad \frac{3x-3}{4}\ge 2$$
$$3x-3\le -8 \quad\text{or}\quad 3x-3\ge 8$$
$$3x\le -5 \quad\text{or}\quad 3x\ge 11$$
$$x\le -\frac{5}{3} \quad\text{or}\quad x\ge \frac{11}{3}$$
$$\left(-\infty,-\frac{5}{3}\right]\cup\left[\frac{11}{3},\infty\right)$$

11. $y=-2x+3$

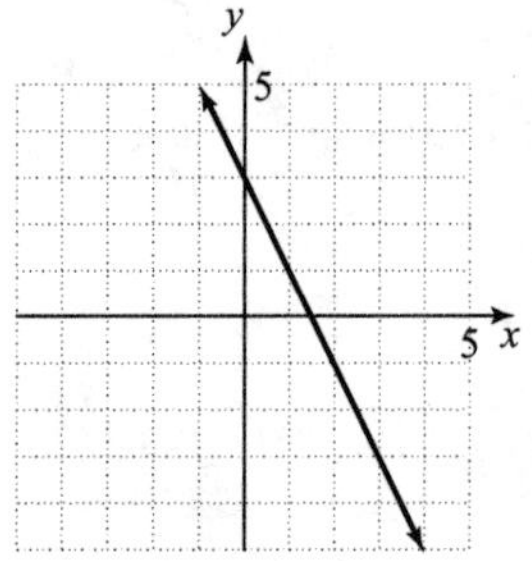

12. $y=-x+3$

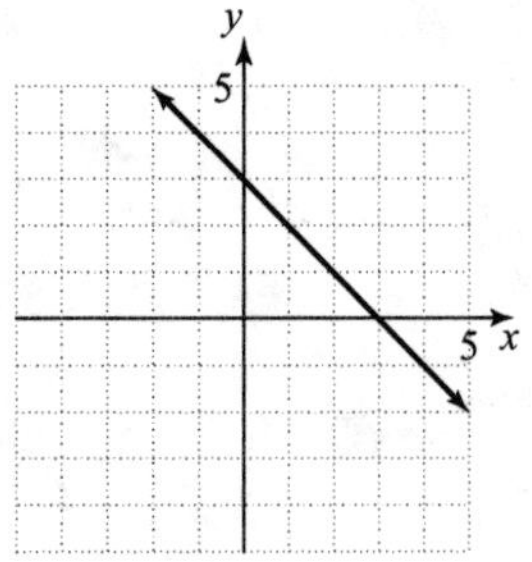

13. a. Function

b. Not a function

c. Function

14. $f(x) = -x^2 + 3x - 2$

a. $f(0) = -(0)^2 + 3(0) - 2 = -2$

b. $f(-3) = -(-3)^2 + 3(-3) - 2$
$= -9 - 9 - 2$
$= -20$

c. $f\left(\frac{1}{3}\right) = -\left(\frac{1}{3}\right)^2 + 3\left(\frac{1}{3}\right) - 2$
$= -\frac{1}{9} + 1 - 2$
$= -\frac{1}{9} + \frac{9}{9} - \frac{18}{9}$
$= \frac{-1 + 9 - 18}{9}$
$= -\frac{10}{9}$

15. $3x + 4y = 12$
Let $y = 0$.
$3x + 4(0) = 12$
$3x = 12$
$x = 4$
Plot (4, 0).
Let $x = 0$.
$3(0) + 4y = 12$
$4y = 12$
$y = 3$
Plot (0, 3).

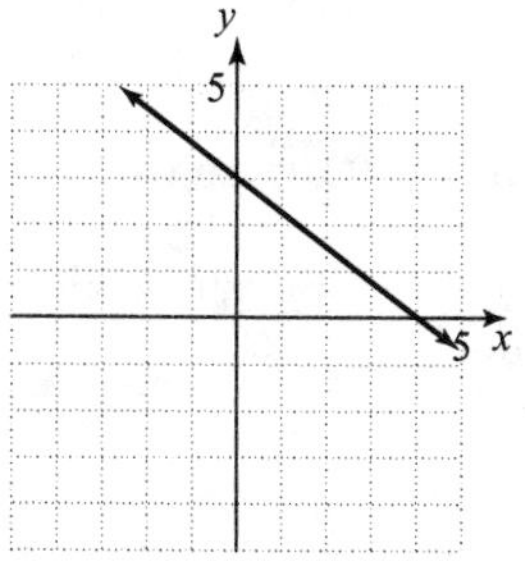

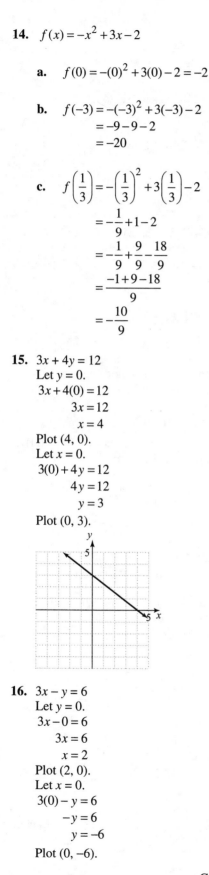

16. $3x - y = 6$
Let $y = 0$.
$3x - 0 = 6$
$3x = 6$
$x = 2$
Plot (2, 0).
Let $x = 0$.
$3(0) - y = 6$
$-y = 6$
$y = -6$
Plot (0, –6).

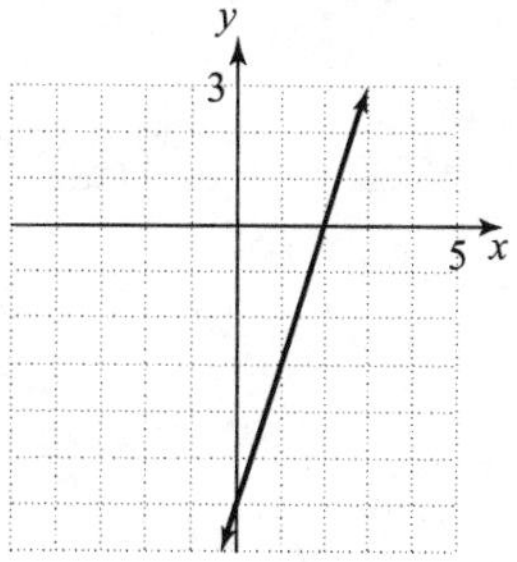

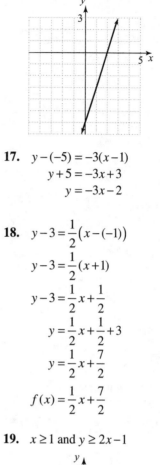

17. $y - (-5) = -3(x - 1)$
$y + 5 = -3x + 3$
$y = -3x - 2$

18. $y - 3 = \frac{1}{2}(x - (-1))$
$y - 3 = \frac{1}{2}(x + 1)$
$y - 3 = \frac{1}{2}x + \frac{1}{2}$
$y = \frac{1}{2}x + \frac{1}{2} + 3$
$y = \frac{1}{2}x + \frac{7}{2}$
$f(x) = \frac{1}{2}x + \frac{7}{2}$

19. $x \geq 1$ and $y \geq 2x - 1$

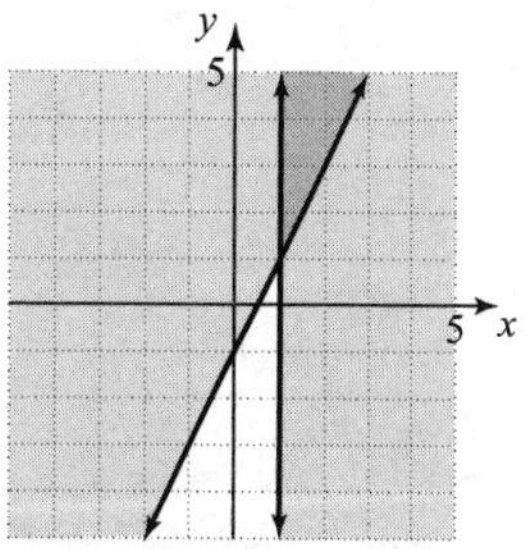

The solution region is the overlap, which has darker shading, along with its boundary.

20. $2x + y \leq 4$ or $y > 2$

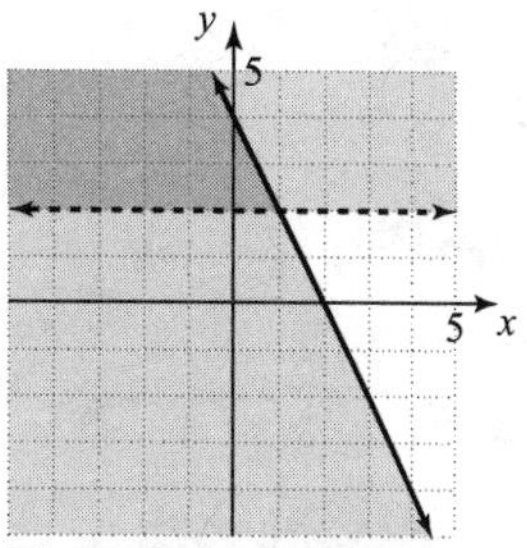

The solution region consists of all shaded regions and the boundary where the boundary line is solid.

21. $\begin{cases} 3x-2y=10 & (1) \\ 4x-3y=15 & (2) \end{cases}$

Multiply E1 by –4 and E2 by 3, and add.

$$\begin{aligned} -12x+8y &= -40 \\ 12x-9y &= 45 \\ \hline -y &= 5 \\ y &= -5 \end{aligned}$$

Replace y with –5 in E1.

$$\begin{aligned} 3x-2(-5) &= 10 \\ 3x+10 &= 10 \\ 3x &= 0 \\ x &= 0 \end{aligned}$$

The solution is (0, –5).

22. $\begin{cases} -2x+3y=6 & (1) \\ 3x-\ \ y=5 & (2) \end{cases}$

Solve E2 for y:

$y = 3x - 5$

Replace y with $3x-5$ in E1.

$$\begin{aligned} -2x+3(3x-5) &= 6 \\ -2x+9x-15 &= 6 \\ 7x &= 21 \\ x &= 3 \end{aligned}$$

Replace x with 3 in the equation $y = 3x - 5$.

$y = 3(3)-5 = 9-5 = 4$

The solution is (3, 4).

23. $\begin{cases} 2x-4y+8z=2 & (1) \\ -x-3y+\ \ z=11 & (2) \\ x-2y+4z=0 & (3) \end{cases}$

Add E2 and E3.

$-5y+5z=11 \quad (4)$

Multiply E2 by 2 and add to E1.

$$\begin{aligned} -2x-6y+2z &= 22 \\ 2x-4y+8z &= 2 \\ \hline -10y+10z &= 24 \quad (5) \end{aligned}$$

Solve the new system.

$\begin{cases} -5y+\ 5z=11 & (4) \\ -10y+10z=24 & (5) \end{cases}$

Multiply E4 by –2 and add to E5.

$$\begin{aligned} 10y-10z &= -22 \\ -10y+10z &= 24 \\ \hline 0 &= 2, \text{ which is impossible.} \end{aligned}$$

The solution is $\varnothing$.

24. $\begin{cases} 2x-2y+4z=6 & (1) \\ -4x-\ \ y+\ \ z=-8 & (2) \\ 3x-\ \ y+\ \ z=6 & (3) \end{cases}$

Multiply E2 by –1 and add to E3.

$$\begin{aligned} 4x+y-z &= 8 \\ 3x-y+z &= 6 \\ \hline 7x \qquad &= 14 \\ x &= 2 \end{aligned}$$

Multiply E2 by –2 and add to E1.

$$\begin{aligned} 8x+2y-2z &= 16 \\ 2x-2y+4z &= 6 \\ \hline 10x \qquad +2z &= 22 \quad \text{or} \quad 5x+z=11 \end{aligned}$$

Replace x with 2 in the equation $5x + z = 11$.

$$\begin{aligned} 5(2)+z &= 11 \\ 10+z &= 11 \\ z &= 1 \end{aligned}$$

Replace x with 2 and z with 1 in E3.

$$\begin{aligned} 3(2)-y+1 &= 6 \\ 7-y &= 6 \\ -y &= -1 \\ y &= 1 \end{aligned}$$

The solution is (2, 1, 1).

25. Let x = measure of the smallest angle,
y = measure of the largest angle and,
z = measure of the remaining angle.

$\begin{cases} x+y+z=180 & (1) \\ y=x+80 & (2) \\ z=x+10 & (3) \end{cases}$

Substitute $x+80$ for y and $x+10$ for z in E1.

$$\begin{aligned} x+(x+80)+(x+10) &= 180 \\ 3x+90 &= 180 \\ 3x &= 90 \\ x &= 30 \end{aligned}$$

Replace x with 30 in E2 and E3.

$y = 30+80 = 110$

$z = 30+10 = 40$

The angles measure 30°, 110°, and 40°.

26. Let x = the price of a ream of paper and
y = the price of a box of manila folders.

$\begin{cases} 3x+2y=21.90 & (1) \\ 5x+\ \ y=24.25 & (2) \end{cases}$

Multiply E2 by – 2 and add to E1.

$$\begin{aligned} -10x-2y &= -48.50 \\ 3x+2y &= 21.90 \\ \hline -7x \qquad &= -26.60 \\ x &= 3.80 \end{aligned}$$

Replace x with 3.80 in E2.

$$\begin{aligned} 5(3.80)+y &= 24.25 \\ 19+y &= 24.25 \\ y &= 5.25 \end{aligned}$$

A ream of paper cost \$3.80 and a box of manila folders cost \$5.25.

27. $\begin{cases} x+2y+z=2 \\ -2x-y+2z=5 \\ x+3y-2z=-8 \end{cases}$

$\left[\begin{array}{ccc|c} 1 & 2 & 1 & 2 \\ -2 & -1 & 2 & 5 \\ 1 & 3 & -2 & -8 \end{array}\right]$

Multiply R1 by 2 and add to R2.
Multiply R1 by –1 and add to R3.

$\left[\begin{array}{ccc|c} 1 & 2 & 1 & 2 \\ 0 & 3 & 4 & 9 \\ 0 & 1 & -3 & -10 \end{array}\right]$

Interchange R2 and R3.

$\left[\begin{array}{ccc|c} 1 & 2 & 1 & 2 \\ 0 & 1 & -3 & -10 \\ 0 & 3 & 4 & 9 \end{array}\right]$

Multiply R2 by –3 and add to R3.

$\left[\begin{array}{ccc|c} 1 & 2 & 1 & 2 \\ 0 & 1 & -3 & -10 \\ 0 & 0 & 13 & 39 \end{array}\right]$

Divide R3 by 13.

$\left[\begin{array}{ccc|c} 1 & 2 & 1 & 2 \\ 0 & 1 & -3 & -10 \\ 0 & 0 & 1 & 3 \end{array}\right]$

This corresponds to $\begin{cases} x+2y+z=2 \\ y-3z=-10 \\ z=3. \end{cases}$

$$\begin{aligned} y-3z&=-10 \\ y-3(3)&=-10 \\ y-9&=-10 \\ y&=-1 \end{aligned} \quad \text{and so} \quad \begin{aligned} x+2(-1)+(3)&=2 \\ x-2+3&=2 \\ x+1&=2 \\ x&=1 \end{aligned}$$

The solution is (1, –1, 3).

28. $\begin{cases} x+y+z=9 \\ 2x-2y+3z=2 \\ -3x+y-z=1 \end{cases}$

$\left[\begin{array}{ccc|c} 1 & 1 & 1 & 9 \\ 2 & -2 & 3 & 2 \\ -3 & 1 & -1 & 1 \end{array}\right]$

Multiply R1 by –2 and add to R2.
Multiply R1 by 3 and add to R3.

$\left[\begin{array}{ccc|c} 1 & 1 & 1 & 9 \\ 0 & -4 & 1 & -16 \\ 0 & 4 & 2 & 28 \end{array}\right]$

Divide R2 by –4.

$\left[\begin{array}{ccc|c} 1 & 1 & 1 & 9 \\ 0 & 1 & -\frac{1}{4} & 4 \\ 0 & 4 & 2 & 28 \end{array}\right]$

Multiply R2 by –4 and add to R3.

$\left[\begin{array}{ccc|c} 1 & 1 & 1 & 9 \\ 0 & 1 & -\frac{1}{4} & 4 \\ 0 & 0 & 3 & 12 \end{array}\right]$

Divide R3 by 3.

$\left[\begin{array}{ccc|c} 1 & 1 & 1 & 9 \\ 0 & 1 & -\frac{1}{4} & 4 \\ 0 & 0 & 1 & 4 \end{array}\right]$

This corresponds to $\begin{cases} x+y+z=9 \\ y-\frac{1}{4}z=4 \\ z=4 \end{cases}$

$$\begin{aligned} y-\frac{1}{4}(4)&=4 \\ y-1&=4 \\ y&=5 \end{aligned} \quad \text{and so} \quad \begin{aligned} x+5+4&=9 \\ x+9&=9 \\ x&=0 \end{aligned}$$

The solution is (0, 5, 4).

29. a. $7^0=1$

b. $-7^0=-1\cdot 7^0=-1\cdot 1=-1$

c. $(2x+5)^0=1$

d. $2x^0=2\cdot x^0=2\cdot 1=2$

30. a. $$\begin{aligned} 2^{-2}+3^{-1}&=\frac{1}{2^2}+\frac{1}{3} \\ &=\frac{1}{4}+\frac{1}{3} \\ &=\frac{3}{12}+\frac{4}{12} \\ &=\frac{7}{12} \end{aligned}$$

b. $-6a^0=-6\cdot a^0=-6\cdot 1=-6$

c. $\frac{x^{-5}}{x^{-2}}=x^{-5-(-2)}=x^{-3}=\frac{1}{x^3}$

31. a. $$\begin{aligned} x^{-b}(2x^b)^2&=\frac{2^2(x^b)^2}{x^b} \\ &=\frac{4x^{2b}}{x^b} \\ &=4x^{2b-b} \\ &=4x^b \end{aligned}$$

b. $\dfrac{(y^{3a})^2}{y^{a-6}} = \dfrac{y^{6a}}{y^{a-6}} = y^{6a-(a-6)} = y^{5a+6}$

32. a. $3x^{4a}(4x^{-a})^2 = 3x^{4a} \cdot 16x^{-2a}$
$= 48x^{4a+(-2a)}$
$= 48x^{2a}$

b. $\dfrac{(y^{4b})^3}{y^{2b-3}} = \dfrac{y^{12b}}{y^{2b-3}}$
$= y^{12b-(2b-3)}$
$= y^{12b-2b+3}$
$= y^{10b+3}$

33. a. $3x^2$ has degree = 2.

b. $-2^3x^5 = -8x^5$ has degree = 5.

c. y has degree = 1.

d. $12x^2yz^3$ has degree = 2 + 1 + 3 = 6.

e. 5.27 has degree = 0.

34. $(2x^2+8x-3)-(2x-7)$
$= 2x^2+8x-3-2x+7$
$= 2x^2+6x+4$

35. $[3+(2a+b)]^2$
$= 3^2+2(3)(2a+b)+(2a+b)^2$
$= 9+6(2a+b)+(4a^2+2(2a)b+b^2)$
$= 9+12a+6b+4a^2+4ab+b^2$

36. $[4+(3x-y)]^2$
$= 4^2+2(4)(3x-y)+(3x-y)^2$
$= 16+8(3x-y)+(9x^2-2(3x)y+y^2)$
$= 16+24x-8y+9x^2-6xy+y^2$

37. $ab-6a+2b-12 = a(b-6)+2(b-6)$
$= (b-6)(a+2)$

38. $xy+2x-5y-10 = x(y+2)-5(y+2)$
$= (y+2)(x-5)$

39. $2n^2-38n+80 = 2(n^2-19n+40)$

40. $6x^2-x-35 = (2x-5)(3x+7)$

41. $x^2+4x+4-y^2 = (x^2+4x+4)-y^2$
$= (x+2)^2-y^2$
$= [(x+2)+y][(x+2)-y]$
$= (x+2+y)(x+2-y)$

42. $4x^2-4x+1-9y^2$
$= (4x^2-4x+1)-9y^2$
$= (2x-1)^2-(3y)^2$
$= [(2x-1)+3y][(2x-1)-3y]$
$= (2x-1+3y)(2x-1-3y)$

43. $(x+2)(x-6) = 0$
$x+2=0$ or $x-6=0$
$x=-2$ or $x=6$
The solutions are –2 and 6.

44. $2x(3x+1)(x-3) = 0$
$x=0$ or $3x+1=0$ or $x-3=0$
$3x=-1$ or $x=3$
$x=-\dfrac{1}{3}$

The solutions are $-\dfrac{1}{3}$, 0, 3.

45. a. $\dfrac{2x^2}{10x^3-2x^2} = \dfrac{2x^2}{2x^2(5x-1)} = \dfrac{1}{5x-1}$

b. $\dfrac{9x^2+13x+4}{8x^2+x-7} = \dfrac{(9x+4)(x+1)}{(8x-7)(x+1)} = \dfrac{9x+4}{8x-7}$

46. a. Domain: $(-\infty, \infty)$; Range: $[-4, \infty)$

b. x-intercepts: (–2, 0), (2, 0)
y-intercept: (0, –4)

c. There is no such point.

d. The point with the least y-value is (0, –4).

e. –2, 2

f. x-values between $x = -2$ and $x = 2$

g. The solutions are –2 and 2.

47. $\dfrac{5k}{k^2-4}-\dfrac{2}{k^2+k-2}$

$=\dfrac{5k}{(k+2)(k-2)}-\dfrac{2}{(k+2)(k-1)}$

$=\dfrac{5k(k-1)-2(k-2)}{(k+2)(k-2)(k-1)}$

$=\dfrac{5k^2-5k-2k+4}{(k+2)(k-2)(k-1)}$

$=\dfrac{5k^2-7k+4}{(k+2)(k-2)(k-1)}$

48. $\dfrac{5a}{a^2-4}-\dfrac{3}{2-a}=\dfrac{5a}{(a+2)(a-2)}+\dfrac{3}{a-2}$

$=\dfrac{5a+3(a+2)}{(a+2)(a-2)}$

$=\dfrac{5a+3a+6}{(a+2)(a-2)}$

$=\dfrac{8a+6}{(a+2)(a-2)}$

$=\dfrac{2(4a+3)}{(a+2)(a-2)}$

49. $\dfrac{3}{x}-\dfrac{x+21}{3x}=\dfrac{5}{3}$

$3x\left(\dfrac{3}{x}-\dfrac{x+21}{3x}\right)=3x\left(\dfrac{5}{3}\right)$

$9-(x+21)=5x$

$9-x-21=5x$

$-x-12=5x$

$-12=6x$

$-2=x$

The solution is –2.

50. $\dfrac{3x-4}{2x}=-\dfrac{8}{x}$

$x(3x-4)=-8(2x)$

$3x^2-4x=-16x$

$3x^2+12x=0$

$3x(x+4)=0$

$3x=0$ or $x+4=0$

$x=0$ or $x=-4$

Discard the answer 0 as extraneous. The solution is –4.

Chapter 7

Section 7.1 Practice Exercises

1. a. $\sqrt{49}=7$ because $7^2=49$ and 7 is not negative.

b. $\sqrt{\frac{0}{1}}=\sqrt{0}=0$ because $0^2=0$ and 0 is not negative.

c. $\sqrt{\frac{16}{81}}=\frac{4}{9}$ because $\left(\frac{4}{9}\right)^2=\frac{16}{81}$ and $\frac{4}{9}$ is not negative.

d. $\sqrt{0.64}=0.8$ because $(0.8)^2=0.64$.

e. $\sqrt{z^8}=z^4$ because $(z^4)^2=z^8$.

f. $\sqrt{16b^4}=4b^2$ because $(4b^2)^2=16b^4$.

g. $-\sqrt{36}=-6$. The negative in front of the radical indicates the negative square root of 36.

h. $\sqrt{-36}$ is not a real number.

2. $\sqrt{45}\approx 6.708$
Since $36<45<49$, then $\sqrt{36}<\sqrt{45}<\sqrt{49}$, or $6<\sqrt{45}<7$. The approximation is between 6 and 7 and thus is reasonable.

3. a. $\sqrt[3]{-1}=-1$ because $(-1)^3=-1$.

b. $\sqrt[3]{27}=3$ because $3^3=27$.

c. $\sqrt[3]{\frac{27}{64}}=\frac{3}{4}$ because $\left(\frac{3}{4}\right)^3=\frac{27}{64}$.

d. $\sqrt[3]{x^{12}}=x^4$ because $(x^4)^3=x^{12}$.

e. $\sqrt[3]{-8x^3}=-2x$ because $(-2x)^3=-8x^3$.

4. a. $\sqrt[4]{10{,}000}=10$ because $10^4=10{,}000$ and 10 is positive.

b. $\sqrt[5]{-1}=-1$ because $(-1)^5=-1$.

c. $-\sqrt{81}=-9$ because -9 is the opposite of $\sqrt{81}$.

d. $\sqrt[4]{-625}$ is not a real number. There is no real number that, when raised to the fourth power, is -625.

e. $\sqrt[3]{27x^9}=3x^3$ because $(3x^3)^3=27x^9$.

5. a. $\sqrt{(-4)^2}=|-4|=4$

b. $\sqrt{x^{14}}=|x^7|$

c. $\sqrt[4]{(x+7)^4}=|x+7|$

d. $\sqrt[3]{(-7)^3}=-7$

e. $\sqrt[5]{(3x-5)^5}=3x-5$

f. $\sqrt{49x^2}=7|x|$

g. $\sqrt{x^2+16x+64}=\sqrt{(x+8)^2}=|x+8|$

6. $f(x)=\sqrt{x+5},\ g(x)=\sqrt[3]{x-3}$

a. $f(11)=\sqrt{11+5}=\sqrt{16}=4$

b. $f(-1)=\sqrt{-1+5}=\sqrt{4}=2$

c. $g(11)=\sqrt[3]{11-3}=\sqrt[3]{8}=2$

d. $g(-6)=\sqrt[3]{-6-3}=\sqrt[3]{-9}$

7. $h(x)=\sqrt{x+2}$
Find the domain.
$x+2\geq 0$
$x\geq -2$
The domain of $h(x)$ is $\{x\mid x\geq -2\}$.

x	$h(x)=\sqrt{x+2}$
-2	0
-1	1
1	$\sqrt{1+2}=\sqrt{3}\approx 1.7$
2	2
7	3

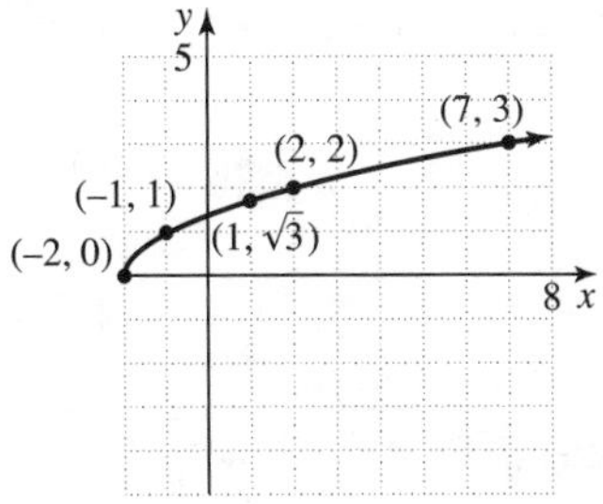

8. $f(x)=\sqrt[3]{x}-4$
The domain is the set of all real numbers.

x	$f(x)=\sqrt[3]{x}-4$
0	-4
1	-3
-1	-5
6	$\sqrt[3]{6}-4\approx 1.8-4=-2.2$
-6	$\sqrt[3]{-6}-4\approx -1.8-4=-5.8$
8	-2
-8	-6

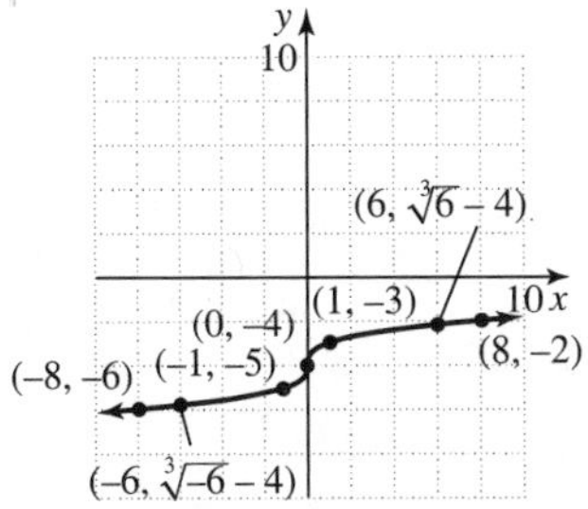

Vocabulary, Readiness & Video Check 7.1

1. In the expression $\sqrt[n]{a}$, the n is called the index, the $\sqrt{\ }$ is called the radical sign, and a is called the radicand.

2. If $\sqrt{a}$ is the positive square root of a, $a \neq 0$, then $-\sqrt{a}$ is the negative square root of a.

3. The square root of a negative number is not a real number.

4. Numbers such as 1, 4, 9, and 25 are called perfect squares, whereas numbers such as 1, 8, 27, and 125 are called perfect cubes.

5. The domain of the function $f(x)=\sqrt{x}$ is $[0, \infty)$.

6. The domain of the function $f(x)=\sqrt[3]{x}$ is $(-\infty, \infty)$.

7. If $f(16) = 4$, the corresponding ordered pair is $(16, 4)$.

8. If $g(-8) = -2$, the corresponding ordered pair is $(-8, -2)$.

9. Divide the index into each exponent in the radicand.

10. Find the nearest perfect squares both less than and greater than the radicand. The square root of the radicand falls between the square roots of these two perfect squares.

11. The square root of a negative number is not a real number, but the cube root of a negative number is a real number.

12. The even root of a negative number is not a real number.

13. For odd roots, there's only one root/answer whether the radicand is positive or negative, so absolute value bars aren't needed.

14. Since the variable x is the radicand of a square root and the square root of a negative number is not a real number, then we know that x cannot be negative.

Exercise Set 7.1

1. $\sqrt{100}=10$ because $10^2=100$.

3. $\sqrt{\frac{1}{4}}=\frac{1}{2}$ because $\left(\frac{1}{2}\right)^2=\frac{1}{4}$.

5. $\sqrt{0.0001} = 0.01$ because $(0.01)^2 = 0.0001$.

7. $-\sqrt{36} = -1 \cdot \sqrt{36} = -1 \cdot 6 = -6$

9. $\sqrt{x^{10}} = x^5$ because $(x^5)^2 = x^{10}$.

11. $\sqrt{16y^6} = 4y^3$ because $(4y^3)^2 = 16y^6$.

13. $\sqrt{7} \approx 2.646$
Since $4 < 7 < 9$, then $\sqrt{4} < \sqrt{7} < \sqrt{9}$, or $2 < \sqrt{7} < 3$. The approximation is between 2 and 3 and thus is reasonable.

15. $\sqrt{38} \approx 6.164$
Since $36 < 38 < 49$, then $\sqrt{36} < \sqrt{38} < \sqrt{49}$, or $6 < \sqrt{38} < 7$. The approximation is between 6 and 7 and thus is reasonable.

17. $\sqrt{200} \approx 14.142$
Since $196 < 200 < 225$, then $\sqrt{196} < \sqrt{200} < \sqrt{225}$, or $14 < \sqrt{200} < 15$. The approximation is between 14 and 15 and thus is reasonable.

19. $\sqrt[3]{64} = 4$ because $4^3 = 64$.

21. $\sqrt[3]{\frac{1}{8}} = \frac{1}{2}$ because $\left(\frac{1}{2}\right)^3 = \frac{1}{8}$.

23. $\sqrt[3]{-1} = -1$ because $(-1)^3 = -1$.

25. $\sqrt[3]{x^{12}} = x^4$ because $(x^4)^3 = x^{12}$.

27. $\sqrt[3]{-27x^9} = -3x^3$ because $(-3x^3)^3 = -27x^9$.

29. $-\sqrt[4]{16} = -2$ because $2^4 = 16$.

31. $\sqrt[4]{-16}$ is not a real number. There is no real number that, when raised to the fourth power, is -16.

33. $\sqrt[5]{-32} = -2$ because $(-2)^5 = -32$.

35. $\sqrt[5]{x^{20}} = x^4$ because $(x^4)^5 = x^{20}$.

37. $\sqrt[6]{64x^{12}} = 2x^2$ because $(2x^2)^6 = 64x^{12}$.

39. $\sqrt{81x^4} = 9x^2$ because $(9x^2)^2 = 81x^4$.

41. $\sqrt[4]{256x^8} = 4x^2$ because $(4x^2)^4 = 256x^8$.

43. $\sqrt{(-8)^2} = |-8| = 8$

45. $\sqrt[3]{(-8)^3} = -8$

47. $\sqrt{4x^2} = |2x| = 2|x|$

49. $\sqrt[3]{x^3} = x$

51. $\sqrt{(x-5)^2} = |x-5|$

53. $\sqrt{x^2+4x+4} = \sqrt{(x+2)^2} = |x+2|$

55. $-\sqrt{121} = -11$

57. $\sqrt[3]{8x^3} = 2x$

59. $\sqrt{y^{12}} = y^6$

61. $\sqrt{25a^2b^{20}} = 5ab^{10}$

63. $\sqrt[3]{-27x^{12}y^9} = -3x^4y^3$

65. $\sqrt[4]{a^{16}b^4} = a^4b$

67. $\sqrt[5]{-32x^{10}y^5} = -2x^2y$

69. $\sqrt{\frac{25}{49}} = \frac{5}{7}$

71. $\sqrt{\frac{x^{20}}{4y^2}} = \frac{x^{10}}{2y}$

73. $-\sqrt[3]{\frac{z^{21}}{27x^3}} = -\frac{z^7}{3x}$

75. $\sqrt[4]{\frac{x^4}{16}} = \frac{x}{2}$

77. $f(x) = \sqrt{2x+3}$
$f(0) = \sqrt{2(0)+3} = \sqrt{3}$

79. $g(x) = \sqrt[3]{x-8}$
$g(7) = \sqrt[3]{7-8} = \sqrt[3]{-1} = -1$

81. $g(x) = \sqrt[3]{x-8}$
$g(-19) = \sqrt[3]{-19-8} = \sqrt[3]{-27} = -3$

83. $f(x) = \sqrt{2x+3}$
$f(2) = \sqrt{2(2)+3} = \sqrt{7}$

85. $f(x) = \sqrt{x}+2$
$x \ge 0$
Domain: $[0, \infty)$

x	$f(x) = \sqrt{x}+2$
0	$\sqrt{0}+2 = 2$
1	$\sqrt{1}+2 = 3$
3	$\sqrt{3}+2 \approx 3.7$
4	$\sqrt{4}+2 = 4$

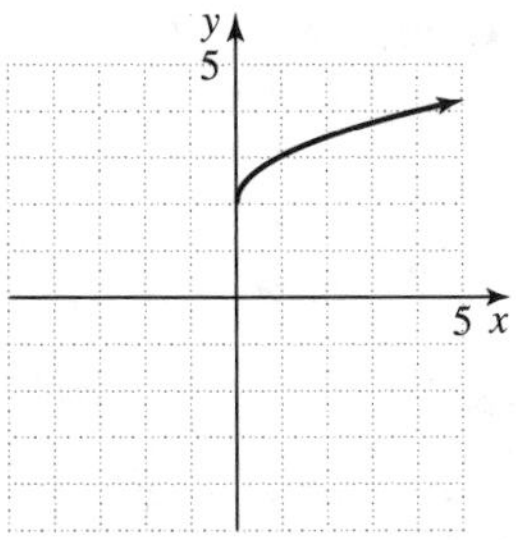

87. $f(x) = \sqrt{x-3}$
$x-3 \ge 0$
$x \ge 3$
Domain: $[3, \infty)$

x	$f(x) = \sqrt{x-3}$
3	$\sqrt{3-3} = \sqrt{0} = 0$
4	$\sqrt{4-3} = \sqrt{1} = 1$
7	$\sqrt{7-3} = \sqrt{4} = 2$
12	$\sqrt{12-3} = \sqrt{9} = 3$

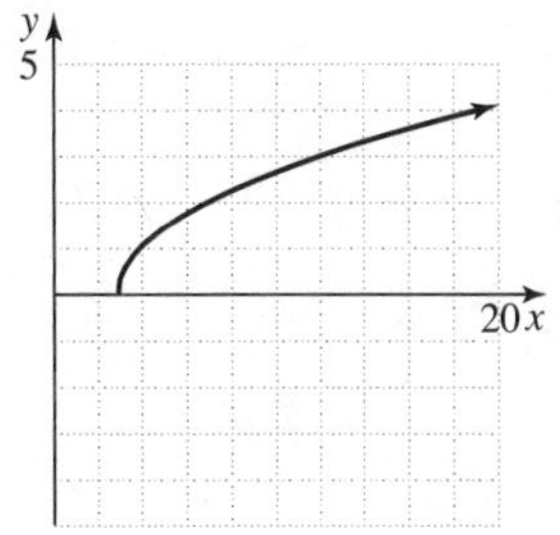

89. $f(x) = \sqrt[3]{x}+1$
Domain: $(-\infty, \infty)$

x	$f(x) = \sqrt[3]{x}+1$
−4	$\sqrt[3]{-4}+1 \approx -0.6$
−1	$\sqrt[3]{-1}+1 = 0$
0	$\sqrt[3]{0}+1 = 1$
1	$\sqrt[3]{1}+1 = 2$
4	$\sqrt[3]{4}+1 \approx 2.6$

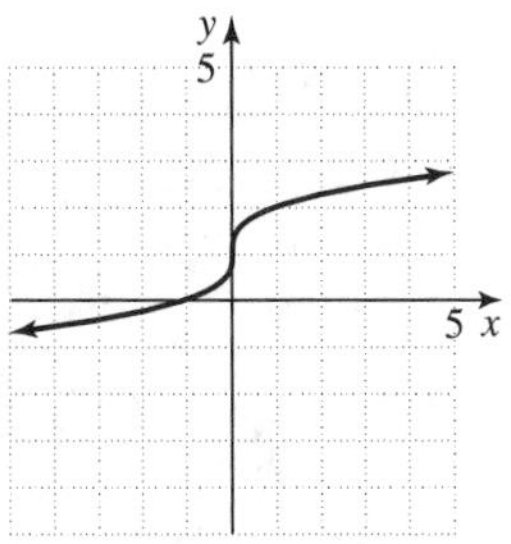

91. $g(x) = \sqrt[3]{x-1}$
Domain: $(-\infty, \infty)$

x	$g(x) = \sqrt[3]{x-1}$
1	$\sqrt[3]{1-1} = \sqrt[3]{0} = 0$
2	$\sqrt[3]{2-1} = \sqrt[3]{1} = 1$
0	$\sqrt[3]{0-1} = \sqrt[3]{-1} = -1$
9	$\sqrt[3]{9-1} = \sqrt[3]{8} = 2$
−7	$\sqrt[3]{-7-1} = \sqrt[3]{-8} = -2$

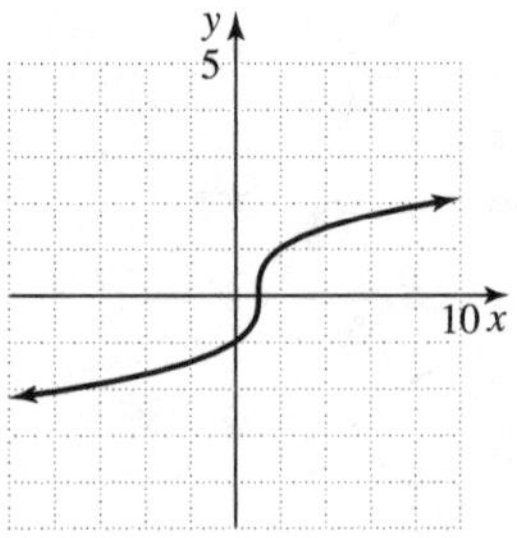

93. $(-2x^3y^2)^5 = (-2)^5 x^{3\cdot 5} y^{2\cdot 5} = -32x^{15}y^{10}$

95. $(-3x^2y^3z^5)(20x^5y^7) = -3(20)x^{2+5}y^{3+7}z^5$
$= -60x^7y^{10}z^5$

97. $\dfrac{7x^{-1}y}{14(x^5y^2)^{-2}} = \dfrac{7x^{-1}y}{14x^{-10}y^{-4}} = \dfrac{x^9y^5}{2}$

99. $\sqrt{-17}$ is not a real number.

101. $\sqrt[10]{-17}$ is not a real number.

103. The radical that is not a real number is $\sqrt{-10}$, choice **d**.

105. The radical that simplifies to −3 is $\sqrt[3]{-27}$, choice **d**.

107. $144 < 160 < 169$ so $\sqrt{144} < \sqrt{160} < \sqrt{169}$, or $12 < \sqrt{160} < 13$. Thus $\sqrt{160}$ is between 12 and 13. Therefore, the answer is **b.**

109. $\sqrt{30} \approx 5$, $\sqrt{10} \approx 3$, and $\sqrt{90} \approx 10$ so $P = \sqrt{30} + \sqrt{10} + \sqrt{90} \approx 5 + 3 + 10 = 18$. Therefore, the answer is **b.**

111. answers may vary

113. $B = \sqrt{\dfrac{hw}{3131}} = \sqrt{\dfrac{66 \cdot 135}{3131}}$
$= \sqrt{\dfrac{8910}{3131}}$
≈ 1.69 sq meters

115. $v = \sqrt{\dfrac{2Gm}{r}} = \sqrt{\dfrac{2(6.67 \times 10^{-11})(5.97 \times 10^{24})}{6.37 \times 10^6}}$
$= \sqrt{\dfrac{2(3.98199 \times 10^{14})}{6.37 \times 10^6}}$
$= \sqrt{\dfrac{7.96398 \times 10^{14}}{6.37 \times 10^6}}$
$\approx \sqrt{125{,}023{,}233.9}$
$\approx 11{,}181$

The escape velocity is 11,181 meters per second.

117. answers may vary

119. $f(x) = \sqrt{x} + 2$

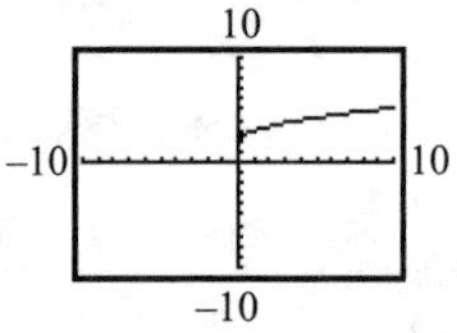

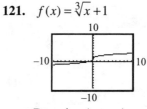

Domain: $[0, \infty)$

121. $f(x) = \sqrt[3]{x} + 1$

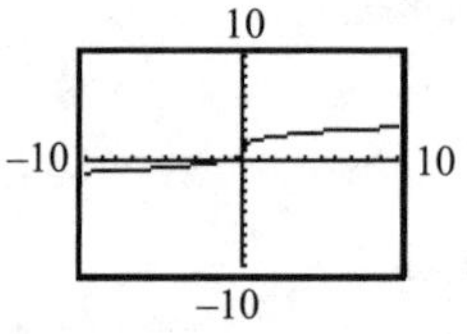

Domain: $(-\infty, \infty)$

Section 7.2 Practice Exercises

1. a. $36^{1/2} = \sqrt{36} = 6$

b. $1000^{1/3} = \sqrt[3]{1000} = 10$

c. $x^{1/3} = \sqrt[3]{x}$

d. $1^{1/4} = \sqrt[4]{1} = 1$

e. $-64^{1/2} = -\sqrt{64} = -8$

f. $(125x^9)^{1/3} = \sqrt[3]{125x^9} = 5x^3$

g. $3x^{1/4} = 3\sqrt[4]{x}$

2. a. $16^{3/2} = \left(\sqrt{16}\right)^3 = 4^3 = 64$

b. $-1^{3/5} = -\left(\sqrt[5]{1}\right)^3 = -(1)^3 = -1$

c. $-(81)^{3/4} = -\left(\sqrt[4]{81}\right)^3 = -(3)^3 = -27$

d. $\left(\frac{1}{25}\right)^{3/2} = \left(\sqrt{\frac{1}{25}}\right)^3 = \left(\frac{1}{5}\right)^3 = \frac{1}{125}$

e. $(3x+2)^{5/9} = \sqrt[9]{(3x+2)^5}$

3. a. $9^{-3/2} = \frac{1}{9^{3/2}} = \frac{1}{\left(\sqrt{9}\right)^3} = \frac{1}{3^3} = \frac{1}{27}$

b. $(-64)^{-2/3} = \frac{1}{(-64)^{2/3}} = \frac{1}{\left(\sqrt[3]{-64}\right)^2} = \frac{1}{(-4)^2} = \frac{1}{16}$

4. a. $y^{2/3} \cdot y^{8/3} = y^{(2/3+8/3)} = y^{10/3}$

b. $x^{3/5} \cdot x^{1/4} = x^{3/5+1/4} = x^{12/20+5/20} = x^{17/20}$

c. $\frac{9^{2/7}}{9^{9/7}} = 9^{2/7-9/7} = 9^{-7/7} = 9^{-1} = \frac{1}{9}$

d. $b^{4/9} \cdot b^{-2/9} = b^{4/9+(-2/9)} = b^{2/9}$

e.
$$\begin{aligned}\frac{\left(3x^{1/4}y^{-2/3}\right)^4}{x^4y} &= \frac{3^4(x^{1/4})^4(y^{-2/3})^4}{x^4y}\\ &= \frac{81xy^{-8/3}}{x^4y}\\ &= 81x^{1-4}y^{-8/3-3/3}\\ &= 81x^{-3}y^{-11/3}\\ &= \frac{81}{x^3y^{11/3}}\end{aligned}$$

5. a.
$$\begin{aligned}x^{3/5}(x^{1/3}-x^2) &= x^{3/5}x^{1/3} - x^{3/5}x^2\\ &= x^{(3/5+1/3)} - x^{(3/5+2)}\\ &= x^{(9/15+5/15)} - x^{(3/5+10/5)}\\ &= x^{14/15} - x^{13/5}\end{aligned}$$

b.
$$\begin{aligned}&(x^{1/2}+6)(x^{1/2}-2)\\ &= x^{2/2} - 2x^{1/2} + 6x^{1/2} - 12\\ &= x + 4x^{1/2} - 12\end{aligned}$$

6.
$$\begin{aligned}2x^{-1/5} - 7x^{4/5} &= (x^{-1/5})(2) - (x^{-1/5})(7x^{5/5})\\ &= x^{-1/5}(2-7x)\end{aligned}$$

7. a. $\sqrt[9]{x^3} = x^{3/9} = x^{1/3} = \sqrt[3]{x}$

b. $\sqrt[4]{36} = 36^{1/4} = (6^2)^{1/4} = 6^{2/4} = 6^{1/2} = \sqrt{6}$

c.
$$\begin{aligned}\sqrt[8]{a^4b^2} &= (a^4b^2)^{1/8}\\ &= a^{4/8}b^{2/8}\\ &= a^{2/4}b^{1/4}\\ &= (a^2b)^{1/4}\\ &= \sqrt[4]{a^2b}\end{aligned}$$

8. a.
$$\begin{aligned}\sqrt[3]{x} \cdot \sqrt[4]{x} &= x^{1/3} \cdot x^{1/4}\\ &= x^{1/3+1/4}\\ &= x^{4/12+3/12}\\ &= x^{7/12}\\ &= \sqrt[12]{x^7}\end{aligned}$$

b.
$$\begin{aligned}\frac{\sqrt[3]{y}}{\sqrt[5]{y}} &= \frac{y^{1/3}}{y^{1/5}}\\ &= y^{1/3-1/5}\\ &= y^{5/15-3/15}\\ &= y^{2/15}\\ &= \sqrt[15]{y^2}\end{aligned}$$

c.
$$\begin{aligned}\sqrt[3]{5} \cdot \sqrt{3} &= 5^{1/3} \cdot 3^{1/2}\\ &= 5^{2/6} \cdot 3^{3/6}\\ &= (5^2 \cdot 3^3)^{1/6}\\ &= \sqrt[6]{5^2 \cdot 3^3}\\ &= \sqrt[6]{675}\end{aligned}$$

Vocabulary, Readiness & Video Check 7.2

1. It is true that $9^{-1/2}$ is a positive number.

2. It is false that $9^{-1/2}$ is a whole number.

3. It is true that $\frac{1}{a^{-m/n}} = a^{m/n}$ (where $a^{m/n}$ is a nonzero real number).

4. To simplify $x^{2/3} \cdot x^{1/5}$, <u>add</u> the exponents; a.

5. To simplify $(x^{2/3})^{1/5}$, <u>multiply</u> the exponents; c.

6. To simplify $\frac{x^{2/3}}{x^{1/5}}$, <u>subtract</u> the exponents; b.

7. $-\sqrt[5]{3x}$

8. The numerator is the power; the denominator is the index.

9. A negative fractional exponent will move a base from the numerator to the <u>denominator</u> with the fractional exponent becoming <u>positive</u>.

10. If applying the product rule of exponents, you <u>add</u> the exponents. If applying the quotient rule of exponents, you <u>subtract</u> the exponents. If applying the power rule of exponents, you <u>multiply</u> the exponents.

11. Write the radical using an equivalent fractional exponent form, simplify the fraction, then write as a radical again.

Exercise Set 7.2

1. $49^{1/2} = \sqrt{49} = 7$

3. $27^{1/3} = \sqrt[3]{27} = 3$

5. $\left(\frac{1}{16}\right)^{1/4} = \sqrt[4]{\frac{1}{16}} = \frac{1}{2}$

7. $169^{1/2} = \sqrt{169} = 13$

9. $2m^{1/3} = 2\sqrt[3]{m}$

11. $(9x^4)^{1/2} = \sqrt{9x^4} = 3x^2$

13. $(-27)^{1/3} = \sqrt[3]{-27} = -3$

15. $-16^{1/4} = -\sqrt[4]{16} = -2$

17. $16^{3/4} = \left(\sqrt[4]{16}\right)^3 = 2^3 = 8$

19. $(-64)^{2/3} = \left(\sqrt[3]{-64}\right)^2 = (-4)^2 = 16$

21. $(-16)^{3/4} = \left(\sqrt[4]{-16}\right)^3$ is not a real number.

23. $(2x)^{3/5} = \sqrt[5]{(2x)^3}$ or $\left(\sqrt[5]{2x}\right)^3$

25. $(7x+2)^{2/3} = \sqrt[3]{(7x+2)^2}$ or $\left(\sqrt[3]{7x+2}\right)^2$

27. $\left(\frac{16}{9}\right)^{3/2} = \left(\sqrt{\frac{16}{9}}\right)^3 = \left(\frac{4}{3}\right)^3 = \frac{64}{27}$

29. $8^{-4/3} = \frac{1}{8^{4/3}} = \frac{1}{\left(\sqrt[3]{8}\right)^4} = \frac{1}{2^4} = \frac{1}{16}$

31. $(-64)^{-2/3} = \frac{1}{(-64)^{2/3}} = \frac{1}{\left(\sqrt[3]{-64}\right)^2} = \frac{1}{(-4)^2} = \frac{1}{16}$

33. $(-4)^{-3/2} = \frac{1}{(-4)^{3/2}} = \frac{1}{\left(\sqrt{-4}\right)^3}$ is not a real number.

35. $x^{-1/4} = \frac{1}{x^{1/4}}$

37. $\frac{1}{a^{-2/3}} = a^{2/3}$

39. $\frac{5}{7x^{-3/4}} = \frac{5x^{3/4}}{7}$

41. $a^{2/3}a^{5/3} = a^{2/3+5/3} = a^{7/3}$

43. $x^{-2/5} \cdot x^{7/5} = x^{-\frac{2}{5}+\frac{7}{5}} = x^{5/5} = x$

45. $3^{1/4} \cdot 3^{3/8} = 3^{\frac{1}{4}+\frac{3}{8}} = 3^{\frac{2}{8}+\frac{3}{8}} = 3^{5/8}$

47. $\frac{y^{1/3}}{y^{1/6}} = y^{\frac{1}{3}-\frac{1}{6}} = y^{\frac{2}{6}-\frac{1}{6}} = y^{1/6}$

49. $(4u^2)^{3/2} = 4^{3/2}u^{2(3/2)}$
$= \left(\sqrt{4}\right)^3 u^3$
$= 2^3u^3$
$= 8u^3$

51. $\dfrac{b^{1/2}b^{3/4}}{-b^{1/4}} = -b^{\frac{1}{2}+\frac{3}{4}-\frac{1}{4}} = -b^{\frac{2}{4}+\frac{3}{4}-\frac{1}{4}} = -b^1 = -b$

53. $\dfrac{(x^3)^{1/2}}{x^{7/2}} = \dfrac{x^{3/2}}{x^{7/2}}$
$= x^{3/2-7/2}$
$= x^{-2}$
$= \dfrac{1}{x^2}$

55. $\dfrac{(3x^{1/4})^3}{x^{1/12}} = \dfrac{3^3x^{3/4}}{x^{1/12}}$
$= 27x^{\frac{3}{4}-\frac{1}{12}}$
$= 27x^{\frac{9}{12}-\frac{1}{12}}$
$= 27x^{8/12}$
$= 27x^{2/3}$

57. $\dfrac{(y^3z)^{1/6}}{y^{-1/2}z^{1/3}} = \dfrac{y^{3/6}z^{1/6}}{y^{-1/2}z^{1/3}}$
$= y^{3/6-(-1/2)}z^{1/6-1/3}$
$= y^{1/2+1/2}z^{1/6-2/6}$
$= y^1z^{-1/6}$
$= \dfrac{y}{z^{1/6}}$

59. $\dfrac{(x^3y^2)^{1/4}}{(x^{-5}y^{-1})^{-1/2}} = \dfrac{x^{3/4}y^{2/4}}{x^{5/2}y^{1/2}}$
$= x^{\frac{3}{4}-\frac{5}{2}}y^{\frac{2}{4}-\frac{1}{2}}$
$= x^{\frac{3}{4}-\frac{10}{4}}y^{\frac{1}{2}-\frac{1}{2}}$
$= x^{-7/4}$
$= \dfrac{1}{x^{7/4}}$

61. $y^{1/2}(y^{1/2} - y^{2/3}) = y^{1/2}y^{1/2} - y^{1/2}y^{2/3}$
$= y^{1/2+1/2} - y^{1/2+2/3}$
$= y^1 - y^{7/6}$
$= y - y^{7/6}$

63. $x^{2/3}(x-2) = x \cdot x^{2/3} - 2x^{2/3}$
$= x^{1+2/3} - 2x^{2/3}$
$= x^{5/3} - 2x^{2/3}$

65. $(2x^{1/3}+3)(2x^{1/3}-3) = (2x^{1/3})^2 - 3^2$
$= 2^2(x^{1/3})^2 - 9$
$= 4x^{2/3} - 9$

67. $x^{8/3} + x^{10/3} = x^{8/3}(1) + x^{8/3}(x^{2/3})$
$= x^{8/3}(1+x^{2/3})$

69. $x^{2/5} - 3x^{1/5} = x^{1/5}(x^{1/5}) - x^{1/5}(3)$
$= x^{1/5}(x^{1/5} - 3)$

71. $5x^{-1/3} + x^{2/3} = x^{-1/3}(5) + x^{-1/3}(x^{3/3})$
$= x^{-1/3}(5+x)$

73. $\sqrt[6]{x^3} = x^{3/6} = x^{1/2} = \sqrt{x}$

75. $\sqrt[6]{4} = 4^{1/6} = (2^2)^{1/6} = 2^{1/3} = \sqrt[3]{2}$

77. $\sqrt[4]{16x^2} = (16x^2)^{1/4}$
$= 16^{1/4}x^{2/4} = 2x^{1/2} = 2\sqrt{x}$

79. $\sqrt[8]{x^4y^4} = (x^4y^4)^{1/8}$
$= x^{4/8}y^{4/8}$
$= x^{1/2}y^{1/2}$
$= (xy)^{1/2}$
$= \sqrt{xy}$

81. $\sqrt[12]{a^8b^4} = a^{8/12}b^{4/12}$
$= a^{2/3}b^{1/3}$
$= (a^2b)^{1/3}$
$= \sqrt[3]{a^2b}$

83. $\sqrt[4]{(x+3)^2} = (x+3)^{2/4} = (x+3)^{1/2} = \sqrt{x+3}$

85. $\sqrt[3]{y}\cdot\sqrt[5]{y^2} = y^{1/3}\cdot y^{2/5}$
$= y^{\frac{1}{3}+\frac{2}{5}}$
$= y^{\frac{5}{15}+\frac{6}{15}}$
$= y^{11/15}$
$= \sqrt[15]{y^{11}}$

87. $\dfrac{\sqrt[3]{b^2}}{\sqrt[4]{b}} = \dfrac{b^{2/3}}{b^{1/4}} = b^{\frac{2}{3}-\frac{1}{4}} = b^{\frac{8}{12}-\frac{3}{12}} = b^{5/12} = \sqrt[12]{b^5}$

89. $\sqrt[3]{x}\cdot\sqrt[4]{x}\cdot\sqrt[8]{x^3} = x^{1/3}\cdot x^{1/4}\cdot x^{3/8}$
$= x^{8/24}\cdot x^{6/24}\cdot x^{9/24}$
$= x^{23/24}$
$= \sqrt[24]{x^{23}}$

91. $\dfrac{\sqrt[3]{a^2}}{\sqrt[6]{a}} = \dfrac{a^{2/3}}{a^{1/6}}$
$= a^{\frac{2}{3}-\frac{1}{6}} = a^{\frac{4}{6}-\frac{1}{6}} = a^{3/6} = a^{1/2} = \sqrt{a}$

93. $\sqrt{3}\cdot\sqrt[3]{4} = 3^{1/2}\cdot 4^{1/3}$
$= 3^{3/6}\cdot 4^{2/6}$
$= (3^3\cdot 4^2)^{1/6}$
$= (27\cdot 16)^{1/6}$
$= (432)^{1/6}$
$= \sqrt[6]{432}$

95. $\sqrt[5]{7}\cdot\sqrt[3]{y} = 7^{1/5}\cdot y^{1/3}$
$= 7^{3/15}\cdot y^{5/15}$
$= (7^3\cdot y^5)^{1/15}$
$= (343y^5)^{1/15}$
$= \sqrt[15]{343y^5}$

97. $\sqrt{5r}\cdot\sqrt[3]{s} = (5r)^{1/2}\cdot s^{1/3}$
$= (5r)^{3/6}\cdot s^{2/6}$
$= [(5r)^3\cdot s^2]^{1/6}$
$= (125r^3s^2)^{1/6}$
$= \sqrt[6]{125r^3s^2}$

99. $75 = 25\cdot 3$ where 25 is a perfect square.

101. $48 = 4\cdot 12$ or $16\cdot 3$ where both 4 and 16 are perfect squares.

103. $16 = 8\cdot 2$ where 8 is a perfect cube.

105. $54 = 27\cdot 2$ where 27 is a perfect cube.

107. $4^{1/2} = 2$, A

109. $(-4)^{1/2}$ is not a real number, C

111. $-8^{1/3} = -2$, B

113. $B(w) = 70w^{3/4}$
$B(60) = 70(60)^{3/4} \approx 1509$
The BMR is 1509 calories.

115. 2015 is 5 years after 2010.
$f(x) = 236x^{1/20}$
$f(5) = 236(5)^{1/20} \approx 255.8$
The model predicts 255.8 million subscriptions in 2015.

117. answers may vary

119. $\square\cdot a^{2/3} = a^{3/3}$
$\square = \dfrac{a^{3/3}}{a^{2/3}}$
$\square = a^{3/3-2/3}$
$\square = a^{1/3}$

121. $\dfrac{\square}{x^{-2/5}} = x^{3/5}$
$x^{-2/5}\left(\dfrac{\square}{x^{-2/5}}\right) = x^{3/5}\cdot x^{-2/5}$
$\square = x^{3/5-2/5}$
$\square = x^{1/5}$

123. $8^{1/4} \approx 1.6818$

125. $18^{3/5} \approx 5.6645$

127. $\dfrac{\sqrt{t}}{\sqrt{u}} = \dfrac{t^{1/2}}{u^{1/2}}$

Section 7.3 Practice Exercises

1. a. $\sqrt{5}\cdot\sqrt{7}=\sqrt{5\cdot 7}=\sqrt{35}$

b. $\sqrt{13}\cdot\sqrt{z}=\sqrt{13z}$

c. $\sqrt[4]{125}\cdot\sqrt[4]{5}=\sqrt[4]{125\cdot 5}=\sqrt[4]{625}=5$

d. $\sqrt[3]{5y}\cdot\sqrt[3]{3x^2}=\sqrt[3]{5y\cdot 3x^2}=\sqrt[3]{15x^2y}$

e. $\sqrt{\frac{5}{m}}\cdot\sqrt{\frac{t}{2}}=\sqrt{\frac{5}{m}\cdot\frac{t}{2}}=\sqrt{\frac{5t}{2m}}$

2. a. $\sqrt{\frac{36}{49}}=\frac{\sqrt{36}}{\sqrt{49}}=\frac{6}{7}$

b. $\sqrt{\frac{z}{16}}=\frac{\sqrt{z}}{\sqrt{16}}=\frac{\sqrt{z}}{4}$

c. $\sqrt[3]{\frac{125}{8}}=\frac{\sqrt[3]{125}}{\sqrt[3]{8}}=\frac{5}{2}$

d. $\sqrt[4]{\frac{5}{81x^8}}=\frac{\sqrt[4]{5}}{\sqrt[4]{81x^8}}=\frac{\sqrt[4]{5}}{3x^2}$

3. a. $\sqrt{98}=\sqrt{49\cdot 2}=\sqrt{49}\cdot\sqrt{2}=7\sqrt{2}$

b. $\sqrt[3]{54}=\sqrt[3]{27\cdot 2}=\sqrt[3]{27}\cdot\sqrt[3]{2}=3\sqrt[3]{2}$

c. The largest perfect square factor of 35 is 1, so $\sqrt{35}$ cannot be simplified further.

d. $\sqrt[4]{243}=\sqrt[4]{81\cdot 3}=\sqrt[4]{81}\cdot\sqrt[4]{3}=3\sqrt[4]{3}$

4. a. $\sqrt{36z^7}=\sqrt{36z^6\cdot z}=\sqrt{36z^6}\cdot\sqrt{z}=6z^3\sqrt{z}$

b.
$$\begin{aligned}\sqrt[3]{32p^4q^7}&=\sqrt[3]{8\cdot 4\cdot p^3\cdot p\cdot q^6\cdot q}\\&=\sqrt[3]{8p^3q^6\cdot 4pq}\\&=\sqrt[3]{8p^3q^6}\cdot\sqrt[3]{4pq}\\&=2pq^2\sqrt[3]{4pq}\end{aligned}$$

c.
$$\begin{aligned}\sqrt[4]{16x^{15}}&=\sqrt[4]{16\cdot x^{12}\cdot x^3}\\&=\sqrt[4]{16x^{12}}\cdot\sqrt[4]{x^3}\\&=2x^3\sqrt[4]{x^3}\end{aligned}$$

5. a. $\frac{\sqrt{80}}{\sqrt{5}}=\sqrt{\frac{80}{5}}=\sqrt{16}=4$

b.
$$\begin{aligned}\frac{\sqrt{98z}}{3\sqrt{2}}&=\frac{1}{3}\cdot\sqrt{\frac{98z}{2}}\\&=\frac{1}{3}\cdot\sqrt{49z}\\&=\frac{1}{3}\cdot\sqrt{49}\cdot\sqrt{z}\\&=\frac{1}{3}\cdot 7\cdot\sqrt{z}\\&=\frac{7}{3}\sqrt{z}\end{aligned}$$

c.
$$\begin{aligned}\frac{5\sqrt[3]{40x^5y^7}}{\sqrt[3]{5y}}&=5\cdot\sqrt[3]{\frac{40x^5y^7}{5y}}\\&=5\cdot\sqrt[3]{8x^5y^6}\\&=5\cdot\sqrt[3]{8x^3y^6\cdot x^2}\\&=5\cdot\sqrt[3]{8x^3y^6}\cdot\sqrt[3]{x^2}\\&=5\cdot 2xy^2\cdot\sqrt[3]{x^2}\\&=10xy^2\sqrt[3]{x^2}\end{aligned}$$

d.
$$\begin{aligned}\frac{3\sqrt[5]{64x^9y^8}}{\sqrt[5]{x^{-1}y^2}}&=3\cdot\sqrt[5]{\frac{64x^9y^8}{x^{-1}y^2}}\\&=3\cdot\sqrt[5]{64x^{10}y^6}\\&=3\cdot\sqrt[5]{32\cdot x^{10}\cdot y^5\cdot 2\cdot y}\\&=3\cdot\sqrt[5]{32x^{10}y^5}\cdot\sqrt[5]{2y}\\&=3\cdot 2x^2y\cdot\sqrt[5]{2y}\\&=6x^2y\sqrt[5]{2y}\end{aligned}$$

6. Let $(x_1, y_1)=(-3, 7)$ and $(x_2, y_2)=(-2, 3)$.

$$\begin{aligned}d&=\sqrt{(x_2-x_1)^2+(y_2-y_1)^2}\\&=\sqrt{[-2-(-3)]^2+(3-7)^2}\\&=\sqrt{(1)^2+(-4)^2}\\&=\sqrt{1+16}\\&=\sqrt{17}\approx 4.123\end{aligned}$$

The distance between the two points is exactly $\sqrt{17}$ units, or approximately 4.123 units.

7. Let $(x_1, y_1) = (5, -2)$ and $(x_2, y_2) = (8, -6)$.

$$\text{midpoint} = \left(\frac{x_1+x_2}{2}, \frac{y_1+y_2}{2}\right) = \left(\frac{5+8}{2}, \frac{-2+(-6)}{2}\right) = \left(\frac{13}{2}, \frac{-8}{2}\right) = \left(\frac{13}{2}, -4\right)$$

The midpoint of the segment is $\left(\frac{13}{2}, -4\right)$.

Vocabulary, Readiness & Video Check 7.3

1. The midpoint of a line segment is a point exactly halfway between the two endpoints of the line segment.
2. The distance between two points is a distance, measured in units.
3. The distance formula is $d = \sqrt{(x_2-x_1)^2 + (y_2-y_1)^2}$.
4. The midpoint formula is $\left(\frac{x_1+x_2}{2}, \frac{y_1+y_2}{2}\right)$.
5. the indexes must be the same
6. If you see that simplifying can be done by separating a fraction radical into separate numerator and denominator radicands or by combining separate numerator and denominator radicands under one radical.
7. The power must be 1. Any even power is a perfect square and will leave no factor in the radicand; any higher odd power can have an even power factored from it, leaving one factor remaining in the radicand.
8. Be careful of signs since you're dealing with subtraction.
9. The x-value of the midpoint is the average of the x-values of the endpoints and the y-value of the midpoint is the average of the y-values of the endpoints.

Exercise Set 7.3

1. $\sqrt{7}\cdot\sqrt{2} = \sqrt{7\cdot 2} = \sqrt{14}$

3. $\sqrt[4]{8}\cdot\sqrt[4]{2} = \sqrt[4]{8\cdot 2} = \sqrt[4]{16} = 2$

5. $\sqrt[3]{4}\cdot\sqrt[3]{9} = \sqrt[3]{4\cdot 9} = \sqrt[3]{36}$

7. $\sqrt{2}\cdot\sqrt{3x} = \sqrt{2\cdot 3x} = \sqrt{6x}$

9. $\sqrt{\frac{7}{x}}\cdot\sqrt{\frac{2}{y}} = \sqrt{\frac{7}{x}\cdot\frac{2}{y}} = \sqrt{\frac{14}{xy}}$

11. $\sqrt[4]{4x^3}\cdot\sqrt[4]{5} = \sqrt[4]{4x^3\cdot 5} = \sqrt[4]{20x^3}$

13. $\sqrt{\frac{6}{49}} = \frac{\sqrt{6}}{\sqrt{49}} = \frac{\sqrt{6}}{7}$

15. $\sqrt{\frac{2}{49}} = \frac{\sqrt{2}}{\sqrt{49}} = \frac{\sqrt{2}}{7}$

17. $\sqrt[4]{\frac{x^3}{16}} = \frac{\sqrt[4]{x^3}}{\sqrt[4]{16}} = \frac{\sqrt[4]{x^3}}{2}$

19. $\sqrt[3]{\frac{4}{27}} = \frac{\sqrt[3]{4}}{\sqrt[3]{27}} = \frac{\sqrt[3]{4}}{3}$

21. $\sqrt[4]{\frac{8}{x^8}} = \frac{\sqrt[4]{8}}{\sqrt[4]{x^8}} = \frac{\sqrt[4]{8}}{x^2}$

23. $$\sqrt[3]{\frac{2x}{81y^{12}}} = \frac{\sqrt[3]{2x}}{\sqrt[3]{81y^{12}}} = \frac{\sqrt[3]{2x}}{\sqrt[3]{27y^{12}}\cdot\sqrt[3]{3}} = \frac{\sqrt[3]{2x}}{3y^4\sqrt[3]{3}}$$

25. $\sqrt{\frac{x^2y}{100}} = \frac{\sqrt{x^2y}}{\sqrt{100}} = \frac{\sqrt{x^2}\sqrt{y}}{10} = \frac{x\sqrt{y}}{10}$

27. $\sqrt{\frac{5x^2}{4y^2}} = \frac{\sqrt{5x^2}}{\sqrt{4y^2}} = \frac{\sqrt{5}\sqrt{x^2}}{2y} = \frac{x\sqrt{5}}{2y}$

29. $-\sqrt[3]{\frac{z^7}{27x^3}} = -\frac{\sqrt[3]{z^7}}{\sqrt[3]{27x^3}} = -\frac{\sqrt[3]{z^6\cdot z}}{3x} = -\frac{z^2\sqrt[3]{z}}{3x}$

31. $\sqrt{32} = \sqrt{16\cdot 2} = \sqrt{16}\cdot\sqrt{2} = 4\sqrt{2}$

33. $\sqrt[3]{192}=\sqrt[3]{64\cdot 3}=\sqrt[3]{64}\cdot\sqrt[3]{3}=4\sqrt[3]{3}$

35. $5\sqrt{75}=5\sqrt{25\cdot 3}=5\sqrt{25}\cdot\sqrt{3}=5(5)\sqrt{3}=25\sqrt{3}$

37. $\sqrt{24}=\sqrt{4\cdot 6}=\sqrt{4}\cdot\sqrt{6}=2\sqrt{6}$

39. $\sqrt{100x^5}=\sqrt{100x^4\cdot x}=\sqrt{100x^4}\cdot\sqrt{x}=10x^2\sqrt{x}$

41. $\sqrt[3]{16y^7}=\sqrt[3]{8y^6\cdot 2y}=\sqrt[3]{8y^6}\cdot\sqrt[3]{2y}=2y^2\sqrt[3]{2y}$

43. $\sqrt[4]{a^8b^7}=\sqrt[4]{a^8b^4\cdot b^3}=\sqrt[4]{a^8b^4}\cdot\sqrt[4]{b^3}=a^2b\sqrt[4]{b^3}$

45. $\sqrt{y^5}=\sqrt{y^4\cdot y}=\sqrt{y^4}\cdot\sqrt{y}=y^2\sqrt{y}$

47. $\sqrt{25a^2b^3}=\sqrt{25a^2b^2\cdot b}$
$=\sqrt{25a^2b^2}\cdot\sqrt{b}$
$=5ab\sqrt{b}$

49. $\sqrt[5]{-32x^{10}y}=\sqrt[5]{-32x^{10}\cdot y}$
$=\sqrt[5]{-32x^{10}}\cdot\sqrt[5]{y}$
$=-2x^2\sqrt[5]{y}$

51. $\sqrt[3]{50x^{14}}=\sqrt[3]{x^{12}\cdot 50x^2}$
$=\sqrt[3]{x^{12}}\cdot\sqrt[3]{50x^2}$
$=x^4\sqrt[3]{50x^2}$

53. $-\sqrt{32a^8b^7}=-\sqrt{16a^8b^6\cdot 2b}$
$=-\sqrt{16a^8b^6}\cdot\sqrt{2b}$
$=-4a^4b^3\sqrt{2b}$

55. $\sqrt{9x^7y^9}=\sqrt{9x^6y^8\cdot xy}$
$=\sqrt{9x^6y^8}\cdot\sqrt{xy}$
$=3x^3y^4\sqrt{xy}$

57. $\sqrt[3]{125r^9s^{12}}=5r^3s^4$

59. $\sqrt[4]{32x^{12}y^5}=\sqrt[4]{16x^{12}y^4}\cdot\sqrt[4]{2y}=2x^3y\sqrt[4]{2y}$

61. $\frac{\sqrt{14}}{\sqrt{7}}=\sqrt{\frac{14}{7}}=\sqrt{2}$

63. $\frac{\sqrt[3]{24}}{\sqrt[3]{3}}=\sqrt[3]{\frac{24}{3}}=\sqrt[3]{8}=2$

65. $\frac{5\sqrt[4]{48}}{\sqrt[4]{3}}=5\sqrt[4]{\frac{48}{3}}=5\sqrt[4]{16}=5(2)=10$

67. $\frac{\sqrt{x^5y^3}}{\sqrt{xy}}=\sqrt{\frac{x^5y^3}{xy}}=\sqrt{x^4y^2}=x^2y$

69. $\frac{8\sqrt[3]{54m^7}}{\sqrt[3]{2m}}=8\sqrt[3]{\frac{54m^7}{2m}}$
$=8\sqrt[3]{27m^6}$
$=8(3m^2)$
$=24m^2$

71. $\frac{3\sqrt{100x^2}}{2\sqrt{2x^{-1}}}=\frac{3}{2}\sqrt{\frac{100x^2}{2x^{-1}}}$
$=\frac{3}{2}\sqrt{50x^3}$
$=\frac{3}{2}\sqrt{25x^2\cdot 2x}$
$=\frac{3}{2}(5x)\sqrt{2x}$
$=\frac{15x}{2}\sqrt{2x}$

73. $\frac{\sqrt[4]{96a^{10}b^3}}{\sqrt[4]{3a^2b^3}}=\sqrt[4]{\frac{96a^{10}b^3}{3a^2b^3}}$
$=\sqrt[4]{32a^8}$
$=\sqrt[4]{16a^8\cdot 2}$
$=2a^2\sqrt[4]{2}$

75. $\frac{\sqrt[5]{64x^{10}y^3}}{\sqrt[5]{2x^3y^{-7}}}=\sqrt[5]{\frac{64x^{10}y^3}{2x^3y^{-7}}}$
$=\sqrt[5]{32x^7y^{10}}$
$=\sqrt[5]{32x^5y^{10}}\cdot\sqrt[5]{x^2}$
$=2xy^2\sqrt[5]{x^2}$

77. (5, 1), (8, 5)
$$\begin{aligned} d &= \sqrt{(8-5)^2+(5-1)^2} \\ &= \sqrt{3^2+4^2} \\ &= \sqrt{9+16} \\ &= \sqrt{25} \\ &= 5 \text{ units} \end{aligned}$$

79. (–3, 2), (1, –3)
$$\begin{aligned} d &= \sqrt{[1-(-3)]^2+(-3-2)^2} \\ &= \sqrt{4^2+(-5)^2} \\ &= \sqrt{16+25} \\ &= \sqrt{41} \approx 6.403 \text{ units} \end{aligned}$$

81. (–9, 4), (–8, 1)
$$\begin{aligned} d &= \sqrt{[-8-(-9)]^2+(1-4)^2} \\ &= \sqrt{1^2+(-3)^2} \\ &= \sqrt{1+9} \\ &= \sqrt{10} \approx 3.162 \text{ units} \end{aligned}$$

83. $\left(0, -\sqrt{2}\right), \left(\sqrt{3}, 0\right)$
$$\begin{aligned} d &= \sqrt{\left(\sqrt{3}-0\right)^2+\left[0-\left(-\sqrt{2}\right)\right]^2} \\ &= \sqrt{\left(\sqrt{3}\right)^2+\left(\sqrt{2}\right)^2} \\ &= \sqrt{3+2} \\ &= \sqrt{5} \approx 2.236 \text{ units} \end{aligned}$$

85. (1.7, –3.6), (–8.6, 5.7)
$$\begin{aligned} d &= \sqrt{(-8.6-1.7)^2+[5.7-(-3.6)]^2} \\ &= \sqrt{(-10.3)^2+(9.3)^2} \\ &= \sqrt{192.58} \approx 13.877 \text{ units} \end{aligned}$$

87. (6, –8), (2, 4)
$$\left(\frac{6+2}{2}, \frac{-8+4}{2}\right) = \left(\frac{8}{2}, \frac{-4}{2}\right) = (4, -2)$$
The midpoint of the segment is (4, –2).

89. (–2, –1), (–8, 6)
$$\left(\frac{-2+(-8)}{2}, \frac{-1+6}{2}\right) = \left(\frac{-10}{2}, \frac{5}{2}\right) = \left(-5, \frac{5}{2}\right)$$
The midpoint of the segment is $\left(-5, \frac{5}{2}\right)$.

91. (7, 3), (–1, –3)
$$\left(\frac{7+(-1)}{2}, \frac{3+(-3)}{2}\right) = \left(\frac{6}{2}, \frac{0}{2}\right) = (3, 0)$$
The midpoint of the segment is (3, 0).

93. $\left(\frac{1}{2}, \frac{3}{8}\right), \left(-\frac{3}{2}, \frac{5}{8}\right)$
$$\left(\frac{\frac{1}{2}+\left(-\frac{3}{2}\right)}{2}, \frac{\frac{3}{8}+\frac{5}{8}}{2}\right) = \left(\frac{-1}{2}, \frac{1}{2}\right)$$
The midpoint of the segment is $\left(-\frac{1}{2}, \frac{1}{2}\right)$.

95. $\left(\sqrt{2}, 3\sqrt{5}\right), \left(\sqrt{2}, -2\sqrt{5}\right)$
$$\begin{aligned} \left(\frac{\sqrt{2}+\sqrt{2}}{2}, \frac{3\sqrt{5}+\left(-2\sqrt{5}\right)}{2}\right) &= \left(\frac{2\sqrt{2}}{2}, \frac{\sqrt{5}}{2}\right) \\ &= \left(\sqrt{2}, \frac{\sqrt{5}}{2}\right) \end{aligned}$$
The midpoint of the segment is $\left(\sqrt{2}, \frac{\sqrt{5}}{2}\right)$.

97. (4.6, –3.5), (7.8, –9.8)
$$\begin{aligned} \left(\frac{4.6+7.8}{2}, \frac{-3.5+(-9.8)}{2}\right) &= \left(\frac{12.4}{2}, \frac{-13.3}{2}\right) \\ &= (6.2, -6.65) \end{aligned}$$
The midpoint of the segment is (6.2, –6.65).

99. $6x + 8x = (6 + 8)x = 14x$

101.
$$\begin{aligned} (2x+3)(x-5) &= 2x^2-10x+3x-15 \\ &= 2x^2-7x-15 \end{aligned}$$

103. $9y^2-8y^2=(9-8)y^2=1y^2=y^2$

105. $-3(x+5)=-3x-3(5)=-3x-15$

107.
$$\begin{aligned} (x-4)^2 &= x^2-2(x)(4)+4^2 \\ &= x^2-8x+16 \end{aligned}$$

109. The statement $\sqrt[n]{a}\cdot\sqrt[n]{b}=\sqrt[n]{ab}$ is <u>true</u>.

111. The statement $\sqrt[3]{7}\cdot\sqrt{11}=\sqrt{77}$ is <u>false</u>.

113. The statement $\frac{\sqrt[n]{a}}{\sqrt[n]{b}}=\sqrt[n]{\frac{a}{b}}$ is <u>true</u>.

115. $\frac{\sqrt[3]{64}}{\sqrt{64}}=\frac{4}{8}=\frac{1}{2}$

117. $\sqrt[5]{x^{35}}=x^7$

119. $\sqrt[4]{a^{12}b^4c^{20}}=a^3bc^5$

121. $\sqrt[3]{z^{32}}=\sqrt[3]{z^{30}\cdot z^2}=\sqrt[3]{z^{30}}\cdot\sqrt[3]{z^2}=z^{10}\sqrt[3]{z^2}$

123.
$$\begin{aligned}\sqrt[7]{q^{17}r^{40}s^7}&=\sqrt[7]{q^{14}\cdot q^3\cdot r^{35}\cdot r^5\cdot s^7}\\&=\sqrt[7]{q^{14}r^{35}s^7\cdot q^3r^5}\\&=q^2r^5s\sqrt[7]{q^3r^5}\end{aligned}$$

125. $r=\sqrt{\frac{A}{4\pi}}=\sqrt{\frac{32.17}{4\pi}}\approx\sqrt{2.56}=1.6$

The radius of a standard zorb is 1.6 meters.

127. $A=\pi r\sqrt{r^2+h^2}$

a.
$$\begin{aligned}A&=\pi(4)\sqrt{4^2+3^2}\\&=4\pi\sqrt{16+9}\\&=4\pi\sqrt{25}\\&=4\pi(5)\\&=20\pi\end{aligned}$$

The area is 20π square centimeters.

b.
$$\begin{aligned}A&=\pi(6.8)\sqrt{(6.8)^2+(7.2)^2}\\&=6.8\pi\sqrt{46.24+51.84}\\&=6.8\pi\sqrt{98.08}\\&\approx211.57\end{aligned}$$

The area is approximately 211.57 square feet.

Section 7.4 Practice Exercises

1. a. $3\sqrt{17}+5\sqrt{17}=(3+5)\sqrt{17}=8\sqrt{17}$

b. $7\sqrt[3]{5z}-12\sqrt[3]{5z}=(7-12)\sqrt[3]{5z}=-5\sqrt[3]{5z}$

c. $3\sqrt{2}+5\sqrt[3]{2}$

This expression cannot be simplified since $3\sqrt{2}$ and $5\sqrt[3]{2}$ do not contain like radicals.

2. a.
$$\begin{aligned}\sqrt{24}+3\sqrt{54}&=\sqrt{4\cdot6}+3\sqrt{9\cdot6}\\&=\sqrt{4}\cdot\sqrt{6}+3\cdot\sqrt{9}\cdot\sqrt{6}\\&=2\cdot\sqrt{6}+3\cdot3\cdot\sqrt{6}\\&=2\sqrt{6}+9\sqrt{6}\\&=11\sqrt{6}\end{aligned}$$

b.
$$\begin{aligned}&\sqrt[3]{24}-4\sqrt[3]{81}+\sqrt[3]{3}\\&=\sqrt[3]{8}\cdot\sqrt[3]{3}-4\cdot\sqrt[3]{27}\cdot\sqrt[3]{3}+\sqrt[3]{3}\\&=2\cdot\sqrt[3]{3}-4\cdot3\cdot\sqrt[3]{3}+\sqrt[3]{3}\\&=2\sqrt[3]{3}-12\sqrt[3]{3}+\sqrt[3]{3}\\&=-9\sqrt[3]{3}\end{aligned}$$

c.
$$\begin{aligned}&\sqrt{75x}-3\sqrt{27x}+\sqrt{12x}\\&=\sqrt{25}\cdot\sqrt{3x}-3\cdot\sqrt{9}\cdot\sqrt{3x}+\sqrt{4}\cdot\sqrt{3x}\\&=5\cdot\sqrt{3x}-3\cdot3\cdot\sqrt{3x}+2\cdot\sqrt{3x}\\&=5\sqrt{3x}-9\sqrt{3x}+2\sqrt{3x}\\&=-2\sqrt{3x}\end{aligned}$$

d.
$$\begin{aligned}\sqrt{40}+\sqrt[3]{40}&=\sqrt{4}\cdot\sqrt{10}+\sqrt[3]{8}\cdot\sqrt[3]{5}\\&=2\sqrt{10}+2\sqrt[3]{5}\end{aligned}$$

e.
$$\begin{aligned}\sqrt[3]{81x^4}+\sqrt[3]{3x^4}&=\sqrt[3]{27x^3}\cdot\sqrt[3]{3x}+\sqrt[3]{x^3}\cdot\sqrt[3]{3x}\\&=3x\sqrt[3]{3x}+x\sqrt[3]{3x}\\&=4x\sqrt[3]{3x}\end{aligned}$$

3. a.
$$\begin{aligned}\frac{\sqrt{28}}{3}-\frac{\sqrt{7}}{4}&=\frac{2\sqrt{7}}{3}-\frac{\sqrt{7}}{4}\\&=\frac{2\sqrt{7}\cdot4}{3\cdot4}-\frac{\sqrt{7}\cdot3}{4\cdot3}\\&=\frac{8\sqrt{7}}{12}-\frac{3\sqrt{7}}{12}\\&=\frac{5\sqrt{7}}{12}\end{aligned}$$

b.
$$\begin{aligned}\sqrt[3]{\frac{6y}{64}}+3\sqrt[3]{6y}&=\frac{\sqrt[3]{6y}}{\sqrt[3]{64}}+3\sqrt[3]{6y}\\&=\frac{\sqrt[3]{6y}}{4}+3\sqrt[3]{6y}\\&=\frac{\sqrt[3]{6y}}{4}+\frac{3\sqrt[3]{6y}\cdot4}{4}\\&=\frac{\sqrt[3]{6y}}{4}+\frac{12\sqrt[3]{6y}}{4}\\&=\frac{13\sqrt[3]{6y}}{4}\end{aligned}$$

4. a. $\sqrt{5}(2+\sqrt{15}) = \sqrt{5}(2)+\sqrt{5}(\sqrt{15})$
$= 2\sqrt{5}+\sqrt{5\cdot 15}$
$= 2\sqrt{5}+\sqrt{5\cdot 5\cdot 3}$
$= 2\sqrt{5}+5\sqrt{3}$

b. $(\sqrt{2}-\sqrt{5})(\sqrt{6}+2)$
$= \sqrt{2}\cdot\sqrt{6}+\sqrt{2}\cdot 2-\sqrt{5}\cdot\sqrt{6}-\sqrt{5}\cdot 2$
$= \sqrt{2\cdot 2\cdot 3}+2\sqrt{2}-\sqrt{30}-2\sqrt{5}$
$= 2\sqrt{3}+2\sqrt{2}-\sqrt{30}-2\sqrt{5}$

c. $(3\sqrt{z}-4)(2\sqrt{z}+3)$
$= 3\sqrt{z}(2\sqrt{z})+3\sqrt{z}(3)-4(2\sqrt{z})-4(3)$
$= 6\cdot z+9\sqrt{z}-8\sqrt{z}-12$
$= 6z+\sqrt{z}-12$

d. $(\sqrt{6}-3)^2 = (\sqrt{6}-3)(\sqrt{6}-3)$
$= \sqrt{6}(\sqrt{6})-\sqrt{6}(3)-3(\sqrt{6})-3(-3)$
$= 6-3\sqrt{6}-3\sqrt{6}+9$
$= 6-6\sqrt{6}+9$
$= 15-6\sqrt{6}$

e. $(\sqrt{5x}+3)(\sqrt{5x}-3)$
$= \sqrt{5x}\cdot\sqrt{5x}-3\sqrt{5x}+3\sqrt{5x}-3\cdot 3$
$= 5x-9$

f. $(\sqrt{x+2}+3)^2 = (\sqrt{x+2})^2+2\cdot\sqrt{x+2}\cdot 3+3^2$
$= x+2+6\sqrt{x+2}+9$
$= x+11+6\sqrt{x+2}$

Vocabulary, Readiness & Video Check 7.4

1. The terms $\sqrt{7}$ and $\sqrt[3]{7}$ are unlike terms.

2. The terms $\sqrt[3]{x^2y}$ and $\sqrt[3]{yx^2}$ are like terms.

3. The terms $\sqrt[3]{abc}$ and $\sqrt[3]{cba}$ are like terms.

4. The terms $2x\sqrt{5}$ and $2x\sqrt{10}$ are unlike terms.

5. $2\sqrt{3}+4\sqrt{3} = \underline{6\sqrt{3}}$

6. $5\sqrt{7}+3\sqrt{7} = \underline{8\sqrt{7}}$

7. $8\sqrt{x}-\sqrt{x} = \underline{7\sqrt{x}}$

8. $3\sqrt{y}-\sqrt{y} = \underline{2\sqrt{y}}$

9. $7\sqrt[3]{x}+\sqrt[3]{x} = \underline{8\sqrt[3]{x}}$

10. $8\sqrt[3]{z}+\sqrt[3]{z} = \underline{9\sqrt[3]{z}}$

11. Sometimes you can't see that there are like radicals until you simplify, so you may incorrectly think you cannot add or subtract if you don't simplify first.

12. The square root of a positive number times the square root of the same positive number is that positive number.

Exercise Set 7.4

1. $\sqrt{8}-\sqrt{32} = \sqrt{4\cdot 2}-\sqrt{16\cdot 2}$
$= \sqrt{4}\cdot\sqrt{2}-\sqrt{16}\cdot\sqrt{2}$
$= 2\sqrt{2}-4\sqrt{2}$
$= -2\sqrt{2}$

3. $2\sqrt{2x^3}+4x\sqrt{8x} = 2\sqrt{x^2\cdot 2x}+4x\sqrt{4\cdot 2x}$
$= 2\sqrt{x^2}\cdot\sqrt{2x}+4x\sqrt{4}\cdot\sqrt{2x}$
$= 2x\sqrt{2x}+4x(2)\sqrt{2x}$
$= 2x\sqrt{2x}+8x\sqrt{2x}$
$= 10x\sqrt{2x}$

5. $2\sqrt{50}-3\sqrt{125}+\sqrt{98}$
$= 2\sqrt{25\cdot 2}-3\sqrt{25\cdot 5}+\sqrt{49\cdot 2}$
$= 2\sqrt{25}\cdot\sqrt{2}-3\sqrt{25}\cdot\sqrt{5}+\sqrt{49}\cdot\sqrt{2}$
$= 2(5)\sqrt{2}-3(5)\sqrt{5}+7\sqrt{2}$
$= 10\sqrt{2}-15\sqrt{5}+7\sqrt{2}$
$= 17\sqrt{2}-15\sqrt{5}$

7. $\sqrt[3]{16x}-\sqrt[3]{54x} = \sqrt[3]{8\cdot 2x}-\sqrt[3]{27\cdot 2x}$
$= \sqrt[3]{8}\cdot\sqrt[3]{2x}-\sqrt[3]{27}\cdot\sqrt[3]{2x}$
$= 2\sqrt[3]{2x}-3\sqrt[3]{2x}$
$= -\sqrt[3]{2x}$

9. $\sqrt{9b^3}-\sqrt{25b^3}+\sqrt{49b^3}$
$= \sqrt{9b^2\cdot b}-\sqrt{25b^2\cdot b}+\sqrt{49b^2\cdot b}$
$= \sqrt{9b^2}\cdot\sqrt{b}-\sqrt{25b^2}\cdot\sqrt{b}+\sqrt{49b^2}\cdot\sqrt{b}$
$= 3b\sqrt{b}-5b\sqrt{b}+7b\sqrt{b}$
$= 5b\sqrt{b}$

11. $\frac{5\sqrt{2}}{3}+\frac{2\sqrt{2}}{5}=\frac{5(5\sqrt{2})+3(2\sqrt{2})}{3(5)}$
$=\frac{25\sqrt{2}+6\sqrt{2}}{15}$
$=\frac{31\sqrt{2}}{15}$

13. $\sqrt[3]{\frac{11}{8}}-\frac{\sqrt[3]{11}}{6}=\frac{\sqrt[3]{11}}{\sqrt[3]{8}}-\frac{\sqrt[3]{11}}{6}$
$=\frac{\sqrt[3]{11}}{2}-\frac{\sqrt[3]{11}}{6}$
$=\frac{3\sqrt[3]{11}-\sqrt[3]{11}}{6}$
$=\frac{2\sqrt[3]{11}}{6}$
$=\frac{\sqrt[3]{11}}{3}$

15. $\frac{\sqrt{20x}}{9}+\sqrt{\frac{5x}{9}}=\frac{\sqrt{4\cdot 5x}}{9}+\frac{\sqrt{5x}}{\sqrt{9}}$
$=\frac{2\sqrt{5x}}{9}+\frac{\sqrt{5x}}{3}$
$=\frac{2\sqrt{5x}+3\sqrt{5x}}{9}$
$=\frac{5\sqrt{5x}}{9}$

17. $7\sqrt{9}-7+\sqrt{3}=7(3)-7+\sqrt{3}$
$=21-7+\sqrt{3}$
$=14+\sqrt{3}$

19. $2+3\sqrt{y^2}-6\sqrt{y^2}+5=2+3y-6y+5$
$=7-3y$

21. $3\sqrt{108}-2\sqrt{18}-3\sqrt{48}$
$=3\sqrt{36\cdot 3}-2\sqrt{9\cdot 2}-3\sqrt{16\cdot 3}$
$=3\sqrt{36}\cdot\sqrt{3}-2\sqrt{9}\cdot\sqrt{2}-3\sqrt{16}\cdot\sqrt{3}$
$=3(6)\sqrt{3}-2(3)\sqrt{2}-3(4)\sqrt{3}$
$=18\sqrt{3}-6\sqrt{2}-12\sqrt{3}$
$=6\sqrt{3}-6\sqrt{2}$

23. $-5\sqrt[3]{625}+\sqrt[3]{40}=-5\sqrt[3]{125\cdot 5}+\sqrt[3]{8\cdot 5}$
$=-5(5)\sqrt[3]{5}+2\sqrt[3]{5}$
$=-25\sqrt[3]{5}+2\sqrt[3]{5}$
$=-23\sqrt[3]{5}$

25. $a^3\sqrt{9ab^3}-\sqrt{25a^7b^3}+\sqrt{16a^7b^3}$
$=a^3\sqrt{9b^2\cdot ab}-\sqrt{25a^6b^2\cdot ab}+\sqrt{16a^6b^2\cdot ab}$
$=a^3\cdot 3b\sqrt{ab}-5a^3b\sqrt{ab}+4a^3b\sqrt{ab}$
$=3a^3b\sqrt{ab}-5a^3b\sqrt{ab}+4a^3b\sqrt{ab}$
$=2a^3b\sqrt{ab}$

27. $5y\sqrt{8y}+2\sqrt{50y^3}=5y\sqrt{4\cdot 2y}+2\sqrt{25y^2\cdot 2y}$
$=5y(2)\sqrt{2y}+2(5y)\sqrt{2y}$
$=10y\sqrt{2y}+10y\sqrt{2y}$
$=20y\sqrt{2y}$

29. $\sqrt[3]{54xy^3}-5\sqrt[3]{2xy^3}+y\sqrt[3]{128x}$
$=\sqrt[3]{27y^3\cdot 2x}-5\sqrt[3]{y^3\cdot 2x}+y\sqrt[3]{64\cdot 2x}$
$=3y\sqrt[3]{2x}-5y\sqrt[3]{2x}+4y\sqrt[3]{2x}$
$=2y\sqrt[3]{2x}$

31. $6\sqrt[3]{11}+8\sqrt{11}-12\sqrt{11}=6\sqrt[3]{11}-4\sqrt{11}$

33. $-2\sqrt[4]{x^7}+3\sqrt[4]{16x^7}-x\sqrt[4]{x^3}$
$=-2\sqrt[4]{x^4\cdot x^3}+3\sqrt[4]{16x^4\cdot x^3}-x\sqrt[4]{x^3}$
$=-2x\sqrt[4]{x^3}+3(2x)\sqrt[4]{x^3}-x\sqrt[4]{x^3}$
$=-2x\sqrt[4]{x^3}+6x\sqrt[4]{x^3}-x\sqrt[4]{x^3}$
$=3x\sqrt[4]{x^3}$

35. $\frac{4\sqrt{3}}{3}-\frac{\sqrt{12}}{3}=\frac{4\sqrt{3}}{3}-\frac{\sqrt{4\cdot 3}}{3}$
$=\frac{4\sqrt{3}-2\sqrt{3}}{3}$
$=\frac{2\sqrt{3}}{3}$

37. $\frac{\sqrt[3]{8x^4}}{7}+\frac{3x\sqrt[3]{x}}{7}=\frac{\sqrt[3]{8x^3\cdot x}}{7}+\frac{3x\sqrt[3]{x}}{7}$
$=\frac{2x\sqrt[3]{x}+3x\sqrt[3]{x}}{7}$
$=\frac{5x\sqrt[3]{x}}{7}$

39. $\sqrt{\frac{28}{x^2}}+\sqrt{\frac{7}{4x^2}}=\frac{\sqrt{28}}{\sqrt{x^2}}+\frac{\sqrt{7}}{\sqrt{4x^2}}$
$=\frac{\sqrt{4\cdot 7}}{x}+\frac{\sqrt{7}}{2x}$
$=\frac{2\sqrt{7}}{x}+\frac{\sqrt{7}}{2x}$
$=\frac{2(2\sqrt{7})+\sqrt{7}}{2x}$
$=\frac{4\sqrt{7}+\sqrt{7}}{2x}$
$=\frac{5\sqrt{7}}{2x}$

41. $\sqrt[3]{\frac{16}{27}}-\frac{\sqrt[3]{54}}{6}=\frac{\sqrt[3]{8\cdot 2}}{\sqrt[3]{27}}-\frac{\sqrt[3]{27\cdot 2}}{6}$
$=\frac{2\sqrt[3]{2}}{3}-\frac{3\sqrt[3]{2}}{6}$
$=\frac{2(2\sqrt[3]{2})-3\sqrt[3]{2}}{6}$
$=\frac{4\sqrt[3]{2}-3\sqrt[3]{2}}{6}$
$=\frac{\sqrt[3]{2}}{6}$

43. $-\frac{\sqrt[3]{2x^4}}{9}+\sqrt[3]{\frac{250x^4}{27}}=-\frac{\sqrt[3]{x^3\cdot 2x}}{9}+\frac{\sqrt[3]{125x^3\cdot 2x}}{\sqrt[3]{27}}$
$=\frac{-x\sqrt[3]{2x}}{9}+\frac{5x\sqrt[3]{2x}}{3}$
$=\frac{-x\sqrt[3]{2x}+3(5x\sqrt[3]{2x})}{9}$
$=\frac{-x\sqrt[3]{2x}+15x\sqrt[3]{2x}}{9}$
$=\frac{14x\sqrt[3]{2x}}{9}$

45. $P=2\sqrt{12}+\sqrt{12}+2\sqrt{27}+3\sqrt{3}$
$=2\sqrt{4\cdot 3}+\sqrt{4\cdot 3}+2\sqrt{9\cdot 3}+3\sqrt{3}$
$=2(2)\sqrt{3}+2\sqrt{3}+2(3)\sqrt{3}+3\sqrt{3}$
$=4\sqrt{3}+2\sqrt{3}+6\sqrt{3}+3\sqrt{3}$
$=15\sqrt{3}$ inches

47. $\sqrt{7}(\sqrt{5}+\sqrt{3})=\sqrt{7}\sqrt{5}+\sqrt{7}\sqrt{3}$
$=\sqrt{35}+\sqrt{21}$

49. $(\sqrt{5}-\sqrt{2})^2=(\sqrt{5})^2-2\sqrt{5}\sqrt{2}+(\sqrt{2})^2$
$=5-2\sqrt{10}+2$
$=7-2\sqrt{10}$

51. $\sqrt{3x}(\sqrt{3}-\sqrt{x})=\sqrt{3x}\sqrt{3}-\sqrt{3x}\sqrt{x}$
$=\sqrt{9x}-\sqrt{3x^2}$
$=3\sqrt{x}-x\sqrt{3}$

53. $(2\sqrt{x}-5)(3\sqrt{x}+1)$
$=2\sqrt{x}(3\sqrt{x})+2\sqrt{x}\cdot 1-5(3\sqrt{x})-5(1)$
$=6x+2\sqrt{x}-15\sqrt{x}-5$
$=6x-13\sqrt{x}-5$

55. $(\sqrt[3]{a}-4)(\sqrt[3]{a}+5)$
$=\sqrt[3]{a}(\sqrt[3]{a})+\sqrt[3]{a}\cdot 5-4\sqrt[3]{a}-4(5)$
$=\sqrt[3]{a^2}+5\sqrt[3]{a}-4\sqrt[3]{a}-20$
$=\sqrt[3]{a^2}+\sqrt[3]{a}-20$

57. $6(\sqrt{2}-2)=6\sqrt{2}-6(2)=6\sqrt{2}-12$

59. $\sqrt{2}(\sqrt{2}+x\sqrt{6})=\sqrt{2}\sqrt{2}+\sqrt{2}(x\sqrt{6})$
$=2+x\sqrt{12}$
$=2+x\sqrt{4\cdot 3}$
$=2+2x\sqrt{3}$

61. $(2\sqrt{7}+3\sqrt{5})(\sqrt{7}-2\sqrt{5})$
$=2\sqrt{7}\sqrt{7}+2\sqrt{7}(-2\sqrt{5})+3\sqrt{5}\sqrt{7}+3\sqrt{5}(-2\sqrt{5})$
$=2(7)-4\sqrt{35}+3\sqrt{35}-6(5)$
$=14-\sqrt{35}-30$
$=-16-\sqrt{35}$

63. $(\sqrt{x}-y)(\sqrt{x}+y)=(\sqrt{x})^2-y^2=x-y^2$

65. $(\sqrt{3}+x)^2=(\sqrt{3})^2+2\sqrt{3}\cdot x+x^2$
$=3+2x\sqrt{3}+x^2$

67. $\left(\sqrt{5x}-2\sqrt{3x}\right)\left(\sqrt{5x}-3\sqrt{3x}\right)$
$=\left(\sqrt{5x}\right)^2-\sqrt{5x}\left(3\sqrt{3x}\right)-2\sqrt{3x}\left(\sqrt{5x}\right)$
$-2\sqrt{3x}\left(-3\sqrt{3x}\right)$
$=5x-3x\sqrt{15}-2x\sqrt{15}+6\cdot 3x$
$=23x-5x\sqrt{15}$

69. $\left(\sqrt[3]{4}+2\right)\left(\sqrt[3]{2}-1\right)$
$=\sqrt[3]{4}\left(\sqrt[3]{2}\right)+\sqrt[3]{4}\cdot(-1)+2\sqrt[3]{2}+2(-1)$
$=\sqrt[3]{8}-\sqrt[3]{4}+2\sqrt[3]{2}-2$
$=2-\sqrt[3]{4}+2\sqrt[3]{2}-2$
$=2\sqrt[3]{2}-\sqrt[3]{4}$

71. $\left(\sqrt[3]{x}+1\right)\left(\sqrt[3]{x^2}-\sqrt[3]{x}+1\right)$
$=\sqrt[3]{x}\left(\sqrt[3]{x^2}\right)-\sqrt[3]{x}\left(\sqrt[3]{x}\right)+\sqrt[3]{x}(1)$
$+1\left(\sqrt[3]{x^2}\right)-1\left(\sqrt[3]{x}\right)+1(1)$
$=\sqrt[3]{x^3}-\sqrt[3]{x^2}+\sqrt[3]{x}+\sqrt[3]{x^2}-\sqrt[3]{x}+1$
$=x+1$

73. $\left(\sqrt{x-1}+5\right)^2=\left(\sqrt{x-1}\right)^2+2\sqrt{x-1}\cdot 5+5^2$
$=(x-1)+10\sqrt{x-1}+25$
$=x+10\sqrt{x-1}+24$

75. $\left(\sqrt{2x+5}-1\right)^2=\left(\sqrt{2x+5}\right)^2-2\sqrt{2x+5}\cdot 1+1^2$
$=(2x+5)-2\sqrt{2x+5}+1$
$=2x-2\sqrt{2x+5}+6$

77. $\dfrac{2x-14}{2}=\dfrac{2(x-7)}{2}=x-7$

79. $\dfrac{7x-7y}{x^2-y^2}=\dfrac{7(x-y)}{(x+y)(x-y)}=\dfrac{7}{x+y}$

81. $\dfrac{6a^2b-9ab}{3ab}=\dfrac{3ab(2a-3)}{3ab}=2a-3$

83. $\dfrac{-4+2\sqrt{3}}{6}=\dfrac{2\left(-2+\sqrt{3}\right)}{6}=\dfrac{-2+\sqrt{3}}{3}$

85. $P=2l+2w$
$=2\left(3\sqrt{20}\right)+2\left(\sqrt{125}\right)$
$=6\sqrt{4\cdot 5}+2\sqrt{25\cdot 5}$
$=6(2)\sqrt{5}+2(5)\sqrt{5}$
$=12\sqrt{5}+10\sqrt{5}$
$=22\sqrt{5}$ feet
$A=lw$
$=\left(3\sqrt{20}\right)\left(\sqrt{125}\right)$
$=3\sqrt{4\cdot 5}\sqrt{25\cdot 5}$
$=3(2)\sqrt{5}\cdot 5\sqrt{5}$
$=30\cdot 5$
$=150$ square feet

87. a. $\sqrt{3}+\sqrt{3}=2\sqrt{3}$

b. $\sqrt{3}\cdot\sqrt{3}=\sqrt{9}=3$

c. answers may vary

89. $\left(\sqrt{2}+\sqrt{3}-1\right)^2$
$=\left[\left(\sqrt{2}+\sqrt{3}\right)-1\right]^2$
$=\left(\sqrt{2}+\sqrt{3}\right)^2-2\left(\sqrt{2}+\sqrt{3}\right)+1^2$
$=\left(\sqrt{2}\right)^2+2\sqrt{2}\sqrt{3}+\left(\sqrt{3}\right)^2-2\sqrt{2}-2\sqrt{3}+1$
$=2+2\sqrt{6}+3-2\sqrt{2}-2\sqrt{3}+1$
$=6+2\sqrt{6}-2\sqrt{2}-2\sqrt{3}$

91. answers may vary

Section 7.5 Practice Exercises

1. a. $\dfrac{5}{\sqrt{3}}=\dfrac{5\cdot\sqrt{3}}{\sqrt{3}\cdot\sqrt{3}}=\dfrac{5\sqrt{3}}{3}$

b. $\dfrac{3\sqrt{25}}{\sqrt{4x}}=\dfrac{3(5)}{2\sqrt{x}}=\dfrac{15}{2\sqrt{x}}=\dfrac{15\cdot\sqrt{x}}{2\sqrt{x}\cdot\sqrt{x}}=\dfrac{15\sqrt{x}}{2x}$

c. $\sqrt[3]{\dfrac{2}{9}}=\dfrac{\sqrt[3]{2}}{\sqrt[3]{9}}=\dfrac{\sqrt[3]{2}\cdot\sqrt[3]{3}}{\sqrt[3]{3^2}\cdot\sqrt[3]{3}}=\dfrac{\sqrt[3]{6}}{3}$

2. $\sqrt{\dfrac{3z}{5y}}=\dfrac{\sqrt{3z}}{\sqrt{5y}}=\dfrac{\sqrt{3z}\cdot\sqrt{5y}}{\sqrt{5y}\cdot\sqrt{5y}}=\dfrac{\sqrt{15yz}}{5y}$

3. $\dfrac{\sqrt[3]{z^2}}{\sqrt[3]{27x^4}} = \dfrac{\sqrt[3]{z^2}}{\sqrt[3]{27x^3}\cdot\sqrt[3]{x}}$
$= \dfrac{\sqrt[3]{z^2}}{3x\sqrt[3]{x}}$
$= \dfrac{\sqrt[3]{z^2}\cdot\sqrt[3]{x^2}}{3x\sqrt[3]{x}\cdot\sqrt[3]{x^2}}$
$= \dfrac{\sqrt[3]{z^2x^2}}{3x\sqrt[3]{x^3}}$
$= \dfrac{\sqrt[3]{x^2z^2}}{3x^2}$

4. a. $\dfrac{5}{3\sqrt{5}+2} = \dfrac{5(3\sqrt{5}-2)}{(3\sqrt{5}+2)(3\sqrt{5}-2)}$
$= \dfrac{5(3\sqrt{5}-2)}{(3\sqrt{5})^2-2^2}$
$= \dfrac{5(3\sqrt{5}-2)}{45-4}$
$= \dfrac{5(3\sqrt{5}-2)}{41}$

b. $\dfrac{\sqrt{2}+5}{\sqrt{3}-\sqrt{5}} = \dfrac{(\sqrt{2}+5)(\sqrt{3}+\sqrt{5})}{(\sqrt{3}-\sqrt{5})(\sqrt{3}+\sqrt{5})}$
$= \dfrac{\sqrt{2}\sqrt{3}+\sqrt{2}\sqrt{5}+5\sqrt{3}+5\sqrt{5}}{(\sqrt{3})^2-(\sqrt{5})^2}$
$= \dfrac{\sqrt{6}+\sqrt{10}+5\sqrt{3}+5\sqrt{5}}{3-5}$
$= \dfrac{\sqrt{6}+\sqrt{10}+5\sqrt{3}+5\sqrt{5}}{-2}$

c. $\dfrac{3\sqrt{x}}{2\sqrt{x}+\sqrt{y}} = \dfrac{3\sqrt{x}(2\sqrt{x}-\sqrt{y})}{(2\sqrt{x}+\sqrt{y})(2\sqrt{x}-\sqrt{y})}$
$= \dfrac{6\sqrt{x^2}-3\sqrt{xy}}{(2\sqrt{x})^2-(\sqrt{y})^2}$
$= \dfrac{6x-3\sqrt{xy}}{4x-y}$

5. $\dfrac{\sqrt{32}}{\sqrt{80}} = \dfrac{\sqrt{16\cdot 2}}{\sqrt{16\cdot 5}} = \dfrac{4\sqrt{2}}{4\sqrt{5}} = \dfrac{\sqrt{2}}{\sqrt{5}} = \dfrac{\sqrt{2}\cdot\sqrt{2}}{\sqrt{5}\cdot\sqrt{2}} = \dfrac{2}{\sqrt{10}}$

6. $\dfrac{\sqrt[3]{5b}}{\sqrt[3]{2a}} = \dfrac{\sqrt[3]{5b}\cdot\sqrt[3]{25b^2}}{\sqrt[3]{2a}\cdot\sqrt[3]{25b^2}} = \dfrac{\sqrt[3]{125b^3}}{\sqrt[3]{50ab^2}} = \dfrac{5b}{\sqrt[3]{50ab^2}}$

7. $\dfrac{\sqrt{x}-3}{4} = \dfrac{(\sqrt{x}-3)(\sqrt{x}+3)}{4(\sqrt{x}+3)}$
$= \dfrac{(\sqrt{x})^2-(3)^2}{4(\sqrt{x}+3)}$
$= \dfrac{x-9}{4(\sqrt{x}+3)}$

Vocabulary, Readiness & Video Check 7.5

1. The conjugate of $a + b$ is $a - b$.
2. The process of writing an equivalent expression, but without a radical in the denominator, is called rationalizing the denominator.
3. The process of writing an equivalent expression, but without a radical in the numerator, is called rationalizing the numerator.
4. To rationalize the denominator of $\dfrac{5}{\sqrt{3}}$, we multiply by $\dfrac{\sqrt{3}}{\sqrt{3}}$.
5. To write an equivalent expression without a radical in the denominator.
6. Using the FOIL order to multiply, the Outer product and the Inner product are opposites and they will subtract out.
7. No, except for the fact you're working with numerators, the process is the same.

Exercise Set 7.5

1. $\dfrac{\sqrt{2}}{\sqrt{7}} = \dfrac{\sqrt{2}\cdot\sqrt{7}}{\sqrt{7}\cdot\sqrt{7}} = \dfrac{\sqrt{14}}{\sqrt{49}} = \dfrac{\sqrt{14}}{7}$

3. $\sqrt{\dfrac{1}{5}} = \dfrac{\sqrt{1}}{\sqrt{5}} = \dfrac{1\cdot\sqrt{5}}{\sqrt{5}\cdot\sqrt{5}} = \dfrac{\sqrt{5}}{5}$

5. $\sqrt{\dfrac{4}{x}} = \dfrac{\sqrt{4}}{\sqrt{x}} = \dfrac{2\cdot\sqrt{x}}{\sqrt{x}\cdot\sqrt{x}} = \dfrac{2\sqrt{x}}{\sqrt{x^2}} = \dfrac{2\sqrt{x}}{x}$

7. $\frac{4}{\sqrt[3]{3}} = \frac{4 \cdot \sqrt[3]{9}}{\sqrt[3]{3} \cdot \sqrt[3]{9}} = \frac{4\sqrt[3]{9}}{\sqrt[3]{27}} = \frac{4\sqrt[3]{9}}{3}$

9. $\frac{3}{\sqrt{8x}} = \frac{3 \cdot \sqrt{2x}}{\sqrt{8x} \cdot \sqrt{2x}} = \frac{3\sqrt{2x}}{\sqrt{16x^2}} = \frac{3\sqrt{2x}}{4x}$

11. $\frac{3}{\sqrt[3]{4x^2}} = \frac{3 \cdot \sqrt[3]{2x}}{\sqrt[3]{4x^2} \cdot \sqrt[3]{2x}} = \frac{3\sqrt[3]{2x}}{\sqrt[3]{8x^3}} = \frac{3\sqrt[3]{2x}}{2x}$

13. $\frac{9}{\sqrt{3a}} = \frac{9 \cdot \sqrt{3a}}{\sqrt{3a} \cdot \sqrt{3a}} = \frac{9\sqrt{3a}}{3a} = \frac{3\sqrt{3a}}{a}$

15. $\frac{3}{\sqrt[3]{2}} = \frac{3 \cdot \sqrt[3]{4}}{\sqrt[3]{2} \cdot \sqrt[3]{4}} = \frac{3\sqrt[3]{4}}{\sqrt[3]{8}} = \frac{3\sqrt[3]{4}}{2}$

17. $\frac{2\sqrt{3}}{\sqrt{7}} = \frac{2\sqrt{3} \cdot \sqrt{7}}{\sqrt{7} \cdot \sqrt{7}} = \frac{2\sqrt{21}}{\sqrt{49}} = \frac{2\sqrt{21}}{7}$

19. $\sqrt{\frac{2x}{5y}} = \frac{\sqrt{2x}}{\sqrt{5y}} = \frac{\sqrt{2x} \cdot \sqrt{5y}}{\sqrt{5y} \cdot \sqrt{5y}} = \frac{\sqrt{10xy}}{5y}$

21. $\sqrt[3]{\frac{3}{5}} = \frac{\sqrt[3]{3}}{\sqrt[3]{5}} \cdot \frac{\sqrt[3]{25}}{\sqrt[3]{25}} = \frac{\sqrt[3]{75}}{5}$

23.
$$\sqrt{\frac{3x}{50}} = \frac{\sqrt{3x}}{\sqrt{50}} = \frac{\sqrt{3x}}{5\sqrt{2}} = \frac{\sqrt{3x} \cdot \sqrt{2}}{5\sqrt{2} \cdot \sqrt{2}} = \frac{\sqrt{6x}}{5 \cdot 2} = \frac{\sqrt{6x}}{10}$$

25. $\frac{1}{\sqrt{12z}} = \frac{1}{\sqrt{4 \cdot 3z}} = \frac{1}{2\sqrt{3z}} \cdot \frac{\sqrt{3z}}{\sqrt{3z}} = \frac{\sqrt{3z}}{6z}$

27. $\frac{\sqrt[3]{2y^2}}{\sqrt[3]{9x^2}} = \frac{\sqrt[3]{2y^2} \cdot \sqrt[3]{3x}}{\sqrt[3]{9x^2} \cdot \sqrt[3]{3x}} = \frac{\sqrt[3]{6xy^2}}{3x}$

29. $\sqrt[4]{\frac{81}{8}} = \frac{\sqrt[4]{81}}{\sqrt[4]{8}} = \frac{3 \cdot \sqrt[4]{2}}{\sqrt[4]{8} \cdot \sqrt[4]{2}} = \frac{3\sqrt[4]{2}}{\sqrt[4]{16}} = \frac{3\sqrt[4]{2}}{2}$

31. $\sqrt[4]{\frac{16}{9x^7}} = \frac{\sqrt[4]{16}}{\sqrt[4]{9x^7}} = \frac{2 \cdot \sqrt[4]{9x}}{\sqrt[4]{9x^7} \cdot \sqrt[4]{9x}} = \frac{2\sqrt[4]{9x}}{\sqrt[4]{81x^8}} = \frac{2\sqrt[4]{9x}}{3x^2}$

33.
$$\frac{5a}{\sqrt[5]{8a^9b^{11}}} = \frac{5a \cdot \sqrt[5]{4ab^4}}{\sqrt[5]{8a^9b^{11}} \cdot \sqrt[5]{4ab^4}} = \frac{5a\sqrt[5]{4ab^4}}{\sqrt[5]{32a^{10}b^{15}}} = \frac{5a\sqrt[5]{4ab^4}}{2a^2b^3} = \frac{5\sqrt[5]{4ab^4}}{2ab^3}$$

35. The conjugate of $\sqrt{2}+x$ is $\sqrt{2}-x$.

37. The conjugate of $5-\sqrt{a}$ is $5+\sqrt{a}$.

39. The conjugate of $-7\sqrt{5}+8\sqrt{x}$ is $-7\sqrt{5}-8\sqrt{x}$.

41.
$$\frac{6}{2-\sqrt{7}} = \frac{6(2+\sqrt{7})}{(2-\sqrt{7})(2+\sqrt{7})} = \frac{6(2+\sqrt{7})}{2^2-(\sqrt{7})^2} = \frac{6(2+\sqrt{7})}{4-7} = \frac{6(2+\sqrt{7})}{-3} = -2(2+\sqrt{7})$$

43.
$$\frac{-7}{\sqrt{x}-3} = \frac{-7(\sqrt{x}+3)}{(\sqrt{x}-3)(\sqrt{x}+3)} = \frac{-7(\sqrt{x}+3)}{(\sqrt{x})^2-(3)^2} = \frac{-7(\sqrt{x}+3)}{x-9} \text{ or } \frac{7(\sqrt{x}+3)}{9-x}$$

45. $\dfrac{\sqrt{2}-\sqrt{3}}{\sqrt{2}+\sqrt{3}}=\dfrac{\left(\sqrt{2}-\sqrt{3}\right)\left(\sqrt{2}-\sqrt{3}\right)}{\left(\sqrt{2}+\sqrt{3}\right)\left(\sqrt{2}-\sqrt{3}\right)}$

$=\dfrac{\left(\sqrt{2}\right)^2-2\sqrt{2}\sqrt{3}+\left(\sqrt{3}\right)^2}{\left(\sqrt{2}\right)^2-\left(\sqrt{3}\right)^2}$

$=\dfrac{2-2\sqrt{6}+3}{2-3}$

$=\dfrac{5-2\sqrt{6}}{-1}$

$=-5+2\sqrt{6}$

47. $\dfrac{\sqrt{a}+1}{2\sqrt{a}-\sqrt{b}}$

$=\dfrac{\left(\sqrt{a}+1\right)\left(2\sqrt{a}+\sqrt{b}\right)}{\left(2\sqrt{a}-\sqrt{b}\right)\left(2\sqrt{a}+\sqrt{b}\right)}$

$=\dfrac{\sqrt{a}\cdot 2\sqrt{a}+\sqrt{a}\sqrt{b}+1\cdot 2\sqrt{a}+1\cdot\sqrt{b}}{\left(2\sqrt{a}\right)^2-\left(\sqrt{b}\right)^2}$

$=\dfrac{2a+\sqrt{ab}+2\sqrt{a}+\sqrt{b}}{4a-b}$

49. $\dfrac{8}{1+\sqrt{10}}=\dfrac{8\left(1-\sqrt{10}\right)}{\left(1+\sqrt{10}\right)\left(1-\sqrt{10}\right)}$

$=\dfrac{8\left(1-\sqrt{10}\right)}{1^2-\left(\sqrt{10}\right)^2}$

$=\dfrac{8\left(1-\sqrt{10}\right)}{1-10}$

$=-\dfrac{8\left(1-\sqrt{10}\right)}{9}$

51. $\dfrac{\sqrt{x}}{\sqrt{x}+\sqrt{y}}=\dfrac{\sqrt{x}\left(\sqrt{x}-\sqrt{y}\right)}{\left(\sqrt{x}+\sqrt{y}\right)\left(\sqrt{x}-\sqrt{y}\right)}$

$=\dfrac{\sqrt{x}\left(\sqrt{x}-\sqrt{y}\right)}{\left(\sqrt{x}\right)^2-\left(\sqrt{y}\right)^2}$

$=\dfrac{\sqrt{x}\left(\sqrt{x}-\sqrt{y}\right)}{x-y}$

$=\dfrac{\sqrt{x}\sqrt{x}-\sqrt{x}\sqrt{y}}{x-y}$

$=\dfrac{x-\sqrt{xy}}{x-y}$

53. $\dfrac{2\sqrt{3}+\sqrt{6}}{4\sqrt{3}-\sqrt{6}}=\dfrac{\left(2\sqrt{3}+\sqrt{6}\right)\left(4\sqrt{3}+\sqrt{6}\right)}{\left(4\sqrt{3}-\sqrt{6}\right)\left(4\sqrt{3}+\sqrt{6}\right)}$

$=\dfrac{8\cdot 3+2\sqrt{18}+4\sqrt{18}+6}{\left(4\sqrt{3}\right)^2-\left(\sqrt{6}\right)^2}$

$=\dfrac{30+6\sqrt{18}}{16\cdot 3-6}$

$=\dfrac{30+6(3)\sqrt{2}}{42}$

$=\dfrac{30+18\sqrt{2}}{42}$

$=\dfrac{6\left(5+3\sqrt{2}\right)}{42}$

$=\dfrac{5+3\sqrt{2}}{7}$

55. $\sqrt{\dfrac{5}{3}}=\dfrac{\sqrt{5}}{\sqrt{3}}=\dfrac{\sqrt{5}\cdot\sqrt{5}}{\sqrt{3}\cdot\sqrt{5}}=\dfrac{\sqrt{25}}{\sqrt{15}}=\dfrac{5}{\sqrt{15}}$

57. $\sqrt{\dfrac{18}{5}}=\dfrac{\sqrt{18}}{\sqrt{5}}$

$=\dfrac{\sqrt{9}\cdot\sqrt{2}}{\sqrt{5}}$

$=\dfrac{3\sqrt{2}}{\sqrt{5}}$

$=\dfrac{3\sqrt{2}\cdot\sqrt{2}}{\sqrt{5}\cdot\sqrt{2}}$

$=\dfrac{3\cdot 2}{\sqrt{10}}$

$=\dfrac{6}{\sqrt{10}}$

59. $\frac{\sqrt{4x}}{7}=\frac{2\sqrt{x}}{7}=\frac{2\sqrt{x}\cdot\sqrt{x}}{7\cdot\sqrt{x}}=\frac{2\sqrt{x^2}}{7\sqrt{x}}=\frac{2x}{7\sqrt{x}}$

61. $\frac{\sqrt[3]{5y^2}}{\sqrt[3]{4x}}=\frac{\sqrt[3]{5y^2}\cdot\sqrt[3]{5^2y}}{\sqrt[3]{4x}\cdot\sqrt[3]{5^2y}}=\frac{\sqrt[3]{5^3y^3}}{\sqrt[3]{100xy}}=\frac{5y}{\sqrt[3]{100xy}}$

63. $\sqrt{\frac{2}{5}}=\frac{\sqrt{2}}{\sqrt{5}}=\frac{\sqrt{2}\cdot\sqrt{2}}{\sqrt{5}\cdot\sqrt{2}}=\frac{\sqrt{4}}{\sqrt{10}}=\frac{2}{\sqrt{10}}$

65. $\frac{\sqrt{2x}}{11}=\frac{\sqrt{2x}\cdot\sqrt{2x}}{11\cdot\sqrt{2x}}=\frac{\sqrt{4x^2}}{11\sqrt{2x}}=\frac{2x}{11\sqrt{2x}}$

67. $\sqrt[3]{\frac{7}{8}}=\frac{\sqrt[3]{7}}{\sqrt[3]{8}}$
$=\frac{\sqrt[3]{7}}{2}$
$=\frac{\sqrt[3]{7}\cdot\sqrt[3]{7^2}}{2\cdot\sqrt[3]{7^2}}$
$=\frac{\sqrt[3]{7^3}}{2\sqrt[3]{49}}$
$=\frac{7}{2\sqrt[3]{49}}$

69. $\frac{\sqrt[3]{3x^5}}{10}=\frac{\sqrt[3]{x^3\cdot 3x^2}}{10}$
$=\frac{x\sqrt[3]{3x^2}}{10}$
$=\frac{x\sqrt[3]{3x^2}\cdot\sqrt[3]{3^2x}}{10\cdot\sqrt[3]{3^2x}}$
$=\frac{x\sqrt[3]{3^3x^3}}{10\sqrt[3]{9x}}$
$=\frac{x\cdot 3x}{10\sqrt[3]{9x}}$
$=\frac{3x^2}{10\sqrt[3]{9x}}$

71. $\sqrt{\frac{18x^4y^6}{3z}}=\frac{\sqrt{18x^4y^6}}{\sqrt{3z}}$
$=\frac{\sqrt{9x^4y^6\cdot 2}}{\sqrt{3z}}$
$=\frac{3x^2y^3\sqrt{2}}{\sqrt{3z}}$
$=\frac{3x^2y^3\sqrt{2}\cdot\sqrt{2}}{\sqrt{3z}\cdot\sqrt{2}}$
$=\frac{3x^2y^3\cdot 2}{\sqrt{6z}}$
$=\frac{6x^2y^3}{\sqrt{6z}}$

73. $\frac{2-\sqrt{11}}{6}=\frac{(2-\sqrt{11})(2+\sqrt{11})}{6(2+\sqrt{11})}$
$=\frac{4-11}{12+6\sqrt{11}}$
$=\frac{-7}{12+6\sqrt{11}}$

75. $\frac{2-\sqrt{7}}{-5}=\frac{(2-\sqrt{7})(2+\sqrt{7})}{-5(2+\sqrt{7})}$
$=\frac{4-7}{-5(2+\sqrt{7})}$
$=\frac{-3}{-5(2+\sqrt{7})}$
$=\frac{3}{5(2+\sqrt{7})}$
$=\frac{3}{10+5\sqrt{7}}$

77. $\frac{\sqrt{x}+3}{\sqrt{x}}=\frac{(\sqrt{x}+3)(\sqrt{x}-3)}{\sqrt{x}(\sqrt{x}-3)}$
$=\frac{\sqrt{x^2}-9}{\sqrt{x^2}-3\sqrt{x}}$
$=\frac{x-9}{x-3\sqrt{x}}$

79. $\frac{\sqrt{2}-1}{\sqrt{2}+1}=\frac{(\sqrt{2}-1)(\sqrt{2}+1)}{(\sqrt{2}+1)(\sqrt{2}+1)}$
$=\frac{\sqrt{4}-1}{\sqrt{4}+2\sqrt{2}+1}$
$=\frac{2-1}{2+2\sqrt{2}+1}$
$=\frac{1}{3+2\sqrt{2}}$

81. $\frac{\sqrt{x}+1}{\sqrt{x}-1}=\frac{(\sqrt{x}+1)(\sqrt{x}-1)}{(\sqrt{x}-1)(\sqrt{x}-1)}$
$=\frac{\sqrt{x^2}-1}{\sqrt{x^2}-2\sqrt{x}+1}$
$=\frac{x-1}{x-2\sqrt{x}+1}$

83. $2x-7=3(x-4)$
$2x-7=3x-12$
$-x-7=-12$
$-x=-5$
$x=5$
The solution is 5.

85. $(x-6)(2x+1)=0$
$x-6=0$ or $2x+1=0$
$x=6$ or $2x=-1$
$x=-\frac{1}{2}$
The solutions are $-\frac{1}{2}$, 6.

87. $x^2-8x=-12$
$x^2-8x+12=0$
$(x-6)(x-2)=0$
$x-6=0$ or $x-2=0$
$x=6$ or $x=2$
The solutions are 2, 6.

89. $r=\sqrt{\frac{A}{4\pi}}$
$=\frac{\sqrt{A}}{\sqrt{4\pi}}$
$=\frac{\sqrt{A}}{2\sqrt{\pi}}$
$=\frac{\sqrt{A}\cdot\sqrt{\pi}}{2\sqrt{\pi}\cdot\sqrt{\pi}}$
$=\frac{\sqrt{A\pi}}{2\pi}$

91. a. $\frac{\sqrt{5y^3}\cdot\sqrt{12x^3}}{\sqrt{12x^3}\cdot\sqrt{12x^3}}=\frac{\sqrt{60x^3y^3}}{\sqrt{(12x^3)^2}}$
$=\frac{2xy\sqrt{15xy}}{12x^3}$
$=\frac{y\sqrt{15xy}}{6x^2}$

b. $\frac{\sqrt{5y^3}\cdot\sqrt{3x}}{\sqrt{12x^3}\cdot\sqrt{3x}}=\frac{\sqrt{15xy^3}}{\sqrt{36x^4}}=\frac{y\sqrt{15xy}}{6x^2}$

c. answers may vary

93. $\frac{9}{\sqrt[3]{5}}=\frac{9}{\sqrt[3]{5}}\cdot\frac{\sqrt[3]{25}}{\sqrt[3]{25}}=\frac{9\sqrt[3]{25}}{\sqrt[3]{125}}=\frac{9\sqrt[3]{25}}{5}$
The smallest number is $\sqrt[3]{25}$.

95. answers may vary

97. answers may vary

Integrated Review

1. $\sqrt{81}=9$ because $9^2=81$.

2. $\sqrt[3]{-8}=-2$ because $(-2)^3=-8$.

3. $\sqrt[4]{\frac{1}{16}}=\frac{1}{2}$ because $\left(\frac{1}{2}\right)^4=\frac{1}{16}$.

4. $\sqrt{x^6}=x^3$ because $(x^3)^2=x^6$.

5. $\sqrt[3]{y^9}=y^3$ because $(y^3)^3=y^9$.

6. $\sqrt{4y^{10}}=2y^5$ because $(2y^5)^2=4y^{10}$.

7. $\sqrt[5]{-32y^5} = -2y$ because $(-2y)^5 = -32y^5$.

8. $\sqrt[4]{81b^{12}} = 3b^3$ because $(3b^3)^4 = 81b^{12}$.

9. $36^{1/2} = \sqrt{36} = 6$

10. $(3y)^{1/4} = \sqrt[4]{3y}$

11. $64^{-2/3} = \dfrac{1}{\left(\sqrt[3]{64}\right)^2} = \dfrac{1}{4^2} = \dfrac{1}{16}$

12. $(x+1)^{3/5} = \sqrt[5]{(x+1)^3}$

13. $y^{-1/6} \cdot y^{7/6} = y^{-\frac{1}{6}+\frac{7}{6}} = y^{6/6} = y$

14.
$$\begin{aligned}\frac{(2x^{1/3})^4}{x^{5/6}} &= 16x^{4/3}x^{-5/6}\\ &= 16x^{\frac{8}{6}-\frac{5}{6}}\\ &= 16x^{3/6}\\ &= 16x^{1/2}\end{aligned}$$

15. $\dfrac{x^{1/4}x^{3/4}}{x^{-1/4}} = x^{\frac{1}{4}+\frac{3}{4}+\frac{1}{4}} = x^{5/4}$

16. $4^{1/3} \cdot 4^{2/5} = 4^{\frac{1}{3}+\frac{2}{5}} = 4^{\frac{5}{15}+\frac{6}{15}} = 4^{11/15}$

17. $\sqrt[3]{8x^6} = (8x^6)^{1/3} = (2^3x^6)^{1/3} = 2^{3/3}x^{6/3} = 2x^2$

18.
$$\begin{aligned}\sqrt[12]{a^9b^6} &= (a^9b^6)^{1/12}\\ &= a^{9/12}b^{6/12}\\ &= a^{3/4}b^{1/2}\\ &= a^{3/4}b^{2/4}\\ &= (a^3b^2)^{1/4}\\ &= \sqrt[4]{a^3b^2}\end{aligned}$$

19. $\sqrt[4]{x} \cdot \sqrt{x} = x^{1/4} \cdot x^{1/2} = x^{\frac{1}{4}+\frac{2}{4}} = x^{3/4} = \sqrt[4]{x^3}$

20.
$$\begin{aligned}\sqrt{5} \cdot \sqrt[3]{2} &= 5^{1/2} \cdot 2^{1/3}\\ &= 5^{3/6} \cdot 2^{2/6}\\ &= (5^3 \cdot 2^2)^{1/6}\\ &= \sqrt[6]{5^3 \cdot 2^2}\\ &= \sqrt[6]{500}\end{aligned}$$

21. $\sqrt{40} = \sqrt{4}\sqrt{10} = 2\sqrt{10}$

22. $\sqrt[4]{16x^7y^{10}} = \sqrt[4]{16x^4y^8}\sqrt[4]{x^3y^2} = 2xy^2\sqrt[4]{x^3y^2}$

23. $\sqrt[3]{54x^4} = \sqrt[3]{27x^3}\sqrt[3]{2x} = 3x\sqrt[3]{2x}$

24. $\sqrt[5]{-64b^{10}} = \sqrt[5]{-32b^{10}}\sqrt[5]{2} = -2b^2\sqrt[5]{2}$

25. $\sqrt{5} \cdot \sqrt{x} = \sqrt{5x}$

26. $\sqrt[3]{8x} \cdot \sqrt[3]{8x^2} = \sqrt[3]{64x^3} = 4x$

27.
$$\begin{aligned}\frac{\sqrt{98y^6}}{\sqrt{2y}} &= \sqrt{\frac{98y^6}{2y}}\\ &= \sqrt{49y^5}\\ &= \sqrt{49y^4} \cdot \sqrt{y}\\ &= 7y^2\sqrt{y}\end{aligned}$$

28.
$$\begin{aligned}\frac{\sqrt[4]{48a^9b^3}}{\sqrt[4]{ab^3}} &= \sqrt[4]{\frac{48a^9b^3}{ab^3}}\\ &= \sqrt[4]{48a^8}\\ &= \sqrt[4]{16a^8} \cdot \sqrt[4]{3}\\ &= 2a^2\sqrt[4]{3}\end{aligned}$$

29.
$$\begin{aligned}\sqrt{20} - \sqrt{75} + 5\sqrt{7} &= \sqrt{4}\sqrt{5} - \sqrt{25}\sqrt{3} + 5\sqrt{7}\\ &= 2\sqrt{5} - 5\sqrt{3} + 5\sqrt{7}\end{aligned}$$

30.
$$\begin{aligned}\sqrt[3]{54y^4} - y\sqrt[3]{16y} &= \sqrt[3]{27y^3}\sqrt[3]{2y} - y\sqrt[3]{8}\sqrt[3]{2y}\\ &= 3y\sqrt[3]{2y} - 2y\sqrt[3]{2y}\\ &= y\sqrt[3]{2y}\end{aligned}$$

31. $\sqrt{3}\left(\sqrt{5} - \sqrt{2}\right) = \sqrt{3}\sqrt{5} - \sqrt{3}\sqrt{2} = \sqrt{15} - \sqrt{6}$

32.
$$\begin{aligned}\left(\sqrt{7} + \sqrt{3}\right)^2 &= \left(\sqrt{7}\right)^2 + 2\sqrt{7}\sqrt{3} + \left(\sqrt{3}\right)^2\\ &= 7 + 2\sqrt{21} + 3\\ &= 10 + 2\sqrt{21}\end{aligned}$$

33.
$$\begin{aligned}\left(2x - \sqrt{5}\right)\left(2x + \sqrt{5}\right) &= (2x)^2 - \left(\sqrt{5}\right)^2\\ &= 4x^2 - 5\end{aligned}$$

34. $\left(\sqrt{x+1}-1\right)^2=\left(\sqrt{x+1}\right)^2-2\left(\sqrt{x+1}\right)+1^2$
$=x+1-2\sqrt{x+1}+1$
$=x+2-2\sqrt{x+1}$

35. $\sqrt{\frac{7}{3}}=\frac{\sqrt{7}}{\sqrt{3}}=\frac{\sqrt{7}}{\sqrt{3}}\cdot\frac{\sqrt{3}}{\sqrt{3}}=\frac{\sqrt{21}}{3}$

36. $\frac{5}{\sqrt[3]{2x^2}}=\frac{5}{\sqrt[3]{2x^2}}\cdot\frac{\sqrt[3]{4x}}{\sqrt[3]{4x}}=\frac{5\sqrt[3]{4x}}{\sqrt[3]{8x^3}}=\frac{5\sqrt[3]{4x}}{2x}$

37. $\frac{\sqrt{3}-\sqrt{7}}{2\sqrt{3}+\sqrt{7}}$
$=\frac{\sqrt{3}-\sqrt{7}}{2\sqrt{3}+\sqrt{7}}\cdot\frac{\left(2\sqrt{3}-\sqrt{7}\right)}{\left(2\sqrt{3}-\sqrt{7}\right)}$
$=\frac{\sqrt{3}\left(2\sqrt{3}\right)-\sqrt{3}\sqrt{7}-\sqrt{7}\left(2\sqrt{3}\right)+\sqrt{7}\sqrt{7}}{\left(2\sqrt{3}\right)^2-\left(\sqrt{7}\right)^2}$
$=\frac{6-\sqrt{21}-2\sqrt{21}+7}{12-7}$
$=\frac{13-3\sqrt{21}}{5}$

38. $\sqrt{\frac{7}{3}}=\frac{\sqrt{7}}{\sqrt{3}}=\frac{\sqrt{7}}{\sqrt{3}}\cdot\frac{\sqrt{7}}{\sqrt{7}}=\frac{7}{\sqrt{21}}$

39. $\sqrt[3]{\frac{9y}{11}}=\frac{\sqrt[3]{9y}}{\sqrt[3]{11}}=\frac{\sqrt[3]{9y}}{\sqrt[3]{11}}\cdot\frac{\sqrt[3]{3y^2}}{\sqrt[3]{3y^2}}=\frac{\sqrt[3]{27y^3}}{\sqrt[3]{31y^2}}=\frac{3y}{\sqrt[3]{33y^2}}$

40. $\frac{\sqrt{x}-2}{\sqrt{x}}=\frac{\sqrt{x}-2}{\sqrt{x}}\cdot\frac{\sqrt{x}+2}{\sqrt{x}+2}$
$=\frac{\left(\sqrt{x}\right)^2-2^2}{\sqrt{x}\sqrt{x}+2\sqrt{x}}$
$=\frac{x-4}{x+2\sqrt{x}}$

Section 7.6 Practice Exercises

1. $\sqrt{3x-5}=7$
$\left(\sqrt{3x-5}\right)^2=7^2$
$3x-5=49$
$3x=54$
$x=18$

Check:
$\sqrt{3x-5}=7$
$\sqrt{3(18)-5}\stackrel{?}{=}7$
$\sqrt{54-5}\stackrel{?}{=}7$
$\sqrt{49}\stackrel{?}{=}7$
$7=7$
The solution is 18.

2. $\sqrt{16x-3}-4x=0$
$\sqrt{16x-3}=4x$
$\left(\sqrt{16x-3}\right)^2=(4x)^2$
$16x-3=16x^2$
$0=16x^2-16x+3$
$0=(4x-1)(4x-3)$
$4x-1=0$ or $4x-3=0$
$x=\frac{1}{4}$ or $x=\frac{3}{4}$

Check $\frac{1}{4}$:
$\sqrt{16\cdot\frac{1}{4}-3}-4\left(\frac{1}{4}\right)\stackrel{?}{=}0$
$\sqrt{4-3}-1\stackrel{?}{=}0$
$\sqrt{1}-1\stackrel{?}{=}0$
$1-1\stackrel{?}{=}0$
$0=0$

Check $\frac{3}{4}$:
$\sqrt{16\cdot\frac{3}{4}-3}-4\left(\frac{3}{4}\right)\stackrel{?}{=}0$
$\sqrt{12-3}-4\left(\frac{3}{4}\right)\stackrel{?}{=}0$
$\sqrt{9}-3\stackrel{?}{=}0$
$3-3\stackrel{?}{=}0$
$0=0$
The solutions are $\frac{1}{4}$ and $\frac{3}{4}$.

3. $\sqrt[3]{x-2}+1=3$
$\sqrt[3]{x-2}=2$
$\left(\sqrt[3]{x-2}\right)^3=2^3$
$x-2=8$
$x=10$

Check:

$\sqrt[3]{x-2}+1=3$

$\sqrt[3]{10-2}+1\stackrel{?}{=}3$

$\sqrt[3]{8}+1\stackrel{?}{=}3$

$2+1=3$

The solution is 10.

4. $\sqrt{16+x}=x-4$

$\left(\sqrt{16+x}\right)^2=(x-4)^2$

$16+x=x^2-8x+16$

$x^2-9x=0$

$x(x-9)=0$

$x=0$ or $x-9=0$

$x=9$

Check 0:

$\sqrt{16+x}=x-4$

$\sqrt{16+0}\stackrel{?}{=}0-4$

$\sqrt{16}\stackrel{?}{=}-4$

$4\neq-4$

Check 9:

$\sqrt{16+x}=x-4$

$\sqrt{16+9}\stackrel{?}{=}9-4$

$\sqrt{25}\stackrel{?}{=}5$

$5=5$

0 does not check, so the only solution is 9.

5. $\sqrt{8x+1}+\sqrt{3x}=2$

$\sqrt{8x+1}=2-\sqrt{3x}$

$\left(\sqrt{8x+1}\right)^2=\left(2-\sqrt{3x}\right)^2$

$8x+1=4-4\sqrt{3x}+3x$

$4\sqrt{3x}=3-5x$

$\left(4\sqrt{3x}\right)^2=(3-5x)^2$

$16(3x)=9-30x+25x^2$

$25x^2-78x+9=0$

$(25x-3)(x-3)=0$

$25x-3=0$ or $x-3=0$

$x=\frac{3}{25}$ or $x=3$

Check $\frac{3}{25}$:

$\sqrt{8x+1}+\sqrt{3x}=2$

$\sqrt{8\left(\frac{3}{25}\right)+1}+\sqrt{3\left(\frac{3}{25}\right)}\stackrel{?}{=}2$

$\sqrt{\frac{24}{25}+\frac{25}{25}}+\sqrt{\frac{9}{25}}\stackrel{?}{=}2$

$\sqrt{\frac{49}{25}}+\sqrt{\frac{9}{25}}\stackrel{?}{=}2$

$\frac{7}{5}+\frac{3}{5}\stackrel{?}{=}2$

$\frac{10}{5}=2$

Check 3:

$\sqrt{8x+1}+\sqrt{3x}=2$

$\sqrt{8(3)+1}+\sqrt{3(3)}\stackrel{?}{=}2$

$\sqrt{25}+\sqrt{9}\stackrel{?}{=}2$

$5+3\neq2$

3 does not check, so the only solution is $\frac{3}{25}$.

6. $a^2+b^2=c^2$

$a^2+6^2=12^2$

$a^2+36=144$

$a^2=108$

$a=\pm\sqrt{108}=\pm\sqrt{36\cdot3}=\pm6\sqrt{3}$

Since a is a length, we will use the positive value only. The unknown leg is $6\sqrt{3}$ meters long.

7. Consider the base of the tank, and the plastic divider in the diagonal. Use the Pythagorean theorem to find l.

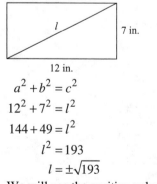

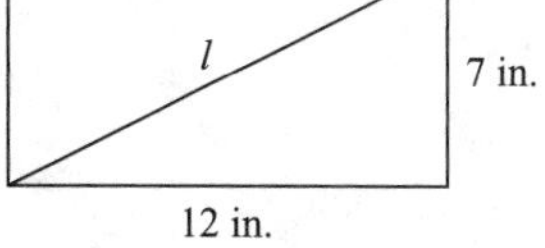

$a^2+b^2=c^2$

$12^2+7^2=l^2$

$144+49=l^2$

$l^2=193$

$l=\pm\sqrt{193}$

We will use the positive value because l represents length. The divider must be $\sqrt{193}\approx13.89$ inches long.

Graphing Calculator Explorations

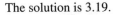

1.

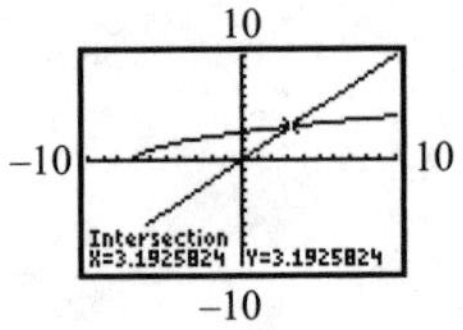

The solution is 3.19.

2.

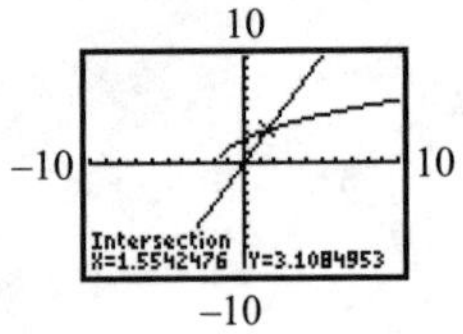

The solution is 1.55.

3.

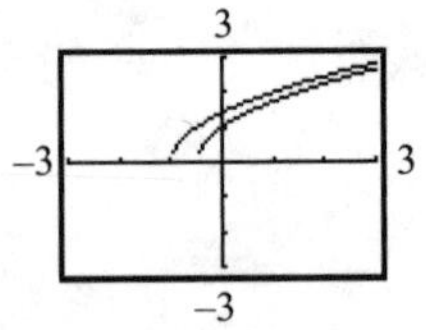

There is no solution. The solution set is ∅.

4.

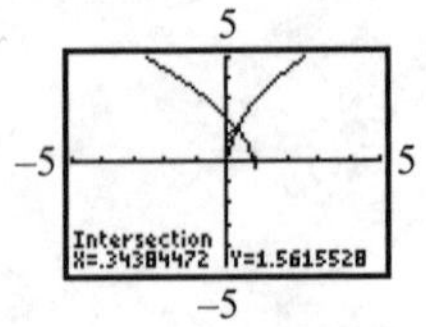

The solution is 0.34.

5.

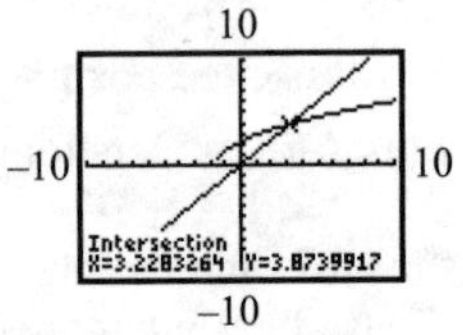

The solution is 3.23.

6.

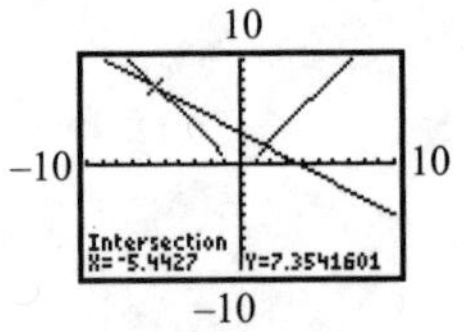

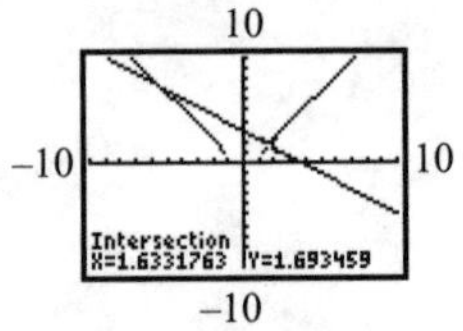

The solutions are −5.44 and 7.35.

Vocabulary, Readiness & Video Check 7.6

1. A proposed solution that is not a solution of the original equation is called an <u>extraneous solution</u>.

2. The Pythagorean theorem states that $a^2+b^2=c^2$ where a and b are the lengths of the <u>legs</u> of a <u>right</u> triangle and c is the length of the <u>hypotenuse</u>.

3. The square of $x - 5$, or $(x-5)^2 = \underline{x^2-10x+25}$.

4. The square of $4-\sqrt{7x}$, or $\left(4-\sqrt{7x}\right)^2 = \underline{16-8\sqrt{7x}+7x}$.

5. Applying the power rule can result in an equation with more solutions than the original equation, so you need to check all proposed solutions in the original equation.

6. The Pythagorean theorem works for a right triangle only and the side opposite the right angle is the hypotenuse, which is c in the formula $a^2+b^2=c^2$.

7. Our answer is either a positive square root of a value or a negative square root of a value. We're looking for a length, which must be positive, so our answer must be the positive square root.

Exercise Set 7.6

1.

$$\sqrt{2x}=4$$
$$\left(\sqrt{2x}\right)^2=4^2$$
$$2x=16$$
$$x=8$$

The solution is 8.

3.

$$\sqrt{x-3}=2$$
$$\left(\sqrt{x-3}\right)^2=2^2$$
$$x-3=4$$
$$x=7$$

The solution is 7.

5. $\sqrt{2x}=-4$

No solution since a principle square root does not yield a negative number, ∅.

7. $\sqrt{4x-3}-5=0$
$\sqrt{4x-3}=5$
$\left(\sqrt{4x-3}\right)^2=5^2$
$4x-3=25$
$4x=28$
$x=7$
The solution is 7.

9. $\sqrt{2x-3}-2=1$
$\sqrt{2x-3}=3$
$\left(\sqrt{2x-3}\right)^2=3^2$
$2x-3=9$
$2x=12$
$x=6$
The solution is 6.

11. $\sqrt[3]{6x}=-3$
$\left(\sqrt[3]{6x}\right)^3=(-3)^3$
$6x=-27$
$x=-\frac{27}{6}=-\frac{9}{2}$
The solution is $-\frac{9}{2}$.

13. $\sqrt[3]{x-2}-3=0$
$\sqrt[3]{x-2}=3$
$\left(\sqrt[3]{x-2}\right)^3=3^3$
$x-2=27$
$x=29$
The solution is 29.

15. $\sqrt{13-x}=x-1$
$\left(\sqrt{13-x}\right)^2=(x-1)^2$
$13-x=x^2-2x+1$
$0=x^2-x-12$
$0=(x-4)(x+3)$
$x-4=0$ or $x+3=0$
$x=4$ or $x=-3$
We discard –3 as extraneous. The solution is 4.

17. $x-\sqrt{4-3x}=-8$
$x+8=\sqrt{4-3x}$
$(x+8)^2=\left(\sqrt{4-3x}\right)^2$
$x^2+16x+64=4-3x$
$x^2+19x+60=0$
$(x+4)(x+15)=0$
$x+4=0$ or $x+15=0$
$x=-4$ or $x=-15$
We discard –15 as extraneous. The solution is –4.

19. $\sqrt{y+5}=2-\sqrt{y-4}$
$\left(\sqrt{y+5}\right)^2=\left(2-\sqrt{y-4}\right)^2$
$y+5=4-4\sqrt{y-4}+(y-4)$
$y+5=y-4\sqrt{y-4}$
$5=-4\sqrt{y-4}$
$5^2=\left(-4\sqrt{y-4}\right)^2$
$25=16(y-4)$
$25=16y-64$
$89=16y$
$\frac{89}{16}=y$
We discard $\frac{89}{16}$ as extraneous. There is no solution, ∅.

21. $\sqrt{x-3}+\sqrt{x+2}=5$
$\sqrt{x-3}=5-\sqrt{x+2}$
$\left(\sqrt{x-3}\right)^2=\left(5-\sqrt{x+2}\right)^2$
$x-3=25-10\sqrt{x+2}+(x+2)$
$x-3=27-10\sqrt{x+2}+x$
$-30=-10\sqrt{x+2}$
$3=\sqrt{x+2}$
$3^2=\left(\sqrt{x+2}\right)^2$
$9=x+2$
$7=x$
The solution is 7.

23. $\sqrt{3x-2}=5$

$$\left(\sqrt{3x-2}\right)^2=5^2$$
$$3x-2=25$$
$$3x=27$$
$$x=9$$

The solution is 9.

25. $-\sqrt{2x}+4=-6$

$$10=\sqrt{2x}$$
$$10^2=\left(\sqrt{2x}\right)^2$$
$$100=2x$$
$$50=x$$

The solution is 50.

27. $\sqrt{3x+1}+2=0$

$$\sqrt{3x+1}=-2$$

No solution since a principle square root does not yield a negative number, $\varnothing$.

29. $\sqrt[4]{4x+1}-2=0$

$$\sqrt[4]{4x+1}=2$$
$$\left(\sqrt[4]{4x+1}\right)^4=2^4$$
$$4x+1=16$$
$$4x=15$$
$$x=\frac{15}{4}$$

The solution is $\frac{15}{4}$.

31. $\sqrt{4x-3}=7$

$$\left(\sqrt{4x-3}\right)^2=7^2$$
$$4x-3=49$$
$$4x=52$$
$$x=13$$

The solution is 13.

33. $\sqrt[3]{6x-3}-3=0$

$$\sqrt[3]{6x-3}=3$$
$$\left(\sqrt[3]{6x-3}\right)^3=3^3$$
$$6x-3=27$$
$$6x=30$$
$$x=5$$

The solution is 5.

35. $\sqrt[3]{2x-3}-2=-5$

$$\sqrt[3]{2x-3}=-3$$
$$\left(\sqrt[3]{2x-3}\right)^3=(-3)^3$$
$$2x-3=-27$$
$$2x=-24$$
$$x=-12$$

The solution is –12.

37. $\sqrt{x+4}=\sqrt{2x-5}$

$$\left(\sqrt{x+4}\right)^2=\left(\sqrt{2x-5}\right)^2$$
$$x+4=2x-5$$
$$-x=-9$$
$$x=9$$

The solution is 9.

39. $x-\sqrt{1-x}=-5$

$$x+5=\sqrt{1-x}$$
$$(x+5)^2=\left(\sqrt{1-x}\right)^2$$
$$x^2+10x+25=1-x$$
$$x^2+11x+24=0$$
$$(x+8)(x+3)=0$$
$$x+8=0 \quad \text{or} \quad x+3=0$$
$$x=-8 \quad \text{or} \quad x=-3$$

We discard –8 as extraneous. The solution is –3.

41. $\sqrt[3]{-6x-1}=\sqrt[3]{-2x-5}$

$$\left(\sqrt[3]{-6x-1}\right)^3=\left(\sqrt[3]{-2x-5}\right)^3$$
$$-6x-1=-2x-5$$
$$-4x=-4$$
$$x=1$$

The solution is 1.

43. $\sqrt{5x-1}-\sqrt{x}+2=3$

$$\sqrt{5x-1}=\sqrt{x}+1$$
$$\left(\sqrt{5x-1}\right)^2=\left(\sqrt{x}+1\right)^2$$
$$5x-1=x+2\sqrt{x}+1$$
$$4x-2=2\sqrt{x}$$
$$2x-1=\sqrt{x}$$
$$(2x-1)^2=\left(\sqrt{x}\right)^2$$
$$4x^2-4x+1=x$$
$$4x^2-5x+1=0$$
$$(4x-1)(x-1)=0$$
$$4x-1=0 \text{ or } x-1=0$$
$$4x=1 \text{ or } x=1$$
$$x=\frac{1}{4}$$

We discard $\frac{1}{4}$ as extraneous. The solution is 1.

45. $\sqrt{2x-1}=\sqrt{1-2x}$

$$\left(\sqrt{2x-1}\right)^2=\left(\sqrt{1-2x}\right)^2$$
$$2x-1=1-2x$$
$$4x=2$$
$$x=\frac{2}{4}=\frac{1}{2}$$

The solution is $\frac{1}{2}$.

47. $\sqrt{3x+4}-1=\sqrt{2x+1}$

$$\sqrt{3x+4}=\sqrt{2x+1}+1$$
$$\left(\sqrt{3x+4}\right)^2=\left(\sqrt{2x+1}+1\right)^2$$
$$3x+4=(2x+1)+2\sqrt{2x+1}+1$$
$$3x+4=2x+2+2\sqrt{2x+1}$$
$$x+2=2\sqrt{2x+1}$$
$$(x+2)^2=\left(2\sqrt{2x+1}\right)^2$$
$$x^2+4x+4=4(2x+1)$$
$$x^2+4x+4=8x+4$$
$$x^2-4x=0$$
$$x(x-4)=0$$
$$x=0 \text{ or } x-4=0$$
$$x=4$$

The solutions are 0 and 4.

49. $\sqrt{y+3}-\sqrt{y-3}=1$

$$\sqrt{y+3}=1+\sqrt{y-3}$$
$$\left(\sqrt{y+3}\right)^2=\left(1+\sqrt{y-3}\right)^2$$
$$y+3=1+2\sqrt{y-3}+(y-3)$$
$$y+3=-2+2\sqrt{y-3}+y$$
$$5=2\sqrt{y-3}$$
$$(5)^2=\left(2\sqrt{y-3}\right)^2$$
$$25=4(y-3)$$
$$25=4y-12$$
$$37=4y$$
$$\frac{37}{4}=y$$

The solution is $\frac{37}{4}$.

51. Let c = length of the hypotenuse.

$$6^2+3^2=c^2$$
$$36+9=c^2$$
$$45=c^2$$
$$\sqrt{45}=\sqrt{c^2}$$
$$\sqrt{9\cdot 5}=c$$
$$3\sqrt{5}=c \text{ so } c=3\sqrt{5} \text{ feet}$$

53. Let b = length of the unknown leg.

$$3^2+b^2=7^2$$
$$9+b^2=49$$
$$b^2=40$$
$$\sqrt{b^2}=\sqrt{40}$$
$$b=\sqrt{4\cdot 10}$$
$$b=2\sqrt{10} \text{ meters}$$

55. Let b = length of the unknown leg.

$$9^2+b^2=\left(11\sqrt{5}\right)^2$$
$$81+b^2=121\cdot 5$$
$$81+b^2=605$$
$$b^2=524$$
$$\sqrt{b^2}=\sqrt{524}$$
$$b=\sqrt{4\cdot 131}$$
$$b=2\sqrt{131}\approx 22.9 \text{ meters}$$

57. Let c = length of the hypotenuse.

$$7^2+(7.2)^2=c^2$$
$$49+51.84=c^2$$
$$100.84=c^2$$
$$100.84=\sqrt{c^2}$$
$$c=\sqrt{100.84}\approx 10.0 \text{ mm}$$

59. Let x = amount of cable needed.

$$15^2+8^2=x^2$$
$$225+64=x^2$$
$$289=x^2$$
$$\sqrt{289}=\sqrt{x^2}$$
$$17=x$$

Thus, 17 feet of cable is needed.

61. Let c = length of the ladder.

$$12^2+5^2=c^2$$
$$144+25=c^2$$
$$169=c^2$$
$$\sqrt{169}=\sqrt{c^2}$$
$$13=c$$

A 13-foot ladder is needed.

63.

$$r=\sqrt{\frac{A}{4\pi}}$$
$$1080=\sqrt{\frac{A}{4\pi}}$$
$$(1080)^2=\left(\sqrt{\frac{A}{4\pi}}\right)^2$$
$$1{,}166{,}400=\frac{A}{4\pi}$$
$$14{,}657{,}415\approx A$$

The surface area is approximately 14,657,415 square miles.

65.

$$v=\sqrt{2gh}$$
$$80=\sqrt{2(32)h}$$
$$(80)^2=\left(\sqrt{64h}\right)^2$$
$$6400=64h$$
$$100=h$$

The object fell 100 feet.

67.

$$S=2\sqrt{I}-9$$
$$11=2\sqrt{I}-9$$
$$20=2\sqrt{I}$$
$$10=\sqrt{I}$$
$$10^2=\left(\sqrt{I}\right)^2$$
$$100=I$$

The estimated IQ is 100.

69.

$$P=2\pi\sqrt{\frac{l}{32}}$$
$$=2\pi\sqrt{\frac{2}{32}}$$
$$=2\pi\sqrt{\frac{1}{16}}$$
$$=2\pi\left(\frac{1}{4}\right)$$
$$=\frac{\pi}{2}\text{ sec}\approx 1.57\text{ sec}$$

71.

$$P=2\pi\sqrt{\frac{l}{32}}$$
$$4=2\pi\sqrt{\frac{l}{32}}$$
$$\frac{4}{2\pi}=\sqrt{\frac{l}{32}}$$
$$\left(\frac{2}{\pi}\right)^2=\left(\sqrt{\frac{l}{32}}\right)^2$$
$$\frac{4}{\pi^2}=\frac{l}{32}$$
$$l=32\left(\frac{4}{\pi^2}\right)\approx 12.97\text{ feet}$$

73. answers may vary

75.

$$s=\frac{1}{2}(6+10+14)=\frac{1}{2}(30)=15$$
$$A=\sqrt{s(s-a)(s-b)(s-c)}$$
$$=\sqrt{15(15-6)(15-10)(15-14)}$$
$$=\sqrt{15(9)(5)(1)}$$
$$=\sqrt{675}$$
$$=\sqrt{225\cdot 3}$$
$$=15\sqrt{3}\text{ sq mi}\approx 25.98\text{ sq mi.}$$

77. answers may vary

79.
$$D(h) = 111.7\sqrt{h}$$
$$80 = 111.7\sqrt{h}$$
$$\frac{80}{111.7} = \sqrt{h}$$
$$\left(\frac{80}{111.7}\right)^2 = \left(\sqrt{h}\right)^2$$
$$0.5129483389 \approx h$$
$$h \approx 0.51 \text{ km}$$

81. Function; no vertical line intersects the graph more than one time.

83. Function; no vertical line intersects the graph more than one time.

85. Not a function; the y-axis is an example of a vertical line that intersects the graph more than one time.

87. $$\frac{\frac{x}{6}}{\frac{2x}{3}+\frac{1}{2}} = \frac{\left(\frac{x}{6}\right)6}{\left(\frac{2x}{3}+\frac{1}{2}\right)6} = \frac{x}{4x+3}$$

89. $$\frac{\frac{z}{5}+\frac{1}{10}}{\frac{z}{20}-\frac{z}{5}} = \frac{\left(\frac{z}{5}+\frac{1}{10}\right)20}{\left(\frac{z}{20}-\frac{z}{5}\right)20}$$
$$= \frac{4z+2}{z-4z}$$
$$= \frac{4z+2}{-3z}$$
$$= -\frac{4z+2}{3z}$$

91.
$$\sqrt{5x-1}+4 = 7$$
$$\sqrt{5x-1} = 3$$
$$\left(\sqrt{5x-1}\right)^2 = 3^2$$
$$5x-1 = 9$$
$$5x = 10$$
$$x = 2$$

93.
$$\sqrt{\sqrt{x+3}+\sqrt{x}} = \sqrt{3}$$
$$\left(\sqrt{\sqrt{x+3}+\sqrt{x}}\right)^2 = \left(\sqrt{3}\right)^2$$
$$\sqrt{x+3}+\sqrt{x} = 3$$
$$\sqrt{x+3} = 3-\sqrt{x}$$
$$\left(\sqrt{x+3}\right)^2 = \left(3-\sqrt{x}\right)^2$$
$$x+3 = 9-6\sqrt{x}+x$$
$$-6 = -6\sqrt{x}$$
$$(-6)^2 = \left(-6\sqrt{x}\right)^2$$
$$36 = 36x$$
$$1 = x$$

95. a. answers may vary

b. answers may vary

97. $3\sqrt{x^2-8x} = x^2-8x$

Let $t = x^2-8x$. Then
$$3\sqrt{t} = t$$
$$\left(3\sqrt{t}\right)^2 = t^2$$
$$9t = t^2$$
$$0 = t^2-9t$$
$$0 = t(t-9)$$
$t = 0$ or $t = 9$

Replace t with x^2-8x.

$x^2-8x = 0$ or $x^2-8x = 9$

$x(x-8) = 0$ | $x^2-8x-9 = 0$

$x = 0$ or $x = 8$ | $(x-9)(x+1) = 0$

$x = 9$ or $x = -1$

The solutions are −1, 0, 8, and 9.

99. $7-(x^2-3x) = \sqrt{(x^2-3x)+5}$

Let $t = x^2-3x$. Then
$$7-t = \sqrt{t+5}$$
$$(7-t)^2 = \left(\sqrt{t+5}\right)^2$$
$$49-14t+t^2 = t+5$$
$$t^2-15t+44 = 0$$
$$(t-11)(t-4) = 0$$
$t = 11$ or $t = 4$

Replace t with x^2-3x.

$x^2-3x=11$ or $x^2-3x=4$

$x^2-3x-11=0$ $\quad$ $x^2-3x-4=0$

Can't factor $\quad$ $(x-4)(x+1)=0$

$x=4$ or $x=-1$

The solutions are –1 and 4.

Section 7.7 Practice Exercises

1. a. $\sqrt{-4}=\sqrt{-1\cdot 4}=\sqrt{-1}\cdot\sqrt{4}=i\cdot 2$, or $2i$

b. $\sqrt{-7}=\sqrt{-1(7)}=\sqrt{-1}\cdot\sqrt{7}=i\sqrt{7}$

c.
$$\begin{aligned}-\sqrt{-18}&=-\sqrt{-1\cdot 18}\\&=-\sqrt{-1}\cdot\sqrt{9\cdot 2}\\&=-i\cdot 3\sqrt{2}\\&=-3i\sqrt{2}\end{aligned}$$

2. a.
$$\begin{aligned}\sqrt{-5}\cdot\sqrt{-6}&=i\sqrt{5}\left(i\sqrt{6}\right)\\&=i^2\sqrt{30}\\&=-1\sqrt{30}\\&=-\sqrt{30}\end{aligned}$$

b. $\sqrt{-9}\cdot\sqrt{-1}=3i\cdot i=3i^2=3(-1)=-3$

c.
$$\begin{aligned}\sqrt{125}\cdot\sqrt{-5}&=5\sqrt{5}\left(i\sqrt{5}\right)\\&=5i\left(\sqrt{5}\sqrt{5}\right)\\&=5i(5)\\&=25i\end{aligned}$$

d. $\dfrac{\sqrt{-27}}{\sqrt{3}}=\dfrac{i\sqrt{27}}{\sqrt{3}}=i\sqrt{9}=3i$

3. a.
$$\begin{aligned}(3-5i)+(-4+i)&=(3-4)+(-5+1)i\\&=-1-4i\end{aligned}$$

b.
$$\begin{aligned}4i-(3-i)&=4i-3+i\\&=-3+(4+1)i\\&=-3+5i\end{aligned}$$

c.
$$\begin{aligned}(-5-2i)-(-8)&=-5-2i+8\\&=(-5+8)-2i\\&=3-2i\end{aligned}$$

4. a. $-4i\cdot 5i=-20i^2=-20(-1)=20=20+0i$

b.
$$\begin{aligned}5i(2+i)&=5i\cdot 2+5i\cdot i\\&=10i+5i^2\\&=10i+5(-1)\\&=10i-5\\&=-5+10i\end{aligned}$$

c.
$$\begin{aligned}(2+3i)(6-i)&=2(6)-2(i)+3i(6)-3i(i)\\&=12-2i+18i-3i^2\\&=12+16i-3(-1)\\&=12+16i+3\\&=15+16i\end{aligned}$$

d.
$$\begin{aligned}(3-i)^2&=(3-i)(3-i)\\&=3(3)-3(i)-3(i)+i^2\\&=9-6i+(-1)\\&=8-6i\end{aligned}$$

e.
$$\begin{aligned}(9+2i)(9-2i)&=9(9)-9(2i)+2i(9)-2i(2i)\\&=81-18i+18i-4i^2\\&=81-4(-1)\\&=81+4\\&=85\\&=85+0i\end{aligned}$$

5. a.
$$\begin{aligned}\frac{4-i}{3+i}&=\frac{(4-i)(3-i)}{(3+i)(3-i)}\\&=\frac{4(3)-4(i)-3(i)+i^2}{3^2-i^2}\\&=\frac{12-7i-1}{9+1}\\&=\frac{11-7i}{10}\\&=\frac{11}{10}-\frac{7i}{10}\text{ or }\frac{11}{10}-\frac{7}{10}i\end{aligned}$$

b.
$$\begin{aligned}\frac{5}{2i}&=\frac{5(-2i)}{2i(-2i)}\\&=\frac{-10i}{-4i^2}\\&=\frac{-10i}{-4(-1)}\\&=\frac{-10i}{4}\\&=\frac{-5i}{2}\\&=0-\frac{5i}{2}\text{ or }0-\frac{5}{2}i\end{aligned}$$

6. a. $i^9 = i^4 \cdot i^4 \cdot i = 1 \cdot 1 \cdot i = i$

b. $i^{16} = (i^4)^4 = 1^4 = 1$

c. $i^{34} = i^{32} \cdot i^2 = (i^4)^8 \cdot i^2 = 1^8(-1) = -1$

d. $i^{-24} = \dfrac{1}{i^{24}} = \dfrac{1}{(i^4)^6} = \dfrac{1}{(1)^6} = \dfrac{1}{1} = 1$

Vocabulary, Readiness & Video Check 7.7

1. A <u>complex</u> number is one that can be written in the form $a + bi$ where a and b are real numbers.
2. In the complex number system, i denotes the <u>imaginary unit</u>.
3. $i^2 = \underline{-1}$
4. $i = \underline{\sqrt{-1}}$
5. A complex number, $a + bi$, is a <u>real</u> number if $b = 0$.
6. A complex number, $a + bi$, is a <u>pure imaginary</u> number if $a = 0$ and $b \neq 0$.
7. The product rule for radicals; you need to first simplify each separate radical and have nonnegative radicands before applying the product rule.
8. combining like terms; i is *not* a variable, but a constant, $\sqrt{-1}$
9. The fact that $i^2 = -1$.
10. using conjugates to rationalize denominators with two terms
11. $i, i^2 = -1, i^3 = -i, i^4 = 1$

Exercise Set 7.7

1. $\sqrt{-81} = \sqrt{-1 \cdot 81} = \sqrt{-1}\sqrt{81} = 9i$

3. $\sqrt{-7} = \sqrt{-1 \cdot 7} = \sqrt{-1}\sqrt{7} = i\sqrt{7}$

5. $-\sqrt{16} = -4$

7. $\sqrt{-64} = \sqrt{-1 \cdot 64} = \sqrt{-1}\sqrt{64} = 8i$

9. $\sqrt{-24} = \sqrt{-1 \cdot 24} = \sqrt{-1}\sqrt{4 \cdot 6} = i \cdot 2\sqrt{6} = 2i\sqrt{6}$

11. $-\sqrt{-36} = -\sqrt{-1 \cdot 36} = -\sqrt{-1}\sqrt{36} = -i \cdot 6 = -6i$

13. $$\begin{aligned} 8\sqrt{-63} &= 8\sqrt{-1 \cdot 63} \\ &= 8\sqrt{-1}\sqrt{9 \cdot 7} \\ &= 8i \cdot 3\sqrt{7} \\ &= 24i\sqrt{7} \end{aligned}$$

15. $-\sqrt{54} = -\sqrt{9 \cdot 6} = -3\sqrt{6} = -3\sqrt{6} + 0i$

17. $$\begin{aligned} \sqrt{-2} \cdot \sqrt{-7} &= i\sqrt{2} \cdot i\sqrt{7} \\ &= i^2\sqrt{14} \\ &= (-1)\sqrt{14} \\ &= -\sqrt{14} \end{aligned}$$

19. $$\begin{aligned} \sqrt{-5} \cdot \sqrt{-10} &= i\sqrt{5} \cdot i\sqrt{10} \\ &= i^2\sqrt{50} \\ &= (-1)\sqrt{25 \cdot 2} \\ &= -5\sqrt{2} \end{aligned}$$

21. $\sqrt{16} \cdot \sqrt{-1} = 4i$

23. $\dfrac{\sqrt{-9}}{\sqrt{3}} = \dfrac{i\sqrt{9}}{\sqrt{3}} = i\sqrt{\dfrac{9}{3}} = i\sqrt{3}$

25. $\dfrac{\sqrt{-80}}{\sqrt{-10}} = \dfrac{i\sqrt{80}}{i\sqrt{10}} = \sqrt{\dfrac{80}{10}} = \sqrt{8} = \sqrt{4 \cdot 2} = 2\sqrt{2}$

27. $$\begin{aligned} (4-7i)+(2+3i) &= (4+2)+(-7+3)i \\ &= 6+(-4)i \\ &= 6-4i \end{aligned}$$

29. $$\begin{aligned} (6+5i)-(8-i) &= 6+5i-8+i \\ &= (6-8)+(5+1)i \\ &= -2+6i \end{aligned}$$

31. $$\begin{aligned} 6-(8+4i) &= 6-8-4i \\ &= (6-8)-4i \\ &= -2-4i \end{aligned}$$

33. $-10i \cdot -4i = 40i^2 = 40(-1) = -40 = -40 + 0i$

35. $$\begin{aligned} 6i(2-3i) &= 12i-18i^2 \\ &= 12i-18(-1) \\ &= 18+12i \end{aligned}$$

37. $(\sqrt{3}+2i)(\sqrt{3}-2i)$
$= \sqrt{3}\cdot\sqrt{3}-\sqrt{3}\cdot 2i+\sqrt{3}\cdot 2i-4i^2$
$= 3-4(-1)+0i$
$= 3+4+0i$
$= 7+0i$

39. $(4-2i)^2 = (4-2i)(4-2i)$
$= 16-4\cdot 2i-4\cdot 2i+4i^2$
$= 16-8i-8i+4(-1)$
$= 16-16i-4$
$= 12-16i$

41. $\frac{4}{i}=\frac{4(-i)}{i(-i)}=\frac{-4i}{-i^2}=\frac{-4i}{-(-1)}=-4i=0-4i$

43. $\frac{7}{4+3i}=\frac{7(4-3i)}{(4+3i)(4-3i)}$
$=\frac{28-21i}{4^2-9i^2}$
$=\frac{28-21i}{16+9}$
$=\frac{28-21i}{25}$
$=\frac{28}{25}-\frac{21}{25}i$

45. $\frac{3+5i}{1+i}=\frac{(3+5i)(1-i)}{(1+i)(1-i)}$
$=\frac{3-3i+5i-5i^2}{1^2-i^2}$
$=\frac{3+2i+5}{1+1}$
$=\frac{8+2i}{2}$
$=\frac{8}{2}+\frac{2}{2}i$
$=4+i$

47. $\frac{5-i}{3-2i}=\frac{(5-i)(3+2i)}{(3-2i)(3+2i)}$
$=\frac{15+10i-3i-2i^2}{3^2-4i^2}$
$=\frac{15+7i+2}{9+4}$
$=\frac{17+7i}{13}$
$=\frac{17}{13}+\frac{7}{13}i$

49. $(7i)(-9i)=-63i^2=-63(-1)=63=63+0i$

51. $(6-3i)-(4-2i)=6-3i-4+2i=2-i$

53. $-3i(-1+9i)=3i-27i^2$
$=3i-27(-1)$
$=27+3i$

55. $\frac{4-5i}{2i}=\frac{4-5i}{2i}\cdot\frac{-2i}{-2i}$
$=\frac{-8i+10i^2}{-4i^2}$
$=\frac{-10-8i}{4}$
$=\frac{-10}{4}-\frac{8}{4}i$
$=-\frac{5}{2}-2i$

57. $(4+i)(5+2i)=20+8i+5i+2i^2$
$=20+13i+2(-1)$
$=20+13i-2$
$=18+13i$

59. $(6-2i)(3+i)=18+6i-6i-2i^2$
$=18+2+0i$
$=20+0i$

61. $(8-3i)+(2+3i)=8-3i+2+3i=10+0i$

63. $(1-i)(1+i)=1+i-i-i^2=1+1+0i=2+0i$

65. $\frac{16+15i}{-3i}=\frac{(16+15i)(3i)}{-3i(3i)}$
$=\frac{48i+45i^2}{-9i^2}$
$=\frac{-45+48i}{9}$
$=\frac{-45}{9}+\frac{48}{9}i$
$=-5+\frac{16}{3}i$

67. $(9+8i)^2=9^2+2(9)(8i)+(8i)^2$
$=81+144i+64i^2$
$=81+144i-64$
$=17+144i$

69. $\frac{2}{3+i}=\frac{2(3-i)}{(3+i)(3-i)}$
$=\frac{6-2i}{3^2-i^2}$
$=\frac{6-2i}{9+1}$
$=\frac{6-2i}{10}$
$=\frac{6}{10}-\frac{2}{10}i$
$=\frac{3}{5}-\frac{1}{5}i$

71. $(5-6i)-4i=5-6i-4i=5-10i$

73. $\frac{2-3i}{2+i}=\frac{(2-3i)(2-i)}{(2+i)(2-i)}$
$=\frac{4-2i-6i+3i^2}{2^2-i^2}$
$=\frac{4-8i-3}{4+1}$
$=\frac{1-8i}{5}$
$=\frac{1}{5}-\frac{8}{5}i$

75. $(2+4i)+(6-5i)=2+4i+6-5i=8-i$

77. $\left(\sqrt{6}+i\right)\left(\sqrt{6}-i\right)=\left(\sqrt{6}\right)^2-i^2$
$=6-(-1)$
$=6+1$
$=7$
$=7+0i$

79. $4(2-i)^2=4(2^2-2\cdot 2i+i^2)$
$=4(4-4i-1)$
$=4(3-4i)$
$=12-16i$

81. $i^8=(i^4)^2=1^2=1$

83. $i^{21}=i^{20}\cdot i=(i^4)^5\cdot i=1^5\cdot i=i$

85. $i^{11}=i^8\cdot i^3=(i^4)^2\cdot i^3=1^2\cdot(-i)=-i$

87. $i^{-6}=\frac{1}{i^6}=\frac{1}{i^4\cdot i^2}=\frac{1}{1\cdot(-1)}=-1$

89. $(2i)^6=2^6i^6=64i^4\cdot i^2=64(1)(-1)=-64$

91. $(-3i)^5=(-3)^5i^5=-243i^4\cdot i=-243(1)i=-243i$

93. $x+50^\circ+90^\circ=180^\circ$
$x+140^\circ=180^\circ$
$x=40^\circ$

95.

1	1	−6	3	−4
		1	−5	−2
	1	−5	−2	−6

Answer: $x^2-5x-2-\frac{6}{x-1}$

97. 5 people

99. 5 + 9 = 14 people

101. $\frac{5\text{ people}}{30\text{ people}}=\frac{1}{6}\approx 0.1666$
About 16.7% of the people reported an average checking balance of \$201 to \$300.

103. $i^3-i^4=-i-1=-1-i$

105. $i^6+i^8=i^4\cdot i^2+(i^4)^2$
$=1(-1)+1^2$
$=-1+1$
$=0$
$=0+0i$

107. $2+\sqrt{-9}=2+i\sqrt{9}=2+3i$

109. $\frac{6+\sqrt{-18}}{3}=\frac{6+i\sqrt{9\cdot 2}}{3}$
$=\frac{6+3i\sqrt{2}}{3}$
$=\frac{6}{3}+\frac{3\sqrt{2}}{3}i$
$=2+i\sqrt{2}$

111. $\frac{5-\sqrt{-75}}{10} = \frac{5-i\sqrt{25\cdot 3}}{10}$
$= \frac{5-5i\sqrt{3}}{10}$
$= \frac{5}{10} - \frac{5\sqrt{3}}{10}i$
$= \frac{1}{2} - \frac{\sqrt{3}}{2}i$

113. answers may vary

115. $\left(8-\sqrt{-3}\right)-\left(2+\sqrt{-12}\right)$
$= 8 - i\sqrt{3} - 2 - 2i\sqrt{3}$
$= 6 - 3i\sqrt{3}$

117. $x^2 + 4 = 0$
$(2i)^2 + 4 = 0$
$4i^2 + 4 = 0$
$4(-1) + 4 = 0$
$-4 + 4 = 0$, which is true.
Yes, $2i$ is a solution.

Chapter 7 Vocabulary Check

1. The conjugate of $\sqrt{3}+2$ is $\sqrt{3}-2$.
2. The principal square root of a positive number a is written as $\sqrt{a}$.
3. The process of writing a radical expression as an equivalent expression but without a radical in the denominator is called rationalizing the denominator.
4. The imaginary unit, written i, is the number whose square is -1.
5. The cube root of a number is written as $\sqrt[3]{a}$.
6. In the notation $\sqrt[n]{a}$, n is called the index and a is called the radicand.
7. Radicals with the same index and the same radicand are called like radicals.
8. A complex number is a number that can be written in the form $a + bi$, where a and b are real numbers.
9. The distance formula is $d = \sqrt{(x_2 - x_1)^2 + (y_2 - y_1)^2}$.
10. The midpoint formula is $\left(\frac{x_1 + x_2}{2}, \frac{y_1 + y_2}{2}\right)$.

Chapter 7 Review

1. $\sqrt{81} = 9$ because $9^2 = 81$.
2. $\sqrt[4]{81} = 3$ because $3^4 = 81$.
3. $\sqrt[3]{-8} = -2$ because $(-2)^3 = -8$.
4. $\sqrt[4]{-16}$ is not a real number.
5. $-\sqrt{\frac{1}{49}} = -\frac{1}{7}$ because $\left(\frac{1}{7}\right)^2 = \frac{1}{49}$.
6. $\sqrt{x^{64}} = x^{32}$ because $(x^{32})^2 = x^{32\cdot 2} = x^{64}$.
7. $-\sqrt{36} = -6$ because $6^2 = 36$.
8. $\sqrt[3]{64} = 4$ because $4^3 = 64$.
9. $\sqrt[3]{-a^6b^9} = \sqrt[3]{-1}\sqrt[3]{a^6}\sqrt[3]{b^9}$
$= -1a^2b^3$
$= -a^2b^3$
10. $\sqrt{16a^4b^{12}} = \sqrt{16}\sqrt{a^4}\sqrt{b^{12}} = 4a^2b^6$
11. $\sqrt[5]{32a^5b^{10}} = \sqrt[5]{32}\sqrt[5]{a^5}\sqrt[5]{b^{10}} = 2ab^2$
12. $\sqrt[5]{-32x^{15}y^{20}} = \sqrt[5]{-32}\sqrt[5]{x^{15}}\sqrt[5]{y^{20}} = -2x^3y^4$
13. $\sqrt{\frac{x^{12}}{36y^2}} = \frac{\sqrt{x^{12}}}{\sqrt{36y^2}} = \frac{x^6}{6y}$
14. $\sqrt[3]{\frac{27y^3}{z^{12}}} = \frac{\sqrt[3]{27y^3}}{\sqrt[3]{z^{12}}} = \frac{3y}{z^4}$
15. $\sqrt{(-x)^2} = |-x|$
16. $\sqrt[4]{(x^2-4)^4} = |x^2-4|$

17. $\sqrt[3]{(-27)^3} = -27$

18. $\sqrt[5]{(-5)^5} = -5$

19. $-\sqrt[5]{x^5} = -x$

20. $-\sqrt[3]{x^3} = -x$

21. $\sqrt[4]{16(2y+z)^4} = 2|2y+z|$

22. $\sqrt{25(x-y)^2} = 5|x-y|$

23. $\sqrt[5]{y^5} = y$

24. $\sqrt[6]{x^6} = |x|$

25. a. $f(x) = \sqrt{x} + 3$
$f(0) = \sqrt{0} + 3 = 0 + 3 = 3$
$f(9) = \sqrt{9} + 3 = 3 + 3 = 6$

b. $f(x) = \sqrt{x} + 3$
$x \geq 0$
Domain: $[0, \infty)$

c.

x	0	1	4	9
$f(x)$	3	4	5	6

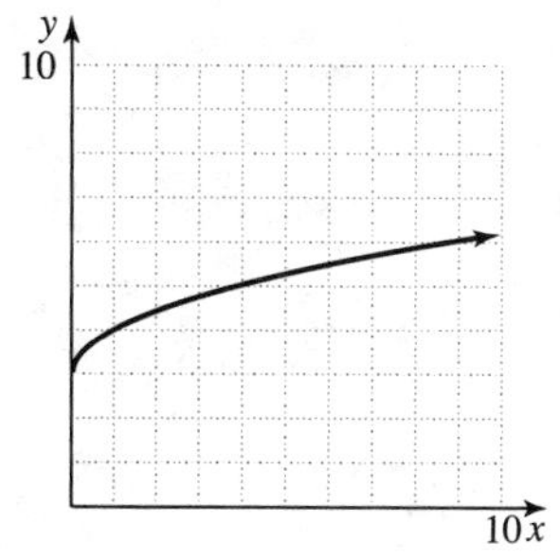

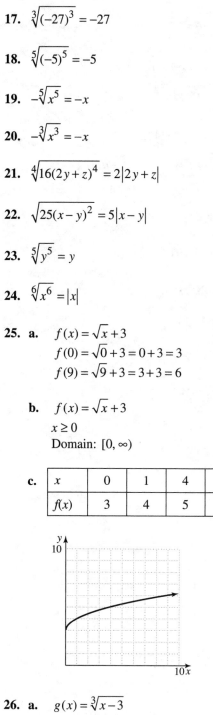

26. a. $g(x) = \sqrt[3]{x-3}$
$g(11) = \sqrt[3]{11-3} = \sqrt[3]{8} = 2$
$g(20) = \sqrt[3]{20-3} = \sqrt[3]{17}$

b. $g(x) = \sqrt[3]{x-3}$
Domain: $(-\infty, \infty)$

c.

x	−5	2	3	4	11
$g(x)$	−2	−1	0	1	2

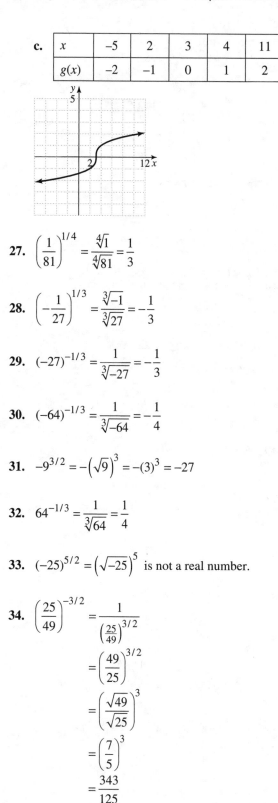

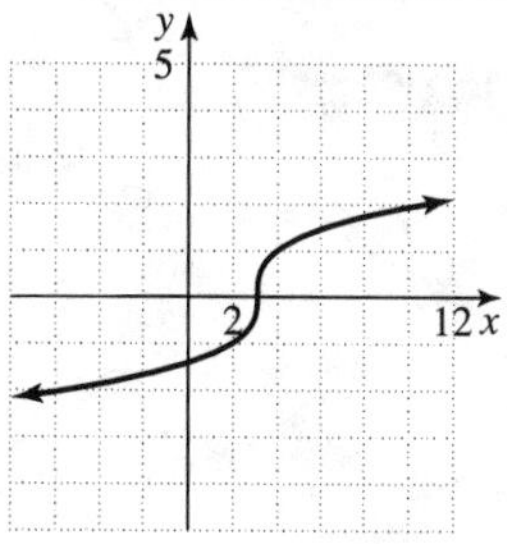

27. $\left(\frac{1}{81}\right)^{1/4} = \frac{\sqrt[4]{1}}{\sqrt[4]{81}} = \frac{1}{3}$

28. $\left(-\frac{1}{27}\right)^{1/3} = \frac{\sqrt[3]{-1}}{\sqrt[3]{27}} = -\frac{1}{3}$

29. $(-27)^{-1/3} = \frac{1}{\sqrt[3]{-27}} = -\frac{1}{3}$

30. $(-64)^{-1/3} = \frac{1}{\sqrt[3]{-64}} = -\frac{1}{4}$

31. $-9^{3/2} = -\left(\sqrt{9}\right)^3 = -(3)^3 = -27$

32. $64^{-1/3} = \frac{1}{\sqrt[3]{64}} = \frac{1}{4}$

33. $(-25)^{5/2} = \left(\sqrt{-25}\right)^5$ is not a real number.

34.
$$\begin{aligned}\left(\frac{25}{49}\right)^{-3/2} &= \frac{1}{\left(\frac{25}{49}\right)^{3/2}} \\ &= \left(\frac{49}{25}\right)^{3/2} \\ &= \left(\frac{\sqrt{49}}{\sqrt{25}}\right)^3 \\ &= \left(\frac{7}{5}\right)^3 \\ &= \frac{343}{125}\end{aligned}$$

35. $\left(\frac{8}{27}\right)^{-2/3} = \frac{1}{\left(\frac{8}{27}\right)^{2/3}}$
$= \left(\frac{27}{8}\right)^{2/3}$
$= \left(\frac{\sqrt[3]{27}}{\sqrt[3]{8}}\right)^2$
$= \left(\frac{3}{2}\right)^2$
$= \frac{9}{4}$

36. $\left(-\frac{1}{36}\right)^{-1/4}$ is not a real number.

37. $\sqrt[3]{x^2} = x^{2/3}$

38. $\sqrt[5]{5x^2y^3} = 5^{1/5}x^{2/5}y^{3/5}$

39. $y^{4/5} = \sqrt[5]{y^4}$

40. $5(xy^2z^5)^{1/3} = 5\sqrt[3]{xy^2z^5}$

41. $(x+2)^{-1/3} = \frac{1}{\sqrt[3]{x+2}}$

42. $(x+2y)^{-1/2} = \frac{1}{\sqrt{x+2y}}$

43. $a^{1/3}a^{4/3}a^{1/2} = a^{2/6}a^{8/6}a^{3/6} = a^{13/6}$

44. $\frac{b^{1/3}}{b^{4/3}} = b^{\frac{1}{3}-\frac{4}{3}} = b^{-3/3} = b^{-1} = \frac{1}{b}$

45. $(a^{1/2}a^{-2})^3 = a^{3/2}a^{-6} = a^{\frac{3}{2}-\frac{12}{2}} = a^{-9/2} = \frac{1}{a^{9/2}}$

46. $(x^{-3}y^6)^{1/3} = x^{-3/3}y^{6/3} = x^{-1}y^2 = \frac{y^2}{x}$

47. $\left(\frac{b^{3/4}}{a^{-1/2}}\right)^8 = \frac{b^{24/4}}{a^{-8/2}} = \frac{b^6}{a^{-4}} = a^4b^6$

48. $\frac{x^{1/4}x^{-1/2}}{x^{2/3}} = x^{\frac{1}{4}-\frac{1}{2}-\frac{2}{3}} = x^{\frac{3}{12}-\frac{6}{12}-\frac{8}{12}} = x^{-\frac{11}{12}} = \frac{1}{x^{11/12}}$

49. $\left(\frac{49c^{5/3}}{a^{-1/4}b^{5/6}}\right)^{-1} = \left(\frac{a^{-1/4}b^{5/6}}{49c^{5/3}}\right) = \frac{b^{5/6}}{49a^{1/4}c^{5/3}}$

50. $a^{-1/4}(a^{5/4} - a^{9/4}) = a^{-\frac{1}{4}+\frac{5}{4}} - a^{-\frac{1}{4}+\frac{9}{4}}$
$= a^{\frac{4}{4}} - a^{\frac{8}{4}}$
$= a - a^2$

51. $\sqrt{20} \approx 4.472$

52. $\sqrt[3]{-39} \approx -3.391$

53. $\sqrt[4]{726} \approx 5.191$

54. $56^{1/3} \approx 3.826$

55. $-78^{3/4} \approx -26.246$

56. $105^{-2/3} \approx 0.045$

57. $\sqrt[3]{2} \cdot \sqrt{7} = 2^{1/3} \cdot 7^{1/2}$
$= 2^{2/6} \cdot 7^{3/6}$
$= \sqrt[6]{2^2 \cdot 7^3}$
$= \sqrt[6]{1372}$

58. $\sqrt[3]{3} \cdot \sqrt[4]{x} = 3^{1/3}x^{1/4}$
$= 3^{4/12} \cdot x^{3/12}$
$= \sqrt[12]{3^4x^3}$
$= \sqrt[12]{81x^3}$

59. $\sqrt{3} \cdot \sqrt{8} = \sqrt{24} = \sqrt{4}\sqrt{6} = 2\sqrt{6}$

60. $\sqrt[3]{7y} \cdot \sqrt[3]{x^2z} = \sqrt[3]{7x^2yz}$

61. $\frac{\sqrt{44x^3}}{\sqrt{11x}} = \sqrt{\frac{44x^3}{11x}} = \sqrt{4x^2} = 2x$

62. $\frac{\sqrt[4]{a^6b^{13}}}{\sqrt[4]{a^2b}} = \sqrt[4]{\frac{a^6b^{13}}{a^2b}} = \sqrt[4]{a^4b^{12}} = ab^3$

63. $\sqrt{60} = \sqrt{4}\sqrt{15} = 2\sqrt{15}$

64. $-\sqrt{75} = -\sqrt{25}\sqrt{3} = -5\sqrt{3}$

65. $\sqrt[3]{162} = \sqrt[3]{27}\sqrt[3]{6} = 3\sqrt[3]{6}$

66. $\sqrt[3]{-32} = \sqrt[3]{-8}\sqrt[3]{4} = -2\sqrt[3]{4}$

67. $\sqrt{36x^7} = \sqrt{36x^6}\sqrt{x} = 6x^3\sqrt{x}$

68. $\sqrt[3]{24a^5b^7} = \sqrt[3]{8a^3b^6}\sqrt[3]{3a^2b^1} = 2ab^2\sqrt[3]{3a^2b}$

69. $\sqrt{\dfrac{p^{17}}{121}} = \dfrac{\sqrt{p^{16}}\sqrt{p}}{\sqrt{121}} = \dfrac{p^8\sqrt{p}}{11}$

70. $\sqrt[3]{\dfrac{y^5}{27x^6}} = \dfrac{\sqrt[3]{y^3}\sqrt[3]{y^2}}{\sqrt[3]{27x^6}} = \dfrac{y\sqrt[3]{y^2}}{3x^2}$

71. $\sqrt[4]{\dfrac{xy^6}{81}} = \dfrac{\sqrt[4]{y^4}\sqrt[4]{xy^2}}{\sqrt[4]{81}} = \dfrac{y\sqrt[4]{xy^2}}{3}$

72. $\sqrt{\dfrac{2x^3}{49y^4}} = \dfrac{\sqrt{x^2}\sqrt{2x}}{\sqrt{49y^4}} = \dfrac{x\sqrt{2x}}{7y^2}$

73. $r = \sqrt{\dfrac{A}{\pi}}$
$= \sqrt{\dfrac{25}{\pi}}$
$= \dfrac{\sqrt{25}}{\sqrt{\pi}}$
$= \dfrac{5}{\sqrt{\pi}}$ meters or $\dfrac{5\sqrt{\pi}}{\pi}$ meters

74. $r = \sqrt{\dfrac{A}{\pi}} = \sqrt{\dfrac{104}{\pi}} = 5.75$ inches

75. $(x_1, y_1) = (-6, 3), (x_2, y_2) = (8, 4)$
$d = \sqrt{(x_2 - x_1)^2 + (y_2 - y_1)^2}$
$= \sqrt{(8+6)^2 + (4-3)^2}$
$= \sqrt{196+1}$
$= \sqrt{197}$ units ≈ 14.036 units

76. $(x_1, y_1) = (-4, -6), (x_2, y_2) = (-1, 5)$
$d = \sqrt{(x_2 - x_1)^2 + (y_2 - y_1)^2}$
$= \sqrt{(-1+4)^2 + (5+6)^2}$
$= \sqrt{9+121}$
$= \sqrt{130}$ units ≈ 11.402 units

77. $(x_1, y_1) = (-1, 5), (x_2, y_2) = (2, -3)$
$d = \sqrt{(x_2 - x_1)^2 + (y_2 - y_1)^2}$
$= \sqrt{(2+1)^2 + (-3-5)^2}$
$= \sqrt{9+64}$
$= \sqrt{73}$ units ≈ 8.544 units

78. $(x_1, y_1) = \left(-\sqrt{2}, 0\right), (x_2, y_2) = \left(0, -4\sqrt{6}\right)$
$d = \sqrt{(x_2 - x_1)^2 + (y_2 - y_1)^2}$
$= \sqrt{\left(0+\sqrt{2}\right)^2 + \left(-4\sqrt{6}-0\right)^2}$
$= \sqrt{2+96}$
$= \sqrt{98}$
$= 7\sqrt{2}$ units ≈ 9.899 units

79. $(x_1, y_1) = \left(-\sqrt{5}, -\sqrt{11}\right)$,
$(x_2, y_2) = \left(-\sqrt{5}, -3\sqrt{11}\right)$
$d = \sqrt{(x_2 - x_1)^2 + (y_2 - y_1)^2}$
$= \sqrt{\left(-\sqrt{5}+\sqrt{5}\right)^2 + \left(-3\sqrt{11}+\sqrt{11}\right)^2}$
$= \sqrt{0+44}$
$= \sqrt{44}$
$= 2\sqrt{11}$ units ≈ 6.633 units

80. $(x_1, y_1) = (7.4, -8.6), (x_2, y_2) = (-1.2, 5.6)$
$d = \sqrt{(-1.2-7.4)^2 + (5.6+8.6)^2}$
$= \sqrt{(-8.6)^2 + (14.2)^2}$
$= \sqrt{73.96+201.64}$
$= \sqrt{275.6}$ units ≈ 16.601 units

81. $(x_1, y_1) = (2, 6), (x_2, y_2) = (-12, 4)$

$$\begin{aligned}\text{midpoint} &= \left(\frac{x_1+x_2}{2}, \frac{y_1+y_2}{2}\right)\\ &= \left(\frac{2-12}{2}, \frac{6+4}{2}\right)\\ &= \left(\frac{-10}{2}, \frac{10}{2}\right)\\ &= (-5, 5)\end{aligned}$$

82. $(x_1, y_1) = (-6, -5), (x_2, y_2) = (-9, 7)$

$$\begin{aligned}\text{midpoint} &= \left(\frac{x_1+x_2}{2}, \frac{y_1+y_2}{2}\right)\\ &= \left(\frac{-6-9}{2}, \frac{-5+7}{2}\right)\\ &= \left(-\frac{15}{2}, 1\right)\end{aligned}$$

83. $(x_1, y_1) = (4, -6), (x_2, y_2) = (-15, 2)$

$$\begin{aligned}\text{midpoint} &= \left(\frac{x_1+x_2}{2}, \frac{y_1+y_2}{2}\right)\\ &= \left(\frac{4-15}{2}, \frac{-6+2}{2}\right)\\ &= \left(-\frac{11}{2}, -2\right)\end{aligned}$$

84. $(x_1, y_1) = \left(0, -\frac{3}{8}\right), (x_2, y_2) = \left(\frac{1}{10}, 0\right)$

$$\begin{aligned}\text{midpoint} &= \left(\frac{x_1+x_2}{2}, \frac{y_1+y_2}{2}\right)\\ &= \left(\frac{0+\frac{1}{10}}{2}, \frac{-\frac{3}{8}+0}{2}\right)\\ &= \left(\frac{1}{20}, -\frac{3}{16}\right)\end{aligned}$$

85. $(x_1, y_1) = \left(\frac{3}{4}, -\frac{1}{7}\right), (x_2, y_2) = \left(-\frac{1}{4}, -\frac{3}{7}\right)$

$$\begin{aligned}\text{midpoint} &= \left(\frac{\frac{3}{4}-\frac{1}{4}}{2}, \frac{-\frac{1}{7}-\frac{3}{7}}{2}\right)\\ &= \left(\frac{\frac{1}{2}}{2}, \frac{-\frac{11}{7}}{2}\right)\\ &= \left(\frac{1}{4}, -\frac{2}{7}\right)\end{aligned}$$

86. $(x_1, y_1) = \left(\sqrt{3}, -2\sqrt{6}\right), (x_2, y_2) = \left(\sqrt{3}, -4\sqrt{6}\right)$

$$\begin{aligned}\text{midpoint} &= \left(\frac{x_1+x_2}{2}, \frac{y_1+y_2}{2}\right)\\ &= \left(\frac{\sqrt{3}+\sqrt{3}}{2}, \frac{-2\sqrt{6}-4\sqrt{6}}{2}\right)\\ &= \left(\frac{2\sqrt{3}}{2}, \frac{-6\sqrt{6}}{2}\right)\\ &= \left(\sqrt{3}, -3\sqrt{6}\right)\end{aligned}$$

87. $\sqrt{20}+\sqrt{45}-7\sqrt{5} = 2\sqrt{5}+3\sqrt{5}-7\sqrt{5} = -2\sqrt{5}$

88. $x\sqrt{75x}-\sqrt{27x^3} = 5x\sqrt{3x}-3x\sqrt{3x} = 2x\sqrt{3x}$

89. $\sqrt[3]{128}+\sqrt[3]{250} = 4\sqrt[3]{2}+5\sqrt[3]{2} = 9\sqrt[3]{2}$

90. $3\sqrt[4]{32a^5}-a\sqrt[4]{162a} = 6a\sqrt[4]{2a}-3a\sqrt[4]{2a} = 3a\sqrt[4]{2a}$

91. $\frac{5}{\sqrt{4}}+\frac{\sqrt{3}}{3} = \frac{5}{2}+\frac{\sqrt{3}}{3} = \frac{15}{6}+\frac{2\sqrt{3}}{6} = \frac{15+2\sqrt{3}}{6}$

92.
$$\begin{aligned}\sqrt{\frac{8}{x^2}}-\sqrt{\frac{50}{16x^2}} &= \frac{2\sqrt{2}}{x}-\frac{5\sqrt{2}}{4x}\\ &= \frac{8\sqrt{2}-5\sqrt{2}}{4x}\\ &= \frac{3\sqrt{2}}{4x}\end{aligned}$$

93.
$$\begin{aligned}2\sqrt{50}-3\sqrt{125}+\sqrt{98} &= 10\sqrt{2}-15\sqrt{5}+7\sqrt{2}\\ &= 17\sqrt{2}-15\sqrt{5}\end{aligned}$$

94.
$$\begin{aligned}&2a\sqrt[4]{32b^5}-3b\sqrt[4]{162a^4b}+\sqrt[4]{2a^4b^5}\\ &= 4ab\sqrt[4]{2b}-9ab\sqrt[4]{2b}+ab\sqrt[4]{2b}\\ &= -4ab\sqrt[4]{2b}\end{aligned}$$

95.
$$\begin{aligned}\sqrt{3}\left(\sqrt{27}-\sqrt{3}\right) &= \sqrt{3}\sqrt{27}-\sqrt{3}\sqrt{3}\\ &= \sqrt{81}-\sqrt{9}\\ &= 9-3\\ &= 6\end{aligned}$$

96. $\left(\sqrt{x}-3\right)^2 = \left(\sqrt{x}\right)^2-(2)(3)\sqrt{x}+9 = x-6\sqrt{x}+9$

97. $(\sqrt{5}-5)(2\sqrt{5}+2)$
$= \sqrt{5}(2\sqrt{5})+2\sqrt{5}-10\sqrt{5}-10$
$= 10-8\sqrt{5}-10$
$= -8\sqrt{5}$

98. $(2\sqrt{x}-3\sqrt{y})(2\sqrt{x}+3\sqrt{y}) = (2\sqrt{x})^2-(3\sqrt{y})^2$
$= 4x-9y$

99. $(\sqrt{a}+3)(\sqrt{a}-3) = (\sqrt{a})^2-(3)^2 = a-9$

100. $(\sqrt[3]{a}+2)^2 = (\sqrt[3]{a})^2+2(2)(\sqrt[3]{a})+2^2$
$= \sqrt[3]{a^2}+4\sqrt[3]{a}+4$

101. $(\sqrt[3]{5x}+9)(\sqrt[3]{5x}-9) = (\sqrt[3]{5x})^2-9^2$
$= \sqrt[3]{25x^2}-81$

102. $(\sqrt[3]{a}+4)\left(\sqrt[3]{a^2}-4\sqrt[3]{a}+16\right)$
$= \sqrt[3]{a}\sqrt[3]{a^2}-\sqrt[3]{a}(4\sqrt[3]{a})+16\sqrt[3]{a}+4\sqrt[3]{a^2}-16\sqrt[3]{a}+64$
$= a-4\sqrt[3]{a^2}+4\sqrt[3]{a^2}+64$
$= a+64$

103. $\frac{3}{\sqrt{7}} = \frac{3}{\sqrt{7}}\cdot\frac{\sqrt{7}}{\sqrt{7}} = \frac{3\sqrt{7}}{7}$

104. $\sqrt{\frac{x}{12}} = \frac{\sqrt{x}}{\sqrt{12}} = \frac{\sqrt{x}}{2\sqrt{3}}\cdot\frac{\sqrt{3}}{\sqrt{3}} = \frac{\sqrt{3x}}{6}$

105. $\frac{5}{\sqrt[3]{4}} = \frac{5}{\sqrt[3]{4}}\cdot\frac{\sqrt[3]{2}}{\sqrt[3]{2}} = \frac{5\sqrt[3]{2}}{2}$

106. $\sqrt{\frac{24x^5}{3y}} = \frac{\sqrt{24x^5}}{\sqrt{3y}}$
$= \frac{2x^2\sqrt{6x}}{\sqrt{3y}}\cdot\frac{\sqrt{3y}}{\sqrt{3y}}$
$= \frac{2x^2\sqrt{9xy}}{3y}$
$= \frac{6x^2\sqrt{2xy}}{3y}$
$= \frac{2x^2\sqrt{2xy}}{y}$

107. $\sqrt[3]{\frac{15x^6y^7}{z^2}} = \frac{\sqrt[3]{15x^6y^7}}{\sqrt[3]{z^2}}$
$= \frac{x^2y^2\sqrt[3]{15y}}{\sqrt[3]{z^2}}\cdot\frac{\sqrt[3]{z}}{\sqrt[3]{z}}$
$= \frac{x^2y^2\sqrt[3]{15yz}}{z}$

108. $\sqrt[4]{\frac{81}{8x^{10}}} = \frac{\sqrt[4]{81}}{\sqrt[4]{8x^{10}}} = \frac{3}{x^2\sqrt[4]{8x^2}}\cdot\frac{\sqrt[4]{2x^2}}{\sqrt[4]{2x^2}} = \frac{3\sqrt[4]{2x^2}}{2x^3}$

109. $\frac{3}{\sqrt{y}-2} = \frac{3}{\sqrt{y}-2}\cdot\frac{\sqrt{y}+2}{\sqrt{y}+2} = \frac{3\sqrt{y}+6}{y-4}$

110. $\frac{\sqrt{2}-\sqrt{3}}{\sqrt{2}+\sqrt{3}} = \frac{\sqrt{2}-\sqrt{3}}{\sqrt{2}+\sqrt{3}}\cdot\frac{\sqrt{2}-\sqrt{3}}{\sqrt{2}-\sqrt{3}}$
$= \frac{2-2\sqrt{6}+3}{2-3}$
$= -5+2\sqrt{6}$

111. $\frac{\sqrt{11}}{3} = \frac{\sqrt{11}}{3}\cdot\frac{\sqrt{11}}{\sqrt{11}} = \frac{11}{3\sqrt{11}}$

112. $\sqrt{\frac{18}{y}} = \frac{\sqrt{18}}{\sqrt{y}} = \frac{3\sqrt{2}}{\sqrt{y}}\cdot\frac{\sqrt{2}}{\sqrt{2}} = \frac{6}{\sqrt{2y}}$

113. $\frac{\sqrt[3]{9}}{7} = \frac{\sqrt[3]{9}}{7}\cdot\frac{\sqrt[3]{3}}{\sqrt[3]{3}} = \frac{3}{7\sqrt[3]{3}}$

114. $\sqrt{\frac{24x^5}{3y^2}} = \frac{\sqrt{24x^5}}{\sqrt{3y^2}}$
$= \frac{2x^2\sqrt{6x}}{y\sqrt{3}} \cdot \frac{\sqrt{6x}}{\sqrt{6x}}$
$= \frac{12x^3}{3y\sqrt{2x}}$
$= \frac{4x^3}{y\sqrt{2x}}$

115. $\sqrt[3]{\frac{xy^2}{10z}} = \frac{\sqrt[3]{xy^2}}{\sqrt[3]{10z}} = \frac{\sqrt[3]{xy^2}}{\sqrt[3]{10z}} \cdot \frac{\sqrt[3]{x^2y}}{\sqrt[3]{x^2y}} = \frac{xy}{\sqrt[3]{10x^2yz}}$

116. $\frac{\sqrt{x}+5}{-3} = \frac{\sqrt{x}+5}{-3} \cdot \frac{\sqrt{x}-5}{\sqrt{x}-5} = \frac{x-25}{-3\sqrt{x}+15}$

117. $\sqrt{y-7} = 5$
$y-7 = 25$
$y = 32$
The solution is 32.

118. $\sqrt{2x}+10 = 4$
$\sqrt{2x} = -6$
No solution since a principal square root does not yield a negative number. The solution set is $\varnothing$.

119. $\sqrt[3]{2x-6} = 4$
$\left(\sqrt[3]{2x-6}\right)^3 = 4^3$
$2x-6 = 64$
$2x = 70$
$x = 35$
The solution is 35.

120. $\sqrt{x+6} = \sqrt{x+2}$
$\left(\sqrt{x+6}\right)^2 = \left(\sqrt{x+2}\right)^2$
$x+6 = x+2$
$6 = 2$ False
The solution set is $\varnothing$.

121. $2x-5\sqrt{x} = 3$
$2x-3 = 5\sqrt{x}$
$4x^2-12x+9 = 25x$
$4x^2-37x+9 = 0$
$(4x-1)(x-9) = 0$
$4x-1 = 0$ or $x-9 = 0$
$4x = 1$ $\quad x = 9$
$x = \frac{1}{4}$

We discard the $\frac{1}{4}$ as extraneous, leaving $x = 9$ as the only solution. The solution is 9.

122. $\sqrt{x+9} = 2+\sqrt{x-7}$
$x+9 = 4+(2)2\sqrt{x-7}+x-7$
$12 = 4\sqrt{x-7}$
$3 = \sqrt{x-7}$
$9 = x-7$
$16 = x$
The solution is 16.

123. $a^2+b^2 = c^2$
$3^2+3^2 = c^2$
$9+9 = c^2$
$18 = c^2$
$\sqrt{18} = c$
$3\sqrt{2} = c$
The unknown length is $3\sqrt{2}$ centimeters.

124. $a^2+b^2 = c^2$
$7^2+\left(8\sqrt{3}\right)^2 = c^2$
$49+192 = c^2$
$241 = c^2$
$\sqrt{241} = c$
The unknown length is $\sqrt{241}$ feet.

125. $a^2+b^2 = c^2$
$a^2+40^2 = 65^2$
$a^2+1600 = 4225$
$a^2 = 2625$
$a = 51.2$
The width of the pond is 51.2 feet.

126.
$$\begin{aligned} a^2+b^2 &= c^2 \\ 3^2+3^2 &= c^2 \\ 9+9 &= c^2 \\ 18 &= c^2 \\ \sqrt{18} &= c \\ 3\sqrt{2} \text{ or } 4.24 &= c \end{aligned}$$
The length is 4.24 feet.

127. $\sqrt{-8}=\sqrt{-1}\cdot\sqrt{8}=2i\sqrt{2}=0+2i\sqrt{2}$

128. $-\sqrt{-6}=-\sqrt{-1}\cdot\sqrt{6}=-i\sqrt{6}=0-i\sqrt{6}$

129.
$$\begin{aligned} \sqrt{-4}+\sqrt{-16} &= \sqrt{-1}\sqrt{4}+\sqrt{-1}\sqrt{16} \\ &= 2i+4i \\ &= 6i \\ &= 0+6i \end{aligned}$$

130.
$$\begin{aligned} \sqrt{-2}\cdot\sqrt{-5} &= \sqrt{-1}\cdot\sqrt{2}\cdot\sqrt{-1}\cdot\sqrt{5} \\ &= i\sqrt{2}\cdot i\sqrt{5} \\ &= i^2\sqrt{10} \\ &= -\sqrt{10} \\ &= -\sqrt{10}+0i \end{aligned}$$

131. $(12-6i)+(3+2i)=(12+3)+(-6+2)i=15-4i$

132.
$$\begin{aligned} (-8-7i)-(5-4i) &= (-8-5)+[-7-(-4)]i \\ &= -13-3i \end{aligned}$$

133. $(2i)^6=2^6\cdot i^6=64\cdot(-1)=-64$

134. $(3i)^4=3^4\cdot i^4=81\cdot 1=81$

135. $-3i(6-4i)=-18i+12(-1)=-12-18i$

136. $(3+2i)(1+i)=3+3i+2i-2=1+5i$

137.
$$\begin{aligned} (2-3i)^2 &= 2^2-2(2)(3i)+(3i)^2 \\ &= 4-12i-9 \\ &= -5-12i \end{aligned}$$

138. $\left(\sqrt{6}-9i\right)\left(\sqrt{6}+9i\right)=\left(\sqrt{6}\right)^2-(9i)^2=6+81=87$

139. $\dfrac{2+3i}{2i}=\dfrac{2+3i}{2i}\cdot\dfrac{-2i}{-2i}=\dfrac{-4i+6}{4}=\dfrac{3}{2}-i$

140. $\dfrac{1+i}{-3i}=\dfrac{1+i}{-3i}\cdot\dfrac{3i}{3i}=\dfrac{3i-3}{9}=\dfrac{-1+i}{3}=-\dfrac{1}{3}+\dfrac{1}{3}i$

141. $\sqrt[3]{x^3}=x$

142. $\sqrt{(x+2)^2}=|x+2|$

143. $-\sqrt{100}=-10$

144. $\sqrt[3]{-x^{12}y^3}=-x^4y$

145. $\sqrt[4]{\dfrac{y^{20}}{16x^{12}}}=\dfrac{\sqrt[4]{y^{20}}}{\sqrt[4]{16x^{12}}}=\dfrac{y^5}{2x^3}$

146. $9^{1/2}=\sqrt{9}=3$

147. $64^{-1/2}=\dfrac{1}{64^{1/2}}=\dfrac{1}{\sqrt{64}}=\dfrac{1}{8}$

148. $\left(\dfrac{27}{64}\right)^{-2/3}=\left(\dfrac{64}{27}\right)^{2/3}=\left(\sqrt[3]{\dfrac{64}{27}}\right)^2=\left(\dfrac{4}{3}\right)^2=\dfrac{16}{9}$

149.
$$\begin{aligned} \frac{(x^{2/3}x^{-3})^3}{x^{-1/2}} &= \frac{x^{6/3}x^{-9}}{x^{-1/2}} \\ &= x^{2-9+\frac{1}{2}} \\ &= x^{-13/2} \\ &= \frac{1}{x^{13/2}} \end{aligned}$$

150. $\sqrt{200x^9}=10x^4\sqrt{2x}$

151. $\sqrt{\dfrac{3n^3}{121m^{10}}}=\dfrac{\sqrt{3n^3}}{\sqrt{121m^{10}}}=\dfrac{n\sqrt{3n}}{11m^5}$

152.
$$\begin{aligned} &3\sqrt{20}-7x\sqrt[3]{40}+3\sqrt[3]{5x^3} \\ &= 3\sqrt{4}\sqrt{5}-7x\sqrt[3]{8}\sqrt[3]{54}+3\sqrt[3]{x^3}\sqrt[3]{5} \\ &= 6\sqrt{5}-14x\sqrt[3]{5}+3x\sqrt[3]{5} \\ &= 6\sqrt{5}-11x\sqrt[3]{5} \end{aligned}$$

153.
$$\begin{aligned} \left(2\sqrt{x}-5\right)^2 &= \left(2\sqrt{x}\right)^2-2(5)\left(2\sqrt{x}\right)+5^2 \\ &= 4x-20\sqrt{x}+25 \end{aligned}$$

154. $(x_1, y_1) = (-3, 5), (x_2, y_2) = (-8, 9)$

$$\begin{aligned} d &= \sqrt{(x_2 - x_1)^2 + (y_2 - y_1)^2} \\ &= \sqrt{(-8+3)^2 + (9-5)^2} \\ &= \sqrt{(-5)^2 + (4)^2} \\ &= \sqrt{25+16} \\ &= \sqrt{41} \end{aligned}$$

The distance is $\sqrt{41}$ units.

155. $(x_1, y_1) = (-3, 8), (x_2, y_2) = (11, 24)$

$$\begin{aligned} \text{midpoint} &= \left(\frac{x_1 + x_2}{2}, \frac{y_1 + y_2}{2}\right) \\ &= \left(\frac{-3+11}{2}, \frac{8+24}{2}\right) \\ &= \left(\frac{8}{2}, \frac{32}{2}\right) \\ &= (4, 16) \end{aligned}$$

156. $\frac{7}{\sqrt{13}} = \frac{7}{\sqrt{13}} \cdot \frac{\sqrt{13}}{\sqrt{13}} = \frac{7\sqrt{13}}{13}$

157. $\frac{2}{\sqrt{x}+3} = \frac{2}{\sqrt{x}+3} \cdot \frac{\sqrt{x}-3}{\sqrt{x}-3} = \frac{2\sqrt{x}-6}{x-9}$

158. $\sqrt{x} + 2 = x$

$$\begin{aligned} \sqrt{x} &= x - 2 \\ \left(\sqrt{x}\right)^2 &= (x-2)^2 \\ x &= x^2 - 4x + 4 \\ 0 &= x^2 - 5x + 4 \\ 0 &= (x-4)(x-1) \end{aligned}$$

$x - 4 = 0$ or $x - 1 = 0$

$x = 4$ $\quad x = 1$

Discard the extraneous solution $x = 1$. The solution is 4.

159. $\sqrt{2x-1} + 2 = x$

$$\begin{aligned} \sqrt{2x-1} &= x - 2 \\ \left(\sqrt{2x-1}\right)^2 &= (x-2)^2 \\ 2x - 1 &= x^2 - 4x + 4 \\ 0 &= x^2 - 6x + 5 \\ 0 &= (x-5)(x-1) \end{aligned}$$

$x - 5 = 0$ or $x - 1 = 0$

$x = 5$ or $x = 1$

Discard the extraneous solution $x = 1$. The solution is 5.

Chapter 7 Getting Ready for the Test

1. $\sqrt{16} = \sqrt{4^2} = 4$

$\sqrt{-16}$ is not a real number.

$\sqrt[3]{64} = \sqrt[3]{4^3} = 4$

$\sqrt[3]{-64} = \sqrt[3]{(-4)^3} = -4$

Choice D is correct.

2. $\sqrt{16} = \sqrt{4^2} = 4$

$\sqrt[3]{16} = \sqrt[3]{2^4} = \sqrt[3]{2^3} \cdot \sqrt[3]{2} = 2\sqrt[3]{2}$

$\sqrt{64} = \sqrt{8^2} = 8$

$\sqrt[3]{64} = \sqrt[3]{4^3} = 4$

Choice B is correct.

3. $\sqrt{5}$ is a real number, since $5 \geq 0$; A.

4. $\sqrt{-11}$ is not a real number, since $-11 < 0$; B.

5. $\sqrt[3]{-9}$ is a real number, since the index (3) is odd; A.

6. $-\sqrt{17}$ is the opposite of $\sqrt{17}$, which is a real number since $17 \geq 0$; A.

7. $25^{1/2} = \sqrt{25} = 5$; A

8. $(-25)^{1/2} = \sqrt{-25}$ is not a real number; C.

9. $(-125)^{1/3} = \sqrt[3]{-125} = -5$, since $(-5)^3 = -125$; B.

10. $-25^{1/2} = -(25^{1/2}) = -\sqrt{25} = -5$; B.

11. $\sqrt{5} \cdot \sqrt{10} = \sqrt{5 \cdot 10} = \sqrt{50} = \sqrt{25 \cdot 2} = 5\sqrt{2}$

The statement is false; B.

12. $\sqrt[3]{4} \cdot \sqrt[3]{9} = \sqrt[3]{4 \cdot 9} = \sqrt[3]{36}$

The statement is true; A.

13. $\frac{\sqrt{12}}{\sqrt{6}} = \sqrt{\frac{12}{6}} = \sqrt{2}$

The statement is true; A.

14. $\frac{\sqrt[3]{10}}{\sqrt[3]{4}} = \sqrt[3]{\frac{10}{4}} = \sqrt[3]{\frac{5}{2}}$

The statement is false; B.

15. $\sqrt{2}+\sqrt{3}$ cannot be simplified since $\sqrt{2}$ and $\sqrt{3}$ are not like radicals (the radicands are different). The statement is false; B.

16. $\sqrt[3]{7}+\sqrt{7}$ cannot be simplified since $\sqrt[3]{7}$ and $\sqrt{7}$ are not like radicals (the indices are different). The statement is false; B.

17. $x^{1/2}+x^{1/2}=1x^{1/2}+1x^{1/2}=(1+1)x^{1/2}=2x^{1/2}$

$x^{1/2}\cdot x^{1/2}=x^{\frac{1}{2}+\frac{1}{2}}=x^1=x$

$(x^{1/2})^2=x^{\frac{1}{2}\cdot 2}=x^1=x$

Since both B and C simplify to x, choice D is correct.

18. To rationalize the numerator of $\dfrac{\sqrt{x}-3}{\sqrt{x}}$ we multiply by 1 in the form of a fraction with the conjugate of $\sqrt{x}-3$ as both the numerator and the denominator, or $\dfrac{\sqrt{x}+3}{\sqrt{x}+3}$; C.

19.
$$\begin{aligned}\sqrt{x+1}&=3+\sqrt{x-1}\\ \left(\sqrt{x+1}\right)^2&=\left(3+\sqrt{x-1}\right)^2\\ x+1&=3^2+2\cdot 3\cdot\sqrt{x-1}+\left(\sqrt{x-1}\right)^2\\ x+1&=9+6\sqrt{x+1}+(x-1)\end{aligned}$$

Choice B is correct.

20.
$$\begin{aligned}(5-2i)^2&=(5-2i)(5-2i)\\ &=5^2-2\cdot 5\cdot 2i+(2i)^2\\ &=25-20i+4i^2\\ &=25-20i+4(-1)\\ &=25-20i-4\\ &=21-20i\end{aligned}$$

Choice D is correct.

21. $\dfrac{3+\sqrt{-9}}{3}=\dfrac{3+i\sqrt{9}}{3}=\dfrac{3+3i}{3}=\dfrac{3(1+i)}{3}=1+i$; A

Chapter 7 Test

1. $\sqrt{216}=\sqrt{36\cdot 6}=6\sqrt{6}$

2. $-\sqrt[4]{x^{64}}=-x^{16}$

3. $\left(\dfrac{1}{125}\right)^{1/3}=\dfrac{1}{125^{1/3}}=\dfrac{1}{\sqrt[3]{125}}=\dfrac{1}{5}$

4. $\left(\dfrac{1}{125}\right)^{-1/3}=\dfrac{1}{\left(\frac{1}{125}\right)^{1/3}}=\dfrac{1}{\frac{1}{5}}=5$

5.
$$\begin{aligned}\left(\frac{8x^3}{27}\right)^{2/3}&=\frac{(8x^3)^{2/3}}{27^{2/3}}\\ &=\frac{\left(\sqrt[3]{8x^3}\right)^2}{\left(\sqrt[3]{27}\right)^2}\\ &=\frac{(2x)^2}{3^2}\\ &=\frac{4x^2}{9}\end{aligned}$$

6. $\sqrt[3]{-a^{18}b^9}=\sqrt[3]{-1a^{18}b^9}=(-1)a^6b^3=-a^6b^3$

7.
$$\begin{aligned}\left(\frac{64c^{4/3}}{a^{-2/3}b^{5/6}}\right)^{1/2}&=\left(\frac{64a^{2/3}c^{4/3}}{b^{5/6}}\right)^{1/2}\\ &=\frac{64^{1/2}(a^{2/3})^{1/2}(c^{4/3})^{1/2}}{(b^{5/6})^{1/2}}\\ &=\frac{\sqrt{64}a^{1/3}c^{2/3}}{b^{5/12}}\\ &=\frac{8a^{1/3}c^{2/3}}{b^{5/12}}\end{aligned}$$

8.
$$\begin{aligned}a^{-2/3}(a^{5/4}-a^3)&=a^{-2/3}a^{5/4}-a^{-2/3}a^3\\ &=a^{-\frac{2}{3}+\frac{5}{4}}-a^{-\frac{2}{3}+3}\\ &=a^{-\frac{8}{12}+\frac{15}{12}}-a^{-\frac{2}{3}+\frac{9}{3}}\\ &=a^{7/12}-a^{7/3}\end{aligned}$$

9. $\sqrt[4]{(4xy)^4}=|4xy|=4|xy|$

10. $\sqrt[3]{(-27)^3}=-27$

11. $\sqrt{\dfrac{9}{y}}=\dfrac{\sqrt{9}}{\sqrt{y}}=\dfrac{3}{\sqrt{y}}=\dfrac{3\cdot\sqrt{y}}{\sqrt{y}\cdot\sqrt{y}}=\dfrac{3\sqrt{y}}{y}$

12. $\frac{4-\sqrt{x}}{4+2\sqrt{x}}=\frac{4-\sqrt{x}}{2(2+\sqrt{x})}$
$=\frac{(4-\sqrt{x})(2-\sqrt{x})}{2(2+\sqrt{x})(2-\sqrt{x})}$
$=\frac{8-4\sqrt{x}-2\sqrt{x}+x}{2\left[2^2-(\sqrt{x})^2\right]}$
$=\frac{8-6\sqrt{x}+x}{2(4-x)}$ or $\frac{8-6\sqrt{x}+x}{8-2x}$

13. $\frac{\sqrt[3]{ab}}{\sqrt[3]{ab^2}}=\sqrt[3]{\frac{ab}{ab^2}}$
$=\sqrt[3]{\frac{1}{b}}$
$=\frac{1}{\sqrt[3]{b}}$
$=\frac{1\cdot\sqrt[3]{b^2}}{\sqrt[3]{b}\cdot\sqrt[3]{b^2}}$
$=\frac{\sqrt[3]{b^2}}{b}$

14. $\frac{\sqrt{6}+x}{8}=\frac{(\sqrt{6}+x)(\sqrt{6}-x)}{8(\sqrt{6}-x)}$
$=\frac{(\sqrt{6})^2-x^2}{8(\sqrt{6}-x)}$
$=\frac{6-x^2}{8(\sqrt{6}-x)}$

15. $\sqrt{125x^3}-3\sqrt{20x^3}=\sqrt{25x^2\cdot 5x}-3\sqrt{4x^2\cdot 5x}$
$=5x\sqrt{5x}-3\cdot 2x\sqrt{5x}$
$=5x\sqrt{5x}-6x\sqrt{5x}$
$=-x\sqrt{5x}$

16. $\sqrt{3}(\sqrt{16}-\sqrt{2})=\sqrt{3}(4-\sqrt{2})$
$=4\sqrt{3}-\sqrt{3}\sqrt{2}$
$=4\sqrt{3}-\sqrt{6}$

17. $(\sqrt{x}+1)^2=(\sqrt{x})^2+2\sqrt{x}+1^2$
$=x+2\sqrt{x}+1$

18. $(\sqrt{2}-4)(\sqrt{3}+1)=\sqrt{2}\sqrt{3}+1\cdot\sqrt{2}-4\sqrt{3}-4$
$=\sqrt{6}+\sqrt{2}-4\sqrt{3}-4$

19. $(\sqrt{5}+5)(\sqrt{5}-5)=(\sqrt{5})^2-5^2$
$=5-25$
$=-20$

20. $\sqrt{561}\approx 23.685$

21. $386^{-2/3}\approx 0.019$

22. $x=\sqrt{x-2}+2$
$x-2=\sqrt{x-2}$
$(x-2)^2=(\sqrt{x-2})^2$
$x^2-4x+4=x-2$
$x^2-5x+6=0$
$(x-2)(x-3)=0$
$x=2$ or $x=3$
The solutions are 2 and 3.

23. $\sqrt{x^2-7}+3=0$
$\sqrt{x^2-7}=-3$
No solution exists since the principle square root of a number is not negative, ∅.

24. $\sqrt[3]{x+5}=\sqrt[3]{2x-1}$
$(\sqrt[3]{x+5})^3=(\sqrt[3]{2x-1})^3$
$x+5=2x-1$
$-x=-6$
$x=6$
The solution is 6.

25. $\sqrt{-2}=i\sqrt{2}=0+i\sqrt{2}$

26. $-\sqrt{-8}=-i\sqrt{4\cdot 2}=-2i\sqrt{2}=0-2i\sqrt{2}$

27. $(12-6i)-(12-3i)=12-6i-12+3i=0-3i$

28. $(6-2i)(6+2i)=6^2-(2i)^2$
$=36-4i^2$
$=36+4$
$=40$
$=40+0i$

29. $(4+3i)^2 = 4^2 + 2\cdot 4\cdot 3i + (3i)^2$
$= 16 + 24i + 9i^2$
$= 16 + 24i - 9$
$= 7 + 24i$

30. $\frac{1+4i}{1-i} = \frac{(1+4i)(1+i)}{(1-i)(1+i)}$
$= \frac{1+i+4i+4i^2}{1^2 - i^2}$
$= \frac{1+5i-4}{1-(-1)}$
$= \frac{-3+5i}{2}$
$= -\frac{3}{2} + \frac{5}{2}i$

31. $x^2 + x^2 = 5^2$
$2x^2 = 25$
$x^2 = \frac{25}{2}$
$\sqrt{x^2} = \sqrt{\frac{25}{2}}$
$x = \frac{5}{\sqrt{2}} = \frac{5\cdot\sqrt{2}}{\sqrt{2}\cdot\sqrt{2}} = \frac{5\sqrt{2}}{2}$

32. $g(x) = \sqrt{x+2}$
$x + 2 \ge 0$
$x \ge -2$
Domain: $[-2, \infty)$

x	−2	−1	2	7
$g(x)$	0	1	2	3

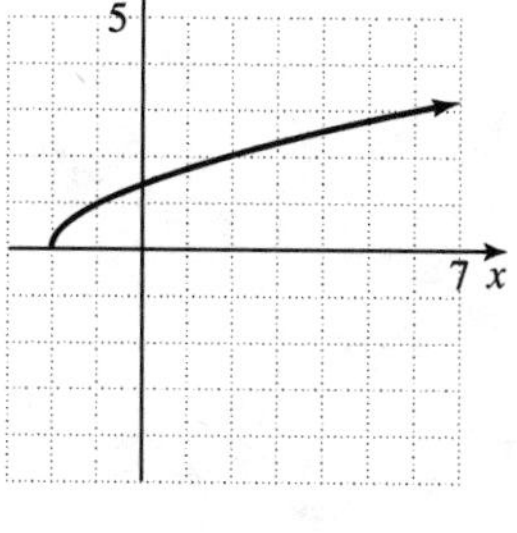

33. $(x_1, y_1) = (-6, 3), (x_2, y_2) = (-8, -7)$
$d = \sqrt{(-8-(-6))^2 + (-7-3)^2}$
$= \sqrt{(-2)^2 + (-10)^2}$
$= \sqrt{4+100}$
$= \sqrt{104}$
$= \sqrt{4\cdot 26}$
$= 2\sqrt{26}$
The distance is $2\sqrt{26}$ units.

34. $(x_1, y_1) = \left(-2\sqrt{5}, \sqrt{10}\right),$
$(x_2, y_2) = \left(-\sqrt{5}, 4\sqrt{10}\right)$
$d = \sqrt{(x_2 - x_1)^2 + (y_2 - y_1)^2}$
$= \sqrt{\left(-\sqrt{5} + 2\sqrt{5}\right)^2 + \left(4\sqrt{10} - \sqrt{10}\right)^2}$
$= \sqrt{\left(\sqrt{5}\right)^2 + \left(3\sqrt{10}\right)^2}$
$= \sqrt{5+90}$
$= \sqrt{95}$
The distance is $\sqrt{95}$ units.

35. $(x_1, y_1) = (-2, -5), (x_2, y_2) = (-6, 12)$
midpoint $= \left(\frac{x_1+x_2}{2}, \frac{y_1+y_2}{2}\right)$
$= \left(\frac{-2-6}{2}, \frac{-5+12}{2}\right)$
$= \left(-\frac{8}{2}, \frac{7}{2}\right)$
$= \left(-4, \frac{7}{2}\right)$

36. $(x_1, y_1) = \left(-\frac{2}{3}, -\frac{1}{5}\right), (x_2, y_2) = \left(-\frac{1}{3}, \frac{4}{5}\right)$
midpoint $= \left(\frac{x_1+x_2}{2}, \frac{y_1+y_2}{2}\right)$
$= \left(\frac{-\frac{2}{3} - \frac{1}{3}}{2}, \frac{-\frac{1}{5} + \frac{4}{5}}{2}\right)$
$= \left(\frac{-\frac{3}{3}}{2}, \frac{\frac{3}{5}}{2}\right)$
$= \left(-\frac{1}{2}, \frac{3}{10}\right)$

37. $V(r)=\sqrt{2.5r}$
$V(300)=\sqrt{2.5(300)}=\sqrt{750}\approx 27$ mph

38. $V(r)=\sqrt{2.5r}$
$30=\sqrt{2.5r}$
$30^2=\left(\sqrt{2.5r}\right)^2$
$900=2.5r$
$r=\frac{900}{2.5}=360$ feet

Chapter 7 Cumulative Review

1. a. $3xy-2xy+5-7+xy=2xy-2$

b. $7x^2+3-5(x^2-4)=7x^2+3-5x^2+20$
$=2x^2+23$

c. $(2.1x-5.6)-(-x-5.3)=2.1x-5.6+x+5.3$
$=3.1x-0.3$

d. $\frac{1}{2}(4a-6b)-\frac{1}{3}(9a+12b-1)+\frac{1}{4}$
$=2a-3b-3a-4b+\frac{1}{3}+\frac{1}{4}$
$=-a-7b+\frac{4}{12}+\frac{3}{12}$
$=-a-7b+\frac{7}{12}$

2. a. $2(x-3)+(5x+3)=2x-6+5x+3$
$=7x-3$

b. $4(3x+2)-3(5x-1)=12x+8-15x+3$
$=-3x+11$

c. $7x+2(x-7)-3x=7x+2x-14-3x$
$=6x-14$

3. $\frac{x+5}{2}+\frac{1}{2}=2x-\frac{x-3}{8}$
$8\left(\frac{x+5}{2}+\frac{1}{2}\right)=8\left(2x-\frac{x-3}{8}\right)$
$4(x+5)+4=16x-(x-3)$
$4x+20+4=16x-x+3$
$4x+24=15x+3$
$-11x=-21$
$x=\frac{21}{11}$

4. $\frac{a-1}{2}+a=2-\frac{2a+7}{8}$
$8\left(\frac{a-1}{2}+a\right)=8\left(2-\frac{2a+7}{8}\right)$
$4(a-1)+8a=16-(2a+7)$
$4a-4+8a=16-2a-7$
$12a-4=9-2a$
$14a=13$
$a=\frac{13}{14}$

5. Let x = the sales needed.
$600+0.20x>1500$
$0.20x>900$
$x>\frac{900}{0.20}$
$x>4500$
The salesperson needs sales of at least \$4500.

6. Let r = their average speed.
$t_{\text{going}}+t_{\text{returning}}=4.5$ hr
$\frac{121.5}{r}+\frac{121.5}{r}=4.5$
$\frac{243}{r}=4.5$
$243=4.5r$
$r=\frac{243}{4.5}=54$
Their average speed was 54 mph.

7. $2|x|+25=23$
$2|x|=-2$
$|x|=-1$, which is impossible.
There is no solution, or the solution set is ∅.

8. $|3x-2|+5=5$
$|3x-2|=0$
$3x-2=0$
$3x=2$
$x=\frac{2}{3}$

9. $\left|\frac{x}{3}-1\right|-7\geq -5$

$\left|\frac{x}{3}-1\right|\geq 2$

$\frac{x}{3}-1\leq -2$ or $\frac{x}{3}-1\geq 2$

$\frac{x}{3}\leq -1$ or $\frac{x}{3}\geq 3$

$x\leq -3$ or $x\geq 9$

$(-\infty, -3]\cup[9, \infty)$

10. $\left|\frac{x}{2}-1\right|\leq 0$

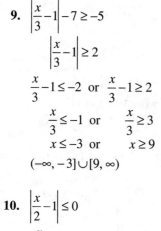

$\frac{x}{2}-1=0$

$\frac{x}{2}=1$

$x=2$

11. $y=|x|$

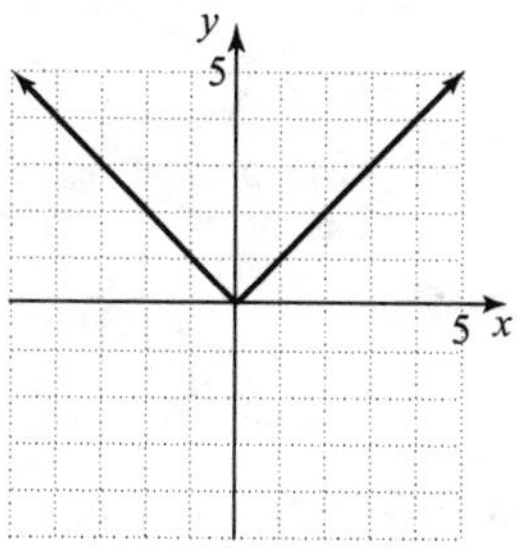

12. $y=|x-2|$

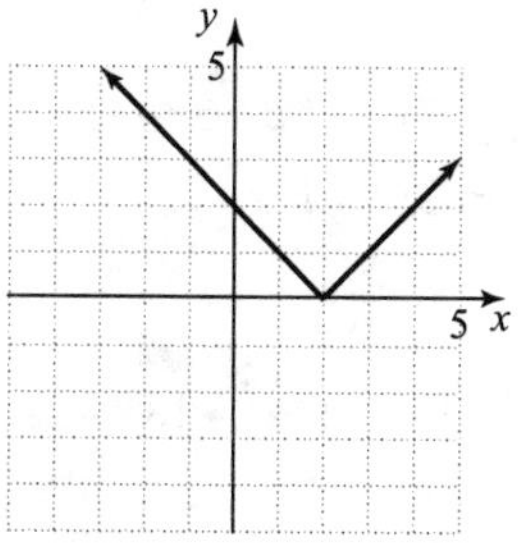

13. a. Domain: {2, 0, 3}
Range: {3, 4, –1}

b. Domain: {–4, –3, –2, –1, 0, 1, 2, 3}
Range: {1}

c. Domain: {El Paso, Virginia Beach, Atlanta, Syracuse, Anaheim}
Range: {448, 345, 674, 145}

14. a. Domain: $(-\infty, 0]$, Range: $(-\infty, \infty)$
not a function

b. Domain: $(-\infty, \infty)$, Range: $(-\infty, \infty)$
function

c. Domain: $(-\infty, -2]\cup[2, \infty)$
Range: $(-\infty, \infty)$
not a function

15. $y=-3$
This is a horizontal line passing through (0, –3).

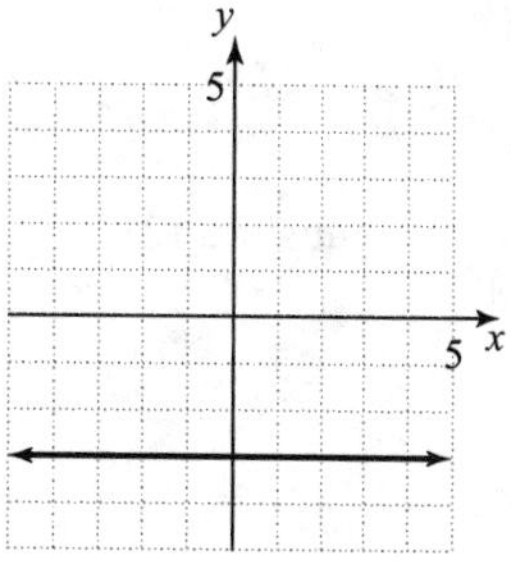

16. $f(x)=-2$
This is a horizontal line passing through (0, –2).

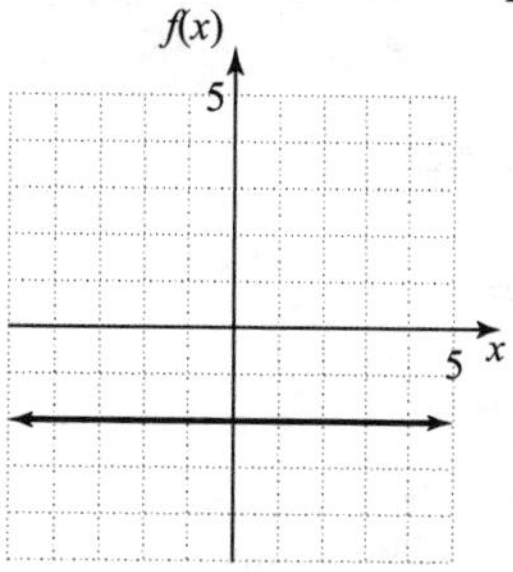

17. $x=-5$ is a vertical line. The slope is undefined.

18. $y=-3$ is a horizontal line. The slope is 0.

19. $\begin{cases}-\frac{x}{6}+\frac{y}{2}=\frac{1}{2}\\ \frac{x}{3}-\frac{y}{6}=-\frac{3}{4}\end{cases}$ or $\begin{cases}-x+3y=3 & (1)\\ 4x-2y=-9 & (2)\end{cases}$

Solve equation (1) for x.

$-x+3y=3$

$3y-3=x$

Replace x with $3y-3$ in equation (2).

$4(3y-3)-2y=-9$

$12y-12-2y=-9$

$10y-12=-9$

$10y=3$

$y=\frac{3}{10}$

Substitute $\frac{3}{10}$ for y in $x = 3y - 3$.

$x = 3\left(\frac{3}{10}\right) - 3 = \frac{9}{10} - \frac{30}{10} = -\frac{21}{10}$

The solution is $\left(-\frac{21}{10}, \frac{3}{10}\right)$.

20. $\begin{cases} \frac{x}{6} - \frac{y}{2} = 1 \\ \frac{x}{3} - \frac{y}{4} = 2 \end{cases}$ or $\begin{cases} x - 3y = 6 & (1) \\ 4x - 3y = 24 & (2) \end{cases}$

Solve equation (1) for x.

$x - 3y = 6$
$x = 3y + 6$

Replace x with $3y + 6$ in equation (2).

$4(3y+6) - 3y = 24$
$12y + 24 - 3y = 24$
$9y + 24 = 24$
$9y = 0$
$y = 0$

Substitute 0 for y in $x = 3y + 6$.

$x = 3(0) + 6 = 0 + 6 = 6$

The solution is (6, 0).

21. a. $2^2 \cdot 2^5 = 2^{2+5} = 2^7$

b. $x^7 x^3 = x^{7+3} = x^{10}$

c. $y \cdot y^2 \cdot y^4 = y^{1+2+4} = y^7$

22. Let x = number of tee-shirts and y = number of shorts.

$\begin{cases} x + y = 9 & (1) \\ 3.50x + 4.25y = 33.75 & (2) \end{cases}$

Solve equation (1) for y.

$x + y = 9$
$y = 9 - x$

Substitute $9 - x$ for y in equation (2).

$3.50x + 4.25(9 - x) = 33.75$
$3.50x + 38.25 - 4.25x = 33.75$
$-0.75x + 38.25 = 33.75$
$-0.75x = -4.5$
$x = \frac{-4.5}{-0.75} = 6$

Replace x with 6 in $y = 9 - x$.

$y = 9 - 6 = 3$

Nana bought 6 shirts and 3 shorts.

23. $\frac{2000 \times 0.000021}{700} = \frac{(2\times10^3)\times(2.1\times10^{-5})}{7\times10^2}$
$= \frac{2(2.1)}{7}\times10^{3+(-5)-2}$
$= 0.6\times10^{-4}$
$= (6\times10^{-1})\times10^{-4}$
$= 6\times10^{-5}$

24. $\frac{0.0000035\times4000}{0.28} = \frac{(3.5\times10^{-6})\times(4\times10^3)}{2.8\times10^{-1}}$
$= \frac{3.5\times4}{2.8}\times10^{-6+3-(-1)}$
$= 5\times10^{-2}$

25. $P(x) = 3x^2 - 2x - 5$

a. $P(1) = 3(1)^2 - 2(1) - 5$
$= 3(1) - 2(1) - 5$
$= 3 - 2 - 5$
$= -4$

b. $P(-2) = 3(-2)^2 - 2(-2) - 5$
$= 3(4) - (-4) - 5$
$= 12 + 4 - 5$
$= 11$

26. $(5x^2 - 3x + 6) + (4x^2 + 5x - 3) - (2x - 5)$
$= 5x^2 - 3x + 6 + 4x^2 + 5x - 3 - 2x + 5$
$= 9x^2 + 8$

27. a. $(x+3)(2x+5) = 2x^2 + 5x + 6x + 15$
$= 2x^2 + 11x + 15$

b. $(2x-3)(5x^2 - 6x + 7)$
$= 10x^3 - 12x^2 + 14x - 15x^2 + 18x - 21$
$= 10x^3 - 27x^2 + 32x - 21$

28. a. $(y-2)(3y+4) = 3y^2 + 4y - 6y - 8$
$= 3y^2 - 2y - 8$

b. $(3y-1)(2y^2 + 3y - 1)$
$= 6y^3 + 9y^2 - 3y - 2y^2 - 3y + 1$
$= 6y^3 + 7y^2 - 6y + 1$

29. $20x^3y = 2^2 \cdot 5x^3y$
$10x^2y^2 = 2 \cdot 5x^2y^2$
$35x^3 = 5 \cdot 7x^3$
$\text{GCF} = 5x^2$

30. $x^3 - x^2 + 4x - 4 = (x^3 - x^2) + (4x - 4)$
$= x^2(x-1) + 4(x-1)$
$= (x-1)(x^2+4)$

31. a. $\frac{x^3+8}{2+x} = \frac{x^3+2^3}{x+2}$
$= \frac{(x+2)(x^2-2x+4)}{x+2}$
$= x^2 - 2x + 4$

b. $\frac{2y^2+2}{y^3-5y^2+y-5} = \frac{2(y^2+1)}{y^2(y-5)+1(y-5)}$
$= \frac{2(y^2+1)}{(y-5)(y^2+1)}$
$= \frac{2}{y-5}$

32. a. $\frac{a^3-8}{2-a} = \frac{a^3-2^3}{2-a}$
$= \frac{(a-2)(a^2+2a+4)}{-1(a-2)}$
$= -1(a^2+2a+4)$
$= -a^2 - 2a - 4$

b. $\frac{3a^2-3}{a^3+5a^2-a-5} = \frac{3(a^2-1)}{a^2(a+5)-1(a+5)}$
$= \frac{3(a^2-1)}{(a+5)(a^2-1)}$
$= \frac{3}{a+5}$

33. a. $\frac{2}{x^2y} + \frac{5}{3x^3y} = \frac{2 \cdot 3x}{x^2y \cdot 3x} + \frac{5}{3x^3y}$
$= \frac{6x+5}{3x^3y}$

b. $\frac{3}{x+2} + \frac{2x}{x-2} = \frac{3(x-2)+2x(x+2)}{(x+2)(x-2)}$
$= \frac{3x-6+2x^2+4x}{(x+2)(x-2)}$
$= \frac{2x^2+7x-6}{(x+2)(x-2)}$

c. $\frac{2x-6}{x-1} - \frac{4}{1-x} = \frac{2x-6}{x-1} + \frac{4}{x-1}$
$= \frac{2x-2}{x-1}$
$= \frac{2(x-1)}{x-1}$
$= 2$

34. a. $\frac{3}{xy^2} - \frac{2}{3x^2y} = \frac{3 \cdot 3x}{xy^2 \cdot 3x} - \frac{2 \cdot y}{3x^2y \cdot y}$
$= \frac{9x-2y}{3x^2y^2}$

b. $\frac{5x}{x+3} - \frac{2x}{x-3} = \frac{5x(x-3)-2x(x+3)}{(x+3)(x-3)}$
$= \frac{5x^2-15x-2x^2-6x}{(x+3)(x-3)}$
$= \frac{3x^2-21x}{(x+3)(x-3)}$
or $\frac{3x(x-7)}{(x+3)(x-3)}$

c. $\frac{x}{x-2} - \frac{5}{2-x} = \frac{x}{x-2} + \frac{5}{x-2} = \frac{x+5}{x-2}$

35. a. $\frac{\frac{5x}{x+2}}{\frac{10}{x-2}} = \frac{5x}{x+2} \cdot \frac{x-2}{10} = \frac{x(x-2)}{2(x+2)}$

b. $\frac{\frac{x}{y^2}+\frac{1}{y}}{\frac{y}{x^2}+\frac{1}{x}} = \frac{\left(\frac{x}{y^2}+\frac{1}{y}\right)x^2y^2}{\left(\frac{y}{x^2}+\frac{1}{x}\right)x^2y^2}$
$= \frac{x^3+x^2y}{y^3+xy^2}$
$= \frac{x^2(x+y)}{y^2(y+x)}$
$= \frac{x^2}{y^2}$

36. a. $\dfrac{\frac{y-2}{16}}{\frac{2y+3}{12}} = \dfrac{y-2}{16} \cdot \dfrac{12}{2y+3} = \dfrac{3(y-2)}{4(2y+3)}$

b.
$$\frac{\frac{x}{16}-\frac{1}{x}}{1-\frac{4}{x}} = \frac{\left(\frac{x}{16}-\frac{1}{x}\right)16x}{\left(1-\frac{4}{x}\right)16x}$$
$$= \frac{x^2-16}{16x-64}$$
$$= \frac{(x+4)(x-4)}{16(x-4)}$$
$$= \frac{x+4}{16}$$

37.
$$\frac{10x^3-5x^2+20x}{5x} = \frac{10x^3}{5x} - \frac{5x^2}{5x} + \frac{20x}{5x}$$
$$= 2x^2 - x + 4$$

38.
$$\begin{array}{r} x^2 \qquad +3 \\ x-2\overline{\smash{)}x^3-2x^2+3x-6} \\ \underline{x^3-2x^2} \qquad\quad \\ 3x-6 \\ \underline{3x-6} \\ 0 \end{array}$$

Answer: x^2+3

39.
$$\begin{array}{r|rrrr} 3 & 2 & -1 & -13 & 1 \\ & & 6 & 15 & 6 \\ \hline & 2 & 5 & 2 & 7 \end{array}$$

Answer: $2x^2+5x+2+\dfrac{7}{x-3}$

40.
$$\begin{array}{r|rrrr} 3 & 4 & -12 & -1 & 12 \\ & & 12 & 0 & -3 \\ \hline & 4 & 0 & -1 & 9 \end{array}$$

Answer: $4y^2-1+\dfrac{9}{y-3}$

41.
$$\frac{x+6}{x-2} = \frac{2(x+2)}{x-2}$$
$$(x-2)\left(\frac{x+6}{x-2}\right) = (x-2)\left[\frac{2(x+2)}{x-2}\right]$$
$$x+6 = 2(x+2)$$
$$x+6 = 2x+4$$
$$-x = -2$$
$$x = 2$$

which we discard as extraneous. There is no solution, ∅.

42.
$$\begin{aligned}
\frac{28}{9-a^2} &= \frac{2a}{a-3}+\frac{6}{a+3}\\
\frac{28}{-(a^2-9)} &= \frac{2a}{a-3}+\frac{6}{a+3}\\
\frac{-28}{(a+3)(a-3)} &= \frac{2a}{a-3}+\frac{6}{a+3}\\
(a+3)(a-3)\cdot\frac{-28}{(a+3)(a-3)} &= (a+3)(a-3)\cdot\left(\frac{2a}{a-3}+\frac{6}{a+3}\right)\\
-28 &= 2a(a+3)+6(a-3)\\
-28 &= 2a^2+6a+6a-18\\
0 &= 2a^2+12a+10\\
0 &= 2(a^2+6a+5)\\
0 &= 2(a+5)(a+1)
\end{aligned}$$

$a=-5$ or $a=-1$

The solutions are –5 and –1.

43.
$$\begin{aligned}
\frac{1}{x}+\frac{1}{y} &= \frac{1}{z}\\
xyz\left(\frac{1}{x}+\frac{1}{y}\right) &= xyz\left(\frac{1}{z}\right)\\
yz+xz &= xy\\
yz &= xy-xz\\
yz &= x(y-z)\\
x &= \frac{yz}{y-z}
\end{aligned}$$

44.
$$\begin{aligned}
A &= \frac{h(a+b)}{2}\\
2A &= h(a+b)\\
2A &= ah+bh\\
2A-bh &= ah\\
\frac{2A-bh}{h} &= a
\end{aligned}$$

45.
$$\begin{aligned}
u &= \frac{k}{w}\\
3 &= \frac{k}{5}\\
k &= 3(5)=15\\
u &= \frac{15}{w}
\end{aligned}$$

46.
$$\begin{aligned}
y &= kx\\
0.51 &= k(3)\\
k &= \frac{0.51}{3}=0.17\\
y &= 0.17x
\end{aligned}$$

47. a. $16^{-3/4} = \frac{1}{16^{3/4}} = \frac{1}{\left(\sqrt[4]{16}\right)^3} = \frac{1}{(2)^3} = \frac{1}{8}$

b. $(-27)^{-2/3} = \frac{1}{(-27)^{2/3}}$
$= \frac{1}{\left(\sqrt[3]{-27}\right)^2}$
$= \frac{1}{(-3)^2}$
$= \frac{1}{9}$

48. a. $81^{-3/4} = \frac{1}{81^{3/4}} = \frac{1}{\left(\sqrt[4]{81}\right)^3} = \frac{1}{(3)^3} = \frac{1}{27}$

b. $(-125)^{-2/3} = \frac{1}{(-125)^{2/3}}$
$= \frac{1}{\left(\sqrt[3]{-125}\right)^2}$
$= \frac{1}{(-5)^2}$
$= \frac{1}{25}$

49. $\frac{\sqrt{x}+2}{5} = \frac{\left(\sqrt{x}+2\right)\left(\sqrt{x}-2\right)}{5\left(\sqrt{x}-2\right)}$
$= \frac{\left(\sqrt{x}\right)^2 - 2^2}{5\left(\sqrt{x}-2\right)}$
$= \frac{x-4}{5\left(\sqrt{x}-2\right)}$

50. a. $\sqrt{36a^3} - \sqrt{144a^3} + \sqrt{4a^3}$
$= \sqrt{36a^2 \cdot a} - \sqrt{144a^2 \cdot a} + \sqrt{4a^2 \cdot a}$
$= 6a\sqrt{a} - 12a\sqrt{a} + 2a\sqrt{a}$
$= -4a\sqrt{a}$

b. $\sqrt[3]{128ab^3} - 3\sqrt[3]{2ab^3} + b\sqrt[3]{16a}$
$= \sqrt[3]{64b^3 \cdot 2a} - 3\sqrt[3]{b^3 \cdot 2a} + b\sqrt[3]{8 \cdot 2a}$
$= 4b\sqrt[3]{2a} - 3b\sqrt[3]{2a} + 2b\sqrt[3]{2a}$
$= 3b\sqrt[3]{2a}$

c. $\frac{\sqrt[3]{81}}{10} + \sqrt[3]{\frac{192}{125}} = \frac{\sqrt[3]{27 \cdot 3}}{10} + \frac{\sqrt[3]{192}}{\sqrt[3]{125}}$
$= \frac{3\sqrt[3]{3}}{10} + \frac{\sqrt[3]{64 \cdot 3}}{5}$
$= \frac{3\sqrt[3]{3}}{10} + \frac{4\sqrt[3]{3}}{5}$
$= \frac{3\sqrt[3]{3}}{10} + \frac{4\sqrt[3]{3} \cdot 2}{5 \cdot 2}$
$= \frac{11\sqrt[3]{3}}{10}$

Chapter 8

Section 8.1 Practice Exercises

1. $x^2 = 32$

$x = \pm\sqrt{32}$

$x = \pm 4\sqrt{2}$

Check:

Let $x = 4\sqrt{2}$.

$x^2 = 32$

$\left(4\sqrt{2}\right)^2 \stackrel{?}{=} 32$

$16 \cdot 2 \stackrel{?}{=} 32$

$32 = 32$ True

Let $x = -4\sqrt{2}$.

$x^2 = 32$

$\left(-4\sqrt{2}\right)^2 \stackrel{?}{=} 32$

$16 \cdot 2 \stackrel{?}{=} 32$

$32 = 32$ True

The solutions are $4\sqrt{2}$ and $-4\sqrt{2}$, or the solution set is $\left\{-4\sqrt{2}, 4\sqrt{2}\right\}$.

2. First we get the squared variable alone on one side of the equation.

$5x^2 - 50 = 0$

$5x^2 = 50$

$x^2 = 10$

$x = \pm\sqrt{10}$

The solutions are $\sqrt{10}$ and $-\sqrt{10}$, or the solution set is $\left\{-\sqrt{10}, \sqrt{10}\right\}$.

3. $(x+3)^2 = 20$

$x + 3 = \pm\sqrt{20}$

$x + 3 = \pm 2\sqrt{5}$

$x = -3 \pm 2\sqrt{5}$

Check:

$(x+3)^2 = 20$

$\left(-3 + 2\sqrt{5} + 3\right)^2 \stackrel{?}{=} 20$

$\left(2\sqrt{5}\right)^2 \stackrel{?}{=} 20$

$4 \cdot 5 \stackrel{?}{=} 20$

$20 = 20$ True

$(x+3)^2 = 20$

$\left(-3 - 2\sqrt{5} + 3\right)^2 \stackrel{?}{=} 20$

$\left(-2\sqrt{5}\right)^2 \stackrel{?}{=} 20$

$4 \cdot 5 \stackrel{?}{=} 20$

$20 = 20$ True

The solutions are $-3 + 2\sqrt{5}$ and $-3 - 2\sqrt{5}$.

4. $(5x-2)^2 = -9$

$5x - 2 = \pm\sqrt{-9}$

$5x - 2 = \pm 3i$

$5x = 2 \pm 3i$

$x = \dfrac{2 \pm 3i}{5}$

The solutions are $\dfrac{2+3i}{5}$ and $\dfrac{2-3i}{5}$.

5. $b^2 + 4b = 3$

Add the square of half the coefficient of b to both sides.

$b^2 + 4b + \left(\dfrac{4}{2}\right)^2 = 3 + \left(\dfrac{4}{2}\right)^2$

$b^2 + 4b + 4 = 7$

$(b+2)^2 = 7$

$b + 2 = \pm\sqrt{7}$

$b = -2 \pm \sqrt{7}$

The solutions are $-2 + \sqrt{7}$ and $-2 - \sqrt{7}$.

6. $p^2 - 3p + 1 = 0$

Subtract 1 from both sides.

$p^2 - 3p = -1$

Add the square of half the coefficient of p to both sides.

$p^2 - 3p + \left(\dfrac{-3}{2}\right)^2 = -1 + \left(\dfrac{-3}{2}\right)^2$

$p^2 - 3p + \dfrac{9}{4} = -1 + \dfrac{9}{4} = \dfrac{5}{4}$

$\left(p - \dfrac{3}{2}\right)^2 = \dfrac{5}{4}$

$p - \dfrac{3}{2} = \pm\dfrac{\sqrt{5}}{2}$

$p = \dfrac{3 \pm \sqrt{5}}{2}$

The solutions are $\dfrac{3+\sqrt{5}}{2}$ and $\dfrac{3-\sqrt{5}}{2}$.

7. $3x^2 - 12x + 1 = 0$
Divide both sides by 3.

$$3x^2 - 12x + 1 = 0$$
$$x^2 - 4x + \frac{1}{3} = 0$$
$$x^2 - 4x = -\frac{1}{3}$$

Find the square of half of −4.

$$\left(\frac{-4}{2}\right)^2 = (-2)^2 = 4$$

Add 4 to both sides of the equation.

$$x^2 - 4x + 4 = -\frac{1}{3} + 4$$
$$(x-2)^2 = -\frac{1}{3} + \frac{12}{3} = \frac{11}{3}$$
$$x - 2 = \pm\sqrt{\frac{11}{3}} = \pm\frac{\sqrt{33}}{3}$$
$$x = \frac{6}{3} \pm \frac{\sqrt{33}}{3} = \frac{6 \pm \sqrt{33}}{3}$$

The solutions are $\frac{6+\sqrt{33}}{3}$ and $\frac{6-\sqrt{33}}{3}$.

8. $2x^2 - 5x + 7 = 0$

$$2x^2 - 5x = -7$$
$$x^2 - \frac{5}{2}x = -\frac{7}{2}$$

Since $\frac{1}{2}\left(-\frac{5}{2}\right) = -\frac{5}{4}$ and $\left(-\frac{5}{4}\right)^2 = \frac{25}{16}$, we add $\frac{25}{16}$ to both sides of the equation.

$$x^2 - \frac{5}{2}x + \frac{25}{16} = -\frac{7}{2} + \frac{25}{16}$$
$$\left(x - \frac{5}{4}\right)^2 = -\frac{56}{16} + \frac{25}{16} = -\frac{31}{16}$$
$$x - \frac{5}{4} = \pm\sqrt{-\frac{31}{16}}$$
$$x = \frac{5}{4} \pm \frac{i\sqrt{31}}{4} = \frac{5 \pm i\sqrt{31}}{4}$$

The solutions are $\frac{5 \pm i\sqrt{31}}{4}$ and $\frac{5 - i\sqrt{31}}{4}$.

9. $A = P(1+r)^t$; $A = 5618$, $P = 5000$, $t = 2$

$$A = P(1+r)^t$$
$$5618 = 5000(1+r)^2$$
$$1.1236 = (1+r)^2$$
$$\pm\sqrt{1.1236} = 1 + r$$
$$-1 \pm 1.06 = r$$
$$0.06 = r \text{ or } -2.06 = r$$

The rate cannot be negative, so we reject −2.06.

Check: $A = 5000(1+0.06)^2$
$= 5000(1.06)^2$
$= 5000 \cdot 1.1236$
$= 5618$

The interest rate is 6% compounded annually.

Graphing Calculator Explorations

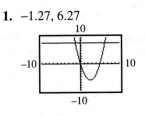

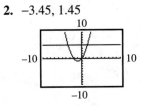

1. −1.27, 6.27

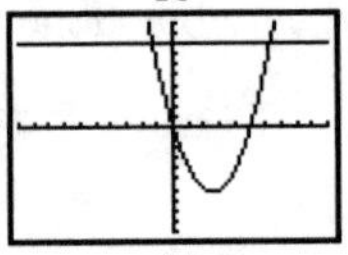

2. −3.45, 1.45

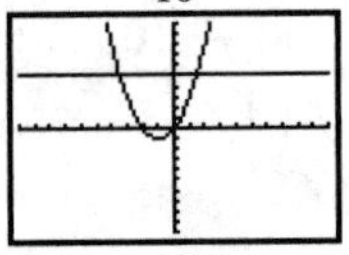

3. −1.10, 0.90

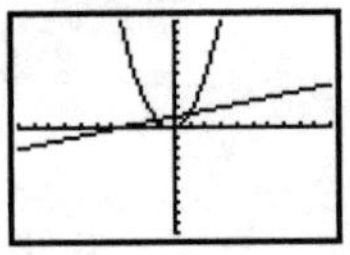

4. −1.54, 1.94

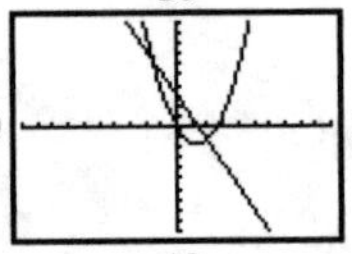

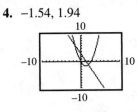

5. No real solutions, or ∅

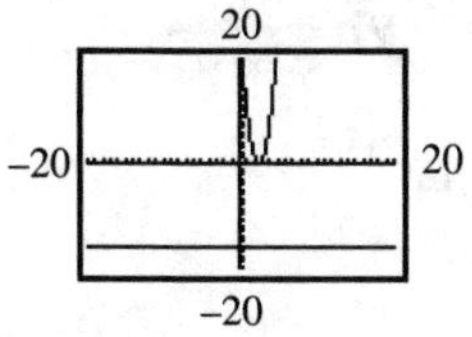

6. answers may vary

Vocabulary, Readiness & Video Check 8.1

1. By the square root property, if *b* is a real number, and $a^2 = b$, then $a = \underline{\pm\sqrt{b}}$.

2. A quadratic equation can be written in the form $ax^2 + bx + c = 0$, $a \neq 0$.

3. The process of writing a quadratic equation so that one side is a perfect square trinomial is called completing the square.

4. A perfect square trinomial is one that can be factored as a binomial squared.

5. To solve $x^2 + 6x = 10$ by completing the square, add 9 to both sides.

6. To solve $x^2 + bx = c$ by completing the square, add $\underline{\left(\frac{b}{2}\right)^2}$ to both sides.

7. We need a quantity shown squared by itself on one side of the equation. The only quantity squared is *x*, so divide both sides by 2 before applying the square root property.

8. The coefficient of y^2 is 3. To use the completing the square method, the coefficient of the squared variable must be 1, so we first divide through by 3.

9. We're looking for an interest rate so a negative value does not make sense.

Exercise Set 8.1

1. $x^2 = 16$

$x = \pm\sqrt{16}$

$x = \pm 4$

The solutions are −4 and 4.

3. $x^2 - 7 = 0$

$x^2 = 7$

$x = \pm\sqrt{7}$

The solutions are $\sqrt{7}$ and $-\sqrt{7}$.

5. $x^2 = 18$

$x = \pm\sqrt{18}$

$x = \pm\sqrt{9 \cdot 2}$

$x = \pm 3\sqrt{2}$

The solutions are $3\sqrt{2}$ and $-3\sqrt{2}$.

7. $3z^2 - 30 = 0$

$3z^2 = 30$

$z^2 = 10$

$z = \pm\sqrt{10}$

The solutions are $\sqrt{10}$ and $-\sqrt{10}$.

9. $(x+5)^2 = 9$

$x + 5 = \pm\sqrt{9}$

$x + 5 = \pm 3$

$x = -5 \pm 3$

$x = -8$ or $x = -2$

The solutions are −8 and −2.

11. $(z-6)^2 = 18$

$z - 6 = \pm\sqrt{18}$

$z - 6 = \pm 3\sqrt{2}$

$z = 6 \pm 3\sqrt{2}$

The solutions are $6 + 3\sqrt{2}$ and $6 - 3\sqrt{2}$.

13. $(2x-3)^2 = 8$

$2x - 3 = \pm\sqrt{8}$

$2x - 3 = \pm 2\sqrt{2}$

$2x = 3 \pm 2\sqrt{2}$

$x = \frac{3 \pm 2\sqrt{2}}{2}$

The solutions are $\frac{3+2\sqrt{2}}{2}$ and $\frac{3-2\sqrt{2}}{2}$.

15. $x^2 + 9 = 0$

$x^2 = -9$

$x = \pm\sqrt{-9}$

$x = \pm 3i$

The solutions are $3i$ and $-3i$.

17. $x^2-6=0$
$x^2=6$
$x=\pm\sqrt{6}$
The solutions are $\sqrt{6}$ and $-\sqrt{6}$.

19. $2z^2+16=0$
$2z^2=-16$
$z^2=-8$
$z=\pm\sqrt{-8}$
$z=\pm i\sqrt{8}$
$z=\pm 2i\sqrt{2}$
The solutions are $2i\sqrt{2}$ and $-2i\sqrt{2}$.

21. $(3x-1)^2=-16$
$3x-1=\pm\sqrt{-16}$
$3x-1=\pm 4i$
$3x=1\pm 4i$
$x=\dfrac{1\pm 4i}{3}$
The solutions are $\dfrac{1+4i}{3}$ and $\dfrac{1-4i}{3}$.

23. $(z+7)^2=5$
$z+7=\pm\sqrt{5}$
$z=-7\pm\sqrt{5}$
The solutions are $-7+\sqrt{5}$ and $-7-\sqrt{5}$.

25. $(x+3)^2+8=0$
$(x+3)^2=-8$
$x+3=\pm\sqrt{-8}$
$x+3=\pm i\sqrt{8}$
$x+3=\pm 2i\sqrt{2}$
$x=-3\pm 2i\sqrt{2}$
The solutions are $-3+2i\sqrt{2}$ and $-3-2i\sqrt{2}$.

27. $x^2+16x+\left(\dfrac{16}{2}\right)^2=x^2+16x+64$
$=(x+8)^2$

29. $z^2-12z+\left(\dfrac{-12}{2}\right)^2=z^2-12z+36$
$=(z-6)^2$

31. $p^2+9p+\left(\dfrac{9}{2}\right)^2=p^2+9p+\dfrac{81}{4}$
$=\left(p+\dfrac{9}{2}\right)^2$

33. $x^2+x+\left(\dfrac{1}{2}\right)^2=x^2+x+\dfrac{1}{4}$
$=\left(x+\dfrac{1}{2}\right)^2$

35. $x^2+8x=-15$
$x^2+8x+\left(\dfrac{8}{2}\right)^2=-15+16$
$x^2+8x+16=1$
$(x+4)^2=1$
$x+4=\pm\sqrt{1}$
$x=-4\pm 1$
$x=-5$ or $x=-3$
The solutions are -5 and -3.

37. $x^2+6x+2=0$
$x^2+6x=-2$
$x^2+6x+\left(\dfrac{6}{2}\right)^2=-2+9$
$x^2+6x+9=7$
$(x+3)^2=7$
$x+3=\pm\sqrt{7}$
$x=-3\pm\sqrt{7}$
The solutions are $-3+\sqrt{7}$ and $-3-\sqrt{7}$.

39.

$$x^2+x-1=0$$
$$x^2+x=1$$
$$x^2+x+\left(\frac{1}{2}\right)^2=1+\frac{1}{4}$$
$$x^2+x+\frac{1}{4}=\frac{5}{4}$$
$$\left(x+\frac{1}{2}\right)^2=\frac{5}{4}$$
$$x+\frac{1}{2}=\pm\sqrt{\frac{5}{4}}$$
$$x=-\frac{1}{2}\pm\frac{\sqrt{5}}{2}=\frac{-1\pm\sqrt{5}}{2}$$

The solutions are $\frac{-1+\sqrt{5}}{2}$ and $\frac{-1-\sqrt{5}}{2}$.

41.

$$x^2+2x-5=0$$
$$x^2+2x=5$$
$$x^2+2x+\left(\frac{2}{2}\right)^2=5+1$$
$$x^2+2x+1=6$$
$$(x+1)^2=6$$
$$x+1=\pm\sqrt{6}$$
$$x=-1\pm\sqrt{6}$$

The solutions are $-1+\sqrt{6}$ and $-1-\sqrt{6}$.

43.

$$y^2+y-7=0$$
$$y^2+y=7$$
$$y^2+y+\left(\frac{1}{2}\right)^2=7+\frac{1}{4}$$
$$y^2+y+\frac{1}{4}=\frac{29}{4}$$
$$\left(y+\frac{1}{2}\right)^2=\frac{29}{4}$$
$$y+\frac{1}{2}=\pm\sqrt{\frac{29}{4}}$$
$$y=-\frac{1}{2}\pm\frac{\sqrt{29}}{2}=\frac{-1\pm\sqrt{29}}{2}$$

The solutions are $\frac{-1+\sqrt{29}}{2}$ and $\frac{-1-\sqrt{29}}{2}$.

45.

$$3p^2-12p+2=0$$
$$p^2-4p+\frac{2}{3}=0$$
$$p^2-4p=-\frac{2}{3}$$
$$p^2-4p+\left(\frac{-4}{2}\right)^2=-\frac{2}{3}+4$$
$$(p-2)^2=\frac{10}{3}$$
$$p-2=\pm\sqrt{\frac{10}{3}}$$
$$p-2=\pm\frac{\sqrt{10}\cdot\sqrt{3}}{\sqrt{3}\cdot\sqrt{3}}$$
$$p-2=\pm\frac{\sqrt{30}}{3}$$
$$p=2\pm\frac{\sqrt{30}}{3}=\frac{6\pm\sqrt{30}}{3}$$

The solutions are $\frac{6+\sqrt{30}}{3}$ and $\frac{6-\sqrt{30}}{3}$.

47.

$$4y^2-2=12y$$
$$4y^2-12y-2=0$$
$$y^2-3y-\frac{1}{2}=0$$
$$y^2-3y=\frac{1}{2}$$
$$y^2-3y+\left(\frac{-3}{2}\right)^2=\frac{1}{2}+\frac{9}{4}$$
$$y^2-3y+\frac{9}{4}=\frac{11}{4}$$
$$\left(y-\frac{3}{2}\right)^2=\frac{11}{4}$$
$$y-\frac{3}{2}=\pm\sqrt{\frac{11}{4}}$$
$$y=\frac{3}{2}\pm\frac{\sqrt{11}}{2}=\frac{3\pm\sqrt{11}}{2}$$

The solutions are $\frac{3+\sqrt{11}}{2}$ and $\frac{3-\sqrt{11}}{2}$.

49.

$$2x^2+7x=4$$
$$x^2+\frac{7}{2}x=2$$
$$x^2+\frac{7}{2}x+\left(\frac{\frac{7}{2}}{2}\right)^2=2+\frac{49}{16}$$
$$x^2+\frac{7}{2}x+\frac{49}{16}=\frac{81}{16}$$
$$\left(x+\frac{7}{4}\right)^2=\frac{81}{16}$$
$$x+\frac{7}{4}=\pm\sqrt{\frac{81}{16}}$$
$$x=-\frac{7}{4}\pm\frac{9}{4}=\frac{-7\pm9}{4}$$
$$x=-4 \text{ or } \frac{1}{2}$$

The solutions are -4 and $\frac{1}{2}$.

51.

$$x^2+8x+1=0$$
$$x^2+8x=-1$$
$$x^2+8x+\left(\frac{8}{2}\right)^2=-1+16$$
$$x^2+8x+16=15$$
$$(x+4)^2=15$$
$$x+4=\pm\sqrt{15}$$
$$x=-4\pm\sqrt{15}$$

The solutions are $-4+\sqrt{15}$ and $-4-\sqrt{15}$.

53.

$$3y^2+6y-4=0$$
$$y^2+2y-\frac{4}{3}=0$$
$$y^2+2y=\frac{4}{3}$$
$$y^2+2y+\left(\frac{2}{2}\right)^2=\frac{4}{3}+1$$
$$y^2+2y+1=\frac{7}{3}$$
$$(y+1)^2=\frac{7}{3}$$
$$y+1=\pm\sqrt{\frac{7}{3}}$$
$$y+1=\pm\frac{\sqrt{7}\cdot\sqrt{3}}{\sqrt{3}\cdot\sqrt{3}}$$
$$y+1=\pm\frac{\sqrt{21}}{3}$$
$$y=-1\pm\frac{\sqrt{21}}{3}=\frac{-3\pm\sqrt{21}}{3}$$

The solutions are $\frac{-3+\sqrt{21}}{3}$ and $\frac{-3-\sqrt{21}}{3}$.

55.

$$2x^2-3x-5=0$$
$$x^2-\frac{3}{2}x-\frac{5}{2}=0$$
$$x^2-\frac{3}{2}x=\frac{5}{2}$$
$$x^2-\frac{3}{2}x+\left(\frac{\frac{3}{2}}{2}\right)^2=\frac{5}{2}+\frac{9}{16}$$
$$x^2-\frac{3}{2}x+\frac{9}{16}=\frac{49}{16}$$
$$\left(x-\frac{3}{4}\right)^2=\frac{49}{16}$$
$$x-\frac{3}{4}=\pm\sqrt{\frac{49}{16}}$$
$$x=\frac{3}{4}\pm\frac{7}{4}=\frac{3\pm7}{4}$$
$$x=-1 \text{ or } \frac{5}{2}$$

The solutions are -1 and $\frac{5}{2}$.

57.
$$y^2+2y+2=0$$
$$y^2+2y=-2$$
$$y^2+2y+\left(\frac{2}{2}\right)^2=-2+1$$
$$y^2+2y+1=-1$$
$$(y+1)^2=-1$$
$$y+1=\pm\sqrt{-1}$$
$$y=-1\pm i$$
The solutions are $-1+i$ and $-1-i$.

59.
$$y^2+6y-8=0$$
$$y^2+6y=8$$
$$y^2+6y+\left(\frac{6}{2}\right)^2=8+9$$
$$y^2+6y+9=17$$
$$(y+3)^2=17$$
$$y+3=\pm\sqrt{17}$$
$$y=-3\pm\sqrt{17}$$
The solutions are $-3+\sqrt{17}$ and $-3-\sqrt{17}$.

61.
$$2a^2+8a=-12$$
$$a^2+4a=-6$$
$$a^2+4a+\left(\frac{4}{2}\right)^2=-6+4$$
$$a^2+4a+4=-2$$
$$(a+2)^2=-2$$
$$a+2=\pm\sqrt{-2}$$
$$a+2=\pm i\sqrt{2}$$
$$a=-2\pm i\sqrt{2}$$
The solutions are $-2+i\sqrt{2}$ and $-2-i\sqrt{2}$.

63.
$$5x^2+15x-1=0$$
$$x^2+3x-\frac{1}{5}=0$$
$$x^2+3x=\frac{1}{5}$$
$$x^2+3x+\left(\frac{3}{2}\right)^2=\frac{1}{5}+\frac{9}{4}$$
$$x^2+3x+\frac{9}{4}=\frac{49}{20}$$
$$\left(x+\frac{3}{2}\right)^2=\frac{49}{20}$$
$$x+\frac{3}{2}=\pm\sqrt{\frac{49}{20}}$$
$$x+\frac{3}{2}=\pm\frac{7}{\sqrt{20}}$$
$$x+\frac{3}{2}=\pm\frac{7}{2\sqrt{5}}$$
$$x+\frac{3}{2}=\pm\frac{7\cdot\sqrt{5}}{2\sqrt{5}\cdot\sqrt{5}}$$
$$x+\frac{3}{2}=\pm\frac{7\sqrt{5}}{10}$$
$$x=-\frac{3}{2}\pm\frac{7\sqrt{5}}{10}=\frac{-15\pm7\sqrt{5}}{10}$$
The solutions are $\frac{-15+7\sqrt{5}}{10}$ and $\frac{-15-7\sqrt{5}}{10}$.

65.
$$2x^2 - x + 6 = 0$$
$$x^2 - \frac{1}{2}x + 3 = 0$$
$$x^2 - \frac{1}{2}x = -3$$
$$x^2 - \frac{1}{2}x + \left(\frac{-\frac{1}{2}}{2}\right)^2 = -3 + \frac{1}{16}$$
$$x^2 - \frac{1}{2}x + \frac{1}{16} = -\frac{47}{16}$$
$$\left(x - \frac{1}{4}\right)^2 = -\frac{47}{16}$$
$$x - \frac{1}{4} = \pm\sqrt{-\frac{47}{16}}$$
$$x - \frac{1}{4} = \pm i\frac{\sqrt{47}}{4}$$
$$x = \frac{1}{4} \pm i\frac{\sqrt{47}}{4} = \frac{1 \pm i\sqrt{47}}{4}$$
The solutions are $\frac{1+i\sqrt{47}}{4}$ and $\frac{1-i\sqrt{47}}{4}$.

67.
$$x^2 + 10x + 28 = 0$$
$$x^2 + 10x = -28$$
$$x^2 + 10x + \left(\frac{10}{2}\right)^2 = -28 + 25$$
$$(x+5)^2 = -3$$
$$x + 5 = \pm\sqrt{-3}$$
$$x = -5 \pm i\sqrt{3}$$
The solutions are $-5 + i\sqrt{3}$ and $-5 - i\sqrt{3}$.

69.
$$z^2 + 3z - 4 = 0$$
$$z^2 + 3z = 4$$
$$z^2 + 3z + \left(\frac{3}{2}\right)^2 = 4 + \frac{9}{4}$$
$$z^2 + 3z + \frac{9}{4} = \frac{25}{4}$$
$$\left(z + \frac{3}{2}\right)^2 = \frac{25}{4}$$
$$z + \frac{3}{2} = \pm\sqrt{\frac{25}{4}}$$
$$z = -\frac{3}{2} \pm \frac{5}{2} = \frac{-3 \pm 5}{2}$$
$$z = -4 \text{ or } 1$$
The solutions are –4 and 1.

71.
$$2x^2 - 4x = -3$$
$$x^2 - 2x = -\frac{3}{2}$$
$$x^2 - 2x + \left(\frac{-2}{2}\right)^2 = -\frac{3}{2} + 1$$
$$x^2 - 2x + 1 = -\frac{1}{2}$$
$$(x-1)^2 = -\frac{1}{2}$$
$$x - 1 = \pm\sqrt{-\frac{1}{2}}$$
$$x - 1 = \pm i\frac{1}{\sqrt{2}}$$
$$x - 1 = \pm i\frac{1 \cdot \sqrt{2}}{\sqrt{2} \cdot \sqrt{2}}$$
$$x - 1 = \pm i\frac{\sqrt{2}}{2}$$
$$x = 1 \pm i\frac{\sqrt{2}}{2} = \frac{2 \pm i\sqrt{2}}{2}$$
The solutions are $\frac{2+i\sqrt{2}}{2}$ and $\frac{2-i\sqrt{2}}{2}$.

73.
$$3x^2 + 3x = 5$$
$$x^2 + x = \frac{5}{3}$$
$$x^2 + x + \left(\frac{1}{2}\right)^2 = \frac{5}{3} + \frac{1}{4}$$
$$x^2 + x + \frac{1}{4} = \frac{23}{12}$$
$$\left(x + \frac{1}{2}\right)^2 = \frac{23}{12}$$
$$x + \frac{1}{2} = \pm\sqrt{\frac{23}{12}}$$
$$x + \frac{1}{2} = \pm\frac{\sqrt{23}}{2\sqrt{3}}$$
$$x + \frac{1}{2} = \pm\frac{\sqrt{23} \cdot \sqrt{3}}{2\sqrt{3} \cdot \sqrt{3}}$$
$$x + \frac{1}{2} = \pm\frac{\sqrt{69}}{6}$$
$$x = -\frac{1}{2} \pm \frac{\sqrt{69}}{6} = \frac{-3 \pm \sqrt{69}}{6}$$
The solutions are $\frac{-3+\sqrt{69}}{6}$ and $\frac{-3-\sqrt{69}}{6}$.

75. $A = P(1+r)^t$
$4320 = 3000(1+r)^2$
$\frac{4320}{3000} = (1+r)^2$
$1.44 = (1+r)^2$
$\pm\sqrt{1.44} = 1+r$
$\pm 1.2 = 1+r$
$-1 \pm 1.2 = r$
$-2.2 = r$ or $0.2 = r$
Rate cannot be negative, so the rate is $r = 0.2 = 20\%$.

77. $A = P(1+r)^t$
$16,224 = 15,000(1+r)^2$
$\frac{16,224}{15,000} = (1+r)^2$
$\pm\sqrt{1.0816} = 1+r$
$\pm 1.04 = 1+r$
$-1 \pm 1.04 = r$
$0.04 = r$ or $-2.04 = r$
Rate cannot be negative, so the rate is $r = 0.04$, or 4%.

79. $s(t) = 16t^2$
$1483 = 16t^2$
$t^2 = \frac{1483}{16}$
$t = \pm\sqrt{\frac{1483}{16}}$
$t \approx 9.63$ or -9.63 (disregard)
It would take 9.63 seconds.

81. $s(t) = 16t^2$
$610 = 16t^2$
$\frac{610}{16} = t^2$
$\pm\sqrt{\frac{610}{16}} = t^2$
$\pm\frac{\sqrt{610}}{4} = t$

The time cannot be negative, so $-\frac{\sqrt{610}}{4}$ is rejected.

It takes the object $\frac{\sqrt{610}}{4} \approx 6.17$ seconds to fall from the top to the base of the dam.

83. Let x be the length of one side.
$x^2 = 225$
$x = \pm\sqrt{225}$
$x = \pm 15$
The length cannot be negative, so -15 is rejected. The dimensions are 15 feet by 15 feet.

85. $a^2 + b^2 = c^2$
$x^2 + x^2 = 20^2$
$2x^2 = 400$
$x^2 = 200$
$x = \pm\sqrt{200}$
$x = \pm 10\sqrt{2}$
The length cannot be negative, so $-10\sqrt{2}$ is rejected. The length of each leg is $10\sqrt{2}$ centimeters.

87. $\frac{1}{2} - \sqrt{\frac{9}{4}} = \frac{1}{2} - \frac{3}{2} = -\frac{2}{2} = -1$

89. $\frac{6+4\sqrt{5}}{2} = \frac{6}{2} + \frac{4\sqrt{5}}{2} = 3 + 2\sqrt{5}$

91. $\frac{3-9\sqrt{2}}{6} = \frac{3}{6} - \frac{9\sqrt{2}}{6} = \frac{1}{2} - \frac{3\sqrt{2}}{2} = \frac{1-3\sqrt{2}}{2}$

93. $\sqrt{b^2 - 4ac} = \sqrt{(4)^2 - 4(2)(-1)}$
$= \sqrt{16+8}$
$= \sqrt{24}$
$= 2\sqrt{6}$

95. $\sqrt{b^2 - 4ac} = \sqrt{(-1)^2 - 4(3)(-2)}$
$= \sqrt{1+24}$
$= \sqrt{25}$
$= 5$

97. The solutions of $(x+1)^2 = -1$ are complex, but not real numbers.

99. The solutions of $3z^2 = 10$ are real numbers.

101. The solutions of $(2y-5)^2 + 7 = 3$ are complex, but not real numbers.

103. $x^2 + \_\_ + 16$

$$\left(\frac{b}{2}\right)^2 = 16$$
$$\frac{b}{2} = \pm\sqrt{16}$$
$$\frac{b}{2} = \pm 4$$
$$b = \pm 8$$

Answer: $\pm 8x$

105. $z^2 + \_\_ + \frac{25}{4}$

$$\left(\frac{b}{2}\right)^2 = \frac{25}{4}$$
$$\frac{b}{2} = \pm\sqrt{\frac{25}{4}}$$
$$\frac{b}{2} = \pm\frac{5}{2}$$
$$b = \pm 5$$

Answer: $\pm 5z$

107. answers may vary

109. compound interest is preferable; answers may vary

111.
$$p = -x^2 + 47$$
$$11 = -x^2 + 47$$
$$x^2 = 36$$
$$x = \pm\sqrt{36}$$
$$x = \pm 6$$

Demand cannot be negative. Therefore, the demand is 6 thousand scissors.

Section 8.2 Practice Exercises

1. $3x^2 - 5x - 2 = 0$

$a = 3, b = -5, c = -2$

$$x = \frac{-b \pm \sqrt{b^2 - 4ac}}{2a}$$
$$= \frac{-(-5) \pm \sqrt{(-5)^2 - 4(3)(-2)}}{2(3)}$$
$$= \frac{5 \pm \sqrt{25 + 24}}{6}$$
$$= \frac{5 \pm \sqrt{49}}{6}$$
$$= \frac{5 \pm 7}{6}$$

$$x = \frac{5+7}{6} = \frac{12}{6} = 2 \text{ or } x = \frac{5-7}{6} = \frac{-2}{6} = -\frac{1}{3}$$

The solutions are $-\frac{1}{3}$ and 2, or the solution set is $\left\{-\frac{1}{3}, 2\right\}$.

2. $3x^2 - 8x = 2$

Write in standard form.

$3x^2 - 8x - 2 = 0$

$a = 3, b = -8, c = -2$

$$x = \frac{-b \pm \sqrt{b^2 - 4ac}}{2a}$$
$$= \frac{-(-8) \pm \sqrt{(-8)^2 - 4(3)(-2)}}{2(3)}$$
$$= \frac{8 \pm \sqrt{64 + 24}}{6}$$
$$= \frac{8 \pm \sqrt{88}}{6}$$
$$= \frac{8 \pm 2\sqrt{22}}{6}$$
$$= \frac{4 \pm \sqrt{22}}{3}$$

The solutions are $\frac{4+\sqrt{22}}{3}$ and $\frac{4-\sqrt{22}}{3}$, or the solution set is $\left\{\frac{4+\sqrt{22}}{3}, \frac{4-\sqrt{22}}{3}\right\}$.

3. $\frac{1}{8}x^2 - \frac{1}{4}x - 2 = 0$

Multiply both sides of the equation by 8.

$$8\left(\frac{1}{8}x^2 - \frac{1}{4}x - 2\right) = 8 \cdot 0$$
$$x^2 - 2x - 16 = 0$$

Substitute $a = 1$, $b = -2$, and $c = -16$ into the quadratic formula and simplify.

$$x = \frac{-(-2) \pm \sqrt{(-2)^2 - 4(1)(-16)}}{2(1)}$$
$$= \frac{2 \pm \sqrt{4+64}}{2}$$
$$= \frac{2 \pm \sqrt{68}}{2}$$
$$= \frac{2 \pm 2\sqrt{17}}{2}$$
$$= 1 \pm \sqrt{17}$$

The solutions are $1+\sqrt{17}$ or $1-\sqrt{17}$.

4. $x = -2x^2 - 2$
The equation in standard form is $2x^2 + x + 2 = 0$. Thus, let $a = 2$, $b = 1$, and $c = 2$ in the quadratic formula.

$$x = \frac{-1 \pm \sqrt{1^2 - 4(2)(2)}}{2(2)}$$
$$= \frac{-1 \pm \sqrt{1-16}}{4}$$
$$= \frac{-1 \pm \sqrt{-15}}{4}$$
$$= \frac{-1 \pm i\sqrt{15}}{4}$$

The solutions are $\frac{-1+i\sqrt{15}}{4}$ and $\frac{-1-i\sqrt{15}}{4}$.

5. a. $x^2 - 6x + 9 = 0$
In $x^2 - 6x + 9$, $a = 1$, $b = -6$, and $c = 9$. Thus,
$b^2 - 4ac = (-6)^2 - 4(1)(9) = 36 - 36 = 0$
Since $b^2 - 4ac = 0$, this equation has one real solution.

b. $x^2 - 3x - 1 = 0$
In this equation, $a = 1$, $b = -3$, and $c = -1$.
$b^2 - 4ac = (-3)^2 - 4(1)(-1) = 9 + 4 = 13 > 0$
Since $b^2 - 4ac$ is positive, this equation has two real solutions.

c. $7x^2 + 11 = 0$
In this equation, $a = 7$, $b = 0$, and $c = 11$.
$b^2 - 4ac = 0^2 - 4(7)(11) = -308 < 0$
Since $b^2 - 4ac$ is negative, this equation has two complex but not real solutions.

6. By the Pythagorean theorem, we have

$$x^2 + (x+3)^2 = 15^2$$
$$x^2 + x^2 + 6x + 9 = 225$$
$$2x^2 + 6x - 216 = 0$$
$$x^2 + 3x - 108 = 0$$

Here, $a = 1$, $b = 3$, and $c = -108$. By the quadratic formula,

$$x = \frac{-3 \pm \sqrt{3^2 - 4(1)(-108)}}{2(1)}$$
$$= \frac{-3 \pm \sqrt{9+432}}{2}$$
$$= \frac{-3 \pm \sqrt{441}}{2}$$
$$= \frac{-3 \pm 21}{2}$$

$$x = \frac{-3+21}{2} = \frac{18}{2} = 9 \text{ or}$$
$$x = \frac{-3-21}{2} = \frac{-24}{2} = -12$$

The length can't be negative, so reject −12. The distance along the sidewalk is
$x + (x + 3) = 2x + 3 = 2(9) + 3 = 18 + 3 = 21$ feet
A person can save 21 − 15 = 6 feet by cutting across the lawn.

7. $h = -16t^2 + 20t + 45$
At the ground, $h = 0$.
$0 = -16t^2 + 20t + 45$
Here, $a = -16$, $b = 20$, and $c = 45$. By the quadratic formula,

$$t = \frac{-20 \pm \sqrt{20^2 - 4(-16)(45)}}{2(-16)}$$
$$= \frac{-20 \pm \sqrt{400+2880}}{-32}$$
$$= \frac{-20 \pm \sqrt{3280}}{-32}$$
$$= \frac{20 \pm \sqrt{16 \cdot 205}}{32}$$
$$= \frac{20 \pm 4\sqrt{205}}{32}$$
$$= \frac{5 \pm \sqrt{205}}{8}$$

$$t = \frac{5+\sqrt{205}}{8} \approx 2.4 \text{ or } t = \frac{5-\sqrt{205}}{8} \approx -1.2$$

Since the time won't be negative, we reject −1.2. The rocket will strike the ground 2.4 seconds after launch.

Vocabulary, Readiness & Video Check 8.2

1. The quadratic formula is $x = \dfrac{-b \pm \sqrt{b^2 - 4ac}}{2a}$.

2. For $2x^2 + x + 1 = 0$, if $a = 2$, then $b = \underline{1}$ and $c = \underline{1}$.

3. For $5x^2 - 5x - 7 = 0$, if $a = 5$, then $b = \underline{-5}$ and $c = \underline{-7}$.

4. For $7x^2 - 4 = 0$, if $a = 7$, then $b = \underline{0}$ and $c = \underline{-4}$.

5. For $x^2 + 9 = 0$, if $c = 9$, then $a = \underline{1}$ and $b = \underline{0}$.

6. The correct simplified form of $\dfrac{5 \pm 10\sqrt{2}}{5}$ is $1 \pm 2\sqrt{2}$. The answer is **c**.

7. **a.** Yes, in order to make sure we have correct values for *a*, *b*, and *c*.

   b. No; clearing fractions makes the work less tedious, but it's not a necessary step.

8. The discriminant is the radicand in the quadratic formula and can be used to find the number and type of solutions of a quadratic equation without solving the equation. To use the discriminant, the quadratic equation needs to be written in standard form.

9. With applications, we need to make sure we answer the question(s) asked. Here we're asked how much distance is saved, so once the dimensions of the triangle are known, further calculations are needed to answer this question and solve the problem.

Exercise Set 8.2

1. $m^2 + 5m - 6 = 0$
   $a = 1, b = 5, c = -6$
   $$m = \frac{-5 \pm \sqrt{(5)^2 - 4(1)(-6)}}{2(1)} = \frac{-5 \pm \sqrt{25 + 24}}{2} = \frac{-5 \pm \sqrt{49}}{2} = \frac{-5 \pm 7}{2} = -6 \text{ or } 1$$
   The solutions are −6 and 1.

3. $2y = 5y^2 - 3$
   $5y^2 - 2y - 3 = 0$
   $a = 5, b = -2, c = -3$
   $$y = \frac{2 \pm \sqrt{(-2)^2 - 4(5)(-3)}}{2(5)} = \frac{2 \pm \sqrt{4 + 60}}{10} = \frac{2 \pm \sqrt{64}}{10} = \frac{2 \pm 8}{10} = -\frac{3}{5} \text{ or } 1$$
   The solutions are $-\frac{3}{5}$ and 1.

5. $x^2 - 6x + 9 = 0$
   $a = 1, b = -6, c = 9$
   $$x = \frac{6 \pm \sqrt{(-6)^2 - 4(1)(9)}}{2(1)} = \frac{6 \pm \sqrt{36 - 36}}{2} = \frac{6 \pm \sqrt{0}}{2} = \frac{6}{2} = 3$$
   The solution is 3.

7. $x^2+7x+4=0$

$a=1, b=7, c=4$

$$x=\frac{-7\pm\sqrt{(7)^2-4(1)(4)}}{2(1)}=\frac{-7\pm\sqrt{49-16}}{2}=\frac{-7\pm\sqrt{33}}{2}$$

The solutions are $\frac{-7+\sqrt{33}}{2}$ and $\frac{-7-\sqrt{33}}{2}$.

9. $8m^2-2m=7$

$8m^2-2m-7=0$

$a=8, b=-2, c=-7$

$$m=\frac{2\pm\sqrt{(-2)^2-4(8)(-7)}}{2(8)}=\frac{2\pm\sqrt{4+224}}{16}=\frac{2\pm\sqrt{228}}{16}=\frac{2\pm\sqrt{4\cdot 57}}{16}=\frac{2\pm 2\sqrt{57}}{16}=\frac{1\pm\sqrt{57}}{8}$$

The solutions are $\frac{1+\sqrt{57}}{8}$ and $\frac{1-\sqrt{57}}{8}$.

11. $3m^2-7m=3$

$3m^2-7m-3=0$

$a=3, b=-7, c=-3$

$$m=\frac{7\pm\sqrt{(-7)^2-4(3)(-3)}}{2(3)}=\frac{7\pm\sqrt{49+36}}{6}=\frac{7\pm\sqrt{85}}{6}$$

The solutions are $\frac{7+\sqrt{85}}{6}$ and $\frac{7-\sqrt{85}}{6}$.

13. $\frac{1}{2}x^2-x-1=0$

$x^2-2x-2=0$

$a=1, b=-2, c=-2$

$$x=\frac{2\pm\sqrt{(-2)^2-4(1)(-2)}}{2(1)}=\frac{2\pm\sqrt{4+8}}{2}=\frac{2\pm\sqrt{12}}{2}=\frac{2\pm 2\sqrt{3}}{2}=1\pm\sqrt{3}$$

The solutions are $1+\sqrt{3}$ and $1-\sqrt{3}$.

15. $\frac{2}{5}y^2+\frac{1}{5}y=\frac{3}{5}$

$2y^2+y-3=0$

$a=2, b=1, c=-3$

$$y=\frac{-1\pm\sqrt{(1)^2-4(2)(-3)}}{2(2)}=\frac{-1\pm\sqrt{1+24}}{4}=\frac{-1\pm\sqrt{25}}{4}=\frac{-1\pm 5}{4}=-\frac{3}{2}\text{ or }1$$

The solutions are $-\frac{3}{2}$ and 1.

17. $\frac{1}{3}y^2=y+\frac{1}{6}$

$\frac{1}{3}y^2-y-\frac{1}{6}=0$

$2y^2-6y-1=0$

$a=2, b=-6, c=-1$

$$y=\frac{6\pm\sqrt{(-6)^2-4(2)(-1)}}{2(2)}$$
$$=\frac{6\pm\sqrt{36+8}}{4}$$
$$=\frac{6\pm\sqrt{44}}{4}$$
$$=\frac{6\pm2\sqrt{11}}{4}$$
$$=\frac{3\pm\sqrt{11}}{2}$$

The solutions are $\frac{3+\sqrt{11}}{2}$ and $\frac{3-\sqrt{11}}{2}$.

19. $x^2+5x=-2$
$$x^2+5x+2=0$$
$a=1, b=5, c=2$
$$x=\frac{-5\pm\sqrt{(5)^2-4(1)(2)}}{2(1)}$$
$$=\frac{-5\pm\sqrt{25-8}}{2}$$
$$=\frac{-5\pm\sqrt{17}}{2}$$

The solutions are $\frac{-5+\sqrt{17}}{2}$ and $\frac{-5-\sqrt{17}}{2}$.

21. $(m+2)(2m-6)=5(m-1)-12$
$$2m^2-6m+4m-12=5m-5-12$$
$$2m^2-7m+5=0$$
$a=2, b=-7, c=5$
$$m=\frac{7\pm\sqrt{(-7)^2-4(2)(5)}}{2(2)}$$
$$=\frac{7\pm\sqrt{49-40}}{4}$$
$$=\frac{7\pm\sqrt{9}}{4}$$
$$=\frac{7\pm3}{4}$$
$$=1 \text{ or } \frac{5}{2}$$

The solutions are 1 and $\frac{5}{2}$.

23. $x^2+6x+13=0$
$a=1, b=6, c=13$
$$x=\frac{-6\pm\sqrt{(6)^2-4(1)(13)}}{2(1)}$$
$$=\frac{-6\pm\sqrt{36-52}}{2}$$
$$=\frac{-6\pm\sqrt{-16}}{2}$$
$$=\frac{-6\pm4i}{2}$$
$$=-3\pm2i$$

The solutions are $-3+2i$ and $-3-2i$.

25. $(x+5)(x-1)=2$
$$x^2+4x-5=2$$
$$x^2+4x-7=0$$
$a=1, b=4, c=-7$
$$x=\frac{-4\pm\sqrt{(4)^2-4(1)(-7)}}{2(1)}$$
$$=\frac{-4\pm\sqrt{16+28}}{2}$$
$$=\frac{-4\pm\sqrt{44}}{2}$$
$$=\frac{-4\pm2\sqrt{11}}{2}$$
$$=-2\pm\sqrt{11}$$

The solutions are $-2+\sqrt{11}$ and $-2-\sqrt{11}$.

27. $6=-4x^2+3x$
$$4x^2-3x+6=0$$
$a=4, b=-3, c=6$
$$x=\frac{3\pm\sqrt{(-3)^2-4(4)(6)}}{2(4)}$$
$$=\frac{3\pm\sqrt{9-96}}{8}$$
$$=\frac{3\pm\sqrt{-87}}{8}$$
$$=\frac{3\pm i\sqrt{87}}{8}$$

The solutions are $\frac{3+i\sqrt{87}}{8}$ and $\frac{3-i\sqrt{87}}{8}$.

29. $$\frac{x^2}{3} - x = \frac{5}{3}$$
$$x^2 - 3x = 5$$
$$x^2 - 3x - 5 = 0$$
$$a = 1,\ b = -3,\ c = -5$$
$$x = \frac{3 \pm \sqrt{(-3)^2 - 4(1)(-5)}}{2(1)} = \frac{3 \pm \sqrt{9+20}}{2} = \frac{3 \pm \sqrt{29}}{2}$$

The solutions are $\frac{3+\sqrt{29}}{2}$ and $\frac{3-\sqrt{29}}{2}$.

31. $$10y^2 + 10y + 3 = 0$$
$$a = 10,\ b = 10,\ c = 3$$
$$y = \frac{-10 \pm \sqrt{(10)^2 - 4(10)(3)}}{2(10)} = \frac{-10 \pm \sqrt{100-120}}{20} = \frac{-10 \pm \sqrt{-20}}{20} = \frac{-10 \pm i\sqrt{4 \cdot 5}}{20} = \frac{-10 \pm 2i\sqrt{5}}{20} = \frac{-5 \pm i\sqrt{5}}{10}$$

The solutions are $\frac{-5+i\sqrt{5}}{10}$ and $\frac{-5-i\sqrt{5}}{10}$.

33. $$x(6x+2) = 3$$
$$x(6x+2) - 3 = 0$$
$$6x^2 + 2x - 3 = 0$$
$$a = 6,\ b = 2,\ c = -3$$
$$x = \frac{-2 \pm \sqrt{(2)^2 - 4(6)(-3)}}{2(6)} = \frac{-2 \pm \sqrt{4+72}}{12} = \frac{-2 \pm \sqrt{76}}{12} = \frac{-2 \pm \sqrt{4 \cdot 19}}{12} = \frac{-2 \pm 2\sqrt{19}}{12} = \frac{-1 \pm \sqrt{19}}{6}$$

The solutions are $\frac{-1+\sqrt{19}}{6}$ and $\frac{-1-\sqrt{19}}{6}$.

35. $$\frac{2}{5}y^2 + \frac{1}{5}y + \frac{3}{5} = 0$$
$$2y^2 + y + 3 = 0$$
$$a = 2,\ b = 1,\ c = 3$$
$$y = \frac{-1 \pm \sqrt{(1)^2 - 4(2)(3)}}{2(2)} = \frac{-1 \pm \sqrt{1-24}}{4} = \frac{-1 \pm \sqrt{-23}}{4} = \frac{-1 \pm i\sqrt{23}}{4}$$

The solutions are $\frac{-1+i\sqrt{23}}{4}$ and $\frac{-1-i\sqrt{23}}{4}$.

37. $$\frac{1}{2}y^2 = y - \frac{1}{2}$$
$$y^2 = 2y - 1$$
$$y^2 - 2y + 1 = 0$$
$$a = 1,\ b = -2,\ c = 1$$
$$y = \frac{2 \pm \sqrt{(-2)^2 - 4(1)(1)}}{2(1)} = \frac{2 \pm \sqrt{4-4}}{2} = \frac{2 \pm \sqrt{0}}{2} = \frac{2}{2} = 1$$

The solution is 1.

39. $(n-2)^2 = 2n$
$n^2 - 4n + 4 = 2n$
$n^2 - 6n + 4 = 0$
$a = 1, b = -6, c = 4$

$$n = \frac{6 \pm \sqrt{(-6)^2 - 4(1)(4)}}{2(1)} = \frac{6 \pm \sqrt{36-16}}{2} = \frac{6 \pm \sqrt{20}}{2} = \frac{6 \pm 2\sqrt{5}}{2} = 3 \pm \sqrt{5}$$

The solutions are $3+\sqrt{5}$ and $3-\sqrt{5}$.

41. $x^2 - 5 = 0$
$a = 1, b = 0, c = -5$
$b^2 - 4ac = 0^2 - 4(1)(-5) = 20 > 0$
Therefore, there are two real solutions.

43. $4x^2 + 12x = -9$
$4x^2 - 12x + 9 = 0$
$a = 4, b = -12, c = 9$

$$b^2 - 4ac = (-12)^2 - 4(4)(9) = 144 - 144 = 0$$

Therefore, there is one real solution.

45. $3x = -2x^2 + 7$
$2x^2 + 3x - 7 = 0$
$a = 2, b = 3, c = -7$

$$b^2 - 4ac = 3^2 - 4(2)(-7) = 9 + 56 = 65 > 0$$

Therefore, there are two real solutions.

47. $6 = 4x - 5x^2$
$5x^2 - 4x + 6 = 0$
$a = 5, b = -4, c = 6$

$$b^2 - 4ac = (-4)^2 - 4(5)(6) = 16 - 120 = -104 < 0$$

Therefore, there are two complex but not real solutions.

49. $9x - 2x^2 + 5 = 0$
$-2x^2 + 9x + 5 = 0$
$a = -2, b = 9, c = 5$

$$b^2 - 4ac = 9^2 - 4(-2)(5) = 81 + 40 = 121 > 0$$

Therefore, there are two real solutions.

51. $(x+8)^2 + x^2 = 36^2$
$(x^2 + 16x + 64) + x^2 = 1296$
$2x^2 + 16x - 1232 = 0$
$a = 2, b = 16, c = -1232$

$$x = \frac{-16 \pm \sqrt{(16)^2 - 4(2)(-1232)}}{2(2)} = \frac{-16 \pm \sqrt{10{,}112}}{4}$$

$x \approx 21$ or $x \approx -29$ (disregard)
$x + (x+8) = 21 + 21 + 8 = 50$
$50 - 36 = 14$
They save about 14 feet of walking distance.

53. Let x = length of leg. Then
$x + 2$ = length of hypotenuse.
$x^2 + x^2 = (x+2)^2$
$2x^2 = x^2 + 4x + 4$
$x^2 - 4x - 4 = 0$
$a = 1, b = -4, c = -4$

$$x = \frac{4 \pm \sqrt{(-4)^2 - 4(1)(-4)}}{2(1)} = \frac{4 \pm \sqrt{32}}{2} = \frac{4 \pm 4\sqrt{2}}{2}$$

$= 2 \pm 2\sqrt{2}$ (disregard the negative)
$= 2 + 2\sqrt{2}$

The sides measure $\left(2+2\sqrt{2}\right)$ cm, $\left(2+2\sqrt{2}\right)$ cm, and $\left(4+2\sqrt{2}\right)$ cm.

55. Let x = width; then $x + 10$ = length.

Area = length · width

$400 = (x+10)x$

$0 = x^2 + 10x - 400$

$a = 1, b = 10, c = -400$

$$x = \frac{-10 \pm \sqrt{(10)^2 - 4(1)(-400)}}{2(1)} = \frac{-10 \pm \sqrt{1700}}{2} = \frac{-10 \pm 10\sqrt{17}}{2} = -5 \pm 5\sqrt{17}$$

Disregard the negative length. The width is $\left(-5+5\sqrt{17}\right)$ ft and the length is $\left(5+5\sqrt{17}\right)$ ft.

57. a. Let x = length.

$x^2 + x^2 = 100^2$

$2x^2 - 10{,}000 = 0$

$a = 2, b = 0, c = -10{,}000$

$$x = \frac{0 \pm \sqrt{(0)^2 - 4(2)(-10{,}000)}}{2(2)} = \frac{\pm\sqrt{80{,}000}}{4} = \frac{\pm 200\sqrt{2}}{4} = \pm 50\sqrt{2}$$

Disregard the negative length. The side measures $50\sqrt{2}$ meters.

b. Area $= s^2$

$$= \left(50\sqrt{2}\right)^2 = 2500(2) = 5000$$

The area is 5000 square meters.

59. Let w = width; then $w + 1.1$ = height.

Area = length · width

$1439.9 = (w+1.1)w$

$0 = w^2 + 1.1w - 1439.9$

$a = 1, b = 1.1, c = -1439.9$

$$w = \frac{-1.1 \pm \sqrt{(1.1)^2 - 4(1)(-1439.9)}}{2(1)} = \frac{-1.1 \pm \sqrt{5760.81}}{2}$$

$= 37.4$ or -38.5 (disregard)

Its width is 37.4 ft and its height is 38.5 ft.

61. Let h = height. Then $2h + 4$ = base.

Area $= \frac{1}{2}$ base · height

$42 = \frac{1}{2}(2h+4)h$

$42 = h^2 + 2h$

$0 = h^2 + 2h - 42$

$a = 1, b = 2, c = -42$

$$h = \frac{-2 \pm \sqrt{(2)^2 - 4(1)(-42)}}{2(1)} = \frac{-2 \pm \sqrt{172}}{2} = \frac{-2 \pm 2\sqrt{43}}{2}$$

$= -1 \pm \sqrt{43}$ (disregard the negative)

base $= 2\left(-1+\sqrt{43}\right) + 4 = 2 + 2\sqrt{43}$

Height: $\left(-1+\sqrt{43}\right)$ cm

Base: $\left(2+2\sqrt{43}\right)$ cm

63. $h = -16t^2 + 20t + 1100$

$0 = -16t^2 + 20t + 1100$

$a = -16, b = 20, c = 1100$

$$t = \frac{-20 \pm \sqrt{(20)^2 - 4(-16)(1100)}}{2(-16)} = \frac{-20 \pm \sqrt{70{,}800}}{-32}$$

≈ 8.9 or -7.7 (disregard)

It will take about 8.9 seconds.

65. $h=-16t^2-20t+180$

$0=-16t^2-20t+180$

$a=-16$, $b=-20$, $c=180$

$$t=\frac{20\pm\sqrt{(-20)^2-4(-16)(180)}}{2(-16)}$$

$$=\frac{20\pm\sqrt{11,920}}{-32}$$

≈ 2.8 or -4.0 (disregard)

It will take about 2.8 seconds.

67. $\sqrt{5x-2}=3$

$\left(\sqrt{5x-2}\right)^2=3^2$

$5x-2=9$

$5x=11$

$x=\frac{11}{5}$

69. $\frac{1}{x}+\frac{2}{5}=\frac{7}{x}$

$5x\left(\frac{1}{x}+\frac{2}{5}\right)=5x\left(\frac{7}{x}\right)$

$5+2x=35$

$2x=30$

$x=15$

71. $x^4+x^2-20=(x^2+5)(x^2-4)$

$=(x^2+5)(x+2)(x-2)$

73. $z^4-13z^2+36=(z^2-9)(z^2-4)$

$=(z+3)(z-3)(z+2)(z-2)$

75. $x^2=-10$

$x^2+10=0$

$a=1$, $b=0$, $c=10$

The correct substitution is **b**.

77. answers may vary

79. $2x^2-6x+3=0$

$a=2$, $b=-6$, $c=3$

$$x=\frac{6\pm\sqrt{(-6)^2-4(2)(3)}}{2(2)}$$

$$=\frac{6\pm\sqrt{12}}{4}$$

≈ 0.6 or 2.4

81. from Sunday to Monday

83. Wednesday

85. $f(x)=3x^2-18x+56$

$f(4)=3(4)^2-18(4)+56=32$

This answers appears to agree with the graph.

87. $f(x)=5x^2+55x+29$

a. 2015 is 20 years after 1995, so $x=20$.

$f(20)=5(20)^2+55(20)+29=3129$

The number of Internet users worldwide in 2015 was 3129 million.

b. Let $f(x)=4500$ and solve for x.

$4500=5x^2+55x+29$

$0=5x^2+55x-4471$

$a=5$, $b=55$, $c=-4471$

$$x=\frac{-b\pm\sqrt{b^2-4ac}}{2a}$$

$$x=\frac{-55\pm\sqrt{55^2-4(5)(-4471)}}{2(5)}$$

$$=\frac{-55\pm\sqrt{92,445}}{10}$$

$x\approx 24.9$ or $x\approx -35.9$

Disregard the negative, since we are not concerned with the past.

$1995+24.9=2019.9$

The number of Internet users worldwide is predicted to be 4500 million in the year 2019.

89. $f(x)=0.2x^2+5.1x+45.3$

a. 2015 is 45 years after 1970, so $x=45$.

$f(45)=0.2(45)^2+5.1(45)+45.3=679.8$

\$679.8 billion was spent in U.S. restaurants in 2015.

b. Let $f(x) = 400$ and solve for x.

$$400 = 0.2x^2 + 5.1x + 45.3$$
$$0 = 0.2x^2 + 5.1x - 354.7$$
$$0 = 2x^2 + 51x - 3547$$
$$a = 2,\ b = 51,\ c = -3547$$
$$x = \frac{-b \pm \sqrt{b^2 - 4ac}}{2a}$$
$$x = \frac{-51 \pm \sqrt{51^2 - 4(2)(-3547)}}{2(2)} = \frac{-51 \pm \sqrt{30{,}977}}{4}$$

$x \approx 31.3$ or $x \approx -56.8$

Disregard the negative value in the context of the application.

$1970 + 31 = 2001$

$400 billion was spent in U.S. restaurants in 2001.

c. Let $f(x) = 1000$ and solve for x.

$$1000 = 0.2x^2 + 5.1x + 45.3$$
$$0 = 0.2x^2 + 5.1x - 954.7$$
$$0 = 2x^2 + 51x - 9547$$
$$a = 2,\ b = 51,\ c = -9547$$
$$x = \frac{-b \pm \sqrt{b^2 - 4ac}}{2a}$$
$$x = \frac{-51 \pm \sqrt{51^2 - 4(2)(-9547)}}{2(2)} = \frac{-51 \pm \sqrt{78{,}977}}{4}$$

$x \approx 57.5$ or $x \approx -83.0$

Disregard the negative value in the context of the application.

$1970 + 57.5 = 2027.5$

The amount spent in U.S. restaurants is predicted to be $1000 billion in 2027.

91. $$\frac{-b + \sqrt{b^2 - 4ac}}{2a} + \frac{-b - \sqrt{b^2 - 4ac}}{2a} = \frac{-b + \sqrt{b^2 - 4ac} - b - \sqrt{b^2 - 4ac}}{2a} = \frac{-2b}{2a} = -\frac{b}{a}$$

93. $3x^2 - \sqrt{12}x + 1 = 0$

$a = 3,\ b = -\sqrt{12},\ c = 1$

$$x = \frac{\sqrt{12} \pm \sqrt{\left(-\sqrt{12}\right)^2 - 4(3)(1)}}{2(3)} = \frac{\sqrt{12} \pm \sqrt{12 - 12}}{6} = \frac{\sqrt{4 \cdot 3} \pm \sqrt{0}}{6} = \frac{2\sqrt{3}}{6} = \frac{\sqrt{3}}{3}$$

The solution is $\frac{\sqrt{3}}{3}$.

95. $x^2 + \sqrt{2}x + 1 = 0$

$a = 1,\ b = \sqrt{2},\ c = 1$

$$x = \frac{-\sqrt{2} \pm \sqrt{\left(\sqrt{2}\right)^2 - 4(1)(1)}}{2(1)} = \frac{-\sqrt{2} \pm \sqrt{2 - 4}}{2} = \frac{-\sqrt{2} \pm \sqrt{-2}}{2} = \frac{-\sqrt{2} \pm i\sqrt{2}}{2}$$

The solutions are $\frac{-\sqrt{2} + i\sqrt{2}}{2}$ and $\frac{-\sqrt{2} - i\sqrt{2}}{2}$.

97. $2x^2 - \sqrt{3}x - 1 = 0$

$a = 2,\ b = -\sqrt{3},\ c = -1$

$$x = \frac{\sqrt{3} \pm \sqrt{\left(-\sqrt{3}\right)^2 - 4(2)(-1)}}{2(2)} = \frac{\sqrt{3} \pm \sqrt{3 + 8}}{4} = \frac{\sqrt{3} \pm \sqrt{11}}{4}$$

The solutions are $\frac{\sqrt{3} + \sqrt{11}}{4}$ and $\frac{\sqrt{3} - \sqrt{11}}{4}$.

99. Exercise 63:

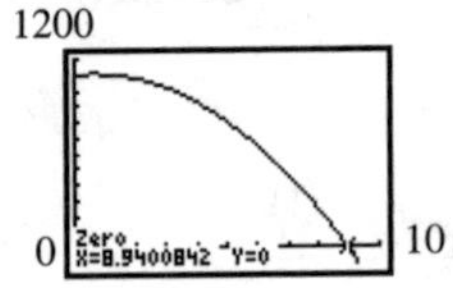

Exercise 65:

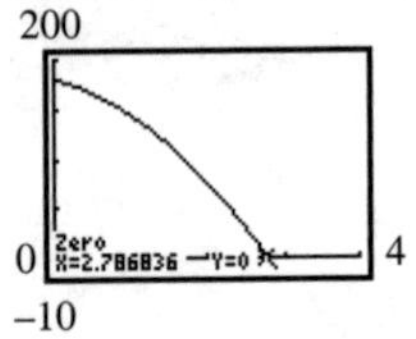

101. $y = 9x - 2x^2 + 5$

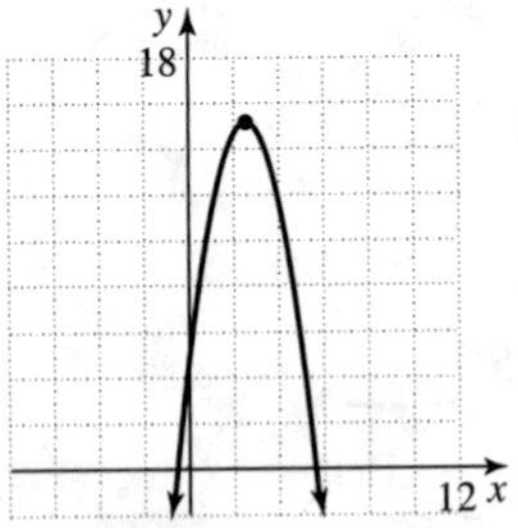

There are two x-intercepts. There are two real solutions.

Section 8.3 Practice Exercises

1. $x - \sqrt{x+1} - 5 = 0$

Get the radical alone on one side of the equation. Then square both sides.

$$x - \sqrt{x+1} - 5 = 0$$
$$x - 5 = \sqrt{x+1}$$
$$(x-5)^2 = x+1$$
$$x^2 - 10x + 25 = x + 1$$
$$x^2 - 11x + 24 = 0$$
$$(x-8)(x-3) = 0$$
$$x - 8 = 0 \quad \text{or} \quad x - 3 = 0$$
$$x = 8 \quad \text{or} \quad x = 3$$

Check:
Let $x = 3$.

$$x - \sqrt{x+1} - 5 = 0$$
$$3 - \sqrt{3+1} - 5 \stackrel{?}{=} 0$$
$$-2 - \sqrt{4} \stackrel{?}{=} 0$$
$$-2 - 2 \stackrel{?}{=} 0$$
$$-4 = 0 \quad \text{False}$$

Let $x = 8$.

$$x-\sqrt{x+1}-5=0$$
$$8-\sqrt{8+1}-5 \stackrel{?}{=} 0$$
$$3-\sqrt{9} \stackrel{?}{=} 0$$
$$3-3 \stackrel{?}{=} 0$$
$$0=0 \quad \text{True}$$

The solution is 8 or the solution set is $\{8\}$.

2. $\dfrac{5x}{x+1}-\dfrac{x+4}{x}=\dfrac{3}{x(x+1)}$

x cannot be either -1 or 0, because these values cause denominators to equal zero. Multiply both sides of the equation by $x(x + 1)$.

$$x(x+1)\left(\frac{5x}{x+1}\right)-x(x+1)\left(\frac{x+4}{x}\right)=x(x+1)\left[\frac{3}{x(x+1)}\right]$$
$$5x^2-(x+1)(x+4)=3$$
$$5x^2-x^2-5x-4=3$$
$$4x^2-5x-7=0$$

Use the quadratic formula with $a = 4$, $b = -5$, and $c = -7$.

$$x=\frac{-(-5)\pm\sqrt{(-5)^2-4(4)(-7)}}{2(4)}=\frac{5\pm\sqrt{25+112}}{8}=\frac{5\pm\sqrt{137}}{8}$$

Neither proposed solution will make denominators 0. The solutions are $\dfrac{5+\sqrt{137}}{8}$ and $\dfrac{5-\sqrt{137}}{8}$ or the solution set is $\left\{\dfrac{5+\sqrt{137}}{8}, \dfrac{5-\sqrt{137}}{8}\right\}$.

3.
$$p^4-7p^2-144=0$$
$$(p^2+9)(p^2-16)=0$$
$$(p^2+9)(p+4)(p-4)=0$$

$p^2+9=0$ or $p+4=0$ or $p-4=0$

$p^2=-9$ $\quad p=-4$ $\quad p=4$

$p=\pm\sqrt{-9}$

$p=\pm 3i$

The solutions are 4, -4, $3i$, and $-3i$.

4. $(x+2)^2-2(x+2)-3=0$

Let $y = x + 2$.

$$y^2-2y-3=0$$
$$(y-3)(y+1)=0$$

$y-3=0$ or $y+1=0$

$y=3$ $\quad y=-1$

Substitute $x + 2$ for y.

$x+2=3$ or $x+2=-1$

$x=1$ $\quad x=-3$

Both 1 and -3 check. The solutions are 1 and -3.

5. $x^{2/3}-5x^{1/3}+4=0$

Let $m=x^{1/3}$.

$m^2-5m+4=0$

$(m-4)(m-1)=0$

$m-4=0$ or $m-1=0$

$m=4$ $\quad$ $m=1$

Since $m=x^{1/3}$, we have

$x^{1/3}=4$ or $x^{1/3}=1$

$x=4^3=64$ $\quad$ $x=1^3=1$

Both 64 and 1 check. The solutions are 64 and 1.

6. Let x = the time in hours it takes Steve to groom all the dogs. Then, $x-1$ = the time it takes Katy to groom all the dogs.

The part of the job completed in one hour by Steve is $\frac{1}{x}$, and the part completed by Katy in one hour is $\frac{1}{x-1}$. In one hour, $\frac{1}{4}$ of the job is completed. We have,

$$\frac{1}{x}+\frac{1}{x-1}=\frac{1}{4}$$

$$4x(x-1)\left(\frac{1}{x}\right)+4x(x-1)\left(\frac{1}{x-1}\right)=4x(x-1)\left(\frac{1}{4}\right)$$

$$4(x-1)+4x=x(x-1)$$

$$4x-4+4x=x^2-x$$

$$0=x^2-9x+4$$

Use the quadratic formula with $a=1$, $b=-9$, and $c=4$.

$$x=\frac{-(-9)\pm\sqrt{(-9)^2-4(1)(4)}}{2(1)}$$

$$x=\frac{9\pm\sqrt{81-16}}{2}=\frac{9\pm\sqrt{65}}{2}$$

$x\approx 8.53$ or $x\approx 0.47$

Since $x-1=0.47-1=-0.53<0$, representing negative time worked, we reject 0.47. It takes Steve $\frac{9+\sqrt{65}}{2}\approx 8.5$ hours and Katy $\frac{9+\sqrt{65}}{2}-1=\frac{7+\sqrt{65}}{2}\approx 7.5$ hours to groom all the dogs when working alone.

7. Let x = the speed driven to Shanghai. Then $x+50$ = the speed driven to Ningbo.

	distance =	*rate* ·	*time*
To Shanghai	36	x	$\frac{36}{x}$
To Ningbo	36	$x+50$	$\frac{36}{x+50}$

The total travel time was 1.3 hours, so

$$\frac{36}{x}+\frac{36}{x+50}=1.3$$

$$x(x+50)\left(\frac{36}{x}\right)+x(x+50)\left(\frac{36}{x+50}\right)=1.3x(x+50)$$

$$36(x+50)+36x=1.3x^2+65x$$

$$36x+1800+36x=1.3x^2+65x$$

$$0=1.3x^2-7x-1800$$

Use the quadratic formula with $a = 1.3$, $b = -7$, and $c = -1800$.

$$x=\frac{-(-7)\pm\sqrt{(-7)^2-4(1.3)(-1800)}}{2(1.3)}=\frac{7\pm\sqrt{9409}}{2.6}$$

$$x=\frac{7+\sqrt{9409}}{2.6}=40 \quad\text{or}\quad x=\frac{7-\sqrt{9409}}{2.6}\approx -34.6$$

The speed is not negative, so reject −34.6. The speed to Shanghai was 40 km/hr and to Ningbo it was 40 + 50 = 90 km/hr.

Vocabulary, Readiness & Video Check 8.3

1. The values we get for the substituted variable are *not* our final answers. Remember to always substitute back to the original variable and solve for it if necessary.

2. The rational equation simplifies to a quadratic equation once you multiply through by the LCD to rid the equation of fractions.

Exercise Set 8.3

1.

$$2x=\sqrt{10+3x}$$

$$4x^2=10+3x$$

$$4x^2-3x-10=0$$

$$(4x+5)(x-2)=0$$

$$4x+5=0 \quad\text{or}\quad x-2=0$$

$$x=-\frac{5}{4} \quad\text{or}\quad x=2$$

Discard $-\frac{5}{4}$. The solution is 2.

3.

$$x-2\sqrt{x}=8$$

$$x-8=2\sqrt{x}$$

$$(x-8)^2=\left(2\sqrt{x}\right)^2$$

$$x^2-16x+64=4x$$

$$x^2-20x+64=0$$

$$(x-16)(x-4)=0$$

$$x-16=0 \quad\text{or}\quad x-4=0$$

$$x=16 \quad\text{or}\quad x=4 \text{ (discard)}$$

The solution is 16.

5. $\sqrt{9x} = x+2$

$\left(\sqrt{9x}\right)^2 = (x+2)^2$

$9x = x^2 + 4x + 4$

$0 = x^2 - 5x + 4$

$0 = (x-4)(x-1)$

$x-4=0$ or $x-1=0$

$x=4$ or $x=1$

The solutions are 1 and 4.

7. $\frac{2}{x} + \frac{3}{x-1} = 1$

Multiply each term by $x(x - 1)$.

$2(x-1)+3x = x(x-1)$

$2x-2+3x = x^2 - x$

$0 = x^2 - 6x + 2$

$x = \frac{6 \pm \sqrt{(-6)^2 - 4(1)(2)}}{2(1)}$

$= \frac{6 \pm \sqrt{28}}{2}$

$= \frac{6 \pm 2\sqrt{7}}{2} = 3 \pm \sqrt{7}$

The solutions are $3+\sqrt{7}$ and $3-\sqrt{7}$.

9. $\frac{3}{x} + \frac{4}{x+2} = 2$

Multiply each term by $x(x + 2)$.

$3(x+2)+4x = 2x(x+2)$

$3x+6+4x = 2x^2 + 4x$

$0 = 2x^2 - 3x - 6$

$x = \frac{3 \pm \sqrt{(-3)^2 - 4(2)(-6)}}{2(2)}$

$= \frac{3 \pm \sqrt{57}}{4}$

The solutions are $\frac{3+\sqrt{57}}{4}$ and $\frac{3-\sqrt{57}}{4}$.

11. $\frac{7}{x^2 - 5x + 6} = \frac{2x}{x-3} - \frac{x}{x-2}$

$\frac{7}{(x-3)(x-2)} = \frac{2x}{x-3} - \frac{x}{x-2}$

Multiply each term by $(x - 3)(x - 2)$.

$7 = 2x(x-2) - x(x-3)$

$7 = 2x^2 - 4x - x^2 + 3x$

$0 = x^2 - x - 7$

$x = \frac{1 \pm \sqrt{(-1)^2 - 4(1)(-7)}}{2(1)}$

$= \frac{1 \pm \sqrt{29}}{2}$

The solutions are $\frac{1+\sqrt{29}}{2}$ and $\frac{2-\sqrt{29}}{2}$.

13. $p^4 - 16 = 0$

$(p^2 - 4)(p^2 + 4) = 0$

$(p+2)(p-2)(p^2+4) = 0$

$p+2=0$ or $p-2=0$ or $p^2+4=0$

$p=-2$ or $p=2$ or $p^2=-4$

$p = \pm\sqrt{-4}$

$p = \pm 2i$

The solutions are –2, 2, –2*i*, and 2*i*.

15. $4x^4 + 11x^2 = 3$

$4x^4 + 11x^2 - 3 = 0$

$(4x^2 - 1)(x^2 + 3) = 0$

$(2x+1)(2x-1)(x^2+3) = 0$

$2x+1=0$ or $2x-1=0$ or $x^2+3=0$

$x = -\frac{1}{2}$ or $x = \frac{1}{2}$ or $x^2 = -3$

$x = \pm\sqrt{-3}$

$x = \pm i\sqrt{3}$

The solutions are $-\frac{1}{2}, \frac{1}{2}, -i\sqrt{3}$, and $i\sqrt{3}$.

17. $z^4 - 13z^2 + 36 = 0$

$(z^2 - 9)(z^2 - 4) = 0$

$(z+3)(z-3)(z+2)(z-2) = 0$

$z=-3,\ z=3,\ z=-2,\ z=2$

The solutions are –3, 3, –2, and 2.

19. $x^{2/3} - 3x^{1/3} - 10 = 0$

Let $y = x^{1/3}$. Then $y^2 = x^{2/3}$ and

$y^2 - 3y - 10 = 0$

$(y-5)(y+2) = 0$

$y-5=0$ or $y+2=0$

$y=5$ or $y=-2$

$x^{1/3}=5$ or $x^{1/3}=-2$

$x=125$ or $x=-8$

The solutions are –8 and 125.

21. $(5n+1)^2+2(5n+1)-3=0$

Let $y = 5n + 1$. Then $y^2=(5n+1)^2$ and

$$\begin{aligned} y^2+2y-3&=0 \\ (y+3)(y-1)&=0 \\ y+3=0 \quad &\text{or} \quad y-1=0 \\ y=-3 \quad &\text{or} \quad y=1 \\ 5n+1=-3 \quad &\text{or} \quad 5n+1=1 \\ 5n=-4 \quad &\text{or} \quad 5n=0 \\ n=-\frac{4}{5} \quad &\text{or} \quad n=0 \end{aligned}$$

The solutions are $-\frac{4}{5}$ and 0.

23. $2x^{2/3}-5x^{1/3}=3$

Let $y=x^{1/3}$. Then $y^2=x^{2/3}$ and

$$\begin{aligned} 2y^2-5y&=3 \\ 2y^2-5y-3&=0 \\ (2y+1)(y-3)&=0 \\ 2y+1=0 \quad &\text{or} \quad y-3=0 \\ y=-\frac{1}{2} \quad &\text{or} \quad y=3 \\ x^{1/3}=-\frac{1}{2} \quad &\text{or} \quad x^{1/3}=3 \\ x=-\frac{1}{8} \quad &\text{or} \quad x=27 \end{aligned}$$

The solutions are $-\frac{1}{8}$ and 27.

25. $1+\frac{2}{3t-2}=\frac{8}{(3t-2)^2}$

$$\begin{aligned} (3t-2)^2+2(3t-2)&=8 \\ (3t-2)^2+2(3t-2)-8&=0 \end{aligned}$$

Let $y=3t-2$. Then $y^2=(3t-2)^2$ and

$$\begin{aligned} y^2+2y-8&=0 \\ (y+4)(y-2)&=0 \\ y+4=0 \quad &\text{or} \quad y-2=0 \\ y=-4 \quad &\text{or} \quad y=2 \\ 3t-2=-4 \quad &\text{or} \quad 3t-2=2 \\ 3t=-2 \quad &\text{or} \quad 3t=4 \\ t=-\frac{2}{3} \quad &\text{or} \quad t=\frac{4}{3} \end{aligned}$$

The solutions are $-\frac{2}{3}$ and $\frac{4}{3}$.

27. $20x^{2/3}-6x^{1/3}-2=0$

Let $y=x^{1/3}$. Then $y^2=x^{2/3}$ and

$$\begin{aligned} 20y^2-6y-2&=0 \\ 2(10y^2-3y-1)&=0 \\ 2(5y+1)(2y-1)&=0 \\ 5y+1=0 \quad &\text{or} \quad 2y-1=0 \\ y=-\frac{1}{5} \quad &\text{or} \quad y=\frac{1}{2} \\ x^{1/3}=-\frac{1}{5} \quad &\text{or} \quad x^{1/3}=\frac{1}{2} \\ x=-\frac{1}{125} \quad &\text{or} \quad x=\frac{1}{8} \end{aligned}$$

The solutions are $\frac{1}{8}$ and $-\frac{1}{125}$.

29. $a^4-5a^2+6=0$

$$\begin{aligned} (a^2-3)(a^2-2)&=0 \\ a^2-3=0 \quad &\text{or} \quad a^2-2=0 \\ a^2=3 \quad &\text{or} \quad a^2=2 \\ a=\pm\sqrt{3} \quad &\text{or} \quad a=\pm\sqrt{2} \end{aligned}$$

The solutions are $-\sqrt{3}, \sqrt{3}, -\sqrt{2}$, and $\sqrt{2}$.

31. $\frac{2x}{x-2}+\frac{x}{x+3}=-\frac{5}{x+3}$

Multiply each term by $(x+3)(x-2)$.

$$\begin{aligned} 2x(x+3)+x(x-2)&=-5(x-2) \\ 2x^2+6x+x^2-2x&=-5x+10 \\ 3x^2+9x-10&=0 \\ x&=\frac{-9\pm\sqrt{(9)^2-4(3)(-10)}}{2(3)} \\ &=\frac{-9\pm\sqrt{201}}{6} \end{aligned}$$

The solutions are $\frac{-9+\sqrt{201}}{6}$ and $\frac{-9-\sqrt{201}}{6}$.

33. $(p+2)^2=9(p+2)-20$

$$(p+2)^2-9(p+2)+20=0$$

Let $x=p+2$. Then $x^2=(p+2)^2$ and

$$\begin{aligned} x^2-9x+20&=0 \\ (x-5)(x-4)&=0 \\ x=5 \quad &\text{or} \quad x=4 \\ p+2=5 \quad &\text{or} \quad p+2=4 \\ p=3 \quad &\text{or} \quad p=2 \end{aligned}$$

The solutions are 2 and 3.

35.
$$2x = \sqrt{11x+3}$$
$$(2x)^2 = \left(\sqrt{11x+3}\right)^2$$
$$4x^2 = 11x+3$$
$$4x^2 - 11x - 3 = 0$$
$$(4x+1)(x-3) = 0$$
$$x = -\frac{1}{4} \text{ (discard) or } x = 3$$
The solution is 3.

37. $x^{2/3} - 8x^{1/3} + 15 = 0$

Let $y = x^{1/3}$. Then $y^2 = x^{2/3}$ and
$$y^2 - 8y + 15 = 0$$
$$(y-5)(y-3) = 0$$
$$y = 5 \quad \text{or} \quad y = 3$$
$$x^{1/3} = 5 \quad \text{or} \quad x^{1/3} = 3$$
$$x = 125 \quad \text{or} \quad x = 27$$
The solutions are 27 and 125.

39.
$$y^3 + 9y - y^2 - 9 = 0$$
$$y(y^2+9) - 1(y^2+9) = 0$$
$$(y^2+9)(y-1) = 0$$
$$y^2 + 9 = 0 \quad \text{or} \quad y - 1 = 0$$
$$y^2 = -9 \quad \text{or} \quad y = 1$$
$$y = \pm\sqrt{-9}$$
$$y = \pm 3i$$
The solutions are 1, –3*i*, and 3*i*.

41. $2x^{2/3} + 3x^{1/3} - 2 = 0$

Let $y = x^{1/3}$. Then $y^2 = x^{2/3}$ and
$$2y^2 + 3y - 2 = 0$$
$$(2y-1)(y+2) = 0$$
$$y = \frac{1}{2} \quad \text{or} \quad y = -2$$
$$x^{1/3} = \frac{1}{2} \quad \text{or} \quad x^{1/3} = -2$$
$$x = \frac{1}{8} \quad \text{or} \quad x = -8$$
The solutions are –8 and $\frac{1}{8}$.

43. $x^{-2} - x^{-1} - 6 = 0$

Let $y = x^{-1}$. Then $y^2 = x^{-2}$ and
$$y^2 - y - 6 = 0$$
$$(y-3)(y+2) = 0$$
$$y = 3 \quad \text{or} \quad y = -2$$
$$x^{-1} = 3 \quad \text{or} \quad x^{-1} = -2$$
$$\frac{1}{x} = 3 \quad \text{or} \quad \frac{1}{x} = -2$$
$$x = \frac{1}{3} \quad \text{or} \quad x = -\frac{1}{2}$$
The solutions are $-\frac{1}{2}$ and $\frac{1}{3}$.

45.
$$x - \sqrt{x} = 2$$
$$x - 2 = \sqrt{x}$$
$$(x-2)^2 = x$$
$$x^2 - 4x + 4 = x$$
$$x^2 - 5x + 4 = 0$$
$$(x-4)(x-1) = 0$$
$x = 4$ or $x = 1$ (discard)

The solution is 4.

47.
$$\frac{x}{x-1} + \frac{1}{x+1} = \frac{2}{x^2-1}$$
$$\frac{x}{x-1} + \frac{1}{x+1} = \frac{2}{(x+1)(x-1)}$$
$$x(x+1) + (x-1) = 2$$
$$x^2 + x + x - 1 = 2$$
$$x^2 + 2x - 3 = 0$$
$$(x+3)(x-1) = 0$$
$x = -3$ or $x = 1$ (discard)

The solution is –3.

49.
$$p^4 - p^2 - 20 = 0$$
$$(p^2-5)(p^2+4) = 0$$
$$p^2 - 5 = 0 \quad \text{or} \quad p^2 + 4 = 0$$
$$p^2 = 5 \quad \text{or} \quad p^2 = -4$$
$$p = \pm\sqrt{5} \quad \text{or} \quad p = \pm 2i$$
The solutions are $-\sqrt{5}, \sqrt{5}, -2i$, and $2i$.

51. $(x+3)(x^2-3x+9)=0$

$x+3=0$ or $x^2-3x+9=0$

$x=-3$ or

$$x=\frac{3\pm\sqrt{(-3)^2-4(1)(9)}}{2(1)}$$

$$=\frac{3\pm\sqrt{-27}}{2}$$

$$=\frac{3\pm 3i\sqrt{3}}{2}$$

The solutions are -3, $\frac{3+3i\sqrt{3}}{2}$, and $\frac{3-3i\sqrt{3}}{2}$.

53.

$$1=\frac{4}{x-7}+\frac{5}{(x-7)^2}$$

$(x-7)^2-4(x-7)-5=0$

Let $y=x-7$. Then $y^2=(x-7)^2$ and

$y^2-4y-5=0$

$(y-5)(y+1)=0$

$y=5$ or $y=-1$

$x-7=5$ or $x-7=-1$

$x=12$ or $x=6$

The solutions are 6 and 12.

55.

$$27y^4+15y^2=2$$

$$27y^4+15y^2-2=0$$

$$(9y^2-1)(3y^2+2)=0$$

$$(3y+1)(3y-1)(3y^2+2)=0$$

$y=-\frac{1}{3}$ or $y=\frac{1}{3}$ or $y^2=-\frac{2}{3}$

$$y=\pm\sqrt{-\frac{2}{3}}$$

$$y=\pm\frac{i\sqrt{6}}{3}$$

The solutions are $-\frac{1}{3}, \frac{1}{3}, -\frac{i\sqrt{6}}{3}$, and $\frac{i\sqrt{6}}{3}$.

57. $x-\sqrt{19-2x}-2=0$

$$x-2=\sqrt{19-2x}$$
$$(x-2)^2=\left(\sqrt{19-2x}\right)^2$$
$$x^2-4x+4=19-2x$$
$$x^2-2x-15=0$$
$$(x-5)(x+3)=0$$
$$x-5=0 \quad \text{or} \quad x+3=0$$
$$x=5 \quad \text{or} \quad x=-3$$

Reject $x=-3$ as an extraneous solution. The solution is 5.

59. Let x be the rate to Tucson. Then $x + 11$ is the rate returning home.

	distance =	*rate* ·	*time*
To Tucson	330	x	$\frac{330}{x}$
Return trip	330	$x + 11$	$\frac{330}{x-11}$

The LCD is $x(x + 11)$.

$$\frac{330}{x}=\frac{330}{x+11}+1$$
$$\frac{330}{x}-\frac{330}{x+11}=1$$
$$x(x+11)\left(\frac{330}{x}\right)-x(x+11)\left(\frac{330}{x+11}\right)=x(x+11)(1)$$
$$330(x+11)-330x=x(x+11)$$
$$330x+3630-330x=x^2+11x$$
$$0=x^2+11x-3630$$
$$0=(x+66)(x-55)$$
$$x+66=0 \quad \text{or} \quad x-55=0$$
$$x=-66 \qquad x=55$$

Reject –66 since rate cannot be negative. The speed to Tucson is 55 mph and the speed returning is $x + 11 = 55 + 11 = 66$ mph.

61. Let x = speed on the first part. Then $x - 1$ = speed on the second part.

$$d=rt \Rightarrow t=\frac{d}{r}$$
$$t_{\text{on first part}}+t_{\text{on second part}}=1\frac{3}{5}$$
$$\frac{3}{x}+\frac{4}{x-1}=\frac{8}{5}$$
$$3\cdot 5(x-1)+4\cdot 5x=8x(x-1)$$
$$15x-15+20x=8x^2-8x$$
$$0=8x^2-43x+15$$
$$0=(8x-3)(x-5)$$

$8x-3=0$ or $x-5=0$

$x=\frac{3}{8}$ or $x=5$

$x-1=4$

Discard $\frac{3}{8}$. Her speeds were 5 mph then 4 mph.

63. Let x = time for hose alone. Then $x-1$ = time for the inlet pipe alone.

$$\frac{1}{x}+\frac{1}{x-1}=\frac{1}{8}$$
$$8(x-1)+8x=x(x-1)$$
$$8x-8+8x=x^2-x$$
$$0=x^2-17x+8$$
$$x=\frac{17\pm\sqrt{(-17)^2-4(1)(8)}}{2(1)}$$
$$=\frac{17\pm\sqrt{257}}{2}$$

$x\approx 0.5$ (discard) or $x\approx 16.5$

$x-1\approx 15.5$

Hose: 16.5 hr; Inlet pipe: 15.5 hr

65. Let x = time for son alone. Then $x-1$ = time for dad alone.

$$\frac{1}{x}+\frac{1}{x-1}=\frac{1}{4}$$
$$4(x-1)+4x=x(x-1)$$
$$4x-4+4x=x^2-x$$
$$0=x^2-9x+4$$
$$x=\frac{9\pm\sqrt{(-9)^2-4(1)(4)}}{2(1)}$$
$$=\frac{9\pm\sqrt{65}}{2}$$

≈ 0.5 (discard) or 8.5

It takes his son about 8.5 hours.

67. Let x = the number.

$$x(x-4)=96$$
$$x^2-4x-96=0$$
$$(x-12)(x+8)=0$$

$x=12$ or $x=-8$

The number is 12 or –8.

69. a. length $=x-3-3=x-6=(x-6)$ in.

b. $V=lwh$

$$300=(x-6)(x-6)\cdot 3$$

c. $300=3(x-6)^2$

$$100=x^2-12x+36$$
$$0=x^2-12x-64$$
$$0=(x-16)(x+4)$$

$x=16$ or $x=-4$ (discard)

The sheet is 16 in. by 16 in.

Check: $V=3(x-6)(x-6)$
$=3(16-6)(16-6)$
$=3(10)(10)$
$=300$ cubic in.

71. Let x = length of the side of the square.

Area $=x^2$

$$920=x^2$$
$$\sqrt{920}=x$$

Adding another radial line to a different corner would yield a right triangle with legs r and hypotenuse x.

$$r^2+r^2=x^2$$
$$2r^2=\left(\sqrt{920}\right)^2$$
$$2r^2=920$$
$$r^2=460$$
$$r=\pm\sqrt{460}=\pm 21.4476$$

Disregard the negative. The smallest radius would be 22 feet.

73. $\frac{5x}{3}+2\le 7$

$$\frac{5x}{3}\le 5$$
$$5x\le 15$$
$$x\le 3$$

$(-\infty, 3]$

75. $\frac{y-1}{15}>-\frac{2}{5}$

$$15\left(\frac{y-1}{15}\right)>15\left(-\frac{2}{5}\right)$$
$$y-1>-6$$
$$y>-5$$

$(-5, \infty)$

77. Domain: $\{x \mid x \text{ is a real number}\}$ or $(-\infty, \infty)$

Range: $\{y \mid y \text{ is a real number}\}$ or $(-\infty, \infty)$

It is a function.

79. Domain: $\{x \mid x \text{ is a real number}\}$ or $(-\infty, \infty)$
Range: $\{y \mid y \geq -1\}$ or $[-1, \infty)$
It is a function.

81.
$$5y^3 - 45y - 5y^2 - 45 = 0$$
$$5y(y^2 + 9) - 5(y^2 + 9) = 0$$
$$(5y - 5)(y^2 + 9) = 0$$
$$5(y - 1)(y^2 + 9) = 0$$
$$y - 1 = 0 \quad \text{or} \quad y^2 + 9 = 0$$
$$y = 1 \qquad y^2 = -9$$
$$y = \pm\sqrt{-9} = \pm 3i$$
The solutions are 1, $3i$, and $-3i$.

83. Let $u = x^{-1}$, then $x^{-2} = u^2$.
$$3x^{-2} - 3x^{-1} - 18 = 0$$
$$3u^2 - 3u - 18 = 0$$
$$3(u^2 - u - 6) = 0$$
$$3(u - 3)(u + 2) = 0$$
$$u - 3 = 0 \quad \text{or} \quad u + 2 = 0$$
$$u = 3 \qquad u = -2$$
$$x^{-1} = 3 \qquad x^{-1} = -2$$
$$x = \frac{1}{3} \qquad x = -\frac{1}{2}$$
The solutions are $\frac{1}{3}$ and $-\frac{1}{2}$.

85.
$$2x^3 = -54$$
$$x^3 = -27$$
$$x^3 + 27 = 0$$
$$(x + 3)(x^2 - 3x + 9) = 0$$
$$x + 3 = 0 \quad \text{or} \quad x^2 - 3x + 9 = 0$$
$$x = -3$$
$$x = \frac{-(-3) \pm \sqrt{(-3)^2 - 4(1)(9)}}{2(1)}$$
$$x = \frac{3 \pm \sqrt{9 - 36}}{2} = \frac{3 \pm \sqrt{-27}}{2}$$
$$x = \frac{3 \pm 3i\sqrt{3}}{2}$$
The solutions are -3, $\frac{3 + 3i\sqrt{3}}{2}$, and $\frac{3 - 3i\sqrt{3}}{2}$.

87. answers may vary

89. a. 3 hours, 5 minutes, and 15 seconds is
$$3 + \frac{5}{60} + \frac{15}{3600} = 3\frac{7}{80} \text{ hours}$$
$$\frac{507.5 \text{ mi}}{3\frac{7}{80} \text{ hr}} \approx 164.37 \ \frac{\text{mi}}{\text{hr}}$$
Joey Logano's average race speed was approximately 164.37 mph.

b. 3 hours, 5 minutes, and 15 seconds is
$3 \cdot 3600 + 5 \cdot 60 + 15 = 11{,}115$ seconds.
$$\frac{11{,}115 \text{ sec}}{203 \text{ laps}} \approx 54.754 \ \frac{\text{sec}}{\text{lap}}$$
Joey Logano's average lap time was approximately 54.754 sec/lap.

c. $54.754 + 1.193 = 55.947$
Dale Earnhart Jr.'s average lap time was approximately 55.947 sec/lap.

d. $$55.947 \ \frac{\text{sec}}{\text{lap}} \cdot 203 \text{ laps} \approx \frac{11{,}357}{3600} \text{ hr}$$
$$\frac{507.5 \text{ mi}}{\frac{11{,}357}{3600} \text{ hr}} \approx 160.87 \ \frac{\text{mi}}{\text{hr}}$$
Dale Earnhart Jr.'s average race speed was approximately 160.87 mph.

Integrated Review

1.
$$x^2 - 10 = 0$$
$$x^2 = 10$$
$$x = \pm\sqrt{10}$$
The solutions are $\sqrt{10}$ and $-\sqrt{10}$.

2.
$$x^2 - 14 = 0$$
$$x^2 = 14$$
$$x = \pm\sqrt{14}$$
The solutions are $\sqrt{14}$ and $-\sqrt{14}$.

3.
$$(x - 1)^2 = 8$$
$$x - 1 = \pm\sqrt{8}$$
$$x - 1 = \pm 2\sqrt{2}$$
$$x = 1 \pm 2\sqrt{2}$$
The solutions are $1 + 2\sqrt{2}$ and $1 - 2\sqrt{2}$.

4. $(x+5)^2 = 12$

$$x+5 = \pm\sqrt{12}$$
$$x+5 = \pm 2\sqrt{3}$$
$$x = -5 \pm 2\sqrt{3}$$

The solutions are $-5+2\sqrt{3}$ and $-5-2\sqrt{3}$.

5. $x^2 + 2x - 12 = 0$

$$x^2 + 2x + \left(\frac{2}{2}\right)^2 = 12 + 1$$
$$x^2 + 2x + 1 = 13$$
$$(x+1)^2 = 13$$
$$x+1 = \pm\sqrt{13}$$
$$x = -1 \pm \sqrt{13}$$

The solutions are $-1+\sqrt{13}$ and $-1-\sqrt{13}$.

6. $x^2 - 12x + 11 = 0$

$$x^2 - 12x + \left(\frac{-12}{2}\right)^2 = -11 + 36$$
$$x^2 - 12x + 36 = 25$$
$$(x-6)^2 = \pm\sqrt{25}$$
$$x - 6 = \pm 5$$
$$x = 6 \pm 5$$
$$x = 1 \text{ or } x = 11$$

The solutions are 1 and 11.

7. $3x^2 + 3x = 5$

$$x^2 + x = \frac{5}{3}$$
$$x^2 + x + \left(\frac{1}{2}\right)^2 = \frac{5}{3} + \frac{1}{4}$$
$$x^2 + x + \frac{1}{4} = \frac{23}{12}$$
$$\left(x + \frac{1}{2}\right)^2 = \frac{23}{12}$$
$$x + \frac{1}{2} = \pm\sqrt{\frac{23}{12}}$$
$$x + \frac{1}{2} = \pm\frac{\sqrt{23}}{2\sqrt{3}}$$
$$x + \frac{1}{2} = \pm\frac{\sqrt{23}\cdot\sqrt{3}}{2\sqrt{3}\cdot\sqrt{3}}$$
$$x + \frac{1}{2} = \pm\frac{\sqrt{69}}{6}$$
$$x = -\frac{1}{2} \pm \frac{\sqrt{69}}{6} = \frac{-3 \pm \sqrt{69}}{6}$$

The solutions are $\frac{-3+\sqrt{69}}{6}$ and $\frac{-3-\sqrt{69}}{6}$.

8. $16y^2 + 16y = 1$

$$y^2 + y = \frac{1}{16}$$
$$y^2 + y + \left(\frac{1}{2}\right)^2 = \frac{1}{16} + \frac{1}{4}$$
$$y^2 + y + \frac{1}{4} = \frac{5}{16}$$
$$\left(y + \frac{1}{2}\right)^2 = \frac{5}{16}$$
$$y + \frac{1}{2} = \pm\sqrt{\frac{5}{16}}$$
$$y + \frac{1}{2} = \pm\frac{\sqrt{5}}{4}$$
$$y = -\frac{1}{2} \pm \frac{\sqrt{5}}{4} = \frac{-2 \pm \sqrt{5}}{4}$$

The solutions are $\frac{-2+\sqrt{5}}{4}$ and $\frac{-2-\sqrt{5}}{4}$.

9. $2x^2 - 4x + 1 = 0$

$a = 2, b = -4, c = 1$

$$x = \frac{4 \pm \sqrt{(-4)^2 - 4(2)(1)}}{2(2)} = \frac{4 \pm \sqrt{8}}{4} = \frac{4 \pm 2\sqrt{2}}{4} = \frac{2 \pm \sqrt{2}}{2}$$

The solutions are $\frac{2+\sqrt{2}}{2}$ and $\frac{2-\sqrt{2}}{2}$.

10. $\frac{1}{2}x^2 + 3x + 2 = 0$

$x^2 + 6x + 4 = 0$

$a = 1, b = 6, c = 4$

$$x = \frac{-6 \pm \sqrt{(6)^2 - 4(1)(4)}}{2(1)} = \frac{-6 \pm \sqrt{20}}{2} = \frac{-6 \pm 2\sqrt{5}}{2} = -3 \pm \sqrt{5}$$

The solutions are $-3+\sqrt{5}$ and $-3-\sqrt{5}$.

11. $x^2 + 4x = -7$

$x^2 + 4x + 7 = 0$

$a = 1, b = 4, c = 7$

$$x = \frac{-4 \pm \sqrt{(4)^2 - 4(1)(7)}}{2(1)} = \frac{-4 \pm \sqrt{-12}}{2} = \frac{-4 \pm i\sqrt{4 \cdot 3}}{2} = \frac{-4 \pm 2i\sqrt{3}}{2} = -2 \pm i\sqrt{3}$$

The solutions are $-2+i\sqrt{3}$ and $-2-i\sqrt{3}$.

12. $x^2 + x = -3$

$x^2 + x + 3 = 0$

$a = 1, b = 1, c = 3$

$$x = \frac{-1 \pm \sqrt{(1)^2 - 4(1)(3)}}{2(1)} = \frac{-1 \pm \sqrt{-11}}{2} = \frac{-1 \pm i\sqrt{11}}{2}$$

The solutions are $\frac{-1+i\sqrt{11}}{2}$ and $\frac{-1-i\sqrt{11}}{2}$.

13. $x^2 + 3x + 6 = 0$

$a = 1, b = 3, c = 6$

$$x = \frac{-3 \pm \sqrt{(3)^2 - 4(1)(6)}}{2(1)} = \frac{-3 \pm \sqrt{-15}}{2} = \frac{-3 \pm i\sqrt{15}}{2}$$

The solutions are $\frac{-3+i\sqrt{15}}{2}$ and $\frac{-3-i\sqrt{15}}{2}$.

14. $2x^2 + 18 = 0$

$2x^2 = -18$

$x^2 = -9$

$x = \pm\sqrt{-9}$

$x = \pm 3i$

The solutions are $3i$ and $-3i$.

15. $x^2 + 17x = 0$

$x(x+17) = 0$

$x = 0$ or $x + 17 = 0$

$x = -17$

$x = 0, -17$

The solutions are 0 and −17.

16. $4x^2-2x-3=0$

$a=4, b=-2, c=-3$

$$x=\frac{2\pm\sqrt{(-2)^2-4(4)(-3)}}{2(4)}=\frac{2\pm\sqrt{52}}{8}=\frac{2\pm2\sqrt{13}}{8}=\frac{1\pm\sqrt{13}}{4}$$

The solutions are $\frac{1+\sqrt{13}}{4}$ and $\frac{1-\sqrt{13}}{4}$.

17. $(x-2)^2=27$

$$\begin{aligned}x-2&=\pm\sqrt{27}\\x-2&=\pm3\sqrt{3}\\x&=2\pm3\sqrt{3}\end{aligned}$$

The solutions are $2+3\sqrt{3}$ and $2-3\sqrt{3}$.

18. $\frac{1}{2}x^2-2x+\frac{1}{2}=0$

$$\begin{aligned}x^2-4x+1&=0\\x^2-4x+\left(\frac{-4}{2}\right)^2&=-1+4\\x^2-4x+4&=3\\(x-2)^2&=3\\x-2&=\pm\sqrt{3}\\x&=2\pm\sqrt{3}\end{aligned}$$

The solutions are $2+\sqrt{3}$ and $2-\sqrt{3}$.

19. $3x^2+2x=8$

$$\begin{aligned}&3x^2+2x-8=0\\&(3x-4)(x+2)=0\\&3x-4=0 \text{ or } x+2=0\\&x=\frac{4}{3} \text{ or } x=-2\end{aligned}$$

The solutions are $\frac{4}{3}$ and -2.

20. $2x^2=-5x-1$

$2x^2+5x+1=0$

$a=2, b=5, c=1$

$$x=\frac{-5\pm\sqrt{(5)^2-4(2)(1)}}{2(2)}=\frac{-5\pm\sqrt{17}}{4}$$

The solutions are $\frac{-5+\sqrt{17}}{4}$ and $\frac{-5-\sqrt{17}}{4}$.

21. $x(x-2)=5$

$$\begin{aligned}x^2-2x&=5\\x^2-2x+\left(\frac{-2}{2}\right)^2&=5+1\\x^2-2x+1&=6\\(x-1)^2&=6\\x-1&=\pm\sqrt{6}\\x&=1\pm\sqrt{6}\end{aligned}$$

The solutions are $1+\sqrt{6}$ and $1-\sqrt{6}$.

22. $x^2-31=0$

$$\begin{aligned}x^2&=31\\x&=\pm\sqrt{31}\end{aligned}$$

The solutions are $\sqrt{31}$ and $-\sqrt{31}$.

23. $5x^2-55=0$

$$\begin{aligned}5x^2&=55\\x^2&=11\\x&=\pm\sqrt{11}\end{aligned}$$

The solutions are $\sqrt{11}$ and $-\sqrt{11}$.

24. $5x^2+55=0$

$$\begin{aligned}5x^2&=-55\\x^2&=-11\\x&=\pm\sqrt{-11}\\x&=\pm i\sqrt{11}\end{aligned}$$

The solutions are $i\sqrt{11}$ and $-i\sqrt{11}$.

25.
$$x(x+5)=66$$
$$x^2+5x=66$$
$$x^2+5x-66=0$$
$$(x+11)(x-6)=0$$
$$x+11=0 \quad \text{or} \quad x-6=0$$
$$x=-11 \quad \text{or} \quad x=6$$
The solutions are −11 and 6.

26. $5x^2+6x-2=0$
$a=5, b=6, c=-2$
$$x=\frac{-6\pm\sqrt{(6)^2-4(5)(-2)}}{2(5)}$$
$$=\frac{-6\pm\sqrt{76}}{10}$$
$$=\frac{-6\pm\sqrt{4\cdot 19}}{10}$$
$$=\frac{-6\pm 2\sqrt{19}}{10}$$
$$=\frac{-3\pm\sqrt{19}}{5}$$
The solutions are $\frac{-3+\sqrt{19}}{5}$ and $\frac{-3-\sqrt{19}}{5}$.

27.
$$2x^2+3x=1$$
$$2x^2+3x-1=0$$
$a=2, b=3, c=-1$
$$x=\frac{-3\pm\sqrt{(3)^2-4(2)(-1)}}{2(2)}$$
$$=\frac{-3\pm\sqrt{17}}{4}$$
The solutions are $\frac{-3+\sqrt{17}}{4}$ and $\frac{-3-\sqrt{17}}{4}$.

28. $x-\sqrt{13-3x}-3=0$
$$x-3=\sqrt{13-3x}$$
$$(x-3)^2=\left(\sqrt{13-3x}\right)^2$$
$$x^2-6x+9=13-3x$$
$$x^2-3x-4=0$$
$$(x-4)(x+1)=0$$
$$x-4=0 \quad \text{or} \quad x+1=0$$
$$x=4 \qquad x=-1$$
The value −1 does not check, so the solution is 4.

29. The LCD is $x(x-2)$.

$$\frac{5x}{x-2}-\frac{x+1}{x}=\frac{3}{x(x-2)}$$
$$x(x-2)\left(\frac{5x}{x-2}\right)-x(x-2)\left(\frac{x+1}{x}\right)=x(x-2)\left[\frac{3}{x(x-2)}\right]$$
$$x(5x)-(x-2)(x+1)=3$$
$$5x^2-(x^2-x-2)=3$$
$$5x^2-x^2+x+2=3$$
$$4x^2+x-1=0$$

$a=4$, $b=1$, $c=-1$

$$x=\frac{-b\pm\sqrt{b^2-4ac}}{2a}$$
$$x=\frac{-1\pm\sqrt{1^2-4(4)(-1)}}{2(4)}$$
$$=\frac{-1\pm\sqrt{1+16}}{8}$$
$$=\frac{-1\pm\sqrt{17}}{8}$$

The solutions are $\frac{-1-\sqrt{17}}{8}$ and $\frac{-1+\sqrt{17}}{8}$.

30. $$a^2+b^2=c^2$$
$$x^2+x^2=20^2$$
$$2x^2=400$$
$$x^2=200$$
$$x=\pm\sqrt{200}$$
$$=\pm10\sqrt{2}\approx14.1421$$

Disregard the negative. A side of the room is $10\sqrt{2}$ feet $\approx$ 14.1 feet.

31. Let x = time for Jack alone. Then $x-2$ = time for Lucy alone.

$$\frac{1}{x}+\frac{1}{x-2}=\frac{1}{4}$$
$$4(x-2)+4x=x(x-2)$$
$$4x-8+4x=x^2-2x$$
$$0=x^2-10x+8$$
$$x=\frac{10\pm\sqrt{(-10)^2-4(1)(8)}}{2(1)}$$
$$=\frac{10\pm\sqrt{68}}{2}$$
$$\approx 9.1 \text{ or } 0.9 \text{ (disregard)}$$
$$x-2=9.1-2=7.1$$

It would take Jack 9.1 hours and Lucy 7.1 hours.

32. Let x = initial speed on treadmill. Then $x + 1$ = speed increased.

$$t_{\text{initial}} + t_{\text{increased}} = \frac{4}{3}$$
$$\frac{5}{x} + \frac{2}{x+1} = \frac{4}{3}$$
$$5 \cdot 3(x+1) + 2 \cdot 3x = 4x(x+1)$$
$$15x + 15 + 6x = 4x^2 + 4x$$
$$0 = 4x^2 - 17x - 15$$
$$0 = (4x+3)(x-5)$$

$x = -\frac{4}{3}$ (disregard) or $x = 5$

$x + 1 = 5 + 1 = 6$

Initial speed: 5 mph

Increased speed: 6 mph

Section 8.4 Practice Exercises

1. $(x - 4)(x + 3) > 0$

Solve the related equation, $(x - 4)(x + 3) = 0$.

$(x - 4)(x + 3) = 0$

$x - 4 = 0$ or $x + 3 = 0$

$x = 4$ $\quad x = -3$

Test points in the three regions separated by $x = 4$ and $x = -3$.

Region	Test Point	$(x - 4)(x + 3) > 0$ Result
A: $(-\infty, -3)$	-4	$(-8)(-1) > 0$ True
B: $(-3, 4)$	0	$(-4)(3) > 0$ False
C: $(4, \infty)$	5	$(1)(8) > 0$ True

The points in regions A and C satisfy the inequality. The numbers 4 and -3 are not included in the solution since the inequality symbol is >. The solution set is $(-\infty, -3) \cup (4, \infty)$.

2. $x^2 - 8x \le 0$

Solve the related equation, $x^2 - 8x = 0$.

$x^2 - 8x = 0$

$x(x - 8) = 0$

$x = 0$ or $x - 8 = 0$

$x = 8$

The numbers 0 and 8 separate the number line into three regions, A, B, and C. Test a point in each region.

Region	Test Point	$x^2 - 8x \le 0$ Result
A: $(-\infty, 0]$	-1	$1 + 8 \le 0$ False
B: $[0, 8]$	1	$1 - 8 \le 0$ True
C: $[8, \infty)$	9	$81 - 72 \le 0$ False

Values in region B satisfy the inequality. The numbers 0 and 8 are included in the solution since the inequality symbol is ≤. The solution set is $[0, 8]$.

3. $(x + 3)(x - 2)(x + 1) \le 0$

Solve $(x + 3)(x - 2)(x + 1) = 0$ by inspection.

$x = -3$ or $x = 2$ or $x = -1$

These separate the number line into four regions. Test points in each region.

Region	Test Point	$(x + 3)(x - 2)(x + 1) \le 0$ Result
A: $(-\infty, -3]$	-4	$(-1)(-6)(-3) \le 0$ True
B: $[-3, -1]$	-2	$(1)(-4)(-1) \le 0$ False
C: $[-1, 2]$	0	$(3)(-2)(1) \le 0$ True
D: $[2, \infty)$	3	$(6)(1)(4) \le 0$ False

The solution set is $(-\infty, -3] \cup [-1, 2]$. We include the numbers -3, -1, and 2 because the inequality symbol is ≤.

4. $\frac{x-5}{x+4} \le 0$

$x+4=0$

$x=-4$

$x=-4$ makes the denominator zero. Solve the related equation $\frac{x-5}{x+4}=0$.

$\frac{x-5}{x+4}=0$

$x-5=0$

$x=5$

Test points in the three regions separated by $x=-4$ and $x=5$.

Region	Test Point	$\frac{x-5}{x+4} \le 0$ Result
A: $(-\infty, -4)$	-5	$\frac{-10}{-1} \le 0$ False
B: $(-4, 5]$	0	$\frac{-5}{4} \le 0$ True
C: $[5, \infty)$	6	$\frac{1}{10} \le 0$ False

The solution set is $(-4, 5]$. The interval includes 5 because 5 satisfies the original inequality. This interval does not include -4, because -4 would make the denominator zero.

5. $\frac{7}{x+3} < 5$

$x+3=0$

$x=-3$

$x=-3$ makes the denominator zero.

Solve $\frac{7}{x+3}=5$.

$$(x+3)\left(\frac{7}{x+3}\right)=5(x+3)$$
$$7=5x+15$$
$$-8=5x$$
$$-\frac{8}{5}=x$$

We use these two solutions to divide the number line into three regions and choose test points.

Region	Test Point	$\frac{7}{x+3} < 5$ Result
A: $(-\infty, -3)$	-4	$\frac{7}{-1} < 5$ True
B: $\left(-3, -\frac{8}{5}\right)$	-2	$\frac{7}{1} < 5$ False
C: $\left(-\frac{8}{5}, \infty\right)$	0	$\frac{7}{3} < 5$ True

The solution set is $(-\infty, -3) \cup \left(-\frac{8}{5}, \infty\right)$.

Vocabulary, Readiness & Video Check 8.4

1. $[-7, 3)$
2. $(-1, 5]$
3. $(-\infty, 0]$
4. $(-\infty, -8]$
5. $(-\infty, -12) \cup [-10, \infty)$
6. $(-\infty, -3] \cup (4, \infty)$
7. We use the solutions to the related equation to divide the number line into regions that either entirely are or entirely are not solution regions; the solutions to the related equation are solutions to the inequality only if the inequality symbol is $\le$ or $\ge$.
8. The solution set cannot include values that make the denominator zero.

Exercise Set 8.4

1. $(x+1)(x+5) > 0$
$x+1=0$ or $x+5=0$
$x=-1$ or $x=-5$

Region	Test Point	$(x+1)(x+5) > 0$ Result
A: $(-\infty, -5)$	-6	$(-5)(-1) > 0$ True
B: $(-5, -1)$	-2	$(-1)(3) > 0$ False
C: $(-1, \infty)$	0	$(1)(5) > 0$ True

Solution: $(-\infty, -5) \cup (-1, \infty)$

3. $(x-3)(x+4) \le 0$
$x-3=0$ or $x+4=0$
$x=3$ or $x=-4$

Region	Test Point	$(x-3)(x+4) \le 0$ Result
A: $(-\infty, -4]$	-5	$(-8)(-1) \le 0$ False
B: $[-4, 3]$	0	$(-3)(4) \le 0$ True
C: $[3, \infty)$	4	$(1)(8) \le 0$ False

Solution: $[-4, 3]$

5. $x^2-7x+10 \le 0$
$(x-5)(x-2) \le 0$
$x-5=0$ or $x-2=0$
$x=5$ or $x=2$

Region	Test Point	$(x-5)(x-2) \le 0$ Result
A: $(-\infty, 2]$	0	$(-5)(-2) \le 0$ False
B: $[2, 5]$	3	$(-2)(1) \le 0$ True
C: $[5, \infty)$	6	$(1)(4) \le 0$ False

Solution: $[2, 5]$

7. $3x^2+16 < -5$
$3x^2+16x+5 < 0$
$(3x+1)(x+5) < 0$
$3x+1=0$ or $x+5=0$
$x=-\frac{1}{3}$ or $x=-5$

Region	Test Point	$(3x+1)(x+5) < 0$ Result
A: $(-\infty, -5)$	-6	$(-17)(-1) < 0$ False
B: $\left(-5, -\frac{1}{3}\right)$	-1	$(-2)(4) < 0$ True
C: $\left(-\frac{1}{3}, \infty\right)$	0	$(1)(5) < 0$ False

Solution: $\left(-5, -\frac{1}{3}\right)$

9. $(x-6)(x-4)(x-2) > 0$
$x-6=0$ or $x-4=0$ or $x-2=0$
$x=6$ or $x=4$ or $x=2$

Region	Test Point	$(x-6)(x-4)(x-2) > 0$ Result
A: $(-\infty, 2)$	0	$(-6)(-4)(-2) > 0$ False
B: $(2, 4)$	3	$(-3)(-1)(1) > 0$ True
C: $(4, 6)$	5	$(-1)(1)(3) > 0$ False
D: $(6, \infty)$	7	$(1)(3)(5) > 0$ True

Solution: $(2, 4) \cup (6, \infty)$

11. $x(x-1)(x+4) \le 0$

$x = 0$ or $x - 1 = 0$ or $x + 4 = 0$

$x = 1$ or $x = -4$

Region	Test Point	$x(x-1)(x+4) \le 0$ Result
A: $(-\infty, -4]$	-5	$-5(-6)(-1) \le 0$ True
B: $[-4, 0]$	-1	$-1(-2)(3) \le 0$ False
C: $[0, 1]$	$\frac{1}{2}$	$\frac{1}{2}\left(-\frac{1}{2}\right)\left(\frac{9}{2}\right) \le 0$ True
D: $[1, \infty)$	2	$2(1)(6) \le 0$ False

Solution: $(-\infty, -4] \cup [0, 1]$

13. $(x^2 - 9)(x^2 - 4) > 0$

$(x+3)(x-3)(x+2)(x-2) > 0$

$x + 3 = 0$ or $x - 3 = 0$ or $x + 2 = 0$ or $x - 2 = 0$

$x = -3$ or $x = 3$ or $x = -2$ or $x = 2$

Region	Test Point	$(x+3)(x-3)(x+2)(x-2) > 0$ Result
A: $(-\infty, -3)$	-4	$(-1)(-7)(-2)(-6) > 0$ True
B: $(-3, -2)$	$-\frac{5}{2}$	$\left(\frac{1}{2}\right)\left(-\frac{11}{2}\right)\left(-\frac{1}{2}\right)\left(-\frac{9}{2}\right) > 0$ False
C: $(-2, 2)$	0	$(3)(-3)(2)(-2) > 0$ True
D: $(2, 3)$	$\frac{5}{2}$	$\left(\frac{11}{2}\right)\left(-\frac{1}{2}\right)\left(\frac{9}{2}\right)\left(\frac{1}{2}\right) > 0$ False
E: $(3, \infty)$	4	$(7)(1)(6)(2) > 0$ True

Solution: $(-\infty, -3) \cup (-2, 2) \cup (3, \infty)$

15. $\dfrac{x+7}{x-2} < 0$

$x+7=0$ or $x-2=0$
$x=-7$ or $x=2$

Region	Test Point	$\dfrac{x+7}{x-2} < 0$ False
A: $(-\infty, -7)$	-8	$\dfrac{-1}{-10} < 0$ False
B: $(-7, 2)$	0	$\dfrac{7}{-2} < 0$ True
C: $(2, \infty)$	3	$\dfrac{10}{1} < 0$ False

Solution: $(-7, 2)$

17. $\dfrac{5}{x+1} > 0$

$x+1=0$
$x=-1$

Region	Test Point	$\dfrac{5}{x+1} > 0$ Result
A: $(-\infty, -1)$	-2	$\dfrac{5}{-1} > 0$ False
B: $(-1, \infty)$	0	$\dfrac{5}{1} > 0$ True

Solution: $(-1, \infty)$

19. $\dfrac{x+1}{x-4} \geq 0$

$x+1=0$ or $x-4=0$
$x=-1$ or $x=4$

Region	Test Point	$\dfrac{x+1}{x-4} \geq 0$ Result
A: $(-\infty, -1]$	-2	$\dfrac{-1}{-6} \geq 0$ True
B: $[-1, 4)$	0	$\dfrac{1}{-4} \geq 0$ False
C: $(4, \infty)$	5	$\dfrac{6}{1} \geq 0$ True

Solution: $(-\infty, -1] \cup (4, \infty)$

21. $\dfrac{3}{x-2} < 4$

The denominator is equal to 0 when $x - 2 = 0$, or $x = 2$.

$$\frac{3}{x-2} = 4$$
$$3 = 4x - 8$$
$$11 = 4x$$
$$\frac{11}{4} = x$$

Region	Test Point	$\dfrac{3}{x-2} < 4$ Result
A: $(-\infty, 2)$	0	$\dfrac{3}{-2} < 4$ True
B: $\left(2, \dfrac{11}{4}\right)$	$\dfrac{5}{2}$	$\dfrac{3}{\frac{1}{2}} = 6 < 4$ False
C: $\left(\dfrac{11}{4}, \infty\right)$	4	$\dfrac{3}{2} < 4$ True

Solution: $(-\infty, 2) \cup \left(\dfrac{11}{4}, \infty\right)$

23. $\dfrac{x^2+6}{5x} \geq 1$

The denominator is equal to 0 when $5x = 0$, or $x = 0$.

$$\frac{x^2+6}{5x} = 1$$
$$x^2+6 = 5x$$
$$x^2-5x+6 = 0$$
$$(x-2)(x-3) = 0$$
$$x-2=0 \quad \text{or} \quad x-3=0$$
$$x=2 \quad \text{or} \quad x=3$$

Region	Test Point	$\frac{x^2+6}{5x} \geq 1$ Result
A: $(-\infty, 0)$	-1	$\frac{7}{-5} \geq 1$ False
B: $(0, 2]$	1	$\frac{7}{5} \geq 1$ True
C: $[2, 3]$	$\frac{5}{2}$	$\frac{\frac{49}{4}}{\frac{25}{2}} = \frac{49}{50} \geq 1$ False
D: $[3, \infty)$	4	$\frac{22}{20} \geq 1$ True

Solution: $(0, 2] \cup [3, \infty)$

25. $\dfrac{x+2}{x-3} < 1$

$x-3=0$

$x = 3$ makes the denominator 0.

$$\frac{x+2}{x-3} = 1$$
$$x+2 = x-3$$
$$2 = -3 \quad \text{False}$$

The related equation has no solution.

Region	Test Point Value	$\frac{x+2}{x-3} < 1$ Result
$(-\infty, 3)$	0	$\frac{2}{-3} < 1$ True
$(3, \infty)$	4	$\frac{6}{1} < 1$ False

Solution: $(-\infty, 3)$

27. $(2x-3)(4x+5) \leq 0$

$$2x-3=0 \quad \text{or} \quad 4x+5=0$$
$$x=\frac{3}{2} \quad \text{or} \quad x=-\frac{5}{4}$$

Region	Test Point	$(2x-3)(4x+5) \leq 0$ Result
A: $\left(-\infty, -\frac{5}{4}\right]$	-2	$(-7)(-3) \leq 0$ False
B: $\left[-\frac{5}{4}, \frac{3}{2}\right]$	0	$(-3)(5) \leq 0$ True
C: $\left[\frac{3}{2}, \infty\right)$	2	$(1)(13) \leq 0$ False

Solution: $\left[-\frac{5}{4}, \frac{3}{2}\right]$

29. $x^2 > x$

$$x^2 - x > 0$$
$$x(x-1) > 0$$
$$x=0 \quad \text{or} \quad x-1=0$$
$$x=1$$

Region	Test Point	$x(x-1) > 0$ Result
A: $(-\infty, 0)$	-1	$-1(-2) > 0$ True
B: $(0, 1)$	$\frac{1}{2}$	$\frac{1}{2}\left(-\frac{1}{2}\right) > 0$ False
C: $(1, \infty)$	2	$2(1) > 0$ True

Solution: $(-\infty, 0) \cup (1, \infty)$

31. $(2x-8)(x+4)(x-6) \le 0$

$2x-8=0$ or $x+4=0$ or $x-6=0$

$x=4$ or $x=-4$ or $x=6$

Region	Test Point	$(2x-8)(x+4)(x-6) \le 0$ Result
A: $(-\infty, -4]$	-5	$(-18)(-1)(-11) \le 0$ True
B: $[-4, 4]$	0	$(-8)(4)(-6) \le 0$ False
C: $[4, 6]$	5	$(2)(9)(-1) \le 0$ True
D: $[6, \infty)$	7	$(6)(11)(1) \le 0$ False

Solution: $(-\infty, -4] \cup [4, 6]$

33. $6x^2 - 5x \ge 6$

$6x^2 - 5x - 6 \ge 0$

$(3x+2)(2x-3) \ge 0$

$3x+2=0$ or $2x-3=0$

$x=-\frac{2}{3}$ or $x=\frac{3}{2}$

Region	Test Point	$(3x+2)(2x-3) \ge 0$ Result
A: $\left(-\infty, -\frac{2}{3}\right]$	-1	$(-1)(-5) \ge 0$ True
B: $\left[-\frac{2}{3}, \frac{3}{2}\right]$	0	$(2)(-3) \ge 0$ False
C: $\left[\frac{3}{2}, \infty\right)$	2	$(8)(1) \ge 0$ True

Solution: $\left(-\infty, -\frac{2}{3}\right] \cup \left[\frac{3}{2}, \infty\right)$

35. $4x^3 + 16x^2 - 9x - 36 > 0$

$4x^2(x+4) - 9(x+4) > 0$

$(x+4)(4x^2 - 9) > 0$

$(x+4)(2x+3)(2x-3) > 0$

$x+4=0$ or $2x+3=0$ or $2x-3=0$

$x=-4$ or $x=-\frac{3}{2}$ or $x=\frac{3}{2}$

Region	Test Point	$(x+4)(2x+3)(2x-3) > 0$
A: $(-\infty, -4)$	-5	$(-1)(-7)(-13) > 0$ False
B: $\left(-4, -\frac{3}{2}\right)$	-3	$(1)(-3)(-9) > 0$ True
C: $\left(-\frac{3}{2}, \frac{3}{2}\right)$	0	$(4)(3)(-3) > 0$ False
D: $\left(\frac{3}{2}, \infty\right)$	4	$(8)(11)(5) > 0$ True

Solution: $\left(-4, -\frac{3}{2}\right) \cup \left(\frac{3}{2}, \infty\right)$

37. $x^4 - 26x^2 + 25 \ge 0$

$(x^2 - 25)(x^2 - 1) \ge 0$

$(x+5)(x-5)(x+1)(x-1) \ge 0$

$x=-5$ or $x=5$ or $x=-1$ or $x=1$

Region	Test Point	$(x+5)(x-5)(x+1)(x-1) \ge 0$ Result
A: $(-\infty, -5]$	-6	$(-1)(-11)(-5)(-7) \ge 0$ True
B: $[-5, -1]$	-2	$(3)(-7)(-1)(-3) \ge 0$ False
C: $[-1, 1]$	0	$(5)(-5)(1)(-1) \ge 0$ True
D: $[1, 5]$	2	$(7)(-3)(3)(1) \ge 0$ False
E: $[5, \infty)$	6	$(11)(1)(7)(5) \ge 0$ True

Solution: $(-\infty, -5] \cup [-1, 1] \cup [5, \infty)$

39. $(2x-7)(3x+5)>0$
$2x-7=0 \quad \text{or} \quad 3x+5=0$
$x=\frac{7}{2} \quad \text{or} \quad x=-\frac{5}{3}$

Region	Test Point	$(2x-7)(3x+5)>0$
A: $\left(-\infty, -\frac{5}{3}\right)$	−2	$(-11)(-1)>0$ True
B: $\left(-\frac{5}{3}, \frac{7}{2}\right)$	0	$(-7)(5)>0$ False
C: $\left(\frac{7}{2}, \infty\right)$	4	$(1)(17)>0$ True

Solution: $\left(-\infty, -\frac{5}{3}\right)\cup\left(\frac{7}{2}, \infty\right)$

41. $\frac{x}{x-10}<0$
$x=0 \quad \text{or} \quad x-10=0$
$x=10$

Region	Test Point	$\frac{x}{x-10}<0$ Result
A: $(-\infty, 0)$	−1	$\frac{-1}{-11}<0$ False
B: $(0, 10)$	5	$\frac{5}{-5}<0$ True
C: $(10, \infty)$	11	$\frac{11}{1}<0$ False

Solution: $(0, 10)$

43. $\frac{x-5}{x+4}\geq 0$
$x-5=0 \quad \text{or} \quad x+4=0$
$x=5 \quad \text{or} \quad x=-4$

Region	Test Point	$\frac{x-5}{x+4} \geq 0$ Result
A: $(-\infty, -4)$	-5	$\frac{-10}{-1} \geq 0$ True
B: $(-4, 5]$	0	$\frac{-5}{4} \geq 0$ False
C: $[5, \infty)$	6	$\frac{1}{10} \geq 0$ True

Solution: $(-\infty, -4) \cup [5, \infty)$

45. $\frac{x(x+6)}{(x-7)(x+1)} \geq 0$

$x = 0$ or $x+6=0$ or $x-7=0$ or $x+1=0$
$x=-6$ or $x=7$ or $x=-1$

Region	Test Point	$\frac{x(x+6)}{(x-7)(x+1)} \geq 0$ Result
A: $(-\infty, -6]$	-7	$\frac{-7(-1)}{(-14)(-6)} \geq 0$ True
B: $[-6, -1)$	-3	$\frac{-3(3)}{(-10)(-2)} \geq 0$ False
C: $(-1, 0]$	$-\frac{1}{2}$	$\frac{-\frac{1}{2}\left(\frac{11}{2}\right)}{\left(-\frac{15}{2}\right)\left(\frac{1}{2}\right)} \geq 0$ True
D: $[0, 7)$	2	$\frac{2(8)}{(-5)(3)} \geq 0$ False
E: $(7, \infty)$	8	$\frac{8(14)}{(1)(9)} \geq 0$ True

Solution: $(-\infty, -6] \cup (-1, 0] \cup (7, \infty)$

47. $\frac{-1}{x-1} > -1$

The denominator is equal to 0 when $x - 1 = 0$, or $x = 1$.

$$\frac{-1}{x-1} = -1$$
$$-1 = -1(x-1)$$
$$-1 = -x+1$$
$$x = 2$$

Region	Test Point	$\frac{-1}{x-1} > -1$ Result
A: $(-\infty, 1)$	0	$\frac{-1}{-1} > -1$ True
B: $(1, 2)$	$\frac{3}{2}$	$\frac{-1}{\frac{1}{2}} = -2 > -1$ False
C: $(2, \infty)$	3	$\frac{-1}{2} > -1$ True

Solution: $(-\infty, 1) \cup (2, \infty)$

49. $\frac{x}{x+4} \le 2$

The denominator is equal to 0 when $x + 4 = 0$, or $x = -4$.

$$\frac{x}{x+4} = 2$$
$$x = 2x+8$$
$$-x = 8$$
$$x = -8$$

Region	Test Point	$\frac{x}{x+4} \le 2$ Result
A: $(-\infty, -8]$	−9	$\frac{-9}{-5} \le 2$ True
B: $[-8, -4)$	−6	$\frac{-6}{-2} \le 2$ False
C: $(-4, \infty)$	0	$\frac{0}{4} \le 2$ True

Solution: $(-\infty, -8] \cup (-4, \infty)$

51. $\frac{z}{z-5} \geq 2z$

The denominator is equal to 0 when $z - 5 = 0$, or $z = 5$.

$$\frac{z}{z-5} = 2z$$
$$z = 2z(z-5)$$
$$z = 2z^2 - 10z$$
$$0 = 2z^2 - 11z$$
$$0 = z(2z-11)$$
$$z = 0 \quad \text{or} \quad 2z - 11 = 0$$
$$z = \frac{11}{2}$$

Region	Test Point	$\frac{z}{z-5} \geq 2z$ Result
A: $(-\infty, 0]$	-1	$\frac{-1}{-6} \geq -2$ True
B: $[0, 5)$	1	$\frac{1}{-4} \geq 2$ False
C: $\left(5, \frac{11}{2}\right]$	$\frac{21}{4}$	$\frac{\left(\frac{21}{4}\right)}{\left(\frac{1}{4}\right)} \geq \frac{21}{2}$ $21 \geq \frac{21}{2}$ True
D: $\left[\frac{11}{2}, \infty\right)$	6	$\frac{6}{1} \geq 12$ False

Solution: $(-\infty, 0] \cup \left(5, \frac{11}{2}\right]$

53. $\frac{(x+1)^2}{5x} > 0$

The denominator is equal to 0 when $5x = 0$, or $x = 0$.

$$\frac{(x+1)^2}{5x} = 0$$
$$(x+1)^2 = 0$$
$$x + 1 = 0$$
$$x = -1$$

Region	Test Point	$\frac{(x+1)^2}{5x} > 0$ Result
A: $(-\infty, -1)$	-2	$\frac{1}{-10} > 0$ False
B: $(-1, 0)$	$-\frac{1}{2}$	$\frac{\left(\frac{1}{4}\right)}{\left(-\frac{5}{2}\right)} > 0$ False
C: $(0, \infty)$	1	$\frac{4}{5} > 0$ True

Solution: $(0, \infty)$

55. $g(x) = |x| + 2$

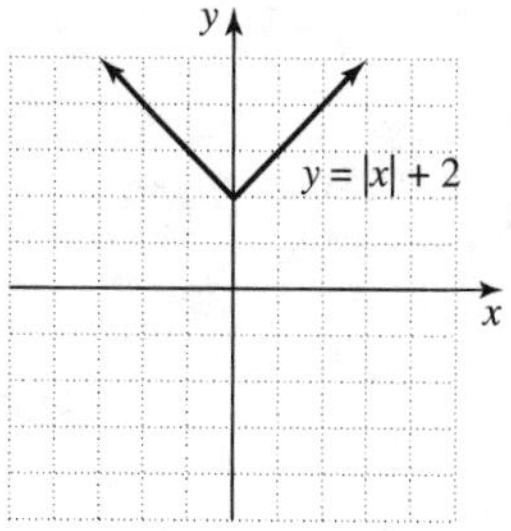

57. $G(x) = |x| - 1$

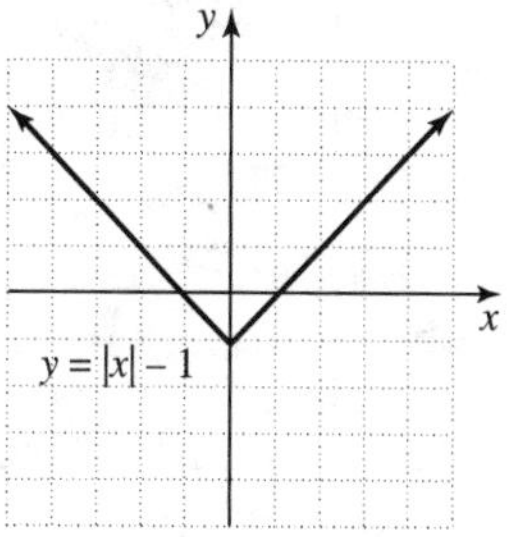

59. $F(x) = x^2 - 3$

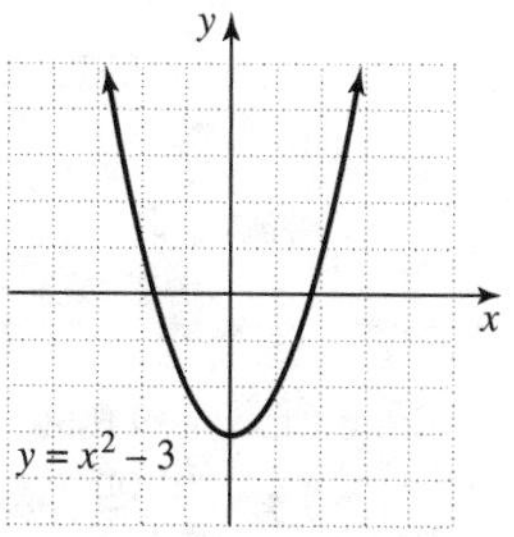

61. $H(x) = x^2 + 1$

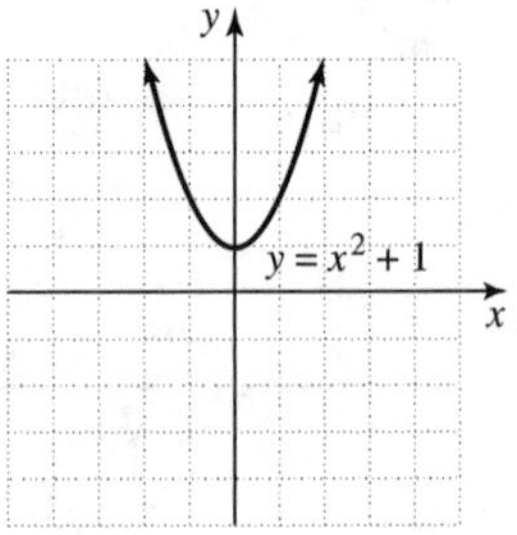

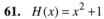

63. answers may vary

65. Let x = the number. Then

$\frac{1}{x}$ = the reciprocal of the number.

$$x - \frac{1}{x} < 0$$
$$\frac{x^2 - 1}{x} < 0$$
$$\frac{(x+1)(x-1)}{x} < 0$$

$x + 1 = 0$ or $x - 1 = 0$ or $x = 0$

$x = -1$ or $x = 1$

Region	Test Point	$\frac{(x+1)(x-1)}{x} < 0$ Result
A: $(-\infty, -1)$	-2	$\frac{(-1)(-3)}{-2} < 0$ True
B: $(-1, 0)$	$-\frac{1}{2}$	$\frac{\left(\frac{1}{2}\right)\left(-\frac{3}{2}\right)}{\left(-\frac{1}{2}\right)} < 0$ False
C: $(0, 1)$	$\frac{1}{2}$	$\frac{\left(\frac{3}{2}\right)\left(-\frac{1}{2}\right)}{\left(\frac{1}{2}\right)} < 0$ True
D: $(1, \infty)$	2	$\frac{(3)(1)}{2} < 0$ False

$(-\infty, -1) \cup (0, 1)$ or any number less than -1 or between 0 and 1 satisfies the conditions.

67. $P(x) = -2x^2 + 26x - 44$

$$-2x^2 + 26x - 44 > 0$$
$$-2(x^2 + 13x - 22) > 0$$
$$-2(x - 11)(x - 2) > 0$$

$x - 11 = 0$ or $x - 2 = 0$

$x = 11$ or $x = 2$

Region	Test Point	$-2(x-11)(x-2) > 0$ Result
A: $(0, 2)$	1	$-2(-10)(-3) > 0$ False
B: $(2, 11)$	3	$-2(-8)(1) > 0$ True
C: $(11, \infty)$	12	$-2(1)(10) > 0$ False

The company makes a profit when x is between 2 and 11.

69.

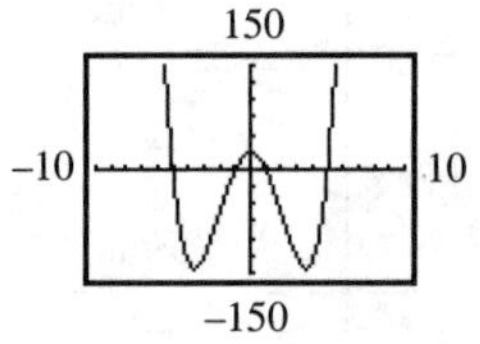

71.

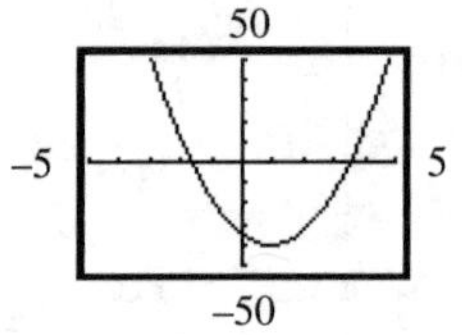

Section 8.5 Practice Exercises

1. $f(x) = x^2$ and $g(x) = x^2 - 4$

Construct a table of values for $f(x)$ and $g(x)$.

x	$f(x) = x^2$	$g(x) = x^2 - 4$
-2	4	0
-1	1	-3
0	0	-4
1	1	-3
2	4	0

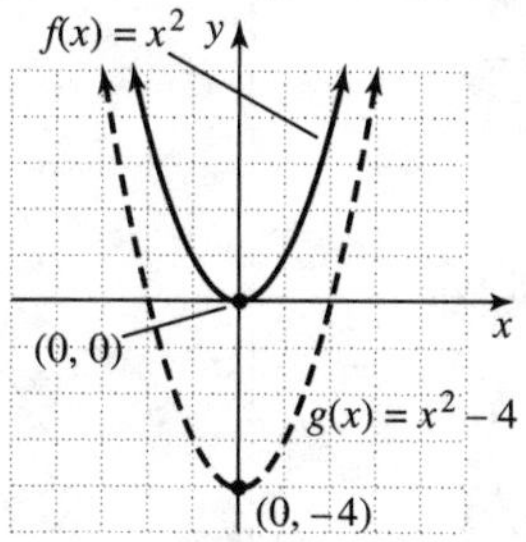

2. a. $f(x) = x^2 - 5$

The graph of $f(x)$ is obtained by shifting the graph of $y = x^2$ downward 5 units.

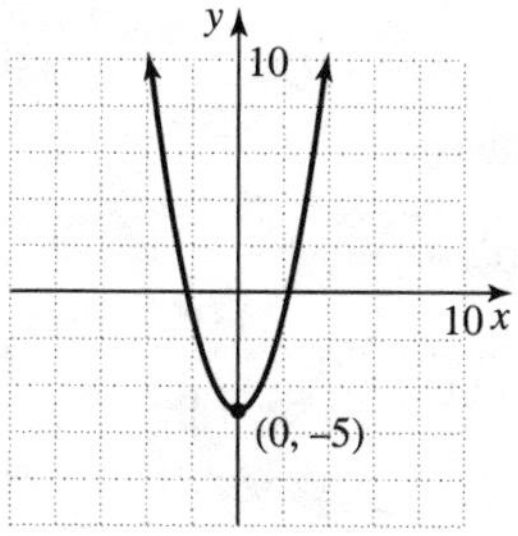

b. $g(x) = x^2 + 3$

The graph of $g(x)$ is obtained by shifting the graph of $y = x^2$ upward 3 units.

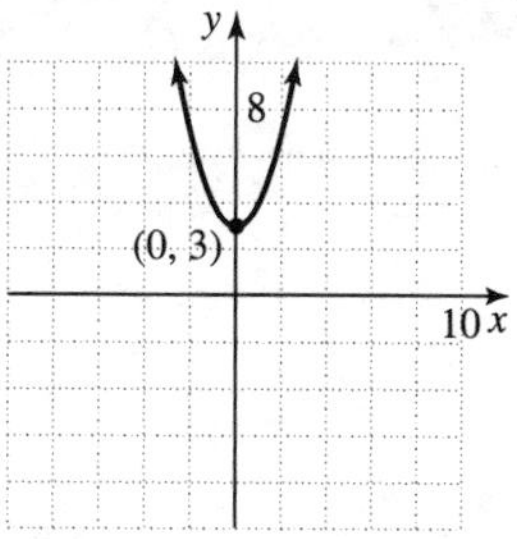

3. $f(x) = x^2$ and $g(x) = (x+6)^2$

Plot points. Notice that the graph of $g(x)$ is the graph of $f(x)$ shifted 6 units to the left.

x	$f(x) = x^2$	x	$g(x) = (x+6)^2$
–2	4	–8	4
–1	1	–7	1
0	0	–6	0
1	1	–5	1
2	4	–4	4

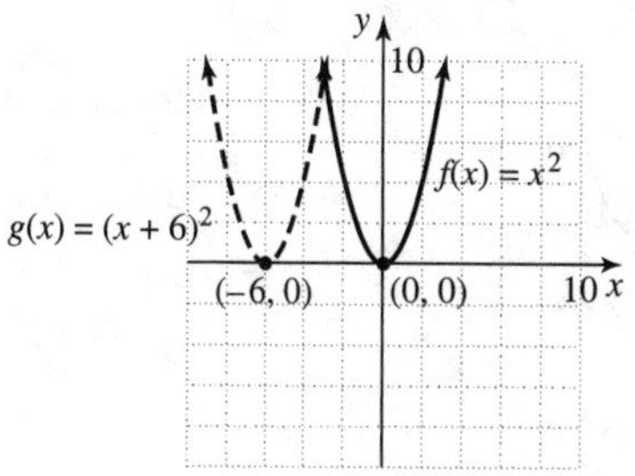

4. a. $G(x) = (x+4)^2$

The graph of $G(x)$ is obtained by shifting the graph of $y = x^2$ to the left 4 units.

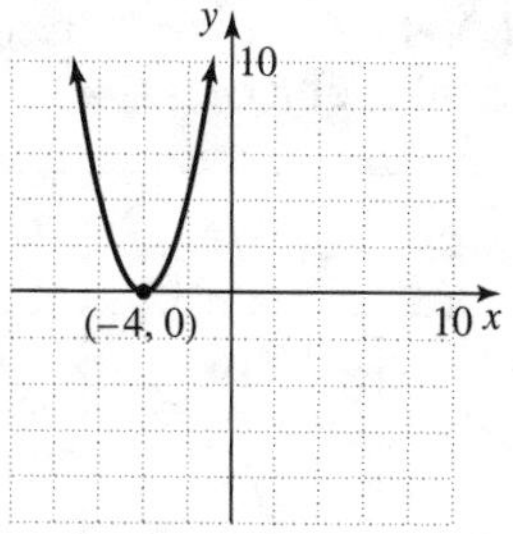

b. $H(x) = (x-7)^2$

The graph of $H(x)$ is obtained by shifting the graph of $y = x^2$ to the right 7 units.

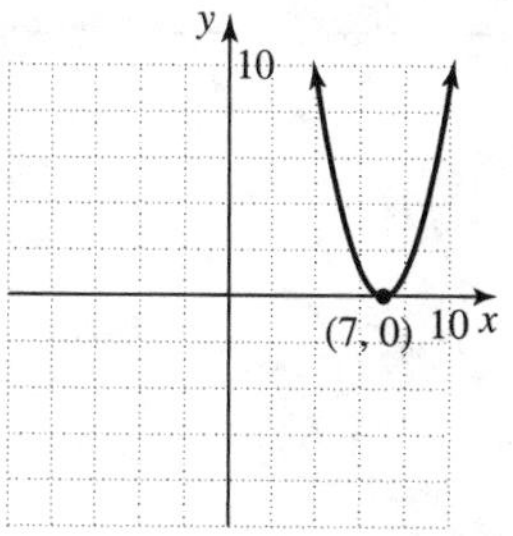

5. $f(x) = (x+2)^2 + 2$

The graph of $f(x)$ is the graph of $y = x^2$ shifted 2 units to the left and 2 units upward. The vertex is then (–2, 2), and the axis of symmetry is $x = -2$.

x	$f(x) = (x+2)^2 + 2$
–4	6
–3	3
–1	3
0	6

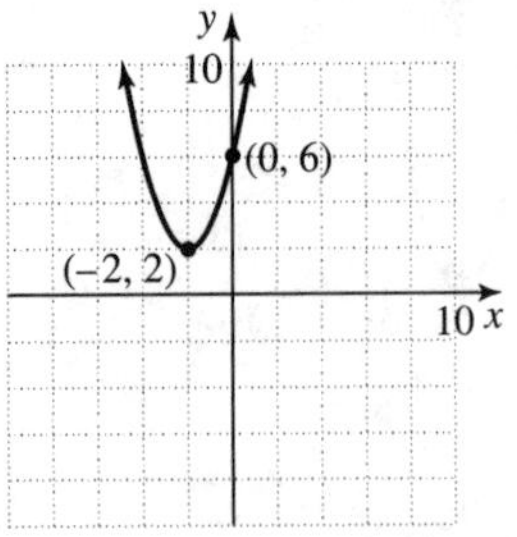

6. $f(x) = x^2$, $g(x) = 4x^2$, and $h(x) = \frac{1}{4}x^2$

Comparing tables of values, we see that for each x-value, the corresponding value of $g(x)$ is four times that of $f(x)$. Similarly, the value of $h(x)$ is one quarter the value of $f(x)$.

x	$f(x) = x^2$	$g(x) = 4x^2$	$h(x) = \frac{1}{4}x^2$
−2	4	16	1
−1	1	4	$\frac{1}{4}$
0	0	0	0
1	1	4	$\frac{1}{4}$
2	4	16	1

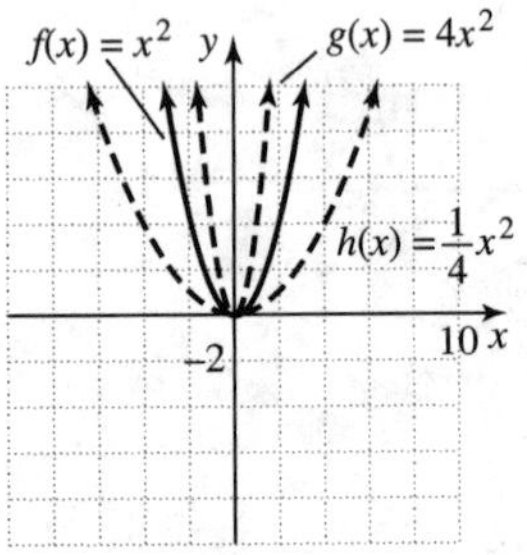

7. $f(x) = -\frac{1}{2}x^2$

Because $a = -\frac{1}{2}$, a negative value, this parabola opens downward. Since $\left|-\frac{1}{2}\right| = \frac{1}{2} < 1$, the parabola is wider than the graph of $y = x^2$. The vertex is (0, 0), and the axis of symmetry is the y-axis.

x	$f(x) = -\frac{1}{2}x^2$
−2	−2
−1	$-\frac{1}{2}$
0	0
1	$-\frac{1}{2}$
2	−2

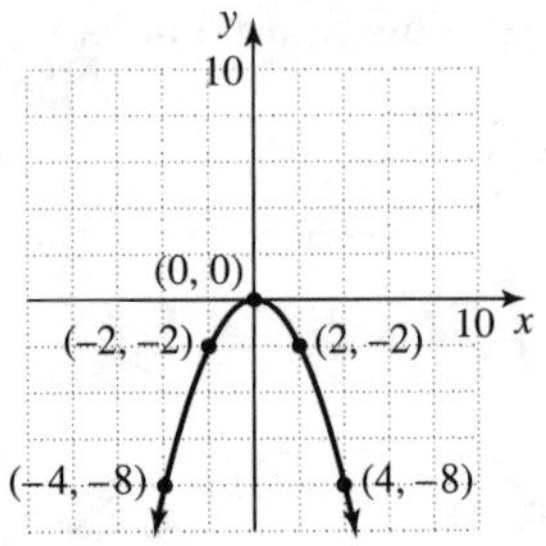

8. $h(x) = \frac{1}{3}(x-4)^2 - 3$

This graph is the same as $y = x^2$ shifted 4 units to the right and 3 units downward, and it is wider because a is $\frac{1}{3}$. The vertex is (4, −3), and the axis of symmetry is $x = 4$.

x	$h(x) = \frac{1}{3}(x-4)^2 - 3$
2	$-\frac{5}{3}$
3	$-\frac{8}{3}$
4	−3
5	$-\frac{8}{3}$
6	$-\frac{5}{3}$

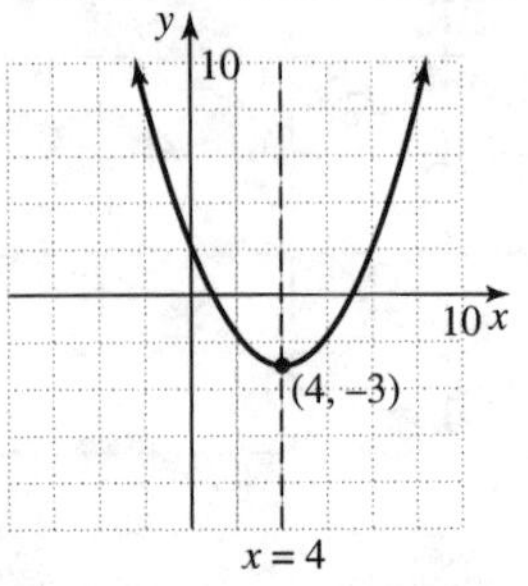

Graphing Calculator Explorations

1.

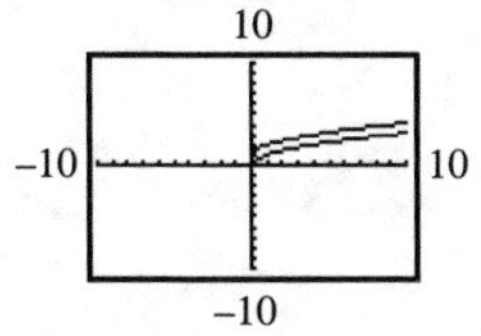

2.

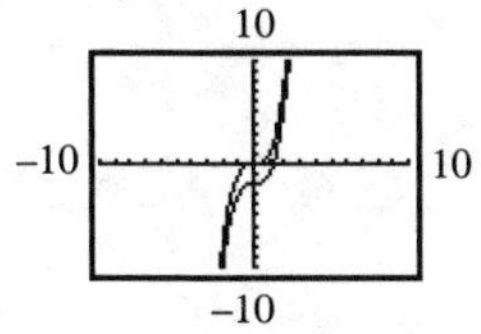

3.

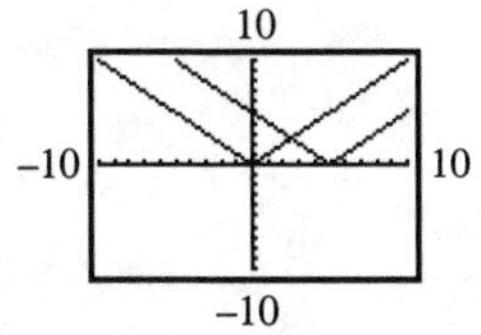

4.

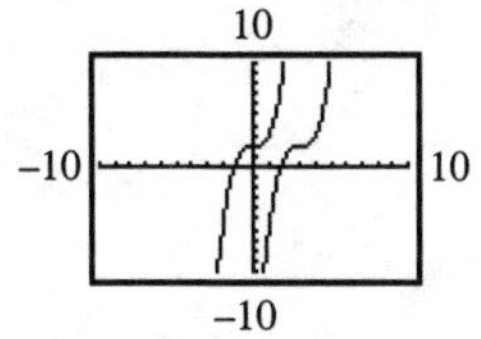

5.

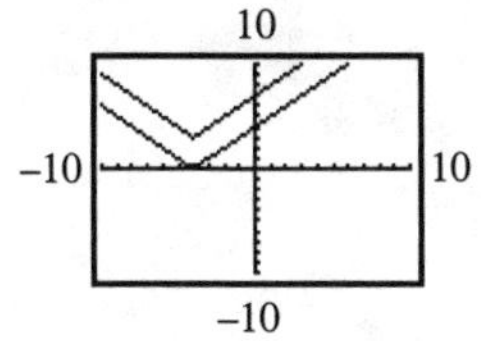

6.

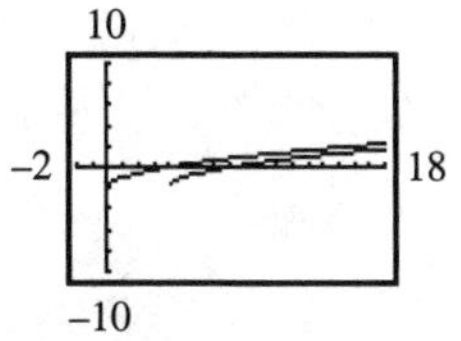

Vocabulary, Readiness & Video Check 8.5

1. A quadratic function is one that can be written in the form $f(x) = ax^2 + bx + c,\ a \neq 0$.

2. The graph of a quadratic function is a parabola opening upward or downward.

3. If $a > 0$, the graph of the quadratic function opens upward.

4. If $a < 0$, the graph of the quadratic function opens downward.

5. The vertex of a parabola is the lowest point if $a > 0$.

6. The vertex of a parabola is the highest point if $a < 0$.

7. $f(x) = x^2$; vertex: (0, 0)

8. $f(x) = -5x^2$; vertex: (0, 0)

9. $g(x) = (x-2)^2$; vertex: (2, 0)

10. $g(x) = (x+5)^2$; vertex: (−5, 0)

11. $f(x) = 2x^2 + 3$; vertex: (0, 3)

12. $h(x) = x^2 - 1$; vertex: (0, −1)

13. $g(x) = (x+1)^2 + 5$; vertex: (−1, 5)

14. $h(x) = (x-10)^2 - 7$; vertex: (10, −7)

15. Graphs of the form $f(x) = x^2 + k$ shift up or down the y-axis k units from $y = x^2$; the y-intercept.

16. Graphs of the form $f(x) = (x-h)^2$ shift right or left on the x-axis h units from $y = x^2$; the x-intercept.

17. The vertex, (h, k) and the axis of symmetry, $x = h$; the basic shape of $y = x^2$ does not change.

18. whether the graph is wider or narrower than $y = x^2$

19. the coordinates of the vertex, whether the graph opens upward or downward, whether the graph is narrower or wider than $y = x^2$, and the graph's axis of symmetry

Exercise Set 8.5

1. $f(x) = x^2 - 1$

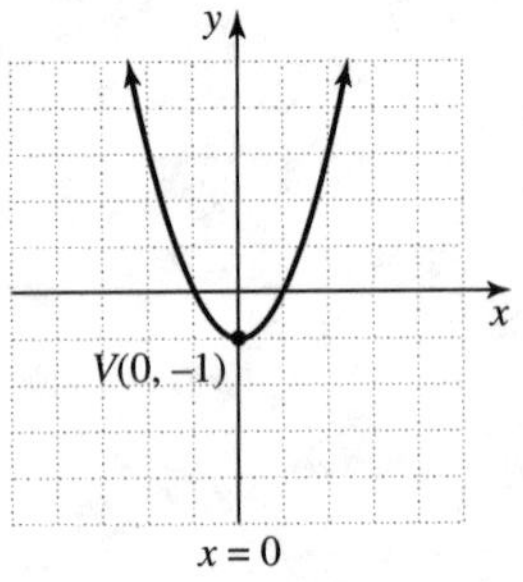

3. $h(x) = x^2 + 5$

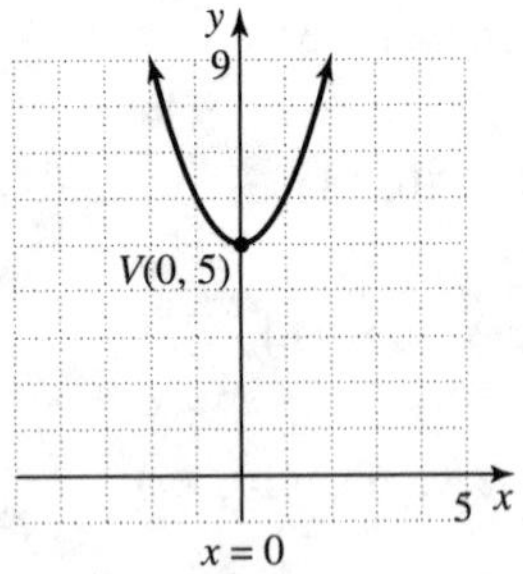

5. $g(x) = x^2 + 7$

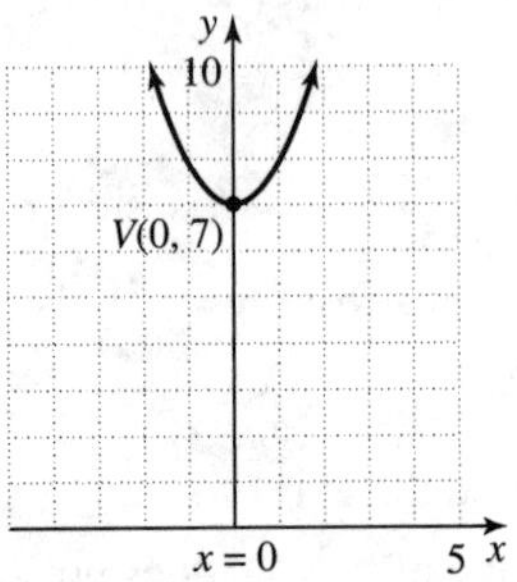

7. $f(x) = (x - 5)^2$

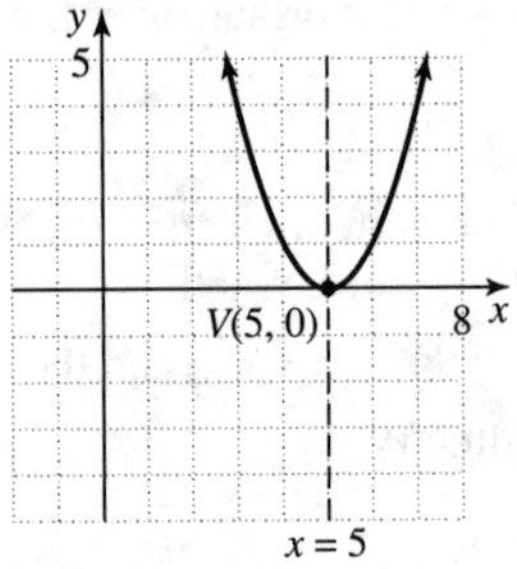

9. $h(x) = (x + 2)^2$

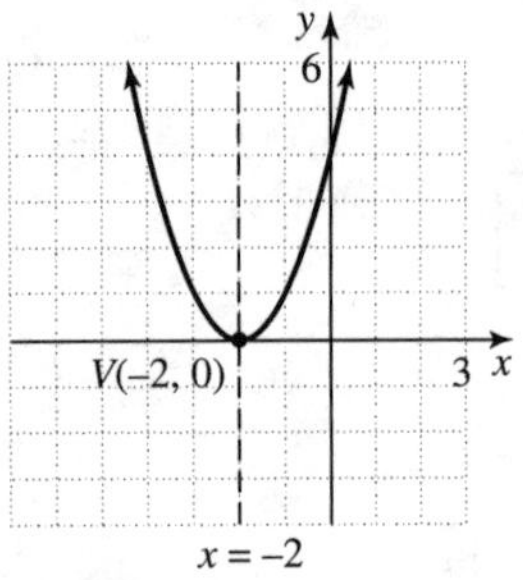

11. $G(x) = (x + 3)^2$

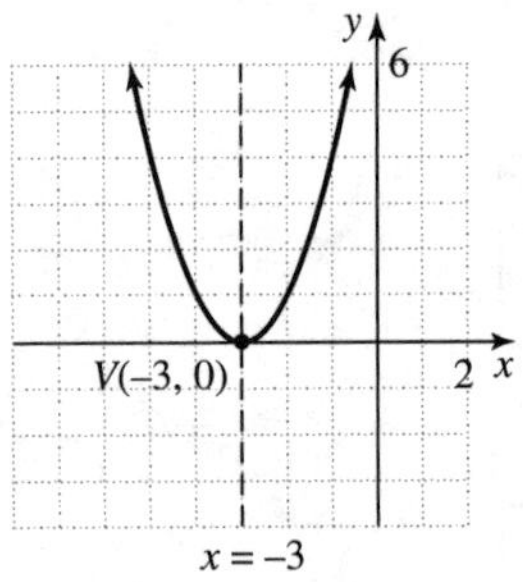

13. $f(x) = (x - 2)^2 + 5$

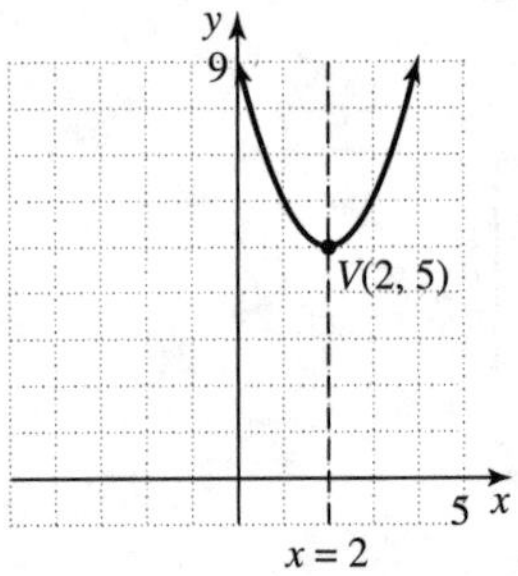

15. $h(x) = (x + 1)^2 + 4$

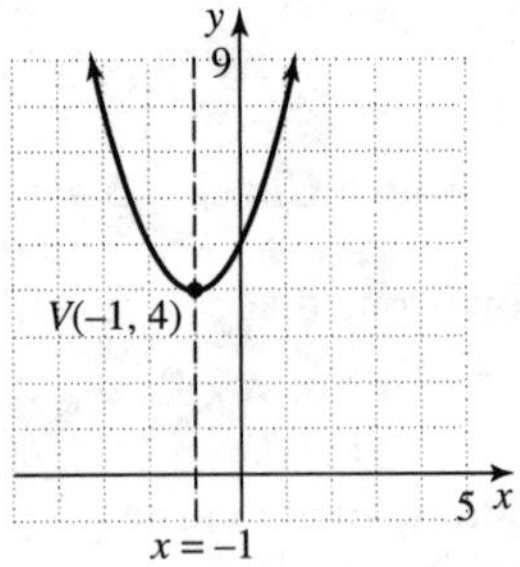

17. $g(x) = (x+2)^2 - 5$

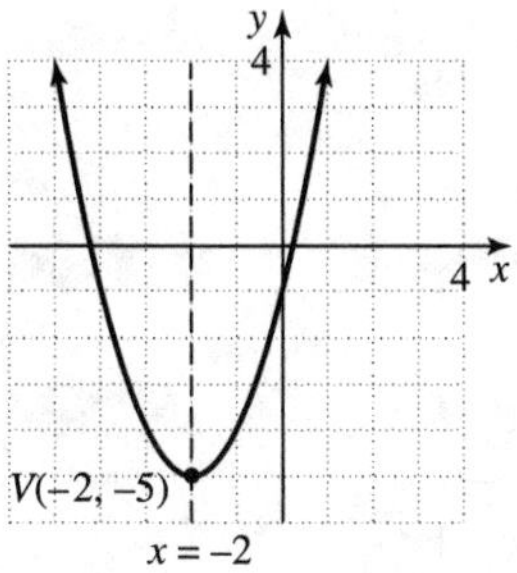

19. $H(x) = 2x^2$

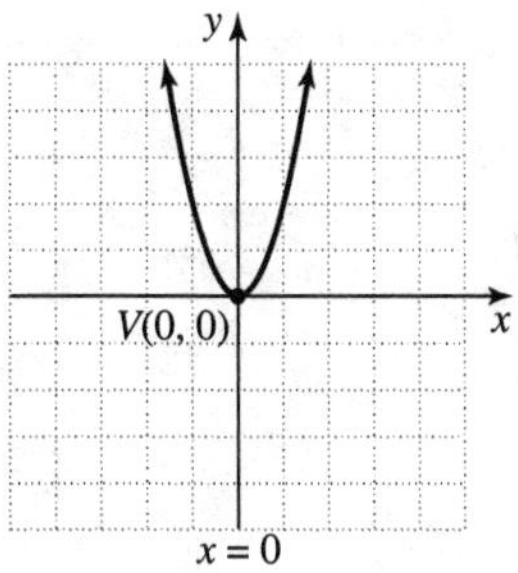

21. $h(x) = \frac{1}{3}x^2$

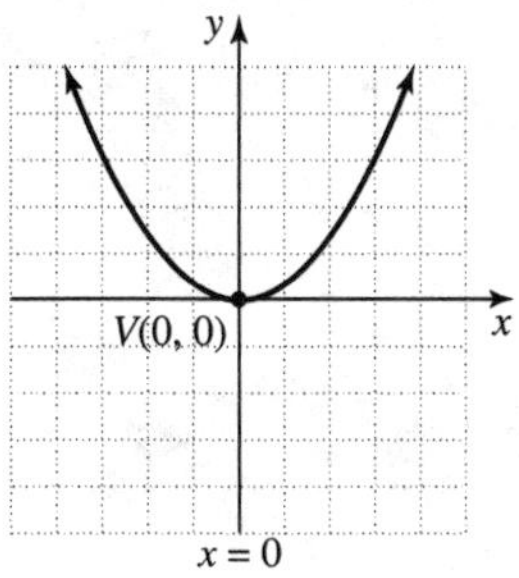

23. $g(x) = -x^2$

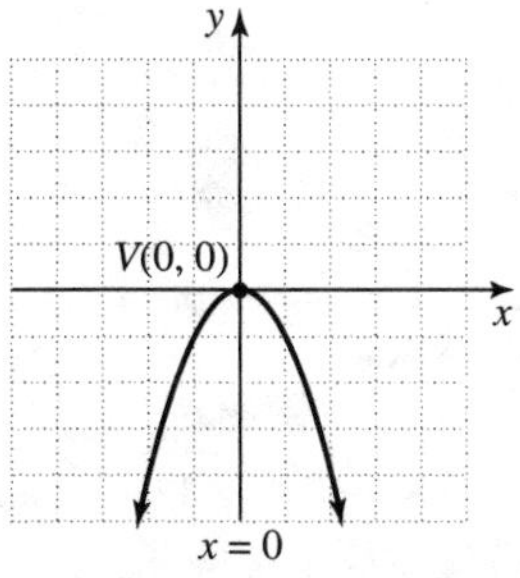

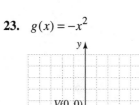

25. $f(x) = 2(x-1)^2 + 3$

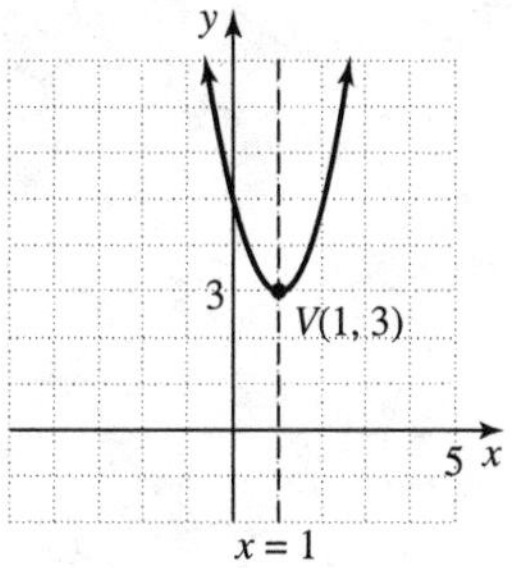

27. $h(x) = -3(x+3)^2 + 1$

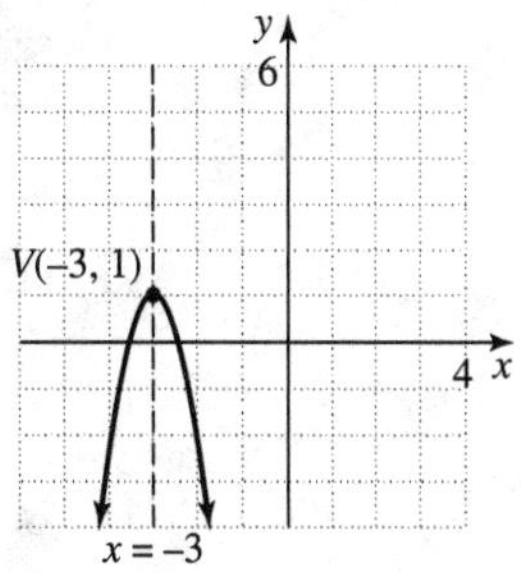

29. $H(x) = \frac{1}{2}(x-6)^2 - 3$

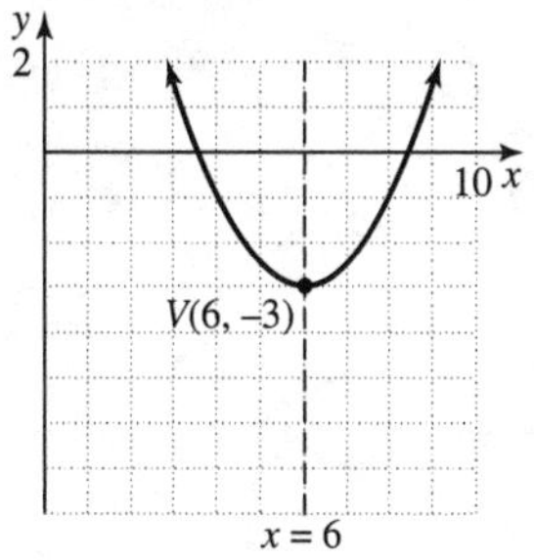

31. $f(x) = -(x-2)^2$

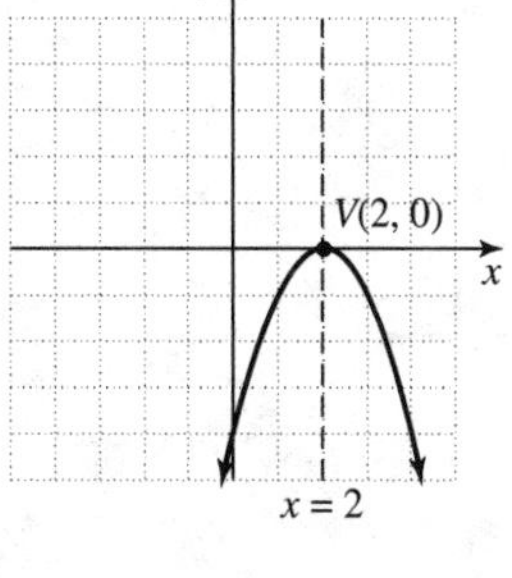

33. $F(x) = -x^2 + 4$

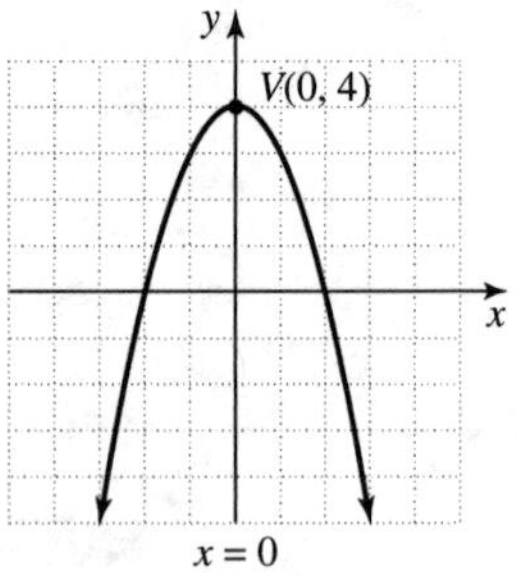

35. $F(x) = 2x^2 - 5$

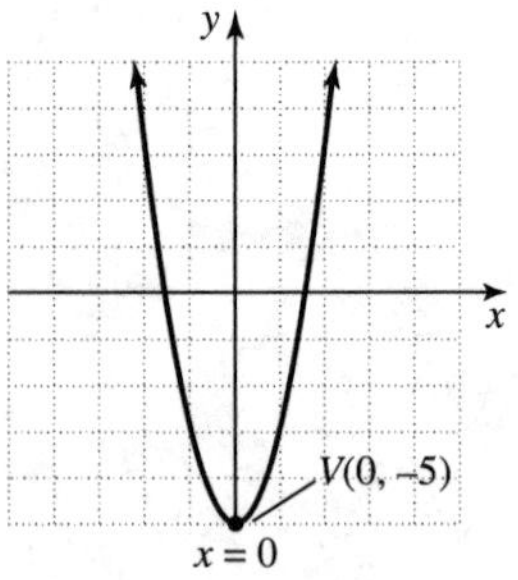

37. $h(x) = (x-6)^2 + 4$

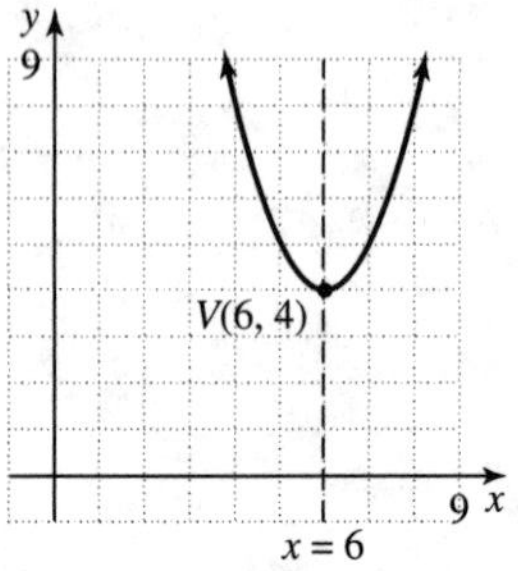

39. $F(x) = \left(x + \frac{1}{2}\right)^2 - 2$

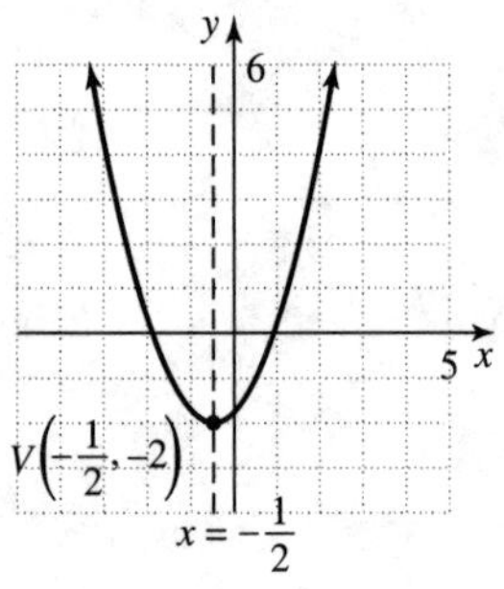

41. $F(x) = \frac{3}{2}(x+7)^2 + 1$

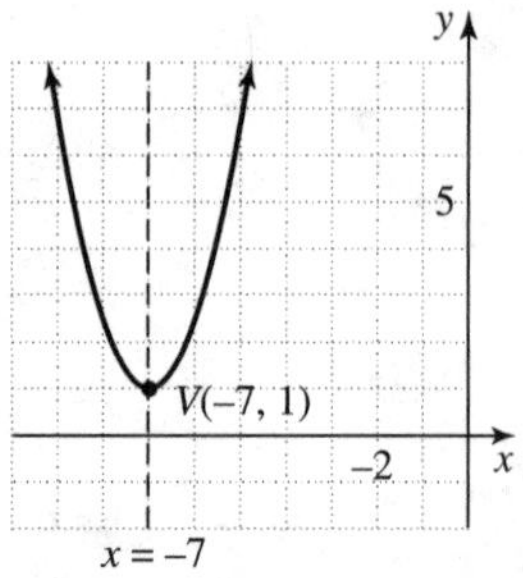

43. $f(x) = \frac{1}{4}x^2 - 9$

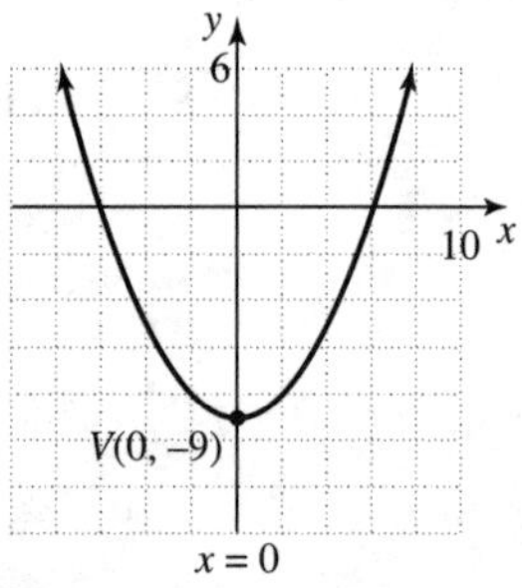

45. $G(x) = 5\left(x + \frac{1}{2}\right)^2$

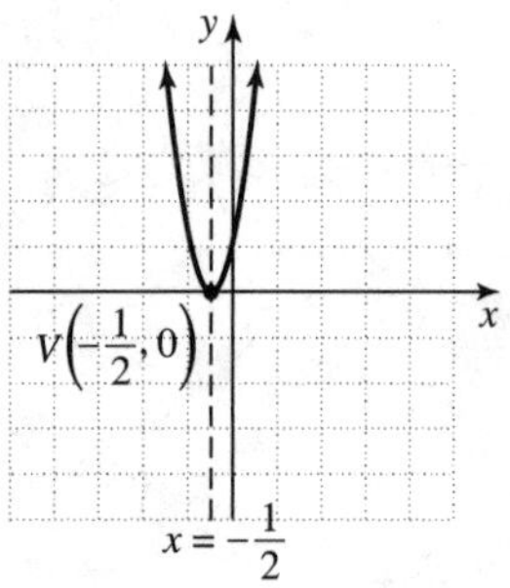

47. $h(x) = -(x-1)^2 - 1$

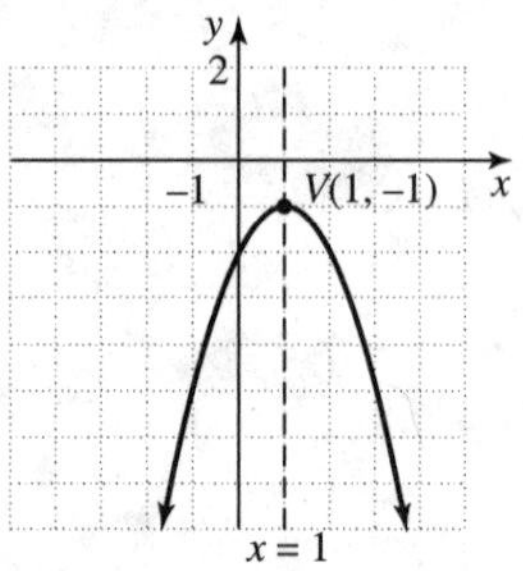

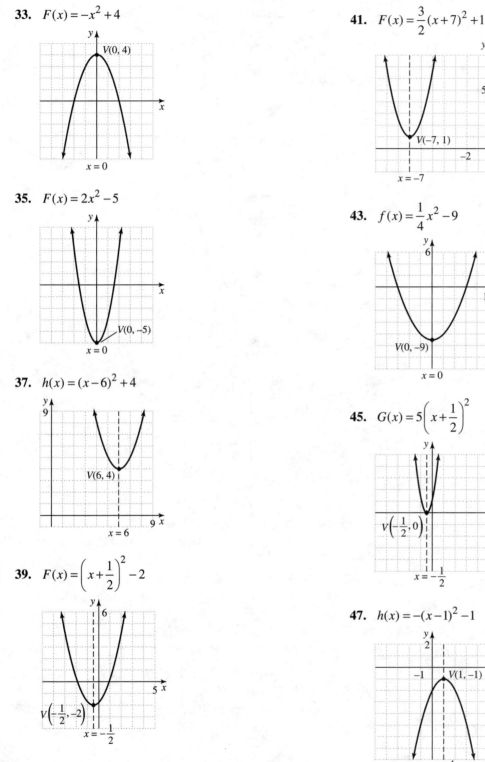

49. $g(x)=\sqrt{3}(x+5)^2+\frac{3}{4}$

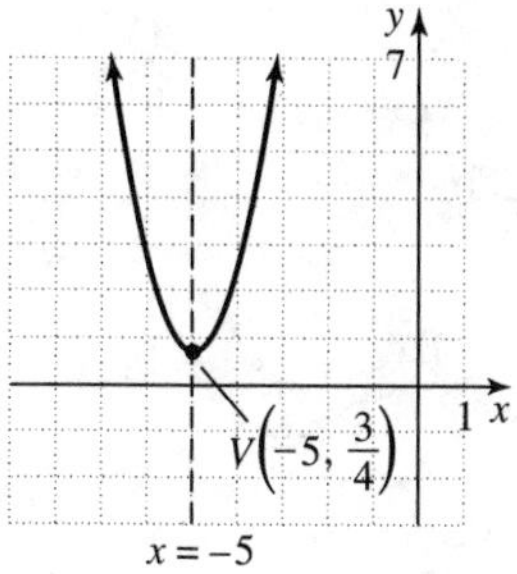

51. $h(x)=10(x+4)^2-6$

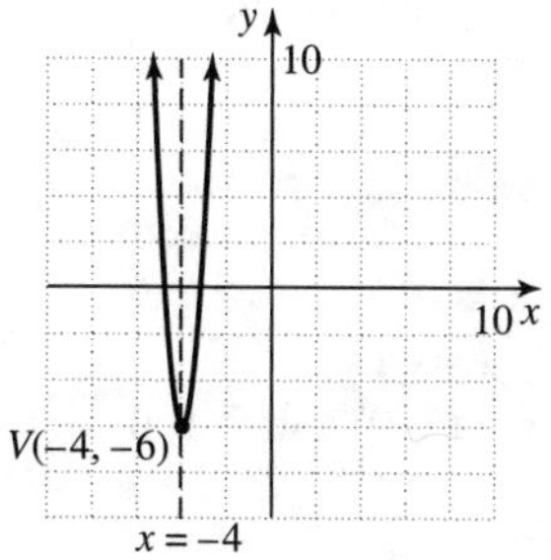

53. $f(x)=-2(x-4)^2+5$

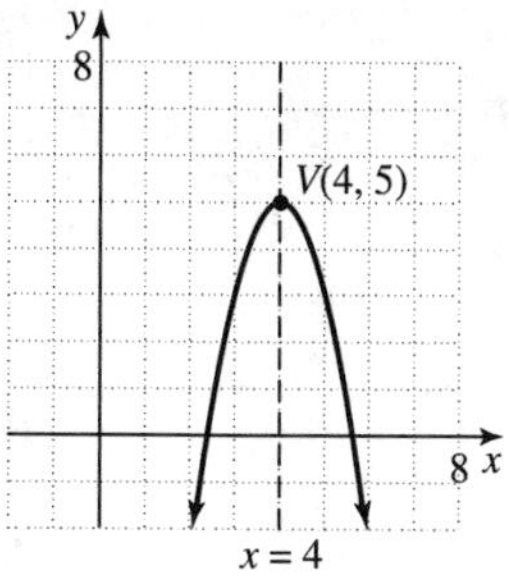

55. x^2+8x

$\left[\frac{1}{2}(8)\right]^2=(4)^2=16$

$x^2+8x+16$

57. z^2-16z

$\left[\frac{1}{2}(-16)\right]^2=(-8)^2=64$

$z^2-16z+64$

59. y^2+y

$\left[\frac{1}{2}(1)\right]^2=\left(\frac{1}{2}\right)^2=\frac{1}{4}$

$y^2+y+\frac{1}{4}$

61.
$$\begin{aligned}x^2+4x&=12\\ x^2+4x+\left(\frac{4}{2}\right)^2&=12+4\\ x^2+4x+4&=16\\ (x+2)^2&=16\\ x+2&=\pm\sqrt{16}\\ x+2&=\pm4\\ x&=-2\pm4\\ x&=-6 \text{ or } 2\end{aligned}$$

63.
$$\begin{aligned}z^2+10z-1&=0\\ z^2+10z&=1\\ z^2+10z+\left(\frac{10}{2}\right)^2&=1+25\\ z^2+10z+25&=26\\ (z+5)^2&=26\\ z+5&=\pm\sqrt{26}\\ z&=-5\pm\sqrt{26}\end{aligned}$$

65.
$$\begin{aligned}z^2-8z&=2\\ z^2-8z+\left(\frac{-8}{2}\right)^2&=2+16\\ z^2-8z+16&=18\\ (z-4)^2&=18\\ z-4&=\pm\sqrt{18}\\ z-4&=\pm3\sqrt{2}\\ z&=4\pm3\sqrt{2}\end{aligned}$$

67. $f(x)=-213(x-0.1)^2+3.6$

$a = -213 < 0$, so $f(x)$ opens downward.
The vertex is (0.1, 3.6). The correct answer is **c**.

69. $f(x)=5(x-2)^2+3$

71. $f(x)=5[x-(-3)]^2+6$
$=5(x+3)^2+6$

73. $y = f(x) + 1$

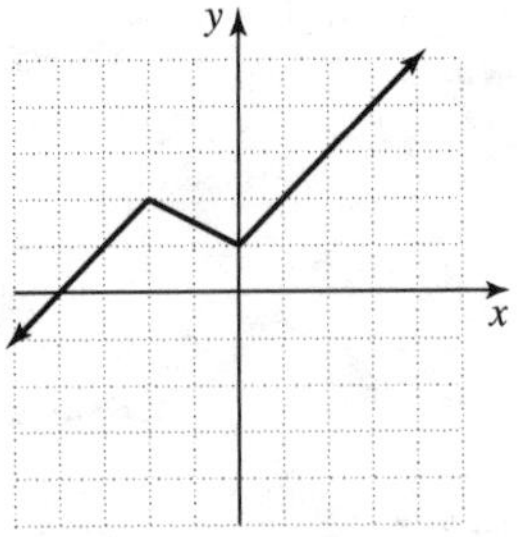

75. $y = f(x-3)$

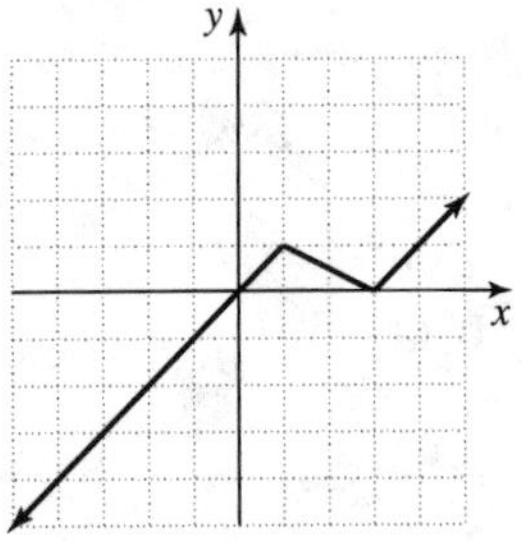

77. $y = f(x+2)+2$

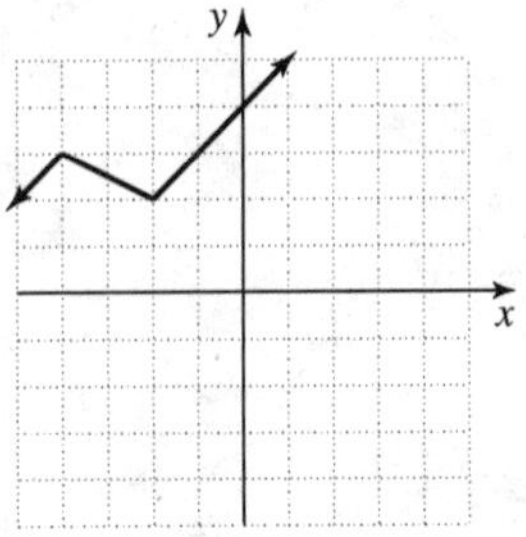

79. $f(x) = 12.5x^2 - 64x + 135$

a. 2013 is 8 years after 2005, so $x = 8$.

$f(8) = 12.5(8)^2 - 64(8) + 135 = 423$

Online retail sales in the United States were \$423 billion in 2013.

b. 2020 is 15 years after 2005, so $x = 15$.

$f(15) = 12.5(15)^2 - 64(15) + 135 = 1987.5$

Online retail sales in the United States are predicted to be \$1987.5 billion in 2020.

Section 8.6 Practice Exercises

1. $g(x) = x^2 - 2x - 3$

Write in the form $y = (x-h)^2 + k$ by completing the square.

$$y = x^2 - 2x - 3$$
$$y + 3 = x^2 - 2x$$
$$y + 3 + \left(\frac{-2}{2}\right)^2 = x^2 - 2x + \left(\frac{-2}{2}\right)^2$$
$$y + 4 = x^2 - 2x + 1$$
$$y = (x-1)^2 - 4$$

The vertex is at (1, –4).
Let $g(x) = 0$.

$0 = x^2 - 2x - 3$
$0 = (x-3)(x+1)$
$x - 3 = 0$ or $x + 1 = 0$
$x = 3$ $\quad x = -1$

The x-intercepts are (3, 0) and (–1, 0).
Let $x = 0$.

$g(0) = 0^2 - 2(0) - 3 = -3$

The y-intercept is (0, –3).

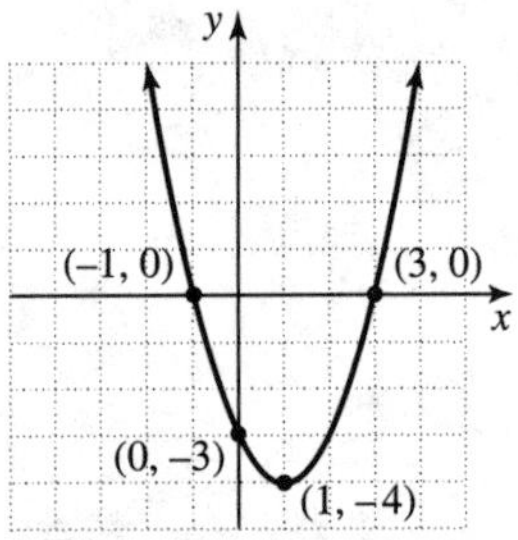

2. $g(x) = 4x^2 + 4x + 3$

Replace $g(x)$ with y and complete the square to write the equation in the form $y = a(x-h)^2 + k$.

$$y = 4x^2 + 4x + 3$$
$$y - 3 = 4x^2 + 4x = 4(x^2 + x)$$
$$y - 3 + 4\left(\frac{1}{2}\right)^2 = 4\left[x^2 + x + \left(\frac{1}{2}\right)^2\right]$$
$$y - 3 + 1 = 4\left(x^2 + x + \frac{1}{4}\right)$$
$$y = 4\left(x + \frac{1}{2}\right)^2 + 2$$

$a = 4$, $h = -\frac{1}{2}$, and $k = 2$.

The parabola opens upward with vertex

$\left(-\frac{1}{2}, 2\right)$, and has an axis of symmetry $x = -\frac{1}{2}$.

Let $x = 0$.

$g(0) = 4(0)^2 + 4(0) + 3 = 3$

The y-intercept is (0, 3). There are no x-intercepts.

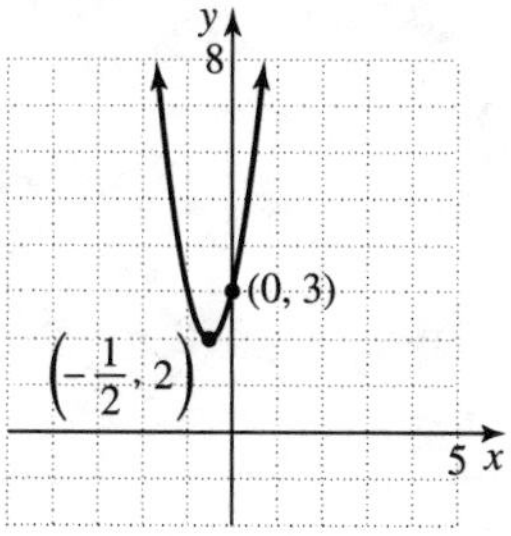

3. $g(x) = -x^2 + 5x + 6$

Write $g(x)$ in the form $a(x-h)^2 + k$ by completing the square. Replace $g(x)$ with y.

$$y = -x^2 + 5x + 6$$
$$y - 6 = -x^2 + 5x$$
$$y - 6 = -1(x^2 - 5x)$$
$$y - 6 - \left(\frac{-5}{2}\right)^2 = -1\left[x^2 - 5x + \left(\frac{-5}{2}\right)^2\right]$$
$$y - 6 - \frac{25}{4} = -1\left(x^2 - 5x + \frac{25}{4}\right)$$
$$y - \frac{49}{4} = -\left(x - \frac{5}{2}\right)^2$$
$$y = -\left(x - \frac{5}{2}\right)^2 + \frac{49}{4}$$

Since $a = -1$, the parabola opens downward with vertex $\left(\frac{5}{2}, \frac{49}{4}\right)$ and axis of symmetry $x = \frac{5}{2}$.

Let $x = 0$.

$y = -0^2 + 5(0) + 6 = 6$

The y-intercept is (0, 6). Let $y = 0$.

$0 = -x^2 + 5x + 6$

$0 = x^2 - 5x - 6$

$0 = (x-6)(x+1)$

$x - 6 = 0$ or $x + 1 = 0$

$x = 6$ $\quad$ $x = -1$

The x-intercepts are (6, 0) and (−1, 0).

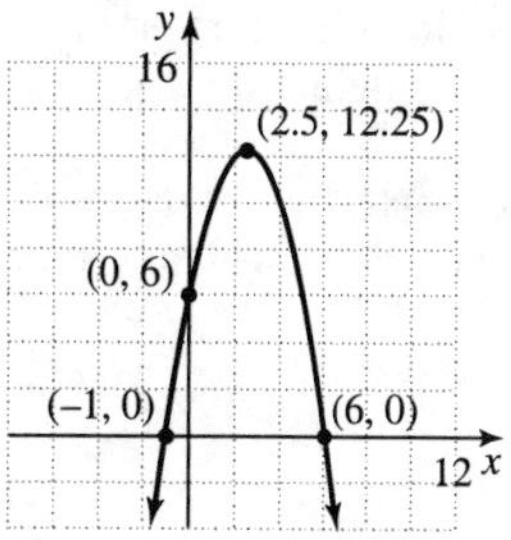

4. $g(x) = x^2 - 2x - 3$

$a = 1$, $b = -2$, and $c = -3$

$\frac{-b}{2a} = \frac{-(-2)}{2(1)} = \frac{2}{2} = 1$

The x-value of the vertex is 1.

$g(1) = 1^2 - 2(1) - 3 = 1 - 2 - 3 = -4$

The vertex is (1, −4).

5. $h(t) = -16t^2 + 24t$

Find the vertex of $h(t)$ to find its maximum value.

$a = -16$, $b = 24$, and $c = 0$

$\frac{-b}{2a} = \frac{-24}{2(-16)} = \frac{3}{4}$

The t-value of the vertex is $\frac{3}{4}$.

$$h\left(\frac{3}{4}\right) = -16\left(\frac{3}{4}\right)^2 + 24\left(\frac{3}{4}\right)$$
$$= -16\left(\frac{9}{16}\right) + 18$$
$$= -9 + 18$$
$$= 9$$

The vertex is $\left(\frac{3}{4}, 9\right)$. Thus, the ball reaches its maximum height of 9 feet in $\frac{3}{4}$ second.

Vocabulary, Readiness & Video Check 8.6

1. If a quadratic function is in the form $f(x) = a(x-h)^2 + k$, the vertex of its graph is $\underline{(h, k)}$.

2. The graph of $f(x) = ax^2 + bx + c$, $a \neq 0$, is a parabola whose vertex has x-value $\underline{\frac{-b}{2a}}$.

3. We can immediately identify the vertex (h, k), whether the parabola opens upward or downward, and know its axis of symmetry; completing the square.

4. This information tells us whether or not the graph has x-intercepts. For example, if the vertex is in quadrant III or IV and the parabola opens downward, then there aren't any x-intercepts and there's no need to go through the steps to locate any.

5. the vertex

Exercise Set 8.6

	Parabola Opens	*Vertex Location*	*Number of x-intercept(s)*	*Number of y-intercept(s)*
1.	up	Q I	0	1
3.	down	Q II	2	1
5.	up	x-axis	1	1
7.	down	Q III	0	
9.	up	Q IV	2	

11. $f(x) = x^2 + 8x + 7$

$-\frac{b}{2a} = \frac{-8}{2(1)} = -4$ and

$$\begin{aligned} f(-4) &= (-4)^2 + 8(-4) + 7 \\ &= 16 - 32 + 7 \\ &= -9 \end{aligned}$$

Thus, the vertex is $(-4, -9)$.

13. $f(x) = -x^2 + 10x + 5$

$-\frac{b}{2a} = \frac{-10}{2(-1)} = 5$ and

$$\begin{aligned} f(5) &= -(5)^2 + 10(5) + 5 \\ &= -25 + 50 + 5 \\ &= 30 \end{aligned}$$

Thus, the vertex is $(5, 30)$.

15. $f(x) = 5x^2 - 10x + 3$

$-\frac{b}{2a} = \frac{-(-10)}{2(5)} = 1$ and

$$\begin{aligned} f(1) &= 5(1)^2 - 10(1) + 3 \\ &= 5 - 10 + 3 \\ &= -2 \end{aligned}$$

Thus, the vertex is $(1, -2)$.

17. $f(x) = -x^2 + x + 1$

$-\frac{b}{2a} = \frac{-1}{2(-1)} = \frac{1}{2}$ and

$$f\left(\frac{1}{2}\right) = -\left(\frac{1}{2}\right)^2 + \left(\frac{1}{2}\right) + 1 = -\frac{1}{4} + \frac{1}{2} + 1 = \frac{5}{4}$$

Thus, the vertex is $\left(\frac{1}{2}, \frac{5}{4}\right)$.

19. $f(x) = x^2 - 4x + 3$

$-\frac{b}{2a} = \frac{-(-4)}{2(1)} = 2$ and

$f(2) = (2)^2 - 4(2) + 3 = -1$

The vertex is (2, –1), so the graph is D.

21. $f(x) = x^2 - 2x - 3$

$-\frac{b}{2a} = \frac{-(-2)}{2(1)} = 1$ and

$f(1) = (1)^2 - 2(1) - 3 = -4$

The vertex is (1, –4), so the graph is B.

23. $f(x) = x^2 + 4x - 5$

$-\frac{b}{2a} = \frac{-4}{2(1)} = -2$ and

$f(-2) = (-2)^2 + 4(-2) - 5 = -9$

Thus, the vertex is (–2, –9).

The graph opens upward ($a = 1 > 0$).

$x^2 + 4x - 5 = 0$

$(x+5)(x-1) = 0$

$x + 5 = 0$ or $x - 1 = 0$

$x = -5$ or $x = 1$

x-intercepts: (–5, 0) and (1, 0).

$f(0) = -5$, so the y-intercept is (0, –5).

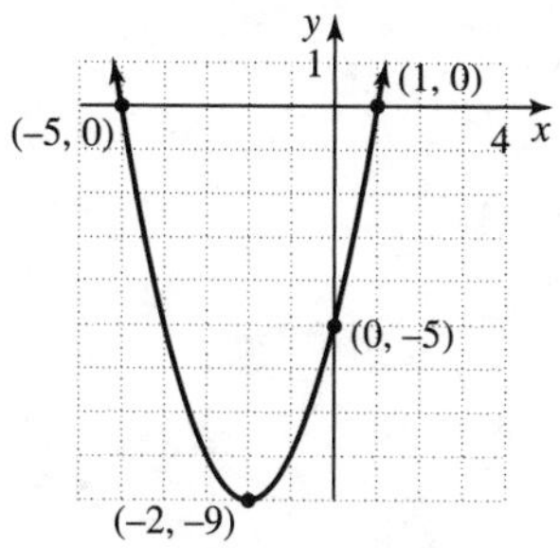

25. $f(x) = -x^2 + 2x - 1$

$-\frac{b}{2a} = \frac{-2}{2(-1)} = 1$ and

$f(1) = -(1)^2 + 2(1) - 1 = 0$

Thus, the vertex is (1, 0).

The graph opens downward ($a = -1 < 0$).

$-x^2 + 2x - 1 = 0$

$x^2 - 2x + 1 = 0$

$(x-1)^2 = 0$

$x - 1 = 0$

$x = 1$

x-intercept: (1, 0).

$f(0) = -1$, so the y-intercept is (0, –1).

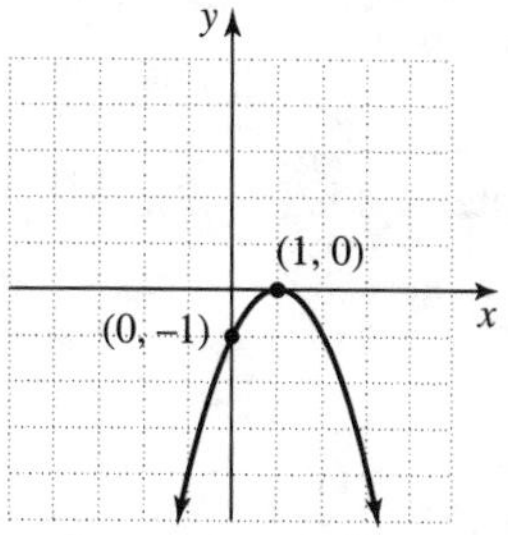

27. $f(x) = x^2 - 4$

$-\frac{b}{2a} = \frac{-0}{2(1)} = 0$ and

$f(0) = (0)^2 - 4 = -4$

Thus, the vertex is (0, –4).

The graph opens upward ($a = 1 > 0$).

$x^2 - 4 = 0$

$(x+2)(x-2) = 0$

$x + 2 = 0$ or $x - 2 = 0$

$x = -2$ or $x = 2$

x-intercepts: (–2, 0) and (2, 0).

$f(0) = -4$, so the y-intercept is (0, –4).

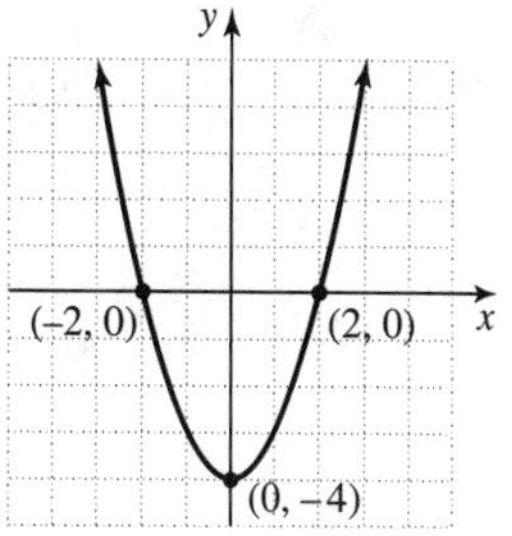

29. $f(x) = 4x^2 + 4x - 3$

$-\frac{b}{2a} = -\frac{4}{2(4)} = -\frac{1}{2}$ and

$f\left(-\frac{1}{2}\right) = 4\left(-\frac{1}{2}\right)^2 + 4\left(-\frac{1}{2}\right) - 3 = -4$

Thus, the vertex is $\left(-\frac{1}{2}, -4\right)$.

The graph opens upward ($a = 4 > 0$).

$4x^2 + 4x - 3 = 0$
$(2x+3)(2x-1) = 0$
$2x+3 = 0$ or $2x-1 = 0$
$x = -\frac{3}{2}$ or $x = \frac{1}{2}$

x-intercepts: $\left(-\frac{3}{2}, 0\right)$ and $\left(\frac{1}{2}, 0\right)$.

$f(0) = -3$, so the *y*-intercept is (0, –3).

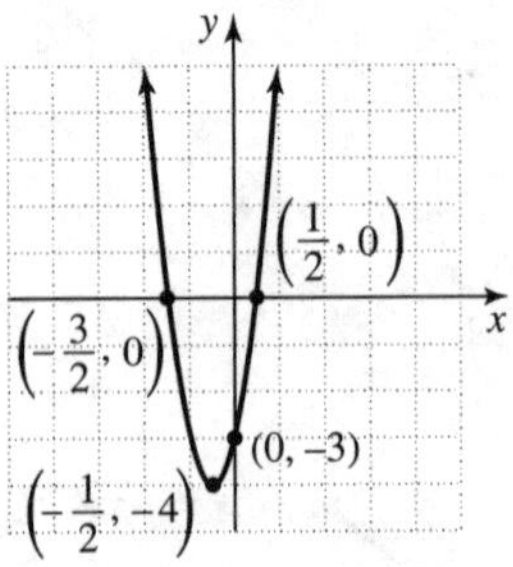

31. $f(x) = \frac{1}{2}x^2 + 4x + \frac{15}{2}$

$-\frac{b}{2a} = -\frac{4}{2\left(\frac{1}{2}\right)} = -\frac{4}{1} = -4$ and

$f(-4) = \frac{1}{2}(-4)^2 + 4(-4) + \frac{15}{2} = -\frac{1}{2}$.

Thus, the vertex is $\left(-4, -\frac{1}{2}\right)$.

The graph opens upward $\left(a = \frac{1}{2} > 0\right)$.

$\frac{1}{2}x^2 + 4x + \frac{15}{2} = 0$
$2\left(\frac{1}{2}x^2 + 4x + \frac{15}{2}\right) = 2(0)$
$x^2 + 8x + 15 = 0$
$(x+5)(x+3) = 0$
$x+5 = 0$ or $x+3 = 0$
$x = -5$ or $x = -3$

x-intercepts: (–5, 0) and (–3, 0).

$f(0) = \frac{15}{2}$, so the *y*-intercept is $\left(0, \frac{15}{2}\right)$.

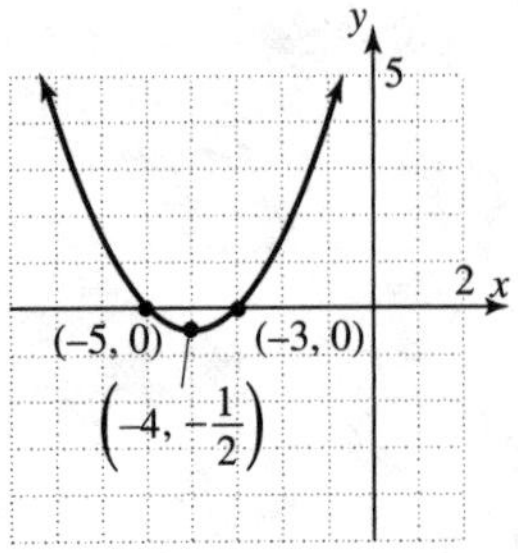

33. $f(x) = x^2 - 6x + 5$
$y = x^2 - 6x + 5$
$y - 5 = x^2 - 6x$
$y - 5 + 9 = x^2 - 6x + 9$
$y + 4 = (x-3)^2$
$y = (x-3)^2 - 4$
$f(x) = (x-3)^2 - 4$

Thus, the vertex is (3, –4).
The graph opens upward ($a = 1 > 0$).

$x^2 - 6x + 5 = 0$
$(x-5)(x-1) = 0$
$x = 5$ or $x = 1$

x-intercepts: (5, 0) and (1, 0).
$f(0) = 5$, so the *y*-intercept is (0, 5).

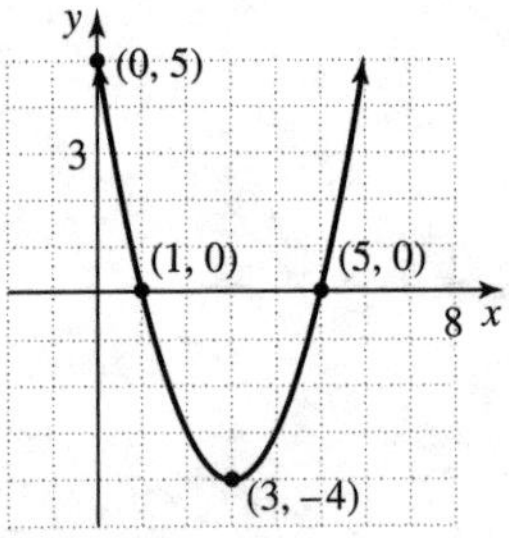

35. $f(x) = x^2 - 4x + 5$
$y = x^2 - 4x + 5$
$y - 5 = x^2 - 4x$
$y - 5 + 4 = x^2 - 4x + 4$
$y - 1 = (x-2)^2$
$y = (x-2)^2 + 1$
$f(x) = (x-2)^2 + 1$

Thus, the vertex is (2, 1).
The graph opens upward ($a = 1 > 0$).

$x^2 - 4x + 5 = 0$

$x = \dfrac{4 \pm \sqrt{(-4)^2 - 4(1)(5)}}{2(1)} = \dfrac{4 \pm \sqrt{-4}}{2}$

which give non-real solutions.

Hence, there are no x-intercepts.

$f(0) = 5$, so the y-intercept is (0, 5).

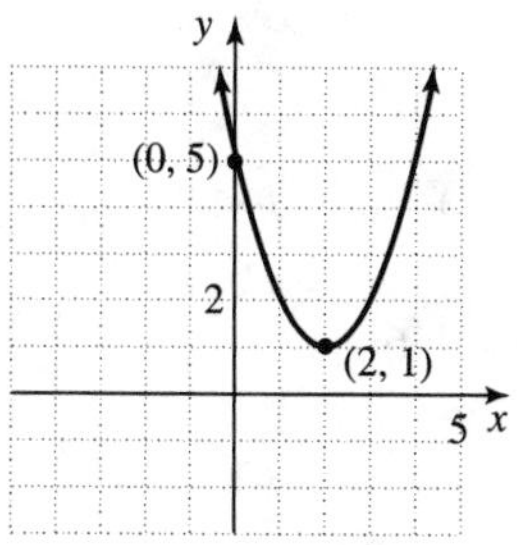

37.

$f(x) = 2x^2 + 4x + 5$

$y = 2x^2 + 4x + 5$

$y - 5 = 2(x^2 + 2x)$

$y - 5 + 2(1) = 2(x^2 + 2x + 1)$

$y - 3 = 2(x+1)^2$

$y = 2(x+1)^2 + 3$

$f(x) = 2(x+1)^2 + 3$

Thus, the vertex is (–1, 3).

The graph opens upward ($a = 2 > 0$).

$2x^2 + 4x + 5 = 0$

$x = \dfrac{-4 \pm \sqrt{(4)^2 - 4(2)(5)}}{2(2)} = \dfrac{-4 \pm \sqrt{-24}}{4}$

which give non-real solutions.

Hence, there are no x-intercepts.

$f(0) = 5$, so the y-intercept is (0, 5).

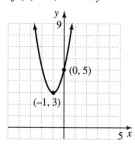

39.

$f(x) = -2x^2 + 12x$

$y = -2(x^2 - 6x)$

$y + [-2(9)] = -2(x^2 - 6x + 9)$

$y - 18 = -2(x-3)^2$

$y = -2(x-3)^2 + 18$

$f(x) = -2(x-3)^2 + 18$

Thus, the vertex is (3, 18).

The graph opens downward ($a = -2 < 0$).

$-2x^2 + 12x = 0$

$-2x(x-6) = 0$

$x = 0$ or $x - 6 = 0$

$x = 6$

x-intercepts: (0, 0) and (6, 0)

$f(0) = 0$, so the y-intercept is (0, 0).

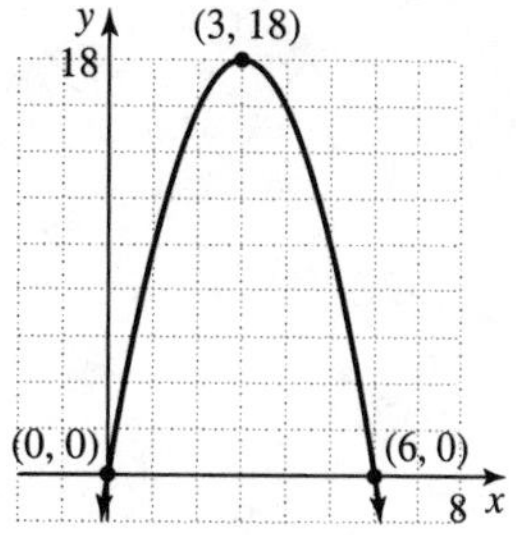

41. $f(x) = x^2 + 1$

$x = -\dfrac{b}{2a} = -\dfrac{0}{2(1)} = 0$

$f(0) = (0)^2 + 1 = 1$

Thus, the vertex is (0, 1).

The graph opens upward ($a = 1 > 0$).

$x^2 + 1 = 0$

$x^2 = -1$

which give non-real solutions.

Hence, there are no x-intercepts.

$f(0) = 1$, so the y-intercept is (0, 1).

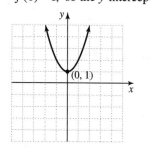

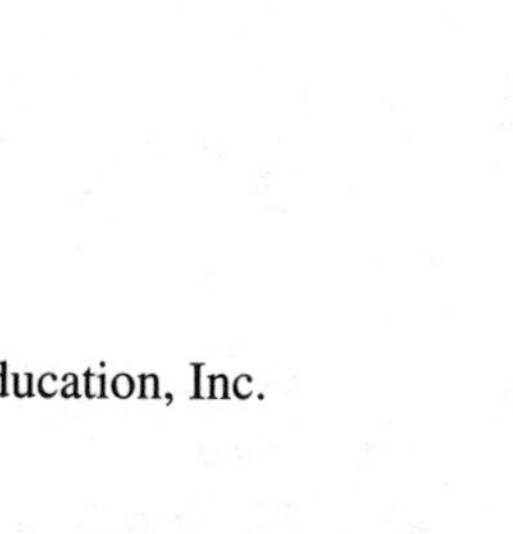

43.
$$\begin{aligned} f(x) &= x^2 - 2x - 15 \\ y &= x^2 - 2x - 15 \\ y + 15 &= x^2 - 2x \\ y + 15 + 1 &= x^2 - 2x + 1 \\ y + 16 &= (x-1)^2 \\ y &= (x-1)^2 - 16 \\ f(x) &= (x-1)^2 - 16 \end{aligned}$$

Thus, the vertex is (1, –16).
The graph opens upward ($a = 1 > 0$).

$x^2 - 2x - 15 = 0$
$(x-5)(x+3) = 0$
$x = 5$ or $x = -3$
x-intercepts: (–3, 0) and (5, 0).
$f(0) = -15$ so the y-intercept is (0, –15).

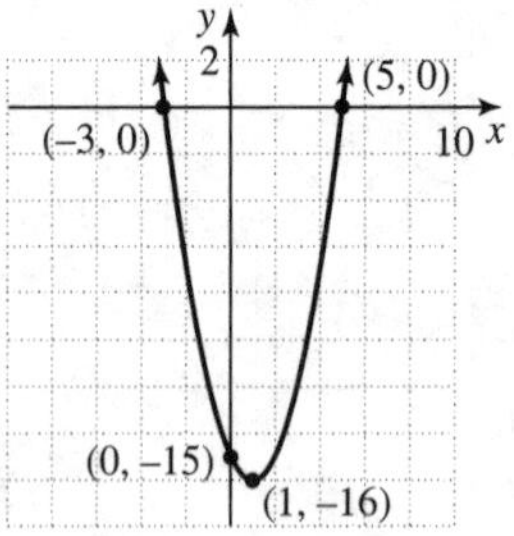

45. $f(x) = -5x^2 + 5x$

$x = -\frac{b}{2a} = \frac{-5}{2(-5)} = \frac{1}{2}$ and

$f\left(\frac{1}{2}\right) = -5\left(\frac{1}{2}\right)^2 + 5\left(\frac{1}{2}\right) = -\frac{5}{4} + \frac{5}{2} = \frac{5}{4}$

Thus, the vertex is $\left(\frac{1}{2}, \frac{5}{4}\right)$.

The graph opens downward ($a = -5 < 0$).

$-5x^2 + 5x = 0$
$-5x(x-1) = 0$
$x = 0$ or $x - 1 = 0$
$x = 1$
x-intercepts: (0, 0) and (1, 0)
$f(0) = 0$, so the y-intercept is (0, 0).

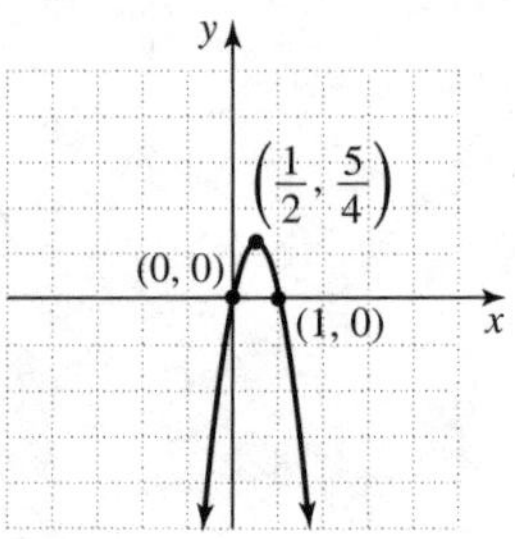

47. $f(x) = -x^2 + 2x - 12$

$x = -\frac{b}{2a} = \frac{-2}{2(-1)} = 1$ and

$f(1) = -(1)^2 + 2(1) - 12 = -11$

Thus, the vertex is (1, –11).
The graph opens downward ($a = -1 < 0$).

$-x^2 + 2x - 12 = 0$
$x^2 - 2x + 12 = 0$

$$x = \frac{2 \pm \sqrt{(-2)^2 - 4(1)(12)}}{2(1)} = \frac{2 \pm \sqrt{-44}}{2}$$

which yields non-real solutions.
Hence, there are no x-intercepts.
$f(0) = -12$ so the y-intercept is (0, –12).

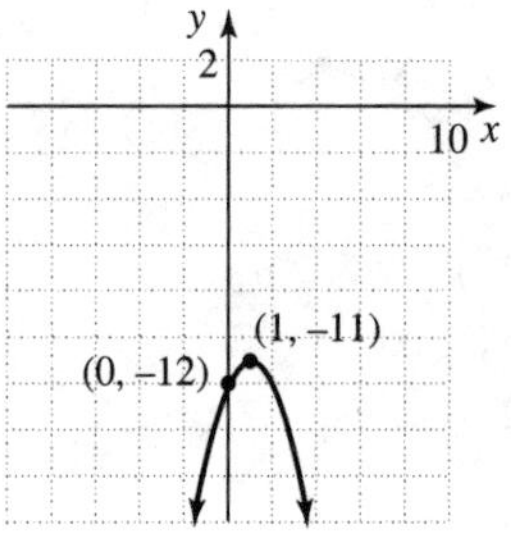

49. $f(x) = 3x^2 - 12x + 15$

$x = -\frac{b}{2a} = \frac{-(-12)}{2(3)} = \frac{12}{6} = 2$ and

$$\begin{aligned} f(2) &= 3(2)^2 - 12(2) + 15 \\ &= 12 - 24 + 15 \\ &= 3 \end{aligned}$$

Thus, the vertex is (2, 3).
The graph opens upward ($a = 3 > 0$).

$3x^2 - 12x + 15 = 0$
$x^2 - 4x + 5 = 0$

$$x = \frac{4 \pm \sqrt{(-4)^2 - 4(1)(5)}}{2(1)} = \frac{4 \pm \sqrt{-4}}{2}$$

which yields non-real solutions.
Hence, there are no x-intercepts.
$f(0) = 15$, so the y-intercept is (0, 15).

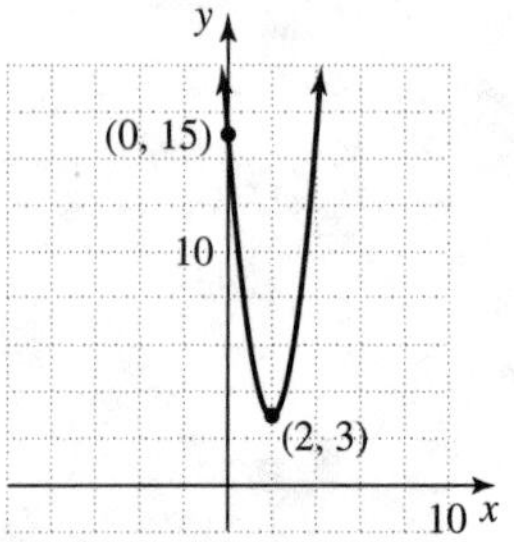

51. $f(x) = x^2 + x - 6$

$x = -\frac{b}{2a} = \frac{-1}{2(1)} = -\frac{1}{2}$ and

$$f\left(-\frac{1}{2}\right) = \left(-\frac{1}{2}\right)^2 + \left(-\frac{1}{2}\right) - 6 = \frac{1}{4} - \frac{1}{2} - 6 = -\frac{25}{4}$$

Thus, the vertex is $\left(-\frac{1}{2}, -\frac{25}{4}\right)$.

The graph opens upward ($a = 1 > 0$).

$x^2 + x - 6 = 0$
$(x+3)(x-2) = 0$
$x = -3$ or $x = 2$

x-intercepts: (–3, 0) and (2, 0).

$f(0) = -6$ so the y-intercept is (0, –6).

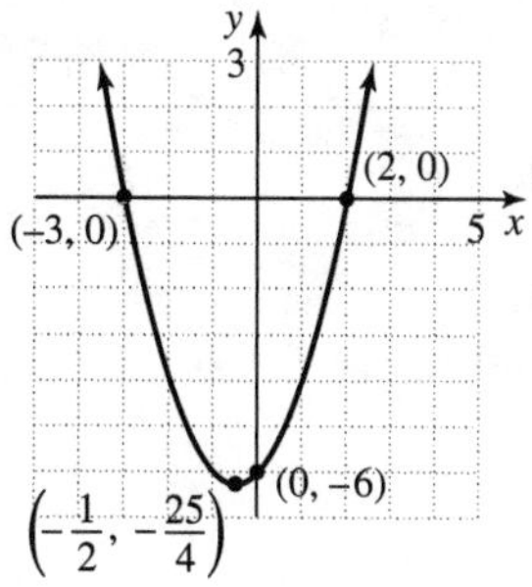

53. $f(x) = -2x^2 - 3x + 35$

$x = -\frac{b}{2a} = \frac{-(-3)}{2(-2)} = -\frac{3}{4}$ and

$$f\left(-\frac{3}{4}\right) = -2\left(-\frac{3}{4}\right)^2 - 3\left(-\frac{3}{4}\right) + 35 = -\frac{9}{8} + \frac{9}{4} + 35 = \frac{289}{8}$$

Thus, the vertex is $\left(-\frac{3}{4}, \frac{289}{8}\right)$.

The graph opens downward ($a = -2 < 0$).

$-2x^2 - 3x + 35 = 0$
$2x^2 + 3x - 35 = 0$
$(2x-7)(x+5) = 0$
$2x - 7 = 0$ or $x + 5 = 0$
$x = \frac{7}{2}$ or $x = -5$

x-intercepts: (–5, 0) and $\left(\frac{7}{2}, 0\right)$.

$f(0) = 35$ so the y-intercept is (0, 35).

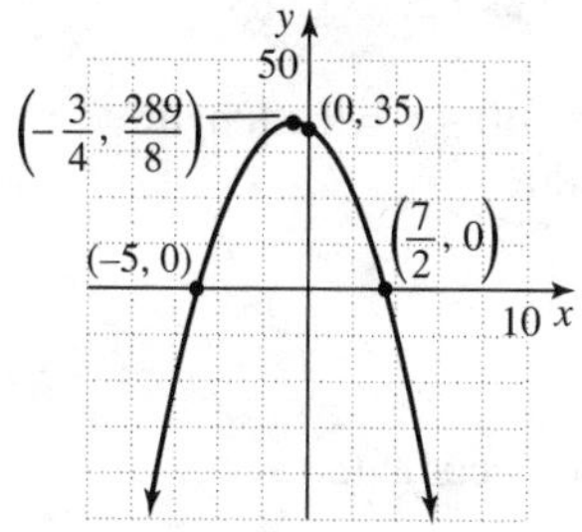

55. $h(t) = -16t^2 + 96t$

$t = -\frac{b}{2a} = \frac{-96}{2(-16)} = \frac{96}{32} = 3$ and

$$h(3) = -16(3)^2 + 96(3) = -144 + 288 = 144$$

The maximum height is 144 feet.

57. $C(x) = 2x^2 - 800x + 92{,}000$

a. $x = -\frac{b}{2a} = \frac{-(-800)}{2(2)} = 200$

200 bicycles are needed to minimize the cost.

b. $C(200) = 2(200)^2 - 800(200) + 92{,}000 = 12{,}000$

The minimum cost is \$12,000.

59. Let x = one number. Then
$60 - x$ = the other number.

$$f(x) = x(60 - x) = 60x - x^2 = -x^2 + 60x$$

The maximum will occur at the vertex.

$x = -\frac{b}{2a} = \frac{-60}{2(-1)} = 30$

$60 - x = 60 - 30 = 30$

The numbers are 30 and 30.

61. Let x = one number. Then
$10 + x$ = the other number.

$$f(x) = x(10 + x) = 10x + x^2 = x^2 + 10x$$

The minimum will occur at the vertex.

$$x = -\frac{b}{2a} = \frac{-10}{2(1)} = -5$$

$$10 + x = 10 + (-5) = 5$$

The numbers are –5 and 5.

63. Let x = width. Then $40 - x$ = the length.

Area = length · width

$$A(x) = (40 - x)x$$
$$= 40x - x^2$$
$$= -x^2 + 40x$$

The maximum will occur at the vertex.

$$x = -\frac{b}{2a} = \frac{-40}{2(-1)} = 20$$

$$40 - x = 40 - 20 = 20$$

The maximum area will occur when the length and width are 20 units each.

65. $f(x) = x^2 + 2$

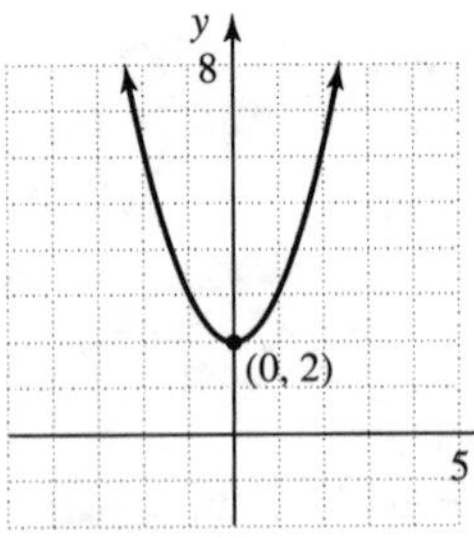

67. $g(x) = x + 2$

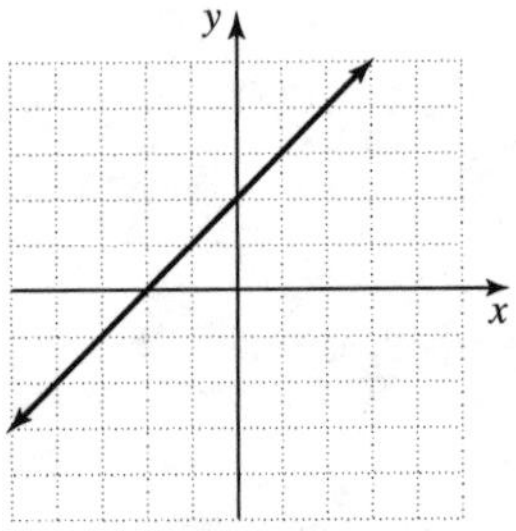

69. $f(x) = (x+5)^2 + 2$

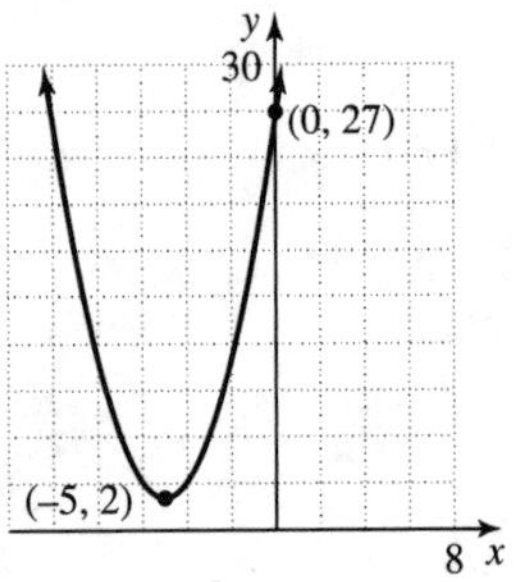

71. $f(x) = 3(x-4)^2 + 1$

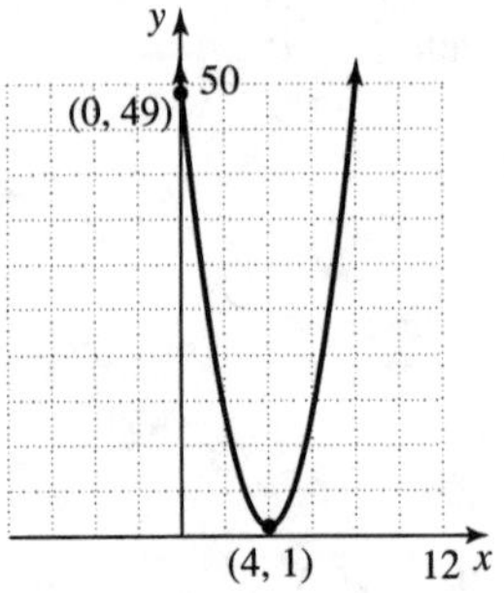

73. $f(x) = -(x-4)^2 + \frac{3}{2}$

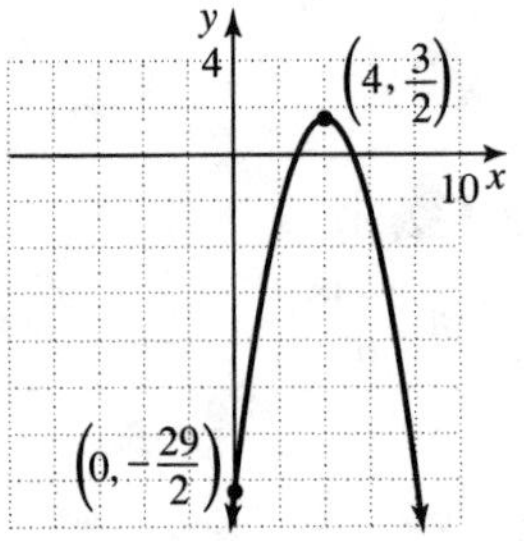

75. $f(x) = 2x^2 - 5$

Since $a = 2 > 0$, the graph opens upward; thus, $f(x)$ has a minimum value.

77. $f(x) = 3 - \frac{1}{2}x^2$

Since $a = -\frac{1}{2} < 0$, the graph opens downward; thus, $f(x)$ has a maximum value.

79. $f(x) = x^2 + 10x + 15$

$$x = -\frac{b}{2a} = \frac{-10}{2(1)} = -5 \text{ and}$$

$$f(-5) = (-5)^2 + 10(-5) + 15 = -10$$

Thus, the vertex is (–5, –10).

The graph opens upward ($a = 1 > 0$).

$f(0) = 15$ so the y-intercept is (0, 15).

$$x^2 + 10x + 15 = 0$$

$$x = \frac{-10 \pm \sqrt{(10)^2 - 4(1)(15)}}{2(1)}$$
$$= \frac{-10 \pm \sqrt{40}}{2}$$
$$\approx -8.2 \text{ or } -1.8$$

The x-intercepts are approximately (–8.2, 0) and (–1.8, 0).

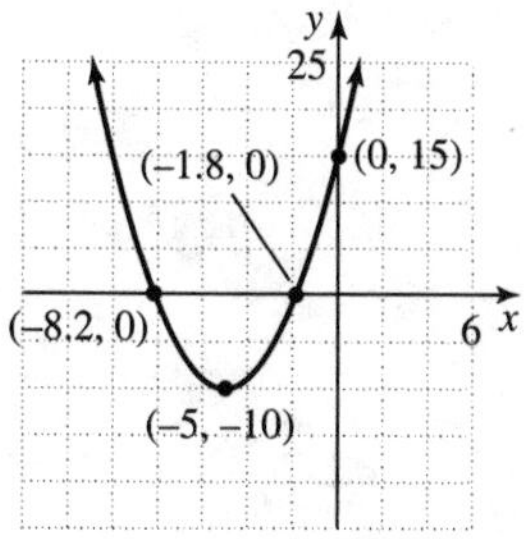

81. $f(x) = 3x^2 - 6x + 7$

$x = -\frac{b}{2a} = \frac{-(-6)}{2(3)} = 1$ and

$f(1) = 3(1)^2 - 6(1) + 7 = 4$

Thus, the vertex is (1, 4).
The graph opens upward ($a = 3 > 0$).
$f(0) = 7$ so the y-intercept is (0, 7).

$3x^2 - 6x + 7 = 0$

$x = \frac{6 \pm \sqrt{(-6)^2 - 4(3)(7)}}{2(3)} = \frac{6 \pm \sqrt{-48}}{6}$

which yields non-real solutions.

Hence, there are no x-intercepts.

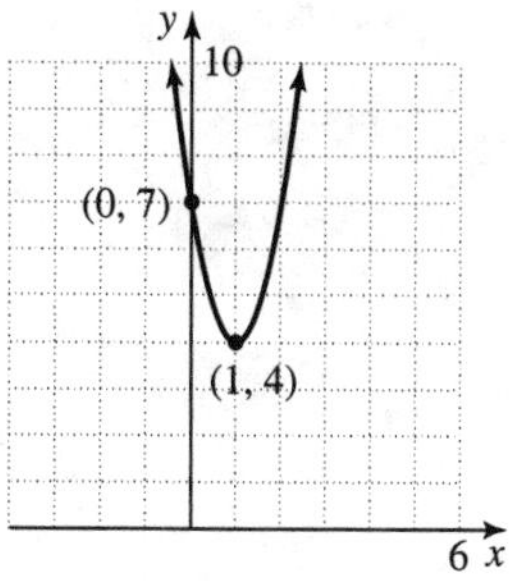

83. $f(x) = 2.3x^2 - 6.1x + 3.2$

$x = \frac{-(-6.1)}{2(2.3)} \approx 1.33$

$f(1.33) \approx -0.84$
minimum ≈ -0.84
Alternative solution:

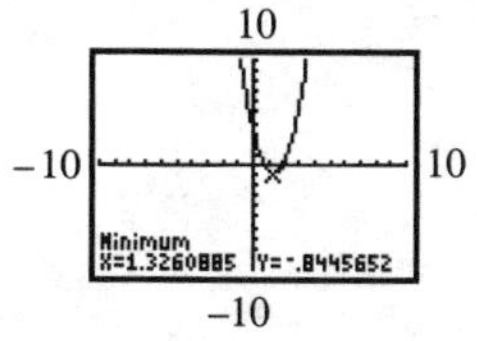

85. $f(x) = -1.9x^2 + 5.6x - 2.7$

$x = \frac{-5.6}{2(-1.9)} \approx 1.47$

$f(1.47) \approx 1.43$
maximum ≈ 1.43

Alternate solution:

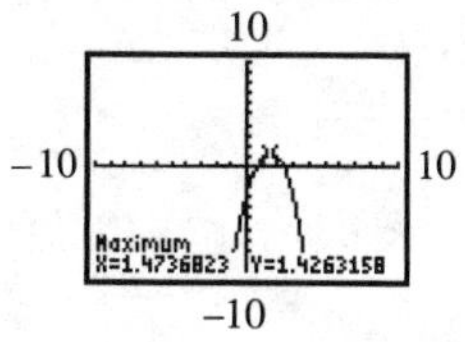

87. a. The function will have a maximum value; answers may vary.

b. $c(x) = -0.4x^2 + 21x + 35$
$a = -0.4$, $b = 21$, $c = 35$

$-\frac{b}{2a} = -\frac{21}{2(-0.4)} = 26.25$

The maximum will be reached 26.25 years after 2009 or in 2009 + 26 = 2035.

c. $c(26.25) = -0.4(26.25)^2 + 21(26.25) + 35$
≈ 310.6

The maximum number of Wi-Fi-enabled cell phones is predicted to be about 311 million.

89.

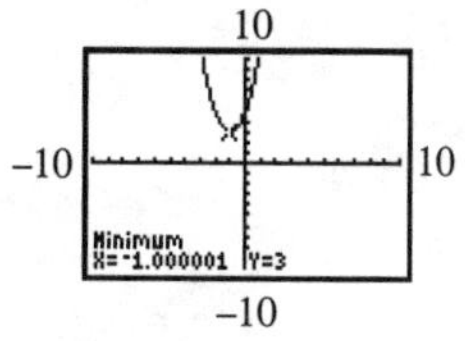

91.

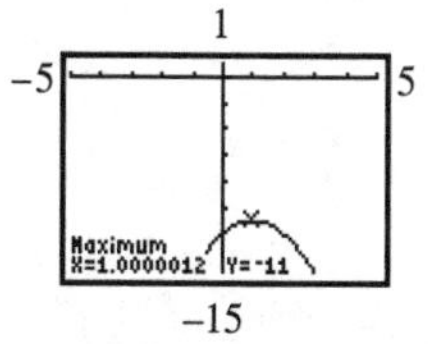

Chapter 8 Vocabulary Check

1. The <u>discriminant</u> helps us find the number and type of solutions of a quadratic equation.

2. If $a^2 = b$, then $a = \underline{\pm\sqrt{b}}$.

3. The graph of $f(x) = ax^2 + bx + c$, where a is not 0, is a parabola whose vertex has x-value $\underline{\frac{-b}{2a}}$.

4. A <u>quadratic inequality</u> is an inequality that can be written so that one side is a quadratic expression and the other side is 0.

5. The process of writing a quadratic equation so that one side is a perfect square trinomial is called completing the square.

6. The graph of $f(x) = x^2 + k$ has vertex (0, k).

7. The graph of $f(x) = (x-h)^2$ has vertex (h, 0).

8. The graph of $f(x) = (x-h)^2 + k$ has vertex (h, k).

9. The formula $x = \frac{-b \pm \sqrt{b^2 - 4ac}}{2a}$ is called the quadratic formula.

10. A quadratic equation is one that can be written in the form $ax^2 + bx + c = 0$ where *a*, *b*, and *c* are real numbers and *a* is not 0.

Chapter 8 Review

1.
$$x^2 - 15x + 14 = 0$$
$$(x-14)(x-1) = 0$$
$$x - 14 = 0 \quad \text{or} \quad x - 1 = 0$$
$$x = 14 \quad \text{or} \quad x = 1$$
The solutions are 1 and 14.

2.
$$7a^2 = 29a + 30$$
$$7a^2 - 29a - 30 = 0$$
$$(7a+6)(a-5) = 0$$
$$7a + 6 = 0 \quad \text{or} \quad a - 5 = 0$$
$$7a = -6 \quad \text{or} \quad a = 5$$
$$a = -\frac{6}{7}$$
The solutions are $-\frac{6}{7}$ and 5.

3.
$$4m^2 = 196$$
$$m^2 = 49$$
$$m = \pm\sqrt{49}$$
$$m = \pm 7$$
The solutions are –7 and 7.

4.
$$(5x-2)^2 = 2$$
$$5x - 2 = \pm\sqrt{2}$$
$$5x = 2 \pm \sqrt{2}$$
$$x = \frac{2 \pm \sqrt{2}}{5}$$
The solutions are $\frac{2+\sqrt{2}}{5}$ and $\frac{2-\sqrt{2}}{5}$.

5.
$$z^2 + 3z + 1 = 0$$
$$z^2 + 3z = -1$$
$$z^2 + 3z + \left(\frac{3}{2}\right)^2 = -1 + \frac{9}{4}$$
$$\left(z + \frac{3}{2}\right)^2 = \frac{5}{4}$$
$$z + \frac{3}{2} = \pm\sqrt{\frac{5}{4}}$$
$$z + \frac{3}{2} = \pm\frac{\sqrt{5}}{2}$$
$$z = -\frac{3}{2} \pm \frac{\sqrt{5}}{2} = \frac{-3 \pm \sqrt{5}}{2}$$
The solutions are $\frac{-3+\sqrt{5}}{2}$ and $\frac{-3-\sqrt{5}}{2}$.

6.
$$(2x+1)^2 = x$$
$$4x^2 + 4x + 1 = x$$
$$4x^2 + 3x = -1$$
$$x^2 + \frac{3}{4}x = -\frac{1}{4}$$
$$x^2 + \frac{3}{4}x + \left(\frac{\frac{3}{4}}{2}\right)^2 = -\frac{1}{4} + \frac{9}{64}$$
$$\left(x + \frac{3}{8}\right)^2 = -\frac{7}{64}$$
$$x + \frac{3}{8} = \pm\sqrt{-\frac{7}{64}}$$
$$x + \frac{3}{8} = \pm\frac{i\sqrt{7}}{8}$$
$$x = -\frac{3}{8} \pm \frac{i\sqrt{7}}{8} = \frac{-3 \pm i\sqrt{7}}{8}$$
The solutions are $\frac{-3+i\sqrt{7}}{8}$ and $\frac{-3-i\sqrt{7}}{8}$.

7.
$$\begin{aligned} A &= P(1+r)^2 \\ 2717 &= 2500(1+r)^2 \\ \frac{2717}{2500} &= (1+r)^2 \\ (1+r)^2 &= 1.0868 \\ 1+r &= \pm\sqrt{1.0868} \\ 1+r &= \pm 1.0425 \\ r &= -1 \pm 1.0425 \\ &= 0.0425 \text{ or } -2.0425 \text{ (disregard)} \end{aligned}$$
The interest rate is 4.25%.

8. Let x = distance traveled.
$$\begin{aligned} a^2+b^2 &= c^2 \\ x^2+x^2 &= (150)^2 \\ 2x^2 &= 22{,}500 \\ x^2 &= 11{,}250 \\ x &= \pm 75\sqrt{2} \approx \pm 106.1 \end{aligned}$$
Disregard the negative. The ships each traveled $75\sqrt{2} \approx 106.1$ miles.

9. Two complex but not real solutions exist.

10. Two real solutions exist.

11. Two real solutions exist.

12. One real solution exists.

13. $x^2 - 16x + 64 = 0$
$a = 1, b = -16, c = 64$
$$\begin{aligned} x &= \frac{16 \pm \sqrt{(-16)^2 - 4(1)(64)}}{2(1)} \\ &= \frac{16 \pm \sqrt{256-256}}{2} \\ &= \frac{16 \pm \sqrt{0}}{2} \\ &= 8 \end{aligned}$$
The solution is 8.

14. $x^2 + 5x = 0$
$a = 1, b = 5, c = 0$
$$\begin{aligned} x &= \frac{-5 \pm \sqrt{(5)^2 - 4(1)(0)}}{2(1)} \\ &= \frac{-5 \pm \sqrt{25}}{2} \\ &= \frac{-5 \pm 5}{2} \\ &= 0 \text{ or } -5 \end{aligned}$$
The solutions are –5 and 0.

15.
$$\begin{aligned} 2x^2 + 3x &= 5 \\ 2x^2 + 3x - 5 &= 0 \end{aligned}$$
$a = 2, b = 3, c = -5$
$$\begin{aligned} x &= \frac{-3 \pm \sqrt{(3)^2 - 4(2)(-5)}}{2(2)} \\ &= \frac{-3 \pm \sqrt{49}}{4} \\ &= \frac{-3 \pm 7}{4} \\ &= 1 \text{ or } -\frac{5}{2} \end{aligned}$$
The solutions are $-\frac{5}{2}$ and 1.

16.
$$\begin{aligned} 9x^2 + 4 &= 2x \\ 9x^2 - 2x + 4 &= 0 \end{aligned}$$
$$\begin{aligned} x &= \frac{2 \pm \sqrt{(-2)^2 - 4(9)(4)}}{2(9)} \\ &= \frac{2 \pm \sqrt{-140}}{18} \\ &= \frac{2 \pm i\sqrt{4 \cdot 35}}{18} \\ &= \frac{2 \pm 2i\sqrt{35}}{18} \\ &= \frac{1 \pm i\sqrt{35}}{9} \end{aligned}$$
The solutions are $\frac{1+i\sqrt{35}}{9}$ and $\frac{1-i\sqrt{35}}{9}$.

17. $6x^2+7=5x$

$$6x^2-5x+7=0$$
$$a=6,\ b=-5,\ c=7$$
$$x=\frac{5\pm\sqrt{(-5)^2-4(6)(7)}}{2(6)}=\frac{5\pm\sqrt{25-168}}{12}=\frac{5\pm\sqrt{-143}}{12}=\frac{5\pm i\sqrt{143}}{12}$$

The solutions are $\frac{5+i\sqrt{143}}{12}$ and $\frac{5-i\sqrt{143}}{12}$.

18. $(2x-3)^2=x$

$$4x^2-12x+9-x=0$$
$$4x^2-13x+9=0$$
$$a=4,\ b=-13,\ c=9$$
$$x=\frac{13\pm\sqrt{(-13)^2-4(4)(9)}}{2(4)}=\frac{13\pm\sqrt{169-144}}{8}=\frac{13\pm\sqrt{25}}{8}=\frac{13\pm5}{8}=\frac{9}{4}\text{ or }1$$

The solutions are 1 and $\frac{9}{4}$.

19. $d(t)=-16t^2+30t+6$

a. $d(1)=-16(1)^2+30(1)+6$
$$=-16+30+6$$
$$=20\text{ feet}$$

b. $-16t^2+30t+6=0$
$$8t^2-15t-3=0$$
$$a=8,\ b=-15,\ c=-3$$
$$t=\frac{15\pm\sqrt{(-15)^2-4(8)(-3)}}{2(8)}=\frac{15\pm\sqrt{225+96}}{16}=\frac{15\pm\sqrt{321}}{16}$$

Disregarding the negative, we have

$$t=\frac{15+\sqrt{321}}{16}\text{ seconds}\approx 2.1\text{ seconds.}$$

20. Let x = length of the legs. Then $x+6$ = length of the hypotenuse.

$$x^2+x^2=(x+6)^2$$
$$2x^2=x^2+12x+36$$
$$x^2-12x-36=0$$
$$a=1,\ b=-12,\ c=-36$$
$$x=\frac{12\pm\sqrt{(-12)^2-4(1)(-36)}}{2(1)}=\frac{12\pm\sqrt{144+144}}{2}=\frac{12\pm\sqrt{144\cdot2}}{2}=\frac{12\pm12\sqrt{2}}{2}=6\pm6\sqrt{2}$$

Disregard the negative. The length of each leg is $\left(6+6\sqrt{2}\right)$ cm.

21.
$$x^3 = 27$$
$$x^3 - 27 = 0$$
$$(x-3)(x^2+3x+9) = 0$$
$$x-3=0 \text{ or } x^2+3x+9=0$$
$$x = 3 \qquad a=1, b=3, c=9$$
$$x = \frac{-3 \pm \sqrt{(3)^2 - 4(1)(9)}}{2(1)} = \frac{-3\pm\sqrt{9-36}}{2} = \frac{-3\pm\sqrt{-27}}{2} = \frac{-3\pm 3i\sqrt{3}}{2}$$

The solutions are 3, $\frac{-3+3i\sqrt{3}}{2}$, and $\frac{-3-3i\sqrt{3}}{2}$.

22.
$$y^3 = -64$$
$$y^3 + 64 = 0$$
$$(y+4)(y^2-4y+16) = 0$$
$$y+4=0 \text{ or } y^2-4y+16=0$$
$$y=-4 \qquad a=1, b=-4, c=16$$
$$y = \frac{4\pm\sqrt{(-4)^2-4(1)(16)}}{2(1)} = \frac{4\pm\sqrt{16-64}}{2} = \frac{4\pm\sqrt{-48}}{2} = \frac{4\pm 4i\sqrt{3}}{2} = 2\pm 2i\sqrt{3}$$

The solutions are –4, $2+2i\sqrt{3}$, and $2-2i\sqrt{3}$.

23.
$$\frac{5}{x} + \frac{6}{x-2} = 3$$
$$x(x-2)\left(\frac{5}{x}+\frac{6}{x-2}\right) = 3x(x-2)$$
$$5(x-2)+6x = 3x^2-6x$$
$$5x-10+6x = 3x^2-6x$$
$$0 = 3x^2-17x+10$$
$$0 = (3x-2)(x-5)$$
$$3x-2=0 \text{ or } x-5=0$$
$$x = \frac{2}{3} \text{ or } x=5$$

The solutions are $\frac{2}{3}$ and 5.

24.
$$x^4-21x^2-100=0$$
$$(x^2-25)(x^2+4)=0$$
$$(x+5)(x-5)(x^2+4)=0$$
$$x+5=0 \text{ or } x-5=0 \text{ or } x^2+4=0$$
$$x=-5 \text{ or } x=5 \text{ or } x^2=-4$$
$$x=\pm 2i$$

The solutions are –5, 5 –2*i*, and 2*i*.

25. $x^{2/3} - 6x^{1/3} + 5 = 0$

Let $y = x^{1/3}$. Then $y^2 = x^{2/3}$ and
$$y^2-6y+5=0$$
$$(y-5)(y-1)=0$$
$$y-5=0 \text{ or } y-1=0$$
$$y=5 \text{ or } y=1$$
$$x^{1/3}=5 \text{ or } x^{1/3}=1$$
$$x=125 \text{ or } x=1$$

The solutions are 1 and 125.

26.
$$5(x+3)^2-19(x+3)=4$$
$$5(x+3)^2-19(x+3)-4=0$$

Let $y = x + 3$. Then $y^2=(x+3)^2$ and
$$5y^2-19y-4=0$$
$$(5y+1)(y-4)=0$$

$$5y+1=0 \quad \text{or} \quad y-4=0$$
$$y=-\frac{1}{5} \quad \text{or} \quad y=4$$
$$x+3=-\frac{1}{5} \quad \text{or} \quad x+3=4$$
$$x=-\frac{16}{5} \quad \text{or} \quad x=1$$

The solutions are $-\frac{16}{5}$ and 1.

27.
$$a^6-a^2=a^4-1$$
$$a^6-a^4-a^2+1=0$$
$$a^4(a^2-1)-1(a^2-1)=0$$
$$(a^2-1)(a^4-1)=0$$
$$(a+1)(a-1)(a^2+1)(a^2-1)=0$$
$$(a+1)(a-1)(a^2+1)(a+1)(a-1)=0$$
$$(a+1)^2(a-1)^2(a^2+1)=0$$
$$(a+1)^2=0 \text{ or } (a-1)^2=0 \text{ or } a^2+1=0$$
$$a+1=0 \text{ or } a-1=0 \text{ or } a^2=-1$$
$$a=-1 \text{ or } a=1 \text{ or } a=\pm i$$

The solutions are –1, 1, –*i*, and *i*.

28. $y^{-2}+y^{-1}=20$
$$\frac{1}{y^2}+\frac{1}{y}=20$$
$$1+y=20y^2$$
$$0=20y^2-y-1$$
$$0=(5y+1)(4y-1)$$
$$5y+1=0 \quad \text{or} \quad 4y-1=0$$
$$y=-\frac{1}{5} \quad \text{or} \quad y=\frac{1}{4}$$

The solutions are $-\frac{1}{5}$ and $\frac{1}{4}$.

29. Let *x* = time for Jerome alone. Then *x* – 1 = time for Tim alone.
$$\frac{1}{x}+\frac{1}{x-1}=\frac{1}{5}$$
$$5(x-1)+5x=x(x-1)$$
$$5x-5+5x=x^2-x$$
$$0=x^2-11x+5$$
$$a=1, b=-11, c=5$$
$$x=\frac{11\pm\sqrt{(-11)^2-4(1)(5)}}{2(1)}$$
$$=\frac{11\pm\sqrt{101}}{2}$$
$$\approx 0.475 \text{ (disregard) or } 10.525$$

Jerome: 10.5 hours
Tim: 9.5 hours

30. Let *x* = the number; then $\frac{1}{x}$ = the reciprocal of the number.
$$x-\frac{1}{x}=-\frac{24}{5}$$
$$5x\left(x-\frac{1}{x}\right)=5x\left(-\frac{24}{5}\right)$$
$$5x^2-5=-24x$$
$$5x^2+24x-5=0$$
$$(5x-1)(x+5)=0$$
$$5x-1=0 \quad \text{or} \quad x+5=0$$
$$x=\frac{1}{5} \quad \text{or} \quad x=-5$$

Disregard the positive value as extraneous. The number is –5.

31.
$$2x^2-50\le 0$$
$$2(x^2-25)\le 0$$
$$2(x+5)(x-5)\le 0$$
$$x+5=0 \quad \text{or} \quad x-5=0$$
$$x=-5 \quad \text{or} \quad x=5$$

Region	Test Point	$2(x+5)(x-5)\le 0$ Result
A: $(-\infty, -5]$	–6	$2(-1)(-11)\le 0$ False
B: $[-5, 5]$	0	$2(5)(-5)\le 0$ True
C: $[5, \infty)$	6	$2(11)(1)\le 0$ False

Solution: [–5, 5]

32.
$$\frac{1}{4}x^2 < \frac{1}{16}$$
$$x^2 < \frac{1}{4}$$
$$x^2 - \frac{1}{4} < 0$$
$$\left(x+\frac{1}{2}\right)\left(x-\frac{1}{2}\right) < 0$$
$$x+\frac{1}{2} = 0 \quad \text{or} \quad x-\frac{1}{2} = 0$$
$$x = -\frac{1}{2} \quad \text{or} \quad x = \frac{1}{2}$$

Region	Test Point	$\left(x+\frac{1}{2}\right)\left(x-\frac{1}{2}\right) < 0$ Result
A: $\left(-\infty, -\frac{1}{2}\right)$	-1	$\left(-\frac{1}{2}\right)\left(-\frac{3}{2}\right) < 0$ False
B: $\left(-\frac{1}{2}, \frac{1}{2}\right)$	0	$\left(\frac{1}{2}\right)\left(-\frac{1}{2}\right) < 0$ True
C: $\left(\frac{1}{2}, \infty\right)$	1	$\left(\frac{3}{2}\right)\left(\frac{1}{2}\right) < 0$ False

Solution: $\left(-\frac{1}{2}, \frac{1}{2}\right)$

33.
$$(x^2 - 4)(x^2 - 25) \le 0$$
$$(x+2)(x-2)(x+5)(x-5) \le 0$$
$$x+2=0 \quad \text{or} \quad x-2=0 \quad \text{or} \quad x+5=0 \quad \text{or} \quad x-5=0$$
$$x=-2 \quad \text{or} \quad x=2 \quad \text{or} \quad x=-5 \quad \text{or} \quad x=5$$

Region	Test Point	$(x^2 - 4)(x^2 - 25) \le 0$ Result
A: $(-\infty, -5)$	-6	$352 \le 0$ False
B: $(-5, -2)$	-3	$-80 \le 0$ True
C: $(-2, 2)$	0	$100 \le 0$ False
D: $(2, 5)$	3	$-80 \le 0$ True
E: $(5, \infty)$	6	$352 \le 0$ False

Solution: $[-5, -2] \cup [2, 5]$

34. $(x^2-16)(x^2-1)>0$
$(x+4)(x-4)(x+1)(x-1)>0$
$x+4=0$ or $x-4=0$ or $x+1=0$ or $x-1=0$
$x=-4$ or $x=4$ or $x=-1$ or $x=1$

Region	Test Point	$(x+4)(x-4)(x+1)(x-1)>0$ Result
A: $(-\infty, -4)$	-5	$(-1)(-9)(-4)(-6)>0$ True
B: $(-4, -1)$	-2	$(2)(-6)(-1)(-3)>0$ False
C: $(-1, 1)$	0	$(4)(-4)(1)(-1)>0$ True
D: $(1, 4)$	2	$(6)(-2)(3)(1)>0$ False
E: $(4, \infty)$	5	$(9)(1)(6)(4)>0$ True

Solution: $(-\infty, -4)\cup(-1, 1)\cup(4, \infty)$

35. $\dfrac{x-5}{x-6}<0$
$x-5=0$ or $x-6=0$
$x=5$ or $x=6$

Region	Test Point	$\dfrac{x-5}{x-6}<0$ Result
A: $(-\infty, 5)$	0	$\dfrac{-5}{-6}<0$ False
B: $(5, 6)$	$\dfrac{11}{2}$	$\dfrac{\frac{1}{2}}{-\frac{1}{2}}<0$ True
C: $(6, \infty)$	7	$\dfrac{2}{1}<0$ False

Solution: $(5, 6)$

36. $\frac{(4x+3)(x-5)}{x(x+6)} > 0$

$4x+3=0,\ x-5=0,\ x=0,\ \text{or } x+6=0$

$x=-\frac{3}{4},\ x=5,\ x=0,\ \text{or } x=-6$

Region	Test Point	$\frac{(4x+3)(x-5)}{x(x+6)} > 0$ Result
A: $(-\infty, -6)$	-7	$\frac{(-25)(-12)}{-7(-1)} > 0$ True
B: $\left(-6, -\frac{3}{4}\right)$	-3	$\frac{(-9)(-8)}{-3(3)} > 0$ False
C: $\left(-\frac{3}{4}, 0\right)$	$-\frac{1}{2}$	$\frac{(1)\left(-\frac{11}{2}\right)}{-\frac{1}{2}\left(\frac{11}{2}\right)} > 0$ True
D: $(0, 5)$	1	$\frac{(7)(-4)}{1(7)} > 0$ False
E: $(5, \infty)$	6	$\frac{(27)(1)}{6(12)} > 0$ True

Solution: $(-\infty, -6) \cup \left(-\frac{3}{4}, 0\right) \cup (5, \infty)$

37. $(x+5)(x-6)(x+2) \le 0$

$x+5=0 \quad \text{or } x-6=0 \quad \text{or } x+2=0$

$x=-5 \quad \text{or} \quad x=6 \quad \text{or} \quad x=-2$

Region	Test Point	$(x+5)(x-6)(x+2) \le 0$ Result
A: $(-\infty, -5]$	-6	$(-1)(-12)(-4) \le 0$ True
B: $[-5, -2]$	-3	$(2)(-9)(-1) \le 0$ False
C: $[-2, 6]$	0	$(5)(-6)(2) \le 0$ True
D: $[6, \infty)$	7	$(12)(1)(9) \le 0$ False

Solution: $(-\infty, -5] \cup [-2, 6]$

38. $x^3+3x^2-25x-75 > 0$

$x^2(x+3)-25(x+3) > 0$

$(x+3)(x^2-25) > 0$

$(x+3)(x+5)(x-5) > 0$

$x+3=0 \quad \text{or } x+5=0 \quad \text{or } x-5=0$

$x=-3 \quad \text{or} \quad x=-5 \quad \text{or} \quad x=5$

Region	Test Point	$(x+3)(x+5)(x-5) > 0$ Result
A: $(-\infty, -5)$	-6	$(-3)(-1)(-11) > 0$ False
B: $(-5, -3)$	-4	$(-1)(1)(-9) > 0$ True
C: $(-3, 5)$	0	$(3)(5)(-5) > 0$ False
D: $(5, \infty)$	6	$(9)(11)(1) > 0$ True

Solution: $(-5, -3) \cup (5, \infty)$

39. $\frac{x^2+4}{3x} \le 1$

The denominator equals 0 when $3x = 0$, or $x = 0$.

$\frac{x^2+4}{3x} = 1$

$x^2+4 = 3x$

$x^2-3x+4=0$

$x=\frac{3\pm\sqrt{(-3)^2-4(1)(4)}}{2(1)}=\frac{3\pm\sqrt{-7}}{2}$

which yields non-real solutions.

Region	Test Point	$\frac{x^2+4}{3x} \le 1$ Result
A: $(-\infty, 0)$	-1	$\frac{5}{-3} \le 1$ True
B: $(0, \infty)$	1	$\frac{5}{3} \le 1$ False

Solution: $(-\infty, 0)$

40. $\dfrac{3}{x-2} > 2$

The denominator is equal to 0 when $x - 2 = 0$, or $x = 2$.

$$\frac{3}{x-2} = 2$$
$$3 = 2(x-2)$$
$$3 = 2x - 4$$
$$7 = 2x$$
$$\frac{7}{2} = x$$

Region	Test Point	$\dfrac{3}{x-2} > 2$ Result
A: $(-\infty, 2)$	0	$\dfrac{3}{-2} > 2$ False
B: $\left(2, \dfrac{7}{2}\right)$	3	$\dfrac{3}{1} > 2$ True
C: $\left(\dfrac{7}{2}, \infty\right)$	5	$\dfrac{3}{3} > 2$ False

Solution: $\left(2, \dfrac{7}{2}\right)$

41. $f(x) = x^2 - 4$

Vertex: (0, –4)

Axis of symmetry: $x = 0$

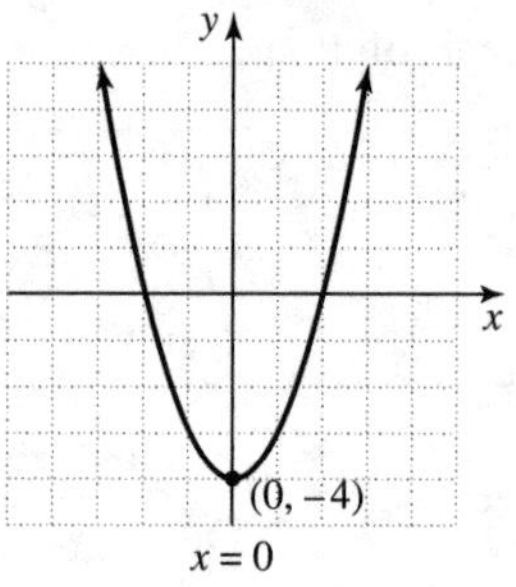

42. $g(x) = x^2 + 7$

Vertex: (0, 7)

Axis of symmetry: $x = 0$

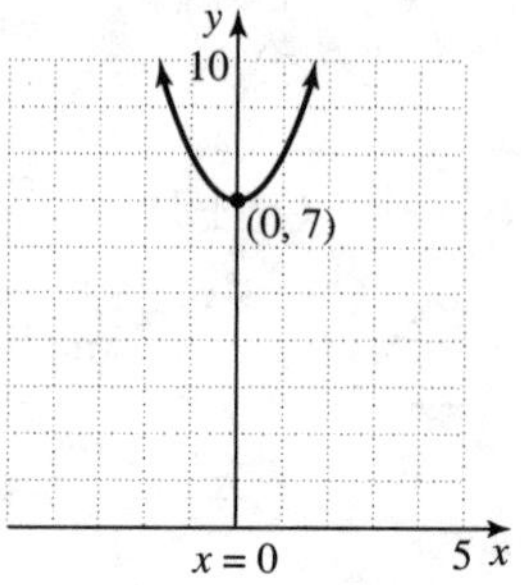

43. $H(x) = 2x^2$

Vertex: (0, 0)

Axis of symmetry: $x = 0$

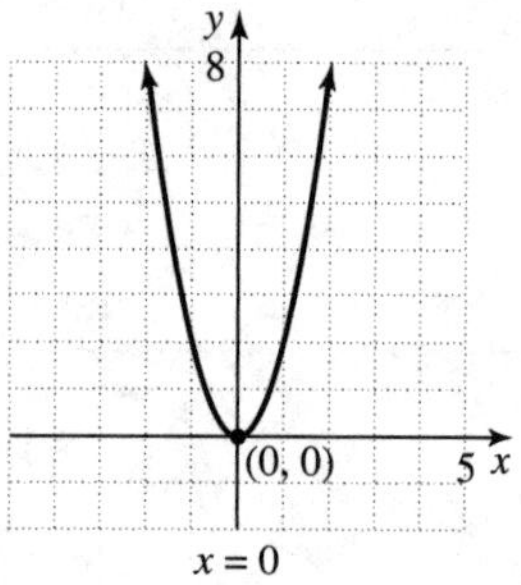

44. $h(x) = -\dfrac{1}{3}x^2$

Vertex: (0, 0)

Axis of symmetry: $x = 0$

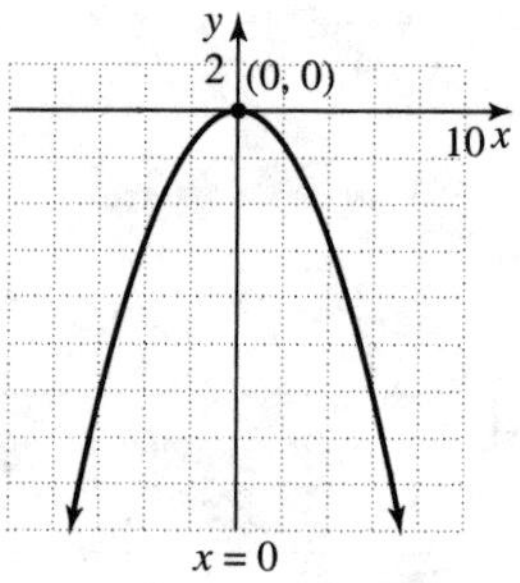

45. $F(x) = (x-1)^2$
Vertex: (1, 0)
Axis of symmetry: $x = 1$

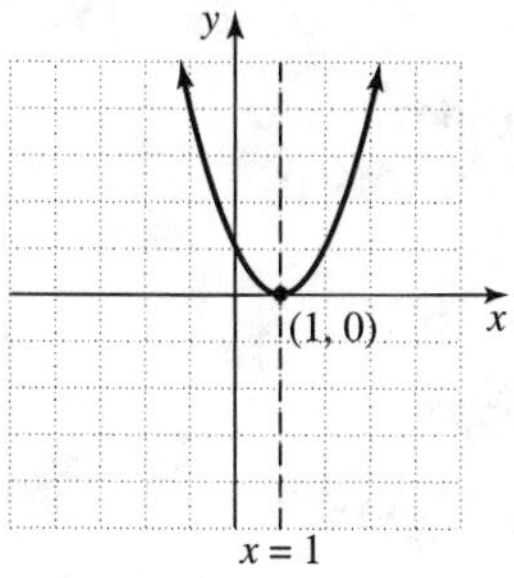

46. $G(x) = (x+5)^2$
Vertex: (–5, 0)
Axis of symmetry: $x = -5$

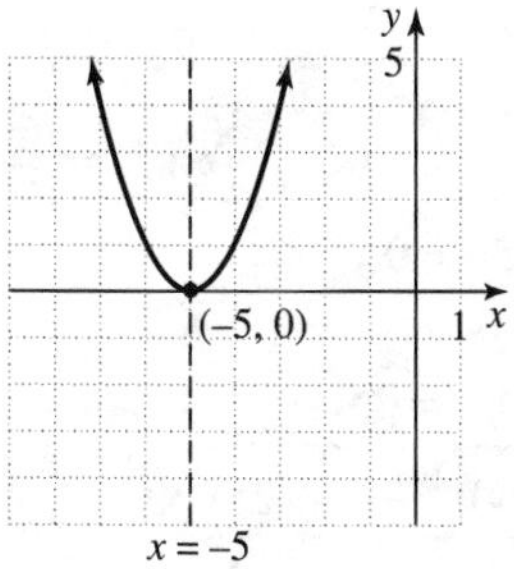

47. $f(x) = (x-4)^2 - 2$
Vertex: (4, –2)
Axis of symmetry: $x = 4$

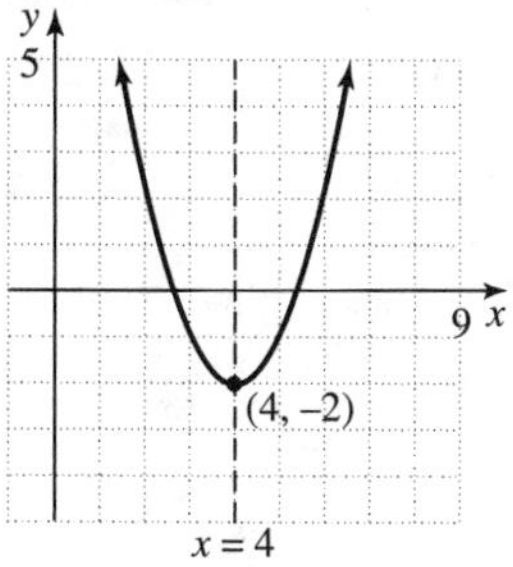

48. $f(x) = -3(x-1)^2 + 1$
Vertex: (1, 1)
Axis of symmetry: $x = 1$

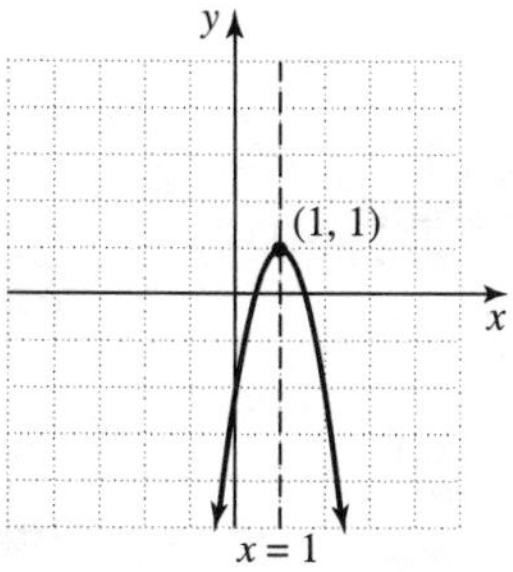

49. $f(x) = x^2 + 10x + 25$

$$x = -\frac{b}{2a} = \frac{-10}{2(1)} = -5$$

$f(-5) = (-5)^2 + 10(-5) + 25 = 0$
Vertex: (–5, 0)

$$x^2 + 10x + 25 = 0$$
$$(x+5)^2 = 0$$
$$x + 5 = 0$$
$$x = -5$$

x-intercept: (–5, 0)
$f(0) = 25$ so the y-intercept is (0, 25).

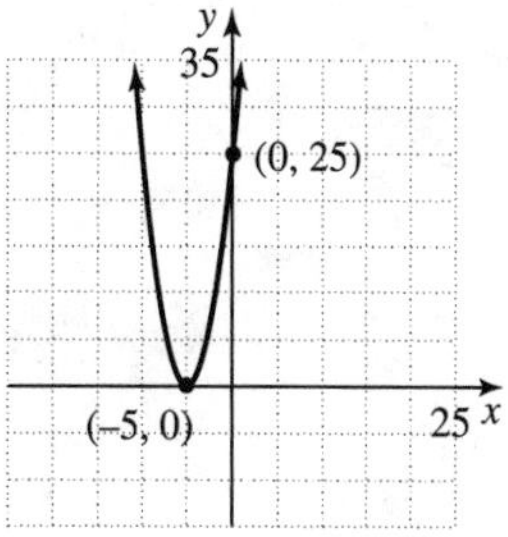

50. $f(x) = -x^2 + 6x - 9$

$$x = -\frac{b}{2a} = \frac{-6}{2(-1)} = 3$$

$f(3) = -(3)^2 + 6(3) - 9 = 0$
Vertex: (3, 0)
x-intercept: (3, 0)
$f(0) = -9$
y-intercept: (0, –9)

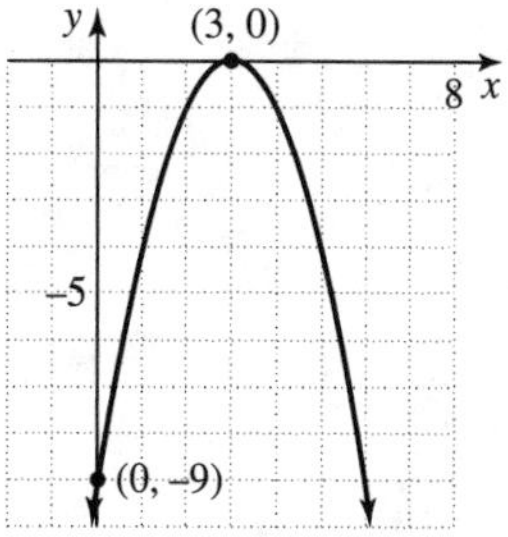

51. $f(x) = 4x^2 - 1$

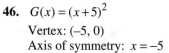

$$x = -\frac{b}{2a} = \frac{-0}{2(4)} = 0$$

$f(0) = 4(0)^2 - 1 = -1$
Vertex: (0, –1)

$$4x^2 - 1 = 0$$
$$(2x+1)(2x-1) = 0$$
$$x = -\frac{1}{2} \text{ or } x = \frac{1}{2}$$

x-intercepts: $\left(-\frac{1}{2}, 0\right), \left(\frac{1}{2}, 0\right)$

$f(0) = -1$

y-intercept: (0, −1)

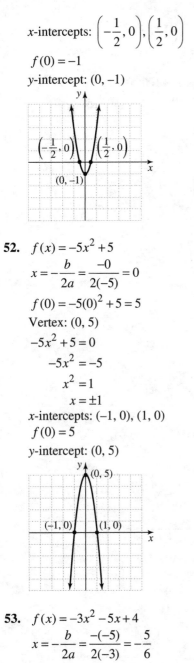

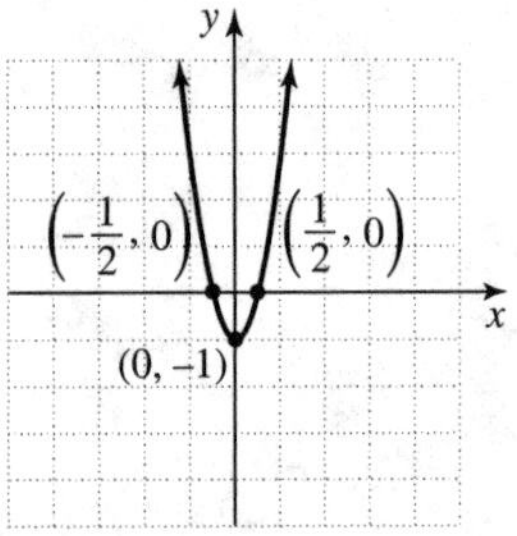

52. $f(x) = -5x^2 + 5$

$x = -\frac{b}{2a} = \frac{-0}{2(-5)} = 0$

$f(0) = -5(0)^2 + 5 = 5$

Vertex: (0, 5)

$-5x^2 + 5 = 0$

$-5x^2 = -5$

$x^2 = 1$

$x = \pm 1$

x-intercepts: (−1, 0), (1, 0)

$f(0) = 5$

y-intercept: (0, 5)

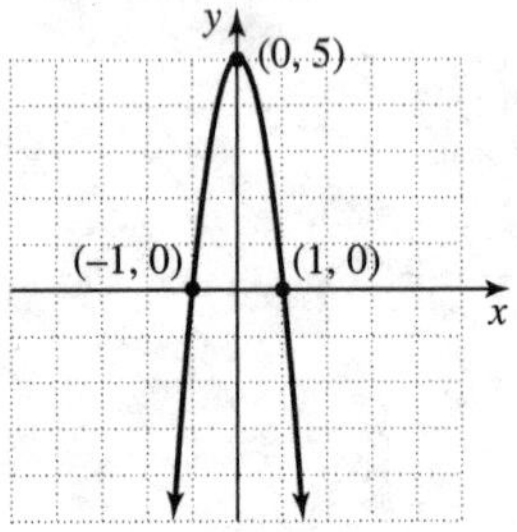

53. $f(x) = -3x^2 - 5x + 4$

$x = -\frac{b}{2a} = \frac{-(-5)}{2(-3)} = -\frac{5}{6}$

$f\left(-\frac{5}{6}\right) = -3\left(-\frac{5}{6}\right)^2 - 5\left(-\frac{5}{6}\right) + 4 = \frac{73}{12}$

Vertex: $\left(-\frac{5}{6}, \frac{73}{12}\right)$

The graph opens downward ($a = -3 < 0$).

$f(0) = 4 \Rightarrow y$-intercept: (0, 4)

$-3x^2 - 5x + 4 = 0$

$x = \frac{5 \pm \sqrt{(-5)^2 - 4(-3)(4)}}{2(-3)}$

$= \frac{5 \pm \sqrt{73}}{-6}$

≈ -2.2573 or 0.5907

x-intercepts: (−2.3, 0), (0.6, 0)

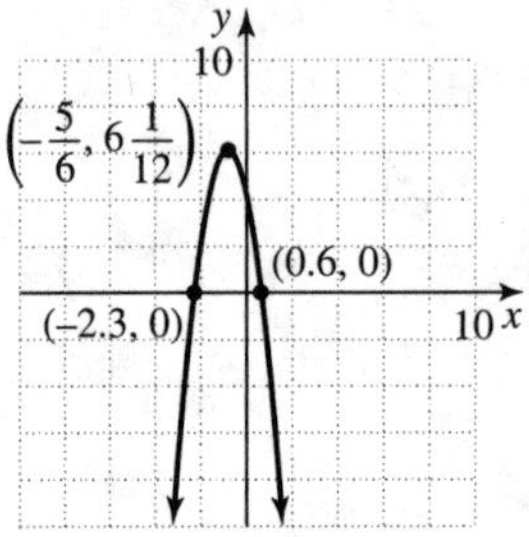

54. $h(t) = -16t^2 + 120t + 300$

a. $350 = -16t^2 + 120t + 300$

$16t^2 - 120t + 50 = 0$

$8t^2 - 60t + 25 = 0$

$a = 8, b = -60, c = 25$

$t = \frac{60 \pm \sqrt{(-60)^2 - 4(8)(25)}}{2(8)}$

$= \frac{60 \pm \sqrt{2800}}{16}$

≈ 0.4 second and 7.1 seconds

b. answers may vary

55. Let x = one number; then

$420 - x$ = the other number.

Let $f(x)$ represent their product.

$f(x) = x(420 - x)$

$= 420x - x^2$

$= -x^2 + 420x$

$x = -\frac{b}{2a} = \frac{-420}{2(-1)} = 210;$

$420 - x = 420 - 210 = 210$

Therefore, the numbers are both 210.

56. $y = a(x-h)^2 + k$

vertex (−3, 7) with $a = -\frac{7}{9}$ gives

$y = -\frac{7}{9}(x+3)^2 + 7$.

57. $x^2 - x - 30 = 0$

$(x+5)(x-6) = 0$

$x+5 = 0$ or $x-6 = 0$

$x = -5$ or $x = 6$

The solutions are –5 and 6.

58. $10x^2 = 3x + 4$

$10x^2 - 3x - 4 = 0$

$(5x-4)(2x+1) = 0$

$5x-4 = 0$ or $2x+1 = 0$

$5x = 4$ or $2x = -1$

$x = \frac{4}{5}$ or $x = -\frac{1}{2}$

The solutions are $-\frac{1}{2}$ and $\frac{4}{5}$.

59. $9y^2 = 36$

$y^2 = 4$

$y = \pm\sqrt{4}$

$y = \pm 2$

The solutions are –2 and 2.

60. $(9n+1)^2 = 9$

$9n+1 = \pm\sqrt{9}$

$9n+1 = \pm 3$

$9n = -1 \pm 3$

$n = \frac{-1\pm 3}{9} = \frac{2}{9}, -\frac{4}{9}$

The solutions are $-\frac{4}{9}$ and $\frac{2}{9}$.

61. $x^2 + x + 7 = 0$

$x^2 + x = -7$

$x^2 + x + \left(\frac{1}{2}\right)^2 = -7 + \frac{1}{4}$

$\left(x + \frac{1}{2}\right)^2 = -\frac{27}{4}$

$x + \frac{1}{2} = \pm\sqrt{-\frac{27}{4}}$

$x + \frac{1}{2} = \pm\frac{i\sqrt{9\cdot 3}}{2}$

$x + \frac{1}{2} = \pm\frac{3i\sqrt{3}}{2}$

$x = -\frac{1}{2} \pm \frac{3i\sqrt{3}}{2} = \frac{-1 \pm 3i\sqrt{3}}{2}$

The solutions are $\frac{-1+3i\sqrt{3}}{2}$ and $\frac{-1-3i\sqrt{3}}{2}$.

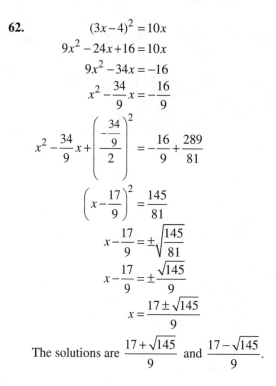

62. $(3x-4)^2 = 10x$

$9x^2 - 24x + 16 = 10x$

$9x^2 - 34x = -16$

$x^2 - \frac{34}{9}x = -\frac{16}{9}$

$x^2 - \frac{34}{9}x + \left(\frac{-\frac{34}{9}}{2}\right)^2 = -\frac{16}{9} + \frac{289}{81}$

$\left(x - \frac{17}{9}\right)^2 = \frac{145}{81}$

$x - \frac{17}{9} = \pm\sqrt{\frac{145}{81}}$

$x - \frac{17}{9} = \pm\frac{\sqrt{145}}{9}$

$x = \frac{17 \pm \sqrt{145}}{9}$

The solutions are $\frac{17+\sqrt{145}}{9}$ and $\frac{17-\sqrt{145}}{9}$.

63. $x^2+11=0$

$a=1, b=0, c=11$

$$x=\frac{0\pm\sqrt{(0)^2-4(1)(11)}}{2(1)}$$
$$=\frac{\pm\sqrt{-44}}{2}$$
$$=\frac{\pm 2i\sqrt{11}}{2}$$
$$=\pm i\sqrt{11}$$

The solutions are $-i\sqrt{11}$ and $i\sqrt{11}$.

64. $x^2+7=0$

$$x^2=-7$$
$$\sqrt{x^2}=\pm\sqrt{-7}$$
$$x=\pm i\sqrt{7}$$

The solutions are $-i\sqrt{7}$ and $i\sqrt{7}$.

65. $(5a-2)^2-a=0$

$$25a^2-20a+4-a=0$$
$$25a^2-21a+4=0$$
$$a=\frac{21\pm\sqrt{(-21)^2-4(25)(4)}}{2(25)}$$
$$=\frac{21\pm\sqrt{441-400}}{50}$$
$$=\frac{21\pm\sqrt{41}}{50}$$

The solutions are $\frac{21+\sqrt{41}}{50}$ and $\frac{21-\sqrt{41}}{50}$.

66. $\frac{7}{8}=\frac{8}{x^2}$

$$7x^2=64$$
$$x^2=\frac{64}{7}$$
$$x=\pm\sqrt{\frac{64}{7}}$$
$$x=\pm\frac{8}{\sqrt{7}}=\pm\frac{8\cdot\sqrt{7}}{\sqrt{7}\cdot\sqrt{7}}=\pm\frac{8\sqrt{7}}{7}$$

The solutions are $-\frac{8\sqrt{7}}{7}$ and $\frac{8\sqrt{7}}{7}$.

67. $x^{2/3}-6x^{1/3}=-8$

$$x^{2/3}-6x^{1/3}+8=0$$

Let $y=x^{1/3}$. Then $y^2=x^{2/3}$ and

$$y^2-6y+8=0$$
$$(y-4)(y-2)=0$$
$$y-4=0 \quad\text{or}\quad y-2=0$$
$$y=4 \quad\text{or}\quad y=2$$
$$x^{1/3}=4 \quad\text{or}\quad x^{1/3}=2$$
$$x=64 \quad\text{or}\quad x=8$$

The solutions are 8 and 64.

68. $(2x-3)(4x+5)\geq 0$

$$2x-3=0 \quad\text{or}\quad 4x+5=0$$
$$x=\frac{3}{2} \quad\text{or}\quad x=-\frac{5}{4}$$

Region	Test Point	$(2x-3)(4x+5)\geq 0$ Result
$A: \left(-\infty, -\frac{5}{4}\right]$	-2	$(-7)(-3)\geq 0$ True
$B: \left[-\frac{5}{4}, \frac{3}{2}\right]$	0	$(-3)(5)\geq 0$ False
$C: \left[\frac{3}{2}, \infty\right)$	3	$(3)(17)\geq 0$ True

Solution: $\left(-\infty, -\frac{5}{4}\right]\cup\left[\frac{3}{2}, \infty\right)$

69. $\dfrac{x(x+5)}{4x-3} \ge 0$

$x = 0$ or $x+5=0$ or $4x-3=0$

$x=-5$ or $x=\dfrac{3}{4}$

Region	Test Point	$\dfrac{x(x+5)}{4x-3} \ge 0$ Result
A: $(-\infty, -5]$	-6	$\dfrac{-6(-1)}{-27} \ge 0$ False
B: $[-5, 0]$	-1	$\dfrac{-1(4)}{-7} \ge 0$ True
C: $\left[0, \dfrac{3}{4}\right)$	$\dfrac{1}{2}$	$\dfrac{\frac{1}{2}\left(\frac{11}{2}\right)}{-1} \ge 0$ False
D: $\left(\dfrac{3}{4}, \infty\right)$	1	$\dfrac{1(6)}{1} \ge 0$ True

Solution: $[-5, 0] \cup \left(\dfrac{3}{4}, \infty\right)$

70. $\dfrac{3}{x-2} > 2$

The denominator is equal to 0 when $x - 2 = 0$, or $x = 2$.

$$\frac{3}{x-2} = 2$$
$$3 = 2(x-2)$$
$$3 = 2x-4$$
$$7 = 2x$$
$$\frac{7}{2} = x$$

Region	Test Point	$\dfrac{3}{x-2} > 2$ Result
A: $(-\infty, 2)$	0	$\dfrac{3}{-2} > 2$ False
B: $\left(2, \dfrac{7}{2}\right)$	3	$\dfrac{3}{1} > 2$ True
C: $\left(\dfrac{7}{2}, \infty\right)$	5	$\dfrac{3}{3} > 2$ False

Solution: $\left(2, \dfrac{7}{2}\right)$

71. $y = -111x^2 + 1960x + 85{,}907$

a. 2020 is 15 years after 2005, so $x = 15$.

$y = -111(15)^2 + 1960(15) + 85{,}907 = 90{,}332$

The passenger traffic at Atlanta's Hartsfield-Jackson International Airport is predicted to be 90,332 thousand passengers in 2020.

b. no; answers may vary

Chapter 8 Getting Ready for the Test

1.
$$x^2 = 8$$
$$x^2 - 8 = 0$$
$$1x^2 + 0x - 8 = 0$$

$a = 1$, $b = 0$, $c = -8$; C

2.
$$\frac{1}{9}x^2 + \frac{1}{3} = x$$
$$\frac{1}{9}x^2 - x + \frac{1}{3} = 0$$
$$9\left(\frac{1}{9}x^2 - x + \frac{1}{3}\right) = 0$$
$$1x^2 - 9x + 3 = 0$$

$a = 1$, $b = -9$, $c = 3$; B

3. $\dfrac{-4\pm\sqrt{-4}}{2}=\dfrac{-4\pm i\sqrt{4}}{2}$
$=\dfrac{-4\pm 2i}{2}$
$=\dfrac{2(-2\pm i)}{2}$
$=-2\pm i$; A

4. $\dfrac{9\pm\sqrt{27}}{3}=\dfrac{9\pm\sqrt{9\cdot 3}}{3}$
$=\dfrac{9\pm 3\sqrt{3}}{3}$
$=\dfrac{3\left(3\pm\sqrt{3}\right)}{3}$
$=3\pm\sqrt{3}$; D

5. To use the completing the square method, the coefficient of the squared term must be 1, so the next step in solving $2x^2+16x=5$ by this method is to divide both sides of the equation by 2.

$2x^2+16x=5$
$x^2+8x=\dfrac{5}{2}$

Choice B is the next correct step.

6. To solve $x^2-7x=5$ by completing the square, the next step is to add the square of $\frac{1}{2}$ of the coefficient of x to both sides. The coefficient of x is −7.

$$\left[\frac{1}{2}(-7)\right]^2=\left(-\frac{7}{2}\right)^2=\frac{49}{4}$$

$x^2-7x=5$
$x^2-7x+\dfrac{49}{4}=5+\dfrac{49}{4}$

Choice D is the next correct step.

7. The difference between the solution set of $(x+5)(x-1)\le 0$ and the solution set of $\dfrac{x+5}{x-1}\le 0$ is that the solution set of $\dfrac{x+5}{x-1}\le 0$ cannot include any value that makes a denominator 0. Thus, the solution set of $\dfrac{x+5}{x-1}\le 0$ is $[-5, 1)$; C.

8. $f(x)=a(x-h)^2+k$
$f(x)=-103(x-20)^2+5.6$

$a=-103$, so the parabola opens downward.
$h=20$, $k=5.6$, so the vertex is (20, 5.6).
The correct description is C.

9. $f(x)=a(x-h)^2+k$
$f(x)=0.5(x+1)^2-3=0.5[x-(-1)]^2+(-3)$

$a=0.5$, so the parabola opens upward.
$h=-1$, $k=-3$, so the vertex is (−1, −3).
The correct description is A.

10.
$$f(x)=3x^2+12x-7$$
$$y=3x^2+12x-7$$
$$y+7=3x^2+12x$$
$$y+7=3(x^2+4x)$$
$$y+7+3(4)=3(x^2+4x+4)$$
$$y+19=3(x+2)^2$$
$$y=3(x+2)^2-19$$
$$f(x)=3(x+2)^2-19$$
$$f(x)=3[x-(-2)]^2+(-19)$$

$h=-2$ and $k=-19$, so the vertex is (−2, −19); B.

11. The x-intercepts occur when $f(x)$ (or y) is 0.

$0=2x^2-x-10$
$0=(2x-5)(x+2)$
$2x-5=0$ or $x+2=0$
$x=\dfrac{5}{2}$ $\quad x=-2$

The x-intercepts are $\left(\dfrac{5}{2}, 0\right)$ and (−2, 0); C.

Chapter 8 Test

1.
$$5x^2-2x=7$$
$$5x^2-2x-7=0$$
$$(5x-7)(x+1)=0$$
$5x-7=0$ or $x+1=0$
$x=\dfrac{7}{5}$ or $x=-1$

The solutions are −1 and $\dfrac{7}{5}$.

2. $(x+1)^2 = 10$

$x+1 = \pm\sqrt{10}$

$x = -1 \pm \sqrt{10}$

The solutions are $-1+\sqrt{10}$ and $-1-\sqrt{10}$.

3. $m^2 - m + 8 = 0$

$a = 1, b = -1, c = 8$

$$m = \frac{1 \pm \sqrt{(-1)^2 - 4(1)(8)}}{2(1)} = \frac{1 \pm \sqrt{1-32}}{2} = \frac{1 \pm \sqrt{-31}}{2} = \frac{1 \pm i\sqrt{31}}{2}$$

The solutions are $\frac{1+i\sqrt{31}}{2}$ and $\frac{1-i\sqrt{31}}{2}$.

4. $u^2 - 6u + 2 = 0$

$a = 1, b = -6, c = 2$

$$u = \frac{-(-6) \pm \sqrt{(-6)^2 - 4(1)(2)}}{2(1)} = \frac{6 \pm \sqrt{36-8}}{2} = \frac{6 \pm \sqrt{28}}{2} = \frac{6 \pm 2\sqrt{7}}{2} = 3 \pm \sqrt{7}$$

The solutions are $3+\sqrt{7}$ and $3-\sqrt{7}$.

5. $7x^2 + 8x + 1 = 0$

$(7x+1)(x+1) = 0$

$7x+1 = 0$ or $x+1 = 0$

$7x = -1$ $\quad$ $x = -1$

$x = -\frac{1}{7}$

The solutions are $-\frac{1}{7}$ and -1.

6. $y^2 - 3y = 5$

$y^2 - 3y - 5 = 0$

$a = 1, b = -3, c = -5$

$$y = \frac{3 \pm \sqrt{(-3)^2 - 4(1)(-5)}}{2(1)} = \frac{3 \pm \sqrt{9+20}}{2} = \frac{3 \pm \sqrt{29}}{2}$$

The solutions are $\frac{3+\sqrt{29}}{2}$ and $\frac{3-\sqrt{29}}{2}$.

7. $\frac{4}{x+2} + \frac{2x}{x-2} = \frac{6}{x^2-4}$

$\frac{4}{x+2} + \frac{2x}{x-2} = \frac{6}{(x+2)(x-2)}$

$4(x-2) + 2x(x+2) = 6$

$4x - 8 + 2x^2 + 4x = 6$

$2x^2 + 8x - 14 = 0$

$x^2 + 4x - 7 = 0$

$a = 1, b = 4, c = -7$

$$x = \frac{-4 \pm \sqrt{(4)^2 - 4(1)(-7)}}{2(1)} = \frac{-4 \pm \sqrt{16+28}}{2} = \frac{-4 \pm \sqrt{44}}{2} = \frac{-4 \pm 2\sqrt{11}}{2} = -2 \pm \sqrt{11}$$

The solutions are $-2+\sqrt{11}$ and $-2-\sqrt{11}$.

8. $x^5 + 3x^4 = x + 3$

$x^5 + 3x^4 - x - 3 = 0$

$x^4(x+3) - 1(x+3) = 0$

$(x+3)(x^4 - 1) = 0$

$(x+3)(x^2+1)(x^2-1) = 0$

$x+3 = 0$ or $x^2+1 = 0$ or $x^2-1 = 0$

$x = -3$ or $x^2 = -1$ or $x^2 = 1$

$x = \pm i$ or $x = \pm 1$

The solutions are –3, –1, 1, –i, and i.

9.
$$\begin{aligned} x^6+1 &= x^4+x^2 \\ x^6-x^4-x^2+1 &= 0 \\ x^4(x^2-1)-(x^2-1) &= 0 \\ (x^4-1)(x^2-1) &= 0 \\ (x^2+1)(x^2-1)(x+1)(x-1) &= 0 \\ (x^2+1)(x+1)^2(x-1)^2 &= 0 \end{aligned}$$

$x^2+1=0$ or $x+1=0$ or $x-1=0$
$x^2=-1$ $\quad x=-1$ $\quad x=1$
$x=\pm i$

The solutions are $-i$, i, -1, and 1.

10. $(x+1)^2-15(x+1)+56=0$

Let $y=x+1$. Then $y^2=(x+1)^2$ and

$y^2-15y+56=0$
$(y-8)(y-7)=0$
$y=8$ or $y=7$
$x+1=8$ or $x+1=7$
$x=7$ or $x=6$

The solutions are 6 and 7.

11.
$$\begin{aligned} x^2-6x &= -2 \\ x^2-6x+\left(\frac{-6}{2}\right)^2 &= -2+9 \\ x^2-6x+9 &= 7 \\ (x-3)^2 &= 7 \\ x-3 &= \pm\sqrt{7} \\ x &= 3\pm\sqrt{7} \end{aligned}$$

The solutions are $3+\sqrt{7}$ and $3-\sqrt{7}$.

12.
$$\begin{aligned} 2a^2+5 &= 4a \\ 2a^2-4a &= -5 \\ a^2-2a &= -\frac{5}{2} \\ a^2-2a+\left(\frac{-2}{2}\right)^2 &= -\frac{5}{2}+1 \\ a^2-2a+1 &= -\frac{3}{2} \\ (a-1)^2 &= -\frac{3}{2} \\ a-1 &= \pm\sqrt{-\frac{3}{2}} = \pm\frac{i\sqrt{3}}{\sqrt{2}} \\ a-1 &= \pm\frac{i\sqrt{6}}{2} \\ a &= 1\pm\frac{i\sqrt{6}}{2} \text{ or } \frac{2\pm i\sqrt{6}}{2} \end{aligned}$$

The solutions are $\frac{2+i\sqrt{6}}{2}$ and $\frac{2-i\sqrt{6}}{2}$.

13.
$$\begin{aligned} 2x^2-7x &> 15 \\ 2x^2-7x-15 &> 0 \\ (2x+3)(x-5) &> 0 \end{aligned}$$

$2x+3=0$ or $x-5=0$
$x=-\frac{3}{2}$ or $x=5$

Region	Test Point	$(2x+3)(x-5)>0$ Result
A: $\left(-\infty, -\frac{3}{2}\right)$	-2	$(-1)(-7)>0$ True
B: $\left(-\frac{3}{2}, 5\right)$	0	$(3)(-5)>0$ False
C: $(5, \infty)$	6	$(15)(1)>0$ True

Solution: $\left(-\infty, -\frac{3}{2}\right)\cup(5, \infty)$

14. $(x^2-16)(x^2-25)\ge 0$

$(x+4)(x-4)(x+5)(x-5)\ge 0$

$x+4=0$ or $x-4=0$ or $x+5=0$ or $x-5=0$

$x=-4$ or $x=4$ or $x=-5$ or $x=5$

Region	Test Point	$(x+4)(x-4)(x+5)(x-5)\ge 0$ Result
A: $(-\infty, -5]$	-6	$(-2)(-10)(-1)(-11)\ge 0$ True
B: $[-5, -4]$	$-\frac{9}{2}$	$\left(-\frac{1}{2}\right)\left(-\frac{17}{2}\right)\left(\frac{1}{2}\right)\left(-\frac{19}{2}\right)\ge 0$ False
C: $[-4, 4]$	0	$(4)(-4)(5)(-5)\ge 0$ True
D: $[4, 5]$	$\frac{9}{2}$	$\left(\frac{17}{2}\right)\left(\frac{1}{2}\right)\left(\frac{19}{2}\right)\left(-\frac{1}{2}\right)\ge 0$ False
E: $[5, \infty)$	6	$(10)(2)(11)(1)\ge 0$ True

Solution: $(-\infty, -5]\cup[-4, 4]\cup[5, \infty)$

15. $\frac{5}{x+3}<1$

The denominator is equal to 0 when $x + 3 = 0$, or $x = -3$.

$\frac{5}{x+3}=1$

$5=x+3$ so $x=2$

Region	Test Point	$\frac{5}{x+3}<1$ Result
A: $(-\infty, -3)$	-4	$\frac{5}{-1}<1$ True
B: $(-3, 2)$	0	$\frac{5}{3}<1$ False
C: $(2, \infty)$	3	$\frac{5}{6}<1$ True

Solution: $(-\infty, -3)\cup(2, \infty)$

16. $\frac{7x-14}{x^2-9} \le 0$

$\frac{7(x-2)}{(x+3)(x-3)} \le 0$

$x-2=0$ or $x+3=0$ or $x-3=0$

$x=2$ or $x=-3$ or $x=3$

Region	Test Point	$\frac{7(x-2)}{(x+3)(x-3)} \le 0$ Result
A: $(-\infty, -3)$	-4	$\frac{7(-6)}{(-1)(-7)} \le 0$ True
B: $(-3, 2]$	0	$\frac{7(-2)}{(3)(-3)} \le 0$ False
C: $[2, 3)$	$\frac{5}{2}$	$\frac{7\left(\frac{1}{2}\right)}{\left(\frac{11}{2}\right)\left(-\frac{1}{2}\right)} \le 0$ True
D: $(3, \infty)$	4	$\frac{7(2)}{(7)(1)} \le 0$ False

Solution: $(-\infty, -3) \cup [2, 3)$

17. $f(x) = 3x^2$

Vertex: (0, 0)

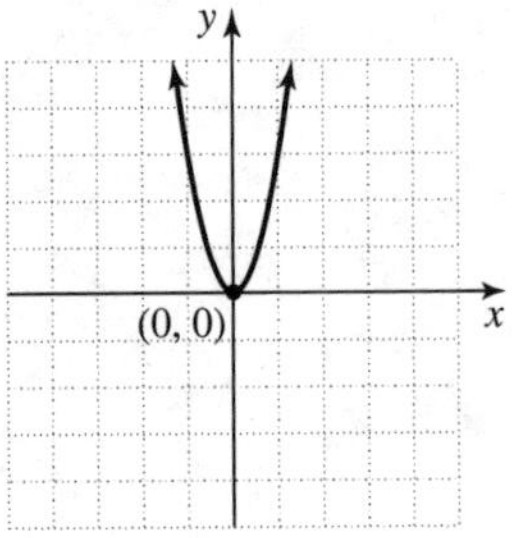

18. $G(x) = -2(x-1)^2 + 5$

Vertex: (1, 5)

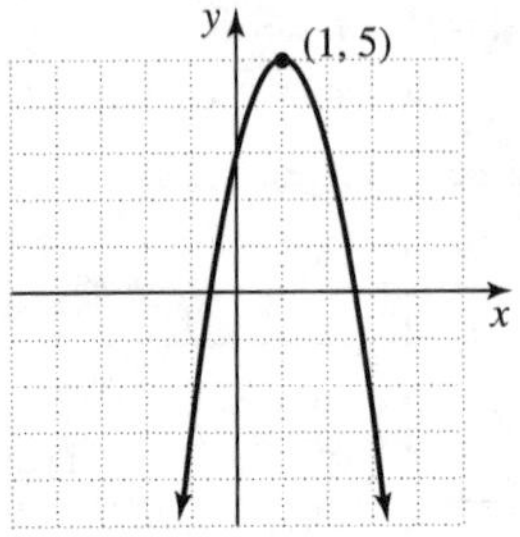

19. $h(x) = x^2 - 4x + 4$

$x = -\frac{b}{2a} = \frac{-(-4)}{2(1)} = 2$

$h(2) = (2)^2 - 4(2) + 4 = 0$

Vertex: (2, 0)

$h(0) = 4 \Rightarrow$ y-intercept: (0, 4)

x-intercept: (2, 0)

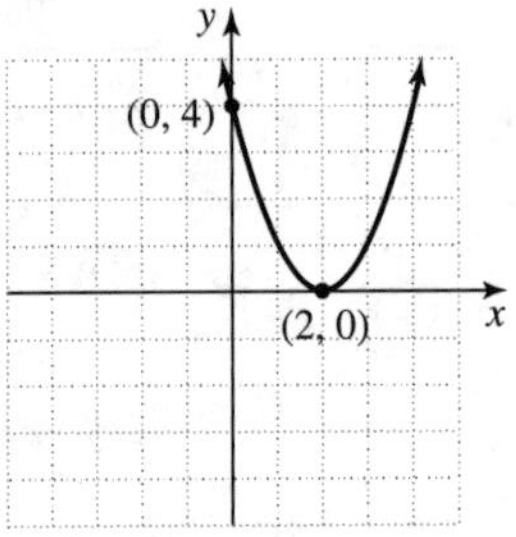

20. $F(x) = 2x^2 - 8x + 9$

$x = -\frac{b}{2a} = \frac{-(-8)}{2(2)} = 2$

$F(2) = 2(2)^2 - 8(2) + 9 = 1$

Vertex: (2, 1)

$F(0) = 9 \Rightarrow$ y-intercept: (0, 9)

$2x^2 - 8x + 9 = 0$

$a = 2, b = -8, c = 9$

$x = \frac{8 \pm \sqrt{(-8)^2 - 4(2)(9)}}{2(2)}$

$= \frac{8 \pm \sqrt{-8}}{4}$

which yields non-real solutons.

Therefore, there are no x-intercepts.

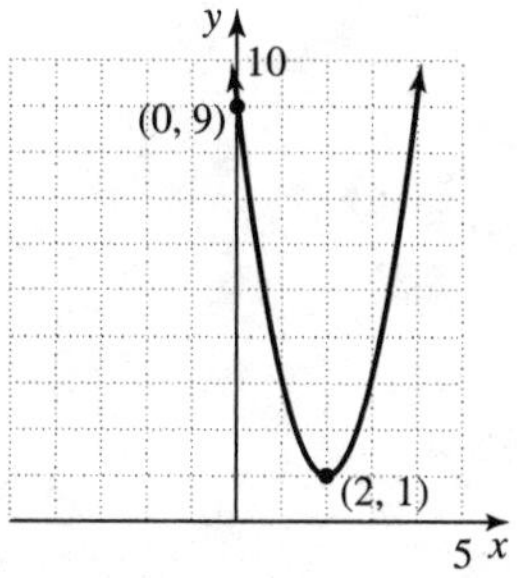

21. Let t = time for Sandy alone. Then $t - 2$ = time for Dave alone.

$$\frac{1}{t}+\frac{1}{t-2}=\frac{1}{4}$$
$$4(t-2)+4t=t(t-2)$$
$$4t-8+4t=t^2-2t$$
$$0=t^2-10t+8$$
$$a=1, b=-10, c=8$$
$$t=\frac{10\pm\sqrt{(-10)^2-4(1)(8)}}{2(1)}$$
$$=\frac{10\pm\sqrt{68}}{2}$$
$$=\frac{10\pm 2\sqrt{17}}{2}$$
$$=5\pm\sqrt{17}$$
$$\approx 9.12 \text{ or } 0.88 \text{ (discard)}$$

It takes her about 9.12 hours.

22. $s(t)=-16t^2+32t+256$

a. $t=-\frac{b}{2a}=\frac{-32}{2(-16)}=1$

$s(1)=-16(1)^2+32(1)+256=272$

Vertex: (1, 272)

The maximum height is 272 feet.

b. $-16t^2+32t+256=0$

$$t^2-2t-16=0$$
$$a=1, b=-2, c=-16$$
$$t=\frac{2\pm\sqrt{(-2)^2-4(1)(-16)}}{2(1)}$$
$$=\frac{2\pm\sqrt{68}}{2}$$
$$=\frac{2\pm 2\sqrt{17}}{2}$$
$$=1\pm\sqrt{17}$$
$$\approx -3.12 \text{ and } 5.12$$

Disregard the negative. The stone will hit the water in about 5.12 seconds.

23.
$$a^2+b^2=c^2$$
$$x^2+(x+8)^2=(20)^2$$
$$x^2+(x^2+16x+64)=400$$
$$2x^2+16x-336=0$$
$$x^2+8x-168=0$$
$$a=1, b=8, c=-168$$
$$x=\frac{-8\pm\sqrt{(8)^2-4(1)(-168)}}{2(1)}$$
$$=\frac{-8\pm\sqrt{736}}{2}$$
$$\approx -17.565 \text{ or } 9.565$$

Disregard the negative.

$x\approx 9.6$

$x+8\approx 9.6+8=17.6$

$17.6+9.6=27.2$

$27.2-20=7.2$

They would save about 7.2 feet.

Chapter 8 Cumulative Review

1. a. $5+y\ge 7$

b. $11\ne z$

c. $20<5-2x$

2. $|3x-2|=-5$ which is impossible. Thus, there is no solution, or $\varnothing$.

3. $m = \frac{5-3}{2-0} = \frac{2}{2} = 1$

Plot the given points and draw a line through them.

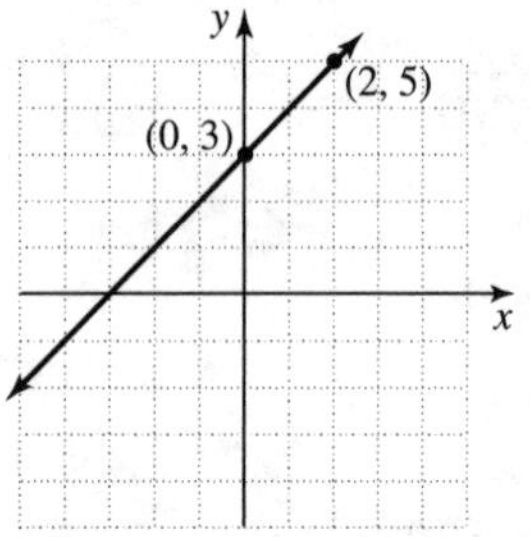

4. $\begin{cases} -6x + y = 5 & (1) \\ 4x - 2y = 6 & (2) \end{cases}$

Multiply E1 by 2 and add to E2.

$$\begin{aligned} -12x + 2y &= 10 \\ 4x - 2y &= 6 \\ \hline -8x \quad &= 16 \\ x &= -2 \end{aligned}$$

Replace x with -2 in E1.

$$\begin{aligned} -6(-2) + y &= 5 \\ 12 + y &= 5 \\ y &= -7 \end{aligned}$$

The solution is $(-2, -7)$.

5. $\begin{cases} x - 5y = -12 & (1) \\ -x + y = 4 & (2) \end{cases}$

Add E1 and E2.

$$\begin{aligned} -4y &= -8 \\ y &= 2 \end{aligned}$$

Replace y with 2 in E1.

$$\begin{aligned} x - 5(2) &= -12 \\ x - 10 &= -12 \\ x &= -2 \end{aligned}$$

The solution is $(-2, 2)$.

6. a. $(a^{-2}bc^3)^{-3} = (a^{-2})^{-3}b^{-3}(c^3)^{-3} = a^6b^{-3}c^{-9} = \frac{a^6}{b^3c^9}$

b. $\left(\frac{a^{-4}b^2}{c^3}\right)^{-2} = \frac{(a^{-4})^{-2}(b^2)^{-2}}{(c^3)^{-2}} = \frac{a^8b^{-4}}{c^{-6}} = \frac{a^8c^6}{b^4}$

c. $\left(\frac{3a^8b^2}{12a^5b^5}\right)^{-2} = \left(\frac{a^3}{4b^3}\right)^{-2} = \frac{(a^3)^{-2}}{4^{-2}(b^3)^{-2}} = \frac{4^2a^{-6}}{b^{-6}} = \frac{16b^6}{a^6}$

7. a. $(2x-7)(3x-4) = 6x^2 - 8x - 21x + 28 = 6x^2 - 29x + 28$

b. $(3x^2 + y)(5x^2 - 2y)$
$= 15x^4 - 6x^2y + 5x^2y - 2y^2$
$= 15x^4 - x^2y - 2y^2$

8. a. $(4a-3)(7a-2) = 28a^2 - 8a - 21a + 6 = 28a^2 - 29a + 6$

b. $(2a+b)(3a-5b)$
$= 6a^2 - 10ab + 3ab - 5b^2$
$= 6a^2 - 7ab - 5b^2$

9. a. $8x^2 + 4 = 4(2x^2 + 1)$

b. $5y - 2z^4$ is a prime polynomial.

c. $6x^2 - 3x^3 + 12x^4 = 3x^2(2 - x + 4x^2)$

10. a. $9x^3 + 27x^2 - 15x = 3x(3x^2 + 9x - 5)$

b. $2x(3y-2) - 5(3y-2)$
$= (3y-2)(2x-5)$

c. $2xy + 6x - y - 3 = 2x(y+3) - 1(y+3) = (y+3)(2x-1)$

11. $x^2 - 12x + 35 = (x-5)(x-7)$

12. $x^2 - 2x - 48 = (x+6)(x-8)$

13. $$\begin{aligned}12a^2x - 12abx + 3b^2x &= 3x(4a^2 - 4ab + b^2) \\ &= 3x(2a-b)^2\end{aligned}$$

14. $$\begin{aligned}2ax^2 - 12axy + 18ay^2 &= 2a(x^2 - 6xy + 9y^2) \\ &= 2a(x-3y)(x-3y) \\ &= 2a(x-3y)^2\end{aligned}$$

15. $$\begin{aligned}3(x^2+4)+5 &= -6(x^2+2x)+13 \\ 3x^2+12+5 &= -6x^2-12x+13 \\ 9x^2+12x+4 &= 0 \\ (3x+2)^2 &= 0 \\ 3x+2 &= 0 \\ 3x &= -2 \\ x &= -\frac{2}{3}\end{aligned}$$

The solutions is $-\frac{2}{3}$.

16. $$\begin{aligned}2(a^2+2)-8 &= -2a(a-2)-5 \\ 2a^2+4-8 &= -2a^2+4a-5 \\ 4a^2-4a+1 &= 0 \\ (2a-1)^2 &= 0 \\ 2a-1 &= 0 \\ 2a &= 1 \\ a &= \frac{1}{2}\end{aligned}$$

The solution is $\frac{1}{2}$.

17. $$\begin{aligned}x^3 &= 4x \\ x^3 - 4x &= 0 \\ x(x^2-4) &= 0 \\ x(x+2)(x-2) &= 0\end{aligned}$$

$x = 0$ or $x+2=0$ or $x-2=0$

$x=-2$ or $x=2$

The solutions are –2, 0, and 2.

18. $f(x) = x^2 + x - 12$

$$x = -\frac{b}{2a} = \frac{-1}{2(1)} = -\frac{1}{2}$$

$$\begin{aligned}f\left(-\frac{1}{2}\right) &= \left(-\frac{1}{2}\right)^2 + \left(-\frac{1}{2}\right) - 12 \\ &= \frac{1}{4} - \frac{1}{2} - 12 \\ &= -\frac{49}{4}\end{aligned}$$

Vertex: $\left(-\frac{1}{2}, -\frac{49}{4}\right)$

$$\begin{aligned}x^2+x-12 &= 0 \\ (x+4)(x-3) &= 0\end{aligned}$$

$x+4=0$ or $x-3=0$

$x=-4$ $\quad$ $x=3$

x-intercepts: (–4, 0), (3, 0)

$f(0) = 0^2 + 0 - 12 = -12$

y-intercept: (0, –12)

19. $$\frac{2x^2}{10x^3-2x^2} = \frac{2x^2}{2x^2(5x-1)} = \frac{1}{5x-1}$$

20. $$\frac{x^2-4x+4}{2-x} = \frac{(x-2)^2}{-(x-2)} = \frac{x-2}{-1} = 2-x$$

21. $$\begin{aligned}&\frac{2x-1}{2x^2-9x-5} + \frac{x+3}{6x^2-x-2} \\ &= \frac{2x-1}{(2x+1)(x-5)} + \frac{x+3}{(2x+1)(3x-2)} \\ &= \frac{(2x-1)(3x-2)+(x+3)(x-5)}{(2x+1)(x-5)(3x-2)} \\ &= \frac{(6x^2-4x-3x+2)+(x^2-5x+3x-15)}{(2x+1)(x-5)(3x-2)} \\ &= \frac{7x^2-9x-13}{(2x+1)(x-5)(3x-2)}\end{aligned}$$

22. $\frac{a+1}{a^2-6a+8}-\frac{3}{16-a^2}$
$=\frac{a+1}{(a-4)(a-2)}-\frac{3}{(4+a)(4-a)}$
$=\frac{a+1}{(a-4)(a-2)}+\frac{3}{(4+a)(a-4)}$
$=\frac{(a+1)(a+4)+3(a-2)}{(a-4)(a-2)(a+4)}$
$=\frac{(a^2+4a+a+4)+3a-6}{(a-4)(a-2)(a+4)}$
$=\frac{a^2+8a-2}{(a-4)(a-2)(a+4)}$

23. $\frac{x^{-1}+2xy^{-1}}{x^{-2}-x^{-2}y^{-1}}=\frac{\frac{1}{x}+\frac{2x}{y}}{\frac{1}{x^2}-\frac{1}{x^2y}}$
$=\frac{\left(\frac{1}{x}+\frac{2x}{y}\right)x^2y}{\left(\frac{1}{x^2}-\frac{1}{x^2y}\right)x^2y}$
$=\frac{xy+2x^3}{y-1}$ or $\frac{x(y+2x^2)}{y-1}$

24. $\frac{(2a)^{-1}+b^{-1}}{a^{-1}+(2b)^{-1}}=\frac{\frac{1}{2a}+\frac{1}{b}}{\frac{1}{a}+\frac{1}{2b}}$
$=\frac{\left(\frac{1}{2a}+\frac{1}{b}\right)2ab}{\left(\frac{1}{a}+\frac{1}{2b}\right)2ab}$
$=\frac{b+2a}{2b+a}$
$=\frac{2a+b}{a+2b}$

25. $\frac{3x^5y^2-15x^3y-x^2y-6x}{x^2y}$
$=\frac{3x^5y^2}{x^2y}-\frac{15x^3y}{x^2y}-\frac{x^2y}{x^2y}-\frac{6x}{x^2y}$
$=3x^3y-15x-1-\frac{6}{xy}$

26.
$$\begin{array}{r}
x^2-6x+8 \\
x+3\overline{\big)\,x^3-3x^2-10x+24} \\
\underline{x^3+3x^2} \\
-6x^2-10x \\
\underline{-6x^2-18x} \\
8x+24 \\
\underline{8x+24} \\
0
\end{array}$$

Answer: x^2-6x+8

27. $P(x)=2x^3-4x^2+5$

a. $P(2)=2(2)^3-4(2)^2+5$
$=2(8)-4(4)+5$
$=16-16+5$
$=5$

b.
$$\begin{array}{r|rrrr}
2 & 2 & -4 & 0 & 5 \\
 & & 4 & 0 & 0 \\
\hline
 & 2 & 0 & 0 & 5
\end{array}$$
Thus, $P(2) = 5$.

28. $P(x)=4x^3-2x^2+3$

a. $P(-2)=4(-2)^3-2(-2)^2+3$
$=4(-8)-2(4)+3$
$=-32-8+3$
$=-37$

b.
$$\begin{array}{r|rrrr}
-2 & 4 & -2 & 0 & 3 \\
 & & -8 & 20 & -40 \\
\hline
 & 4 & -10 & 20 & -37
\end{array}$$
Thus, $P(-2)=-37$.

29. $\frac{4x}{5}+\frac{3}{2}=\frac{3x}{10}$
$10\left(\frac{4x}{5}+\frac{3}{2}\right)=10\left(\frac{3x}{10}\right)$
$2(4x)+5(3)=3x$
$8x+15=3x$
$5x=-15$
$x=-3$
The solution is –3.

30.
$$\frac{x+3}{x^2+5x+6}=\frac{3}{2x+4}-\frac{1}{x+3}$$
$$\frac{x+3}{(x+3)(x+2)}=\frac{3}{2(x+2)}-\frac{1}{x+3}$$
$$2(x+3)=3(x+3)-2(x+2)$$
$$2x+6=3x+9-2x-4$$
$$2x+6=x+5$$
$$x=-1$$

31. Let x = the number.
$$\frac{9-x}{19+x}=\frac{1}{3}$$
$$3(9-x)=1(19+x)$$
$$27-3x=19+x$$
$$-4x=-8$$
$$x=2$$
The number is 2.

32. Let t = time to roof the house together.
$$\frac{1}{24}+\frac{1}{40}=\frac{1}{t}$$
$$120t\left(\frac{1}{24}+\frac{1}{40}\right)=120t\left(\frac{1}{t}\right)$$
$$5t+3t=120$$
$$8t=120$$
$$t=\frac{120}{8}=15$$
It would take them 15 hours to roof the house working together.

33. $y=kx$
$$5=k(30)$$
$$k=\frac{5}{30}=\frac{1}{6} \text{ and } y=\frac{1}{6}x$$

34. $y=\frac{k}{x}$
$$8=\frac{k}{14}$$
$$k=8(14)=112 \text{ and } y=\frac{112}{x}$$

35. a. $\sqrt{(-3)^2}=|-3|=3$

b. $\sqrt{x^2}=|x|$

c. $\sqrt[4]{(x-2)^4}=|x-2|$

d. $\sqrt[3]{(-5)^3}=-5$

e. $\sqrt[5]{(2x-7)^5}=2x-7$

f. $\sqrt{25x^2}=\sqrt{25}\cdot\sqrt{x^2}=5|x|$

g. $\sqrt{x^2+2x+1}=\sqrt{(x+1)^2}=|x+1|$

36. a. $\sqrt{(-2)^2}=|-2|=2$

b. $\sqrt{y^2}=|y|$

c. $\sqrt[4]{(a-3)^4}=|a-3|$

d. $\sqrt[3]{(-6)^3}=-6$

e. $\sqrt[5]{(3x-1)^5}=3x-1$

37. a. $\sqrt[8]{x^4}=x^{4/8}=x^{1/2}=\sqrt{x}$

b.
$$\sqrt[6]{25}=(25)^{1/6}$$
$$=(5^2)^{1/6}=5^{2/6}=5^{1/3}=\sqrt[3]{5}$$

c.
$$\sqrt[4]{r^2s^6}=(r^2s^6)^{1/4}$$
$$=r^{2/4}s^{6/4}$$
$$=r^{1/2}s^{3/2}$$
$$=(rs^3)^{1/2}=\sqrt{rs^3}$$

38. a. $\sqrt[4]{5^2}=5^{2/4}=5^{1/2}=\sqrt{5}$

b. $\sqrt[12]{x^3}=x^{3/12}=x^{1/4}=\sqrt[4]{x}$

c.
$$\sqrt[6]{x^2y^4}=(x^2y^4)^{1/6}$$
$$=x^{2/6}y^{4/6}$$
$$=x^{1/3}y^{2/3}$$
$$=(xy^2)^{1/3}=\sqrt[3]{xy^2}$$

39. a. $\sqrt{25x^3}=\sqrt{25x^2\cdot x}=5x\sqrt{x}$

b.
$$\sqrt[3]{54x^6y^8}=\sqrt[3]{27x^6y^6\cdot 2y^2}$$
$$=3x^2y^2\sqrt[3]{2y^2}$$

c. $\sqrt[4]{81z^{11}}=\sqrt[4]{81z^8\cdot z^3}=3z^2\sqrt[4]{z^3}$

40. a. $\sqrt{64a^5} = \sqrt{64a^4 \cdot a} = 8a^2\sqrt{a}$

b. $\sqrt[3]{24a^7b^9} = \sqrt[3]{8a^6b^9 \cdot 3a} = 2a^2b^3\sqrt[3]{3a}$

c. $\sqrt[4]{48x^9} = \sqrt[4]{16x^8 \cdot 3x} = 2x^2\sqrt[4]{3x}$

41. a. $\dfrac{2}{\sqrt{5}} = \dfrac{2 \cdot \sqrt{5}}{\sqrt{5} \cdot \sqrt{5}} = \dfrac{2\sqrt{5}}{5}$

b. $\dfrac{2\sqrt{16}}{\sqrt{9x}} = \dfrac{2 \cdot 4}{3\sqrt{x}} = \dfrac{8 \cdot \sqrt{x}}{3\sqrt{x} \cdot \sqrt{x}} = \dfrac{8\sqrt{x}}{3x}$

c. $\sqrt[3]{\dfrac{1}{2}} = \dfrac{\sqrt[3]{1}}{\sqrt[3]{2}} = \dfrac{1}{\sqrt[3]{2}} = \dfrac{1 \cdot \sqrt[3]{2^2}}{\sqrt[3]{2} \cdot \sqrt[3]{2^2}} = \dfrac{\sqrt[3]{4}}{2}$

42. a.
$$\begin{aligned}&\left(\sqrt{3}-4\right)\left(2\sqrt{3}+2\right)\\ &= \sqrt{3} \cdot 2\sqrt{3} + 2\sqrt{3} - 4 \cdot 2\sqrt{3} - 4 \cdot 2\\ &= 2(3) + 2\sqrt{3} - 8\sqrt{3} - 8\\ &= 6 - 6\sqrt{3} - 8\\ &= -2 - 6\sqrt{3}\end{aligned}$$

b.
$$\begin{aligned}\left(\sqrt{5}-x\right)^2 &= \left(\sqrt{5}\right)^2 - 2 \cdot \sqrt{5} \cdot x + x^2\\ &= 5 - 2x\sqrt{5} + x^2\end{aligned}$$

c.
$$\begin{aligned}\left(\sqrt{a}+b\right)\left(\sqrt{a}-b\right) &= \left(\sqrt{a}\right)^2 - b^2\\ &= a - b^2\end{aligned}$$

43.
$$\begin{aligned}\sqrt{2x+5} + \sqrt{2x} &= 3\\ \sqrt{2x+5} &= 3 - \sqrt{2x}\\ \left(\sqrt{2x+5}\right)^2 &= \left(3-\sqrt{2x}\right)^2\\ 2x+5 &= 9 - 6\sqrt{2x} + 2x\\ -4 &= -6\sqrt{2x}\\ (-4)^2 &= \left(-6\sqrt{2x}\right)^2\\ 16 &= 36(2x)\\ 16 &= 72x\\ x &= \frac{16}{72} = \frac{2}{9}\end{aligned}$$

The solution is $\dfrac{2}{9}$.

44.
$$\begin{aligned}\sqrt{x-2} &= \sqrt{4x+1} - 3\\ \left(\sqrt{x-2}\right)^2 &= \left(\sqrt{4x+1}-3\right)^2\\ x-2 &= (4x+1) - 6\sqrt{4x+1} + 9\\ 6\sqrt{4x+1} &= 3x+12\\ 2\sqrt{4x+1} &= x+4\\ \left(2\sqrt{4x+1}\right)^2 &= (x+4)^2\\ 4(4x+1) &= x^2+8x+16\\ 16x+4 &= x^2+8x+16\\ 0 &= x^2-8x+12\\ 0 &= (x-6)(x-2)\end{aligned}$$

$x-6=0$ or $x-2=0$

$x=6$ or $x=2$

The solutions are 2 and 6.

45. a.
$$\begin{aligned}\frac{2+i}{1-i} &= \frac{(2+i)\cdot(1+i)}{(1-i)\cdot(1+i)}\\ &= \frac{2+2i+1i+i^2}{1^2-i^2}\\ &= \frac{2+3i-1}{1+1}\\ &= \frac{1+3i}{2}\\ &= \frac{1}{2}+\frac{3}{2}i\end{aligned}$$

b. $\dfrac{7}{3i} = \dfrac{7\cdot(-3i)}{3i\cdot(-3i)} = \dfrac{-21i}{-9i^2} = \dfrac{-21i}{9} = -\dfrac{7}{3}i = 0 - \dfrac{7}{3}i$

46. a.
$$\begin{aligned}3i(5-2i) &= 15i - 6i^2\\ &= 15i + 6\\ &= 6 + 15i\end{aligned}$$

b.
$$\begin{aligned}(6-5i)^2 &= 6^2 - 2(6)(5i) + (5i)^2\\ &= 36 - 60i + 25i^2\\ &= 36 - 60i - 25\\ &= 11 - 60i\end{aligned}$$

c.
$$\begin{aligned}\left(\sqrt{3}+2i\right)\left(\sqrt{3}-2i\right) &= \left(\sqrt{3}\right)^2 - (2i)^2\\ &= 3 - 4i^2\\ &= 3 + 4\\ &= 7\\ &= 7 + 0i\end{aligned}$$

47. $(x+1)^2 = 12$

$x+1 = \pm\sqrt{12}$

$x+1 = \pm 2\sqrt{3}$

$x = -1 \pm 2\sqrt{3}$

The solutions are $-1+2\sqrt{3}$ and $-1-2\sqrt{3}$.

48. $(y-1)^2 = 24$

$y-1 = \pm\sqrt{24}$

$y-1 = \pm 2\sqrt{6}$

$y = 1 \pm 2\sqrt{6}$

The solutions are $1+2\sqrt{6}$ and $1-2\sqrt{6}$.

49. $x - \sqrt{x} - 6 = 0$

Let $y = \sqrt{x}$. Then $y^2 = x$ and

$$y^2 - y - 6 = 0$$
$$(y-3)(y+2) = 0$$

$y-3=0$ or $y+2=0$

$y=3$ or $y=-2$

$\sqrt{x}=3$ or $\sqrt{x}=-2$ (can't happen)

$x=9$

The solution is 9.

50. $m^2 = 4m+8$

$m^2 - 4m - 8 = 0$

$a = 1, b = -4, c = -8$

$$x = \frac{4 \pm \sqrt{(-4)^2 - 4(1)(-8)}}{2(1)}$$
$$= \frac{4 \pm \sqrt{16+32}}{2}$$
$$= \frac{4 \pm \sqrt{48}}{2}$$
$$= \frac{4 \pm 4\sqrt{3}}{2}$$
$$= 2 \pm 2\sqrt{3}$$

The solutions are $2+2\sqrt{3}$ and $2-2\sqrt{3}$.

Chapter 9

Section 9.1 Practice Exercises

1. $f(x) = x + 2$; $g(x) = 3x + 5$

a. $$\begin{aligned}(f+g)(x) &= f(x)+g(x)\\ &= (x+2)+(3x+5)\\ &= 4x+7\end{aligned}$$

b. $$\begin{aligned}(f-g)(x) &= f(x)-g(x)\\ &= (x+2)-(3x+5)\\ &= x+2-3x-5\\ &= -2x-3\end{aligned}$$

c. $$\begin{aligned}(f\cdot g)(x) &= f(x)\cdot g(x)\\ &= (x+2)(3x+5)\\ &= 3x^2+6x+5x+10\\ &= 3x^2+11x+10\end{aligned}$$

d. $\left(\frac{f}{g}\right)(x) = \frac{f(x)}{g(x)} = \frac{x+2}{3x+5}$, where $x \neq -\frac{5}{3}$.

2. $f(x) = x^2+1$; $g(x) = 3x - 5$

a. $(f\circ g)(4) = f(g(4)) = f(7) = 50$
$(g\circ f)(4) = g(f(4)) = g(17) = 46$

b. $$\begin{aligned}(f\circ g)(x) &= f(g(x))\\ &= f(3x-5)\\ &= (3x-5)^2+1\\ &= 9x^2-30x+26\end{aligned}$$
$$\begin{aligned}(g\circ f)(x) &= g(f(x))\\ &= g(x^2+1)\\ &= 3(x^2+1)-5\\ &= 3x^2-2\end{aligned}$$

3. $f(x) = x^2+5$; $g(x) = x+3$

a. $$\begin{aligned}(f\circ g)(x) &= f(g(x))\\ &= f(x+3)\\ &= (x+3)^2+5\\ &= x^2+6x+14\end{aligned}$$

b. $$\begin{aligned}(g\circ f)(x) &= g(f(x))\\ &= g(x^2+5)\\ &= (x^2+5)+3\\ &= x^2+8\end{aligned}$$

4. $f(x) = 3x$; $g(x) = x - 4$; $h(x) = |x|$

a. $F(x) = |x - 4|$
$$\begin{aligned}F(x) &= (h\circ g)(x)\\ &= h(g(x))\\ &= h(x-4)\\ &= |x-4|\end{aligned}$$

b. $G(x) = 3x - 4$
$$\begin{aligned}G(x) &= (g\circ f)(x)\\ &= g(f(x))\\ &= g(3x)\\ &= 3x-4\end{aligned}$$

Vocabulary, Readiness & Video Check 9.1

1. $(f\circ g)(x) = f(g(x))$; C

2. $(f\cdot g)(x) = f(x)\cdot g(x)$; E

3. $(f-g)(x) = f(x) - g(x)$; F

4. $(g\circ f)(x) = g(f(x))$; A

5. $\left(\frac{f}{g}\right)(x) = \frac{f(x)}{g(x)}$, $g(x) \neq 0$; D

6. $(f+g)(x) = f(x) + g(x)$; B

7. You can find $(f+g)(x)$ and then find $(f+g)(2)$ or you can find $f(2)$ and $g(2)$ and then add those results.

8. Yes, sometimes they can be equal.

Exercise Set 9.1

1. a. $(f+g)(x) = (x-7)+(2x+1) = 3x-6$

b. $$\begin{aligned}(f-g)(x) &= (x-7)-(2x+1)\\ &= x-7-2x-1\\ &= -x-8\end{aligned}$$

c. $(f\cdot g)(x) = (x-7)(2x+1) = 2x^2-13x-7$

d. $\left(\dfrac{f}{g}\right)(x)=\dfrac{x-7}{2x+1}$, where $x \neq -\dfrac{1}{2}$.

3. a. $(f+g)(x)=(x^2+1)+5x=x^2+5x+1$

b. $(f-g)(x)=(x^2+1)-5x=x^2-5x+1$

c. $(f\cdot g)(x)=(x^2+1)(5x)=5x^3+5x$

d. $\left(\dfrac{f}{g}\right)(x)=\dfrac{x^2+1}{5x}$, where $x \neq 0$

5. a. $(f+g)(x)=\sqrt[3]{x}+x+5$

b. $(f-g)(x)=\sqrt[3]{x}-(x+5)=\sqrt{x}-x-5$

c. $$\begin{aligned}(f\cdot g)(x)&=\sqrt[3]{x}(x+5)\\&=x\sqrt[3]{x}+5\sqrt[3]{x}\end{aligned}$$

d. $\left(\dfrac{f}{g}\right)(x)=\dfrac{\sqrt[3]{x}}{x+5}$; where $x \neq -5$.

7. a. $(f+g)(x)=-3x+5x^2$ or $5x^2-3x$

b. $(f-g)(x)=-3x-5x^2$ or $-5x^2-3x$

c. $(f\cdot g)(x)=(-3x)(5x^2)=-15x^3$

d. $$\begin{aligned}\left(\frac{f}{g}\right)(x)&=\frac{-3x}{5x^2}\\&=-\frac{3}{5x}, \text{ where } x\neq 0.\end{aligned}$$

9. $$\begin{aligned}(f\circ g)(2)&=f(g(2))\\&=f(-4)\\&=(-4)^2-6(-4)+2\\&=16+24+2\\&=42\end{aligned}$$

11. $$\begin{aligned}(g\circ f)(-1)&=g(f(-1))\\&=g(9)\\&=-2(9)\\&=-18\end{aligned}$$

13. $$\begin{aligned}(g\circ h)(0)&=g(h(0))\\&=g(0)\\&=-2(0)\\&=0\end{aligned}$$

15. $$\begin{aligned}(f\circ g)(x)&=f(g(x))\\&=f(5x)\\&=(5x)^2+1\\&=25x^2+1\end{aligned}$$
$$\begin{aligned}(g\circ f)(x)&=g(f(x))\\&=g(x^2+1)\\&=5(x^2+1)\\&=5x^2+5\end{aligned}$$

17. $$\begin{aligned}(f\circ g)(x)&=f(g(x))\\&=f(x+7)\\&=2(x+7)-3\\&=2x+14-3\\&=2x+11\end{aligned}$$
$$\begin{aligned}(g\circ f)(x)&=g(f(x))\\&=g(2x-3)\\&=(2x-3)+7\\&=2x+4\end{aligned}$$

19. $$\begin{aligned}(f\circ g)(x)&=f(g(x))\\&=f(-2x)\\&=(-2x)^3+(-2x)-2\\&=-8x^3-2x-2\end{aligned}$$
$$\begin{aligned}(g\circ f)(x)&=g(f(x))\\&=g(x^3+x-2)\\&=-2(x^3+x-2)\\&=-2x^3-2x+4\end{aligned}$$

21. $$\begin{aligned}(f\circ g)(x)&=f(g(x))\\&=f(10x-3)\\&=|10x-3|\end{aligned}$$
$$(g\circ f)(x)=g(f(x))=g(|x|)=10|x|-3$$

23. $(f\circ g)(x)=f(g(x))=f(-5x+2)=\sqrt{-5x+2}$
$(g\circ f)(x)=g(f(x))=g\left(\sqrt{x}\right)=-5\sqrt{x}+2$

25. $$\begin{aligned}H(x)&=(g\circ h)(x)\\&=g(h(x))\\&=g(x^2+2)\\&=\sqrt{x^2+2}\end{aligned}$$

27. $F(x) = (h \circ f)(x)$
$= h(f(x))$
$= h(3x)$
$= (3x)^2 + 2$
$= 9x^2 + 2$

29. $G(x) = (f \circ g)(x)$
$= f(g(x))$
$= f\left(\sqrt{x}\right)$
$= 3\sqrt{x}$

31. answers may vary; for example, $g(x) = x + 2$ and $f(x) = x^2$

33. answers may vary; for example, $g(x) = x + 5$ and $f(x) = \sqrt{x} + 2$

35. answers may vary; for example, $g(x) = 2x - 3$ and $f(x) = \frac{1}{x}$

37. $x = y + 2$
$y = x - 2$

39. $x = 3y$
$y = \frac{x}{3}$

41. $x = -2y - 7$
$2y = -x - 7$
$y = -\frac{x+7}{2}$

43. $(f + g)(2) = f(2) + g(2) = 7 + (-1) = 6$

45. $(f \circ g)(2) = f(g(2)) = f(-1) = 4$

47. $(f \cdot g)(7) = f(7) \cdot g(7) = 1 \cdot 4 = 4$

49. $\left(\frac{f}{g}\right)(-1) = \frac{f(-1)}{g(-1)} = \frac{4}{-4} = -1$

51. answers may vary

53. Profit is equal to the revenue minus the cost; $P(x) = R(x) - C(x)$

Section 9.2 Practice Exercises

1. a. $f = \{(4, -3), (3, -4), (2, 7), (5, 0)\}$
f is one-to-one since each y-value corresponds to only one x-value.

b. $g = \{(8, 4), (-2, 0), (6, 4), (2, 6)\}$
g is not one-to-one because the y-value 4 in (8, 4) and (6, 4) corresponds to two different x-values.

c. $h = \{(2, 4), (1, 3), (4, 6), (-2, 4)\}$
h is not one-to-one because the y-value 4 in (2, 4) and (−2, 4) corresponds to two different x-values.

d.

Year	1950	1963	1968	1975	1997	2008
Federal Minimum Wage	\$0.75	\$1.25	\$1.60	\$2.10	\$5.15	\$6.55

This function is one-to-one because each wage corresponds to only one year.

e. The function represented by the graph is not one-to-one because the y-value 2 in (2, 2) and (3, 2) corresponds to two different x-values.

f. The function represented by the diagram is not one-to-one because the score 495 corresponds to two different states.

2. Graphs **a**, **b**, and **c** all pass the vertical line test, so only these graphs are functions. But, of these, only **b** and **c** pass the horizontal line test, so only **b** and **c** are graphs of one-to-one functions.

3. $f = \{(3, 4), (-2, 0), (2, 8), (6, 6)\}$
Switching the coordinates of each ordered pair gives $f^{-1} = \{(4, 3), (0, -2), (8, 2), (6, 6)\}$

4. $f(x) = 6 - x$
Replace $f(x)$ with y.
$y = 6 - x$
Interchange x and y.
$x = 6 - y$
Solve for y.
$x = 6 - y$
$y = 6 - x$
Replace y with $f^{-1}(x)$.
$f^{-1}(x) = 6 - x$

5. $f(x) = 5x + 2$
Replace $f(x)$ with y.
$y = 5x + 2$
Interchange x and y.
$x = 5y + 2$
Solve for y.
$x = 5y + 2$
$x - 2 = 5y$
$\frac{x-2}{5} = y$
Replace y with $f^{-1}(x)$.
$f^{-1}(x) = \frac{x-2}{5}$

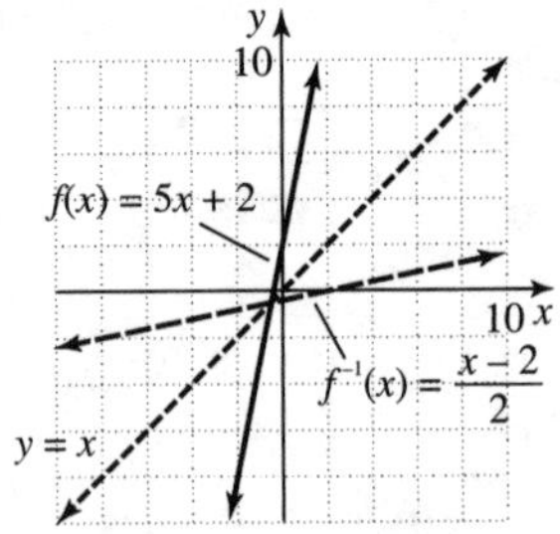

6. a.

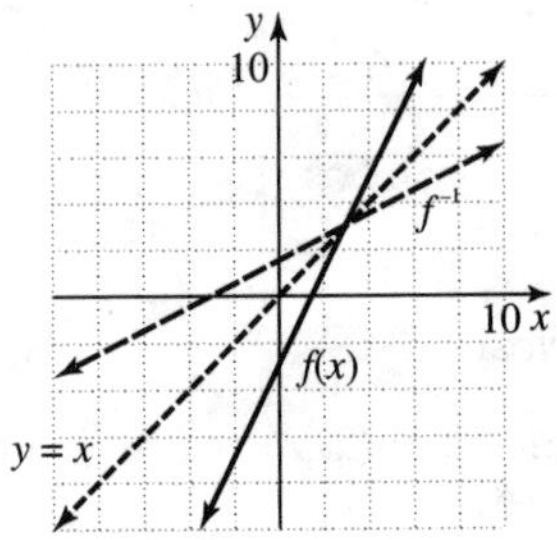

b.

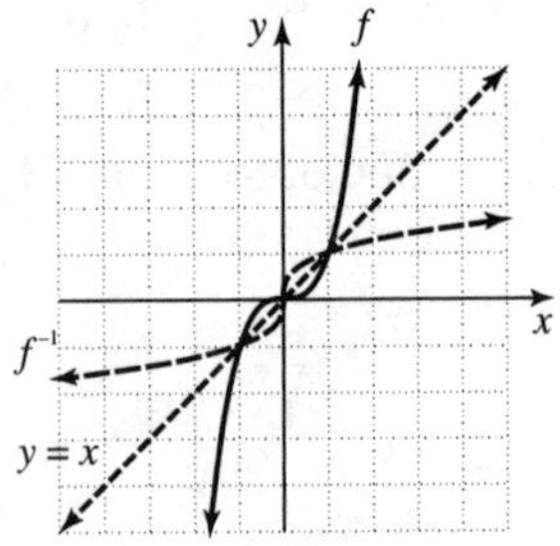

7. $f(x) = 4x - 1;\ f^{-1}(x) = \frac{x+1}{4}$

$$\begin{aligned}(f \circ f^{-1})(x) &= f(f^{-1}(x)) \\ &= f\left(\frac{x+1}{4}\right) \\ &= 4\left(\frac{x+1}{4}\right) - 1 \\ &= x + 1 - 1 \\ &= x\end{aligned}$$

$$\begin{aligned}(f^{-1} \circ f)(x) &= f^{-1}(f(x)) \\ &= f^{-1}(4x-1) \\ &= \frac{(4x-1)+1}{4} \\ &= \frac{4x-1+1}{4} \\ &= \frac{4x}{4} \\ &= x\end{aligned}$$

Since $f \circ f^{-1} = x$ and $f^{-1} \circ f = x$, if $f(x) = 4x - 1$, $f^{-1}(x) = \frac{x+1}{4}$.

Vocabulary, Readiness & Video Check 9.2

1. If $f(2) = 11$, the corresponding ordered pair is (2, 11).

2. If (7, 3) is an ordered pair solution of $f(x)$, and $f(x)$ has an inverse, then an ordered pair solution of $f^{-1}(x)$ is (3, 7).

3. The symbol f^{-1} means the inverse of f.

4. True or false: The function notation $f^{-1}(x)$ means $\frac{1}{f(x)}$. false

5. To tell whether a graph is the graph of a function, use the vertical line test.

6. To tell whether the graph of a function is also a one-to-one function, use the horizontal line test.

7. The graphs of f and f^{-1} are symmetric about the line $y = x$.

8. Two functions are inverse of each other if $(f \circ f^{-1})(x) = x$ and $(f^{-1} \circ f)(x) = x$.

9. Every function must have each x-value correspond to only one y-value. A one-to-one function must also have each y-value correspond to only one x-value.

10. No, a graph must pass the vertical line test to even be a function—a graph must pass *both* the vertical and horizontal line tests to be a one-to-one function.

11. Yes; by the definition of an inverse function.

12. The definition of inverse function tells us that f^{-1} consists of the ordered pairs (y, x) when (x, y) belongs to f. So it makes sense that switching x and y in the equations would result in switching the x and y values in the ordered pairs.

13. Once you know some points of the original equation or graph, you can switch the x's and y's of these points to find points that satisfy the inverse and then graph it. You can also check that the two graphs (the original and the inverse) are symmetric about the line $y = x$.

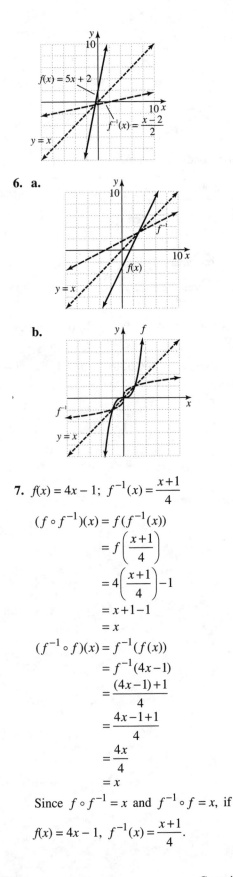

14. You must show that $f(f^{-1}(x))$ and $f^{-1}(f(x))$ both equal x in order to prove they are inverses of each other.

Exercise Set 9.2

1. $f = \{(-1,-1),(1,1),(0,2),(2,0)\}$ is a one-to-one function.
$f^{-1} = \{(-1,-1),(1,1),(2,0),(0,2)\}$

3. $h = \{(10,10)\}$ is a one-to-one function.
$h^{-1} = \{(10,10)\}$

5. $f = \{(11,12),(4,3),(3,4),(6,6)\}$ is a one-to-one function.
$f^{-1} = \{(12,11),(3,4),(4,3),(6,6)\}$

7. This function is not one-to-one because the months February, March, and May have the same output, 5.5.

9. This function is one-to-one.

Rank in Population (input)	1	47	16	25	36	7
State (output)	California	Alaska	Arizona	Louisiana	New Mexico	Ohio

11. $f(x) = x^3 + 2$

a. $f(1) = 1^3 + 2 = 3$

b. $f^{-1}(3) = 1$

13. $f(x) = x^3 + 2$

a. $f(-1) = (-1)^3 + 2 = 1$

b. $f^{-1}(1) = -1$

15. The graph represents a one-to-one function because it passes the horizontal line test.

17. The graph does not represent a one-to-one function because it does not pass the horizontal line test.

19. The graph represents a one-to-one function because it passes the horizontal line test.

21. The graph does not represent a one-to-one function because it does not pass the horizontal line test.

23.
$$f(x) = x + 4$$
$$y = x + 4$$
$$x = y + 4$$
$$y = x - 4$$
$$f^{-1}(x) = x - 4$$

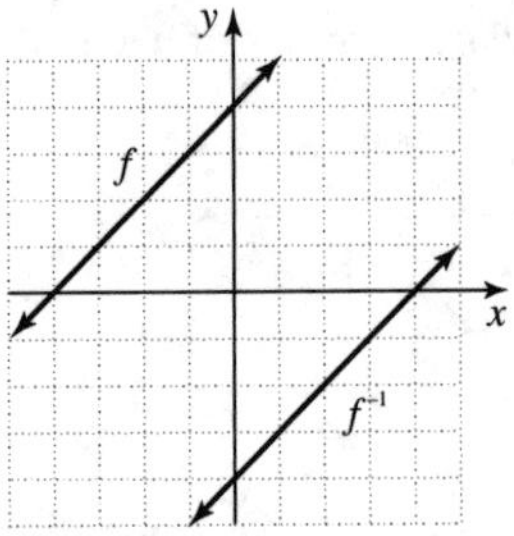

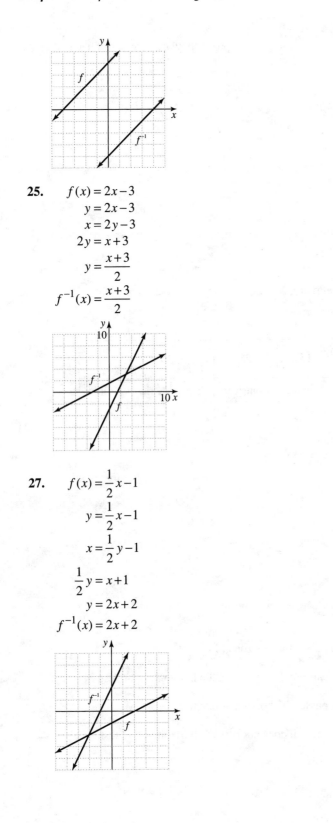

25.
$$
\begin{aligned}
f(x) &= 2x-3\\
y &= 2x-3\\
x &= 2y-3\\
2y &= x+3\\
y &= \frac{x+3}{2}\\
f^{-1}(x) &= \frac{x+3}{2}
\end{aligned}
$$

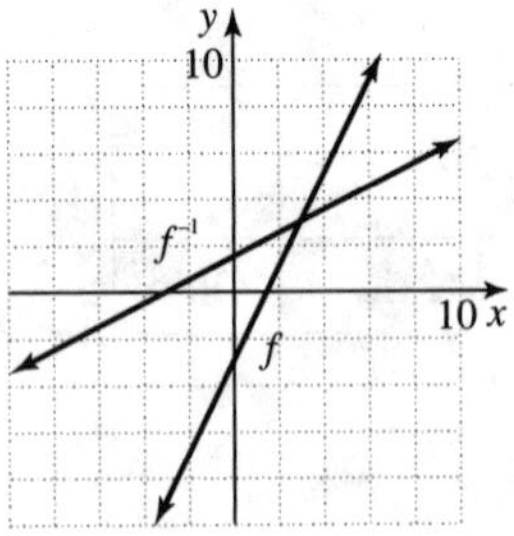

27.
$$
\begin{aligned}
f(x) &= \frac{1}{2}x-1\\
y &= \frac{1}{2}x-1\\
x &= \frac{1}{2}y-1\\
\frac{1}{2}y &= x+1\\
y &= 2x+2\\
f^{-1}(x) &= 2x+2
\end{aligned}
$$

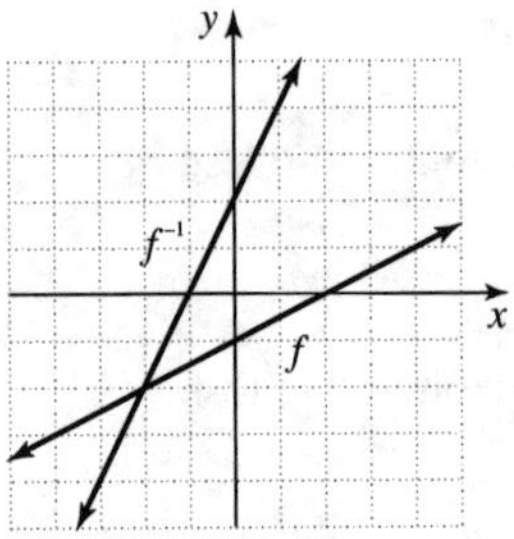

29.
$$
\begin{aligned}
f(x) &= x^3\\
y &= x^3\\
x &= y^3\\
y &= \sqrt[3]{x}\\
f^{-1}(x) &= \sqrt[3]{x}
\end{aligned}
$$

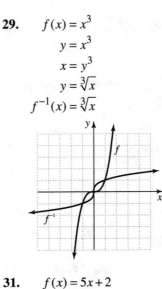

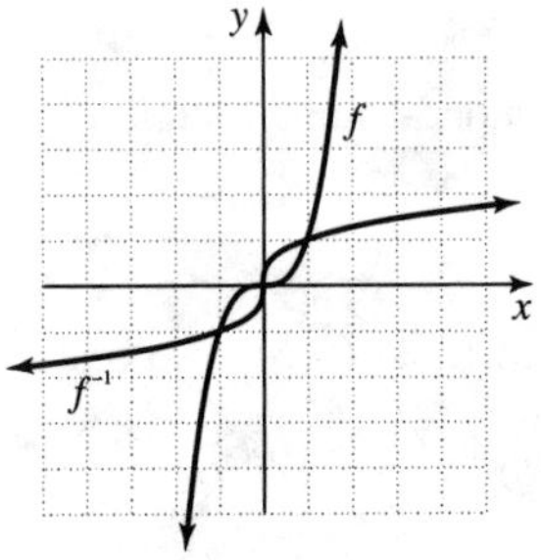

31.
$$
\begin{aligned}
f(x) &= 5x+2\\
y &= 5x+2\\
x &= 5y+2\\
5y &= x-2\\
y &= \frac{x-2}{5}\\
f^{-1}(x) &= \frac{x-2}{5}
\end{aligned}
$$

33.
$$
\begin{aligned}
f(x) &= \frac{x-2}{5}\\
y &= \frac{x-2}{5}\\
x &= \frac{y-2}{5}\\
5x &= y-2\\
y &= 5x+2\\
f^{-1}(x) &= 5x+2
\end{aligned}
$$

35.
$$
\begin{aligned}
f(x) &= \sqrt[3]{x}\\
y &= \sqrt[3]{x}\\
x &= \sqrt[3]{y}\\
x^3 &= y\\
f^{-1}(x) &= x^3
\end{aligned}
$$

37.
$$\begin{aligned} f(x) &= \frac{5}{3x+1} \\ y &= \frac{5}{3x+1} \\ x &= \frac{5}{3y+1} \\ 3y+1 &= \frac{5}{x} \\ 3y &= \frac{5}{x}-1 \\ 3y &= \frac{5-x}{x} \\ y &= \frac{5-x}{3x} \\ f^{-1}(x) &= \frac{5-x}{3x} \end{aligned}$$

39.
$$\begin{aligned} f(x) &= (x+2)^3 \\ y &= (x+2)^3 \\ x &= (y+2)^3 \\ \sqrt[3]{x} &= y+2 \\ \sqrt[3]{x}-2 &= y \\ f^{-1}(x) &= \sqrt[3]{x}-2 \end{aligned}$$

41.

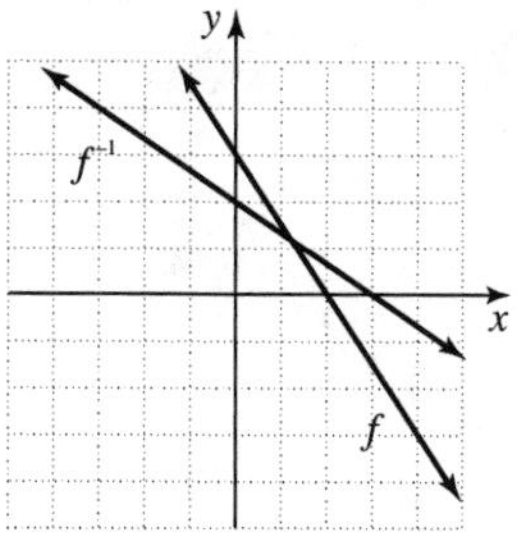

43.

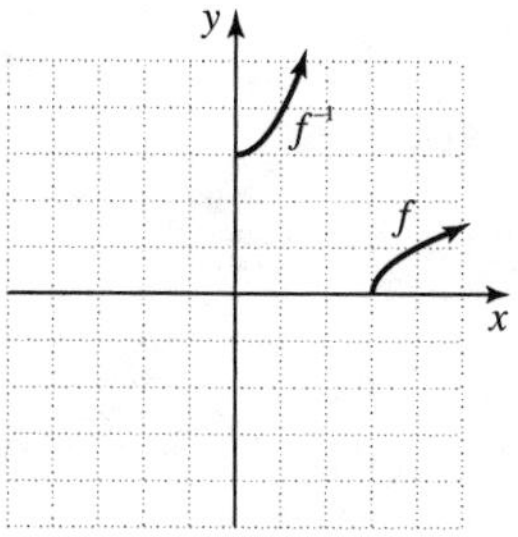

45.

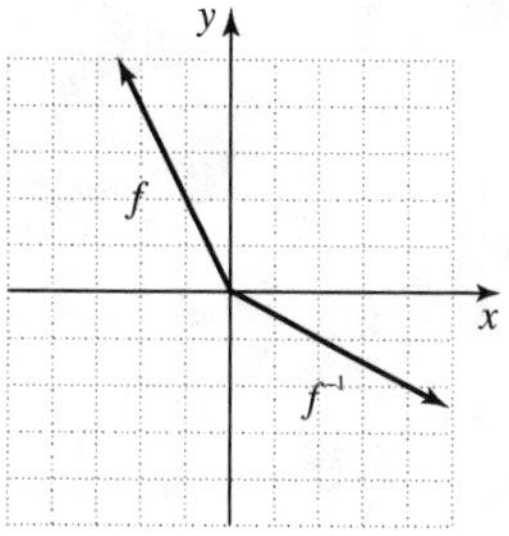

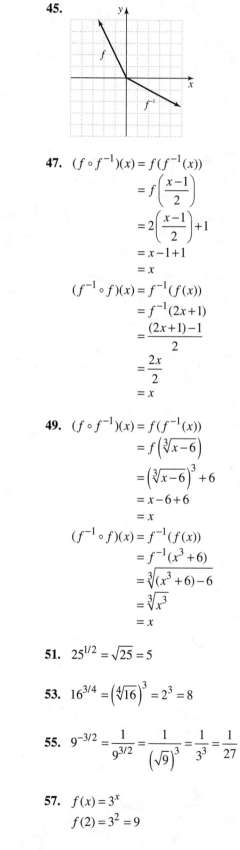

47.
$$\begin{aligned} (f \circ f^{-1})(x) &= f(f^{-1}(x)) \\ &= f\left(\frac{x-1}{2}\right) \\ &= 2\left(\frac{x-1}{2}\right)+1 \\ &= x-1+1 \\ &= x \end{aligned}$$
$$\begin{aligned} (f^{-1} \circ f)(x) &= f^{-1}(f(x)) \\ &= f^{-1}(2x+1) \\ &= \frac{(2x+1)-1}{2} \\ &= \frac{2x}{2} \\ &= x \end{aligned}$$

49.
$$\begin{aligned} (f \circ f^{-1})(x) &= f(f^{-1}(x)) \\ &= f\left(\sqrt[3]{x-6}\right) \\ &= \left(\sqrt[3]{x-6}\right)^3+6 \\ &= x-6+6 \\ &= x \end{aligned}$$
$$\begin{aligned} (f^{-1} \circ f)(x) &= f^{-1}(f(x)) \\ &= f^{-1}(x^3+6) \\ &= \sqrt[3]{(x^3+6)-6} \\ &= \sqrt[3]{x^3} \\ &= x \end{aligned}$$

51. $25^{1/2} = \sqrt{25} = 5$

53. $16^{3/4} = \left(\sqrt[4]{16}\right)^3 = 2^3 = 8$

55. $9^{-3/2} = \frac{1}{9^{3/2}} = \frac{1}{\left(\sqrt{9}\right)^3} = \frac{1}{3^3} = \frac{1}{27}$

57. $f(x) = 3^x$
$f(2) = 3^2 = 9$

59. $f(x) = 3^x$

$f\left(\frac{1}{2}\right) = 3^{1/2} \approx 1.73$

61. $f(2) = 9$

a. (2, 9)

b. (9, 2)

63. a. $\left(-2, \frac{1}{4}\right), \left(-1, \frac{1}{2}\right), (0,1), (1,2), (2,5)$

b. $\left(\frac{1}{4}, -2\right), \left(\frac{1}{2}, -1\right), (1,0), (2,1), (5,2)$

c, d.

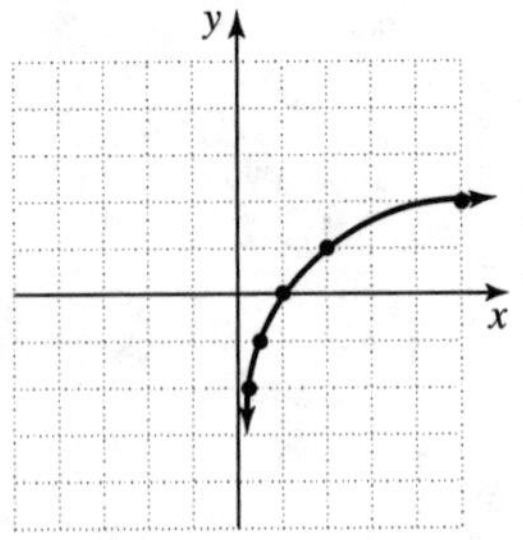

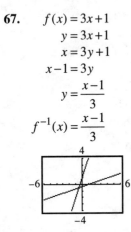

65. answers may vary

67.

$f(x) = 3x + 1$

$y = 3x + 1$

$x = 3y + 1$

$x - 1 = 3y$

$y = \frac{x-1}{3}$

$f^{-1}(x) = \frac{x-1}{3}$

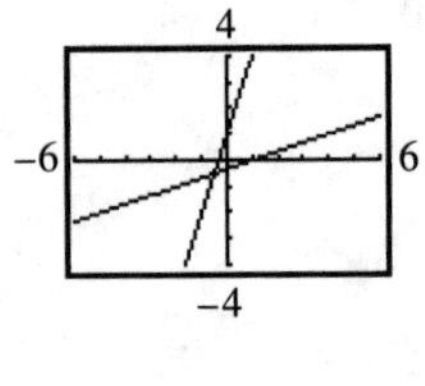

69. $f(x) = \sqrt[3]{x+1}$

$y = \sqrt[3]{x+1}$

$x = \sqrt[3]{y+1}$

$x^3 = y+1$

$y = x^3 - 1$

$f^{-1}(x) = x^3 - 1$

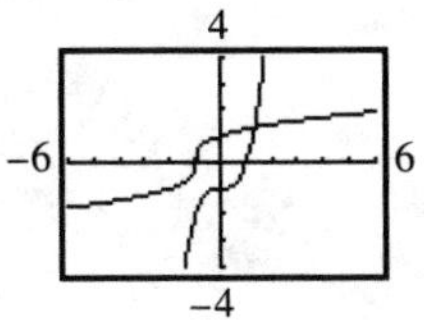

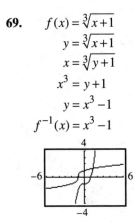

Section 9.3 Practice Exercises

1.

$f(x) = 2^x$	x	0	1	2	3	−1	−2
	$f(x)$	1	2	4	8	$\frac{1}{2}$	$\frac{1}{4}$

$g(x) = 7^x$	x	0	1	2	3	−1	−2
	$g(x)$	1	7	49	343	$\frac{1}{7}$	$\frac{1}{49}$

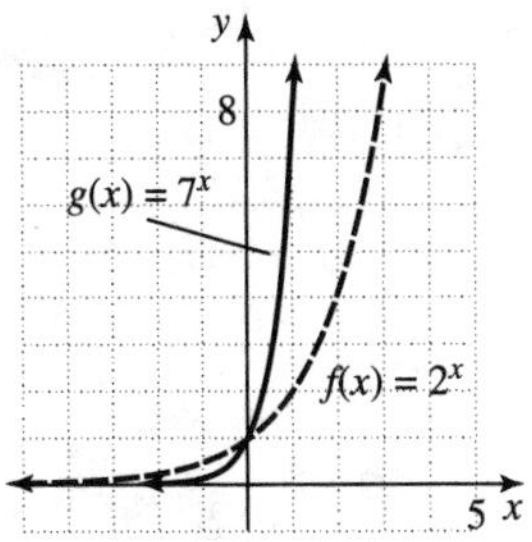

2.

$f(x) = \left(\frac{1}{3}\right)^x$	x	0	1	2	3	−1	−2
	$f(x)$	1	$\frac{1}{3}$	$\frac{1}{9}$	$\frac{1}{27}$	3	9

$g(x) = \left(\frac{1}{5}\right)^x$	x	0	1	2	3	−1	−2
	$g(x)$	1	$\frac{1}{5}$	$\frac{1}{25}$	$\frac{1}{125}$	5	25

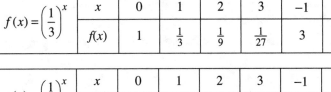

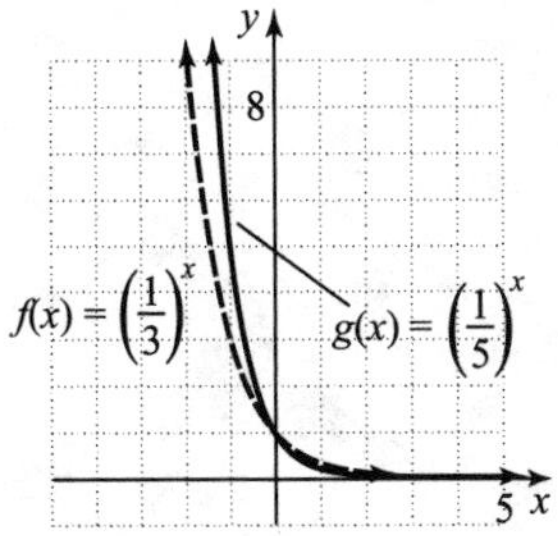

3. $f(x) = 2^{x-3}$

$f(x) = 2^{x-3}$	x	6	5	4	3	2	1	0
	$f(x)$	8	4	2	1	$\frac{1}{2}$	$\frac{1}{4}$	$\frac{1}{8}$

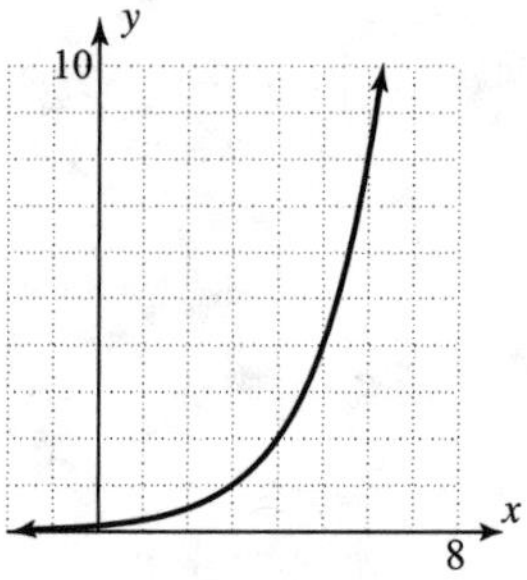

4. a. $3^x = 9$

Write 9 as a power of 3, $9 = 3^2$.

$3^x = 3^2$, thus, $x = 2$.

b. $8^x = 16$

Write 8 and 16 as powers of 2.

$8 = 2^3$ and $16 = 2^4$.

$$\begin{aligned} 8^x &= 16 \\ (2^3)^x &= 2^4 \\ 2^{3x} &= 2^4 \\ 3x &= 4 \\ x &= \frac{4}{3} \end{aligned}$$

c. $125^x = 25^{x-2}$

Write 125 and 25 as powers of 5.

$125 = 5^3$ and $25 = 5^2$.

$$\begin{aligned} 125^x &= 25^{x-2} \\ (5^3)^x &= (5^2)^{x-2} \\ 5^{3x} &= 5^{2x-4} \\ 3x &= 2x-4 \\ x &= -4 \end{aligned}$$

5. $P = \$3000$, $r = 7\% = 0.07$, $n = 2$, and $t = 4$.

$$A = P\left(1+\frac{r}{n}\right)^{nt}$$

$$A = 3000\left(1+\frac{0.07}{2}\right)^{2(4)}$$

$$= 3000(1.035)^8$$

$$\approx 3950.43$$

Thus, the amount owed is approximately \$3950.43.

6. a. $p(n) = 100(2.7)^{-0.05n}$, $n = 2$

$$p(2) = 100(2.7)^{-0.05(2)}$$

$$= 100(2.7)^{-0.1}$$

$$\approx 90.54$$

Thus, approximately 90.54% of the light passes through.

b. $p(n) = 100(2.7)^{-0.05n}$, $n = 10$

$$p(10) = 100(2.7)^{-0.05(10)}$$

$$= 100(2.7)^{-0.5}$$

$$\approx 60.86$$

Thus, approximately 60.86% of the light passes through.

Graphing Calculator Explorations

1.

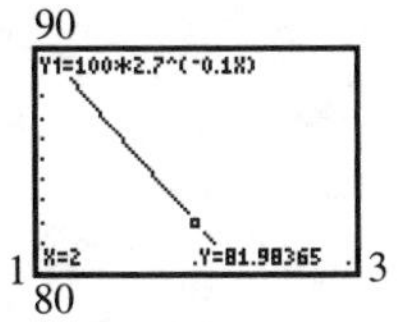

The expected percent after 2 days is 81.98%.

2.

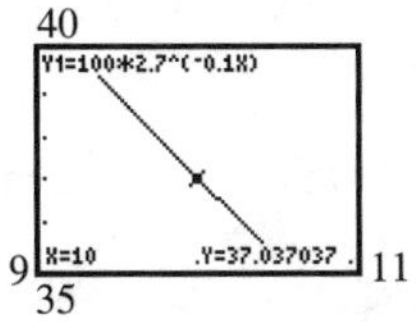

The expected percent after 10 days is 37.04%.

3.

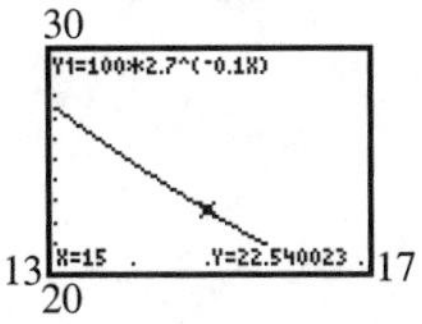

The expected percent after 15 days is 22.54%.

4.

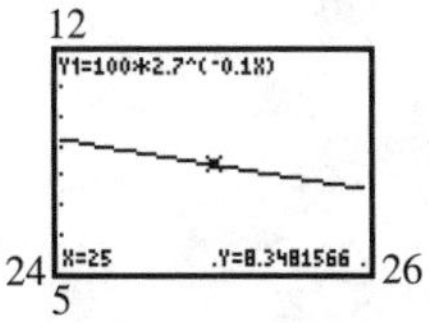

The expected percent after 25 days is 8.35%.

Vocabulary, Readiness & Video Check 9.3

1. A function such as $f(x) = 2^x$ is an exponential function; **C**.

2. If $7^x = 7^y$, then $x = y$; **B**.

3. Yes, the graph passes the vertical line test.

4. Yes, the function passes both the vertical and horizontal line tests.

5. The function has no x-intercept.

6. The function has a y-intercept of (0, 1).

7. The domain of this function, in interval notation, is $(-\infty, \infty)$.

8. The range of this function, in interval notation, is $(0, \infty)$.

9. In a polynomial function, the base is the variable and the exponent is the constant; in an exponential function, the base is the constant and the exponent is the variable.

10. rewrite the equation so the bases are the same

11. $y = 30(0.996)^{101} \approx 20.0$ lb

Exercise Set 9.3

1. $y = 5^x$

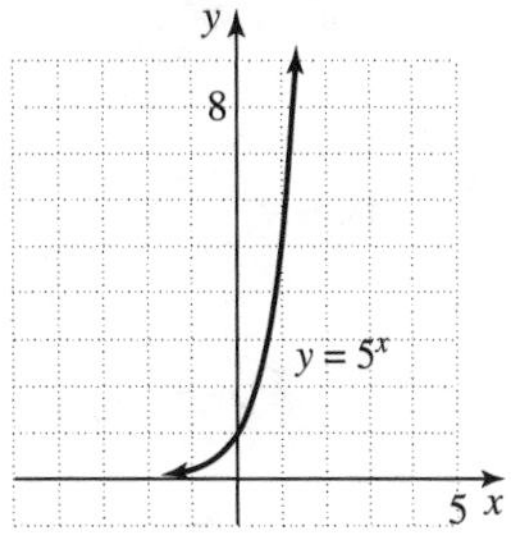

3. $y = 2^x + 1$

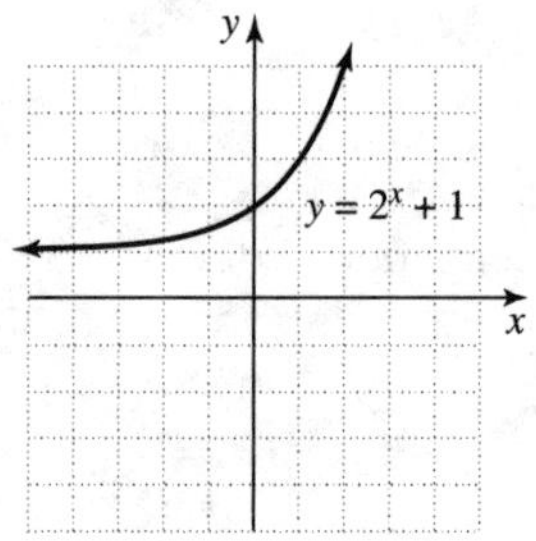

5. $y = \left(\frac{1}{4}\right)^x$

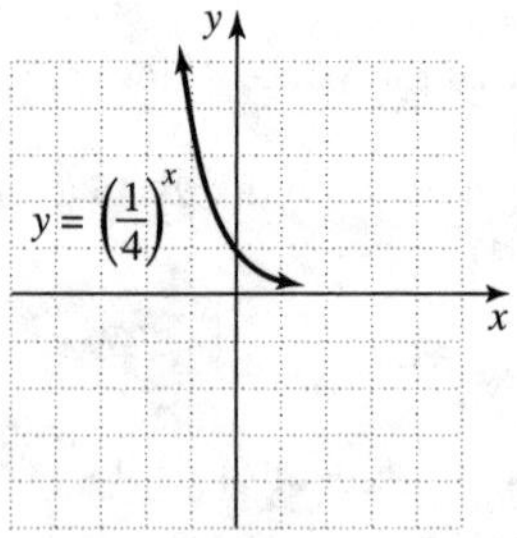

7. $y = \left(\frac{1}{2}\right)^x - 2$

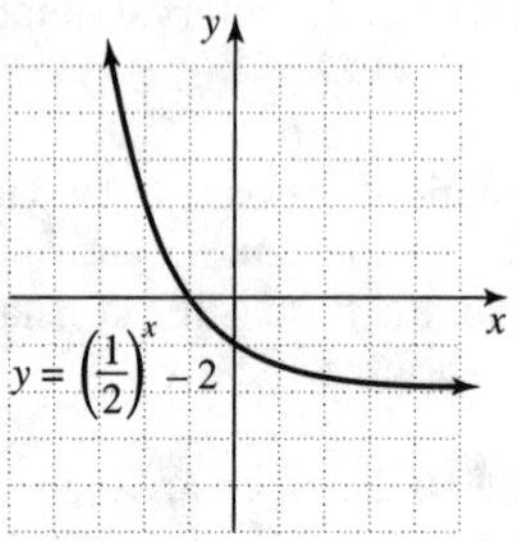

9. $y = -2^x$

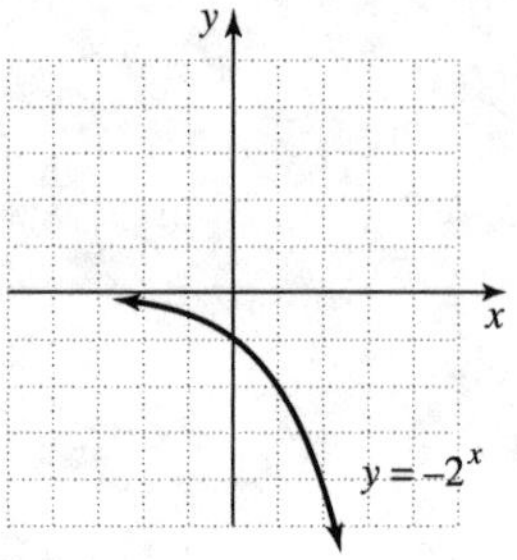

11. $y = -\left(\frac{1}{4}\right)^x$

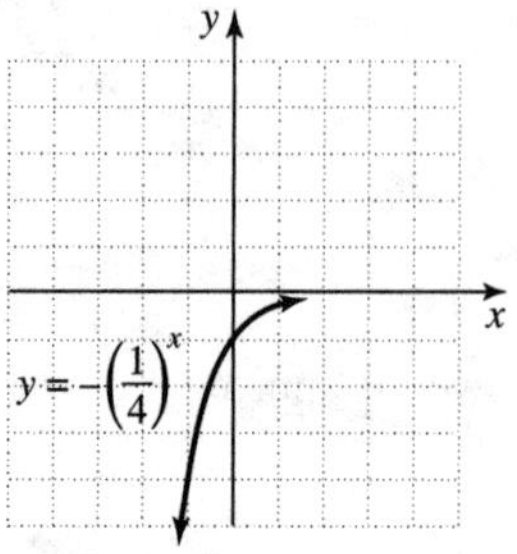

13. $f(x) = 2^{x+1}$

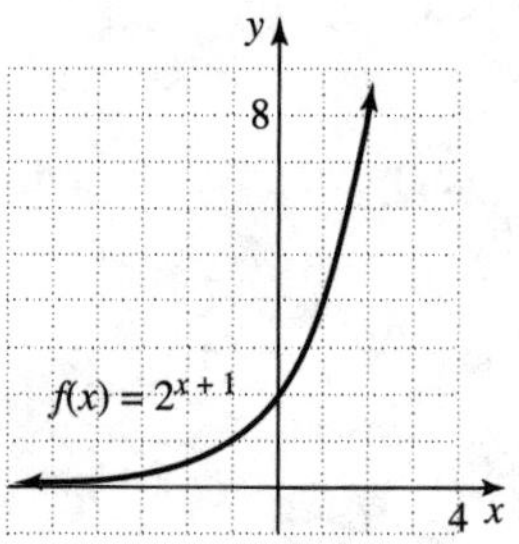

15. $f(x) = 4^{x-2}$

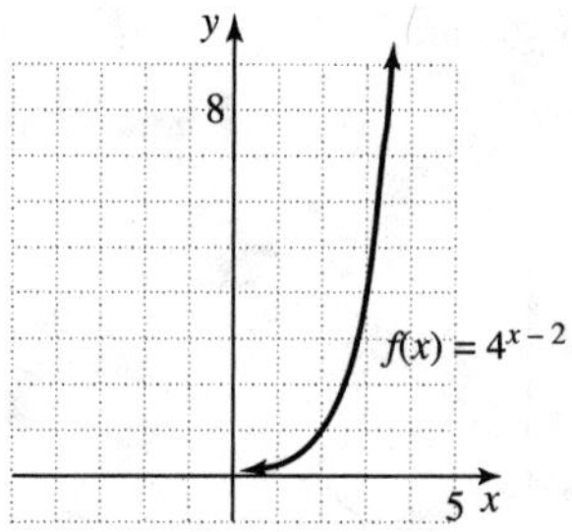

17. C

19. B

21. $3^x = 27$
$3^x = 3^3$
$x = 3$
The solution is 3.

23. $16^x = 8$
$(2^4)^x = 2^3$
$2^{4x} = 2^3$
$4x = 3$
$x = \frac{3}{4}$

The solution is $\frac{3}{4}$.

25. $32^{2x-3} = 2$
$(2^5)^{2x-3} = 2^1$
$2^{10x-15} = 2^1$
$10x - 15 = 1$
$10x = 16$
$x = \frac{8}{5}$
The solution is $\frac{8}{5}$.

27. $\frac{1}{4} = 2^{3x}$
$2^{-2} = 2^{3x}$
$3x = -2$
$x = -\frac{2}{3}$
The solution is $-\frac{2}{3}$.

29. $5^x = 625$
$5^x = 5^4$
$x = 4$
The solution is 4.

31. $4^x = 8$
$(2^2)^x = 2^3$
$2^{2x} = 2^3$
$2x = 3$
$x = \frac{3}{2}$
The solution is $\frac{3}{2}$.

33. $27^{x+1} = 9$
$(3^3)^{x+1} = 3^2$
$3^{3x+3} = 3^2$
$3x + 3 = 2$
$3x = -1$
$x = -\frac{1}{3}$
The solution is $-\frac{1}{3}$.

35. $81^{x-1} = 27^{2x}$
$(3^4)^{x-1} = (3^3)^{2x}$
$3^{4x-4} = 3^{6x}$
$4x - 4 = 6x$
$-4 = 2x$
$x = -2$
The solution is -2.

37. $y = 30(0.996)^x$
$y = 30(0.996)^{50} \approx 24.6$
There will be about 24.6 pounds left after 50 days.

39. $y = 29.2(1.015)^x$

a. 2010 is 5 years after 2005, so = 5.
$y = 29.2(1.015)^5 \approx 31.5$
Consumption of cheese in the United States was 31.5 pounds per person per year in 2010.

b. 2020 is 15 years after 2005, so $x = 15$.
$y = 29.2(1.015)^{15} \approx 36.5$
Consumption of cheese in the United States is predicted to be 36.5 pounds per person per year in 2020.

41. $y = 122.1(1.065)^x$

a. 2003 is 5 years after 1998, so $x = 5$.
$y = 122.1(1.065)^5 \approx 167.3$
Approximately 167.3 thousand American college students studied abroad in 2003.

b. 2018 is 20 years after 1998, so $x = 20$.
$y = 122.1(1.065)^{20} \approx 430.2$
The model predicts that 430.2 thousand Americans will study abroad in 2018.

43. $y = 5723(1.33)^x$
2020 is 13 years after 2007, so $x = 13$.
$y = 5723(1.33)^{13} \approx 233{,}182$
The model predicts 233,182 thousand Netflix subscribers at the beginning of 2020.

45. $y = 200{,}000(1.08)^x,\ x = 13$
$y = 200{,}000(1.08)^{13} \approx 544{,}000$
There will be approximately 544,000 mosquitoes on May 25.

47. $A = P\left(1+\frac{r}{n}\right)^{nt}$

$t = 3$, $P = 6000$, $r = 0.08$, and $n = 12$

$$A = 6000\left(1+\frac{0.08}{12}\right)^{12(3)}$$
$$= 6000\left(1+\frac{0.08}{12}\right)^{36}$$
$$\approx 7621.42$$

Erica would owe \$7621.42 after 3 years.

49. $A = P\left(1+\frac{r}{n}\right)^{nt}$

$P = 2000$

$r = 0.06,\ n = 2,$ and $t = 12$

$$A = 2000\left(1+\frac{0.06}{2}\right)^{2(12)}$$
$$= 2000(1.03)^{24}$$
$$\approx 4065.59$$

Janina has approximately \$4065.59 in her savings account.

51. $5x - 2 = 18$

$5x = 20$

$x = 4$

The solution is 4.

53. $3x - 4 = 3(x+1)$

$3x - 4 = 3x + 3$

$-4 = 3$

This is a false statement. The solution set is ∅.

55. $x^2 + 6 = 5x$

$x^2 - 5x + 6 = 0$

$(x-2)(x-3) = 0$

$x = 2$ or $x = 3$

The solutions are 2 and 3.

57. $2^x = 8$

$2^3 = 8$

$x = 3$

59. $5^x = \frac{1}{5}$

$5^{-1} = \frac{1}{5}$

$x = -1$

61. Since there are no variables in the exponent in $f(x) = 1.5x^2$, it is not an exponential function.

63. Since there are no variables in the exponent in $h(x) = \left(\frac{1}{2}x\right)^2$, it is not an exponential function.

65. $f(x) = 2^{-x}$

$f(1) = 2^{-1} = \frac{1}{2}$

This is graph C.

67. $f(x) = 4^{-x}$

$f(1) = 4^{-1} = \frac{1}{4}$

This is graph D.

69. answers may vary

71. $y = |3^x|$

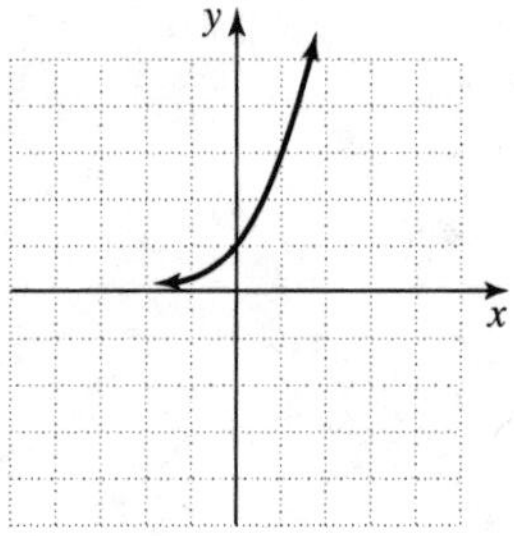

73. $y = 3^{|x|}$

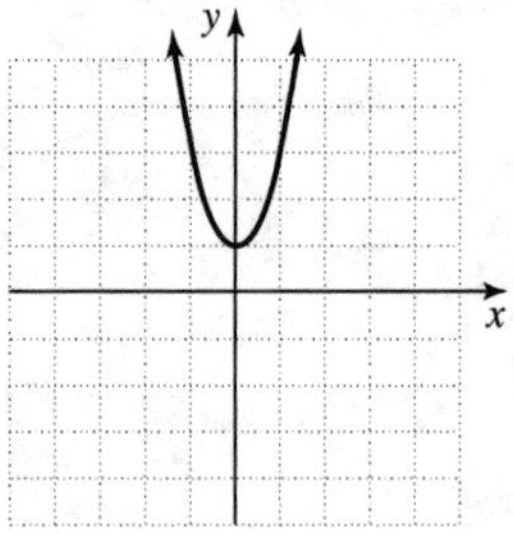

75.

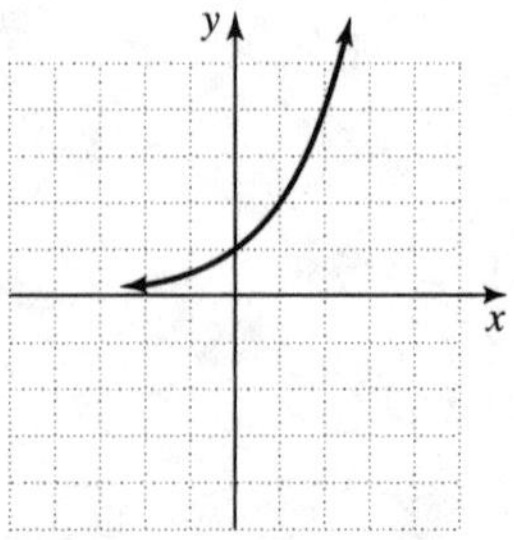

The graphs are the same, since $\left(\frac{1}{2}\right)^{-x} = 2^x$.

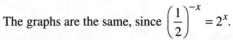

77. The result is the same, 24.55 pounds.

79. $y = 30(0.996)^x$

$y = 30(0.996)^{100} \approx 20.09$

After 100 days, the estimate is 20.09 pounds left.

Section 9.4 Practice Exercises

1. $C = 25{,}000,\ r = 0.12,\ x = 2022 - 2007 = 15$

$y = C(1+r)^x$

$y = 25{,}000(1+0.12)^{15}$

$= 25{,}000(1.12)^{15}$

$\approx 136{,}839$

In 2022, the predicted population is 136,839.

2. $C = 800,\ r = 0.30,\ x = 9$

$y = C(1-r)^x$

$y = 800(1-0.30)^9 = 800(0.70)^9 \approx 32$

After 9 rounds, there are 32 players remaining.

3. $C = 500,\ r = 0.50,\ x = \frac{51}{15} = 3.4$

$y = C(1-r)^x$

$y = 500(1-0.50)^{3.4} = 500(0.50)^{3.4} \approx 47.4$

In 51 years, 47.4 grams of DDT remain.

Vocabulary, Readiness & Video Check 9.4

1. For Example 1, the growth rate is given as 5% per year. Since this is "per year," the number of time intervals is the "number of years," or 8.

2. The number of employees is decreasing and not increasing. It is exponential decay because the decrease is the same percent per year.

3. time intervals = years/half-life; the decay rate is 50% or $\frac{1}{2}$ because half-life is the amount of time it takes half of a substance to decay.

Exercise Set 9.4

	Original Amount	Growth Rate Per Year	Number of Years, x	Final Amount after x Years of Growth
1.	305	5%	8	$y = 305(1+0.05)^8 \approx 451$
3.	2000	11%	41	$y = 2000(1+0.11)^{41} \approx 144{,}302$
5.	17	29%	28	$y = 17(1+0.29)^{28} \approx 21{,}231$

	Original Amount	Decay Rate per Year	Number of Years, x	Final Amount after x Years of Decay
7.	305	5%	8	$y = 305(1-0.05)^8 \approx 202$
9.	10,000	12%	15	$y = 10,000(1-0.12)^{15} \approx 1470$
11.	207,000	32%	25	$y = 207,000(1-0.32)^{25} \approx 13$

13. $C = 500,000,\ r = 0.03,\ x = 12$

$y = C(1+r)^x = 500,000(1+0.03)^{12} \approx 712,880$

In 12 years, the population is predicted to be 712,880.

15. $C = 640,\ r = 0.05,\ x = 10$

$y = C(1-r)^x = 640(1-0.05)^{10} \approx 383$

In 10 years, the number of employees is predicted to be 383.

17. $C = 260,\ r = 0.025,\ x = 10$

$y = C(1+r)^x = 260(1+0.025)^{10} \approx 333$

In 10 years, the predicted number of bison is 333.

19. $C = 5,\ r = 0.15,\ x = 10$

$y = C(1-r)^x = 5(1-0.15)^{10} \approx 1$

After 10 seconds, there will be 1 gram of the isotope.

	Original Amount	Half-Life (in years)	Number of Years	Time intervals, x	Final Amount after x Time Intervals	Is amount reasonable?
21. a.	40	7	14	$\frac{14}{7} = 2$	$40(1-0.5)^2 = 10$	yes
b.	40	7	11	$\frac{11}{7} \approx 1.6$	$40(1-0.5)^{1.6} \approx 13.2$	yes
23.	21	152	500	$\frac{500}{152} \approx 3.3$	$21(1-0.5)^{3.3} \approx 2.1$	yes

25. $C = 30,\ r = 0.5,\ x = \dfrac{250}{96}$

$y = 30(1-0.5)^{250/96}$

$y \approx 4.9$

In 250 years, the predicted amount is 4.9 grams.

27. $2^x = 8$

$2^x = 2^3$

$x = 3$

29. $5^x = \frac{1}{5}$

$5^x = 5^{-1}$

$x = -1$

31. no; answers may vary.

Section 9.5 Practice Exercises

1. a. $\log_3 81 = 4$ means $3^4 = 81$.

b. $\log_5 \frac{1}{5} = -1$ means $5^{-1} = \frac{1}{5}$.

c. $\log_7 \sqrt{7} = \frac{1}{2}$ means $7^{1/2} = \sqrt{7}$.

d. $\log_{13} y = 4$ means $13^4 = y$.

2. a. $4^3 = 64$ means $\log_4 64 = 3$.

b. $6^{1/3} = \sqrt[3]{6}$ means $\log_6 \sqrt[3]{6} = \frac{1}{3}$.

c. $5^{-3} = \frac{1}{125}$ means $\log_5 \frac{1}{125} = -3$.

d. $\pi^7 = z$ means $\log_\pi z = 7$.

3. a. $\log_3 9 = 2$ because $3^2 = 9$.

b. $\log_2 \frac{1}{8} = -3$ because $2^{-3} = \frac{1}{8}$.

c. $\log_{49} 7 = \frac{1}{2}$ because $49^{1/2} = 7$.

4. a. $\log_5 \frac{1}{25} = x$

$\log_5 \frac{1}{25} = x$ means $5^x = \frac{1}{25}$. Solve $5^x = \frac{1}{25}$.

$5^x = \frac{1}{25}$

$5^x = 5^{-2}$

Since the bases are the same, by the uniqueness of b^x, we have that $x = -2$. The solution is -2 or the solution set is $\{-2\}$.

b. $\log_x 8 = 3$

$x^3 = 8$

$x^3 = 2^3$

$x = 2$

c. $\log_6 x = 2$

$6^2 = x$

$36 = x$

d. $\log_{13} 1 = x$

$13^x = 1$

$13^x = 13^0$

$x = 0$

e. $\log_h 1 = x$

$h^x = 1$

$h^x = h^0$

$x = 0$

5. a. From Property 2, $\log_5 5^4 = 4$.

b. From Property 2, $\log_9 9^{-2} = -2$.

c. From Property 3, $6^{\log_6 5} = 5$.

d. From Property 3, $7^{\log_7 4} = 4$.

6. $y = \log_9 x$ means that $9^y = x$. Find some ordered pair solutions that satisfy $9^y = x$.

$x = 9^y$	y
1	0
9	1
$\frac{1}{9}$	-1
$\frac{1}{81}$	-2

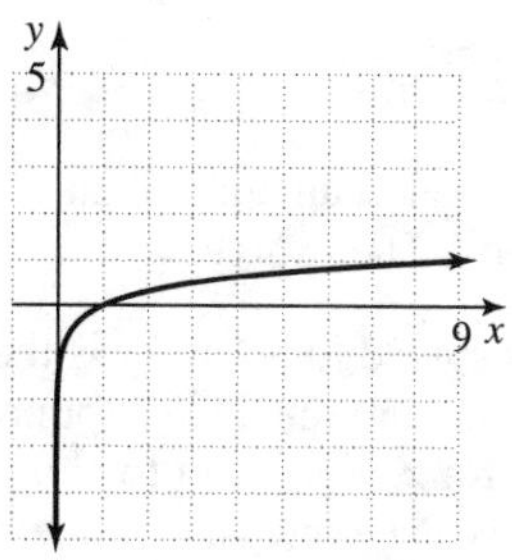

7. $y = \log_{1/4} x$ means that $\left(\frac{1}{4}\right)^y = x$. Find some ordered-pair solutions that satisfy $\left(\frac{1}{4}\right)^y = x$.

$x = \left(\frac{1}{4}\right)^y$	y
1	0
$\frac{1}{4}$	1
4	−1
16	−2

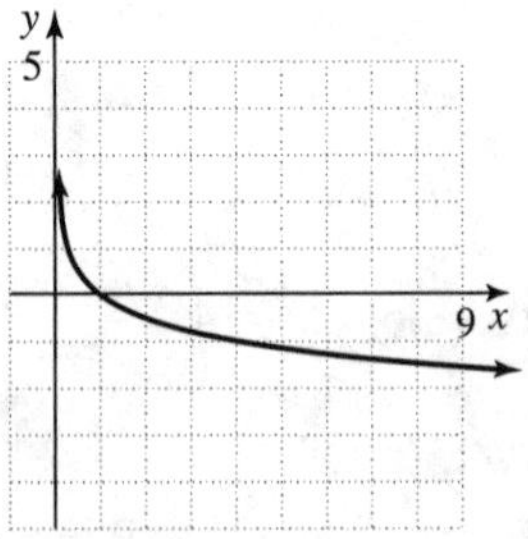

Vocabulary, Readiness & Video Check 9.5

1. A function such as $y = \log_2 x$ is a logarithmic function; **B**.
2. If $y = \log_2 x$, then $2^y = x$; **C**.
3. Yes, the function passes both the horizontal- and vertical-line tests.
4. The function has an x-intercept of (1, 0).
5. The function has no y-intercept.
6. The domain of this function, in interval notation, is $(0, \infty)$.
7. The range of this function, in interval notation, is $(-\infty, \infty)$.
8. Logarithms are exponents.
9. First write the equation as an equivalent exponential equation. Then solve.
10. The exponential is solved for x, and y is the exponent in the equation. Since the exponential equation is solved for x, it is easier to choose a y-value and simplify the expression containing y, which is then the x-value.

Exercise Set 9.5

1. $\log_6 36 = 2$
$6^2 = 36$

3. $\log_3 \frac{1}{27} = -3$
$3^{-3} = \frac{1}{27}$

5. $\log_{10} 1000 = 3$
$10^3 = 1000$

7. $\log_9 x = 4$
$9^4 = x$

9. $\log_\pi \frac{1}{\pi^2} = -2$
$\pi^{-2} = \frac{1}{\pi^2}$

11. $\log_7 \sqrt{7} = \frac{1}{2}$
$7^{1/2} = \sqrt{7}$

13. $\log_{0.7} 0.343 = 3$
$0.7^3 = 0.343$

15. $\log_3 \frac{1}{81} = -4$
$3^{-4} = \frac{1}{81}$

17. $2^4 = 16$
$\log_2 16 = 4$

19. $10^2 = 100$
$\log_{10} 100 = 2$

21. $\pi^3 = x$
$\log_\pi x = 3$

23. $10^{-1} = \frac{1}{10}$
$\log_{10} \frac{1}{10} = -1$

25. $4^{-2} = \frac{1}{16}$

$\log_4 \frac{1}{16} = -2$

27. $5^{1/2} = \sqrt{5}$

$\log_5 \sqrt{5} = \frac{1}{2}$

29. $\log_2 8 = 3$ since $2^3 = 8$.

31. $\log_3 \frac{1}{9} = -2$ since $3^{-2} = \frac{1}{9}$.

33. $\log_{25} 5 = \frac{1}{2}$ since $25^{1/2} = 5$.

35. $\log_{1/2} 2 = -1$ since $\left(\frac{1}{2}\right)^{-1} = 2$.

37. $\log_6 1 = 0$ since $6^0 = 1$.

39. $\log_{10} 100 = \log_{10} 10^2 = 2$

41. $\log_3 81 = \log_3 3^4 = 4$

43. $\log_4 \frac{1}{64} = \log_4 4^{-3} = -3$

45. $\log_3 9 = x$

$3^x = 9$

$3^x = 3^2$

$x = 2$

47. $\log_3 x = 4$

$x = 3^4 = 81$

49. $\log_x 49 = 2$

$x^2 = 49$

$x = \pm 7$

We discard the negative base.
$x = 7$

51. $\log_2 \frac{1}{8} = x$

$2^x = \frac{1}{8}$

$2^x = 2^{-3}$

$x = -3$

53. $\log_3 \frac{1}{27} = x$

$\frac{1}{27} = 3^x$

$3^{-3} = 3^x$

$-3 = x$

55. $\log_8 x = \frac{1}{3}$

$x = 8^{1/3} = 2$

57. $\log_4 16 = x$

$4^x = 16$

$4^x = 4^2$

$x = 2$

59. $\log_{3/4} x = 3$

$\left(\frac{3}{4}\right)^3 = x$

$\frac{27}{64} = x$

61. $\log_x 100 = 2$

$x^2 = 100$

$x = \pm 10$

We discard the negative base.
$x = 10$

63. $\log_2 2^4 = x$

$2^x = 2^4$

$x = 4$

65. $3^{\log_3 5} = x$

$5 = x$

67. $\log_x \frac{1}{7} = \frac{1}{2}$

$x^{1/2} = \frac{1}{7}$

$x = \frac{1}{49}$

69. $\log_5 5^3 = 3$

71. $2^{\log_2 3} = 3$

73. $\log_9 9 = 1$

75. $\log_8 (8)^{-1} = -1$

77. $y = \log_3 x$
$y = 0$:
$\log_3 x = 0$
$x = 3^0 = 1$
(1, 0) is the only x-intercept. No y-intercept exists.

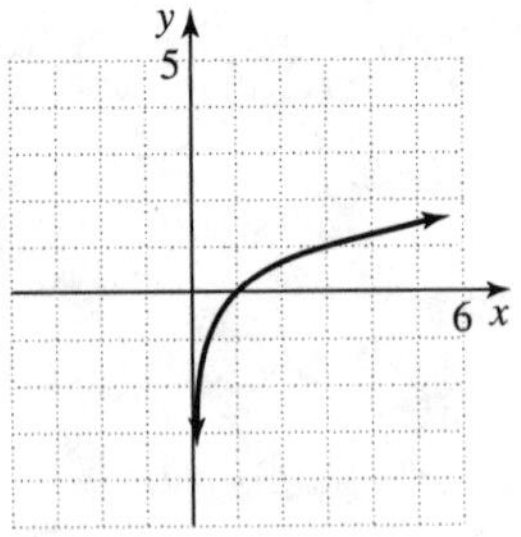

79. $f(x) = \log_{1/4} x$
$y = 0$: $0 = \log_{1/4} x$
$x = \left(\frac{1}{4}\right)^0 = 1$
(1, 0) is the x-intercept. No y-intercept exists.

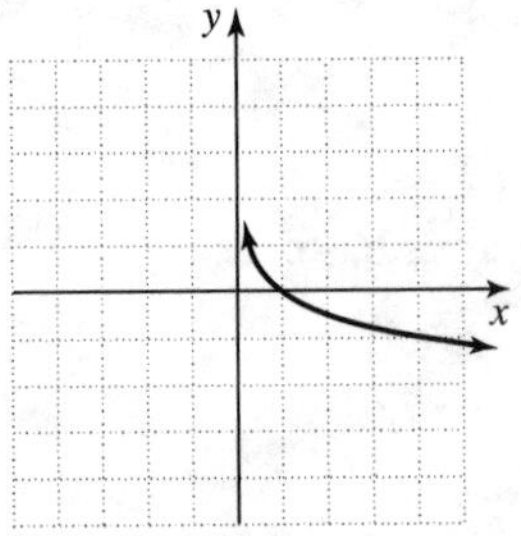

81. $f(x) = \log_5 x$
$y = 0$: $0 = \log_5 x$
$x = 5^0 = 1$
(1, 0) is the x-intercept. No y-intercept exists.

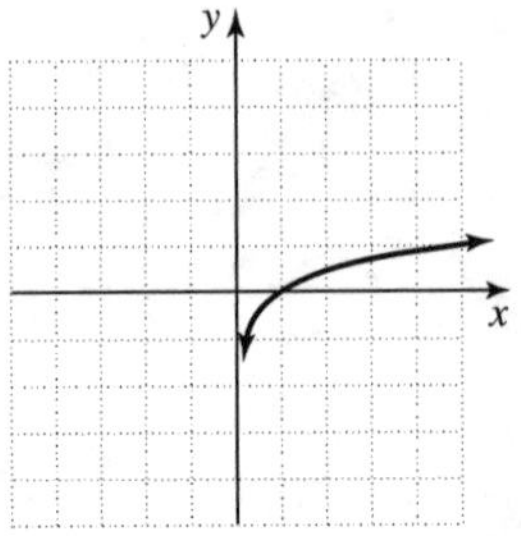

83. $f(x) = \log_{1/16} x$
$y = 0$:
$0 = \log_{1/6} x$
$x = \left(\frac{1}{6}\right)^0 = 1$
(1, 0) is the x-intercept. No y-intercept exists.

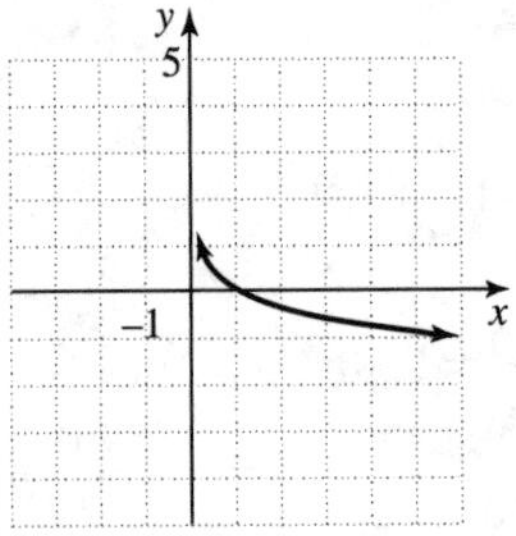

85. $\frac{x+3}{3+x} = \frac{x+3}{x+3} = 1$

87. $\frac{x^2-8x+16}{2x-8} = \frac{(x-4)^2}{2(x-4)} = \frac{x-4}{2}$

89. $\frac{2}{x} + \frac{3}{x^2} = \frac{2x}{x^2} + \frac{3}{x^2} = \frac{2x+3}{x^2}$

91. $\frac{3x}{x+3} + \frac{9}{x+3} = \frac{3x+9}{x+3} = \frac{3(x+3)}{x+3} = 3$

93. $f(x) = \log_5 x$; $f^{-1}(x) = g(x) = 5^x$

a. (2, 25) implies $g(2) = 25$.

b. Since $f^{-1}(x) = g(x)$, (25, 2) is a solution of $f(x)$.

c. (25, 2) implies $f(25) = 2$.

95. answers may vary

97. $\log_7(5x-2)=1$
$$5x-2=7^1$$
$$5x=9$$
$$x=\frac{9}{5}$$

99. $\log_3(\log_5 125)=\log_3(3)=1$

101. $y=4^x$; $y=\log_4 x$

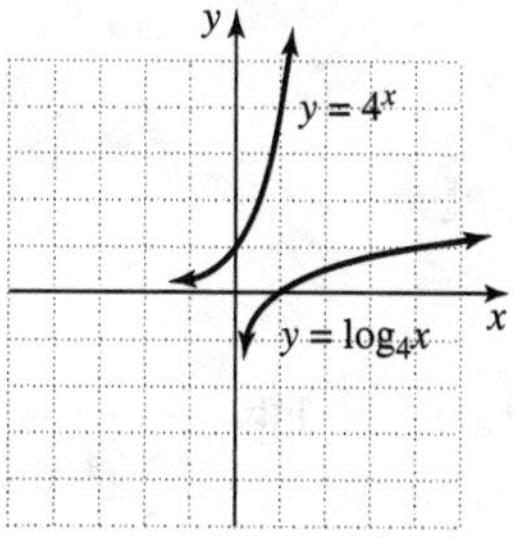

103. $y=\left(\frac{1}{3}\right)^x$; $y=\log_{1/3} x$

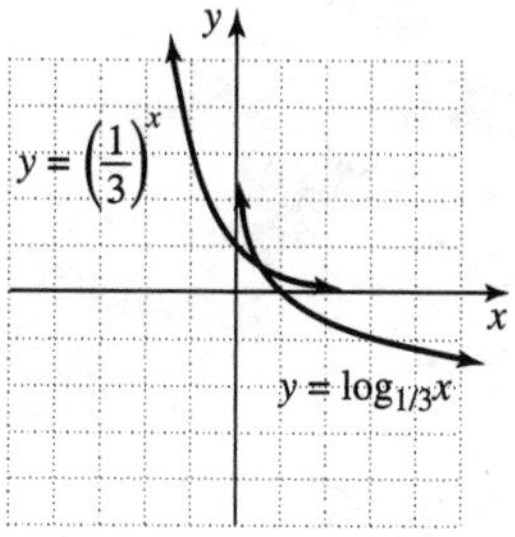

105. answers may vary

107. $\log_{10}(1-k)=\frac{-0.3}{H}$, $H=8$
$$\log_{10}(1-k)=\frac{-0.3}{8}=-0.0375$$
$$1-k=10^{-0.0375}$$
$$1-10^{-0.0375}=k$$
$$k\approx 0.0827$$
The rate of decay is 0.0827.

Section 9.6 Practice Exercises

1. a. $\log_8 5+\log_8 3=\log_8(5\cdot 3)=\log_8 15$

b. $\log_2\frac{1}{3}+\log_2 18=\log_2\left(\frac{1}{3}\cdot 18\right)=\log_2 6$

c. $\log_5(x-1)+\log_5(x+1)=\log_5[(x+1)(x+1)]$
$$=\log_5(x^2-1)$$

2. a. $\log_5 18-\log_5 6=\log_5\frac{18}{6}=\log_5 3$

b. $\log_6 x-\log_6 3=\log_6\frac{x}{3}$

c. $\log_4(x^2+1)-\log_4(x^2+3)=\log_4\frac{x^2+1}{x^2+3}$

3. a. $\log_7 x^8=8\log_7 x$

b. $\log_5\sqrt[4]{7}=\log_5 7^{1/4}=\frac{1}{4}\log_5 7$

4. a. $2\log_5 4+5\log_5 2=\log_5 4^2+\log_5 2^5$
$$=\log_5 16+\log_5 32$$
$$=\log_5(16\cdot 32)$$
$$=\log_5 512$$

b. $2\log_8 x-\log_8(x+3)=\log_8 x^2-\log_8(x+3)$
$$=\log_8\frac{x^2}{x+3}$$

c. $\log_7 12+\log_7 5-\log_7 4$
$$=\log_7(12\cdot 5)-\log_7 4$$
$$=\log_7 60-\log_7 4$$
$$=\log_7\frac{60}{4}$$
$$=\log_7 15$$

5. a. $\log_5\frac{4\cdot 3}{7}=\log_5(4\cdot 3)-\log_5 7$
$$=\log_5 4+\log_5 3-\log_5 7$$

b. $\log_4\frac{a^2}{b^5}=\log_4 a^2-\log_4 b^5$
$$=2\log_4 a-5\log_4 b$$

6. $\log_b 5=0.83$ and $\log_b 3=0.56$

a. $\log_b 15=\log_b(3\cdot 5)$
$$=\log_b 3+\log_b 5$$
$$=0.56+0.83$$
$$=1.39$$

b. $\log_b 25 = \log_b 5^2 = 2\log_b 5 = 2(0.83) = 1.66$

c. $\log_b \sqrt{3} = \log_b 3^{1/2}$
$= \frac{1}{2}\log_b 3$
$= \frac{1}{2}(0.56)$
$= 0.28$

Vocabulary, Readiness & Video Check 9.6

1. $\log_b 12 + \log_b 3 = \log_b(12 \cdot 3) = \log_b \underline{36}$; **a.**

2. $\log_b 12 - \log_b 3 = \log_b \frac{12}{3} = \log_b \underline{4}$; **c.**

3. $7\log_b 2 = \underline{\log_b 2^7}$; **b.**

4. $\log_b 1 = \underline{0}$; **c.**

5. $b^{\log_b x} = \underline{x}$; **a.**

6. $\log_5 5^2 = \underline{2}$; **b.**

7. No, the product property says the logarithm of a product can be written as a sum of logarithms—the expression in Example 2 is a logarithm of a sum.

8. The bases must be the same.

9. Since $\frac{1}{x} = x^{-1}$, this gives us $\log_2 x^{-1}$. Using the power property, we get $-1\log_2 x$ or $-\log_2 x$.

10. From writing logarithms as equivalent exponents and then using the rules for exponents.

Exercise Set 9.6

1. $\log_5 2 + \log_5 7 = \log_5(2 \cdot 7) = \log_5 14$

3. $\log_4 9 + \log_4 x = \log_4 9x$

5. $\log_6 x + \log_6(x+1) = \log_6[x(x+1)]$
$= \log_6(x^2 + x)$

7. $\log_{10} 5 + \log_{10} 2 + \log_{10}(x^2 + 2)$
$= \log_{10}\left[5 \cdot 2\left(x^2 + 2\right)\right]$
$= \log_{10}\left(10x^2 + 20\right)$

9. $\log_5 12 - \log_5 4 = \log_5 \frac{12}{4} = \log_5 3$

11. $\log_3 8 - \log_3 2 = \log_3 \frac{8}{2} = \log_3 4$

13. $\log_2 x - \log_2 y = \log_2 \frac{x}{y}$

15. $\log_2(x^2 + 6) - \log(x^2 + 1) = \log_2 \frac{x^2 + 6}{x^2 + 1}$

17. $\log_3 x^2 = 2\log_3 x$

19. $\log_4 5^{-1} = -\log_4 5$

21. $\log_5 \sqrt{y} = \log_5 y^{1/2} = \frac{1}{2}\log_5 y$

23. $\log_2 5 + \log_2 x^3 = \log_2 5x^3$

25. $3\log_4 2 + \log_4 6 = \log_4 2^3 + \log_4 6$
$= \log_4 8 + \log_4 6$
$= \log_4(8 \cdot 6)$
$= \log_4 48$

27. $3\log_5 x + 6\log_5 z = \log_5 x^3 + \log_5 z^6$
$= \log_5 x^3 z^6$

29. $\log_4 2 + \log_4 10 - \log_4 5 = \log_4(2 \cdot 10) - \log_4 5$
$= \log_4 \frac{20}{5}$
$= \log_4 4$
$= 1$

31. $\log_7 6 + \log_7 3 - \log_7 4 = \log_7(6 \cdot 3) - \log_7 4$
$= \log_7 \frac{18}{4}$
$= \log_7 \frac{9}{2}$

33. $\log_{10} x - \log_{10}(x+1) + \log_{10}(x^2-2)$
$= \log_{10}\dfrac{x}{x+1} + \log_{10}(x^2-2)$
$= \log_{10}\dfrac{x(x^2-2)}{x+1}$
$= \log_{10}\dfrac{x^3-2x}{x+1}$

35. $3\log_2 x + \frac{1}{2}\log_2 x - 2\log_2(x+1)$
$= \log_2 x^3 + \log_2 x^{1/2} - \log_2(x+1)^2$
$= \log_2(x^3 \cdot x^{1/2}) - \log_2(x+1)^2$
$= \log_2 x^{7/2} - \log_2(x+1)^2$
$= \log_2 \dfrac{x^{7/2}}{(x+1)^2}$

37. $2\log_8 x - \frac{2}{3}\log_8 x + 4\log_8 x = \left(2 - \frac{2}{3} + 4\right)\log_8 x$
$= \frac{16}{3}\log_8 x$
$= \log_8 x^{16/3}$

39. $\log_3 \frac{4y}{5} = \log_3 4y - \log_3 5$
$= \log_3 4 + \log_3 y - \log_3 5$

41. $\log_4 \frac{5}{9z} = \log_4 5 - \log_4 9z$
$= \log_4 5 - (\log_4 9 + \log_4 z)$
$= \log_4 5 - \log_4 9 - \log_4 z$

43. $\log_2 \frac{x^3}{y} = \log_2 x^3 - \log_2 y$
$= 3\log_2 x - \log_2 y$

45. $\log_b \sqrt{7x} = \log_b (7x)^{1/2}$
$= \frac{1}{2}\log_b(7x)$
$= \frac{1}{2}[\log_b 7 + \log_b x]$
$= \frac{1}{2}\log_b 7 + \frac{1}{2}\log_b x$

47. $\log_6 x^4 y^5 = \log_6 x^4 + \log_6 y^5$
$= 4\log_6 x + 5\log_6 y$

49. $\log_5 x^3(x+1) = \log_5 x^3 + \log_5(x+1)$
$= 3\log_5 x + \log_5(x+1)$

51. $\log_6 \frac{x^2}{x+3} = \log_6 x^2 - \log_6(x+3)$
$= 2\log_6 x - \log_6(x+3)$

53. $\log_b 15 = \log_b(5\cdot 3)$
$= \log_b 5 + \log_b 3$
$= 0.7 + 0.5$
$= 1.2$

55. $\log_b \frac{5}{3} = \log_b 5 - \log_b 3 = 0.7 - 0.5 = 0.2$

57. $\log_b \sqrt{5} = \log_b 5^{1/2} = \frac{1}{2}\log_b 5 = \frac{1}{2}(0.7) = 0.35$

59. $\log_b 8 = \log_b 2^3 = 3\log_b 2 = 3(0.43) = 1.29$

61. $\log_b \frac{3}{9} = \log_b 3 - \log_b 9$
$= \log_b 3 - \log_b 3^2$
$= \log_b 3 - 2\log_b 3$
$= -\log_b 3$
$= -0.68$

63. $\log_b \sqrt{\frac{2}{3}} = \log_b \left(\frac{2}{3}\right)^{1/2}$
$= \frac{1}{2}\log_b \frac{2}{3}$
$= \frac{1}{2}(\log_b 2 - \log_b 3)$
$= \frac{1}{2}(0.43 - 0.68)$
$= \frac{1}{2}(-0.25)$
$= -0.125$

65. $y = 10^x$ and $y = \log_{10} x$

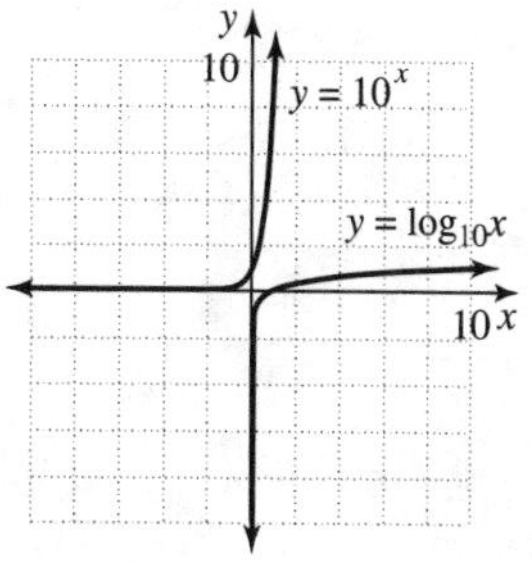

67. $\log_{10} 100 = \log_{10} 10^2 = 2$

69. $\log_7 7^2 = 2$

71. $\log_3 \frac{14}{11} = \log_3 14 - \log_3 11$; **b**

73. $\log_2 x^3 = 3\log_2 x$ is true.

75. $\frac{\log_7 10}{\log_7 5} = \log_7 2$ is false.

77. $\frac{\log_7 x}{\log_7 y} = (\log_7 x) - (\log_7 y)$ is false.

79. $\log_b 8$ equals $\log_b 8 + \log_b 1$ because $\log_b 1 = 0$.

Integrated Review

1. $(f+g)(x) = x - 6 + x^2 + 1 = x^2 + x - 5$

2. $(f-g)(x) = x - 6 - (x^2 + 1) = -x^2 + x - 7$

3. $(f \cdot g)(x) = (x-6)(x^2+1) = x^3 - 6x^2 + x - 6$

4. $\left(\frac{f}{g}\right)(x) = \frac{x-6}{x^2+1}$

5. $(f \circ g)(x) = f(g(x)) = f(3x-1) = \sqrt{3x-1}$

6. $(g \circ f)(x) = g(f(x)) = g\left(\sqrt{x}\right) = 3\sqrt{x} - 1$

7. one-to-one; inverse: $\{(6,-2), (8,4), (-6,2), (3,3)\}$

8. not one-to-one

9. not one-to-one

10. one-to-one

11. not one-to-one

12.
$$f(x) = 3x$$
$$y = 3x$$
$$x = 3y$$
$$y = \frac{x}{3}$$
$$f^{-1}(x) = \frac{x}{3}$$

13.
$$f(x) = x + 4$$
$$y = x + 4$$
$$x = y + 4$$
$$y = x - 4$$
$$f^{-1}(x) = x - 4$$

14.
$$f(x) = 5x - 1$$
$$y = 5x - 1$$
$$x = 5y - 1$$
$$5y = x + 1$$
$$y = \frac{x+1}{5}$$
$$f^{-1}(x) = \frac{x+1}{5}$$

15.
$$f(x) = 3x + 2$$
$$y = 3x + 2$$
$$x = 3y + 2$$
$$3y = x - 2$$
$$y = \frac{x-2}{3}$$
$$f^{-1}(x) = \frac{x-2}{3}$$

16. $y = \left(\frac{1}{2}\right)^x$

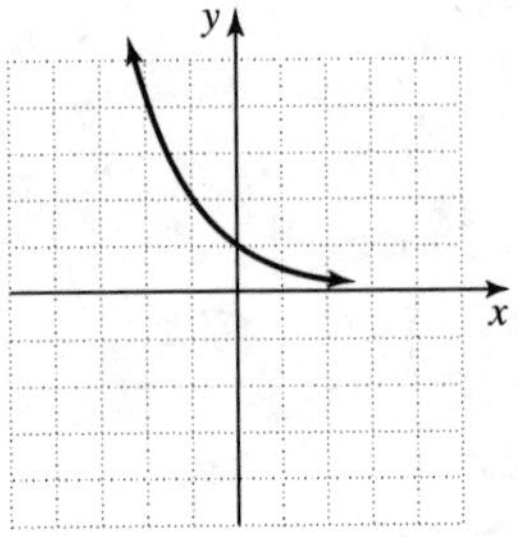

17. $y = 2^x + 1$

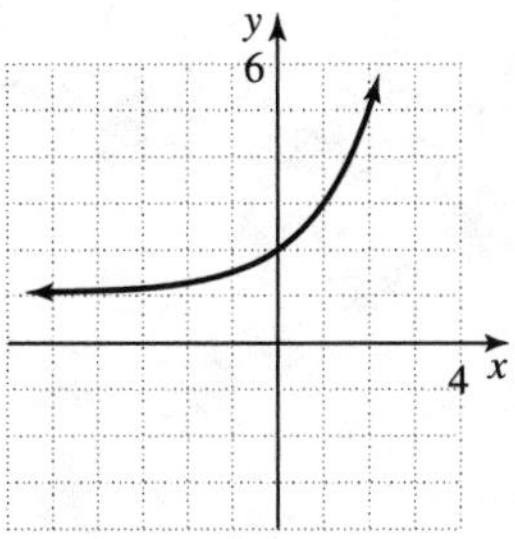

18. $y = \log_3 x$

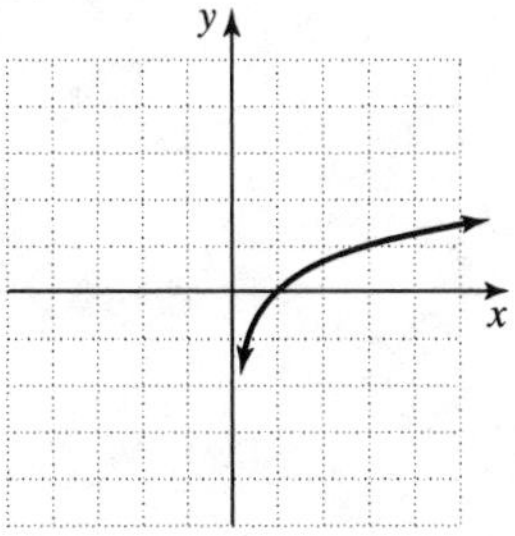

19. $y = \log_{1/3} x$

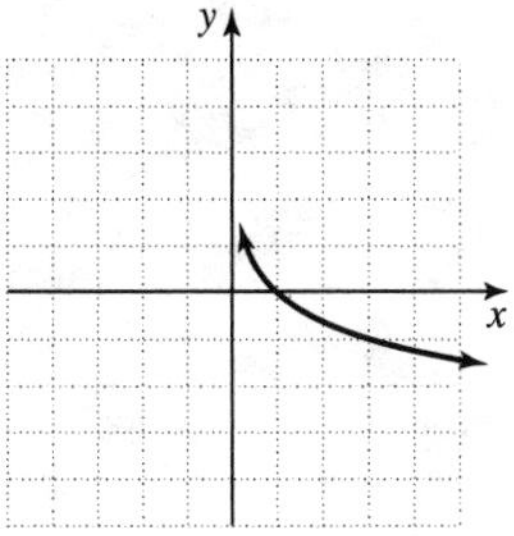

20. $2^x = 8$
$2^x = 2^3$
$x = 3$
The solution is 3.

21. $9 = 3^{x-5}$
$3^2 = 3^{x-5}$
$2 = x - 5$
$7 = x$
The solution is 7.

22. $4^{x-1} = 8^{x+2}$
$(2^2)^{x-1} = (2^3)^{x+2}$
$2^{2x-2} = 2^{3x+6}$
$2x - 2 = 3x + 6$
$-8 = x$
The solution is –8.

23. $25^x = 125^{x-1}$
$(5^2)^x = (5^3)^{x-1}$
$5^{2x} = 5^{3x-3}$
$2x = 3x - 3$
$3 = x$
The solution is 3.

24. $\log_4 16 = x$
$4^x = 16$
$4^x = 4^2$
$x = 2$
The solution is 2.

25. $\log_{49} 7 = x$
$49^x = 7$
$(7^2)^x = 7$
$7^{2x} = 7$
$2x = 1$
$x = \frac{1}{2}$
The solution is $\frac{1}{2}$.

26. $\log_2 x = 5$
$2^5 = x$
$32 = x$
The solution is 32.

27. $\log_x 64 = 3$
$x^3 = 64$
$x^3 = 4^3$
$x = 4$
The solution is 4.

28. $\log_x \frac{1}{125} = -3$
$x^{-3} = \frac{1}{125}$
$x^{-3} = 5^{-3}$
$x = 5$
The solution is 5.

29. $\log_3 x = -2$
$3^{-2} = x$
$x = \frac{1}{3^2} = \frac{1}{9}$
The solution is $\frac{1}{9}$.

30. $5\log_2 x = \log_2 x^5$

31. $x\log_2 5 = \log_2 5^x$

32. $3\log_5 x - 5\log_5 y = \log_5 x^3 - \log_5 y^5 = \log_5 \frac{x^3}{y^5}$

33. $9\log_5 x + 3\log_5 y = \log_5 x^9 + \log_5 y^3$
$= \log_5 x^9 y^3$

34. $\log_2 x + \log_2(x-3) - \log_2(x^2+4)$
$= \log_2[x(x-3)] - \log_2(x^2+4)$
$= \log_2(x^2-3x) - \log_2(x^2+4)$
$= \log_2 \frac{x^2-3x}{x^2+4}$

35. $\log_3 y - \log_3(y+2) + \log_3(y^3+11)$
$= \log_3 \frac{y}{y+2} + \log_3(y^3+11)$
$= \log_3 \frac{y(y^3+11)}{y+2}$
$= \log_3 \frac{y^4+11y}{y+2}$

36. $\log_7 \frac{9x^2}{y} = \log_7 9x^2 - \log_7 y$
$= \log_7 9 + \log_7 x^2 - \log_7 y$
$= \log_7 9 + 2\log_7 x - \log_7 y$

37. $\log_6 \frac{5y}{z^2} = \log_6 5y - \log_6 z^2$
$= \log_6 5 + \log_6 y - 2\log_6 z$

38. $C = 100{,}000$, $r = 6\% = 0.06$, $x = 17 - 1 = 16$
$y = C(1+r)^x$
$= 100{,}000(1+0.06)^{16}$
$\approx 254{,}000$
There will be approximately 254,000 mosquitoes on April 17.

Section 9.7 Practice Exercises

1. To four decimal places, log 15 ≈ 1.1761.

2. a. $\log\frac{1}{100} = \log 10^{-2} = 2$

b. $\log 100{,}000 = \log 10^5 = 5$

c. $\log \sqrt[5]{10} = \log 10^{1/5} = \frac{1}{5}$

d. $\log 0.001 = \log 10^{-3} = -3$

3. $\log x = 3.4$
$x = 10^{3.4}$
$x \approx 2511.8864$

4. $a = 450$ micrometers
$T = 4.2$ seconds
$B = 3.6$
$R = \log\left(\frac{a}{T}\right) + B$
$= \log\left(\frac{450}{4.2}\right) + 3.6$
$\approx 2.0 + 3.6$
$= 5.6$
The earthquake had a magnitude of 5.6 on the Richter scale.

5. To four decimal places, ln 13 ≈ 2.5649.

6. a. $\ln e^4 = 4$

b. $\ln \sqrt[3]{e} = \ln e^{1/3} = \frac{1}{3}$

7. $\ln 5x = 8$
$e^8 = 5x$
$\frac{e^8}{5} = x$
$x = \frac{1}{5}e^8 \approx 596.1916$

8. $P = \$2400$
$r = 6\% = 0.06$
$t = 4$ years
$A = Pe^{rt} = 2400e^{0.06(4)} = 2400e^{0.24} \approx 3051.00$
The total amount of money owed is \$3051.00.

9. $\log_8 5 = \frac{\log 5}{\log 8} \approx \frac{0.6989700043}{0.903089987} \approx 0.773976$

To four decimal places, $\log_8 5 \approx 0.7740$.

Vocabulary, Readiness & Video Check 9.7

1. The base of log 7 is <u>10</u>; **c.**

2. The base of ln 7 is $\underline{e}$; **a**.

3. $\log_{10} 10^7 = \underline{7}$; **b**.

4. $\log_7 1 = \underline{0}$; **d**.

5. $\log_e e^5 = \underline{5}$; **b**.

6. $\ln e^5 = \underline{5}$; **b**.

7. $\log_2 7 = \underline{\frac{\log 7}{\log 2} \text{ or } \frac{\ln 7}{\ln 2}}$; **a and b**.

8. 10

9. The understood base of a common logarithm is 10. If you're finding the common logarithm of a known power of 10, then the common logarithm is the known power of 10.

10. e

11. $\log_b b^x = x$

12. $\frac{\ln 4}{\ln 6}$ or $\frac{\log 4}{\log 6}$; also $\frac{\ln 4}{\ln 6} = \frac{\log 4}{\log 6}$

Exercise Set 9.7

1. $\log 8 \approx 0.9031$

3. $\log 2.31 \approx 0.3636$

5. $\ln 2 \approx 0.6931$

7. $\ln 0.0716 \approx -2.6367$

9. $\log 12.6 \approx 1.1004$

11. $\ln 5 \approx 1.6094$

13. $\log 41.5 \approx 1.6180$

15. $\log 100 = \log 10^2 = 2$

17. $\log \frac{1}{1000} = \log 10^{-3} = -3$

19. $\ln e^2 = 2$

21. $\ln \sqrt[4]{e} = \ln e^{1/4} = \frac{1}{4}$

23. $\log 10^3 = 3$

25. $\ln e^{-7} = -7$

27. $\log 0.0001 = \log 10^{-4} = -4$

29. $\ln \sqrt{e} = \ln e^{1/2} = \frac{1}{2}$

31.
$$\begin{aligned} \ln 2x &= 7 \\ 2x &= e^7 \\ x &= \frac{1}{2}e^7 \approx 548.3166 \end{aligned}$$

33.
$$\begin{aligned} \log x &= 1.3 \\ x &= 10^{1.3} \approx 19.9526 \end{aligned}$$

35.
$$\begin{aligned} \log 2x &= 1.1 \\ 2x &= 10^{1.1} \\ x &= \frac{10^{1.1}}{2} \approx 6.2946 \end{aligned}$$

37.
$$\begin{aligned} \ln x &= 1.4 \\ x &= e^{1.4} \approx 4.0552 \end{aligned}$$

39.
$$\begin{aligned} \ln(3x-4) &= 2.3 \\ 3x-4 &= e^{2.3} \\ 3x &= 4+e^{2.3} \\ x &= \frac{4+e^{2.3}}{3} \approx 4.6581 \end{aligned}$$

41.
$$\begin{aligned} \log x &= 2.3 \\ x &= 10^{2.3} \approx 199.5262 \end{aligned}$$

43.
$$\begin{aligned} \ln x &= -2.3 \\ x &= e^{-2.3} \approx 0.1003 \end{aligned}$$

45.
$$\begin{aligned} \log(2x+1) &= -0.5 \\ 2x+1 &= 10^{-0.5} \\ 2x &= 10^{-0.5} - 1 \\ x &= \frac{10^{-0.5}-1}{2} \approx -0.3419 \end{aligned}$$

47. $\ln 4x = 0.18$
$4x = e^{0.18}$
$x = \dfrac{e^{0.18}}{4} \approx 0.2993$

49. $\log_2 3 = \dfrac{\log 3}{\log 2} \approx 1.5850$

51. $\log_{1/2} 5 = \dfrac{\ln 5}{\ln\left(\frac{1}{2}\right)} \approx -2.3219$

53. $\log_4 9 = \dfrac{\ln 9}{\ln 4} \approx 1.5850$

55. $\log_3\left(\dfrac{1}{6}\right) = \dfrac{\log\left(\frac{1}{6}\right)}{\log 3} \approx -1.6309$

57. $\log_8 6 = \dfrac{\log 6}{\log 8} \approx 0.8617$

59. $R = \log\left(\dfrac{a}{T}\right) + B$, $a = 200$, $T = 1.6$
$B = 2.1$
$R = \log\left(\dfrac{200}{1.6}\right) + 2.1 \approx 4.2$
The earthquake measures 4.2 on the Richter scale.

61. $R = \log\left(\dfrac{a}{T}\right) + B$, $a = 400$, $T = 2.6$
$B = 3.1$
$R = \log\left(\dfrac{400}{2.6}\right) + 3.1 \approx 5.3$
The earthquake measures 5.3 on the Richter scale.

63. $A = Pe^{rt}$, $t = 12$, $P = 1400$, $r = 0.08$
$A = 1400e^{(0.08)12} = 1400e^{0.96} \approx 3656.38$
Dana has \$3656.38 after 12 years.

65. $A = Pe^{rt}$, $t = 4$, $P = 2000$, $r = 0.06$
$A = 2000e^{(0.06)4} = 2000e^{0.24} \approx 2542.50$
Barbara owes \$2542.50 at the end of 4 years.

67. $6x - 3(2 - 5x) = 6$
$6x - 6 + 15x = 6$
$21x = 12$
$x = \dfrac{12}{21} = \dfrac{4}{7}$
The solution is $\dfrac{4}{7}$.

69. $2x + 3y = 6x$
$3y = 4x$
$x = \dfrac{3y}{4}$

71. $x^2 + 7x = -6$
$x^2 + 7x + 6 = 0$
$(x + 6)(x + 1) = 0$
$x + 6 = 0$ or $x + 1 = 0$
$x = -6$ or $x = -1$
The solutions are −6 and −1.

73. $\begin{cases} x + 2y = -4 \\ 3x - y = 9 \end{cases}$
Multiply the second equation by 2, then add.
$x + 2y = -4$
$6x - 2y = 18$
$7x = 14$
$x = 2$
Replace x with 2 in the first equation.
$x + 2y = -4$
$2 + 2y = -4$
$2y = -6$
$y = -3$
The solution is (2, −3).

75. answers may vary

77. ln 50 is larger. answers may vary

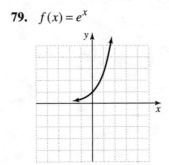

79. $f(x) = e^x$

81. $f(x) = e^{-3x}$

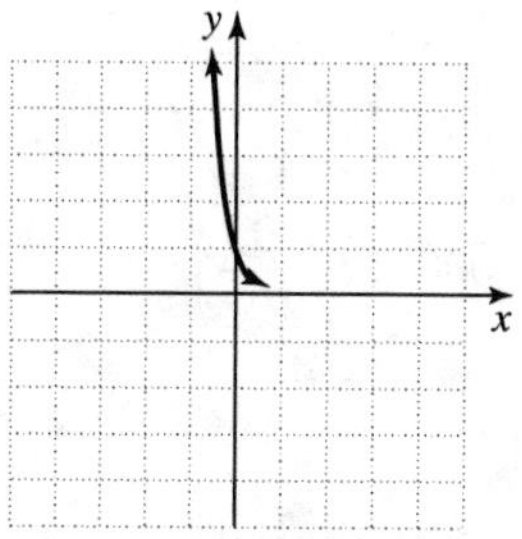

83. $f(x) = e^x + 2$

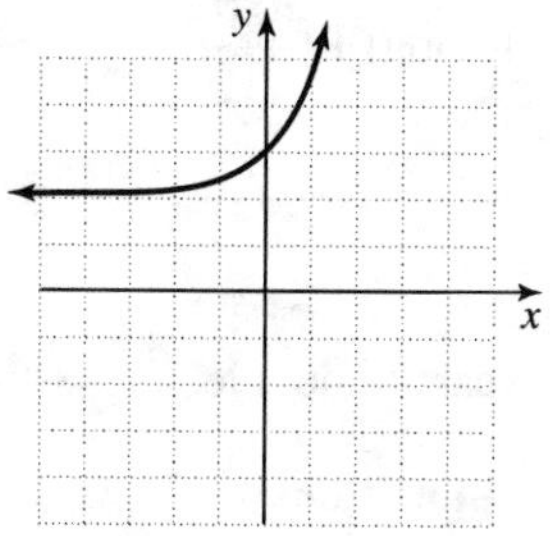

85. $f(x) = e^{x-1}$

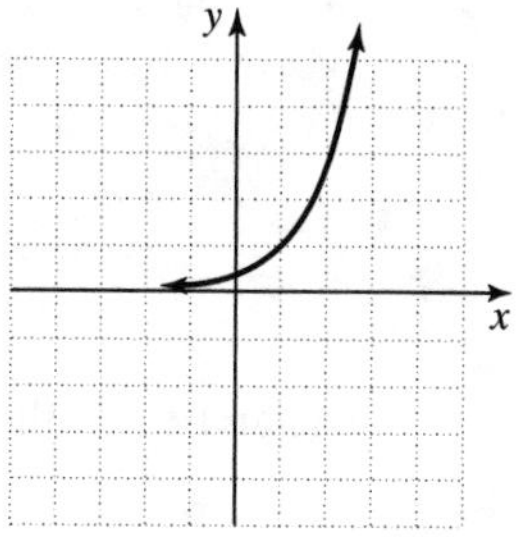

87. $f(x) = 3e^x$

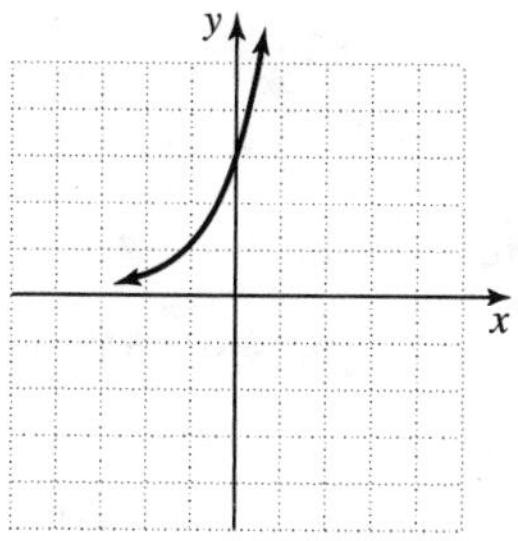

89. $f(x) = \ln x$

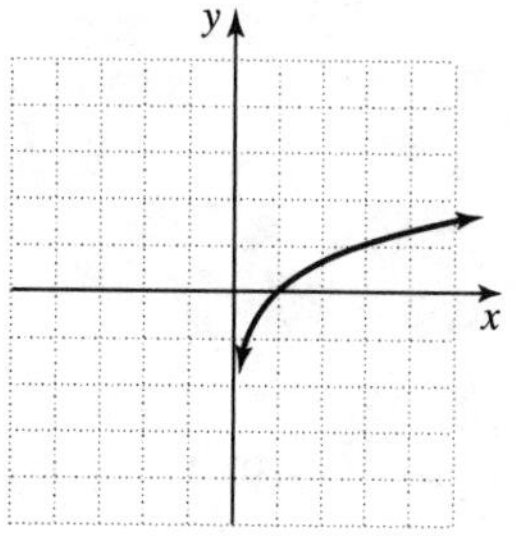

91. $f(x) = -2\log x$

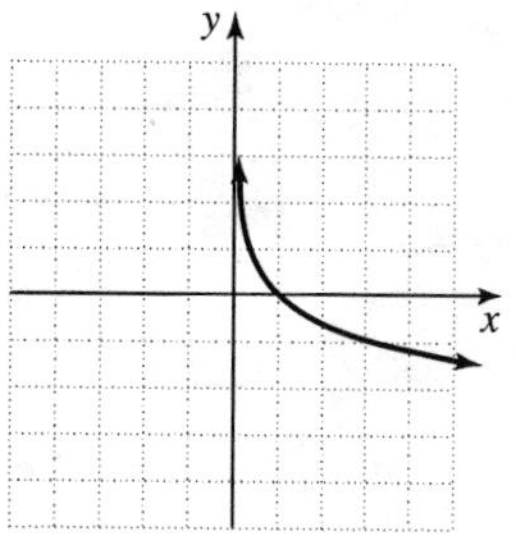

93. $f(x) = \log(x+2)$

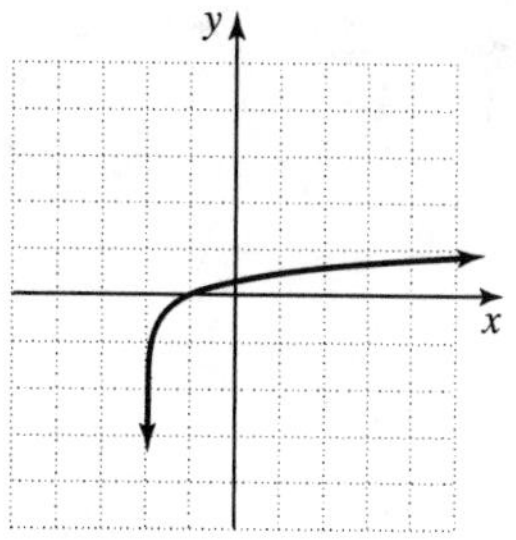

95. $f(x) = \ln x - 3$

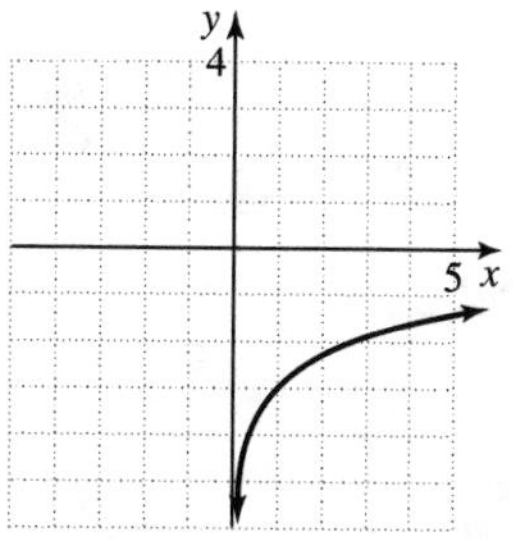

97. $f(x) = e^x$

$f(x) = e^x + 2$

$f(x) = e^x - 3$

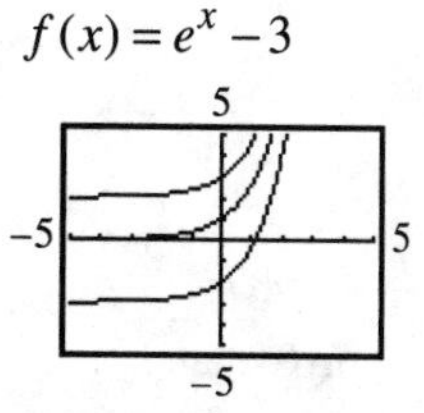

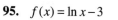

answers may vary

Section 9.8 Practice Exercises

1.
$$5^x = 9$$
$$\log 5^x = \log 9$$
$$x \log 5 = \log 9$$
$$x = \frac{\log 9}{\log 5} \approx 1.3652$$

The solution is $\frac{\log 9}{\log 5}$, or approximately 1.3652.

2. $\log_2(x-1) = 5$
$$2^5 = x-1$$
$$32 = x-1$$
$$33 = x$$

Check: $\log_2(x-1) = 5$
$$\log_2(33-1) \stackrel{?}{=} 5$$
$$\log_2 32 \stackrel{?}{=} 5$$
$$2^5 = 32 \quad \text{True}$$

The solution is 33.

3. $\log_5 x + \log_5(x+4) = 1$
$$\log_5 x(x+4) = 1$$
$$\log_5(x^2+4x) = 1$$
$$5^1 = x^2+4x$$
$$0 = x^2+4x-5$$
$$0 = (x+5)(x-1)$$
$$x+5=0 \quad \text{or} \quad x-1=0$$
$$x=-5 \qquad\qquad x=1$$

Since $\log_5(-5)$ is undefined, –5 is rejected. The solution is 1.

4. $\log(x+3) - \log x = 1$
$$\log\frac{x+3}{x} = 1$$
$$10^1 = \frac{x+3}{x}$$
$$10x = x+3$$
$$9x = 3$$
$$x = \frac{1}{3}$$

The solution is $\frac{1}{3}$.

5. $y_0 = 60;\ t = 3$
$$y = y_0 e^{0.916t}$$
$$y = 60e^{0.916(3)} = 60e^{2.748} \approx 937$$

The population will be approximately 937 rabbits.

6. $P = \$3000;\ r = 7\% = 0.07;\ n = 12;$
$A = 2P = \$6000$
$$A = P\left(1+\frac{r}{n}\right)^{nt}$$
$$6000 = 3000\left(1+\frac{0.07}{12}\right)^{12t}$$
$$2 = \left(1+\frac{0.07}{12}\right)^{12t}$$
$$\log 2 = \log\left(1+\frac{0.07}{12}\right)^{12t}$$
$$\log 2 = 12t\log\left(1+\frac{0.07}{12}\right)$$
$$\frac{\log 2}{12\log\left(1+\frac{0.07}{12}\right)} = t$$
$$9.9 \approx t$$

It takes nearly 10 years to double.

Graphing Calculator Explorations

1. $Y_1 = 5000\left(1+\frac{0.05}{4}\right)^{4x},\ Y_2 = 6000$

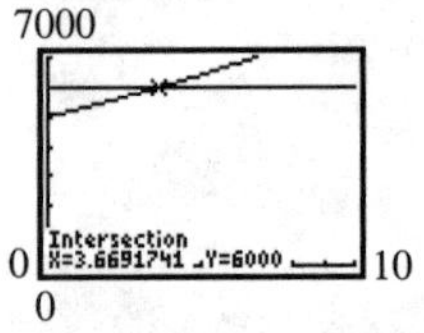

It takes 3.67 years, or 3 years and 8 months.

2. $Y_1 = 1000\left(1+\frac{0.045}{365}\right)^{365x},\ Y_2 = 2000$

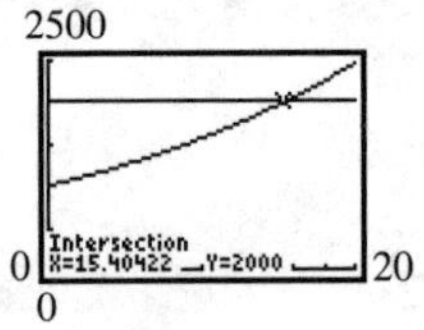

It takes 15.40 years or 15 years and 5 months.

3. $Y_1 = 10{,}000\left(1+\frac{0.06}{12}\right)^{12x},\ Y_2 = 40{,}000$

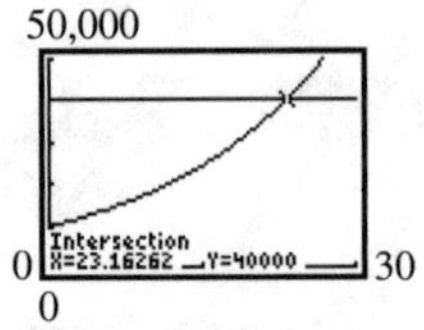

It takes 23.16 years or 23 years and 2 months.

4. $Y_1 = 500\left(1+\frac{0.04}{2}\right)^{2x}$, $Y_2 = 800$

1000

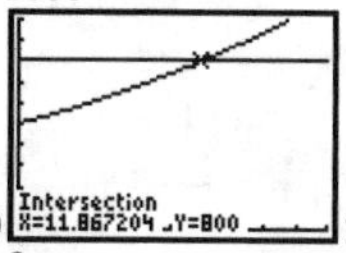

0 20

0

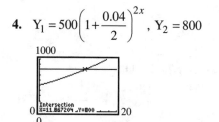

It takes 11.87 years or 11 years and 10 months.

Vocabulary, Readiness & Video Check 9.8

1. ln(4*x* – 2) = ln 3 is the same as $\log_e(4x-2) = \log_e 3$. Therefore, from the logarithm property of equality, we know that 4*x* – 2 = 3.

2. Substituting –8 in the original equation gives us the logarithm of a negative number, which does not exist—we can only take the logarithm of a positive number.

3. $2000 = 1000\left(1+\frac{0.07}{12}\right)^{12\cdot t}$

$t \approx 9.9$

As long as the interest rate and compounding are the same, it takes any amount of money the same time to double.

Exercise Set 9.8

1.
$$3^x = 6$$
$$\log 3^x = \log 6$$
$$x\log 3 = \log 6$$
$$x = \frac{\log 6}{\log 3} \approx 1.6309$$

3.
$$3^{2x} = 3.8$$
$$\log 3^{2x} = \log 3.8$$
$$2x\log 3 = \log 3.8$$
$$x = \frac{\log 3.8}{2\log 3} \approx 0.6076$$

5.
$$2^{x-3} = 5$$
$$\log 2^{x-3} = \log 5$$
$$(x-3)\log 2 = \log 5$$
$$x\log 2 - 3\log 2 = \log 5$$
$$x\log 2 = 3\log 2 + \log 5$$
$$x = \frac{3\log 2 + \log 5}{\log 2}$$
or
$$x = 3 + \frac{\log 5}{\log 2}$$
$$x \approx 5.3219$$

7.
$$9^x = 5$$
$$\log 9^x = \log 5$$
$$x\log 9 = \log 5$$
$$x = \frac{\log 5}{\log 9} \approx 0.7325$$

9.
$$4^{x+7} = 3$$
$$\log 4^{x+7} = \log 3$$
$$(x+7)\log 4 = \log 3$$
$$x\log 4 + 7\log 4 = \log 3$$
$$x\log 4 = \log 3 - 7\log 4$$
$$x = \frac{\log 3 - 7\log 4}{\log 4}$$
or
$$x = \frac{\log 3}{\log 4} - 7$$
$$x \approx -6.2075$$

11.
$$\log_2(x+5) = 4$$
$$x+5 = 2^4$$
$$x+5 = 16$$
$$x = 11$$

13.
$$\log_4 2 + \log_4 x = 0$$
$$\log_4(2x) = 0$$
$$2x = 4^0$$
$$2x = 1$$
$$x = \frac{1}{2}$$

15. $\log_2 6 - \log_2 x = 3$

$$\log_2\left(\frac{6}{x}\right) = 3$$
$$\frac{6}{x} = 2^3$$
$$\frac{6}{x} = 8$$
$$8x = 6$$
$$x = \frac{3}{4}$$

17. $\log_6(x^2 - x) = 1$

$$6^1 = x^2 - x$$
$$0 = x^2 - x - 6$$
$$0 = (x-3)(x+2)$$

$x = 3$ or $x = -2$

19. $\log_4 x + \log_4(x+6) = 2$

$$\log_4 x(x+6) = 2$$
$$x(x+6) = 4^2$$
$$x^2 + 6x = 16$$
$$x^2 + 6x - 16 = 0$$
$$(x+8)(x-2) = 0$$

$x = -8$ or $x = 2$

We discard -8 as extraneous, the solution is 2.

21. $\log_5(x+3) - \log_5 x = 2$

$$\log_5\left(\frac{x+3}{x}\right) = 2$$
$$\frac{x+3}{x} = 5^2$$
$$\frac{x+3}{x} = 25$$
$$x + 3 = 25x$$
$$3 = 24x$$
$$x = \frac{1}{8}$$

23. $7^{3x-4} = 11$

$$\log 7^{3x-4} = \log 11$$
$$(3x-4)\log 7 = \log 11$$
$$3x\log 7 - 4\log 7 = \log 11$$
$$3x\log 7 = 4\log 7 + \log 11$$
$$x = \frac{4\log 7 + \log 11}{3\log 7}$$

or

$$x = \frac{1}{3}\left(4 + \frac{\log 11}{\log 7}\right)$$

$x \approx 1.7441$

25. $\log_4(x^2 - 3x) = 1$

$$x^2 - 3x = 4$$
$$x^2 - 3x - 4 = 0$$
$$(x-4)(x+1) = 0$$

$x = 4$ or $x = -1$

27. $e^{6x} = 5$

$$\ln e^{6x} = \ln 5$$
$$6x = \ln 5$$
$$x = \frac{\ln 5}{6} \approx 0.2682$$

29. $\log_3 x^2 = 4$

$$x^2 = 3^4$$
$$x^2 = 81$$
$$x = \pm 9$$

31. $\ln 5 + \ln x = 0$

$$\ln(5x) = 0$$
$$e^0 = 5x$$
$$1 = 5x$$
$$\frac{1}{5} = x$$

33. $3\log x - \log x^2 = 2$

$$3\log x - 2\log x = 2$$
$$\log x = 2$$
$$x = 10^2$$
$$x = 100$$

35. $\log_4 x - \log_4(2x-3) = 3$

$$\log_4\left(\frac{x}{2x-3}\right) = 3$$
$$\frac{x}{2x-3} = 4^3$$
$$x = 64(2x-3)$$
$$x = 128x - 192$$
$$192 = 127x$$
$$x = \frac{192}{127}$$

37. $\log_2 x + \log_2(3x+1) = 1$

$$\log_2 x(3x+1) = 1$$
$$x(3x+1) = 2$$
$$3x^2 + x - 2 = 0$$
$$(3x-2)(x+1) = 0$$
$$3x-2 = 0 \quad \text{or} \quad x+1 = 0$$
$$x = \frac{2}{3} \quad \text{or} \quad x = -1$$

We discard −1 as extraneous, the solution is $\frac{2}{3}$.

39. $\log_2 x + \log_2(x+5) = 1$

$$\log_2 x(x+5) = 1$$
$$x(x+5) = 2$$
$$x^2 + 5x - 2 = 0$$

$a = 1$, $b = 5$, $c = -2$

$$x = \frac{-5 \pm \sqrt{5^2 - 4(1)(-2)}}{2(1)}$$
$$x = \frac{-5 \pm \sqrt{33}}{2}$$

Discard $\frac{-5-\sqrt{33}}{2}$, the solution is $\frac{-5+\sqrt{33}}{2}$.

41. Let $y_0 = 83$ and $t = 5$.

$$y = y_0 e^{0.043t}$$
$$y = 83e^{0.043(5)} = 83e^{0.215} \approx 103$$

The population is estimated to be 103 wolves in 5 years.

43. Let $y_0 = 13{,}700$ and $t = 4$.

$$y = y_0 e^{-1.044t}$$
$$y = 13{,}700e^{-1.044(4)} \approx 13{,}700e^{-4.176} \approx 210$$

The population of the Cook Islands is predicted to be 210 in 2018.

45. $A = P\left(1+\frac{r}{n}\right)^{nt}$, $P = 600$,

$A = 2(600) = 1200$, $r = 0.07$, $n = 12$

$$1200 = 600\left(1+\frac{0.07}{12}\right)^{12t}$$
$$2 = \left(1+\frac{0.07}{12}\right)^{12t}$$
$$\log 2 = \log\left(1+\frac{0.07}{12}\right)^{12t}$$
$$\log 2 = 12t\log\left(1+\frac{0.07}{12}\right)$$
$$\frac{\log 2}{12\log\left(1+\frac{0.07}{12}\right)} = t$$
$$9.9 \approx t$$

It takes approximately 9.9 years for the \$600 to double.

47. $A = P\left(1+\frac{r}{n}\right)^{nt}$, $P = 1200$,

$A = P + I = 1200 + 200 = 1400$

$r = 0.09$, $n = 4$

$$1400 = 1200\left(1+\frac{0.09}{4}\right)^{4t}$$
$$\frac{7}{6} = (1.0225)^{4t}$$
$$\log\frac{7}{6} = \log 1.0225^{4t}$$
$$\log\frac{7}{6} = 4t\log 1.0225$$
$$t = \frac{\log\frac{7}{6}}{4\log 1.0225}$$
$$t \approx 1.7$$

It would take the investment approximately 1.7 years to earn \$200.

49. $A = P\left(1+\frac{r}{n}\right)^{nt}$, $P = 1000$

$A = 2(1000) = 2000$, $r = 0.08$, $n = 2$

$$2000 = 1000\left(1+\frac{0.08}{2}\right)^{2t}$$
$$2 = (1.04)^{2t}$$
$$\log 2 = \log 1.04^{2t}$$
$$\log 2 = 2t \log 1.04$$
$$t = \frac{\log 2}{2\log 1.04}$$
$$t \approx 8.8$$

It takes 8.8 years to double.

51. $w = 0.00185h^{2.67}$, and $h = 35$

$$w = 0.00185(35)^{2.67} \approx 24.5$$

The expected weight of a boy 35 inches tall is 24.5 pounds.

53. $w = 0.00185h^{2.67}$, and $w = 85$

$$85 = 0.00185h^{2.67}$$
$$\frac{85}{0.00185} = h^{2.67}$$
$$h = \left(\frac{85}{0.00185}\right)^{1/2.67} \approx 55.7$$

The expected height of the boy is 55.7 inches.

55. $P = 14.7e^{-0.21x}$, $x = 1$

$$P = 14.7e^{-0.21(1)}$$
$$= 14.7e^{-0.21}$$
$$\approx 11.9$$

The average atmospheric pressure in Denver is approximately 11.9 pounds per square inch.

57. $P = 14.7e^{-0.21x}$, $P = 7.5$

$$7.5 = 14.7e^{-0.21x}$$
$$\frac{7.5}{14.7} = e^{-0.21x}$$
$$-0.21x = \ln\left(\frac{7.5}{14.7}\right)$$
$$x = -\frac{1}{0.21}\ln\left(\frac{7.5}{14.7}\right) \approx 3.2$$

The elevation of the jet is approximately 3.2 miles.

59. $t = \frac{1}{c}\ln\left(\frac{A}{A-N}\right)$

$$t = \frac{1}{0.09}\ln\left(\frac{75}{75-50}\right)$$
$$t = \frac{1}{0.09}\ln(3)$$
$$t \approx 12.21$$

It will take 12 weeks.

61. $t = \frac{1}{c}\ln\left(\frac{A}{A-N}\right)$

$$t = \frac{1}{0.07}\ln\left(\frac{210}{210-150}\right)$$
$$t = \frac{1}{0.07}\ln(3.5)$$
$$t \approx 17.9$$

It will take 18 weeks.

63. $\frac{x^2 - y + 2z}{3x} = \frac{(-2)^2 - 0 + 2(3)}{3(-2)}$

$$= \frac{4+6}{-6}$$
$$= \frac{10}{-6}$$
$$= -\frac{5}{3}$$

65. $\frac{3z - 4x + y}{x + 2z} = \frac{3(3) - 4(-2) + 0}{-2 + 2(3)} = \frac{9+8}{-2+6} = \frac{17}{4}$

67. $f(x) = 5x + 2$

$$y = 5x + 2$$

$$x = 5y + 2$$
$$\frac{x-2}{5} = y$$
$$f^{-1}(x) = \frac{x-2}{5}$$

69. $y = y_0 e^{0.0015t}$

$y = 9{,}950{,}000,\ y_0 = 9{,}910{,}000$

$$9{,}950{,}000 = 9{,}910{,}000e^{0.0015t}$$

$$\frac{9{,}950{,}000}{9{,}910{,}000} = e^{0.0015t}$$

$$\ln\frac{995}{991} = 0.0015t$$

$$t = \frac{\ln\frac{995}{991}}{0.0015} \approx 2.7$$

The population of Michigan will be 9,950,000 after approximately 2.7 years.

71. answers may vary

73. $Y_1 = e^{0.3x},\ Y_2 = 8$

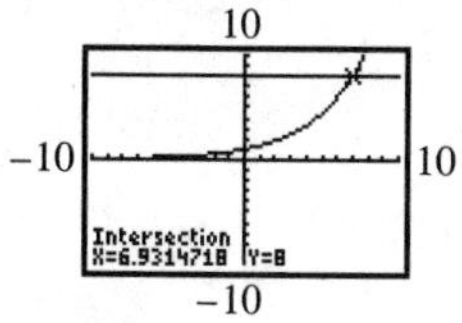

$x \approx 6.93$

75. $Y_1 = 2\log(-5.6x+1.3)+x+1,\ Y_2 = 0$

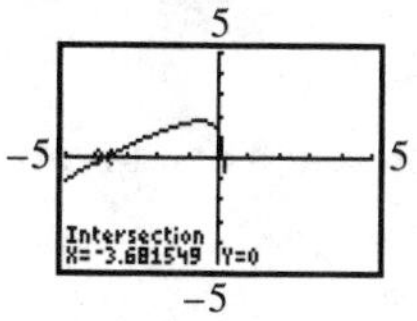

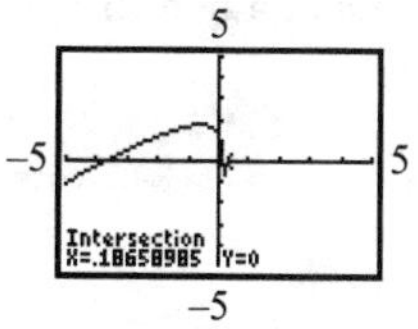

$x \approx -3.68, 0.19$

77. $Y_1 = 7^{3x-4} - 11,\ Y_2 = 0$

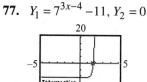

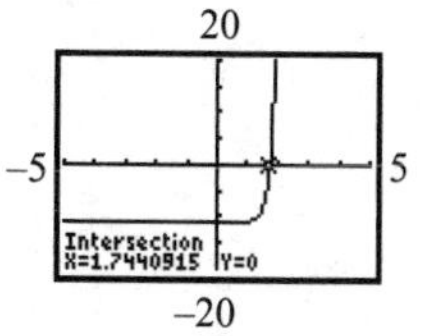

$x \approx 1.74$

79. $Y_1 = \ln 5 + \ln x,\ Y_2 = 0$

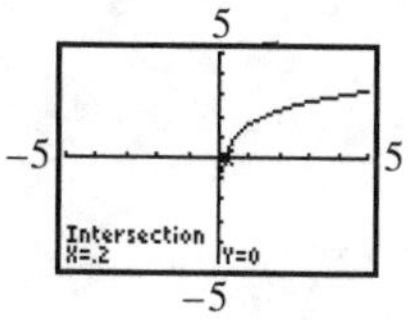

$x = 0.2 = \frac{1}{5}$

Chapter 9 Vocabulary Check

1. For a one-to-one function, we can find its inverse function by switching the coordinates of the ordered pairs of the function.

2. The composition of functions f and g is $(f \circ g)(x) = f(g(x))$.

3. A function of the form $f(x) = b^x$ is called an exponential function if $b > 0$, b is not 1, and x is a real number.

4. The graphs of f and f^{-1} are symmetric about the line $y = x$.

5. Natural logarithms are logarithms to base e.

6. Common logarithms are logarithms to base 10.

7. To see whether a graph is the graph of a one-to-one function, apply the vertical line test to see whether it is a function, and then apply the horizontal line test to see whether it is a one-to-one function.

8. A logarithmic function is a function that can be defined by $f(x) = \log_b x$ where x is a positive real number, b is a constant positive real number, and b is not 1.

9. Half-life is the amount of time it takes for half of the amount of a substance to decay.

10. A quantity that grows or decays by the same percent at regular time periods is said to have exponential growth or decay.

Chapter 9 Review

1.
$$\begin{aligned}(f+g)(x) &= f(x)+g(x)\\ &= (x-5)+(2x+1)\\ &= x-5+2x+1\\ &= 3x-4\end{aligned}$$

2. $(f-g)(x) = f(x)-g(x)$
$= (x-5)-(2x+1)$
$= x-5-2x-1$
$= -x-6$

3. $(f\cdot g)(x) = f(x)\cdot g(x)$
$= (x-5)(2x+1)$
$= 2x^2+x-10x-5$
$= 2x^2-9x-5$

4. $\left(\frac{g}{f}\right)(x) = \frac{g(x)}{f(x)} = \frac{2x+1}{x-5}, x \neq 5$

5. $(f\circ g)(x) = f(g(x))$
$= f(x+1)$
$= (x+1)^2-2$
$= x^2+2x-1$

6. $(g\circ f)(x) = g(f(x))$
$= g(x^2-2)$
$= x^2-2+1$
$= x^2-1$

7. $(h\circ g)(2) = h(g(2)) = h(3) = 3^3-3^2 = 18$

8. $(f\circ f)(x) = f(f(x))$
$= f(x^2-2)$
$= (x^2-2)^2-2$
$= x^4-4x^2+4-2$
$= x^4-4x^2+2$

9. $(f\circ g)(-1) = f(g(-1)) = f(0) = 0^2-2 = -2$

10. $(h\circ h)(2) = h(h(2)) = h(4) = 4^3-4^2 = 48$

11. The function is one-to-one.
$h^{-1} = \{(14,-9),(8,6),(12,-11),(15,15)\}$

12. The function is not one-to-one.

13. The function is one-to-one.

Rank in Housing Starts for 2014 (Input)	4	3	1	2
U.S. Region (Output)	Northeast	Midwest	South	West

14. The function is not one-to-one.

15. $f(x) = \sqrt{x+2}$

a. $f(7) = \sqrt{7+2} = \sqrt{9} = 3$

b. $f^{-1}(3) = 7$

16. $f(x) = \sqrt{x+2}$

a. $f(-1) = \sqrt{-1+2} = \sqrt{1} = 1$

b. $f^{-1}(1) = -1$

17. The graph does not represent a one-to-one function.

18. The graph does not represent a one-to-one function.

19. The graph does not represent a one-to-one function.

20. The graph represents a one-to-one function.

21. $f(x) = x-9$
$y = x-9$
$x = y-9$
$y = x+9$
$f^{-1}(x) = x+9$

22. $f(x) = x+8$
$y = x+8$
$x = y+8$
$y = x-8$
$f^{-1}(x) = x-8$

23.
$$f(x) = 6x + 11$$
$$y = 6x + 11$$
$$x = 6y + 11$$
$$6y = x - 11$$
$$y = \frac{x - 11}{6}$$
$$f^{-1}(x) = \frac{x - 11}{6}$$

24.
$$f(x) = 12x - 1$$
$$y = 12x - 1$$
$$x = 12y - 1$$
$$x + 1 = 12y$$
$$y = \frac{x + 1}{12}$$
$$f^{-1}(x) = \frac{x + 1}{12}$$

25.
$$f(x) = x^3 - 5$$
$$y = x^3 - 5$$
$$x = y^3 - 5$$
$$y^3 = x + 5$$
$$y = \sqrt[3]{x + 5}$$
$$f^{-1}(x) = \sqrt[3]{x + 5}$$

26.
$$f(x) = \sqrt[3]{x + 2}$$
$$y = \sqrt[3]{x + 2}$$
$$x = \sqrt[3]{y + 2}$$
$$x^3 = y + 2$$
$$y = x^3 - 2$$
$$f^{-1}(x) = x^3 - 2$$

27.
$$g(x) = \frac{12x - 7}{6}$$
$$y = \frac{12x - 7}{6}$$
$$x = \frac{12y - 7}{6}$$
$$6x = 12y - 7$$
$$12y = 6x + 7$$
$$y = \frac{6x + 7}{12}$$
$$g^{-1}(x) = \frac{6x + 7}{12}$$

28.
$$r(x) = \frac{13x - 5}{2}$$
$$y = \frac{13x - 5}{2}$$
$$x = \frac{13y - 5}{2}$$
$$2x = 13y - 5$$
$$2x + 5 = 13y$$
$$y = \frac{2x + 5}{13}$$
$$r^{-1}(x) = \frac{2x + 5}{13}$$

29.
$$f(x) = -2x + 3$$
$$y = -2x + 3$$
$$x = -2y + 3$$
$$x - 3 = -2y$$
$$-\frac{x - 3}{2} = y$$
$$f^{-1}(x) = -\frac{x - 3}{2}$$

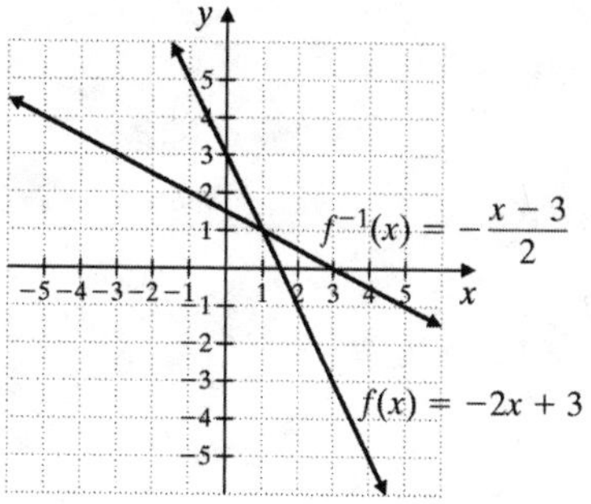

30.
$$f(x) = 5x - 5$$
$$y = 5x - 5$$
$$x = 5y - 5$$
$$x + 5 = 5y$$
$$\frac{x + 5}{5} = y$$
$$f^{-1}(x) = \frac{x + 5}{5}$$

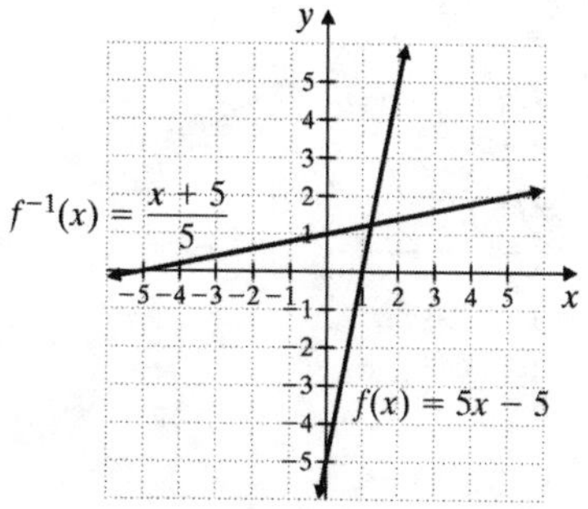

31. $4^x = 64$
$4^x = 4^3$
$x = 3$

32. $3^x = \frac{1}{9}$
$3^x = 3^{-2}$
$x = -2$

33. $2^{3x} = \frac{1}{16}$
$2^{3x} = 2^{-4}$
$3x = -4$
$x = -\frac{4}{3}$

34. $5^{2x} = 125$
$5^{2x} = 5^3$
$2x = 3$
$x = \frac{3}{2}$

35. $9^{x+1} = 243$
$(3^2)^{x+1} = 3^5$
$3^{2x+2} = 3^5$
$2x + 2 = 5$
$2x = 3$
$x = \frac{3}{2}$

36. $8^{3x-2} = 4$
$(2^3)^{3x-2} = 2^2$
$2^{9x-6} = 2^2$
$9x - 6 = 2$
$9x = 8$
$x = \frac{8}{9}$

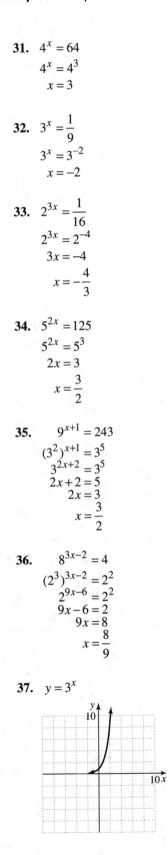

37. $y = 3^x$

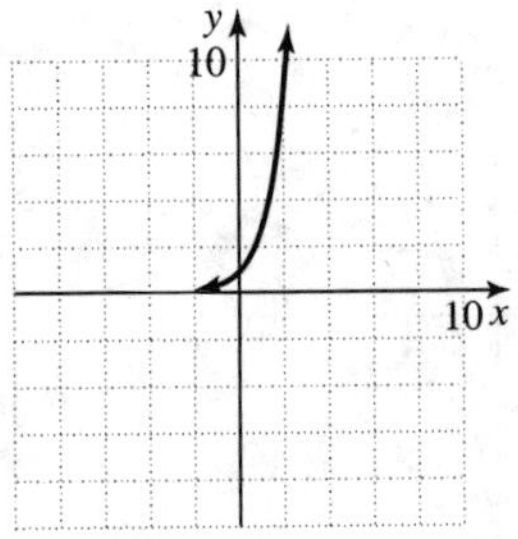

38. $y = \left(\frac{1}{3}\right)^x$

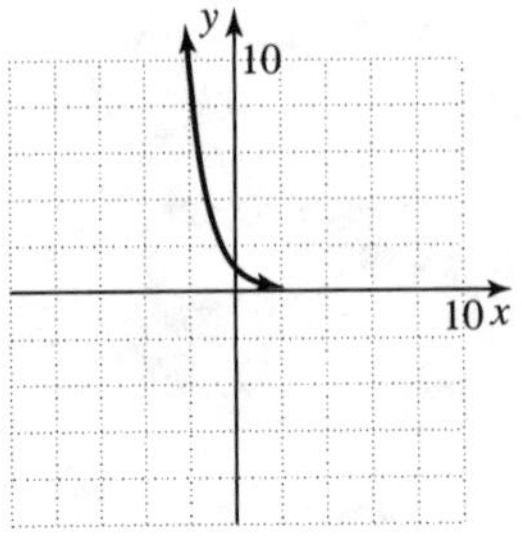

39. $y = 2^{x-4}$

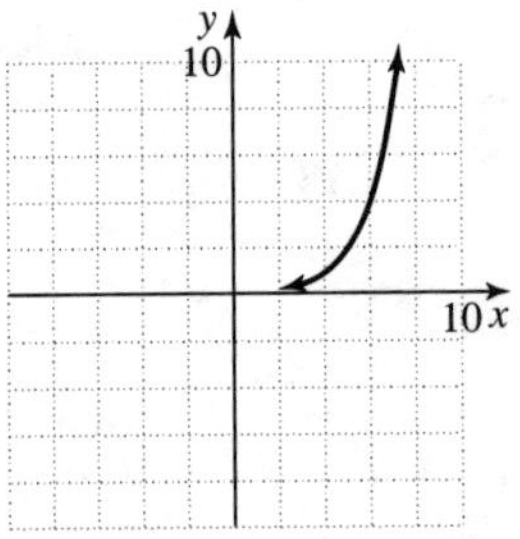

40. $y = 2^x + 4$

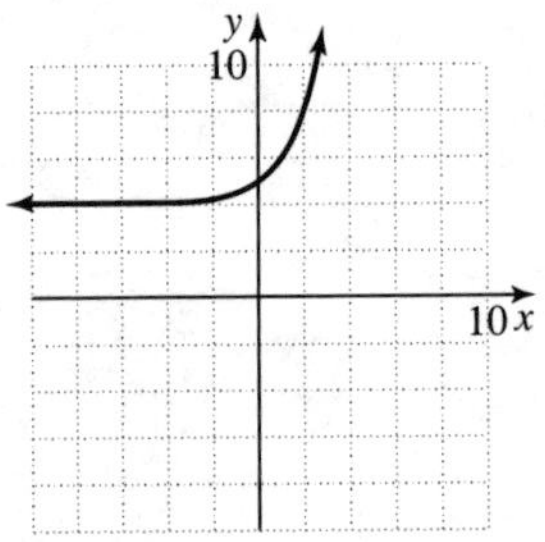

41. $A = P\left(1 + \frac{r}{n}\right)^{nt}$

$A = 1600\left(1 + \frac{0.09}{2}\right)^{(2)(7)}$

$A \approx 2963.11$

The amount accrued is \$2963.11.

42. $A = P\left(1 + \frac{r}{n}\right)^{nt}$

$A = 800\left(1 + \frac{0.07}{4}\right)^{(4)(5)}$

$A \approx 1131.82$

The certificate is worth \$1131.82 at the end of 5 years.

43. $C = 257{,}437,\ r = 0.052,\ x = 2020 - 2010 = 10$

$y = C(1+r)^x$

$y = 257{,}437(1+0.052)^{10} \approx 427{,}394$

Henderson's population is predicted to be 427,394 in 2020.

44. $C = 403{,}971,\ r = 0.069,\ x = 2019 - 2010 = 9$

$y = C(1+r)^x$

$y = 403{,}971(1+0.069)^9 \approx 736{,}461$

Raleigh's population is predicted to be 736,461 in 2019.

45. $C = 1024,\ r = 0.5,\ x = 7$

$y = C(1-r)^x$

$y = 1024(1-0.5)^7$

$y = 8$

After 7 rounds there will be 8 players.

46. $C = 1280,\ r = 0.11,\ x = 6$

$y = C(1-r)^x$

$y = 1280(1-0.11)^6$

$y \approx 636$

The predicted bear population in 6 years is 636.

47. $49 = 7^2$

$\log_7 49 = 2$

48. $2^{-4} = \frac{1}{16}$

$\log_2 \frac{1}{16} = -4$

49. $\log_{1/2} 16 = -4$

$\left(\frac{1}{2}\right)^{-4} = 16$

50. $\log_{0.4} 0.064 = 3$

$0.4^3 = 0.064$

51. $\log_4 x = -3$

$x = 4^{-3} = \frac{1}{64}$

52. $\log_3 x = 2$

$x = 3^2 = 9$

53. $\log_3 1 = x$

$3^x = 1$

$3^x = 3^0$

$x = 0$

54. $\log_4 64 = x$

$4^x = 64$

$4^x = 4^3$

$x = 3$

55. $\log_4 4^5 = x$

$x = 5$

56. $\log_7 7^{-2} = x$

$x = -2$

57. $5^{\log_5 4} = x$

$x = 4$

58. $2^{\log_2 9} = x$

$9 = x$

59. $\log_2(3x-1) = 4$

$3x - 1 = 2^4$

$3x - 1 = 16$

$3x = 17$

$x = \frac{17}{3}$

60. $\log_3(2x+5) = 2$

$2x + 5 = 3^2$

$2x + 5 = 9$

$2x = 4$

$x = 2$

61. $\log_4(x^2 - 3x) = 1$

$x^2 - 3x = 4$

$x^2 - 3x - 4 = 0$

$(x+1)(x-4) = 0$

$x = -1$ or $x = 4$

62. $\log_8(x^2 + 7x) = 1$

$x^2 + 7x = 8$

$x^2 + 7x - 8 = 0$

$(x+8)(x-1) = 0$

$x = -8$ or $x = 1$

63. $y = 2^x$ and $y = \log_2 x$

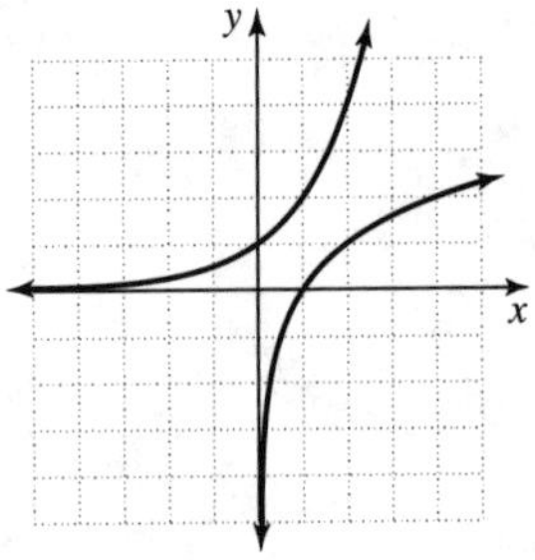

64. $y = \left(\frac{1}{2}\right)^x$ and $y = \log_{1/2} x$

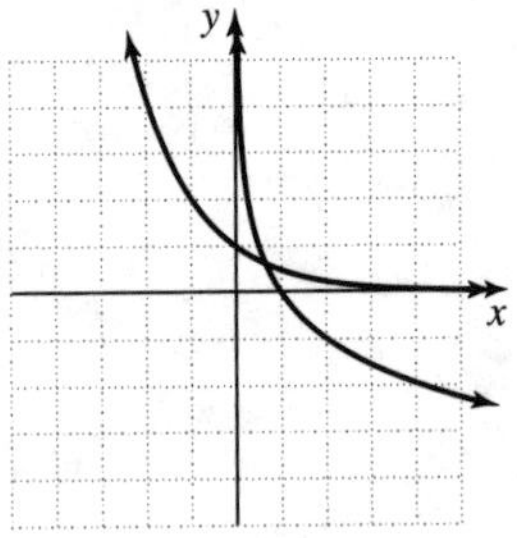

65. $\log_3 8 + \log_3 4 = \log_3(8 \cdot 4) = \log_3 32$

66. $\log_2 6 + \log_2 3 = \log_2(6 \cdot 3) = \log_2 18$

67. $\log_7 15 - \log_7 20 = \log_7 \frac{15}{20} = \log_7 \frac{3}{4}$

68. $\log 18 - \log 12 = \log \frac{18}{12} = \log \frac{3}{2}$

69. $\log_{11} 8 + \log_{11} 3 - \log_{11} 6 = \log_{11} \frac{(8)(3)}{6}$
$= \log_{11} 4$

70. $\log_5 14 + \log_5 3 - \log_5 21$
$= \log_5(14 \cdot 3) - \log_5 21$
$= \log_5 \frac{42}{21}$
$= \log_5 2$

71. $2\log_5 x - 2\log_5(x+1) + \log_5 x$
$= \log_5 x^2 - \log_5(x+1)^2 + \log_5 x$
$= \log_5 \frac{(x^2)(x)}{(x+1)^2}$
$= \log_5 \frac{x^3}{(x+1)^2}$

72. $4\log_3 x - \log_3 x + \log_3(x+2)$
$= 3\log_3 x + \log_3(x+2)$
$= \log_3 x^3 + \log_3(x+2)$
$= \log_3 \left[x^3(x+2)\right]$
$= \log_3(x^4 + 2x^3)$

73. $\log_3 \frac{x^3}{x+2} = \log_3 x^3 - \log_3(x+2)$
$= 3\log_3 x - \log_3(x+2)$

74. $\log_4 \frac{x+5}{x^2} = \log_4(x+5) - \log_4 x^2$
$= \log_4(x+5) - 2\log_4 x$

75. $\log_2 \frac{3x^2 y}{z} = \log_2(3x^2 y) - \log_2 z$
$= \log_2 3 + \log_2 x^2 + \log_2 y - \log_2 z$
$= \log_2 3 + 2\log_2 x + \log_2 y - \log_2 z$

76. $\log_7 \frac{yz^3}{x} = \log_7(yz^3) - \log_7 x$
$= \log_7 y + \log_7 z^3 - \log_7 x$
$= \log_7 y + 3\log_7 z - \log_7 x$

77. $\log_b 50 = \log_b(5)(5)(2)$
$= \log_b(5) + \log_b(5) + \log_b(2)$
$= 0.83 + 0.83 + 0.36$
$= 2.02$

78. $\log_b \frac{4}{5} = \log_b 4 - \log_b 5$
$= \log_b 2^2 - \log_b 5$
$= 2\log_b 2 - \log_b 5$
$= 2(0.36) - 0.83$
$= 0.72 - 0.83$
$= -0.11$

79. $\log 3.6 \approx 0.5563$

80. $\log 0.15 \approx -0.8239$

81. $\ln 1.25 \approx 0.2231$

82. $\ln 4.63 \approx 1.5326$

83. $\log 1000 = \log 10^3 = 3$

84. $\log\frac{1}{10} = \log 10^{-1} = -1$

85. $\ln\frac{1}{e} = \ln e^{-1} = -1$

86. $\ln e^4 = 4$

87. $\ln(2x) = 2$
$$2x = e^2$$
$$x = \frac{e^2}{2}$$

88. $\ln(3x) = 1.6$
$$3x = e^{1.6}$$
$$x = \frac{e^{1.6}}{3}$$

89. $\ln(2x-3) = -1$
$$2x-3 = e^{-1}$$
$$x = \frac{e^{-1}+3}{2}$$

90. $\ln(3x+1) = 2$
$$3x+1 = e^2$$
$$3x = e^2 - 1$$
$$x = \frac{e^2-1}{3}$$

91.
$$\ln\frac{I}{I_0} = -kx$$
$$\ln\frac{0.03I_0}{I_0} = -2.1x$$
$$\ln 0.03 = -2.1x$$
$$\frac{\ln 0.03}{-2.1} = x$$
$$x \approx 1.67$$
The depth is 1.67 millimeters.

92.
$$\ln\frac{I}{I_0} = -kx$$
$$\ln\frac{0.02I_0}{I_0} = -3.2x$$
$$\ln 0.02 = -3.2x$$
$$\frac{\ln 0.02}{-3.2} = x$$
$$x \approx 1.22$$
2% of the original radioactivity will penetrate at a depth of approximately 1.22 millimeters.

93. $\log_5 1.6 = \frac{\log 1.6}{\log 5} \approx 0.2920$

94. $\log_3 4 = \frac{\log 4}{\log 3} \approx 1.2619$

95. $A = Pe^{rt}$
$$A = 1450e^{(0.03)(5)}$$
$$A \approx 1684.66$$
The accrued amount is \$1684.66.

96. $A = Pe^{rt}$
$$A = 940e^{0.04(3)} = 940e^{0.12} \approx 1059.85$$
The investment grows to \$1059.85.

97.
$$3^{2x} = 7$$
$$\log 3^{2x} = \log 7$$
$$2x\log 3 = \log 7$$
$$x = \frac{\log 7}{2\log 3} \approx 0.8856$$

98.
$$6^{3x} = 5$$
$$\log 6^{3x} = \log 5$$
$$3x\log 6 = \log 5$$
$$x = \frac{\log 5}{3\log 6} \approx 0.2994$$

99.

$$\begin{aligned} 3^{2x+1} &= 6 \\ \log 3^{2x+1} &= \log 6 \\ (2x+1)\log 3 &= \log 6 \\ 2x\log 3 + \log 3 &= \log 6 \\ 2x\log 3 &= \log 6 - \log 3 \\ x &= \frac{\log 6 - \log 3}{2\log 3} \end{aligned}$$

or

$$x = \frac{1}{2}\left(\frac{\log 6}{\log 3} - 1\right)$$

$x \approx 0.3155$

100.

$$\begin{aligned} 4^{3x+2} &= 9 \\ \log 4^{3x+2} &= \log 9 \\ (3x+2)\log 4 &= \log 9 \\ 3x\log 4 + 2\log 4 &= \log 9 \\ 3x\log 4 &= \log 9 - 2\log 4 \\ x &= \frac{\log 9 - 2\log 4}{3\log 4} \end{aligned}$$

or

$$x = \frac{1}{3}\left(\frac{\log 9}{\log 4} - 2\right)$$

$x \approx -0.1383$

101.

$$\begin{aligned} 5^{3x-5} &= 4 \\ \log 5^{3x-5} &= \log 4 \\ (3x-5)\log 5 &= \log 4 \\ 3x\log 5 - 5\log 5 &= \log 4 \\ 3x\log 5 &= \log 4 + 5\log 5 \\ x &= \frac{\log 4 + 5\log 5}{3\log 5} \end{aligned}$$

or

$$x = \frac{1}{3}\left(\frac{\log 4}{\log 5} + 5\right)$$

$x \approx 1.9538$

102.

$$\begin{aligned} 8^{4x-2} &= 3 \\ \log 8^{4x-2} &= \log 3 \\ (4x-2)\log 8 &= \log 3 \\ 4x\log 8 - 2\log 8 &= \log 3 \\ 4x\log 8 &= \log 3 + 2\log 8 \\ x &= \frac{\log 3 + 2\log 8}{4\log 8} \end{aligned}$$

or

$$x = \frac{1}{4}\left(\frac{\log 3}{\log 8} + 2\right)$$

$x \approx 0.6321$

103.

$$\begin{aligned} 5^{x-1} &= \frac{1}{2} \\ \log 5^{x-1} &= \log \frac{1}{2} \\ (x-1)\log 5 &= \log \frac{1}{2} \\ x\log 5 - \log 5 &= \log \frac{1}{2} \\ x\log 5 &= \log \frac{1}{2} + \log 5 \\ x &= \frac{\log \frac{1}{2} + \log 5}{\log 5} \end{aligned}$$

or

$$x = -\frac{\log 2}{\log 5} + 1$$

$x \approx 0.5693$

104.

$$\begin{aligned} 4^{x+5} &= \frac{2}{3} \\ \log 4^{x+5} &= \log \frac{2}{3} \\ (x+5)\log 4 &= \log \frac{2}{3} \\ x\log 4 + 5\log 4 &= \log \frac{2}{3} \\ x\log 4 &= \log \frac{2}{3} - 5\log 4 \\ x &= \frac{\log \frac{2}{3} - 5\log 4}{\log 4} \end{aligned}$$

or

$$x = \frac{\log \frac{2}{3}}{\log 4} - 5$$

$x \approx -5.2925$

105. $\log_5 2+\log_5 x=2$

$$\log_5 2x=2$$
$$2x=5^2$$
$$2x=25$$
$$x=\frac{25}{2}$$

106. $\log_3 x+\log_3 10=2$

$$\log_3(10x)=2$$
$$10x=3^2$$
$$10x=9$$
$$x=\frac{9}{10}$$

107. $\log(5x)-\log(x+1)=4$

$$\log\frac{5x}{x+1}=4$$
$$\frac{5x}{x+1}=10^4$$
$$\frac{5x}{x+1}=10{,}000$$
$$5x=10{,}000x+10{,}000$$
$$x=-1.0005$$

no solution, or $\varnothing$

108. $-\log_6(4x+7)+\log_6 x=1$

$$\log_6\frac{x}{4x+7}=1$$
$$\frac{x}{4x+7}=6$$
$$x=6(4x+7)$$
$$x=24x+42$$
$$x=-\frac{42}{23}$$

$-\frac{42}{23}$ is rejected since $\log_6\left(-\frac{42}{23}\right)$ is undefined.

There is no solution, or $\varnothing$.

109. $\log_2 x+\log_2 2x-3=1$

$$\log_2(x\cdot 2x)=4$$
$$2x^2=2^4$$
$$2x^2=16$$
$$x^2=8$$
$$x=\pm 2\sqrt{2}$$

$-2\sqrt{2}$ is rejected since $\log_2\left(-2\sqrt{2}\right)$ is undefined. The solution is $2\sqrt{2}$.

110. $\log_3(x^2-8x)=2$

$$3^2=x^2-8x$$
$$0=x^2-8x-9$$
$$0=(x-9)(x+1)$$

$x=9$ or $x=-1$

111. Let $y_0=27$, $y=425$, $r=11.4\%=0.114$.

$$y=y_0e^{kt}$$
$$425=27e^{(0.114)t}$$
$$\frac{425}{27}=e^{0.114t}$$
$$\ln\frac{425}{27}=0.114t$$
$$\frac{1}{0.114}\ln\frac{425}{27}=t$$
$$24.2\approx t$$

It will take approximately 24.2 years.

112. Let $y_0=66{,}030{,}000$, $y=70{,}000{,}000$, $k=0.005$.

$$y=y_0e^{kt}$$
$$70{,}000{,}000=66{,}030{,}000e^{0.005t}$$
$$\frac{7000}{6603}=e^{0.005t}$$
$$\ln\frac{7000}{6603}=0.005t$$
$$\frac{1}{0.005}\ln\frac{7000}{6603}=t$$
$$11.7\approx t$$

It will take approximately 11.7 years.

113. Let $y_0=23{,}130{,}000$, $y=2(23{,}130{,}000)=46{,}260{,}000$, $k=0.018$.

$$y=y_0e^{kt}$$
$$46{,}260{,}000=23{,}130{,}000e^{0.018t}$$
$$2=e^{0.018t}$$
$$\ln 2=0.018t$$
$$\frac{1}{0.018}\ln 2=t$$
$$38.5\approx t$$

It will take approximately 38.5 years for the population to double.

114. Let $y_0 = 35{,}160{,}000$,
$y = 2(35{,}160{,}000) = 70{,}320{,}000$, $k = 0.012$.

$$y = y_0 e^{kt}$$
$$70{,}320{,}000 = 35{,}160{,}000e^{0.012t}$$
$$2 = e^{0.012t}$$
$$\ln 2 = 0.012t$$
$$\frac{1}{0.012}\ln 2 = t$$
$$57.8 \approx t$$

It will take approximately 57.8 years for the population to double.

115.
$$A = P\left(1+\frac{r}{n}\right)^{nt}$$
$$10{,}000 = 5000\left(1+\frac{0.08}{4}\right)^{4t}$$
$$2 = (1.02)^{4t}$$
$$\log 2 = \log 1.02^{4t}$$
$$\log 2 = 4t\log 1.02$$
$$t = \frac{\log 2}{4\log 1.02} \approx 8.8$$

It will take 8.8 years.

116.
$$A = P\left(1+\frac{r}{n}\right)^{nt}$$
$$10{,}000 = 6000\left(1+\frac{0.06}{12}\right)^{12t}$$
$$\frac{5}{3} = (1.005)^{12t}$$
$$\log\frac{5}{3} = \log 1.005^{12t}$$
$$\log\frac{5}{3} = 12t\log 1.005$$
$$t = \frac{1}{12}\left(\frac{\log\left(\frac{5}{3}\right)}{\log(1.005)}\right) \approx 8.5$$

It was invested for approximately 8.5 years.

117. $Y_1 = e^x$, $Y_2 = 2$

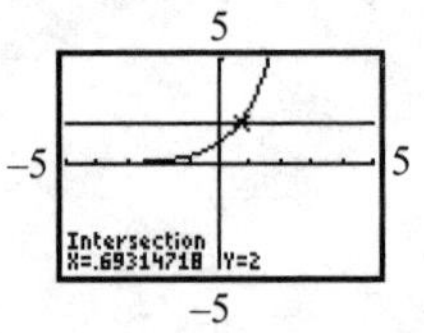

$x \approx 0.69$

118. $Y_1 = 10^{0.3x}$, $Y_2 = 7$

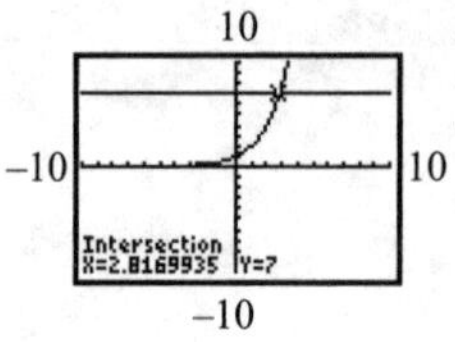

$x \approx 2.82$

119.
$$3^x = \frac{1}{81}$$
$$3^x = 3^{-4}$$
$$x = -4$$

120.
$$7^{4x} = 49$$
$$7^{4x} = 7^2$$
$$4x = 2$$
$$x = \frac{1}{2}$$

121.
$$8^{3x-2} = 32$$
$$(2^3)^{(3x-2)} = 2^5$$
$$2^{9x-6} = 2^5$$
$$9x - 6 = 5$$
$$9x = 11$$
$$x = \frac{11}{9}$$

122.
$$9^{x-2} = 27$$
$$(3^2)^{(x-2)} = 3^3$$
$$3^{2x-4} = 3^3$$
$$2x - 4 = 3$$
$$2x = 7$$
$$x = \frac{7}{2}$$

123.
$$\log_4 4 = x$$
$$4^x = 4^1$$
$$x = 1$$

124.
$$\log_3 x = 4$$
$$3^4 = x$$
$$81 = x$$

125. $\log_5(x^2-4x)=1$

$$5^1 = x^2-4x$$
$$0 = x^2-4x-5$$
$$0 = (x-5)(x+1)$$

$x-5=0$ or $x+1=0$

$x=5$ $\quad x=-1$

Both check, so the solutions are 5 and −1.

126. $\log_4(3x-1)=2$

$$4^2 = 3x-1$$
$$16+1 = 3x$$
$$\frac{17}{3} = x$$

127. $\ln x = -3.2$

$$e^{\ln x} = e^{-3.2}$$
$$x = e^{-3.2}$$

128. $\log_5 x + \log_5 10 = 2$

$$\log_5(10x) = 2$$
$$5^2 = 10x$$
$$\frac{25}{10} = x$$
$$\frac{5}{2} = x$$

129. $\ln x - \ln 2 = 1$

$$\ln\frac{x}{2} = 1$$
$$e^{\ln\frac{x}{2}} = e^1$$
$$\frac{x}{2} = e$$
$$x = 2e$$

130. $\log_6 x - \log_6(4x+7) = 1$

$$\log_6\frac{x}{4x+7} = 1$$
$$6^1 = \frac{x}{4x+7}$$
$$24x+42 = x$$
$$23x = -42$$
$$x = -\frac{42}{23}$$

$-\frac{42}{23}$ is rejected since $\log_6\left(-\frac{42}{23}\right)$ is undefined.

There is no solution, or ∅.

Chapter 9 Getting Ready for the Test

1. $(f+g)(9) = f(9)+g(9) = 6+8 = 14$; B

2. $(f \circ g)(5) = f(g(5)) = f(-3) = -5$; A

3. $(f \cdot g)(-3) = f(-3)\cdot g(-3) = -5 \cdot 5 = -25$; D

4. $\left(\frac{f}{g}\right)(0) = \frac{f(0)}{g(0)} = \frac{4}{-2} = -2$; C

5. If f is a one-to-one function and $f(-2) = 7$, then an ordered pair solution of f is (−2, 7), so an ordered pair solution of f^{-1} is (7, −2); D.

6. If $f(x) = 3^{-x}$, then $f(-1) = 3^{-(-1)} = 3^1 = 3$ so (−1, 3) is a point on the graph; B.

7. If $f(x) = \left(\frac{1}{3}\right)^{-x}$, then

$f(1) = \left(\frac{1}{3}\right)^{-1} = (3^{-1})^{-1} = 3^1 = 3$, so (1, 3) is a point on the graph; D.

8. If $f(x) = 5^{-x}$, then $f(-1) = 5^{-(-1)} = 5^1 = 5$, so (−1, 5) is a point on the graph; C.

9. If $f(x) = \left(\frac{1}{5}\right)^{-x}$, then

$f(1) = \left(\frac{1}{5}\right)^{-1} = (5^{-1})^{-1} = 5^1 = 5$, so (1, 5) is a point on the graph; A.

10. Half of a substance will have decayed after 1 half-life, so one-half of the original amount will be left. Since 32 years is the half-life of the substance, $\frac{1}{2}$ of the original 50 grams, or 25 grams, will remain after 32 years; B.

11. $\log_7\frac{13}{9} = \log_7 13 - \log_7 9$; B

12. $\log_3 2 = \frac{\log 2}{\log 3} = \frac{\ln 2}{\ln 3}$; D

13. $\log_2(a+b)$ cannot be simplified, so the statement is false; B.

14. $\log_5 7^2 = 2\log_5 7$ is true; A.

15. $\dfrac{\log_2 x}{\log_2 y}$ cannot be simplified, so the statement is false; B.

Chapter 9 Test

1. $f(x) = x$ and $g(x) = 2x - 3$

$(f \cdot g)(x) = f(x) \cdot g(x) = x(2x-3) = 2x^2 - 3x$

2. $f(x) = x$ and $g(x) = 2x - 3$

$$\begin{aligned}(f-g)(x) &= f(x) - g(x)\\ &= x-(2x-3)\\ &= -x+3\\ &= 3-x\end{aligned}$$

3. $(f \circ h)(0) = f(h(0)) = f(5) = 5$

4. $(g \circ f)(x) = g(f(x)) = g(x) = x - 7$

5.
$$\begin{aligned}(g \circ h)(x) &= g(h(x))\\ &= g(x^2-6x+5)\\ &= x^2-6x+5-7\\ &= x^2-6x-2\end{aligned}$$

6. $f(x) = 7x - 14,\ f^{-1}(x) = \dfrac{x+14}{7}$

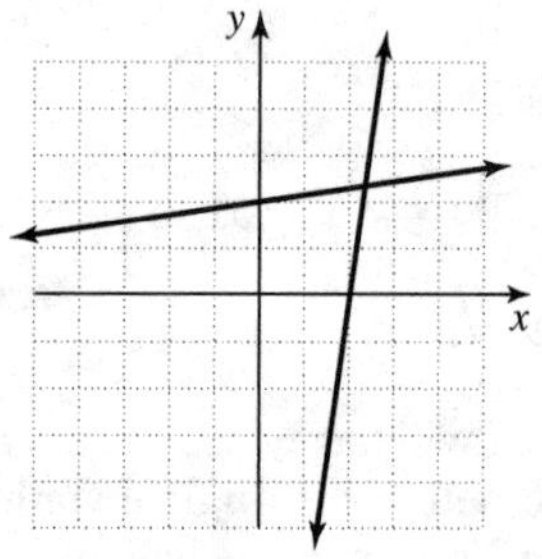

7. The graph represents a one-to-one function.

8. The graph does not represent a one-to-one function.

9. $y = 6 - 2x$ is one-to-one.

$$\begin{aligned}x &= 6-2y\\ 2y &= -x+6\\ y &= \frac{-x+6}{2}\\ f^{-1}(x) &= \frac{-x+6}{2}\end{aligned}$$

10. $f = \{(0,0), (2,3), (-1,5)\}$ is one-to-one.

$f^{-1} = \{(0,0), (3,2), (5,-1)\}$

11. The function is not one-to-one.

12. $\log_3 6 + \log_3 4 = \log_3(6 \cdot 4) = \log_3 24$

13.
$$\begin{aligned}&\log_5 x + 3\log_5 x - \log_5(x+1)\\ &= 4\log_5 x - \log_5(x+1)\\ &= \log_5 x^4 - \log_5(x+1)\\ &= \log_5 \frac{x^4}{x+1}\end{aligned}$$

14.
$$\begin{aligned}\log_6 \frac{2x}{y^3} &= \log_6 2x - \log_6 y^3\\ &= \log_6 2 + \log_6 x - 3\log_6 y\end{aligned}$$

15.
$$\begin{aligned}\log_b\left(\frac{3}{25}\right) &= \log_b 3 - \log_b 25\\ &= \log_b 3 - \log_b 5^2\\ &= \log_b 3 - 2\log_b 5\\ &= 0.79 - 2(1.16)\\ &= -1.53\end{aligned}$$

16. $\log_7 8 = \dfrac{\ln 8}{\ln 7} \approx 1.0686$

17.
$$\begin{aligned}8^{x-1} &= \frac{1}{64}\\ 8^{x-1} &= 8^{-2}\\ x-1 &= -2\\ x &= -1\end{aligned}$$

18.
$$\begin{aligned}3^{2x+5} &= 4\\ \log 3^{2x+5} &= \log 4\\ (2x+5)\log 3 &= \log 4\\ 2x &= \frac{\log 4}{\log 3} - 5\\ x &= \frac{1}{2}\left(\frac{\log 4}{\log 3} - 5\right)\\ x &\approx -1.8691\end{aligned}$$

19.
$$\begin{aligned}\log_3 x &= -2\\ x &= 3^{-2}\\ x &= \frac{1}{9}\end{aligned}$$

20. $\ln\sqrt{e} = x$

$\ln e^{1/2} = x$

$\frac{1}{2} = x$

21. $\log_8(3x-2) = 2$

$3x-2 = 8^2$

$3x-2 = 64$

$3x = 66$

$x = \frac{66}{3} = 22$

22. $\log_5 x + \log_5 3 = 2$

$\log_5(3x) = 2$

$3x = 5^2$

$3x = 25$

$x = \frac{25}{3}$

23. $\log_4(x+1) - \log_4(x-2) = 3$

$\log_4 \frac{x+1}{x-2} = 3$

$\frac{x+1}{x-2} = 4^3$

$\frac{x+1}{x-2} = 64$

$x+1 = 64x-128$

$129 = 63x$

$\frac{129}{63} = x$

$\frac{43}{21} = x$

24. $\ln(3x+7) = 1.31$

$3x+7 = e^{1.31}$

$3x = e^{1.31} - 7$

$x = \frac{e^{1.31}-7}{3} \approx -1.0979$

25. $y = \left(\frac{1}{2}\right)^x + 1$

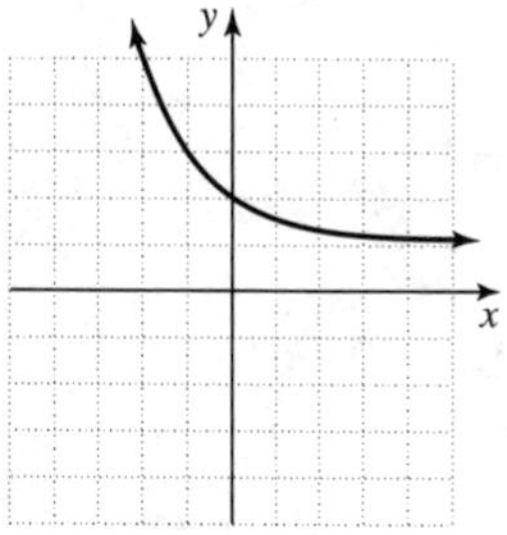

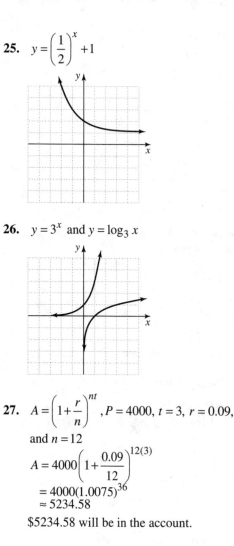

26. $y = 3^x$ and $y = \log_3 x$

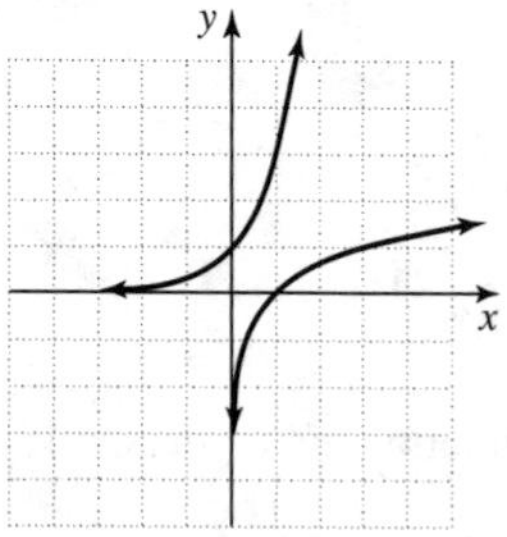

27. $A = \left(1+\frac{r}{n}\right)^{nt}$, $P = 4000$, $t = 3$, $r = 0.09$, and $n = 12$

$A = 4000\left(1+\frac{0.09}{12}\right)^{12(3)}$

$= 4000(1.0075)^{36}$

≈ 5234.58

\$5234.58 will be in the account.

28. $A = \left(1+\frac{r}{n}\right)^{nt}$, $P = 2000$, $A = 3000$

$r = 0.07$, $n = 2$

$3000 = 2000\left(1+\frac{0.07}{2}\right)^{2t}$

$1.5 = (1.035)^{2t}$

$\log 1.5 = \log 1.035^{2t}$

$\log 1.5 = 2t \log 1.035$

$t = \frac{\log 1.5}{2\log 1.035} \approx 5.9$

It would take 6 years.

29. Let $P = 3000$, $t = 10$.
Semiannually: Let $n = 2$, $r = 0.065$.

$$A = P\left(1+\frac{r}{n}\right)^{nt}$$
$$A = 3000\left(1+\frac{0.065}{2}\right)^{2(10)}$$
$$A \approx 5688$$

Monthly: Let $n = 12$, $r = 0.06$.

$$A = 3000\left(1+\frac{0.06}{12}\right)^{12(10)}$$
$$A \approx 5458$$

The better investment is the 6.5% investment by \$5688 − \$5458 = \$230.

30. $C = 150{,}000$, $r = 0.02$, $x = 20$

$$y = C(1-r)^x$$
$$y = 150{,}000(1-0.02)^{20}$$
$$y \approx 100{,}141$$

The predicted population in 20 years is 100,141.

31. $C = 57{,}000$, $r = 0.026$, $x = 5$

$$y = C(1+r)^x$$
$$y = 57{,}000(1+0.026)^5$$
$$y \approx 64{,}805$$

In 5 years, the predicted population is 64,805 animals.

32. $C = 400$, $r = 0.062$, $y = 1000$

$$y = C(1+r)^x$$
$$1000 = 400(1+0.062)^x$$
$$\frac{1000}{400} = 1.062^x$$
$$\ln\left(\frac{1000}{400}\right) = x\ln 1.062$$
$$x = \frac{\ln\left(\frac{1000}{400}\right)}{\ln 1.062}$$
$$x \approx 15$$

It will take about 15 years to reach their goal.

33. Let $t = 0.5$.

$$R(t) = 1.6^{-(1/3)t}$$
$$R(0.5) = 1.6^{-(1/3)(0.5)} \approx 0.92$$

The probability is about 92%.

34. Let $t = 2$.

$$R(t) = 1.6^{-(1/3)(2)} \approx 0.73$$

The probability is about 73%.

Chapter 9 Cumulative Review

1. a. $(-8)(-1) = 8$

b. $-2\left(\frac{1}{6}\right) = -\frac{1}{3}$

c. $-1.2(0.3) = -0.36$

d. $0(-11) = 0$

e. $\frac{1}{5}\left(-\frac{10}{11}\right) = -\frac{2}{11}$

f. $(7)(1)(-2)(-3) = 42$

g. $8(-2)(0) = 0$

2.
$$\frac{1}{3}(x-2) = \frac{1}{4}(x+1)$$
$$4(x-2) = 3(x+1)$$
$$4x-8 = 3x+3$$
$$x = 11$$

3. $y = x^2$

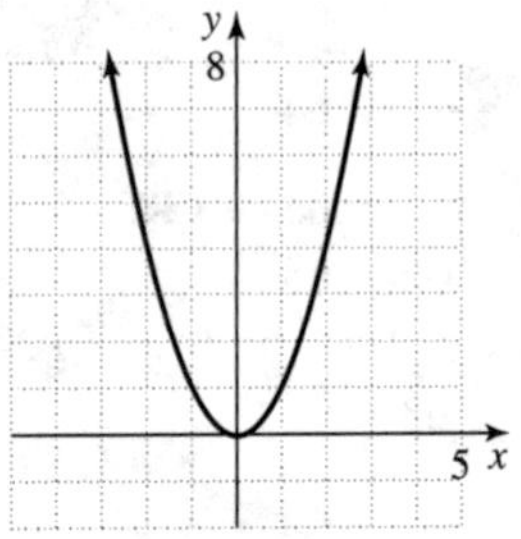

4. $y = f(x) = -3x + 4$, $m = -3$

Perpendicular line: $m = \frac{1}{3}$, through $(-2, 6)$

$$y - y_1 = m(x - x_1)$$
$$y - 6 = \frac{1}{3}[x-(-2)]$$
$$y - 6 = \frac{1}{3}x + \frac{2}{3}$$
$$y = \frac{1}{3}x + \frac{20}{3}$$
$$f(x) = \frac{1}{3}x + \frac{20}{3}$$

5. Equation 2 is twice the opposite of equation 1 and equation 3 is one-half of equation 1. Therefore, the system is dependent. The solution is $\{(x, y, z) | x - 5y - 2z = 6\}$.

6. The angles labeled $y°$ and $(x - 40)°$ are alternate interior angles, so $y = x - 40$. The angles labeled $x°$ and $y°$ are supplementary, so $x + y = 180$.

$$\begin{cases} y = x - 40 \\ x + y = 180 \end{cases}$$

Replace y with $x - 40$ in the second equation.

$$x + (x - 40) = 180$$
$$2x = 220$$
$$x = 110$$

$y = x - 40 = 110 - 40 = 70$

7. a. $\dfrac{x^7}{x^4} = x^{7-4} = x^3$

b. $\dfrac{5^8}{5^2} = 5^{8-2} = 5^6$

c. $\dfrac{20x^6}{4x^5} = \dfrac{20}{4}x^{6-5} = 5x^1 = 5x$

d. $\dfrac{12y^{10}z^7}{14y^8z^7} = \dfrac{12}{14}y^{10-8}z^{7-7} = \dfrac{6}{7}y^2z^0 = \dfrac{6y^2}{7}$

8. a. $(4a^3)^2 = 4^2(a^3)^2 = 16a^6$

b. $\left(-\dfrac{2}{3}\right)^3 = \dfrac{(-2)^3}{3^3} = \dfrac{-8}{27} = -\dfrac{8}{27}$

c. $\left(\dfrac{4a^5}{b^3}\right)^3 = \dfrac{4^3(a^5)^3}{(b^3)^3} = \dfrac{64a^{15}}{b^9}$

d. $\left(\dfrac{3^{-2}}{x}\right)^{-3} = \dfrac{(3^{-2})^{-3}}{x^{-3}} = \dfrac{3^6}{x^{-3}} = 729x^3$

e.
$$\begin{aligned}(a^{-2}b^3c^{-4})^{-2} &= (a^{-2})^{-2}(b^3)^{-2}(c^{-4})^{-2} \\ &= a^4b^{-6}c^8 \\ &= \frac{a^4c^8}{b^6}\end{aligned}$$

9. a.
$$\begin{aligned}C(100) &= \frac{2.6(100) + 10{,}000}{100} \\ &= 102.60\end{aligned}$$

The cost is \$102.60 per disc for 100 discs.

b.
$$\begin{aligned}C(1000) &= \frac{2.6(1000) + 10{,}000}{1000} \\ &= 12.60\end{aligned}$$

The cost is \$12.60 per disc for 1000 discs.

10. a.
$$\begin{aligned}(3x-1)^2 &= (3x)^2 - 2(3x)(1) + 1^2 \\ &= 9x^2 - 6x + 1\end{aligned}$$

b.
$$\begin{aligned}\left(\frac{1}{2}x+3\right)\left(\frac{1}{2}x-3\right) &= \left(\frac{1}{2}x\right)^2 - 3^2 \\ &= \frac{1}{4}x^2 - 9\end{aligned}$$

c.
$$\begin{aligned}(2x-5)(6x+7) &= 12x^2 + 14x - 30x - 35 \\ &= 12x^2 - 16x - 35\end{aligned}$$

11. a. $\dfrac{x}{4} + \dfrac{5x}{4} = \dfrac{6x}{4} = \dfrac{3x}{2}$

b. $\dfrac{5}{7z^2} + \dfrac{x}{7z^2} = \dfrac{5+x}{7z^2}$

c.
$$\begin{aligned}\frac{x^2}{x+7} - \frac{49}{x+7} &= \frac{x^2-49}{x+7} \\ &= \frac{(x+7)(x-7)}{x+7} \\ &= x - 7\end{aligned}$$

d. $\dfrac{x}{3y^2} - \dfrac{x+1}{3y^2} = \dfrac{x-x-1}{3y^2} = -\dfrac{1}{3y^2}$

12.
$$\begin{aligned}&\frac{5}{x-2} + \frac{3}{x^2+4x+4} - \frac{6}{x+2} \\ &= \frac{5}{x-2} + \frac{3}{(x+2)^2} - \frac{6}{x+2} \\ &= \frac{5(x+2)^2 + 3(x-2) - 6(x-2)(x+2)}{(x-2)(x+2)(x+2)} \\ &= \frac{-x^2 + 23x + 38}{(x-2)(x+2)^2}\end{aligned}$$

13.

$$\begin{array}{r} 3x^2+2x+3 \\ x^2-1\overline{)3x^4+2x^3\qquad\quad -8x+6} \\ \underline{3x^4\qquad -3x^2}\qquad\qquad \\ 2x^3+3x^2-8x\qquad \\ \underline{2x^3\qquad\quad -2x}\qquad \\ 3x^2-6x+6 \\ \underline{3x^2\qquad\quad -3} \\ -6x+9 \end{array}$$

Solution: $3x^2+2x+3+\dfrac{-6x+9}{x^2-1}$

14. a. $\dfrac{\frac{a}{5}}{\frac{a-1}{10}}=\dfrac{a}{5}\cdot\dfrac{10}{a-1}=\dfrac{2a}{a-1}$

b. $\dfrac{\frac{3}{2+a}+\frac{6}{2-a}}{\frac{5}{a+2}-\frac{1}{a-2}}=\dfrac{\frac{3}{a+2}-\frac{6}{a-2}}{\frac{5}{a+2}-\frac{1}{a-2}}$

Multiply the numerator and the denominator by $(a+2)(a-2)$.

$$\begin{aligned}\frac{3(a-2)-6(a+2)}{5(a-2)-1(a+2)}&=\frac{3a-6-6a-12}{5a-10-a-2}\\&=\frac{-3a-18}{4a-12}\\&=-\frac{3(a+6)}{4(a-3)}\end{aligned}$$

c. $\dfrac{x^{-1}+y^{-1}}{xy}=\dfrac{\frac{1}{x}+\frac{1}{y}}{xy}=\dfrac{\left(\frac{1}{x}+\frac{1}{y}\right)xy}{(xy)(xy)}=\dfrac{y+x}{x^2y^2}$

15.

$$\begin{aligned}\frac{2x}{2x-1}+\frac{1}{x}&=\frac{1}{2x-1}\\(2x-1)x\left(\frac{2x}{2x-1}+\frac{1}{x}\right)&=(2x-1)x\left(\frac{1}{2x-1}\right)\\2x^2+2x-1&=x\\2x^2+x-1&=0\\(2x-1)(x+1)&=0\end{aligned}$$

$2x-1=0$ or $x+1=0$

$2x=1 \qquad x=-1$

$x=\dfrac{1}{2}$

$x=\dfrac{1}{2}$ makes the denominator $2x-1$ zero, so the only solution is $x=-1$.

16.

$$\begin{aligned}\frac{x^3-8}{x-2}&=\frac{(x-2)(x^2+2x+4)}{(x-2)}\\&=x^2+2x+4\end{aligned}$$

17. Use distance = (rate)(time). Let c be the speed of the current.

	d	r	t
Upstream	72	$30-c$	$1.5t$
Downstream	72	$30+c$	t

Upstream: $72=1.5t(30-c)$ or $t=\dfrac{72}{1.5(30-c)}$

Downstream: $72=t(30+c)$ or $t=\dfrac{72}{30+c}$

$$\begin{aligned}\frac{72}{1.5(30-c)}&=\frac{72}{30+c}\\72(30+c)&=72(1.5)(30-c)\\2160+72c&=3240-108c\\180c&=1080\\c&=6\end{aligned}$$

The speed of the current is 6 miles per hour.

18.

$$\begin{array}{r|rrr} 2 & 8 & -12 & -7 \\ & & 16 & 8 \\ \hline & 8 & 4 & 1 \end{array}$$

Solution: $8x+4+\dfrac{1}{x-2}$

19. a. $\sqrt[4]{81}=\sqrt[4]{3^4}=3$

b. $\sqrt[5]{-243}=\sqrt[5]{(-3)^5}=-3$

c. $-\sqrt{25}=-\sqrt{5^2}=-5$

d. $\sqrt[4]{-81}$ is not a real number.

e. $\sqrt[3]{64x^3}=\sqrt[3]{4^3x^3}=4x$

20. $$\frac{1}{a+5} = \frac{1}{3a+6} - \frac{a+2}{a^2+7a+10}$$
$$\frac{1}{a+5} = \frac{1}{3(a+2)} - \frac{a+2}{(a+2)(a+5)}$$
$$3(a+2) = a+5-3(a+2)$$
$$3a+6 = a+5-3a-6$$
$$5a = -7$$
$$a = -\frac{7}{5}$$

21. a. $\sqrt{x} \cdot \sqrt[4]{x} = x^{1/2} \cdot x^{1/4} = x^{3/4} = \sqrt[4]{x^3}$

b. $\frac{\sqrt{x}}{\sqrt[3]{x}} = \frac{x^{1/2}}{x^{1/3}} = x^{\frac{1}{2}-\frac{1}{3}} = x^{1/6} = \sqrt[6]{x}$

c. $$\sqrt[3]{3} \cdot \sqrt{2} = 3^{1/3} \cdot 2^{1/2}$$
$$= 3^{2/6} \cdot 2^{3/6}$$
$$= 9^{1/6} \cdot 8^{1/6}$$
$$= 72^{1/6}$$
$$= \sqrt[6]{72}$$

22. $$y = kx$$
$$\frac{1}{2} = 12k$$
$$k = \frac{1}{24},\ y = \frac{1}{24}x$$

23. a. $\sqrt{3}\left(5+\sqrt{30}\right) = 5\sqrt{3}+\sqrt{90} = 5\sqrt{3}+3\sqrt{10}$

b. $\left(\sqrt{5}-\sqrt{6}\right)\left(\sqrt{7}+1\right) = \sqrt{35}+\sqrt{5}-\sqrt{42}-\sqrt{6}$

c. $$\left(7\sqrt{x}+5\right)\left(3\sqrt{x}-\sqrt{5}\right)$$
$$= 21x - 7\sqrt{5x} + 15\sqrt{x} - 5\sqrt{5}$$

d. $$\left(4\sqrt{3}-1\right)^2$$
$$= \left(4\sqrt{3}\right)^2 - 2\left(4\sqrt{3}\right)(1) + 1^2$$
$$= 16 \cdot 3 - 8\sqrt{3} + 1$$
$$= 49 - 8\sqrt{3}$$

e. $$\left(\sqrt{2x}-5\right)\left(\sqrt{2x}+5\right) = \left(\sqrt{2x}\right)^2 - 5^2$$
$$= 2x - 25$$

f. $$\left(\sqrt{x-3}+5\right)^2 = \left(\sqrt{x-3}\right)^2 + 2\sqrt{x-3}(5) + 5^2$$
$$= x-3+10\sqrt{x-3}+25$$
$$= x+22+10\sqrt{x-3}$$

24. a. $\sqrt[3]{27} = \sqrt[3]{3^3} = 3$

b. $\sqrt[3]{-27} = \sqrt[3]{(-3)^3} = -3$

c. $\sqrt{\frac{9}{64}} = \sqrt{\left(\frac{3}{8}\right)^2} = \frac{3}{8}$

d. $\sqrt[4]{x^{12}} = x^3$

e. $\sqrt[3]{-125y^6} = -5y^2$

25. $$\frac{\sqrt[4]{x}}{\sqrt[4]{81y^5}} = \frac{\sqrt[4]{x}}{\sqrt[4]{81y^5}} \cdot \frac{\sqrt[4]{y^3}}{\sqrt[4]{y^3}} = \frac{\sqrt[4]{xy^3}}{3y^2}$$

26. a. $a^{1/4}(a^{3/4} - a^{7/4}) = a^{4/4} - a^{8/4} = a - a^2$

b. $$(x^{1/2}-3)(x^{1/2}+5)$$
$$= x^{2/2} + 5x^{1/2} - 3x^{1/2} - 15$$
$$= x + 2x^{1/2} - 15$$

27. $$\sqrt{4-x} = x-2$$
$$\left(\sqrt{4-x}\right)^2 = (x-2)^2$$
$$4-x = x^2-4x+4$$
$$0 = x^2-3x$$
$$0 = x(x-3)$$
$$x = 0 \quad \text{or} \quad x-3 = 0$$
$$x = 3$$

$x = 0$ does not check, so the only solution is $x = 3$.

28. a. $\sqrt{\frac{54}{6}} = \sqrt{9} = 3$

b. $$\frac{\sqrt{108a^2}}{3\sqrt{3}} = \frac{1}{3}\sqrt{\frac{108a^2}{3}}$$
$$= \frac{1}{3}\sqrt{36a^2}$$
$$= \frac{1}{3}(6a)$$
$$= 2a$$

c. $$\frac{3\sqrt[3]{81a^5b^{10}}}{\sqrt[3]{3b^4}} = 3\sqrt[3]{\frac{81a^5b^{10}}{3b^4}} = 3\sqrt[3]{27a^5b^6} = 9ab^2\sqrt[3]{a^2}$$

29. $$3x^2 - 9x + 8 = 0$$
$$x^2 - 3x + \frac{8}{3} = 0$$
$$x^2 - 3x = -\frac{8}{3}$$
$$x^2 - 3x + \left(\frac{-3}{2}\right)^2 = -\frac{8}{3} + \left(\frac{-3}{2}\right)^2$$
$$x^2 - 3x + \frac{9}{4} = -\frac{8}{3} + \frac{9}{4}$$
$$\left(x - \frac{3}{2}\right)^2 = -\frac{5}{12}$$
$$x - \frac{3}{2} = \pm\sqrt{-\frac{5}{12}}$$
$$x - \frac{3}{2} = \pm\frac{i\sqrt{5}}{2\sqrt{3}}$$
$$x - \frac{3}{2} = \pm\frac{i\sqrt{15}}{6}$$
$$x = \frac{3}{2} \pm \frac{i\sqrt{15}}{6} = \frac{9}{6} \pm \frac{i\sqrt{15}}{6} = \frac{9 \pm i\sqrt{15}}{6}$$

The solutions are $\frac{9 + i\sqrt{15}}{6}$ and $\frac{9 - i\sqrt{15}}{6}$.

30. a. $$\frac{\sqrt{20}}{3} + \frac{\sqrt{5}}{4} = \frac{2\sqrt{5}}{3} + \frac{\sqrt{5}}{4} = \frac{8\sqrt{5} + 3\sqrt{5}}{12} = \frac{11\sqrt{5}}{12}$$

b. $$\sqrt[3]{\frac{24x}{27}} - \frac{\sqrt[3]{3x}}{2} = \frac{2\sqrt[3]{3x}}{3} - \frac{\sqrt[3]{3x}}{2} = \frac{4\sqrt[3]{3x} - 3\sqrt[3]{3x}}{6} = \frac{\sqrt[3]{3x}}{6}$$

31. $$\frac{3x}{x-2} - \frac{x+1}{x} = \frac{6}{x(x-2)}$$
$$3x(x) - (x+1)(x-2) = 6$$
$$3x^2 - x^2 + x + 2 = 6$$
$$2x^2 + x - 4 = 0$$
$a = 2$, $b = 1$, $c = -4$
$$x = \frac{-1 \pm \sqrt{1^2 - 4(2)(-4)}}{2(2)} = \frac{-1 \pm \sqrt{33}}{4}$$

32. $$\sqrt[3]{\frac{27}{m^4n^8}} = \frac{\sqrt[3]{27}}{\sqrt[3]{m^4n^8}} = \frac{3}{mn^2\sqrt[3]{mn^2}} = \frac{3 \cdot \sqrt[3]{m^2n}}{mn^2\sqrt[3]{mn^2} \cdot \sqrt[3]{m^2n}} = \frac{3\sqrt[3]{m^2n}}{m^2n^3}$$

33. $x^2 - 4x \le 0$
$x(x-4) = 0$
$x = 0$, $x = 4$

Region	Test Point	$x(x-4) \le 0$	Result
$x < 0$	$x = -1$	$(-1)(-5) \le 0$	False
$0 < x < 4$	$x = 2$	$2(-2) \le 0$	True
$x > 4$	$x = 5$	$5(1) \le 0$	False

Solution: [0, 4]

34. $$c^2 = a^2 + b^2$$
$$8^2 = 4^2 + b^2$$
$$64 = 16 + b^2$$
$$48 = b^2$$
$$\pm 4\sqrt{3} = b$$
$b > 0$ so the length is $4\sqrt{3}$ inches.

35. $F(x) = (x-3)^2 + 1$

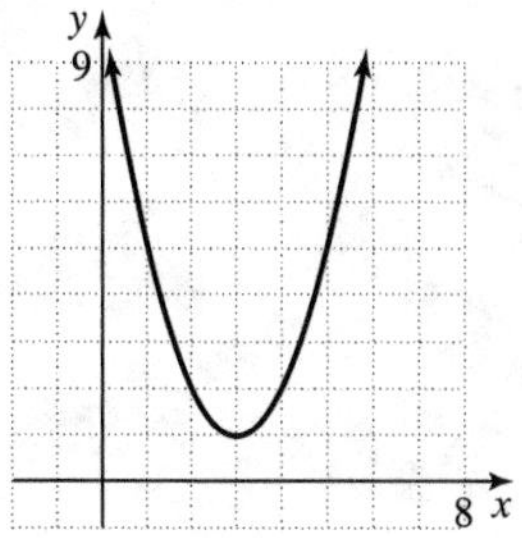

36. **a.** $i^8 = (i^2)^4 = (-1)^4 = 1$

b. $i^{21} = i(i^{20}) = i$

c. $i^{42} = i^2(i^{40}) = i^2 = -1$

d. $i^{-13} = \frac{1}{i^{13}} = \frac{1}{i(i^{12})} = \frac{1}{i} = \frac{i}{i^2} = -i$

37. $f(x) = x - 1$ and $g(x) = 2x - 3$

a.
$$\begin{aligned}(f+g)(x) &= f(x) + g(x)\\ &= x - 1 + 2x - 3\\ &= 3x - 4\end{aligned}$$

b.
$$\begin{aligned}(f-g)(x) &= f(x) - g(x)\\ &= x - 1 - (2x - 3)\\ &= -x + 2\end{aligned}$$

c.
$$\begin{aligned}(f \cdot g)(x) &= f(x) \cdot g(x)\\ &= (x-1)(2x-3)\\ &= 2x^2 - 5x + 3\end{aligned}$$

d. $\left(\frac{f}{g}\right)(x) = \frac{f(x)}{g(x)} = \frac{x-1}{2x-3}$, where $x \neq \frac{3}{2}$.

38.
$$\begin{aligned}4x^2 + 8x - 1 &= 0\\ x^2 + 2x - \frac{1}{4} &= 0\\ x^2 + 2x &= \frac{1}{4}\\ x^2 + 2x + \left(\frac{2}{2}\right)^2 &= \frac{1}{4} + \left(\frac{2}{2}\right)^2\\ x^2 + 2x + 1 &= \frac{1}{4} + 1\\ (x+1)^2 &= \frac{5}{4}\\ x + 1 &= \pm\sqrt{\frac{5}{4}}\\ x + 1 &= \pm\frac{\sqrt{5}}{2}\\ x &= -1 \pm \frac{\sqrt{5}}{2}\\ &= \frac{-2 \pm \sqrt{5}}{2}\end{aligned}$$

The solutions are $\frac{-2+\sqrt{5}}{2}$ and $\frac{-2-\sqrt{5}}{2}$.

39.
$$\begin{aligned}f(x) &= x + 3\\ y &= x + 3\\ x &= y + 3\\ y &= x - 3\\ f^{-1}(x) &= x - 3\end{aligned}$$

40.
$$\begin{aligned}\left(x - \frac{1}{2}\right)^2 &= \frac{x}{2}\\ x^2 - x + \frac{1}{4} &= \frac{1}{2}x\\ x^2 - \frac{3}{2}x + \frac{1}{4} &= 0\\ 4x^2 - 6x + 1 &= 0\end{aligned}$$

$a = 4$, $b = -6$, $c = 1$

$$\begin{aligned}x &= \frac{-(-6) \pm \sqrt{(-6)^2 - 4(4)(1)}}{2(4)}\\ &= \frac{6 \pm \sqrt{20}}{8}\\ &= \frac{6 \pm 2\sqrt{5}}{8}\\ &= \frac{3 \pm \sqrt{5}}{4}\end{aligned}$$

The solutions are $\frac{3+\sqrt{5}}{4}$ and $\frac{3-\sqrt{5}}{4}$.

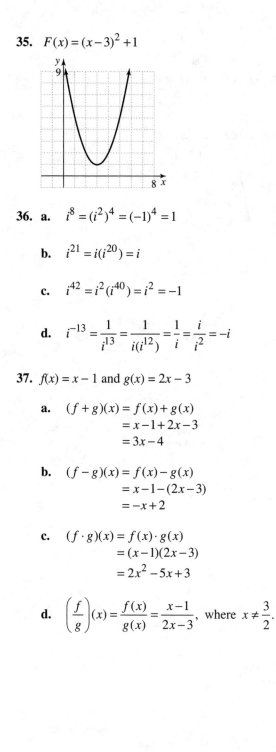

41. a. $\log_4 16 = \log_4 4^2 = 2$

b. $\log_{10} \frac{1}{10} = \log_{10} 10^{-1} = -1$

c. $\log_9 3 = \log_9 9^{1/2} = \frac{1}{2}$

42. $f(x) = -(x+1)^2 + 1$

Vertex: (−1, 1)

Axis of symmetry: $x = -1$

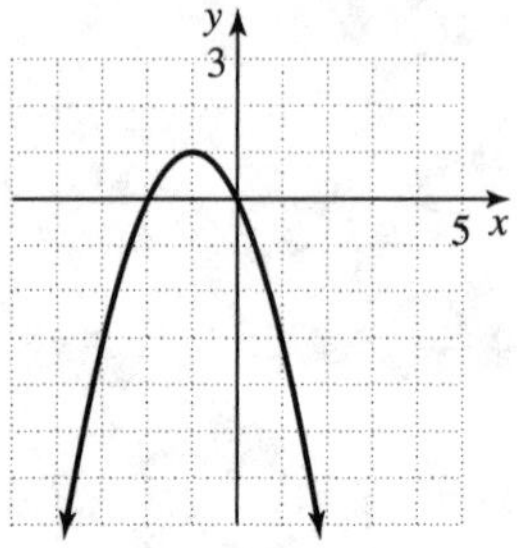

Chapter 10

Section 10.1 Practice Exercises

1. $x = \frac{1}{2}y^2$; $a = \frac{1}{2}$, $h = 0$, $k = 0$; vertex: (0, 0)

x	y
2	−2
$\frac{1}{2}$	−1
0	0
$\frac{1}{2}$	1
2	2

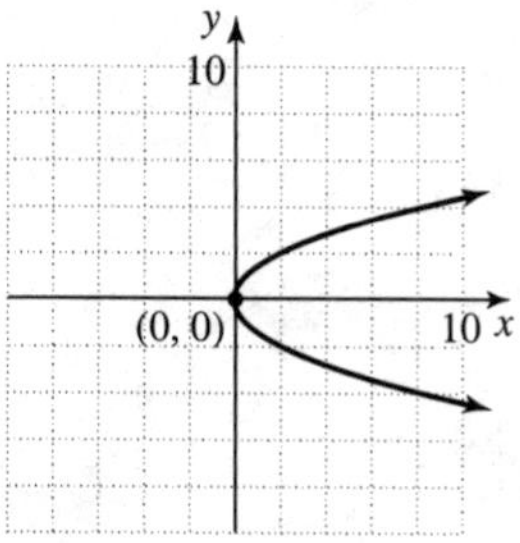

2. $x = -2(y+4)^2 - 1$; $a = -2$, $h = -1$, $k = -4$;
vertex: (−1, −4)

x	y
−9	−6
3	−5
−1	−4
−3	−3
−9	−2

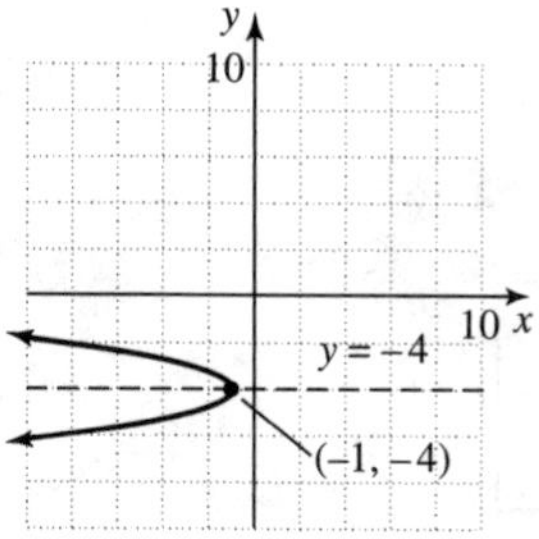

3.
$$y = -x^2 + 4x + 6$$
$$y - 6 = -x^2 + 4x$$
$$y - 6 = -(x^2 - 4x)$$
$$y - 6 - (+4) = -(x^2 - 4x + 4)$$
$$y - 10 = -(x-2)^2$$
$$y = -(x-2)^2 + 10$$

$a = -1$, $h = 2$, $k = 10$
vertex: (2, 10)

x	y
−1	1
0	6
1	9
2	10
3	9
4	6
5	1

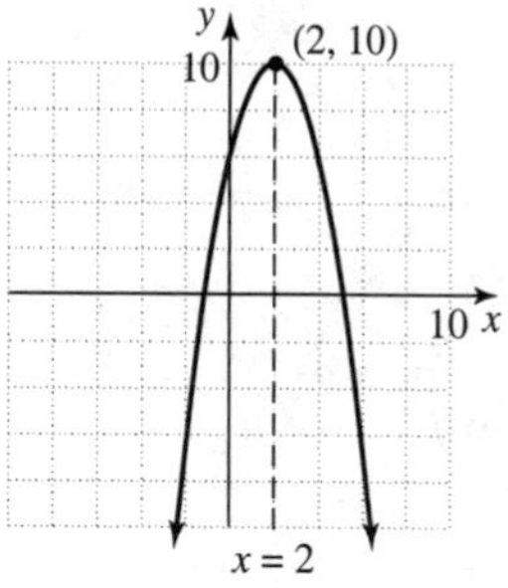

4. $x = 3y^2 + 6y + 4$
Find the vertex.

$$y = \frac{-b}{2a} = \frac{-6}{2(3)} = -1$$

$$x = 3(-1)^2 + 6(-1) + 4 = 3 - 6 + 4 = 1$$

vertex: (1, −1)
The axis of symmetry is the line $y = -1$.
Since $a > 0$, the parabola opens to the right.

$$x = 3(0)^2 + 6(0) + 4 = 4$$

The x-intercept is (4, 0).

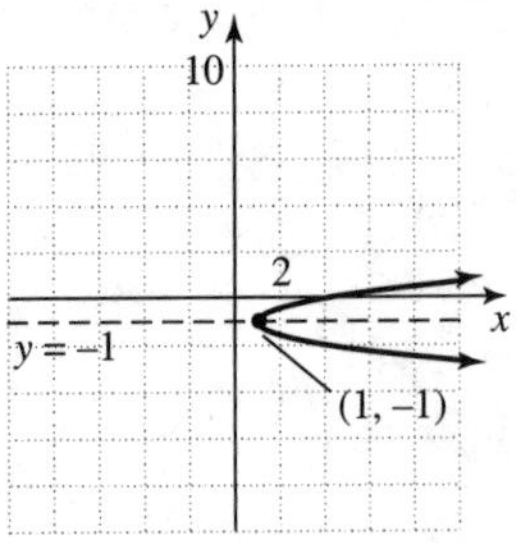

5.

$$x^2 + y^2 = 25$$
$$(x-0)^2 + (y-0)^2 = 5^2$$

center: (0, 0); radius = 5

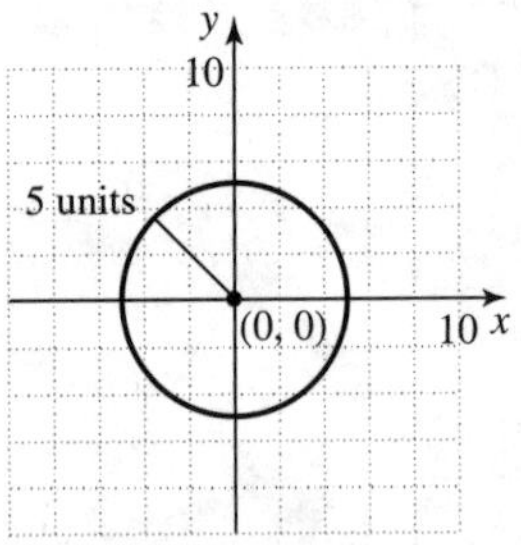

6. $(x-3)^2 + (y+2)^2 = 4$

$h = 3,\ k = -2,\ r = \sqrt{4} = 2$

center: (3, −2)

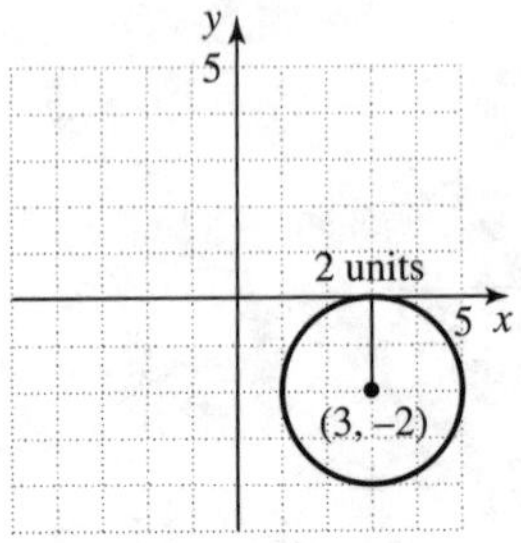

7.

$$x^2 + y^2 + 6x - 2y = 6$$
$$(x^2 + 6x) + (y^2 - 2y) = 6$$
$$(x^2 + 6x + 9) + (y^2 - 2y + 1) = 6 + 9 + 1$$
$$(x+3)^2 + (y-1)^2 = 16$$

Center: (−3, 1); radius $= \sqrt{16} = 4$

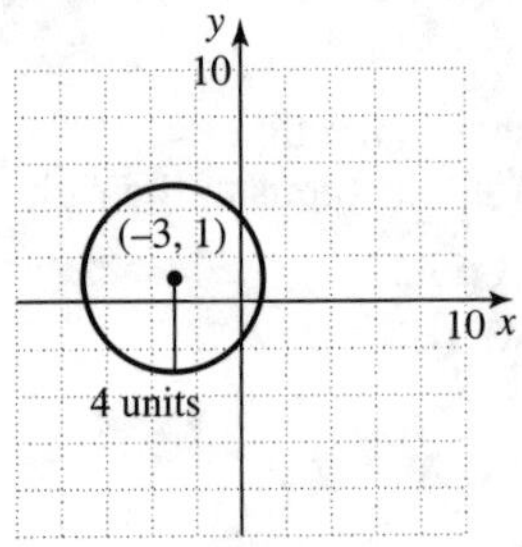

8. Center: (−2, −5); radius = 9

$(x-h)^2 + (y-k)^2 = r^2$

$h = -2,\ k = -5$, and $r = 9$.

The equation is $(x+2)^2 + (y+5)^2 = 81$.

Graphing Calculator Explorations

1. $x^2 + y^2 = 55$

$$y^2 = 55 - x^2$$
$$y = \pm\sqrt{55 - x^2}$$

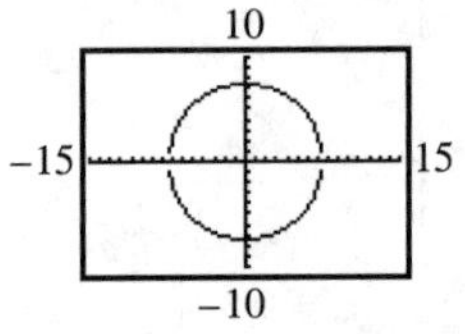

2. $x^2 + y^2 = 20$

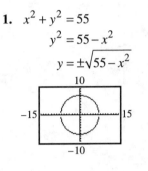

$$y^2 = 20 - x^2$$
$$y = \pm\sqrt{20 - x^2}$$

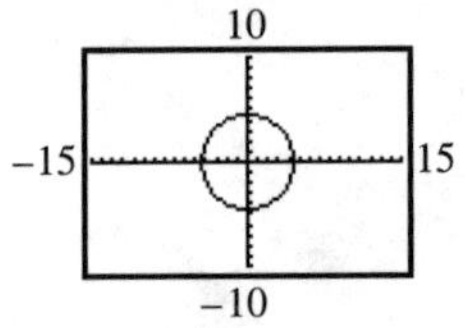

3. $5x^2 + 5y^2 = 50$

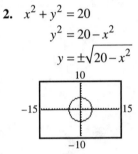

$$5y^2 = 50 - 5x^2$$
$$y^2 = 10 - x^2$$
$$y = \pm\sqrt{10 - x^2}$$

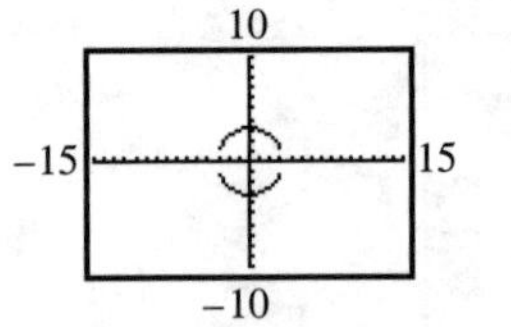

4. $6x^2 + 6y^2 = 105$

$$6y^2 = 105 - 6x^2$$
$$y^2 = 17.5 - x^2$$
$$y = \pm\sqrt{17.5 - x^2}$$

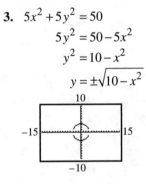

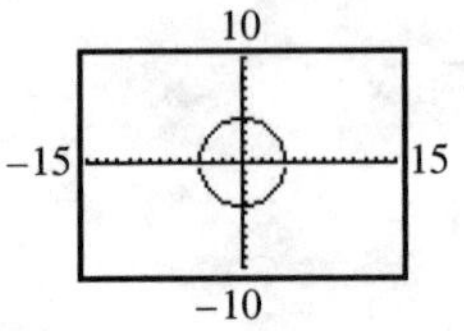

5. $2x^2+2y^2-34=0$

$$2y^2=34-2x^2$$
$$y^2=17-x^2$$
$$y=\pm\sqrt{17-x^2}$$

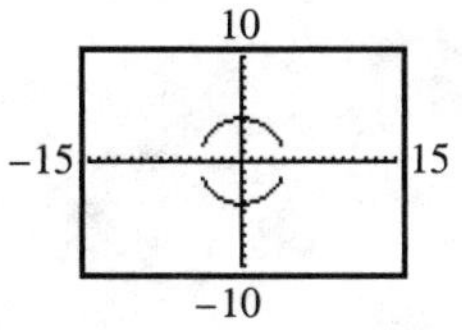

6. $4x^2+4y^2-48=0$

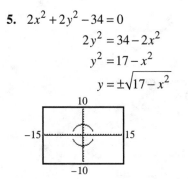

$$4y^2=48-4x^2$$
$$y^2=12-x^2$$
$$y=\pm\sqrt{12-x^2}$$

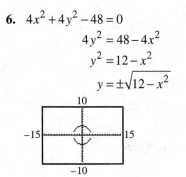

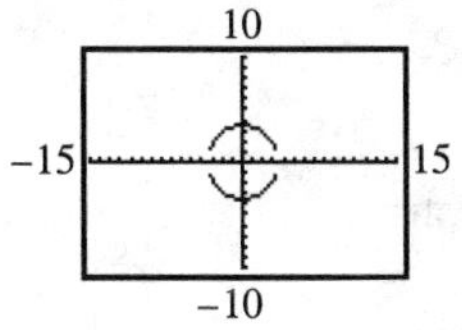

7. $7x^2+7y^2-89=0$

$$7y^2=89-7x^2$$
$$y^2=\frac{89-7x^2}{7}$$
$$y=\pm\sqrt{\frac{89-7x^2}{7}}$$

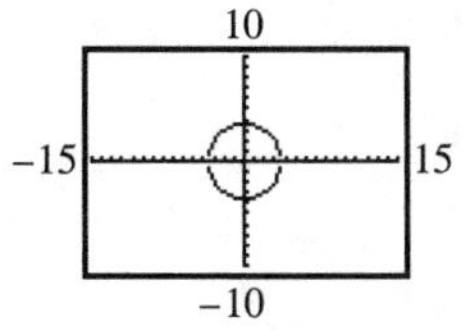

8. $3x^2+3y^2-35=0$

$$3y^2=35-3x^2$$
$$y^2=\frac{35-3x^2}{3}$$
$$y=\pm\sqrt{\frac{35-3x^2}{3}}$$

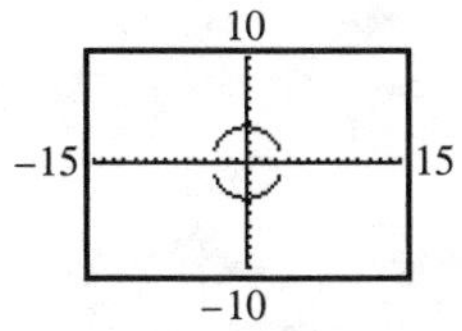

Vocabulary, Readiness & Video Check 10.1

1. The circle, parabola, ellipse, and hyperbola are called the <u>conic sections</u>.

2. For a parabola that opens upward the lowest point is the <u>vertex</u>.

3. A <u>circle</u> is the set of all points in a plane that are the same distance from a fixed point. The fixed point is called the <u>center</u>.

4. The midpoint of a diameter of a circle is the <u>center</u>.

5. The distance from the center of a circle to any point of the circle is called the <u>radius</u>.

6. Twice a circle's radius is its <u>diameter</u>.

7. No, their graphs don't pass the vertical line test.

8. $x^2+y^2=r^2$

9. The formula for the standard form of a circle identifies the center and radius, so you just need to substitute these values into this formula and simplify.

10. Since the standard form of a circle involves a squared binomial for both x and y, we need to complete the square on both x and y.

Exercise Set 10.1

1. $y=x^2-7x+5$; $a=1$, upward

3. $x=-y^2-y+2$; $a=-1$, to the left

5. $y=-x^2+2x+1$; $a=-1$, downward

7. $x=3y^2$

   $x=3(y-0)^2+0$

   $V(0, 0)$; opens to the right

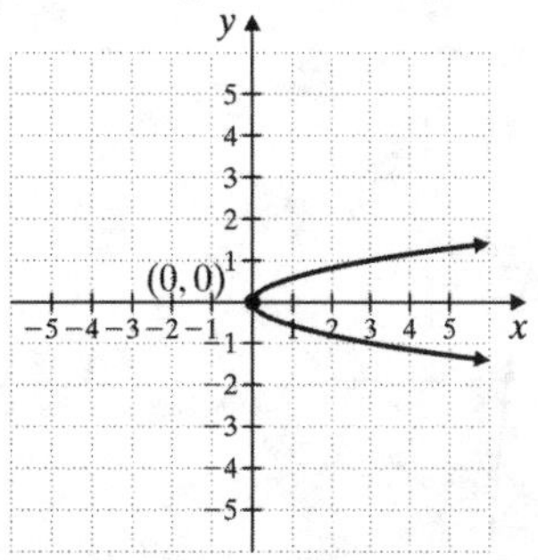

9. $x = -2y^2$

$x = -2(y-0)^2 + 0$

$V(0, 0)$; opens to the left

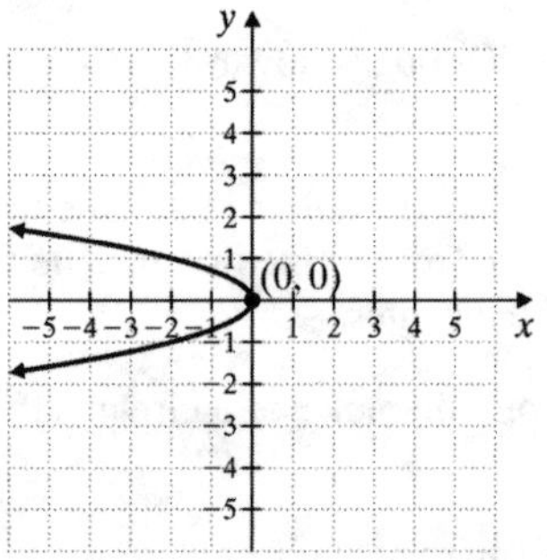

11. $y = -4x^2 = -4(x-0)^2 + 0$

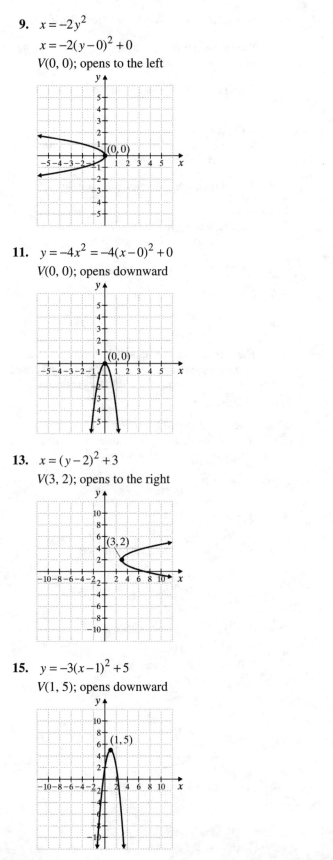

$V(0, 0)$; opens downward

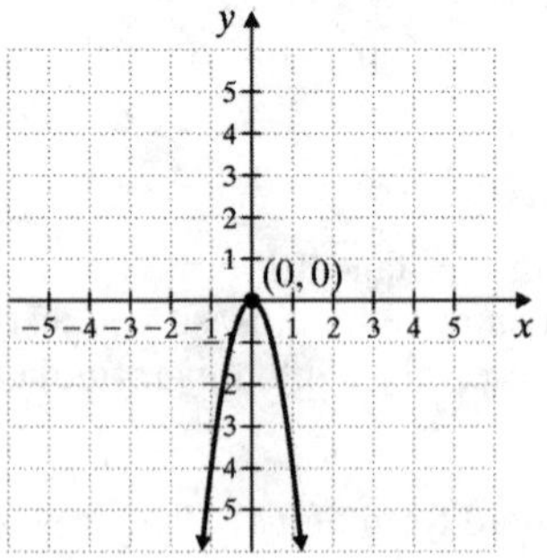

13. $x = (y-2)^2 + 3$

$V(3, 2)$; opens to the right

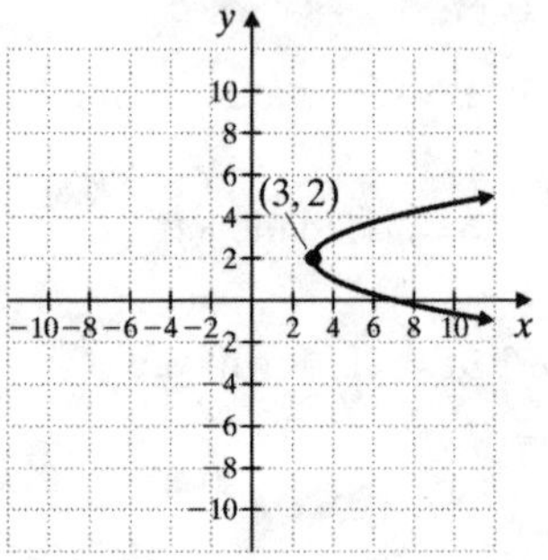

15. $y = -3(x-1)^2 + 5$

$V(1, 5)$; opens downward

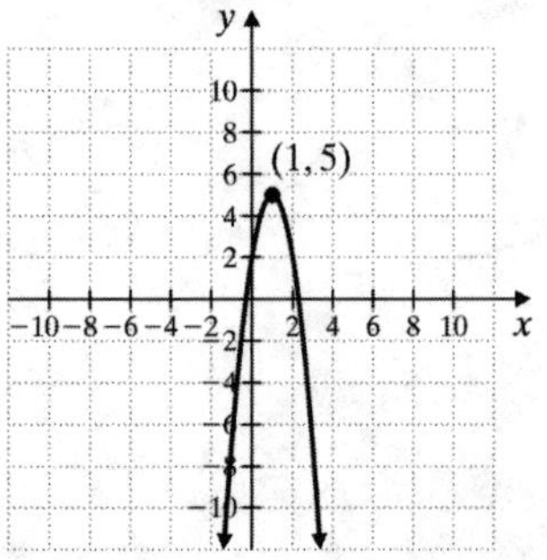

17. $x = y^2 + 6y + 8$

$x = y^2 + 6y + 9 + 8 - 9$

$x = (y+3)^2 - 1$

$V(-1, -3)$; opens to the right

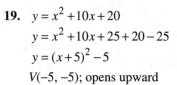

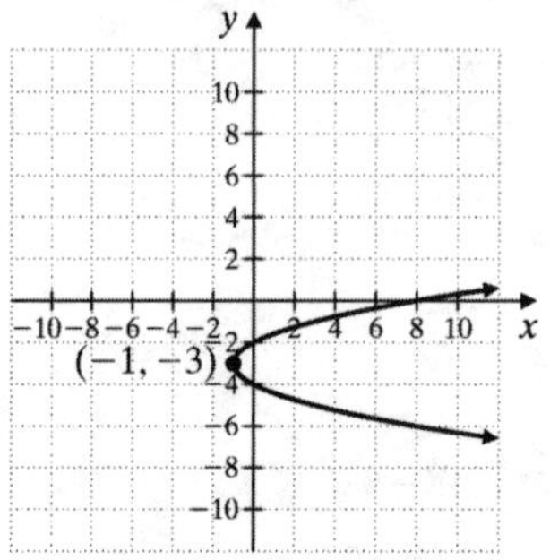

19. $y = x^2 + 10x + 20$

$y = x^2 + 10x + 25 + 20 - 25$

$y = (x+5)^2 - 5$

$V(-5, -5)$; opens upward

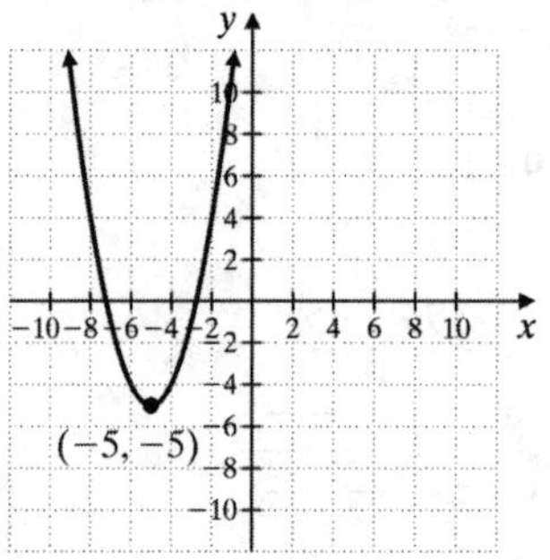

21. $x = -2y^2 + 4y + 6$

$x = -2(y^2 - 2y) + 6$

$x = -2(y^2 - 2y + 1) + 6 + 2$

$x = -2(y-1)^2 + 8$

$V(8, 1)$; opens to the left

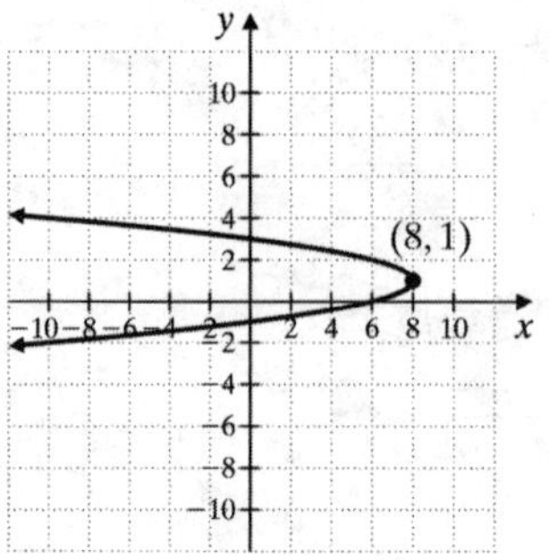

23. $x^2 + y^2 = 9$

$(x-0)^2 + (y-0)^2 = 3^2$

$C(0, 0)$ and $r = 3$

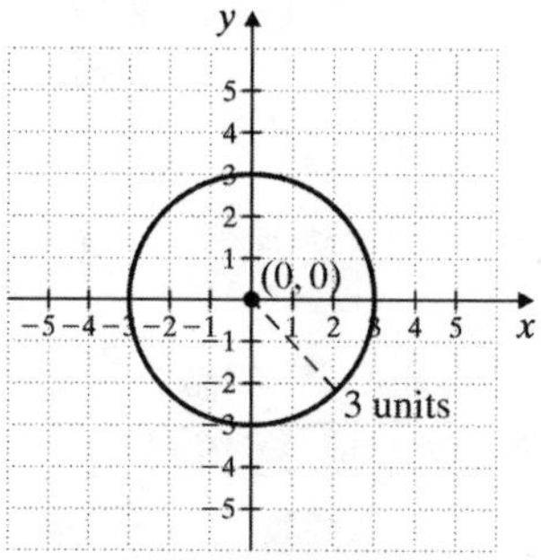

25. $x^2 + (y-2)^2 = 1$

$(x-0)^2 + (y-2)^2 = 1^2$

$C(0, 2)$ and $r = 1$

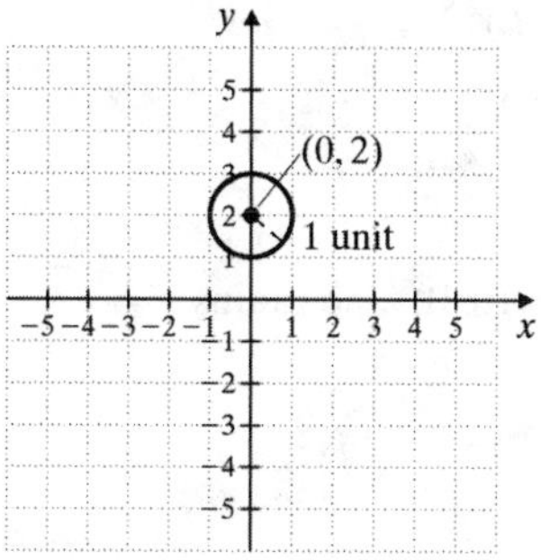

27. $(x-5)^2 + (y+2)^2 = 1$

$(x-5)^2 + (y+2)^2 = 1^2$

$C(5, -2)$ and $r = 1$

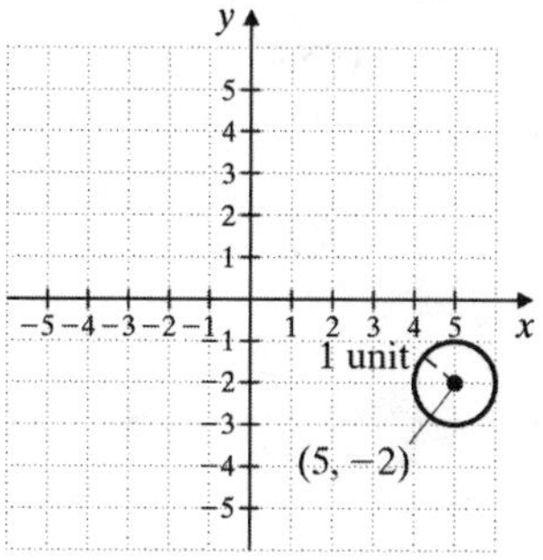

29. $x^2 + y^2 + 6y = 0$

$x^2 + y^2 + 6y + 9 = 0 + 9$

$(x-0)^2 + (y+3)^2 = 3^2$

$C(0, -3)$ and $r = 3$

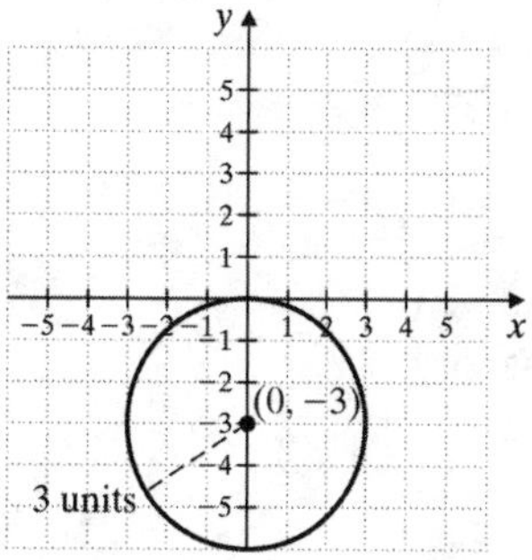

31. $x^2 + y^2 + 2x - 4y = 4$

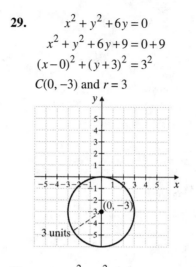

$x^2 + 2x + 1 + y^2 - 4y + 4 = 4 + 1 + 4$

$(x+1)^2 + (y-2)^2 = 9$

$(x+1)^2 + (y-2)^2 = 3^2$

$C(-1, 2)$ and $r = 3$

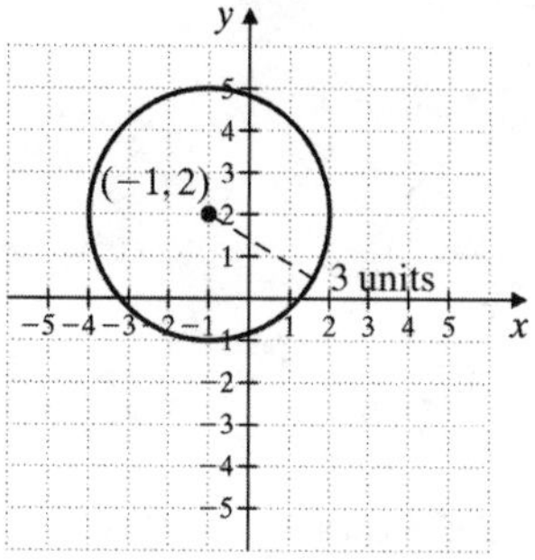

33. $(x+2)^2 + (y-3)^2 = 7$

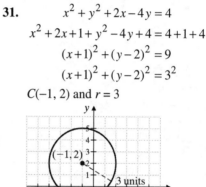

$(x+2)^2 + (y-3)^2 = \left(\sqrt{7}\right)^2$

$C(-2, 3)$ and $r = \sqrt{7}$

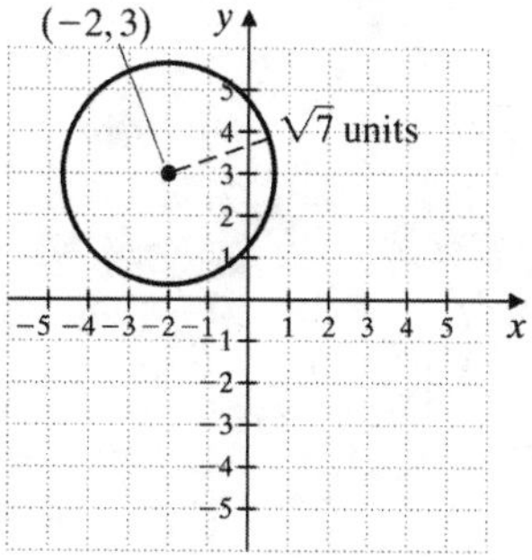

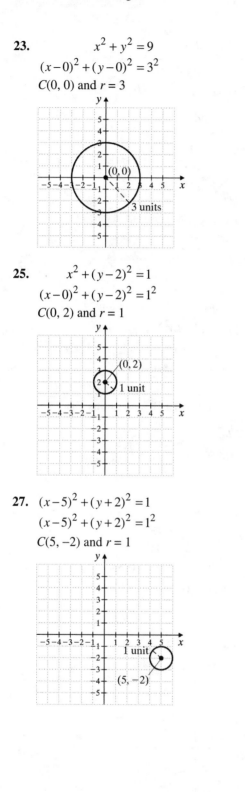

35. $x^2+y^2-4x-8y-2=0$

$x^2-4x+4+y^2-8y+16=2+4+16$

$(x-2)^2+(y-4)^2=22$

$(x-2)^2+(y-4)^2=\left(\sqrt{22}\right)^2$

$C(2, 4)$ and $r=\sqrt{22}$

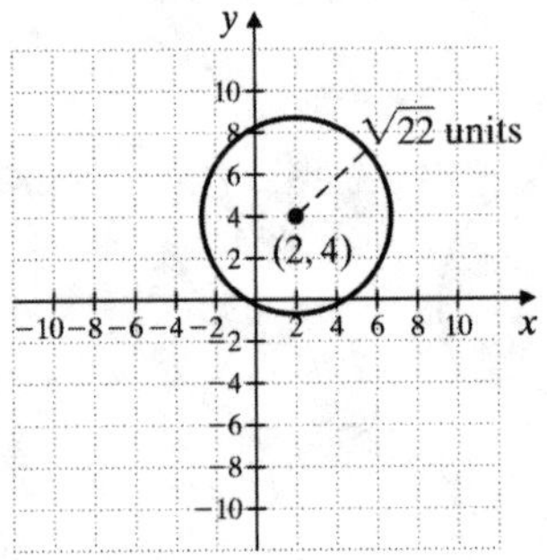

37. $3x^2+3y^2=75$

$x^2+y^2=25$

$(x-0)^2+(y-0)^2=5^2$

$C(0, 0)$ and $r=5$

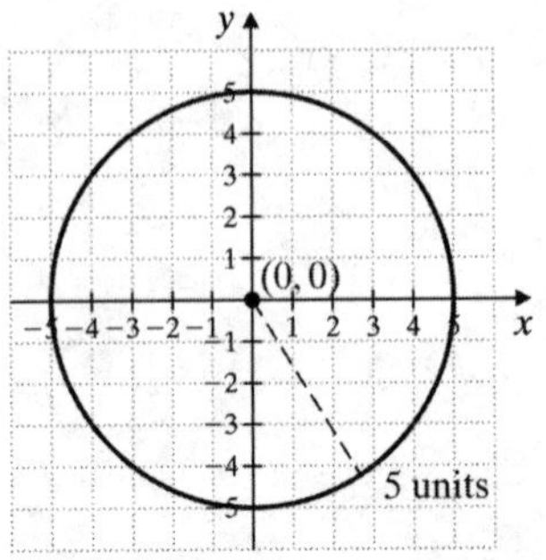

39. $6(x-4)^2+6(y-1)^2=24$

$(x-4)^2+(y-1)^2=4$

$(x-4)^2+(y-1)^2=2^2$

$C(4, 1)$ and $r=2$

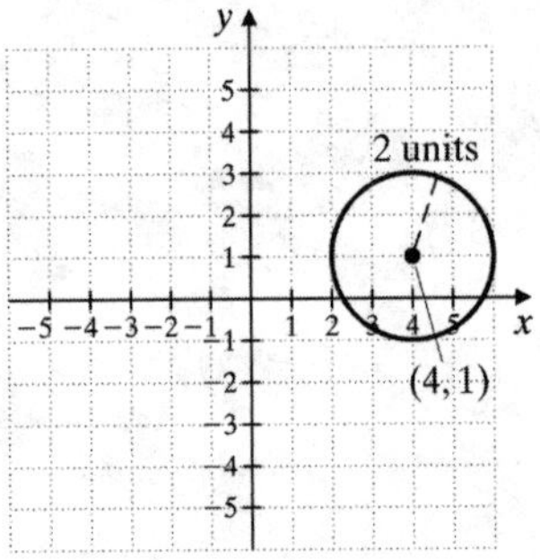

41. $4(x+1)^2+4(y-3)^2=12$

$(x+1)^2+(y-3)^2=3$

$(x+1)^2+(y-3)^2=\left(\sqrt{3}\right)^2$

$C(-1, 3)$ and $r=\sqrt{3}$

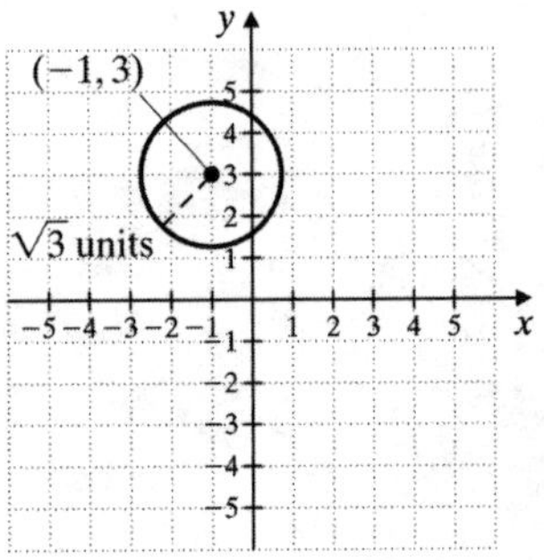

43. Center $(h, k)=(2, 3)$ and radius $r=6$.

$(x-h)^2+(y-k)^2=r^2$

$(x-2)^2+(y-3)^2=6^2$

$(x-2)^2+(y-3)^2=36$

45. Center $(h, k)=(0, 0)$ and radius $r=\sqrt{3}$.

$(x-h)^2+(y-k)^2=r^2$

$(x-0)^2+(y-0)^2=\left(\sqrt{3}\right)^2$

$x^2+y^2=3$

47. Center $(h, k)=(-5, 4)$ and radius $r=3\sqrt{5}$.

$(x-h)^2+(y-k)^2=r^2$

$[x-(-5)]^2+(y-4)^2=\left(3\sqrt{5}\right)^2$

$(x+5)^2+(y-4)^2=45$

49. $x=y^2-3$

$x=(y-0)^2-3$

Vertex: $(-3, 0)$

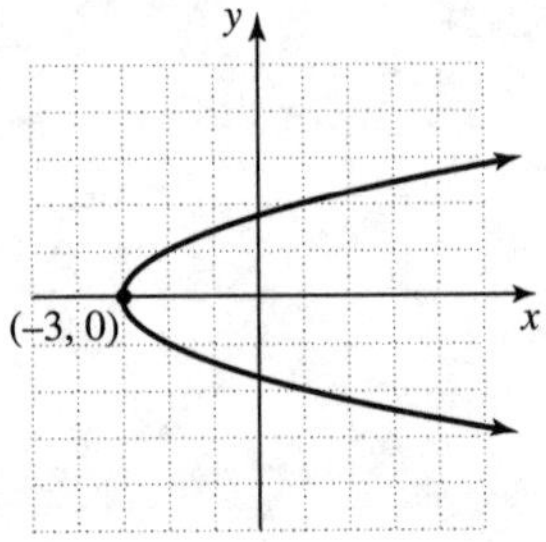

51. $y = (x-2)^2 - 2$

Vertex: (2, –2)

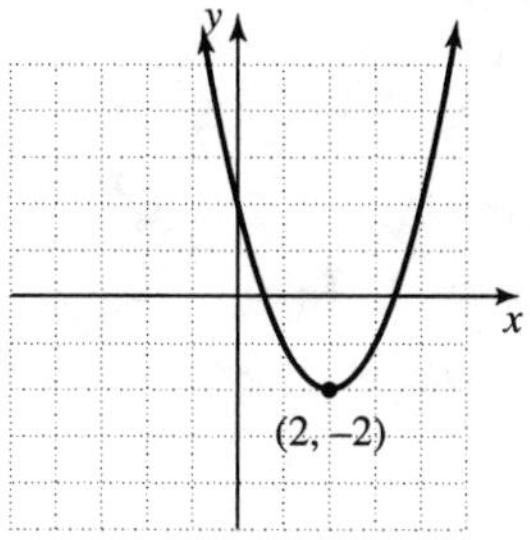

53. $x^2 + y^2 = 1$

Center: (0, 0), radius $r = \sqrt{1} = 1$

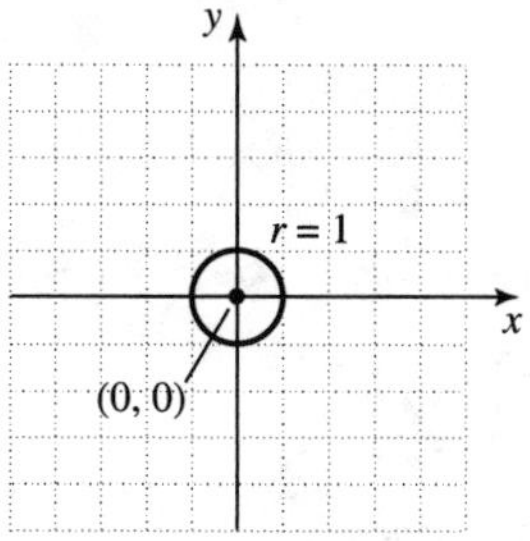

55. $x = (y+3)^2 - 1$

Vertex: (–1, –3)

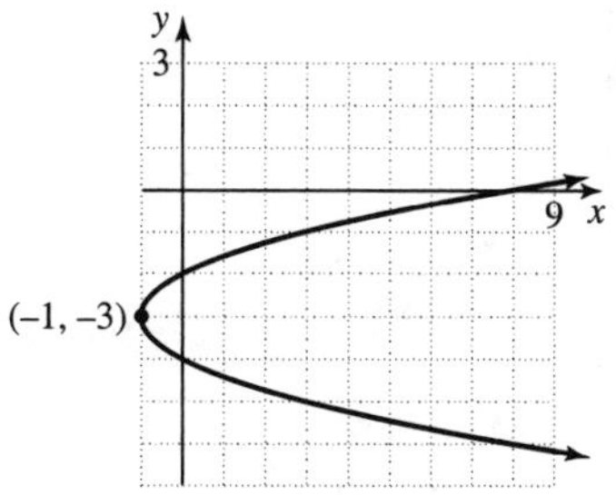

57. $(x-2)^2 + (y-2)^2 = 16$

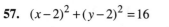

Center: (2, 2), radius $r = \sqrt{16} = 4$

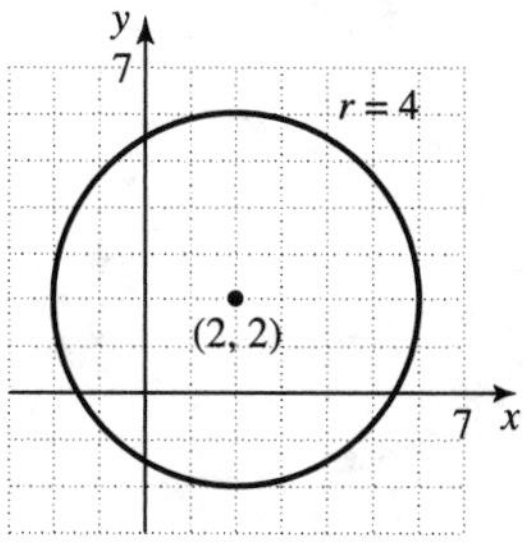

59. $x = -(y-1)^2$

Vertex: (0, 1)

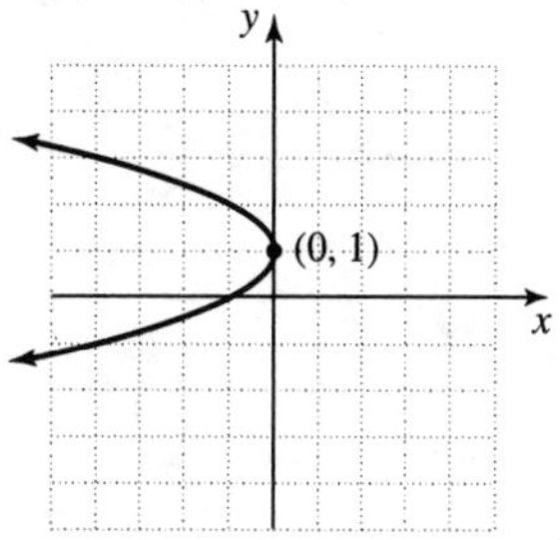

61. $(x-4)^2 + y^2 = 7$

Center: (4, 0), radius $r = \sqrt{7}$

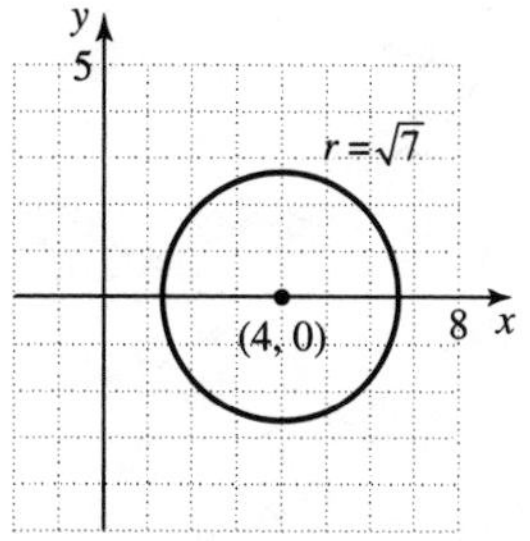

63. $y = 5(x+5)^2 + 3$

Vertex: (–5, 3)

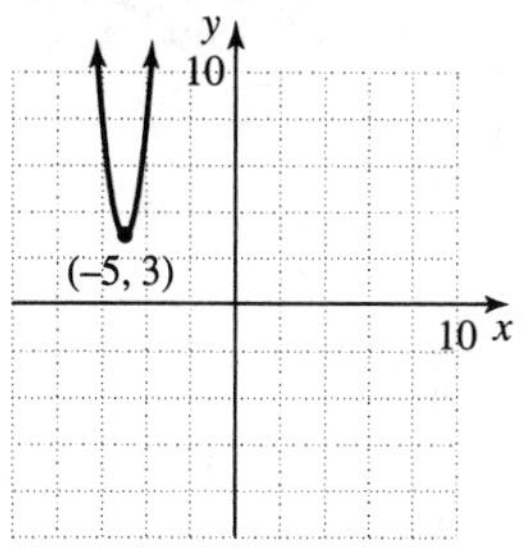

65.

$$\frac{x^2}{8} + \frac{y^2}{8} = 2$$

$$8\left(\frac{x^2}{8} + \frac{y^2}{8}\right) = 8(2)$$

$$x^2 + y^2 = 16$$

Center: (0, 0), radius $r = \sqrt{16} = 4$

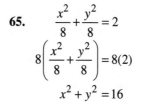

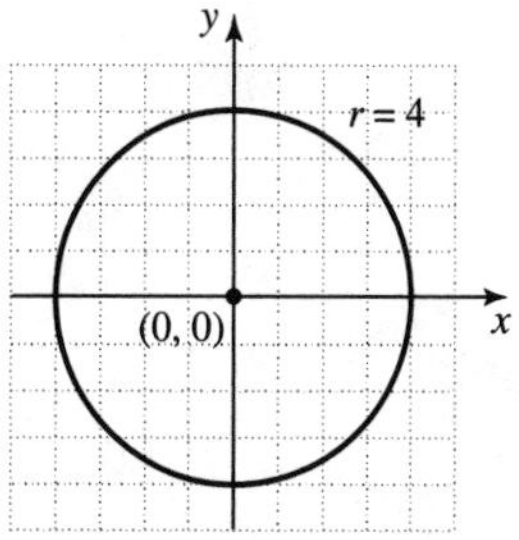

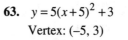

67.

$$\begin{aligned}
y &= x^2 + 7x + 6 \\
y - 6 &= x^2 + 7x \\
y - 6 + \frac{49}{4} &= x^2 + 7x + \frac{49}{4} \\
y + \frac{25}{4} &= \left(x + \frac{7}{2}\right)^2 \\
y &= \left(x + \frac{7}{2}\right)^2 - \frac{25}{4}
\end{aligned}$$

Vertex: $\left(-\frac{7}{2}, -\frac{25}{4}\right)$

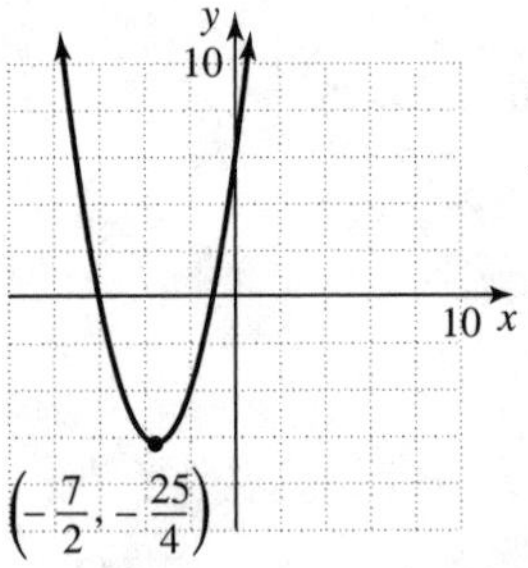

69.

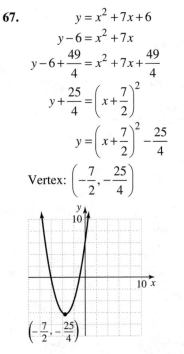

$$\begin{aligned}
x^2 + y^2 + 2x + 12y - 12 &= 0 \\
(x^2 + 2x) + (y^2 + 12y) &= 12 \\
(x^2 + 2x + 1) + (y^2 + 12y + 36) &= 12 + 1 + 36 \\
(x+1)^2 + (y+6)^2 &= 49
\end{aligned}$$

Center: (–1, –6), radius $r = \sqrt{49} = 7$

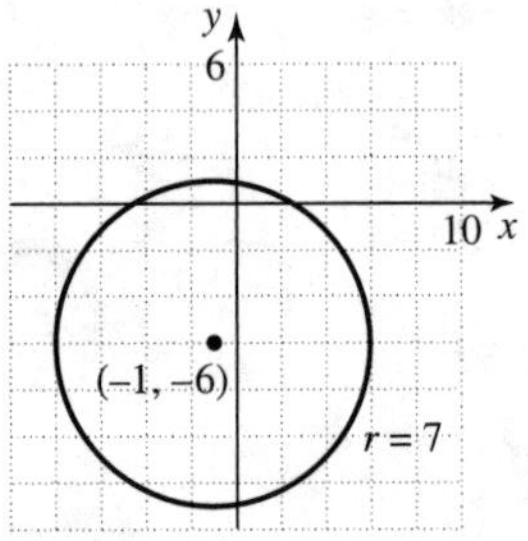

71.

$$\begin{aligned}
x &= y^2 + 8y - 4 \\
x + 4 &= y^2 + 8y \\
x + 4 + 16 &= y^2 + 8y + 16 \\
x + 20 &= (y+4)^2 \\
x &= (y+4)^2 - 20
\end{aligned}$$

Vertex: (–20, –4)

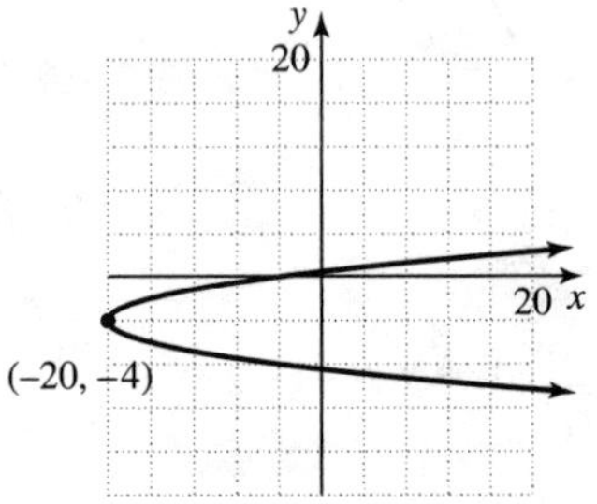

73.

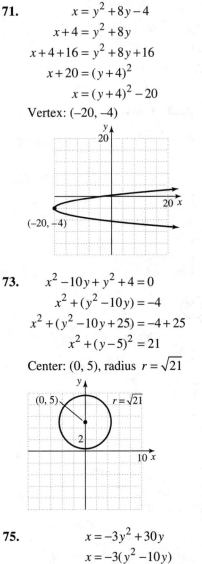

$$\begin{aligned}
x^2 - 10y + y^2 + 4 &= 0 \\
x^2 + (y^2 - 10y) &= -4 \\
x^2 + (y^2 - 10y + 25) &= -4 + 25 \\
x^2 + (y-5)^2 &= 21
\end{aligned}$$

Center: (0, 5), radius $r = \sqrt{21}$

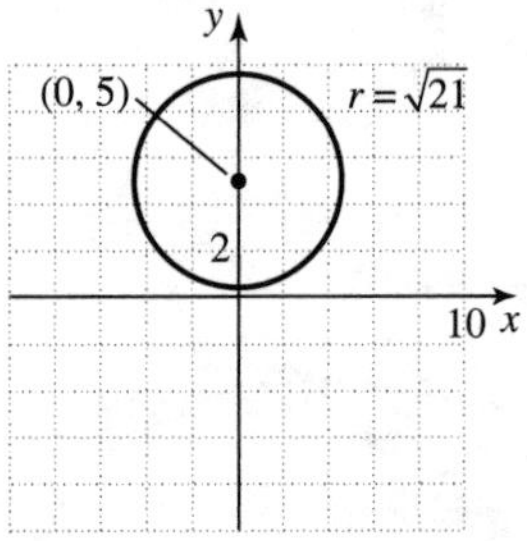

75.

$$\begin{aligned}
x &= -3y^2 + 30y \\
x &= -3(y^2 - 10y) \\
x + [-3(25)] &= -3(y^2 - 10y + 25) \\
x - 75 &= -3(y-5)^2 \\
x &= -3(y-5)^2 + 75
\end{aligned}$$

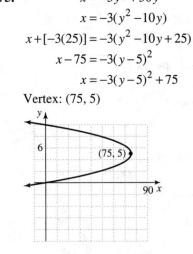

Vertex: (75, 5)

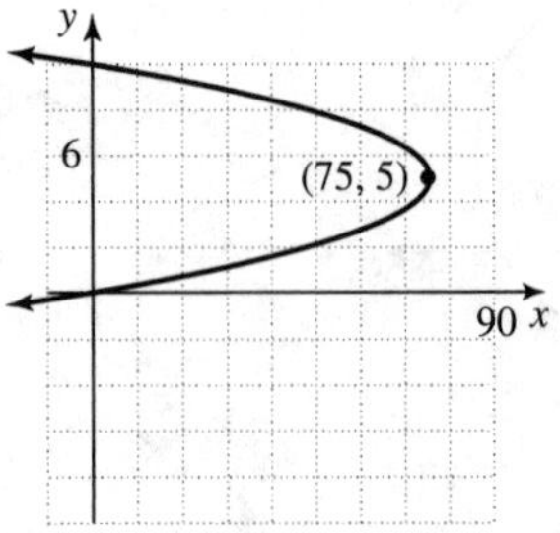

77. $5x^2 + 5y^2 = 25$
$x^2 + y^2 = 5$
Center: (0, 0), radius $r = \sqrt{5}$

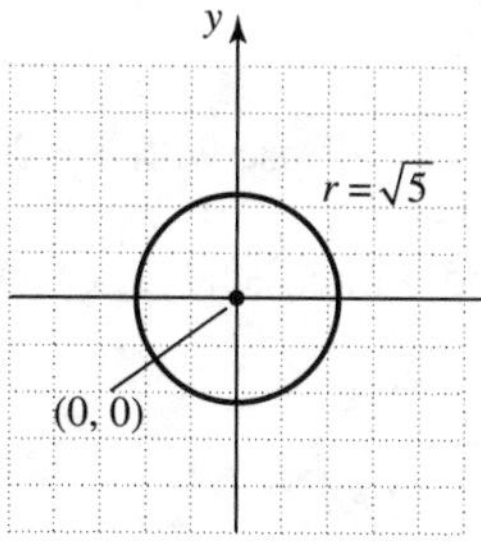

79.
$$y = 5x^2 - 20x + 16$$
$$y - 16 = 5(x^2 - 4x)$$
$$y - 16 + 5(4) = 5(x^2 - 4x + 4)$$
$$y + 4 = (x-2)^2$$
$$y = (x-2)^2 - 4$$
Vertex: (2, –4)

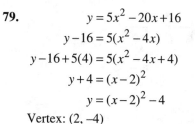

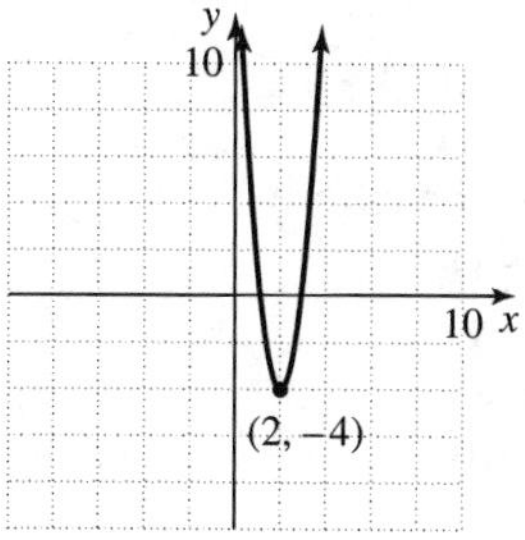

81. $y = 2x + 5$

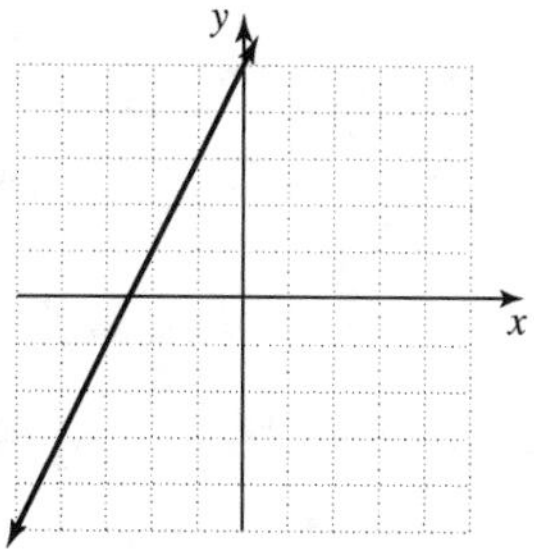

83. $y = 3$

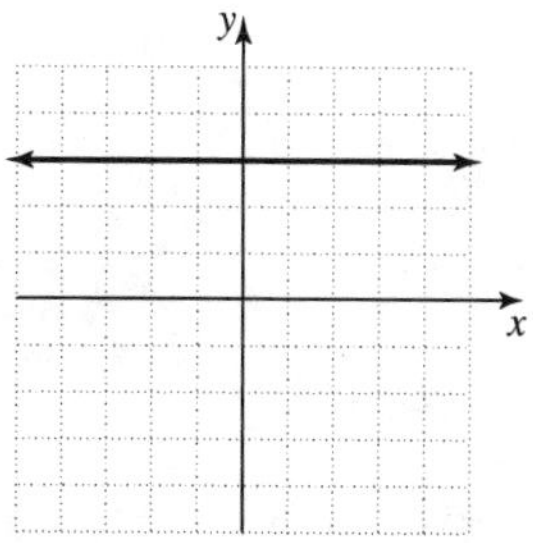

85. $\frac{1}{\sqrt{3}} = \frac{1 \cdot \sqrt{3}}{\sqrt{3} \cdot \sqrt{3}} = \frac{\sqrt{3}}{3}$

87. $\frac{4\sqrt{7}}{\sqrt{6}} = \frac{4\sqrt{7} \cdot \sqrt{6}}{\sqrt{6} \cdot \sqrt{6}} = \frac{4\sqrt{42}}{6} = \frac{2\sqrt{42}}{3}$

89. The vertex is (1, –5).

91. **a.** radius $= \frac{1}{2}$(diameter)
$= \frac{1}{2}$(33 meters)
$= 16.5$ meters

b. circumference = π(diameter)
= π(33 meters)
≈ 103.67 meters

c. $\frac{103.67}{30} \approx 3.5$ meters apart

d. center: (0, 16.5)

e. $(x-0)^2 + (y-16.5)^2 = 16.5^2$
$x^2 + (y-16.5)^2 = 16.5^2$

93. **a.** The radius was one-half of the diameter, or 125 feet.

b. 264 – 250 = 14
The wheel was 14 feet above the ground.

c. 125 + 14 = 139
The center of the wheel was 139 feet from the ground.

d. From the drawing, the center was at (0, 139).

e. $(x-0)^2 + (y-139)^2 = 125^2$
$x^2 + (y-139)^2 = 125^2$

95. answers may vary

97. Using A as (0, 0), point B is at (3, 1) and point C is at (19, 13).

$$\begin{aligned} d(B, C) &= \sqrt{(x_2 - x_1)^2 + (y_2 - y_1)^2} \\ &= \sqrt{(19-3)^2 + (13-1)^2} \\ &= \sqrt{16^2 + 12^2} \\ &= \sqrt{256 + 144} \\ &= \sqrt{400} \\ &= 20 \end{aligned}$$

The distance across the lake is 20 meters.

99. $5x^2 + 5y^2 = 25$

$$\begin{aligned} 5y^2 &= 25 - 5x^2 \\ y^2 &= 5 - x^2 \\ y &= \pm\sqrt{5 - x^2} \end{aligned}$$

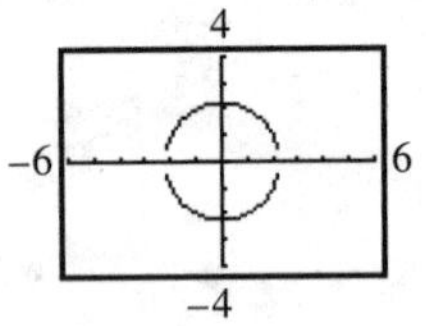

101. $y = 5x^2 - 20x + 16$

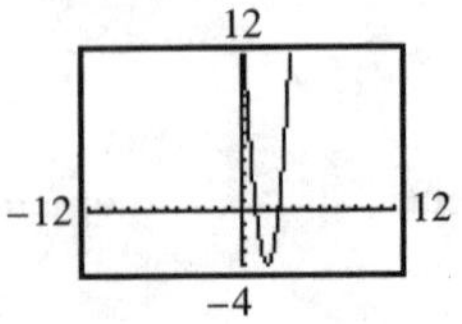

Section 10.2 Practice Exercises

1. $\dfrac{x^2}{25} + \dfrac{y^2}{4} = 1$

The equation is an ellipse with $a = 5$ and $b = 2$. The center is (0, 0). The x-intercepts are (5, 0) and (−5, 0). The y-intercepts are (2, 0) and (−2, 0).

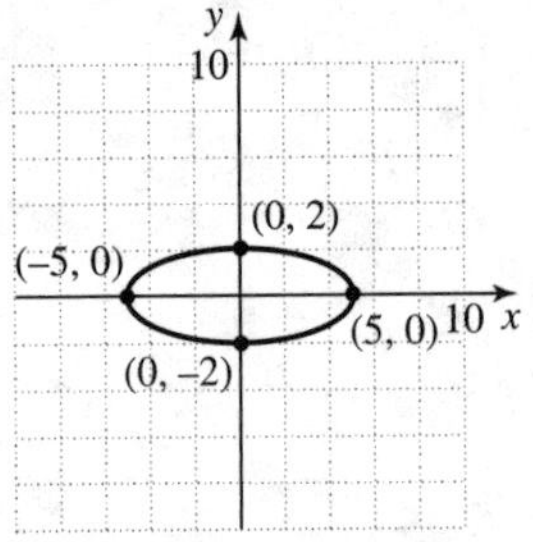

2. $9x^2 + 4y^2 = 36$

$$\begin{aligned} \frac{9x^2}{36} + \frac{4y^2}{36} &= \frac{36}{36} \\ \frac{x^2}{4} + \frac{y^2}{9} &= 1 \end{aligned}$$

This is an equation of an ellipse with $a = 2$ and $b = 3$. The ellipse has center (0, 0), x-intercepts (2, 0) and (−2, 0), and y-intercepts (3, 0) and (−3, 0).

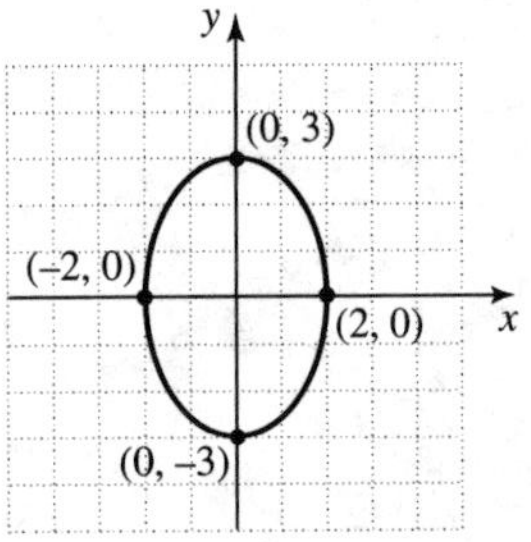

3. $\dfrac{(x-4)^2}{49} + \dfrac{(y+1)^2}{81} = 1$

This ellipse has center (4, −1).
$a = 7$ and $b = 9$.
Find four points on the ellipse.
(4 + 7, −1) = (11, −1)
(4 − 7, −1) = (−3, −1)
(4, −1 + 9) = (4, 8)
(4, −1 − 9) = (4, −10)

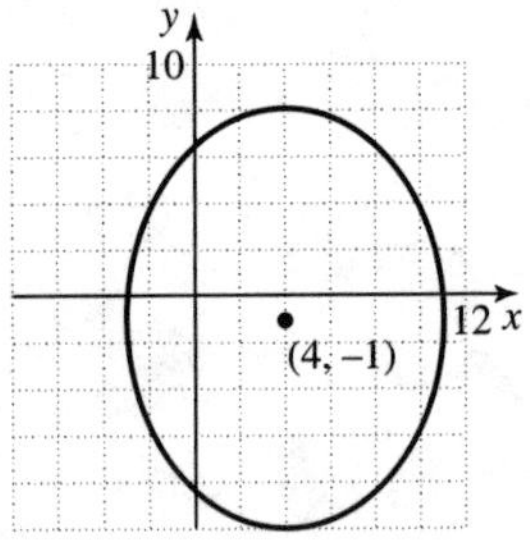

4. $\frac{x^2}{9} - \frac{y^2}{16} = 1$

This is a hyperbola with $a = 3$ and $b = 4$. It has center (0, 0) and x-intercepts (3, 0) and (−3, 0). The asymptotes pass through (3, 4), (3, −4), (−3, 4), and (−3, −4).

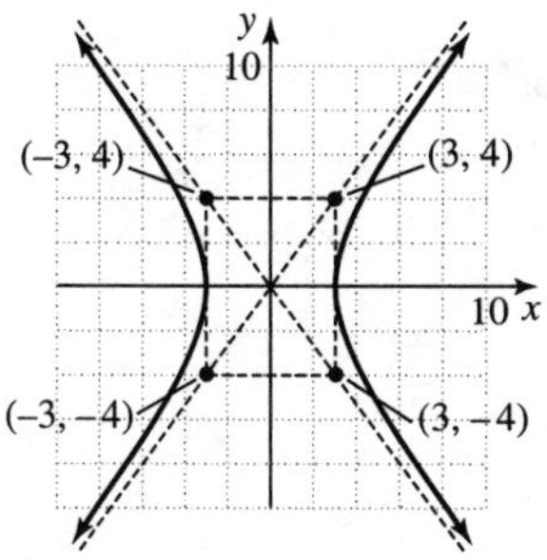

5. $9y^2 - 25x^2 = 225$

$$\frac{9y^2}{225} - \frac{25x^2}{225} = \frac{225}{225}$$

$$\frac{y^2}{25} - \frac{x^2}{9} = 1$$

This is a hyperbola with $a = 3$ and $b = 5$. The center is at (0, 0) with y-intercepts (0, 5) and (0, −5). The asymptotes pass through (3, 5), (3, −5), (−3, 5), and (−3, −5).

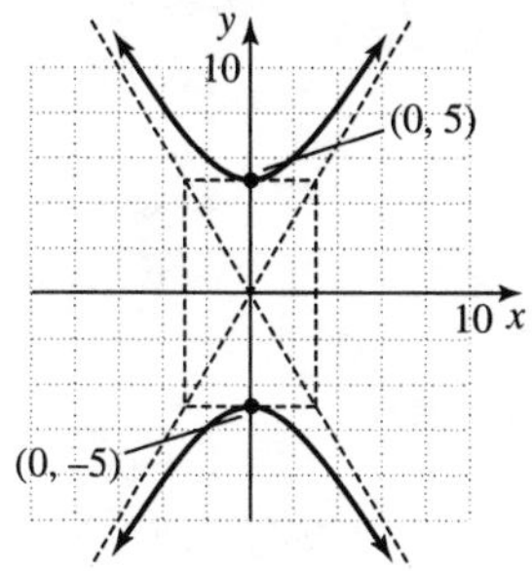

Graphing Calculator Explorations

1. $10x^2 + y^2 = 32$

$$y^2 = 32 - 10x^2$$

$$y = \pm\sqrt{32 - 10x^2}$$

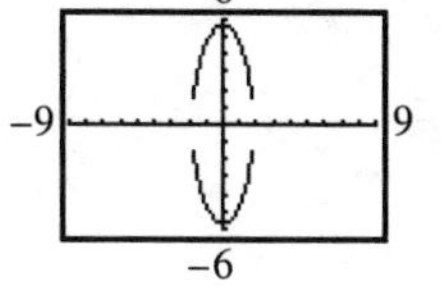

2. $x^2 + 6y^2 = 35$

$$6y^2 = 35 - x^2$$

$$y^2 = \frac{35 - x^2}{6}$$

$$y = \pm\sqrt{\frac{35 - x^2}{6}}$$

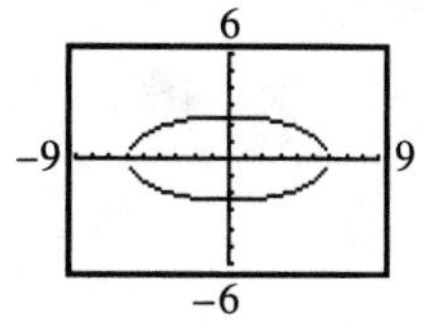

3. $20x^2 + 5y^2 = 100$

$$5y^2 = 100 - 20x^2$$

$$y^2 = 20 - 4x^2$$

$$y = \pm\sqrt{20 - 4x^2}$$

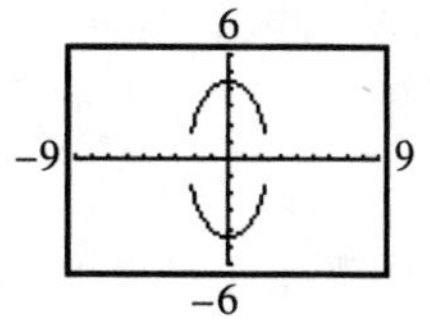

4. $4y^2 + 12x^2 = 48$

$$4y^2 = 48 - 12x^2$$

$$y^2 = 12 - 3x^2$$

$$y = \pm\sqrt{12 - 3x^2}$$

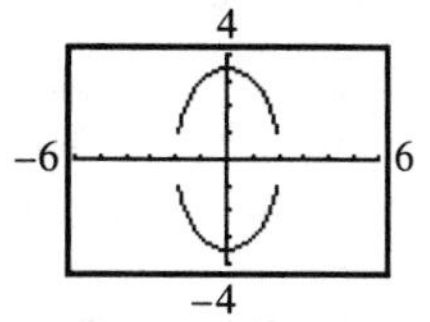

5. $7.3x^2 + 15.5y^2 = 95.2$

$$15.5y^2 = 95.2 - 7.3x^2$$

$$y^2 = \frac{95.2 - 7.3x^2}{15.5}$$

$$y = \pm\sqrt{\frac{95.2 - 7.3x^2}{15.5}}$$

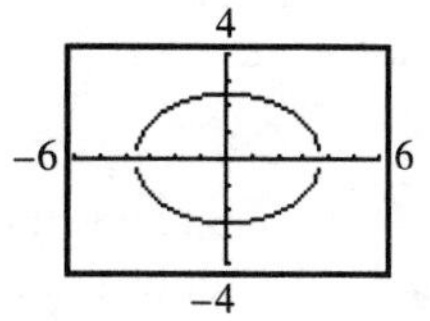

6. $18.8x^2 + 36.1y^2 = 205.8$

$$36.1y^2 = 205.8 - 18.8x^2$$

$$y^2 = \frac{205.8 - 18.8x^2}{36.1}$$

$$y = \pm\sqrt{\frac{205.8 - 18.8x^2}{36.1}}$$

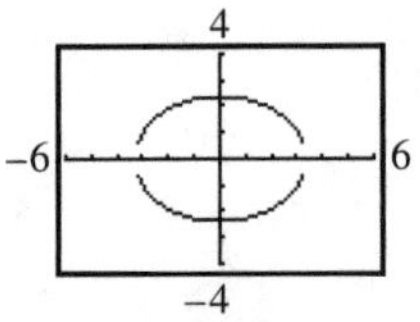

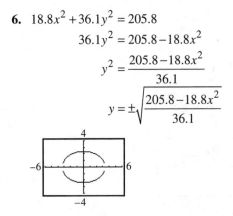

Vocabulary, Readiness & Video Check 10.2

1. A <u>hyperbola</u> is the set of points in a plane such that the absolute value of the differences of their distances from two fixed points is constant.

2. An <u>ellipse</u> is the set of points in a plane such that the sum of their distances from two fixed points is constant.

3. The two fixed points are each called a <u>focus</u>.

4. The point midway between the foci is called the <u>center</u>.

5. The graph of $\frac{x^2}{a^2} - \frac{y^2}{b^2} = 1$ is a <u>hyperbola</u> with center <u>(0, 0)</u> and <u>x</u>-intercepts of <u>(*a*, 0) and (−*a*, 0)</u>.

6. The graph of $\frac{x^2}{a^2} + \frac{y^2}{b^2} = 1$ is an <u>ellipse</u> with center <u>(0, 0)</u> and *x*-intercepts of <u>(*a*, 0) and (−*a*, 0)</u>.

7. *a* and *b* give us the location of 4 intercepts: $(a, 0)$, $(-a, 0)$, $(0, b)$ and $(0, -b)$ for $\frac{x^2}{a^2} + \frac{y^2}{b^2} = 1$ with center (0, 0). For Example 2, the values of *a* and *b* give us 4 points of the graph, just not intercepts. Here we move a distance of *a* units horizontally to the left and right of the center and *b* units above and below the center.

8. We use these points to draw asymptotes (also not part of the graph) which help us draw the correct shape of the hyperbola. The graph of a hyperbola gets closer and closer to the asymptotes without crossing them.

Exercise Set 10.2

1. $\frac{x^2}{16} + \frac{y^2}{4} = 1$ is an ellipse.

3. $x^2 - 5y^2 = 3$ is a hyperbola.

5. $-\frac{y^2}{25} + \frac{x^2}{36} = 1$ or

$\frac{x^2}{36} - \frac{y^2}{25} = 1$ is a hyperbola.

7. $\frac{x^2}{4} + \frac{y^2}{25} = 1$

$\frac{x^2}{2^2} + \frac{y^2}{5^2} = 1$

Center: (0, 0)
x-intercepts: (−2, 0), (2, 0)
y-intercepts: (0, −5), (0, 5)

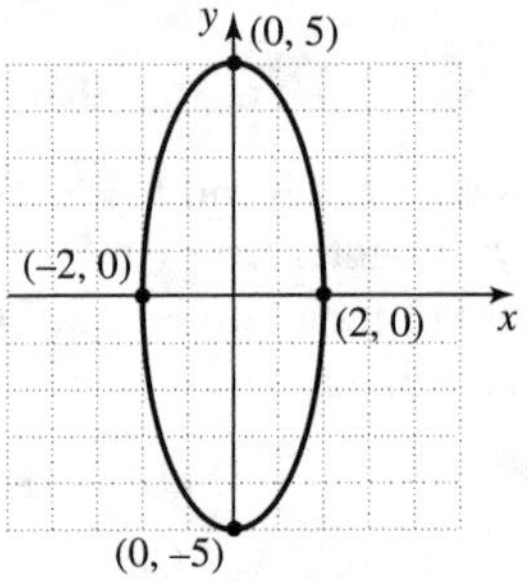

9. $\frac{x^2}{9} + y^2 = 1$

$\frac{x^2}{3^2} + \frac{y^2}{1^2} = 1$

Center: (0, 0)
x-intercepts: (−3, 0), (3, 0)
y-intercepts: (0, −1), (0, 1)

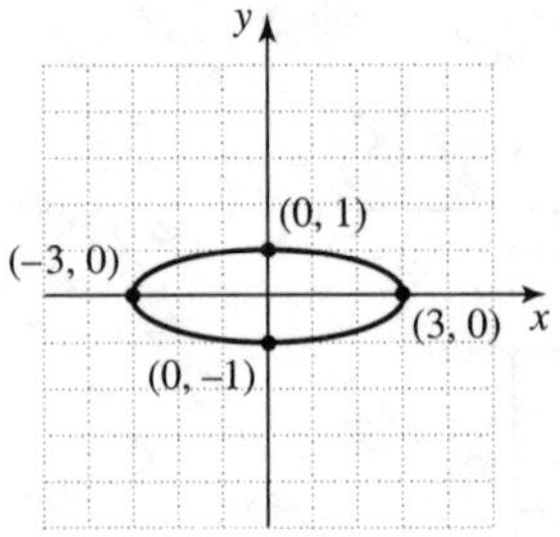

11. $9x^2 + y^2 = 36$

$$\frac{x^2}{4} + \frac{y^2}{36} = 1$$

$$\frac{x^2}{2^2} + \frac{y^2}{6^2} = 1$$

Center: (0, 0)
x-intercepts: (–2, 0), (2, 0)
y-intercepts: (0, –6), (0, 6)

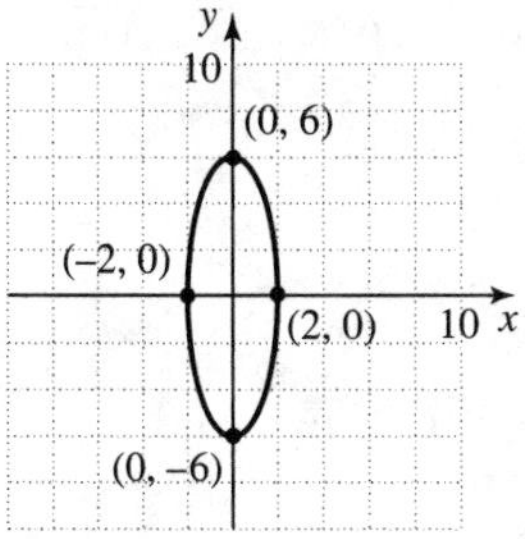

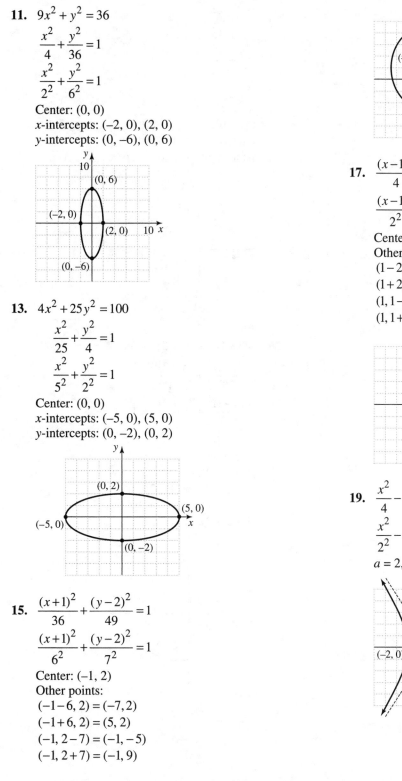

13. $4x^2 + 25y^2 = 100$

$$\frac{x^2}{25} + \frac{y^2}{4} = 1$$

$$\frac{x^2}{5^2} + \frac{y^2}{2^2} = 1$$

Center: (0, 0)
x-intercepts: (–5, 0), (5, 0)
y-intercepts: (0, –2), (0, 2)

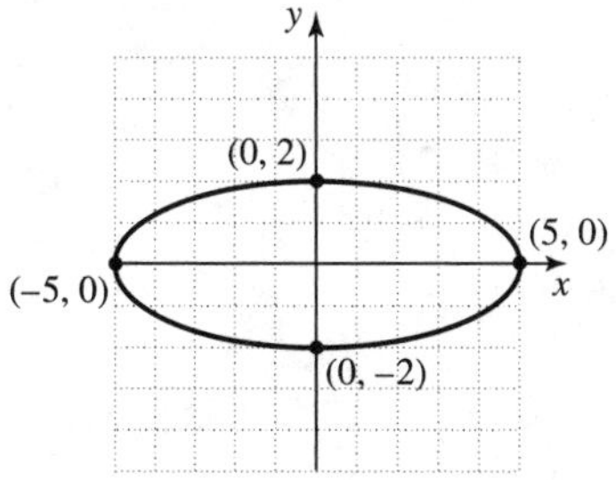

15. $\frac{(x+1)^2}{36} + \frac{(y-2)^2}{49} = 1$

$$\frac{(x+1)^2}{6^2} + \frac{(y-2)^2}{7^2} = 1$$

Center: (–1, 2)
Other points:
$(-1-6, 2) = (-7, 2)$
$(-1+6, 2) = (5, 2)$
$(-1, 2-7) = (-1, -5)$
$(-1, 2+7) = (-1, 9)$

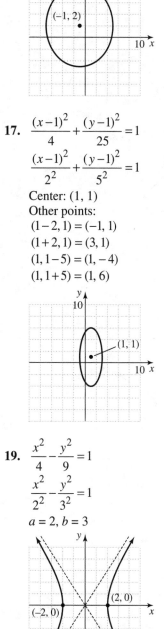

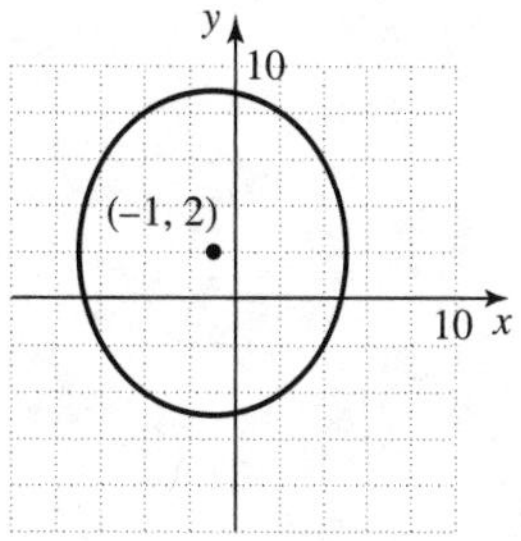

17. $\frac{(x-1)^2}{4} + \frac{(y-1)^2}{25} = 1$

$$\frac{(x-1)^2}{2^2} + \frac{(y-1)^2}{5^2} = 1$$

Center: (1, 1)
Other points:
$(1-2, 1) = (-1, 1)$
$(1+2, 1) = (3, 1)$
$(1, 1-5) = (1, -4)$
$(1, 1+5) = (1, 6)$

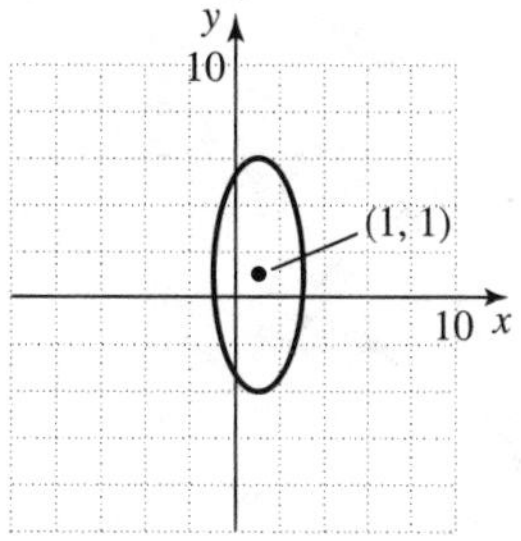

19. $\frac{x^2}{4} - \frac{y^2}{9} = 1$

$$\frac{x^2}{2^2} - \frac{y^2}{3^2} = 1$$

$a = 2, b = 3$

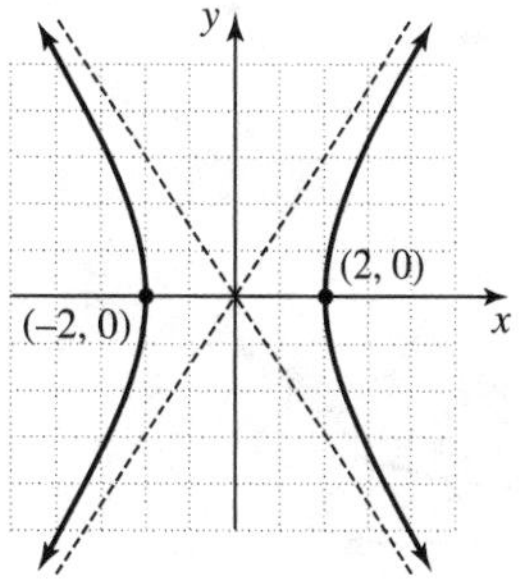

21. $\frac{y^2}{25}-\frac{x^2}{16}=1$

$\frac{y^2}{5^2}-\frac{x^2}{4^2}=1$

$a=4,\ b=5$

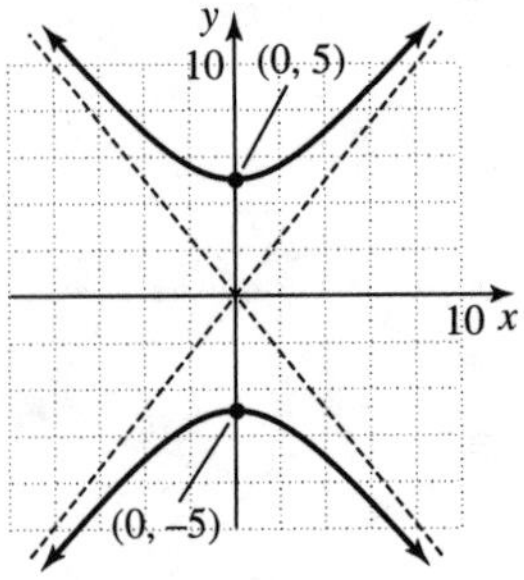

23. $x^2-4y^2=16$

$\frac{x^2}{16}-\frac{y^2}{4}=1$

$\frac{x^2}{4^2}-\frac{y^2}{2^2}=1$

$a=4,\ b=2$

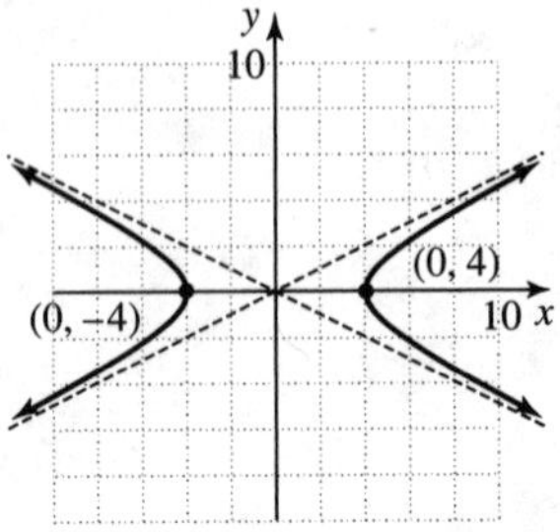

25. $16y^2-x^2=16$

$\frac{y^2}{1}-\frac{x^2}{16}=1$

$\frac{y^2}{1^2}-\frac{x^2}{4^2}=1$

$a=4,\ b=1$

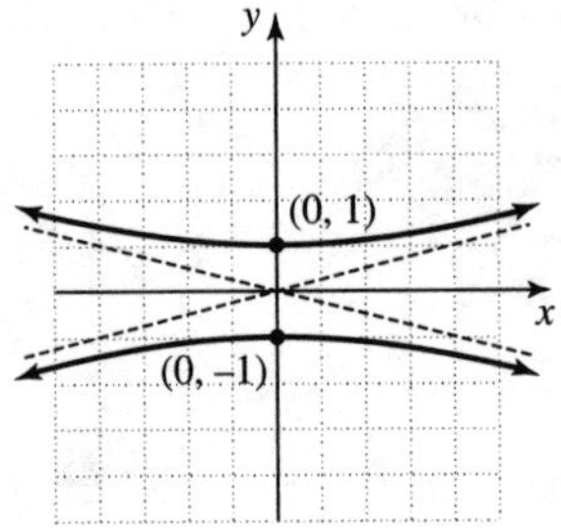

27. $\frac{y^2}{36}=1-x^2$

$x^2+\frac{y^2}{36}=1$

$\frac{x^2}{1^2}+\frac{y^2}{6^2}=1$

$C(0, 0)$

x-intercepts: (–1, 0), (1, 0)

y-intercepts: (0, –6), (0, 6)

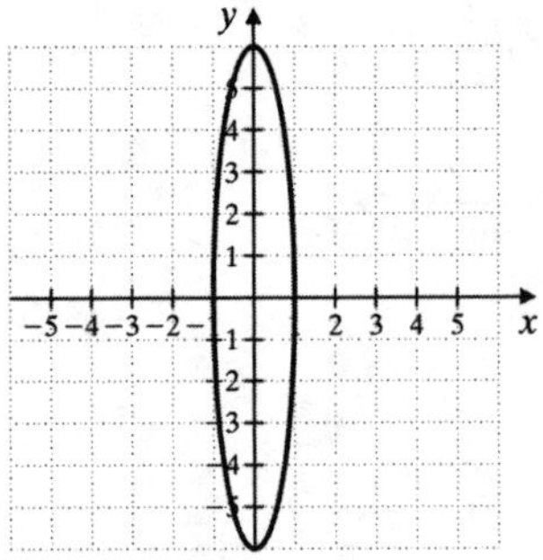

29. $4(x-1)^2+9(y+2)^2=36$

$\frac{(x-1)^2}{9}+\frac{(y+2)^2}{4}=1$

$\frac{(x-1)^2}{3^2}+\frac{(y+2)^2}{2^2}=1$

$C(1, -2)$

other points:

(1 – 3, –2), or (–2, –2)

(1 + 3, –2), or (4, –2)

(1, –2 – 2), or (1, –4)

(1, –2 + 2), or (1, 0)

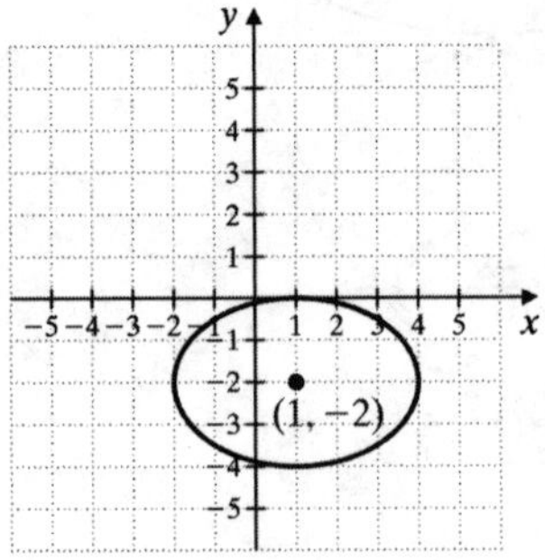

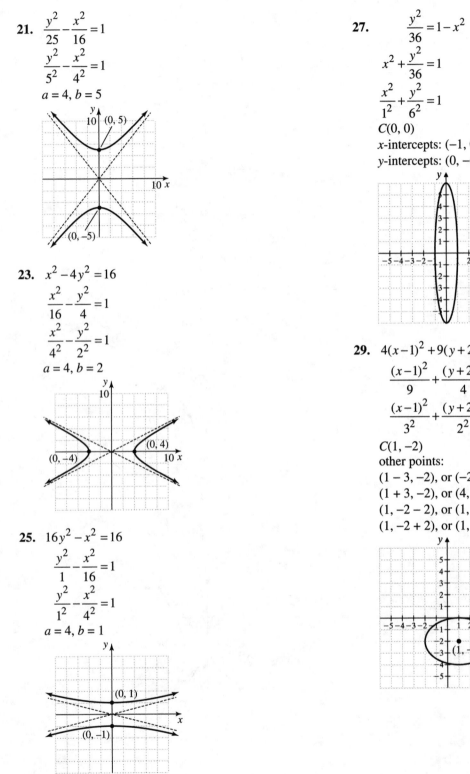

31. $8x^2 + 2y^2 = 32$

$$\frac{8x^2}{32} + \frac{2y^2}{32} = 1$$

$$\frac{x^2}{4} + \frac{y^2}{16} = 1$$

$$\frac{x^2}{2^2} + \frac{y^2}{4^2} = 1$$

$C(0, 0)$

x-intercepts: $(-2, 0)$, $(2, 0)$

y-intercepts: $(0, -4)$, $(0, 4)$

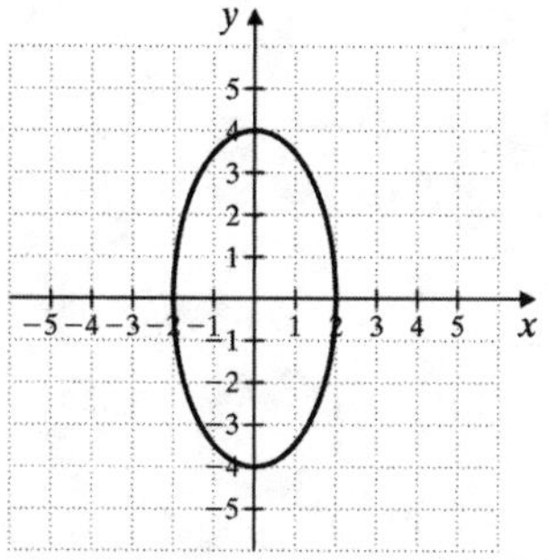

33. $25x^2 - y^2 = 25$

$$x^2 - \frac{y^2}{25} = 1$$

$$\frac{x^2}{1^2} - \frac{y^2}{5^2} = 1$$

$a = 1, b = 5$

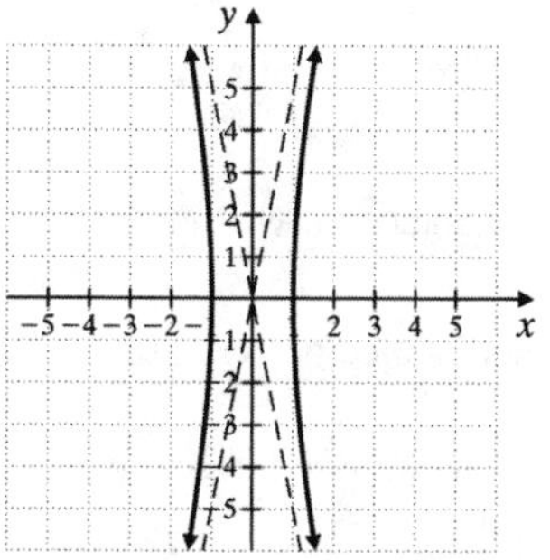

35. $(x-7)^2 + (y-2)^2 = 4$

Circle; center (7, 2), radius $r = \sqrt{4} = 2$

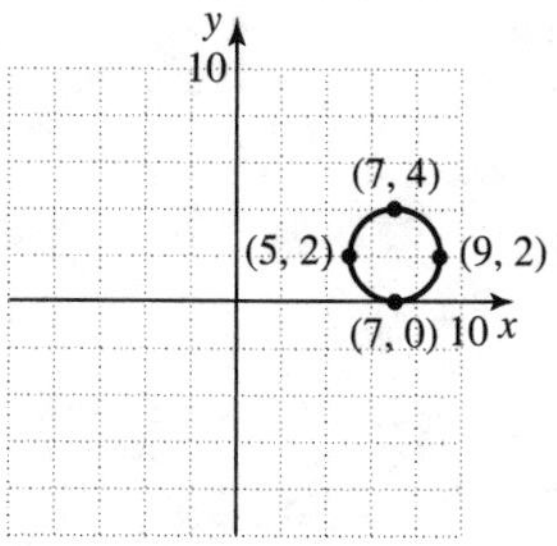

37. $y = x^2 + 12x + 36$

Parabola; $x = \frac{-b}{2a} = \frac{-12}{2(1)} = -6$

$y = (-6)^2 + 12(-6) + 36 = 0$

Vertex: (–6, 0), opens upward

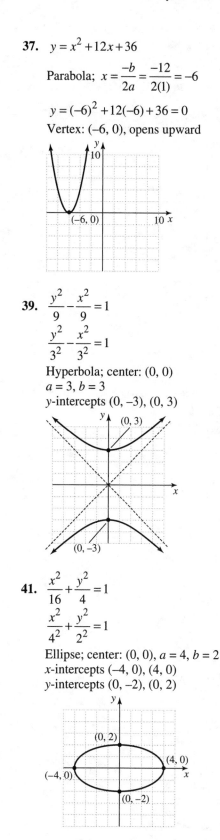

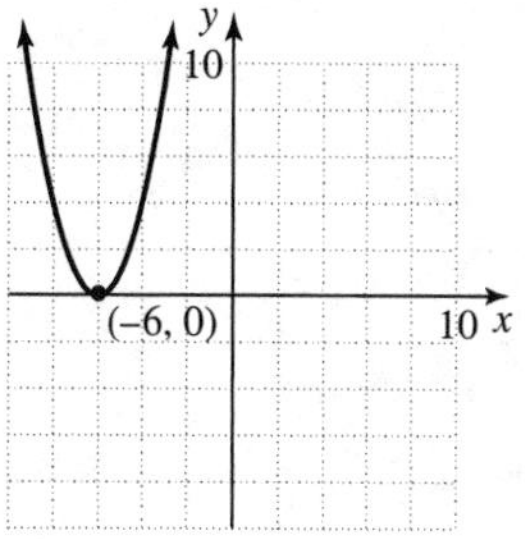

39. $\frac{y^2}{9} - \frac{x^2}{9} = 1$

$$\frac{y^2}{3^2} - \frac{x^2}{3^2} = 1$$

Hyperbola; center: (0, 0)

$a = 3, b = 3$

y-intercepts (0, –3), (0, 3)

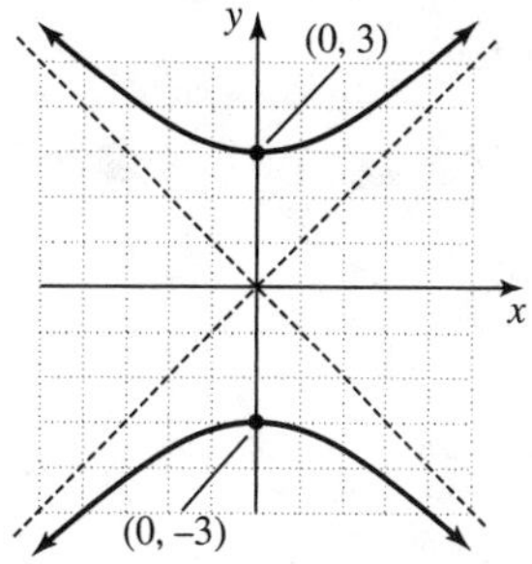

41. $\frac{x^2}{16} + \frac{y^2}{4} = 1$

$$\frac{x^2}{4^2} + \frac{y^2}{2^2} = 1$$

Ellipse; center: (0, 0), $a = 4, b = 2$

x-intercepts (–4, 0), (4, 0)

y-intercepts (0, –2), (0, 2)

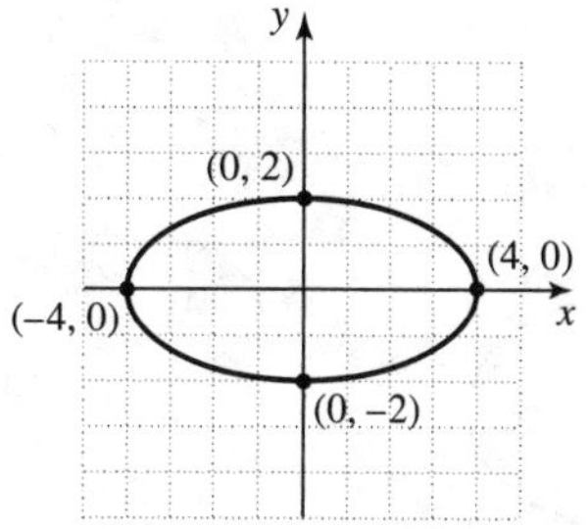

43. $x = y^2 + 4y - 1$

Parabola: $y = \frac{-b}{2a} = \frac{-4}{2(1)} = -2$

$x = (-2)^2 + 4(-2) - 1 = -5$

Vertex: (–5, –2), opens to the right.

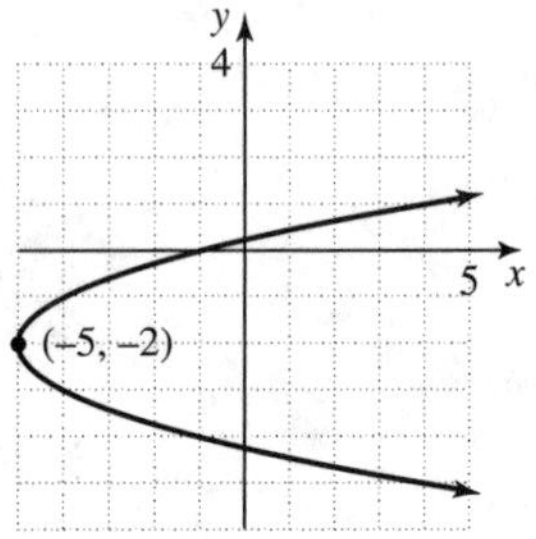

45. $9x^2 - 4y^2 = 36$

$\frac{x^2}{4} - \frac{y^2}{9} = 1$

$\frac{x^2}{2^2} - \frac{y^2}{3^2} = 1$

Hyperbola: center = (0, 0)
$a = 2, b = 3$
x-intercepts: (2, 0), (–2, 0)

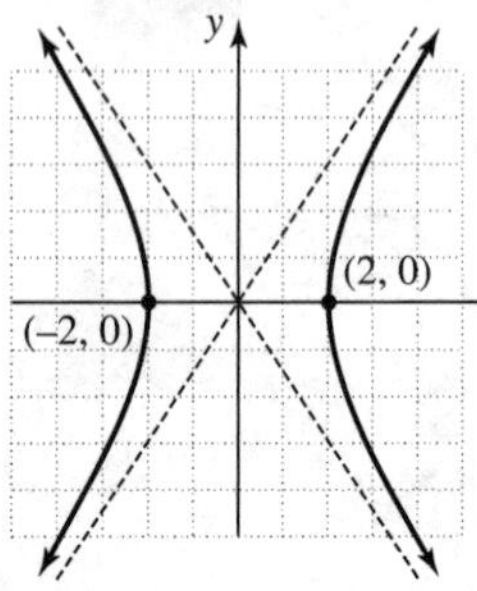

47. $\frac{(x-1)^2}{49} + \frac{(y+2)^2}{25} = 1$

$\frac{(x-1)^2}{7^2} + \frac{(y+2)^2}{5^2} = 1$

Ellipse; center: (1, –2)
$a = 7, b = 5$

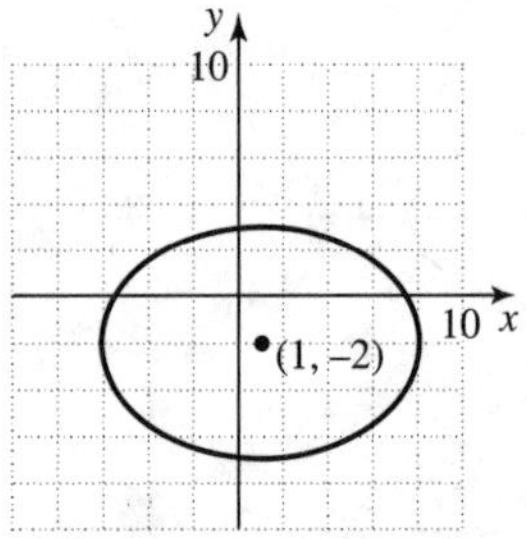

49. $\left(x + \frac{1}{2}\right)^2 + \left(y - \frac{1}{2}\right)^2 = 1$

Circle; center: $\left(-\frac{1}{2}, \frac{1}{2}\right)$, radius $r = 1$

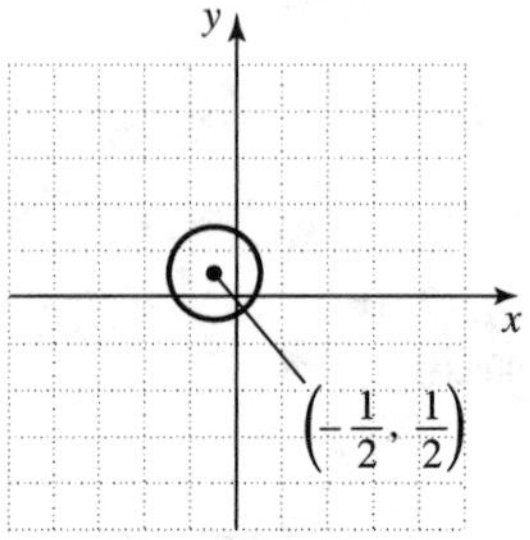

51. $(2x^3)(-4x^2) = -8x^5$

53. $-5x^2 + x^2 = -4x^2$

55. $\frac{x^2}{16} + \frac{y^2}{25} = 1$

$\sqrt{16} = 4$, so the distance between the x-intercepts is 4 + 4 = 8 units.
$\sqrt{25} = 5$, so the distance between the y-intercepts is 5 + 5 = 10 units.
The distance between the y-intercepts is longer by 10 – 8 = 2 units.

57. $4x^2 + y^2 = 16$

$\frac{x^2}{4} + \frac{y^2}{16} = 1$

$\sqrt{4} = 2$, so the distance between the x-intercepts is 2 + 2 = 4 units.
$\sqrt{16} = 4$, so the distance between the y-intercepts is 4 + 4 = 8 units.
The distance between the y-intercepts is longer by 8 – 4 = 4 units.

59. answers may vary

61. Circles: B, F
Ellipses: C, E, H
Hyperbolas: A, D, G

63. A: $c^2 = 36 + 13 = 49;\ c = \sqrt{49} = 7$

B: $c^2 = 4 - 4 = 0;\ c = \sqrt{0} = 0$

C: $c^2 = |25 - 16| = 9;\ c = \sqrt{9} = 3$

D: $c^2 = 39 + 25 = 64;\ c = \sqrt{64} = 8$

E: $c^2 = |81-17| = 64;\ c = \sqrt{64} = 8$

F: $c^2 = |36-36| = 0;\ c = \sqrt{0} = 0$

G: $c^2 = 65+16 = 81;\ c = \sqrt{81} = 9$

H: $c^2 = |144-140| = 4;\ c = \sqrt{4} = 2$

65. A: $e = \frac{7}{6}$

B: $e = \frac{0}{2} = 0$

C: $e = \frac{3}{5}$

D: $e = \frac{8}{5}$

E: $e = \frac{8}{9}$

F: $e = \frac{0}{6} = 0$

G: $e = \frac{9}{4}$

H: $e = \frac{2}{12} = \frac{1}{6}$

67. They are equal to 0.

69. answers may vary

71. $a = 130{,}000{,}000 \Rightarrow a^2 = (130{,}000{,}000)^2 = 1.69\times10^{16}$

$b = 125{,}000{,}000 \Rightarrow b^2 = (125{,}000{,}000)^2 = 1.5625\times10^{16}$

Thus, the equation is

$$\frac{x^2}{1.69\times10^{16}} + \frac{y^2}{1.5625\times10^{16}} = 1.$$

73. $9x^2 + 4y^2 = 36$

$$4y^2 = 36 - 9x^2$$

$$y^2 = \frac{36-9x^2}{4}$$

$$y = \pm\sqrt{\frac{36-9x^2}{4}} = \pm\frac{\sqrt{36-9x^2}}{2}$$

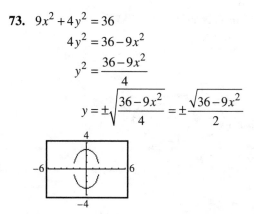

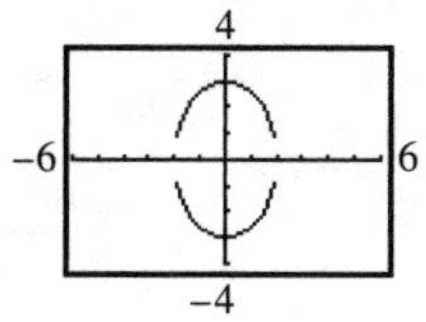

75. $\frac{(x-1)^2}{4} - \frac{(y+1)^2}{25} = 1$

Center: (1, −1)

$a = 2, b = 5$

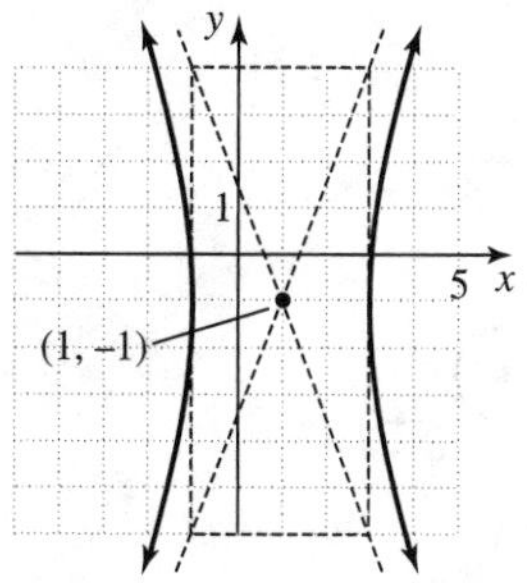

77. $\frac{y^2}{16} - \frac{(x+3)^2}{9} = 1$

Center: (−3, 0)

$a = 3, b = 4$

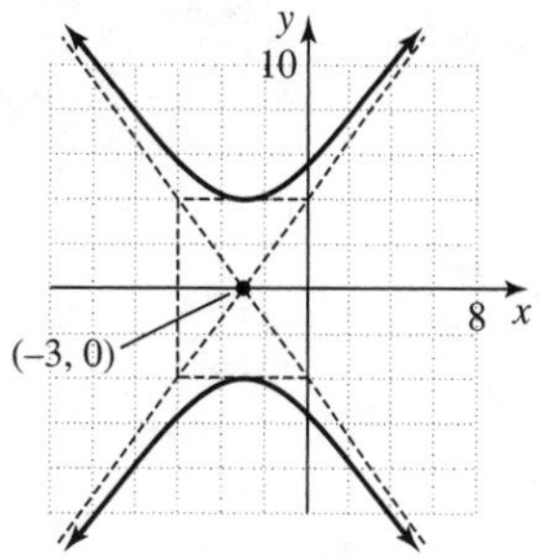

79. $\frac{(x+5)^2}{16} - \frac{(y+2)^2}{25} = 1$

Center: (−5, −2)

$a = 4, b = 5$

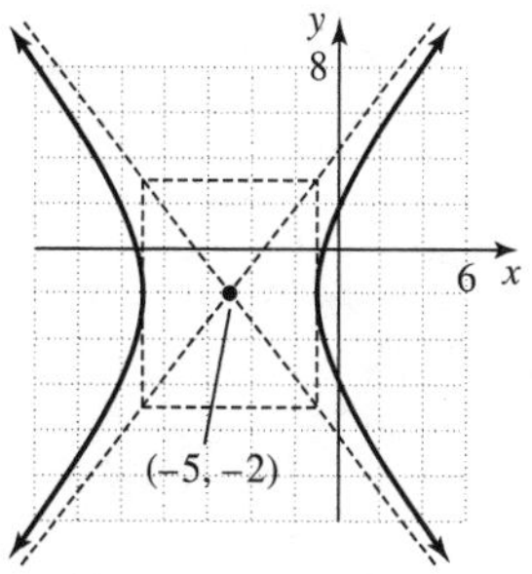

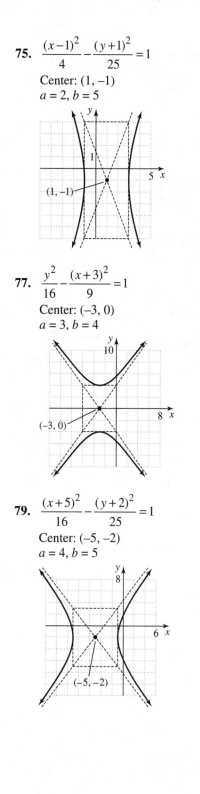

Integrated Review

1. $(x-7)^2+(y-2)^2=4$
Circle; center: (7, 2),
radius: $r=\sqrt{4}=2$

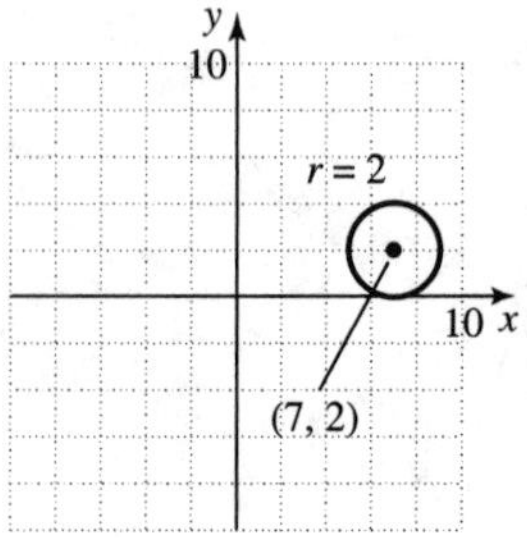

2. $y=x^2+4$
Parabola; vertex: (0, 4)

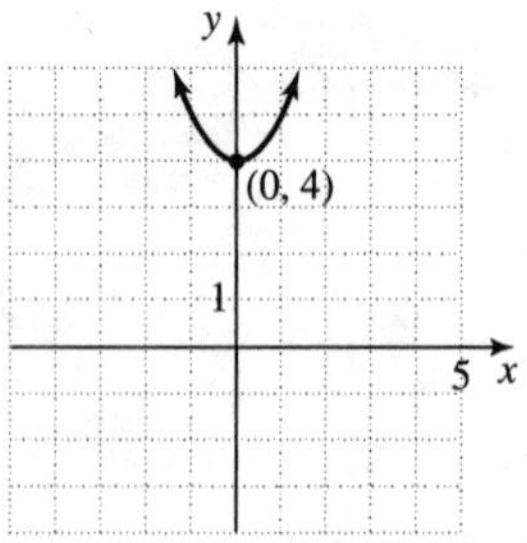

3. $y=x^2+12x+36$

Parabola; $x=\dfrac{-b}{2a}=\dfrac{-12}{2(1)}=-6$

$y=(-6)^2+12(-6)+36=0$
Vertex: (–6, 0)

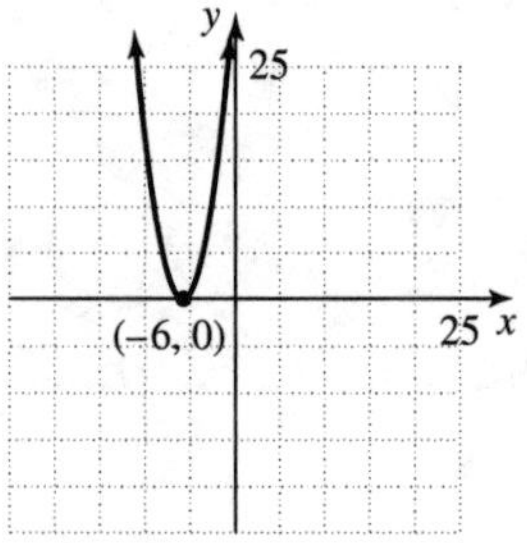

4. $\dfrac{x^2}{4}+\dfrac{y^2}{9}=1$
Ellipse; center: (0, 0)
$a=2$, $b=3$

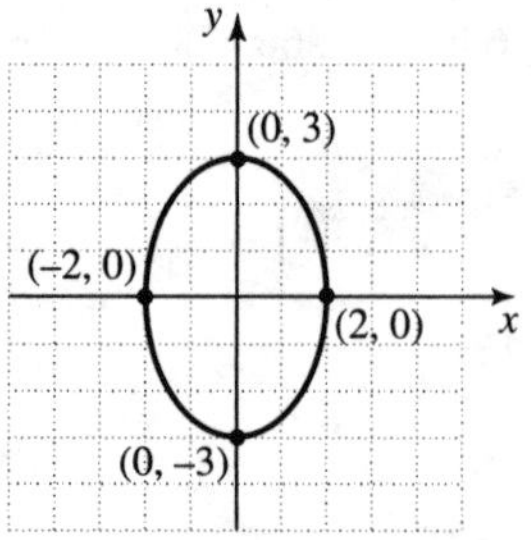

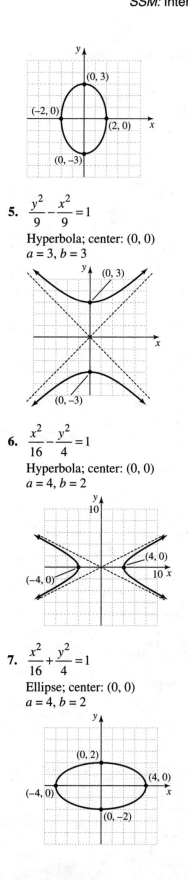

5. $\dfrac{y^2}{9}-\dfrac{x^2}{9}=1$
Hyperbola; center: (0, 0)
$a=3$, $b=3$

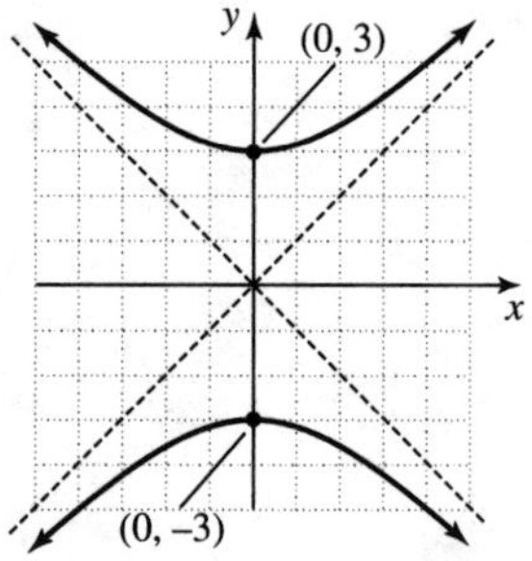

6. $\dfrac{x^2}{16}-\dfrac{y^2}{4}=1$
Hyperbola; center: (0, 0)
$a=4$, $b=2$

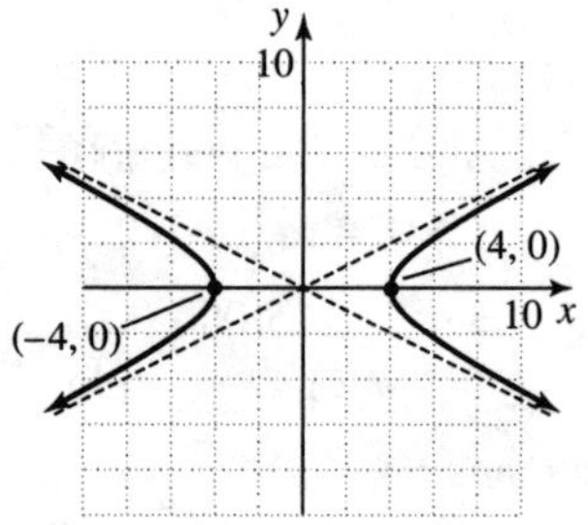

7. $\dfrac{x^2}{16}+\dfrac{y^2}{4}=1$
Ellipse; center: (0, 0)
$a=4$, $b=2$

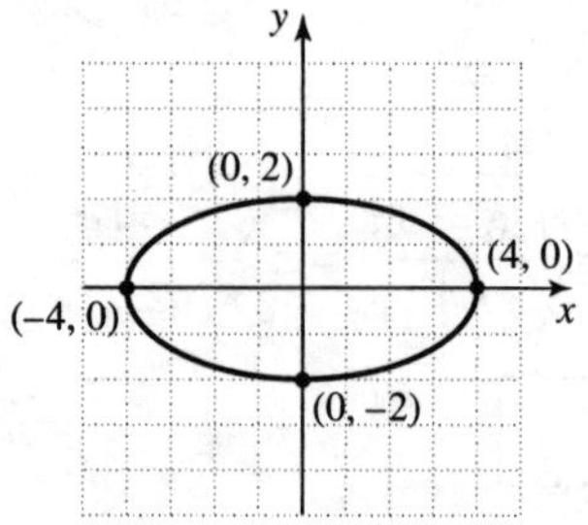

8. $x^2 + y^2 = 16$
Circle; center: (0, 0)
radius: $r = \sqrt{16} = 4$

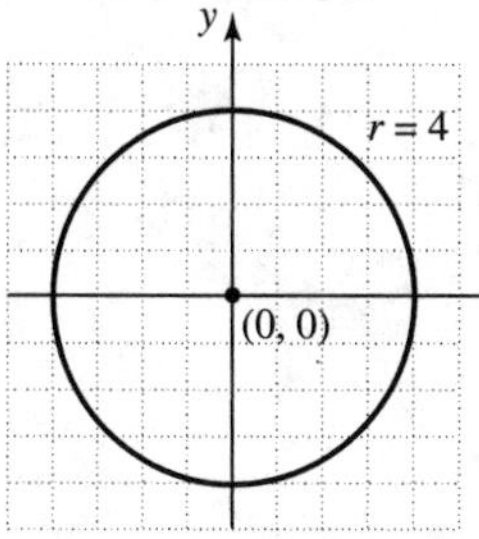

9. $x = y^2 + 4y - 1$

Parabola; $y = \frac{-b}{2a} = \frac{-4}{2(1)} = -2$

$x = (-2)^2 + 4(-2) - 1 = -5$
Vertex: (–5, –2)

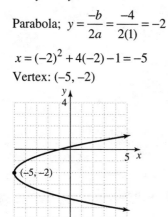

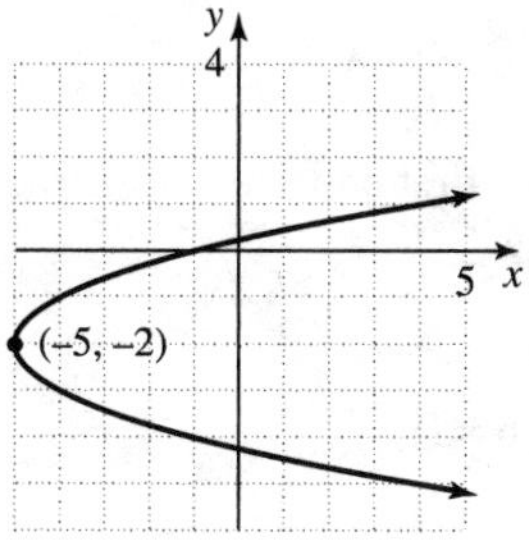

10. $x = -y^2 + 6y$

Parabola; $y = \frac{-b}{2a} = \frac{-6}{2(-1)} = 3$

$x = -(3)^2 + 6(3) = 9$
Vertex: (9, 3)

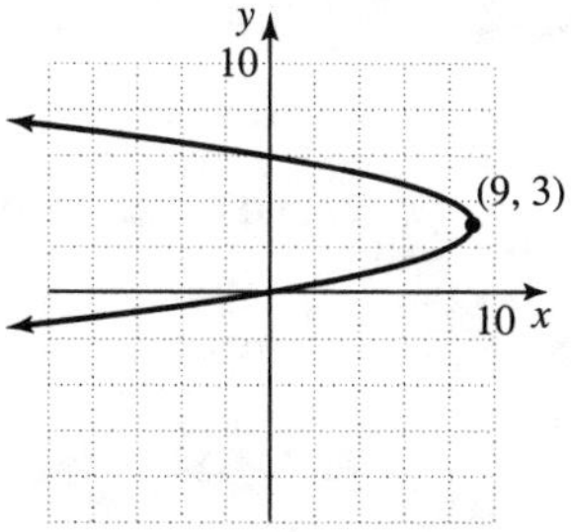

11. $9x^2 - 4y^2 = 36$

$\frac{x^2}{4} - \frac{y^2}{9} = 1$

Hyperbola; center: (0, 0)
$a = 2, b = 3$

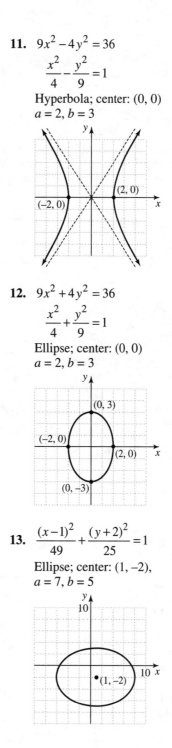

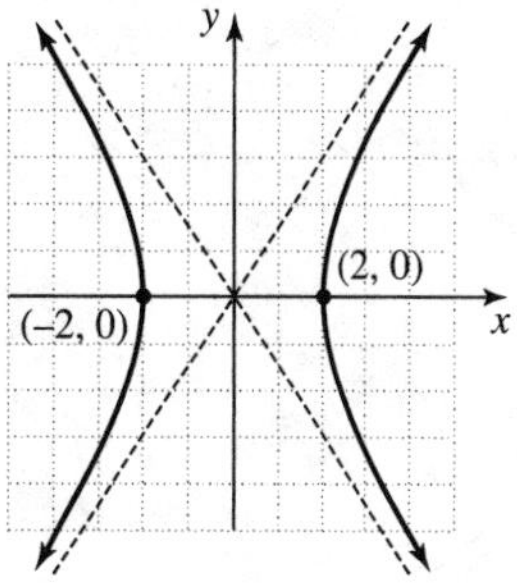

12. $9x^2 + 4y^2 = 36$

$\frac{x^2}{4} + \frac{y^2}{9} = 1$

Ellipse; center: (0, 0)
$a = 2, b = 3$

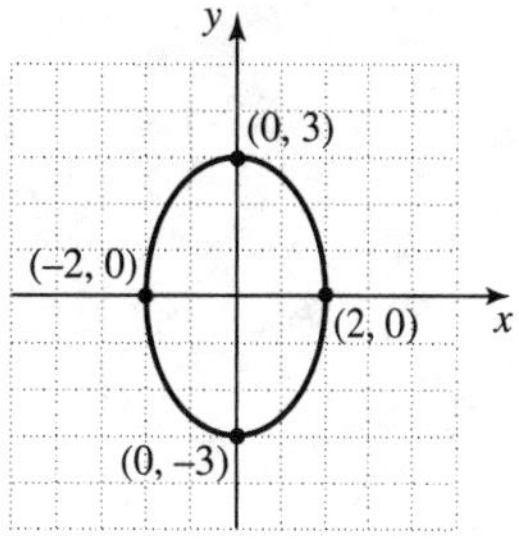

13. $\frac{(x-1)^2}{49} + \frac{(y+2)^2}{25} = 1$

Ellipse; center: (1, –2),
$a = 7, b = 5$

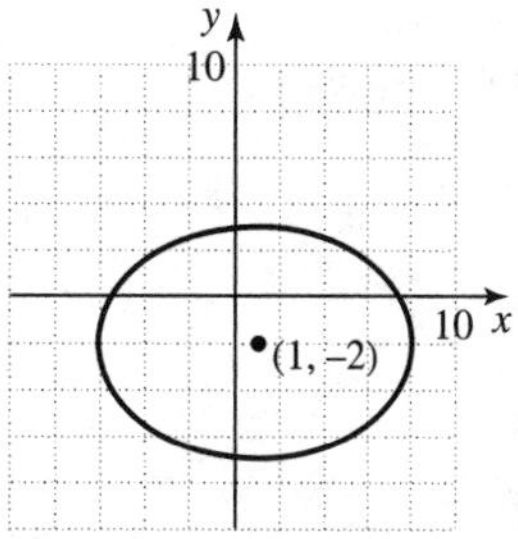

14. $y^2 = x^2 + 16$

$y^2 - x^2 = 16$

$\frac{y^2}{16} - \frac{x^2}{16} = 1$

Hyperbola; center: (0, 0)

$a = 4, b = 4$

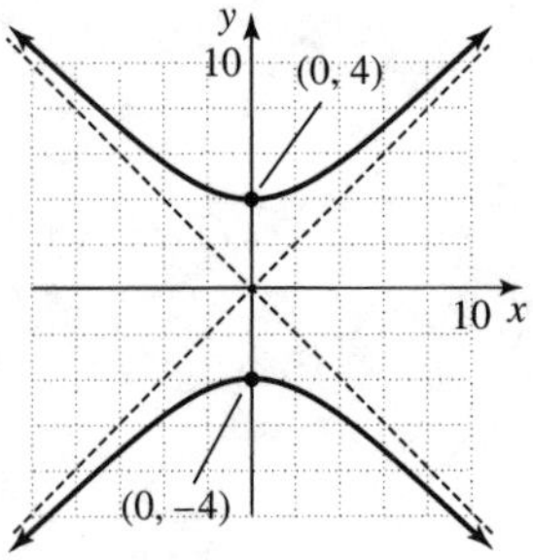

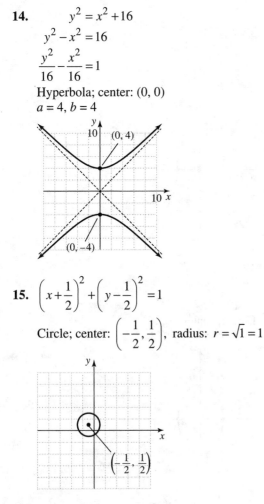

15. $\left(x+\frac{1}{2}\right)^2 + \left(y-\frac{1}{2}\right)^2 = 1$

Circle; center: $\left(-\frac{1}{2}, \frac{1}{2}\right)$, radius: $r = \sqrt{1} = 1$

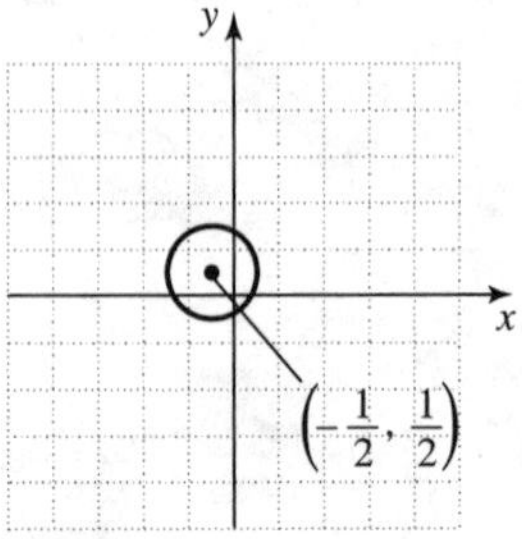

Section 10.3 Practice Exercises

1. $\begin{cases} x^2 - 4y = 4 \\ x + y = -1 \end{cases}$

Solve $x + y = -1$ for y.

$y = -x - 1$

Replace y with $-x - 1$ in the first equation and solve for x.

$x^2 - 4(-x-1) = 4$

$x^2 + 4x + 4 = 4$

$x^2 + 4x = 0$

$x(x+4) = 0$

$x = 0$ or $x = -4$

Let $x = 0$, $y = -0 - 1 = -1$

Let $x = -4$, $y = -(-4) - 1 = 3$

The solutions are (0, −1) and (−4, 3).

2. $\begin{cases} y = -\sqrt{x} \\ x^2 + y^2 = 20 \end{cases}$

Substitute $-\sqrt{x}$ for y in the second equation.

$x^2 + \left(-\sqrt{x}\right)^2 = 20$

$x^2 + x = 20$

$x^2 + x - 20 = 0$

$(x+5)(x-4) = 0$

$x = -5$ or $x = 4$

Let $x = -5$.

$y = -\sqrt{-5}$ Not a real number

Let $x = 4$.

$y = -\sqrt{4} = -2$

The solution is (4, −2).

3. $\begin{cases} x^2 + y^2 = 9 \\ x - y = 5 \end{cases}$

Solve the second equation for x.

$x = y + 5$

Let $x = y + 5$ in the first equation.

$(y+5)^2 + y^2 = 9$

$y^2 + 10y + 25 + y^2 = 9$

$2y^2 + 10y + 16 = 0$

$y^2 + 5y + 8 = 0$

By the quadratic formula,

$$y = \frac{-5 \pm \sqrt{5^2 - 4(1)(8)}}{2(1)} = \frac{-5 \pm \sqrt{-7}}{2}$$

$\sqrt{-7}$ is not a real number. There is no real solution, or ∅.

4. $\begin{cases} x^2 + 4y^2 = 16 \\ x^2 - y^2 = 1 \end{cases}$

Add the opposite of the second equation to the first.

$x^2 + 4y^2 = 16$

$-x^2 + y^2 = -1$

$0 + 5y^2 = 15$

$y^2 = 3$

$y = \pm\sqrt{3}$

Let $y=\sqrt{3}$.

$$x^2-\left(\sqrt{3}\right)^2=1$$
$$x^2-3=1$$
$$x^2=4$$
$$x=\pm 2$$

Let $y=-\sqrt{3}$.

$$x^2-\left(-\sqrt{3}\right)^2=1$$
$$x^2-3=1$$
$$x^2=4$$
$$x=\pm 2$$

The solutions are $\left(2, \sqrt{3}\right)$, $\left(2, -\sqrt{3}\right)$, $\left(-2, \sqrt{3}\right)$, and $\left(-2, -\sqrt{3}\right)$.

Vocabulary, Readiness & Video Check 10.3

1. Solving for y would either introduce tedious fractions (2nd equation) or a square root (1st equation) into the calculations.

2. When you multiply the left side of the equation by this number, do not forget to also multiply the right side.

Exercise Set 10.3

1. $\begin{cases} x^2+y^2=25 & (1) \\ 4x+3y=0 & (2) \end{cases}$

Solve E2 for y.

$$3y=-4x$$
$$y=-\frac{4x}{3}$$

Substitute into E1.

$$x^2+\left(-\frac{4x}{3}\right)^2=25$$
$$x^2+\frac{16x^2}{9}=25$$
$$9\left(x^2+\frac{16x^2}{9}\right)=9(25)$$
$$9x^2+16x^2=225$$
$$25x^2=225$$
$$x^2=9$$
$$x=\pm\sqrt{9}=\pm 3$$
$$x=3: y=-\frac{4(3)}{3}=-4$$
$$x=-3: y=-\frac{4(-3)}{3}=4$$

The solutions are (3, –4) and (–3, 4).

3. $\begin{cases} x^2+4y^2=10 & (1) \\ y=x & (2) \end{cases}$

Substitute x for y in E1.

$$x^2+4x^2=10$$
$$5x^2=10$$
$$x^2=2$$
$$x=\pm\sqrt{2}$$

Substitute these values into E2.

$$x=\sqrt{2}: y=x=\sqrt{2}$$
$$x=-\sqrt{2}: y=x=-\sqrt{2}$$

The solutions are $\left(\sqrt{2}, \sqrt{2}\right)$ and $\left(-\sqrt{2}, -\sqrt{2}\right)$.

5. $\begin{cases} y^2=4-x & (1) \\ x-2y=4 & (2) \end{cases}$

Solve E2 for x.

$$x=2y+4$$

Substitute into E1.

$$y^2=4-(2y+4)$$
$$y^2=-2y$$
$$y^2+2y=0$$
$$y(y+2)=0$$
$$y=0 \text{ or } y+2=0$$
$$y=-2$$

Substitute these values into the equation $x = 2y + 4$.

$$y=0: x=2(0)+4=4$$
$$y=-2: x=2(-2)+4=0$$

The solutions are (4, 0) and (0, –2).

7. $\begin{cases} x^2+y^2=9 & (1) \\ 16x^2-4y^2=64 & (2) \end{cases}$

Multiply E1 by 4 and add to E2.

$$4x^2+4y^2=36$$
$$16x^2-4y^2=64$$
$$20x^2=100$$
$$x^2=5$$
$$x=\pm\sqrt{5}$$

Substitute 5 for x^2 into E1.

$$5+y^2=9$$
$$y^2=4$$
$$y=\pm 2$$

The solutions are $\left(-\sqrt{5}, -2\right)$, $\left(-\sqrt{5}, 2\right)$, $\left(\sqrt{5}, -2\right)$, and $\left(\sqrt{5}, 2\right)$.

9. $\begin{cases} x^2+2y^2=2 & (1) \\ x-\ \ y=2 & (2) \end{cases}$

Solve E2 for x: $x=y+2$

Substitute into E1.

$$(y+2)^2+2y^2=2$$
$$y^2+4y+4+2y^2=2$$
$$3y^2+4y+2=0$$
$$y=\frac{-4\pm\sqrt{(4)^2-4(3)(2)}}{2(3)}=\frac{-4\pm\sqrt{-8}}{6}$$

There are no real solutions. The solution is ∅.

11. $\begin{cases} y=x^2-3 & (1) \\ 4x-y=6 & (2) \end{cases}$

Substitute x^2-3 for y in E2.

$$4x-(x^2-3)=6$$
$$4x-x^2+3=6$$
$$0=x^2-4x+3$$
$$0=(x-3)(x-1)$$
$$x-3=0 \text{ or } x-1=0$$
$$x=3 \text{ or } x=1$$

Substitute these values into E1.

$x=3: y=(3)^2-3=6$

$x=1: y=(1)^2-3=-2$

The solutions are (3, 6) and (1, –2).

13. $\begin{cases} y=x^2 & (1) \\ 3x+y=10 & (2) \end{cases}$

Substitute x^2 for y in E2.

$$3x+x^2=10$$
$$x^2+3x-10=0$$
$$(x+5)(x-2)=0$$
$$x+5=0 \text{ or } x-2=0$$
$$x=-5 \text{ or } x=2$$

Substitute these values into E1.

$x=-5: y=(-5)^2=25$

$x=2: y=(2)^2=4$

The solutions are (–5, 25) and (2, 4).

15. $\begin{cases} y=2x^2+1 & (1) \\ x+y=-1 & (2) \end{cases}$

Substitute $2x^2+1$ for y in E2.

$$x+2x^2+1=-1$$
$$2x^2+x+2=0$$
$$x=\frac{-1\pm\sqrt{(1)^2-4(2)(2)}}{2(2)}=\frac{-1\pm\sqrt{-15}}{4}$$

There are no real solutions. The solution is ∅.

17. $\begin{cases} y=x^2-4 & (1) \\ y=x^2-4x & (2) \end{cases}$

Substitute x^2-4 for y in E2.

$$x^2-4=x^2-4x$$
$$-4=-4x$$
$$1=x$$

Substitute this value into E1.

$y=(1)^2-4=-3$

The solution is (1, –3).

19. $\begin{cases} 2x^2+3y^2=14 & (1) \\ -x^2+\ \ y^2=3 & (2) \end{cases}$

Multiply E2 by 2 and add to E1.

$$\begin{array}{r} 2x^2+3y^2=14 \\ -2x^2+2y^2=6 \\ \hline 5y^2=20 \end{array}$$
$$y^2=4$$
$$y=\pm 2$$

Substitute 4 for y^2 into E2.

$$-x^2+4=3$$
$$-x^2=-1$$
$$x^2=1$$
$$x=\pm 1$$

The solutions are (–1, –2), (–1, 2), (1, –2), and (1, 2).

21. $\begin{cases} x^2+y^2=1 & (1) \\ x^2+(y+3)^2=4 & (2) \end{cases}$

Multiply E1 by –1 and add to E2.

$$\begin{aligned} -x^2-y^2&=-1 \\ x^2+(y+3)^2&=4 \\ \hline (y+3)^3-y^2&=3 \\ y^2+6y+9-y^2&=3 \\ 6y&=-6 \\ y&=-1 \end{aligned}$$

Replace y with –1 in E1.

$$\begin{aligned} x^2+(-1)^2&=1 \\ x^2&=0 \\ x&=0 \end{aligned}$$

The solution is (0, –1).

23. $\begin{cases} y=x^2+2 & (1) \\ y=-x^2+4 & (2) \end{cases}$

Add E1 and E2.

$$\begin{aligned} y&=x^2+2 \\ y&=-x^2+4 \\ \hline 2y&=6 \\ y&=3 \end{aligned}$$

Substitute this value into E1.

$$\begin{aligned} 3&=x^2+2 \\ 1&=x^2 \\ \pm1&=x \end{aligned}$$

The solutions are (–1, 3) and (1, 3).

25. $\begin{cases} 3x^2+y^2=9 & (1) \\ 3x^2-y^2=9 & (2) \end{cases}$

Add E1 and E2.

$$\begin{aligned} 3x^2+y^2&=9 \\ 3x^2-y^2&=9 \\ \hline 6x^2&=18 \\ x^2&=3 \\ x&=\pm\sqrt{3} \end{aligned}$$

Substitute 3 for x^2 in E1.

$$\begin{aligned} 3(3)+y^2&=9 \\ y^2&=0 \\ y&=0 \end{aligned}$$

The solutions are $\left(-\sqrt{3},0\right)$, $\left(\sqrt{3},0\right)$.

27. $\begin{cases} x^2+3y^2=6 & (1) \\ x^2-3y^2=10 & (2) \end{cases}$

Solve E2 for x^2: $x^2=3y^2+10$.

Substitute into E1.

$$\begin{aligned} (3y^2+10)+3y^2&=6 \\ 6y^2&=-4 \\ y^2&=-\frac{2}{3} \end{aligned}$$

There are no real solutions. The solution is ∅.

29. $\begin{cases} x^2+y^2=36 & (1) \\ y=\frac{1}{6}x^2-6 & (2) \end{cases}$

Solve E1 for x^2: $x^2=36-y^2$.

Substitute into E2.

$$\begin{aligned} y&=\frac{1}{6}(36-y^2)-6 \\ y&=6-\frac{1}{6}y^2-6 \\ 6y&=-y^2 \\ y^2+6y&=0 \\ y(y+6)&=0 \\ y=0 \text{ or } y&=-6 \end{aligned}$$

Substitute these values into the equation $x^2=36-y^2$.

$$\begin{aligned} y=0: x^2&=36-(0)^2 \\ x^2&=36 \\ x&=\pm6 \\ y=-6: x^2&=36-(6)^2 \\ x^2&=0 \\ x&=0 \end{aligned}$$

The solutions are (–6, 0), (6, 0) and (0, –6).

31. $\begin{cases} y=\sqrt{x} \\ x^2+y^2=12 \end{cases}$

Substitute.

$$\begin{aligned} x^2+\left(\sqrt{x}\right)^2&=12 \\ x^2+x&=12 \\ x^2+x-12&=0 \\ (x+4)(x-3)&=0 \end{aligned}$$

$x+4=0$ or $x-3=0$

$x=-4$ $\quad$ $x=3$

$x=-4$: $y=\sqrt{-4}$

$x = 3$: $y = \sqrt{3}$

Since $\sqrt{-4}$ is not a real number, the only solution is $\left(3, \sqrt{3}\right)$.

33. $x > -3$

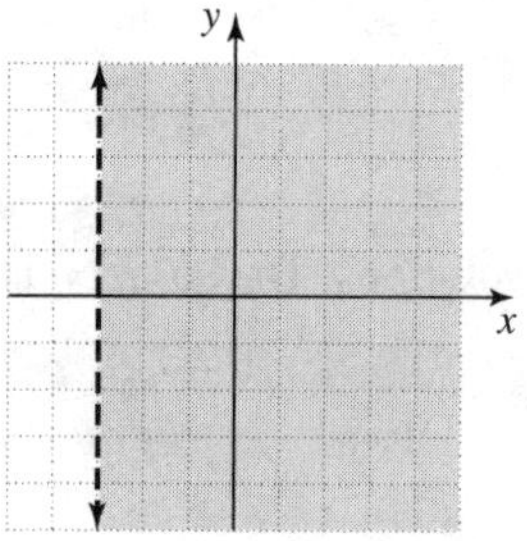

35. $y < 2x - 1$

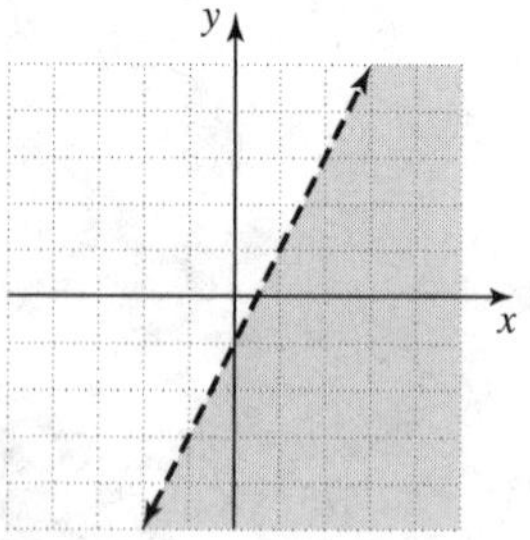

37. $P = x + (2x - 5) + (5x - 20) = (8x - 25)$ inches

39. $P = 2(x^2 + 3x + 1) + 2(x^2)$
$= 2x^2 + 6x + 2 + 2x^2$
$= (4x^2 + 6x + 2)$ meters

41. answers may vary

43. There are 0, 1, 2, 3, or 4 possible real solutions. answers may vary

45. Let x and y represent the numbers.

$$\begin{cases} x^2 + y^2 = 130 \\ x^2 - y^2 = 32 \end{cases}$$

Add the equations.

$$\begin{aligned} x^2 + y^2 &= 130 \\ x^2 - y^2 &= 32 \\ \hline 2x^2 \quad &= 162 \\ x^2 &= 81 \\ x &= \pm 9 \end{aligned}$$

Replace x^2 with 81 in the first equation.

$$\begin{aligned} 81 + y^2 &= 130 \\ y^2 &= 49 \\ y &= \pm 7 \end{aligned}$$

The numbers are –9 and –7, –9 and 7, 9 and –7, and 9 and 7.

47. Let x and y be the length and width.

$$\begin{cases} xy = 285 \\ 2x + 2y = 68 \end{cases}$$

Solve the first equation for y: $y = \dfrac{285}{x}$.

Substitute into the second equation.

$$\begin{aligned} 2x + 2\left(\frac{285}{x}\right) &= 68 \\ x + \frac{285}{x} &= 34 \\ x^2 + 285 &= 34x \end{aligned}$$

$x^2 - 34x + 285 = 0$
$(x - 19)(x - 15) = 0$
$x = 19$ or $x = 15$

Using $x = 19$, $y = \dfrac{285}{x} = \dfrac{285}{19} = 15$.

Using $x = 15$, $y = \dfrac{285}{x} = \dfrac{285}{15} = 19$.

The dimensions are 19 cm by 15 cm.

49. $$\begin{cases} p = -0.01x^2 - 0.2x + 9 \\ p = 0.01x^2 - 0.1x + 3 \end{cases}$$

Substitute.

$$\begin{aligned} -0.01x^2 - 0.2x + 9 &= 0.01x^2 - 0.1x + 3 \\ 0 &= 0.02x^2 + 0.1x - 6 \\ 0 &= x^2 + 5x - 300 \\ 0 &= (x + 20)(x - 15) \end{aligned}$$

$x + 20 = 0$ or $x - 15 = 0$
$x = -20$ or $x = 15$

Disregard the negative.

$p = -0.01(15)^2 - 0.2(15) + 9$
$p = 3.75$

The equilibrium quantity is 15,000 compact discs, and the corresponding price is $3.75.

51. $\begin{cases} x^2 + 4y^2 = 10 \\ y = x \end{cases}$

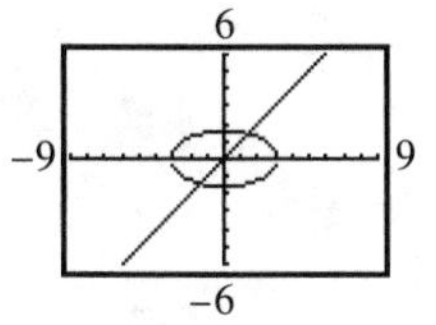

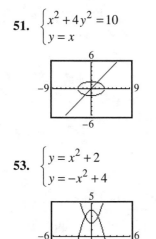

53. $\begin{cases} y = x^2 + 2 \\ y = -x^2 + 4 \end{cases}$

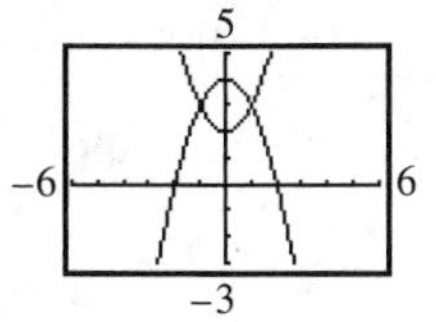

Section 10.4 Practice Exercises

1. $\dfrac{x^2}{36} + \dfrac{y^2}{16} \geq 1$

First graph the ellipse $\dfrac{x^2}{36} + \dfrac{y^2}{16} = 1$ as a solid curve. Choose (0, 0) as a test point.

$$\frac{x^2}{36} + \frac{y^2}{16} \geq 1$$
$$\frac{0^2}{36} + \frac{0^2}{16} \geq 1$$
$$0 \geq 1 \quad \text{False}$$

The solution set is the region that does not contain (0, 0).

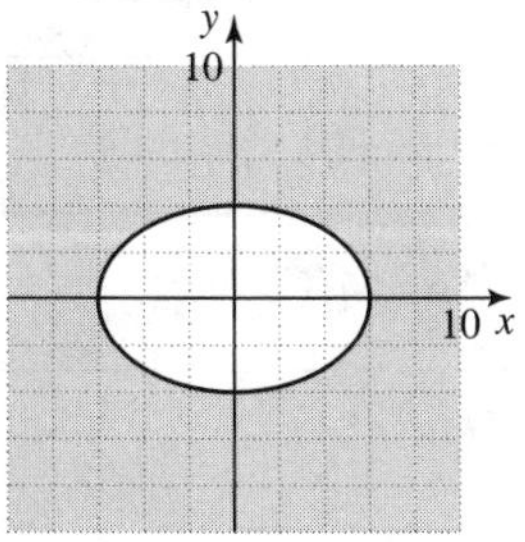

2. $16y^2 > 9x^2 + 144$

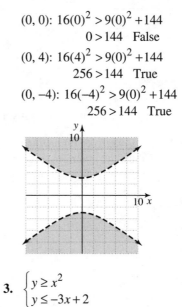

The related equation is $16y^2 = 9x^2 + 144.$

$$16y^2 - 9x^2 = 144$$
$$\frac{y^2}{9} - \frac{x^2}{16} = 1$$

Graph the hyperbola as a dashed curve.
Choose (0, 0), (0, 4), and (0, –4) as test points.

(0, 0): $16(0)^2 > 9(0)^2 + 144$
$0 > 144$ False

(0, 4): $16(4)^2 > 9(0)^2 + 144$
$256 > 144$ True

(0, –4): $16(-4)^2 > 9(0)^2 + 144$
$256 > 144$ True

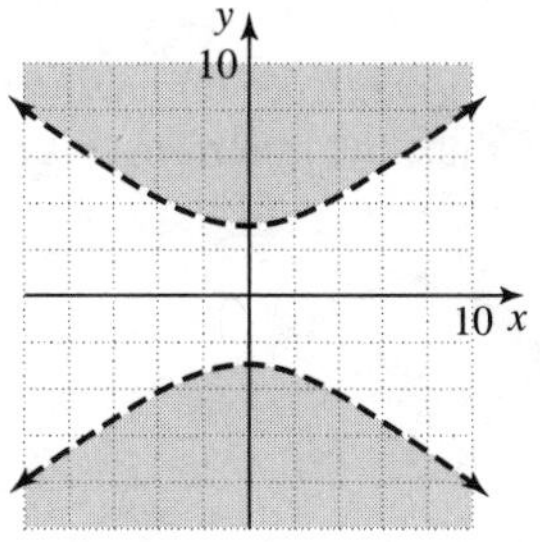

3. $\begin{cases} y \geq x^2 \\ y \leq -3x + 2 \end{cases}$

Solve the related system $\begin{cases} y = x^2 \\ y = -3x + 2 \end{cases}$.

Substitute –3*x* + 2 for *y* in the first equation.

$$x^2 = -3x + 2$$
$$x^2 + 3x - 2 = 0$$
$$x = \frac{-3 \pm \sqrt{3^2 - 4(1)(-2)}}{2(1)}$$
$$= \frac{-3 \pm \sqrt{17}}{2}$$
$$\approx 0.56 \text{ or } -3.56$$

$y = -3x + 2 \approx -3(0.56) + 2 = 0.32$
$y \approx -3(-3.56) + 2 = 12.68$

The points of intersection are approximately (0.56, 0.32) and (–3.56, 12.68).

Graph $y = x^2$ and $y = -3x + 2$ as solid curves.

The region of the solution set is above the parabola but below the line.

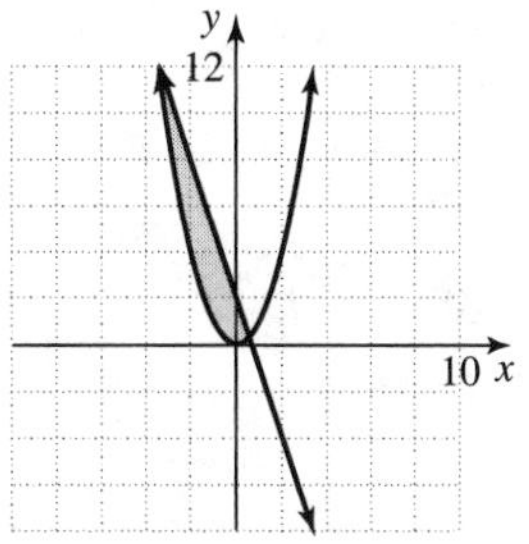

4. $\begin{cases} x^2 + y^2 < 16 \\ \dfrac{x^2}{4} - \dfrac{y^2}{9} < 1 \\ y < x + 3 \end{cases}$

Graph $x^2 + y^2 = 16$, $\dfrac{x^2}{4} - \dfrac{y^2}{9} = 1$, and $y = x + 3$.

The test point (0, 0) gives true statements for all three inequalities; thus, the innermost region is the solution set.

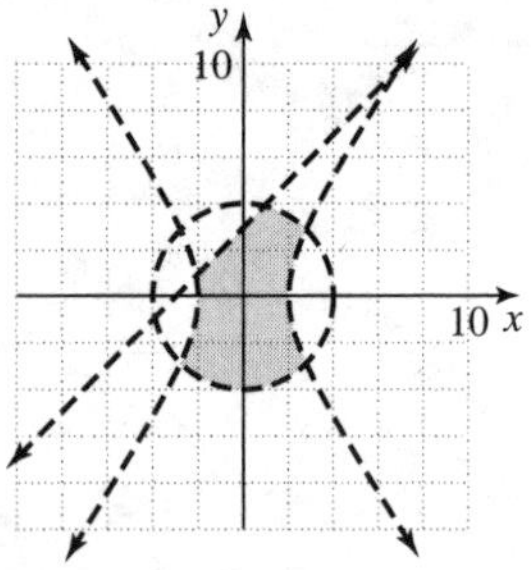

Vocabulary, Readiness & Video Check 10.4

1. For both, we graph the related equation to find the boundary and sketch it as a solid boundary for ≤ or ≥ and a dashed boundary for < or >; also we choose a test point (or test points) not on the boundary and shade that region if the test point is a solution of the original inequality.

2. A circle within a circle (either circle solid or dashed) where the inner circle is shaded inside and the outer circle is shaded outside; also, two non-intersecting circles (either circle solid or dashed), both shaded inside just to name a few examples.

Exercise Set 10.4

1. $y < x^2$

First graph the parabola as a dashed curve.

Test Point	$y < x^2$; Result
(0, 1)	$1 < 0^2$; False

Shade the region which does not contain (0, 1).

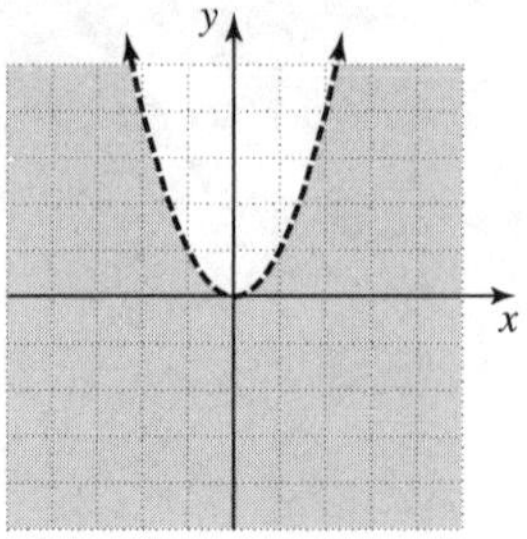

3. $x^2 + y^2 \geq 16$

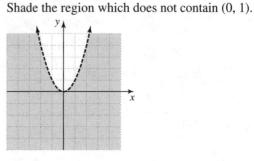

First graph the circle as a solid curve.

Test Point	$x^2 + y^2 \geq 16$; Result
(0, 0)	$0^2 + 0^2 \geq 16$; False

Shade the region which does not contain (0, 0).

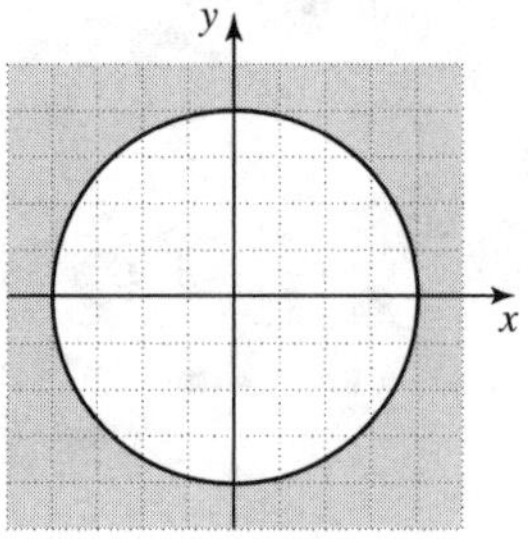

5. $\dfrac{x^2}{4} - y^2 < 1$

First graph the hyperbola as a dashed curve.

Test Point	$\frac{x^2}{4} - y^2 < 1$; Result
(−4, 0)	$\frac{(-4)^2}{4} - 0^2 < 1$; False
(0, 0)	$\frac{(0)^2}{4} - 0^2 < 1$; True
(4, 0)	$\frac{(4)^2}{4} - 0^2 < 1$; False

Shade the region containing (0, 0).

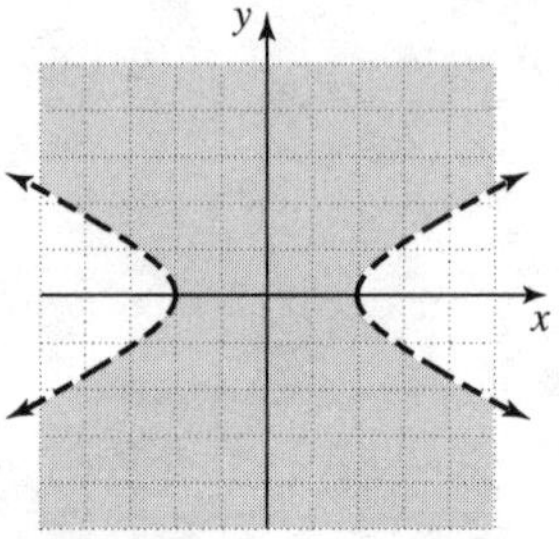

7. $y > (x-1)^2 - 3$

First graph the parabola as a dashed curve.

Test Point	$y > (x-1)^2 - 3$; Result
(0, 0)	$0 > (0-1)^2 - 3$; True

Shade the region containing (0, 0).

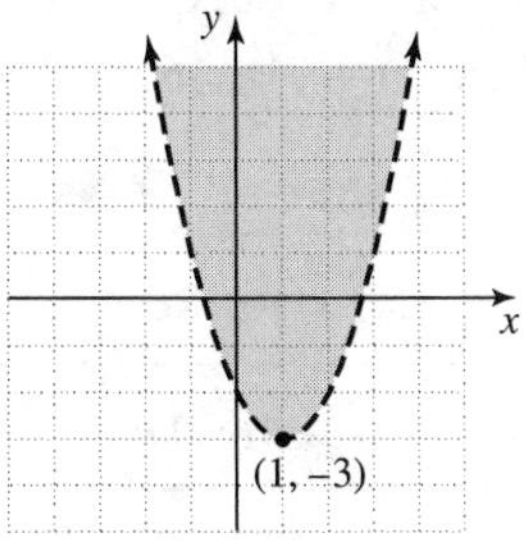

9. $x^2 + y^2 \le 9$

First graph the circle as a solid curve.

Test Point	$x^2 + y^2 \le 9$; Result
(0, 0)	$0^2 + 0^2 \le 9$; True

Shade the region containing (0, 0).

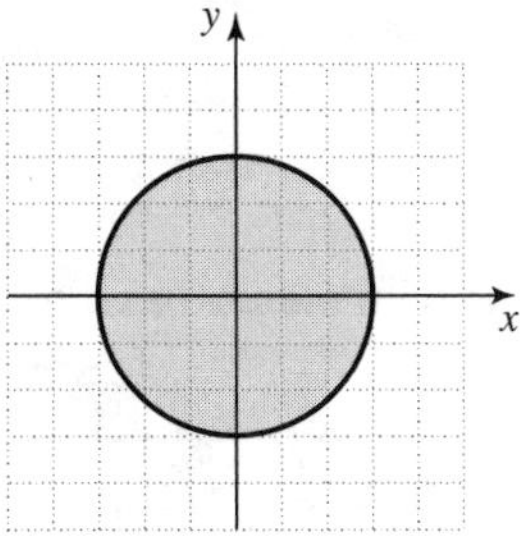

11. $y > -x^2 + 5$

First graph the parabola as a dashed curve.

Test Point	$y > -x^2 + 5$; Result
(0, 0)	$0 > -(0)^2 + 5$; False

Shade the region which does not contain (0, 0).

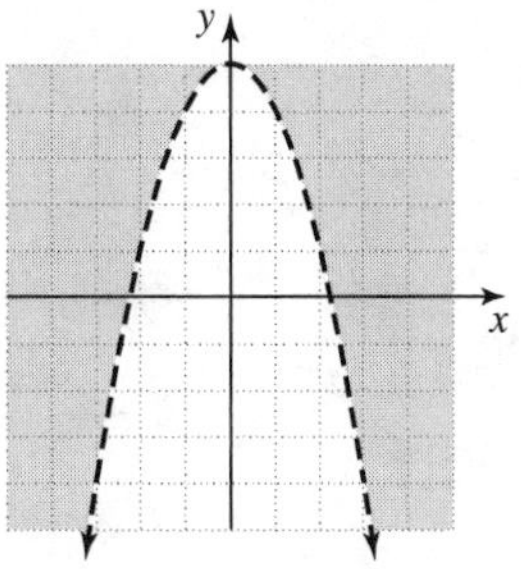

13. $\dfrac{x^2}{4} + \dfrac{y^2}{9} \le 1$

First graph the ellipse as a solid curve.

Test Point	$\frac{x^2}{4} + \frac{y^2}{9} \le 1$; Result
(0, 0)	$\frac{(0)^2}{4} + \frac{(0)^2}{9} \le 1$; True

Shade the region containing (0, 0).

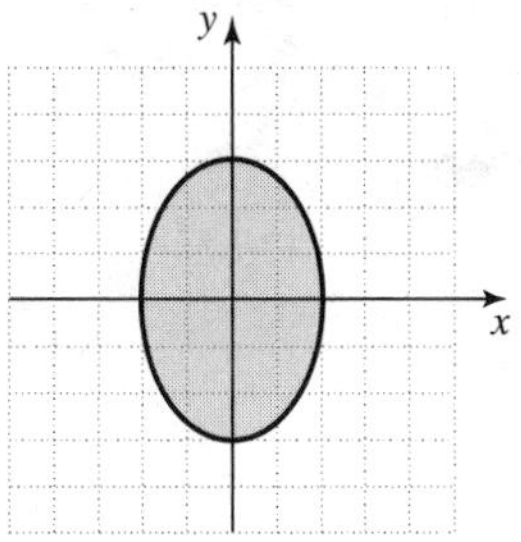

15. $\dfrac{y^2}{4} - x^2 \le 1$

First graph the hyperbola as solid curves.

Test Point	$\frac{y^2}{4} - x^2 \le 1$; Result
(0, −4)	$\frac{(-4)^2}{4} - 0^2 \le 1$; False
(0, 0)	$\frac{(0)^2}{4} - 0^2 \le 1$; True
(0, 4)	$\frac{(4)^2}{4} - 0^2 \le 1$; False

Shade the region containing (0, 0).

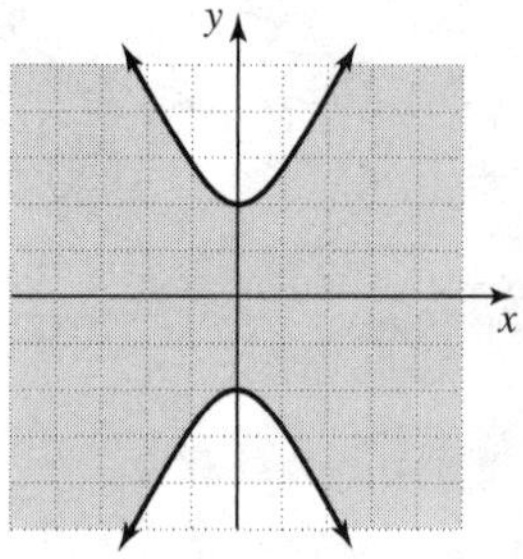

17. $y < (x-2)^2 + 1$

First graph the parabola as a dashed curve.

Test Point	$y < (x-2)^2 + 1$; Result
(0, 0)	$0 < (0-2)^2 + 1$; True

Shade the region containing (0, 0).

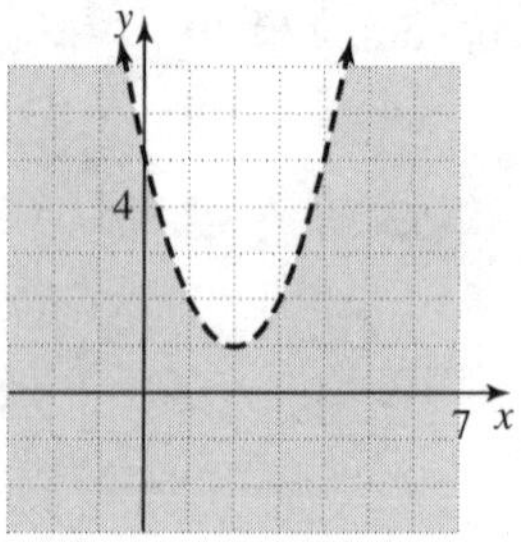

19. $y \le x^2 + x - 2$

First graph the parabola as a solid curve.

Test Point	$y \le x^2 + x - 2$; Result
(0, 0)	$0 \le (0)^2 + (0) - 2$; False

Shade the region which does not contain (0, 0).

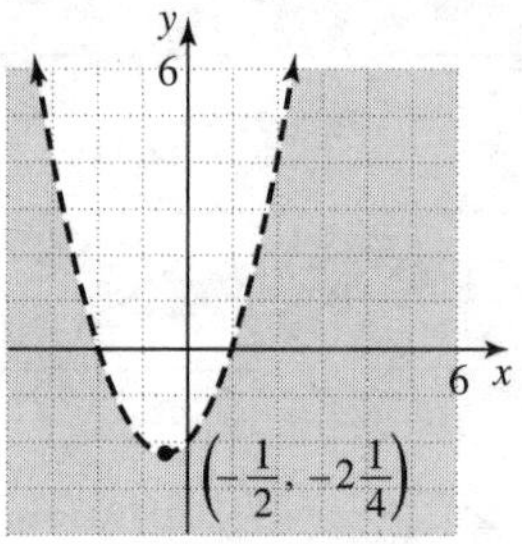

21. $\begin{cases} 4x + 3y \ge 12 \\ x^2 + y^2 < 16 \end{cases}$

First graph $4x + 3y = 12$ as a solid line.

Test Point	$4x + 3y \ge 12$; Result
(0, 0)	$4(0) + 3(0) \ge 12$; False

Shade the region which does not contain (0, 0). Next, graph the circle $x^2 + y^2 = 16$ as a dashed curve.

Test Point	$x^2 + y^2 < 16$; Result
(0, 0)	$0^2 + 0^2 < 16$; True

Shade the region containing (0, 0). The solution to the system is the intersection.

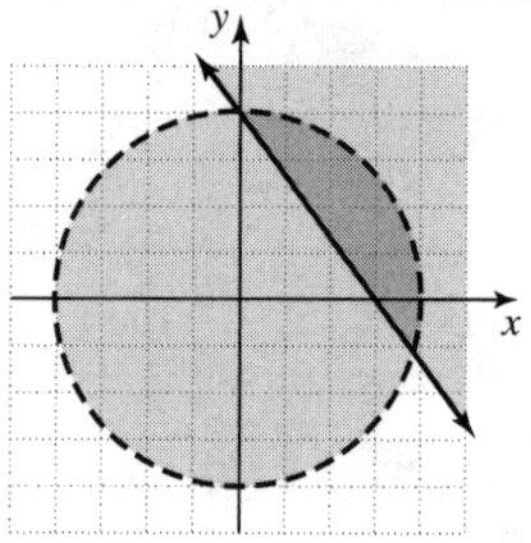

23. $\begin{cases} x^2 + y^2 \le 9 \\ x^2 + y^2 \ge 1 \end{cases}$

First graph the circle with radius 3 as a solid curve.

Test Point	$x^2 + y^2 \le 9$; Result
(0, 0)	$0^2 + 0^2 \le 9$; True

Shade the region containing (0, 0). Next, graph the circle with 1 as a dashed curve.

Test Point	$x^2 + y^2 \ge 1$; Result
(0, 0)	$0^2 + 0^2 \ge 1$; False

Shade the region which does not contain (0, 0). The solution to the system is the intersection.

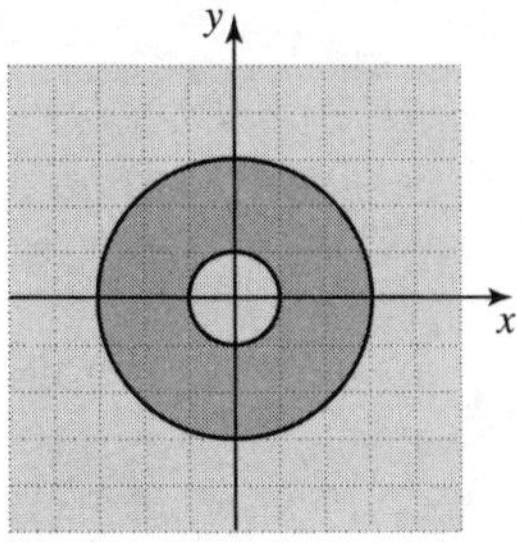

25. $\begin{cases} y > x^2 \\ y \geq 2x+1 \end{cases}$

First graph the parabola as a dashed curve.

Test Point	$y > x^2$; Result
(0, 1)	$1 > 0^2$; True

Shade the region containing (0, 1). Next, graph $y = 2x+1$ as a solid line.

Test Point	$y \geq 2x + 1$; Result
(0, 0)	$0 \geq 2(0) + 1$; False

Shade the region which does not contain (0, 0). The solution to the system is the intersection.

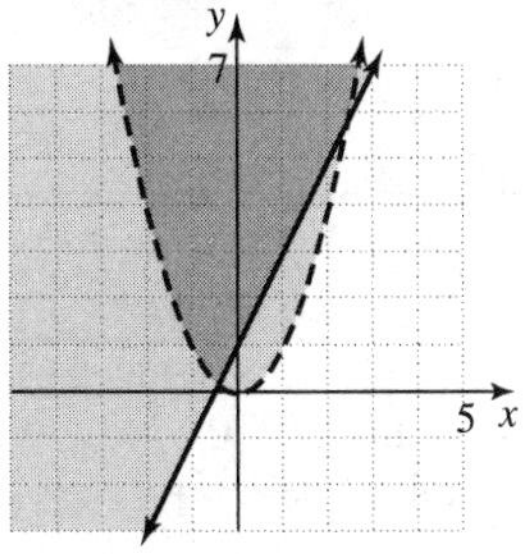

27. $\begin{cases} x^2 + y^2 > 9 \\ y > x^2 \end{cases}$

First graph the circle as a dashed curve.

Test Point	$x^2 + y^2 > 9$; Result
(0, 0)	$0^2 + 0^2 > 9$; False

Shade the region which does not contain (0, 0). Next, graph the parabola as a dashed curve.

Test Point	$y > x^2$; Result
(0, 1)	$1 > 0^2$; True

Shade the region containing (0, 1). The solution to the system is the intersection.

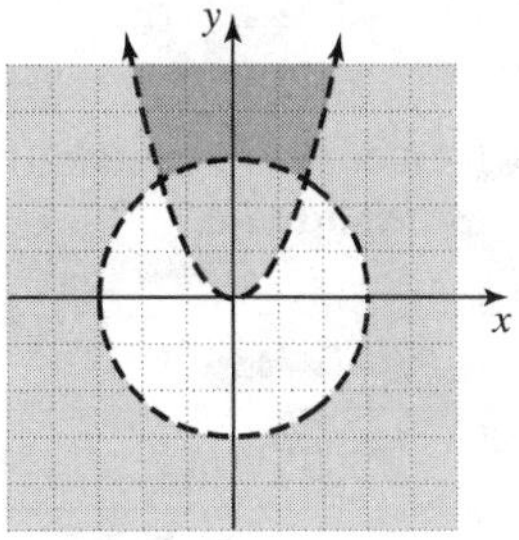

29. $\begin{cases} \dfrac{x^2}{4} + \dfrac{y^2}{9} \geq 1 \\ x^2 + y^2 \geq 4 \end{cases}$

First graph the ellipse as a solid curve.

Test Point	$\frac{x^2}{4} + \frac{y^2}{9} \geq 1$; Result
(0, 0)	$\frac{0^2}{4} + \frac{0^2}{9} \geq 1$; False

Shade the region which does not contain (0, 0). Next, graph the circle as a solid curve.

Test Point	$x^2 + y^2 \geq 4$; Result
(0, 0)	$0^2 + 0^2 \geq 4$; False

Shade the region which does not contain (0, 0). The solution to the system is the intersection.

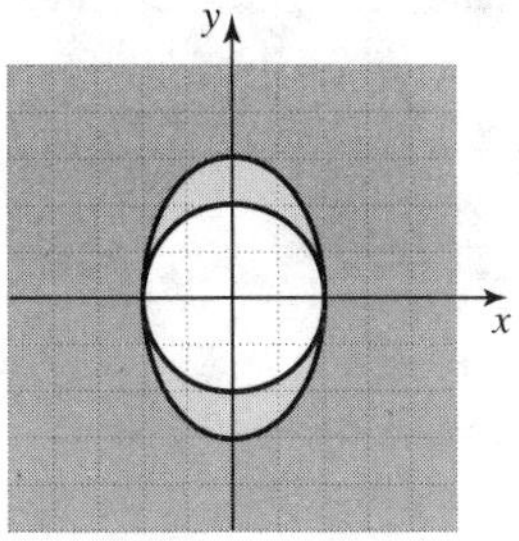

31. $\begin{cases} x^2 - y^2 \geq 1 \\ y \geq 0 \end{cases}$

First graph the hyperbola as solid curves.

Test Point	$x^2 - y^2 \geq 1$; Result
(−2, 0)	$(-2)^2 - 0^2 \geq 1$; True
(0, 0)	$0^2 - 0^2 \geq 1$; False
(2, 0)	$2^2 - 0^2 \geq 1$; True

Shade the region which does not contain (0, 0). Next, graph $y = 0$ as a solid line.

Test Point	$y > 0$; Result
(0, 1)	$1 \geq 0$; True

Shade the region containing (0, 1). The solution to the system is the intersection.

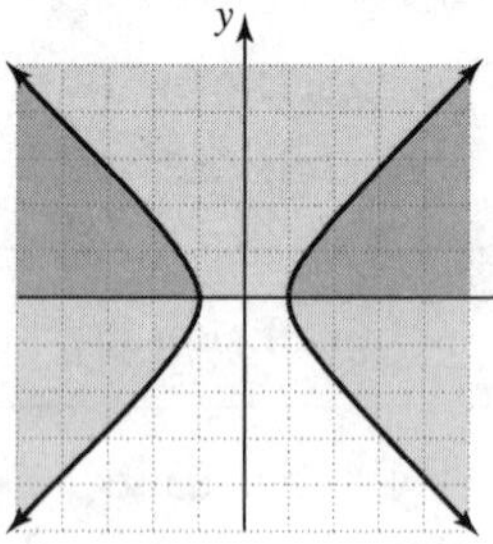

33. $\begin{cases} x + y \geq 1 \\ 2x + 3y < 1 \\ x > -3 \end{cases}$

First graph $x + y = 1$ as a solid line.

Test Point	$x + y \geq 1$; Result
(0, 0)	$0 + 0 \geq 1$; False

Shade the region which does not contain (0, 0). Next, graph $2x + 3y = 1$ as a dashed line.

Test Point	$2x + 3y < 1$; Result
(0, 0)	$2(0)\,1 + 3(0) < 1$; True

Shade the region containing (0, 0). Now graph the line $x = -3$ as a dashed line.

Test Point	$x > -3$; Result
(0, 0)	$0 > -3$; True

Shade the region containing (0, 0). The solution to the system is the intersection.

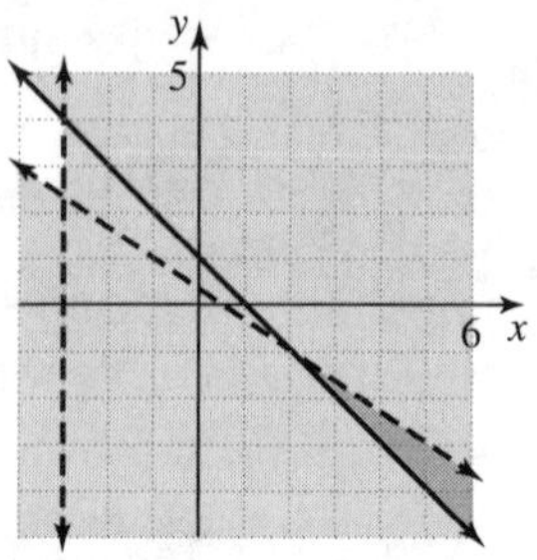

35. $\begin{cases} x^2 - y^2 < 1 \\ \dfrac{x^2}{16} + y^2 \leq 1 \\ x \geq -2 \end{cases}$

First graph the hyperbola as dashed curves.

Test Point	$x^2 - y^2 < 1$; Result
(−2, 0)	$(-2)^2 - 0^2 < 1$; False
(0, 0)	$0^2 - 0^2 < 1$; True
(2, 0)	$2^2 - 0^2 < 1$; False

Shade the region containing (0, 0). Next, graph the ellipse as a solid curve.

Test Point	$\frac{x^2}{16} + y^2 \leq 1$; Result
(0, 0)	$\frac{0^2}{16} + 0^2 \leq 1$; True

Shade the region containing (0, 0). Now graph the line $x = -2$ as a solid line.

Test Point	$x \geq -2$; Result
(0, 0)	$0 \geq -2$; True

Shade the region containing (0, 0). The solution to the system is the intersection.

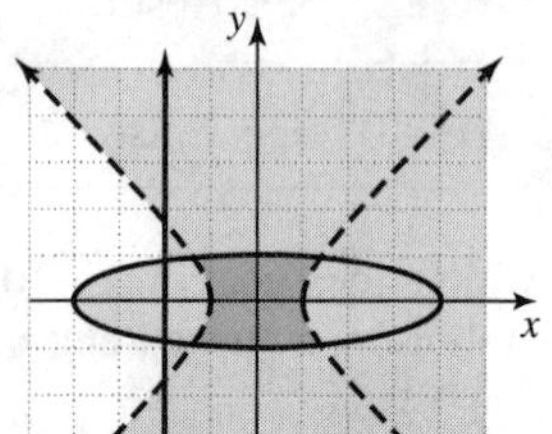

37. This is not a function because a vertical line can cross the graph in more than one place.

39. This is a function because a vertical line can cross the graph in no more than one place.

41. $f(x) = 3x^2 - 2$
$f(-1) = 3(-1)^2 - 2 = 3 - 2 = 1$

43. $f(x) = 3x^2 - 2$
$f(a) = 3(a)^2 - 2 = 3a^2 - 2$

45. answers may vary

47. $\begin{cases} y \le x^2 \\ y \ge x + 2 \\ x \ge 0 \\ y \ge 0 \end{cases}$

First graph $y = x^2$ as a solid curve.

Test Point	$y \le x^2$; Result
(0, 1)	$1 \le 0^2$; False

Shade the region which does not contain (0, 1). Next, graph $y = x + 2$ as a solid line.

Test Point	$y \ge x + 2$; Result
(0, 0)	$0 \ge 0 + 2$; False

Shade the region which does not contain (0, 0). Next graph the line $x = 0$ as a solid line, and shade to the right. Now graph the line $y = 0$ as a solid line, and shade above. The solution to the system is the intersection.

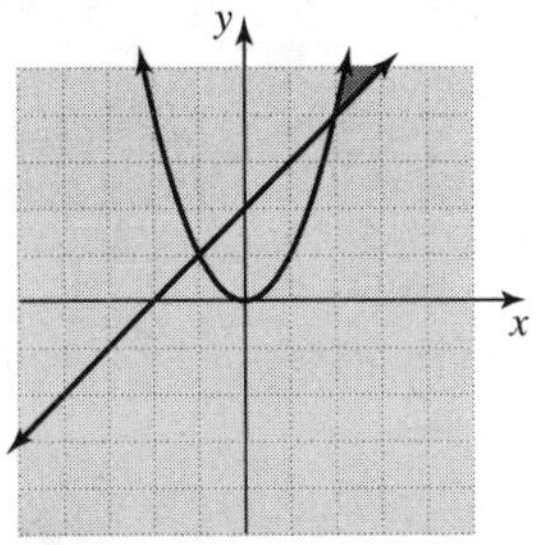

Chapter 10 Vocabulary Check

1. A circle is the set of all points in a plane that are the same distance from a fixed point, called the center.

2. A nonlinear system of equations is a system of equations at least one of which is not linear.

3. An ellipse is the set of points in a plane such that the sum of the distances of those points from two fixed points is a constant.

4. In a circle, the distance from the center to a point of the circle is called its radius.

5. A hyperbola is the set of points in a plane such that the absolute value of the difference of the distance from two fixed points is constant.

6. The circle, parabola, ellipse, and hyperbola are called the conic sections.

7. For a parabola that opens upward, the lowest point is the vertex.

8. Twice a circle's radius is its diameter.

Chapter 10 Review

1. center (–4, 4), radius 3
$[x-(-4)]^2 + (y-4)^2 = 3^2$
$(x+4)^2 + (y-4)^2 = 9$

2. center (5, 0), radius 5
$(x-5)^2 + (y-0)^2 = 5^2$
$(x-5)^2 + y^2 = 25$

3. center (–7, –9), radius $\sqrt{11}$
$[x-(-7)]^2 + [y-(-9)]^2 = \left(\sqrt{11}\right)^2$
$(x+7)^2 + (y+9)^2 = 11$

4. center (0, 0), radius $\frac{7}{2}$
$(x-0)^2 + (y-0)^2 = \left(\frac{7}{2}\right)^2$
$x^2 + y^2 = \frac{49}{4}$

5. $x^2 + y^2 = 7$

Circle; center (0, 0), radius $r = \sqrt{7}$

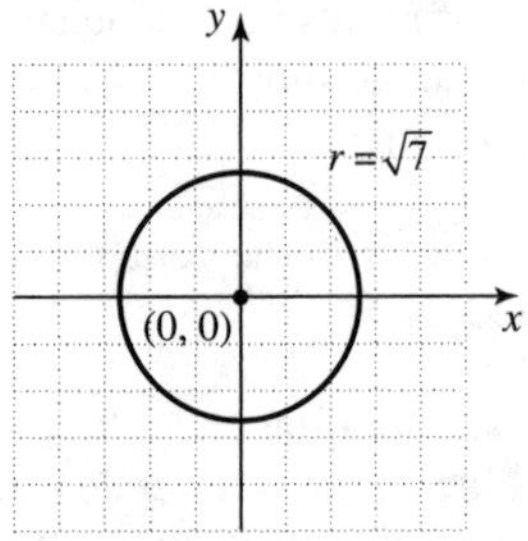

6. $x = 2(y-5)^2 + 4$

Parabola; vertex: (4, 5)

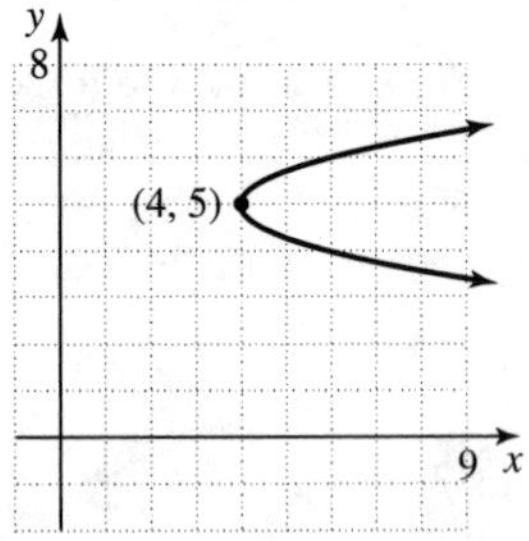

7. $x = -(y+2)^2 + 3$

Parabola; vertex: (3, –2)

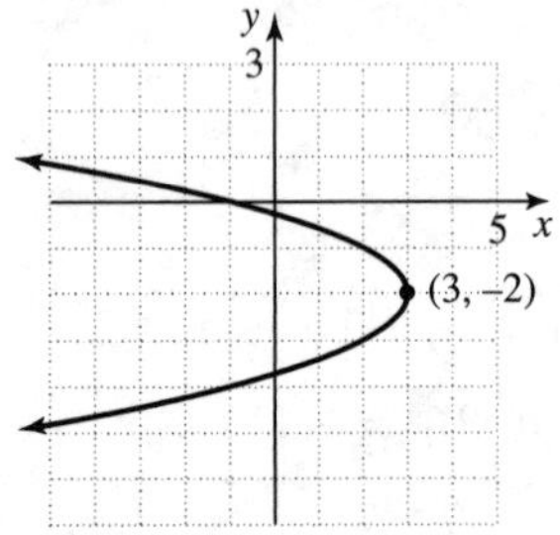

8. $(x-1)^2 + (y-2)^2 = 4$

Circle; center (1, 2), radius $r = \sqrt{4} = 2$

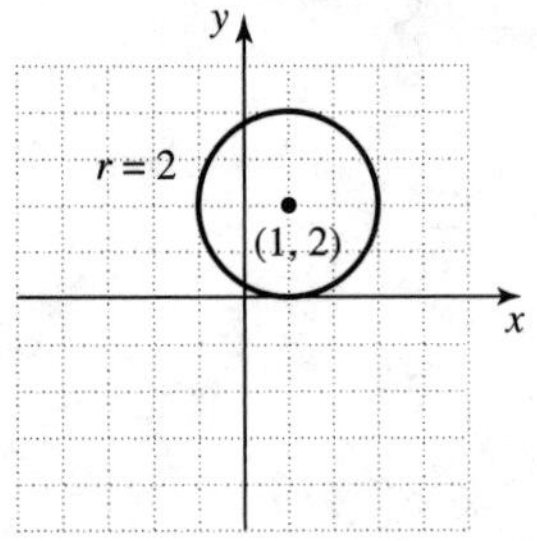

9. $y = -x^2 + 4x + 10$

Parabola; $x = \dfrac{-b}{2a} = \dfrac{-4}{2(-1)} = 2$

$y = -(2)^2 + 4(2) + 10 = 14$

Vertex: (2, 14)

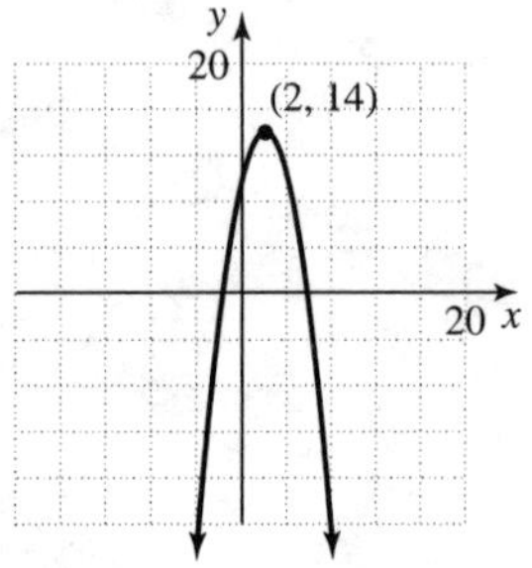

10. $x = -y^2 - 4y + 6$

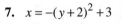

Parabola; $y = \dfrac{-b}{2a} = \dfrac{-(-4)}{2(-1)} = -2$

$x = -(-2)^2 - 4(-2) + 6 = 10$

Vertex: (10, –2)

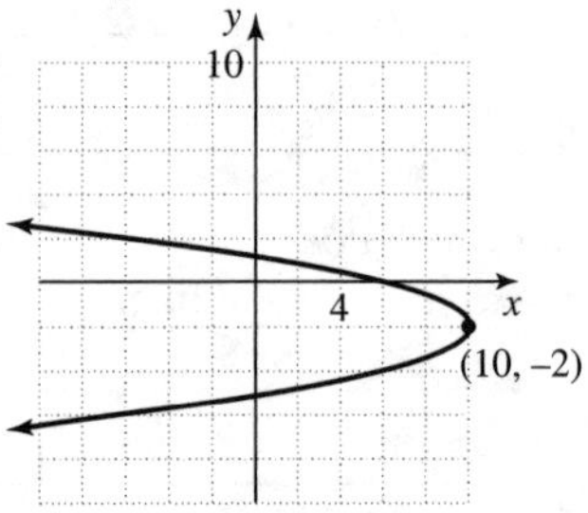

11. $x = \dfrac{1}{2}y^2 + 2y + 1$

Parabola; $y = \dfrac{-b}{2a} = \dfrac{-2}{2\left(\frac{1}{2}\right)} = -2$

$x = \dfrac{1}{2}(-2)^2 + 2(-2) + 1 = -1$

Vertex: (–1, –2)

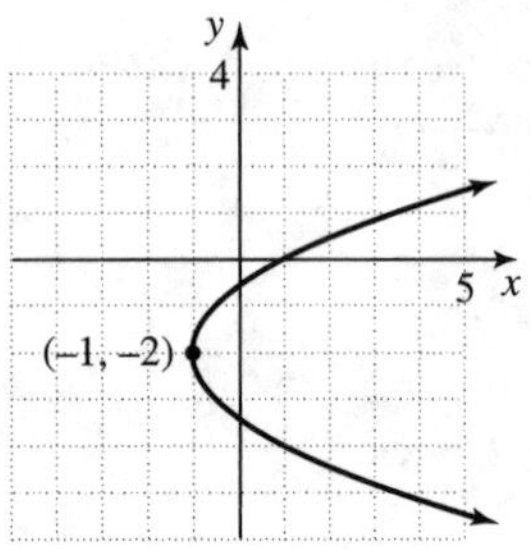

12. $y = -3x^2 + \frac{1}{2}x + 4$

Parabola; $x = \frac{-b}{2a} = \frac{-\frac{1}{2}}{2(-3)} = \frac{1}{12}$

$$y = -3\left(\frac{1}{12}\right)^2 + \frac{1}{2}\left(\frac{1}{12}\right) + 4 = \frac{193}{48}$$

Vertex: $\left(\frac{1}{12}, \frac{193}{48}\right)$

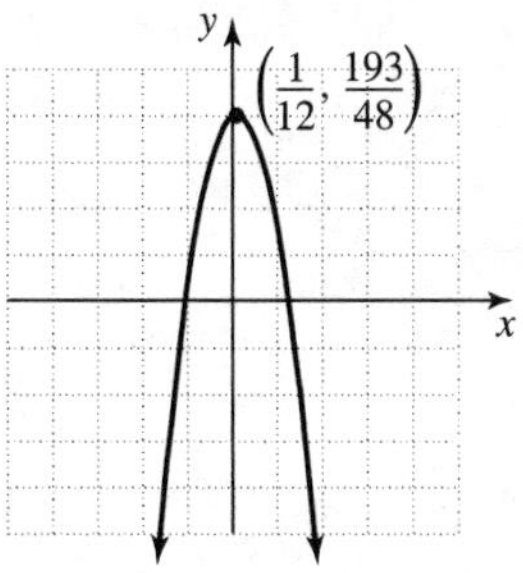

13.

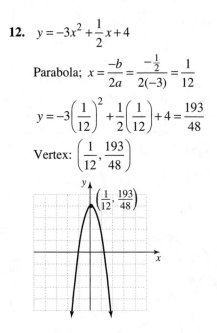

$$x^2 + y^2 + 2x + y = \frac{3}{4}$$
$$(x^2 + 2x) + (y^2 + y) = \frac{3}{4}$$
$$(x^2 + 2x + 1) + \left(y^2 + y + \frac{1}{4}\right) = \frac{3}{4} + 1 + \frac{1}{4}$$
$$(x+1)^2 + \left(y + \frac{1}{2}\right)^2 = 2$$

Circle; center $\left(-1, -\frac{1}{2}\right)$, radius $r = \sqrt{2}$

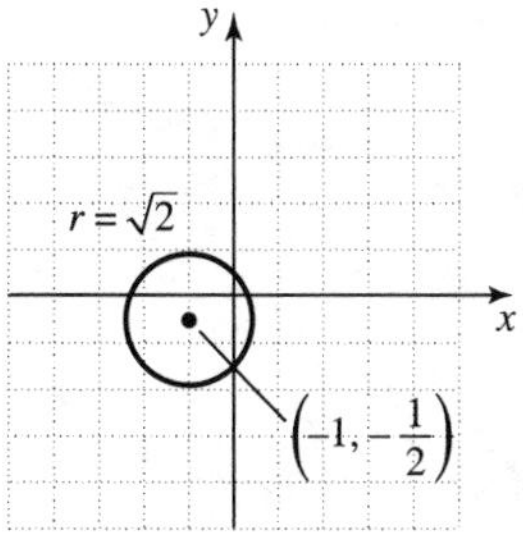

14.

$$x^2 + y^2 - 3y = \frac{7}{4}$$
$$x^2 + \left(y^2 - 3y + \frac{9}{4}\right) = \frac{7}{4} + \frac{9}{4}$$
$$x^2 + \left(y - \frac{3}{2}\right)^2 = 4$$

Circle; center $\left(0, \frac{3}{2}\right)$, radius $r = \sqrt{4} = 2$

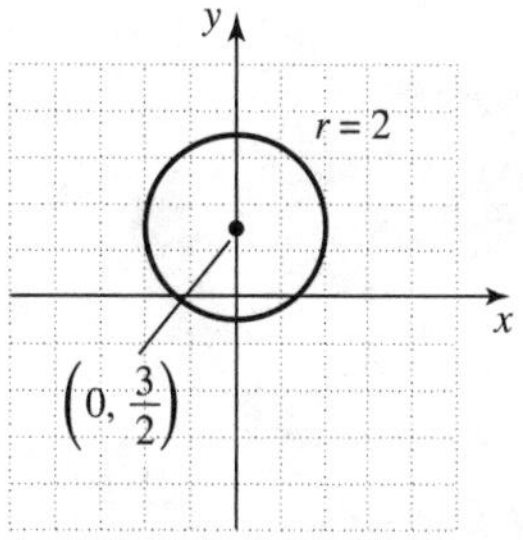

15.

$$4x^2 + 4y^2 + 16x + 8y = 1$$
$$(x^2 + 4x) + (y^2 + 2y) = \frac{1}{4}$$
$$(x^2 + 4x + 4) + (y^2 + 2y + 1) = \frac{1}{4} + 4 + 1$$
$$(x+2)^2 + (y+1)^2 = \frac{21}{4}$$

Circle; center (–2, –1), radius $r = \sqrt{\frac{21}{4}} = \frac{\sqrt{21}}{2}$

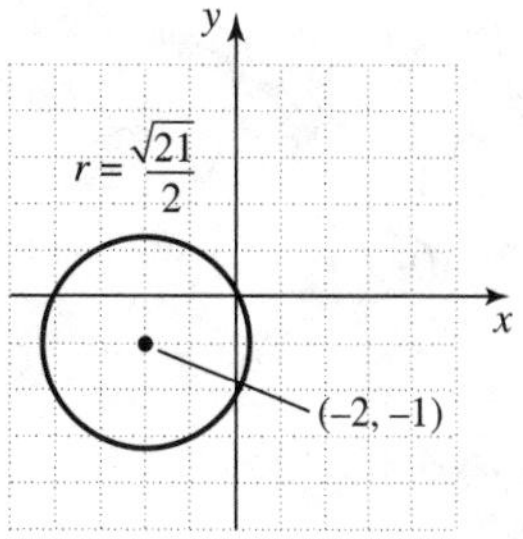

16.

$$3x^2 + 3y^2 + 18x - 12y = -12$$
$$x^2 + y^2 + 6x - 4y = -4$$
$$x^2 + 6x + 9 + y^2 - 4y + 4 = -4 + 9 + 4$$
$$(x+3)^2 + (y-2)^2 = 9$$
$$(x+3)^2 + (y-2)^2 = 3^2$$

$C(-3, 2)$, $r = 3$

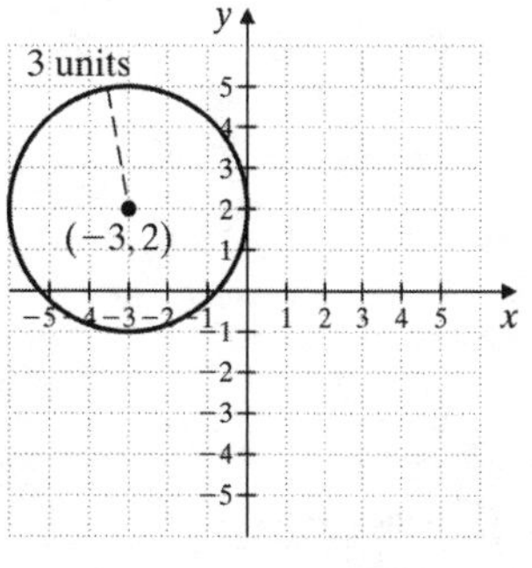

17. $x^2 - \dfrac{y^2}{4} = 1$

$\dfrac{x^2}{1^2} - \dfrac{y^2}{2^2} = 1$

$a = 1, b = 2$

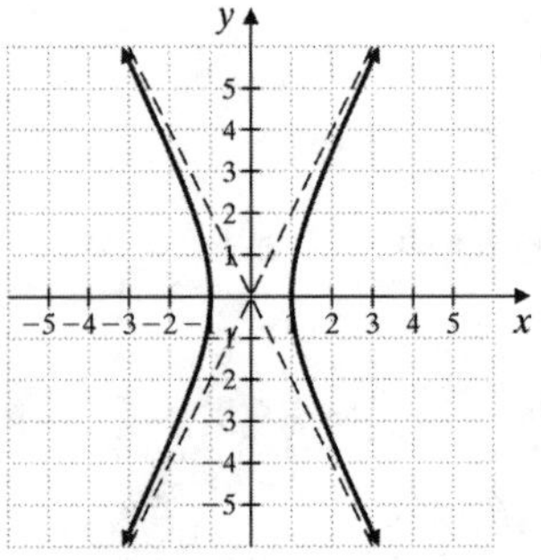

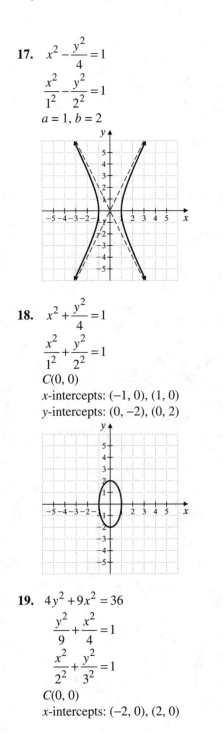

18. $x^2 + \dfrac{y^2}{4} = 1$

$\dfrac{x^2}{1^2} + \dfrac{y^2}{2^2} = 1$

$C(0, 0)$

x-intercepts: $(-1, 0), (1, 0)$

y-intercepts: $(0, -2), (0, 2)$

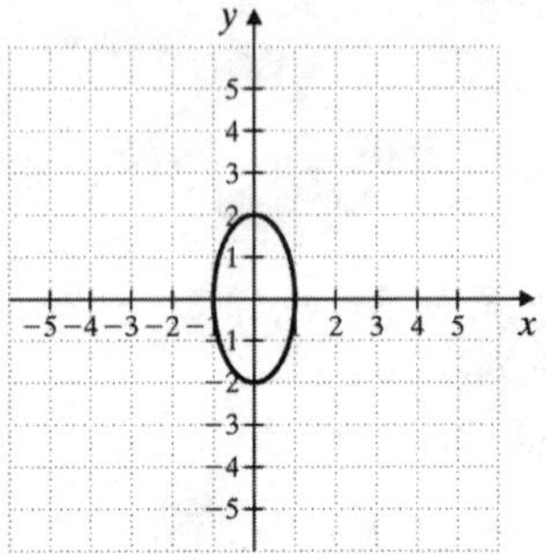

19. $4y^2 + 9x^2 = 36$

$\dfrac{y^2}{9} + \dfrac{x^2}{4} = 1$

$\dfrac{x^2}{2^2} + \dfrac{y^2}{3^2} = 1$

$C(0, 0)$

x-intercepts: $(-2, 0), (2, 0)$

y-intercepts: $(0, -3), (0, 3)$

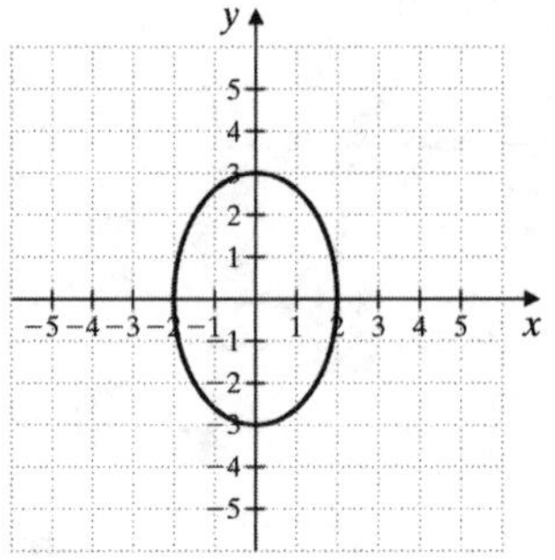

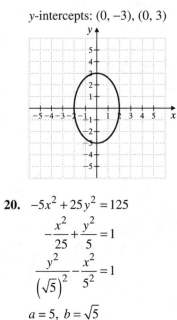

20. $-5x^2 + 25y^2 = 125$

$-\dfrac{x^2}{25} + \dfrac{y^2}{5} = 1$

$\dfrac{y^2}{\left(\sqrt{5}\right)^2} - \dfrac{x^2}{5^2} = 1$

$a = 5,\ b = \sqrt{5}$

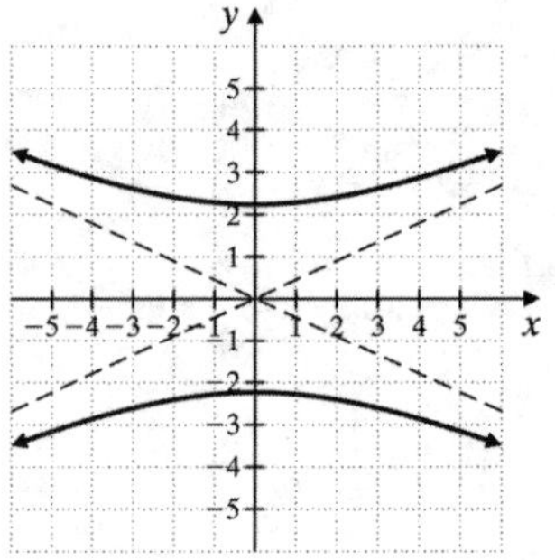

21. $x^2 - y^2 = 1$

$\dfrac{x^2}{1^2} - \dfrac{y^2}{1^2} = 1$

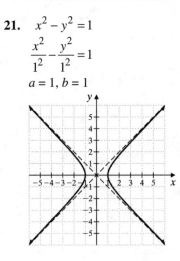

$a = 1, b = 1$

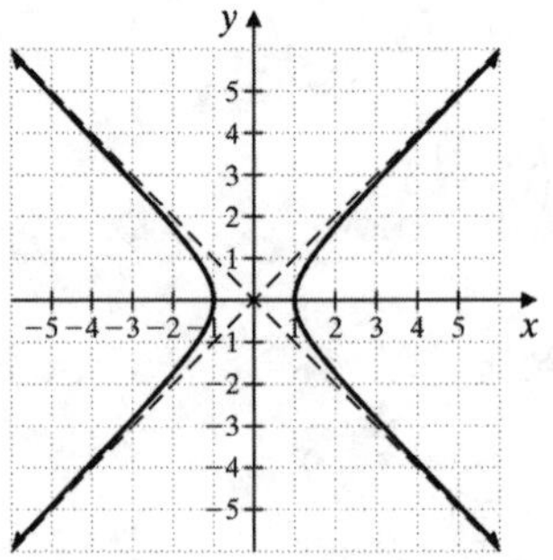

22. $\frac{(x+3)^2}{9}+\frac{(y-4)^2}{25}=1$

$\frac{(x+3)^2}{3^2}+\frac{(y-4)^2}{5^2}=1$

$C(-3, 4)$
other points:
$(-3 + 3, 4)$, or $(0, 4)$
$(-3 - 3, 4)$, or $(-6, 4)$
$(-3, 4 - 5)$, or $(-3, -1)$
$(-3, 4 + 5)$, or $(-3, 9)$

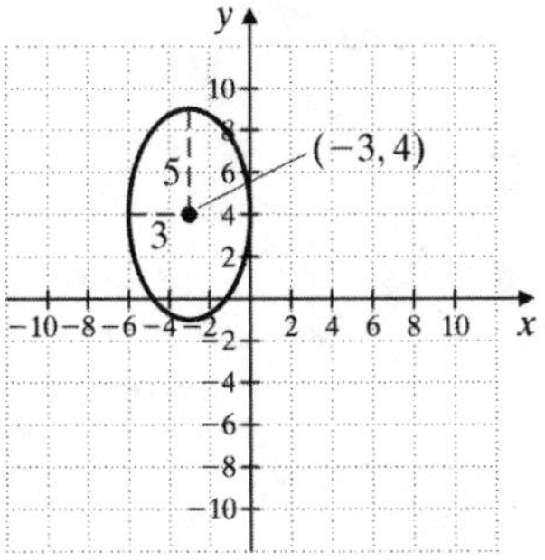

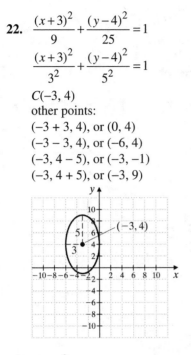

23. $y = x^2 + 9$
$V(0, 9)$
The parabola opens upward.

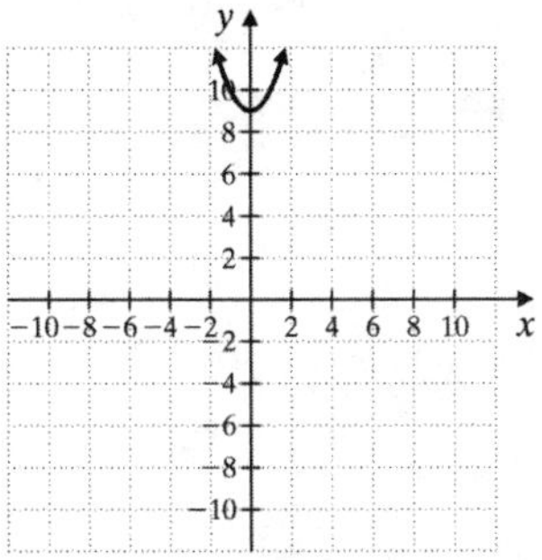

24. $36y^2 - 49x^2 = 1764$

$\frac{y^2}{49}-\frac{x^2}{36}=1$

$\frac{y^2}{7^2}-\frac{x^2}{6^2}=1$

$a = 6, b = 7$

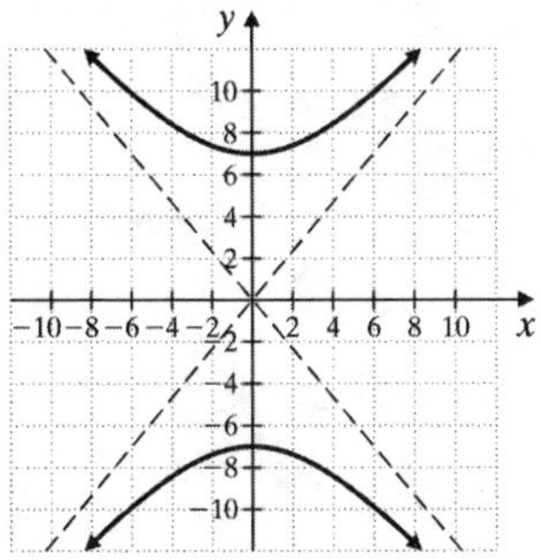

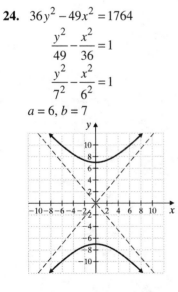

25. $x = 4y^2 - 16$
$V(-16, 0)$
The parabola opens to the right.

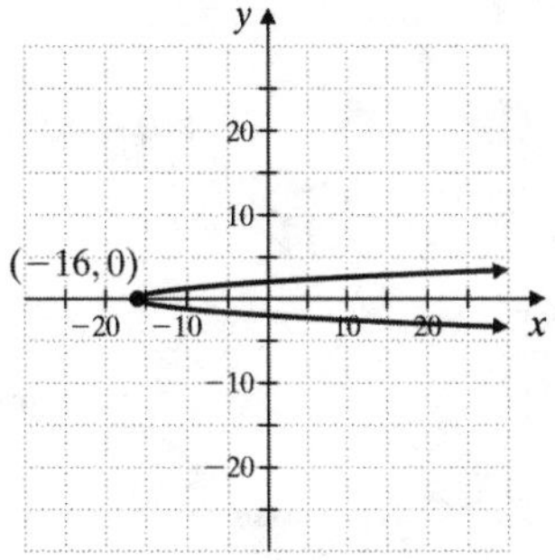

26. $y = x^2 + 4x + 6$

$y = (x^2 + 4x + 4) + 6 - 4$

$y = (x + 2)^2 + 2$

$V(-2, 2)$
The parabola opens upward.

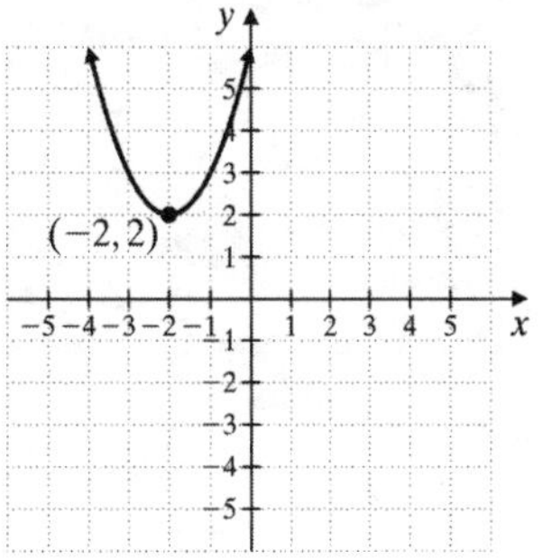

27. $y^2 + 2(x-1)^2 - 8 = 0$

$y^2 + 2(x-1)^2 = 8$

$\frac{y^2}{8}+\frac{(x-1)^2}{4}=1$

$\frac{y^2}{\left(\sqrt{8}\right)^2}+\frac{(x-1)^2}{2^2}=1$

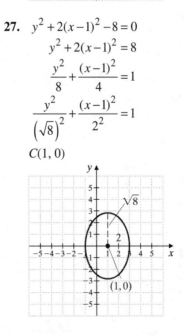

$C(1, 0)$

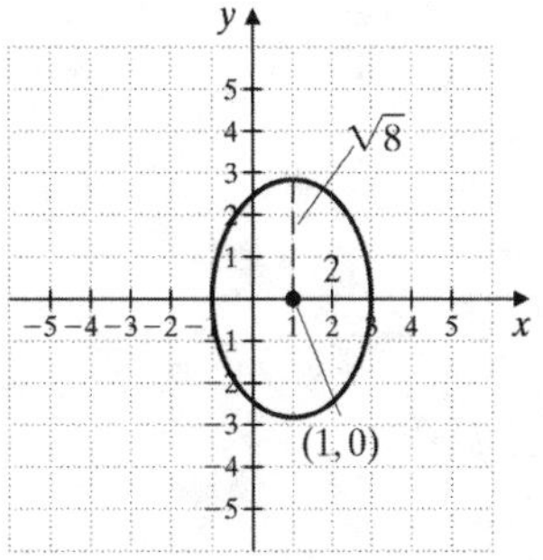

28. $x - 4y = y^2$

$$x = y^2 + 4y$$
$$x = y^2 + 4y + 4 - 4$$
$$x = (y+2)^2 - 4$$

$V(-4, -2)$

The parabola opens to the right.

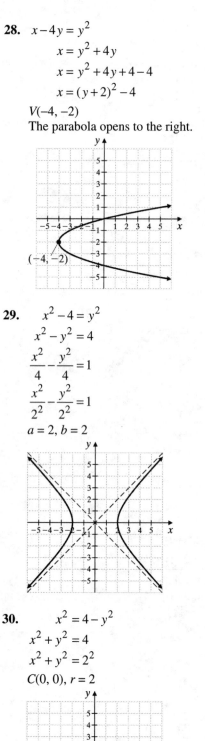

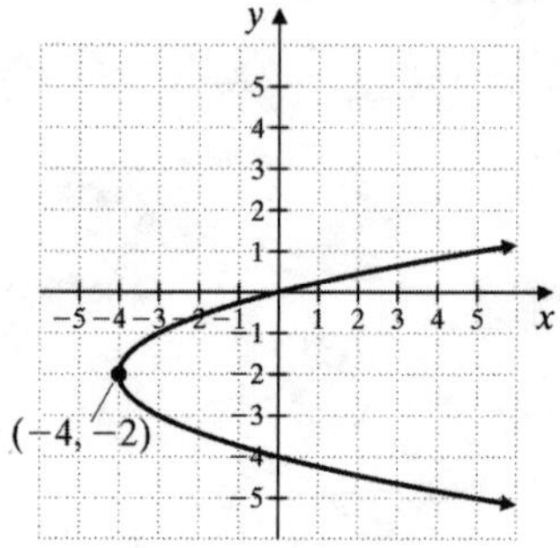

29. $x^2 - 4 = y^2$

$$x^2 - y^2 = 4$$
$$\frac{x^2}{4} - \frac{y^2}{4} = 1$$
$$\frac{x^2}{2^2} - \frac{y^2}{2^2} = 1$$

$a = 2$, $b = 2$

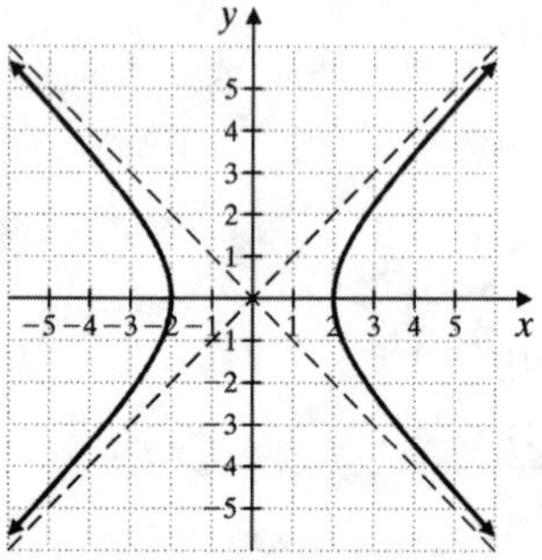

30. $x^2 = 4 - y^2$

$$x^2 + y^2 = 4$$
$$x^2 + y^2 = 2^2$$

$C(0, 0)$, $r = 2$

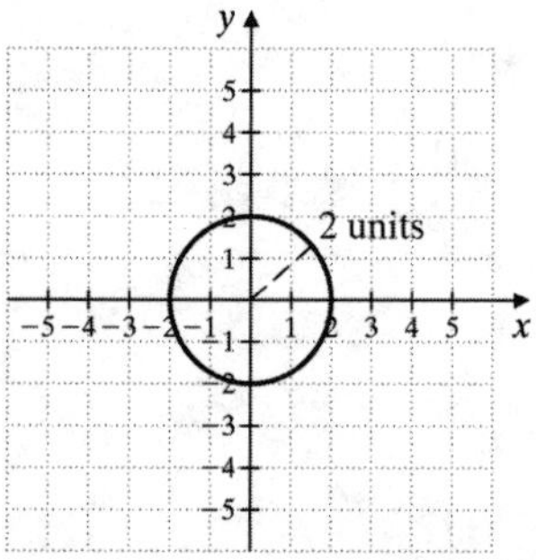

31. $36y^2 = 576 + 16x^2$

$$36y^2 - 16x^2 = 576$$
$$\frac{y^2}{16} - \frac{x^2}{36} = 1$$
$$\frac{y^2}{4^2} - \frac{x^2}{6^2} = 1$$

$a = 6$, $b = 4$

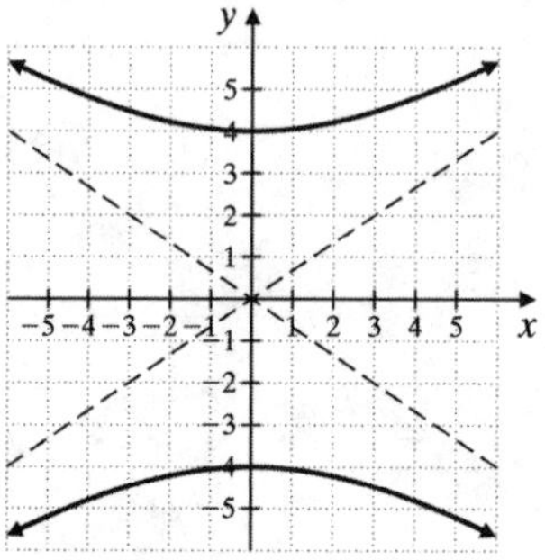

32. $3(x-7)^2 + 3(y+4)^2 = 1$

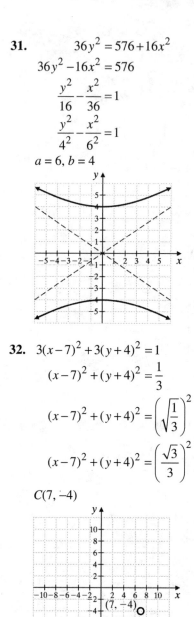

$$(x-7)^2 + (y+4)^2 = \frac{1}{3}$$
$$(x-7)^2 + (y+4)^2 = \left(\sqrt{\frac{1}{3}}\right)^2$$
$$(x-7)^2 + (y+4)^2 = \left(\frac{\sqrt{3}}{3}\right)^2$$

$C(7, -4)$

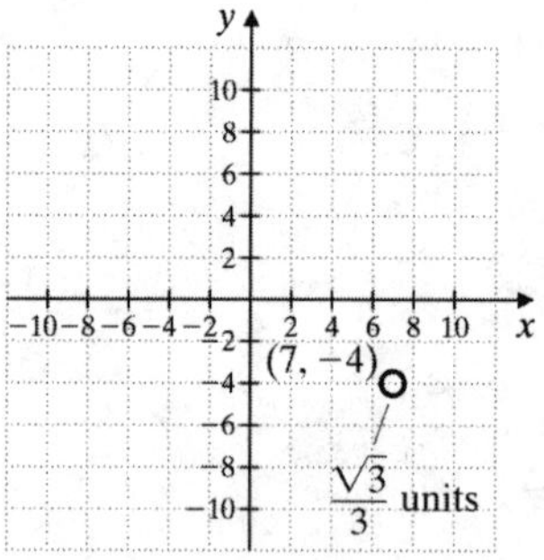

33. $\begin{cases} y = 2x-4 \\ y^2 = 4x \end{cases}$

Substitute.

$$\begin{aligned} y^2 &= 4x \\ (2x-4)^2 &= 4x \\ 4x^2 - 16x + 16 &= 4x \\ 4x^2 - 16x + 16 &= 4x \\ 4x^2 - 20x + 16 &= 0 \\ 4(x^2 - 5x + 4) &= 0 \\ 4(x-4)(x-1) &= 0 \end{aligned}$$

$x - 4 = 0$ or $x - 1 = 0$

$x = 4$ $\quad$ $x = 1$

$x = 4$: $y = 2(4) - 4 = 4$

$x = 1$: $y = 2(1) - 4 = -2$

The solutions are (1, −2) and (4, 4).

34. $\begin{cases} x^2 + y^2 = 4 \\ x - y = 4 \end{cases}$

Solve equation 2 for *x*.

$x = 4 + y$

Substitute.

$$\begin{aligned} x^2 + y^2 &= 4 \\ (4+y)^2 + y^2 &= 4 \\ 16 + 8y + y^2 + y^2 &= 4 \\ 2y^2 + 8y + 12 &= 0 \end{aligned}$$

$a = 2$, $b = 8$, $c = 12$

$$y = \frac{-8 \pm \sqrt{8^2 - 4(2)(12)}}{2(2)} = \frac{-8 \pm \sqrt{-32}}{4}$$

Since $\sqrt{-32}$ is not a real number, there is no solution. The solution set is ∅.

35. $\begin{cases} y = x+2 \\ y = x^2 \end{cases}$

Substitute.

$$\begin{aligned} x + 2 &= x^2 \\ 0 &= x^2 - x - 2 \\ 0 &= (x-2)(x+1) \end{aligned}$$

$x - 2 = 0$ or $x + 1 = 0$

$x = 2$ $\quad$ $x = -1$

$x = 2$: $y = 2^2 = 4$

$x = -1$: $y = (-1)^2 = 1$

The solutions are (−1, 1) and (2, 4).

36. $\begin{cases} 4x - y^2 = 0 \\ 2x^2 + y^2 = 16 \end{cases}$

Add the equations.

$$\begin{aligned} 2x^2 + 4x &= 16 \\ 2x^2 + 4x - 16 &= 0 \\ 2(x^2 + 2x - 8) &= 0 \\ 2(x+4)(x-2) &= 0 \end{aligned}$$

$x + 4 = 0$ or $x - 2 = 0$

$x = -4$ $\quad$ $x = 2$

$x = -4$: $4(-4) - y^2 = 0$

$$\begin{aligned} -16 &= y^2 \\ \pm\sqrt{-16} &= y \end{aligned}$$

$x = 2$: $4(2) - y^2 = 0$

$$\begin{aligned} 8 &= y^2 \\ \pm\sqrt{8} &= y \\ \pm 2\sqrt{2} &= y \end{aligned}$$

Since $\sqrt{-16}$ is not a real number, the solutions are $\left(2, 2\sqrt{2}\right)$ and $\left(2, -2\sqrt{2}\right)$.

37. $\begin{cases} x^2 + 4y^2 = 16 \\ x^2 + y^2 = 4 \end{cases}$

Multiply equation 2 by −1. Add the results.

$$\begin{aligned} x^2 + 4y^2 &= 16 \\ -x^2 - y^2 &= -4 \\ \hline 3y^2 &= 12 \\ y^2 &= 4 \\ y &= \pm\sqrt{4} \\ y &= \pm 2 \end{aligned}$$

$y = -2$: $x^2 + (-2)^2 = 4$

$$\begin{aligned} x^2 &= 0 \\ x &= 0 \end{aligned}$$

$y = 2$: $x^2 + (2)^2 = 4$

$$\begin{aligned} x^2 &= 0 \\ x &= 0 \end{aligned}$$

The solutions are (0, 2) and (0, −2).

38. $\begin{cases} x^2 + 2y = 9 \\ 5x - 2y = 5 \end{cases}$

Add the equations.

$$\begin{aligned} x^2 + 5x &= 14 \\ x^2 + 5x - 14 &= 0 \\ (x+7)(x-2) &= 0 \end{aligned}$$

$$x+7=0 \quad \text{or} \quad x-2=0$$
$$x=-7 \qquad\qquad x=2$$
$$x=-7:\ 5(-7)-2y=5$$
$$-2y=40$$
$$y=-20$$
$$x=2:\ 5(2)-2y=5$$
$$-2y=-5$$
$$y=\frac{5}{2}$$

The solutions are $\left(2, \frac{5}{2}\right)$ and $(-7, -20)$.

39. $\begin{cases} y=3x^2+5x-4 \\ y=3x^2-x+2 \end{cases}$

Substitute.
$$3x^2+5x-4=3x^2-x+2$$
$$5x-4=-x+2$$
$$6x=6$$
$$x=1$$
$$x=1:\ y=3(1)^2+5(1)-4=4$$

The solution is $(1, 4)$.

40. $\begin{cases} x^2-3y^2=1 \\ 4x^2+5y^2=21 \end{cases}$

Solve equation 1 for x^2.
$$x^2=1+3y^2$$
Substitute.
$$4(1+3y^2)+5y^2=21$$
$$4+12y^2+5y^2=21$$
$$17y^2=17$$
$$y^2=1$$
$$y=\pm1$$
$$y=-1:\ x^2=1+3(-1)^2$$
$$x^2=4$$
$$x=\pm2$$
$$y=1:\ x^2=1+3(1)^2$$
$$x^2=4$$
$$x=\pm2$$

The solutions are $(-2, -1)$, $(-2, 1)$, $(2, -1)$, and $(2, 1)$.

41. Let x be the width and y be the length.

$\begin{cases} xy=150 \\ 2x+2y=50 \end{cases}$

Solve equation 2 for x.
$$2x+2y=50$$
$$x+y=25$$
$$x=25-y$$
Substitute.
$$xy=150$$
$$(25-y)y=150$$
$$25y-y^2=150$$
$$0=y^2-25y+150$$
$$0=(y-15)(y-10)$$
$$y-15=0 \quad \text{or} \quad y-10=0$$
$$y=15 \qquad\qquad y=10$$

$y = 15$: $x = 25 - 15 = 10$
$y = 10$: $x = 25 - 10 = 15$

The dimensions are length = 15 feet and width = 10 feet.

42. An ellipse and a hyperbola can intersect at a maximum of 4 points. Therefore, there are a maximum of 4 real number solutions.

43. $y \le -x^2+3$

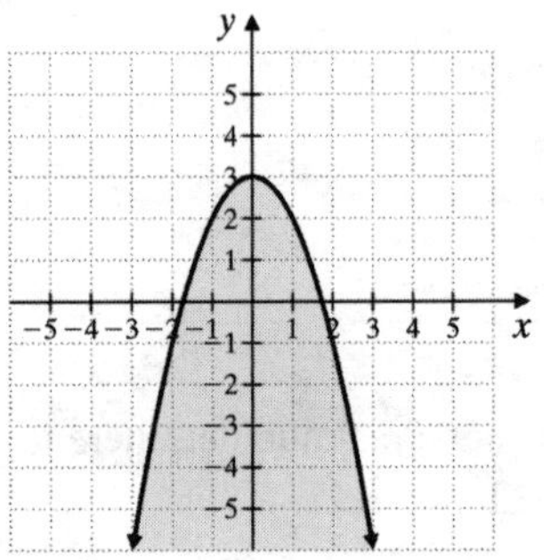

44. $x \le y^2-1$

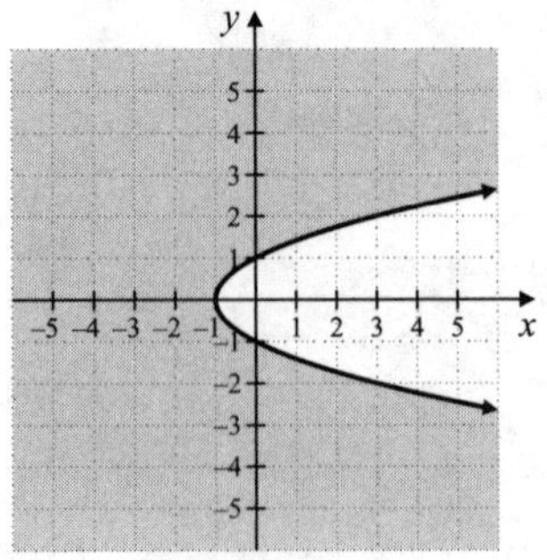

45. $x^2 + y^2 < 9$

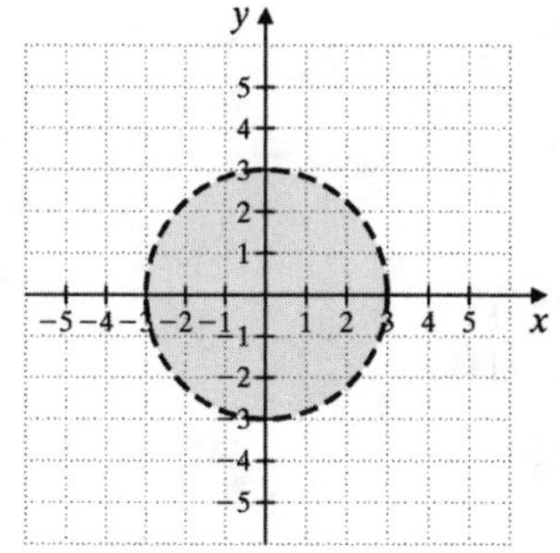

46. $\frac{x^2}{4} + \frac{y^2}{9} \geq 1$

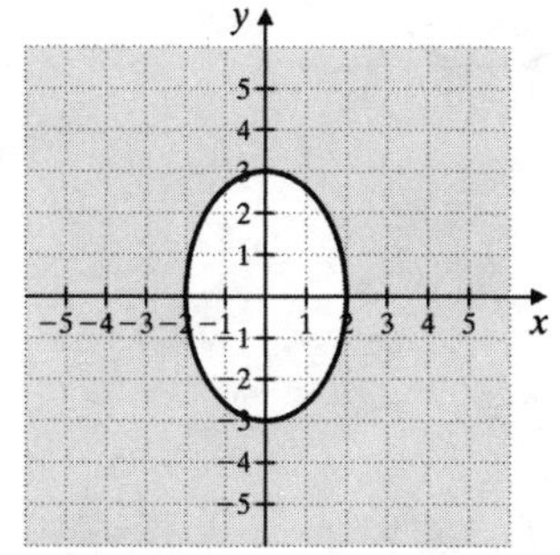

47. $\begin{cases} 3x + 4y \leq 12 \\ x - 2y > 6 \end{cases}$

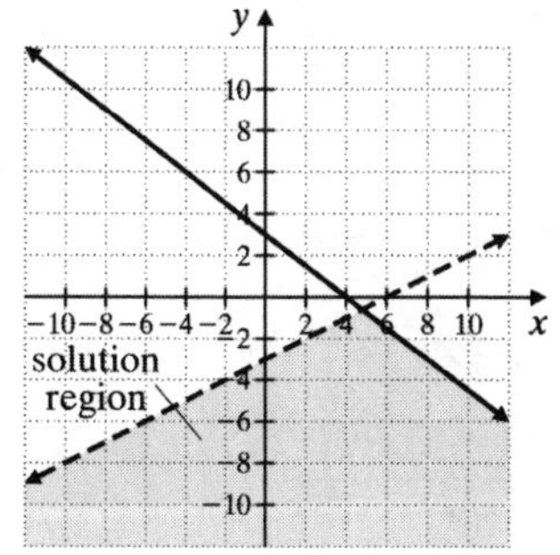

48. $\begin{cases} x^2 + y^2 \leq 16 \\ x^2 + y^2 \geq 4 \end{cases}$

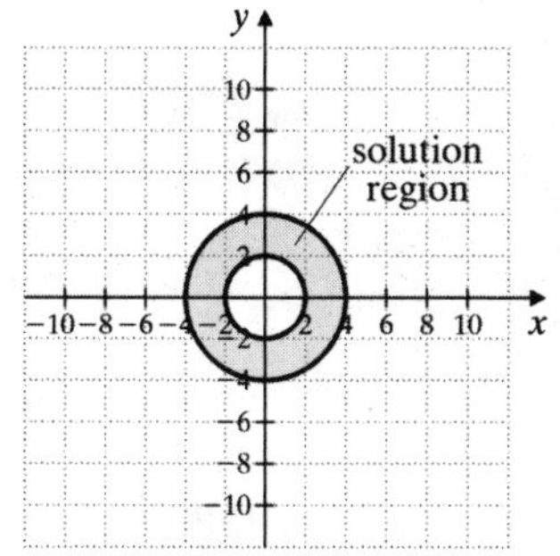

49. $\begin{cases} x^2 + y^2 < 4 \\ x^2 - y^2 \leq 1 \end{cases}$

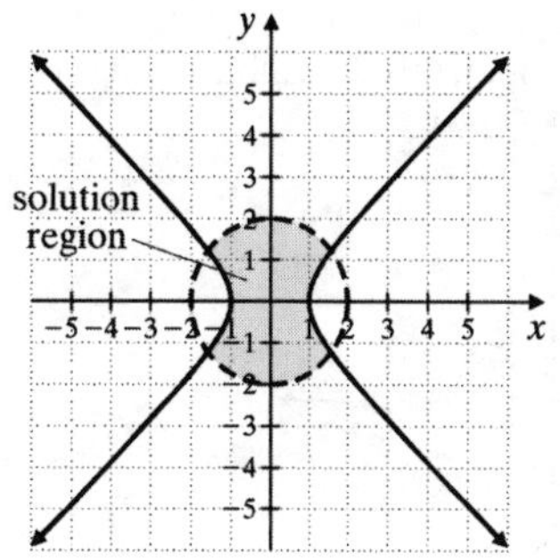

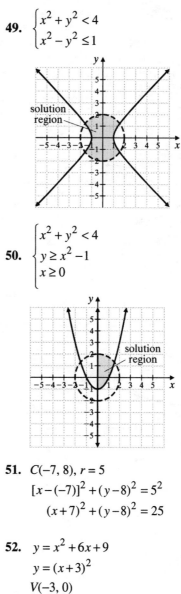

50. $\begin{cases} x^2 + y^2 < 4 \\ y \geq x^2 - 1 \\ x \geq 0 \end{cases}$

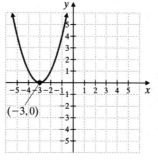

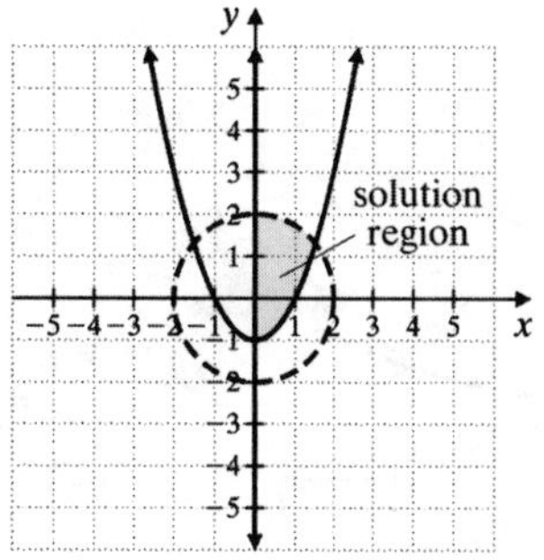

51. $C(-7, 8)$, $r = 5$

$$[x - (-7)]^2 + (y - 8)^2 = 5^2$$
$$(x + 7)^2 + (y - 8)^2 = 25$$

52. $y = x^2 + 6x + 9$

$y = (x + 3)^2$

$V(-3, 0)$

The parabola opens upward.

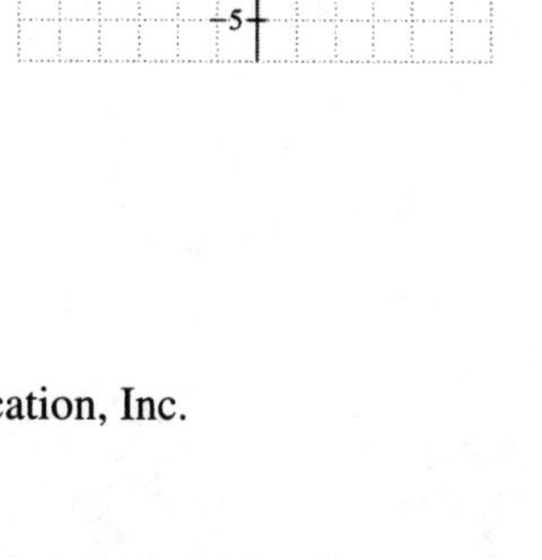

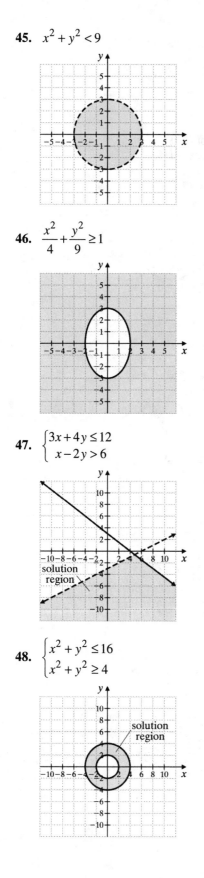

53. $x = y^2 + 6y + 9$

$x = (y+3)^2$

$V(0, -3)$

The parabola opens to the right.

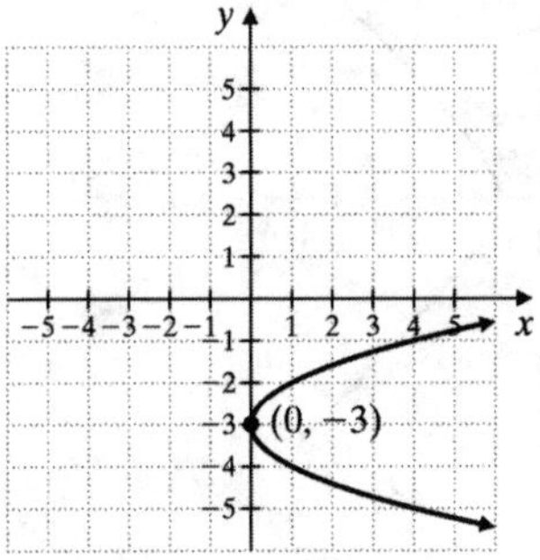

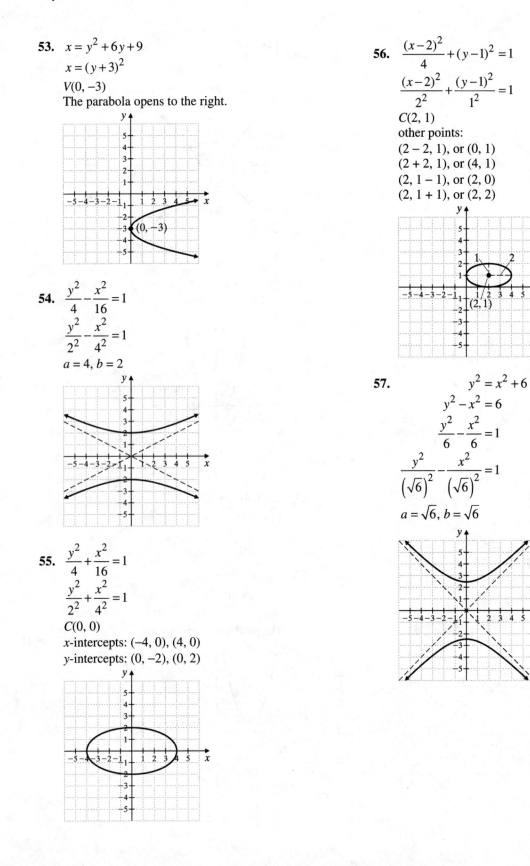

54. $\dfrac{y^2}{4} - \dfrac{x^2}{16} = 1$

$\dfrac{y^2}{2^2} - \dfrac{x^2}{4^2} = 1$

$a = 4, b = 2$

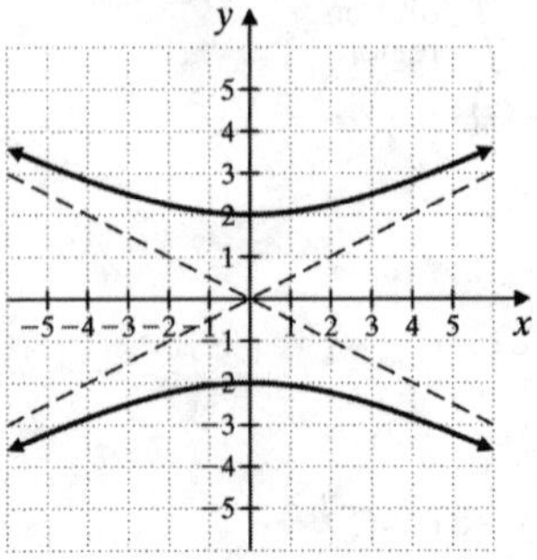

55. $\dfrac{y^2}{4} + \dfrac{x^2}{16} = 1$

$\dfrac{y^2}{2^2} + \dfrac{x^2}{4^2} = 1$

$C(0, 0)$

x-intercepts: $(-4, 0)$, $(4, 0)$

y-intercepts: $(0, -2)$, $(0, 2)$

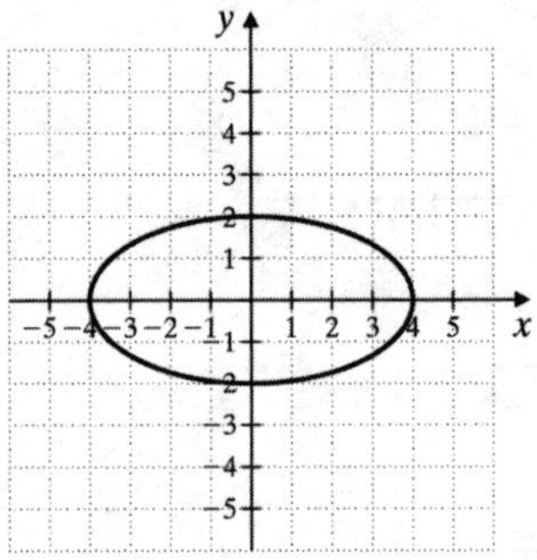

56. $\dfrac{(x-2)^2}{4} + (y-1)^2 = 1$

$\dfrac{(x-2)^2}{2^2} + \dfrac{(y-1)^2}{1^2} = 1$

$C(2, 1)$

other points:

$(2 - 2, 1)$, or $(0, 1)$

$(2 + 2, 1)$, or $(4, 1)$

$(2, 1 - 1)$, or $(2, 0)$

$(2, 1 + 1)$, or $(2, 2)$

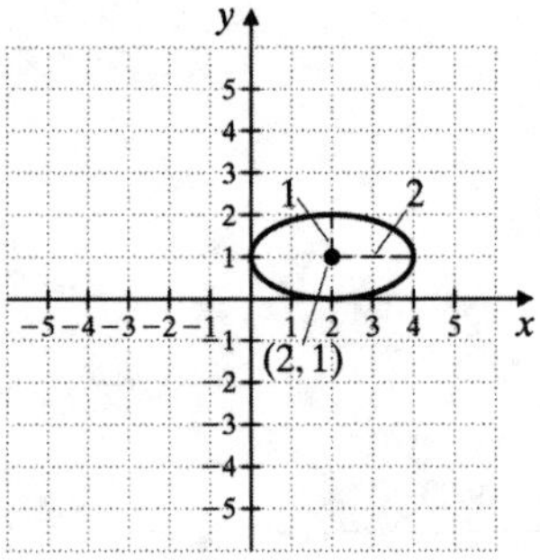

57. $y^2 = x^2 + 6$

$y^2 - x^2 = 6$

$\dfrac{y^2}{6} - \dfrac{x^2}{6} = 1$

$\dfrac{y^2}{\left(\sqrt{6}\right)^2} - \dfrac{x^2}{\left(\sqrt{6}\right)^2} = 1$

$a = \sqrt{6}, b = \sqrt{6}$

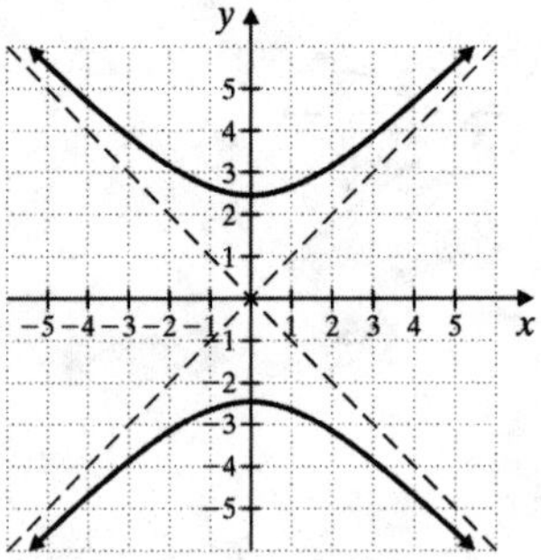

58. $y^2+(x-2)^2=10$

$(x-2)^2+y^2=\left(\sqrt{10}\right)^2$

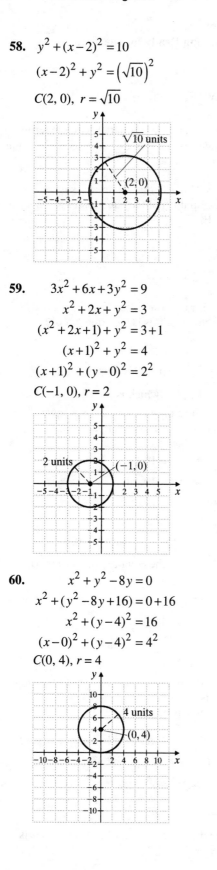

$C(2,\ 0),\ r=\sqrt{10}$

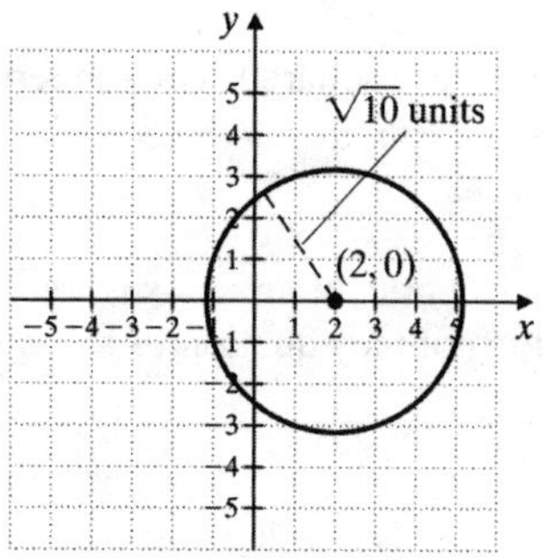

59. $3x^2+6x+3y^2=9$

$x^2+2x+y^2=3$

$(x^2+2x+1)+y^2=3+1$

$(x+1)^2+y^2=4$

$(x+1)^2+(y-0)^2=2^2$

$C(-1,\ 0),\ r=2$

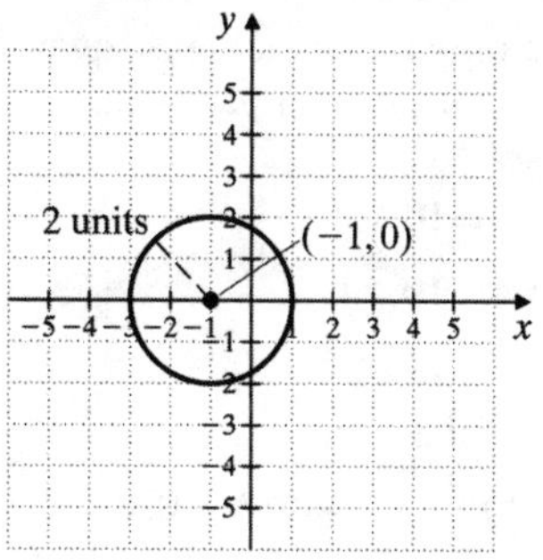

60. $x^2+y^2-8y=0$

$x^2+(y^2-8y+16)=0+16$

$x^2+(y-4)^2=16$

$(x-0)^2+(y-4)^2=4^2$

$C(0,\ 4),\ r=4$

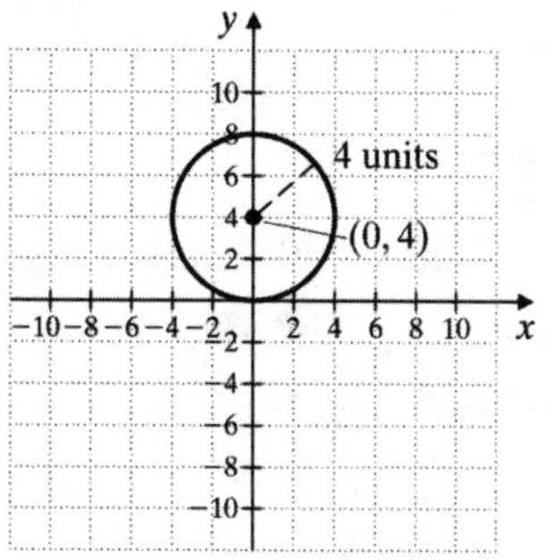

61. $6(x-2)^2+9(y+5)^2=36$

$\dfrac{(x-2)^2}{6}+\dfrac{(y+5)^2}{4}=1$

$\dfrac{(x-2)^2}{\left(\sqrt{6}\right)^2}+\dfrac{(y+5)^2}{2^2}=1$

$C(2,\ -5)$

other points:

$\left(2-\sqrt{6},\ -5\right)$

$\left(2+\sqrt{6},\ -5\right)$

(2, −5 − 2), or (2, −7)

(2, −5 + 2), or (2, −3)

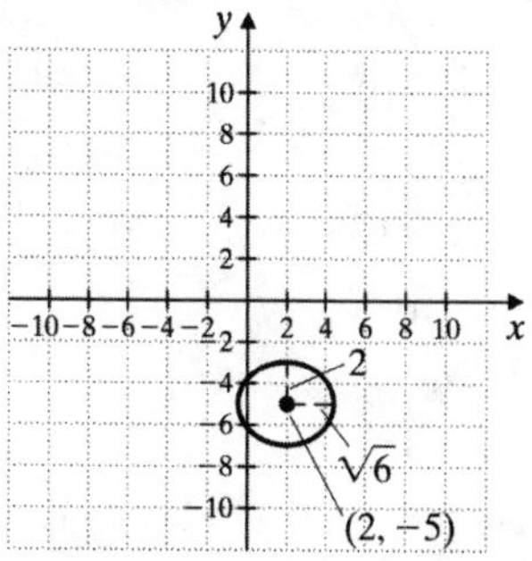

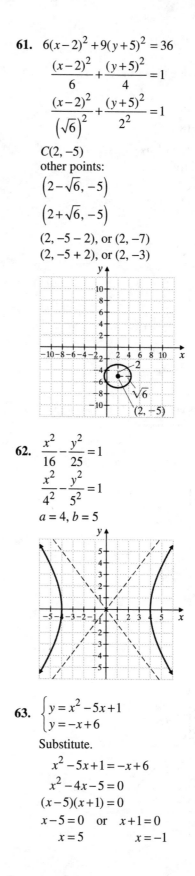

62. $\dfrac{x^2}{16}-\dfrac{y^2}{25}=1$

$\dfrac{x^2}{4^2}-\dfrac{y^2}{5^2}=1$

$a=4,\ b=5$

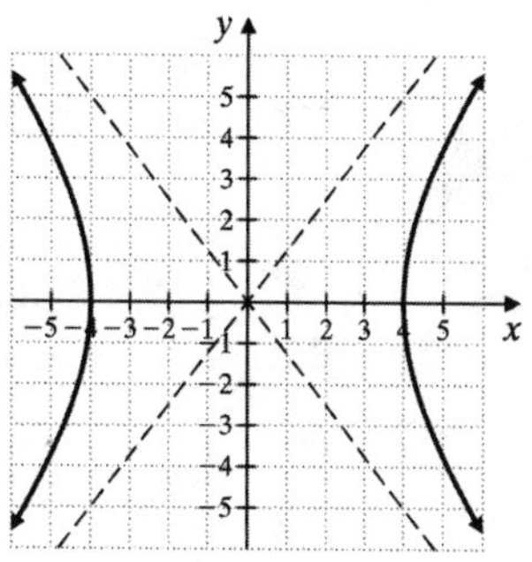

63. $\begin{cases} y=x^2-5x+1 \\ y=-x+6 \end{cases}$

Substitute.

$x^2-5x+1=-x+6$

$x^2-4x-5=0$

$(x-5)(x+1)=0$

$x-5=0$ or $x+1=0$

$x=5 \qquad x=-1$

$x = 5$: $y = -5 + 6 = 1$
$x = -1$: $y = -(-1) + 6 = 7$
The solutions are (5, 1) and (−1, 7).

64. $\begin{cases} x^2 + y^2 = 10 \\ 9x^2 + y^2 = 18 \end{cases}$

Multiply equation 1 by −1. Add the results.

$$\begin{aligned} -x^2 - y^2 &= -10 \\ 9x^2 + y^2 &= 18 \\ \hline 8x^2 \quad &= 8 \\ x^2 &= 1 \\ x &= \pm\sqrt{1} \\ x &= \pm 1 \end{aligned}$$

$x = -1$: $(-1)^2 + y^2 = 10$
$y^2 = 9$
$y = \pm 3$

$x = 1$: $(1)^2 + y^2 = 10$
$y^2 = 9$
$y = \pm 3$

The solutions are (−1, 3), (−1, −3), (1, 3), and (1, −3).

65. $x^2 - y^2 < 1$

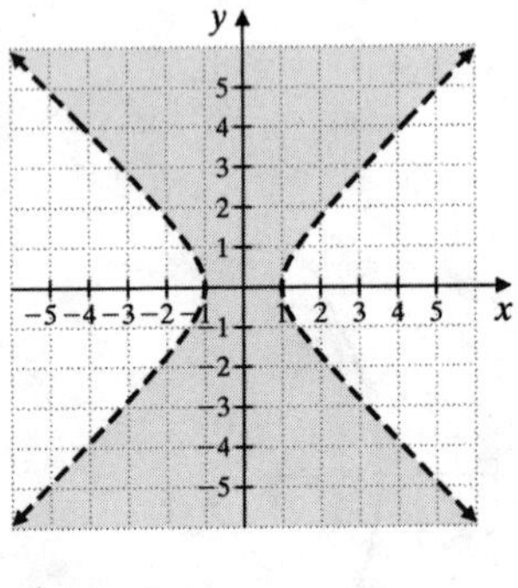

66. $\begin{cases} y > x^2 \\ x + y \ge 3 \end{cases}$

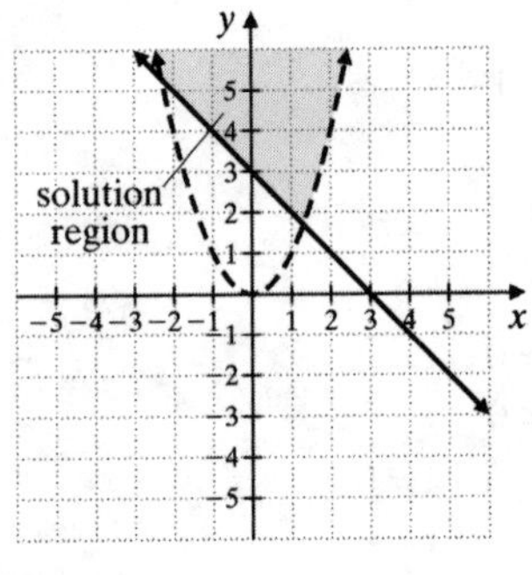

Chapter 10 Getting Ready for the Test

1. $6x^2 + 3y^2 = 24$

$$\frac{x^2}{4} + \frac{y^2}{8} = 1$$

Since both x and y are squared with the terms being added, and $a \ne b$, this is the equation of an ellipse; C.

2. $6x + 3y = 24$ does not contain any squared variables, so it is not the equation of a conic section; E.

3. $6x^2 - 3y^2 = 24$

$$\frac{x^2}{4} - \frac{y^2}{8} = 1$$

Since both x and y are squared with the terms being subtracted, this is the equation of a hyperbola; D.

4. $x = 5(y-2)^2 + 3$ is in the form $x = a(y-k)^2 + h$ which is the equation of a parabola, since only one variable is squared; A. Note that this is a horizontal parabola.

5.
$$\begin{aligned} x^2 + y^2 - 4y &= 10 \\ x^2 + y^2 - 4y + 4 &= 10 + 4 \\ x^2 + (y-2)^2 &= 14 \end{aligned}$$

The equation has the form $(x-h)^2 + (y-k)^2 = r^2$ with $h = 0$, $k = 2$, and $r = \sqrt{14}$. This is the equation of a circle; B.

6.
$$\begin{aligned} y &= x^2 + 2x \\ y &= x^2 + 2x + 1 - 1 \\ y &= (x+1)^2 - 1 \end{aligned}$$

This is in the form $y = a(x-h)^2 + k$, which is the equation of a parabola, since only one variable is squared; A.
Note that this is a vertical parabola.

7. The equation of a circle with center (h, k) and radius r is $(x-h)^2 + (y-k)^2 = r^2$.

$$(x+2)^2 + y^2 = 22$$
$$[x-(-2)]^2 + (y-0)^2 = \left(\sqrt{22}\right)^2$$

The center is $(h, k) = (-2, 0)$ and the radius is $r = \sqrt{22}$; D.

8. The equation of a horizontal parabola has the form $x = a(y-k)^2 + h,$ where (h, k) is the vertex.
In $x = 3(y+2)^2 + 1 = 3[y-(-2)]^2 + 1,$ $a = 3,$ $h = 1,$ and $k = -2.$ The vertex is $(h, k) = (1, -2)$ and since $a = 3 > 0,$ the parabola opens to the right; B.

9. $\frac{x^2}{100} + \frac{y^2}{64} = 1$
$\frac{x^2}{10^2} + \frac{y^2}{8^2} = 1$
The graph of this equation has y-intercepts of (0, 8) and (0, −8). The distance between these points is 8 + 8 = 16 units; D.

10. $x^2 - 4y^2 = 36$
$\frac{x^2}{36} - \frac{y^2}{9} = 1$
$\frac{x^2}{6^2} - \frac{y^2}{3^2} = 1$
The graph of this equation has x-intercepts (6, 0) and (−6, 0). The distance between these points is 6 + 6 = 12 units; C.

11. The graph of $x^2 + y^2 = 1$ is a circle centered at the origin with radius $\sqrt{1} = 1,$ which is the set of all points that are 1 unit from the origin. The graph of $x^2 + y^2 = 9$ is a circle centered at the origin with radius $\sqrt{9} = 3,$ which is the set of all points that are 3 units from the origin. Since no point can be both 1 unit and 3 units from the origin, the graphs have no points of intersection; A.
Alternative solution:
Subtract the equations.

$$\begin{array}{r} x^2 + y^2 = 1 \\ -\,(x^2 + y^2 = 9) \\ \hline 0 = -8 \end{array} \quad \text{False}$$

The nonlinear system of equations has no solution, so the graphs of the equations do not intersect; A.

12. A line and a circle may intersect in a maximum of 2 points; C.

13. A line and a parabola may intersect in a maximum of 2 points; C.

14. A circle and an ellipse may intersect in a maximum of 4 points; E.

15. An ellipse and a hyperbola may intersect in a maximum of 4 points; E.

16. An ellipse and an ellipse may intersect in a maximum of 4 points; E.

17. Two distinct lines may intersect in a maximum of 1 point; B.

Chapter 10 Test

1. $x^2 + y^2 = 36$
Circle; center: (0, 0), radius $r = \sqrt{36} = 6$

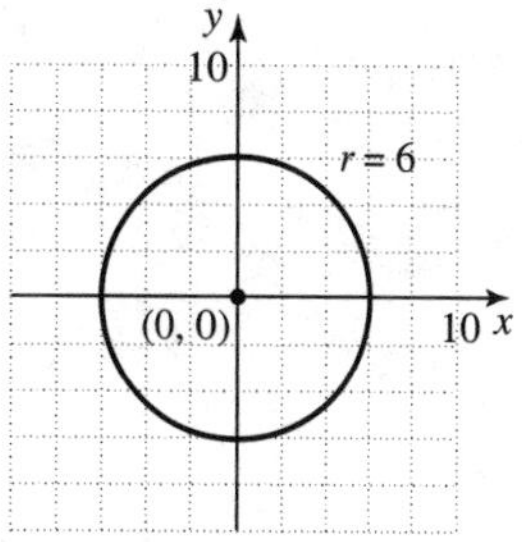

2. $x^2 - y^2 = 36$
$\frac{x^2}{36} - \frac{y^2}{36} = 1$
Hyperbola; center: (0, 0), $a = 6,$ $b = 6$

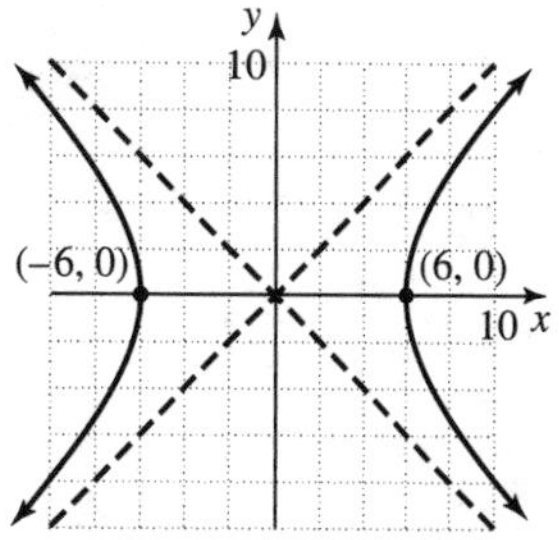

3. $16x^2 + 9y^2 = 144$
$\frac{x^2}{9} + \frac{y^2}{16} = 1$
Ellipse; center: (0, 0), $a = 3,$ $b = 4$

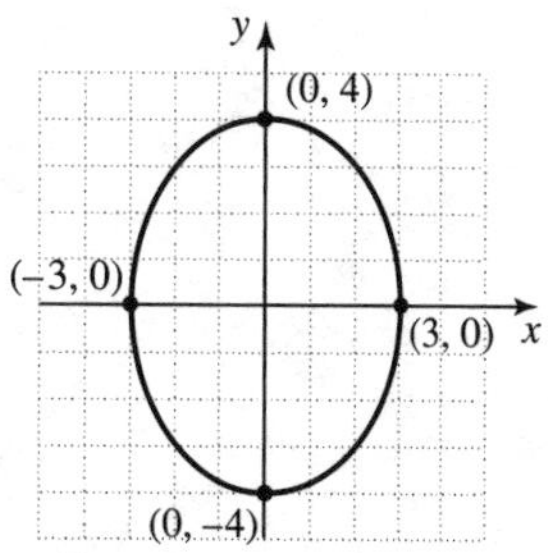

4. $y = x^2 - 8x + 16$

$y = (x-4)^2$

Parabola; vertex: (4, 0)

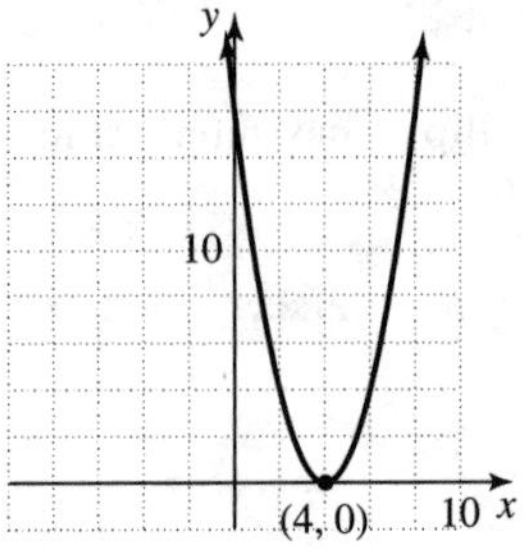

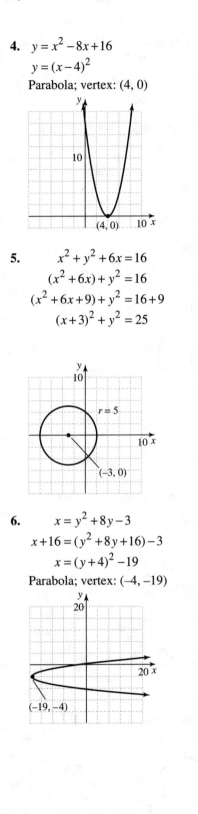

5.
$$x^2 + y^2 + 6x = 16$$
$$(x^2 + 6x) + y^2 = 16$$
$$(x^2 + 6x + 9) + y^2 = 16 + 9$$
$$(x+3)^2 + y^2 = 25$$

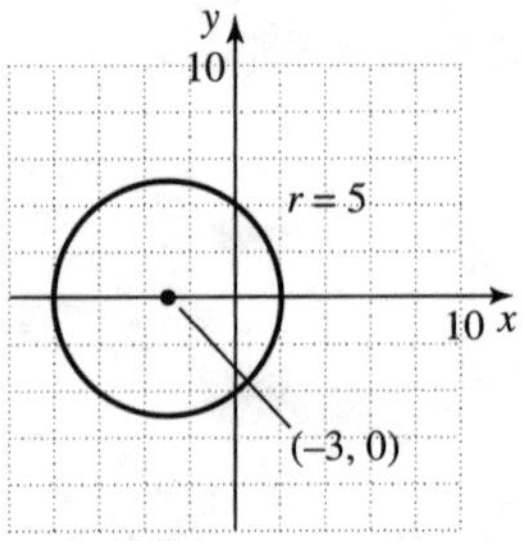

6.
$$x = y^2 + 8y - 3$$
$$x + 16 = (y^2 + 8y + 16) - 3$$
$$x = (y+4)^2 - 19$$

Parabola; vertex: (–4, –19)

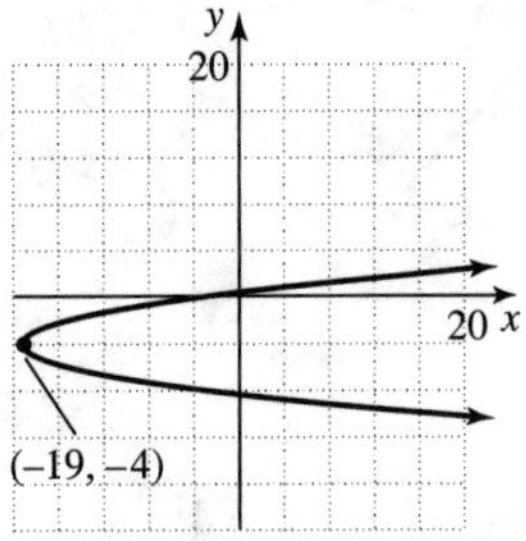

7. $\dfrac{(x-4)^2}{16} + \dfrac{(y-3)^2}{9} = 1$

Ellipse: center: (4, 3), $a = 4$, $b = 3$

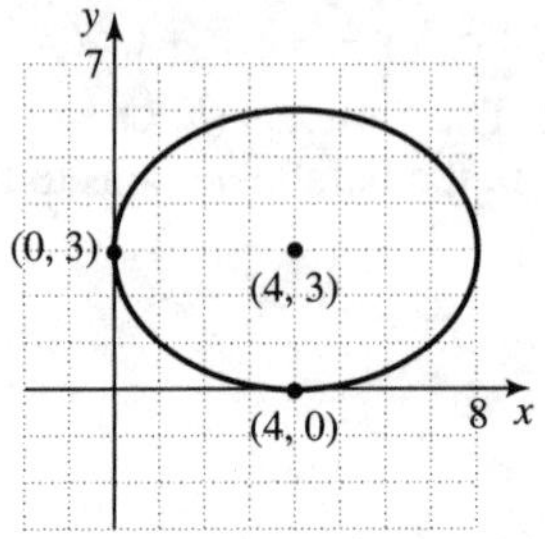

8. $y^2 - x^2 = 1$

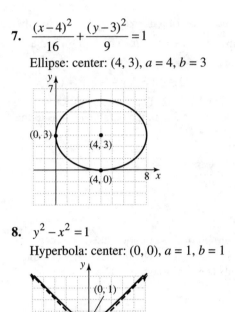

Hyperbola: center: (0, 0), $a = 1$, $b = 1$

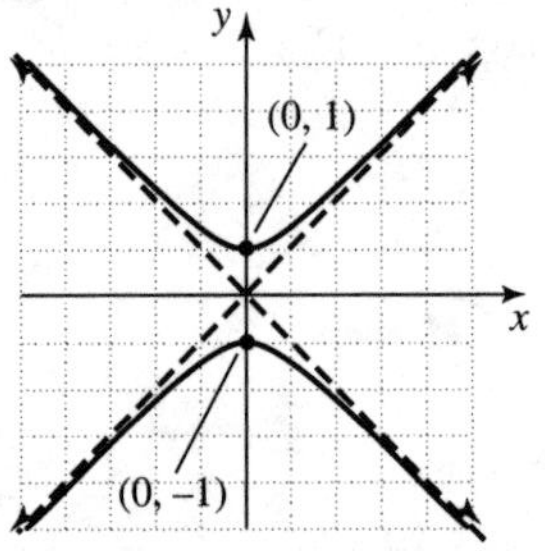

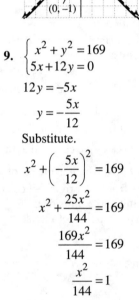

9. $\begin{cases} x^2 + y^2 = 169 \\ 5x + 12y = 0 \end{cases}$

$12y = -5x$

$y = -\dfrac{5x}{12}$

Substitute.

$$x^2 + \left(-\frac{5x}{12}\right)^2 = 169$$
$$x^2 + \frac{25x^2}{144} = 169$$
$$\frac{169x^2}{144} = 169$$
$$\frac{x^2}{144} = 1$$

$x^2 = 144$ so $x = \pm 12$.

Substitute back.

$x = 12$: $y = -\dfrac{5}{12}(12) = -5$

$x = -12$: $y = -\dfrac{5}{12}(-12) = 5$

The solutions are (12, –5) and (–12, 5).

10. $\begin{cases} x^2 + y^2 = 26 \\ x^2 - 2y^2 = 23 \end{cases}$

Multiply the second equation by −1. Add the results.

$$\begin{aligned} x^2 + y^2 &= 26 \\ -x^2 + 2y^2 &= -23 \\ \hline 3y^2 &= 3 \\ y^2 &= 1 \\ y &= \pm 1 \end{aligned}$$

$y = -1$: $x^2 + (-1)^2 = 26$
$x^2 = 25$
$x = \pm 5$

$y = 1$: $x^2 + 1^2 = 26$
$x^2 = 25$
$x = \pm 5$

The solutions are (−5, −1), (−5, 1), (5, −1), and (5, 1).

11. $\begin{cases} y = x^2 - 5x + 6 \\ y = 2x \end{cases}$

Substitute.

$$\begin{aligned} x^2 - 5x + 6 &= 2x \\ x^2 - 7x + 6 &= 0 \\ (x-6)(x-1) &= 0 \end{aligned}$$

$x - 6 = 0$ or $x - 1 = 0$
$x = 6$ $\quad$ $x = 1$

$x = 6$: $y = 2(6) = 12$
$x = 1$: $y = 2(1) = 2$

The solutions are (6, 12) and (1, 2).

12. $\begin{cases} x^2 + 4y^2 = 5 \\ y = x \end{cases}$

Substitute.

$$\begin{aligned} x^2 + 4x^2 &= 5 \\ 5x^2 &= 5 \\ x^2 &= 1 \\ x &= \pm 1 \end{aligned}$$

$x = -1$: $y = -1$
$x = 1$: $y = 1$

The solutions are (−1, −1) and (1, 1).

13. $\begin{cases} 2x + 5y \ge 10 \\ y \ge x^2 + 1 \end{cases}$

First graph $\begin{cases} 2x + 5y = 10 \\ y = x^2 + 1 \end{cases}$ or

$\begin{cases} y = -\dfrac{2}{5}x + 2 \\ y = 1 \cdot (x-0)^2 + 1 \end{cases}$

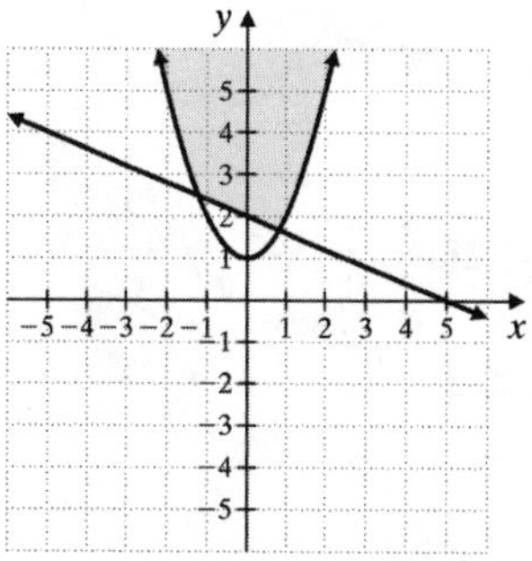

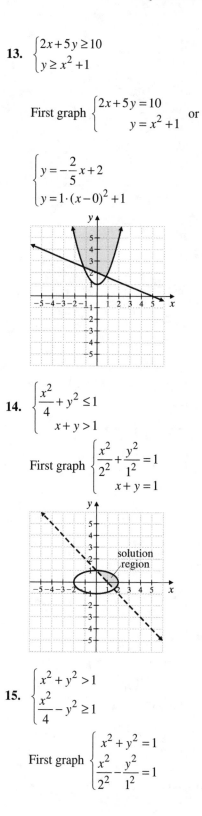

14. $\begin{cases} \dfrac{x^2}{4} + y^2 \le 1 \\ x + y > 1 \end{cases}$

First graph $\begin{cases} \dfrac{x^2}{2^2} + \dfrac{y^2}{1^2} = 1 \\ x + y = 1 \end{cases}$

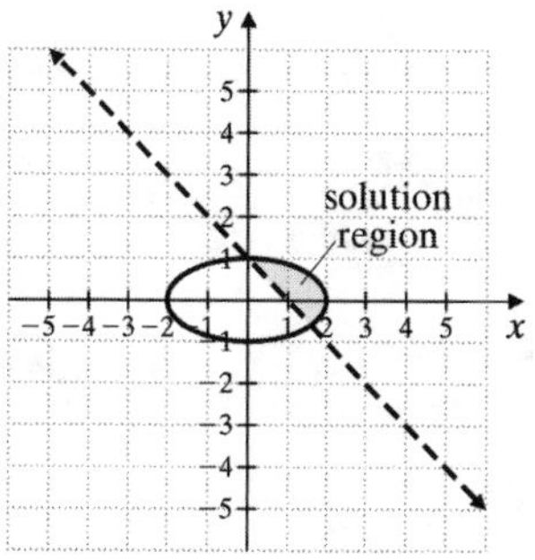

15. $\begin{cases} x^2 + y^2 > 1 \\ \dfrac{x^2}{4} - y^2 \ge 1 \end{cases}$

First graph $\begin{cases} x^2 + y^2 = 1 \\ \dfrac{x^2}{2^2} - \dfrac{y^2}{1^2} = 1 \end{cases}$

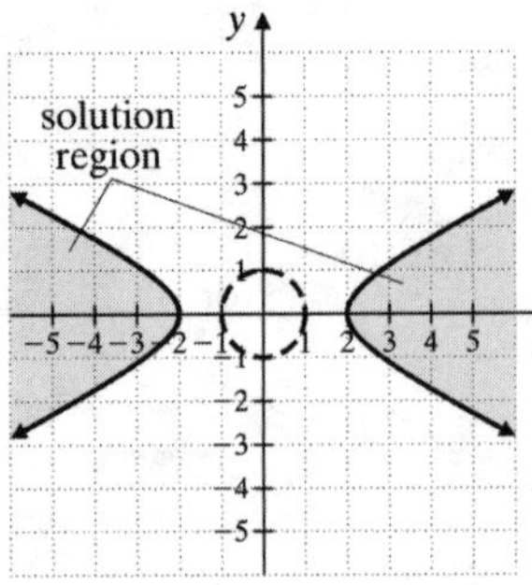

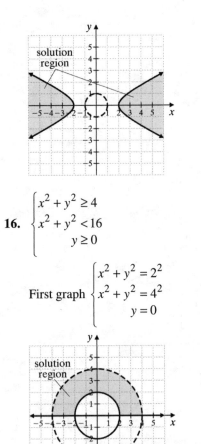

16. $\begin{cases} x^2 + y^2 \geq 4 \\ x^2 + y^2 < 16 \\ y \geq 0 \end{cases}$

First graph $\begin{cases} x^2 + y^2 = 2^2 \\ x^2 + y^2 = 4^2 \\ y = 0 \end{cases}$

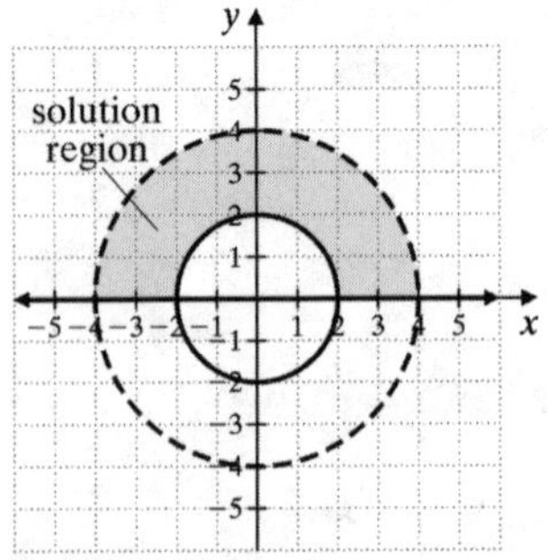

17. Graph B; vertex in second quadrant, opens to the right.

18. $100x^2 + 225y^2 = 22{,}500$

$$\frac{x^2}{225} + \frac{y^2}{100} = 1$$

$$\frac{x^2}{15^2} + \frac{y^2}{10^2} = 1$$

Height = 10 feet
Width = 2(15) = 30 feet

Chapter 10 Cumulative Review

1. $4 \cdot (9y) = (4 \cdot 9)y = 36y$

2. $3x + 4 > 1 \quad and \quad 2x - 5 \leq 9$

$3x > -3 \quad and \quad 2x \leq 14$

$x > -1 \quad and \quad x \leq 7$

$-1 < x \leq 7$

$(-1, 7]$

3. $x = -2y$

$y = -\frac{1}{2}x$

x	$y = -\frac{1}{2}x$
–2	1
0	0
4	–2

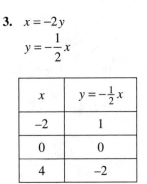

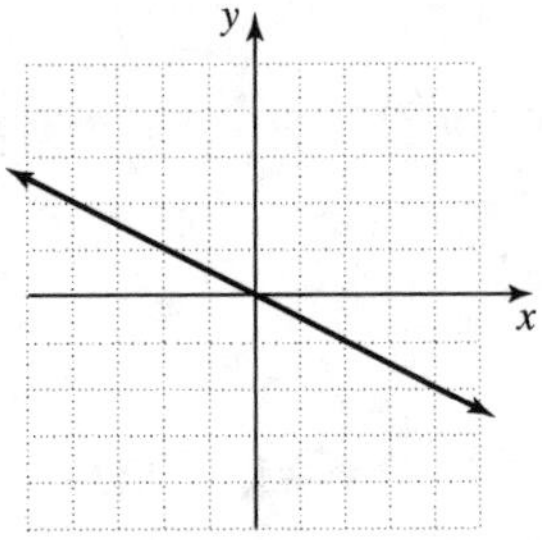

4. (3, 2), (1, –4)

$$m = \frac{-4 - 2}{1 - 3} = \frac{-6}{-2} = 3$$

5. $\begin{cases} 3x + \frac{y}{2} = 2 \quad (1) \\ 6x + y = 5 \quad (2) \end{cases}$

Multiply E1 by –2 and add to E2.

$-6x - y = -4$

$6x + y = 5$

$0 = 1$

This is a false statement. Therefore, the solution is $\varnothing$.

6. Let x = speed of one plane. Then $x + 25$ = speed of the other plane.

$d_{\text{plane 1}} + d_{\text{plane 2}} = 650$ miles

$2x + 2(x + 25) = 650$

$2x + 2x + 50 = 650$

$4x = 600$

$x = 150$

$x + 25 = 150 + 25 = 175$

The planes are traveling at 150 mph and 175 mph.

7. a. $(5x^2)^3 = 5^3(x^2)^3 = 125x^6$

b. $\left(\frac{2}{3}\right)^3 = \frac{2^3}{3^3} = \frac{8}{27}$

c. $\left(\frac{3p^4}{q^5}\right)^2 = \frac{3^2(p^4)^2}{(q^5)^2} = \frac{9p^8}{q^{10}}$

d. $\left(\frac{2^{-3}}{y}\right)^{-2} = \frac{(2^{-3})^{-2}}{y^{-2}}$
$= 2^6 y^2$
$= 64y^2$

e. $(x^{-5}y^2z^{-1})^7 = (x^{-5})^7(y^2)^7(z^{-1})^7$
$= x^{-35}y^{14}z^{-7}$
$= \frac{y^{14}}{x^{35}z^7}$

8. a. $\frac{4^8}{4^3} = 4^{8-3} = 4^5$

b. $\frac{y^{11}}{y^5} = y^{11-5} = y^6$

c. $\frac{32x^7}{4x^6} = \frac{32}{4}x^{7-6} = 8x$

d. $\frac{18a^{12}b^6}{12a^8b^6} = \frac{18}{12}a^{12-8}b^{6-6} = \frac{3}{2}a^4b^0 = \frac{3a^4}{2}$

9.
$2x^2 = \frac{17}{3}x + 1$
$3(2x^2) = 3\left(\frac{17}{3}x + 1\right)$
$6x^2 = 17x + 3$
$6x^2 - 17x - 3 = 0$
$(6x+1)(x-3) = 0$
$6x + 1 = 0$ or $x - 3 = 0$
$6x = -1$ or $x = 3$
$x = -\frac{1}{6}$

The solutions are $-\frac{1}{6}$ and 3.

10. a. $3y^2 + 14y + 15 = (3y+5)(y+3)$

b. $20a^5 + 54a^4 + 10a^3$
$= 2a^3(10a^2 + 27a + 5)$
$= 2a^3(2a+5)(5a+1)$

c. $(y-3)^2 - 2(y-3) - 8$

Let $u = y - 3$. Then $u^2 = (y-3)^2$ and
$u^2 - 2u - 8 = (u-4)(u+2)$
$= [(y-3)-4][(y-3)+2]$
$= (y-7)(y-1)$

11. $\frac{7}{x-1} + \frac{10x}{x^2-1} - \frac{5}{x+1}$
$= \frac{7}{x-1} + \frac{10x}{(x+1)(x-1)} - \frac{5}{x+1}$
$= \frac{7(x+1) + 10x - 5(x-1)}{(x+1)(x-1)}$
$= \frac{7x + 7 + 10x - 5x + 5}{(x+1)(x-1)}$
$= \frac{12x + 12}{(x+1)(x-1)}$
$= \frac{12(x+1)}{(x+1)(x-1)}$
$= \frac{12}{x-1}$

12. $\frac{2}{3a-15} - \frac{a}{25-a^2}$
$= \frac{2}{3(a-5)} + \frac{a}{a^2-25}$
$= \frac{2}{3(a-5)} + \frac{a}{(a+5)(a-5)}$
$= \frac{2(a+5) + 3a}{3(a+5)(a-5)}$
$= \frac{2a + 10 + 3a}{3(a+5)(a-5)}$
$= \frac{5a + 10}{3(a+5)(a-5)}$
$= \frac{5(a+2)}{3(a+5)(a-5)}$

13. a. $\frac{\frac{2x}{27y^2}}{\frac{6x^2}{9}} = \frac{2x}{27y^2} \cdot \frac{9}{6x^2} = \frac{1}{3y^2} \cdot \frac{1}{3x} = \frac{1}{9xy^2}$

b. $\frac{\frac{5x}{x+2}}{\frac{10}{x-2}} = \frac{5x}{x+2} \cdot \frac{x-2}{10} = \frac{x(x-2)}{2(x+2)}$

c. $$\frac{\frac{x}{y^2}+\frac{1}{y}}{\frac{y}{x^2}+\frac{1}{x}}=\frac{\left(\frac{x}{y^2}+\frac{1}{y}\right)x^2y^2}{\left(\frac{y}{x^2}+\frac{1}{x}\right)x^2y^2}=\frac{x^3+x^2y}{y^3+xy^2}=\frac{x^2(x+y)}{y^2(y+x)}=\frac{x^2}{y^2}$$

14. a. $$(a^{-1}-b^{-1})^{-1}=\left(\frac{1}{a}-\frac{1}{b}\right)^{-1}=\left(\frac{b-a}{ab}\right)^{-1}=\frac{ab}{b-a}$$

b. $$\frac{2-\frac{1}{x}}{4x-\frac{1}{x}}=\frac{\left(2-\frac{1}{x}\right)x}{\left(4x-\frac{1}{x}\right)x}=\frac{2x-1}{4x^2-1}=\frac{2x-1}{(2x+1)(2x-1)}=\frac{1}{2x+1}$$

15.
$$\begin{array}{r}
2x-5 \\
x+2\overline{)2x^2-\ x-10} \\
\underline{2x^2+4x} \\
-5x-10 \\
\underline{-5x-10} \\
0
\end{array}$$

Answer: $2x - 5$

16.
$$\frac{2}{x+3}=\frac{1}{x^2-9}-\frac{1}{x-3}$$
$$\frac{2}{x+3}=\frac{1}{(x+3)(x-3)}-\frac{1}{x-3}$$
$$2(x-3)=1-1(x+3)$$
$$2x-6=1-x-3$$
$$2x-6=-x-2$$
$$3x=4$$
$$x=\frac{4}{3}$$

17.
$$\begin{array}{r|rrrrrrr}
4 & 4 & -25 & 35 & 0 & 17 & 0 & 0 \\
 & & 16 & -36 & -4 & -16 & 4 & 16 \\
\hline
 & 4 & -9 & -1 & -4 & 1 & 4 & 16
\end{array}$$

Thus, $P(4) = 16$.

18.
$$y=\frac{k}{x}$$
$$3=\frac{k}{\frac{2}{3}}$$
$$k=3\left(\frac{2}{3}\right)=2$$

Thus, the equation is $y=\frac{2}{x}$.

19.
$$\frac{2x}{x-3}+\frac{6-2x}{x^2-9}=\frac{x}{x+3}$$
$$\frac{2x}{x-3}+\frac{-2(x-3)}{(x+3)(x-3)}=\frac{x}{x+3}$$
$$\frac{2x}{x-3}-\frac{2}{x+3}=\frac{x}{x+3}$$
$$\frac{2x}{x-3}=\frac{x}{x+3}+\frac{2}{x+3}$$
$$\frac{2x}{x-3}=\frac{x+2}{x+3}$$
$$2x(x+3)=(x+2)(x-3)$$
$$2x^2+6x=x^2-x-6$$
$$x^2+7x+6=0$$
$$(x+6)(x+1)=0$$
$$x+6=0 \quad\text{or}\quad x+1=0$$
$$x=-6 \quad\text{or}\quad x=-1$$

The solutions are –6 and –1.

20. a. $\sqrt[5]{-32}=-2$ because $(-2)^5=-32$.

b. $\sqrt[4]{625}=5$ because $5^4=625$.

c. $-\sqrt{36}=-6$ because $6^2=36$.

d. $-\sqrt[3]{-27x^3}=-(-3x)=3x$

e. $\sqrt{144y^2}=12y$

21. Let t = time it will take together.

$$\frac{1}{4}+\frac{1}{5}=\frac{1}{t}$$
$$20t\left(\frac{1}{4}+\frac{1}{5}\right)=20t\left(\frac{1}{t}\right)$$
$$5t+4t=20$$
$$9t=20$$
$$t=\frac{20}{9}=2\frac{2}{9}$$

It will take them $2\frac{2}{9}$ hours. No, they can not finish before the movie starts.

22. **a.** $\frac{\sqrt{32}}{\sqrt{4}}=\sqrt{\frac{32}{4}}=\sqrt{8}=\sqrt{4\cdot 2}=2\sqrt{2}$

b.
$$\frac{\sqrt[3]{240y^2}}{5\sqrt[3]{3y^{-4}}}=\frac{1}{5}\sqrt[3]{\frac{240y^2}{3y^{-4}}}$$
$$=\frac{1}{5}\sqrt[3]{80y^6}$$
$$=\frac{1}{5}\sqrt[3]{8y^6\cdot 10}$$
$$=\frac{2y^3\sqrt[3]{10}}{5}$$

c.
$$\frac{\sqrt[5]{64x^9y^2}}{\sqrt[5]{2x^2y^{-8}}}=\sqrt[5]{\frac{64x^9y^2}{2x^2y^{-8}}}$$
$$=\sqrt[5]{32x^7y^{10}}$$
$$=\sqrt[5]{32x^5y^{10}\cdot x^2}$$
$$=2xy^2\sqrt[5]{x^2}$$

23. **a.** $\sqrt[3]{1}=1$

b. $\sqrt[3]{-64}=-4$

c. $\sqrt[3]{\frac{8}{125}}=\frac{\sqrt[3]{8}}{\sqrt[3]{125}}=\frac{2}{5}$

d. $\sqrt[3]{x^6}=x^2$

e. $\sqrt[3]{-27x^9}=-3x^3$

24. **a.**
$$\sqrt{5}\left(2+\sqrt{15}\right)=2\sqrt{5}+\sqrt{5}\cdot\sqrt{15}$$
$$=2\sqrt{5}+\sqrt{75}$$
$$=2\sqrt{5}+5\sqrt{3}$$

b.
$$\left(\sqrt{3}-\sqrt{5}\right)\left(\sqrt{7}-1\right)$$
$$=\sqrt{3}\cdot\sqrt{7}-\sqrt{3}\cdot 1-\sqrt{5}\cdot\sqrt{7}+\sqrt{5}\cdot 1$$
$$=\sqrt{21}-\sqrt{3}-\sqrt{35}+\sqrt{5}$$

c.
$$\left(2\sqrt{5}-1\right)^2=\left(2\sqrt{5}\right)^2-2\cdot 2\sqrt{5}\cdot 1+1^2$$
$$=4(5)-4\sqrt{5}+1$$
$$=21-4\sqrt{5}$$

d.
$$\left(3\sqrt{2}+5\right)\left(3\sqrt{2}-5\right)=\left(3\sqrt{2}\right)^2-5^2$$
$$=9(2)-25$$
$$=18-25$$
$$=-7$$

25. **a.**
$$z^{2/3}(z^{1/3}-z^5)=z^{2/3+1/3}-z^{2/3+5}$$
$$=z^{3/3}-z^{2/3+15/3}$$
$$=z-z^{17/3}$$

b.
$$(x^{1/3}-5)(x^{1/3}+2)$$
$$=x^{1/3}\cdot x^{1/3}+2x^{1/3}-5x^{1/3}-5(2)$$
$$=x^{2/3}-3x^{1/3}-10$$

26.
$$\frac{-2}{\sqrt{3}+3}=\frac{-2\left(\sqrt{3}-3\right)}{\left(\sqrt{3}+3\right)\left(\sqrt{3}-3\right)}$$
$$=\frac{-2\left(\sqrt{3}-3\right)}{\left(\sqrt{3}\right)^2-3^2}$$
$$=\frac{-2\left(\sqrt{3}-3\right)}{3-9}$$
$$=\frac{-2\left(\sqrt{3}-3\right)}{-6}$$
$$=\frac{\sqrt{3}-3}{3}$$

27. **a.** $\frac{\sqrt{20}}{\sqrt{5}}=\sqrt{\frac{20}{5}}=\sqrt{4}=2$

b. $\frac{\sqrt{50x}}{2\sqrt{2}}=\frac{1}{2}\sqrt{\frac{50x}{2}}=\frac{1}{2}\sqrt{25x}=\frac{5\sqrt{x}}{2}$

c. $\dfrac{7\sqrt[3]{48x^4y^8}}{\sqrt[3]{6y^2}} = 7\sqrt[3]{\dfrac{48x^4y^8}{6y^2}}$
$= 7\sqrt[3]{8x^4y^6}$
$= 7\sqrt[3]{8x^3y^6 \cdot x}$
$= 7 \cdot 2xy^2\sqrt[3]{x}$
$= 14xy^2\sqrt[3]{x}$

d. $\dfrac{2\sqrt[4]{32a^8b^6}}{\sqrt[4]{a^{-1}b^2}} = 2\sqrt[4]{\dfrac{32a^8b^6}{a^{-1}b^2}}$
$= 2\sqrt[4]{32a^9b^4}$
$= 2\sqrt[4]{16a^8b^4 \cdot 2a}$
$= 2 \cdot 2a^2b\sqrt[4]{2a}$
$= 4a^2b\sqrt[4]{2a}$

28. $\sqrt{2x-3} = x-3$
$\left(\sqrt{2x-3}\right)^2 = (x-3)^2$
$2x-3 = x^2-6x+9$
$0 = x^2-8x+12$
$0 = (x-6)(x-2)$
$x-6=0$ or $x-2=0$
$x=6$ or $x=2$
Discard 2 as an extraneous solution. The solution is 6.

29. a. $\dfrac{\sqrt{45}}{4} - \dfrac{\sqrt{5}}{3} = \dfrac{3\sqrt{5}}{4} - \dfrac{\sqrt{5}}{3}$
$= \dfrac{9\sqrt{5}-4\sqrt{5}}{12}$
$= \dfrac{5\sqrt{5}}{12}$

b. $\sqrt[3]{\dfrac{7x}{8}} + 2\sqrt[3]{7x} = \dfrac{\sqrt[3]{7x}}{2} + 2\sqrt[3]{7x}$
$= \dfrac{\sqrt[3]{7x}}{2} + \dfrac{4\sqrt[3]{7x}}{2}$
$= \dfrac{5\sqrt[3]{7x}}{2}$

30. $9x^2-6x=-4$
$9x^2-6x+4=0$
$a=9, b=-6, c=4$
$b^2-4ac = (-6)^2-4(9)(4)$
$= 36-144$
$= -108$
two complex but not real solutions

31. $\sqrt{\dfrac{7x}{3y}} = \dfrac{\sqrt{7x}}{\sqrt{3y}} = \dfrac{\sqrt{7x}\cdot\sqrt{3y}}{\sqrt{3y}\cdot\sqrt{3y}} = \dfrac{\sqrt{21xy}}{3y}$

32. $\dfrac{4}{x-2} - \dfrac{x}{x+2} = \dfrac{16}{x^2-4}$
$\dfrac{4}{x-2} - \dfrac{x}{x+2} = \dfrac{16}{(x+2)(x-2)}$
$4(x+2)-x(x-2) = 16$
$4x+8-x^2+2x = 16$
$0 = x^2-6x+8$
$0 = (x-4)(x-2)$
$x-4=0$ or $x-2=0$
$x=4$ or $x=2$
Discard the solution 2 as extraneous. The solution is 4.

33. $\sqrt{2x-3} = 9$
$\left(\sqrt{2x-3}\right)^2 = 9^2$
$2x-3 = 81$
$2x = 84$
$x = 42$
The solution is 42.

34. $x^3+2x^2-4x>8$
$x^3+2x^2-4x-8>0$
$x^2(x+2)-4(x+2)>0$
$(x+2)(x^2-4)>0$
$(x+2)(x+2)(x-2)>0$
$(x+2)^2(x-2)>0$
$(x+2)^2=0$ or $x-2=0$
$x+2=0$ or $x=2$
$x=-2$

Region	Test Point	$(x+2)^2(x-2) > 0$ Result
A: $(-\infty, -2)$	-3	$(-1)^2(-5) > 0$ False
B: $(-2, 2)$	0	$(2)^2(-2) > 0$ False
C: $(2, \infty)$	3	$(5)^2(1) > 0$ True

Solution: $(2, \infty)$

35. a. $i^7 = i^4 \cdot i^3 = 1 \cdot (-i) = -i$

b. $i^{20} = (i^4)^5 = 1^5 = 1$

c. $i^{46} = i^{44} \cdot i^2 = (i^4)^{11} \cdot (-1) = 1^{11}(-1) = -1$

d. $i^{-12} = \frac{1}{i^{12}} = \frac{1}{(i^4)^3} = \frac{1}{1^3} = 1$

36. $f(x) = (x+2)^2 - 1$

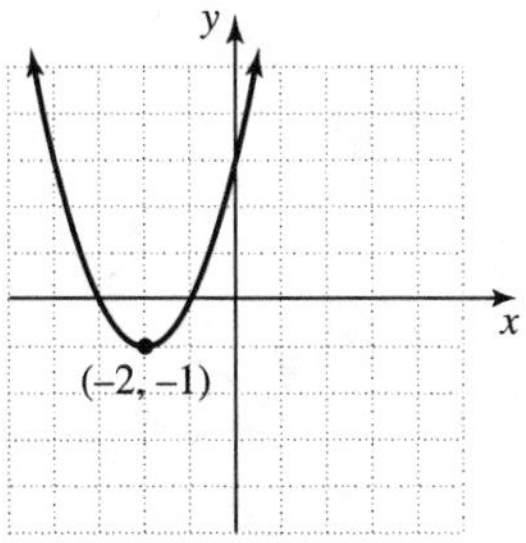

37.
$$p^2 + 2p = 4$$
$$p^2 + 2p + \left(\frac{2}{2}\right)^2 = 4 + 1$$
$$p^2 + 2p + 1 = 5$$
$$(p+1)^2 = 5$$
$$p + 1 = \pm\sqrt{5}$$
$$p = -1 \pm \sqrt{5}$$
The solutions are $-1+\sqrt{5}$ and $-1-\sqrt{5}$.

38. $f(x) = -x^2 - 6x + 4$
The maximum will occur at the vertex.
$$x = \frac{-b}{2a} = \frac{-(-6)}{2(-1)} = -3$$
$$f(-3) = -(-3)^2 - 6(-3) + 4 = 13$$
The maximum value is 13.

39.
$$\frac{1}{4}m^2 - m + \frac{1}{2} = 0$$
$$4\left(\frac{1}{4}m^2 - m + \frac{1}{2}\right) = 4(0)$$
$$m^2 - 4m + 2 = 0$$
$$a = 1, b = -4, c = 2$$
$$m = \frac{-(-4) \pm \sqrt{(-4)^2 - 4(1)(2)}}{2(1)}$$
$$= \frac{4 \pm \sqrt{16-8}}{2}$$
$$= \frac{4 \pm \sqrt{8}}{2}$$
$$= \frac{4 \pm 2\sqrt{2}}{2}$$
$$= 2 \pm \sqrt{2}$$
The solutions are $2+\sqrt{2}$ and $2-\sqrt{2}$.

40.
$$f(x) = \frac{x+1}{2}$$
$$y = \frac{x+1}{2}$$
$$x = \frac{y+1}{2}$$
$$2x = y + 1$$
$$2x - 1 = y$$
$$f^{-1}(x) = 2x - 1$$

41.
$$p^4 - 3p^2 - 4 = 0$$
$$(p^2 - 4)(p^2 + 1) = 0$$
$$(p+2)(p-2)(p^2+1) = 0$$
$$p + 2 = 0 \text{ or } p - 2 = 0 \text{ or } p^2 + 1 = 0$$
$$p = -2 \text{ or } p = 2 \text{ or } p^2 = -1$$
$$p = \pm i$$
The solutions are –2, 2, $-i$, and i.

42. $f(x) = x^2 - 3x + 2$
$g(x) = -3x + 5$

a. $$\begin{aligned}(f \circ g)(x) &= f[g(x)] \\ &= f(-3x+5) \\ &= (-3x+5)^2 - 3(-3x+5) + 2 \\ &= 9x^2 - 30x + 25 + 9x - 15 + 2 \\ &= 9x^2 - 21x + 12\end{aligned}$$

b. $$\begin{aligned}(f \circ g)(-2) &= f[g(-2)] \\ &= f[-3(-2)+5] \\ &= f(11) \\ &= (11)^2 - 3(11) + 2 \\ &= 121 - 33 + 2 \\ &= 90\end{aligned}$$

c. $$\begin{aligned}(g \circ f)(x) &= g[f(x)] \\ &= g(x^2 - 3x + 2) \\ &= -3(x^2 - 3x + 2) + 5 \\ &= -3x^2 + 9x - 6 + 5 \\ &= -3x^2 + 9x - 1\end{aligned}$$

d. $$\begin{aligned}(g \circ f)(5) &= g[f(5)] \\ &= g[(5)^2 - 3(5) + 2] \\ &= g(12) \\ &= -3(12) + 5 \\ &= -36 + 5 \\ &= -31\end{aligned}$$

43. $\dfrac{x+2}{x-3} \le 0$
$x + 2 = 0$ or $x - 3 = 0$
$x = -2$ or $x = 3$

Region	Test Point	$\dfrac{x+2}{x-3} \le 0$ Result
A: $(-\infty, -2)$	-3	$\dfrac{-1}{-6} \le 0$; False
B: $(-2, 3)$	0	$\dfrac{2}{-3} \le 0$; True
C: $(3, \infty)$	4	$\dfrac{6}{1} \le 0$; False

Solution: $[-2, 3)$

44. $4x^2 + 9y^2 = 36$
$$\frac{x^2}{9} + \frac{y^2}{4} = 1$$
Ellipse: center (0, 0), $a = 3$, $b = 2$

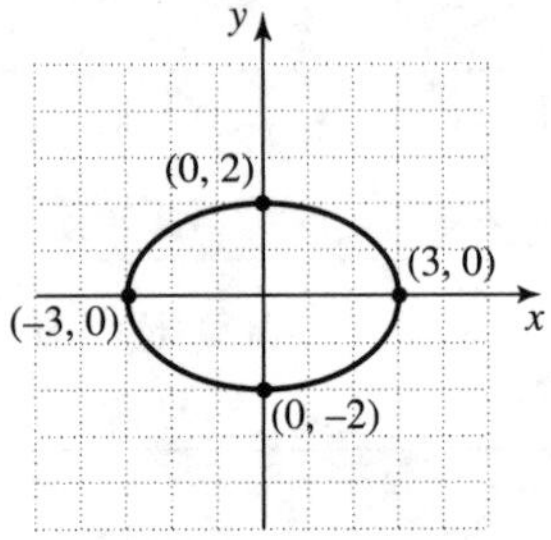

45. $g(x) = \dfrac{1}{2}(x+2)^2 + 5$
Vertex: $(-2, 5)$, axis: $x = -2$

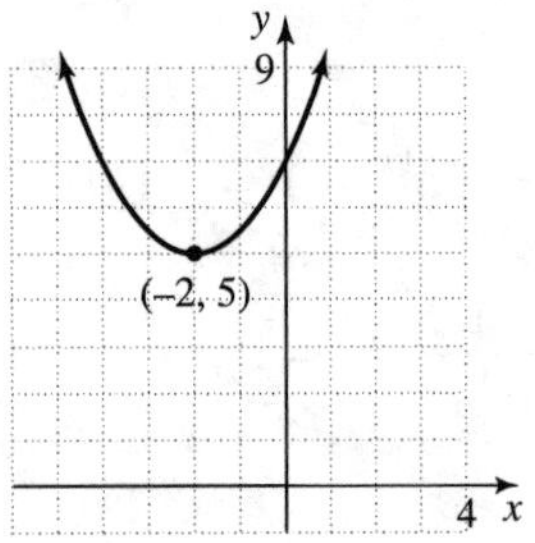

46. a. $$\begin{aligned}64^x &= 4 \\ (4^2)^x &= 4 \\ 4^{2x} &= 4 \\ 2x &= 1 \\ x &= \frac{1}{2}\end{aligned}$$

b. $$\begin{aligned}125^{x-3} &= 25 \\ (5^3)^{x-3} &= 5^2 \\ 5^{3x-9} &= 5^2 \\ 3x - 9 &= 2 \\ 3x &= 11 \\ x &= \frac{11}{3}\end{aligned}$$

c. $$\begin{aligned}\frac{1}{81} &= 3^{2x} \\ 3^{-4} &= 3^{2x} \\ -4 &= 2x \\ -\frac{4}{2} &= x \\ -2 &= x\end{aligned}$$

47. $f(x) = x^2 - 4x - 12$

$x = \frac{-b}{2a} = \frac{-(-4)}{2(1)} = 2$

$f(2) = (2)^2 - 4(2) - 12 = -16$

Vertex: (2, –16)

48. $\begin{cases} x + 2y < 8 \\ y \geq x^2 \end{cases}$

First, graph $x + 2y = 8$ as a dashed line.

Test Point	$x + 2y < 8$; Result
(0, 0)	$0 + 2(0) < 8$; True

Shade the region containing (0, 0). Next, graph the parabola $y = x^2$ as a solid curve.

Test Point	$y \geq x^2$; Result
(0, 1)	$1 \geq 0^2$; True

Shade the region containing (0, 1). The solution to the system is the intersection.

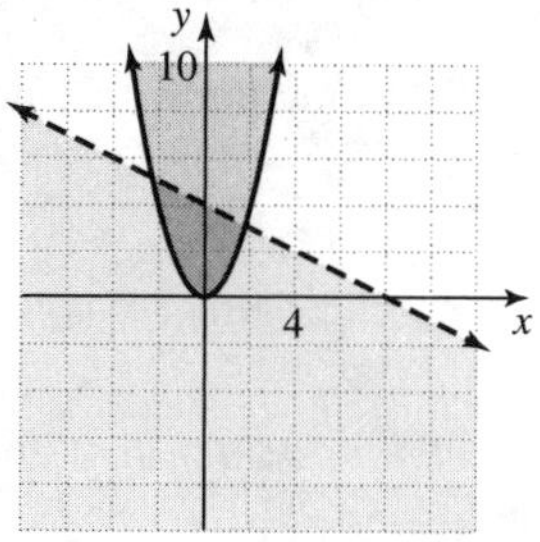

49. (2, –5), (1, –4)

$d = \sqrt{[-4-(-5)]^2 + (1-2)^2}$

$= \sqrt{1^2 + (-1)^2}$

$= \sqrt{2} \approx 1.414$

50. $\begin{cases} x^2 + y^2 = 36 & (1) \\ y = x + 6 & (2) \end{cases}$

Substitute $x + 6$ for y in E1.

$x^2 + (x+6)^2 = 36$

$x^2 + (x^2 + 12x + 36) = 36$

$2x^2 + 12x = 0$

$2x(x+6) = 0$

$2x = 0$ or $x + 6 = 0$

$x = 0$ or $x = -6$

Use these values in E2 to find y.

$x = 0: y = 0 + 6 = 6$

$x = -6: y = -6 + 6 = 0$

The solutions are (0, 6) and (–6, 0).

Chapter 11

Section 11.1 Practice Exercises

1. $a_n = 5+n^2$

$a_1 = 5+1^2 = 5+1 = 6$

$a_2 = 5+2^2 = 5+4 = 9$

$a_3 = 5+3^2 = 5+9 = 14$

$a_4 = 5+4^2 = 5+16 = 21$

$a_5 = 5+5^2 = 5+25 = 30$

Thus, the first five terms of the sequence are 6, 9, 14, 21, and 30.

2. $a_n = \frac{(-1)^n}{5n}$

a. $a_1 = \frac{(-1)^1}{5(1)} = -\frac{1}{5}$

b. $a_4 = \frac{(-1)^4}{5(4)} = \frac{1}{20}$

c. $a_{30} = \frac{(-1)^{30}}{5(30)} = \frac{1}{150}$

d. $a_{19} = \frac{(-1)^{19}}{5(19)} = -\frac{1}{95}$

3. a. 1, 3, 5, 7, ...
These numbers are the first four odd natural numbers, so a general term might be $a_n = 2n-1$.

b. 3, 9, 27, 81, ...
These numbers are all powers of 3 ($3 = 3^1$, $9 = 3^2$, $27 = 3^3$, and $81 = 3^4$), so a general term might be $a_n = 3^n$.

c. $\frac{1}{2}, \frac{2}{3}, \frac{3}{4}, \frac{4}{5}, \ldots$
The numerators are the first four natural numbers and each denominator is one greater than the numerator, so a general term might be $a_n = \frac{n}{n+1}$.

d. $-\frac{1}{2}, -\frac{1}{3}, -\frac{1}{4}, -\frac{1}{5}, \ldots$
The denominators are consecutive natural numbers beginning with 2 and each term is negative, so a general term might be

$a_n = -\frac{1}{n+1}$.

4. $v_n = 3950(0.8)^n$

$v_3 = 3950(0.8)^3$
$= 3950(0.512)$
$= 2022.4$

The value of the copier after three years is $2022.40.

Vocabulary, Readiness & Video Check 11.1

1. The *n*th term of the sequence a_n is called the general term.

2. A finite sequence is a function whose domain is {1, 2, 3, 4, ..., *n*} where *n* is some natural number.

3. An infinite sequence is a function whose domain is {1, 2, 3, 4, ...}.

4. $a_n = 7^n$

$a_1 = 7^1 = 7$

5. $a_n = \frac{(-1)^n}{n}$

$a_1 = \frac{(-1)^1}{1} = -1$

6. $a_n = (-1)^n \cdot n^4$

$a_1 = (-1)^1 \cdot 1^4 = -1$

7. A sequence is a function whose domain is the set of natural numbers. We use a_n to mean the general term of a sequence.

8. If the negative is inside the parentheses, such as $(-2)^n$, it is also raised to the power. Since a_2 and a_4 would then have even powers, they would then be positive terms—but all original terms are negative.

9. $a_9 = 0.10(2)^{9-1} = \$25.60$

Exercise Set 11.1

1. $a_n = n+4$

$a_1 = 1+4 = 5$
$a_2 = 2+4 = 6$
$a_3 = 3+4 = 7$
$a_4 = 4+4 = 8$
$a_5 = 5+4 = 9$

Thus, the first five terms of the sequence $a_n = n+4$ are 5, 6, 7, 8, 9.

3. $a_n = (-1)^n$

$a_1 = (-1)^1 = -1$
$a_2 = (-1)^2 = 1$
$a_3 = (-1)^3 = -1$
$a_4 = (-1)^4 = 1$
$a_5 = (-1)^5 = -1$

Thus, the first five terms of the sequence $a_n = (-1)^n$ are −1, 1, −1, 1, −1.

5. $a_n = \dfrac{1}{n+3}$

$a_1 = \dfrac{1}{1+3} = \dfrac{1}{4}$
$a_2 = \dfrac{1}{2+3} = \dfrac{1}{5}$
$a_3 = \dfrac{1}{3+3} = \dfrac{1}{6}$
$a_4 = \dfrac{1}{4+3} = \dfrac{1}{7}$
$a_5 = \dfrac{1}{5+3} = \dfrac{1}{8}$

Thus, the first five terms of the sequence $a_n = \dfrac{1}{n+3}$ are $\dfrac{1}{4}, \dfrac{1}{5}, \dfrac{1}{6}, \dfrac{1}{7}, \dfrac{1}{8}$.

7. $a_n = 2n$

$a_1 = 2(1) = 2$
$a_2 = 2(2) = 4$
$a_3 = 2(3) = 6$
$a_4 = 2(4) = 8$
$a_5 = 2(5) = 10$

Thus, the first five terms of the sequence $a_n = 2n$ are 2, 4, 6, 8, 10.

9. $a_n = -n^2$

$a_1 = -1^2 = -1$
$a_2 = -2^2 = -4$
$a_3 = -3^2 = -9$
$a_4 = -4^2 = -16$
$a_5 = -5^2 = -25$

Thus, the first five terms of the sequence $a_n = n^2$ are −1, −4, −8, −16, −25.

11. $a_n = 2^n$

$a_1 = 2^1 = 2$
$a_2 = 2^2 = 4$
$a_3 = 2^3 = 8$
$a_4 = 2^4 = 16$
$a_5 = 2^5 = 32$

Thus, the first five terms of the sequence $a_n = 2^n$ are 2, 4, 8, 16, 32.

13. $a_n = 2n+5$

$a_1 = 2(1)+5 = 2+5 = 7$
$a_2 = 2(2)+5 = 4+5 = 9$
$a_3 = 2(3)+5 = 6+5 = 11$
$a_4 = 2(4)+5 = 8+5 = 13$
$a_5 = 2(5)+5 = 10+5 = 15$

Thus, the first five terms of the sequence $a_n = 2n+5$ are 7, 9, 11, 13, 15.

15. $a_n = (-1)^n n^2$

$a_1 = (-1)^1(1)^2 = -1(1) = -1$
$a_2 = (-1)^2(2)^2 = 1(4) = 4$
$a_3 = (-1)^3(3)^2 = -1(9) = -9$
$a_4 = (-1)^4(4)^2 = 1(16) = 16$
$a_5 = (-1)^5(5)^2 = -1(25) = -25$

Thus, the first five terms of the sequence $a_n = (-1)^n n^2$ are −1, 4, −9, 16, −25.

17. $a_n = 3n^2$

$a_5 = 3(5)^2 = 3(25) = 75$

19. $a_n = 6n-2$

$a_{20} = 6(20)-2 = 120-2 = 118$

21. $a_n = \dfrac{n+3}{n}$

$a_{15} = \dfrac{15+3}{15} = \dfrac{18}{15} = \dfrac{6}{5}$

23. $a_n = (-3)^n$

$a_6 = (-3)^6 = 729$

25. $a_n = \dfrac{n-2}{n+1}$

$a_6 = \dfrac{6-2}{6+1} = \dfrac{4}{7}$

27. $a_n = \dfrac{(-1)^n}{n}$

$a_8 = \dfrac{(-1)^8}{8} = \dfrac{1}{8}$

29. $a_n = -n^2 + 5$

$a_{10} = -10^2 + 5 = -100 + 5 = -95$

31. $a_n = \dfrac{(-1)^n}{n+6}$

$a_{19} = \dfrac{(-1)^{19}}{19+6} = -\dfrac{1}{25}$

33. 3, 7, 11, 15, or 4(1) – 1, 4(2) – 1, 4(3) – 1, 4(4) – 1. In general, $a_n = 4n - 1$.

35. –2, –4, –8, –16, or –2, -2^2, -2^3, -2^4

In general, $a_n = -2^n$.

37. $\dfrac{1}{3}, \dfrac{1}{9}, \dfrac{1}{27}, \dfrac{1}{81}$, or $\dfrac{1}{3}, \dfrac{1}{3^2}, \dfrac{1}{3^3}, \dfrac{1}{3^4}$

In general, $a_n = \dfrac{1}{3^n}$.

39. $a_n = 32n - 16$

$a_2 = 32(2) - 16 = 64 - 16 = 48$ ft

$a_3 = 32(3) - 16 = 96 - 16 = 80$ ft

$a_4 = 32(4) - 16 = 128 - 16 = 112$ ft

41. 0.10, 0.20, 0.40, or 0.10, 0.10(2), $0.10(2)^2$

In general, $a_n = 0.10(2)^{n-1}$

$a_{14} = 0.10(2)^{13} = \$819.20$

43. $a_n = 75(2)^{n-1}$

$a_6 = 75(2)^5 = 75(32) = 2400$ cases

$a_1 = 75(2)^0 = 75(1) = 75$ cases

45. $a_n = \dfrac{1}{2}a_{n-1}$ for $n > 1, a_1 = 800$

In 2000, $n = 1$ and $a_1 = 800$.

In 2001, $n = 2$ and $a_2 = \dfrac{1}{2}(800) = 400$.

In 2002, $n = 3$ and $a_3 = \dfrac{1}{2}(400) = 200$.

In 2003, $n = 4$ and $a_4 = \dfrac{1}{2}(200) = 100$.

In 2004, $n = 5$ and $a_5 = \dfrac{1}{2}(100) = 50$.

The population estimate for 2004 is 50 sparrows. Continuing the sequence:

In 2005, $n = 6$ and $a_6 = \dfrac{1}{2}(50) = 25$.

In 2006, $n = 7$ and $a_7 = \dfrac{1}{2}(25) \approx 12$.

In 2007, $n = 8$ and $a_8 = \dfrac{1}{2}(12) = 6$.

In 2008, $n = 9$ and $a_9 = \dfrac{1}{2}(6) = 3$.

In 2009, $n = 10$ and $a_{10} = \dfrac{1}{2}(3) \approx 1$.

In 2010, $n = 11$ and $a_{11} = \dfrac{1}{2}(1) \approx 0$.

The population is estimated to have become extinct in 2010.

47. $f(x) = (x-1)^2 + 3$

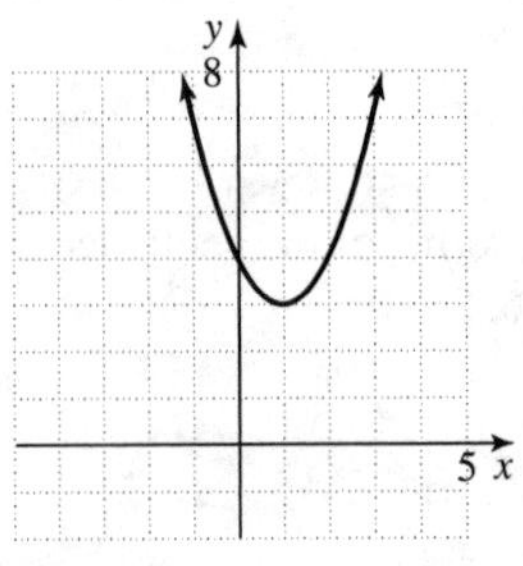

49. $f(x) = 2(x+4)^2 + 2$

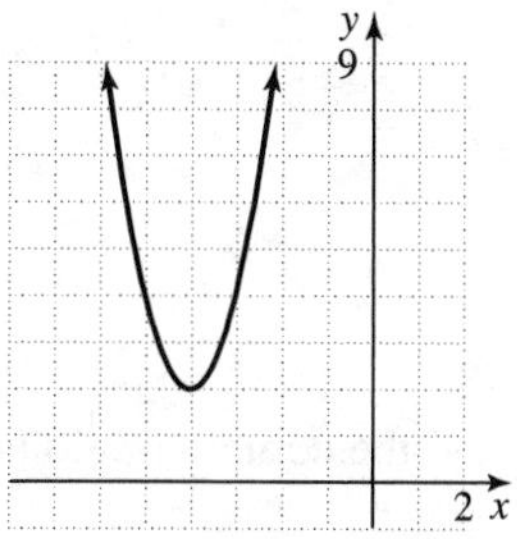

51. $(-4, -1)$ and $(-7, -3)$

$$\begin{aligned} d &= \sqrt{[-7-(-4)]^2 + [-3-(-1)]^2} \\ &= \sqrt{(-7+4)^2 + (-3+1)^2} \\ &= \sqrt{(-3)^2 + (-2)^2} \\ &= \sqrt{9+4} \\ &= \sqrt{13} \text{ units} \end{aligned}$$

53. $(2, -7)$ and $(-3, -3)$

$$\begin{aligned} d &= \sqrt{(-3-2)^2 + [-3-(-7)]^2} \\ &= \sqrt{(-5)^2 + (-3+7)^2} \\ &= \sqrt{(-5)^2 + (4)^2} \\ &= \sqrt{25+16} \\ &= \sqrt{41} \text{ units} \end{aligned}$$

55. $a_n = \dfrac{1}{\sqrt{n}}$

$$\begin{aligned} a_1 &= \frac{1}{\sqrt{1}} = \frac{1}{1} = 1 \\ a_2 &= \frac{1}{\sqrt{2}} \approx 0.7071 \\ a_3 &= \frac{1}{\sqrt{3}} \approx 0.5774 \\ a_4 &= \frac{1}{\sqrt{4}} = \frac{1}{2} = 0.5 \\ a_5 &= \frac{1}{\sqrt{5}} \approx 0.4472 \end{aligned}$$

Thus, the first five terms of the sequence $a_n = \dfrac{1}{\sqrt{n}}$ are 1, 0.7071, 0.5774, 0.5, 0.4472.

57. $a_n = \left(1 + \dfrac{1}{n}\right)^n$

$$\begin{aligned} a_1 &= \left(1+\frac{1}{1}\right)^1 = (2)^1 = 2 \\ a_2 &= \left(1+\frac{1}{2}\right)^2 = \left(\frac{3}{2}\right)^2 = 2.25 \\ a_3 &= \left(1+\frac{1}{3}\right)^3 = \left(\frac{4}{3}\right)^3 \approx 2.3704 \\ a_4 &= \left(1+\frac{1}{4}\right)^4 = \left(\frac{5}{4}\right)^4 \approx 2.4414 \\ a_5 &= \left(1+\frac{1}{5}\right)^5 = \left(\frac{6}{5}\right)^5 \approx 2.4883 \end{aligned}$$

Thus, the first five terms of the sequence $a_n = \left(1+\dfrac{1}{n}\right)^n$ are 2, 2.25, 2.3704, 2.4414, 2.4883.

Section 11.2 Practice Exercises

1. $a_1 = 4$

$a_2 = 4 + 5 = 9$

$a_3 = 9 + 5 = 14$

$a_4 = 14 + 5 = 19$

$a_5 = 19 + 5 = 24$

The first five terms are 4, 9, 14, 19, 24.

2. a. $a_n = a_1 + (n-1)d$

Here, $a_1 = 2$ and $d = -3$.

$a_n = 2 + (n-1)(-3) = 2 - 3n + 3 = 5 - 3n$

b. $a_n = 5 - 3n$

$a_{12} = 5 - 3 \cdot 12 = 5 - 36 = -31$

3. Since the sequence is arithmetic, the ninth term is $a_9 = a_1 + (9-1)d = a_1 + 8d$.

a_1 is the first term of the sequence, so $a_1 = 3$. d is the constant difference, so $d = a_2 - a_1 = 9 - 3 = 6$. Thus, $a_9 = a_1 + 8d = 3 + 8 \cdot 6 = 51$.

4. We need to find a_1 and d. The given facts, $a_3 = 23$ and $a_8 = 63$, lead to a system of linear equations.

$$\begin{cases} a_3 = a_1 + (3-1)d \\ a_8 = a_1 + (8-1)d \end{cases} \text{ or } \begin{cases} 23 = a_1 + 2d \\ 63 = a_1 + 7d \end{cases}$$

We solve the system by elimination. Multiply both sides of the second equation by -1.

$$\begin{cases} 23 = a_1 + 2d \\ -1(63) = -1(a_1 + 7d) \end{cases} \text{ or } \begin{cases} 23 = a_1 + 2d \\ -63 = -a_1 - 7d \end{cases}$$
$$-40 = -5d$$
$$8 = d$$

To find a_1, let $d = 8$ in $23 = a_1 + 2d$.

$23 = a_1 + 2(8)$
$23 = a_1 + 16$
$7 = a_1$

Thus, $a_1 = 7$ and $d = 8$, so
$a_n = 7 + (n-1)(8) = 7 + 8n - 8 = -1 + 8n$ and
$a_6 = -1 + 8 \cdot 6 = 47$.

5. The first term, a_1, is 57,000, and d is 2200.
$a_n = 57,000 + (n-1)(2200)$
$= 54,800 + 2200n$
$a_3 = 54,800 + 2200 \cdot 3 = 61,400$
The salary for the third year is $61,400.

6. $a_1 = 8$
$a_2 = 8(-3) = -24$
$a_3 = -24(-3) = 72$
$a_4 = 72(-3) = -216$
The first four terms are 8, −24, 72, and −216.

7. $a_n = a_1 r^{n-1}$

Here, $a_1 = 64$ and $r = \frac{1}{4}$.

Evaluate a_n for $n = 7$.

$a_7 = 64\left(\frac{1}{4}\right)^{7-1}$
$= 64\left(\frac{1}{4}\right)^6$
$= 64\left(\frac{1}{4096}\right)$
$= \frac{1}{64}$

8. Since the sequence is geometric and $a_1 = -3$, the seventh term must be $a_1 r^{7-1}$, or $-3r^6$. r is the common ratio of terms, so r must be $\frac{6}{-3}$, or −2.

$a_7 = -3r^6$
$a_7 = -3(-2)^6 = -192$

9. Notice that $\frac{27}{4} \div \frac{9}{2} = \frac{3}{2}$, so $r = \frac{3}{2}$.

$a_2 = a_1\left(\frac{3}{2}\right)^{2-1}$
$\frac{9}{2} = a_1\left(\frac{3}{2}\right)^1$, or $a_1 = 3$

The first term is 3, and the common ratio is $\frac{3}{2}$.

10. Since the culture is reduced by one-half each day, the population sizes are modeled by a geometric sequence. Here, $a_1 = 4800$ and $r = \frac{1}{2}$.

$a_n = a_1 r^{n-1} = 4800\left(\frac{1}{2}\right)^{n-1}$

$a_7 = 4800\left(\frac{1}{2}\right)^{7-1} = 75$

The bacterial culture should measure 75 units at the beginning of day 7.

Vocabulary, Readiness & Video Check 11.2

1. A geometric sequence is one in which each term (after the first) is obtained by multiplying the preceding term by a constant r. The constant r is called the common ratio.

2. An arithmetic sequence is one in which each term (after the first) differs from the preceding term by a constant amount d. The constant d is called the common difference.

3. The general term of an arithmetic sequence is $a_n = a_1 + (n-1)d$ where a_1 is the first term and d is the common difference.

4. The general term of a geometric sequence is $a_n = a_1 r^{n-1}$ where a_1 is the first term and r is the common ratio.

5. If there is a common difference between each term and its preceding term in a sequence, it's an arithmetic sequence.

6. An arithmetic sequence has a common difference between terms—you add the same number to go from term to term. A geometric sequence has a common ratio between terms—you multiply by the same number to go from term to term.

Exercise Set 11.2

1. $a_n = a_1 + (n-1)d$

$a_1 = 4;\ d = 2$
$a_1 = 4$
$a_2 = 4 + (2-1)2 = 6$
$a_3 = 4 + (3-1)2 = 8$
$a_4 = 4 + (4-1)2 = 10$
$a_5 = 4 + (5-1)2 = 12$

The first five terms are 4, 6, 8, 10, 12.

3. $a_n = a_1 + (n-1)d$

$a_1 = 6,\ d = -2$
$a_1 = 6$
$a_2 = 6 + (2-1)(-2) = 4$
$a_3 = 6 + (3-1)(-2) = 2$
$a_4 = 6 + (4-1)(-2) = 0$
$a_5 = 6 + (5-1)(-2) = -2$

The first five terms are 6, 4, 2, 0, −2.

5. $a_n = a_1 r^{n-1}$

$a_1 = 1,\ r = 3$
$a_1 = 1(3)^{1-1} = 1$
$a_2 = 1(3)^{2-1} = 3$
$a_3 = 1(3)^{3-1} = 9$
$a_4 = 1(3)^{4-1} = 27$
$a_5 = 1(3)^{5-1} = 81$

The first five terms are 1, 3, 9, 27, 81.

7. $a_n = a_1 r^{n-1}$

$a_1 = 48,\ r = \frac{1}{2}$

$a_1 = 48\left(\frac{1}{2}\right)^{1-1} = 48$

$a_2 = 48\left(\frac{1}{2}\right)^{2-1} = 24$

$a_3 = 48\left(\frac{1}{2}\right)^{3-1} = 12$

$a_4 = 48\left(\frac{1}{2}\right)^{4-1} = 6$

$a_5 = 48\left(\frac{1}{2}\right)^{5-1} = 3$

The first five terms are 48, 24, 12, 6, 3.

9. $a_n = a_1 + (n-1)d$

$a_1 = 12,\ d = 3$
$a_n = 12 + (n-1)3$
$a_8 = 12 + 7(3) = 12 + 21 = 33$

11. $a_n = a_1 r^{n-1}$

$a_1 = 7,\ d = -5$
$a_n = a_1 r^{n-1}$
$a_4 = 7(-5)^3 = 7(-125) = -875$

13. $a_n = a_1 + (n-1)d$

$a_1 = -4,\ d = -4$
$a_n = -4 + (n-1)(-4)$
$a_{15} = -4 + 14(-4) = -4 - 56 = -60$

15. 0, 12, 24

$a_1 = 0$ and $d = 12$
$a_n = 0 + (n-1)12$
$a_9 = 8(12) = 96$

17. 20, 18, 16

$a_1 = 20$ and $d = -2$
$a_n = 20 + (n-1)(-2)$
$a_{25} = 20 + 24(-2) = 20 - 48 = -28$

19. 2, −10, 50

$a_1 = 2$ and $r = -5$
$a_n = 2(-5)^{n-1}$
$a_5 = 2(-5)^4 = 2(625) = 1250$

21. $a_4 = 19,\ a_{15} = 52$

$\begin{cases} a_4 = a_1 + (4-1)d \\ a_{15} = a_1 + (15-1)d \end{cases}$ or

$\begin{cases} 19 = a_1 + 3d \\ 52 = a_1 + 14d \end{cases}$

$\begin{cases} -19 = -a_1 - 3d \\ 52 = a_1 + 14d \end{cases}$

Adding yields 33 = 11d or d = 3. Then $a_1 = 19 - 3(3) = 10$.

$$\begin{aligned} a_n &= 10 + (n-1)3 \\ &= 10 + 3n - 3 \\ &= 7 + 3n \end{aligned}$$

and $a_8 = 7 + 3(8)$
$= 7 + 24$
$= 31$

23. $a_2 = -1,\ a_4 = 5$

$\begin{cases} a_2 = a_1 + (2-1)d \\ a_4 = a_1 + (4-1)d \end{cases}$ or

$\begin{cases} -1 = a_1 + d \\ 5 = a_1 + 3d \end{cases}$

$\begin{cases} 1 = -a_1 - d \\ 5 = a_1 + 3d \end{cases}$

Adding yields 6 = 2*d* or *d* = 3. Then

$a_1 = -1 - 3 = -4.$

$$\begin{aligned} a_n &= -4 + (n-1)3 \\ &= -4 + 3n - 3 \\ &= -7 + 3n \end{aligned}$$

and $a_9 = -7 + 3(9)$
$= -7 + 27$
$= 20$

25. $a_2 = -\frac{4}{3}$ and $a_3 = \frac{8}{3}$

Notice that $\frac{8}{3} \div \frac{-4}{3} = \frac{8}{3} \cdot -\frac{3}{4} = -2$, so $r = -2$.

Then

$a_2 = a_1(-2)^{2-1}$

$-\frac{4}{3} = a_1(-2)$

$\frac{2}{3} = a_1.$

The first term is $\frac{2}{3}$ and the common ratio is -2.

27. answers may vary

29. 2, 4, 6 is an arithmetic sequence.

$a_1 = 2$ and $d = 2$

31. 5, 10, 20 is a geometric sequence.

$a_1 = 5$ and $r = 2$

33. $\frac{1}{2}, \frac{1}{10}, \frac{1}{50}$ is a geometric sequence.

$a_1 = \frac{1}{2}$ and $r = \frac{1}{5}$

35. $x, 5x, 25x$ is a geometric sequence.

$a_1 = x$ and $r = 5$

37. $p, p+4, p+8$ is an arithmetic sequence.

$a_1 = p$ and $d = 4$

39. $a_1 = 14$ and $d = \frac{1}{4}$

$a_n = 14 + (n-1)\frac{1}{4}$

$a_{21} = 14 + 20\left(\frac{1}{4}\right) = 14 + 5 = 19$

41. $a_1 = 3$ and $r = -\frac{2}{3}$

$a_n = 3\left(-\frac{2}{3}\right)^{n-1}$

$a_4 = 3\left(-\frac{2}{3}\right)^3 = 3\left(-\frac{8}{27}\right) = -\frac{8}{9}$

43. $\frac{3}{2}, 2, \frac{5}{2}, \ldots$

$a_1 = \frac{3}{2}$ and $d = \frac{1}{2}$

$a_n = \frac{3}{2} + (n-1)\frac{1}{2}$

$a_{15} = \frac{3}{2} + 14\left(\frac{1}{2}\right) = \frac{17}{2}$

45. $24, 8, \frac{8}{3}, \ldots$

$a_1 = 24$ and $r = \frac{1}{3}$

$a_n = 24\left(\frac{1}{3}\right)^{n-1}$

$a_6 = 24\left(\frac{1}{3}\right)^5 = 24\left(\frac{1}{243}\right) = \frac{8}{81}$

47. $a_3 = 2,\ a_{17} = -40$

$\begin{cases} a_3 = a_1 + (3-1)d \\ a_{17} = a_1 + (17-1)d \end{cases}$ or

$\begin{cases} 2 = a_1 + 2d \\ -40 = a_1 + 16d \end{cases}$

$\begin{cases} -2 = -a_1 - 2d \\ -40 = a_1 + 16d \end{cases}$

Adding yields $-42 = 14d$ or $d = -3$. Then

$a_1 = 2 - 2(-3) = 8.$

$a_n = 8 + (n-1)(-3) = 8 - 3n + 3 = 11 - 3n$

and

$a_{10} = 11 - 3(10) = 11 - 30 = -19$

49. $54, 58, 62$

$a_1 = 54$ and $d = 4$

$a_n = 54 + (n-1)4$

$a_{20} = 54 + 19(4) = 54 + 76 = 130$

The general term of the sequence is $a_n = 4n + 50$. There are 130 seats in the twentieth row.

51. $a_1 = 6$ and $r = 3$

$a_n = 6(3)^{n-1} = 2 \cdot 3 \cdot (3)^{n-1} = 2(3)^n$

The general term of the sequence is $a_n = 6(3)^{n-1}$ or $a_n = 2(3)^n$.

53. $a_1 = 486$ and $r = \dfrac{1}{3}$

Initial height $= a_1 = 486\left(\dfrac{1}{3}\right)^{1-1} = 486$

Rebound $1 = a_2 = 486\left(\dfrac{1}{3}\right)^{2-1} = 162$

Rebound $2 = a_3 = 486\left(\dfrac{1}{3}\right)^{3-1} = 54$

Rebound $3 = a_4 = 486\left(\dfrac{1}{3}\right)^{4-1} = 18$

Rebound $4 = a_5 = 486\left(\dfrac{1}{3}\right)^{5-1} = 6$

The first five terms of the sequence are 486, 162, 54, 18, 6.

The general term is $a_n = 486\left(\dfrac{1}{3}\right)^{n-1}$ or $a_n = \dfrac{486}{3^{n-1}}$. Since $a_6 = 2$ and $a_7 = \dfrac{2}{3}$, a_7 is the first term less than 1. Since a_7 corresponds to the 6th bounce, it takes 6 bounces for the ball to rebound less than 1 foot.

55. $a_1 = 4000$ and $d = 125$

$a_n = 4000 + (n-1)125$ or

$a_n = 3875 + 125n$

$a_{12} = 4000 + 11(125) = 5375$

His salary for his last month of training is \$5375.

57. $a_1 = 400$ and $r = \dfrac{1}{2}$

12 hours = 4(3 hours), so we seek the fourth term after a_1, namely a_5.

$a_n = a_1 r^{n-1}$

$a_5 = 400\left(\dfrac{1}{2}\right)^4 = \dfrac{400}{16} = 25$

25 grams of the radioactive material remain after 12 hours.

59. $$\frac{1}{3(1)} + \frac{1}{3(2)} + \frac{1}{3(3)} = \frac{1}{3} + \frac{1}{6} + \frac{1}{9} = \frac{6}{18} + \frac{3}{18} + \frac{2}{18} = \frac{11}{18}$$

61. $3^0 + 3^1 + 3^2 + 3^3 = 1 + 3 + 9 + 27 = 40$

63. $$\frac{8-1}{8+1} + \frac{8-2}{8+2} + \frac{8-3}{8+3} = \frac{7}{9} + \frac{6}{10} + \frac{5}{11} = \frac{770}{990} + \frac{594}{990} + \frac{450}{990} = \frac{1814}{990} = \frac{907}{495}$$

65. $a_1 = \$11,782.40$

$r = 0.5$

$a_2 = (11,782.40)(0.5) = \$5891.20$

$a_3 = (5891.20)(0.5) = \$2945.60$

$a_4 = (2945.60)(0.5) = \$1472.80$

The first four terms of the sequence are \$11,782.40, \$5891.20, \$2945.60, \$1472.80.

67. $a_1 = 19.652$ and $d = -0.034$

$a_2 = 19.652 - 0.034 = 19.618$

$a_3 = 19.618 - 0.034 = 19.584$

$a_4 = 19.584 - 0.034 = 19.550$

69. answers may vary

Section 11.3 Practice Exercises

1. a. $$\sum_{i=0}^{4} \frac{i-3}{4} = \frac{0-3}{4} + \frac{1-3}{4} + \frac{2-3}{4} + \frac{3-3}{4} + \frac{4-3}{4} = \left(-\frac{3}{4}\right) + \left(-\frac{2}{4}\right) + \left(-\frac{1}{4}\right) + 0 + \frac{1}{4} = -\frac{5}{4} \text{ or } -1\frac{1}{4}$$

b. $\sum_{i=2}^{5} 3^i = 3^2 + 3^3 + 3^4 + 3^5$
$= 9 + 27 + 81 + 243$
$= 360$

2. a. Since the difference of each term and the preceding term is 5, the terms correspond to the first six terms of the arithmetic sequence $a_n = 5 + (n-1)5 = 5n$. Thus, in summation notation,

$$5 + 10 + 15 + 20 + 25 + 30 = \sum_{i=1}^{6} 5i.$$

b. Since each term is the product of the preceding term and $\frac{1}{5}$, these terms correspond to the first four terms of the geometric sequence $a_n = \frac{1}{5}\left(\frac{1}{5}\right)^{n-1} = \left(\frac{1}{5}\right)^n$.

In summation notation,

$$\frac{1}{5} + \frac{1}{25} + \frac{1}{125} + \frac{1}{625} = \sum_{i=1}^{4}\left(\frac{1}{5}\right)^i.$$

3. $S_4 = \sum_{i=1}^{4} \frac{2+3i}{i^2}$
$= \frac{2+3\cdot 1}{1^2} + \frac{2+3\cdot 2}{2^2} + \frac{2+3\cdot 3}{3^2} + \frac{2+3\cdot 4}{4^2}$
$= \frac{5}{1} + \frac{8}{4} + \frac{11}{9} + \frac{14}{16}$
$= 5 + 2 + \frac{11}{9} + \frac{7}{8}$
$= \frac{655}{72}$ or $9\frac{7}{72}$

4. $S_5 = \sum_{i=1}^{5} i(2i-1)$
$= 1(2\cdot 1 - 1) + 2(2\cdot 2 - 1) + 3(2\cdot 3 - 1) + 4(2\cdot 4 - 1) + 5(2\cdot 5 - 1)$
$= 1 + 6 + 15 + 28 + 45$
$= 95$

There are 95 plants after 5 years.

Vocabulary, Readiness & Video Check 11.3

1. A series is an <u>infinite</u> series if it is the sum of all the terms of an infinite sequence.

2. A series is a <u>finite</u> series if it is the sum of a finite number of terms.

3. A shorthand notation for denoting a series when the general term of the sequence is known is called <u>summation</u> notation.

4. In the notation $\sum_{i=1}^{7}(5i-2)$, the Σ is the Greek uppercase letter <u>sigma</u> and the i is called the <u>index of summation</u>.

5. The sum of the first n terms of a sequence is a finite series known as a <u>partial sum</u>.

6. For the notation in Exercise 4 above, the beginning value of i is <u>1</u> and the ending value of i is <u>7</u>.

7. sigma/sum, index of summation, beginning value of i, ending value of i, and general term of the sequence

8. the sum of the first 7 terms of the sequence

Exercise Set 11.3

1. $\sum_{i=1}^{4}(i-3) = (1-3) + (2-3) + (3-3) + (4-3)$
$= -2 + (-1) + 0 + 1$
$= -2$

3. $\sum_{i=4}^{7}(2i+4) = [2(4)+4] + [2(5)+4] + [2(6)+4] + [2(7)+4]$
$= 12 + 14 + 16 + 18$
$= 60$

5. $\sum_{i=2}^{4}(i^2-3) = (2^2-3) + (3^2-3) + (4^2-3)$
$= 1 + 6 + 13$
$= 20$

7. $\sum_{i=1}^{3}\left(\frac{1}{i+5}\right) = \frac{1}{1+5} + \frac{1}{2+5} + \frac{1}{3+5}$
$= \frac{1}{6} + \frac{1}{7} + \frac{1}{8}$
$= \frac{28}{168} + \frac{24}{168} + \frac{21}{168}$
$= \frac{73}{168}$

9. $\sum_{i=1}^{3} \frac{1}{6i} = \frac{1}{6(1)} + \frac{1}{6(2)} + \frac{1}{6(3)}$
$= \frac{1}{6} + \frac{1}{12} + \frac{1}{18}$
$= \frac{6+3+2}{36}$
$= \frac{11}{36}$

11. $\sum_{i=2}^{6} 3i = 3(2)+3(3)+3(4)+3(5)+3(6)$
$= 6+9+12+15+18$
$= 60$

13. $\sum_{i=3}^{5} i(i+2) = 3(3+2)+4(4+2)+5(5+2)$
$= 15+24+35$
$= 74$

15. $\sum_{i=1}^{5} 2^i = 2^1+2^2+2^3+2^4+2^5$
$= 2+4+8+16+32$
$= 62$

17. $\sum_{i=1}^{4} \frac{4i}{i+3} = \frac{4(1)}{1+3} + \frac{4(2)}{2+3} + \frac{4(3)}{3+3} + \frac{4(4)}{4+3}$
$= 1 + \frac{8}{5} + 2 + \frac{16}{7}$
$= \frac{105}{35} + \frac{56}{35} + \frac{80}{35}$
$= \frac{241}{35}$

19. $1 + 3 + 5 + 7 + 9$
$a_1 = 1,\ d = 2$
$a_n = 1+(n-1)2 = 2n-1$
$\sum_{i=1}^{5}(2i-1)$

21. $4+12+36+108 = 4+4(3)+4(3)^2+4(3)^3$
$= \sum_{i=1}^{4} 4(3)^{i-1}$

23. $12+9+6+3+0+(-3)$
$a_1 = 12,\ d = -3$
$a_n = 12+(n-1)(-3) = -3n+15$
$\sum_{i=1}^{6}(-3i+15)$

25. $12+4+\frac{4}{3}+\frac{4}{9} = \frac{4}{3^{-1}} + \frac{4}{3^0} + \frac{4}{3} + \frac{4}{3^2}$
$= \sum_{i=1}^{4} \frac{4}{3^{i-2}}$

27. $1+4+9+16+25+36+49$
$= 1^2+2^2+3^2+4^2+5^2+6^2+7^2$
$= \sum_{i=1}^{7} i^2$

29. $a_n = (n+2)(n-5)$
$S_2 = \sum_{i=1}^{2}(i+2)(i-5)$
$= (1+2)(1-5)+(2+2)(2-5)$
$= 3(-4)+4(-3)$
$= -12-12$
$= -24$

31. $a_n = (-1)^n$
$S_6 = \sum_{i=1}^{6}(-1)^i$
$= (-1)^1+(-1)^2+(-1)^3+(-1)^4+(-1)^5$
$+(-1)^6$
$= -1+1+(-1)+1+(-1)+1$
$= 0$

33. $a_n = (n+3)(n+1)$
$S_4 = \sum_{i=1}^{4}(i+3)(i+1)$
$= (1+3)(1+1)+(2+3)(2+1)+(3+3)(3+1)$
$+(4+3)(4+1)$
$= 4(2)+5(3)+6(4)+7(5)$
$= 8+15+24+35$
$= 82$

35. $a_n = -2n$

$$S_4 = \sum_{i=1}^{4}(-2i) = -2(1)+(-2)(2)+(-2)(3)+(-2)(4) = -2-4-6-8 = -20$$

37. $a_n = -\frac{n}{3}$

$$S_3 = \sum_{i=1}^{3} -\frac{i}{3} = -\frac{1}{3}-\frac{2}{3}-\frac{3}{3} = -2$$

39. 1, 2, 3,…,10

$a_n = n$

$$S_{10} = \sum_{i=1}^{10} i = 1+2+3+\ldots+10 = 55$$

A total of 55 trees were planted.

41. $a_n = 6 \cdot 2^{n-1}$

$$S_5 = \sum_{i=1}^{5} 6 \cdot 2^{i-1} = 6(2)^0+6(2)^1+6(2)^2+6(2)^3+6(2)^4 = 6+12+24+48+96 = 186$$

Adding in the original 6 units, there will be 192 fungus units at the end of the fifth day.

43. $a_4 = (4+1)(4+3) = 5(7) = 35$

In the fourth year, 35 species were born.

$$S_4 = \sum_{i=1}^{4}(i+1)(i+3) = 2(4)+3(5)+4(6)+5(7) = 8+15+24+12 = 82$$

There were 82 species born in the first four years.

45. $a_n = (n+1)(n+2)$

$a_4 = (4+1)(4+2) = 5(6) = 30$

30 opossums were killed in the fourth month.

$$S_4 = \sum_{i=1}^{4}(i+1)(i+2) = 2(3)+(3)(4)+(4)(5)+(5)(6) = 6+12+20+30 = 68$$

68 opossums were killed in the four months.

47. $a_n = 100(0.5)^n$

$a_4 = 100(0.5)^4 = 6.25$

The decay in the fourth year is 6.25 pounds.

$$S_4 = \sum_{i=1}^{4} 100(0.5)^i = 100(0.5)^1+100(0.5)^2+100(0.5)^3+100(0.5)^4 = 100(0.5)+100(0.25)+100(0.125)+100(0.0625) = 50+25+12.5+6.25 = 93.75$$

The decay over the four years is 93.75 pounds.

49. $a_1 = 40$ and $r = \frac{4}{5}$

$$a_5 = 40\left(\frac{4}{5}\right)^4 = 16.384$$

The length of the fifth swing is approximately 16.4 inches.

$$S_5 = \sum_{i=1}^{5} 40\left(\frac{4}{5}\right)^{i-1} = 40\left(\frac{4}{5}\right)^0+40\left(\frac{4}{5}\right)^1+40\left(\frac{4}{5}\right)^2+40\left(\frac{4}{5}\right)^3+40\left(\frac{4}{5}\right)^4 = 40+32+25.6+20.48+16.384 = 134.464$$

The pendulum swings about 134.5 inches in five swings.

51. $\frac{5}{1-\frac{1}{2}} = \frac{5}{\frac{1}{2}} = 5 \cdot \frac{2}{1} = 10$

53. $\frac{\frac{1}{3}}{1-\frac{1}{10}} = \frac{\frac{1}{3}}{\frac{9}{10}} = \frac{1}{3} \cdot \frac{10}{9} = \frac{10}{27}$

55. $\frac{3(1-2^4)}{1-2} = \frac{3(1-16)}{-1} = \frac{3(-15)}{-1} = \frac{-45}{-1} = 45$

57. $\frac{10}{2}(3+15)=\frac{10}{2}(18)=\frac{180}{2}=90$

59. a. $\sum_{i=1}^{7}(i+i^2)$
$=(1+1^2)+(2+2^2)+(3+3^2)+(4+4^2)$
$+(5+5^2)+(6+6^2)+(7+7^2)$
$=2+6+12+20+30+42+56$

b. $\sum_{i=1}^{7}i+\sum_{i=1}^{7}i^2$
$=(1+2+3+4+5+6+7)$
$+(1+4+9+16+25+36+49)$

c. answers may vary

d. true; answers may vary

Integrated Review

1. $a_n=n-3$
$a_1=1-3=-2$
$a_2=2-3=-1$
$a_3=3-3=0$
$a_4=4-3=1$
$a_5=5-3=2$
Therefore, the first five terms are −2, −1, 0, 1, 2.

2. $a_n=\frac{7}{1+n}$
$a_1=\frac{7}{1+1}=\frac{7}{2}$
$a_2=\frac{7}{1+2}=\frac{7}{3}$
$a_3=\frac{7}{1+3}=\frac{7}{4}$
$a_4=\frac{7}{1+4}=\frac{7}{5}$
$a_5=\frac{7}{1+5}=\frac{7}{6}$
The first five terms are $\frac{7}{2}, \frac{7}{3}, \frac{7}{4}, \frac{7}{5},$ and $\frac{7}{6}$.

3. $a_n=3^{n-1}$
$a_1=3^{1-1}=3^0=1$
$a_2=3^{2-1}=3^1=3$
$a_3=3^{3-1}=3^2=9$
$a_4=3^{4-1}=3^3=27$
$a_5=3^{5-1}=3^4=81$
The first five terms are 1, 3, 9, 27, and 81.

4. $a_n=n^2-5$
$a_1=1^2-5=1-5=-4$
$a_2=2^2-5=4-5=-1$
$a_3=3^2-5=9-5=4$
$a_4=4^2-5=16-5=11$
$a_5=5^2-5=25-5=20$
The first five terms are −4, −1, 4, 11, and 20.

5. $(-2)^n$; a_6
$a_6=(-2)^6=64$

6. $-n^2+2$; a_4
$a_4=-(4)^2+2=-16+2=-14$

7. $\frac{(-1)^n}{n}$; a_{40}
$a_{40}=\frac{(-1)^{40}}{40}=\frac{1}{40}$

8. $\frac{(-1)^n}{2n}$; a_{41}
$a_{41}=\frac{(-1)^{41}}{2(41)}=\frac{-1}{82}=-\frac{1}{82}$

9. $a_1=7$; $d=-3$
$a_1=7$
$a_2=7-3=4$
$a_3=4-3=1$
$a_4=1-3=-2$
$a_5=-2-3=-5$
The first five terms are 7, 4, 1, −2, −5.

10. $a_1 = -3;\ r = 5$

$a_1 = -3$
$a_2 = -3(5) = -15$
$a_3 = -15(5) = -75$
$a_4 = -75(5) = -375$
$a_5 = -375(5) = -1875$

The first five terms are −3, −15, −75, −375, −1875.

11. $a_1 = 45;\ r = \frac{1}{3}$

$a_1 = 45$
$a_2 = 45\left(\frac{1}{3}\right) = 15$
$a_3 = 15\left(\frac{1}{3}\right) = 5$
$a_4 = 5\left(\frac{1}{3}\right) = \frac{5}{3}$
$a_5 = \frac{5}{3}\left(\frac{1}{3}\right) = \frac{5}{9}$

The first five terms are 45, 15, 5, $\frac{5}{3}$, $\frac{5}{9}$.

12. $a_1 = -12;\ d = 10$

$a_1 = -12$
$a_2 = -12 + 10 = -2$
$a_3 = -2 + 10 = 8$
$a_4 = 8 + 10 = 18$
$a_5 = 18 + 10 = 28$

The first five terms are −12, −2, 8, 18, 28.

13. $a_1 = 20;\ d = 9$

$a_n = a_1 + (n-1)d$

$$\begin{aligned} a_{10} &= 20 + (10-1)9 \\ &= 20 + 81 \\ &= 101 \end{aligned}$$

14. $a_1 = 64;\ r = \frac{3}{4}$

$a_n = a_1 r^{n-1}$

$$\begin{aligned} a_6 &= 64\left(\frac{3}{4}\right)^{6-1} \\ &= 64\left(\frac{3}{4}\right)^5 \\ &= 64\left(\frac{243}{1024}\right) \\ &= \frac{243}{16} \end{aligned}$$

15. $a_1 = 6;\ r = \frac{-12}{6} = -2$

$a_n = a_1 r^{n-1}$

$a_7 = 6(-2)^{7-1} = 6(-2)^6 = 6(64) = 384$

16. $a_1 = -100;\ d = -85 - (-100) = 15$

$a_n = a_1 + (n-1)d$

$$\begin{aligned} a_{20} &= -100 + (20-1)(15) \\ &= -100 + (19)(15) \\ &= -100 + 285 \\ &= 185 \end{aligned}$$

17. $a_4 = -5,\ a_{10} = -35$

$a_n = a_1 + (n-1)d$

$$\begin{cases} a_4 = a_1 + (4-1)d \\ a_{10} = a_1 + (10-1)d \end{cases}$$

$$\begin{cases} -5 = a_1 + 3d \\ -35 = a_1 + 9d \end{cases}$$

Multiply eq. 2 by −1, then add the equations.

$$\begin{cases} -5 = a_1 + 3d \\ (-1)(-35) = -1(a_1 + 9d) \end{cases}$$

$$\begin{cases} -5 = a_1 + 3d \\ 35 = -a_1 - 9d \end{cases}$$

$30 = -6d$
$-5 = d$

To find a_1, let $d = -5$ in

$-5 = a_1 + 3d$
$-5 = a_1 + 3(-5)$
$10 = a_1$

Thus, $a_1 = 10$ and $d = -5$, so

$a_n = 10 + (n-1)(-5) = -5n + 15$

$a_5 = -5(5) + 15 = -10$

18. $a_4 = 1;\ a_7 = \frac{1}{8}$

$a_n = a_1 r^{n-1}$

$a_4 = a_1 r^{4-1}$ so $1 = a_1 r^3$

$a_7 = a_1 r^{7-1}$ so $\frac{1}{8} = a_1 r^6$

Since $a_1 r^6 = (a_1 r^3) r^3$, $\frac{1}{8} = 1 \cdot r^3$ and $r = \frac{1}{2}$.

$a_5 = a_4 \cdot r$ so $a_5 = 1 \cdot \frac{1}{2} = \frac{1}{2}$

19. $\sum_{i=1}^{4} 5i = 5(1) + 5(2) + 5(3) + 5(4)$
$= 5 + 10 + 15 + 20$
$= 50$

20. $\sum_{i=1}^{7} (3i+2)$
$= (3(1)+2) + (3(2)+2) + (3(3)+2) + (3(4)+2) + (3(5)+2) + (3(6)+2) + (3(7)+2)$
$= 5 + 8 + 11 + 14 + 17 + 20 + 23$
$= 98$

21. $\sum_{i=3}^{7} 2^{i-4}$
$= 2^{3-4} + 2^{4-4} + 2^{5-4} + 2^{6-4} + 2^{7-4}$
$= 2^{-1} + 2^0 + 2^1 + 2^2 + 2^3$
$= \frac{1}{2} + 1 + 2 + 4 + 8$
$= 15\frac{1}{2}$
$= \frac{31}{2}$

22. $\sum_{i=2}^{5} \frac{i}{i+1} = \frac{2}{2+1} + \frac{3}{3+1} + \frac{4}{4+1} + \frac{5}{5+1}$
$= \frac{2}{3} + \frac{3}{4} + \frac{4}{5} + \frac{5}{6}$
$= \frac{61}{20}$

23. $S_3 = \sum_{i=1}^{3} i(i-4)$
$= 1(1-4) + 2(2-4) + 3(3-4)$
$= -3 - 4 - 3$
$= -10$

24. $S_{10} = \sum_{i=1}^{10} (-1)^i (i+1)$
$= (-1)^1(1+1) + (-1)^2(2+1) + (-1)^3(3+1) + (-1)^4(4+1) + (-1)^5(5+1) + (-1)^6(6+1) + (-1)^7(7+1) + (-1)^8(8+1) + (-1)^9(9+1) + (-1)^{10}(10+1)$
$= -2 + 3 - 4 + 5 - 6 + 7 - 8 + 9 - 10 + 11$
$= 5$

Section 11.4 Practice Exercises

1. 2, 9, 16, 23, 30, ...
Use the formula for S_n of an arithmetic sequence, replacing n with 5, a_1 with 2, and a_n with 30.

$S_n = \frac{n}{2}(a_1 + a_n)$

$S_5 = \frac{5}{2}(2+30) = \frac{5}{2}(32) = 80$

2. Because 1, 2, 3, …, 50 is an arithmetic sequence, use the formula for S_n with $n = 50$, $a_1 = 1$, and $a_n = 50$.

$S_n = \frac{n}{2}(a_1 + a_n)$

$S_5 = \frac{50}{2}(1+50) = 25(51) = 1275$

3. The list 6, 7, …, 15 is the first 10 terms of an arithmetic sequence. Use the formula for S_n with $n = 10$, $a_1 = 6$, and $a_n = 15$.

$S_{10} = \frac{10}{2}(6+15) = 5(21) = 105$

There are a total of 105 blocks of ice.

4. 32, 8, 2, $\frac{1}{2}$, $\frac{1}{8}$

Use the formula for the partial sum S_n of the terms of a geometric sequence. Here, $n = 5$, the first term $a_1 = 32$, and the common ratio $r = \frac{1}{4}$.

$$S_n = \frac{a_1(1-r^n)}{1-r}$$
$$S_5 = \frac{32\left[1-\left(\frac{1}{4}\right)^5\right]}{1-\frac{1}{4}}$$
$$= \frac{32\left(1-\frac{1}{1024}\right)}{\frac{3}{4}}$$
$$= \frac{32-\frac{1}{32}}{\frac{3}{4}}$$
$$= \frac{\frac{1023}{32}}{\frac{3}{4}}$$
$$= \frac{1023}{32} \cdot \frac{4}{3}$$
$$= \frac{341}{8} \text{ or } 42\frac{5}{8}$$

5. The donations are modeled by the first seven terms of a geometric sequence. Evaluate S_n when $n = 7$, $a_1 = 250{,}000$, and $r = 0.8$.
$$S_7 = \frac{250{,}000[1-(0.8)^7]}{1-0.8} = 987{,}856$$
The total amount donated during the seven years is \$987,856.

6. $7, \frac{7}{4}, \frac{7}{16}, \frac{7}{64}, \ldots$

For this geometric sequence $r = \frac{1}{4}$. Since $|r| < 1$, use the formula for S_∞ of a geometric sequence with $a_1 = 7$ and $r = \frac{1}{4}$.
$$S_\infty = \frac{a_1}{1-r} = \frac{7}{1-\frac{1}{4}} = \frac{7}{\frac{3}{4}} = \frac{28}{3} \text{ or } 9\frac{1}{3}$$

7. The ball travels both up and down before it comes to rest. Each direction of motion can be modeled by an infinite geometric sequence. For the downward direction, the first term, a_1, is 36 and the common ratio is 0.96. Since $|r| < 1$, we may use the formula for S_∞.
$$S_\infty = \frac{a_1}{1-r} = \frac{36}{1-0.96} = \frac{36}{0.04} = 900$$
For the upward direction, each bounce except the initial drop of 36 inches is traveled both upward and downward, so the upward distance is $900 - 36 = 864$ inches.
The ball travels a total distance of 900 in. + 864 in. = 1764 in.

Vocabulary, Readiness & Video Check 11.4

1. Each term after the first is 5 more than the preceding term; the sequence is <u>arithmetic</u>.

2. Each term after the first is 2 times the preceding term; the sequence is <u>geometric</u>.

3. Each term after the first is −3 times the preceding term; the sequence is <u>geometric</u>.

4. Each term after the first is 2 more than the preceding term; the sequence is <u>arithmetic</u>.

5. Each term after the first is 7 more than the preceding term; the sequence is <u>arithmetic</u>.

6. Each term after the first is −1 times the preceding term; the sequence is <u>geometric</u>.

7. Use the general term formula from Section 11.2 for the general term of an arithmetic sequence: $a_n = a_1 + (n-1)d$.

8. It would be a sequence in which every number is the same since you would multiply each term by 1 to get the next term; to find a partial sum of n terms, just multiply the first term by n since that term, which just repeats, will be added n times.

9. The common ratio r is 3 for this sequence so that $|r| \geq 1$, or $|3| \geq 1$; S_∞ doesn't exist if $|r| \geq 1$.

Exercise Set 11.4

1. 1, 3, 5, 7, ...
$d = 2$; $a_6 = 1 + (6-1)(2) = 11$
$$S_6 = \frac{6}{2}(1+11) = 3(12) = 36$$

3. 4, 12, 36, ...
$a_1 = 4$, $r = 3$, $n = 5$
$$S_5 = \frac{4(1-3^5)}{1-3} = 484$$

5. 3, 6, 9, ...
$d = 3$; $a_6 = 3 + (6-1)(3) = 18$
$$S_6 = \frac{6}{2}(3+18) = 3(21) = 63$$

7. $2, \frac{2}{5}, \frac{2}{25}, \ldots$

$a_1 = 2,\ r = \frac{1}{5},\ n = 4$

$S_4 = \frac{2\left[1-\left(\frac{1}{5}\right)^4\right]}{1-\frac{1}{5}} = \frac{\frac{1248}{625}}{\frac{4}{5}} = \frac{312}{125}$

9. $1, 2, 3, \ldots, 10$

The first term is 1 and the tenth term is 10.

$S_{10} = \frac{10}{2}(1+10) = 5(11) = 55$

11. $1, 2, 3, 7$

The first term is 1 and the fourth term is 7.

$S_4 = \frac{4}{2}(1+7) = 2(8) = 16$

13. 12, 6, 3, ...

$a_1 = 12,\ r = \frac{1}{2}$

$S_\infty = \frac{12}{1-\frac{1}{2}} = \frac{12}{\frac{1}{2}} = 12 \cdot \frac{2}{1} = 24$

15. $\frac{1}{10}, \frac{1}{100}, \frac{1}{1000}, \ldots$

$a_1 = \frac{1}{10},\ r = \frac{1}{10}$

$S_\infty = \frac{\frac{1}{10}}{1-\frac{1}{10}} = \frac{\frac{1}{10}}{\frac{9}{10}} = \frac{1}{10} \cdot \frac{10}{9} = \frac{1}{9}$

17. $-10, -5, -\frac{5}{2}, \ldots$

$a_1 = -10,\ r = \frac{1}{2}$

$S_\infty = \frac{-10}{1-\frac{1}{2}} = \frac{-10}{\frac{1}{2}} = -10 \cdot \frac{2}{1} = -20$

19. $2, -\frac{1}{4}, \frac{1}{32}, \ldots$

$a_1 = 2,\ r = -\frac{1}{8}$

$S_\infty = \frac{2}{1-\left(-\frac{1}{8}\right)} = \frac{2}{\frac{9}{8}} = 2 \cdot \frac{8}{9} = \frac{16}{9}$

21. $\frac{2}{3}, -\frac{1}{3}, \frac{1}{6}, \ldots$

$a_1 = \frac{2}{3},\ r = -\frac{1}{2}$

$S_\infty = \frac{\frac{2}{3}}{1-\left(-\frac{1}{2}\right)} = \frac{\frac{2}{3}}{\frac{3}{2}} = \frac{2}{3} \cdot \frac{2}{3} = \frac{4}{9}$

23. $-4, 1, 6, \ldots, 41$

The first term is –4 and the tenth term is 41.

$S_{10} = \frac{10}{2}(-4+41) = 5(37) = 185$

25. $3, \frac{3}{2}, \frac{3}{4}, \ldots$

$a_1 = 3,\ r = \frac{1}{2},\ n = 7$

$S_7 = \frac{3\left[1-\left(\frac{1}{2}\right)^7\right]}{1-\frac{1}{2}} = \frac{381}{64}$

27. $-12, 6, -3, \ldots$

$a_1 = -12,\ r = -\frac{1}{2},\ n = 5$

$S_5 = \frac{-12\left[1-\left(-\frac{1}{2}\right)^5\right]}{1-\left(-\frac{1}{2}\right)} = -\frac{33}{4}$ or -8.25

29. $\frac{1}{2}, \frac{1}{4}, 0, \ldots, -\frac{17}{4}$

The first term is $\frac{1}{2}$ and the twentieth term is $-\frac{17}{4}$.

$S_{20} = \frac{20}{2}\left(\frac{1}{2} - \frac{17}{4}\right) = 10\left(\frac{-15}{4}\right) = -\frac{75}{2}$

31. $a_1 = 8,\ r = -\frac{2}{3},\ n = 3$

$S_3 = \frac{8\left[1-\left(-\frac{2}{3}\right)^3\right]}{1-\left(-\frac{2}{3}\right)} = \frac{56}{9}$

33. The first five terms are 4000, 3950, 3900, 3850, 3800.

$a_1 = 4000,\ d = -50,\ n = 12$

$a_{12} = 4000 + 11(-50) = 3450$

3450 cars will be sold in month 12.

$S_{12} = \frac{12}{2}(4000 + 3450) = 44,700$

44,700 cars will be sold in the first year.

35. Firm *A*:

The first term is 22,000 and the tenth term is 31,000.

$$S_{10} = \frac{10}{2}(22,000 + 31,000) = \$265,000$$

Firm *B*:

The first term is 20,000 and the tenth term is 30,800.

$$S_{10} = \frac{10}{2}(20,000 + 30,800) = \$254,000$$

Thus, Firm *A* is making the more profitable offer.

37. $a_1 = 30,000,\ r = 1.10,\ n = 4$

$a_4 = 30,000(1.10)^{4-1} = 39,930$

She made \$39,930 during her fourth year of business.

$$S_4 = \frac{30,000(1 - 1.10^4)}{1 - 1.10} = 139,230$$

She made \$139,230 during the first four years of business.

39. $a_1 = 30,\ r = 0.9,\ n = 5$

$a_5 = 30(0.9)^{5-1} = 19.683$

Approximately 20 minutes to assemble the first computer.

$$S_5 = \frac{30(1 - 0.9^5)}{1 - 0.9} = 122.853$$

Approximately 123 minutes to assemble the first 5 computers.

41. $a_1 = 20,\ r = \frac{4}{5}$

$$S_\infty = \frac{20}{1 - \frac{4}{5}} = 100$$

We double the number (to account for the flight up as well as down) and subtract 20 (since the first bounce was preceded by only a downward flight). Thus, the ball travels 2(100) – 20 = 180 feet.

43. Player *A*:

The first term is 1 and the ninth term is 9.

$S_9 = \frac{9}{2}(1 + 9) = 45$ points

Player *B*:

The first term is 10 and the sixth term is 15.

$S_6 = \frac{6}{2}(10 + 15) = 75$ points

45. The first term is 200 and the twentieth is 200 – 19(5) = 105.

$S_{20} = \frac{20}{2}(200 + 105) = 3050$

Thus, \$3050 rent is paid for 20 days during the holiday rush.

47. $a_1 = 0.01,\ r = 2,\ n = 30$

$$S_3 = \frac{0.01\left[1 - 2^{30}\right]}{1 - 2} = 10,737,418.23$$

He would pay \$10,737,418.23 in room and board for the 30 days.

49. $6 \cdot 5 \cdot 4 \cdot 3 \cdot 2 \cdot 1 = 720$

51. $\frac{3 \cdot 2 \cdot 1}{2 \cdot 1} = \frac{3 \cdot \cancel{2} \cdot \cancel{1}}{\cancel{2} \cdot \cancel{1}} = 3$

53. $(x + 5)^2 = x^2 + 2 \cdot x \cdot 5 + 5^2 = x^2 + 10x + 25$

55.
$$\begin{aligned}(2x - 1)^3 &= (2x - 1)^2(2x - 1)\\ &= (4x^2 - 4x + 1)(2x - 1)\\ &= 8x^3 - 4x^2 + 2x - 8x^2 + 4x - 1\\ &= 8x^3 - 12x^2 + 6x - 1\end{aligned}$$

57.
$$\begin{aligned}0.88\bar{8} &= 0.8 + 0.08 + 0.008 + \cdots\\ &= \frac{8}{10} + \frac{8}{100} + \frac{8}{1000} + \cdots\end{aligned}$$

This is a geometric series with $a_1 = \frac{8}{10},\ r = \frac{1}{10}$.

$$S_\infty = \frac{\frac{8}{10}}{1 - \frac{1}{10}} = \frac{\frac{8}{10}}{\frac{9}{10}} = \frac{8}{10} \cdot \frac{10}{9} = \frac{8}{9}$$

59. answers may vary

Section 11.5 Practice Exercises

1. $(p+r)^7$

The $n = 7$ row of Pascal's triangle is

1 7 21 35 35 21 7 1

Using the $n = 7$ row of Pascal's triangle as the coefficients, $(p + r)^7$ can be expanded as

$p^7 + 7p^6r + 21p^5r^2 + 35p^4r^3 + 35p^3r^4 + 21p^2r^5 + 7pr^6 + r^7$

2. a. $\frac{6!}{7!} = \frac{6\cdot5\cdot4\cdot3\cdot2\cdot1}{7\cdot6\cdot5\cdot4\cdot3\cdot2\cdot1} = \frac{1}{7}$

b.
$$\begin{aligned}\frac{8!}{4!2!} &= \frac{8\cdot7\cdot6\cdot5\cdot4!}{4!\cdot2\cdot1}\\ &= \frac{8\cdot7\cdot6\cdot5}{2\cdot1}\\ &= 4\cdot7\cdot6\cdot5\\ &= 840\end{aligned}$$

c. $\frac{5!}{4!1!} = \frac{5\cdot4\cdot3\cdot2\cdot1}{4\cdot3\cdot2\cdot1\cdot1} = 5$

d. $\frac{9!}{9!0!} = \frac{9!}{9!\cdot1} = 1$

3. $(a+b)^9$

Let $n = 9$ in the binomial formula.

$$\begin{aligned}(a+b)^9 &= a^9 + \frac{9}{1!}a^8b + \frac{9\cdot8}{2!}a^7b^2 + \frac{9\cdot8\cdot7}{3!}a^6b^3 + \frac{9\cdot8\cdot7\cdot6}{4!}a^5b^4 + \frac{9\cdot8\cdot7\cdot6\cdot5}{5!}a^4b^5 + \frac{9\cdot8\cdot7\cdot6\cdot5\cdot4}{6!}a^3b^6\\ &\quad + \frac{9\cdot8\cdot7\cdot6\cdot5\cdot4\cdot3}{7!}a^2b^7 + \frac{9\cdot8\cdot7\cdot6\cdot5\cdot4\cdot3\cdot2}{8!}ab^8 + b^9\\ &= a^9 + 9a^8b + 36a^7b^2 + 84a^6b^3 + 126a^5b^4 + 126a^4b^5 + 84a^3b^6 + 36a^2b^7 + 9ab^8 + b^9\end{aligned}$$

4. $(a+5b)^3$

Replace b with $5b$ in the binomial formula.

$$\begin{aligned}(a+5b)^3 &= a^3 + \frac{3}{1!}a^2(5b) + \frac{3\cdot2}{2!}a(5b)^2 + (5b)^3\\ &= a^3 + 3a^2(5b) + 3a(25b^2) + 125b^3\\ &= a^3 + 15a^2b + 75ab^2 + 125b^3\end{aligned}$$

5. $(3x-2y)^3$

Let $a = 3x$ and $b = -2y$ in the binomial formula.

$$\begin{aligned}(3x-2y)^3 &= (3x)^3 + \frac{3}{1!}(3x)^2(-2y) + \frac{3\cdot2}{2!}(3x)(-2y)^2 + (-2y)^3\\ &= 27x^3 + 3(9x^2)(-2y) + 3(3x)(4y^2) - 8y^3\\ &= 27x^3 - 54x^2y + 36xy^2 - 8y^3\end{aligned}$$

6. $(x-4y)^{11}$

Use the formula with $n = 11$, $a = x$, $b = -4y$, and $r + 1 = 7$. Notice that, since $r + 1 = 7$, $r = 6$.

$$\begin{aligned}\frac{n!}{r!(n-r)!}a^{n-r}b^{r} &= \frac{11!}{6!5!}x^{5}(-4y)^{6}\\ &= 462x^{5}(4096y^{6})\\ &= 1{,}892{,}352x^{5}y^{6}\end{aligned}$$

Vocabulary, Readiness & Video Check 11.5

1. $0! = \underline{1}$

2. $1! = \underline{1}$

3. $4! = 4 \cdot 3 \cdot 2 \cdot 1 = \underline{24}$

4. $2! = 2 \cdot 1 = \underline{2}$

5. $3!0! = 3 \cdot 2 \cdot 1 \cdot 1 = \underline{6}$

6. $0!2! = 1 \cdot 2 \cdot 1 = \underline{2}$

7. Pascal's triangle gives you the coefficients of the terms of the expanded binomial; also, the power tells you how many terms the expansion has (1 more than the power on the binomial).

8. $4 \cdot 3 \cdot 2 \cdot 1 = 24$; $0! = 1$

9. The theorem is in terms of $(a+b)^n$, so if your binomial is in the form $(a-b)^n$, then remember to think of it as $(a+(-b))^n$, so your second term is $-b$.

10. We are using the formula for the $(r + 1)$st term in a binomial—so if $r + 1 = 4$ then $r = 3$.

Exercise Set 11.5

1. $(m+n)^3 = m^3 + 3m^2n + 3mn^2 + n^3$

3. $(c+d)^5 = c^5 + 5c^4d + 10c^3d^2 + 10c^2d^3 + 5cd^4 + d^5$

5.
$$\begin{aligned}(y-x)^5 &= [y+(-x)]^5\\ &= y^5 - 5y^4x + 10y^3x^2 - 10y^2x^3 + 5yx^4 - x^5\end{aligned}$$

7. answers may vary

9. $\frac{8!}{7!} = \frac{8 \cdot 7!}{7!} = 8$

11. $\frac{7!}{5!} = \frac{7 \cdot 6 \cdot 5!}{5!} = 7 \cdot 6 = 42$

13. $\frac{10!}{7!2!} = \frac{10 \cdot 9 \cdot 8 \cdot 7!}{7!2!} = \frac{10 \cdot 9 \cdot 8}{2 \cdot 1} = 360$

15. $\frac{8!}{6!0!} = \frac{8\cdot7\cdot6!}{6!1} = 8\cdot7 = 56$

17. Let $n = 7$ in the binomial theorem.

$$(a+b)^7 = a^7 + \frac{7}{1!}a^6b + \frac{7\cdot6}{2!}a^5b^2 + \frac{7\cdot6\cdot5}{3!}a^4b^3 + \frac{7\cdot6\cdot5\cdot4}{4!}a^3b^4 + \frac{7\cdot6\cdot5\cdot4\cdot3}{5!}a^2b^5 + \frac{7\cdot6\cdot5\cdot4\cdot3\cdot2}{6!}ab^6 + b^7$$
$$= a^7 + 7a^6b + 21a^5b^2 + 35a^4b^3 + 35a^3b^4 + 21a^2b^5 + 7ab^6 + b^7$$

19. Let $b = 2b$ and $n = 5$ in the binomial theorem.

$$(a+2b)^5 = a^5 + \frac{5}{1!}a^4(2b) + \frac{5\cdot4}{2!}a^3(2b)^2 + \frac{5\cdot4\cdot3}{3!}a^2(2b)^3 + \frac{5\cdot4\cdot3\cdot2}{4!}a(2b)^4 + (2b)^5$$
$$= a^5 + 10a^4b + 40a^3b^2 + 80a^2b^3 + 80ab^4 + 32b^5$$

21. Let $a = q$, $b = r$, and $n = 9$ in the binomial theorem.

$$(q+r)^9 = q^9 + \frac{9}{1!}q^8r + \frac{9\cdot8}{2!}q^7r^2 + \frac{9\cdot8\cdot7}{3!}q^6r^3 + \frac{9\cdot8\cdot7\cdot6}{4!}q^5r^4 + \frac{9\cdot8\cdot7\cdot6\cdot5}{5!}q^4r^5 + \frac{9\cdot8\cdot7\cdot6\cdot5\cdot4}{6!}q^3r^6$$
$$+ \frac{9\cdot8\cdot7\cdot6\cdot5\cdot4\cdot3}{7!}q^2r^7 + \frac{9\cdot8\cdot7\cdot6\cdot5\cdot4\cdot3\cdot2}{8!}qr^8 + r^9$$
$$= q^9 + 9q^8r + 36q^7r^2 + 84q^6r^3 + 126q^5r^4 + 126q^4r^5 + 84q^3r^6 + 36q^2r^7 + 9qr^8 + r^9$$

23. Let $a = 4a$ and $n = 5$ in the binomial theorem.

$$(4a+b)^5 = (4a)^5 + \frac{5}{1!}(4a)^4b + \frac{5\cdot4}{2!}(4a)^3b^2 + \frac{5\cdot4\cdot3}{3!}(4a)^2b^3 + \frac{5\cdot4\cdot3\cdot2}{4!}(4a)b^4 + b^5$$
$$= 1024a^5 + 1280a^4b + 640a^3b^2 + 160a^2b^3 + 20ab^4 + b^5$$

25. Let $a = 5a$, $b = -2b$, and $n = 4$ in the binomial theorem.

$$(5a-2b)^4 = (5a)^4 + \frac{4}{1!}(5a)^3(-2b) + \frac{4\cdot3}{2!}(5a)^2(-2b)^2 + \frac{4\cdot3\cdot2}{3!}(5a)(-2b)^3 + (-2b)^4$$
$$= 625a^4 - 1000a^3b + 600a^2b^2 - 160ab^3 + 16b^4$$

27. Let $a = 2a$, $b = 3b$, and $n = 3$ in the binomial theorem.

$$(2a+3b)^3 = (2a)^3 + \frac{3}{1!}(2a)^2(3b) + \frac{3\cdot2}{2!}(2a)(3b)^2 + (3b)^3$$
$$= 8a^3 + 36a^2b + 54ab^2 + 27b^3$$

29. Let $a = x$, $b = 2$, and $n = 5$ in the binomial theorem.

$$(x+2)^5 = x^5 + \frac{5}{1!}x^4(2) + \frac{5\cdot4}{2!}x^3(2)^2 + \frac{5\cdot4\cdot3}{3!}x^2(2)^3 + \frac{5\cdot4\cdot3\cdot2}{4!}x(2)^4 + (2)^5$$
$$= x^5 + 10x^4 + 40x^3 + 80x^2 + 80x + 32$$

31. 5th term of $(c-d)^5$ corresponds to $r = 4$:

$$\frac{5!}{4!(5-4)!}c^{5-4}(-d)^4 = 5cd^4$$

33. 8th term of $(2c+d)^7$ corresponds to $r = 7$:

$$\frac{7!}{7!(7-7)!}(2c)^{7-7}(d)^7 = d^7$$

35. 4th term of $(2r-s)^5$ corresponds to $r=3$:

$$\frac{5!}{3!(5-3)!}(2r)^{5-3}(-s)^3 = -40r^2s^3$$

37. 3rd term of $(x+y)^4$ corresponds to $r=2$: $\frac{4!}{2!(4-2)!}(x)^{4-2}(y)^2 = 6x^2y^2$

39. 2nd term of $(a+3b)^{10}$ corresponds to $r=1$: $\frac{10!}{1!(10-1)!}(a)^{10-1}(3b)^1 = 30a^9b$

41. $f(x)=|x|$

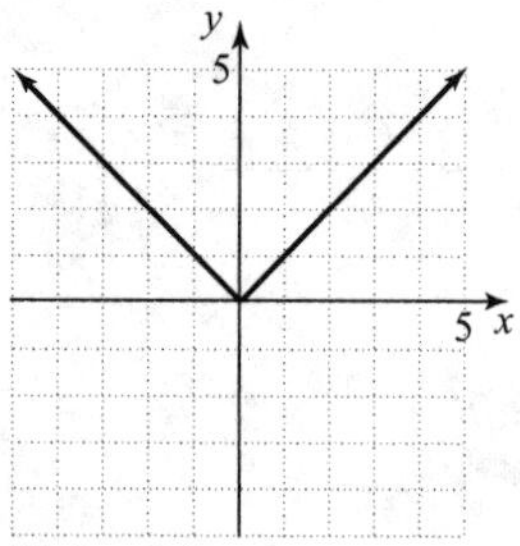

Not one-to-one

43. $H(x)=2x+3$

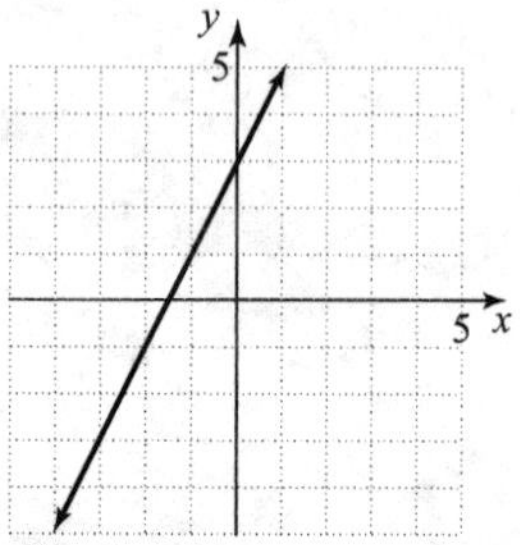

One-to-one

45. $f(x)=x^2+3$

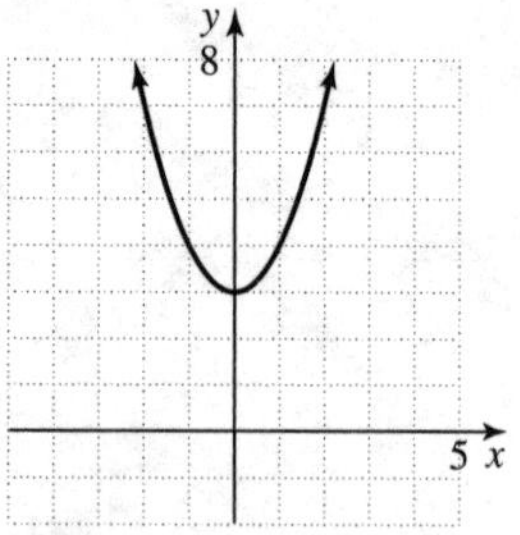

Not one-to-one

47. $(\sqrt{x}+\sqrt{3})^5$

Use the binomial theorem with $n=5$, $a=\sqrt{x}$, , and $b=\sqrt{3}$.

$$\begin{aligned}\left(\sqrt{x}+\sqrt{3}\right)^5 &= \left(\sqrt{x}\right)^5+5\left(\sqrt{x}\right)^4\left(\sqrt{3}\right)+10\left(\sqrt{x}\right)^3\left(\sqrt{3}\right)^2+10\left(\sqrt{x}\right)^2\left(\sqrt{3}\right)^3+5\left(\sqrt{x}\right)\left(\sqrt{3}\right)^4+\left(\sqrt{3}\right)^5\\ &= x^2\sqrt{x}+5\sqrt{3}x^2+30x\sqrt{x}+30\sqrt{3}x+45\sqrt{x}+9\sqrt{3}\end{aligned}$$

49. $\binom{9}{5} = \frac{9!}{5!(9-5)!}$
$= \frac{9!}{5!4!}$
$= \frac{9 \cdot 8 \cdot 7 \cdot 6 \cdot 5 \cdot 4 \cdot 3 \cdot 2 \cdot 1}{(5 \cdot 4 \cdot 3 \cdot 2 \cdot 1) \cdot (4 \cdot 3 \cdot 2 \cdot 1)}$
$= 126$

51. $\binom{8}{2} = \frac{8!}{2!(8-2)!}$
$= \frac{8!}{2!6!}$
$= \frac{8 \cdot 7 \cdot 6 \cdot 5 \cdot 4 \cdot 3 \cdot 2 \cdot 1}{(2 \cdot 1) \cdot (6 \cdot 5 \cdot 4 \cdot 3 \cdot 2 \cdot 1)}$
$= 28$

53. answers may vary

Chapter 11 Vocabulary Check

1. A finite sequence is a function whose domain is the set of natural numbers {1, 2, 3, ..., *n*}, where *n* is some natural number.
2. The factorial of *n*, written *n*!, is the product of the first *n* consecutive natural numbers.
3. An infinite sequence is a function whose domain is the set of natural numbers.
4. A geometric sequence is a sequence in which each term (after the first) is obtained by multiplying the preceding term by a constant amount *r*. The constant *r* is called the common ratio of the sequence.
5. A sum of the terms of a sequence is called a series.
6. The *n*th term of the sequence a_n is called the general term.
7. An arithmetic sequence is a sequence in which each term (after the first) differs from the preceding term by a constant amount *d*. The constant *d* is called the common difference of the sequence.
8. A triangular array of the coefficients of the terms of the expansions of $(a+b)^n$ is called Pascal's triangle.

Chapter 11 Review

1. $a_n = -3n^2$
$a_1 = -3(1)^2 = -3$
$a_2 = -3(2)^2 = -12$
$a_3 = -3(3)^2 = -27$
$a_4 = -3(4)^2 = -48$
$a_5 = -3(5)^2 = -75$

2. $a_n = n^2 + 2n$
$a_1 = 1^2 + 2(1) = 3$
$a_2 = 2^2 + 2(2) = 8$
$a_3 = 3^2 + 2(3) = 15$
$a_4 = 4^2 + 2(4) = 24$
$a_5 = 5^2 + 2(5) = 35$

3. $a_n = \frac{(-1)^n}{100}$
$a_{100} = \frac{(-1)^{100}}{100} = \frac{1}{100}$

4. $a_n = \frac{2n}{(-1)^n}$
$a_{50} = \frac{2(50)}{(-1)^{50}} = 100$

5. $\frac{1}{6 \cdot 1}, \frac{1}{6 \cdot 2}, \frac{1}{6 \cdot 3}, \ldots$
In general, $a_n = \frac{1}{6n}$.

6. $-1, 4, -9, 16, \ldots$
$a_n = (-1)^n n^2$

7. $a_n = 32n - 16$
$a_5 = 32(5) - 16 = 144$ feet
$a_6 = 32(6) - 16 = 176$ feet
$a_7 = 32(7) - 16 = 208$ feet

8. Since the measure is 80 at the end of the first day is 80, the measures at the ends of the first 5 days are:
80
$2 \cdot 80 = 160$
$2 \cdot 160 = 320$
$2 \cdot 320 = 640$
$2 \cdot 640 = 1280$

The general terms for the measure at the end of the *n*th day is $80 \cdot 2^{n-1}$.

$$80 \cdot 2^{n-1} \geq 10{,}000$$
$$2^{n-1} \geq 125$$

Since $2^6 = 64$ and $2^7 = 128$, $n - 1 = 7$, so $n = 8$. It will take 8 days before the yeast culture measures at least 10,000.

9. 2010: $a_1 = 97{,}000$

2011: $a_2 = 97{,}000(1.5) = 145{,}500$

2012: $a_3 = 145{,}500(1.5) = 218{,}250$

2013: $a_4 = 218{,}250(1.5) = 327{,}375$

2014: $a_5 = 327{,}375(1.5) = 491{,}062.5 \approx 491{,}063$

In 2014, there were approximately 491,063 acres of pine trees newly infested with pine beetles.

10. $a_n = 50 + (n-1)8$

$a_1 = 50$
$a_2 = 50 + 8 = 58$
$a_3 = 50 + 2(8) = 66$
$a_4 = 50 + 3(8) = 74$
$a_5 = 50 + 4(8) = 82$
$a_6 = 50 + 5(8) = 90$
$a_7 = 50 + 6(8) = 98$
$a_8 = 50 + 7(8) = 106$
$a_9 = 50 + 8(8) = 114$
$a_{10} = 50 + 9(8) = 122$

There are 122 seats in the tenth row.

11. $a_1 = -2,\ r = \frac{2}{3}$

$a_1 = -2$

$$a_2 = -2\left(\frac{2}{3}\right) = -\frac{4}{3}$$
$$a_3 = \left(-\frac{4}{3}\right)\left(\frac{2}{3}\right) = -\frac{8}{9}$$
$$a_4 = \left(-\frac{8}{9}\right)\left(\frac{2}{3}\right) = -\frac{16}{27}$$
$$a_5 = \left(-\frac{16}{27}\right)\left(\frac{2}{3}\right) = -\frac{32}{81}$$

12. $a_n = 12 + (n-1)(-1.5)$

$a_1 = 12$
$a_2 = 12 + (1)(-1.5) = 10.5$
$a_3 = 12 + 2(-1.5) = 9$
$a_4 = 12 + 3(-1.5) = 7.5$
$a_5 = 12 + 4(-1.5) = 6$

13. $a_n = -5 + (n-1)4$

$a_{30} = -5 + (30-1)4 = 111$

14. $a_n = 2 + (n-1)\frac{3}{4}$

$$a_{11} = 2 + 10\left(\frac{3}{4}\right) = \frac{19}{2}$$

15. 12, 7, 2,...

$a_1 = 12,\ d = -5,\ n = 20$

$a_{20} = 12 + (20-1)(-5) = -83$

16. $a_n = a_1 r^{n-1},\ a_1 = 4,\ r = \frac{3}{2}$

$$a_6 = 4\left(\frac{3}{2}\right)^{6-1} = \frac{243}{8}$$

17. $a_4 = 18,\ a_{20} = 98$

$$\begin{cases} a_4 = a_1 + (4-1)d \\ a_{20} = a_1 + (20-1)d \end{cases}$$
$$\begin{cases} 18 = a_1 + 3d \\ 98 = a_1 + 19d \end{cases}$$
$$\begin{cases} -18 = -a_1 - 3d \\ 98 = a_1 + 19d \end{cases}$$

Adding yields $80 = 16d$ or $d = 5$.
Then $a_1 = 18 - 3(5) = 3$.

18. $a_3 = -48,\ a_4 = 192$

$$r = \frac{a_4}{a_3} = \frac{192}{-48} = -4$$
$$a_3 = a_1 r^{3-1}$$
$$-48 = a_1(-4)^2$$
$$-48 = 16a_1$$
$$-3 = a_1$$
$$r = -4,\ a_1 = -3$$

19. $\frac{3}{10}, \frac{3}{10^2}, \frac{3}{10^3}, \ldots$

In general, $a_n = \frac{3}{10^n}$

20. 50, 58, 66, ...

$a_n = 50 + (n-1)8$ or $a_n = 42 + 8n$

21. $\frac{8}{3}, 4, 6, \ldots$

Geometric; $a_1 = \frac{8}{3}$,

$r = \frac{4}{\frac{8}{3}} = 4 \cdot \frac{3}{8} = \frac{12}{8} = \frac{3}{2}$

22. $-10.5, -6.1, -1.7$
Arithmetic; $a_1 = -10.5$,
$d = -6.1 - (-10.5) = 4.4$

23. $7x, -14x, 28x$

Geometric; $a_1 = 7x$, $r = -2$

24. $3x^2, 3x^2 + 5, 3x^2 + 10$

Arithmetic: $a_1 = 3x^2$

$d = 3x^2 + 5 - 3x^2 = 5$

25. $a_1 = 8$, $r = 0.75$

$a_1 = 8$
$a_2 = 8(0.75) = 6$
$a_3 = 8(0.75)^2 = 4.5$
$a_4 = 8(0.75)^3 \approx 3.4$
$a_5 = 8(0.75)^4 \approx 2.5$
$a_6 = 8(0.75)^5 \approx 1.9$

Yes, a ball that rebounds to a height of 2.5 feet after the fifth bounce is good, since $2.5 \geq 1.9$.

26. $a_1 = 25$, $d = -4$
$a_n = a_1 + (n-1)d$
$a_n = 25 + (n-1)(-4) = 29 - 4n$
$a_7 = 25 + 6(-4) = 1$
Continuing the progression as far as possible leaves 1 can in the top row.

27. $a_1 = 1$, $r = 2$

$a_n = 2^{n-1}$

$a_{10} = 2^9 = 512$

$a_{30} = 2^{29} = 536{,}870{,}912$

You save \$512 on the tenth day and \$536,870,912 on the thirtieth day.

28. $a_n = a_1 r^{n-1}$, $a_1 = 30$, $r = 0.7$

$a_5 = 30(0.7)^4 = 7.203$

The length is 7.203 inches on the fifth swing.

29. $a_1 = 900$, $d = 150$

$a_n = 900 + (n-1)150 = 150n + 750$
$a_6 = 900 + (6-1)150 = 1650$

Her salary is \$1650 per month at the end of training.

30. $\frac{1}{512}, \frac{1}{256}, \frac{1}{128}, \ldots$

first fold: $a_1 = \frac{1}{256}$, $r = 2$

$a_{15} = \frac{1}{256}(2)^{15-1} = 64$

After 15 folds, the thickness is 64 inches.

31.
$$\begin{aligned}\sum_{i=1}^{5}(2i-1) &= [2(1)-1]+[2(2)-1]+[2(3)-1] \\ &\quad +[2(4)-1]+[2(5)-1] \\ &= 1+3+5+7+9 \\ &= 25\end{aligned}$$

32.
$$\begin{aligned}\sum_{i=1}^{5} i(i+2) &= 1(1+2)+2(2+2)+3(3+2) \\ &\quad +4(4+2)+5(5+2) \\ &= 3+8+15+24+35 \\ &= 85\end{aligned}$$

33.
$$\begin{aligned}\sum_{i=2}^{4}\frac{(-1)^i}{2i} &= \frac{(-1)^2}{2(2)}+\frac{(-1)^3}{2(3)}+\frac{(-1)^4}{2(4)} \\ &= \frac{1}{4}-\frac{1}{6}+\frac{1}{8} \\ &= \frac{5}{24}\end{aligned}$$

34.
$$\begin{aligned}\sum_{i=3}^{5}5(-1)^{i-1} &= 5(-1)^{3-1}+5(-1)^{4-1}+5(-1)^{5-1} \\ &= 5(1)+5(-1)+5(1) \\ &= 5-5+5 \\ &= 5\end{aligned}$$

35.
$$\begin{aligned}&1+3+9+27+81+243 \\ &= 3^0+3^1+3^2+3^3+3^4+3^5 \\ &= \sum_{i=1}^{6}3^{i-1}\end{aligned}$$

36. $6+2+(-2)+(-6)+(-10)+(-14)+(-18)$

$a_1 = 6,\ d = -4$

$a_n = 6+(n-1)(-4)$

$$\sum_{i=1}^{7}[6+(i-1)(-4)]$$

37. $$\frac{1}{4}+\frac{1}{16}+\frac{1}{64}+\frac{1}{256}=\frac{1}{4^1}+\frac{1}{4^2}+\frac{1}{4^3}+\frac{1}{4^4}=\sum_{i=1}^{4}\frac{1}{4^i}$$

38. $$1+\left(-\frac{3}{2}\right)+\frac{9}{4}=\left(-\frac{3}{2}\right)^0+\left(-\frac{3}{2}\right)^1+\left(-\frac{3}{2}\right)^2=\sum_{i=1}^{3}\left(-\frac{3}{2}\right)^{i-1}$$

39. $a_1 = 20,\ r = 2$

$a_n = 20(2)^n$ represents the number of yeast, where n represents the number of 8-hour periods. Since $48 = 6(8)$ here, $n = 6$.

$a_6 = 20(2)^6 = 1280$

There are 1280 yeast after 48 hours.

40. $a_n = n^2 + 2n - 1$

$a_4 = (4)^2 + 2(4) - 1 = 23$

$$S_4 = \sum_{i=1}^{4}(i^2+2i-1)$$
$$= (1+2-1)+(4+4-1)+(9+6-1)+(16+8-1)$$
$$= 46$$

23 cranes born in the fourth year and 46 cranes born in the first four years.

41. For Job *A*: $a_1 = 39{,}500,\ d = 2200;$

$a_5 = 39{,}500+(5-1)2200 = \$48{,}330$

For Job *B*: $a_1 = 41{,}000,\ d = 1400$

$a_5 = 41{,}000+(5-1)1400 = \$46{,}600$

For the fifth year, Job *A* has a higher salary.

42. $a_n = 200(0.5)^n$

$a_3 = 200(0.5)^3 = 25$

$$S_3 = \sum_{i=1}^{3}200(0.5)^i$$
$$= 200(0.5)+200(0.5)^2+200(0.5)^3$$
$$= 175$$

25 kilograms decay in the third year and 175 kilograms decay in the first three years.

43. $a_n = (n-3)(n+2)$

$$S_4 = \sum_{i=1}^{4}(i-3)(i+2)$$
$$= (1-3)(1+2)+(2-3)(2+2)+(3-3)(3+2)+(4-3)(4+2)$$
$$= -6-4+0+6$$
$$= -4$$

44. $a_n = n^2$

$$S_6 = \sum_{i=1}^{6}i^2$$
$$= (1)^2+(2)^2+(3)^2+(4)^2+(5)^2+(6)^2$$
$$= 91$$

45. $a_n = -8+(n-1)3 = 3n-11$

$$S_5 = \sum_{i=1}^{5}(3i-11)$$
$$= [3(1)-11]+[3(2)-11]+[3(3)-11]+[3(4)-11]+[3(5)-11]$$
$$= -8-5-2+1+4$$
$$= -10$$

46. $a_n = 5(4)^{n-1}$

$$S_3 = \sum_{i=1}^{3}5(4)^{i-1} = 5(4)^0+5(4)^1+5(4)^2 = 105$$

47. 15, 19, 23, ...

$a_1 = 15,\ d = 4,\ a_6 = 15+(6-1)4 = 35$

$$S_6 = \frac{6}{2}[15+35] = 150$$

48. $5, -10, 20, \ldots$

$a_1 = 5,\ r = -2$

$$S_n = \frac{a_1(1-r^n)}{1-r}$$

$$S_9 = \frac{5(1-(-2)^9)}{1-(-2)} = 855$$

49. $a_1 = 1,\ d = 2,\ n = 30,\ a_{30} = 1 + (30-1)2 = 59$

$S_{30} = \frac{30}{2}[1+59] = 900$

50. 7, 14, 21, 28, ...

$a_n = 7 + (n-1)7$
$a_{20} = 7 + (20-1)7 = 140$
$S_{20} = \frac{20}{2}(7+140) = 1470$

51. 8, 5, 2, ...

$a_1 = 8,\ d = -3,\ n = 20$
$a_{20} = 8 + (20-1)(-3) = -49$
$S_{20} = \frac{20}{2}[8+(-49)]$
$= -410$

52. $\frac{3}{4}, \frac{9}{4}, \frac{27}{4}, \ldots$

$a_1 = \frac{3}{4},\ r = 3$

$S_8 = \frac{\frac{3}{4}(1-3^8)}{1-3} = 2460$

53. $a_1 = 6,\ r = 5$

$S_4 = \frac{6(1-5^4)}{1-5} = 936$

54. $a_1 = -3,\ d = -6$

$a_n = -3 + (n-1)(-6)$
$a_{100} = -3 + (100-1)(-6) = -597$
$S_{100} = \frac{100}{2}(-3+(-597)) = -30{,}000$

55. $5, \frac{5}{2}, \frac{5}{4}, \ldots$

$a_1 = 5,\ r = \frac{1}{2}$

$S_\infty = \frac{5}{1-\frac{1}{2}} = 10$

56. $18, -2, \frac{2}{9}, \ldots$

$a_1 = 18,\ r = -\frac{1}{9}$

$S_\infty = \frac{18}{1+\frac{1}{9}} = \frac{81}{5}$

57. $-20, -4, -\frac{4}{5}, \ldots$

$a_1 = -20,\ r = \frac{1}{5}$

$S_\infty = \frac{-20}{1-\frac{1}{5}} = -25$

58. 0.2, 0.02, 0.002,...

$a_1 = 0.2 = \frac{1}{5},\ r = \frac{1}{10}$

$S_\infty = \frac{\frac{1}{5}}{1-\frac{1}{10}} = \frac{2}{9}$

59. $a_1 = 20{,}000,\ r = 1.15,\ n = 4$

$a_4 = 20{,}000(1.15)^{4-1} = 30{,}418$

$S_4 = \frac{20{,}000(1-1.15^4)}{1-1.15} = 99{,}868$

He earned $30,418 during the fourth year and $99,868 over the four years.

60. $a_n = 40(0.8)^{n-1}$

$a_4 = 40(0.8)^{4-1} = 20.48$

$S_4 = \frac{40(1-0.8^4)}{1-0.8} = 118.08$

He takes 20 minutes to assemble the fourth television and 118 minutes to assemble the first four televisions.

61. $a_1 = 100, d = -7, n = 7$

$a_7 = 100 + (7-1)(-7) = 58$

$S_7 = \frac{7}{2}(100+58) = 553$

The rent for the seventh day is $58 and the rent for 7 days is $553.

62. $a_1 = 15,\ r = 0.8$

$$S_\infty = \frac{15}{1-0.8} = 75 \text{ feet downward}$$

$a_1 = 15(0.8) = 12,\ r = 0.8$

$$S_\infty = \frac{12}{1-0.8} = 60 \text{ feet upward}$$

The total distance is 135 feet.

63. 1800, 600, 200,...

$a_1 = 1800,\ r = \frac{1}{3},\ n = 6$

$$S_6 = 1800\frac{\left(1-\left(\frac{1}{3}\right)^6\right)}{1-\frac{1}{3}} \approx 2696$$

Approximately 2696 mosquitoes were killed during the first six days after the spraying.

64. 1800, 600, 200, ...

For which n is $a_n < 1$?

$$a_n = 1800\left(\frac{1}{3}\right)^{n-1} < 1$$

$$\left(\frac{1}{3}\right)^{n-1} < \frac{1}{1800}$$

$$(n-1)\log\frac{1}{3} < \log\frac{1}{1800}$$

$$(n-1)\log 3^{-1} < \log 1800^{-1}$$

$$(n-1)(-\log 3) < -\log 1800$$

$$n-1 > \frac{-\log 1800}{-\log 3}$$

$$n > 1 + \frac{\log 1800}{\log 3}$$

$$n > 7.8$$

No longer effective on the 8th day

$$S_8 = \frac{1800\left(1-\left(\frac{1}{3}\right)^8\right)}{1-\frac{1}{3}} \approx 2700$$

About 2700 mosquitoes were killed.

65. $0.55\overline{5} = 0.5 + 0.05 + 0.005 + \cdots$

$a_1 = 0.5,\ r = 0.1$

$$S_\infty = \frac{0.5}{1-0.1} = \frac{0.5}{0.9} = \frac{5}{9}$$

66. 27, 30, 33, ...
$a_n = 27 + (n-1)(3)$
$a_{20} = 27 + (20-1)(3) = 84$
$S_{20} = \frac{20}{2}(27+84) = 1110$
There are 1110 seats in the theater.

67. $(x+z)^5 = x^5 + 5x^4z + 10x^3z^2 + 10x^2z^3 + 5xz^4 + z^5$

68. $$\begin{aligned}(y-r)^6 &= y^6 + 6y^5(-r) + 15y^4(-r)^2 + 20y^3(-r)^3 + 15y^2(-r)^4 + 6y(-r)^5 + (-r)^6\\ &= y^6 - 6y^5r + 15y^4r^2 - 20y^3r^3 + 15y^2r^4 - 6yr^5 + r^6\end{aligned}$$

69. $$\begin{aligned}(2x+y)^4 &= (2x)^4 + 4(2x)^3y + 6(2x)^2y^2 + 4(2x)y^3 + y^4\\ &= 16x^4 + 32x^3y + 24x^2y^2 + 8xy^3 + y^4\end{aligned}$$

70. $$\begin{aligned}(3y-z)^4 &= (3y)^4 + 4(3y)^3(-z) + 6(3y)^2(-z)^2 + 4(3y)(-z)^3 + (-z)^4\\ &= 81y^4 - 108y^3z + 54y^2z^2 - 12yz^3 + z^4\end{aligned}$$

71. $$\begin{aligned}(b+c)^8 &= b^8 + \frac{8}{1!}b^7c + \frac{8\cdot 7}{2!}b^6c^2 + \frac{8\cdot 7\cdot 6}{3!}b^5c^3 + \frac{8\cdot 7\cdot 6\cdot 5}{4!}b^4c^4 + \frac{8\cdot 7\cdot 6\cdot 5\cdot 4}{5!}b^3c^5\\ &\quad + \frac{8\cdot 7\cdot 6\cdot 5\cdot 4\cdot 3}{6!}b^2c^6 + \frac{8\cdot 7\cdot 6\cdot 5\cdot 4\cdot 3\cdot 2}{7!}bc^7 + c^8\\ &= b^8 + 8b^7c + 28b^6c^2 + 56b^5c^3 + 70b^4c^4 + 56b^3c^5 + 28b^2c^6 + 8bc^7 + c^8\end{aligned}$$

72. $$\begin{aligned}(x-w)^7 &= x^7 + \frac{7}{1!}x^6(-w) + \frac{7\cdot 6}{2!}x^5(-w)^2 + \frac{7\cdot 6\cdot 5}{3!}x^4(-w)^3 + \frac{7\cdot 6\cdot 5\cdot 4}{4!}x^3(-w)^4 + \frac{7\cdot 6\cdot 5\cdot 4\cdot 3}{5!}x^2(-w)^5\\ &\quad + \frac{7\cdot 6\cdot 5\cdot 4\cdot 3\cdot 2}{6!}x(-w)^6 + (-w)^7\\ &= x^7 - 7x^6w + 21x^5w^2 - 35x^4w^3 + 35x^3w^4 - 21x^2w^5 + 7xw^6 - w^7\end{aligned}$$

73. $$\begin{aligned}(4m-n)^4 &= (4m)^4 + \frac{4}{1!}(4m)^3(-n) + \frac{4\cdot 3}{2!}(4m)^2(-n)^2 + \frac{4\cdot 3\cdot 2}{3!}(4m)(-n)^3 + (-n)^4\\ &= 256m^4 - 256m^3n + 96m^2n^2 - 16mn^3 + n^4\end{aligned}$$

74. $$\begin{aligned}(p-2r)^5 &= p^5 + \frac{5}{1!}p^4(-2r) + \frac{5\cdot 4}{2!}p^3(-2r)^2 + \frac{5\cdot 4\cdot 3}{3!}p^2(-2r)^3 + \frac{5\cdot 4\cdot 3\cdot 2}{4!}p(-2r)^4 + (-2r)^5\\ &= p^5 - 10p^4r + 40p^3r^2 - 80p^2r^3 + 80pr^4 - 32r^5\end{aligned}$$

75. The 4th term corresponds to $r = 3$.
$$\frac{7!}{3!(7-3)!}a^{7-3}b^3 = 35a^4b^3$$

76. The 11th term corresponds to $r = 10$.
$$\frac{10!}{10!0!}y^{10-10}(2z)^{10} = 1024z^{10}$$

77. $\sum_{i=1}^{4} i^2(i+1) = 1^2(1+1)+2^2(2+1)+3^2(3+1)+4^2(4+1)$

$$= 1(2)+4(3)+9(4)+16(5)$$
$$= 2+12+36+80$$
$$= 130$$

78. 14, 8, 2

$a_1 = 14,\ d = 8 - 14 = -6$

$a_n = a_1 + (n-1)d = 14 - 6(n-1) = 20 - 6n$

$a_{15} = 20 - 6(15) = 20 - 90 = -70$

79. 27, 9, 3, 1

$a_1 = 27,\ r = \frac{9}{27} = \frac{1}{3}$

$S_\infty = \frac{a_1}{1-r} = \frac{27}{1-\frac{1}{3}} = \frac{27}{\frac{2}{3}} = 27\left(\frac{3}{2}\right) = 40.5$

80. $(2x-3)^4 = (2x)^4 + 4(2x)^3(-3) + 6(2x)^2(-3)^2 + 4(2x)(-3)^3 + (-3)^4$

$$= 16x^4 - 96x^3 + 216x^2 - 216x + 81$$

Chapter 11 Getting Ready for the Test

1. $a_n = -n^2$

$a_3 = -(3)^2 = -9$; B

2. 6, 9, 12

Each term after the first is 3 more than the preceding term. This is an arithmetic sequence with $a_1 = 6$ and $d = 3$; C.

3. 6, 9, $\frac{27}{2}$

Each term after the first is $\frac{3}{2}$ times the preceding term. This is a geometric sequence with $a_1 = 6$ and $r = \frac{3}{2}$; B.

4. 6, 3, 0

Each term after the first is 3 less than the preceding term. This is an arithmetic sequence with $a_1 = 6$ and $d = -3$; D.

5. 6, 3, $\frac{3}{2}$

Each term after the first is $\frac{1}{2}$ times the preceding term. This is a geometric sequence with $a_1 = 6$ and $r = \frac{1}{2}$; A.

6. $\sum_{i=3}^{5} (i^2 - 1) = (3^2 - 1) + (4^2 - 1) + (5^2 - 1)$

$$= 8 + 15 + 24$$
$$= 47$$

Choice A is correct.

7. $a_n = 12 + (n-1)5$

$a_1 = 12 + (1-1)5 = 12$

$a_{20} = 12 + (20-1)5 = 107$

$S_n = \frac{n}{2}(a_1 + a_n)$

$S_{20} = \frac{20}{2}(12 + 107) = 10(119) = 1190$

Choice C is correct.

8. $a_n = 3(2)^{n-1}$

$a_1 = 3(2)^{1-1} = 3;\ r = 2$

$S_n = \frac{a_1(1-r^n)}{1-r}$

$S_{10} = \frac{3(1-2^{10})}{1-2} = \frac{3(1-1024)}{-1} = 3069$

Choice A is correct.

9. $a_n = 3\left(\frac{1}{2}\right)^{n-1}$

$a_1 = 3\left(\frac{1}{2}\right)^{1-1} = 3;\ r = \frac{1}{2}$

Since $|r| = \left|\frac{1}{2}\right| = \frac{1}{2} < 1$, we can find S_∞.

$S_\infty = \frac{a_1}{1-r} = \frac{3}{1-\frac{1}{2}} = \frac{3}{\frac{1}{2}} = 6$

Choice D is correct.

10. $\frac{10 \cdot 9 \cdot 8 \cdot 7 \cdot 6}{5!} = \frac{10 \cdot 9 \cdot 8 \cdot 7 \cdot 6}{5 \cdot 4 \cdot 3 \cdot 2 \cdot 1}$

$= \frac{\cancel{5} \cdot 2 \cdot 9 \cdot \cancel{4} \cdot 2 \cdot 7 \cdot \cancel{3} \cdot \cancel{2}}{\cancel{5} \cdot \cancel{4} \cdot \cancel{3} \cdot \cancel{2} \cdot 1}$

$= 2 \cdot 9 \cdot 2 \cdot 7$

$= 252$

Choice B is correct.

Chapter 11 Test

1. $a_n = \frac{(-1)^n}{n+4}$

$a_1 = \frac{(-1)^1}{1+4} = -\frac{1}{5}$

$a_2 = \frac{(-1)^2}{2+4} = \frac{1}{6}$

$a_3 = \frac{(-1)^3}{3+4} = -\frac{1}{7}$

$a_4 = \frac{(-1)^4}{4+4} = \frac{1}{8}$

$a_5 = \frac{(-1)^5}{5+4} = -\frac{1}{9}$

2. $a_n = 10 + 3(n-1)$

$a_{80} = 10 + 3(80-1) = 247$

3. $\frac{2}{5}, \frac{2}{25}, \frac{2}{125}, \ldots$

In general, $a_n = \frac{2}{5}\left(\frac{1}{5}\right)^{n-1}$ or $a_n = \frac{2}{5^n}$.

4. $-9, 18, -27, 36, \ldots$

$9(-1)^1, 9(-1)^2(2), 9(-1)^3(3), 9(-1)^4(4), \ldots$

In general, $9(-1)^n n = (-1)^n 9n$.

5. $a_n = 5(2)^{n-1}, S_5 = \frac{5(1-2^5)}{1-2} = 155$

6. $a_n = 18 + (n-1)(-2)$

$a_1 = 18,\ a_{30} = 18 + (30-1)(-2) = -40$

$S_{30} = \frac{30}{2}[18 - 40] = -330$

7. $a_1 = 24,\ r = \frac{1}{6}$

$S_\infty = \frac{24}{1-\frac{1}{6}} = \frac{144}{5}$

8. $\frac{3}{2}, -\frac{3}{4}, \frac{3}{8}, \ldots$

$a_1 = \frac{3}{2},\ r = -\frac{1}{2}$

$S_\infty = \frac{\frac{3}{2}}{1-\left(-\frac{1}{2}\right)} = 1$

9. $\sum_{i=1}^{4} i(i-2) = 1(1-2)+2(2-2)+3(3-2)+4(4-2)$
$= -1+0+3+8-20+40-80$
$= 10$

10. $\sum_{i=2}^{4} 5(2)^i(-1)^{i-1} = 5(2)^2(-1)^{2-1} + 5(2)^3(-1)^{3-1} + 5(2)^4(-1)^{4-1} = -20+40-80 = -60$

11. $(a-b)^6 = a^6 - 6a^5b + 15a^4b^2 - 20a^3b^3 + 15a^2b^4 - 6ab^5 + b^6$

12. $(2x+y)^5 = (2x)^5 + \frac{5}{1!}(2x)^4y + \frac{5\cdot 4}{2!}(2x)^3y^2 + \frac{5\cdot 4\cdot 3}{3!}(2x)^2y^3 + \frac{5\cdot 4\cdot 3\cdot 2}{4!}(2x)y^4 + y^5$
$= 32x^5 + 80x^4y + 80x^3y^2 + 40x^2y^3 + 10xy^4 + y^5$

13. $a_n = 250 + 75(n-1)$

$a_{10} = 250 + 75(10-1) = 925$

There were 925 people in the town at the beginning of the tenth year.

$a_1 = 250 + 75(1-1) = 250$

There were 250 people in the town at the beginning of the first year.

14. 1, 3, 5, ...

$a_1 = 1,\ d = 2,\ n = 8$

$a_8 = 1 + (8-1)2 = 15$

$1+3+5+7+9+11+13+15$

$S_8 = \frac{8}{2}[1+15] = 64$

There were 64 shrubs planted in the 8 rows.

15. $a_1 = 80,\ r = \frac{3}{4},\ n = 4$

$a_4 = 80\left(\frac{3}{4}\right)^{4-1} = 33.75$

The arc length is 33.75 cm on the 4th swing.

$S_4 = \frac{80\left(1-\left(\frac{3}{4}\right)^4\right)}{1-\frac{3}{4}} = 218.75$

The total of the arc lengths is 218.75 cm for the first 4 swings.

16. $a_1 = 80,\ r = \frac{3}{4}$

$S_\infty = \frac{80}{1-\frac{3}{4}} = 320$

The total of the arc lengths is 320 cm before the pendulum comes to rest.

17. 16, 48, 80,...

$a_{10} = 16 + (10-1)32 = 304$

He falls 304 feet during the 10th second.

$S_{10} = \frac{10}{2}[16+304] = 1600$

He falls 1600 feet during the first 10 seconds.

18. $0.42\overline{42} = 0.42 + 0.0042 + \ 0.000042$

$a_1 = 0.42 = \frac{42}{100},\ r = 0.01 = \frac{1}{100}$

$S_\infty = \frac{\frac{42}{100}}{1-\frac{1}{100}} = \frac{42}{100} \cdot \frac{100}{99} = \frac{14}{33}$

Thus, $0.42\overline{42} = \frac{14}{33}$.

Chapter 11 Cumulative Review

1. a. $\frac{20}{-4} = -5$

b. $\frac{-9}{-3} = 3$

c. $-\frac{3}{8} \div 3 = -\frac{3}{8} \cdot \frac{1}{3} = -\frac{1}{8}$

d. $\frac{-40}{10} = -4$

e. $\frac{-1}{10} \div \frac{-2}{5} = \frac{1}{10} \cdot \frac{5}{2} = \frac{1}{4}$

f. $\frac{8}{0}$ is undefined.

2. a. $3a - (4a+3) = 3a - 4a - 3 = -a - 3$

b. $(5x-3) + (2x+6) = 7x + 3$

c. $4(2x-5) - 3(5x+1) = 8x - 20 - 15x - 3$
$= -7x - 23$

3. Let x = the original price, then

$x - 0.08x = 2162$
$0.92x = 2162$
$x = 2350$

The original price is \$2350.

4. Let x = the price before taxes, then

$x + 0.06x = 344.50$
$1.06x = 344.50$
$x = 325$

The price before taxes was \$325.

5. $3y - 2x = 7$
$3y = 2x + 7$
$y = \frac{1}{3}(2x+7)$
$y = \frac{2x}{3} + \frac{7}{3}$

6. If the line is to be parallel, then the slope has to be the same as the slope of the given line.

Therefore, $m = \frac{3}{2}$.

$(y-(-2)) = \frac{3}{2}(x-3)$
$y + 2 = \frac{3}{2}(x-3)$
$y = \frac{3}{2}x - \frac{13}{2}$
$f(x) = \frac{3}{2}x - \frac{13}{2}$

7. a. $(3x^6)(5x) = 3 \cdot 5x^{6+1} = 15x^7$

b. $(-2.4x^3p^2)(4xp^{10}) = -2.4 \cdot 4x^{3+1}p^{2+10}$
$= -9.6x^4p^{12}$

8. $y^3 + 5y^2 - y = 5$
$y^3 + 5y^2 - y - 5 = 0$
$(y^3 + 5y^2) + (-y - 5) = 0$
$y^2(y+5) - 1(y+5) = 0$
$(y^2 - 1)(y+5) = 0$
$(y+1)(y-1)(y+5) = 0$
$y = -5, -1, 1$

9.
$$\begin{array}{r|rrrrr} -2 & 1 & -2 & -11 & 0 & 34 \\ & & -2 & 8 & 6 & -12 \\ \hline & 1 & -4 & -3 & 6 & 22 \end{array}$$

Answer: $x^3 - 4x^2 - 3x + 6 + \dfrac{22}{x+2}$

10.
$$\begin{aligned} &\frac{5}{3a-6} - \frac{a}{a-2} + \frac{3+2a}{5a-10} \\ &= \frac{5}{3(a-2)} - \frac{a}{a-2} + \frac{3+2a}{5(a-2)} \\ &= \frac{5\cdot 5 - 3\cdot 5a + 3(3+2a)}{3\cdot 5(a-2)} \\ &= \frac{25-15a+9+6a}{15(a-2)} \\ &= \frac{34-9a}{15(a-2)} \end{aligned}$$

11. a. $\sqrt{50} = \sqrt{2}\sqrt{25} = 5\sqrt{2}$

b. $\sqrt[3]{24} = \sqrt[3]{8}\sqrt[3]{3} = 2\sqrt[3]{3}$

c. $\sqrt{26} = \sqrt{26}$

d. $\sqrt[4]{32} = \sqrt[4]{16}\sqrt[4]{2} = 2\sqrt[4]{2}$

12.
$$\begin{aligned} \sqrt{3x+6} - \sqrt{7x-6} &= 0 \\ \sqrt{3x+6} &= \sqrt{7x-6} \\ \left(\sqrt{3x+6}\right)^2 &= \left(\sqrt{7x-6}\right)^2 \\ 3x+6 &= 7x-6 \\ -4x &= -12 \\ x &= 3 \end{aligned}$$

13.
$$\begin{aligned} 2420 &= 2000(1+r)^2 \\ \frac{2420}{2000} &= (1+r)^2 \\ \frac{121}{100} &= (1+r)^2 \\ \pm\sqrt{\frac{121}{100}} &= 1+r \\ \pm\frac{11}{10} &= 1+r \\ -1 \pm \frac{11}{10} &= r \end{aligned}$$

Discard the negative value.

$r = -1 + \dfrac{11}{10} = \dfrac{1}{10} = 0.10$

The interest rate is 10%.

14. a. $\sqrt[3]{\dfrac{4}{3x}} = \dfrac{\sqrt[3]{4}}{\sqrt[3]{3x}} = \left(\dfrac{\sqrt[3]{9x^2}}{\sqrt[3]{9x^2}}\right) = \dfrac{\sqrt[3]{36x^2}}{3x}$

b.
$$\begin{aligned} \frac{\sqrt{2}+1}{\sqrt{2}-1} &= \frac{\sqrt{2}+1}{\sqrt{2}-1} \cdot \left(\frac{\sqrt{2}+1}{\sqrt{2}+1}\right) \\ &= \frac{2+2\sqrt{2}+1}{2-1} \\ &= 3+2\sqrt{2} \end{aligned}$$

15.
$$\begin{aligned} (x-3)^2 - 3(x-3) - 4 &= 0 \\ x^2 - 6x + 9 - 3x + 9 - 4 &= 0 \\ x^2 - 9x + 14 &= 0 \\ (x-2)(x-7) &= 0 \\ x &= 2, 7 \end{aligned}$$

16.
$$\begin{aligned} \frac{10}{(2x+4)^2} - \frac{1}{2x+4} &= 3 \\ 10 - (2x+4) &= 3(2x+4)^2 \\ 10 - 2x - 4 &= 3(4x^2 + 16x + 16) \\ -2x + 6 &= 12x^2 + 48x + 48 \\ 12x^2 + 50x + 42 &= 0 \\ 6x^2 + 25x + 21 &= 0 \\ (6x+7)(x+3) &= 0 \\ x &= -\frac{7}{6}, -3 \end{aligned}$$

17.
$$\begin{aligned} \frac{5}{x+1} &< -2 \\ x+1 &= 0 \\ x &= -1 \end{aligned}$$

Solve $\dfrac{5}{x+1} = -2.$

$$\begin{aligned} (x+1)\frac{5}{x+1} &= (x+1)(-2) \\ 5 &= -2x-2 \\ 7 &= -2x \\ -\frac{7}{2} &= x \end{aligned}$$

Region	Test Point	$\frac{5}{x+1} < -2$; Result
$\left(-\infty, -\frac{7}{2}\right)$	$x = -6$	$\frac{5}{-5} < -2$; False
$\left(-\frac{7}{2}, -1\right)$	$x = -2$	$\frac{5}{-1} < -2$; True
$(-1, \infty)$	$x = 4$	$\frac{5}{5} < -2$; False

The solution set is $\left(-\frac{7}{2}, -1\right)$.

18. $f(x) = (x+2)^2 - 6$
Axis of symmetry: $x = -2$
vertex: $(-2, -6)$

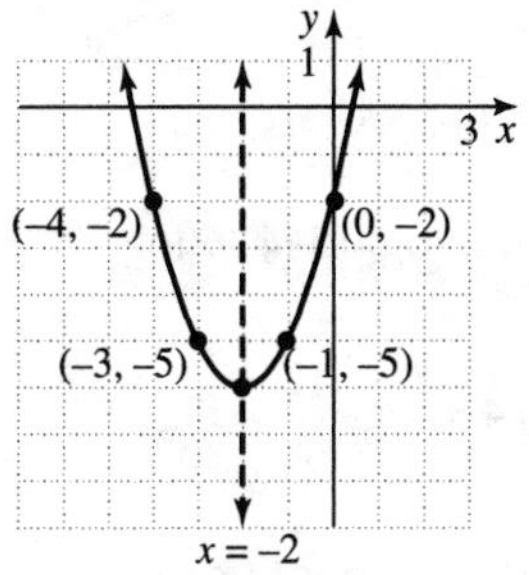

19. $f(t) = -16t^2 + 20t$
The maximum height occurs at the vertex.
$$t = \frac{-20}{2(-16)} = \frac{5}{8}$$
$$f\left(\frac{5}{8}\right) = -16\left(\frac{5}{8}\right)^2 + 20\left(\frac{5}{8}\right) = \frac{25}{4}$$
The maximum height of $\frac{25}{4}$ feet occurs at $\frac{5}{8}$ second.

20. $f(x) = x^2 + 3x - 18$
$a = 1, b = 3, c = -18$
$$x = \frac{-3}{2(1)} = -\frac{3}{2}$$
$$f\left(-\frac{3}{2}\right) = \left(-\frac{3}{2}\right)^2 + 3\left(\frac{3}{2}\right) - 18 = -\frac{81}{4}$$
The vertex is $\left(-\frac{3}{2}, -\frac{81}{4}\right)$.

21. a. $(f \circ g)(2) = f(g(2)) = f(5) = 5^2 = 25$
$(g \circ f)(2) = g(f(2)) = g(4) = 4 + 3 = 7$

b. $(f \circ g)(x) = f(x+3)$
$= (x+3)^2$
$= x^2 + 6x + 9$
$(g \circ f)(x) = g(x^2) = x^2 + 3$

22. $f(x) = -2x + 3$
$y = -2x + 3$
$x = -2y + 3$
$x - 3 = -2y$
$\frac{x-3}{-2} = y$
$f^{-1}(x) = -\frac{x-3}{2}$ or $f^{-1}(x) = \frac{3-x}{2}$

23. $f^{-1} = \{(1,0), (7,-2), (-6,3), (4,4)\}$

24. a. $(f \circ g)(2) = f(g(2)) = f(3) = 3^2 - 2 = 7$
$(g \circ f)(2) = g(f(2)) = g(2) = 2 + 1 = 3$

b. $(f \circ g)(x) = f(x+1)$
$= (x+1)^2 - 2$
$= x^2 + 2x - 1$
$(g \circ f)(x) = g(x^2 - 2) = x^2 - 2 + 1 = x^2 - 1$

25. a. $2^x = 16$
$2^x = 2^4$
$x = 4$

b. $9^x = 27$
$(3^2)^x = 3^3$
$2x = 3$
$x = \frac{3}{2}$

c. $4^{x+3} = 8^x$
$(2^2)^{x+3} = (2^3)^x$
$2^{2x+6} = 2^{3x}$
$2x + 6 = 3x$
$x = 6$

26. a. $\log_2 32 = x$
$$2^x = 32$$
$$2^x = 2^5$$
$$x = 5$$

b. $\log_4 \frac{1}{64} = x$
$$4^x = \frac{1}{64}$$
$$4^x = 4^{-3}$$
$$x = -3$$

c. $\log_{1/2} x = 5$
$$\left(\frac{1}{2}\right)^5 = x$$
$$x = \frac{1}{32}$$

27. a. $\log_3 3^2 = 2$

b. $\log_7 7^{-1} = -1$

c. $5^{\log_5 3} = 3$

d. $2^{\log_2 6} = 6$

28. a. $4^x = 64$
$$\left(2^2\right)^x = 2^6$$
$$2x = 6$$
$$x = 3$$

b. $8^x = 32$
$$\left(2^3\right)^x = 2^5$$
$$3x = 5$$
$$x = \frac{5}{3}$$

c. $9^{x+4} = 243^x$
$$(3^2)^{x+4} = (3^5)^x$$
$$3^{2x+8} = 3^{5x}$$
$$2x + 8 = 5x$$
$$8 = 3x$$
$$x = \frac{8}{3}$$

29. a. $\log_{11} 10 + \log_{11} 3 = \log_{11}(10 \cdot 3) = \log_{11} 30$

b. $\log_3 \frac{1}{2} + \log_3 12 = \log_3\left(\frac{1}{2} \cdot 12\right) = \log_3 6$

c. $\log_2(x+2) + \log_2 x = \log_2[(x+2)x]$
$$= \log_2(x^2 + 2x)$$

30. a. $\log 100{,}000 = \log_{10} 10^5 = 5$

b. $\log 10^{-3} = \log_{10} 10^{-3} = -3$

c. $\ln \sqrt[5]{e} = \ln e^{1/5} = \frac{1}{5}$

d. $\ln e^4 = 4$

31. $A = Pe^{rt}$
$$A = 1600e^{0.09(5)} \approx 2509.30$$
\$2509.30 is owed after 5 years.

32. a. $\log_6 5 + \log_6 4 = \log_6(5 \cdot 4) = \log_6 20$

b. $\log_8 12 - \log_8 4 = \log_8 \frac{12}{4} = \log_8 3$

c. $2\log_2 x + 3\log_2 x - 2\log_2(x-1)$
$$= 5\log_2 x - \log_2(x-1)^2$$
$$= \log_2 x^5 - \log_2(x-1)^2$$
$$= \log_2 \frac{x^5}{(x-1)^2}$$

33. $3^x = 7$
$$\log 3^x = \log 7$$
$$x \log 3 = \log 7$$
$$x = \frac{\log 7}{\log 3} \approx 1.7712$$

34. $10{,}000 = 5000\left(1 + \frac{0.02}{4}\right)^{4t}$
$$2 = (1.005)^{4t}$$
$$\ln 2 = \ln 1.005^{4t}$$
$$\ln 2 = 4t \ln(1.005)$$
$$t = \frac{\ln 2}{4 \ln 1.005} \approx 34.7$$
It takes 34.7 years.

35. $\log_4(x-2)=2$

$$4^2 = x-2$$
$$x-2=16$$
$$x=18$$

36. $\log_4 10 - \log_4 x = 2$

$$\log_4 \frac{10}{x} = 2$$
$$4^2 = \frac{10}{x}$$
$$16 = \frac{10}{x}$$
$$16x = 10$$
$$x = \frac{5}{8}$$

37. $\frac{x^2}{16} - \frac{y^2}{25} = 1$

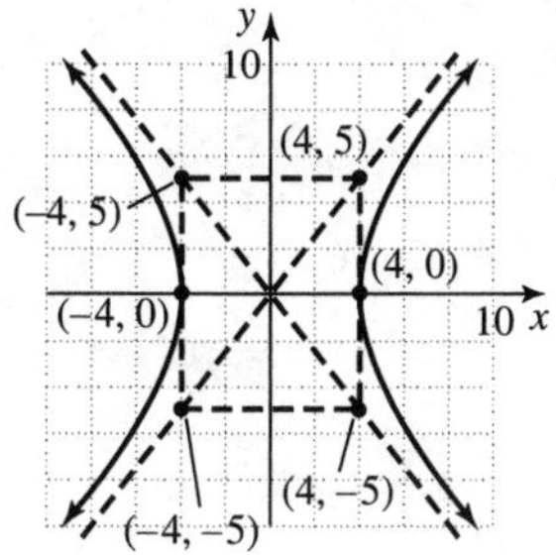

38. (8, 5), (–2, 4)

$$d = \sqrt{(-2-8)^2 + (4-5)^2} = \sqrt{101} \text{ units}$$

39. $\begin{cases} y = \sqrt{x} \\ x^2 + y^2 = 6 \end{cases}$

Replace y with $\sqrt{x}$ in the first equation.

$$(x)^2 + \left(\sqrt{x}\right)^2 = 6$$
$$x^2 + x - 6 = 0$$
$$(x+3)(x-2) = 0$$
$$x = -3 \text{(discard) or } x = 2$$
$$x = 2: \ y = \sqrt{x} = \sqrt{2}$$
$$\left(2, \sqrt{2}\right)$$

40. $\begin{cases} x^2 + y^2 = 36 \\ x - y = 6 \Rightarrow x = y + 6 \end{cases}$

Replace x with $y + 6$ in the first equation.

$$(y+6)^2 + y^2 = 36$$
$$2y^2 + 12y = 0$$
$$2y(y+6) = 0$$

$y = 0$ or $y = -6$

$x = 0 + 6 = 6$ $\quad x = -6 + 6 = 0$

$(0, -6); (6, 0)$

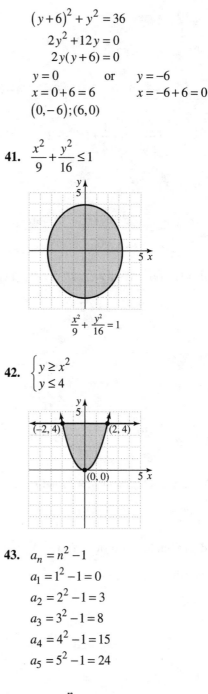

41. $\frac{x^2}{9} + \frac{y^2}{16} \le 1$

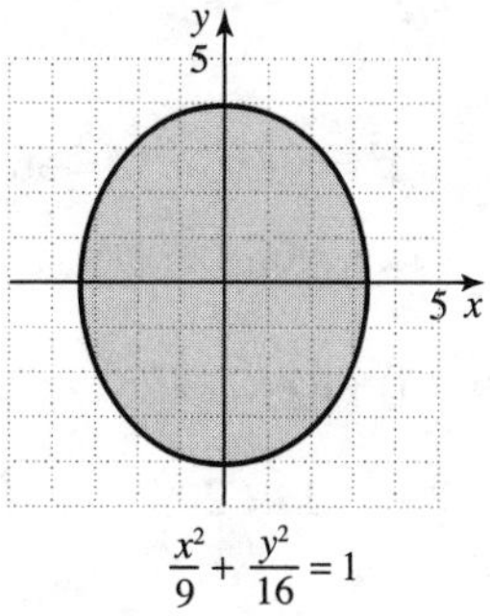

42. $\begin{cases} y \ge x^2 \\ y \le 4 \end{cases}$

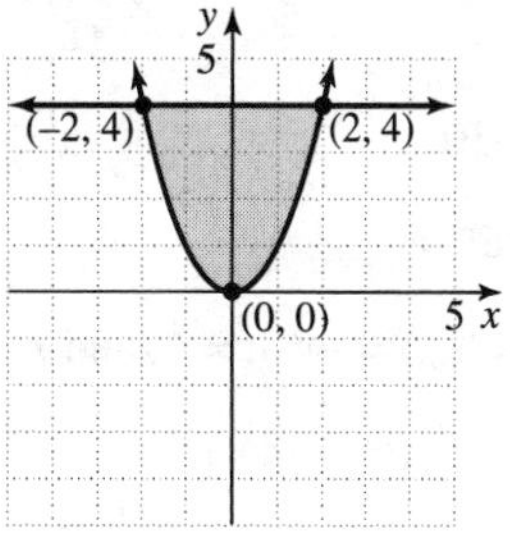

43. $a_n = n^2 - 1$

$$a_1 = 1^2 - 1 = 0$$
$$a_2 = 2^2 - 1 = 3$$
$$a_3 = 3^2 - 1 = 8$$
$$a_4 = 4^2 - 1 = 15$$
$$a_5 = 5^2 - 1 = 24$$

44. $a_n = \frac{n}{n+4}$

$$a_8 = \frac{8}{8+4} = \frac{8}{12} = \frac{2}{3}$$

45. $a_1 = 2, \ d = 9 - 2 = 7$

$$a_{11} = 2 + (11-1)(7) = 72$$

46. $a_1 = 2,\ r = \frac{10}{2} = 5$

$a_6 = 2(5)^{6-1} = 2(5)^5 = 6250$

47. a. $\sum_{i=0}^{6} \frac{i-2}{2} = \frac{0-2}{2} + \frac{1-2}{2} + \frac{2-2}{2} + \frac{3-2}{2} + \frac{4-2}{2} + \frac{5-2}{2} + \frac{6-2}{2}$

$= -1 - \frac{1}{2} + 0 + \frac{1}{2} + 1 + \frac{3}{2} + 2$

$= \frac{7}{2} \text{ or } 3\frac{1}{2}$

b. $\sum_{i=3}^{5} 2^i = 2^3 + 2^4 + 2^5 = 8 + 16 + 32 = 56$

48. a. $\sum_{i=0}^{4} i(i+1) = 0(0+1) + 1(1+1) + 2(2+1) + 3(3+1) + 4(4+1)$

$= 0 + 2 + 6 + 12 + 20$

$= 40$

b. $\sum_{i=0}^{3} 2^i = 2^0 + 2^1 + 2^2 + 2^3 = 1 + 2 + 4 + 8 = 15$

49. $a_1 = 1,\ a_{30} = 30$

$S_n = \frac{n}{2}(a_1 + a_n) = \frac{30}{2}(1 + 30) = 465$

50. $(x-y)^6$ where $a = x$, $b = -y$, $n = 6$, and $r = 2$.

$\frac{6!}{2!(6-2)!} x^{6-2} y^2 = 15x^4y^2$

The third term in the expansion of $(x-y)^6$ is $15x^4y^2$.

Appendices

Appendix A.3 Exercise Set

1. $V = lwh$
$= (6 \text{ in.})(4 \text{ in.})(3 \text{ in.})$
$= 72 \text{ cu in.}$
$SA = 2lh + 2wh + 2lw$
$= 2(6 \text{ in.})(3 \text{ in.}) + 2(4 \text{ in.})(3 \text{ in.}) + 2(6 \text{ in.})(4 \text{ in.})$
$= 36 \text{ sq in.} + 24 \text{ sq in.} + 48 \text{ sq in.}$
$= 108 \text{ sq in.}$

3. $V = s^3 = (8 \text{ cm})^3 = 512 \text{ cu cm}$

$SA = 6s^2 = 6(8 \text{ cm})^2 = 384 \text{ sq cm}$

5. $V = \frac{1}{3}\pi r^2 h$
$= \frac{1}{3}\pi(2 \text{ yd})^2(3 \text{ yd})$
$= 4\pi \text{ cu yd}$
$\approx 4\left(\frac{22}{7}\right) \text{ cu yd}$
$\approx \frac{88}{7} \text{ cu yd}$
$\approx 12\frac{4}{7} \text{ cu yd}$

$SA = \pi r\sqrt{r^2 + h^2} + \pi r^2$
$= \pi(2 \text{ yd})\sqrt{(2 \text{ yd})^2 + (3 \text{ yd})^2} + \pi(2 \text{ yd})^2$
$= \pi(2 \text{ yd})\left(\sqrt{13} \text{ yd}\right) + \pi(4 \text{ sq yd})$
$= \left(2\sqrt{13} + 4\right)\pi \text{ sq yd}$
$\approx 3.14\left(2\sqrt{13} + 4\right) \text{ sq yd}$
$\approx 35.20 \text{ sq yd}$

7. $V = \frac{4}{3}\pi r^3$
$= \frac{4}{3}\pi(5 \text{ in.})^3$
$= \frac{500}{3}\pi \text{ cu in.}$
$\approx \frac{500}{3}\left(\frac{22}{7}\right) \text{ cu in.}$
$\approx 523\frac{17}{21} \text{ cu in.}$

$$\begin{aligned} SA &= 4\pi r^2 \\ &= 4\pi(5\text{ in.})^2 \\ &= 100\pi\text{ sq in.} \\ &\approx 100\left(\frac{22}{7}\right)\text{ sq in.} \\ &\approx 314\frac{2}{7}\text{ sq in.} \end{aligned}$$

9. $V = \frac{1}{3}s^2h = \frac{1}{3}(6\text{ cm})^2(4\text{ cm}) = 48\text{ cu cm}$

$$\begin{aligned} SA &= B + \frac{1}{2}pl \\ &= (6\text{ cm})^2 + \frac{1}{2}(24\text{ cm})(5\text{ cm}) \\ &= 36\text{ sq cm} + 60\text{ sq cm} \\ &= 96\text{ sq cm} \end{aligned}$$

11. $V = s^3$

$$\begin{aligned} &= \left(1\frac{1}{3}\text{ in.}\right)^3 \\ &= \left(\frac{4}{3}\text{ in.}\right)^3 \\ &= \frac{64}{27}\text{ cu in.} \\ &= 2\frac{10}{27}\text{ cu in.} \end{aligned}$$

13. $SA = 2lh + 2wh + 2lw$

$$\begin{aligned} &= 2(2\text{ ft})(1.4\text{ ft}) + 2(2\text{ ft})(3\text{ ft}) + 2(1.4\text{ ft})(3\text{ ft}) \\ &= 5.6\text{ sq ft} + 12\text{ sq ft} + 8.4\text{ sq ft} \\ &= 26\text{ sq ft} \end{aligned}$$

15. $V = \frac{1}{3}s^2h$

$$\begin{aligned} &= \frac{1}{3}(5\text{ in.})^2(1.3\text{ in.}) \\ &= \frac{1}{3}(25\text{ sq in.})\left(\frac{13}{10}\text{ in.}\right) \\ &= \frac{65}{6}\text{ cu in.} \\ &= 10\frac{5}{6}\text{ cu in.} \end{aligned}$$

17. $V = \frac{1}{3}s^2h = \frac{1}{3}(12\text{ cm})^2(20\text{ cm}) = 960\text{ cu cm}$

19. $SA = 4\pi r^2 = 4\pi(7\text{ in.})^2 = 196\pi\text{ sq in.}$

21. $V = (2\text{ ft})\left(2\frac{1}{2}\text{ ft}\right)\left(1\frac{1}{2}\text{ ft}\right) = 7\frac{1}{2}\text{ cu ft}$

23. $V = \frac{1}{3}\pi r^2 h$

$$\begin{aligned} &\approx \frac{1}{3}\left(\frac{22}{7}\right)(2\text{ cm})^2(3\text{ cm}) \\ &= \frac{88}{7}\text{ cu cm} \\ &\approx 12\frac{4}{7}\text{ cu cm} \end{aligned}$$

Appendix B Exercise Set

1. $f(x) = 3|x|$
Find and plot ordered-pair solutions.

x	$f(x) = 3\|x\|$
−1	$3\|-1\| = 3$
0	$3\|0\| = 0$
1	$3\|1\| = 3$

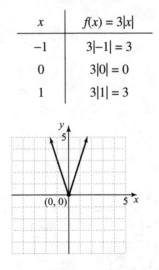

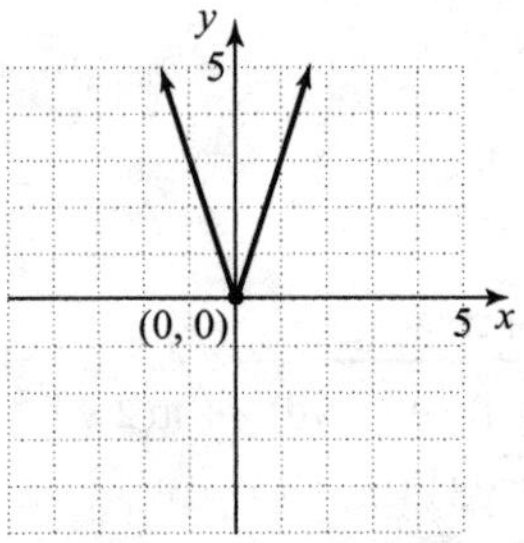

3. $f(x) = \frac{1}{4}|x|$

Find and plot ordered-pair solutions.

x	$f(x) = \frac{1}{4}\|x\|$
−4	1
0	0
4	1

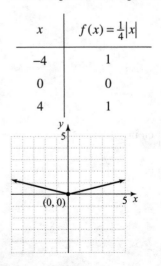

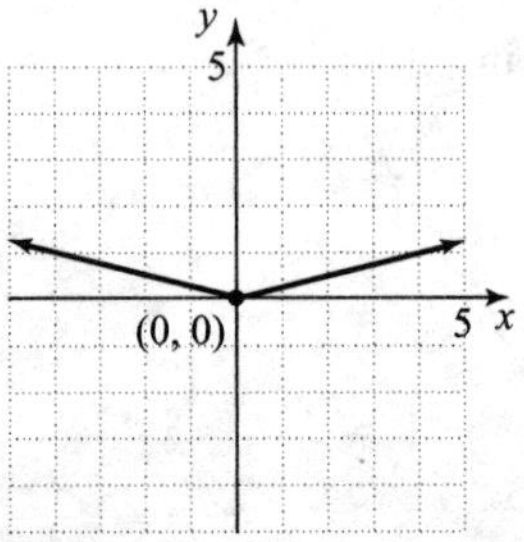

5. $g(x) = 2|x| + 3$
Write in the form $g(x) = a|x - h| + k$.
$g(x) = 2|x - 0| + 3$

- vertex is $(h, k) = (0, 3)$
- since $a > 0$, V-shape opens up
- since $|a| = |2| = 2 > 1$, the graph is narrower than $y = |x|$

x	$g(x) = 2\lvert x\rvert + 3$
−1	$2\lvert -1\rvert + 3 = 5$
0	$2\lvert 0\rvert + 3 = 3$
1	$2\lvert 1\rvert + 3 = 5$

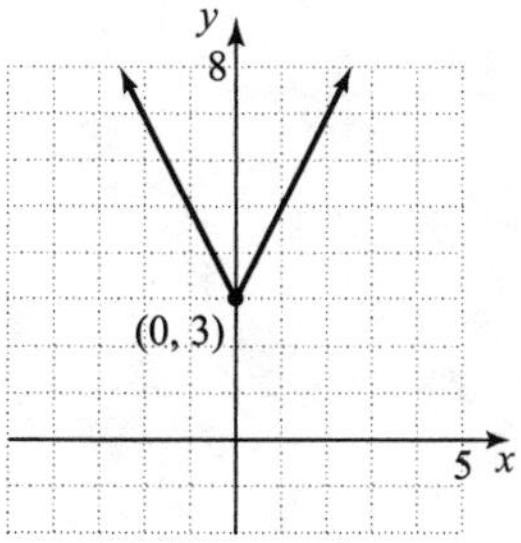

7. $h(x) = -\frac{1}{2}|x|$

- vertex is $(h, k) = (0, 0)$
- since $a < 0$, V-shape opens down
- since $|a| = \left|-\frac{1}{2}\right| = \frac{1}{2} < 1,$ the graph is wider than $y = |x|$

x	$h(x) = -\frac{1}{2}\lvert x\rvert$
−2	$-\frac{1}{2}\lvert -2\rvert = -1$
0	$-\frac{1}{2}\lvert 0\rvert = 0$
2	$-\frac{1}{2}\lvert 2\rvert = -1$

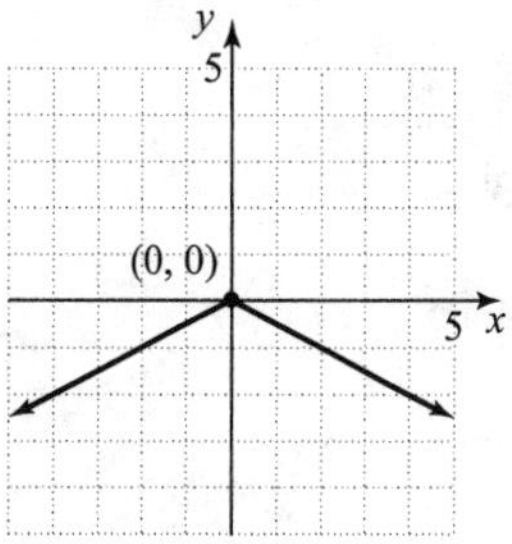

9. $f(x) = 4|x - 1|$
Write in the form $f(x) = a|x - h| + k$.
$f(x) = 4|x - 1| + 0$

- vertex is $(h, k) = (1, 0)$
- since $a > 0$, V-shape opens up
- since $|a| = |4| = 4 > 1$, the graph is narrower than $y = |x|$

x	$f(x) = 4\lvert x - 1\rvert$
−2	$4\lvert -2 - 1\rvert = 12$
0	$4\lvert 0 - 1\rvert = 4$
2	$4\lvert 2 - 1\rvert = 4$

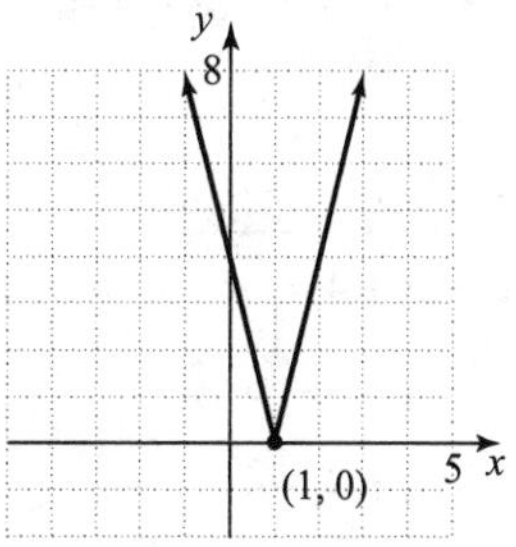

11. $g(x) = -\frac{1}{3}|x| - 2$
Write in the form $g(x) = a|x - h| + k$.
$g(x) = -\frac{1}{3}|x - 0| + (-2)$

- vertex is $(h, k) = (0, -2)$
- since $a < 0$, V-shape opens down
- since $|a| = \left|-\frac{1}{3}\right| = \frac{1}{3} < 1,$ the graph is wider than $y = |x|$

x	$g(x) = -\frac{1}{3}\lvert x\rvert - 2$
-3	$-\frac{1}{3}\lvert -3\rvert - 2 = -3$
0	$-\frac{1}{3}\lvert 0\rvert - 2 = -2$
3	$-\frac{1}{3}\lvert 3\rvert - 2 = -3$

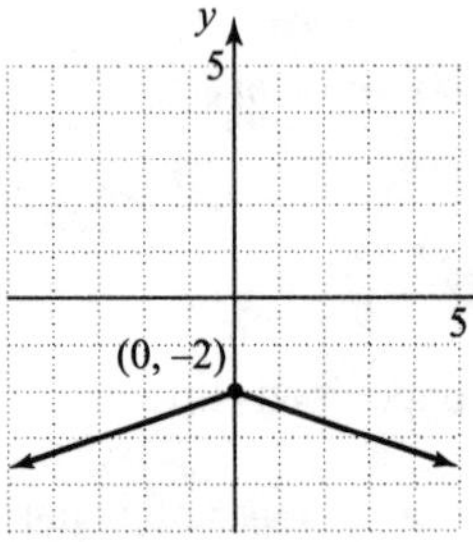

13. $f(x) = -2\lvert x - 3\rvert + 4$
This function is already written in the form $f(x) = a\lvert x - h\rvert + k$.

- vertex is $(h, k) = (3, 4)$
- since $a < 0$, V-shape opens down
- since $\lvert a\rvert = \lvert -2\rvert = 2 > 1$, the graph is narrower than $y = \lvert x\rvert$

x	$f(x) = -2\lvert x - 3\rvert + 4$
-1	$-2\lvert -1 - 3\rvert + 4 = -4$
1	$-2\lvert 1 - 3\rvert + 4 = 0$
4	$-2\lvert 4 - 3\rvert + 4 = 2$

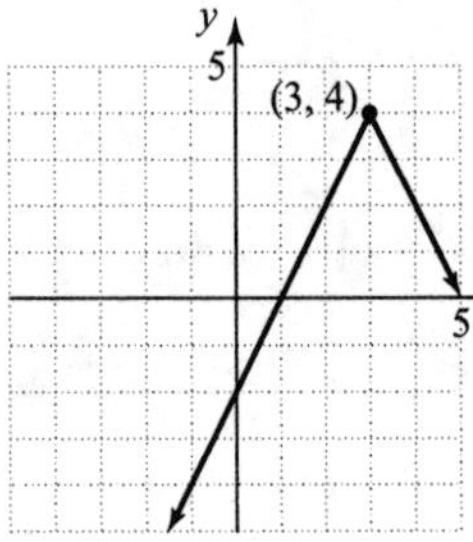

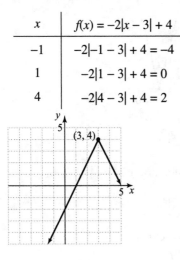

15. $f(x) = \frac{2}{3}\lvert x + 2\rvert - 5$
Write in the form $f(x) = a\lvert x - h\rvert + k$.
$f(x) = \frac{2}{3}\lvert x - (-2)\rvert + (-5)$

- vertex is $(h, k) = (-2, -5)$
- since $a > 0$, V-shape opens up
- since $\lvert a\rvert = \left\lvert \frac{2}{3} \right\rvert = \frac{2}{3} < 1$, the graph is wider than $y = \lvert x\rvert$

x	$f(x) = \frac{2}{3}\lvert x + 2\rvert - 5$
-5	$\frac{2}{3}\lvert -5 + 2\rvert - 5 = -3$
-2	$\frac{2}{3}\lvert -2 + 2\rvert - 5 = -5$
1	$\frac{2}{3}\lvert 1 + 2\rvert - 5 = -3$

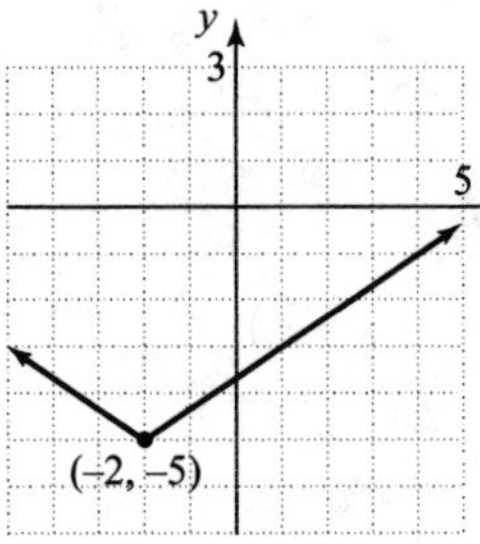

Appendix C

Vocabulary and Readiness Check Appendix C

1. $\begin{vmatrix} 7 & 2 \\ 0 & 8 \end{vmatrix} = 56$

2. $\begin{vmatrix} 6 & 0 \\ 1 & 2 \end{vmatrix} = 12$

3. $\begin{vmatrix} -4 & 2 \\ 0 & 8 \end{vmatrix} = -32$

4. $\begin{vmatrix} 5 & 0 \\ 3 & -5 \end{vmatrix} = -25$

5. $\begin{vmatrix} -2 & 0 \\ 3 & -10 \end{vmatrix} = 20$

6. $\begin{vmatrix} -1 & 4 \\ 0 & -18 \end{vmatrix} = 18$

Appendix C Exercise Set

1. $\begin{vmatrix} 3 & 5 \\ -1 & 7 \end{vmatrix} = ad - bc = 3(7) - 5(-1) = 21 + 5 = 26$

3. $\begin{vmatrix} 9 & -2 \\ 4 & -3 \end{vmatrix} = ad - bc$
$$= 9(-3) - (-2)(4)$$
$$= -27 + 8$$
$$= -19$$

5. $\begin{vmatrix} -2 & 9 \\ 4 & -18 \end{vmatrix} = ad - bc$
$$= -2(-18) - 9(4)$$
$$= 36 - 36$$
$$= 0$$

7. $\begin{vmatrix} \frac{3}{4} & \frac{5}{2} \\ -\frac{1}{6} & \frac{7}{3} \end{vmatrix} = ad - bc$
$$= \left(\frac{3}{4}\right)\left(\frac{7}{3}\right) - \left(\frac{5}{2}\right)\left(-\frac{1}{6}\right)$$
$$= \frac{7}{4} + \frac{5}{12}$$
$$= \frac{21}{12} + \frac{5}{12}$$
$$= \frac{26}{12}$$
$$= \frac{13}{6}$$

9. $\begin{cases} 2y - 4 = 0 \\ x + 2y = 5 \end{cases}$ or $\begin{cases} 0x + 2y = 4 \\ 1x + 2y = 5 \end{cases}$

$D = \begin{vmatrix} 0 & 2 \\ 1 & 2 \end{vmatrix} = 0 - 2 = -2$

$D_x = \begin{vmatrix} 4 & 2 \\ 5 & 2 \end{vmatrix} = 8 - 10 = -2$

$D_y = \begin{vmatrix} 0 & 4 \\ 1 & 5 \end{vmatrix} = 0 - 4 = -4$

$x = \frac{D_x}{D} = \frac{-2}{-2} = 1$

$y = \frac{D_y}{D} = \frac{-4}{-2} = 2$

The solution is (1, 2).

11. $\begin{cases} 3x + y = 1 \\ 2y = 2 - 6x \end{cases}$ or $\begin{cases} 3x + 1y = 1 \\ 6x + 2y = 2 \end{cases}$

$D = \begin{vmatrix} 3 & 1 \\ 6 & 2 \end{vmatrix} = 6 - 6 = 0$

Since $D = 0$, Cramer's Rule cannot be used. Notice that equation (2) is equation (1) multiplied by 2.
The solution is $\{(x, y) | 3x + y = 1\}$.

13. $\begin{cases} 5x - 2y = 27 \\ -3x + 5y = 18 \end{cases}$

$D = \begin{vmatrix} 5 & -2 \\ -3 & 5 \end{vmatrix} = 25 - 6 = 19$

$D_x = \begin{vmatrix} 27 & -2 \\ 18 & 5 \end{vmatrix} = 135 + 36 = 171$

$D_y = \begin{vmatrix} 5 & 27 \\ -3 & 18 \end{vmatrix} = 90 + 81 = 171$

$x = \frac{D_x}{D} = \frac{171}{19} = 9$ and $y = \frac{D_y}{D} = \frac{171}{19} = 9$

The solution is (9, 9).

15. $\begin{cases} 2x - 5y = 4 \\ x + 2y = -7 \end{cases}$

$D = \begin{vmatrix} 2 & -5 \\ 1 & 2 \end{vmatrix} = 4 + 5 = 9$

$D_x = \begin{vmatrix} 4 & -5 \\ -7 & 2 \end{vmatrix} = 8 - 35 = -27$

$D_y = \begin{vmatrix} 2 & 4 \\ 1 & -7 \end{vmatrix} = -14 - 4 = -18$

$x = \frac{D_x}{D} = \frac{-27}{9} = -3$ and $y = \frac{D_y}{D} = \frac{-18}{9} = -2$

The solution is (−3, −2).

17. $\begin{cases} \frac{2}{3}x - \frac{3}{4}y = -1 \\ -\frac{1}{6}x + \frac{3}{4}y = \frac{5}{2} \end{cases}$

$D = \begin{vmatrix} \frac{2}{3} & -\frac{3}{4} \\ -\frac{1}{6} & \frac{3}{4} \end{vmatrix} = \frac{1}{2} - \frac{1}{8} = \frac{3}{8}$

$D_x = \begin{vmatrix} -1 & -\frac{3}{4} \\ \frac{5}{2} & \frac{3}{4} \end{vmatrix} = -\frac{3}{4} + \frac{15}{8} = \frac{9}{8}$

$D_y = \begin{vmatrix} \frac{2}{3} & -1 \\ -\frac{1}{6} & \frac{5}{2} \end{vmatrix} = \frac{10}{6} - \frac{1}{6} = \frac{9}{6}$

$x = \frac{D_x}{D} = \frac{\frac{9}{8}}{\frac{3}{8}} = 3$ and $y = \frac{D_y}{D} = \frac{\frac{9}{6}}{\frac{3}{8}} = 4$

The solution is (3, 4).

19. Expand by first row.

$$\begin{vmatrix} 2 & 1 & 0 \\ 0 & 5 & -3 \\ 4 & 0 & 2 \end{vmatrix} = 2\begin{vmatrix} 5 & -3 \\ 0 & 2 \end{vmatrix} - 1\begin{vmatrix} 0 & -3 \\ 4 & 2 \end{vmatrix} + 0\begin{vmatrix} 0 & 5 \\ 4 & 0 \end{vmatrix}$$
$$= 2(10-0) - 1(0+12) + 0$$
$$= 20 - 12$$
$$= 8$$

21. Expand by third column.

$$\begin{vmatrix} 4 & -6 & 0 \\ -2 & 3 & 0 \\ 4 & -6 & 1 \end{vmatrix} = 0\begin{vmatrix} -2 & 3 \\ 4 & -6 \end{vmatrix} - 0\begin{vmatrix} 4 & -6 \\ 4 & -6 \end{vmatrix} + 1\begin{vmatrix} 4 & -6 \\ -2 & 3 \end{vmatrix}$$
$$= 0 - 0 + 1(12-12)$$
$$= 0$$

23. Expand by first row.

$$\begin{vmatrix} 1 & 0 & 4 \\ 1 & -1 & 2 \\ 3 & 2 & 1 \end{vmatrix} = 1\begin{vmatrix} -1 & 2 \\ 2 & 1 \end{vmatrix} - 0\begin{vmatrix} 1 & 2 \\ 3 & 1 \end{vmatrix} + 4\begin{vmatrix} 1 & -1 \\ 3 & 2 \end{vmatrix}$$
$$= 1(-1-4) - 0 + 4(2+3)$$
$$= -5 + 20$$
$$= 15$$

25. Expand by first row.

$$\begin{vmatrix} 3 & 6 & -3 \\ -1 & -2 & 3 \\ 4 & -1 & 6 \end{vmatrix} = 3\begin{vmatrix} -2 & 3 \\ -1 & 6 \end{vmatrix} - 6\begin{vmatrix} -1 & 3 \\ 4 & 6 \end{vmatrix} - 3\begin{vmatrix} -1 & -2 \\ 4 & -1 \end{vmatrix}$$
$$= 3(-12+3) - 6(-6-12) - 3(1+8)$$
$$= -27 + 108 - 27$$
$$= 54$$

27. $\begin{cases} 3x \quad\quad + z = -1 \\ -x - 3y + z = 7 \\ \quad\quad 3y + z = 5 \end{cases}$

$$D = \begin{vmatrix} 3 & 0 & 1 \\ -1 & -3 & 1 \\ 0 & 3 & 1 \end{vmatrix} = 3\begin{vmatrix} -3 & 1 \\ 3 & 1 \end{vmatrix} - 0\begin{vmatrix} -1 & 1 \\ 0 & 1 \end{vmatrix} + 1\begin{vmatrix} -1 & -3 \\ 0 & 3 \end{vmatrix}$$
$$= 3(-3-3) - 0 + 1(-3-0)$$
$$= -18 - 3$$
$$= -21$$

$$D_x = \begin{vmatrix} -1 & 0 & 1 \\ 7 & -3 & 1 \\ 5 & 3 & 1 \end{vmatrix}$$
$$= -1\begin{vmatrix} -3 & 1 \\ 3 & 1 \end{vmatrix} - 0\begin{vmatrix} 7 & 1 \\ 5 & 1 \end{vmatrix} + 1\begin{vmatrix} 7 & -3 \\ 5 & 3 \end{vmatrix}$$
$$= -1(-3-3) - 0 + 1|21+15|$$
$$= 6 + 36$$
$$= 42$$

$$D_y = \begin{vmatrix} 3 & -1 & 1 \\ -1 & 7 & 1 \\ 0 & 5 & 1 \end{vmatrix} = 3\begin{vmatrix} 7 & 1 \\ 5 & 1 \end{vmatrix} + 1\begin{vmatrix} -1 & 1 \\ 5 & 1 \end{vmatrix} + 0\begin{vmatrix} -1 & 1 \\ 7 & 1 \end{vmatrix}$$
$$= 3(7-5) + 1(-1-5) + 0$$
$$= 6 - 6$$
$$= 0$$

$$D_z = \begin{vmatrix} 3 & 0 & -1 \\ -1 & -3 & 7 \\ 0 & 3 & 5 \end{vmatrix}$$
$$= 3\begin{vmatrix} -3 & 7 \\ 3 & 5 \end{vmatrix} - 0\begin{vmatrix} -1 & 7 \\ 0 & 5 \end{vmatrix} - 1\begin{vmatrix} -1 & -3 \\ 0 & 3 \end{vmatrix}$$
$$= 3(-15-21) - 0 - 1(-3-0)$$
$$= -108 + 3$$
$$= -105$$

$$x = \frac{D_x}{D} = \frac{42}{-21} = -2, \quad y = \frac{D_y}{D} = \frac{0}{-21} = 0,$$
$$z = \frac{D_z}{D} = \frac{-105}{-21} = 5$$

The solution is (−2, 0, 5).

29. $\begin{cases} x + y + z = 8 \\ 2x - y - z = 10 \\ x - 2y + 3z = 22 \end{cases}$

$$D = \begin{vmatrix} 1 & 1 & 1 \\ 2 & -1 & -1 \\ 1 & -2 & 3 \end{vmatrix}$$
$$= 1\begin{vmatrix} -1 & -1 \\ -2 & 3 \end{vmatrix} - 1\begin{vmatrix} 2 & -1 \\ 1 & 3 \end{vmatrix} + 1\begin{vmatrix} 2 & -1 \\ 1 & -2 \end{vmatrix}$$
$$= 1(-3-2) - 1(6+1) + 1(-4+1)$$
$$= -5 - 7 - 3$$
$$= -15$$

$$D_x = \begin{vmatrix} 8 & 1 & 1 \\ 10 & -1 & -1 \\ 22 & -2 & 3 \end{vmatrix}$$
$$= 8\begin{vmatrix} -1 & -1 \\ -2 & 3 \end{vmatrix} - 10\begin{vmatrix} 1 & 1 \\ -2 & 3 \end{vmatrix} + 22\begin{vmatrix} 1 & 1 \\ -1 & -1 \end{vmatrix}$$
$$= 8(-3-2) - 10(3+2) + 22(-1+1)$$
$$= -40 - 50 + 0$$
$$= -90$$

$$D_y = \begin{vmatrix} 1 & 8 & 1 \\ 2 & 10 & -1 \\ 1 & 22 & 3 \end{vmatrix}$$
$$= 1\begin{vmatrix} 10 & -1 \\ 22 & 3 \end{vmatrix} - 8\begin{vmatrix} 2 & -1 \\ 1 & 3 \end{vmatrix} + 1\begin{vmatrix} 2 & 10 \\ 1 & 22 \end{vmatrix}$$
$$= 1(30+22) - 8(6+1) + 1(44-10)$$
$$= 52 - 56 + 34$$
$$= 30$$

$$D_z=\begin{vmatrix}1&1&8\\2&-1&10\\1&-2&22\end{vmatrix}$$
$$=1\begin{vmatrix}-1&10\\-2&22\end{vmatrix}-1\begin{vmatrix}2&10\\1&22\end{vmatrix}+8\begin{vmatrix}2&-1\\1&-2\end{vmatrix}$$
$$=1(-22+20)-1(44-10)+8(-4+1)$$
$$=-2-34-24$$
$$=-60$$

$$x=\frac{D_x}{D}=\frac{-90}{-15}=6,\quad y=\frac{D_y}{D}=\frac{30}{-15}=-2,$$
$$z=\frac{D_z}{D}=\frac{-60}{-15}=4$$

The solution is (6, −2, 4).

31. $\begin{cases}2x+2y+z=1\\-x+y+2z=3\\x+2y+4z=0\end{cases}$

$$D=\begin{vmatrix}2&2&1\\-1&1&2\\1&2&4\end{vmatrix}=2\begin{vmatrix}1&2\\2&4\end{vmatrix}-2\begin{vmatrix}-1&2\\1&4\end{vmatrix}+1\begin{vmatrix}-1&1\\1&2\end{vmatrix}$$
$$=2(4-4)-2(-4-2)+1(-2-1)$$
$$=0+12-3$$
$$=9$$

$$D_x=\begin{vmatrix}1&2&1\\3&1&2\\0&2&4\end{vmatrix}=1\begin{vmatrix}1&2\\2&4\end{vmatrix}-3\begin{vmatrix}2&1\\2&4\end{vmatrix}+0$$
$$=1(4-4)-3(8-2)$$
$$=-18$$

$$D_y=\begin{vmatrix}2&1&1\\-1&3&2\\1&0&4\end{vmatrix}=2\begin{vmatrix}3&2\\0&4\end{vmatrix}-1\begin{vmatrix}-1&2\\1&4\end{vmatrix}+1\begin{vmatrix}-1&3\\1&0\end{vmatrix}$$
$$=2(12-0)-1(-4-2)+1(0-3)$$
$$=24+6-3$$
$$=27$$

$$D_z=\begin{vmatrix}2&2&1\\-1&1&3\\1&2&0\end{vmatrix}=2\begin{vmatrix}1&3\\2&0\end{vmatrix}-2\begin{vmatrix}-1&3\\1&0\end{vmatrix}+1\begin{vmatrix}-1&1\\1&2\end{vmatrix}$$
$$=2(0-6)-2(0-3)+1(-2-1)$$
$$=-12+6-3$$
$$=-9$$

$$x=\frac{D_x}{D}=\frac{-18}{9}=-2,\quad y=\frac{D_y}{D}=\frac{27}{9}=3,$$
$$z=\frac{D_z}{D}=\frac{-9}{9}=-1$$

The solution is (−2, 3, −1).

33. $\begin{cases}x-2y+z=-5\\3y+2z=4\\3x-y=-2\end{cases}$

$$D=\begin{vmatrix}1&-2&1\\0&3&2\\3&-1&0\end{vmatrix}=1\begin{vmatrix}3&2\\-1&0\end{vmatrix}+2\begin{vmatrix}0&2\\3&0\end{vmatrix}+1\begin{vmatrix}0&3\\3&-1\end{vmatrix}$$
$$=1(0+2)+2(0-6)+1(0-9)$$
$$=2-12-9$$
$$=-19$$

$$D_x=\begin{vmatrix}-5&-2&1\\4&3&2\\-2&-1&0\end{vmatrix}$$
$$=-5\begin{vmatrix}3&2\\-1&0\end{vmatrix}+2\begin{vmatrix}4&2\\-2&0\end{vmatrix}+1\begin{vmatrix}4&3\\-2&-1\end{vmatrix}$$
$$=-5(0+2)+2(0+4)+1(-4+6)$$
$$=-10+8+2$$
$$=0$$

$$D_y=\begin{vmatrix}1&-5&1\\0&4&2\\3&-2&0\end{vmatrix}=1\begin{vmatrix}4&2\\-2&0\end{vmatrix}+5\begin{vmatrix}0&2\\3&0\end{vmatrix}+1\begin{vmatrix}0&4\\3&-2\end{vmatrix}$$
$$=1(0+4)+5(0-6)+1(0-12)$$
$$=4-30-12$$
$$=-38$$

$$D_z=\begin{vmatrix}1&-2&-5\\0&3&4\\3&-1&-2\end{vmatrix}$$
$$=1\begin{vmatrix}3&4\\-1&-2\end{vmatrix}+2\begin{vmatrix}0&4\\3&-2\end{vmatrix}-5\begin{vmatrix}0&3\\3&-1\end{vmatrix}$$
$$=1(-6+4)+2(0-12)-5(0-9)$$
$$=-2-24+45$$
$$=19$$

$$x=\frac{D_x}{D}=\frac{0}{-19}=0,\quad y=\frac{D_y}{D}=\frac{-38}{-19}=2,$$
$$z=\frac{D_z}{D}=\frac{19}{-19}=-1$$

The solution is (0, 2, −1).

35. $\begin{vmatrix}1&x\\2&7\end{vmatrix}=-3$

$$(1)(7)-x\cdot 2=-3$$
$$7-2x=-3$$
$$-2x=-10$$
$$x=5$$

37. 0; answers may vary

Appendix D

Viewing Window and Interpreting Window Settings Exercise Set

1. Yes, since every coordinate is between −10 and 10.

3. No, since −11 is less than −10.

5. Answers may vary. Any values such that Xmin < −90, Ymin < −80, Xmax > 55, and Ymax > 80.

7. Answers may vary. Any values such that Xmin < −11, Ymin < −5, Xmax > 7, and Ymax > 2.

9. Answers may vary. Any values such that Xmin < 50, Ymin < −50, Xmax > 200, and Ymax > 200.

11. Xmin = −12 Ymin = −12
Xmax = 12 Ymax = 12
Xscl = 3 Yscl = 3

13. Xmin = −9 Ymin = −12
Xmax = 9 Ymax = 12
Xscl = 1 Yscl = 2

15. Xmin = −10 Ymin = −25
Xmax = 10 Ymax = 25
Xscl = 2 Yscl = 5

17. Xmin = −10 Ymin = −30
Xmax = 10 Ymax = 30
Xscl = 1 Yscl = 3

19. Xmin = −20 Ymin = −30
Xmax = 30 Ymax = 50
Xscl = 5 Yscl = 10

Graphing Equations and Square Viewing Window Exercise Set

1. Setting A:

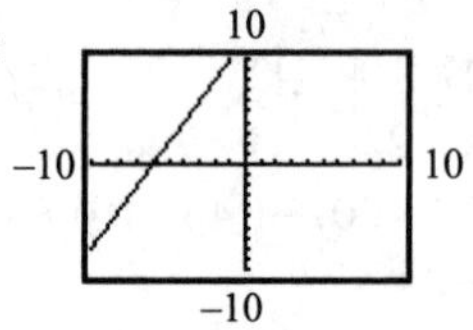

Setting B:

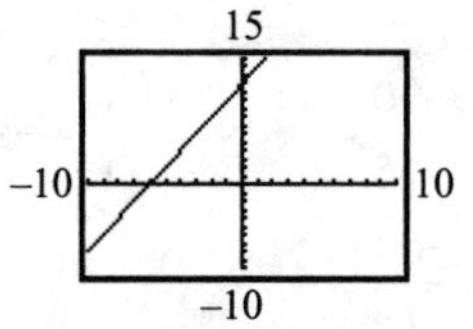

Setting B shows all intercepts.

3. Setting A:

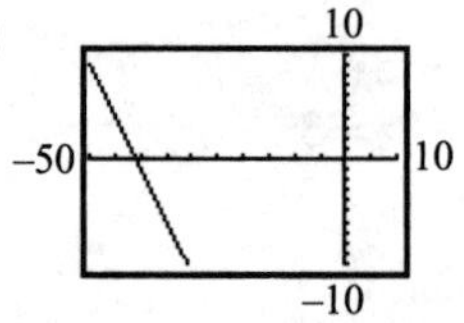

Setting B:

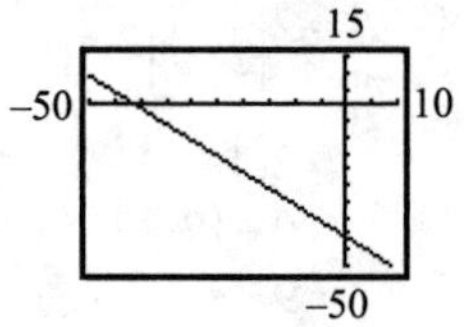

Setting B shows all intercepts.

5. Setting A:

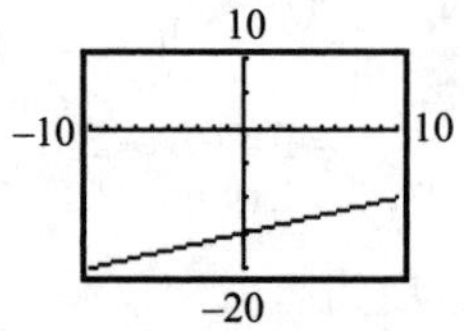

Setting B:

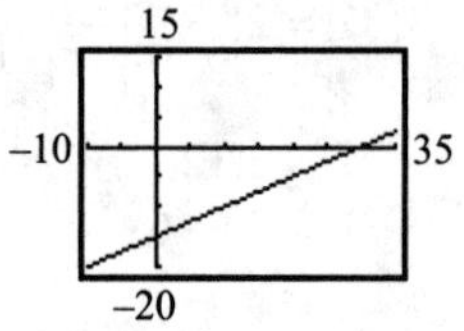

Setting B shows all intercepts.

7. $3x = 5y$

$$y = \frac{3}{5}x$$

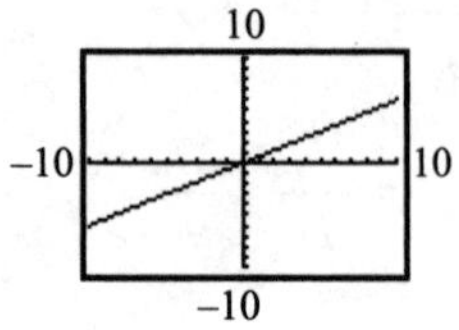

9. $9x - 5y = 30$

$$-5y = -9x + 30$$

$$y = \frac{9}{5}x - 6$$

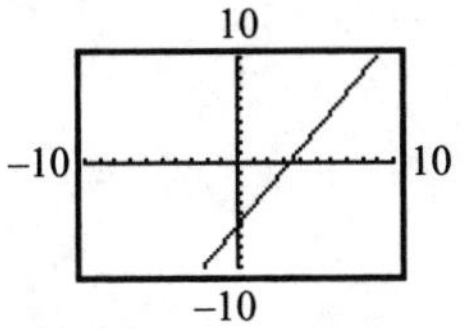

11. $y = -7$

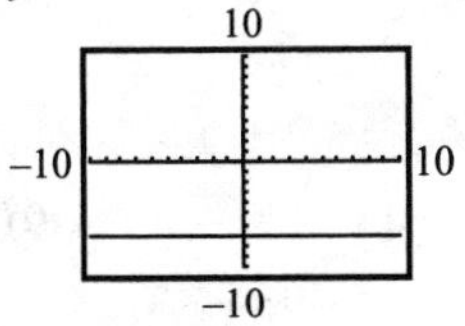

13. $x + 10y = -5$

$$10y = -x - 5$$

$$y = -\frac{1}{10}x - \frac{1}{2}$$

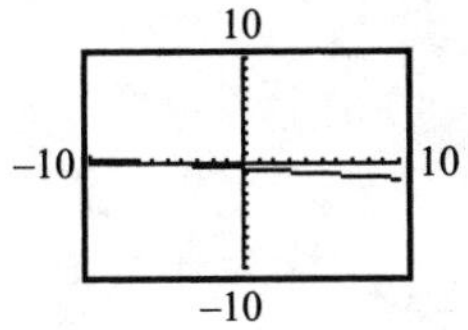

15. $y = \sqrt{x}$

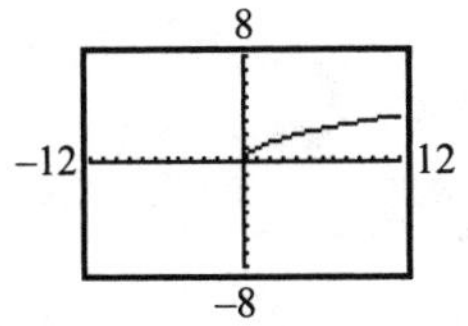

17. $y = x^2 + 2x + 1$

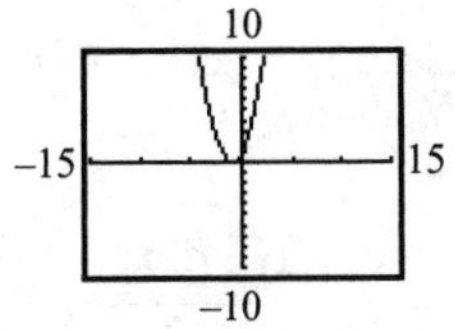

19. $y = |x|$

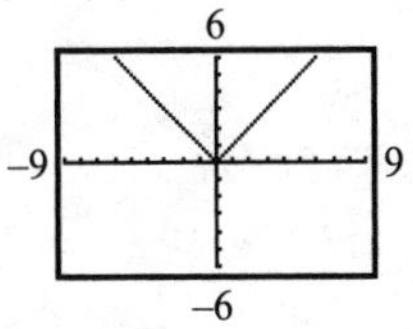

21. $x + 2y = 30$

$$2y = -x + 30$$

$$y = -\frac{1}{2}x + 15$$

Standard window:

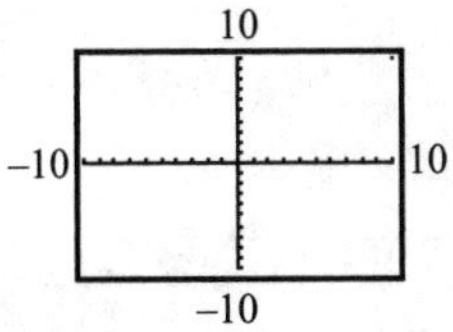

Adjusted window:

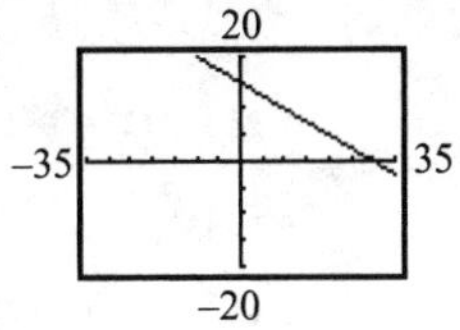

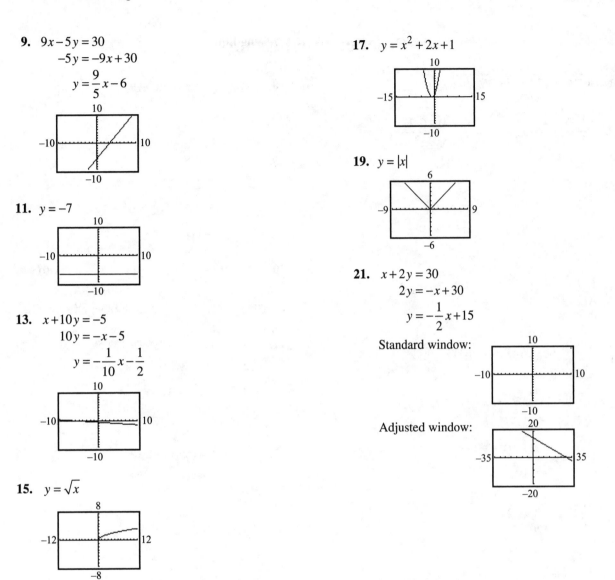

Practice Final Exam

1. $\sqrt{216} = \sqrt{36 \cdot 6} = \sqrt{36} \cdot \sqrt{6} = 6\sqrt{6}$

2. $$\frac{(4-\sqrt{16})-(-7-20)}{-2(1-4)^2} = \frac{(4-4)-(-27)}{-2(-3)^2} = \frac{0+27}{-2(9)} = \frac{27}{-18} = -\frac{3}{2}$$

3. $\left(\frac{1}{125}\right)^{-1/3} = 125^{1/3} = \sqrt[3]{125} = \sqrt[3]{5^3} = 5$

4. $(-9x)^{-2} = \frac{1}{(-9x)^2} = \frac{1}{(-9)^2 x^2} = \frac{1}{81x^2}$

5. $$\frac{\frac{5}{x}-\frac{7}{3x}}{\frac{9}{8x}-\frac{1}{x}} = \frac{24x\left(\frac{5}{x}-\frac{7}{3x}\right)}{24x\left(\frac{9}{8x}-\frac{1}{x}\right)} = \frac{24x\left(\frac{5}{x}\right)-24x\left(\frac{7}{3x}\right)}{24x\left(\frac{9}{8x}\right)-24x\left(\frac{1}{x}\right)} = \frac{24\cdot 5-8\cdot 7}{3.9-24} = \frac{120-56}{27-24} = \frac{64}{3}$$

6. $$\frac{6^{-1}a^2b^{-3}}{3^{-2}a^{-5}b^2} = \frac{3^2 a^{2-(-5)}b^{-3-2}}{6^1} = \frac{9a^7b^{-5}}{6} = \frac{3a^7}{2b^5}$$

7. $$\left(\frac{64c^{4/3}}{a^{-2/3}b^{5/6}}\right)^{1/2} = \frac{64^{1/2}c^{\frac{4}{3}\cdot\frac{1}{2}}}{a^{-\frac{2}{3}\cdot\frac{1}{2}}b^{\frac{5}{6}\cdot\frac{1}{2}}} = \frac{8c^{2/3}}{a^{-1/3}b^{5/12}} = \frac{8a^{1/3}c^{2/3}}{b^{5/12}}$$

8. $$\begin{aligned} 3x^2y-27y^3 &= 3y(x^2-9y^2) \\ &= 3y[x^2-(3y)^2] \\ &= 3y(x+3y)(x-3y) \end{aligned}$$

9. $$\begin{aligned} 16y^3-2 &= 2(8y^3-1) \\ &= 2[(2y)^3-1^3] \\ &= 2(2y-1)[(2y)^2+2y\cdot 1+1^2] \\ &= 2(2y-1)(4y^2+2y+1) \end{aligned}$$

10. $$\begin{aligned} x^2y-9y-3x^2+27 &= y(x^2-9)-3(x^2-9) \\ &= (x^2-9)(y-3) \\ &= (x+3)(x-3)(y-3) \end{aligned}$$

11. $$\begin{aligned} &(4x^3y-3x-4)-(9x^3y+8x+5) \\ &= 4x^3y-3x-4-9x^3y-8x-5 \\ &= 4x^3y-9x^3y-3x-8x-4-5 \\ &= -5x^3y-11x-9 \end{aligned}$$

12. $$\begin{aligned} (6m+n)^2 &= (6m)^2+2(6m)(n)+n^2 \\ &= 36m^2+12mn+n^2 \end{aligned}$$

13. $$\begin{aligned} &(2x-1)(x^2-6x+4) \\ &= 2x(x^2-6x+4)-1(x^2-6x+4) \\ &= 2x^3-12x^2+8x-x^2+6x-4 \\ &= 2x^3-13x^2+14x-4 \end{aligned}$$

14. $$\frac{3x^2-12}{x^2+2x-8} \div \frac{6x+18}{x+4} = \frac{3x^2-12}{x^2+2x-8}\cdot\frac{x+4}{6x+18} = \frac{3(x^2-4)}{(x+4)(x-2)}\cdot\frac{x+4}{6(x+3)} = \frac{3(x+2)(x-2)(x+4)}{(x+4)(x-2)\cdot 6(x+3)} = \frac{x+2}{2(x+3)}$$

15. $\dfrac{2x^2+7}{2x^4-18x^2}-\dfrac{6x+7}{2x^4-18x^2}=\dfrac{(2x^2+7)-(6x+7)}{2x^4-18x^2}$

$$=\frac{2x^2+7-6x-7}{2x^4-18x^2}$$
$$=\frac{2x^2-6x}{2x^2(x^2-9)}$$
$$=\frac{2x(x-3)}{2x^2(x+3)(x-3)}$$
$$=\frac{1}{x(x+3)}$$

16. $\dfrac{3}{x^2-x-6}+\dfrac{2}{x^2-5x+6}$

$$=\frac{3}{(x-3)(x+2)}+\frac{2}{(x-2)(x-3)}$$
$$=\frac{3(x-2)}{(x-3)(x+2)(x-2)}+\frac{2(x+2)}{(x-2)(x-3)(x+2)}$$
$$=\frac{3(x-2)+2(x+2)}{(x-3)(x+2)(x-2)}$$
$$=\frac{3x-6+2x+4}{(x-3)(x+2)(x-2)}$$
$$=\frac{5x-2}{(x-3)(x+2)(x-2)}$$

17. $\sqrt{125x^3}-3\sqrt{20x^3}=\sqrt{25x^2\cdot 5x}-3\sqrt{4x^2\cdot 5x}$

$$=5x\sqrt{5x}-3\cdot 2x\sqrt{5x}$$
$$=5x\sqrt{5x}-6x\sqrt{5x}$$
$$=(5x-6x)\sqrt{5x}$$
$$=-x\sqrt{5x}$$

18. $\left(\sqrt{5}+5\right)\left(\sqrt{5}-5\right)=\left(\sqrt{5}\right)^2-5^2=5-25=-20$

19.

$$\begin{array}{r|l} & 2x^2-x-2 \\ \hline 2x+1 & 4x^3+0x^2-5x+0 \\ & \underline{4x^3+2x^2} \\ & \quad -2x^2-5x \\ & \quad \underline{-2x^2\ \ -x} \\ & \qquad\quad -4x+0 \\ & \qquad\quad \underline{-4x-2} \\ & \qquad\qquad\ \ 2 \end{array}$$

$$(4x^3-5x)\div(2x+1)=2x^2-x-2+\frac{2}{2x+1}$$

20.
$$\begin{aligned} 9(x+2)&=5[11-2(2-x)+3] \\ 9x+18&=5(11-4+2x+3) \\ 9x+18&=5(2x+10) \\ 9x+18&=10x+50 \\ 9x-10x&=50-18 \\ -x&=32 \\ x&=-32 \end{aligned}$$

21.
$$|6x-5|-3=-2$$
$$|6x-5|=1$$
$6x-5=-1$ or $6x-5=1$

$6x=4$ or $6x=6$

$x=\dfrac{4}{6}$ or $x=1$

$x=\dfrac{2}{3}$

Both solutions check.

22.
$$\begin{aligned} 3n(7n-20)&=96 \\ 21n^2-60n&=96 \\ 21n^2-60n-96&=0 \\ 3(7n^2-20n-32)&=0 \\ 3(7n+8)(n-4)&=0 \end{aligned}$$
$7n+8=0$ or $n-4=0$

$n=-\dfrac{8}{7}$ or $n=4$

Both solutions check.

23.
$$\begin{aligned} -3<2(x-3)&\le 4 \\ -3<2x-6&\le 4 \\ -3+6<2x-6+6&\le 4+6 \\ 3<2x&\le 10 \\ \frac{3}{2}<\frac{2x}{2}&\le\frac{10}{2} \\ \frac{3}{2}<x&\le 5 \end{aligned}$$
$$\left(\frac{3}{2}, 5\right]$$

24. $|3x+1|>5$

$3x+1<-5$ or $3x+1>5$

$3x<-6$ $\quad 3x>4$

$x<-2$ $\quad x>\dfrac{4}{3}$

$$(-\infty, -2)\cup\left(\frac{4}{3}, \infty\right)$$

25. $\frac{x^2+8}{x}-1=\frac{2(x+4)}{x}$

$$x\left(\frac{x^2+8}{x}-1\right)=x\left(\frac{2(x+4)}{x}\right)$$
$$x^2+8-x=2(x+4)$$
$$x^2-x+8=2x+8$$
$$x^2-3x=0$$
$$x(x-3)=0$$
$$x=0 \quad \text{or} \quad x-3=0$$
$$x=3$$

The only solution is 3.

26. $y^2-3y=5$

$$y^2-3y-5=0$$
$$y=\frac{-(-3)\pm\sqrt{(-3)^2-4(1)(-5)}}{2(1)}$$
$$y=\frac{3\pm\sqrt{9+20}}{2}$$
$$y=\frac{3\pm\sqrt{29}}{2}$$

27. $x=\sqrt{x-2}+2$

$$x-2=\sqrt{x-2}$$
$$(x-2)^2=\left(\sqrt{x-2}\right)^2$$
$$x^2-4x+4=x-2$$
$$x^2-5x+6=0$$
$$(x-2)(x-3)=0$$
$$x-2=0 \quad \text{or} \quad x-3=0$$
$$x=2 \quad \text{or} \quad x=3$$

28. $2x^2-7x>15$

$$2x^2-7x-15>0$$
$$(2x+3)(x-5)>0$$
$$2x+3=0 \quad \text{or} \quad x-5=0$$
$$x=-\frac{3}{2} \quad \text{or} \quad x=5$$

Region	Test Point	$2x^2-7x>15$ Result
$\left(-\infty, -\frac{3}{2}\right)$	-3	$39>15$ True
$\left(-\frac{3}{2}, 5\right)$	0	$0>15$ False
$(5, \infty)$	6	$30>15$ True

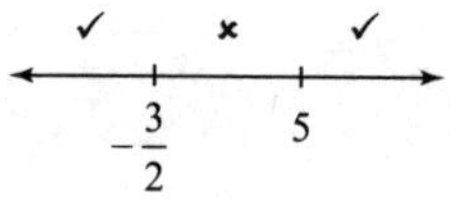

$$\left(-\infty, -\frac{3}{2}\right)\cup(5, \infty)$$

29. $\begin{cases} \frac{x}{2}+\frac{y}{4}=-\frac{3}{4} \\ x+\frac{3}{4}y=-4 \end{cases}$

$$\begin{cases} -8\left(\frac{x}{2}+\frac{y}{4}\right)=-8\left(-\frac{3}{4}\right) \\ 4\left(x+\frac{3}{4}y\right)=4(-4) \end{cases}$$
$$-4x-2y=6$$
$$\underline{4x+3y=-16}$$
$$y=-10$$

Let $y=-10$ in the second equation.

$$x+\frac{3}{4}y=-4$$
$$x+\frac{3}{4}(-10)=-4$$
$$x-\frac{15}{2}=-\frac{8}{2}$$
$$x=\frac{7}{2}$$

The solution is $\left(\frac{7}{2}, -10\right)$.

30. $4x+6y=7$

$$6y=-4x+7$$
$$y=-\frac{2}{3}x+\frac{7}{6}$$

slope $=-\frac{2}{3}$, y-intercept $\left(0, \frac{7}{6}\right)$

Plot points: $\left(0, \frac{7}{6}\right), \left(\frac{7}{4}, 0\right), \left(-3, \frac{19}{6}\right)$

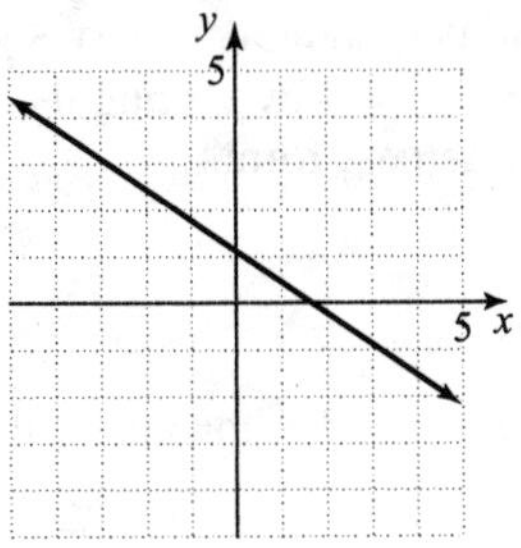

31. $2x - y > 5$
Graph $2x - y = 5$ with a dashed line because the inequality is >.

x	$2x - y = 5$	y
0	$2(0) - y = 5$	−5
1	$2(1) - y = 5$	−3
3	$2(3) - y = 5$	1

Test point (0, 0): $2(0) - 0 \overset{?}{>} 5$

$0 > 5$ False

Shade the region that does not include (0, 0).

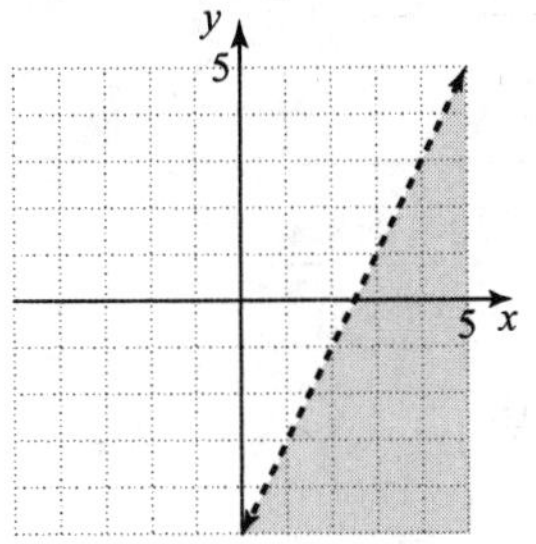

32. $y = -3$
The graph of $y = -3$ is a horizontal line with a y-intercept of (0, −3).

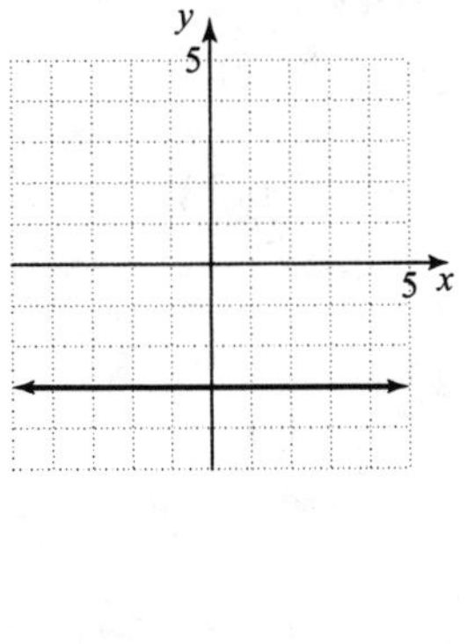

33. $g(x) = -|x + 2| - 1$

x	$g(x) = -\lvert x + 2\rvert - 1$	$g(x)$
−5	$-\lvert -5 + 2\rvert - 1 = -4$	−4
−4	$-\lvert -4 + 2\rvert - 1 = -3$	−3
−3	$-\lvert -3 + 2\rvert - 1 = -2$	−2
−2	$-\lvert -2 + 2\rvert - 1 = -1$	−1
−1	$-\lvert -1 + 2\rvert - 1 = -2$	−2
0	$-\lvert 0 + 2\rvert - 1 = -3$	−3
1	$-\lvert 1 + 2\rvert - 1 = -4$	−4

Domain: All real numbers, $(-\infty, \infty)$
Range: $(-\infty, -1]$

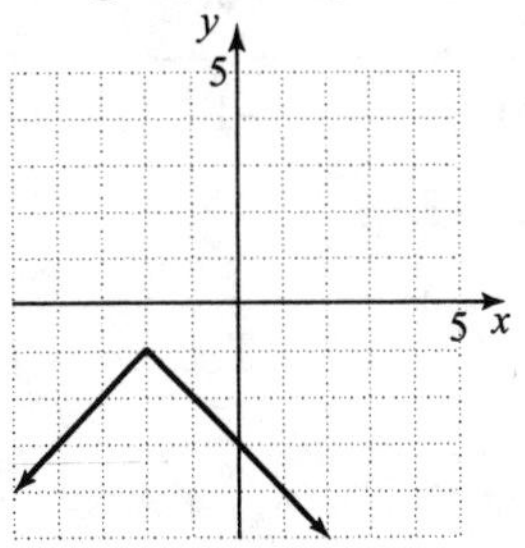

34. $h(x) = x^2 - 4x + 4$
x-intercept: Let $h(x) = 0$ and solve for x.

$$0 = x^2 - 4x + 4$$
$$0 = (x - 2)^2$$
$$x - 2 = 0$$
$$x = 2$$

x-intercept: (2, 0)
y-intercept: Let $x = 0$.

$$h(0) = 0^2 - 4(0) + 4 = 4$$

y-intercept: (0, 4)
x-coordinate of vertex:

$$-\frac{b}{2a} = -\frac{-4}{2(1)} = 2$$

vertex: (2, 0)

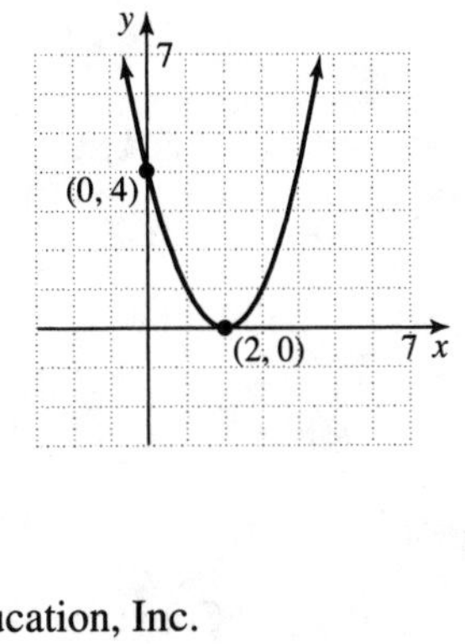

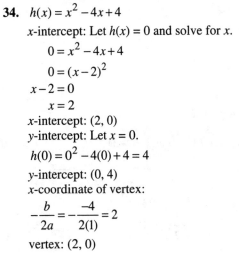

35. $f(x)=\begin{cases}-\frac{1}{2}x & \text{if } x\le 0\\ 2x-3 & \text{if } x>0\end{cases}$

If $x \le 0$

x	$-\frac{1}{2}x$	$f(x)$
0	$-\frac{1}{2}(0)$	0
−2	$-\frac{1}{2}(-2)$	1
−4	$-\frac{1}{2}(-4)$	2

If $x > 0$

x	$2x-3$	$f(x)$
1	2(1) − 3	−1
2	2(2) − 3	1
3	2(3) − 3	3

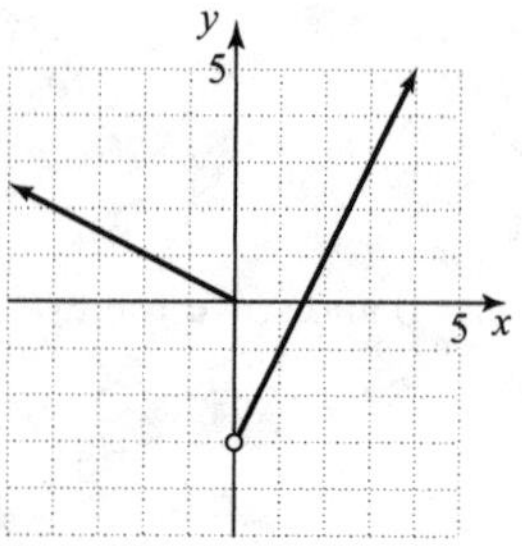

Domain: $(-\infty, \infty)$
Range: $(-3, \infty)$

36. through (4, −2) and (6, −3)

$$\text{slope} = m = \frac{y_2-y_1}{x_2-x_1} = \frac{-3-(-2)}{6-4} = \frac{-1}{2}$$

$$y-y_1 = m(x-x_1)$$
$$y-(-2) = -\frac{1}{2}(x-4)$$
$$y+2 = -\frac{1}{2}x+2$$
$$y = -\frac{1}{2}x$$
$$f(x) = -\frac{1}{2}x$$

37. through (−1, 2) and perpendicular to $3x - y = 4$
Find the slope of $3x - y = 4$ by writing the equation in slope-intercept form.

$$3x-y=4$$
$$-y=-3x+4$$
$$y=3x-4$$

The slope is 3. The slope of a line perpendicular to this line is $-\frac{1}{3}$.

Substitute $m=-\frac{1}{3}$ and $(x_1, y_1)=(-1, 2)$ in the equation:

$$y-y_1=m(x-x_1)$$
$$y-2=-\frac{1}{3}[x-(-1)]$$
$$y-2=-\frac{1}{3}(x+1)$$
$$y-2=-\frac{1}{3}x-\frac{1}{3}$$
$$y=-\frac{1}{3}x+\frac{5}{3}$$
$$f(x)=-\frac{1}{3}x+\frac{5}{3}$$

38. $(x_1, y_1)=(-6, 3);\ (x_2, y_2)=(-8, -7)$

$$d=\sqrt{(x_2-x_1)^2+(y_2-y_1)^2}$$
$$=\sqrt{[-8-(-6)]^2+(-7-3)^2}$$
$$=\sqrt{(-2)^2+(-10)^2}$$
$$=\sqrt{4+100}$$
$$=\sqrt{104}$$
$$=2\sqrt{26} \text{ units}$$

39. $(x_1, y_1)=(-2, -5);\ (x_2, y_2)=(-6, 12)$

$$\text{midpoint}=\left(\frac{x_1+x_2}{2}, \frac{y_1+y_2}{2}\right)$$
$$=\left(\frac{-2+(-6)}{2}, \frac{-5+12}{2}\right)$$
$$=\left(\frac{-8}{2}, \frac{7}{2}\right)$$
$$=\left(-4, \frac{7}{2}\right)$$

40. $\sqrt{\frac{9}{y}}=\frac{\sqrt{9}}{\sqrt{y}}=\frac{\sqrt{9}}{\sqrt{y}}\cdot\frac{\sqrt{y}}{\sqrt{y}}=\frac{\sqrt{9}\cdot\sqrt{y}}{\sqrt{y}\cdot\sqrt{y}}=\frac{3\sqrt{y}}{y}$

41. $\frac{4-\sqrt{x}}{4+2\sqrt{x}} = \frac{4-\sqrt{x}}{4+2\sqrt{x}} \cdot \frac{4-2\sqrt{x}}{4-2\sqrt{x}}$

$= \frac{(4-\sqrt{x})(4-2\sqrt{x})}{(4+2\sqrt{x})(4-2\sqrt{x})}$

$= \frac{16-12\sqrt{x}+2x}{16-4x}$

$= \frac{2(8-6\sqrt{x}+x)}{2(8-2x)}$

$= \frac{8-6\sqrt{x}+x}{8-2x}$

42. Let x = the number of people employed as registered nurses in 2012. Then $x + 0.19x = 3{,}240{,}000$.

$x+0.19x = 3,240,000$

$1.19x = 3,240,000$

$x \approx 2,723,000$

There were 2,723,000 people employed as registered nurses in 2012.

43. Subtract the area of the small square from the area of the large square.

$x^2 - (2y)^2 = (x+2y)(x-2y)$

44. Let x = the number.

$(x+1)\cdot\left(2\cdot\frac{1}{x}\right) = \frac{12}{5}$

$\frac{2}{x}(x+1) = \frac{12}{5}$

$5x\left[\frac{2}{x}(x+1)\right] = 5x\left(\frac{12}{5}\right)$

$10(x+1) = x\cdot 12$

$10x+10 = 12x$

$10 = 2x$

$5 = x$

The number is 5.

45. $W = \frac{k}{V}$

Find k by substituting $W = 20$ and $V = 12$.

$20 = \frac{k}{12}$

$240 = k$

Write the inverse relation equation.

$W = \frac{240}{V}$

Let $V = 15$ and find W.

$W = \frac{240}{15}$

$W = 16$

46. Use the Pythagorean theorem.

$c^2 = a^2 + b^2$

$20^2 = x^2 + (x+8)^2$

$400 = x^2 + x^2 + 16x + 64$

$0 = 2x^2 + 16x - 336$

$0 = 2(x^2 + 8x - 168)$

$x = \frac{-8 \pm \sqrt{8^2 - 4(1)(-168)}}{2(1)}$

$x = \frac{-8 \pm \sqrt{736}}{2}$

$x \approx -17.6$ or $x \approx 9.6$

Discard a negative distance.

$x + 8 + x = 9.6 + 8 + 9.6 = 27.2$

$27.2 - 20 = 7.2$

A person saves about 7.2 feet.

47. a. Find the vertex.

$s(t) = -16t^2 + 32t + 256$

t-value: $\frac{-b}{2a} = \frac{-32}{2(-16)} = 1$

$s(t)$-value:

$s(1) = -16(1)^2 + 32(1) + 256 = 272$

The maximum height is 272 feet.

b. Let $s(t) = 0$ and solve for t.

$0 = -16t^2 + 32t + 256$

$0 = -16(t^2 - 2t - 16)$

$t = \frac{-(-2) \pm \sqrt{(-2)^2 - 4(1)(-16)}}{2(1)}$

$t = \frac{2 \pm \sqrt{68}}{2}$

$t = \frac{2 \pm 2\sqrt{17}}{2}$

$t = 1 \pm \sqrt{17}$

$t \approx -3.12$ or $t \approx 5.12$

Discard a negative time.

The stone will hit the water in approximately 5.12 seconds.

48. Let x = amount of 10% solution to add to mixture.

solution	amount of solution	amount of fructose
10%	x	$0.10x$
20%	$20 - x$	$0.20(20 - x)$
17.5%	20	$0.175(20)$

$$0.10x + 0.20(20 - x) = 0.175(20)$$
$$0.10x + 4 - 0.20x = 3.5$$
$$-0.10x = -0.5$$
$$x = 5$$

$20 - x = 15$

Therefore, mix 5 gallons of 10% solution with 15 gallons of 20% solution.

49. $-\sqrt{-8} = -\sqrt{4\cdot(-1)\cdot 2}$
$$= -\sqrt{4}\cdot\sqrt{-1}\cdot\sqrt{2}$$
$$= -2i\sqrt{2}$$
$$= 0 - 2i\sqrt{2}$$

50. $(12 - 6i) - (12 - 3i) = 12 - 6i - 12 + 3i$
$$= 12 - 12 - 6i + 3i$$
$$= 0 - 3i$$

51. $(4 + 3i)^2 = (4 + 3i)(4 + 3i)$
$$= 16 + 12i + 12i + 9i^2$$
$$= 16 + 24i - 9$$
$$= 7 + 24i$$

52. $\dfrac{1+4i}{1-i} = \dfrac{1+4i}{1-i}\cdot\dfrac{1+i}{1+i}$
$$= \frac{(1+4i)(1+i)}{(1-i)(1+i)}$$
$$= \frac{1+5i+4i^2}{1-i^2}$$
$$= \frac{1+5i-4}{1-(-1)}$$
$$= \frac{-3+5i}{2}$$
$$= -\frac{3}{2} + \frac{5}{2}i$$

53. $g(x) = x - 7$ and $h(x) = x^2 - 6x + 5$

$(g \circ h)(x) = (x^2 - 6x + 5) - 7 = x^2 - 6x - 2$

54. $f(x) = 6 - 2x$ is a one-to-one function since there is only one $f(x)$ value for each x-value.

Inverse:

$y = 6 - 2x \quad \Rightarrow \quad x = 6 - 2y$
$$x + 2y = 6$$
$$2y = -x + 6$$
$$y = \frac{-x+6}{2}$$
$$f^{-1}(x) = \frac{-x+6}{2}$$

55. $\log_5 x + 3\log_5 x - \log_5(x+1)$
$$= \log_5 x + \log_5 x^3 - \log_5(x+1)$$
$$= \log_5 x\cdot x^3 - \log_5(x+1)$$
$$= \log_5 x^4 - \log_5(x+1)$$
$$= \log_5 \frac{x^4}{x+1}$$

56. $8^{x-1} = \dfrac{1}{64}$
$$(2^3)^{x-1} = \frac{1}{2^6}$$
$$2^{3(x-1)} = 2^{-6}$$
$$3(x-1) = -6$$
$$3x - 3 = -6$$
$$3x = -3$$
$$x = -1$$

57. $3^{2x+5} = 4$
$$\log 3^{2x+5} = \log 4$$
$$(2x+5)\log 3 = \log 4$$
$$2x + 5 = \frac{\log 4}{\log 3}$$
$$2x = \frac{\log 4}{\log 3} - 5$$
$$x = \frac{1}{2}\left(\frac{\log 4}{\log 3} - 5\right)$$
$$x \approx -1.8691$$

58. $\log_8(3x - 2) = 2$
$$8^2 = 3x - 2$$
$$64 = 3x - 2$$
$$66 = 3x$$
$$22 = x$$

59. $\log_4(x+1)-\log_4(x-2)=3$

$$\log_4\frac{x+1}{x-2}=3$$
$$4^3=\frac{x+1}{x-2}$$
$$64=\frac{x+1}{x-2}$$
$$64(x-2)=x+1$$
$$64x-128=x+1$$
$$63x=129$$
$$x=\frac{129}{63}=\frac{43}{21}$$

60. $\ln\sqrt{e}=x$

$$\ln e^{1/2}=x$$
$$\frac{1}{2}\ln e=x$$
$$\frac{1}{2}=x$$

61. $y=\left(\frac{1}{2}\right)^x+1$

x	$\left(\frac{1}{2}\right)^x+1$	y
−3	$\left(\frac{1}{2}\right)^{-3}+1=9$	9
−2	$\left(\frac{1}{2}\right)^{-2}+1=5$	5
−1	$\left(\frac{1}{2}\right)^{-1}+1=3$	3
0	$\left(\frac{1}{2}\right)^{0}+1=2$	2
1	$\left(\frac{1}{2}\right)^{1}+1=1\frac{1}{2}$	$1\frac{1}{2}$
2	$\left(\frac{1}{2}\right)^{2}+1=1\frac{1}{4}$	$1\frac{1}{4}$
3	$\left(\frac{1}{2}\right)^{3}+1=1\frac{1}{8}$	$1\frac{1}{8}$

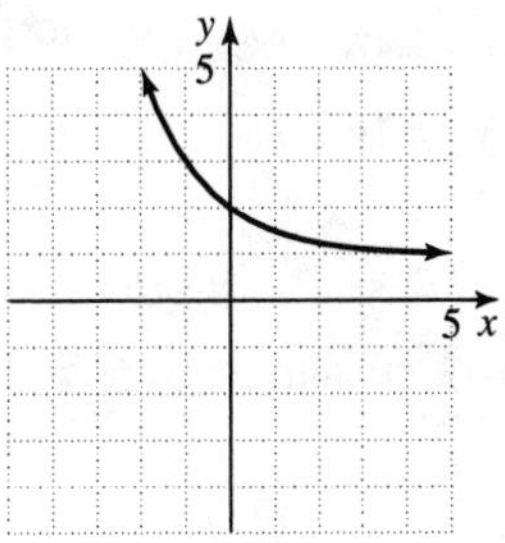

62. Use $y=C(1+r)^x$ with $C = 57{,}000$, $r = 2.6\% = 0.026$, and $x = 5$.

$y=57{,}000(1+0.026)^5\approx 64{,}805$

In 5 years, there will be 64,805 prairie dogs.

63. $x^2-y^2=36$

$x^2-y^2=6^2$

hyperbola, with x-intercepts (−6, 0), (6, 0)

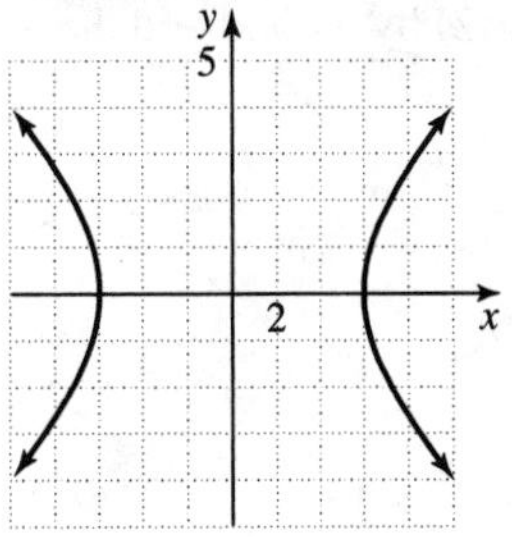

64. $16x^2+9y^2=144$

$$\frac{16x^2}{144}+\frac{9y^2}{144}=\frac{144}{144}$$
$$\frac{x^2}{9}+\frac{y^2}{16}=1$$

Ellipse, x-intercepts (−3, 0), (3, 0)
y-intercepts (0, −4), (0, 4)

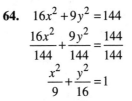

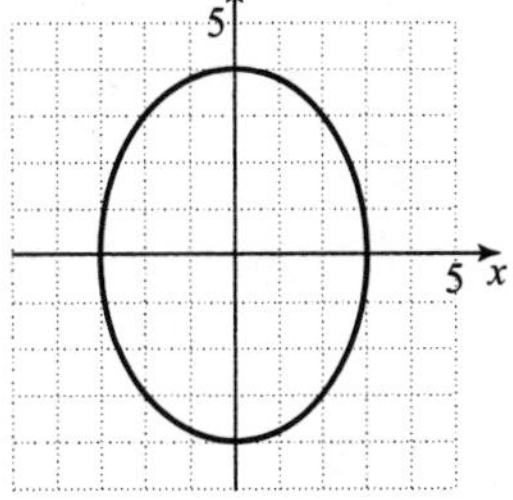

65.
$$x^2 + y^2 + 6x = 16$$
$$(x^2 + 6x + 9) + y^2 = 16 + 9$$
$$(x+3)^2 + y^2 = 25$$
$$[x-(-3)]^2 + (y-0)^2 = 5^2$$
circle with center (−3, 0) and radius 5

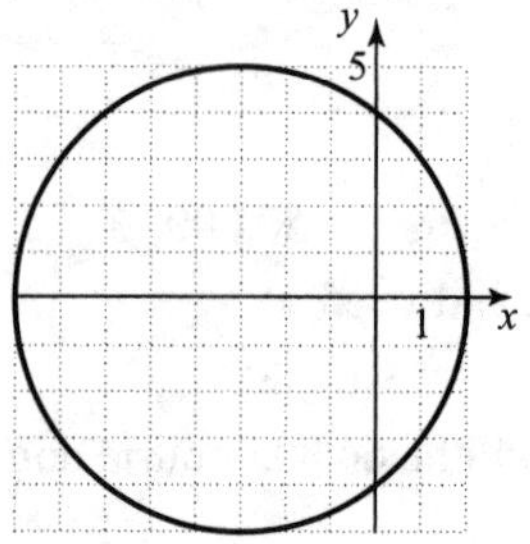

66. $\begin{cases} x^2 + y^2 = 26 \\ x^2 - 2y^2 = 23 \end{cases}$

Multiply equation (2) by −1 and add the equations.
$$\begin{array}{r} x^2 + y^2 = 26 \\ -x^2 + 2y^2 = -23 \\ \hline 3y^2 = 3 \\ y^2 = 1 \\ y = \pm 1 \end{array}$$
Substitute $y = -1$ and $y = 1$ into equation (1).
$$x^2 + (-1)^2 = 26$$
$$x^2 = 25$$
$$x = \pm 5$$
$$x^2 + 1^2 = 26$$
$$x^2 = 25$$
$$x = \pm 5$$
The solutions are (−5, −1), (−5, 1), (5, −1), (5, 1).

67. $a_n = \dfrac{(-1)^n}{n+4}$
$$a_1 = \frac{(-1)^1}{1+4} = -\frac{1}{5}$$
$$a_2 = \frac{(-1)^2}{2+4} = \frac{1}{6}$$
$$a_3 = \frac{(-1)^3}{3+4} = -\frac{1}{7}$$
$$a_4 = \frac{(-1)^4}{4+4} = \frac{1}{8}$$
$$a_5 = \frac{(-1)^5}{5+4} = -\frac{1}{9}$$
The first five terms are $-\frac{1}{5}, \frac{1}{6}, -\frac{1}{7}, \frac{1}{8}, -\frac{1}{9}$.

68. $a_n = 5(2)^{n-1}$
$$a_1 = 5(2)^{1-1} = 5(2)^0 = 5$$
$r = 2$
$n = 5$
$$S_n = \frac{a_1(1-r^n)}{1-r}$$
$$S_5 = \frac{5(1-2^5)}{1-2} = \frac{5(1-32)}{-1} = 155$$

69. Sequence $\frac{3}{2}, -\frac{3}{4}, \frac{3}{8}, \ldots$
$$a_1 = \frac{3}{2}, \ r = -\frac{1}{2}$$
$$S_\infty = \frac{a_1}{1-r} = \frac{\frac{3}{2}}{1-\left(-\frac{1}{2}\right)} = \frac{\frac{3}{2}}{\frac{3}{2}} = 1$$

70.
$$\sum_{i=1}^{4} i(i-2) = 1(1-2) + 2(2-2) + 3(3-2) + 4(4-2)$$
$$= 1(-1) + 2(0) + 3(1) + 4(2)$$
$$= -1 + 0 + 3 + 8$$
$$= 10$$

71. $(2x+y)^5 = \binom{5}{0}(2x)^5 + \binom{5}{1}(2x)^4(y) + \binom{5}{2}(2x)^3(y)^2 + \binom{5}{3}(2x)^2(y)^3 + \binom{5}{4}(2x)^1(y)^4 + \binom{5}{5}y^5$
$= 2^5x^5 + 5 \cdot 2^4x^4y + 10 \cdot 2^3x^3y^2 + 10 \cdot 2^2x^2y^3 + 5 \cdot 2xy^4 + y^5$
$= 32x^5 + 80x^4y + 80x^3y^2 + 40x^2y^3 + 10xy^4 + y^5$